MANUFACTURING ENGINEERING HANDBOOK

MANUFACTURING ENGINEERING HANDBOOK

Hwaiyu Geng, CMfgE, PE **Editor in Chief**

Project Manager, Hewlett-Packard Company
Palo Alto, California

McGRAW-HILL

New York Chicago San Francisco Lisbon London Madrid
Mexico City Milan New Delhi San Juan Seoul
Singapore Sydney Toronto

The McGraw-Hill Companies

Library of Congress Cataloging-in-Publication Data

Manufacturing engineering handbook/Hwaiyu Geng, editor in chief.
p. cm.
ISBN 0-07-139825-2
1. Production engineering. I. Geng, Hwaiyu.

TS176.M3615 2004
658.5—dc22 2004049949

1 2 3 4 5 6 7 8 9 0 DOC/DOC 0 9 8 7 6 5 4

ISBN 0-07-139825-2

The sponsoring editor for this book was Kenneth McCombs and the production supervisor was Pamela Pelton. The art director for the cover was Handel Low. It was set in HBI Design in Times Roman by International Typesetting and Composition.

Printed and bound by RR Donnelley.

This book was printed on acid-free paper.

To our mothers who cradle the world

CONTENTS

Part 2 Manufacturing Automation and Technologies

Part 3 Heat Treating, Hot Working, and Metalforming

Part 4 Metalworking, Moldmaking, and Machine Design

CONTRIBUTORS

Lawrence S. Aft, PE *Aft Systems, Inc., Atlanta, Georgia* (CHAPS 4, 53)
Bruce J. Andersen, CPIM *Richard Muther and Associates, Inc., Marietta, Georgia,* (CHAP 52)
Takashi Asano *Japan Management Consultants, Inc., Cincinnati, Ohio,* (CHAP 51)
Gary D. Baldwin *Kennametal University, Latrobe, Pennsylvania* (CHAP 27)
Ann M. Ball *Milacron, Inc., CIMCOOL Global Industrial Fluids, Cincinnati, Ohio* (CHAP 33)
Steve Benedict *Com Tal Machine & Engineering, Inc., St. Paul, Minnesota* (CHAP 47)
Dennis Berry *SKD Automotive Group, Troy, Michigan,* (CHAP 23)
Charles E. Boyer *ABB Inc., Fort Collins, Colorado* (CHAP 45)
Jay Boyle *Murietta, Georgia* (CHAP 4)
Kenneth W. Chase *Brigham Young University, Provo, Utah* (CHAP 7)
Nicholas P. Cheremisinoff *Princeton Energy Resources International, LLC, Rockville, Maryland* (CHAP 61)
Irvan Christy *CoCreate Software, Fort Collins, Colorado* (CHAP 8)
Kevin D. Creehan, PhD *Virginia Polytechnic Institute and State University, Blacksburg, Virginia* (CHAP 60)
David Curry, PhD, CHFP *Packer Engineering, Napperville, Illinois* (CHAP 58)
Peter Dewhurst *University of Rhode Island, Kingston, Rhode Island* (CHAP 2)
David J. Dunn *F.L.D. Enterprises, Aurora, Ohio* (CHAP 50)
William E. Fillmore, PE *Richard Muther and Associates, Inc., Marietta, Georgia* (CHAP 52)
Gerald A. Fleischer *University of Southern California, Los Angeles, California* (CHAP 54)
Leslie W. Flott *Summitt Process Consultant, Inc., Wabash, Indiana* (CHAP 48)
Thomas O. Floyd *Carboloy, Inc., Warren, Michigan* (CHAP 28)
Keith Frantz *Cimnet, Inc., Robesonia, Pennsylvania* (CHAP 44)
Kevin Gaudette, Maj, USAF *Indiana University, Bloomington, Indiana* (CHAP 55)
Howard A. Greis *Kinefac Corporation, Worcester, Massachussetts* (CHAP 22)
Todd Grimm *T.A. Grimm & Associates, Edgewood, Kentucky* (CHAP 5)
Chunghun Ha *Texas A&M University, College Station, Texas* (CHAP 13)
H. Lee Hales *Richard Muther and Associates, Inc., Marietta, Georgia* (CHAP 52)
Charles Harrell *Brigham Young University, Provo, Utah* (CHAP 10)
Daniel H. Herring *The Herring Group, Inc., Elmhurst, Illinois* (CHAP 18)
Mark J. Jackson *Tennessee Technological University, Cookeville, Tennessee* (CHAP 31)
F. Robert Jacobs *Indiana University, Bloomington, Indiana* (CHAP 55)
Mark Johnson *Tapmatic Corporation, Post Falls, Idaho* (CHAP 29)
Ali Khosravi Kamrani, PhD *University of Houston, Houston, Texas* (CHAP 17)

Albert V. Karvelis, PhD, PE *Packer Engineering, Naperville, Illinois* (CHAP 58)

James Kaszynski *Boehler Uddeholm, Rolling Meadows, Illinois* (CHAP 39)

Ian M. Kay *Cast Metals, Inc., American Foundry Society, Inc., Des Plaines, Illinois* (CHAPS 16, 19)

Way Kuo, PhD *University of Tennessee, Knoxville, Tennessee* (CHAP 13)

Chaman Lall, PhD *Metal Powder Products Company, Westfield, Indiana* (CHAP 20)

Gisbert Ledvon *Lincolnshire, Illinois* (CHAP 37)

V. Jorge Leon *Texas A&M University, College Station, Texas* (CHAP 15)

Arthur F. Lubiarz *NACHI America, Inc., Macomb, Michigan* (CHAP 30)

Goetz Marczinski, Dr. Ing. *CIMSOURCE Software Company, Ann Arbor, Michigan* (CHAP 16)

Steve Marek *Lucas-Milhaupt, Inc., Cudahy, Wisconsin* (CHAP 25)

David D. McCorry *Kaltenbach, Inc., Columbus, Indiana* (CHAP 32)

Leonard Migliore *Coherent, Inc., Santa Monica, California* (CHAP 35)

Duane K. Miller, PhD *Lincoln Electric Company, Cleveland, Ohio* (CHAP 21)

Ilya Mirman *SolidWorks Corporation, Concord, Massachussetts* (CHAP 9)

Roderick A. Munro *RAM Q Universe, Inc., Reno, Nevada* (CHAP 57)

John H. Olsen, PhD *OMAX Corporation, Kent, Washington* (CHAP 38)

Joseph F. Otero, CVS *Pratt & Whitney, Springfield, Massachussetts* (CHAP 3)

Todd Park *Athenahealth, Waltham, Massachussetts* (CHAP 1)

Sheila R. Poling *Pinnacle Partners, Inc., Oak Ridge, Tennessee* (CHAP 56)

V. Rajendra Prasad *Texas A&M University, College Station, Texas* (CHAP 13)

Jerry G. Scherer *GE Fanuc Product Development, Charlottesville, Virginia* (CHAP 14)

Ronald Schildge *Eitel Presses, Inc., Orwigsburg, Pennsylvania* (CHAP 24)

Kenneth F. Smith *Mayfran International, Cleveland, Ohio* (CHAP 43)

Andreas Somogyi *Rockwell Automation, Mayfield Heights, Ohio* (CHAP 11)

Paul Spink, BSHE, CMTSE *Mori Seiki, USA, Inc., Irving, Texas* (CHAP 12)

Vijay Srinivasan, PhD *IBM Corporation/Columbia University, New York, New York* (CHAP 6)

Eric Stange *Tools for Bending, Denver, Colorado* (CHAP 26)

Fred G. Steil *D-M-E Company, Madison Heights, Michigan* (CHAP 40)

Mal Sudhakar *Mikron Bostomatic Corporation, Holliston, Massachussetts* (CHAP 41)

Peter M. Sweeney *Bijur Lubricating Corporation, Morrisville, North Carolina* (CHAP 42)

Rodger Talbert *R. Talbert Consulting, Inc., Grand Rapids, Michigan* (CHAP 49)

Atsushi Terada *JMA Consultants America, Inc., Arlington Heights, Illinois* (CHAP 59)

Sophronia Ward, PhD *Pinnacle Partners, Inc., Oak Ridge, Tennessee* (CHAP 56)

Y. Lawrence Yao *Columbia University, New York, New York* (CHAP 34)

Wenwu Zhang *General Electric Global Research Center, Schenectady, New York* (CHAP 34)

Nello Zuech *Vision Systems International, Inc., Yardley, Pennsylvania* (CHAP 46)

Jerry Zybko *LEISTER Technologies, LLC, Itasca, Illinois* (CHAP 36)

PREFACE

Whether as an engineer, manager, researcher, professor, or student, we are all facing increasing challenges in a cross-functional manufacturing environment. For each problem, we must identify the givens, the unknowns, feasible solutions, and how to validate each of these. How can we best apply technical knowledge to assemble a proposal, to lead a project, or to support the team?

Our challenges may include designing manufacturing processes for new products, improving manufacturing yield, implementing automated manufacturing and production facilities, and establishing quality and safety programs. A good understanding of how manufacturing engineering works, as well as how it relates to other departments, will enable one to plan, design, and implement projects more effectively.

The goal of the *Manufacturing Engineering Handbook* is to provide readers with the essential tools needed for working in manufacturing engineering for problem solving, for establishing manufacturing processes, and for improving existing production lines in an enterprise. This *Handbook* embraces both conventional and emerging manufacturing tools and processes used in the automotive, aerospace, and defense industries and their supply chain industries.

The *Handbook* is organized into six major parts. These six parts comprise 61 chapters. In general, each chapter includes three components principles, operational considerations, and references. The principles are the fundamentals of a technology and its application. Operational considerations provide useful tips for planning, implementing, and controlling manufacturing processes. The references are a list of relevant books, technical papers, and websites for additional reading.

Part 1 of the *Handbook* gives background information on e-manufacturing. Tools for product development and design are introduced. Part 2 covers conventional and emerging manufacturing automation and technologies that are useful for planning and designing a manufacturing process. Part 3 offers fundamentals on heat-treating, hot-working, and metal-forming. Part 4 discusses major metalworking processes, briefly reviews moldmaking, and describes machine design fundamentals. Part 5 covers essential assembling operations including robotics, machine vision, automated assembly, and surface preparation. Part 6 reviews useful tools, processes, and considerations when planning, designing, and implementing a new or existing manufacturing process.

The *Handbook* covers topics ranging from product development, manufacturing automation, and technologies, to manufacturing process systems. Manufacturing industry engineers, managers, researchers, teachers, students, and others will find this to be a useful and enlightening resource because it covers the breadth and depth of manufacturing engineering. The *Manufacturing Engineering Handbook* is the most comprehensive single-source guide ever published in its field.

HWAIYU GENG, CMFGE, P.E.

ACKNOWLEDGMENTS

The *Manufacturing Engineering Handbook* is a collective representation of an international community of scientists and professionals. Over 60 authors have contributed to this book. Many others from both industry and academia offered their suggestions and advice while I prepared and organized the book. I would like to thank the contributors who took time from their busy schedules and personal lives to share their wisdom and valuable experiences. Special thanks and appreciation go to the following individuals, companies, societies, and institutes for their contributions and/or for granting permission for the use of copyrighted materials: Jane Gaboury, Institute of Industrial Engineers; Lew Gedansky, Project Management Institute; Ian Kay, Cast Metals Institute; Larry aft, Aft Systems; Vijay Srinivasan, IBM; Duane Miller, Lincoln Electric; Howard Greis, Kinefac Corporation; Fred Steil, D-M-E Company; Takashi Asano, Lean Manufacturing; David Curry, Packer Engineering; Gary Baldwin, Kennametal University; Lawrence Yao, Columbia University; Way Kuo, University of Tennessee; Gerald Fleischer, University of Southern California; Ken Chase, Brigham Young University; and Ken McComb, McGraw-Hill Company. I would also like to thank the production staff at ITC and McGraw-Hill, whose "can do" spirit and teamwork were instrumental in producing this book. My special thanks to my wife, Limei, and to my daughters, Amy and Julie, for their support and encouragement while I was preparing this book.

HWAIYU GENG, CMFGE, P.E.

MANUFACTURING ENGINEERING HANDBOOK

P · A · R · T · 1

PRODUCT DEVELOPMENT AND DESIGN

CHAPTER 1
E-MANUFACTURING

Todd Park
Athenahealth, Inc.
Waltham, Massachusetts

1.1 INTRODUCTION

In the past decade, so much ink has been spilled (not to mention blood and treasure) on the concepts of e-business and e-manufacturing that it has been extraordinarily difficult to separate hope from hype. If the early pronouncements from e-seers were to be believed, the Internet was destined to become a force of nature that, within only a few years, would transform manufacturers and manufacturing processes beyond all recognition. Everyone—customers, suppliers, management, line employees, machines, etc.—would be on-line, and fully integrated. It would be a grand alignment—one that would convert a customer's every e-whim into perfectly realized product, with all customer communication and transactions handled via the web, products designed collaboratively with customers on-line, all the right inputs delivered in exactly the right quantities at exactly the right millisecond (cued, of course, over the web), machines in production across the planet conversing with each other in a web-enabled symphony of synchronization, and total process transparency of all shop floors to the top floor, making managers omniscient gods of a brave new manufacturing universe.

These initial predictions now seem overly rosy at best, yet it is far too easy (and unfair) to dismiss e-business and e-manufacturing as fads in the same category as buying pet food and barbeque grills over the Internet. Gartner Group has estimated that as of 2001, only 1 percent of U.S. manufacturers had what could be considered full-scale e-manufacturing implementations. By 2006, U.S. Dept. of Commerce has estimated that almost half of the U.S. workforce will be employed by industries that are either major producers or intensive users of information technology products and services. The most successful e-companies, it turns out, have not been companies with ".com" emblazoned after their name, but, rather, traditional powerhouses like Intel and General Electric, who have led the way on everything from selling goods and services over the web to Internet-enabled core manufacturing processes. Perhaps most startlingly, the U.S. Bureau of Labor Statistics has projected that the rise of e-manufacturing could potentially equal or even exceed the impact of steam and electricity on industrial productivity. The Bureau recently concluded that the application of computers and early uses of the Internet in the supply chain had been responsible for a 3-percent point increase in annual U.S. manufacturing productivity growth, to 5 percent, during the 1973–1999 timeframe. The Bureau then projected that the rise of e-manufacturing could build upon those gains by boosting productivity growth by another two percentage points to an astounding 7 percent per year.[1] In fact, many analysts have pointed to e-manufacturing as the next true paradigm shift in manufacturing processes—albeit one that will take a long time to fulfill, but one that will ultimately be so pervasive that the term "e-manufacturing" will eventually become synonymous with manufacturing itself.

The purpose of this chapter is not to teach you everything there is to know about e-business and e-manufacturing. The field is moving too rapidly for any published compendium of current technologies and techniques to be valid for any relevant length of time. Rather, this chapter aims to introduce you to the core notions of e-business, the core principles of e-manufacturing, and to give you a simple operational and strategic framework which you can utilize to evaluate and pursue the application of "e" to the manufacturing process.

1.2 WHAT IS E-MANUFACTURING?

As is common with new phenomena, there is currently a good deal of semantic confusion around the words "e-business" and "e-manufacturing." Let us therefore start with some simple working definitions. "E-business" is defined most cogently and accurately as *the application of the Internet to business*. Somewhat confusingly, "e-business" is sometimes characterized as synonymous with "e-commerce," which is more narrowly defined as the buying and selling of things on the Internet. In my view, "e-commerce" is just one subset of "e-business"—and one, though it dominated e-business-related headlines during the go-go nineties, will ultimately be one of the less important applications of the Internet to business. Far more important than whether one can buy things on the Internet is the question of whether the Internet, like electricity and other fundamental technologies, can actually change (a) *the fundamental customer value produced by business* and (b) *the efficiency via which that value can be produced.*

This is where "e-manufacturing" comes in. E-manufacturing can be most cogently and generally described as *the application of the Internet to manufacturing*. Let us first say what it is not, for the sake of analytical clarity: e-manufacturing as a discipline is not the same thing as production automation or the so-called "digital factory." The application of computing technology to the factory floor is its own phenomenon, and can be pursued wholly independently of any use of the Internet. That being said, while e-manufacturing is not the same thing as production automation, it is perfectly complementary to the idea of production automation—an additional strategy and approach that can turbocharge the value produced by the application of technology to the factory. Business 2.0 has memorably defined e-manufacturing as "the marriage of the digital factory and the Internet."[1] What, then, are the dynamics of this marriage, and where specifically does it add value?

1.3 WHERE, WHEN, AND HOW CAN MANUFACTURING ENGINEERS APPLY E-MANUFACTURING?

There are literally hundreds of different frameworks that have been created and promulgated to describe e-manufacturing and how and where it can be usefully applied. If one were to seek a common thread of collectively exhaustive truth that runs through all of these frameworks, it would be the following: everyone agrees that e-manufacturing can impact both (a) the fundamental customer value produced by the manufacturing process and (b) the core efficiency of that process itself (Fig. 1.1).

1.3.1 Impacting Customer Value

The business of manufacturing has always been a guessing game. What do customers want? So therefore, what should we produce? One of the most profound implications of the Internet for manufacturing is its potential ability to deliver upon an objective that has been a Holy Grail of sorts for manufacturers since the beginning of manufacturing: build exactly what the customer wants, exactly when the customer wants it. This concept has been clothed in many different terms: "collaborative product commerce," "collaborative product development," "mass customization," "adaptive manufacturing," "c-manufacturing," "made-to-order manufacturing," etc. All of these refer to the same basic concept: utilizing the Internet, a customer (or salesperson or distributor representing the customer) electronically communicates his or her preferences, up to and including jointly designing the end product with the manufacturer. This specification is then delivered to the factory floor, where the customer's vision is made into reality.

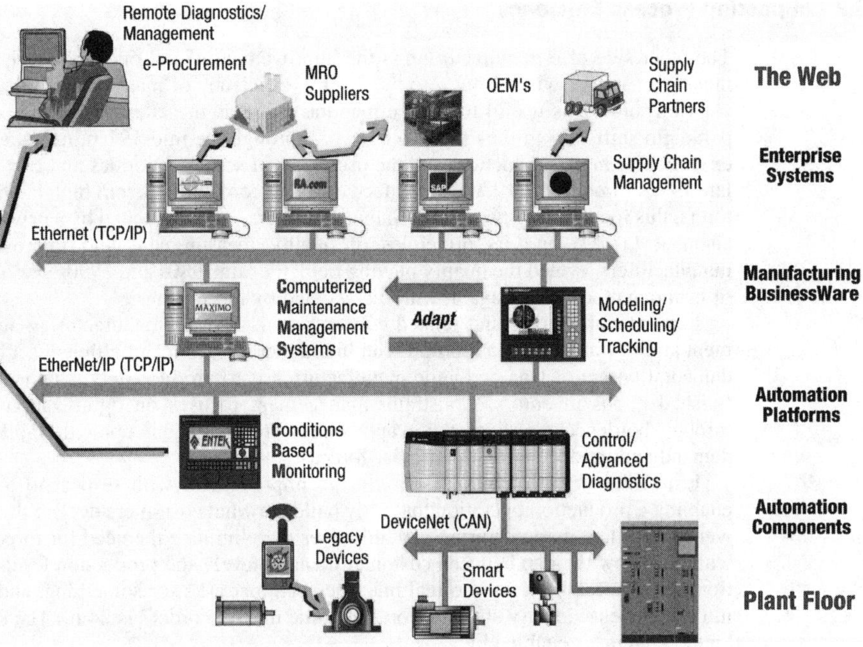

FIGURE 1.1 e-business and e-manufacturing. (*Courtesy of Rockwell Automation, Inc.*)

The simplest and earliest examples of this "made-to-order" approach have been in the technology industry, where companies like Dell have pioneered approaches such as allowing customers on their websites to customize PCs they are ordering. These applications have been facilitated by the relative simplicity of the end product, with a fairly limited number of parameters against which customers express preferences, and where product components can be manufactured to an intermediate step, with the manufacturing process being completed when the customer actually communicates a customized product order.

However, the "made-to-order" approach is now being applied to far more complicated product businesses. Perhaps the most famous example is Cutler-Hammer, a major manufacturer of panel boards, motor control centers, and other complex assemblies. Cutler-Hammer has built and deployed a proprietary system called Bid Manager, which allows customers from literally thousands of miles away to easily configure custom designs of items as complicated as a motor control center—down to the specific placement of switches, circuit breakers, etc.—with the assistance of a powerful rules engine and alerts that ensure correct design. The design, once completed and transmitted by a customer, is then processed by Bid Manager, which then, often within minutes of the transmittal of the order, instructs machines and people on Cutler-Hammer factory floors to build the product the customer wants. Cutler-Hammer has reported that it processes over 60,000 orders per year via Bid Manager, and that this comprehensive e-manufacturing application has increased Cutler-Hammer's market share for configured products by 15 percent, sales of larger assemblies by 20 percent, productivity by 35 percent, and had a dramatic impact on profitability and quality.[1]

While Bid Manager is an example of a proprietary software package, there are a rapidly expanding number of generally available software tools and approaches available to help make customer integration into manufacturing a reality, and utilizable by an ever-broadening range of manufacturers. New technologies such as XML enable seamless integration of customer-facing, web-based e-commerce, and product configuration applications with the software that powers "digital factories." The end result is to step closer and closer to the ideal of a "customer-driven" business where what the customer wants is exactly what the customer gets, with unprecedented flexibility and speed.

1.3.2 Impacting Process Efficiency

The other side of e-manufacturing is the improvement of not only the level of precision of fulfillment of customer wishes, but also the level of efficiency of manufacturing processes. For a historical viewpoint, it is useful to view e-manufacturing as the latest in a series of production process paradigm shifts. From the era of Henry Ford through the mid-1970s, manufacturers focused on the execution of mass production, and the principles of scale economies and cost efficiencies. From the late 1970s through the 1980s, in the face of rising competition from high quality Japanese manufacturers, this focus, at least among U.S. manufacturers, was succeeded by a new one: total quality management (TQM) and its principles of quality measurement and improvement. As American manufacturers leveled the quality playing field, the late 1980s and 1990s saw a new focus: the notion of lean manufacturing—again, with the way led by the Japanese.[2]

Lean manufacturing and related concepts such as agile manufacturing and constraint management aim to transform mass production into a more flexible and efficient set of processes. The fundamental notion of lean and agile manufacturing is to produce only what is required with minimal finished goods inventory. Constraint management focuses on optimization of flow of materials through bottlenecks. All of these approaches are dependent upon the ability to forecast future demand and produce to that particular forecast.[3]

E-manufacturing enables a step change improvement with respect to lean manufacturing by enabling a production operation that truly builds to what customers declare they want and when they want it. While e-manufacturing certainly does not eliminate the need for forecasting, it does significantly narrow the gap between customer demand levels and production levels by bringing the relationship between the two into realtime: the customer asks for something, and the factory produces it. (The current industry standard for "real-time make to order" is 24 h.) The development of such a "pull" system[3] is enabled by

1. The implementation of the "digital factory"—i.e., the use of integrated information systems such as Manufacturing Execution Software (MES) to coordinate production scheduling, quality, SCADA/HMI systems for data collection and machine and operator interface control, maintenance, and warehouse/inventory management.[4]
2. The connection of the "digital factory" not only to e-commerce and product design applications that face the customer, as described earlier, but to (1) Enterprise Resource Planning (ERP) systems that need information from the factory to understand how to manage the flow of resources within the enterprise to feed that factory and (2) to the external supply chain via Internet-based communications tools that allow suppliers to understand what is required from them and where and when to deliver it. These external supply chain connectivity applications may also contain an auction or procurement exchange component, via which a manufacturer may electronically array suppliers in competition with one another in order to get the best real-time deal.

The implementation of e-manufacturing infrastructure, if executed properly, can generate benefits of 25 to 60 percent in inventory reduction, 30 to 45 percent in cycle time reduction, 17 to 55 percent in WIP reduction, and 35 to 55 percent in paperwork reduction.[4] While each implementation situation has its own specific dynamics, this is certainly the order of magnitude of targeted statistics for which one should strive.

1.3.3 Where It All Comes Together: Information Synthesis and Transparency

While impacting customer value and process efficiency are the two primary axes of e-manufacturing programs, the strategic level where e-manufacturing may ultimately have the most impact is in the realm of information synthesis and understanding. If one properly structures one's e-manufacturing infrastructure, with an emphasis not only on automation and connectivity but also on reportability—i.e., the systematic capture and organization in realtime of key workflow and operational data—then a critical additional benefit can be realized from putting one's customers, factory, and supply chain on an interconnected electronic foundation: the fact that information from across this infrastructure can be retrieved and synthesized electronically—and allow, for the first time, managers to have visibility across the extended manufacturing enterprise. Reports from research houses such as Forrester have repeatedly asserted that poor visibility into the shop floor, into the mind of the customer, and into the state of the supply chain are the biggest problems facing manufacturing management.[5] E-manufacturing's most important consequence may be the lifting of the fog of war that currently clouds even advanced manufacturing operations that don't have the benefit of comprehensive, real-time information measurement and synthesis. One cannot manage what one cannot measure and see, and e-manufacturing—again, if implemented with reportability as well as connectivity in mind—can help enormously with the ability to see across the manufacturing value chain.

1.4 WHAT IS THE FUTURE OF E-MANUFACTURING?

A realistic projection of the future of e-manufacturing would simultaneously take into account the very real power of the innovations embodied in the e-manufacturing paradigm, while also noting the fundamental difficulties of changing any manufacturing culture. While the good news is that approaches and technologies have finally arrived that can help make e-manufacturing a reality, and that companies across multiple industries have made enormous gains through e-manufacturing, it is nevertheless the case that an e-manufacturing implementation remains an exercise in organizational change as much as technological change—and organizational change is never easy. However, there is much to be gained from careful analysis of one's manufacturing enterprise, and applying the frameworks of e-manufacturing to see where value can be produced. It is a concept that may have as much impact on manufacturing value as the notions of mass production, TQM, and lean manufacturing have had, and it is certainly beneficial for every manufacturing engineer to be knowledgeable about its fundamental principles and goals.

REFERENCES

1. Bylinsky, Gene. "The E-Factory Catches On," *Business 2.0*, July 2001.
2. O'Brien, Kevin. "Value-Chain Report: Next-Generation Manufacturing," *Industry Week*, September 10, 2001.
3. Tompkins, James. "E-Manufacturing: Made to Order," *IQ Magazine*, July/August 2001.
4. Software Toolbox, Inc. and Unifi Technology Group. "Building the Infrastructure for e-Manufacturing." 2000.
5. *Manufacturing Deconstructed*, Forrester Research, July 2000.

CHAPTER 2
DESIGN FOR MANUFACTURE AND ASSEMBLY

Peter Dewhurst
University of Rhode Island
Kingston, Rhode Island

2.1 INTRODUCTION

This chapter describes the process of analyzing product designs in order to identify design changes which will improve assembly and manufacturing efficiency. The process consists of three main steps: design for assembly (DFA), selection of materials and processes, and design for individual part manufacture (DFM). The process of applying these three analysis steps is referred to as design for manufacture and assembly (DFMA). Case studies are presented in the chapter to show that DFMA can produce dramatic cost reductions coupled with substantial quality improvements.

2.1.1 Changes in the Product Development Process

A complete change has taken place in the process of product development over the past decade. The seeds of this change were planted in the early 1980s with two separate developments which were to come together over a period of several years. The first of these seeds was a redefinition of the expected outcome of the activity of design for manufacture. The redefinition arose in major part from a National Science Foundation funded research program at the University of Massachusetts (UMASS).[1] This work formed the basis for a Design for Manufacture Research Center at the University of Rhode Island (URI) which has been in existence since 1985. Research at URI over the past decades[2,3,4,5] has changed the process, which has become known as DFMA (Design for Manufacture and Assembly), from tables and lookup charts to interactive computer software used throughout the world.[6] The process of DFMA is now well established in industrial product development.[7,8,9,10]

The second change started in the early 1980s with the recognition by a few U.S. corporations that product design was simply too important to be entrusted to design engineers working in isolation. This led to the establishment of new procedures for product development in which product performance and the required manufacturing procedures for the product are considered together from the earliest concept stages of a new design. This process was gradually adopted in the development of consumer products where the title of *simultaneous engineering* or *concurrent engineering* was usually given to it. The main ingredient of simultaneous or concurrent engineering is the establishment of cross-functional product development teams which encompass the knowledge and expertise necessary to ensure that all the requirements of a new product are addressed. These requirements are usually defined to be that the product should meet customer performance requirements and should be efficient to manufacture in order to meet both cost and quality goals. The core product development team then comprises personnel from marketing, design engineering, and industrial and manufacturing engineering. By the end of the 1980s simultaneous engineering had become synonymous with design for manufacture and assembly and had become widely adopted across U.S. Industry.[11]

Simultaneous engineering is now the accepted method of product development. It has been stated that manufacturers of discrete goods who do not practice simultaneous engineering will be unlikely to survive in today's competitive global markets. This widespread adoption of simultaneous engineering has increased the need, in product development, for formal methods of design for manufacture so that manufacturing efficiency measures can be obtained early in the design process. In this way the manufacturing representatives of the team become empowered in the decision making process and design choices are not based solely on performance comparisons which can be readily quantified with CAE tools. Also, when looking critically at the product development procedure, with a view to changing to simultaneous engineering, many corporations had come to the realization that the bulk of the cost of a new product is locked in place from the earliest concept stage of design. Thus, if manufacturing cost is not assessed in these early stages then it is often too late during detailed design execution to have any major effect on final product cost.

2.1.2 The Traditional Practice of Design for Manufacture

The term, design for manufacture (DFM), is often applied to a process of using rules or guidelines to assist in the design of individual parts for efficient processing. In this form, DFM has been practiced for decades and the rule sets have often been made available to designers through company specific design manuals. An excellent recent example of this approach to DFM provides a compilation of rules for a large number of processes, provided by experts for each of the process methods.[12] Such rule sets are usually accompanied by information on material stock form availability, on the problems of achieving given tolerance and surface finish values, and information on the application of different coatings and surface treatments. Such information is clearly invaluable to designer teams who can make very costly decisions about the design of individual parts if these are made without regard to the capabilities and limitations of the required manufacturing processes. However, if DFM rules are used as the main principles to guide a new design in the direction of manufacturing efficiency then the result will usually be very unsatisfactory. The reason is that in order to achieve individual part manufacturing efficiency, the direction will invariably be one of individual part simplicity. This might take the form of sheet metal parts for which all of the bends can be produced simultaneously in a simple bending tool, or die castings which can be produced without the need for any mechanisms in the die, or powder metal parts which have the minimum number of different levels, and so on. Figure 2.1 is an illustration taken from a DFM Industrial Handbook,[13] in which the manufacture and spot welding of two simple sheet metal parts is recommended instead of the more complex single cross-shaped part. Such advice is invariably bad. The end result of this guidance toward individual part simplicity will often be a product with an unnecessarily large number of individual functional parts, with a corresponding large number of interfaces between parts, and a large number of associated items for spacing, supporting, connecting, and securing. At the assembly level, as opposed to the manufactured part level, the resulting product will often be very far from optimal with respect to total cost or reliability.

2.1.3 The New Approach to Design for Manufacture and Assembly (DFMA)

The alternative approach to part-focused DFM is to concentrate initially on the structure of the product and try to reach team consensus on the design structure which is likely to minimize cost when total parts manufacturing costs, assembly cost, and other cost sources are considered. The other cost sources may include cost of rework of faulty products, and costs associated with manufacturing support such as purchasing, documentation, and inventory. In addition, the likely costs of warranty service

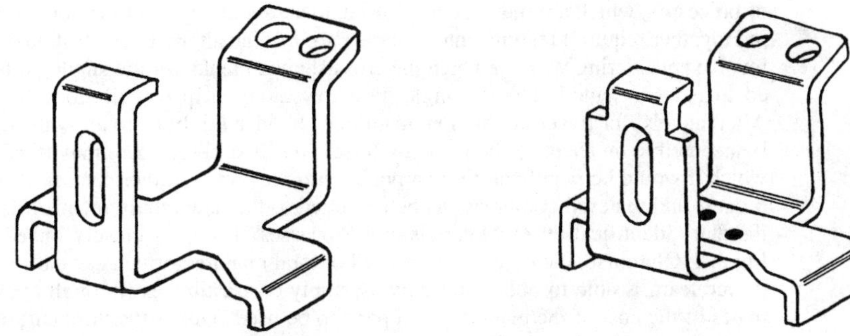

FIGURE 2.1 Single part and two part spot weld design.

and support may be included if procedures are in place to quantify these costs at the early design stage. In this chapter we will focus our discussion on manufacturing and assembly cost reduction using the process of DFMA. We will also address the likely product quality benefits which arise from the application of the DFMA process.

Design for manufacture and assembly uses design for assembly (DFA) as the primary vehicle for decision making as to the structural form of the proposed product. The DFMA method can be represented by the flow chart shown in Fig. 2.2. The three upper blocks in Fig. 2.2 represent the main iteration loop in the process of identifying the optimal product structure. This iteration process stops when the team reaches some consensus as to the best product structure coupled with the wisest choices of processes and associated materials to be used for the manufactured parts. In this iteration process DFA is the starting point and can be viewed as the driving activity. The process ends when the DFA analysis results are seen to represent a robust structure for the product which it is believed can be assembled efficiently. In Fig. 2.2 the activity of DFA is also referred to as *Product Simplification*. This is because DFA naturally guides the design team in the direction of part count reduction. DFA challenges the product development team to reduce the time and cost required to assemble the product. Clearly, a powerful way to achieve this result is to reduce the number of parts which must be put together in the assembly process. This often leads to a review of the capabilities

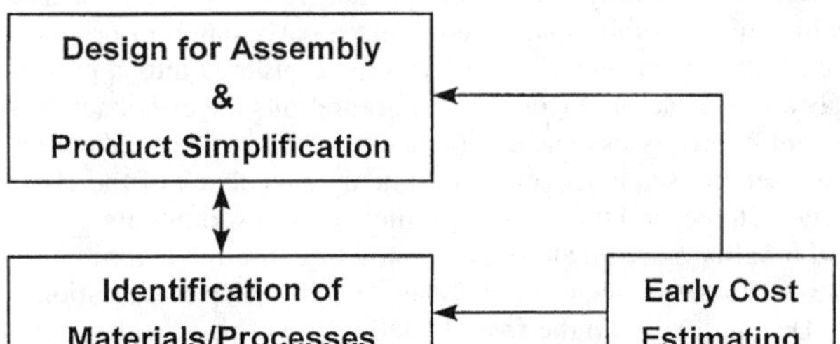

FIGURE 2.2 The design for manufacture and assembly process.

of processes, which are intended to be used, to assess the possibilities for combining parts, or bringing together required features into a single part. For the alternatives illustrated in Fig. 2.1, this would involve considering ways in which the cross-shaped blanks for the single-part design might be nested together, oriented at a 45° angle, in a row along strip or coil stock for stamping operations. Alternatively, for lower production quantities, the idea might be to array them overlapping as closely as possible on sheets to be made by turret press working. There may of course be other features which it would be beneficial to incorporate into the part, and then the search might expand to other process and material combinations better suited to the new requirements. It is through such scenarios that "Identification of Materials and Processes" becomes closely linked to DFA as shown in Fig.2. 2. Of course, the objective is control of total manufacturing cost and so it is important that the design team is able to obtain not only assembly cost estimates (through DFA) but also quick estimates of the cost of the manufactured parts to be used. This is the third part of the iteration process referred to in Fig. 2.2 as "Early Cost Estimating." The need for early cost estimating procedures in product development cannot be overstressed. In many organizations the cost of proposed new parts cannot be obtained until detailed designs are available. This is invariably too late in the design process to consider radical changes in design, particularly if the changes require different process selections. It is for this reason that the writer has been involved, with his colleagues, in developing economic models of processes which could be applied with only the limited amount of information available at the sketch stage of design.[5]

The final stage in the DFMA process, as illustrated in Fig. 2.2, is *component design for efficient processing*. This is equivalent to the historical approach to DFM described above. However, as part of the DFMA process it is intended that it should occur only after the important decisions regarding product structure and process choices have been fully explored. In the case of the single cross-shaped sheet metal part shown in Fig. 2.1, this might involve ensuring that the most economical choice of alloy and gage thickness have been made; adjusting the shape to facilitate closer nesting of the part, on the sheet or along the strip, to reduce scrap; questioning the direction of the bend above the punched slot for easier forming; checking that the tolerances on linear dimensions and angles are within standard press-working capabilities; increasing the profile radii of the bends if necessary so as not to exceed the ductility of the selected material; and so on.

2.2 DESIGN FOR ASSEMBLY

2.2.1 The Role of DFA in Product Simplification

Design for assembly is a systematic analysis procedure for assessing assembly difficulties and for estimating assembly times or costs in the early stages of product design. Assembly difficulties which are identified during a DFA analysis are translated into appropriate time or cost penalties and in this way a single measure (time or cost) represents the efficiency of the proposed design for the required assembly processes. The design team is then able to make adjustments to the design of parts or to the assembly sequence and get immediate feedback of the effect of such changes on assembly efficiency. However, DFA is also a vehicle for questioning the relationship between the parts in a design and for attempting to simplify the structure through combinations of parts or features, through alternative choices of securing methods, or through spatial relationship changes.

Dealing first with the spatial relationship of parts within a proposed product structure, parts often exist in a design solely because of the chosen relative position of other items. For example, separate bracket supports may be required for two items which could be supported on the same bracket if they were moved into closer proximity. Alternatively, and much more commonly, parts may exist just as connectors between items which have been separated arbitrarily in the product structure. Such connectors may be used to transmit signals, electrical power, gasses, fluids, or forces for motions. For example, in the exploded view of a pressure recorder device[5] illustrated in Fig. 2.3, the tube assembly comprising five items (one copper tube, two nuts, and two seals) is required simply because of

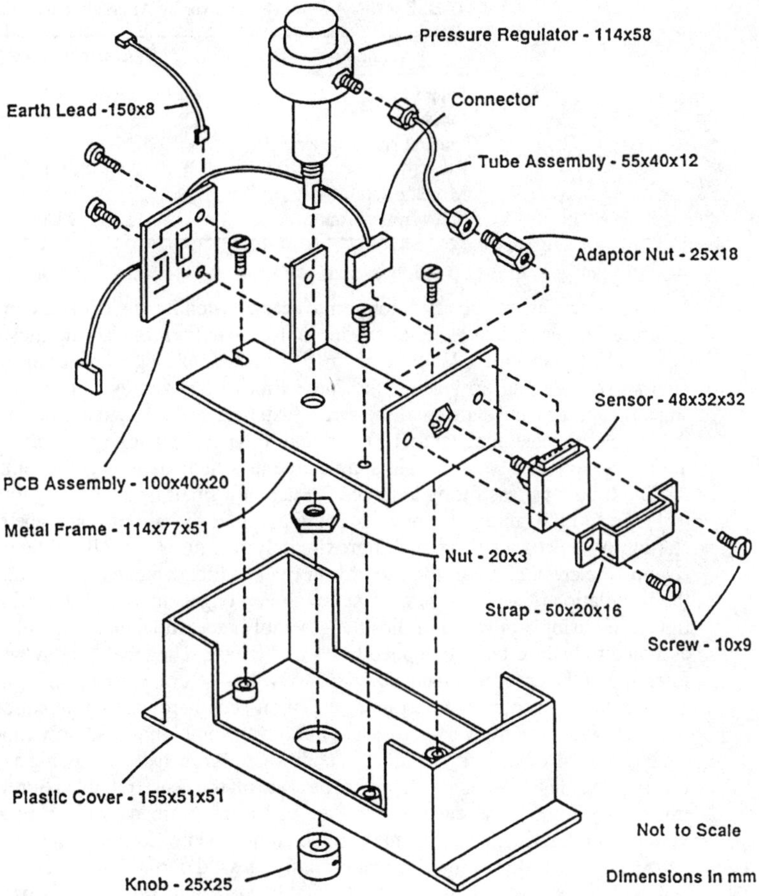

FIGURE 2.3 Existing design of a pressure recorder.

the decision to mount the sensor onto the metal frame rather than to secure it directly to the casting of the pressure regulator. Also, the electrical lead labeled *connector* is needed because the sensor and the printed circuit board (PCB) assembly have been mounted onto opposite sides of the metal frame.

An important role of DFA is to assist in the determination of the most efficient fastening methods for necessary interfaces between separate items in a design. This is an important consideration since separate fasteners are often the most labor intensive group of items when considering mechanical assembly work. For the pressure recorder assembly in Fig. 2.3, for example, approximately 47 percent of the assembly time is spent on the insertion and tightening of separate screws and nuts. To reduce the assembly cost of dealing with separate fasteners, fastening methods which are an integral part of functional items should be considered. For plastic molded parts, well-designed snap fits of various types can often provide reliable high-quality fastening arrangements which are extremely efficient for product assembly.[14,15] Less commonly, sheet metal parts might be made from spring steel to incorporate integral fastening features with savings in assembly cost more than sufficient to offset the increase in material cost. Alternatively, metal parts may be made with projections for riveting or forming of permanent joints, or they may be press-fitted together or may contain threads for screw fastening.

TABLE 2.1 Assembly of Cover by Alternative Methods

Method	Assembly time (s)
Snap fit	4.1
Press fit	7.3
Integral screw fastener	11.5
Rivet (4)	36.1
Machine screw (4)	40.5
Screw/washer/nut (4)	73.8

It is worth paying specific attention to screw fastening since it is the most widely used method of securing in mechanical assemblies, and unfortunately, it is also the most inefficient one. Table 2.1 gives industry-average DFA time estimates[6] for assembling a cover part to a base using a variety of alternative fastening arrangements. These times do not allow for any difficulties in the assembly steps except for hand starting of the screw fasteners as discussed below. The cover is assumed to have the largest dimension equal to 100 mm and to be easy to align and self-locating. The snap-fit operation is assumed to be with rigid snap elements which are engaged simultaneously.

The time for press fitting assumes the use of a small bench press with foot pedal or similar control. The time for integral screw fastening assumes the use of a jar-type cover, the need for careful starting of the cover thread and approximately five turns for full tightening. The estimated time for riveting the cover assumes the use of four rivets which are hand-loaded into a power tool. Finally for the installation of separate screws, screw thread engagement is assumed to be by hand followed by tightening using a power tool. For the assembly method labeled "Machine Screw (4)," 4 screws are assumed to be inserted into tapped holes in the base. The label "Screw/washer/nut (4)," refers to the fastening of the cover to a flange on the base with 4 screws inserted through the flange and one washer and nut fastened to each. Of course, the often-seen hardware combinations of two flatwashers and one lock washer with every screw and nut have even higher assembly times.

It should also be mentioned that in addition to screw fastening being a relatively inefficient securing method, it is also recognized to be one of the least reliable. Screws can be cross-threaded, improperly torqued because of burrs or malformed threads, or can become loosened because of vibrations in service. Experiments conducted by Loctite Corporation show that transverse vibrations across a screw fastened interface can rapidly unwind correctly torqued screws, even with any of the different types of standard lock washers.[16] In mechanical design it is preferable to consider screws or nuts as threaded "unfasteners" and to avoid their use in situations where joints do not need to be separated in service.

Avoiding separate connectors and separate fasteners wherever possible in a design does not ensure that the design has an optimum part count. To force the design team to consider every possibility for reducing the number of separate manufactured parts in an assembly, the BDI DFA method[6] challenges each part according to three simple criteria. These are applied to each part in turn as the DFA analysis steps through the assembly sequence. The criteria are intended to be a catalyst to brain storming of ideas for consolidation or elimination of parts. As each part is considered, the part is allowed to be defined as necessarily separate, if with respect to parts already assembled it:

1. Must be made of different material for some fundamental performance-related reason, or must be isolated from parts for which the same material could be used.
2. Must move in rigid-body fashion, involving rotation or translation not possible with flexure of an integrated part.
3. Must be separate for reasons of assembly; i.e., combination of the part with any others already assembled would make it impossible to position parts in their required locations.

If a part does not satisfy at least one of these three criteria then it is considered to be a candidate for elimination. The design team is then expected to discuss possibilities and document ideas for eliminating the part from the design. In this way the results of a DFA analysis include a list of possibilities for product structure simplification, in addition to the estimates of assembly time and cost. The design team can then edit the DFA structure file to incorporate all or selected ideas from the list, update the part cost estimates, and develop a full cost estimate comparison for the revised design.

2.2.2 The DFA Time-Standard System

The DFA procedure utilizes a database of standard times for handling and insertion based on part size and symmetry, and on the securing method to be used. In addition, appropriate penalties are added for difficulties in handling or inserting items during assembly. The difficulties included in the DFA analysis procedure are those which incur a significant time penalty on the assembly processes. Avoidance of these difficulties thus represents the essence of good detail design for assembly. These can be presented as a set of rules, divided into the two categories of part handling and part insertion as listed below.

1. Handling

- Design parts so that they do not nest or interlock (tangle) together when in bulk.
- Avoid flexible parts which do not maintain their shape under their own weight.
- Avoid sharp edges or points on parts which are to be handled manually.
- Make parts as symmetrical as possible.
- If parts are not symmetrical then ensure that the asymmetry is obvious.

2. Insertion

- If parts are not secured immediately on insertion then ensure that mating surfaces hold the part in the correct position and orientation during subsequent operations.
- Design mating parts with lips, leads, tapers, chamfers, etc., so that alignment is easy.
- Limit forces required for insertion of parts in manual assembly.[17]
- Choose clearances between parts so that jamming cannot occur during insertion. The required clearance for a given part can be established from the part thickness, hole or recess dimensions, and the coefficient of friction between the mating surfaces.[6]
- Select directions of insertion to minimize the need for reorienting the partially built assembly as assembly proceeds.
- For manual assembly ensure that the assembler can see the mating surfaces or edges for ease of location of the parts to be inserted.
- Ensure adequate access for the part, for the assembly worker's hand or fingers, or for the assembly tool if one is required.

The last three insertion rules are often satisfied by designing a product so that all parts are added vertically—so called Z-axis assembly. However, it should be noted that Z-axis assembly is much more important for assembly automation than for manual assembly. With the former, vertical insertions can be performed by simpler, less expensive, and more reliable devices.

Assembly problems of the types listed above are identified during a DFA analysis. At the same time, parts are classified as only for fastening or connecting, or they are assessed, according to the three criteria, for independent existence. The results of the analysis are presented on a DFA worksheet, the rows of which provide information on each assembly step. Figure 2.4 shows the DFA worksheet for the pressure recorder assembly illustrated in Fig. 2.3. It can be seen that 24 steps are involved in final assembly of the pressure recorder with an estimated assembly time of 215 s. This is useful information to have at an early stage of assembly design. However, it is important to be able to interpret the data with respect to the goal of assembly efficiency.

At the detail level, we can review the times for the individual assembly steps and compare them to an ideal benchmark value. From the DFA time-standard database it can be determined that the average time per assembly step for bench assembly of items which present no assembly difficulties (all are easy to grasp, align, and insert with simple motions and small forces) is approximately 3 s. With this value in mind we can identify inefficient assembly steps on a DFA worksheet. For example, in

Item name	Number of items/ operations	Operation time, sec.	Minimum part count
Pressure regulator	1	3.5	1
Metal frame	1	7.4	1
Nut	1	9.1	0
Reorient	1	9.0	—
Sensor	1	8.4	1
Strap	1	8.3	0
Screw	2	19.6	0
Adapter nut	1	12.0	0
Tube assembly	1	7.0	0
Nut tighten	2	15.1	—
PCB assembly	1	12.1	1
Screw	2	19.6	0
Connector	1	6.9	0
Reorient	1	9.0	0
Knob	1	8.4	1
Set screw tighten	1	5.0	—
Plastic cover	1	8.4	0
Reorient	1	9.0	—
Screw	3	36.9	0
Total	24	214.7	5

FIGURE 2.4 DFA worksheet for the pressure recorder.

Fig. 2.4, the first operation of placing the pressure regulator in the work fixture is an efficient one since it takes only 3.5 s. However, the next step of adding the metal frame to the pressure regulator is obviously inefficient, taking more than twice the benchmark time value. The problem with this item is the lack of any alignment features to fix the required angular orientation, and the need to hold down the item during the following operation. At the bottom of the worksheet, it can be seen that the three screws which secure the frame to the cover represent the most inefficient of the assembly steps. The problem here is not only difficulty of alignment of the screws with the nonlocated frame and cover, but also the restricted access of the screws against the deep end walls of the cover and frame. If a separate cover and frame, secured by three screws, is the best design solution, then it would be an easy matter to put locating projections in the bottom of the injection molded cover and move the screw locations adjacent to the side-wall cutouts for easier access. Attention to such details for ease of assembly is inexpensive at the early design phase and DFA can be used to quantify the likely assembly cost benefits. However, as will be discussed later, the benefits of efficient, operator-frustration-free, assembly steps are likely to have the even more important benefit of improvements in product quality.

2.2.3 The DFA Index

The above discussion of the assembly of the frame and the cover was preceded by the qualifier that detail improvements should be made if these separate items represent components of a "best design solution." A measure of the overall quality of the proposed design for assembly is obtained by using the numbers in the right-hand column of Fig. 2.4. These are obtained during DFA analysis by scoring only those items whose function is other than just fastening or connecting and which satisfy one of the three criteria for separate parts listed above. The summation of these values then gives a total which is regarded as the theoretical minimum part count. For the pressure recorder this value is five. The reverse of this value is that 19 of the 24 assembly steps are considered to be candidates for elimination. Ideas for actual elimination of these steps would have been documented during the DFA process.

For example, it can be seen that the plastic cover (Fig. 2.3) was identified as an elimination candidate. This is because, with respect to the metal frame, it does not have to be made of a different material, does not have to move, and a combined cover and frame would not make it impossible to assemble the other necessary items in the product. Of course, this does not mean that a combined cover and frame part must be made from the same material as the metal frame. A more sensible choice in this case would be an engineering resin so that an injection molded structural cover can have features sufficient to support the pressure recorder, the PCB assembly, and the sensor, items supported by the metal frame in the initial design proposal.

The minimum part count can be used to determine a DFA Index[6] which includes not just the assessment of possible part-count redundancy, but also the assembly difficulties in the design being analyzed. This is defined as

$$\text{DFA Index} = \frac{N_m \times t_m}{t_a} \times 100 \tag{2.1}$$

where N_m = theoretical minimum number of parts
t_m = minimum assembly time per part
t_a = estimated total assembly time

For the pressure recorder, this gives

$$\text{DFA Index} = \frac{5 \times 3}{214.7} \times 100 = 7.0$$

Since the ideal design for assembly would have a minimum number of items and no assembly difficulties, the DFA Index for such a design would be 100. The score of 7.0 for the pressure recorder, on a scale of 0 to 100, clearly identifies the need for substantial assembly efficiency improvements. If the required production volumes are sufficient to justify large custom tooling investments then we could envision a design comprising only a structural cover, a pressure regulator, a sensor, a PCB, and a control knob. This would require a custom die cast body on the pressure regulator with an inlet boss and screw thread to match the screw connector on the sensor. The PCB could then connect directly to the sensor, and the structural cover could contain supports and snap features to fasten itself to matching steps or undercuts on the die cast body of the pressure regulator and to secure the PCB. A push-on knob would then complete the assembly. Assuming these five items were easy to assemble, then this would comprise the ideal design for assembly. If it is not possible to justify manufacture of a custom pressure regulator then the design must accommodate the nonmatching screw threads on the purchased pressure regulator and sensor. Also, the only method of securing the regulator to the structural cover would be with a separate nut as in the existing design. These compromises from the "ideal" design lead to a product structure which might be as shown in Fig. 2.5. It can be seen that the structural plastic cover has an extensive rib structure to provide the required stiffness. It also has three internal undercuts, labeled *Board Snaps*, into which the PCB will be snapped during connection to the sensor. A DFA worksheet for this new design is given in Fig. 2.6.

Comparison of this with Fig. 2.4 shows that the estimated assembly time has decreased by 60 percent from the original design and the DFA Index has increased from 7 to 19. Also the number of parts has been reduced dramatically from 18 to 7, and the number of separate assembly operations has reduced from 6 to 3. The likely positive effects of this reduction of assembly operations, in addition to the decrease in assembly cost, will be discussed after considering other case studies.

2.2.4 DFA Case Studies

Two positive additional benefits of a DFA product redesign can be seen in a case study from Texas Instruments.[18] The original design of a gun sight mechanism is shown in Fig. 2.7 and the redesign after DFA analysis is illustrated in Fig. 2.8. In this case the original design was actually in production

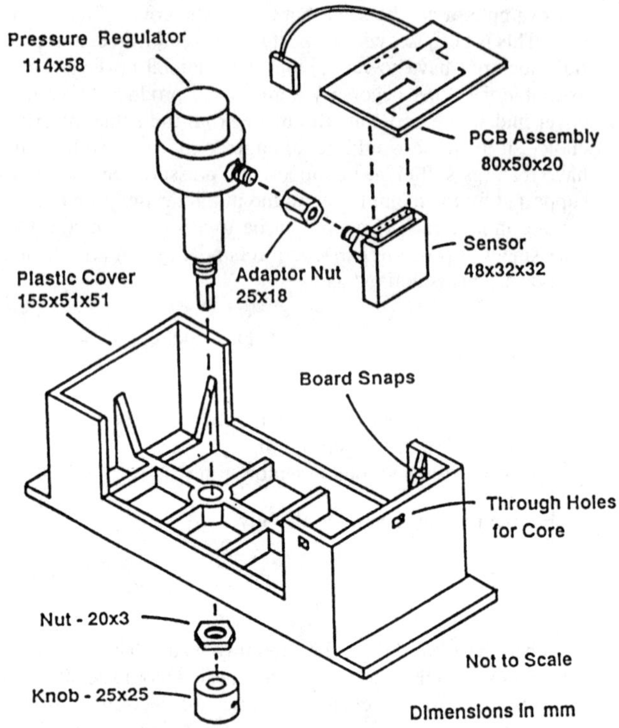

FIGURE 2.5 DFA redesign of the pressure recorder.

Item name	Number of items/ operations	Operation time, sec.	Minimum part count
Pressure regulator	1	3.5	1
Plastic cover	1	7.4	1
Nut	1	9.1	0
Knob	1	8.4	1
Set screw tighten	1	5.0	—
Reorient	1	9.0	—
Apply tape	1	12.0	—
Adapter nut	1	12.0	0
Sensor	1	9.9	1
PCB assembly	1	7.6	1
Total	10	83.9	5

FIGURE 2.6 DFA worksheet for the redesigned pressure recorder.

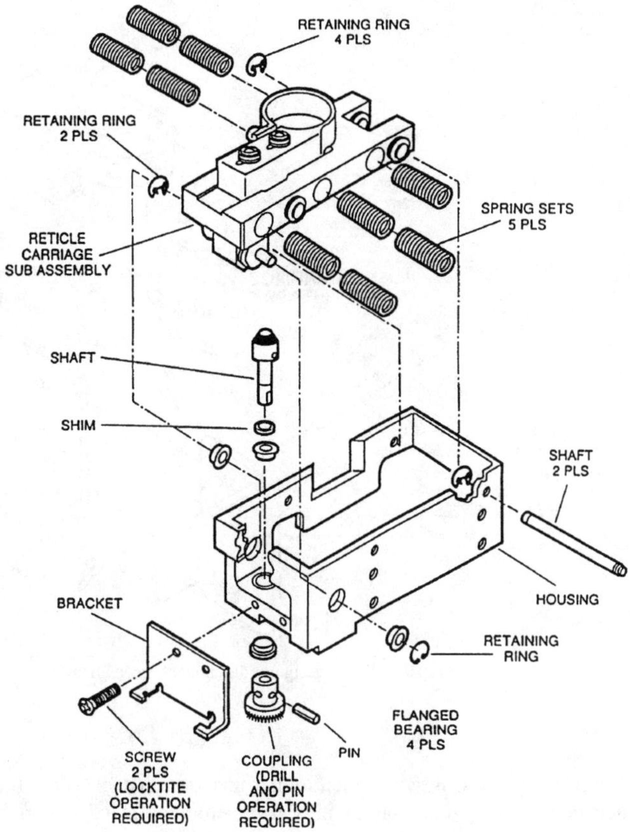

FIGURE 2.7 Original design of a gun sight mechanism.

and yet the advantages of the redesign were so great that manufacture was changed to the new design. The part count reduction is even more dramatic than in the pressure recorder example above. The effect of applying the minimum-parts criteria during analysis of the existing design can be considered for the case of the compression springs. When the first spring to be inserted into the carriage subassembly is considered it satisfies the need for a different material than exists in items already assembled. However, the next eight springs do not have to be made from different material than already present (in the first spring), do not have to move in rigid body fashion, and do not have to be separate for assembly purposes. This may lead the design team to consider a single custom spring formed from spring steel wire or stamped and formed from spring steel sheet, or to consider ways of simply eliminating one or more of the standard compression springs. It can be seen in Fig. 2.8 that the latter approach prevailed with a resulting design containing only two springs.

Table 2.2 shows the benefits of the redesigned gun sight mechanism.[18] The reduction of assembly time by 84.7 percent represents the intended achievement of the DFA analysis. However, it can be seen that a much larger saving has been obtained in part manufacturing time–8.98 h. reduction compared to 1.82 h. saved in assembly time. This result is typical of designs with greatly simplified structures resulting from DFA application. While a few parts may often become individually more expensive, this is usually more than offset by the savings from elimination of other items.

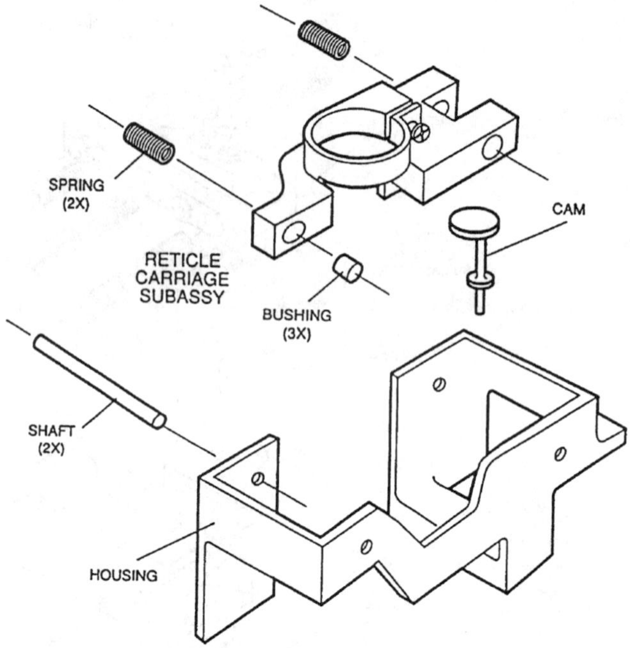

FIGURE 2.8 Redesign of a gun sight mechanism.

At this point it is worth mentioning that the savings from elimination of items in a simplified design go far beyond the savings from elimination of materials and manufacturing processes. Eliminated parts also remove associated costs for purchasing, inventory, quality control, documentation, production control, and scheduling. Savings in these overhead functions can often outweigh the reduction in direct manufacturing and assembly costs.

Table 2.3 shows the benefits of DFA implementation obtained from 94 case studies published in the literature.[19] The numbers in the second column of the table refer to the total number of references to each particular benefit in the 94 cases. Perhaps the most important indirect benefit listed in the table is the reduction of assembly defects. Unfortunately this was measured and reported in only three of the 94 case studies. However one of these cases produced some profound results on the effect of assembly times and efficiency on defect rates and this will be discussed in the next section.

TABLE 2.2 Benefits of DFA Redesign of Gun Sight Mechanism

Attribute	Original design	Redesign	Improvement (%)
Assembly time (h)	2.15	0.33	85
Number of different parts	24	8	67
Total number of parts	47	12	75
Total number of operations	58	13	78
Part manufacturing time (h)	12.63	3.65	71
Weight (lb)	0.48	0.26	46

TABLE 2.3 DFA Results from 94 Published Case Studies

Category	No. of cases	Average reduction (%)
Part count	80	53
Assembly time	49	61
Product cost	21	50
Assembly cost	17	45
Assembly operations	20	53
Separate fasteners	15	67
Labor costs	8	42
Manufacturing cycle	6	58
Weight	6	31
Assembly tools	5	69
Part cost	4	45
Unique parts	4	59
Material cost	4	32
Manufacturing process steps	3	45
No. of suppliers	4	51
Assembly defects	3	68
Cost savings per year	6	$1,283,000

2.3 ASSEMBLY QUALITY

Design for assembly has been used by Motorola Inc. since the mid 1980s to simplify products and reduce assembly costs. In 1991 they reported the results of a DFA redesign of the motor vehicle adapter for their family of two-way professional hand-held radios.[20] Their benchmarking of competitors' electronic products indicated a best-in-class DFA Index value, as given by Eq. (2.1), of 50 percent, and they evaluated many different concepts to reach that goal. The final design had 78 percent fewer parts than their previous vehicle adapter and an 87 percent reduction in assembly time. They also measured the assembly defect rates of the new design in production and compared the results to defect rates for the old design. The result was a reduction of 80 percent in defect rates per part, roughly equivalent to the percentage part count reduction. However, combining the 78 percent reduction in part count with an 80 percent reduction in assembly defects per part gives a startling 95.6 percent reduction in assembly defects per product. Encouraged by this result, the Motorola engineers surveyed a number of products which had been analyzed using DFA and produced a correlation between assembly defects per part and the DFA Index as shown in Fig. 2.9. This clearly shows a strong relationship between assembly quality and the DFA Index values.

This Motorola data was subsequently analyzed independently by other researchers[21] to produce an even more powerful relationship for use in early design evaluation. These researchers postulated that since assembly time can be related to increasing difficulty of assembly operations then the probability of an assembly error may also be a function of assembly operation time. In the study it was reported that 50 combinations of defect rates to assembly characteristics were tested for meaningful correlation. Of these, the variation of average assembly defect rate per operation with average DFA time estimate per operation showed the strongest linear correlation, with correlation coefficient $r = 0.94$. The actual data is shown illustrated in Fig. 2.10. The equation of the regression line is given by

$$D_i = 0.0001(t_i - 3.3) \tag{2.2}$$

where D_i = average probability of assembly defect per operation
t_i = average assembly time per operation

As mentioned earlier, the average assembly time for small parts, which presents no assembly difficulties, is approximately 3 s from the DFA time-standard database. Thus Eq. (2.2) can be interpreted

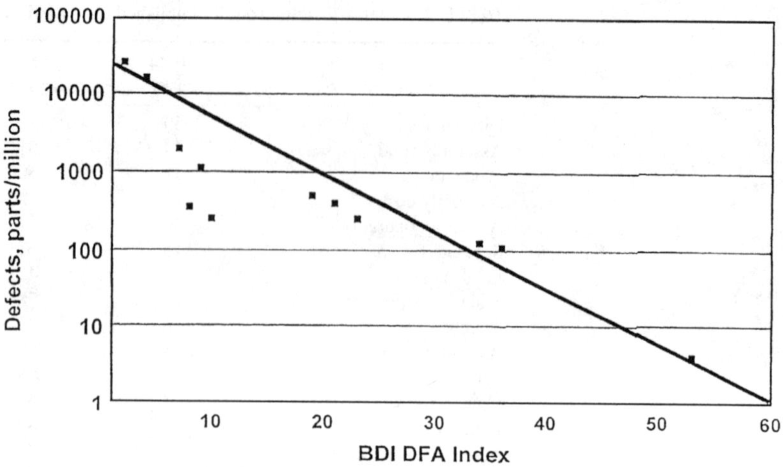

FIGURE 2.9 Relationship between assembly defects and the DFA index.

as an estimated assembly defect rate of 0.0001, or 1 in 10000, for every second of extra time associated with difficulties of assembly.

For a product requiring n assembly operations, the probability of one or more assembly defects is therefore

$$D_a = 1 - (1 - 0.0001(t_i - 3.3))^n \tag{2.3}$$

This relationship can be applied very easily in the early stages of design to compare the possible assembly reject rates of alternative design ideas. This can provide powerful directional guidance for

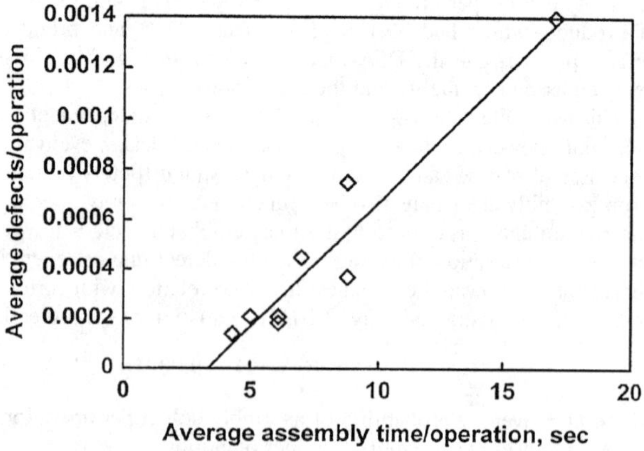

FIGURE 2.10 Relationship between assembly defects and average assembly time per operation.

product quality improvements, since it is becoming widely accepted that faulty assembly steps are more often the reason for production defects than part variability.[22]

For the pressure recorder example, the existing design has an average DFA assembly time per operation of 8.95 s for a total of 34 operations; see Fig. 2.4. Applying Eq. (2.3) then gives an estimated probability of a defective assembly as 0.13, or 13 per 1000. For the redesigned pressure recorder, the number of operations is 10 with an average time of 8.39 s, and the likely number of defective assemblies is predicted to be five per thousand; a likely quality improvement of 60 percent from the original design. This could be improved further by considering detail design improvements to reduce the average operation time from the still-high value of 8.39 s. This might involve adding features to make the plastic cover to make it self-locating when placed on the pressure regulator, using a counter-bored nut for easy alignment, and using an adhesive applicator for thread sealing/locking instead of the tape application. The effect of such changes on the likely defect rate could be tested by making appropriate changes in DFA and reapplying Eq. (2.3).

Finally it should be noted that the above calculations can readily be extended to include assembly defects due to part variability.[21]

2.4 CHOICE OF MATERIALS AND PROCESSES

It has long been recognized that product designers often consider only a very few of the wide range of material and process combinations which are available for product design.[23] Much of the reason for this stems from the personal responsibility for lifetime successful performance of the product, which rests with the design team. This, coupled with an often too-short design time, leads designers to choose the processes with which they are comfortable and familiar. Only if a particular design requirement cannot be satisfied by these familiar processes will the design team be forced to explore the wider range of process possibilities. In this way design is too often the art of the possible, and lower cost manufacturing solutions are overlooked.

A system which would guide a product design team to make wise choices of material and process combinations at the earliest stages of design would be of tremendous value. Unfortunately little progress has been made in this important area. Some work was carried out in the early 1980s in development of a system called MAPS for Material and Process Selection.[24] This was a FORTRAN-based mainframe program for selection of primary shape forming processes based on part size, geometry classification, and performance requirements. The system had two major limitations. First, it did not allow for the stock form availability of materials. For example, an input that the desired part should be circular and prismatic in shape would be likely to produce wire or bar drawing as a possible process. Second, it did not allow for the possibility of a sequence of processes to satisfy the desired requirements. Thus secondary machining operations could satisfy the tolerance or surface finish requirements not possible by casting, or coatings could be used satisfactorily on a material excluded because of corrosion resistance requirements, and so on. Later attempts to overcome these limitations were made with a PC-based system using a commercial relational database.[25] This system called CAMPS (Computer-Aided Material and Process Selection) allowed what-if games to be played with shape and performance requirements with immediate feedback on process and material possibilities. However, linking of the system with automatic process sequence generation was never achieved satisfactorily.

Experience with the CAMPS system has led the writer to the belief that specifying both material performance and geometric shape requirements for such a system is too constraining. Often at the end of this process CAMPS would only suggest a very few, often obvious, candidates. A preferable approach, in the writer's view, if a wider range of alternative possibilities is sought, is to concentrate first on just material performance requirements. This approach can often produce surprising material candidates and the identification of associated processes will lead to geometric shapes which are different than might initially have been chosen. Selection of material can be based on fundamental material properties such as yield stress, fracture toughness, Young's modulus, and so on. For example, assume that wall stiffness is important in the design of a part. The design team would then know that the value of Young's modulus will be important. However, the minimum acceptable value of Young's modulus cannot be determined until the cross-sectional area or wall thickness of the loaded region has been defined. This in turn may depend upon the material cost, acceptable weight, or capabilities of the selected process. One way to proceed with this problem is to utilize derived material properties which more closely match the design requirement. If the part wall is to be subjected to bending movements and low weight is a design requirement, then a defined parameter which represents bending stiffness per weight would be useful for selection purposes. Such defined parameters have been used by Dieter[26] and Ashby[27] for material selection purposes. Reference to the book by Ashby can be made to verify that the defined property for bending stiffness per weight is given by

$$P_1 = \frac{E^{1/3}}{\rho} \tag{2.4}$$

where E = Young's modulus
ρ = material density

Materials with a high value of P_1 can then be investigated further with regard to shape possibilities and comparative costs. However, if minimum material cost is really the more important consideration, then the defined property for bending stiffness per unit cost simply becomes

$$P_2 = \frac{E^{1/3}}{\rho C_m} \tag{2.5}$$

where C_m = material cost per unit weight

Materials with a high value of P_2 could than be compared with respect to possible shapes and weights.

Work by the writer has been concerned with transforming the important derived parameters in mechanical design, such as the two given above, onto common 0 to 100 scales.[28] This allows for easy concept design selection without regard to the units to be used for subsequent design calculations.

Irrespective of the material selection criteria, cost is invariably important and it cannot be obtained without considering both material cost and the effect of material selection on processing cost. For this reason, early cost estimating is the key to design for manufacture. The ability to produce cost estimates must be available from the earliest sketch stages. Unfortunately in many manufacturing organizations reliable cost information is not available until detailed drawings have been submitted to manufacturing or to vendors for formal quotes. This makes it impossible to consider the numerous alternatives which may be necessary to arrive at a low cost solution. As an example of this process the design of the structural cover for the pressure recorder will be considered.

The initial proposal for the cover design is illustrated in Fig. 2.5. The important decisions to be made with regard to the cost of the cover are the choice of the thermoplastic to be used and the detailed design of the features. For the former, it is a relatively easy matter to estimate the volume of material required for alternative polymers and thus find the lowest cost material selections. However, if this is carried out independently of the effect on processing cost then the least cost solution is certainly not assured.

Assume that it is deemed necessary to have the main wall of the cover equivalent in stiffness to the 20-gage (0.91-mm) low carbon steel of the frame in the initial design (Fig. 2.3). From simple bending theory this requires wall thickness values proportional to the cube root of the Young's modulus values of the alternative materials. Using this relationship, a low-cost polymer choice such as high-impact polystyrene would require a main wall thickness of 4.8 mm, while the more expensive engineering thermoplastic choice of glass-reinforced polycarbonate would require a wall thickness of only 3.3 mm. Thus the volume of a polycarbonate cover would be approximately 45 percent of the volume of a high-impact polystyrene one. However since glass-filled polycarbonate is about four times as expensive per unit volume as high-impact polystyrene, based on just material cost, polystyrene would be the obvious choice. However, if we consider the effect of the material choice on the injection molding cycle time then the selection is not so obvious. Mold-filling and mold-opening and -closing times are unlikely to be significantly different for the two material choices. However, the cooling time in the mold is proportional to the square of the part wall thickness and inversely proportional to the material thermal diffusivity.[29,30] Using typical injection, mold, and ejection temperatures, and thermal diffusivity values for the two polymers, the cooling time in the mold for a polypropylene cover is likely to be 41 s compared to only 17 s for a glass-filled polycarbonate cover. It now becomes a question of machine rate to determine if the reduced cycle time will more than compensate for higher material cost. Such trade-offs are common in material selection. Alternative material choices may affect material removal times, molding or forming cycle times, press sizes and therefore press rates, die cost, or die life, and so on. The most economical material choice, just like the most economical process choice can only be determined through the use of process models which can provide accurate early cost assessments.

2.5 DETAIL DESIGN FOR MANUFACTURE

The details of each part design for ease of manufacture can have a substantial effect on the cost of individual items. A study at Rolls-Royce in the UK[31] was carried out on parts which were currently being manufactured by the company, to identify any opportunities for cost reduction which had been missed. Of all of the avoidable costs which were identified in this study, 30 percent of them would have resulted from changes in the detail design of parts. Thus the final DFM checks on part design should not be forgotten, even though, as noted earlier, any detail design changes for easier manufacture should not unduly compromise an efficient structure for the product. This should be determined with the large picture of total manufacturing cost, assembly cost, and product quality in mind.

Taking the structural cover for the pressure recorder as a further example, the decision to include snap-fit features into the cover was justified for the resulting savings in assembly cost. However, the proposed design of these snap-fit features may possibly be improved. The concept of undercuts in the tapered ribs (gussets) as shown in the Fig. 2.5 will require extra moving cores in the mold in order to prevent the part from becoming die locked when it solidifies. With holes through the side walls corresponding to the undercuts, as shown in the figure, the cores can move outwards on slideways. The mold for the proposed design would require three of these slideway-mounted cores—so-called core pulls. The need for these core pulls could be avoided if the undercuts were separated from the underside board supports and if small holes were permissible in the face of the cover for core pins to protrude directly from the mold cavity to the undercut surfaces.[15] This small change could save an estimated 140 h of mold-making time, with a corresponding mold cost reduction of approximately $7,000 at current U.S. mold-making rates.[32] In addition slots could be molded alongside the undercuts to produce cantilever elements. Through appropriate choice of cantilever length and width, this would allow much better control of assembly forces than would be possible with the sidewall distortion of the proposed design.[15]

2.6 CONCLUDING COMMENTS

Effective design for manufacture must include recognition of the fact that assembly is part of the manufacturing process of a product. Even though assembly represents the final steps of manufacture there is great advantage to be gained by considering it first in design assessment. The result of this will be a drive toward simplicity of product structure with wide ranging benefits in every activity from material or parts procurement to reliability and customer satisfaction.

REFERENCES

1. Boothroyd, G., "Design for Economic Manufacture," *Annals of CIRP*, Vol. 28, No.1, 1979.
2. Dewhurst, P., and G. Boothroyd, "Design for Assembly in Action," *Assembly Engineering*, 1987.
3. Boothroyd, G., and P. Dewhurst, "Early Cost Estimating in Product Design," *Journal of Manufacturing Systems*, Vol. 7, No. 3, 1988, p. 183.
4. Boothroyd, G., P. Dewhurst, and W.A. Knight, "Research Program on the Selection of Materials and Processes for Component Parts," *Int. Journal of Advanced Manufacturing Technology*, Vol. 6, 1991.
5. Boothroyd, G., P. Dewhurst, and W.A. Knight, *Product Design for Manufacture and Assembly*, Marcel Dekker, New York, 1994.
6. *Design for Manufacture and Assembly Software*, Boothroyd Dewhurst, Wakefield, RI, 1985–present.
7. Pugh, S., *Total Design*, Addison-Wesley, Reading, MA, 1991.
8. Kobe, G., "DFMA at Cadillac," *Automotive Industries Magazine*, May 1992.
9. Ulrich, K.T., and S.D. Eppinger, *Product Design and Development*, McGraw-Hill, New York, 1995.
10. Ashley, S., "Cutting costs and time with DFMA," *Mechanical Engineering Magazine*, March 1995.
11. Allen, C.W. ed., *Simultaneous Engineering:Integrating Manufacturing and Design*, SME, Dearborn, MI, 1990.
12. Bralla, G.J., *Handbook of Product Design for Manufacturing*, McGraw-Hill, New York, 1986.
13. Pahl, G., and W. Beitz, *Engineering Design; A Systematic Approach*, Springer, London, 1996.
14. *G.E. Plastics Engineering Thermoplastic Design Guide*, Pittsfield, MA, 1997.
15. *Plastic snap-fit joints*, Bayer Corporation, Pittsburgh, PA, 1992.
16. *Loctite Worldwide Design Handbook*, Loctite North America, Rocky Hill, CT, 1996.
17. *Ergonomics Design Guidelines*, Version 3.0, Auburn Engineers, Auburn, AL, 1997.
18. "Designing for Manufacture and Assembly," *Industry Week*, September 4, 1989.
19. *A Compilation of Published Case Studies on the Application of DFMA*, Boothroyd Dewhurst, Wakefield, RI, 1997.
20. Branan, W., "DFA Cuts Assembly Defects by 80%," *Appliance Manufacturer*, November 1991.
21. Barkan, P., and C.M. Hinckley, "The Benefits and Limitations of Structured Design Methodologies," *Manufacturing Review*, Vol. 6, No. 3 (September 1993).
22. Hinckley, C.M., "The Quality Question," *Assembly*, November 1997.
23. Bishop, R., "Huge Gaps in Designers' Knowledge Revealed," *Eureka* (UK), October 1985.
24. Dargie, P.P., K. Parmeshwar, and W.R.D. Wilson, "MAPS-1: Computer-Aided Design System for Preliminary Material and Manufacturing Process Selection," *ASME Transactions*, Vol. 104 (January 1982).
25. Shea, C., and P. Dewhurst, "Computer-Aided Material and Process Selection," *Proc. 4th Int. Conference on DFMA*, Newport, RI, June 1989.
26. Dieter, G., *Engineering Design*, McGraw-Hill, New York, 1983.
27. Ashby, M.F., *Materials Selection in Mechanical Design*, Pergamon Press, Elmsford, NY, 1992.
28. Dewhurst, P., and C.R. Reynolds, "A Novel Procedure for the Selection of Materials in Concept Design," *J. Materials Engineering and Performance*, Vol. 6, No. 3 (June 1997).
29. Ballman, P., and R. Shusman, "Easy way to calculate injection molding set-up time," *Modern Plastics*, 1959.
30. Yu, Chi J., and J.E. Sunderland, "Determination of Ejection Temperature and Cooling Time in Injection Molding," *Polymer Engineering and Science*, Vol. 32, No. 3, 1992.
31. Corbett, J., "Design for Economic Manufacture," *Annals of CIRP*, Vol. 35, No. 1, 1986.

CHAPTER 3
VALUE ENGINEERING AND MANAGEMENT

Joseph Otero, CVS
Pratt & Whitney, UTC
Springfield, Massachusetts

3.1 OVERVIEW

The following topics are covered in this chapter:

Value Engineering (Section 3.2).

Value Management and Its Value Methodology (Section 3.3).

Phases of Value Methodology. (Section 3.4) Each of the phases and their purpose are briefly examined.

Organizing to Manage Value. (Section 3.5) This section shares recommendations on organizing an effective value management office.

Conclusions (Section 3.6).

3.2 VALUE ENGINEERING*

In reporting the death of Silicon Valley cofounder William Hewlett in 2001, the news media were quick to acknowledge the unique corporate culture he and David Packard created in 1939. Their business philosophy, called the "HP Way," is a people-oriented approach with decentralized decision making and management by objective. The tenets of the Hewlett-Packard philosophy are respect for the individual, contribution to customer and community, integrity, teamwork, and innovation. To a value engineer these are familiar characteristics embodied in the value methodology. They represent the way value practitioners view their work and help explain why the value process for solving problems is so successful.

Value engineering (VE) is often misunderstood. Even though VE enjoys a half-century of history as a successful technique for improving the value of projects, products, and processes, there remains a vague understanding in the engineering community of what VE is and what it can accomplish. The history of value improvement work dates back to the 1940s when Lawrence Miles, working for

* "Understanding Value Engineering" by Roger B. Sperling, CVS. Copyright 2001 by IIE Solutions. Reprinted with minor editing by permission of IIE Solutions. Minor edits have been incorporated. None affect the meaning of the original article and are clearly reflected by brackets ([]) to enclose additions and by triple periods (...) to mark omissions.

General Electric, developed value analysis. Miles' concept evolved out of the need to redesign GE's products because of shortages of critical raw materials during World War II. The U.S. military then named the process *value engineering*, embracing it in their quest to eliminate unnecessary costs of defense systems. Expanding use of VE in the public and private sectors followed in the United States and abroad. Mandated VE studies now save billions of dollars of public funds and corporate VE programs assure the competitive edge in the private sector.

The search for better value is based on the VE job plan . . . an organized, step-by-step problem-solving methodology. This systematic process, beginning with the . . . information [phase], is the same regardless of the item under study. It is carefully designed to analyze functions of a project, product, or process before moving to the [idea generation] and evaluative phases. The final . . . phases, [development, reporting, and implementation] . . . complete the protocol. All phases must be completed in an orderly way to achieve optimum results.

3.2.1 Value Has a Definition

The definition of value,

$$\text{Value} = \frac{\text{Function}}{\text{Cost}},$$

is a key to understanding value engineering. Improving value means enhancing function, reducing cost, or both. Therefore, it's necessary to consider the function of an item—what its purpose is—before value improvements are suggested.

For example, when studying a mousetrap for cost reduction, suggestions for making minor modifications to the existing mousetrap (e.g., use a plastic base) can be made. However, after value analyzing the function of the mousetrap—to kill mice—alternative ways to kill mice (e.g., use poison) can be suggested. Clearly, these are two different creative thinking paths: The first leads to small changes while the latter has the potential for large changes. The unique approach of function analysis is the essential precursor to the search for creative alternatives. Understanding what work an item is intended to do must precede the search for better value alternatives.

This is what makes VE unique and gives it the power to achieve surprising value improvements.

Failure to understand the functional approach of VE leads to a false conclusion that VE is merely a cost-cutting exercise. Unfortunately, many studies are conducted in the name of value engineering in which the function analysis phase of the VE job plan is omitted. This overenthusiastic leap from the information phase to the [idea generation] . . . phase (skipping the function analysis phase) defeats the very goal of value studies, which is to improve value, not just cost. Table 3.1 ("Wastewater diversion facility") illustrates this point. In the information phase of this study, the team received instructions from the design manager not to change the project's location. But by moving the facility to a new location, it was possible for the team to more than double the capacity of the system for the same cost and within the same footprint. Management was pleased and surprised that VE worked so well because expectations for this initial VE study were low—only minor cost-cutting ideas had been anticipated.

3.2.2 A Team Process

Value studies rely on the synergy of teams to solve a common problem. Typically, mixed-discipline teams, with some members having prior knowledge of the item under study and some without, are used in value studies. The individual strengths of every team member are melded into a dynamic team that achieves sometimes startling results. Trained and certified team facilitators work with diverse teams and stimulate synergist behavior that allows them to find solutions that may have been overlooked.

The VE process ensures that the ideas of each team member are considered objectively. When ideas are suggested for improving value, they are faithfully recorded without prejudice for later evaluation. This suspension of evaluation is what allows value teams to generate many new ideas; not all

TABLE 3.1 Example VE Study No. 1: Wastewater Diversion Facility

Description of Project: Tankage and controls to allow retention and treatment of potentially hazardous wastewater prior to discharging it to city wastewater treatment plant.

VE Study Design: An in-house team facilitator worked with an in-house team of engineers, an architect, and one technical consultant.

Original Concept: Horizontal tanks, 50,000 gallons capacity, below ground level in a pit with piping and instrumentation.

VE Alternative Concept: Vertical tanks, 120,000 gallons capacity, mounted at ground level with piping and instrumentation.

Advantages: More than double the capacity for the same project cost without increasing the "footprint" of the facility.

Disadvantages: No significant cost savings (but capacity increased); concern about odors at neighboring buildings.

Results: The VE alternative concept was adopted (increase of capacity welcomed); objections of close "neighbors" overcome by assurances odors would be controlled.

of them are of equal value but they are honored equally since a lesser idea can lead to a greater idea. The relative values of all ideas are determined in the evaluative phase by a careful judgment process in which each idea is given a fair evaluation against specific stakeholder criteria.

Outside value consultants often are needed to augment in-house resources. They can provide technical experts to sit on value teams and trained team facilitators. Where proprietary designs are being studied, in-house staff is used exclusively. However, consultants are often needed to train the people who will be invited to form value teams and then facilitate them.

Table 3.2 ("Design process for transportation systems") illustrates how two consultants, one a team member and the other the facilitator, helped a value team of in-house design professionals achieve significant improvements to their own familiar design process. The state highway design procedure under review was lengthy and complex. The consultant had worked on contracts for the agency and had a view from outside the organization. He was able to make suggestions for improvement that were developed into viable alternatives to shorten the processing of designs. The value

TABLE 3.2 Example VE Study No. 2: Design Process for Transportation Systems

Description of Process: State transportation departments' design delivery system was complex and lengthy, as executed in 12 regional offices throughout the state.

VE Study Design: A team of in-house project engineers and project managers—plus one consultant design manager—was led by a consultant team facilitator in one regional office.

Original Concept: The bottlenecks in the process for developing a design were not clearly understood; no remedies were apparent to reduce delays in putting projects out to bid.

VE Alternative Concept: The VE team applied the VE tools to the design process to identify the critical functions and problem areas; several dozen alternatives were developed to give specific remedies for shortening project delivery time.

Advantages: Bottlenecks and redundancies were identified and specific solutions were developed in detail, involving several different departments.

Disadvantages: Acceptance of the VE alternatives required extensive briefings to obtain the "buy-in" from the many departments involved.

Results: Many of the VE alternatives were adopted in the regional office sponsoring the VE study and some were adopted statewide, trimming project delivery time by one month, improving accountability, and leveling the playing field with the private sector.

TABLE 3.3 Example VE Study No. 3: Manufacturing of Electronic Circuit Boards

Description of Product: Printed circuit board for a temperature controller in a commercial appliance was losing market share to new domestic and foreign competitors.

VE Study Design: A team of in-house engineers and procurement officers was led by two consultant team facilitators (no other outside assistance on proprietary design).

Original Concept: Printed circuit board with eleven components was assembled in eight manufacturing steps; "move" and "wait" times were excessive.

VE Alternative Concept: Analysis of the component costs led to alternatives for procurement, and a study of the manufacturing processes revised the layout of the assembly line.

Advantages: Component prices were reduced to a small degree and the assembly time was reduced to a large degree.

Disadvantages: Plant layout had to be changed to achieve estimated savings in "move" and "wait" times.

Results: Cost of components was reduced and cost of manufacture was reduced to reach the goal for a return to profitable, competitive pricing.

methods allowed in-house staff to accept ideas from the private sector to enhance its process. Many schedule-shortening changes were adopted.

The most frequently asked question about VE is: What is the best time to conduct a value improvement study? The answer—anytime. However, the trend is to do VE sooner rather than later. The use of VE after the original design concept is nearly ready for release is prone to develop antagonisms between the stakeholders and the VE team. It is preferable to use VE sooner in the development process, allowing the design team and the value team to work in concert to explore—in functional terms—what the project, product, or process is intended to serve and generate a wide range of alternative concepts. VE is an excellent way to sharpen the scope of work on ill-defined projects.

Table 3.3 ("Manufacturing electronic circuit boards") illustrates how the familiar assembly line operations for an electronic circuit board can be analyzed with VE to reduce component costs and manufacturing time. This study was not conducted at the early development stage of the circuit board but after it had been in production for some time. The purpose of the value study was to find value improvements to help regain market share for a highly competitive commercial appliance. Redesign of the assembly line to reduce move and wait times resulted from this study.

3.2.3 Success

Successful application of VE requires a commitment from top management and a dedicated staff to manage the studies. Without willingness by managers—both in the private and public sectors—to support the training of staff in value methods and to nurture the administration of an organized value program, the benefits of VE cannot be realized.

A full-time VE coordinator is the individual who organizes VE study teams and monitors their performance. The coordinator reports to the consultant on the performance of the team and summarizes the results of each study to the project manager.

Annual summaries of implemented VE study results are elevated to management, and VE successes are publicized to the organization.

Written descriptions of the VE process are inadequate to convey the energy and excitement that is inherent in value teams as they work to improve the value of projects, products, and processes. One needs to be part of a VE team to experience the value methodology and to become infected with the value ethic.

The value methodology fully embodies the five tenets of the HP Way:

- *Respect.* VE honors the ideas of its team members.
- *Contribution.* VE results in improvements to the benefit of owners and society.
- *Integrity.* VE maintains the integrity of the owner's projects.
- *Teamwork.* VE relies on synergistic teams to produce surprising results.
- *Innovation.* VE develops alternatives from carefully evaluated creative ideas.

3.2.4 A Half-Century of Value Engineering

1940s. Lawrence D. Miles, an electrical engineer, developed value analysis (VA) as a tool for replacing scarce materials during World War II in General Electric's manufactured products. New materials resulted in lower cost and improved performance, giving birth to the discipline of VA.

1950s. Value analysis—the study of functions of an item and its associated costs—was codified as a creative team process to stimulate the elimination of unnecessary costs. Its use expanded to the U.S. Navy's Bureau of Ships to analyze designs before construction, and it became known as value engineering (VE). The Society of American Value Engineers (SAVE) was founded in 1958.

1960s. The U.S. Department of Defense applied VE to military systems; VE expanded to military construction projects through the Navy Facilities Engineering Command, the Army Corps of Engineers, and commercial manufacturing in the United States. VE was embraced internationally in Japan, Australia, Great Britain, Italy, and Canada.

1970s. The Environmental Protection Agency began requiring VE for wastewater facilities valued at more than $10 million. Public building services began requiring it for construction management. The U.S. Department of Transportation encouraged voluntary use of VE by state departments of transportation. Private-sector use expanded to communications, manufacturing, automobiles, chemicals, building products, shipping, and design and construction projects.

1980s. VA and VE applications grew nationally and internationally to include hardware and software; systems and procedures; buildings; highways; infrastructure; water and wastewater facilities; and commercial, government, and military facilities. There was increased use of VE early in the life of projects and products, which refined scopes of work and budgets.

1990s. The U.S. Office of Management and Budget required government-wide use of VA and VE on large, federally funded projects. The National Highway System Designation Act required VE on transportation projects valued at more than $25 million. SAVE International, "The Value Society," adopted its new name in 1998, with members in 35 countries.

2000s. The future of VE is bright as practitioners and applications expand worldwide.

3.3 VALUE MANAGEMENT AND ITS VALUE METHODOLOGY

What It Is. Value management is centered around a process, called the value methodology (sometimes referred to as the value engineering job plan or the value analysis job plan), that examines the functions of goods and services to deliver essential functions in the most profitable manner. Value management (hereafter referred to occasionally as VM) is what its name implies—managing the value related to projects in a company or agency.

Where to Use It. In manufacturing, the value methodology can be employed to improve products and processes, to design new manufacturing facilities or improve existing ones, and to design or improve business processes that support manufacturing. Furthermore, use of the value methodology extends beyond the manufacturing arena and is employed in construction and service industries. Indeed value management "can be applied wherever cost and/or performance improvement is desired. That improvement can be measured in terms of monetary aspects and/or other critical factors such as productivity, quality, time, energy, environmental impact, and durability. VM can beneficially be applied to virtually all areas of human endeavor."*

When to Use It. The best value is achieved by employing the value methodology early in a project, in the early planning or design stages, before capital equipment and tooling are locked in, and while there is flexibility to implement the choices of highest value.

A solid value management program employs parts of the value methodology in determining customer needs and expectations. It then generates ideas to address those needs and wants, and puts the best ideas into packages of implementation plans.

3.4 PHASES OF VALUE METHODOLOGY

Value management employs the value methodology, which consists of the following sequential steps, called phases:

- Information phase
- Function analysis phase
- Idea generation phase
- Evaluation phase
- Development phase
- Reporting phase
- Implementation phase

3.4.1 Value Study

Employing the first six steps or phases is a creative problem-solving effort called a *value study*. A value study is typically done by a team of several people representing all of the stakeholders—people or organizations that can affect the outcome or are impacted by it—in a project, and is led by a facilitator trained in the value methodology. More about the team makeup is discussed in Section 3.5. The purpose of each phase is laid out in the following sections.

3.4.2 Information Phase

The purpose of the information phase is to frame and focus a value study. The information phase creates a framework for a value study team to work within for the remainder of the study. To this end the information phase is designed to clearly define the problem that the study team will strive to resolve in the remainder of the study, to identify issues that surround the problem, and to gather the information necessary to effectively execute the study. The following actions are accomplished in the information phase:

* Value Methodology Standard. SAVE International, 1998. www.value-eng.org.

Identify Team Members. These are the people that represent those who affect or are affected by the problem that will be addressed by the team. Details are covered in Section 3.5.

Secure Management Approval and Support. Get management okay before launching the Information Phase, then confirm their buy-in and support at the end of the information phase and before starting the function analysis phase.

Gather Data. Gather data necessary to substantiate the problem, measure current state, and yardstick potential solutions.

Identify the Problem. Several different exercises may be employed to accomplish this. The facilitator will employ the exercise(s) best suited to the situation.

Set Goals and/or Objectives. Set goals and/or objectives that if reached solve the problem.

Identify Potential Barriers. Identify potential barriers to achieving the goals and objectives.

Identify Metrics. Identify the metrics by which to evaluate current state and measure the value of solutions proposed to solve the problem. The metrics are determined from the customer's perspective. Indeed, when possible, the customer is involved directly in determining measures of merit. What does this mean from a practical standpoint for manufacturing? It means that manufacturers make decisions that are based on what its customers deem to be of value. Some manufacturing engineers think their management is the customer. This is not correct. The purchaser of their product or the downstream user of the product is the customer. For example, if a manufacturer makes shovels, the person who buys the shovel to dig holes is the customer for whom value decisions must be made (not the vice president, or the store that sells the shovels).

3.4.3 Function Analysis Phase

This phase separates the value methodology from all other methodologies and problem solving vehicles. It regularly causes teams to do what is rarely achieved by other well-established means: it engages both analytical and creative processes simultaneously. Think of it this way: refining the scope of the problem is convergent thinking; expanding the number of possible answers is divergent thinking. Function analysis does both.

So, what is function analysis? It is breaking a process or product into discrete functions that represent what is happening, why it is happening, and how it all happens. For example, a match has a basic function to "generate heat." It has a dependent function that answers how heat is generated—"ignite fuel." That function is accomplished by another "how" function—"create friction." See Fig. 3.1.

Notice that each function is represented as two words when possible—an active verb and a measurable noun. What is the power of this process? A few strengths begin to exhibit themselves upon examination of this simple set of functions. First, technical and nontechnical people alike readily understand each simple set of words; this common understanding helps the team members—who represent a multitude of technical and nontechnical disciplines—to build a common frame of reference for communicating within the team. Furthermore, the sets of words promote viewing the product—a match in this example—from different perspectives. We all know that friction ignites the fuel. Usually rubbing the head of the match over a rough surface generates friction. Rubbing a rotating wheel against a stationary flint can also generate friction. This can generate sparks that ignite the fuel, which in the case of the match is wood or paper. Notice that we started with a traditional explanation of a match and naturally progressed to a nontraditional way to create friction. In others words we asked ourselves, "how else can we ignite fuel?" Now suppose we ask ourselves "what different fuels might we use instead of wood or paper?" Could we use butane? If so, and we combine the butane fuel with our idea of a rotating wheel and flint to create friction, we now have a cigarette lighter instead of a match. Notice that the set of functions is still valid; yet, we have moved beyond a match. Suppose we now ask, "are there other ways to ignite fuel besides creating friction?" The answer is yes. An electric spark can ignite fuel. This can be achieved by a piezoelectric device or a small capacitor switch and battery. So we can

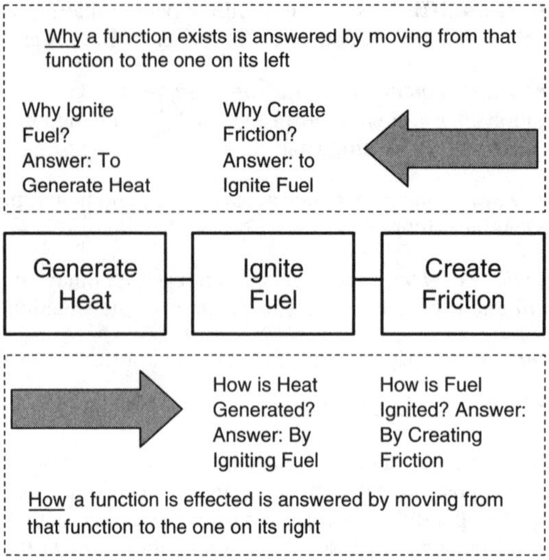

FIGURE 3.1 Three related functions for a wooden or paper match. Functions are related in both left and right directions, with the interpretation as illustrated.

invent an electronic cigarette lighter. But we have still limited ourselves to a flame. Now suppose we ask if there are other ways to generate heat besides igniting fuel. Again, the answer is yes. For example, we can use electric resistance—pass current through a coil—to generate heat; now we have an automobile cigarette lighter. (And who says all of these devices have as their purpose to light cigarettes? We can think of many uses, such as starting a campfire, or igniting fuel in an enclosed cylinder—like in an automobile or truck engine.) This simple set of functions illustrates that function analysis helps a team to change the way it looks at a product or process and to think of different ways to execute the functions, or even replace existing functions. Hence, function analysis helps create divergent thinking.

Remember that it also aids in convergent thinking. This is accomplished by analyzing the functions relative to a metric. For example, we could determine the cost (a metric) of the function "create friction." Then we compare the worth of that function against the worth of other functions of the product. In a value study we typically select the functions of least worth as most in need of a better value solution. This is a different way to select items to brainstorm for potential improvements than the traditional approach of creating a Pareto of the costs of the components or features. Incidentally, even this convergent approach helps to create divergent thinking.

Function Analysis Models. Functions clustered in related sets are called function models. Figure 3.2 shows an example of a simplified technical function model of an overhead projector. This kind of model is called a FAST (function analysis system technique) model.

Notice that functions on the left are more abstract than those on the right. These abstract functions help to create divergent thinking. They imply the question, "what ways can functions on the left be accomplished other than by performing the functions on the right?"

Notice that a third direction for interpreting the relationships of the functions is represented in addition to HOW and WHY; this is the WHEN direction. Functions linked by vertical lines have logical relationships, but are not linked in both How and Why directions. For example Fig. 3.2 shows that when the overhead project performs the function "emit light," heat is dissipated. This function, like many "when" functions, typically is one of the highest cost functions of an overhead projector, since it is usually accomplished with the use of a fan. In fact a Pareto of the cost of components would

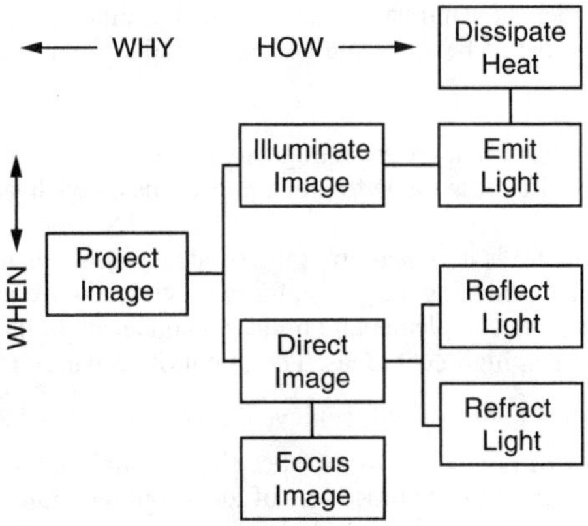

FIGURE 3.2 Simplified FAST (function analysis system technique) model of an overhead projector.

highlight the cost of a fan; and traditional cost cutting would look for ways to make the fan cheaper. Function analysis implies a more abstract question: How may heat be dissipated at a lower cost? At least one company now sells an overhead projector that has no fan, thus eliminating that expense; to dissipate heat, the bulb is mounted where air can circulate freely from bottom to top, thus encouraging natural convection to cool the bulb and dissipate heat. That style projector is also quieter than conventional designs, since without a fan it makes no noise. This example is typical of function analysis results: high value solutions emerge when we challenge the traditional ways functions are performed, especially when we think abstractly about what a function does.

This chapter touches only lightly on function analysis and the building and interpreting of FAST models. Much more information on the process of analyzing functions can be found in reference literature listed at the end of this chapter.

3.4.4 Idea Generation Phase

The purpose of the idea generation phase is to develop a large quantity of ideas for performing each function selected for study. This is a creative process. Effort is made to suspend judgment and curtail discussion. The team brainstorms and/or uses other idea generation techniques.

3.4.5 Evaluation Phase

Many ideas, typically hundreds, are generated in the idea generation phase. This creates a challenge: how does a team reduce this large set of ideas to the ones that have the best potential of satisfying the study objectives, and do it within the time constraints of a program? A properly executed evaluation phase solves this challenge.

Sorting and Sifting Through Ideas. To best utilize the time of a study team, several idea-sorting filters are used. The first filter is very coarse and eliminates most of the poor ideas. Succeeding filters simultaneously eliminate additional ideas that don't measure up and begin to refine remaining ideas. This process, like the other phases, is best facilitated by an expert—at least until all team members have a working command of the process.

3.4.6 Development Phase

A fuzzy line exists between the evaluation phase and the development phase, since filtering out of ideas that do not lead to the objectives of the study continues in this phase and refining of ideas begins in the evaluation phase. There are, however, distinct activities that occur in the development phase:

Expand Surviving Ideas. Remaining ideas, the ones that have survived the filtering process so far, are expanded, usually into one-page summaries of each surviving idea. These summaries typically include a brief description of the idea in contrast to the current baseline, a list of assumptions required for the idea to work, a short list of benefits and concerns, and a rough estimate of costs to implement the idea.

Assess Surviving Ideas. The whole team in a group setting usually assesses the summaries, though individuals may write the summaries. Effort is made to reach consensus on each summary.

Group Ideas into Scenarios. A scenario is a logical grouping of ideas whose intent is to resolve, or contribute to resolving, a problem and to achieving the goals of a value study. For example, one scenario might include only low-risk, low-cost, short-implementation ideas, another might be of moderate-risk ideas, and another might be of high-risk, high-cost ideas. The intent of scenarios is twofold:

- One reason for scenarios is to show a path that will achieve the goals of the value study.
- Another reason is to show what will not work. Remember that example of low-cost, low-risk, short-implementation ideas? Management likes those sets of ideas, but they rarely reach the goals by themselves. Showing this set and its impacts—or lack thereof—to management convinces them that they need to invest a bit more money and let the team take more risk.

Evaluate Scenarios. After ideas are clustered into scenarios, they are evaluated for their net impact—both positive and negative—on the value of the product or process under study.

Determine Nonrecurring Expenses For manufacturing, non-recurring expenses are usually important. The non-recurring expense of each scenario is estimated.

Build a Business Case for Best Scenario(s) The best looking scenarios can sell themselves on a solid business case.

Validate Scenarios. Some teams take a couple of weeks to validate the numbers they arrive at during the intense sessions of a value study. This process increases their credibility. They need to do a few things:

- Make certain that their assumptions are correct.
- Fill in missing data gaps.
- Alter assessments as needed.
- Alter conclusions as needed.

3.4.7 Reporting Phase

The value study team reports to their management. The team gives a brief description of the objectives of the study, summarizes the results of the study and asks for the management commitment, funds, and manpower to implement the recommendations of the study.

3.4.8 Implementation Phase

The team employs the necessary resources to implement the recommendations of the value study. They deliver progress reports and report successes to their management and to the value management group.

3.5 ORGANIZING TO MANAGE VALUE

A value management program that successfully carries out value studies and sees them through to implementation is not a set of accidents. It is a deliberate and carefully managed effort. This section covers two topics. The first topic enumerates several keys to success. The second topic outlines how to set up a value management program.

3.5.1 Keys to Success

Key elements of an effective value management program include the following:

An Executive Champion. The best champion is the top executive of the company. If this is not possible, the best alternative is the executive responsible for the program for which value studies are targeted. The need for an executive champion cannot be overemphasized. Many well conceived value management programs have failed or fallen way below their potential due lack of executive sponsorship.

Value Study Teams. Carefully chosen value study teams are key to the successful outcome of a study. Missing key contributors or stakeholders can have disastrous effects not only on the quality of the study itself, but also on the ability to implement the recommendations of the team. To this end, a few guidelines are useful.

- *Represent all disciplines* that can impact the problem the team will focus on. Do not leave out key players unless the reason is compelling.
- *Select individuals who have decision-making authority* or are given it by their management for the purpose of the value study.
- *Select individuals of approximately equal authority.* Big mismatches in job classification nearly always adversely impact the ability of team members to work closely together.
- *Pick individuals who are open-minded and positive.* A "wild card" member fits this bill. The "wild card" is a technical expert without a direct stake in the outcome of the value study. Other members of the team should also have a reputation for being open-minded.
- *Limit the size of the team.* Teams of five to eight members are effective. Smaller teams tend to lack broad experience. Larger teams tend to be inhibited. If larger teams must be used, there are ways to address the team size—such as splitting it into small groups for many activities during the value study.
- *Okay the team makeup with the facilitator.* Let him or her know of any unusual challenges so that the facilitator can offer guidance and prepare appropriately.

An Implementation Team. Ideally, the same people will be in the value study team and implementation team. This, however, rarely happens. As much continuity as possible is recommended. Integrated product teams (IPT's) are good cores for value studies and for implementation teams.

A Training Program. The better the training received by members of value study teams, the better they will perform in a value study. As a minimum, training of team members and their managers consists of a two hour overview of the methodology and value management process. In-house facilitators are a great boon to a value management program and will require training, coaching, and mentoring.

Good Facilitators. Facilitators have a tremendous impact on the quality of the outcome of a value study. Their impact may be difficult to measure since a good facilitator strives to help the team believe that they are responsible for the success of a value study. Good facilitators have a thorough command of the value methodology, have excellent communication skills, and employ outstanding interpersonal and team dynamics skills.

SAVE International, "the value society," has a list of facilitation consultants on their website: www.value-eng.org

SAVE International has members in 35 countries. It delivers training via conferences, workshops, literature, and networking. It also has a professional certification program culminating in the highest level of professional certification in the value methodology: certified value specialist (CVS). Select a consultant with manufacturing and management background to 1) help facilitate, 2) train team members and managers in the value methodology, and 3) advise in setting up an in-house value management program.

A Reporting Process. Results of value studies must be measured and reported regularly. Data that show success drive effective decisions.

3.5.2 Setting up a Value Management Program

Employ the keys to success. See the preceding section.

Select a Consultant/Advisor. Follow the guidelines found in the paragraphs headed "Good Facilitator."

Select the First Training Workshop. It should be a challenge that has a high chance of success. Choose a process or product that needs to have the cost lowered and whose outcome can be readily measured.

Establish an In-house Staff. The staff needs a manager to sell the methodology and coordinate value studies, an engineer or financial expert to track implementation effectiveness, and eventually will also include one or more in-house facilitators. If working on government contracts, hire someone—either full-time or as a consultant—who has solid experience with VECP's (value engineering change proposals) in manufacturing (experience in construction VECP's is of little or no use in manufacturing).

Report to the Chief Executive. The higher the value management organization is placed in the company, the less likely it is that it will get pushed aside. Because the value methodology can be utilized throughout the company, it makes sense to place it in the organization where the entire company can use it.

Continue with a Few Cost Reduction Value Studies. Value Management is most effective in new processes and new products. But contrasting the changes in cost it brings about by being able to compare it with known costs helps to sell it internally. Do not under any circumstances call yourselves cost cutters, nor permit others to give you this title. Why? For one thing, cost cutters are among the first ones cut when the budget ax falls.

Evolve Toward Using the Value Methodology to Design New Processes and New Products. This is where value management can shine the brightest.

Monitor Progress. The effectiveness of value studies must be measured. Measured activities are taken seriously and energy is applied to them. Measured results are also the best marketing tool.

Track Implementation. The value studies will generate hundreds, perhaps thousands of ideas with great potential, but the ideas are of no value unless they are implemented.

Report Implementation. Report to your executive champion and your chain of command if they are not one and the same person. See that the reports are widely distributed.

Trumpet Successes. Success, as shown in the hard data in your reports, is one of the strongest allies of a value management program.

3.6 CONCLUSIONS

The value methodology—a creative problem solving process—is the powerful core of effective value management. The core of the value methodology is function analysis—analyzing goods and services to deliver key functions in the most profitable manner.

There are several keys to successfully creating and implementing a program or office dedicated to enhancing the value of products and services. Using the keys will result in a company delivering products that are highly valued by customers and profitable for the business.

BIBLIOGRAPHY

Kaufman, J. Jerry, *Value Engineering for the Practitioner*, North Carolina State University, 1985.

Kaufman, J. Jerry, "Value Management (Creating Competitive Advantage)," Crisp Management Library of Crisp Publications, 1998.

King, Thomas R., *Value Engineering Theory and Practice,* The Lawrence D. Miles Value Foundation, 2000.

Mudge, Arthur E., *Value Engineering: A Systematic Approach*, J. Pohl Associates, 1989.

Value Methodology Standard, SAVE International, 1998. SAVE International. found on SAVE International's website: www.value-eng.org.

Woodhead, Roy and James McCuish, *Achieving Results: How to Create Value*, Thomos Telford Limited, 2002

CHAPTER 4
QUALITY FUNCTION DEPLOYMENT AND DESIGN OF EXPERIMENTS

Lawrence S. Aft
Aft Systems, Inc.
Roswell, Georgia

Jay Boyle
Marietta, Georgia

4.1 INTRODUCTION—QUALITY FUNCTION DEVELOPMENT

A key component of all the quality improvement processes is recognizing the customer and meeting and exceeding customer requirements. Not surprisingly, quality function deployment (QFD) began more than 30 years ago in Japan as a quality system focused on delivering products and services that satisfy customers. To efficiently deliver value to customers it is necessary to listen to the voice of the customer throughout the product or service development. The late Drs. Shigeru Mizuno and Yoji Akao, and other quality experts in Japan developed the tools and techniques and organized them into a comprehensive system to assure quality and customer satisfaction in new products and services.*

QFD links the needs of the customer (end user) with design, development, engineering, manufacturing, and service functions. It helps organizations seek out both spoken and unspoken needs, translate these into actions and designs, and focus various business functions toward achieving this common goal. QFD empowers organizations to exceed normal expectations and provide a level of unanticipated excitement that generates value.† "QFD uses a series of interlocking matrices that translates customer needs into product and process characteristics."

QFD is:

1. Understanding customer requirements
2. Quality systems thinking + psychology + knowledge/epistemology
3. Maximizing positive quality that adds value
4. Comprehensive quality system for customer satisfaction
5. Strategy to stay ahead of the game

* See Ref. 6.
† http://www.qfdi.org/

In QFD, product development translates customer expectations on function requirements into specific engineering and quality characteristics.[5] Quality function deployment has four phases. Phase 1 gathers the voice of the customer, puts it in words accurately understood by the producing organizations and analyzes it versus the capability and strategic plans of the organizations. Phase 2 identifies the area of priority breakthrough that will result in dramatic growth in market share for the producer. Phase 3 represents the breakthrough to new technology. Phase 4 represents the production of the new product and new technology at the highest possible quality standards.*

The following is one of the classic QFD examples. In the early 1980s International Harvester and Komatsu ended a partnering relationship. Since International Harvester had owned all the patents, Komatsu had to develop 11 new heavy equipment models in the short period of 24 months.

Komatsu engineers went out to the field to watch and observe the actual use of the equipment. They observed the discomfort and toil of the operator. As they studied this it became clear that two improvement areas might be the comfort of the driver in the cab and reducing the effort to shift the vehicle, since it was constantly going back and forth.

In the case of the cab, Komatsu engineers reworked the window structure so that there was a clearer view in all directions. They put in air conditioning that would stand up in a dusty environment. They made a seat that was comfortable to sit in for long periods of time. In the case of the shifting they looked into electronic shifting. They considered twelve different approaches. After considerable testing, they chose the one that would be the most reliable and easy to use.

When Komatsu introduced its new line of heavy trucks, it was met with great enthusiasm. Because of its ease of use, it led to higher productivity and driver preference. Soon Komatsu became a dominant force in the heavy truck business, a position it maintained for over a decade.

4.2 METHODOLOGY

QFD uses a series of matrices to document information collected and developed and represent the team's plan for a product. The QFD methodology is based on a systems engineering approach consisting of the following general steps:†

1. Derive top-level product requirements or technical characteristics from customer needs (product planning matrix).
2. Develop product concepts to satisfy these requirements.
3. Evaluate product concepts to select the optimum one (concept selection matrix).
4. Partition system concept or architecture into subsystems or assemblies and flow-down higher level requirements or technical characteristics to these subsystems or assemblies.
5. Derive lower level product requirements (assembly or part characteristics) and specifications from subsystem/assembly requirements (assembly/part deployment matrix).
6. For critical assemblies or parts, flow-down lower level product requirements (assembly or part characteristics) to process planning.
7. Determine manufacturing process steps to meet these assembly or part characteristics.
8. Based in these process steps, determine set-up requirements, process controls and quality controls to assure achievement of these critical assembly or part characteristics.

The following methodology has been suggested for implementing QFD.

The following steps are important in QFD. However, there is a very specific process that should be followed when building the House of Quality—a complex graphical tool that is essentially a product planning matrix (see Fig. 4.1). These steps are provided as an introduction.††

* GOAL QPC Web Site

† http://www.npd-solutions.com/bok.html

†† http://egweb.mines.edu/eggn491/lecture/qfd/

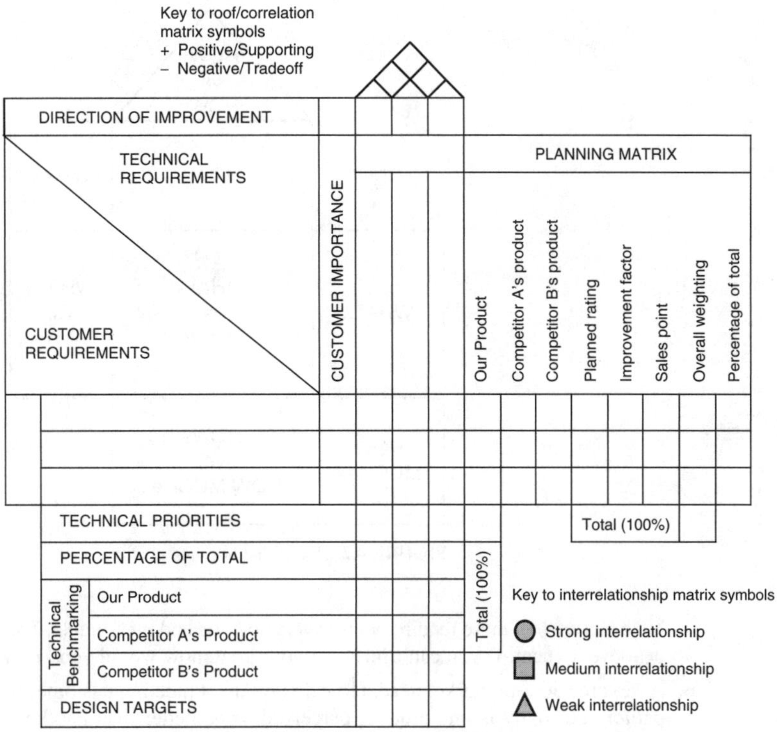

FIGURE 4.1 The expanded house of quality. (*http://www.proactdev.com/pages/ehoq.htm.*)

1. Listen to the voice of the customer. What, specifically, is important to our customers? For example, if we were trying to build the perfect cup of coffee, the customer requirements might include flavor; served warm but not too hot; ability to hold without burning the fingers; inexpensive; served quickly. These customer requirements are moved into the appropriate room in the House of Quality. Customer requirements can be gathered through a variety of sources including focus groups, interviews and calls to customer service centers, or customer complaints. (Additionally, these items can be used in the development of a future satisfaction survey.)
2. Rank the customer requirements in terms of importance. If you can't focus on all attributes, consider those which are most important.
3. Figure out how you will measure customer requirements by translating customer requirements into design requirements. To continue our example, "served warm but not too hot" would be measured by service temperature, "ability to hold without burning the fingers" would be measured by outside cup temperature, "inexpensive" would be measured by price. Note that each of these measurements use a variable scale, are specific and controllable, and are nonconstraining (which means we leave as many options open as possible). Although it seems that we could not measure "flavor" using these requirements, it can be measured by a panel of experts. Especially important in this step is to avoid specific product attributes of current products. Again, these design requirements are moved into the appropriate room in the House of Quality.
4. Rate the design attributes in terms of organizational difficulty. It is very possible that some attributes are in direct conflict. For example, increasing service temperature will conflict with cup temperature.
5. Determine the target values for the design requirements. It is very important that these target values be identified through research, not simply applied arbitrarily or based on current product attributes.

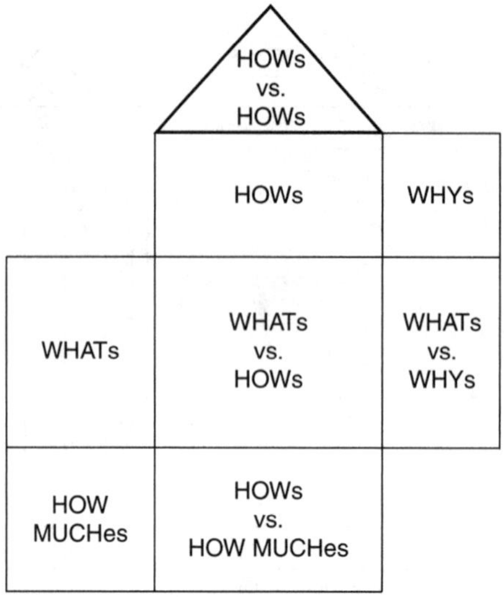

FIGURE 4.2 Expanded house of quality.

Note that each of these requirements is very specific and measurable. This is very important for product development. If you can't measure your goal, how would you know if you've achieved it?

6. Assess the current marketplace. How do you do at meeting customer requirements? How do competitors do? Why is one product perceived to be better than another? This can be completed in many ways—through customer surveys, traditional market research, panel discussions, reverse engineering, getting the designers out to sample the competitor's products, etc.

The most important thing to know about QFD is that it is a systematic way to ensure that customer requirements drive the design process. QFD ensures that customer requirements are met through the use of the House of Quality. The general format for the House of Quality is shown here in Fig. 4.1.*

As seen in Fig. 4.2, this "Expanded House of Quality" consists of multiple "rooms." Four of the rooms form the basic axes of the house. These are lists of "WHATs," "HOWs," "WHYs," and "HOW MUCHes." Four of the rooms consist of relationships between these lists. A brief explanation of each room is in order.

4.2.1 Whats

This is a list of what the customer wants or what is to be achieved. When the "Expanded House of Quality" is used with end user requirements, these would be customer statements about what they want to see in the product.

Hint: A common problem is that a lot of customers tend to state their requirements in terms of a possible solution. It is important that you understand the true requirement rather than accepting customer statements at face value.

4.2.2 Hows

This is a list of what your company can measure and control in order to ensure that you are going to satisfy the customer's requirements. Typically, the entries on this list are parameters for which a

* http://www.gsm.mq.edu.au/cmit/hoq/

means of measurement and a measurable target value can be established. Sometimes HOWs are also known as quality characteristics or design requirements.

Hint: It is best to try to keep these entries as concept-independent as possible. Failure to do this will lock you into a particular design solution that will almost never be what you would arrive at if you do QFD correctly. For example, if you were developing the lock for a car door you might be tempted to define HOWs such as "key insert force" and "key turn torque." These both imply that the lock will be key actuated. You will have immediately eliminated concepts such as combination locks that might have security and cost advantages for your particular application. A better HOW might be "Lock/Unlock Work" which could be measured for both key-operated or combination-operated locks.

4.2.3 Whys

Conceptually, this is a list that describes the current market. It is a way of explaining why this product needs to exist. It indicates what data will be used to prioritize the list of WHATs. Commonly included are lists of the customer groups your product must satisfy and their importance relative to each other. Also included are lists of products that will compete with yours in the marketplace.

4.2.4 How Muches

This list is used to specify how much of each HOW is required to satisfy the WHATs. Commonly it contains a listing of the products on which testing will be performed. This testing helps establish realistic target values for the HOWs. It also includes entries where the priority of each of the HOWs can be established. In general, WHYs and HOW MUCHes are very similar. WHYs lead to the importance of the WHATs while HOW MUCHes document and refine the importance of the HOWs.

4.2.5 Whats vs. Hows

This is a relationship matrix that correlates what the customer wants from a product and how the company can meet those requirements. It is the core matrix of QFD. Relationships within this matrix are usually defined using a strong, medium, weak, or none scale. If a HOW is a strong measure of compliance with a WHAT, then the WHAT and HOW are strongly correlated. Similarly, if a HOW provides no indication as to whether your product complies with the WHAT, there is probably no relationship. Filling and analyzing this matrix will likely take a large portion of the time you spend in QFD meetings.

4.2.6 Whats vs. Whys

This is a relationship matrix that is used to prioritize the WHATs based upon market information. Usually, the data in this matrix consist of ratings on how important different customer groups perceive each of the WHATs to be. Ratings of how well competitive products are perceived to meet each of the WHATs can also be included here. Averaging the stated importance ratings and factoring in where your product is perceived relative to your competition helps establish the overall importance of each WHAT.

4.2.7 Hows vs. How Muches

This is a relationship matrix that helps you decide what the next step in the project should be. Typically, this matrix includes calculated values which identify the relative importance of each of the HOWs. Also included is information about how your competition performs relative to each of the HOWs. This information can lead you to establish realistic and measurable target values which, if met, will ensure that you meet the customer's requirements.

4.2.8 Hows vs. Hows

This matrix forms the roof of the "Expanded House of Quality" and gives it its name. It is used to identify the interactions between different HOWs. The relationships in this matrix are rated as strong positive, positive, negative, strong negative, and none. If two HOWs help each other meet their target values, they are rated as positive or strong positive. If meeting one HOW's target value makes it harder or impossible to meet another HOW's target, those two HOWs are rated with a negative or strong negative relationship.

4.3 QFD SUMMARY

Quality function deployment and the house of quality serve as a living document and a source of ready reference for related products, processes, and future improvements. Their purpose is to serve as a method for strengthening communications and tearing down internal and external walls. "Through customer needs and competitive analysis, QFD helps to identify the critical technical components that require change. Issues are addressed that may never have surfaced before. These critical issues are then driven … to identify the critical parts, manufacturing operations, and quality control measures needed to produce a product that fulfills both customer needs and producer needs within a shorter development cycle time."* Tools such as designed experiments assist in the improvement of processes to meet those needs.

4.4 INTRODUCTION—DESIGN OF EXPERIMENTS (DOE)

Sir Ronald Fisher invented experimental design in the early 1920s to provide better results in agricultural experiments. Farmers wanted to know how to better control the planting-growing process. Much like industrial processes, there are many variables that affect the output, such as seed, soil, temperature, sunlight, moisture, and fertilizer. Obviously these factors interact, but how much of each is optimum, and which has the most effect and in what proportions? Using DOE gave new insight into plant growth. The DOE technique has also been used for many years in food and drug industry research.

DOE is an experimental test technique that identifies and quantifies process variables that have the most effect on process output variation. Many variables can then be changed and tested at the same time, reducing the cost of testing. Common manufacturing processes including casting, forming, injection molding, and thread making have been improved significantly using DOE. There have also been applications in marketing and telecommunications. It has also accelerated the product design cycle when used in conjunction with concurrent engineering. DOE is a strategic competitive weapon providing more reliable products, reducing concept-to-market time, and lowering life cycle costs.

Design of experiments is a planned, structured observation of two or more process input variables and their effect on the output variables under study. The objective is to select important input variables, known as factors, and their levels that will optimize the average output response levels and variability. These experiments can provide process managers the data on selecting input variables that will make the output less sensitive (robust) to the process and product operational environments.[3]

4.5 STATISTICAL METHODS INVOLVED

The structured process of DOE requires data collection and, based on analysis, draws conclusions that help improve the performance of a process. In order to begin a study of DOE one must first study or review statistical tools for analyzing data. The statistical methods used to properly analyze the data are called *descriptive and inferential statistics*.

* Tapke, Muller, Johnson, and Sieck, "House of Quality—Steps in Understanding the House of Quality," IE 361

Descriptive statistics describe data in terms of averages, variances, appearances, and distributions. Descriptive statistics are used to determine the parameters of a population and/or the statistics of a sample. These are used in statistical inference to forecast and predict hypothesized outcomes. *Statistical inference* is based on the belief that small samples drawn from processes can be used to estimate or approximate the populations from which they are drawn. It is founded on the concept that all sample measurements will vary. The key point is that the existence of sampling variation means that any one sample cannot be relied upon to always give an adequate decision. The statistical approach analyzes the results of the sample, taking into account the possible sampling variation that could occur.

4.5.1 Definitions

Experiments can have a wide variety of objectives, and the best strategy depends on the objective. In some experiments the objective is to find the most important variables affecting a quality characteristic. Design of experiments is the plan for conducting such experiments.

Recall from the history presented above, the first experimental designs were used in agricultural settings. A plot of land was marked off into different strips to test different fertilizers as a test of the various brands. The experimenters felt that other effects (or factors) such as rain, sun, and soil conditions would be the same (or could be controlled) in each strip, so that the only effect would be due to the fertilizers (fertilizer as a whole is one *factor*). The word *treatment* used in DOE comes from treating the strips with various brands (each brand is also called a *level*) of fertilizer. When the crop *yield* (or also called *response*) data were put into a matrix form, the treatment results were put into the columns.

Of course, the experimenters eventually wanted to study more than one factor. They subdivided the strips into rows, called *blocks*, to plant different types of crops. Now they were testing two factors, fertilizer brands and crops, within a single experiment. When the crop yield data were put into a matrix form, the treatments results were still in the *columns* while the blocks were in the *rows*. As the experimenters went to more than two factors, new experimental design techniques were developed which were called *factorial* designs.

4.6 OBJECTIVES OF EXPERIMENTAL DESIGNS

The variable thought to affect the response and thought to be controlled by the experimenter is called a *factor*. The various settings of these factors are called *levels*. The combination of a factor with one of a factor's levels defines a *treatment*. The output readings or yields, obtained by some relative measuring procedure, are called the dependent, or *response* variables.

In the fertilizer example, each brand is compared for its ability in growing crops side-by-side in a measured field. The variable under investigation (fertilizer) is the single *factor*. Each brand of fertilizer is a *treatment* within that factor. Treatments are also called *levels* of a factor. For the case with the factor being fertilizer, the three levels of fertilizer (or three treatments) could be brands A, B, and C.

The word level is also used when describing variations within a treatment. Not only can the factor have levels, each treatment (a level) can be subdivided into levels. In this case the factor is fertilizer, a treatment level is brand A, and the amount of brand A fertilizer to be used could be subdivided into two further levels such as 100 and 120 lb/acre.

Factors may be qualitative (different brands of fertilizer) or quantitative (amounts of fertilizer). We can make quantitative factors such as the 100- and 120-lb amounts of fertilizer into qualitative ones by coding them into settings called *low* or *high* amounts.

Some experiments have a *fixed effects* model, i.e., the treatments being investigated represent all levels of concern to the investigator, e.g., three brands of fertilizers. Other experiments have a *random effects* model, i.e., the levels chosen are just a sample from a larger population, e.g., two spreader settings controlling the amount of fertilizer used.

4.6.1 Selection of an Experimental Design

The choice of an experimental design depends on the objectives of the study and the number of factors to be investigated. One-factor experiments require analysis of variance (ANOVA) techniques. Two-factor experiments can be analyzed with either ANOVA or *factorial* techniques; however, ANOVA must be used when studies include more than three levels of the factors. Two or more factors with two or three levels per factor are studied using factorial techniques.

4.7 ANOVA-BASED EXPERIMENTAL DESIGNS

Fisher's pioneering work on DOE involved using an analysis technique called ANOVA. As mentioned earlier, ANOVA techniques study the variation between the total responses compared to the variation of responses within each factor. The ANOVA studies are augmented with results attained from applying multiple regression techniques to the yield (responses) data. Using the regression, we can form prediction equations that model the study responses obtained.

All of the ANOVA experimental design methods are essentially tests of hypothesis. A hypothesis test is used to determine the equivalence of the multiple means (average of each level) for each factor. The general null and alternate hypotheses statements for each factor are of the form:

$$H_0: \mu_{\mathrm{I}} = \mu_{\mathrm{II}} = \mu_{\mathrm{III}} = \cdots = \mu_k \quad \text{and} \quad H_1\text{: At least one mean is different}$$

The ANOVA study results in a test statistic per factor analyzed with a hypothesis test to determine the significance of each factor.

4.7.1 Single-Factor Design or Completely Randomized Design

In the *classical* design all factors are fixed except the one under investigation. The one factor could be fertilizer from our earlier analogy with three brands as the factor levels or treatments. Thus a total of nine tests could be run: three tests for each of the three brands. The rainfall, time of harvest, temperature, and all other factors are held equivalent (controlled) for each treatment. The major drawback to this method is that the conclusions about the brands would apply only to the specific conditions run in the experiment. The table shown below is a way to visualize this design where the numbers are not the response values but a simple numbering of the total number of tests that are to be performed.

I	II	III
1	2	3
4	5	6
7	8	9

The single factor design is also known as the *completely randomized* design. This naming is because the nine tests that are to be completed are performed in a completely random fashion. This will randomize any random variation in fixed factors (e.g., water, sunshine, and temperature at the time of each test).

I	II	III
3	8	1
6	2	9
4	7	5

In ANOVA terminology, this would be called a *one-way analysis of variance* since all of the studied variation in responses is contained only in the columns (treatments).

4.7.2 Calculations for Single-Factor ANOVA Tables

Single Factor			
Treatment 1	Treatment 2	...	Treatment k
$y_{1,1}$	$y_{1,2}$	...	$y_{1,k}$
$y_{2,1}$	$y_{2,2}$	...	$y_{2,k}$
...	$y_{3,2}$	...	...
$y_{n1,1}$	...	...	$y_{nk,k}$
	$y_{n2,2}$		
$T_k = \sum_{i=1}^{n_k} y_k$	$T_2 = \sum_{i=1}^{n_2} y_{i,2}$	...	$T_k = \sum_{j=1}^{n_k} y_{k,j}$
n_1	n_2	...	n_k

Some sources give and use a different set of equations for balanced (same number of observations in each treatment) and unbalanced (different number of observations in each treatment) designs. We will use the one set of equations given below for both designs. As the yields (responses) are captured from the tests, they are recorded in a matrix. The equations reference elements of the yield matrix are shown below. When there are a variable number of responses (y's) in any treatment (column), this is the unbalanced design. The balanced design is simply when every treatment has the same number of responses (i.e., the same number of rows per treatment such that $n_1 = n_2 = \cdots = n_k$).

Constants from the inputs of the design:

k = number of treatments

n_j = number of samples in the jth treatment

N = total sample size = $n_1 + n_2 + \cdots + n_k$

$y_{i,j}$ = yield in row i and column j

Calculations made from yield matrix:

$$T_m = \text{sample total of } m\text{th treatment} = \sum_{i=1}^{n_m} y_{i,m}$$

$$\sum y = \text{overall sample total} = \sum_{j=1}^{k}\sum_{i=1}^{n_j} y_{i,j} \text{ or } = \sum_{j=1}^{k} T_j$$

$$\sum y^2 = \text{sum of squares of all } n \text{ samples} = \sum_{j=1}^{k}\sum_{i=1}^{n_j} y_{i,j}^2$$

Sum of square calculations:

$$\text{SST} = \text{sum of squares for treatments} = \sum_{j=1}^{k} \frac{T_j^2}{n_j} - \frac{(\Sigma y)^2}{N}$$

$$\text{TSS} = \text{total sum of square} = \left(\sum y^2\right) - \frac{(\Sigma y)^2}{N}$$

$$\text{SSE} = \text{sum of squares for error} = \text{TSS} - \text{SST} - \text{SSB}$$

4.7.3 Single-Factor ANOVA Table

Values of k, N, SST, SSE and TSS from the above equations are used to complete the ANOVA table shown below. In this table, the sums of square (SS) terms are converted to variances (MS terms) and the F-test is the ratio of the MS terms. Some minor computations are necessary to complete the table.

Source	*df*	SS	MS	*F*-test
Treatment	$k - 1$	SST	MST = SST/(k – 1)	F = MST/MSE
Error	$N - k$	SSE	MSE = SSE/(N – k)	
Total	$N - 1$	TSS		

Using a hypothesis test, the F-test value of the treatment determines the significance of the factor (i.e., whether the treatment's means are equivalent to each other).

4.7.4 Two-Factor Design or Randomized Block Design

The next design recognizes a second factor, called *blocks* (e.g., crops A, B, and C, where A is acorn squash, B is beans, and C is corn). Both the original factor with its treatments and the added factor, the blocks, are studied. Again, data for each response must be collected in a completely randomized fashion.

In the *randomized block design* each block (row) is a crop (acorn squash, beans, corn) and the fertilizer brands are the treatments (columns). Each brand, crop combination is tested in random order. This guards against any possible bias due to the order in which the brands and crops are used.

Fertilizer Brands⇒ Crops⇓	I	II	III
A	3	8	1
B	6	2	9
C	5	7	5

The randomized block design has advantages in the subsequent data analysis and conclusions. First, from the same nine observations, a hypothesis test can be run to compare brands and a separate hypothesis test run to compare crops. Second, the conclusions concerning brands apply for the three crops and vice versa, thus providing conclusions over a wider range of conditions.

The ANOVA terminology calls the randomized block design *two-way analysis of variance* since the studied variations in responses are contained both in the columns (treatments) and in the rows (blocks).

4.7.5 Calculations for Two-Factor ANOVA Tables

The generalized matrix for two-factor solutions is shown below. As with the one-factor equations, the two way matrix and elements are referenced by the two-way equations.

		Treatment 1	Treatment 2	…	Treatment k	Totals
Blocks	1	$y_{1,1}$	$y_{1,2}$	…	$y_{1,k}$	$B_1 = \sum_{j=1}^{k} y_{1,j}$
	2	$y_{2,1}$	$y_{2,2}$	…	$y_{2,k}$	$B_2 = \sum_{j=1}^{k} y_{2,j}$
	⋮	⋮	⋮	…	⋮	⋮
	b	$y_{n,1}$	$y_{n,2}$	…	$y_{n,k}$	$B_n = \sum_{j=1}^{k} y_{n,j}$
		$T_1 = \sum_{i=1}^{b} y_{i,1}$	$T_2 = \sum_{i=1}^{b} y_{i,2}$	…	$T_k = \sum_{j=1}^{b} y_{k,j}$	

All columns are size $n = b$ and all rows are size k.

The two-factor equations are shown below. In addition to calculating the total sum of squares (TSS) for all yields (responses) and the sum of squares for all treatments (SST), the sum of squares for all blocks (SSB) must be calculated. This time the sum of squares of the error (SSE) is the difference between TSS and both SST and SSB.

Constants from the design:

K = number of treatments

N = total sample size = bk

b = number of blocks

$y_{i,j}$ = yield in row i and column j

Calculations from the yield matrix:

$$T_m = \text{sample total of } m\text{th treatment} = \sum_{i=1}^{b} y_{i,m}$$

$$B_m = \text{sample total of } m\text{th block} = \sum_{j=1}^{k} y_{m,j}$$

$$\sum y = \text{overall sample total} = \sum_{j=1}^{k}\sum_{i=1}^{n_j} y_{i,j} \text{ or } = \sum_{j=1}^{k} T_j$$

$$\sum y = \text{sum of squares of all } n \text{ samples} = \sum_{j=1}^{k}\sum_{i=1}^{n_j} y_{i,j}^2$$

Sum of square calculations:

$$\text{SST} = \text{sum of squares for treatments} = \frac{1}{b}\sum_{j=1}^{k} T_j^2 - \frac{(\Sigma y)^2}{N}$$

$$\text{SSB} = \text{sum of squares for blocks} = \frac{1}{k}\sum_{i=1}^{b} B_j^2 - \frac{(\Sigma y)^2}{N}$$

$$\text{TSS} = \text{total sum of square} = \left(\sum\right)^2 - \frac{(\Sigma y)^2}{N}$$

$$\text{SSE} = \text{sum of squares for error} = \text{TSS} - \text{SST} - \text{SSB}$$

4.7.6 Two-Factor ANOVA Table

In the two factor ANOVA table, there is an additional row calculation added for the sum of squares for the blocks (SSB). As with single factors, the SS terms are converted to variances (MST, MSB, and MSE terms). Values of k, b, N, SST, SSB, SSE and TSS from the above equations are used to complete the ANOVA table shown below. This time two hypothesis tests are required: one for treatments and one for blocks.

Source	*df*	SS	MS	*F*-test
Treatment	$k - 1$	SST	MST = SST/(k – 1)	F = MST/MSE
Blocks	$b - 1$	SSB	MSB = SSB/(b – 1)	F = MSB/MSE
Error	$(b - 1)(k - 1)$	SSE	MSE = SSE/(N – k)	
Total	$bk - 1$	TSS		

4.7.7 Two Factor with Interaction Design

The last issue to consider is interaction between the two factors. A new term, *interaction,* is defined as the effect of mixing, in this case, of specific fertilizer brands with specific crops. There may be a possibility that under a given set of conditions something "strange" happens when there are interactions among the factors. The *two factor with interaction* design investigates not only the two main factors but the possible interaction between them. In this design each test is repeated (replicated) for every combination of the main factors. In our agricultural example with three replications using the three brands of fertilizer and three crops, we have 3 × 3 × 3 or 27 possibilities (responses). Separate tests of hypothesis can be run to evaluate the main factors and the possible interaction.

	I	I	II	III
Hi	---	---	---	---
Med	---	---	---	---
Lo	---	---	---	---

The two-factor design with interaction is also known as *two-way analysis of variance with replications.*

4.7.8 Calculations for Two Factor with Interaction ANOVA Tables

To show the yield matrix for this situation, the term *replication* must be defined. Replication is where yields for all treatments and blocks are observed for "*r*" times. The generalized matrix is shown on the below:

		Factor A					
		1	2	3	. . .	a	Totals
		---	---	---		---	
		---	---	---		---	
		---	---	---		---	
	1	$T_{1,1}$	$T_{1,2}$	$T_{1,3}$	. . .	$T_{1,a}$	B_1
		---	---	---		---	
		---	---	---		---	
		---	---	---		---	
	2	$T_{2,1}$	$T_{2,2}$	$T_{2,3}$	. . .	$T_{2,a}$	B_2
		---	---	---		---	
Factor B		---	---	---		---	
		---	---	---		---	
	3	$T_{3,1}$	$T_{3,2}$	$T_{3,3}$	. . .	$T_{3,a}$	B_3
		⋮	⋮	⋮		⋮	⋮
		---	---	---		---	
		---	---	---		---	
		---	---	---		---	
	b	$T_{b,1}$	$T_{b,2}$	$T_{b,3}$	. . .	$T_{b,a}$	B_b
	Totals	A_1	A_2	A_3	. . .	A_a	Σy

The associated calculations for the above matrix are shown below:
Input from the Design:

a = Factor A Treatments b = Factor B Blocks r = Number of replications
N = Total Sample Size = abr $y_{i,j,m}$ = yield in row i column j and observation m

Calculations from the yield matrix:

$$T_{i,j} = \text{Replication Total of } i, j\text{th Treatment} = \sum_{m=1}^{m} y_{i,j,m}$$

$$B_m = \text{Sample Total of } m\text{th Block} = \sum_{j=1}^{a} T_{m,j}$$

$$\sum Y = \text{Overall sample total} = \sum_{j=1}^{a} A_j \text{ or} = \sum_{i=1}^{b} T_i$$

$$\sum Y^2 = \text{Sum of squares of all N samples} = \sum_{m=1}^{r}\sum_{j=1}^{a}\sum_{i=1}^{b} y_{i,j,m}^2$$

Sum of Squares Calculations:

$$\text{SST} = \text{Sum of Squares for all Treatments} = \sum_{j=1}^{a}\sum_{i=1}^{b}\frac{T_{i,j}^2}{r} - \frac{(\Sigma Y)^2}{N}$$

$$\text{SS(A)} = \text{Sum of Squares for Factor A} = \sum_{i=1}^{a}\frac{A_i^2}{br} - \frac{(\Sigma Y)^2}{N}$$

$$\text{SS(B)} = \text{Sum of Squares for Factor B} = \sum_{j=1}^{b}\frac{B_j^2}{ar} - \frac{(\Sigma Y)^2}{N}$$

$$\text{SS(A}\times\text{B)} = \text{Sum of Squares for Interaction A}\times\text{B} = \text{SST} - \text{SS(A)} - \text{SS(B)}$$

$$\text{TSS} = \text{Total Sum of Square} = \left(\sum Y\right)^2 - \frac{(\Sigma Y)^2}{N}$$

$$\text{SSE} = \text{Sum of Squares for Error} = \text{TSS} - \text{SST} - \text{SS(A)} - \text{SS(B)} - \text{SS(A}\times\text{B)}$$

4.7.9 Two Factor with Interaction ANOVA Table

The ANOVA table for two factors with interaction is shown below. This time hypothesis tests are required for A, B, and AB to determine significance.

Source	*df*	SS	MS	*F*-test
Treatments	$ab - 1$	SST		
Factor A	$a - 1$	SS(A)	MS(A) = SS(A)/(a – 1)	F = MS(A)/MSE
Factor B	$b - 1$	SS(B)	MSB = SS(B)/(b – 1)	F = MS(B)/MSE
Factor A × B	$(a - 1)(b - 1)$	SS(A × B)	MS(A × B) = SS(A × B)/(a – 1)(b – 1)	F = MS(A × B)/MSE
Error	$ab(r - 1)$	SSE	MSE = SSE/ab(r – 1)	
Total	$N - 1$	TSS		

4.7.10 Factorial Base Experimental Designs

Factorial solutions use a matrix approach that can study multiple factors, interactions, and levels. A *full factorial* design gets responses at every level for each factor. The following table shows a full factorial design for studying four factors—tire compounds, road temperatures, tire pressures, and vehicle types. The design tests all four factors at each of the two levels, in every possible combination.

Factor	Levels	
Compounding formula	X	Y
Road temperature	75	80
Tire pressure	28	34
Vehicle type	I	II

This is called a 2^4 design, since there are two levels and four factors. The "2" is for the two levels and the "4" is for the factors. The general form is 2^k, where the k is the number of factors. To test this model using a full factorial design, a minimum of sixteen (2^4) different experimental runs are needed.

4.7.11 Full Factorial of Two Factors at Two Levels (2^2) Design

This table shows the complete layout of a two factor with two levels (2^2) experiment. Factor A could be road temperature and factor B could be tire pressure. The yields of interest could be tire wear.

Run (Trial)	Factors	
1	75	28
2	80	28
3	75	34
4	80	34

4.7.12 Full Factorial 2^2 Design with Two Replications

When only one trial is performed for each combination, statistical analysis is difficult and may be misleading. When the experiment is replicated it is possible to perform a test of hypothesis on the effects to determine which, if any, are statistically significant. In this example we will use the twice-replicated runs ($r = 2$) for the experiment.

When there is more than one replication, it is necessary to compute the mean of each run. The variance will also be useful.

Run	Factor A	Factor B	Y_1	Y_2	Average	s^2
1	75	28	55	56	55.5	0.5
2	80	28	70	69	69.5	0.5
3	75	34	65	71	68.0	18.0
4	80	34	45	47	46.0	2.0

The results indicate that run 2 gets the best average yield and has a low variance. Run 3 has a slightly lower average but a larger variance. Common sense would indicate that the experiments should use the settings of run 2 for the best results.

Taking a graphical approach, look first at factor A. When low settings are used the average yield is 61.75—the two averaged yields where A is set low are 55.5 and 68.0, which is an average of 61.75.

Similarly for the high setting A the average yield is 57.75. That means that as factor A increases—goes from the low setting to the high setting—then the magnitude changes (decreases) by a magnitude of –4.0. This is called the effect of factor A. See factor A chart below. A similar graph is plotted for the factor B effect, which is –5.5.

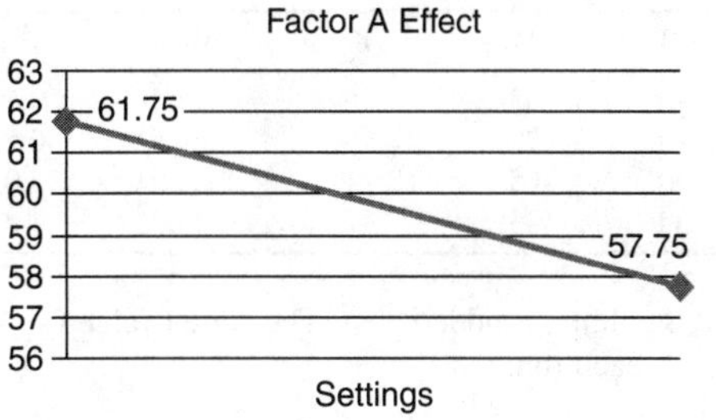

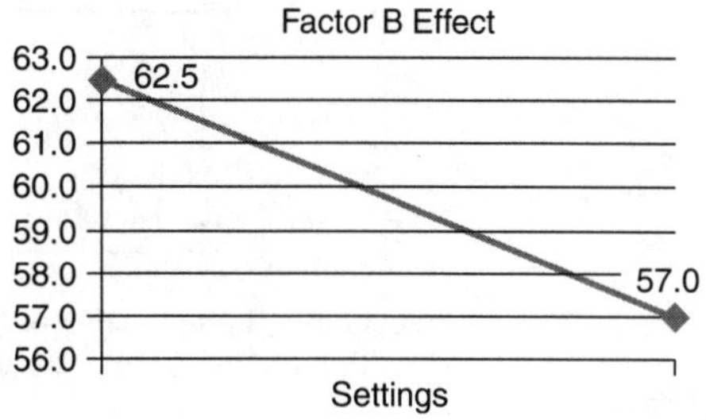

The figure below shows the interaction of the factors. If the high setting for factor A is chosen, and factor B is set low, the least fire wear is achived. Note that varying only one factor at a time would have missed the large interaction between A and B.

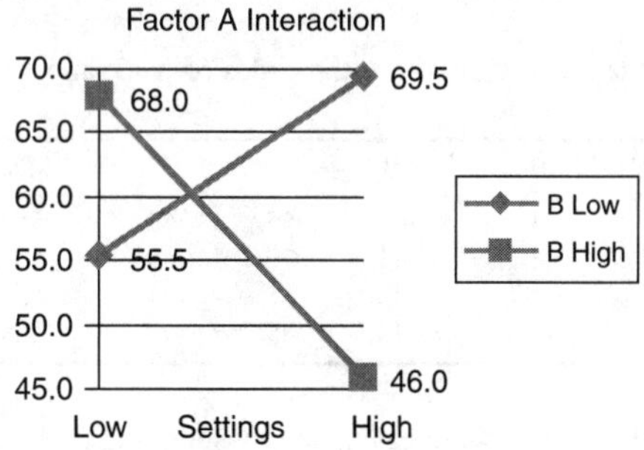

4.7.13 Full Factorial 2^2 Design—Linear Equation Model

Graphical analysis does not provide a complete picture. We now need to investigate a statistical method of analysis. The graphical analysis, especially when coupled with a statistical analysis, allows the experimenter to get reasonable ideas as to the behavior of the process.

Returning to the two-factor (2^2) example we will assume a linear modeling equation of:

$$\hat{y} = b_0 + b_1\text{A} + b_2\text{B}.$$

The b_1A and b_2B terms are called the main effects. The coefficients b_1 and b_2 are the slopes of factors A and B respectively. The term b_3AB is the interaction effect and coefficient b_3 is the slope of the interaction AB.

As an aid in the statistical analysis we will use codes instead of the words low and high. A *standard coding* terminology substitutes a "–1" for each low setting. A "+1" is used for a high setting. For the factor "tire pressure" in the table, the 28 would be the low setting and the 34 would be the high setting. In the "coded" design they would be specified –1 and +1, respectively. Note that in each coded column there is the same number of pluses as there are minuses for each run. This means that all factors at all levels are sampled equally, and that the design is balanced. This order is sometimes called the standard order.

4.7.14 Full Factorial 2^2 Design Calculations

Returning to the previous example, the statistical calculations are initiated by adding the "codes," columns, and rows to the matrix array. The first step is to rewrite using the codes for high and low values.

Run	A	B	Y_1	Y_2	Mean	s^2
1	−1	−1	55	56	55.5	0.5
2	+1	−1	70	69	69.5	0.5
3	−1	+1	65	71	68.0	18.0
4	+1	+1	45	47	46.0	2.0

A column for interaction or the AB column is added next. The coded value for AB is determined by multiplying the A times B codes for each run.

Run	A	B	AB	Y_1	Y_2	Mean	s^2
1	−1	−1	+1	55	56	55.5	0.5
2	+1	−1	−1	70	69	69.5	0.5
3	−1	+1	−1	65	71	68.0	18.0
4	+1	+1	+1	45	47	46.0	2.0

Then the mean yield times the codes of A, B, and AB, are multiplied, respectively.

Run	A	B	AB	Y_1	Y_2	Mean	s^2	A	B	AB
1	−1	−1	+1	55	56	55.5	0.5	−55.5	−55.5	+55.5
2	+1	−1	−1	70	69	69.5	0.5	+69.5	−69.5	−69.5
3	−1	+1	−1	65	71	68.0	18.0	−68.0	+68.0	−68.0
4	+1	+1	+1	45	47	46.0	2.0	+46.0	+46.0	+46.0

Next the contrasts are calculated by summing each run value in columns.

Run	A	B	AB	Y_1	Y_2	Mean	s^2	A	B	AB
1	−1	−1	+1	55	56	55.5	0.5	−55.5	−55.5	+55.5
2	+1	−1	−1	70	69	69.5	0.5	+69.5	−69.5	−69.5
3	−1	+1	−1	65	71	68.0	18.0	−68.0	+68.0	−68.0
4	+1	+1	+1	45	47	46.0	2.0	+46.0	+46.0	+46.0
Contrasts						239	21.0	−8.0	−11.0	−36.0

The effects are the contrasts divided by 2^{k-1} or $2^{2-1} = 2$ in this case. Note that the means and variance columns do not have effects.

Run	A	B	AB	Y_1	Y_2	Mean	s^2	A	B	AB
1	−1	−1	+1	55	56	55.5	0.5	−55.5	−55.5	+55.5
2	+1	−1	−1	70	69	69.5	0.5	+69.5	−69.5	−69.5
3	−1	+1	−1	65	71	68.0	18.0	−68.0	+68.0	−68.0
4	+1	+1	+1	45	47	46.0	2.0	+46.0	+46.0	+46.0
Contrasts						239	21.0	−8.0	−11.0	−36.0
Effects						N/A	N/A	−4.0	−5.5	−18.0

The calculations in the row labeled "Effects" are nothing more than what we saw in the effect plots earlier. Again, the largest effect is the interaction effect, as shown by the −18 in the AB column, which is much more significant than the effect of A or B.

The final addition to the table is the coefficients. The effects sum for A, B, and AB divided by two. For the means, it is a divided by $2^k = 4$. Nothing applies to the variance column.

Run	A	B	AB	Y_1	Y_2	Mean	s^2	A	B	AB
1	−1	−1	+1	55	56	55.5	0.5	−55.5	−55.5	+55.5
2	+1	−1	−1	70	69	69.5	0.5	+69.5	−69.5	−69.5
3	−1	+1	−1	65	71	68.0	18.0	−68.0	+68.0	−68.0
4	+1	+1	+1	45	47	46.0	2.0	+46.0	+46.0	+46.0
Contrasts						239	21.0	−8.0	−11.0	−36.0
Effects						N/A	N/A	−4.0	−5.5	−18.0
Coefficients						59.75	N/A	−2.0	−2.75	−9.0

The coefficients are used to predict the yield for any setting of the factors. However, only significant factors are used in this equation. The purpose of this example is to conduct a "*t*-test" on each of the coefficients. First, the standard errors S_e and S_β are found as follows:

$$\text{Pooled variance} = \text{SSE} = s_p^2 = (\text{sum of the run variances}) = 21$$

$$\text{Standard error} = s_e = \sqrt{\frac{s_p^2}{2^k}} = 2.291$$

$$\text{Standard error of the coefficients} = s_\beta = \sqrt{\frac{s_e^2}{r \times 2^k}} = \sqrt{\frac{2.291^2}{2 \times 2^2}} = 0.81$$

(Where k = number of factors = 2, r = number replications = 2)

Then the hypothesis test for each coefficient is stated as:

H_{0A}: coefficient of A = 0 $\quad$ H_{1A}: It is significant

H_{0B}: coefficient of B = 0 $\quad$ H_{1B}: It is significant

H_{0AB}: coefficient of AB = 0 $\quad$ H_{1AB}: It is significant

In each case the test statistic is calculated using the relationship:

$$t_{\text{test}} = [\text{coefficient}/s_\beta]$$

The three test statistics are calculated.

$$t_A = [-2.0/0.81] = -2.47 \qquad t_B = [-2.75/0.81] = -3.40 \qquad t_{AB} = [-9.0/0.81] = -11.11$$

The degrees of freedom are determined from $(r - 1) \times 2^k = (2 - 1) \times 2^2 = 1 \times 4 = 4$

The "*t*" table two-sided critical value, when $\alpha = .05$ and $df = 4$, is 2.776. Comparing the calculated *t* values with the table values, the coefficients for factor B and the interaction AB are significant.

The resulting linear predicting relationship for yields is:

$$\hat{y} = 59.75 - 2.75\text{B} - 9.0\text{AB}$$

4.7.15 Observations

- The full factorial experiment can be extended for 2 levels of three factors (2^3) and larger experiments. The number of interactions grows dramatically as the size of the experiment increases, though.
- Experiments are not really done in the order listed. They are performed in a random fashion that causes resetting of the variables and thus reduces bias in the results. Remember the way they are listed is called the standard order.

4.7.16 Full Factorial Three Factor with Two Levels (2^3) Designs

Instead of explaining a build up of steps starting with graphical methods and then going onto statistical analysis of a 2^3 design, this section combines the steps. The layout of 2^3 design for a milling machine's power consumption (wattage) for three factors [speed (A), pressure (B), and angle (C)] is as follows:

Factor	Levels	
A	200	600
B	Low	High
C	20	28

The null hypotheses are that there is no significant difference in any factor, regardless of the setting, and that there is no significant interaction. Appropriate alternative hypotheses are that there is a significant difference in a factor based on the setting and that there is significant interaction.

4.7.17 Graphical and Statistical Analysis of a Full Factorial 2^3 Design

Testing the above model using a full factorial design will require eight separate runs or trials. Performing two replications ($r = 2$) of the above design and adding the coding resulted in the following matrix showing the two sets of yields (watts used) from the runs.

Run	Factors			A	B	C	Y_1	Y_2
1	200	Low	20	−1	−1	−1	221	311
2	600	Low	20	1	−1	−1	325	435
3	200	High	20	−1	1	−1	354	348
4	600	High	20	1	1	−1	552	472
5	200	Low	28	−1	−1	1	440	453
6	600	Low	28	1	−1	1	406	377
7	200	High	28	−1	1	1	605	500
8	600	High	28	1	1	1	392	419

Beginning below the additional columns and rows will be added to the table so that the results can be analyzed.

Run	Factors			A	B	C	AB	AC	BC	ABC	Y_1	Y_2
1	200	Low	20	−1	−1	−1	1	1	1	−1	221	311
2	600	Low	20	1	−1	−1	−1	−1	1	1	325	435
3	200	High	20	−1	1	−1	−1	1	−1	1	354	348
4	600	High	20	1	1	−1	1	−1	−1	−1	552	472
5	200	Low	28	−1	−1	1	1	−1	−1	1	440	453
6	600	Low	28	1	−1	1	−1	1	−1	−1	406	377
7	200	High	28	−1	1	1	−1	−1	1	−1	605	500
8	600	High	28	1	1	1	1	1	1	1	392	419

Columns of the table were removed for space considerations.

Run	Average	S2		A	B	C	AB	AC	BC	ABC
1	266	4050		−266	−266	−266	266	266	266	−266
2	380	6050		380	−380	−380	−380	−380	380	380
3	351	18		−351	351	−351	−351	351	−351	351
4	512	3200		512	512	−512	512	−512	−512	−512
5	446.5	84.5		−446.5	−446.5	446.5	446.5	−446.5	−446.5	446.5
6	391.5	420.5		391.5	−391.5	391.5	−391.5	391.5	−391.5	−391.5
7	552.5	5512.5		−552.5	552.5	552.5	−552.5	−552.5	552.5	−552.5
8	405.5	364.5		405.5	405.5	406	405.5	405.5	405.5	405.5

The contrasts, effects and coefficients are summed. The contrasts, except the one for yield, are divided by 2^{k-1} to calculate the effects. Coefficients are calculated by dividing the effects by two, except the coefficient for yield is its contrast divided by 2^3 or 8.

Run	Average	S2		A	B	C	AB	AC	BC	ABC
1	266	4050		−266	−266	−266	266	266	266	−266
2	380	6050		380	−380	−380	−380	−380	380	380
3	351	18		−351	351	−351	−351	351	−351	351
4	512	3200		512	512	−512	512	−512	−512	−512
5	446.5	84.5		−446.5	−446.5	446.5	446.5	−446.5	−446.5	446.5
6	391.5	420.5		391.5	−391.5	391.5	−391.5	391.5	−391.5	−391.5
7	552.5	5512.5		−552.5	552.5	552.5	−552.5	−552.5	552.5	−552.5
8	405.5	364.5		405.5	405.5	405.5	405.5	405.5	405.5	405.5
	3305	19700	Contrasts	73	337	287	−45	−477	−97	−139
	N/A	s_p = SSE	Effects	18.25	84.25	71.75	−11.25	−119.25	−24.25	−34.75
Bo	413.125	N/A	Coefficients	9.125	42.125	35.875	−5.625	−59.625	−12.125	−17.375
TSS	134587.8		SS_i	1332.25	28392.25	20592.25	506.25	56882.25	2352.25	4830.25
	sigma = s_e = 49.624		t_{test}	0.73554	3.395563	2.89177	−0.453	−4.80618	−0.9774	−1.4005
	S_{beta} 12.406			B1	B2	B3	B4	B5	B6	B7

The same equations listed earlier for 2^2 apply and they are:

$$\text{Pooled variance} = \text{SSE} = s_p^2 = (\text{sum of the run variances}) = 19700$$

$$\text{Standard error} = s_e = \sqrt{\frac{s_p^2}{2^k}} = 49.6236$$

$$\text{Standard error of the coefficients} = s_\beta = \sqrt{\frac{s_e^2}{r \times 2^k}} = \sqrt{\frac{49.6236^2}{2 \times 2^3}} = 12.4059$$

(Where k = number of factors = 2, r = number replications = 2)

$$\text{Sum square of each effect} = \text{SS}_i = \frac{r \times (\text{Contrasts}_i)^2}{2^k} \quad \text{for example SSA}= \frac{r \times 73^2}{2^3} = 1332.25$$

The test values for the t's are calculated from: $t_{\text{test}} = \text{Coefficient}_i/s_\beta$ for example the B_1 test value is calculated as follows = 9.125/12.4059 = 0.73554.

The critical value for a t-test at the .05 level with 8 df is 2.306, comparing each of the test values to the critical results in significant coefficients for main effect B, main effect C, and interaction effect AC.

The ANOVA table can be constructed from the sum squares calculated above.

ANOVA Source	*df*	SS	MS	F	sig F	Regression Significant?
Regression	7	114887.75	16412.536	6.664989	0.0079	yes
Error	8	19700	2462.5			
Total	15	134587.75				

Remember that sum square of the regression is calculated by adding up the seven main and interaction effects sum squares.

$$\text{SSR} = \text{SSA} + \text{SSB} + \text{SSC} + \text{SSAB} + \text{SSAC} + \text{SSBC} + \text{SSABC}$$

Total sum square, or TSS, is the numerator term for the variance of all of the yields and is equal to sum square of regression plus sum square of error

$$(\text{TSS} = \text{SSR} + \text{SSE})$$

Another statistics to look at is the coefficient of correlation R. Its value is simply calculated from

$$R = \text{SSR/TSS} = 0.9239 \text{ and so } R^2 = 0.8536$$

4.7.18 Additional Thoughts on Factorial Designs

Higher order full factorial designs, such as 2^4 could be run having four factors at two levels each. It would require 16 trials to collect all the data. For 2^k designs there are 2^{k-1} columns of main and interaction effects to code and calculate. This means for four factors a total of 15 columns are needed to account for the main, 2nd, 3rd, and 4th order effects.

Higher order interactions, such as ABC, ABCD, or ABCDE, are rarely important. As a general rule, avoid assuming that all 2nd order interactions have no meaning in order to perform a statistical analysis. While higher order interactions are rarely important or significant, second order interactions are frequently significant.

Analysis techniques are available that consider only part of the full factorial runs and they are called fractional factorials. Also available are 2^{k-p} factorial designs which use fractional designs combined with placing additional factors in the third order and above interactions. The 2^{k-p} is a way to add factors without requiring the addition runs necessary in full factorial designs. Whole books are devoted to such designs.

Situations calling for three levels call for a 3^k design. The calculations change from the above 2^k technique and require an additional centering run. A three factor with 3 levels would require 3 or 27 runs plus the centering run.

REFERENCES

1. American Supplier Institute, "An Introduction to QFD Seminar," 1977.
2. Bossert, J.L., *Quality Function Deployment: A Practitioner's Approach,* ASQ Quality Press, Milwaukee, 1991.
3. Cartin, T., and A. Jacoby, *A Review of Managing Quality and a Primer for the Certified Quality Manager Exam,* ASQ Quality Press, Milwaukee, 1997.
4. Gryna, F., *Quality Planning and Analysis,* 4th ed., McGraw-Hill, New York, 2001, p. 336.
5. Juran, J., *Quality Control Handbook,* 4th ed., McGraw-Hill, New York, 1988, p. 13.13.
6. Mazur, G., "QFD for Service Industries," *Fifth Symposium on Quality Function Deployment,* Novi, Michigan, June 1993, p. 1.

USEFUL WEBSITES

http://www.qfdi.org/

http://www.npd-solutions.com/bok.html

http://egweb.mines.edu/eggn491/lecture/qfd/

http://www.gsm.mq.edu.au/cmit/hoq/

http://www.proactdev.com/pages/ehoq.htm

CHAPTER 5
RAPID PROTOTYPING, TOOLING, AND MANUFACTURING

Todd Grimm
T. A. Grimm & Associates, Inc.
Edgewood, Kentucky

5.1 INTRODUCTION

5.1.1 Growth and Challenge

Rapid prototyping came to light in the late 1980s. Since the delivery of the first system, the scope of applications and breadth of use have swelled. Rapid prototyping is used in virtually every industry that produces mechanical components.

As presented in the *Wohlers Report* by Wohlers Associates, rapid prototyping is nearly a billion dollar industry with two dozen system vendors that have installed more than 8000 machines around the globe.* With the growth in the application of the technology for the development of prototypes, other applications have come to light, namely rapid tooling and rapid manufacturing.

Best known by the original technology, *stereolithography,* rapid prototyping now has numerous methodologies and processes. The common element of these technologies is that they derive speed in the construction of complex geometry through the additive nature of the process. Neither subtractive nor formative, rapid prototyping constructs designs without the use of molds or machine tools.

While the industry has had an exceptional track record, it is not without its challenges. The general consensus is that less than 20 percent of the design and product development community use rapid prototyping. In the manufacturing and manufacturing engineering disciplines, the level of use is far less.

The obstacles that rapid prototyping faces are not unique. As with any new technology, there is a resistance to change and a reluctance to work through the challenges of a developing technology. However, there are other factors that are unique to this industry. Since rapid prototyping requires 3D digital definition of the part, its growth rate is limited to that of CAD solid modeling, an application that is far from being used by the majority of design professionals. Additionally, rapid prototyping has been burdened with a negative perception that the parts are "brittle." While true many years ago, this is no longer an appropriate generalization. Yet, many use the belief that rapid prototypes are brittle to justify not evaluating or using the technology.

While rapid prototyping may not pose a competitive threat to those who do not use it, many who have implemented the technology have discovered powerful advantages in applications that range from product development to manufacturing to sales and marketing.

* Terry Wohlers, *Wohlers Report 2002*, Wohlers Associates, Inc., Fort Collins, CO, www.wohlersassociates.com

5.1.2 Widespread Applications and Powerful Results

Rapid prototyping's impact reaches far and wide. There is diversity in the application of rapid prototyping in terms of the disciplines that use it, the processes that benefit from it, the industries that employ it, and the products that are better because of it. The common element of all of these applications is that rapid prototyping has been a tool that makes the process faster, the product better, and the cost lower.

Industrial design, engineering, manufacturing, and sales and marketing are just some of the disciplines that have applied rapid prototyping. The processes to which each has applied rapid prototyping match the breadth of these disciplines. A small sampling includes conceptualization, form, fit, function analysis, tooling patterns, tool design, tool building, sales presentations, and marketing materials.

Every industry that makes metal or plastic parts has used rapid prototyping. Aerospace, automotive, consumer products, electronics, toys, power tools, industrial goods, and durable goods are some of the commonly referenced industries. With each successful application, the list grows. The technology is now applied to medical modeling, biomedical development, orthodontics, and custom jewelry manufacturing. Rapid prototyping is so pervasive that it would be unlikely that any individual could go about a daily routine without using a product that has in some way benefited from rapid prototyping (Fig. 5.1).

The list of products is too large to attempt to capture in a few words. Yet, some of the most exciting are those where rapid prototypes have actually taken flight in fighter aircraft and space vehicles. Equally impressive is that both NASCAR and Indy racing use rapid prototyping to win races. And finally, rapid prototyping has even been used as a presurgical planning tool for the separation of conjoined twins.

While the design of an injection mold or foundry tool may not represent life or death, as would surgical tools in the case of conjoined twins, the sophistication of today's products merits the use of rapid prototyping. The constant emphasis on "better, faster, and cheaper" demands new tools, processes, and ways of thinking.

FIGURE 5.1 From small 3D printers to large format devices such as the FDM Maxum pictured here, rapid prototyping systems offer a wide range of functionality, price, and performance. (*Photo courtesy of Stratasys, Inc.*)

5.1.3 A Tool for Change

Faced with economic challenges and global competition, the way business is done is changing. Organizations around the globe need to drive costs out of the process and product while enhancing quality and reducing time to market. Those that shoulder the burden of these requirements and initiatives find themselves with more work to do, fewer resources, and crushing deadlines. To cope or excel in this environment, the way business is done has to change.

Although this change will come in many forms, two key elements are collaboration and innovation. Design engineering and manufacturing engineering need to eliminate the barriers between the departments. Rather than "throwing a design over the wall," design and manufacturing should communicate early in the process. This communication will produce a better product at less cost and in less time. To innovate, change is required, and this change demands that nothing be taken for granted, that no process is sacred. New methods, processes, and procedures are required in the highly competitive business environment.

Rapid prototyping is ideally positioned as a tool for change. Quickly creating complex geometry with minimal labor, the advantages are obvious. Yet, to realize the full potential, rapid prototyping should be adopted by all functions within an organization. It cannot be designated as merely a design tool. Manufacturing needs to find ways to benefit from the technology, and it should demand access to this tool. This is also true for all other departments: operations, sales, marketing, and even executive management. When adopted throughout the organization, rapid prototyping can be a catalyst to powerful and lasting change.

5.2 TECHNOLOGY OVERVIEW

5.2.1 Rapid Prototyping Defined

To accurately discuss and describe rapid prototyping, it must first be defined. As the term *rapid prototyping* has become common, it has taken on many meanings. In a time of increased competitive pressure, many organizations have added rapid prototyping to their repertoire of products and services without any investment in this class of technology. Since the opposite of rapid is slow, and since no one wants to be viewed as a slow prototyper, many have adopted this term for processes that are quick but not truly rapid prototyping. Everything from machining to molding is now described as a rapid prototyping process.

For this text, rapid prototyping is used in the context of its original meaning, which was coined with the commercial release of the first stereolithography systems.

Rapid prototyping. A collection of technologies that, directly driven by CAD data, produce physical models and parts in an additive fashion.

In simpler terms, rapid prototyping is a digital tool that grows parts on a layer-by-layer basis without machining, molding, or casting.

To an even greater extent, rapid tooling and rapid manufacturing are subject to multiple definitions. Once again, if the process is completed quickly, many will describe it as either rapid tooling or rapid manufacturing. For this text, these processes are defined as follows:

Rapid tooling. The production of tools, molds, or dies—directly or indirectly—from a rapid prototyping technology.

Rapid manufacturing. The production of end use parts—directly or indirectly—from a rapid prototyping technology.

Direct means that the actual tool (or tool insert) or sellable part is produced on the rapid prototyping system. *Indirect* means that there is a secondary process between the output of the rapid prototyping system and the final tool or sellable part.

5.2.2 The Rapid Prototyping Process

While the various rapid prototyping technologies each have their own unique methodology and process, there are common elements that apply, at least in part, to each of these technologies. Although not specifically stated in the following process description, the unique factors of rapid prototyping, when compared to other manufacturing processes, are that there is minimal labor required and there is little requirement for thorough consideration of part design or construction techniques.

CAD. Rapid prototyping requires unambiguous, three-dimensional digital data as its input. Therefore, the starting point of any rapid prototyping process is the creation of a 3D CAD database. This file may be constructed as either a surfaced model or a solid model.

It is important to note that the quality of the model is critical to rapid prototyping. Construction techniques and design shortcuts that can be accommodated by other manufacturing processes may not be appropriate for rapid prototyping. The CAD file must be watertight (no gaps, holes, or voids), and geometry must not overlap. What looks good to the eye may produce poor results as a rapid prototype.

The CAD data is used to generate the STL file.

STL File Generation. The STL file is a neutral file format designed such that any CAD system can feed data to the rapid prototyping process. All commercial systems in use today can produce an STL file.

The STL file is an approximation of the geometry in the CAD file. Using a mesh of triangular elements, the bounding surfaces of the CAD file are represented in a simple file that denotes the coordinates of each vertex of the triangle.

When exporting the STL file, the goal is to balance model quality and file size. This is done by dictating the allowable deviation between the model's surface and the face of the triangle. Although there are various terms for this deviation—chord height, facet deviation, and others—each CAD system allows the user to identify the allowable gap between the triangle face and the part surface. With smaller deviation, accuracy improves.

In general, a facet deviation of 0.001 to 0.002 in. is sufficient for the rapid prototyping process. For the average CAD file, this automated process takes only a few minutes.

File Verification. Both the original CAD model and the STL generator can yield defects in the STL file that will impact the quality of the prototype or prevent its use. Common defects include near flat triangles, noncoincident triangles, and gaps between triangles.

To resolve these corruptions, there are software tools that diagnose and repair STL files. These tools are available as both third party software applications and as an integral component of the rapid prototyping system's preprocessing software.

For most files, the verification software will repair the STL file so that it is ready for building. In some cases, however, modification may be required on the original CAD model. This is often the result of poor CAD modeling techniques.

For the average file, verification and repair take just a few minutes.

File Processing. To prepare the STL files for building, there are several steps required. While this process may appear to be time consuming, it typically takes as little as a few minutes and no more than an hour. The steps in file processing include part orientation, support structure generation, part placement, slicing, and build file creation.

Part Orientation. Careful consideration of the orientation of the prototype is important in balancing the part quality and machine time. In most rapid prototyping systems, the height of the part has a significant impact on build time. As height increases, build times get longer. However, this must be balanced with part quality since the accuracy, surface finish, and feature definitions may vary by the plane in which they are located.

Support Structures. All but a few rapid prototyping systems require support structures. Supports serve two functions; rigidly attaching the part to the build platen and supporting any overhanging geometry. The generation of support structures is an automated process and, like file verification, it can be done with third party software or vendor-supplied tools. While advanced users may invest additional time to modify and locate the support structures, in most cases the operation will be performed satisfactorily without user intervention.

Part Placement. For operational efficiency and productivity, it is important to pack the platen with parts. Individual STL files (with their supports) are placed within the footprint of the build platen. The individual files are tightly packed in the build envelope so that any system-defined minimum spacing between parts can be preserved.

Slicing. With a completed build layout, the STL files are sliced into thin, horizontal cross sections. It is these individual cross sections that define the layer thickness and the "tool path" for each cross section.

Within system-allowed parameters, the slice thickness (layer thickness) is specified. As with part orientation, the layer thickness is defined with consideration of both build time and part quality. Increasing the number of layers in the part increases the build time. Yet, finer layers create smoother surface finishes by minimizing the stairstepping effect that results from the 2-axis operation of rapid prototyping.

Commercially available systems currently offer layer thicknesses of 0.0005 to 0.020 in.

Build File Creation. Each rapid prototyping system offers user-defined build parameters that are specified by the system's and the material's operating parameters. In some cases, the geometry of the part will dictate parameters that affect the quality of the prototype. After the parameters are specified, the build file is created and sent to the rapid prototyping machine.

Part Construction. Perhaps one of the biggest advantages of rapid prototyping is that the operation is unattended. With few exceptions, rapid prototyping systems operate 24 h a day without labor. The only labor required of all rapid prototyping systems is for machine preparation, build launch, and the removal of the prototypes upon completion.

To prepare for the build, material is added. And in some cases, the system is allowed to reach operating temperature.

Starting with the bottommost layer, the rapid prototyping device solidifies the part geometry. The build platform then moves downward, and the process is repeated. This continues until the last, uppermost layer is created. In effect, rapid prototyping is much like a $2\,^1/_2$-axis machining operation that adds material instead of removing it.

The length of time to construct prototypes varies dramatically by system, operating parameters, and build volume. While a large, thick-walled part could take 3 days or more, most machine runs will range from $^1/_2$ h to 48 h.

Part Cleaning. In general, rapid prototyping requires the removal of any excess material and support structures. All other aspects are process dependent and may include items such as post-curing, chemical stripping, bead blasting, or water jetting. Although hard to generalize, this process typically takes a minimum of 15 min and as much as 4 h.

As interest grows in the application of rapid prototyping in a desktop or office environment, this stage in the process is the limiting factor. The process can be messy and may require equipment and chemicals that are suited only for a shop environment.

Part Benching. Dependent on the application of the prototype, additional part preparation and finishing may be required. This is especially true when a mold-ready or paint-ready surface is desired. This process can take hours, if not days, to achieve the desired result. And, it is the one process that requires a significant amount of manual labor.

5.3 THE BENEFITS OF RAPID PROTOTYPING

In describing the rapid prototyping process, the major benefits of the technology have been revealed. At the highest level, these strengths relate to time, cost, quality, and capability. As with the use of any prototype or prototype tool, the product itself will be delivered faster, better, and at less cost. The key benefits of rapid prototyping come at the operational level where the prototyping process is executed faster with less expense and higher quality when compared to other prototyping and manufacturing techniques.

5.3.1 Time

From the name of the technology, it is obvious that the major benefit of rapid prototyping is speed. But, this advantage is generally considered only with respect to the machine time to build the prototype. The greater benefit is that the total cycle time is greatly reduced, offering convenience and efficiency in every step of the process. For the total cycle time, from data receipt to part delivery, rapid prototyping can be much faster than other manufacturing processes. This results from a combination of factors: automated processes, unattended operation, process simplification, queue minimization, and geometry insensitivity.

With rapid prototyping, it is possible to receive data at 4:30 p.m. and deliver parts the next morning. Without multiple shifts or overtime, this would be nearly impossible for other processes. Rapid prototyping virtually eliminates the time to prepare data for construction. With automated, push-button data processing, there is no need for hours of design and tool path generation. Once build data is generated, the job can be started at the end of the business day since rapid prototyping machines can be operated around the clock with no staffing. The speed and efficiency of the process decreases total cycle time and increases throughput and productivity.

Rapid prototyping also eliminates the time associated with machine setup, fixturing, mold building, and other steps in conventional processes. By eliminating these steps, delivery time is further decreased and operational efficiencies are gained. CNC machining requires that labor, materials, and machine time are available simultaneously, which can create a work-in-progress backlog. Rapid prototyping simplifies the scheduling process and minimizes work-in-progress since only data and machine time need to be available. And finally, rapid prototyping is fast for even the most complex geometries. Where, in conventional processes, a simple undercut can add hours if not days to the manufacturing process, this design feature has no impact on the time for rapid prototyping.

Even with all of these time advantages, rapid prototyping is not the fastest process for all parts. For simple, straightforward geometry, such as a basic profile with a pocket or two, CNC machining is often the fastest technology. The key difference is that rapid prototyping is fast for even the most complex designs.

5.3.2 Cost

While rapid prototyping may be much more expensive than a CNC mill in terms of initial capital expense and operating expense, the fully burdened hourly cost of rapid prototyping can be less than that for CNC. Rapid prototyping utilization is measured as a percent of all hours in the year, not work days or shift hours. With this item alone, rapid prototyping can gain a threefold advantage over an operation with a single shift. As mentioned in the discussion of time, rapid prototyping requires much less labor, especially from skilled craftspeople. This can also offer a threefold advantage.

In a shop where labor is 20 percent of the machining cost, the combination of increased utilization and reduced labor can give rapid prototyping a fivefold advantage. In other words, rapid prototyping could be five times more expensive in terms of purchase price, operating expense, and material cost and still be cheaper than CNC machining on a cost per hour basis.

5.3.3 Quality

In most areas, CNC machining has the quality advantage over rapid prototyping. But there are a few exceptions.

Rapid prototyping can produce some features that affect quality and are not available when machining. For example, rapid prototyping can produce sharp inside corners and high aspect ratio features, such as deep, narrow channels or tall, narrow ribs. To match this ability, a CNC machined part would require secondary operations, often manual operations or time consuming processes that could impact quality, time, or cost.

5.3.4 Capability

A key to the benefits listed above is that rapid prototyping is insensitivity to complexity. Eliminating the need for material removal or material molding and casting, the additive nature of rapid prototyping proves to be both efficient and accommodating. No matter how complicated or challenging the design, rapid prototyping can produce it quickly and cost effectively.

With few limitations as to what is possible, rapid prototyping promotes creativity and innovation. For both parts and tools, rapid prototyping allows designers and manufacturing engineers to experiment and try new approaches that were previously unthinkable or impossible.

5.4 APPLICATION OF RAPID PROTOTYPING, TOOLING, AND MANUFACTURING

5.4.1 Design

For design engineering, rapid prototyping is a powerful tool for conceptualization, form and fit review, functional analysis, and pattern generation. These applications are also relevant to the manufacturing process.

Evaluating and understanding a component for tool design can be somewhat difficult when one is presented with engineering drawings or 3D CAD data. A physical model offers quick, clear, and concise definition of the component's design, which leads to easier and faster visualization of the required tool design. With this clear understanding, accurate estimates of time, cost, and challenges can be offered. Also, the manufacturing team can readily participate in a collaborative effort to modify the part design to improve tooling in terms of quality, time, cost, and service life. And finally, the rapid prototype makes a great visual aid for the manufacturing team as a tool is being designed and constructed.

As a pattern generator or tool builder, rapid prototyping can allow manufacturing a functional analysis before any chips are put on the floor. Injection molding, for example, presents many challenges: knit lines, mold filling, shrinkage, gate location, ejector location, slides, lifters, sinks. Using rapid prototyping early to build a short run prototype tool allows each of these factors to be evaluated and changed before an investment is made in production tooling. This minimizes the risk of rework delays and expense.

As the product design progresses, it seems that the demand for parts swells. This usually happens well before short-run or production tooling has even been started. Rapid prototyping can satisfy this demand by offering limited quantities of product before tooling has begun. This leads to the next application, rapid tooling.

5.4.2 Tooling

In the quest to reduce cost and time in the construction of prototype, short-run, and production tooling, many have looked to rapid prototyping as a solution. While some have had great success with rapid tooling, its application has been limited. As advances in machined tooling have driven out cost and time, rapid tooling's advantages have lessened, and its limitations remain unchanged.

The most often used technologies for rapid tooling allow the production of tooling inserts in metal. However, secondary operations are required to deliver the accuracy and surface finish demanded of a tool. When added to the process, rapid tooling often offers only a slight time and cost advantage over machined tooling.

With some innovation and research, rapid tooling holds promise as a solution that greatly reduces cycle time in the manufacturing process. Since rapid prototyping is insensitive to complexity, it can produce tooling that offers nonlinear cooling channels. Conformal cooling follows a convoluted path to efficiently remove heat from the tool. In doing so, molding cycle times may be greatly reduced.

Currently, there is research into the use of gradient materials. This concept envisions the construction of the tooling insert in multiple materials. The graduated placement of multiple materials

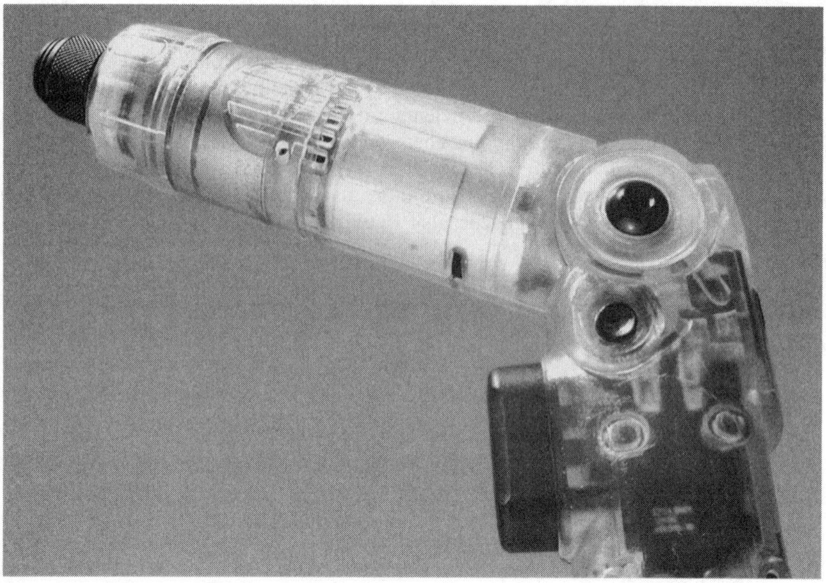

FIGURE 5.2 A prototype of a new drill, made in DSM Somos WaterClear resin, offers insight into the assembly without injection molding of clear polycarbonate parts. (*Photo courtesy of DSM Somos.*)

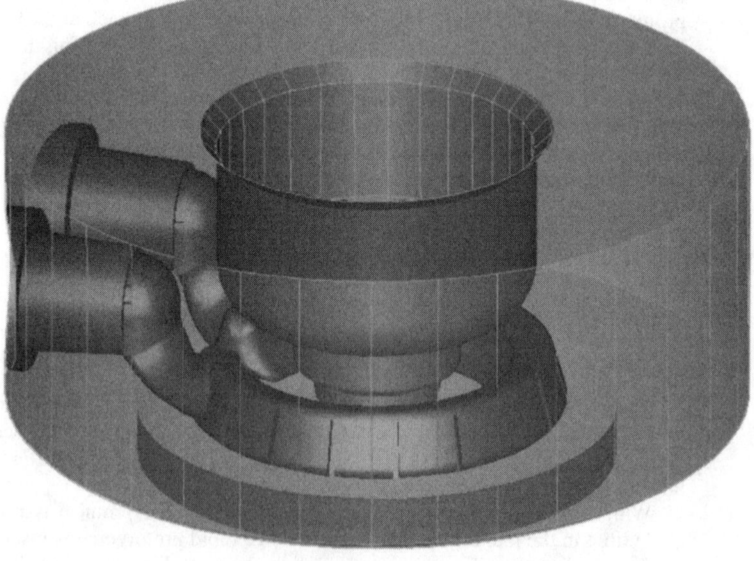

FIGURE 5.3 The additive nature of rapid prototyping processes, such as those used by MoldFusion, offers unprecedented design freedom for nonlinear, conformal cooling channels. (*Photo courtesy of D-M-E Company.*)

can simultaneously address strength, surface wear, weight, and heat dissipation. Tools constructed with gradient materials would increase efficiency and tool life while driving down cycle time. However, this concept is yet to be made viable since no rapid prototyping system offers a method for the application of multiple materials in the same machine run.

5.4.3 Manufacturing

Today few consider rapid prototyping as a viable option for manufacturing end-use product. Many view it as a possibility well into the future. However, necessity and innovation have already yielded beneficial rapid manufacturing applications.

There are few industries or applications that are required to meet specifications as stringent as those applied to military aircraft and space vehicles. So, some find it surprising, even amazing, that rapid prototyping has already been used in fighter aircraft, the space shuttle, and the space station. Fully qualified for flight, the rapid-manufactured parts have yielded time and cost savings. For the limited number of units in production, tooling and molding were much more expensive and time consuming.

Coming down to earth, rapid prototyping has been applied to other products with extremely low production volumes, such as racecars. Both directly and indirectly, rapid prototyping is used to construct metal and plastic components for Indy cars and NASCARs. In this fast-paced environment where every ounce of weight reduction is critical, race teams have found that rapid prototyping allows them to quickly realize production parts that improve performance

Obviously, these representative examples are unique. Each has production runs measured in tens, not tens of thousands, and each faces design challenges that are not common in the typical consumer or industrial product. Yet, these everyday applications can also benefit from rapid manufacturing. Innovative applications are emerging every day as companies consider the advantages and possibilities rather than the obstacles and risks.

As more companies explore the opportunities, and as the technology develops into a suitable manufacturing process, rapid manufacturing will grow beyond a niche application to become a routinely used solution.

5.4.4 Cutting Across Departments

In many companies, a compartmentalized, departmental deployment of rapid prototyping exists. While powerful in this segmented way, those that see the maximum benefit are those that use rapid prototyping as a cross-functional, collaborative tool.

As many realize, designers, especially new designers, often create products with little consideration of how they will be made and at what expense. Rapid prototyping can be used to foster collaboration and communication between design and manufacturing. While it is true that rapid prototyping can construct parts that cannot be manufactured, this is a problem only for those who do not use rapid prototyping to intercede before it is too late. With a physical model, designers and manufacturing engineers can clearly communicate intent, requirements, and challenges early in the design process. In doing so, designs can be modified to eliminate costly, time-consuming, or difficult manufacturing processes.

Once a design is complete, the rapid prototype often is discarded or placed on a shelf as a trophy. A better approach is to pass it on. As tool design is being considered, manufacturing processes evaluated, or dies are being cut, it is helpful to have the physical part to review and evaluate. Rather than making critical decisions from engineering drawings or 3D CAD data, reviewing the physical part often offers clarity and eliminates assumptions.

5.5 ECONOMIC JUSTIFICATION

Justifying the implementation of rapid prototyping can be difficult. The challenge arises for two reasons: cost of acquisition and operation and the difficulty of showing a return on investment. As a result, many current users of rapid prototyping are innovators and risk takers, those who downplay the expense while stressing the possible gains and opportunities. In a sense, many current users have justified the acquisition on a subjective or operational basis.

With system prices of $20,000 to $800,000 and total start up cost of $30,000 to $1.5 million, many organizations believe that rapid prototyping is out of their reach. When annual operating expenses are added—materials, maintenance, power, labor, training and support—the financial justification becomes even more difficult. This is compounded by the difficulty of justifying the technology on factors that are hard to measure: cost avoidance, quality improvement, and time reduction. Unless these items can be made measurable and quantifiable, a financial justification is unlikely to succeed.

Since an expenditure of this size will require management approval, quantifying the financial gain in terms of sales revenue or profit is often well received. Following is a hypothetical example of this tactic:

Total acquisition and first year expense for rapid prototyping: $750,000

- *Projected new product sales (year 1):* $50,000,000
- *Projected gross profit:* $30,000,000
- *Daily gross profit:* $120,000

Break-even point: 6.25 days [expense ($750,000) ÷ daily gross profit ($120,000/day)]

Time-to-market (historical): 9 months

Justification: Rapid prototyping needs to improve time to market by 3.47 percent (if only one new product). At a minimum, a 2-week reduction is expected for each product, yielding a 5.55 percent reduction.

Of course, there are many softer justifications for rapid prototyping. Additional justification may include efficiency, throughput, fully burdened hourly cost, and enhanced capability. One factor that should not be ignored is the reduction in demand for labor and skilled craftsmen. Companies are leaner today, and the pool of skilled trades people continues to shrink. Rapid prototyping could be justified as a tool that supports workforce reduction initiatives or as a solution to the difficulty in hiring skilled CNC machinists.

5.6 IMPLEMENTATION AND OPERATION

5.6.1 In-House or Outsource

As with any process that requires capital investment, operational overhead, and staffing, the first decision is to choose between acquisition and operation of a rapid prototyping system and outsourcing the work to a qualified service bureau. The justification for either approach will be made with consideration of system utilization, expected benefits, and total expense.

Rapid prototyping can be difficult to evaluate since there is a limited body of publicly available information. This is most evident when trying to determine the cost of operation and limitations of the technology. With dozens of available systems and limited information, it can be challenging to select the most appropriate for the current needs. For this reason, many companies elect to use service bureaus prior to a system purchase. The use of the service bureau allows the evaluation of multiple technologies and materials with minimal risk. It also establishes a baseline upon which financial projections can be made.

Should the implementation of a rapid prototyping system be justified, many find that they still require outsourced prototypes to support users' demands. There are three reasons for this outsourcing strategy. First, it is not economically sound to have the available capacity to meet peak demands. Doing so means that on most days the system will be idle or underutilized. In this case, the service bureau is used to provide capacity when demand outstrips supply. Second, for some rapid prototyping systems, carrying multiple materials can be expensive, and material conversion can be time consuming. Rather than bearing the expense of carrying all possible materials and impacting efficiency and productivity with downtime for conversion, parts are outsourced when desired material properties cannot be satisfied with the in-house material inventory. Third, it is unlikely that one technology can address all applications. It is best to implement a technology that addresses the majority of the demands while outsourcing the balance of work to service bureaus that possess the desired alternative technologies.

5.6.2 Implementing Rapid Prototyping

Independent of the decision between in-house operations or outsource purchases, there are key elements to the successful application of rapid prototyping. Like any other manufacturing technology, a strong operation is built upon education, organization, process, measurement, and management. Without these elements, the full impact of the technology will not be realized.

Beyond training, processes, and management, an in-house implementation of a rapid prototyping system has three areas of consideration: front-end systems, rapid prototyping systems, and back-end operations.

5.6.3 Technology Implementation

Front-End Systems. To deliver the speed and responsiveness expected of rapid prototyping, the implementation of front-end systems must address both process and computing needs. The areas that are addressed are the receipt, management, and processing of data for the rapid prototyping builds.

Rapid prototyping is fast-paced and subject to frequent change. Therefore, a process to coordinate, manage, and schedule the operation is vital. The schedule will be dynamic, often changing many times a day. So, a process for submitting work and managing the schedule needs to be devised. Also, multiple revisions to each part's design should be expected, and this requires a strategy for revision control and data archival needs to be developed.

The installation of front-end systems is relatively straightforward, where many elements are common to other information technology (IT) projects. Dependent on typical part size, STL files can get quite large, so it is important to establish a local area network that can transmit large files rapidly. It is also important to consider wide area network demands if data will originate from outside the rapid prototyping operations facility. This would require an FTP server with large bandwidth.

Checklist of implementation elements:

- ○ Data communications
  - □ FTP communications
  - □ Local area network
  - □ Computer servers and workstations
- ○ Data handling
  - □ Revision control
  - □ Archival
- ○ Data preparation
  - □ STL generation
  - □ File processing
- ○ Scheduling
  - □ Order receipt
  - □ Job scheduling
  - □ Order confirmation

RP Systems. Implementing the hardware for a rapid prototyping operation requires some advanced planning. Prior to the delivery of the system, facility modifications are often required. Most rapid prototyping systems are best suited to a controlled lab environment, not the shop floor or the office area. In constructing the lab, considerations include HVAC, isolation from (or for) airborne contaminants, and electricity. For some systems, supply lines for gases or water may also be required. Also allot space for material inventory, tools, and supporting equipment in the lab.

Prior to the installation, practices and procedures should be created for material storage, handling, and disposal. For those materials that are not treated as hazardous, the procedures may focus only on proper handling, disposal, or reclamation of materials. For systems that use materials that are considered hazardous, employee safety procedures should be created and corporate policies and governmental regulations should be reviewed.

In all cases, the equipment vendor will be an important information source for facility, safety, and equipment requirements. This information will be offered in advance of system delivery so that the facility is ready for the installation.

Checklist of implementation elements:

- ○ Facilities
  - □ Space allocation and modification
  - □ Electricity
  - □ Uninterruptible power supplies
  - □ Environmental control (HVAC)
  - □ Ventilation
  - □ Isolation of airborne contaminants
- ○ Installation
  - □ Set-up
  - □ Calibration
  - □ Testing
- ○ Maintenance
  - □ Routine preventative maintenance
  - □ Routine system calibration
  - □ Repairs
- ○ Materials
  - □ Material selection (may be third party)
  - □ Inventory control
  - □ Waste disposal
- ○ Safety
  - □ Equipment (gloves, respirators)
  - □ Handling and operation procedures

Back-End Operations. The post-build requirements of the rapid prototyping systems vary greatly. However, the one common element is that no technology produces a part that is ready for use directly from the machine. In general, there are two components to consider during the implementation, cleaning, and benching.

The considerations for back-end operations are similar to those for any model shop environment. In fact, if a model shop exists, the implementation may require only the addition of a few pieces of specialized equipment.

Rapid prototypes require cleaning after being removed from the system. For most processes, this entails removal of excess material (resin or powder) that coats the part's surface and the removal of support structures. The system vendor will recommend the appropriate equipment, which may include part washers, solvent tanks, or downdraft tables.

Benching is the most labor-dependent operation in the rapid prototyping process. For every system, supplying a prototype, pattern, or tool with the desired level of finish will require some degree of benching. This process will require facility modification for workstations, solvent baths, debris isolation, and possibly paint booths. Additionally, an inventory of supplies and tools will be needed.

These operations generate waste and contaminants, so thought should be given to the disposal of wastes (some considered hazardous), safety, and isolation of airborne contaminants.

Checklist of implementation elements:

- ○ Facilities
  - □ Debris isolation
  - □ Ventilation
  - □ Workstations
  - □ Lighting
- ○ Equipment
  - □ Solvent tanks
  - □ Hand tools
  - □ Downdraft tables
  - □ Paint booths
  - □ Shop equipment
    - ▷ Mills, drills, lathes
  - □ Bead blaster
  - □ Ovens
- ○ Supplies
  - □ Consumables
    - ▷ Adhesives
    - ▷ Solvents and other chemical agents
    - ▷ Sand paper
    - ▷ Primer
    - ▷ Paint
  - □ Packing materials
- ○ Waste
  - □ Disposal
    - ▷ Waste hauler
    - ▷ Procedures
  - □ Regulatory controls
- ○ Safety
  - □ Equipment
    - ▷ Gloves
    - ▷ Safety glasses or shields
    - ▷ Respirators
  - □ Handling and operation procedures

5.7 SYSTEM SELECTION: HARDWARE AND SOFTWARE

With a myriad of processes and technologies, selecting the right rapid prototyping system requires a thorough evaluation. As with other manufacturing tools, each system has both strengths and weaknesses. A successful selection is one where these attributes are fitting for the majority, but not all, of the intended applications.

For most operations, a single technology will not satisfy all the user demands. Many companies that have integrated rapid prototyping within the product development process have implemented multiple technologies.

5.7.1 Hardware

To evaluate the rapid prototyping technologies, several areas should be considered. These are:

1. Desired applications for the prototypes
2. Physical properties of the prototypes
3. Operational considerations
4. Total investment
   a. Initial acquisition and implementation
   b. Annual operating expense

While it will be fairly easy to discover the strengths of each technology, determining the limitations and operational constraints may prove difficult. Finding this information will require investigation: talking with users, attending conferences, and possibly seeking outside assistance.

To begin the selection, first define all the potential applications of the technology. For example, will the system be applied to conceptualization, form, fit and function analysis, pattern generation, tool design and creation, or rapid manufacturing? The second step is to take these applications and list the requirements for each. These considerations could include accuracy, material properties, or physical size. With this list of requirements, the evaluation of the available technologies can begin.

As a starting point, a sample criteria listing is offered below.

- Physical properties
  Available materials
  Material properties
  Accuracy
  Surface finish
  Feature definition
  Machineability
  Environmental resistance
  Maximum size
- Operational constraints
  Build times and total cycle time
  Support structures
  Throughput
  Staffing
  Secondary operations
  Facility modification
  Material selection
- Total investment
  Initial expense
  - System price
  - Facility modification
  - Material inventory
  - Training
  - Supporting equipment and facilities
    Front end and back end

  Annual expense
  - Labor
  - Materials
  - Maintenance and repair
  - Consumables
    Lasers, extrusion tips, print heads
  - Waste disposal
  - Electricity (power)
  - Insurance

For manufacturing engineering, it is critical that a representative play a role in the identification of requirements and the selection of the system. Most organizations approach rapid prototyping as a design engineering tool. As such, the requirements are driven by design engineering, which means that manufacturing engineering may have a system selected without consideration of its needs.

5.7.2 Software

Software selection is simple and straightforward when compared to the hardware evaluation. The key software components for rapid prototyping are 3D CAD and rapid prototyping preprocessing tools. Without an existing 3D CAD system implementation, it is unwise to consider a rapid prototyping system. The process of selection, implementation, and transition to 3D CAD technology is a major undertaking, so software selection for 3D CAD is simplified because it must be executed prior to a rapid prototyping evaluation.

While there are a limited number of software preprocessing tools to consider, it is recommended that this evaluation be delayed until after the successful implementation of the rapid prototyping system. Each system offers the fundamental tools to prepare STL files, so a functioning rapid prototyping operation is possible without additional software. After the hardware implementation, the true needs for preprocessing software are discovered. With this information, a software evaluation can commence.

5.8 WHAT THE FUTURE HOLDS

Rapid prototyping, tooling, and manufacturing will continue to develop in the coming years. Some of this development will be expected, but much of it will be a surprise as it originates from innovative ideas and applications.

5.8.1 User Demands

In forums of rapid prototyping users, there are four frequent and common requests. These include material development, simplified operations, cost reductions, and improvements in accuracy and repeatability.

To varying degrees, each of these areas has been addressed by today's rapid prototyping suppliers. There have been significant improvements in material properties. For example, stereolithography is no longer the technology that builds "brittle" prototypes. Likewise, there are systems available in the $30,000 price range, and these offer very simple installation, implementation, and operation. Yet, users continue to demand more. In the coming years, advancements will be made in these areas to further address the requests of the user base.

However, the most significant developments are likely to be realized in two distinct application areas, desktop systems and manufacturing tools.

5.8.2 Desktop Device or Production Tool?

Rapid prototyping systems vendors have been drawn to the two extremes of the application spectrum. Due to the higher demand, attention is shifting from high-end prototyping tools to low-cost concept modeling devices and high-end manufacturing systems. For every prototype, there is the potential for dozens of concept models and hundreds, if not thousands, of end-use parts. With this potential demand, rapid prototyping manufacturers are targeting these applications with expectations of significant increases in users.

Today's systems, generally speaking, are not ideally suited for either application. So, much of the future research and development will be focused on the necessary enhancements that will make the systems fitting for these two application areas.

Desktop Devices. Desktop rapid prototyping is not the best description for the low-end concept modeling market. Although some systems may, in the future, become small enough for the desktop, it is more likely that reasonably small, office-friendly systems become the standard. The three terms in use today that best describe this configuration are concept modelers, office modelers, and 3D printers.

In the short term, it is unlikely that rapid prototyping will advances to the point of being a low-cost, desktop device. What is more likely is that the technology will develop such that it becomes appropriate for an office environment where it is a shared resource of the engineering and manufacturing departments. To this end, several obstacles will be overcome. These include:

1. *Cost reduction.* Both the purchase price and operating expense must decrease.
2. *Size.* Systems in this category must reduce the physical size to that of an office machine.
3. *Ease of use.* To be practical as a shared device, the systems must be easily installed and operated without vendor training and support.
4. *Noise.* The noise levels must decrease to something that is equivalent to that of a copy machine.
5. *Cleanliness.* All systems produce some dirt, dust, or debris which makes them best suited for a lab or shop environment. This must be remedied.

These advances will be made in the near future, opening a big market for rapid prototyping systems that produce conceptualization tools in the office environment.

Production Tools. Rapid prototyping has had some success as both a rapid tooling and rapid manufacturing solution. But, the successes have been limited in scope and in breadth of application. Future developments will work to augment the systems and technologies to broaden the range of applications and increase system use.

Rapid tooling, once the focus of application development, has had limited success. This is especially true now that conventional processes have stepped up to meet the competitive forces. In general, rapid tooling has deficiencies when compared to machined tooling. And as the efficiency and speed of CAM and machining increases, the limitations grow. However, there are two strengths of rapid prototyping that can set it apart from CNC machining: conformal cooling and gradient materials. Each of these solutions can dramatically affect cycle time during the molding process. Offering cycle time reductions, rapid tooling could become a strong contender for cycle time management applications when improvements in surface finish, accuracy, and material properties are made.

Rapid manufacturing is truly the exciting growth area for the technologies. With an ability to make an economic order quantity of one, the possibilities are endless. However, there are barriers to the growth of this application. Currently, the technology is designed as a prototype tool that lacks the controls of a viable production device. Likewise, most of the available materials are best suited for product development applications. While some rapid prototyping technologies deliver properties that approach or mimic those of an injection molded part, most do not. To lead the way to rapid manufacturing, systems will be redesigned as production devices and new classes of materials, including plastics, composites, and metals, will be developed.

5.9 CONCLUSION

Rapid prototyping is far from being a mainstream, commonly applied tool. However, in time, it will become a mainstay in both the product development and manufacturing processes. Using an additive approach distinguishes this class of technology from all others, allowing it to quickly produce parts that are extremely complex. Yet, this differentiation is not beneficial to all parts and all applications. As a result, rapid prototyping will be one of many options available for the quick, accurate, and cost-effective completion of projects.

FURTHER READING

Burns, Marshall, *Automated Fabrication; Improving Productivity in Manufacturing,* Prentice Hall, Englewood Cliffs, NJ, 1993.

Cooper, Kenneth G., *Rapid Prototyping Technology; Selection and Application,* Marcel Dekker, New York, 2001.

Grimm, Todd A., *User's Guide to Prototyping,* Society of Manufacturing Engineers, Dearborn, MI, 2004.

Hilton, Peter, and Paul Jacobs, eds., *Rapid Tooling; Technologies and Industrial Applications*, Marcel Dekker, New York, 2000.

Jacobs, Paul F., *Stereolithography and Other RP and M Technologies; From Rapid Prototyping to Rapid Tooling,* Society of Manufacturing Engineers, New York, 1995.

Jacobs, Paul F., *Rapid Prototyping & Manufacturing; Fundamentals of Stereolithography,* Society of Manufacturing Engineers, Dearborn, MI, 1992.

Leu, Donald, *Handbook of Rapid Prototyping and Layered Manufacturing,* Academic Press, New York, 2000.

McDonald, J. A., C. J. Ryall, and D. I. Wimpenny, eds., *Rapid Prototyping Casebook,* Professional Engineering Publications, 2001.

Moldmaking Technology, Communication Technologies, Inc., www.moldmakingtechnology.com

Pham, D. T., and S. S. Dimov, *Rapid Manufacturing; The Technologies and Applications of Rapid Prototyping and Rapid Tooling,* Springer Verlag, London, 2001.

Rapid Prototyping Journal, Emerald Journals, www.emeraldinsight.com/rpsv/rpj.htm

Rapid Prototyping Report, Cad/Cam Publishing, Inc., www.cadcamnet.com

Time Compression Technologies (Europe), Rapid News Publications plc, www.time-compression.com

Time Compression Technologies (North America), Communication Technologies, Inc., www.timecompress.com

Wohlers, Terry, *Wohlers Report; Rapid Prototyping & Tooling State of the Industry,* Wohlers Associates, (www.wohlersassociates.com).

INFORMATION RESOURCES

Manufacturers

Actify, Inc., San Francisco, CA, www.actify.com

Deskartes Oy, Helsinki, Finland, www.deskartes.com

DSM Somos, New Castle, DE, www.dsmsomos.com

Materialise GmbH, Leuven, Belgium, Ann Arbor, MI, www.materialise.com

Raindrop Geomagics, Inc., Research Triangle Park, NC, www.geomagic.com

Solid Concepts Inc., Valencia, CA, www.solidconcepts.com

Stratasys, Inc., Eden Prairie, MN, www.stratasy.com

3D Systems, Valencia, CA, www.3dsystems.com

Vantico Inc., East Lansing, MI, www.vantico.com

Z Corporation, Burlington, MA, www.zcorp.com

Associations

Association of Professional Model Makers, Austin, Texas, www.modelmakers.org

Global Alliance of Rapid Prototyping Associations (GARPA), www.garpa.org. including these international associations:

Australia's QMI Solutions Ltd, www.qmisolutions.com.au/

Canadian Association of Rapid Prototyping, Tooling and Manufacturing, www.nrc.ca/imti

Chinese Rapid Forming Technology Committee, www.geocities.com/CollegePark/Lab/8600/rftc.htm

Danish Technological Institute

Finnish Rapid Prototyping Association, ltk.hut.fi/firpa/

French Rapid Prototyping Association, www.art-of-design.com/afpr/

Germany's NC Society, www.ncg.de/
Hong Kong Society for Rapid Prototyping Tooling and Manufacturing, hkumea.hku.hk/~CRPDT/RP&T.html
Italian Rapid Prototyping Association, www.apri-rapid.it/
Japanese Association of Rapid Prototyping Industry, www.rpjp.or.jp/
Association for RP Companies in The Netherlands
Rapid Product Development Association of South Africa, www.garpa.org/members.html#za
Swedish Industrial Network on FFF, www.ivf.se/FFF/fffblad.pdf
UK's Rapid Prototyping and Manufacturing Association, www.imeche.org.uk/manufacturing/rpma/
USA's Rapid Prototyping Association of the Society of Manufacturing Engineers, www.sme.org/rpa
Rapid Prototyping Association of the Society of Manufacturing Engineers, Dearborn, MI, www.sme.org/rpa

Web sites

Rapid Prototyping Home Page, University of Utah, www.cc.utah.edu/~asn8200/rapid.html
Rapid Prototyping Mailing List (RPML), rapid.lpt.fi/rp-ml/
Wohlers Associates, Wohlers Associates, Inc., www.wohlersassociates.com
Worldwide Guide to Rapid Prototyping, Castle Island, home.att.net/~castleisland/

Consultants

Edward Mackenzie Ltd, Derbyshire, UK , www.edwardmackenzie.com
Ennex Corporation, Santa Barbara, CA, www.ennex.com
New Product Dynamics, Portland, OR, www.newproductdynamics.com
T. A. Grimm & Associates, Inc., Edgewood, KY, www.tagrimm.com
Wohlers Associates, Inc. Fort Collins, CO, www.wohlersassociates.com

Events

Euromold, Frankfurt, Germany, www.euromold.de
Moldmaking Expo, Cleveland, OH, www.moldmakingtechnology.com/expo.cfm
Rapid Prototyping & Manufacturing, Dearborn, MI, www.sme.org/rapid
Solid Freeform Fabrication, Austin, TX
Siggraph, San Diego, CA, www.siggraph.org
TCT Exhibition, Manchester, UK, www.time-compress.com

CHAPTER 6
DIMENSIONING AND TOLERANCING

Vijay Srinivasan
IBM Corporation and Columbia University
New York, NY

6.1 OVERVIEW

This chapter deals with some fundamental and practical aspects of dimensioning and tolerancing and their importance to manufacturing engineering. Every engineer should be familiar with the notion of dimensioning a sketch of a part, either by formal training or by intuitive trait. He or she will be less familiar with tolerancing because it is usually not taught as a part of the engineering curriculum. This is unfortunate because dimensioning and tolerancing form the bulk of the engineering documentation in industry. This chapter provides a brief description of this important topic and points to other sources for details.

6.2 INTRODUCTION

Dimensions are numerical values assigned to certain geometric parameters. These are measures of some distances and angles, and are expressed in their appropriate units of measure (e.g., inches, millimeters, degrees, minutes). Classical dimensioning is closely associated with projected views presented in engineering drawings, which may be hand drawn or generated on a computer screen using a computer-aided drafting software system. Modern computer-aided design (CAD) systems are more powerful. They are capable of generating, storing, and transmitting three-dimensional geometric models of parts. Increasing use of such CAD systems has enlarged the scope of dimensioning because we can now assign numerical values to some parameters in a 3D CAD model and treat them as dimensions; alternatively we can query some distance or angle measures of geometric elements in these CAD models and treat them as dimensions.

If we are dealing only with dimensioning, it is possible to live completely in the world of ideal geometric forms. These platonic ideal forms have been studied for nearly 2500 years, and we have a wealth of knowledge of these from which we can develop a very good understanding of dimensioning. The dimensioning information presented in this chapter is condensed from such understanding.

But these ideal forms are never to be found in nature or in man-made artifacts. As Plato himself observed, no circle, however carefully it is drawn, can be perfectly circular. Extending this notion, we observe that no manufactured object has ideal geometric form. Even worse, we notice that no two manufactured objects are geometrically identical. This is due to a fundamental axiom in manufacturing that states that *all manufacturing processes are inherently imprecise and produce parts that vary*. One can try to reduce this variability by applying economic resources, but it can never be completely eliminated. Consequently, this variability is explicitly accommodated in design using tolerances and consciously controlled in production using process controls. While the designer takes the responsibility for specifying tolerances that don't compromise the function of the product, the manufacturing engineer is responsible for correctly interpreting the tolerance specifications and selecting appropriate manufacturing processes to meet these specifications. Both are responsible for keeping the overall cost under control—this can only be achieved by concurrent engineering or early manufacturing involvement in product development. This is often referred to as *design for manufacturability*.

In contrast to dimensioning, we have only limited theoretical understanding of tolerancing. We have the benefit of just a few centuries of practice following the industrial revolution and, more recently, mass production and interchangeability. Some of the best practices found thus far have been codified by national and international standards. These form the basis for what we know and use in tolerancing. As our understanding improves, these tolerancing standards also undergo changes. The tolerancing information found in this chapter provides a snapshot of where we stand in the beginning of the twenty-first century.

6.3 DIMENSIONING INTRINSIC CHARACTERISTICS

A simple, but useful, dimensioning procedure starts with dimensioning simple geometric objects in a part and then dimensioning the relationship among these objects. In this section we focus on dimensioning simple geometric objects such as elementary curves and surfaces. A later section deals with dimensioning the relationship among these objects.

A theory of dimensioning can be developed on the basis of the simple idea of congruence. We consider two geometric objects to be equivalent if they are congruent under rigid motion. Congruence is a valid relationship for defining equivalence because it satisfies the following three properties for an equivalence relation:

1. *Reflexive.* A geometric object A is congruent to itself.
2. *Symmetric.* If A is congruent to B, then B is congruent to A.
3. *Transitive.* If A is congruent to B and B is congruent to C, then A is congruent to C.

So we ask when two curves or surfaces are congruent and use the answer to dimension them as described below.

6.3.1 Dimensioning Elementary Curves

The simplest curve is the (unbounded) straight line. It doesn't have any intrinsic dimension. Next in the hierarchy of complexity are the second-degree curves called conics. Table 6.1 lists all possible conics that can occur in engineering. The last column of this table lists the intrinsic parameters of the associated conic. For example, the semi-major axis a and semi-minor axis b are the intrinsic parameters for an ellipse. These are called intrinsic because they don't change if the ellipse is moved around in space. More formally, we say that intrinsic characteristics are those that remain invariant under rigid motion.

An important special case of the ellipse is the circle, when the semi-major and the semi-minor axes equal the radius. Its intrinsic parameter is the radius, or equivalently, its diameter. All circles that have the same radius are congruent, and hence are equivalent. The only thing that distinguishes one circle from another is its radius. Therefore, we say that the circle belongs to a one-parameter family of curves, where the parameter is the radius. So we can dimension a circle by specifying a numerical value for its radius or diameter.

TABLE 6.1 Intrinsic Dimensions for Conics Are the Numerical Values for the Intrinsic Parameters

	Type	Canonical equation	Intrinsic parameters
Non-degenerate conics	Ellipse	$\frac{x^2}{a^2}+\frac{y^2}{b^2}=1$	a, b
	Special case: Circle	$x^2 + y^2 = a^2$	a = radius
	Hyperbola	$\frac{x^2}{a^2}-\frac{y^2}{b^2}=1$	a, b
	Parabola	$y^2 - 2lx = 0$	l
Degenerate conics	Parallel lines	$x^2 - a^2 = 0$	a = half of the distance between the parallel lines
	Intersecting lines	$\frac{x^2}{a^2}-\frac{y^2}{b^2}=0$	$\tan^{-1}(b/a)$ = half of the angle between the intersecting lines
	Coincident lines	$x^2 = 0$	None

Degenerate conics listed in Table 6.1 correspond to a pair of straight lines in the plane. If they are distinct and parallel, then we just dimension the distance between them. If they are distinct, but intersect, then the angle between them can be dimensioned. If the pair of lines coincide, then there is no intrinsic dimensioning issue.

After conics, the most important types of curves used in engineering are the free-form curves. These include the Bézier and B-spline curves. These can be dimensioned by dimensioning their control polygons. Figure 6.1 shows how a simple Bézier curve can be dimensioned. Alternatively, it can be dimensioned by coordinate dimensioning of its control points.

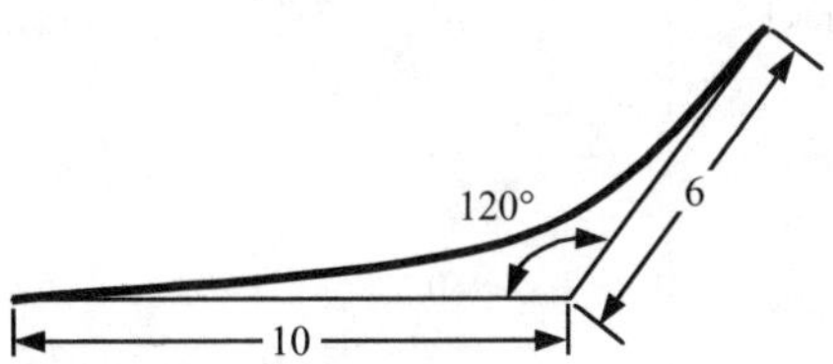

FIGURE 6.1 Dimensioning a second-degree Bézier curve.

6.3.2 Dimensioning Elementary Surfaces

Moving to surfaces, we first note that the simplest surface is an unbounded plane. It doesn't have any intrinsic dimension. Next in the hierarchy of complexity are the second-degree surfaces called quadrics. Table 6.2 lists all possible quadrics that can occur in engineering. The last column in this table lists the intrinsic parameters of these surfaces. The quadrics can be dimensioned by assigning numerical values to these parameters.

Two special cases of the nondegenerate quadrics are of importance because they occur frequently in engineering. A sphere is a special case of an ellipsoid. Its dimension is its radius, or equivalently, its diameter. A (right circular) cylinder is a special case of an elliptic cylinder. Its dimension is its radius, or equivalently, its diameter. Both sphere and cylinder belong to the one-parameter family of surfaces.

The degenerate quadrics in Table 6.2 correspond to the degenerate conics of Table 6.1. Two distinct, parallel planes can be dimensioned by the distance between them. Two distinct, but intersecting, planes can be dimensioned by the angle between them. If the two planes are coincident then there is no intrinsic dimensioning issue.

Armed with just these basic facts, we can justify the dimensioning scheme shown in Fig. 6.2. It is a simple example of a rectangular plate of constant thickness, with a cylindrical hole. We will use this

TABLE 6.2 Intrinsic Dimensions for Quadrics Are the Numerical Values for the Intrinsic Parameters

	Type	Canonical equation	Intrinsic parameters
Non-degenerate quadrics	Ellipsoid	$\frac{x^2}{a^2}+\frac{y^2}{b^2}+\frac{z^2}{c^2}=1$	a, b, c
	Special case: Sphere	$x^2 + y^2 + z^2 = a^2$	a = radius
	Hyperboloid of one sheet	$\frac{x^2}{a^2}+\frac{y^2}{b^2}-\frac{z^2}{c^2}=1$	a, b, c
	Hyperboloid of two sheets	$\frac{x^2}{a^2}+\frac{y^2}{b^2}-\frac{z^2}{c^2}=-1$	a, b, c
	Quadric cone	$\frac{x^2}{a^2}+\frac{y^2}{b^2}-\frac{z^2}{c^2}=0$	a/c, b/c
	Elliptic paraboloid	$\frac{x^2}{a^2}+\frac{y^2}{b^2}-2z=0$	a, b
	Hyperbolic paraboloid	$\frac{x^2}{a^2}-\frac{y^2}{b^2}+2z=0$	a, b
	Elliptic cylinder	$\frac{x^2}{a^2}+\frac{y^2}{b^2}=1$	a, b
	Special case: Cylinder	$x^2 + y^2 = a^2$	a = radius
	Hyperbolic cylinder	$\frac{x^2}{a^2}-\frac{y^2}{b^2}=1$	a, b
	Parabolic cylinder	$y^2 - 2lx = 0$	l
Degenerate quadrics	Parallel planes	$x^2 - a^2 = 0$	a = half of the distance between the parallel planes
	Intersecting planes	$\frac{x^2}{a^2}-\frac{y^2}{b^2}=0$	$\tan^{-1}(b/a)$ = half of the angle between the intersecting planes
	Coincident planes	$x^2 = 0$	None

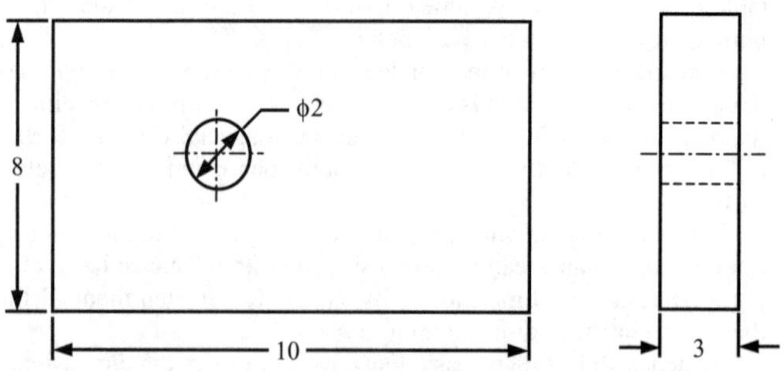

FIGURE 6.2 Example of dimensioning intrinsic characteristics.

example throughout this chapter. The diameter of the cylindrical surface and the distance between two parallel planes (there are three such pairs of parallel planes in this example) have been dimensioned. We justify such dimensioning based on the reasoning about quadrics given above.

After quadrics, the most important types of surfaces used in engineering are the free-form surfaces. These include the Bézier and B-spline surface patches. These can be dimensioned by dimensioning their control nets.

6.4. TOLERANCING INDIVIDUAL CHARACTERISTICS

As indicated in the introduction, tolerancing practice is not as general as dimensioning and is restricted to some well-known cases. In this section we focus on tolerancing individual characteristics. Tolerancing relational characteristics is discussed in a later section.

6.4.1 Size Tolerancing

It might be tempting to think that any intrinsic dimension encountered in the last section can be toleranced by specifying (upper and lower) limits to that dimension. This is not the case. There are only three cases, called features of size, currently permitted in the U.S. standards for size tolerancing. They are (1) a spherical surface, (2) a cylindrical surface, and (3) a set of two opposed elements or opposed parallel surfaces. Interestingly, all these cases appear in Table 6.2, where the sphere and the cylinder make their appearance as special cases of nondegenerate quadrics, and a pair of parallel planes appears as a degenerate quadric. Figure 6.3 illustrates an example of size tolerancing.

How should these size tolerance specifications be interpreted? According to the current U.S. standards, there are two conditions that should be checked. The first is the actual local size at each cross section of the feature of size, which can be checked by the so-called "two-point measurements." For example, Fig. 6.4 shows an exaggerated illustration of the side view of an actual part toleranced in Fig. 6.3. Note that this illustration captures the fact that there are no ideal geometric forms in an actual, manufactured part. To check whether the part conforms to the size tolerancing specified as 3 ± 0.1 in Fig. 6.3, we first check whether all actual local sizes, one of which is shown in Fig. 6.4(*a*), are within the limits of 2.9 and 3.1 units. But this is only the first of the two checks that are necessary. The second check is whether the two actual surface features involved in this size tolerancing do not extend beyond the boundary, also called the envelope, of perfect form at the maximum material condition (MMC). For example, it means that in Fig. 6.4(*b*) the actual surface features should lie between two parallel planes that are 3.1 units apart. These two conditions apply for all three types of features of size mentioned earlier. Note that the envelope requirement can be checked using functional gages. If the second condition, also called the envelope principle, is not required in some application, then a note that PERFECT FORM AT MMC NOT REQD can be specified for that size tolerance.

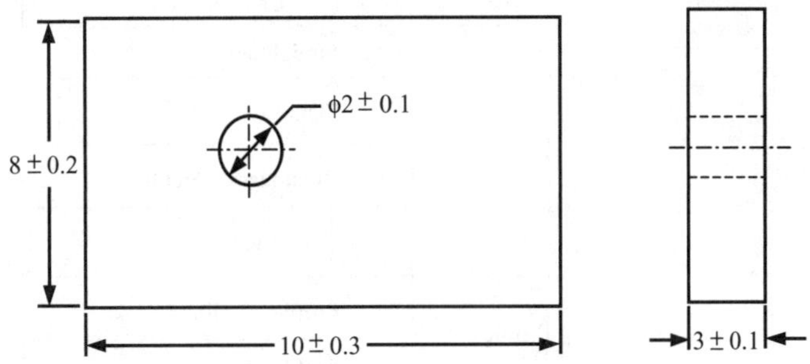

FIGURE 6.3 Example of size tolerancing.

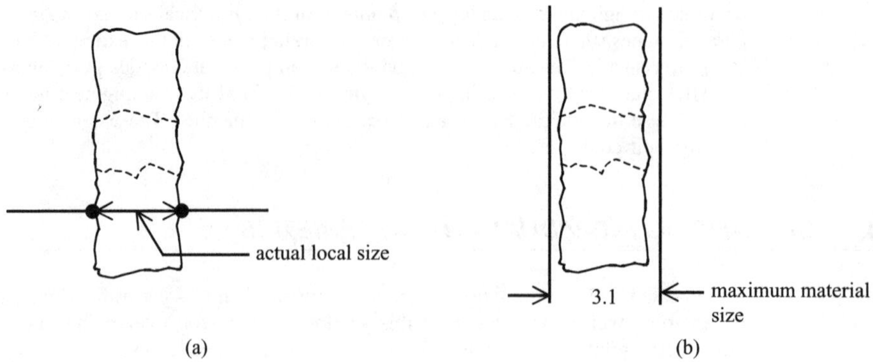

FIGURE 6.4 Example of checking size tolerance for conformance.

A given actual feature of size can have an infinite number of actual local sizes. But it can have only one actual mating size. The actual mating size of the actual feature shown in Fig. 6.4(*b*), for example, is the smallest distance between two parallel planes within which these two actual surface features are contained.

Before leaving size tolerancing, we note that national and international standards committees are constantly examining ways to define it more precisely and to extend its coverage to several cases beyond just the three mentioned earlier.

6.4.2 Form Tolerancing

The fact that an actual, manufactured surface does not possess an ideal geometric form brings into question how far that actual surface can deviate from the ideal form. This is addressed by form tolerancing. Table 6.3 shows the standardized form tolerances for individual characteristics. Of these, four cases are specially designated as form tolerances and these will be examined in this section. The next section covers profile tolerances, which are generalizations of form tolerances.

The four cases of form tolerances in Table 6.3 cover straightness, flatness, roundness, and cylindricity. As the names imply, these form tolerances apply only to those individual features that are nominally defined to be straight line segments, planar patches, circles (full or partial), and finite

TABLE 6.3 Form and Profile Tolerancing Individual Characteristics

Type	Characteristic	Symbol
Form	Straightness	—
	Flatness	⏥
	Roundness (Circularity)	○
	Cylindricity	⌭
Profile	Profile of a line	⌒
	Profile of a surface	⌓

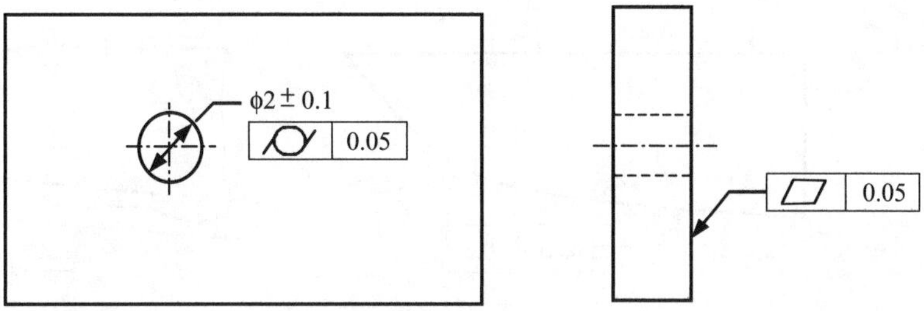

FIGURE 6.5 Example of form tolerancing.

cylinders (full or partial), respectively. Figure 6.5 shows specification of flatness and cylindricity tolerances on an example part.

Figure 6.6 shows how these form tolerance specifications should be interpreted. This is the first instance where we come across the notion of a tolerance zone, which is a very important concept in tolerancing. Figure 6.6(*a*) shows a tolerance zone for cylindricity; it is bounded by two coaxial cylinders whose radial separation is 0.05 units. The actual cylindrical feature should lie within this tolerance zone to conform to the cylindricity specification of Fig 6.5. Note that the radius of the inner or outer cylinder in this tolerance zone is not specified. Only the difference between these radii is specified because cylindricity controls only how close the actual feature is to an ideal cylindrical form and not how big this cylinder is.

Similarly, Fig. 6.6(*b*) shows a tolerance zone for flatness; it is bounded by two parallel planes separated by 0.05 units. The actual planar feature should lie within this tolerance zone to conform to the flatness specification of Fig. 6.5. The tolerance zone can be positioned anywhere in space, as long as it contains the actual planar feature because flatness controls only how close the actual feature is to an ideal planar form.

Straightness and roundness deal with how straight and circular certain geometric elements should be. Roundness, which is also called circularity, may be applied to any circular cross section of a nominally axially symmetric surface. Straightness may be applied to any straight line-segment that can be obtained as a planar cross section of a nominal surface; it may also be applied to the axis of a nominal surface feature.

Given an actual feature, it is possible to define an actual value for its form tolerances. It is the thickness of the thinnest tolerance zone within which the actual feature is contained. For example, the actual value for cylindricity for the actual cylindrical feature in Fig 6.6(*a*) is the smallest radial separation between two coaxial cylinders within which the actual feature lies. Similarly, the actual

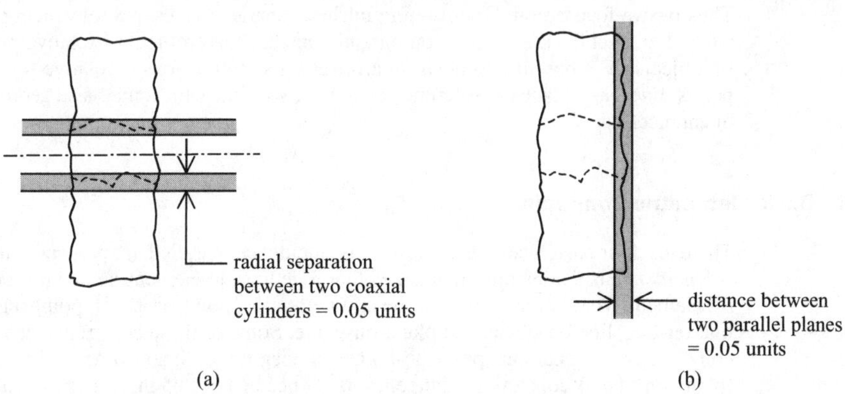

FIGURE 6.6 Example of checking form tolerances for conformance. (*a*) Cylindricity specifies a tolerance zone bounded by two coaxial cylinders. (*b*) Flatness specifies a tolerance zone bounded by two parallel planes.

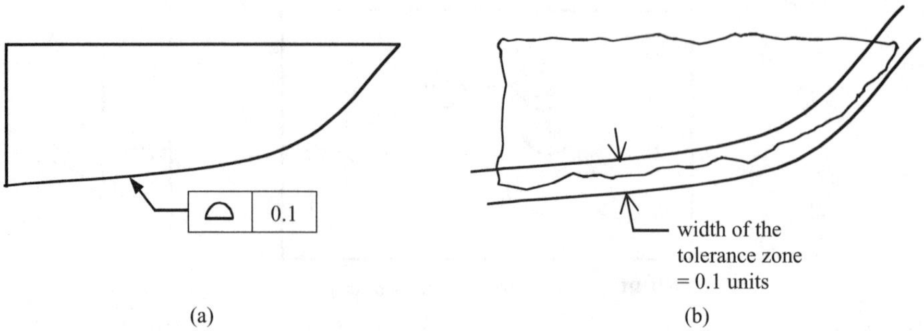

FIGURE 6.7 Example of profile tolerancing.

value for the flatness for the planar feature in Fig 6.6(*b*) is the smallest distance between two parallel planes within which the actual feature lies.

It is easy to see that form tolerancing may be extended to other surfaces such as spheres and cones. These are under consideration in national and international standards committees.

6.4.3 Profile Tolerancing

A simple curve or surface can be subjected to profile tolerancing listed in Table 6.3. Profile of a line actually controls the profile of a curve. Profile tolerancing is a generalization of form tolerancing and it can be applied to any arbitrary curve or surface.

Figure 6.7 (*a*) shows a free-form surface that is under profile tolerancing. The boundary for the tolerance zone in this case, shown in Fig 6.7(*b*), is obtained by taking the nominal free-form surface and offsetting it by 0.05 units in one direction and 0.05 units in the other direction. Actual value for profile tolerancing can be defined by considering the thickness of the thinnest tolerance zone within which the actual feature lies.

Profile tolerancing can also be applied to control relationships among features.

6.5 DIMENSIONING RELATIONAL CHARACTERISTICS

Thus far we focused on dimensioning intrinsic characteristics and tolerancing individual characteristics. Let's get back to dimensioning and consider the problem of relative positioning of geometric objects. It is possible to position arbitrary geometric objects relative to each other using only points, lines, and planes as reference elements. We start with some basic geometric constraints used in engineering.

6.5.1 Basic Geometric Constraints

There are four basic geometric constraints: incidence, parallelism, perpendicularity, and chirality.

Incidence is a constraint that states that a geometric element lies within or on another geometric element. These constraints can be enumerated as point-on-point, point-on-line, point-on-plane, line-on-line, line-on-plane, and plane-on-plane. Some of these constraints can also be referred to as coincidence (for example, point-on-point), concentric, collinear (line-on-line), coaxial, or coplanar (plane-on-plane) constraints. Tangency is a special type of incidence constraint. Note that more than two elements can be involved in an incidence constraint.

Parallelism applies to lines and planes. There can be parallelism constraint between line and line, line and plane, and plane and plane. More than two elements can be involved in a parallelism constraint. *Perpendicularity* also applies to lines and planes. There can be perpendicularity constraint between two lines in a plane, between a line and a plane, and between two planes.

Chirality refers to the left- or right-handedness of an object. It is an important constraint because left- and right-handed objects are usually not interchangeable. For example, a helix has chirality; a right-handed helical thread is not interchangeable with a left-handed thread.

6.5.2 Reference Elements

We start with the observation that any geometric object belongs to one and only one of the seven classes of symmetry listed in Table 6.4. The second column in that table gives us a set of points, lines, and planes that can serve as reference elements for the corresponding row object. These reference element(s) are important because the associated object can be positioned using the reference element, as we will see in what follows.

For a sphere, its center is the reference element. This holds true even for a solid sphere or a set of concentric spheres. For a cylinder, its axis is the reference element. This is also true for a solid cylinder, a cylindrical hole, or a set of coaxial cylinders. For a plane, the reference element is the plane itself. If we have several parallel planes, then any plane parallel to them can serve as a reference element. For a slot or a slab, which is bounded by two parallel planes, it is customary to take their middle (median) plane and treat it as the reference element. For a helical object, such as a screw thread, a helix having the same chirality, axis, and pitch will serve as a reference element. In practice, we just take the axis of the helix and use it to position the helical object.

Any surface of revolution is a revolute object. A cone, for example, belongs to the revolute class. Its reference elements consist of the axis of the cone and the vertex of the cone (which lies on the axis of the cone). An oval-shaped slot belongs to a prismatic class. Its reference elements consist of a plane, which could be the plane of symmetry of the slot, and a line in that plane along the generator of the prism. Finally, if an object does not belong to any of the six described so far, it belongs to the general class. We then need to find a plane, a line in the plane, and a point on the line and treat them as the reference elements for this general object.

The advantage of reference element(s) is that they can be used to dimension the relative positioning of one object with respect to another. For example, the relative position of a sphere and a cylinder can be dimensioned by dimensioning the distance between the center of the sphere and the axis of the cylinder. Similarly, the relative position between two planes, if they are not coplanar, can be dimensioned by the distance between them if they are parallel and by the angle between them if they intersect.

Figure 6.8 shows how a cylindrical hole can be positioned relative to two planes. In this figure, several other relative positions are assumed from the drawing convention. The planar faces indicated as

TABLE 6.4 Classes of Symmetry and the Associated Reference Element(s)

Class	Reference element(s)
Spherical	Center (point)
Cylindrical	Axis (line)
Planar	Plane
Helical	Helix
Revolute	Axis (line) and a point on the axis
Prismatic	Plane and a line on the plane
General	Plane, line, and point

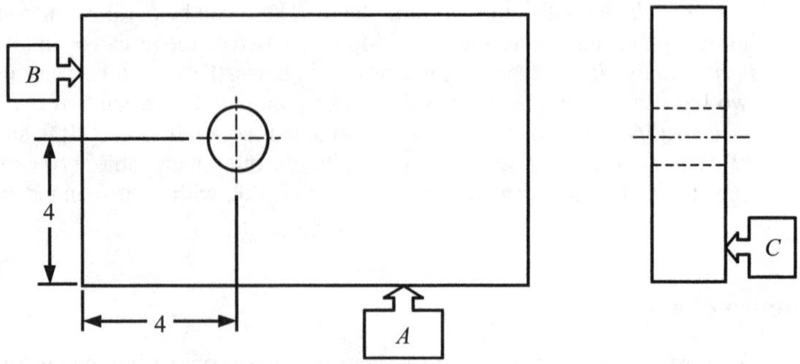

FIGURE 6.8 Example of relative positioning.

A and *B* are assumed to intersect at 90° because that is how they seem to be drawn. Also, the planar face indicated as *C* is assumed to be perpendicular to both *A* and *B*, from the way it is drawn. The axis of the cylindrical hole in Fig. 6.8 is assumed to be perpendicular to *C* for the same reason. Therefore, the axis of this cylindrical face is parallel to both *A* and *B*, and this justifies the dimensional scheme shown in Fig. 6.8 to position the cylinder relative to *A* and *B*.

Table 6.5 shows how many dimensions may be needed to position point, line, and plane relative to each other. The line-line case refers to the general case of skew lines, which may be dimensioned by the shortest distance between the lines and their (signed) twist angle. If the lines lie in a plane, then only one dimension (distance between parallel lines or angle between intersecting lines) is necessary for their relative positioning.

The reference element(s) of Table 6.4 can be used beyond the realm of dimensioning. They will be used to guide the definition of datums in a later section for tolerancing relational characteristics.

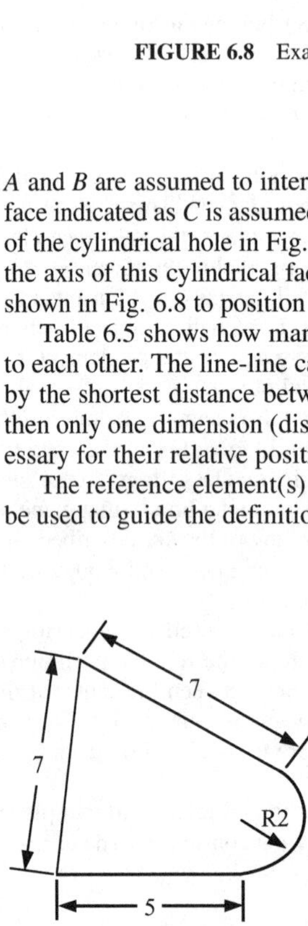

FIGURE 6.9 A sketch with dimensional constraints.

6.5.3 Dimensional Constraints

Figure 6.9 shows a planar sketch with dimensional constraints. The circular arc is tangential to the line segments at both ends. This is a valid dimensioning scheme but drawing such a sketch to scale is not a trivial exercise. Modern CAD systems can handle these dimensional constraints and produce correct figures. Such practices are becoming increasingly popular with designers, especially in the so-called feature-based designs. They specify a set of simultaneous dimensions and geometric constraints, and let the CAD system figure out the rest of the geometric details.

TABLE 6.5 Number of Dimensions Needed for Relative Positioning

	Point	Line	Plane
Point	1	1	1
Line	1	2	1
Plane	1	1	1

TABLE 6.6 Tolerancing Relational Characteristics

Type	Characteristic	Symbol
Orientation	Parallelism	//
	Perpendicularity	⊥
	Angularity	∠
Location	Position	⌖
	Concentricity	◎
	Symmetry	⌯
Runout	Circular runout	↗
	Total runout	⌰
Profile	Profile of a line	⌒
	Profile of a surface	⌓

6.6 TOLERANCING RELATIONAL CHARACTERISTICS

Table 6.6 lists the type of tolerances defined in the U.S. standards to control the relational characteristics of features in a manufactured object. These tolerances control variations in the relative position (location and/or orientation) of features. In specifying such tolerances, at least two features are involved; one of these two is used to define a datum and the other feature is toleranced relative to this datum. It is, therefore, important to understand how datums are defined.

6.6.1 Datums

Recall from an earlier section that simple reference elements can be defined for any geometric object to ease the relative positioning problem. Given a feature on a manufactured object, we first use a fitting procedure to derive an ideal feature that has perfect form (such as a plane, cylinder, or sphere). Then we extract one or more reference elements (such as plane, axis, or center) from these ideal features, which can be used as a single datum or in a datum system. The use of datums will become clear in the next few sections.

6.6.2 Orientation, Location, and Runout

Table 6.6 lists parallelism, perpendicularity, and angularity as the three characteristics under orientation tolerancing. Note that parallelism and perpendicularity are two of the four basic geometric constraints described earlier. Let's look at parallelism tolerancing in some detail.

Figure 6.10 (*a*) illustrates the specification of a parallelism tolerance using the plate with a hole as an example. The face indicated as *C* in Fig. 6.8 is specified as a datum feature in Fig. 6.10(*a*). Its opposite face is nominally parallel to it and this face is now subjected to a parallelism tolerance of 0.025 units with respect to the datum feature *C*. On a manufactured part, such as the one shown in Fig. 6.10(*b*), a datum plane is first fitted to the feature that corresponds to the datum feature *C*.

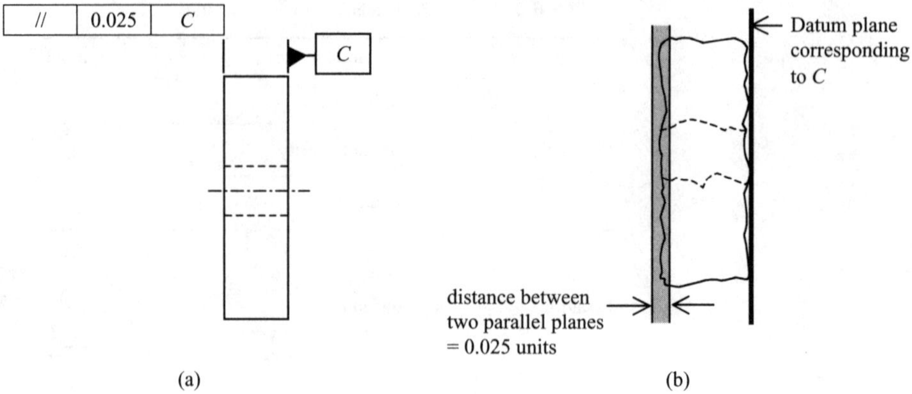

FIGURE 6.10 (*a*) Specification of parallelism tolerance, and (*b*) its interpretation. The shaded tolerance zone is parallel to the established datum plane in (*b*).

We know that a plane will serve as a datum because the datum feature is nominally planar and its reference element is a plane, according to Table 6.4. There are several ways in which this datum plane can be established. One method is to mount this face carefully on a surface plate and use the surface plate as the datum. It is also possible to measure several points on this feature using a coordinate measuring machine (CMM) and apply a mathematical algorithm to fit a plane to these points.

The interpretation of the specified parallelism, illustrated in Fig. 6.10(*b*), is that on a manufactured part the toleranced feature should lie within two parallel planes, which are themselves parallel to the datum plane, that are 0.025 units apart. Note that only the orientation of this tolerance zone, bounded by two parallel planes, is controlled by the datum; it can be located anywhere relative to the datum. On an actual part, the width of the thinnest tolerance zone that satisfies the parallelism constraint and contains the actual feature is the actual value for the parallelism tolerance.

Table 6.6 also lists position, concentricity, and symmetry as the three characteristics under location tolerancing. Let's look at position tolerancing in some detail. Figure 6.11(*a*) illustrates

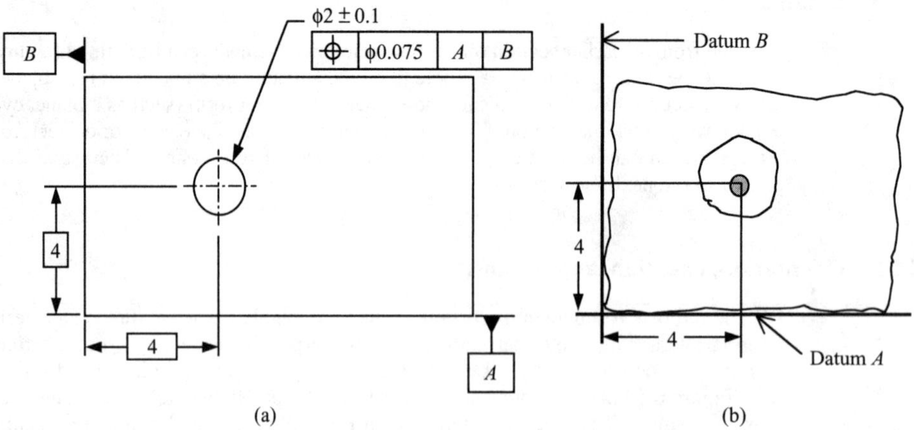

FIGURE 6.11 Example of location tolerancing. The shaded cylindrical tolerance zone in (*b*) has a diameter of 0.075 units.

how the location of a cylindrical hole can be toleranced using a datum system. Given an actual part, such as the one shown in Fig. 6.11(*b*), we first establish the primary datum *A*, which is a plane, by a fitting process or by using a surface plate. Then the secondary datum *B* is established. It is also a plane and it is constrained to be perpendicular to the datum plane *A*. These two datum planes then define an object belonging to the prismatic class (per Table 6.4), and this is the datum system relative to which we will position a cylindrical tolerance zone of diameter 0.075 units, as shown in Fig. 6.11(*b*). The interpretation of the position tolerance is that the axis of the actual cylindrical hole should lie within this tolerance zone. For an actual part, the diameter of the smallest cylindrical tolerance zone within which the axis lies is the actual value for the position tolerance.

Circular and total runouts listed in Table 6.6 are two special relational characteristics that are toleranced when rotating elements are involved. The reader is referred to Refs.1, 2, 3, and 4 for details on the runout and other relational characteristics (e.g., profile tolerancing of relational characteristics) not covered here.

6.6.3 MMC, LMC, and Boundary Conditions

For functional reasons, one may want to specify tolerances on relational characteristics by invoking maximum or least material conditions. Figure 6.12 (*a*) illustrates one such specification using MMC (maximum material condition) on the plate with a hole example. The datums *A* and *B* are established as before. We then construct a virtual boundary, which is a cylinder of diameter 2 – 0.1 – 0.075 = 1.825 units, as shown in Fig. 6.12(*b*). The interpretation of the position tolerancing under MMC is that the actual cylindrical hole should lie outside the virtual boundary. Note that it is possible to verify this on an actual part using a functional gage. In fact, functional gaging is a basic motivator for MMC type specifications. In the example shown, an actual related mating size for an actual part is defined as the largest diameter of the virtual boundary, located at the specified distances from the datum system, which still leaves the actual hole outside of it.

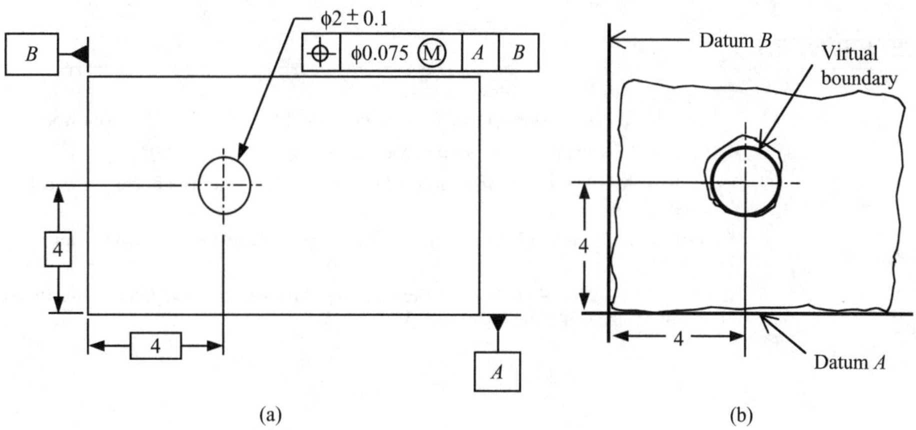

FIGURE 6.12 Example of location tolerancing under MMC. The virtual boundary in (*b*) is a cylinder of diameter 1.825 units.

6.7 MANUFACTURING CONSIDERATIONS

Dimensions and tolerances drive two major concerns for manufacturing engineers. One concern deals with manufacturing process capability and process control and the other deals with inspection issues.

When manufacturing engineers receive a tolerance specification, one of their first tasks is to figure out whether the manufacturing resources at their disposal are capable of meeting the tolerance requirements. Assuming that such capability exists, the next concern is whether these manufacturing tools and processes can be controlled over a long period of time to mass produce these parts still meeting the specification. These concerns are addressed by manufacturing process capability studies and by the use of statistical process control techniques, respectively.

Inspection issues fall under two broad categories. The first is the set of measurement issues associated with manufacturing process diagnosis. Here parts are measured to establish and/or verify the health of the manufacturing processes that produce these parts. The second is the set of measurement issues that arise when we attempt to verify whether the manufactured parts meet the tolerance specifications. In both cases statistical techniques that can address measurement uncertainty and sampling issues (using statistical quality control) should be employed.

6.8 SUMMARY AND FURTHER READING

Dimensioning and tolerancing is a vast topic and this chapter gives only a short summary of this topic. It attempts a balanced treatment of dimensioning and tolerancing. Both intrinsic and relational dimensioning are described. Their tolerancing counterparts are individual and relational tolerancing, which are also described. The McGraw-Hill *Dimensioning and Tolerancing Handbook*[1] is the best reference for detailed information on tolerancing. For details on a theory of dimensioning, the reader is referred to a recent book devoted exclusively to this topic.[2]

ASME Y14.5M-1994[3] is the US standard for dimensioning and tolerancing. ISO 1101[4] is the corresponding international standard. ASME Y14.5.1M-1994[5] gives a mathematical theory of tolerancing. Updated versions of these standards are expected within the next few years and they may address statistical tolerancing in some detail. ISO is currently issuing a series of GPS (Geometrical Product Specification) standards that deal with detailed tolerance specification and verification issues.

REFERENCES

1. P. Drake, Jr, ed., *Dimensioning and Tolerancing Handbook*, McGraw-Hill, New York, 1999.
2. V. Srinivasan, *Theory of Dimensioning*, Marcel Dekker, New York, 2003.
3. ASME Y14.5M-1994, *Dimensioning and Tolerancing,* The American Society of Mechanical Engineers, New York, 1995.
4. ISO 1101-1983, *Technical Drawing—Geometrical Tolerancing,* International Organization for Standardization, Geneva, 1983.
5. ASME Y14.5.1-1994, *Mathematical Definition of Dimensioning and Tolerancing Principles*, The American Society of Mechanical Engineers, New York, 1995.

CHAPTER 7
BASIC TOOLS FOR TOLERANCE ANALYSIS OF MECHANICAL ASSEMBLIES

Ken Chase
Mechanical Engineering Department
Brigham Young University, Provo, Utah

7.1 INTRODUCTION

As manufacturing companies pursue higher quality products, they spend much of their efforts monitoring and controlling variation. Dimensional variation in production parts accumulate or stack up statistically and propagate through an assembly kinematically, causing critical features of the final product to vary. Such variation can cause costly problems during assembly, requiring extensive rework or scrapped parts. It can also cause unsatisfactory performance of the finished product, drastically increasing warranty costs and creating dissatisfied customers.

One of the effective tools for variation management is tolerance analysis. This is a quantitative tool for predicting the accumulation of variation in an assembly by performing a stack-up analysis. It involves the following steps:

1. Identifying the dimensions which chain together to control a critical assembly dimension or feature.
2. The mean, or average, assembly dimension is determined by summing the mean of the dimensions in the chain.
3. The variation in the assembly dimension is estimated by summing the corresponding component variations. This process is called a "stack-up."
4. The predicted assembly variation is compared to the engineering limits to estimate the number of rejects, or nonconforming assemblies.
5. Design or production changes may be made after evaluating the results of the analysis.

If the parts are production parts, actual measured data may be used. This is preferred. However, if the parts are not yet in production, measured data is not available. In that case, the engineer searches for data on similar parts and processes. That failing, he or she may substitute the tolerance on each dimension in place of its variation, assuming that quality controls will keep the individual part variations within tolerance. This substitution is so common in the design stage that the process is generally called *tolerance analysis*.

The four most popular models for tolerance stack-up are shown in Table 7.1. Each has its own advantages and limitations.

TABLE 7.1 Models for Tolerance Stack-Up Analysis in Assemblies

Model	Stack formula	Predicts	Application
Worst Case (WC)	$\sigma_{ASM} = \sum \lvert T_i \rvert$ Not statistical	Extreme limits of variation No rejects permitted	Critical systems Most costly
Statistical (RSS)	$\sigma_{ASM} = \sqrt{\sum \left(\frac{T_i}{3}\right)^2}$	Probable variation Percent rejects	Reasonable estimate Some rejects allowed Less costly
Six Sigma (6s)	$\sigma_{ASM} = \sqrt{\sum \left(\frac{T_i}{3C_P(1-k)}\right)^2}$	Long-term variation Percent rejects	Drift in mean over time is expected High quality levels desired
Measured Data (Meas)	$\sigma_{ASM} = \sqrt{\sum \sigma_i^2}$	Variation using existing part measurements Percent rejects	After parts are made What-if? studies

7.2 COMPARISON OF STACK-UP MODELS

The two most common stack-up models are:

Worst Case (WC). Computes the extreme limits by summing absolute values of the tolerances, to obtain the worst combination of over and undersize parts. If the worst case is within assembly tolerance limits, there will be no rejected assemblies. For given assembly limits, WC will require the tightest part tolerances. Thus, it is the most costly.

Statistical (RSS). Adds variations by root-sum-squares (RSS). Since it considers the statistical probabilities of the possible combinations, the predicted limits are more reasonable. RSS predicts the statistical distribution of the assembly feature, from which percent rejects can be estimated. It can also account for static mean shifts.

As an example, suppose we had an assembly of nine components of equal precision, such that the same tolerance T_i may be assumed for each. The predicted assembly variation would be:

$$\text{WC:}\quad T_{ASM} = \sum \lvert T_i \rvert = 9 \times 0.01 = \pm 0.09$$

$$\text{RSS:}\quad T_{ASM} = \sqrt{\sum T_i^2} = \sqrt{9 \times 0.01^2} = \pm 0.03$$

(± denotes a symmetric range of variation)

Clearly, WC predicts much more variation than RSS. The difference is even greater as the number of component dimensions in the chain increases.

Now, suppose $T_{ASM} = 0.09$ is specified as a design requirement. The stack-up analysis is reversed. The required component tolerances are determined from the assembly tolerance.

$$\text{WC:}\quad T_i = \frac{T_{ASM}}{9} = \frac{0.09}{9} = \pm 0.01$$

$$\text{RSS:}\quad T_i = \frac{T_{ASM}}{\sqrt{9}} = \frac{0.09}{3} = \pm 0.03$$

Here, WC requires much tighter tolerances than RSS to meet an assembly requirement.

7.3 USING STATISTICS TO PREDICT REJECTS

All manufacturing processes produce random variations in each dimension. If you measured each part and kept track of how many are produced at each size, you could make a *frequency plot,* as shown in Fig. 7.1.

Generally, most of the parts will be clustered about the mean or average value, causing the plot to bulge in the middle. The further you get from the mean, the fewer parts will be produced, causing the frequency plot to decrease to zero at the extremes.

A common statistical model used to describe random variations is shown in the figure. It is called a *normal*, or *Gaussian*, distribution. The *mean* μ marks the highest point on the curve and tells how close the process is to the target dimension. The spread of the distribution is expressed by its *standard deviation* σ, which indicates the precision or process capability.

UL and *LL* mark the *upper and lower limits* of size, as set by the design requirements. If *UL* and *LL* correspond to the $\pm 3\sigma$ process capability, as shown, a few parts will be rejected (about 3 per 1000).

Any normal distribution may be converted to a *standard normal,* which has a mean of zero and σ of 1.0. Instead of plotting the frequency versus size, it is plotted in terms of the number of standard deviations from the mean. Standard tables then permit you to determine the fraction of assemblies which will fail to meet the engineering limits. This is accomplished as follows:

1. Perform a tolerance stack-up analysis to calculate the mean and standard deviation of the assembly dimension X, which has design requirements X_{UL} and X_{LL}.
2. Calculate the number of standard deviations from the mean to each limit:

$$Z_{UL} = \frac{X_{UL} - \overline{X}}{\sigma_X} \qquad Z_{LL} = \frac{X_{LL} - \overline{X}}{\sigma_X}$$

   where $\overline{X}$ and σ_X are the mean and standard deviation of the assembly dimension X, and $\overline{Z} = 0$ and $\sigma_Z = 1.0$ are the mean and standard deviation of the transformed distribution curve.
3. Using standard normal tables, look up the fraction of assemblies lying between Z_{LL} and Z_{UL} (the area under the curve). This is the predicted *yield,* or fraction of assemblies which will meet the requirements. The fraction lying outside the limits is (1.0 – yield). These are the predicted *rejects,* usually expressed in parts per million (ppm).

   Note: Standard tables list only positive Z, since the normal distribution is symmetric.

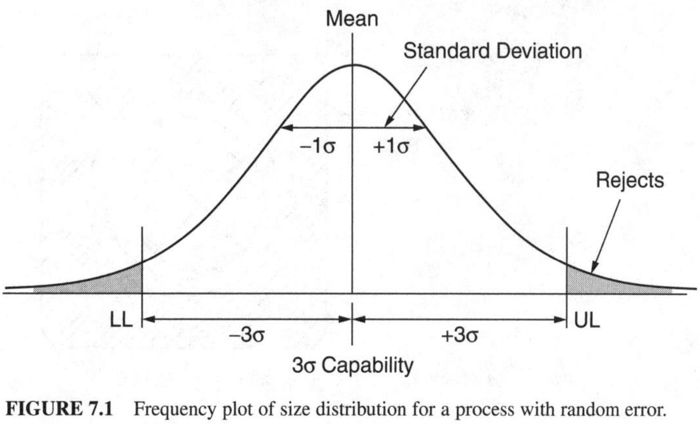

FIGURE 7.1 Frequency plot of size distribution for a process with random error.

TABLE 7.2 Comparative Quality Level vs. Number of Standard Deviations in Z_{LL} and Z_{UL}

Z_{LL} and Z_{UL}	Yield fraction	Rejects per million	Quality level
$\pm 2\sigma$	0.9545	45500	Unacceptable
$\pm 3\sigma$	0.9973	2700	Moderate
$\pm 4\sigma$	0.9999366	63.4	High
$\pm 5\sigma$	0.999999426	0.57	Very high
$\pm 6\sigma$	0.999999998	0.002	Extremely high

Expressing the values Z_{LL} and Z_{UL} in standard deviations provides a nondimensional measure of the quality level of an assembly process. A comparison of the relative quality in terms of the number of σ is presented in Table 7.2.

7.4 PERCENT CONTRIBUTION

Another valuable, yet simple, evaluation tool is the *percent contribution*. By calculating the percent contribution that each variation contributes to the resultant assembly variation, designers and production personnel can decide where to concentrate their quality improvement efforts. The contribution is just the ratio of a component standard deviation to the total assembly standard deviation:

$$\text{WC: } \%\text{Cont} = 100\frac{T_i}{T_{ASM}} \qquad \text{RSS: } \%\text{Cont} = 100\frac{\sigma_i^2}{\sigma_{ASM}^2}$$

7.5 EXAMPLE 1—CYLINDRICAL FIT

A clearance must be maintained between the rotating shaft and bushing shown in Fig. 7.2. The minimum clearance must not be less than 0.002 in. The max is not specified. Nominal dimension and tolerance for each part are given in Table 7.3, below:

The first step involves converting the given dimensions and tolerances to centered dimensions and symmetric tolerances. This is a requirement for statistical tolerance analysis. The resulting centered

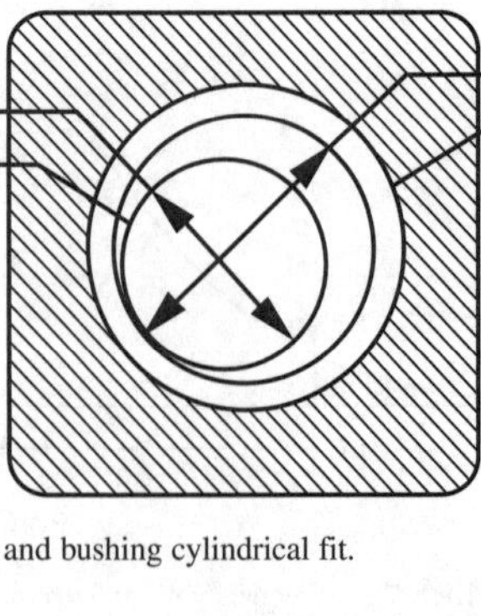

FIGURE 7.2 Shaft and bushing cylindrical fit.

TABLE 7.3 Dimensions and Tolerances—Cylindrical Fit Assembly

Part	Nominal dimension	*LL* tolerance	*UL* tolerance	Centered dimension	Plus/minus tolerance
Bushing *B*	0.75	−0	+.0020	0.7510	±0.0010
Shaft *S*	0.75	−0.0028	−0.0016	0.7478	±0.0006
Clearance *C*				0.0032	±0.0016 WC ±0.00117RSS

dimensions and symmetric tolerances are listed in the last two columns. If you calculate the maximum and minimum dimensions for both cases, you will see that they are equivalent.

The next step is to calculate the mean clearance and variation about the mean. The variation has been calculated both by WC and RSS stackup, for comparison.

$$\text{Mean clearance: } \bar{C} = \bar{B} - \bar{S} = 0.7510 - 0.7478 = 0.0032 \text{ in}$$

(the bar denotes the mean or average value)

$$\text{WC variation: } T_C = |T_B| + |T_S| = 0.0010 + 0.0006 = 0.0016 \text{ in}$$

$$\text{RSS variation: } T_C = \sqrt{T_B^2 + T_S^2} = \sqrt{0.0010^2 + 0.0006^2} = 0.00117 \text{ in}$$

Note that even though *C* is the difference between *B* and *S*, the tolerances are summed. Component tolerances are always summed. You can think of the absolute value canceling the negative sign for WC and the square of the tolerance canceling for RSS.

The predicted range of the clearance is $C = 0.0032 \pm 0.00117$ in (RSS),

or, $C_{max} = 0.0044$, $C_{min} = 0.00203$ in

Note that C_{max} and C_{min} are not absolute limits. They represent the $\pm 3\sigma$ limits of the variation. It is the overall process capability of this assembly process, calculated from the process capabilities of each of the component dimensions in the chain. The tails of the distribution actually extend beyond these limits.

So, how many assemblies will have a clearance less than 0.002 in? To answer this question, we must first calculate Z_{LL} in terms of dimensionless σ units. The corresponding yield is obtained by table lookup in a math table or by using a spreadsheet, such as Microsoft Excel:

$$\sigma_C = \frac{T_C}{3} = 0.00039 \text{ in}$$

$$Z_{LL} = \frac{LL - \bar{C}}{\sigma_C} = \frac{0.002 - 0.0032}{0.00039} = -3.087\sigma$$

The results from Excel are:

Yield = NORMSDIST(Z_{LL}) = 0.998989 Reject fraction = 1.0 − Yield = 0.001011

or, 99.8989 percent good assemblies, 1011 ppm (parts per million) rejects.

Only Z_{LL} was needed, since there was no upper limit specified.

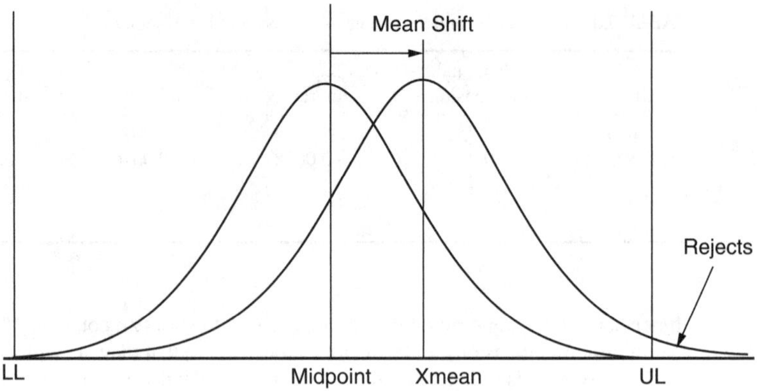

FIGURE 7.3 Normal distribution with a mean shift causes an increase in rejects.

A Z_{LL} magnitude of 3.087σ indicates a moderate quality level is predicted, provided the specified tolerances truly represent the $\pm 3\sigma$ process variations. Figure 7.3 is a plot showing a normal distribution with a mean positive shift. Note the increase in rejects due to the shift.

7.6 HOW TO ACCOUNT FOR MEAN SHIFTS

It is common practice in statistical tolerance analysis to assume that the mean of the distribution is stationary, located at the midpoint between the *LL* and *UL*. This is generally not true. All processes shift with time due to numerous causes, such as tool wear, thermal expansion, drift in the electronic control systems, operator errors, and the like. Other errors cause shifts of a fixed amount, including fixture errors, setup errors, setup differences from batch to batch, material properties differences, etc. A shift in the nominal dimension of any part in the chain can throw the whole assembly off center by a corresponding amount.

When the mean of the distribution shifts off center, it can cause serious problems. More of the tail of the distribution is shoved beyond the limit, increasing the number of rejects. The slope of the curve steepens as you move the mean toward the limit, so the rejects can increase dramatically. Mean shifts can become the dominant source of rejects. No company can afford to ignore them.

There are two kinds of mean shifts that must be considered: static and dynamic. *Static mean shifts* occur once, and affect every part produced thereafter with a fixed error. They cause a fixed shift in the mean of the distribution. *Dynamic mean shifts* occur gradually over time. They may drift in one direction, or back and forth. Over time, large-scale production requires multiple setups, multicavity molds, multiple suppliers, etc. The net result of each dynamic error source is to degrade the distribution, increasing its spread. Thus, more of the tails will be thrust beyond the limits.

To model the effect of static mean shifts, one simply alters the mean value of one or more of the component dimensions. If you have data of actual mean shifts, that is even better. When you calculate the distance from the mean to *LL* and *UL* in σ units, you can calculate the rejects at each limit. That gives you a handle on the problem.

Modeling dynamic mean shifts requires altering the tolerance stackup model. Instead of estimating the standard deviation σ_i of the dimensional tolerances from $T_i = 3\sigma_i$, as in conventional RSS tolerance analysis, a modified form is used to account for higher quality level processes:

$$T_i = 3\,Cp_i\,\sigma_i \qquad \text{where } Cp \text{ is the } \textit{process capability index}$$

$$C_P = \frac{UL - LL}{6\sigma}$$

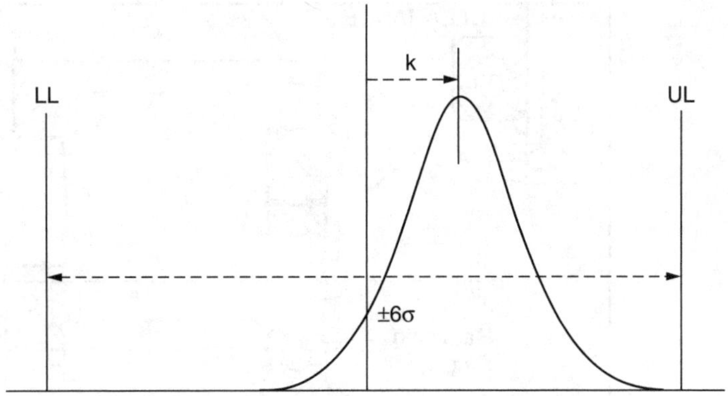

FIGURE 7.4 The six sigma model uses a drift factor k and $\pm 6\sigma$ limits to simulate high quality levels.

If the UL and LL correspond to $\pm 3\sigma$ of the process, then the difference $UL - LL = 6\sigma$, and Cp will be 1.0. Thus, a Cp of 1.0 corresponds to a "moderate quality level" of $\pm 3\sigma$. If the tolerances correspond to $\pm 6\sigma$, $UL - LL = 12\sigma$, and $Cp = 2.0$, corresponding to an "extremely high quality level" of $\pm 6\sigma$.

The Six Sigma model for tolerance stack-up accounts for both high quality and dynamic mean shift by altering the stack-up equation to include the Cp and a drift factor k for each dimension in the chain.

$$\sigma_{ASM} = \sqrt{\sum \left(\frac{T_i}{3C_{Pi}(1-k_i)} \right)^2}$$

As Cp increases, the contribution of that dimension decreases, causing σ_{ASM} to decrease.

The drift factor k measures how much the mean of a distribution has been observed to drift during production. Factor k is a fraction, between 0 and 1.0. Figure 7.4 shows that k corresponds to the shift in the mean as a percent of the tolerance. If there is no data, it is usually set to $k = 0.25$. The effects of these modifications are demonstrated by a comprehensive example.

7.7 EXAMPLE 2—AXIAL SHAFT AND BEARING STACK

The shaft and bearing assembly shown in Fig. 7.5 requires clearance between the shoulder and inner bearing race (see inset) to allow for thermal expansion during operation. Dimensions A through G stack up to control the clearance U. They form a chain of dimensions, indicated by vectors added tip-to-tail in the figure. The chain is 1-D, but the vectors are offset vertically for clarity. The vector chain passes from mating-part to mating-part as it crosses each pair of mating surfaces. Note that all the vectors acting to the right are positive and to the left are negative. By starting the chain on the left side of the clearance and ending at the right, a positive sum indicates a clearance and a negative sum, an interference.

Each dimension is subject to variation. Variations accumulate through the chain, causing the clearance to vary as the resultant of the sum of variations. The nominal and process tolerance limits for each one are listed in Table 7.4 with labels corresponding to the figure.

The design requirement for the clearance U is given below. The upper and lower limits of clearance U are determined by the designer, from performance requirements. Such assembly

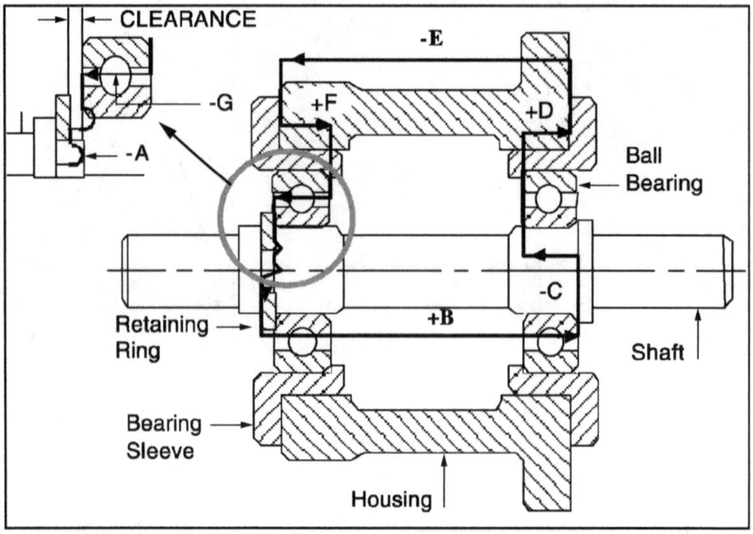

FIGURE 7.5 Example Problem 2: Shaft and bearing assembly. (Fortini, 1967)

requirements are called *key characteristics*. They represent critical assembly features, which affect performance.

Design requirement: Clearance (U) = 0.020 ± 0.015 in

Initial design tolerances for dimensions B, D, E, and F were selected from a tolerance chart, which describes the "natural variation" of the processes by which parts are made (Trucks 1987). It is a bar chart, indicating the range of variation achievable by each process. Also note that the range of variation depends on the nominal size of the part dimension. The tolerances for B, D, E, and F were chosen from the middle of the range of the turning process, corresponding to the nominal size of each. These values are used as a first estimate, since no parts have been made. As the variation analysis progresses, the designer may elect to modify them to meet the design requirements. The bearings, and retaining ring, however, are vendor-supplied. The dimensions and tolerances for A, C, and G are therefore fixed, not subject to modification.

The next step is to calculate the mean clearance and variation about the mean. The variation has been calculated both by WC and RSS stackup, for comparison.

TABLE 7.4 Nominal Dimensions and Tolerances for the Example Problem 2

Part	Dimension	Nominal in	Tolerance in	Process limits	
				Min Tol	Max Tol
Retaining ring	A*	−.0505	±.0015*	*	*
Shaft	B	8.000	±.008	±0.003	±0.020
Bearing	C*	−.5090	±.0025*	*	*
Bearing sleeve	D	.400	±.002	±0.0008	±0.005
Housing	E	−7.705	±.006	±0.0025	±0.0150
Bearing sleeve	F	.400	±.002	±0.0008	±0.005
Bearing	G*	−.5090	±.0025*	*	*

* Vendor-supplied part

Mean clearance:

$$
\begin{aligned}
\overline{U} &= -\overline{A} + \overline{B} - \overline{C} + \overline{D} - \overline{E} + \overline{F} - \overline{G} \\
&= -0.0505 + 8.000 - 0.509 + 0.400 - 7.705 + 0.400 - 0.509 \\
&= 0.0265
\end{aligned}
$$

WC variation:

$$
\begin{aligned}
T_U &= |T_A| + |T_B| + |T_C| + |T_D| + |T_E| + |T_F| + |T_G| \\
&= 0.0015 + 0.008 + 0.0025 + 0.002 + 0.006 + 0.002 + 0.0025 \\
&= 0.0245
\end{aligned}
$$

RSS variation:

$$
\begin{aligned}
T_U &= \sqrt{T_A^2 + T_B^2 + T_C^2 + T_D^2 + T_E^2 + T_F^2 + T_G^2} \\
&= \sqrt{0.0015^2 + 0.008^2 + 0.0025^2 + 0.002^2 + 0.006^2 + 0.002^2 + 0.0025^2} \\
&= 0.01108
\end{aligned}
$$

Parts-per-million rejects:

$$
Z_{UL} = \frac{U_{UL} - \overline{U}}{\sigma_U} = \frac{0.035 - 0.0265}{0.00369} = 2.30\sigma \quad \Rightarrow \quad 10{,}679 \quad PPM\_\text{Rejects}
$$

$$
Z_{LL} = \frac{U_{LL} - \overline{U}}{\sigma_U} = \frac{0.005 - 0.0265}{0.00369} = -5.82\sigma \quad \Rightarrow \quad 0.0030 \quad PPM\_\text{Rejects}
$$

Percent Contribution. The percent contribution has been calculated for all seven dimensions, for both WC and RSS. A plot of the results is shown in Fig. 7.6. RSS is greater because it is the square of the ratio of the variation.

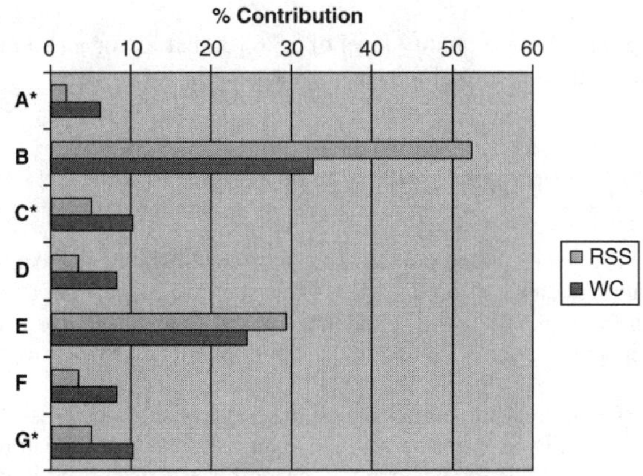

% Contribution	
WC	**RSS**
6.12	1.83
32.65	52.14
10.20	5.09
8.16	3.26
24.49	29.33
8.16	3.26
10.20	5.09

FIGURE 7.6 Percent contribution chart for Example 2.

7.8 CENTERING

The example problem discovered a mean shift of 0.0065 in from the target value, 0.020 in, midway between *LL* and *UL*. The analysis illustrates the effect of the mean shift—a large increase in rejects at the upper limit and reduced rejects at the lower limit. To correct the problem, we must modify one or more nominal values of the dimensions *B*, *D*, *E*, or *F*, since *A*, *C*, and *G* are fixed.

Correcting the problem is more challenging. Simply changing a callout on a drawing to center the mean will not make it happen. The mean value of a single dimension is the average of many produced parts. Machinists cannot tell what the mean is until they have made many parts. They can try to compensate, but it is difficult to know what to change. They must account for tool wear, temperature changes, set up errors, etc. The cause of the problem must be identified and corrected. It may require tooling modifications, changes in the processes, careful monitoring of the target value, a temperature-controlled workplace, adaptive machine controls, etc. Multicavity molds may have to be qualified cavity-by-cavity and modified if needed. It may require careful evaluation of all the dimensions in the chain to see which is most cost effective to modify.

In this case, we have chosen to increase dimension *E* by 0.0065 in, to a value of 7.7115 in. The results are:

Mean	σ_{ASM}	Z_{LL}	Rejects	Z_{UL}	Rejects
0.020 in	0.01108 in	–4.06	24 ppm	4.06	24 ppm

This would be a good solution, if we could successfully hold that mean value by better fixturing, more frequent tool sharpening, statistical process control, and the like.

7.9 ADJUSTING THE VARIANCE

Suppose the mean of the process cannot be controlled sufficiently. In that case, we may choose to adjust the tolerance of one or more dimensions. The largest contributors are dimensions *B* on the shaft and *E* on the housing. We reduce them both to 0.004 in with the results:

Mean	σ_{ASM}	Z_{LL}	Rejects	Z_{UL}	Rejects
0.0265 in	0.00247 in	–8.72	0 ppm	3.45	284 ppm

This corresponds to an effective quality level of $\pm 3.63\sigma$, that is, for a two-tailed, centered distribution having the same number of total rejects (142 at each limit).

7.10 MIXING NORMAL AND UNIFORM DISTRIBUTIONS

Suppose the shaft (Part **B**) is jobbed out to a new shop, with which we have no previous experience. We are uncertain how much variation to expect. How shall we account for this uncertainty? We could do a worst case analysis, but that would penalize the entire assembly for just one part of unknown quality. We could instead resort to a uniform distribution, applied to dimension *B*, leaving the others as normal.

The *uniform* distribution is sometimes called the "equal likelihood" distribution. It is rectangular in shape. There are no tails, as with the normal. Every size between the upper and lower tolerance limits has an equal probability of occurring. The uniform distribution is conservative. It predicts greater variation than the normal, but not as great as worst case.

For a uniform distribution, the tolerance limits are not $\pm 3\sigma$, as they are for the normal. They are equal to $\pm\sqrt{3}\sigma$. Thus, the stackup equation becomes:

$$\sigma_{ASM} = \sqrt{\sum\left(\frac{T_i}{3}\right)^2 + \sum\left(\frac{T_j}{\sqrt{3}}\right)^2}$$

where the first summation is the squares of the σ_i for the normal distributions and the second sum is for the uniform distributions.

For the example problem, dimension *B* has upper and lower limits of 8.008 and 7.992 in, respectively, corresponding to the $\pm\sqrt{3}\sigma$ limits. We assume the assembly distribution has been centered and the only change is that *B* is uniform rather than normal. Substituting the tolerance for *B* in the second summation and the tolerance for each of the other dimensions in the first summation, the results are:

Mean	σ_{ASM}	Z_{LL}	Rejects	Z_{UL}	Rejects
0.020 in	0.00528 in	−2.84	2243 ppm	2.84	2243 ppm

The predicted rejects assume that the resulting distribution of assembly clearance *U* is normal. This is generally true if there are five or more dimensions in the stack. Even if all of the component dimensions were uniform, the resultant would still approximate a normal distribution. However, if one non-normal dimension has a much larger variation than the sum of all the others in the stack, the assembly distribution would be non-normal.

7.11 SIX SIGMA ANALYSIS

Six Sigma analysis accounts for long-term drift in the mean, or dynamic mean shift, in manufactured parts. It uses the process capability index *Cp* and drift factor *k* to simulate the long term spreading of the distribution, as mentioned earlier. In the following, Six Sigma is applied to two models for example Problem 2. The first uses *Cp* = 1.0 for comparison directly with RSS, corresponding to a $\pm 3\sigma$ quality level, with and without drift correction. The second case uses *Cp* = 2.0 for comparison of $\pm 6\sigma$ quality levels with $\pm 3\sigma$. The results are presented in Table 7.5 alongside WC and RSS results for comparison. All centered cases used the modified nominals to center the distribution mean.

TABLE 7. 5 Comparison of Tolerance Analysis Models for Example 2

Model	Mean in	σ_{ASM} in	Z_{LL}/Z_{UL} σ	Rejects ppm	Quality σ
Centered					
WC	0.020	0.00820*	N/A	N/A	N/A
RSS—Uniform	0.020	0.00640	±2.34	19027	2.34
RSS—Normal	0.020	0.00369	±4.06	48	4.06
6Sigma—*Cp* = 1	0.020	0.00492	±3.05	2316	3.05
6Sigma—*Cp* = 2	0.020	0.00246	±6.1	0.0011	6.10
Mean Shift					
RSS—Uniform	0.0265	0.00640	−3.36/1.33	92341	1.68
RSS—Normal	0.0265	0.00369	−5.82/2.30	10679	2.55
6Sigma—*Cp* = 1	0.0265	0.00492	−4.37/1.73	42162	2.03
6Sigma—*Cp* = 2	0.0265	0.00246	−8.73/3.45	278	3.64

*WC has no σ. This is calculated from $T_{ASM}/3$ for comparison with RSS methods.

Noncentered cases used a mean shift of 0.0065 in. The RSS—uniform results were not presented before, as this case applied uniform distributions to all seven dimensions for comparison to WC.

7.12 REMARKS

The foregoing discussion has presented techniques for predicting tolerance stacking, or the accumulation of variation, in mechanical assembly processes. There is quite a wide range of results, depending on the assumptions, available data, and the quality goals involved. As with any analytical modeling, it is wise to verify the results by measurements. When production data become available, values of the mean and standard deviation of the measured dimensions may be substituted into the RSS stack equation. This will give real-world data to benchmark against.

In 1-D stacks, the means do add linearly and standard deviations do add by root-sum-squares, as long as the variations are independent (not correlated). There are tests for correlation, which may be applied. Verification will build confidence in the methods. Experience will improve your assembly modeling skills and help you decide which analytical models are most appropriate for given applications.

There are many topics which have been omitted from this introduction, including:

1. Modeling variable clearances, such as the clearance around a bolt or shaft, which can introduce variation into a chain of dimensions as an input rather than a resultant assembly gap.
2. Treating errors due to human assembly operations, such as positioning parts in a slip-joint before tightening the bolts.
3. Available standards for tolerancing, such as cylindrical fits, or standard parts, like fasteners.
4. How to apply GD&T to tolerance stacks.
5. Tolerance allocation algorithms, which assist in assigning tolerances systematically.
6. When and how to use Monte Carlo Simulation, design of experiments, response surface methodology, and method of system moments for advanced applications.
7. How to treat non-normal distributions, such as skewed distributions.
8. Methods for modeling 2-D and 3-D assembly stacks.
9. CAD-based tolerance analysis tools.

The results here presented were obtained using an Excel spreadsheet called CATS 1-D, which is available as a free download, along with documents, from the ADCATS web site, listed in the References.

For further reading, see below. Additional papers which discuss many of these topics are available on the ADCATS web site.

REFERENCES

Trucks, H. E., *Designing for Economical Production,* 2nd ed., Society of Manufacturing Engineers, Dearborn, Michigan, 1987.

Fortini, E.T., *Dimensioning for Interchangeable Manufacture,* Industrial Press, New York, 1967.

FURTHER READING

ADCATS web site. http://adcats.et.byu.edu

Chase, K. W., J. Gao, and S. P. Magleby, "General 2-D Tolerance Analysis of Mechanical Assemblies with Small Kinematic Adjustments," *J. of Design and Manufacturing,* vol. 5, no. 4, 1995, pp. 263–274.

Chase, K. W., J. Gao, and S. P. Magleby, "Tolerance Analysis of 2-D and 3-D Mechanical Assemblies with Small Kinematic Adjustments," Chap. 5 in *Advanced Tolerancing Techniques,* John Wiley, 1998, pp. 103–137.

Chase, K. W., and W. H. Greenwood, "Design Issues in Mechanical Tolerance Analysis," *Manufacturing Review,* ASME, vol. 1, no. 1, March, 1988, pp. 50–59.

Chase, K. W., S. P. Magleby, and C. G. Glancy, "A Comprehensive System for Computer-Aided Tolerance Analysis of 2-D and 3-D Mechanical Assemblies," *Proceedings of the 5th CIRP Seminar on Computer-Aided Tolerancing,* Toronto, Ontario, April 28–29, 1997.

Chase, K. W., and A. R. Parkinson, "A Survey of Research in the Application of Tolerance Analysis to the Design of Mechanical Assemblies," *Research in Engineering Design,* vol. 3, 1991, pp. 23–37.

Creveling, C. M., *Tolerance Design*, Addison-Wesley, Reading, MA, 1997.

Drake, Paul J. Jr., *Dimensioning and Tolerancing Handbook,* McGraw-Hill, New York, 1999.

Fortini, E. T., *Dimensioning for Interchangeable Manufacture,* Industrial Press, New York, 1967.

Spotts, M. F., *Dimensioning and Tolerancing for Quantity Production,* Prentice-Hall, Englewood Cliffs, New Jersey, 1983.

CHAPTER 8
DESIGN AND MANUFACTURING COLLABORATION

Irvan Christy
Director, CoCreate Software, Inc.
Fort Collins, Colorado

8.1 INTRODUCTION

Imagine that a crisis greets you, the principal manufacturing engineer, at work on Monday morning: The scrap rate level for the new product you're manufacturing indicates that the process has exceeded the control limit. If you can't identify and correct the problem quickly, you're in danger of burning through your profit margin and failing to meet your commitment of just-in-time delivery to the assembly plant. You try to reach the off-site product manager, but he's unavailable. You have to settle for e-mail and voicemail to communicate the problem to him and other key people.

Because of time zone differences, a day passes before you have the product manager's response. He says he asked the process engineer, the tooling engineer, and the product design engineer to send you pertinent data to help identify the problem.

Another day passes before you receive all the faxes of the tooling drawings and other data. You squint at the faxes. The image quality is poor, but faxing was necessary; you and the contractor use different engineering applications and cannot exchange data directly. You think you can make out a change in the product design that would affect the manufacturing process already in place. But you need more information—and you cannot implement any changes until you have sign-off.

Time zone differences, disparate engineering applications, and difficulty getting the right information to the right people prolong the identification of the problem and the formalization of the resulting engineering/production change order by several days. Many phone calls, e-mails, faxes, and process and product tooling modifications later, the change order is finalized. Meanwhile, you've all but exhausted your profit margin, production volume has significantly slowed, and similar issues have arisen for other products.

Now, imagine this instead: A crisis greets you, the principal manufacturing engineer, at work on Monday morning: the scrap rate level for the new product you're manufacturing indicates that the process has exceeded the control limit. If you can't identify and correct the problem quickly, you're in danger of burning through your profit margin and failing to meet your commitment for just-in-time delivery to the assembly plant.

It's time to bring together the troubleshooters and decision makers. You use your browser to access an online collaboration environment where you check the schedules of the product manager, the process engineer, the tooling engineer, and the product design engineer. You see a window of time this morning when everyone is free, so you schedule the meeting and include a brief description of the problem. All invitees receive an automatically generated e-mail containing a URL that links to the meeting.

All project members for this project can access product and process data in a secure, online project data space. Through your browser, you enter the project space now to review data before the meeting. You see data such as a discussion thread, a three-dimensional model, screenshots, notes, bills of materials, and process sheets.

At the scheduled time, you all link to the meeting through your e-mail invitations to examine the product and process data together. You run an application from your coordinate measuring machine to show the real-time defect generation coming off the machine in the plant, and your team note taker captures the information with video, notes, and marked-up screenshots to go into an electronic report.

The group determines the cause of the defect: A product change affecting the process was not communicated to manufacturing. Working together in the collaboration environment, the group reaches a solution. An inexpensive tooling change can be made on the manufacturing floor that will not affect your process or output rate. It will require only minimal time to complete the tooling update.

Your note taker records the needed details, and the decision-makers' sign off. With the needed information captured and the key people and expertise at hand, the team confidently ends the meeting. Your note taker generates an automatic, electronic report and checks it into the project space.

A new, automatically generated message arrives in everyone's e-mail inbox, linking you to the report that formalizes the tooling changes. By this afternoon, manufacturing will be back on schedule.

8.2 COLLABORATIVE ENGINEERING DEFINED

The first scenario may be frustratingly familiar. Fortunately, the latter scenario is actually possible now, thanks to an engineering process known as *collaborative engineering*. Collaborative engineering is a team- and project-centric communication process that occurs throughout a product's life cycle. With the help of technology, the collaborative engineering process incorporates the extended project team members, regardless of their geographical locations, into the process of taking a product from concept to market. The extended team is typically diverse (for instance, original equipment manufacturers, marketers, mechanical engineers, design engineers, suppliers, electrical engineers, tooling engineers, and manufacturing engineers). This process can also include outside experts—people separate from the product team who are consulted for their knowledge in a specific area.

The process also includes in-depth documentation of the issues, decisions, and next steps needed as a product progresses through its life cycle. This documentation is made available to the product team throughout the cycle.

Collaboration itself is nothing new. People have always needed to work together, or collaborate, to achieve a common purpose. Collaborative engineering, on the other hand, began as an extension of the concurrent engineering process, which makes engineering processes within the design and manufacturing process simultaneous when possible. Starting in the mid-1990s, concurrent engineering processes began converging with Internet capabilities, CAD applications, and other technologies to make possible the early form of collaborative engineering.

Initially, use of collaborative engineering was limited. The first to adopt it were technology pioneers—those willing to work around limitations such as prolonged, expensive deployment that required the additional purchase of on-site consulting services, unwieldy user interfaces, security concerns, and lack of integration with other desktop applications. Over time, however, the enabling technology has become increasingly easy to deploy and use. It has also become more reliable and secure. Simultaneously, its price point has plummeted: The depth of functionality that once cost six figures now is available for three figures, making the technology available to a wider range of users. And as the technology has evolved to fit in with engineers' daily work environments, product design and manufacturing groups have adopted the collaborative engineering process on a broader scale.

8.3 WHY USE COLLABORATIVE ENGINEERING?

Product teams that can benefit from the collaborative engineering process usually find that one or more of the following symptoms have affected their end-product quality, profit margins, or time-to-market:

- Product development changes occur without consideration of impact on manufacturing.
- Manufacturing lacks a means to give timely input.
- Design changes require multiple iterations to resolve.
- Product programs lack a consistent way to support the decision-making process for accountability.
- Manufacturing supply chains are geographically dispersed, with cultural and language differences.
- Product reviews:
  - Are disorganized or vary in nature.
  - Incur substantial time- and travel expenses.
  - Lack attendance from key people because of geographic limitations.
  - Fail to identify problems in a timely manner.
  - Fail to address manufacturing feasibility and concerns.

Manufacturing benefits from collaborative engineering in several areas. Some of these are discussed below.

8.3.1 Product Design for Optimal Manufacturability

Early manufacturing input into the product design can result in better, more efficiently manufactured product designs. This early engineering partnership between the original equipment manufacturers (OEM) and the supplier reduces internal manufacturing costs in areas such as production tooling (savings that can be passed on to the customer) and decreases production cycle time. This leaves more manufacturing capital available for investing toward the future with actions such as creating unique technologies or purchasing new equipment to improve manufacturing efficiency.

8.3.2 Access to Project Information Throughout the Product Cycle

Manufacturing engineers can efficiently, accurately communicate online with other project members on-demand. Additionally, easy access to project information itself gives manufacturing more understanding of the product. As a result, manufacturing becomes a true design partner and a valued service provider that can give needed input at any point in the product cycle. For example, a deeper product understanding lets a manufacturing supplier steer project choices toward higher-quality product design, materials, and assembly configurations that result in lower warranty and recall chargebacks from the OEM after the product reaches the marketplace.

8.3.3 Availability of Decision Makers

In virtual meetings, at decision points such as whether a product design is ready to turn over to production, all needed decision makers can attend, with simultaneous access to all product information needed for making informed decisions. With less time spent pursuing needed sign-offs from decision makers, manufacturing can devote more time to the core responsibilities that help it remain competitive.

8.3.4 More Competitive Supply Chains

Clear, in-depth communication and reduced time and travel costs across geographically dispersed supply chains let suppliers submit more competitive bids and increase profit margins.

8.3.5 Improved Accountability

Collaborative technology can provide tools to electronically capture design and manufacturing issues with text, two-dimensional images, markups, three-dimensional data, and video. This creates a detailed audit trail that clearly specifies next steps and individual accountability, saving time and reducing misunderstandings.

8.3.6 Reduced Misunderstandings and Engineering Change Orders

Engineering change orders add manufacturing costs and increase product time to market. Because the collaborative engineering process increases accurate communication and documentation throughout the product cycle, misunderstandings in areas such as design intent and manufacturing instructions are reduced. This results in fewer engineering change orders.

8.3.7 Increased Manufacturing Yields

By the time a product created with collaborative engineering reaches the production stage, it is designed to maximize manufacturing yields.

8.3.8 Increased Competitiveness

All the factors above combine to make products faster, more efficient, and less expensive to manufacture, increasing the competitiveness of the manufacturing provider and the success of the extended product team.

8.4 HOW IT WORKS

Collaborative engineering uses the Internet to provide access to technology that facilitates the exchange of ideas and information needed to move a product from design concept to the assembly plant.

Collaborative engineering encompasses a variety of process-enabling technologies. At its most basic level, it should include the capabilities for some forms of project management and online meetings. The particulars of the technology vary by product. Keep in mind that the purpose behind all the technology discussed in this section is to provide the depth of communication and data access necessary to collaborative engineering. Business value increases as these capabilities increase. Ultimately, through these rich means of communication, designing, sharing, and managing data become a unified collaborative process, as indicated in Fig. 8.1.

This section discusses the range of collaborative engineering technologies and how they contribute to the collaborative engineering process.

8.4.1 Project Management

Project management is central to collaborative engineering. Project management, not to be confused with software applications that simply track employee schedules, lets team members organize and access all data related to a project. This capability in collaborative engineering products ranges from

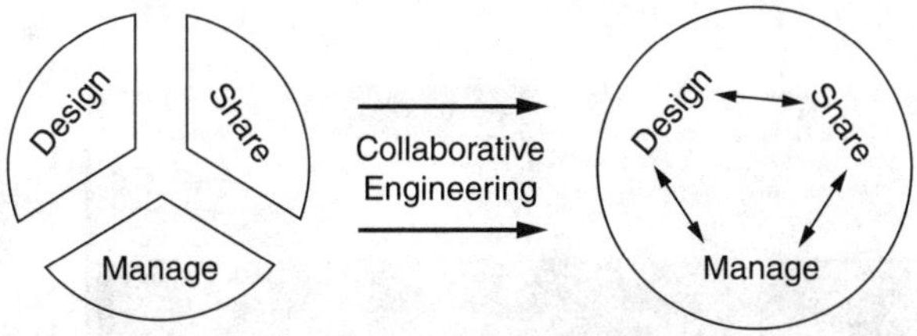

FIGURE 8.1 Unified collaborative process in communication, designing, sharing, and managing data.

none, to nonintegrated, companion products, to full integration with other collaborative technology. Ideally, the project management system is browser-based, meaning that team members both inside and outside of a company network can access it through a Web browser.

Project management tools may also include methods for creating and storing other project information such as discussion threads and electronic reports documenting what occurred in an online meeting. Project members who use these tools can access this project information online, at will, in addition to attending online meetings when needed, as discussed below.

8.4.2 Online Meetings

An online collaborative engineering meeting usually combines simultaneous use of two technologies: a conference call and a meeting held in a virtual, online space. Because the meeting is online, team members from multiple locations come together virtually rather than having to travel. This ease of communication results in more frequent consultations throughout the team and throughout the product cycle. As a consequence, teams find that misunderstandings and mistakes are reduced.

The range of capabilities available in online meeting technology is discussed below.

Application Sharing. Application sharing uses the Internet to let a meeting member show the contents of his or her computer display to others in the meeting. With this technology, the meeting member can share bitmapped images of an application window, such as an FEA analysis tool or a word-processing program. This person can also choose to let another meeting member take control of the application remotely.

Certain online technologies are limited strictly to application-sharing technology. In this case, any meeting attendee can view and, given permission, work in an application installed on one user's computer. However, no tools specific to the collaborative process, such as project management or documentation features, are included.

Because it does not support active, two-way communication, this type of online meeting functions best as a training or presentation mechanism, where one or more meeting members take turns presenting data to the group from their computer desktops.

The richest application-sharing environments for collaborative engineering are fine-tuned for three-dimensional graphics applications such as CAD programs. Ideally, these environments also contain tools designed specifically for collaborative engineering, such as tools to create engineering-style markups, notes, and reports. In this type of environment, meeting attendees work together in a combination of the collaboration-specific environment and any application installed on the desktop of a meeting member, as shown in Fig. 8.2.

Other Integrated Capabilities. Online meeting capabilities may include an integrated tool for instant messaging ("chatting" with other meeting members by exchanging real-time, typed-in messages that are displayed on a common window in the meeting). These integrated capabilities also can include the exchange of real-time voice and video communications.

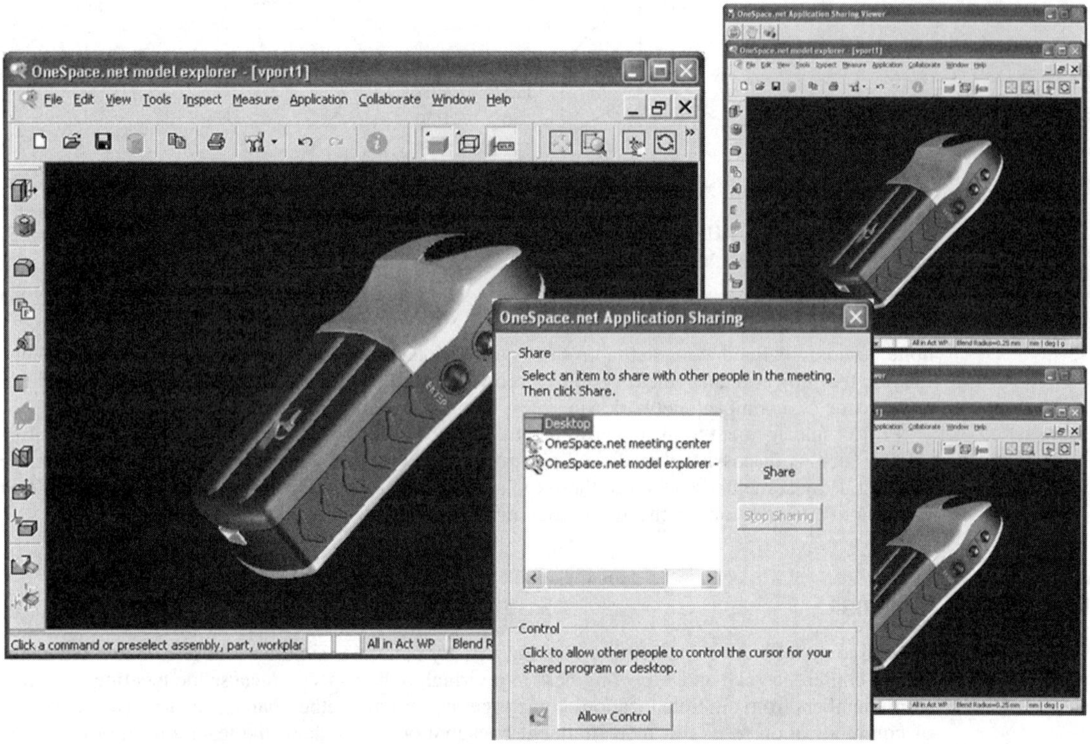

FIGURE 8.2 Meeting attendees use applcation sharing to work together online and avoid travel for a face to face meeting.

Synchronous and Asynchronous Use. Online collaboration technology is often used synchronously, meaning that users attend an online meeting at the same time, usually in conjunction with a conference call. The members examine a variety of data together to facilitate exchange of information and ideas, troubleshooting, decision making, and the like.

For example, a design engineer and a manufacturing engineer meet online synchronously to clarify a set of assembly instructions. The manufacturing engineer has several questions about the instructions. The manufacturing engineer talks these over with the designer while together they inspect the design of the product to be manufactured. They take screenshots and make markups (drawings and text on the screenshot images) as needed, incorporating the images and markups as part of a clear, new set of instructions in their online meeting notes. When both attendees are satisfied that all questions are addressed, they save their meeting data and end the online meeting.

Online collaboration technology also may be used *asynchronously,* meaning that users work in a project and/or meeting space sequentially rather than concurrently. In this case, each team member works in the project and/or meeting space individually. Asynchronous use is particularly helpful when team members work in different parts of the world and need round-the-clock progress on a project.

For example, a project manager in Asia needs to know how proposed design changes to a product's housing would impact the product's electronics design. He accesses the online collaboration technology, placing three-dimensional design data, notes, and marked-up screenshots showing where the change would be made. He saves the electronic summary file in a company repository and sends a request to the electronics engineer in the United States to access this information. While it's night for the project manager, the electronics engineer starts his own work day and inspects the posted data. He documents needed modifications to the electronics design with three-dimensional data, screen shots, and notes, and then saves the electronic summary file. In Asia the next morning, the manager opens the updated electronic summary file to access the electronic engineer's input.

In practice, collaborative engineering teams often use online collaboration technology both synchronously and asynchronously. Asynchronous use lets project members in different time zones make individual contributions, and synchronous use is important for making and documenting final decisions.

Data Viewing and Exchange. Data viewing capabilities vary across collaborative engineering products. Some products let meeting attendees inspect or modify 2-dimensional and/or three-dimensional data during the meeting, while others only let attendees view the data.

Additionally, the products differ in their ability to process large, three-dimensional images, so graphics performance varies widely. In some cases, larger design files cause delays in image transmittal to meeting attendees who thus tend to work more with screen shots prepared prior to the meeting. Other online meeting products process large design files with minimal or no delay.

Finally, interoperability, the ability of two or more systems or components to exchange and use design data without extraordinary effort by the user, varies across online meeting products. Project teams may use an online meeting product that lets users load only three-dimensional designs created in the software parent company's proprietary application. Any other native design data thus requires translation into IGES or STEP files before attendees can load it into the meeting software. This translation can cause geometric data loss and affect model accuracy, impacting manufacturability. Thus, noninteroperable collaborative engineering applications are most useful when all team members work with a single CAD application.

Product teams that use a range of different CAD applications, on the other hand, benefit more from online meeting technology that provides a CAD-neutral environment. This environment accommodates file formats from all CAD applications. Thus, meeting attendees can load data from any CAD application into a common meeting space for viewing, inspection, and modification.

Data Inspection and Modification. Online meeting attendees' power to work with three-dimensional design data also varies widely. Some products permit users only to view three-dimensional data. Others permit detailed geometric inspection of the data, as shown in Fig. 8.3 below. Finally,

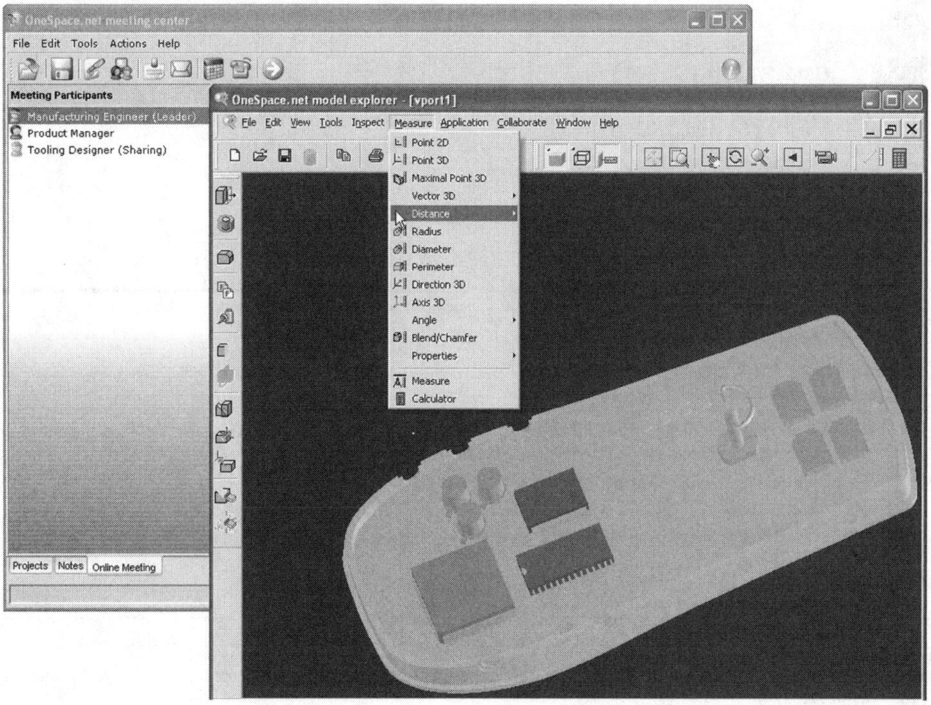

FIGURE 8.3 Online collaboration working with three-dimensional design data.

capability to explore design changes by modifying 3-dimensional data in the meeting space is less common, but also available. Useful for brainstorming and troubleshooting, this lets users experiment with "what if" modification scenarios without needing to use the native CAD application that created a design.

Meeting Documentation. Capturing design issues, team decisions, action items, and other information from the online meeting is integral to the collaborative engineering process. This ability differs by product. When the online meeting product offers no integrated documentation features, meeting users capture meeting data by taking notes and screen captures manually or by using another application to capture them. Ideally, the information is then distributed across the team using an agreed-upon process and format.

Another online meeting technology integrates information-capturing capabilities. Meeting members work directly within the meeting environment to document the discussion and tasks arising from the meeting. Integrated information capture may include abilities to take notes and screen shots, make markups, create a copy of a 3D design to save as a reference file, and generate automatic meeting summary reports. They may also include capability for using a project workspace to store and update the tasks, as shown in Fig. 8.4.

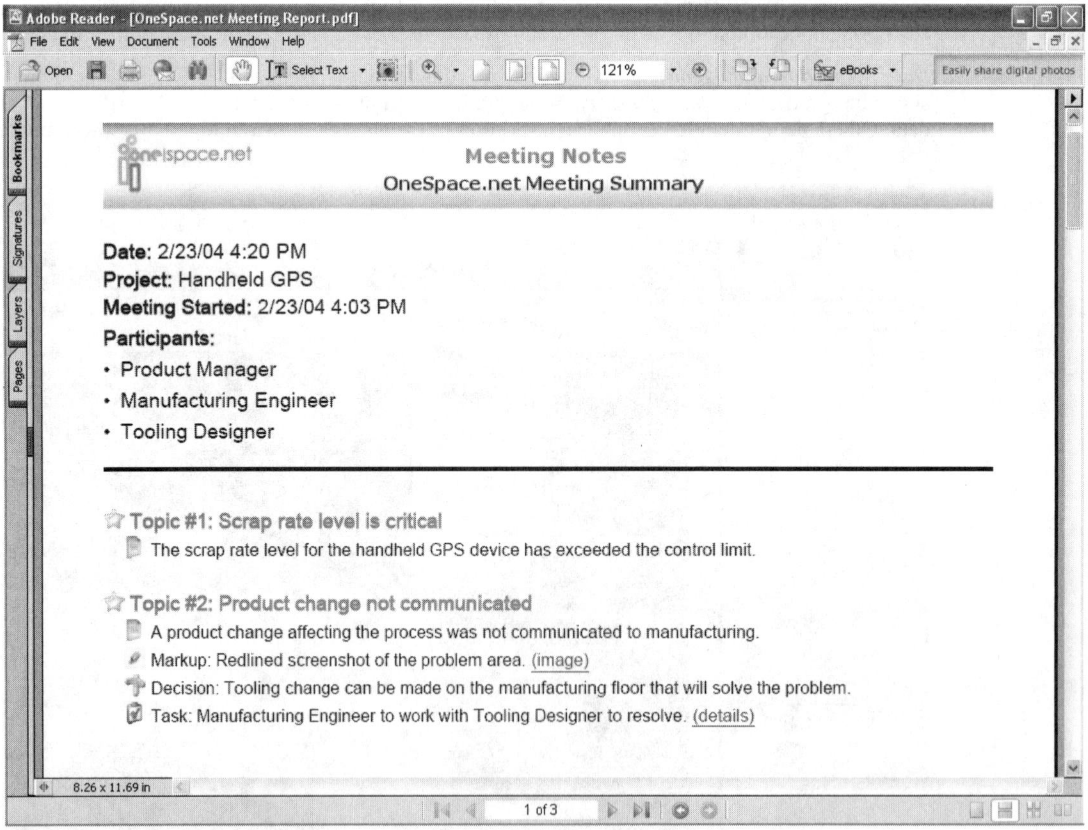

FIGURE 8.4 Team members use collaboration to document project issues, capture markups, assign tasks, and record decisions when problems surface.

8.4.3 Data Protection

Users of collaborative engineering need to consider several aspects of data protection. These are discussed here.

Firewall Protection. In the most secure collaborative engineering products, a firewall protects data in online meetings and in project spaces. The firewall grants access only to trusted systems through a specified port. If a product does not use a firewall, project members should avoid posting sensitive data in the project space or bringing it into an online meeting.

Data Transfer. Data transfer over the Internet must be protected. Some companies, such as those in the defense industry, also want to protect data transferred over their own intranets. The most secure collaborative engineering products use the industry standard for secure data transfer, the secure socket layer (SSL) protocol, which encrypts all the data being transferred over the SSL connection.

Data Persistence. In the most secure collaborative engineering products, data brought into online meetings does not remain in the caches on the computers of other meeting attendees. Otherwise, a meeting participant can unintentionally leave copies of proprietary data with other meeting members.

Application Security. At the most granular level, collaborative engineering technology lets a project manager adjust the level of data access by individual project members. In Fig. 8.3, for instance, a project has three different types of levels of data access ("roles"): project manager, who has both project control and full read and write access to project file; team member, who has full read and write access to project files; and guest, who has read-only access to the files.

8.4.4 Deployment Methods and System Requirements

Deployment methods and system requirements vary widely across collaborative engineering products. Some technologies are available only as ownership purchases where applications, clients, and servers are locally installed and configured within the user's internal network. Other technologies are available only as subscription-based services offered by application server providers (ASPs). Only a few technologies are available by both methods.

To add optimum value to the working environment, collaborative engineering technology should have as many of the following traits as possible:

- *Client size.* Local installations require minimal hard drive space.
- *Memory requirements.* Application performs smoothly with minimal memory requirements.
- *Versatility.* Technology functions on a variety of systems and allows use of multiple data formats.
- *Distribution and deployment.* Application can be made available quickly to any desired collaboration participant with minimal or no support from a company's information technology (IT) group.
- *Security.* Data can be seamlessly protected both with network security measures and within the public Internet. Online meeting data is not cached on local users' machines to persist after the meeting.

8.5 USE MODELS

The collaborative engineering process is becoming integrated into numerous areas of product design and manufacturing. The following examples show how the collaborative engineering process fits into these areas.

8.5.1 Design for Manufacturability and Assembly

How a product will be manufactured is one of the most critical aspects of a new design. Because manufacturing drives much of the product cost, this activity is often outsourced—sometimes overseas. When manufacturing and development sites are geographically separated, collaborative engineering helps close the gap.

Using collaborative engineering technology early in the product cycle, manufacturing engineers can hold online meetings with designers. Manufacturing directs the design toward a more easily manufactured and assembled product, before design specifications are formalized.

Meeting members may want to share data such as Bills of Material, 3D models, manufacturing work instructions, part and assembly drawings, and product test plans and instructions.

Frequent, informal collaborative engineering meetings are recommended at this stage. Product team members should think of online meetings at this point as a way to stop by each other's virtual desks for quick but critical input.

8.5.2 Supplier Management

Collaborative engineering eases supplier management in areas such as supplier bidding and the process of working with offshore suppliers.

Supplier Bidding. Collaborative engineering lets a manufacturing company access tooling, manufacturing data, and other information needed to create an accurate bid. The original equipment manufacturer (OEM) may choose to give a potential supplier selective access to an area of the project repository. Once the supplier reviews the data there, both groups attend an online meeting to clarify product specifications in person. The supplier can later post the bid in the project repository space.

Offshore Suppliers. Distributed supply chains commonly juggle challenges including time zone differences, language barriers, and diverse design environments. Asynchronous use of online collaborative technology accommodates time zone differences and decreases travel needs. Communicating with marked up screenshots, 2-D and 3-D design data, and text eases language barrier problems.

8.5.3 Tooling Design Reviews

Tooling designers and manufacturers are often one and the same. Even when they differ, they must communicate closely in preparation for product production. Tooling problems increase tooling cost and the potential for impact on a product release schedule. Part designers and tooling engineers must review designs for accuracy, discuss trade-offs of the decisions that they make, minimize tooling costs, and ensure the function and life of the tooling.

With collaborative engineering meetings early in a project, tooling engineers give input about creating a design that meshes with tooling capabilities. Meeting decisions are formalized with documentation capabilities in the collaborative technology tools and stored in a project repository.

8.5.4 Project Team Meetings

Throughout a project, collaborative engineering teams hold online meetings to determine next steps to further the project. By attending these meetings throughout the project, a manufacturing representative stays informed of a product's progress toward the production stage. Early on, manufacturing input may be needed to ensure design for manufacturability, as previously discussed. Later, manufacturing will stay informed about changes that could affect the manufacturing schedule.

For instance, a representative from marketing may suggest a design change that the manufacturing engineer knows would add weeks to production. The manufacturing engineer can provide a manufacturing-based perspective about the feasibility of the change so the team can make an informed decision about whether the change is worth the tradeoff.

Manufacturing work instructions can be created during the meetings when appropriate. Meeting members use markup-, notes-, and report-creation tools to capture all data and member input and then store the meeting results in the project repository.

8.5.5 Manufacturing Readiness Reviews

Before releasing a product to production, project members formally review the product's readiness for manufacturing. Manufacturing bears responsibility for the product from here on and must ensure that all the deliverables it needs are complete.

Before holding a formal readiness review online, team members can individually visit the project repository to prepare for the meeting. In the formal review, meeting members meet online to review data including drawings and work instructions, 3D models of parts and tooling, supply chain information, and other documentation that could impact the decision to go to production.

Team members document issues with the note- and markup-creation features in the collaborative technology tool. Manufacturing work instructions are touched up and finalized. When everyone agrees that the product is production-ready, they formalize manufacturing's signoff and capture all information in a meeting report to store in the project repository.

8.5.6 Ad Hoc Problem Solving

As issues that may impact manufacturing arise, the needed parties hold online meetings to troubleshoot and problem-solve. With all data needed for troubleshooting easily at hand, and with quick access to needed decision-makers and experts, the team can often resolve an issue before the need for an engineering change order arises.

8.5.7 Engineering Change Process

Once a product is released to production, changes to the original design may be proposed to improve manufacturability, reduce product cost, or improve product performance or reliability. Manufacturing may initiate an engineering change order request when a change is needed to make a design easier to produce. Another member of the extended project team may also initiate the request. Either way, engineering change orders are subject to approval by numerous parts of a company's organization and require significant analysis, preparation, and documentation.

Online meetings and project repositories let the needed decision makers come together and access the information needed to finalize and formalize engineering change orders.

8.5.8 Data Conversion and Exchange Across Design Applications

To exchange data across CAD systems, companies may resort to using IGES or STEP files. Use of these formats can result in loss of model accuracy, part names, assembly structure, and attributes. In cases where an OEM contracts manufacturing services to an outside provider, manufacturing needs a method to exchange 3D designs accurately.

Using CAD-neutral collaborative technology, the OEM loads a 3D design from any CAD application into an online, collaborative meeting. Once the design is loaded, meeting participants who are given the proper access permissions can save it directly from the online meeting into their own CAD systems as a reference file that contains accurate geometry, part names, assembly structure, and attributes.

8.6 CONCLUSION

The most useful collaborative engineering technology is affordable, easily deployed, integrated seamlessly into the daily working tools and processes of its users, and designed to specifically address product development needs. Such technology enables a level of collaborative engineering that makes manufacturing both more predictable and more profitable. Ease in accessing data, managing supply chains, tracking accountability, and obtaining support for decisions that affect production increases predictability and establishment of manufacturing as a true design partner increases manufacturability and fosters long-term relationships with customers, increasing profitability.

As collaborative engineering technology continues the trend toward integration into the engineer's desktop working environment, it is becoming an increasingly standard part of the design and manufacturing cycle.

P • A • R • T • 2

MANUFACTURING AUTOMATION AND TECHNOLOGIES

CHAPTER 9
CAD/CAM/CAE

Ilya Mirman
SolidWorks Corporation
Concord, Massachusetts

Robert McGill
SolidWorks Corporation
Concord, Massachusetts

9.1 INTRODUCTION

The need for illustrating, visualizing, and documenting mechanical designs prior to production has existed ever since human beings began creating machines, mechanisms, and products. Over the last century, methods for achieving this function have evolved dramatically from the blackboard illustrations of the early twentieth century and from manual drafting systems that were commonplace 50 years ago to today's automated 3D solid modeling software. As computer technology has advanced, so have the tools designers and product engineers use to create and illustrate design concepts. Today, powerful computer hardware and software have supplanted the drafting tables and T-squares of the 1960s and have advanced to the point of playing a pivotal role in not only improving design visualization but also in driving the entire manufacturing process.

9.1.1 What Is CAD?

The advances made over the last 30 years in the use of computers for mechanical design have occurred at a more rapid pace than all the progress in design visualization that preceded the advent of computer technology. When computer hardware and software systems first appeared, the acronym CAD actually represented the term *computer-aided drafting*. That's because the early 2D computer design packages merely automated the manual drafting process. The first 2D CAD packages enabled designers/drafters to produce design drawings and manufacturing documentation more efficiently than the manual drafting of the past. The introduction of 2D drafting packages represented the first widespread migration of engineers to new design tools, and manufacturers readily embraced this technology because of the productivity gains it offered.

The next stage in the evolution of design tools was the move from 2D to 3D design systems (Fig. 9.1). Beginning in the 1990s, this represented the second large migration of engineers to a new design paradigm and the watershed that shifted the meaning of the acronym CAD from "computer-aided drafting" to "computer-aided design." That's because 3D solid modeling removed the emphasis from using computer technology to document or capture a design concept and gave engineers a tool that truly helped them create more innovative designs and manufacture higher quality products.

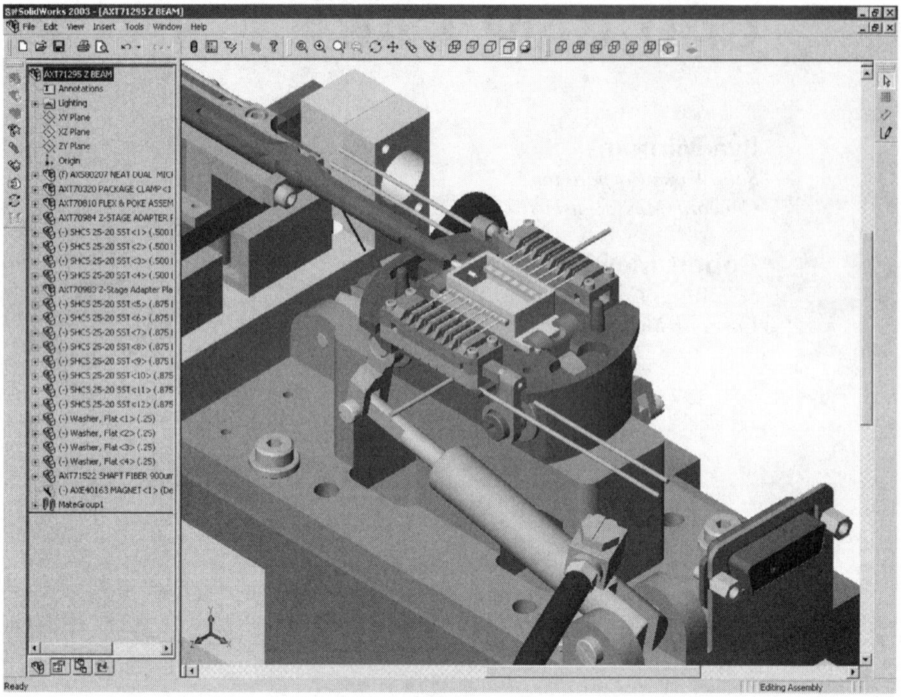

FIGURE 9.1 Modern CAD software facilitates design of high-precision machinery. (*Courtesy of Axsun Technologies and SolidWorks Corporation*).

Instead of having the production of an engineering drawing as the final goal, engineers employ the enhanced design visualization and manipulation capabilities of 3D CAD systems to refine designs, improve products, and create 3D design data, which can be leveraged throughout the product development process.

Yet, not all 3D solid modelers are the same, and since the introduction of 3D CAD systems two decades ago, many advances have been made. The early 3D systems were slow, expensive, based on proprietary hardware, and difficult to use because they frequently required the memorization of manual commands. The introduction of affordable, yet powerful computers and the Windows operating environment gave 3D CAD developers the foundation they needed to create the fast, affordable, and easy-to-use 3D solid modelers of today.

Advances in 3D CAD technology enable the following benefits:

Parametric Design. All features and dimensions are driven off design parameters. When an engineer wants to change the design, he or she simply changes the value of the parameter, and the geometry updates accordingly.

Bidirectional Associativity Means Quick Design Changes. All elements of a solid model (part models, assembly models, detail drawings) are associated in both directions. When a change is made to any of these documents, the change automatically propagates to all associated files.

Intelligent Geometry for Downstream Applications. 3D CAD data support other design and manufacturing functions, such as machining, prototyping, analysis, assembly management, and documentation, without the need to convert or translate files.

Large Assembly Capabilities. 3D CAD technology has the ability to design assemblies and subassemblies, some of which can involve thousands of parts, as well as individual components. When a product design requires large, complex assemblies involving thousands of moving parts, 2D design techniques become labor-intensive and time-consuming. Managing the numerous production-level drawings alone can be tedious. With 3D solid modeling software, managing the accuracy and completeness of assembly production drawings becomes a less costly and more manageable process.

Configurations of Derivative Products. Using design tables, engineers can create varied configurations of products, assemblies, or product families, with varying sizes, dimensions, weights, and capacities from a single solid model, leveraging existing designs and simplifying the development of new product models.

Design Stylization. As more and more products compete on the basis of aesthetics and ergonomics, the need to easily create free-form, organic, and stylized shapes is becoming increasingly important. State-of-the-art CAD systems have the capability of creating complex models, surfaces, and shapes, including curves, blends, fillets, and other unique design features.

Automatic Creation of Drawings. 3D solid modelers can automatically output engineering drawings comprising various views, such as isometric, exploded assembly, detail, and section views, from the base solid model without the need to draw them manually. The dimensions used to create the model can be used to annotate the drawing.

Communicates Design Intent. CAD data is basically a geometrical representation of an engineer's imagination, capturing the engineer's creativity and design intent. With 2D drawings, engineers and manufacturing personnel have to interpret or visualize a flat 2D drawing as a 3D part or assembly. At times, interpreting 2D drawings results in a loss or misinterpretation of the engineer's original design intent, leading to delays and rework. With 3D solid modeling software, design intent is maintained and effectively communicated through the actual 3D representation of the part or assembly, leaving little possibility for misinterpretation.

Assesses Fit and Tolerance Problems. Engineers who design assemblies and subassemblies cannot assess fit and tolerance problems effectively in 2D. Using a 2D layout drawing that shows product components, subassembly interfaces, and working envelopes, engineers cannot fully visualize the 3D fit, interface, and function of assembly components. Often, this results in fit and tolerance problems that go undetected until late in the design cycle, when they become more costly and time-consuming to correct. With 3D solid modeling software, an engineer can assess and address fit and tolerance problems during the initial stage of design.

Minimizes Reliance on Physical Prototyping. Traditionally, physical prototyping is nearly a prerequisite in new product development to detect parts that collide or interfere with one another and ensure that all components have adequate clearances. With 3D solid modeling software, the same objectives can be accomplished on the computer, saving both time and prototype development costs. It is not uncommon for teams that transition from 2D to solids-based design to cut multiple prototype iterations from their typical product development cycles.

Eliminates Lengthy Error Checking. With 2D, most assembly designs require a lengthy, labor-intensive error check of drawings, which itself is prone to error. Checkers spend countless hours checking fit and tolerance dimensions between drawings. This process becomes more complicated when drafters use different dimensioning parameters for parts from the same assembly. The error-checking process takes even more time when redlined drawings are sent back to the designer for corrections and then returned to the checker for final approval. With 3D solid modeling software, there is no need to check drawings because the designer addresses fit and tolerance problems in the model as part of assembly design.

Today's 3D CAD systems are the culmination of more than 30 years of research and development and have demonstrated and proven their value in actual manufacturing environments for more than a decade. 3D systems are much better than their predecessors in capturing and communicating the engineer's original design intent and automating engineering tasks, creating a sound platform on which the entire manufacturing process is based.

9.2 WHAT IS CAM?

Just as the acronym CAD evolved from meaning "computer-aided drafting" to "computer-aided design," the meaning of the acronym CAM has changed from "computer-aided machining" to "computer-aided manufacturing." The basic premise of CAM technology is to leverage product design information (CAD data) to drive manufacturing functions. The development of CAM technology to automate and manage machining, tooling, and mold creation with greater speed and accuracy is intimately linked to the development of CAD technology, which is why the term CAD/CAM is often used as a single acronym.

The introduction of CAM systems allowed manufacturing and tooling engineers to write computer programs to control machine tool operations such as milling and turning. These computer numerically controlled (CNC or NC) programs contain hundreds or thousands of simple commands, much like driving instructions, needed to move the machine tool precisely from one position to the next. These commands are sent to the machine tool's controller to control highly precise stepper motors connected to the machine tool's various axes of travel. CNC control represents a huge improvement over the traditional method of reading a blueprint and manually adjusting the position of a machine tool through hand cranks. The accuracy and repeatability of CNC machining has had a permanent impact on the reliability and quality of today's manufacturing environment.

With the development of 3D CAD solid modeling systems, the interim step of developing computer code to control 3D CAM machining operations has been automated (see Fig. 9.2). Because the data included in solid models represent three-dimensional shapes with complete accuracy, today's CAM systems can directly import 3D solid models and use them to generate the CNC computer code required to control manufacturing operations with an extremely high degree of precision. While

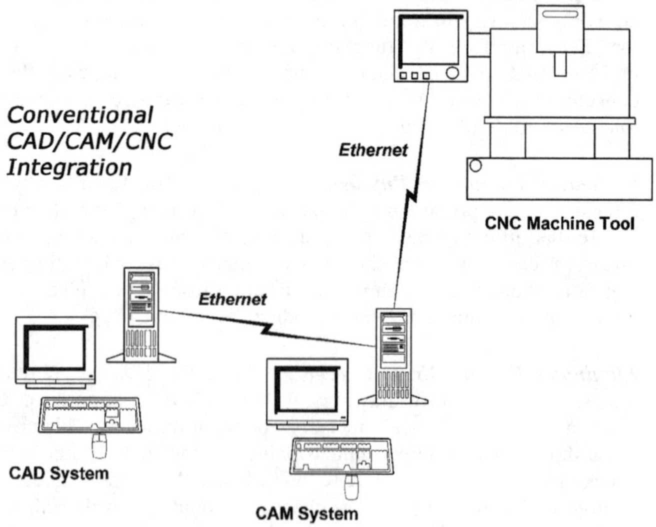

FIGURE 9.2 Conventional CAD/CAM information flow. (*Courtesy Gibbs and Associates.*)

manufacturers initially applied CAM technology for tooling and mass production machining operations, its use has expanded to include other manufacturing processes such as the creation of molds for plastic injection-molding and certain automatic (robotic) assembly operations, all directly from the 3D solid model.

9.2.1 The CNC Machine Tool

The CNC machine tool industry has evolved in response to the increasing sophistication of the CAM products that are on the market. The earliest numerically controlled machines were expensive, special-purpose machines built by aerospace manufacturers to accurately and repeatedly machine the complex contours of airframe components. Today, with the rapid adoption of high-end 3D CAM systems, a wide variety of CNC milling, turning, and EDM machines have been introduced in a range of prices that makes it possible for the smallest shop to own at least one.

Accurate, repeatable machining is not new. CAMs were added to machine tools during the Civil War to speed weapons production. CAM-driven screw machines, also known as Swiss machines, have accurately mass produced small fasteners and watch components since the mid-1800s. Programmable CNC controls and high precision stepper motors have replaced CAMs on modern machine tools making short run, close tolerance parts achievable by the typical job shop.

Three-axis CNC milling machines and two-axis lathes are now commonplace. Inexpensive rotary tables have brought four- and five-axis machining within the reach of many shops. The advent of the CNC Mill Turn or turning center which combines conventional live tooling milling operations with turning operations is again revolutionizing the machining industry. Complex parts can now be completely machined as part of a totally automated, lights-out manufacturing process.

In most modern machine shops, the CAM programmer generates the NC code on a computer away from the shop floor. This computer is connected to one or more CNC machine tools out in the shop via a shielded cable in the same way that you might connect a printer. Each machine tool has its own machine controller which is usually attached directly to the machine. Most of these controllers have proprietary designs specific to the maker of the machine tool but in some cases the controller is a commercially available PC in a hardened cabinet. Shop-floor programming takes advantage of these hardened terminals to allow full CAM programming right at the machine tool.

Since most CAM systems are designed to operate with many different sizes and makes of machine tools, they utilize a machine-specific translator, or post processor, to convert the generic CAM instructions to the low-level, machine-specific code that the machine controller can understand. In this way, the same CAM program can be used to run several different machine tools.

9.2.2 CNC Programming

Whether the machinist is drilling a hole on a CNC milling machine or machining a complex part on a machining center, setting up the machine to run a CAM program is essentially the same. First, the coordinate systems of the CAM system must be accurately matched to the machine tool. This is a one time process and most machine tools offer special verification tools to assist with this step. Then, the operator clamps the material or stock onto the machine tool and aligns it with the machine tool axis. If the stock is accurately represented as a solid in the CAM system, the machining process can begin. If not, the operator can place a special probe in the machine tool and bring the probe in contact with the stock to establish the stock dimensions.

In the language of NC code, machining a straight line is a single command, but machining a circle requires multiple commands to tell the machine to move a small distance in X and Y, based on the accuracy required, until the circle is complete. The CAM system automates the creation of the NC code by using the geometry of the CAD model. To machine a circle, the CAM programmer simply selects the proper tool and the desired depth and indicates the circle in the CAD model he or she wants machined. The CAM system creates the complete NC tool path, from tool selection to the thousands of small incremental steps required to machine the circle. Taking this down to the level of the machine, many modern machine tools include *circular interpolation* as a standard feature of the machine controller.

9.2.3 Generating the Code

The job of the CAM system is to generate the machining processes and the precise path that the tool must follow to accurately reproduce the shape of the 3D solid model without the tool or the machine colliding with the part or with itself. A series of "roughing passes" with large-diameter fast-cutting tools quickly removes most of the stock material. Additional "finishing passes" bring out the final shape of the part (see Fig. 9.3).

Based on the shape of the cutting tool, the finishing passes are a set of curves through which the tool tip must pass. For a round tool, this is a simple offset of the 3D part surface. For a flat or bull-nosed cutter, the offset constantly changes with the 3D shape of the part. Successive curves are generated based on the diameter of the tool, the hardness of the material, and the available cutting power of the machine. The same holds true for the depth of the cut.

Parts with flat faces and pockets with vertical sides can be very accurately machined. Parts with complex three-dimensional surfaces require many multiple passes with progressively smaller tools to achieve an acceptable finish. The difference between the desired surface and the machined surface can often be seen as a series of scallops left behind by a round cutter. If necessary, these scallops can be removed with hand grinding or benching to achieve the desired surface finish.

Holes make up a large part of production machining operations, and so most CAM systems offer special functions for making holes. Hole tables allow the CAM programmer to quickly specify a

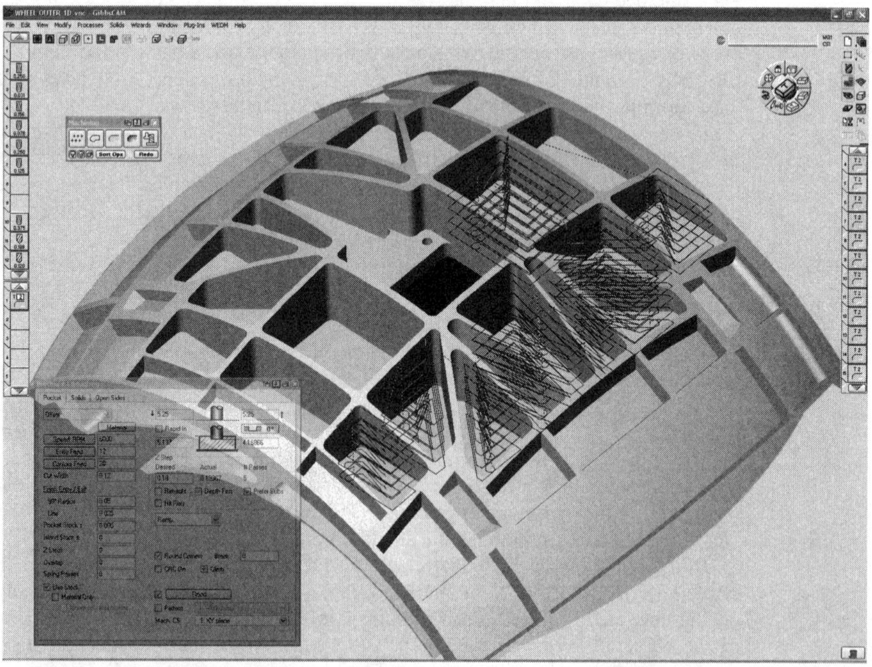

FIGURE 9.3 Computer-aided manufacture of the Mars Rover wheel. (*Courtesy of Next Intent and Gibbs and Associates.*)

pattern of holes to be drilled using the same drilling process. Special operations built into the controllers are available to automatically control chip removal.

9.2.4 Getting Fancy

As the demand for higher quality, faster delivery times, and mass customization keeps accelerating, the demand grows for faster and more accurate machining. Eliminating hand grinding, which is both slow and imprecise, is a priority. Today's high-speed machining (HSM) systems combine conventional 3-axis machining with high travel speeds and small cut depth. These extreme table speeds can impart enormous shock loads to the cutting tool and have prompted new thinking in tool design and material selection. New CAM techniques protect the tool and the machine by looking ahead and slowing the travel speed prior to changes in cut depth and using arcs rather than sharp corners in the tool travel. HSM has enabled huge stamping dies, used to stamp automobile panels, to be machined to final shape with no secondary manual grinding operations.

Five-axis machining is used to reduce part setups or speed the finishing process. Where a three-axis machine can only cut what it can see from directly above the work piece, a five-axis machine can cut from any angle not blocked by the hold-down clamps. Calculating the tool path for five-axis operation is made more difficult by the limitations in the travel of most five-axis heads. Cutting completely around a sphere, for example, requires the CAM code to reposition the head several times to allow the tool to continue cutting around the surface. Machining internal surfaces requires long shank cutters which must be precisely controlled to minimize wobble and chatter. In addition, the extra joints required to provide five-axis movement often result in a machine tool that is less rigid and precise than the equivalent three-axis machine.

9.2.5 Lowering the Cost of Manufacturing

Nearly every product available on store shelves is more reliable and less expensive to manufacture today thanks to the precision and productivity of CNC machine tools. For example, plastic injection molds are automatically manufactured from the 3D solid CAD models. Changes in the part design can be quickly reflected in the mold and the NC toolpath changed accordingly. Creating the CNC program to cut a highly precise mold takes hours where it once took days or even weeks. These same cost savings are being realized with milled parts, turned parts, and sheet metal parts.

9.3 WHAT IS CAE?

CAE stands for *computer-aided engineering* and primarily encompasses two engineering software technologies that manufacturers use in conjunction with CAD to engineer, analyze, and optimize product designs. Design analysis and knowledge-based engineering (KBE) applications help manufacturers to refine design concepts by simulating a product's physical behavior in its operating environment and infusing the designer's knowledge and expertise into the manufacturing process.

Creating a CAD model is one thing, but capturing the knowledge or design intent that went into designing the model and reusing it as part of manufacturing represents the essence of KBE software technology. These applications propagate the designer's process-specific knowledge throughout the design and manufacturing process, leveraging the organization's knowledge to produce consistent quality and production efficiencies.

Design analysis, also frequently referred to as finite element analysis (FEA), is a software technology used to simulate the physical behavior of a design under specific conditions. FEA breaks a solid model down into many small and simple geometric elements (bricks, tetrahedrons) to solve a series of equations formulated around how these elements interact with each other and the external

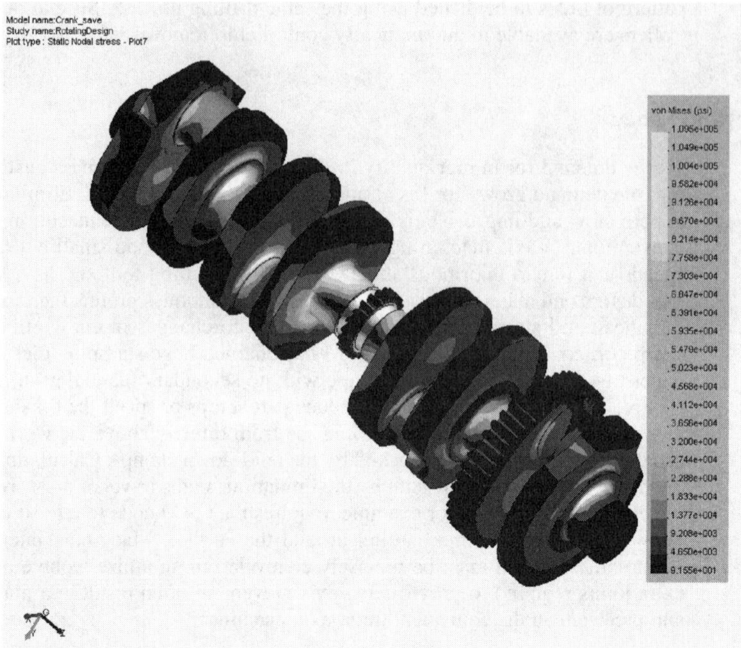

FIGURE 9.4 Example of a structural analysis performed on a crankshaft. *(Courtesy of SolidWorks Corporation.)*

loads. Using this technique, engineers can simulate responses of designs to operating forces and use these results to improve design performance and minimize the need to build physical prototypes (see Fig. 9.4).

Some of the questions FEA helps answer early in the design stage include:

- Will structural stresses or fatigue cause it to break, buckle, or deform?
- Will thermal stresses weaken a component or cause it to fail?
- Will electromagnetic forces cause a system to behave in a manner inconsistent with its intended use?
- How much material can be removed, and where, while still maintaining the required safety factor?

Depending on the component or system, and how it is used, effective design analysis can mean the difference between product success and acceptance, or even life and death. For example, airplane manufacturers use FEA to ensure that aircraft wings can withstand the forces of flight. The more common types of design analyses that are performed include structural, thermal, kinematics (motion), electromagnetic, and fluid dynamics analyses.

Early design analysis software packages were separate applications, often with their own geometric modeling application. Analysts would often have to rebuild a model after they received it from a designer in order to complete the particular type of design analysis required. Today, many analysis systems operate directly on the 3D CAD solid model, and some packages are even integrated directly with 3D CAD software, combining design and analysis within the same application. The benefit of an integrated design analysis package is that a design engineer can optimize the design as part of the conceptual design rather than after the design concept is finished.

9.4 CAD'S INTERACTION WITH OTHER TOOLS

Once a separate application, CAD has steadily evolved to work with a variety of other software applications (Fig. 9.5). Product development is not a single step but rather a continuous process that impacts various departments and functions from the idea stage all the way through the actual introduction of a product. Instead of treating design as a separate, autonomous function, CAD vendors recognize that the value of 3D CAD data extends far beyond conceptual design. By making CAD data compatible with other functions such as manufacturing, documentation, and marketing, CAD developers have accelerated the rate at which information is processed throughout the product development organization and have produced efficiencies and productivity improvements that were unanticipated and unsupported during the early application of CAD technology. Increasingly, CAD data has become the data thread that weaves its way across the extended manufacturing enterprise, accelerating the rate at which information is processed throughout product development.

9.4.1 Interoperability

Throughout the product development process, there are complementary software tools that work with and leverage CAD systems and CAD data. The CAD data leveraged by these tools can largely be split into two categories: visual information (graphical representation, bills of materials, etc.) and precise geometric data for manufacturing operations.

One of the important developments that has produced broad compatibility among software packages and enables the leveraging of the visual information by other applications is the Microsoft object

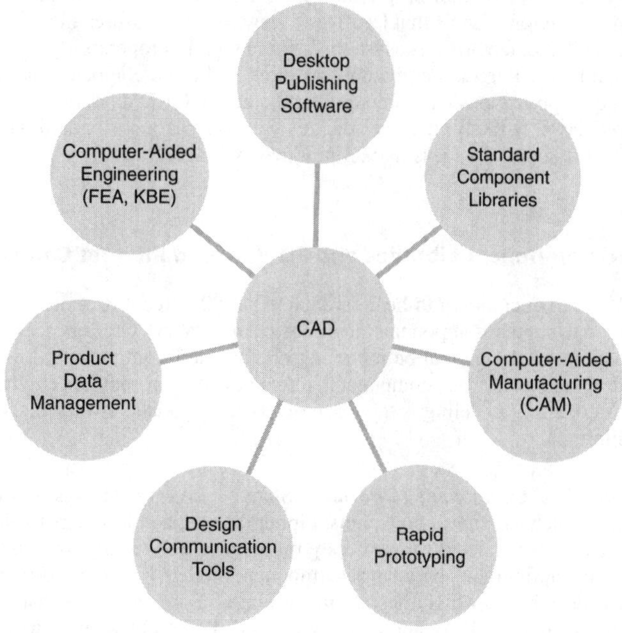

FIGURE 9.5 CAD data drives and interacts with a broad array of tools and technologies. (*Courtesy of SolidWorks Corporation.*)

linking and embedding (OLE) design standard, which treats computer files as objects, enabling them to be embedded inside other applications with links to their specific executable programs. Leveraging OLE, for example, easily enables the documentation department to incorporate isometric views and a bill of materials directly from the CAD drawing into a service manual document.

Reading geometry information from the CAD files is a prerequisite for manufacturing operations such as machining, rapid prototyping, and mold design and fabrication. Although DWG is the de facto file format for sharing 2D data, there is not yet a universally adopted standard format for sharing 3D data. With different solid modeling systems in use around the world, data interoperability, at the time of this writing, can still be a significant productivity drain. Today, there are two approaches to sharing geometry information, each with their own advantages and disadvantages.

Neutral File Formats. There are two general industry data standards that manufacturers use heavily to leverage CAD data for other applications—IGES and STEP. The U.S. Department of Commerce's National Bureau of Standards established the Initial Graphics Exchange Specification (IGES) to facilitate data exchange among CAD systems and other software applications. The Product Data Exchange Specification/Standard for the Exchange of Product Data (PDES/STEP), which is most frequently referred to as STEP, is an international standard developed to achieve the same goal. Although many applications can read these file formats, they inherently require a translation step from the original ("native") CAD file format. This translation step introduces two concerns. First, the translated file may not have the same precision as the source geometry (for example, a circular cutout may be represented by a series of straight lines). Second, feature information that may facilitate machine tool selection (e.g., a standard countersink) may not carry through to the neutral file.

Native CAD Data. By definition, the purest way to share CAD data is to use the file format of the CAD application that generated the data. Some CAD vendors have gone to great lengths to enable other application vendors to work directly with the native file format. The opening up of the CAD system's application programming interface (API) to thousands of developers worldwide, and the proliferation of tools that let anyone view, share, measure, and mark up designs, have proven to facilitate "manufacturing ecosystems" where data interoperability challenges are minimized. And as the solid modeling market matures, as more and more companies are using the same CAD tool, and as more applications can directly read the native CAD data, interoperability challenges will inevitably reduce. It is likely that over the next several years, a standard will emerge among 3D file formats, driven largely by mass market production usage.

9.4.2 Standard Component Libraries and 3D-Powered Internet Catalogs

It is not uncommon to hear estimates that 80 percent or more of a typical design uses off-the-shelf or outsourced components and subsystems. When one considers that some of the least enjoyable aspects of design can be modeling off-the-shelf components, it is no wonder that finding a way to drag-and-drop these components into a new design captures engineers' attention. Indeed, a key benefit of solid modeling is the reuse of existing models. Two primary ways exist to access solid model libraries.

Standard Component Libraries. State-of-the-art CAD systems now include libraries of standard parts such as screws, bolts, nuts, piping fittings, and fasteners to eliminate the time, effort, and quality concerns related to needlessly modeling these readily available parts (Fig. 9.6). These are typically configurable, parametric component models with the additional benefit that they can help drive the design into which they are incorporated. For example, a matching array of appropriately tapped holes can be quickly and easily added to an assembly, which would automatically update when the fastener specifications change.

FIGURE 9.6 Example of bolts dragged and dropped from a fastener library directly into a CAD model. *(Courtesy of SolidWorks Corporation.)*

3D-Powered Internet Catalogs. Some manufacturers and distributors have enhanced their physical catalogs with a web-based selection, configuration, and purchasing system (Fig. 9.7), and others will likely follow suit. Instead of leafing through and ordering from a paper catalog, customers can now peruse the actual 3D models of products and components, configure and view the models they want, download models for inclusion in their designs, and order their selections online. The days when every engineer who wanted to use an off-the-shelf component had to remodel the same part will soon be history. These systems also benefit part suppliers by easing the process of having their parts incorporated—and thus implicitly specified within customer designs.

9.4.3 Design Presentation, Communication, and Documentation Tools

The improved visualization characteristics of 3D CAD models provide new opportunities for using computer visualizations and animations for marketing and product presentation materials. CAD systems can either export graphics file formats directly—for use on web sites, in marketing brochures, or in product catalogs—or are integrated with imaging or graphics manipulation software packages, enabling the production of both web- and print-quality visuals from the original solid model. Similarly, CAD systems can be used in conjunction with animation packages to create moving product visuals such as walkthroughs, flythroughs, and animated product demonstrations. And increasingly, CAD is being leveraged to create compelling visuals for new product proposals very early in the development process, before any physical prototypes have even been fabricated.

The most commonly used desktop productivity applications are part of the Microsoft Office software suite, which includes Microsoft Word (a word processor), Microsoft Excel (a spreadsheet application), and Microsoft PowerPoint (slide/presentation software). Engineers can import drawings or

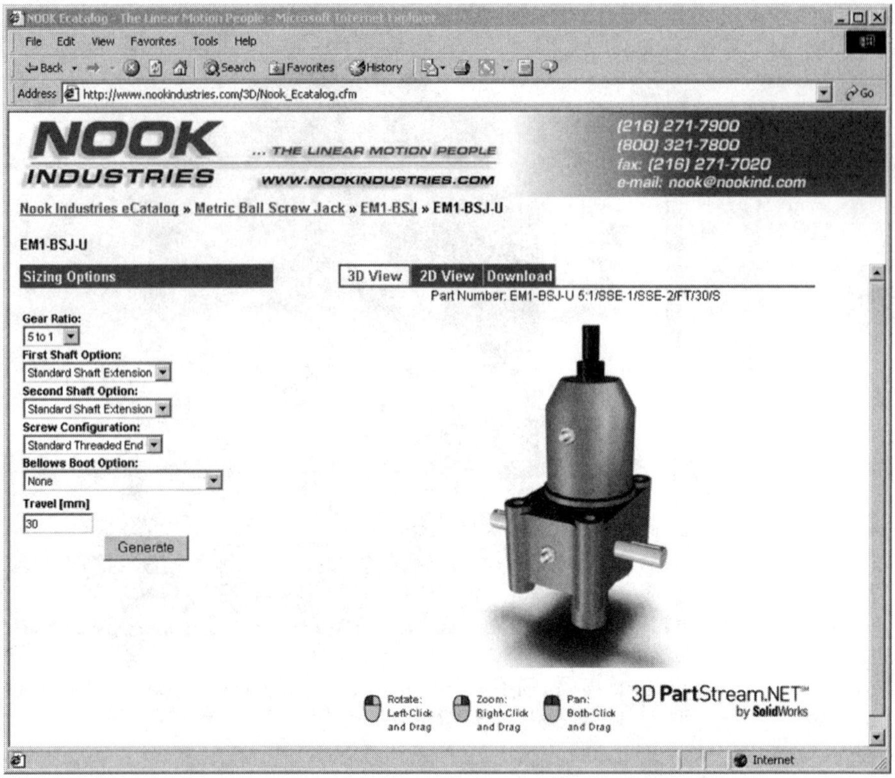

FIGURE 9.7 Example of 3D-enabled web-based catalog offers instant 3D solid models to drag and drop into CAD assemblies. *(Courtesy of Nook Industries.)*

model images directly into Word or PowerPoint for document and presentation preparation related to design reviews. Some CAD systems also leverage Excel for certain functions such as exporting design tables directly into an Excel spreadsheet.

After or as part of product manufacturing, all companies create some sort of product documentation for their customers, whether this documentation takes the form of a basic schematic, product assembly instructions, or a full-blown technical manual. Manufacturers use a variety of software tools to produce and publish this information. FrameMaker, Quicksilver, PageMaker, Quark, Microsoft Word, and WordPerfect are some of the more popular packages for creating and publishing product documentation. In the past, visual illustrations had to be recreated from the original product design data. But now many CAD systems can automatically produce the visual information necessary for product documentation. Automatic generation of exploded assembly views of a product design with numbered balloon labels of individual components is one example of how CAD packages work with documentation publishing packages.

Engineers can also use CAD with web- and email-based communications tools to communicate design information with collaborators, customers, and suppliers, across geographically disparate locations. When all the members of a design effort are not in the same building in the same city, the effectiveness of using documents and presentations for communicating design information becomes compromised. In such instances, communications tools that leverage the Internet can instantly communicate design information anywhere in the world. Some state-of-the-art CAD systems permit the simple emailing of self-viewing files containing 2D and 3D design information with full markup capability or the creation of live web sites with interactive 3D design content (Fig. 9.8).

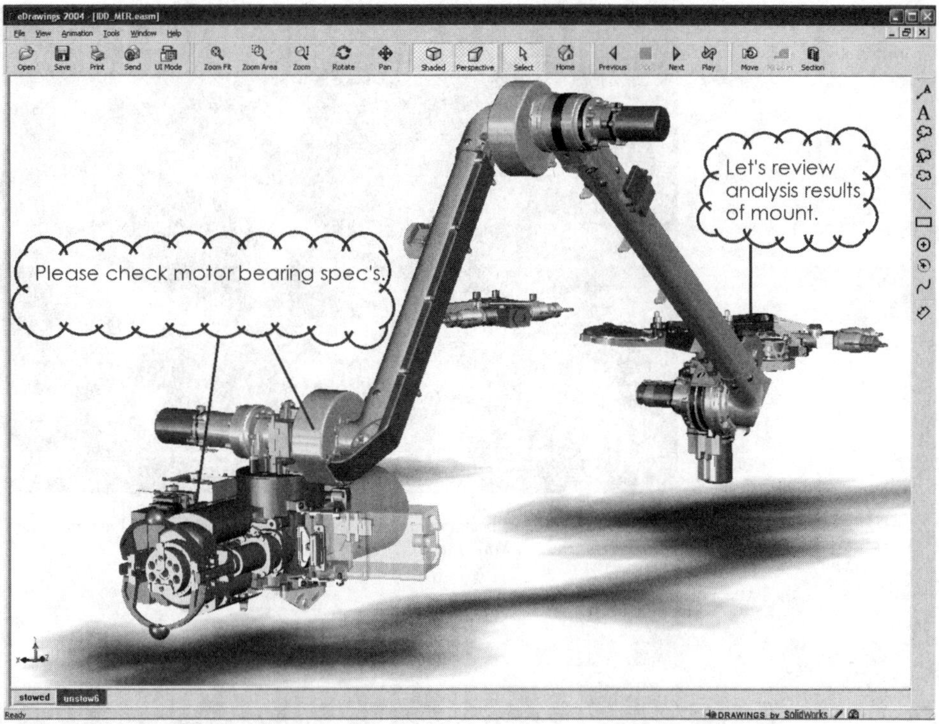

FIGURE 9.8 Self-viewing eDrawing file of the Mars Rover Arm enables remote design reviews between team members. (*Courtesy of Alliance Spacesystems, Inc.*)

9.4.4 Product Data Management (PDM)

Most manufacturers operate their CAD software in conjunction with a product data management (PDM) system. Since solid modeling allows the inclusion of multiple subsystem design files in a new design and the creation of complementary data such as analysis, documentation, and other related electronic data, there is a need to manage the revisions of and relations between a growing body of electronic data. Managing these large amounts of data manually can be not only burdensome but also dangerous because of a lack of rigid security and strict revision control. PDM systems provide the structure for managing, tracking, and securing design data, which can foster design collaboration across engineering teams. Engineers have access to a model's complete history: Who created it? What revision number is it? What assembly does it belong to? What part number has been assigned to it? etc. While many PDM systems are separate applications, some are integrated inside the CAD system, and can provide links to enterprise-wide information systems.

9.4.5 Rapid Prototyping

Rapid prototyping, also known as 3D printing, is a set of technologies that take 3D CAD data as input to specialized machines which quickly make physical prototypes. The machines in this space build the prototypes by applying material in a series of layers, each layer built on its predecessor. This is referred to as an "additive" process, which can be contrasted to milling and turning, which are "subtractive" processes. The method and materials used to construct the layers will vary considerably. Processes include using lasers to cure or fuse material (Stereolithography and Selective Laser

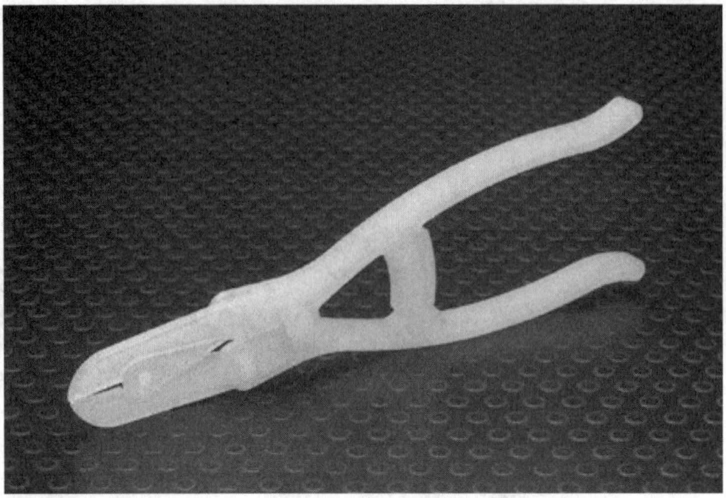

FIGURE 9.9 Rapid prototype created through Fused Deposition Modeling (FDM) technology. (*Courtesy of Xpress3D, Inc.*)

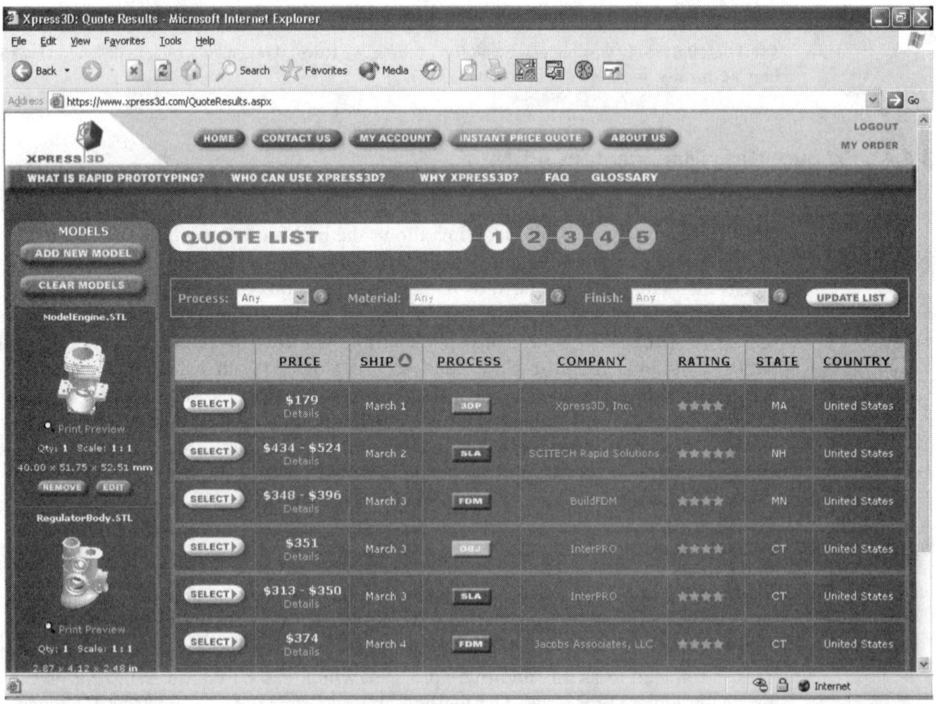

FIGURE 9.10 Instant web-based price quoting for rapid prototypes. (*Courtesy of Xpress3D, Inc.*)

Sintering), drawing the part in 3D with a molten filament of plastic (Fused Deposition Modeling, Fig. 9.9), and inkjet printing. Materials include photosensitive resins; powders made of plaster, nylon, and even metal; and plastics. Each process and material combination has its advantages such as build speed, accuracy, and strength. Prototypes can be built with in-house equipment, or ordered from service bureaus by transmitting 3D CAD data over the internet (Fig. 9.10).

9.5 THE VALUE OF CAD DATA

Increasingly, CAD data has become the thread that weaves its way across the extended manufacturing enterprise, accelerating the rate at which information is processed throughout the product development process. Modern 3D CAD systems extend the value of CAD data to others involved in product development, both internal and external (Fig. 9.11). The value of CAD systems is no longer limited only to helping engineers design products. By improving a designer's capacity for communicating design information with other product development contributors and creating opportunities for leveraging design data, CAD systems add value to the entire manufacturing process and help companies to launch products successfully in a competitive global marketplace.

9.5.1 Manufacturing Engineers

There are two key ways in which manufacturing team members benefit from 3D CAD—better visualization and communication and lower scrap rates.

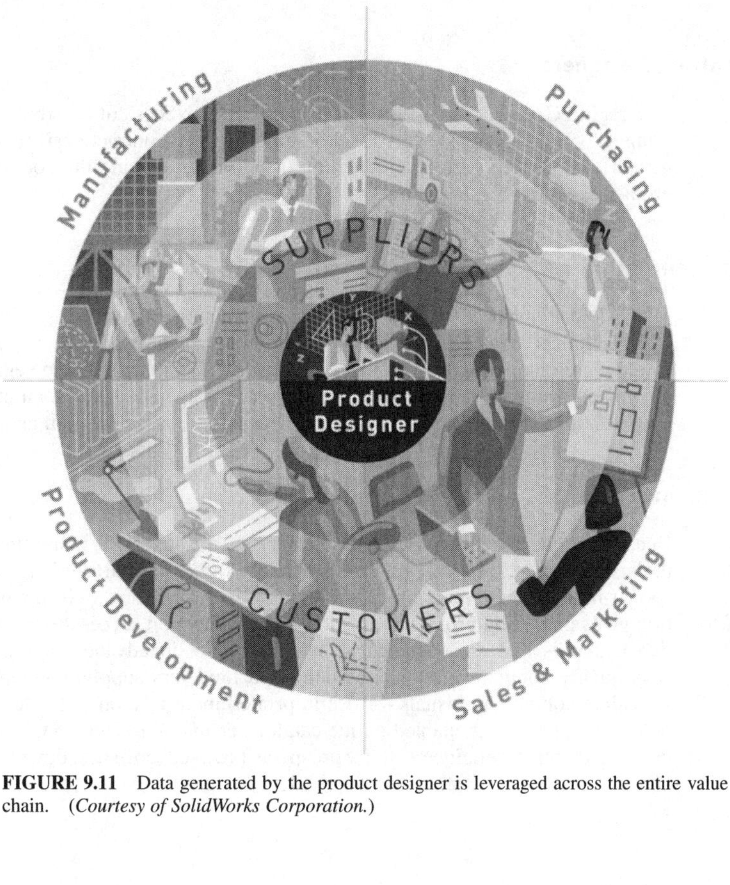

FIGURE 9.11 Data generated by the product designer is leveraged across the entire value chain. (*Courtesy of SolidWorks Corporation.*)

Better Visualization. With access to detail drawings having rich detail, manufacturing engineers can better visualize, and thus impact, new product designs. Furthermore, self-viewing email-enabled design communication tools enable a distributed workforce to review, comment on, and mark up designs early in the design process.

Lower Scrap Rates. Instead of studying 2D engineering drawings and writing computer production control programs in order to develop manufacturing processes and control automated machining and assembly systems—a process which by its very nature can result in errors, misinterpretations, and misunderstandings—manufacturing engineers can import 3D solid models directly into automated CAM systems. In addition to saving time, direct use of CAD models for manufacturing improves quality and reduces the cost of scrap related to design errors or miscommunication of design data. If a manufacturing engineer has a question about a design, the solid model becomes the visual foundation for discussions with design engineers.

9.5.2 Purchasing and Suppliers

CAD data makes the process of obtaining quotes and buying supplied components for use in product designs more clear and efficient. Bidirectional associativity of the CAD system ensures that the Bill of Materials (BOM) on the detail drawing is accurate and up to date. In addition to providing rich detail drawings to suppliers for quotes and orders, purchasing departments can use 3D CAD communication tools to email solid modeling data. This approach saves time for purchasers and eliminates questions and/or confusion over the desired component for the supplier. Furthermore, working directly off the native CAD files will save the suppliers both time and money, savings which can be passed on to the customer.

9.5.3 Contract Designers

Leveraging CAD data extends the reach and effectiveness of external contract designers. By delivering subsystem designs in the form of solid models, contract designers save their clients the effort of converting, translating, or recreating design data, making the contract designer's services more efficient, compatible, and attractive.

9.5.4 Documentation

3D CAD data makes things easier for documentation professionals. Whether the task involves writing a technical manual or laying out an assembly schematic, 3D solid models provide an excellent visual representation of the final product, enabling documentation to begin and finish well in advance of product availability. Automatic generation of drawings, such as an exploded assembly view, and graphics directly from the CAD package save additional illustration and graphics creation tasks.

9.5.5 Marketing

Because solid models provide such rich visual information, marketing professionals can gain an understanding of, and thus impact, a product's overall characteristics and appealing attributes far sooner in the development process. Having this information well in advance of a product introduction gives marketing professionals more time to develop successful marketing campaigns and effective materials that are designed to seed a market in advance of product availability. Access to high-quality computer graphics, which solid modelers supply, complements marketing efforts by providing compelling visuals for use in printed materials, on web sites, and in clear, visually compelling proposals. 3D-enabled online catalogs enable marketers to take advantage of the 3D models developed by their engineers, to let prospects browse, configure, download, and thus incorporate the components and subsystems in new product designs.

9.5.6 Sales

Improved computer hardware and advanced 3D solid modelers combine to create a visual representation of a product that's the next best thing to actually holding the finished product in your hand. Sales professionals can use solid models to demonstrate a soon-to-be-introduced product to potential customers and secure orders in advance of product availability. This capability is especially beneficial to build-to-order companies and manufacturers of large systems, equipment, and mechanisms. Instead of having to build a demo product or prototype, these companies can demonstrate the system on the computer, helping sales professionals move the sales cycle forward and close business sooner.

9.5.7 Analysts

CAD data adds value to design analysts because they no longer have to rebuild models and can perform analyses directly on the original 3D solid model. This development provides additional time for conducting more thorough and often more beneficial analyses of product designs, enabling manufacturers to both reduce costs through more efficient use of material and improve the quality, safety, and effectiveness of their products. Furthermore, analysts using design analysis packages that are integrated inside CAD systems can make suggested design changes directly on the solid model and more effectively collaborate with product designers.

9.5.8 Service and Repair Teams

With consumer and industrial products of even moderate complexity, assembly, repair, and maintenance are nontrivial issues. Having assemblies designed in 3D enables manufacturers to create, with relatively little effort, clear and interactive documentation (Fig. 9.12) for use on the manufacturing

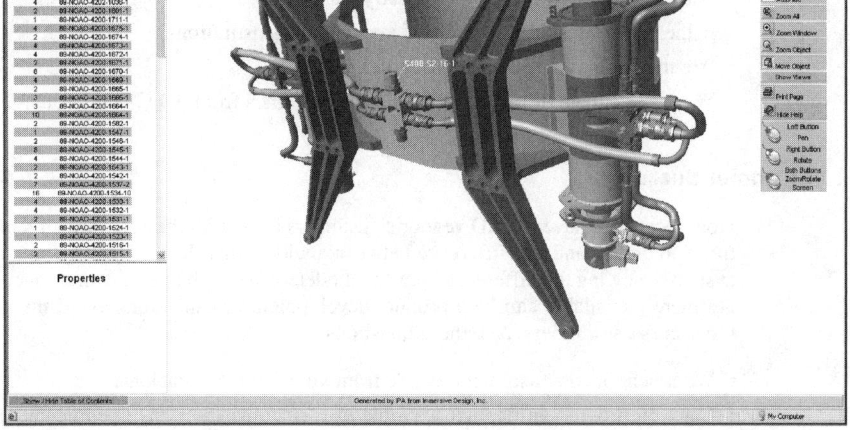

FIGURE 9.12 3D data generated by the product designer is repurposed in an interactive service manual. (*Courtesy of Immersive Design Corporation.*)

floor, service facilities, and online. With the intelligence built into 3D data, working with Bills of Materials, visualizing assembly instructions one step at a time, and keeping track of component and subassembly properties (e.g., cost, weight, part numbers, re-order details) becomes simpler, more enjoyable, and more efficient.

9.6 PLANNING, PURCHASING, AND INSTALLATION

In planning for, evaluating, selecting, purchasing, and installing a 3D solid modeling system, product development organizations face a myriad of options and have different software packages to choose from. What is the appropriate package for a particular manufacturer, and what factors should the product development team consider as part of its preparation for and evaluation of a 3D CAD system? While each manufacturer's needs are different, the factors each should consider to match needs with available solutions are the same. The successful deployment of a 3D CAD system depends upon effective planning, a thorough understanding of both business and technical needs, and the ability to match those needs to CAD capabilities as part of the evaluation of available 3D solid modelers. When planning to acquire a 3D CAD system, manufacturers should consider the following business and technical factors to ensure they acquire a tool that helps rather than hinders their product development efforts.

9.6.1 Company Strength, Market Share, Vision

Manufacturers should assess the CAD vendor's size, position in the industry, commitment to customer support, and vision for the future. The company should be financially secure and strong enough to continue aggressive research and development. The 3D CAD software should be used widely and proven in a manufacturer's industry. A focus on mechanical design provides a greater likelihood of long-term success. Evaluating a CAD vendor is as important as evaluating a solid modeling system. Ask these questions:

- How many 3D CAD licenses has the company sold?
- How many customers are in production with the CAD system?
- Is the company's product the market leader?
- Is the company's product the industry standard?
- Is the company's product taught in educational institutions?
- Are the company's 3D CAD revenues growing?
- What percentage of the company's revenue comes from 3D CAD products?

9.6.2 Customer Success

How successful have a CAD vendor's customers been with the software? It's important for manufacturers to understand the difference between a widely distributed CAD system and a successful CAD system. Knowing the difference between modelers that help manufacturers succeed and modelers that are merely available can help product development organizations avoid the mistakes and replicate the successes of others. Ask these questions:

- What benefits do customers realize from your 3D CAD package?
- Can you provide an explicit example of a customer that has documented a return on its investment (ROI) as a result of using your 3D CAD system?
- Is there a methodology for calculating ROI related to your system?

- Can you provide customer references and/or testimonials?
- Are there extensive training programs available for this CAD system, regardless of geographic location?

9.6.3 Availability of Complementary Products

Another important consideration is the availability of complementary products that extend the capabilities of a 3D solid modeler or provide additional specialized functionality. The availability of complementary products indicates the breadth and depth of the CAD system's use in real-world manufacturing settings. Ask these questions:

- What products are available for extending the capabilities of your core 3D CAD product?
- How mature are the complementary solutions for your 3D CAD system?
- How many users are there for these complementary solutions?
- Does the CAD system have integrated applications for:
  - Design analysis?
  - Computer-aided manufacturing (CAM)?
  - Product data management (PDM)?
  - Fluid flow analysis?
  - Printed circuit board (PCB) design?
  - Photorealistic rendering?
  - Animation?
  - Surfacing functionality?
  - Feature recognition?
  - Tolerance analysis?
  - Mold design?

9.6.4 Product Maturity

A CAD system's history in the marketplace often provides indications of its utility in actual production settings. Just as buying the first model year of a new automobile line is a risky proposition, it can take years and thousands of users to work out problems and performance issues for a CAD system. The maturity of a CAD system is also a mark of the "bugs" that have been resolved and the new functionality that has been added. Ask these questions:

- How many major releases has the CAD software had to date?
- When was the CAD system last updated or revised?
- How often are mid-release service packs, enhancements, and extensions distributed?

9.6.5 Legacy Data Management

Many product development organizations delay the migration to 3D solid modeling because of concerns over large amounts of legacy 2D data, which designers frequently access to design new products. Legacy data can exist in a variety of data formats including 2D and 3D CAD files. When migrating to 3D, manufacturers should consider how they will access and utilize legacy data, and look for a solid modeler with data translation formats and built-in productivity tools for converting 2D and other forms of legacy data to 3D solid models. Ask these questions:

- Can the 3D CAD system import legacy 2D and 3D data such as DXF, DWG, STL, IGES, and STEP files?
- Does the 3D CAD system work with integrated feature recognition software to improve the handling of legacy data?

- Which file types can the CAD system import and export?
- Is the CAD application OLE compliant, providing seamless data exchange with Microsoft Office applications?

9.6.6 Product Innovation

The number and frequency of innovations that a CAD vendor has made is often related to how focused the company is to its customers needs and how well the company listens to its customers. The vendors who are most receptive to their customers are the ones that tend to break new ground in CAD. To assess the level of product innovation in a CAD package, ask these questions:

- Do customers have input into product upgrades? If so, how?
- How innovative have new product enhancements been?
- What CAD innovations has your company been responsible for?
- How many patents and patents pending related to your CAD system does your company have?
- Have you ever been forced to discontinue a product? If so, which ones and why?

9.6.7 Large Assembly Capabilities

Most product design involves assemblies and subassemblies as well as individual parts. Some 3D CAD systems can handle large assemblies involving thousands of parts, and manufacturers should evaluate their assembly design needs and the varying large assembly capabilities of different 3D packages. Ask these questions:

- Does the 3D system support assemblies involving thousands of parts?
- How does the CAD package manage assemblies?
- Does the CAD package support collaboration on an assembly by many individual designers?
- Does the 3D system include built-in tools for assembly design evaluation such as interference checking and collision detection?
- Does the CAD package provide display and modeling techniques for improving computer performance when working with large, complex assemblies?

9.6.8 Configuration Flexibility and Automation

Product developers should also consider whether the automatic configuration of assembly, part, and product variations fits with their needs. Manufacturers that produce families of parts and products with varying sizes, dimensions, weights, and capacities can benefit greatly from the flexibility to configure products automatically from a single original design. Instead of designing variations of an assembly individually, manufacturers whose products vary by nature should look for solid modeling systems that can produce these derivative products or product families automatically. Ask these questions:

- How does the CAD system create similar parts with different dimensions?
- Can I create a family of different parts with varying dimensions from a single part design?
- Can I create different assembly configurations from a single assembly design?

9.6.9 Specialized Capabilities

In addition to base mechanical solid modeling functionality, many manufacturers should consider whether they need a 3D CAD package that offers special features that support specific needs.

Effective solid modelers should offer specialized capabilities that enable productivity gains. Ask these questions:

- Does the CAD system include specialized capabilities or can they be added?
- Can I add specialized functionality for sheet metal design?
- Can I add specialized functionality for designing piping systems?
- Can I add specialized functionality for designed electronics packages?
- Can I add specialized functionality for creating molds?

9.6.10 Visualization and Virtual Prototyping

In evaluating 3D CAD systems, manufacturers should consider a solid modeler's visualization, design evaluation, and animation capabilities and the impact they have on prototyping needs and costs. In addition to minimizing physical prototyping, 3D visualization and animation capabilities can support functions outside the design cycle such as sales, marketing, and customer service. Ask these questions:

- Does the CAD system include full 3D visualization capabilities even during model rotation?
- Does the CAD system permit design evaluation of assembly motion by detecting part collisions and interferences?
- Does the CAD system allow for viewing the internal portions of an assembly?
- Can I animate my assembly model cost-effectively?
- Can I create print-quality visuals and graphics from my CAD package?

9.6.11 Web-Based Communication Tools

The Internet has changed the way much of the world does business, and exploiting the web from a design perspective is an important consideration for companies that are evaluating solid modeling software. Manufacturers should consider whether a 3D CAD package provides web-based communication tools for easily sharing design data with vendors and customers and collaborating with colleagues and partners.

- Does the CAD software provide a means for efficiently emailing design data?
- Does the CAD software provide a means for publishing interactive web sites with 3D solid model content?

9.7 SUCCESSFUL IMPLEMENTATION

While CAD packages differ greatly, the steps a product development organization should take to implement a CAD system are very similar. Basically, the plan should address every functional area in the company that will be impacted by the transition, from design and engineering through manufacturing and information systems. At a minimum, the CAD implementation plan should contain these elements:

- *Standards.* A set of documents that define recommended practices for using the CAD system.
- *Installation.* The set of procedures that define the hardware requirements and how the CAD software is installed and configured.

- *Training.* The set of procedures and schedule for training the user base on how to operate the new CAD system.
- *Legacy data.* The set of procedures for how legacy design data will be managed and reused.
- *Data Management.* The standard for how the company will define, modify, manage, and archive design data created in the new CAD system.
- *Evaluation.* A method for evaluating the effectiveness of the CAD implementation such as an accurate methodology for calculating ROI.

9.7.1 Standards

Manufacturers should collect and publish all documents governing a company's approved design practices either as a printed manual or as an online resource that is available to all users. A standards manual is an important resource for manufacturing companies that are transitioning to a new CAD system and provides a single resource for addressing and resolving user questions. The standards manual should include the following information.

Design and Engineering. This describes the company's standards for engineering models and drawings. Should engineers use a specified company standard or ANSI standards? When working internationally, should the designers use European, British, or ISO standards? This information should address any questions related to how models and drawings are labeled, dimensioned, etc.

Data Exchange. This describes the company's standards for exchanging data. Is there a preferred format for importing and exporting CAD data and for interacting with vendors, customers, or suppliers? Are there approved data exchange methods such as FTP, e-mail, compressed ZIP files, or web communication tools? Are there approved data healing approaches?

Design Communication and Collaboration. This describes the company's standards for collaborating on and communicating design data. Are there certain design and life cycle management requirements that come into play? How does a designer go about setting up a design review or requesting an engineering change? What are the design-for-manufacturing implications such as bending allowances, edge tolerances for punching, corner radius requirements, machining allowances, or CNC download requirements?

CAD-Related Configurations. This describes the standard configuration settings for both computer hardware and the CAD system. What display and performance settings should a designer use? Should users keep the default settings or use company-approved settings for things such as file locations, data backup, revisions, materials, part numbers, drawing numbers, and templates? Does the company have approved templates for drawings? Which fonts, line weights, arrowheads, and units should be used?

Design Practices/Methodologies. This describes methodologies for handling certain types of designs such as assemblies. Should engineers design assemblies from the top down to the component level or from the component level up? What are the standards for tolerances within assemblies?

Sketching. This describes the standards for creating engineering sketches. How will engineering sketches be used? What level of detail is required? Where should dimensions be located? What constitutes a fully defined sketch?

Part Modeling. This describes the standards for creating solid models of parts. How should designers handle models for purchased components? How should designers annotate the model? How should designers prepare the model for interfacing with finite element analysis (FEA) programs? How should engineers use part configurations?

Assembly Modeling. This describes the standards for creating assembly models. What structure should designers use for subassemblies? How should designers check for interferences within an assembly? How should designers apply annotations, notes, and datums to an assembly model? How should engineers use assembly configurations?

Drawings. This describes the standards for creating engineering drawings. What dimension styles should designers use? How should designers handle balloons and annotations? What drawing views are required? Why type of detail, projections, and sections need to be done? What external files should drawings reference?

Legacy Data. This describes the company policy on accessing and reusing legacy design data. How can a designer access legacy data? What are the requirements for accessing legacy data? When can legacy data be used? How should legacy data be imported into the new CAD system?

General/Administrative. This describes the company policy for updating the CAD system. When should the CAD system be updated? What is the procedure for requesting an update? When should a designer develop custom programming for certain capabilities or macros for automating common repetitive tasks?

Education, Training, and Support. This describes the company policy for requesting additional training and technical support. Are there procedures for obtaining additional training? What are the support procedures? Are there internal support guidelines? Is there a procedure for accessing external support services?

9.7.2 Installation

The CAD implementation plan should address the information system needs of the new CAD system.

Computer Hardware. The plan should address the minimum system requirements for a user. What operating system (OS) is required? How much random access memory (RAM) does a user need? What is the minimum CPU (computer processor) that will run the CAD software? What video cards and drivers does the CAD system support?

Printing Hardware. The plan should describe the printers, plotters, and peripherals that designers will use with the CAD system. Does the company need additional printing hardware to support the new CAD system?

Network Hardware and Topology. If the CAD system is used across a computer network, the plan should address any additional network hardware needs. Does new network hardware need to be acquired to support the new CAD system?

9.7.3 Training

The CAD implementation plan should address the level of training that each user should receive and schedule training in the least disruptive and most productive manner. In preparing the training plan, manufacturers should create a detailed training plan for each user.

- *Essentials.* Training that every user will need to operate the CAD system effectively.
- *Advanced part modeling.* Training that only users who are responsible for the design of unique parts will need to operate the CAD system effectively.
- *Advanced assembly modeling.* Training only for users who are responsible for the design of complex assemblies will be needed to operate the CAD system effectively.

- *Specialized modeling*. Training only for users who are responsible for specialized design functions, such as sheet-metal, plastic injection-molded parts, and piping systems, will be needed to operate the CAD system effectively.
- *CAD productivity.* Training in the use of CAD productivity tools such as utilities and feature-recognition software.
- *Programming macro development.* Training on how to leverage the CAD system's application programming interface (API) to develop Visual Basic scripts, C++ coding, and macros for automating frequent tasks.

9.7.4 Legacy Data

The CAD implementation plan should address how the new CAD system will interface with legacy design data, whether 2D or 3D in nature, and establish procedures for how legacy data will be leveraged, managed, and reused. What is the preferred design format for importing legacy data? How should it be saved?

9.7.5 Data Management

All CAD implementation plans should take product data management (PDM) needs into account. The plan should include procedures on how designers will define, modify, revise, update, and archive CAD design data. Will this be done manually, or will the company use an integrated or standalone PDM system? If a new PDM system will be installed as part of the CAD implementation, are there additional training needs? Will the PDM implementation coincide with the CAD transition or take place later?

9.7.6 Evaluation

How can a product development organization determine whether the implementation of a CAD system has been successful? One way to evaluate a CAD system's success is to develop a methodology for comparing product development cycles and design costs against those experienced with the previous CAD system. Have design cycles gotten shorter or longer? Have design costs gone up or down? Have a company's scrap costs increased or decreased? Some CAD vendors provide methodologies and surveys that are designed to calculate a customer's return on investment (ROI), an indication of the success or failure of a new CAD transition. When using vendor-supplied methodologies make sure that the items used for comparison are easily quantifiable. Building an evaluation component into a CAD implementation plan is important for gaining reliable feedback on whether the new CAD system is working.

9.7.7 Start with a Pilot Program

A particularly effective means of ensuring the successful implementation of a CAD system is to start small before going big. Designing and executing a pilot CAD implementation plan at the department or group level is an excellent way to gauge the probable impact and potential success of a CAD transition across the entire product development organization. Simply develop the implementation plan for a single department or group, following the guidelines described above, and evaluate the results of that implementation as it applies to the larger organization as a whole.

Critical questions to ask include: Is the group more productive? Have costs gone down? Can the organization expect similar results company-wide?

9.8 FUTURE CAD TRENDS

Today, 3D CAD systems provide efficient solutions for automating the product development process. Designers, engineers, and manufacturing professionals can now leverage solid modeling data in ways that were unimaginable just a generation ago. CAD technology has matured to the point of providing ample evidence of its usefulness in compressing product design and manufacturing cycles, reducing design and production costs, improving product quality, and sparking design innovation and creativity, all of which combine to make manufacturing concerns more profitable and competitive. As more manufacturers reap the benefits of 3D CAD technology, research and development will continue to push the technology forward and make CAD systems easier to use and deploy.

Anticipating the future course of CAD technology requires a solid understanding of how far the technology has come, how far it still has to go, and what areas hold the greatest potential for advancement. The appeal of CAD technology from a business perspective has always been tied to the ability to create operational efficiencies and foster innovation in product design and manufacturing settings. To provide these benefits, CAD systems must be fast, easy-to-use, robust (in terms of design functionality), portable, and integrated with other design and manufacturing systems. CAD technology has made great strides in recent years in each of these areas and will continue to do so in the years to come.

9.8.1 Performance

CAD systems and the computer hardware they run on have matured greatly since the first cryptically complex, command-driven, UNIX-based drafting and modeling packages. Perhaps the most substantive developments in the evolution of CAD technology to date have been the application of production-quality 3D solid modeling technology to the Windows operating environment and the availability of affordable, high-performance PCs. These developments extended the reach and economic viability of CAD technology for all manufacturers, both large and small. As computing power and graphics display capabilities continue to advance, so will the speed and overall performance of CAD software. Just 10 years ago, engineers could not rotate a large solid model on a computer screen in real time. Now, this type of display performance is an everyday occurrence. Some of today's CAD systems enable designers to put complex assemblies in dynamic motion to check for part collisions and interferences. In the future, we can expect CAD systems to deliver even more power, performing computationally intensive and graphically demanding tasks far faster than they do today.

9.8.2 Ease-of-Use

Although CAD has become significantly easier since the days of command-line or Unix-based systems, the "CAD overhead" associated with product design is still too high. At the same time, CAD systems have also become more capable, which inevitably adds to the software tool's complexity and presents new user interface challenges. It is therefore not unreasonable to expect that at least some of the following promising concepts will be embraced by the mainstream.

Heads-Up User Interface. The notion of a "heads-up UI" has been around for some time and is found in a variety of real-world applications, including fighter jet cockpit controls. The idea is to seamlessly integrate the visual presentation with the controls so that the operator does not need to change focus. Some leading CAD systems have already started to take this approach, minimizing the designer's distraction away from the graphics, and intuitively arranging the controls. Further advancements, such as context-sensitive controls, "wizards" that guide the user, and context-sensitive help and tutorials, are right around the corner.

3D Input Devices. The keyboard-and-mouse paradigm has largely been untouched for 30 years. To take a close look at something in the real world, you pick it up and examine it from every angle. The action is so natural, you don't even think about it. In a similar manner, using a 3D motion controller, the designer can intuitively zoom, pan and rotate a 3D model nearly as naturally as if it were in his or her hands.

3D User Interface. The computer desktop has been 2D for 30 years and may likely enter the third dimension soon; the 3D interface is coming to the Windows operating system, helping to navigate icons, controls, and other objects. For example, imagine that icons and toolbars that are used less often are moved to the background, and ones used more often are moved to the foreground. And there can be some degree of transparency so that you can still see the objects in the back, but they are less prominent.

9.8.3 Further Integration

Functions that used to require separate specialty packages, such as advanced surfacing, on-the-fly engineering analysis, and kinematics studies, will likely become part of basic solid modeling packages in the years to come. Some of the currently available 3D CAD packages are already illustrating this trend, adding specialized functions, such as sheet metal, piping system design, and finite element analysis, to their core packages. Another likely trend is real-time photorealistic rendering of the CAD models, making for a more immersive design experience by taking advantage of graphics cards' increasing computational capabilities.

9.8.4 Interoperability

Some CAD vendors have already addressed some of the obstacles to interoperability by integrating CAD systems with other design and manufacturing functions such as design analysis, rapid prototyping technologies and CAM. This trend is likely to continue in other areas including interoperability across competing CAD systems because whenever CAD data have to be recreated, converted, or translated, duplicated effort and an opportunity for error occur.

9.9 FUTURE CAM TRENDS

Computer-aided manufacturing (CAM) technology that was considered "future trends" only a few years ago is in widespread production use today. High speed machining, palletized operations, and multispindle machining centers are all routinely programmed with modern CAM systems. Where might CAM be headed in the next 10 years?

An area that shows a lot of promise is the integration of tolerance information into the CAD/CAM process. Tolerancing information plays a critical role in determining a part's overall manufacturing strategy, but conventional CAM systems operate from nominal, or nontoleranced geometry. As CAD/CAM progresses from geometric models to complete product models, the CAD system will supply the CAM system with a part's tolerance specification directly, eliminating the need for 2D drawings.

The CAM system could then combine this product model information with knowledge-based manufacturing (KBM) to automate macro planning and toolpath generation (see Fig. 9.13). Ultimately, this knowledge base might be used to create manufacturing-aware design advisors providing feedback to the designer from a manufacturing perspective. This would allow designers to easily evaluate the manufacturing implications of design decisions, resulting in designs that can be manufactured faster, cheaper, and at higher quality.

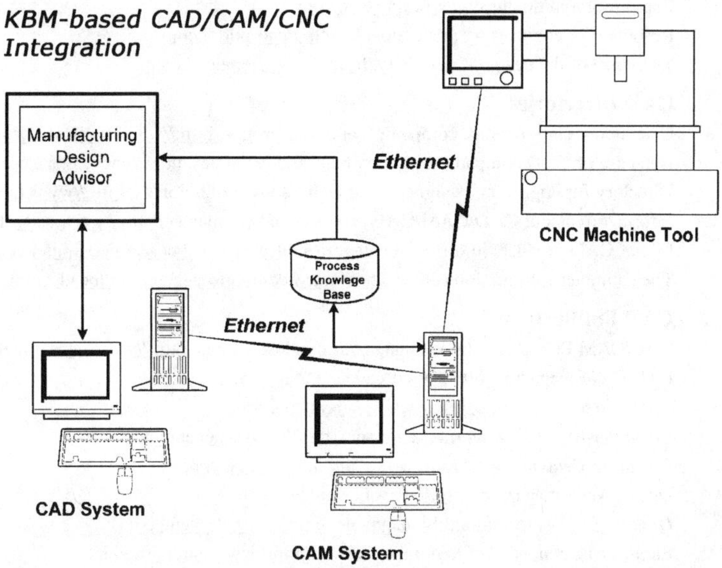

FIGURE 9.13 Future knowledge-based design advisors. (*Courtesy of Gibbs and Associates.*)

9.10 CONCLUSION

CAD technology has come a long way since the early, esoteric, command-driven systems, which required as much if not more of an engineer's attention as the actual process of design, and now helps manufacturers to streamline their design processes, reduce costs, and improve product quality. Today's engineering and manufacturing professionals need a design platform that complements their creativity, innovation, and engineering skills so that they can approach design and manufacturing challenges without distraction.

Today's CAD systems have progressed a great deal toward achieving that goal, requiring less mental energy to run so that an engineer can focus more on bringing better products to market faster. CAD technology operates efficiently on affordable computing hardware. CAD packages are now integrated with more complementary design, manufacturing, and desktop productivity applications and CAD data can now automate many functions across the product development organization. In many ways, 3D solid modeling data have become both the foundation and the "glue" that drive today's efficient, high-quality manufacturing operations.

INFORMATION RESOURCES

CAD Resources

CAD information resource, http://www.cadwire.com.

CAD news, reviews, and information, http://www.cadinfo.net.

Interactive 3D documentation, http://www.immdesign.com.

Mechanical CAD resource, http://www.mcadcafe.com.

Mechanical design software, http://www.solidworks.com.

Rankings for CAD software, http://www.daratech.com.

Rapid prototyping, http://www.xpress3d.com.
Resource for engineering professionals, http://cad-portal.com.
The voice of the design community, http://www.digitalcad.com.

CAD Directories

CAD news, employment, companies, and user groups, http://www.cad-forum.com.
Database of CAD companies, publications, and resources, http://www.3zone.com.
Directory for engineers, designers, and technology professionals, http://www.tenlinks.com.
Information about CAD/CAM/CAE products and companies, http://www.cadontheweb.com.
List of CAD products in different categories, http://www.dmoz.org/computers/CAD/.
The computer information center, http://www.compinfo-center.com/tpcad-t.htm.

CAD Publications

CAD/CAM Publishing: books, magazines, industry links, http://www.cadcamnet.com.
CADENCE magazine, http://www.cadence-mag.com.
CAD Systems magazine, http://www.cadsystems.com.
Computer-Aided Engineering magazine, http://www.caenet.com.
Computer Graphics World magazine, http://www.cgw.com.
Design News magazine, http://www.designnews.com.
Desktop Engineering monthly magazine, http://www.deskeng.com.
Engineering Automation Report newsletter, http://www.eareport.com.
Engineering handbooks online, http://www.engnetbase.com.
Machine Design magazine, http://www.machinedesign.com.
MCAD Vision: mechanical design technology magazine, http://www.mcadvision.com.
Technicom MCAD weekly newsletter: CAD industry analysis, http://www.technicom.com.
Weekly CAD magazine, http://www.upfrontezine.com.

CAD Research

California Polytechnic State University CAD Research Center, http://www.cadrc.calpoly.edu.
Massachusetts Institute of Technology (MIT) CAD Laboratory, http://cadlab.mit.edu.
Purdue University CAD Laboratory, http://www.cadlab.ecn.purdue.edu.
University of California (Berkeley) Design Technology Warehouse, http://www-cad.eecs.berkeley.edu.
University of Southern California Advanced Design Automation Laboratory, http://atrak.usc.edu.
University of Strathclyde (Scotland) CAD Centre, http://www.cad.strat.ac.uk.

Organizations

American Design and Drafting Associations (ADDA), http://www.adda.org.
The American Society of Mechanical Engineers (ASME), http://www.asme.org.
The CAD Society, http://www.cadsociety.org.
The Institute of Electrical & Electronics Engineers (IEEE), http://www.ieee.org.

CHAPTER 10
MANUFACTURING SIMULATION

Charles Harrell
Brigham Young University
Provo, Utah

10.1 INTRODUCTION

"Man is a tool using animal. … Without tools he is nothing, with tools he is all."
—Thomas Carlyle

Computer simulation is becoming increasingly recognized as a quick and effective way to design and improve the operational performance of manufacturing systems. Simulation is essentially a virtual prototyping tool that can answer many of the design questions traditionally requiring the use of hardware and expensive trial and error techniques. Here we describe the use of simulation in the design and operational improvement of manufacturing systems.

The Oxford American Dictionary (1980) defines simulation as a way "to reproduce the conditions of a situation, as by means of a model, for study or testing or training, etc." To analyze a manufacturing system, one might construct a simple flow chart, develop a spreadsheet model, or build a compuer simulation model depending on the complexity of the system and the desired precision in the answer. Flowcharts and spreadsheet models are fine for modeling simple processes with little or no interdependencies or variability. However, for complex processes a computer simulation which is capable of imitating the complex interactions of the system over time is needed. This type of dynamic simulation has been defined by Schriber (1987) as "the modeling of a process or system in such a way that the model mimics the response of the actual system to events that take place over time." Thus, by studying the behavior of the dynamic model we can gain insights into the behavior of the actual system.

In practice, manufacturing simulation is performed using commercial simulation software such as ProModel or AutoMod that have modeling constructs specifically designed for capturing the dynamic behavior of systems. Using the modeling constructs available, the user builds a model that captures the processing logic and constraints of the system being studied. As the model is "run," performance statistics are gathered and automatically summarized for analysis. Modern simulation software provides a realistic, graphical animation of the system being modeled to better visualize how the system behaves under different conditions (see Fig. 10.1).

During the simulation, the user can interactively adjust the animation speed and even make changes to model parameter values to do "what if" analysis on the fly. State-of-the-art simulation technology even provides optimization capability—not that simulation itself optimizes, but scenarios that satisfy defined feasibility constraints can be automatically run and analyzed using special goal-seeking algorithms.

Because simulation accounts for interdependencies and variability, it provides insights into the complex dynamics of a system that cannot be obtained using other analysis techniques. Simulation

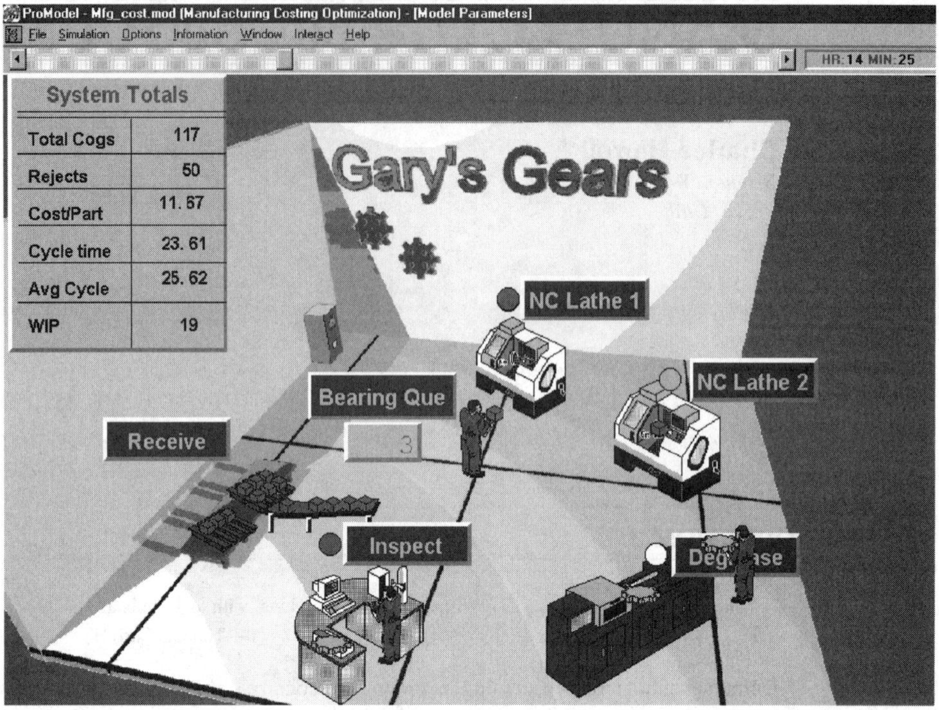

FIGURE 10.1 Simulation provides both visualization and performance statistics.

gives systems planners unlimited freedom to try out different ideas for improvement, risk free—with virtually no cost, no waste of time, and no disruption to the current system. Furthermore, the results are both visual and quantitative with performance statistics automatically reported on all measures of interest.

The procedure for doing simulation follows the scientific method of (1) formulating a hypothesis, (2) setting up an experiment, (3) testing the hypothesis through experimentation, and (4) drawing conclusions about the validity of the hypothesis. In simulation, we formulate a hypothesis about what design or operating policies work best. We then set up an experiment in the form of a simulation model to test the hypothesis. With the model, we conduct multiple replications of the experiment or simulation. Finally, we analyze the simulation results and draw conclusions about our hypothesis. If our hypothesis was correct, we can confidently move ahead in making the design or operational changes (assuming time and other implementation constraints are satisfied). As shown in Fig. 10.2, this process is repeated until we are satisfied with the results.

As can be seen, simulation is essentially an experimentation tool in which a computer model of a new or existing system is created for the purpose of conducting experiments. The model acts as a surrogate for the actual or real-world system. Knowledge gained from experimenting on the model can be transferred to the real system. Thus, when we speak of doing simulation, we are talking about "the process of designing a model of a real system and conducting experiments with this model" (Shannon, 1998). Everyone is aware of the benefits flight simulators provide in training pilots before turning them loose in actual flight. Just as a flight simulator reduces the risk of making costly errors in actual flight, system simulation reduces the risk of having systems that operate inefficiently or that fail to meet minimum performance requirements. Rather than leave design decisions to chance, simulation provides a way to validate whether or not the best decisions are being made. Simulation

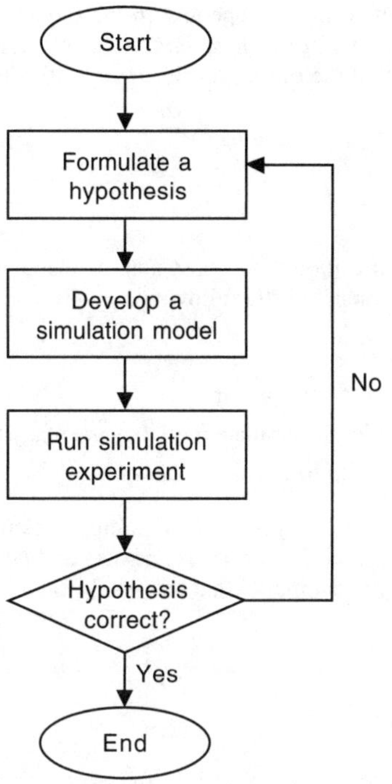

FIGURE 10.2 The process of simulation experimentation.

avoids the time, expense, and disruption associated with traditional trial-and-error techniques.

By now it should be obvious that simulation itself is not a solution tool but rather an evaluation tool. It describes how a defined system will behave; it does not prescribe how it should be designed. Simulation doesn't compensate for one's ignorance of how a system is supposed to operate. Neither does it excuse one from being careful and responsible in handling input data and interpreting output results. Rather than being perceived as a substitute for thinking, simulation should be viewed as an extension of the mind that enables one to understand the complex dynamics of a system.

Simulation promotes a try-it-and-see attitude that stimulates innovation and encourages thinking "outside the box." It helps one get into the system with sticks and beat the bushes to flush out problems and find solutions. It also puts an end to fruitless debates over what solution will work best and by how much. Simulation takes the emotion out of the decision-making process by providing objective evidence that is difficult to refute.

By using a computer to model a system before it is built or to test operating policies before they are actually implemented, many of the pitfalls that are often encountered in the start-up of a new system or the modification of an existing system can be avoided. Improvements that traditionally took months and even years of fine-tuning to achieve can be attained in a matter days or even hours. Because simulation runs in compressed time, weeks of system operation can be simulated in only a few minutes or even seconds.

Even if no problems to a system design are found through simulation, the exercise of developing a model is, in itself, beneficial in that it forces one to think through the operational details of the process. Simulation can work with inaccurate information, but it can't work with incomplete information. If you can't define how the system operates, you won't be able to simulate it. Often solutions present themselves simply by going through the model-building exercise before any simulation run is made. System planners often gloss over the details of how a system will operate and then get tripped up during the implementation phase by all of the loose ends. The expression "the devil is in the details" has definite application to systems planning. Simulation forces decisions on critical details so that they are not left to chance or to the last minute when it may be too late.

10.2 SIMULATION CONCEPTS

To gain a basic understanding of how simulation works, let's look a little more in detail at a few of the key concepts involved in simulation. A simulation is defined using the *modeling constructs* provided by the simulation software. When a model is run, the model definition is converted to a sequence of *events* that are processed in chronological order. As the simulation progresses, *state and statistical variables* are updated to reflect what is happening in the model. To mimic the random behavior that occurs in manufacturing systems, *random variates* are generated from appropriate distributions defined by the user. To ensure that the output results are statistically valid, an appropriate number of *replications* should be run.

10.2.1 Modeling Constructs

Every simulation package provides specific modeling constructs or elements that can be used to build a model. Typically, these elements consist of the following:

- *Entities.* The items being processed.
- *Workstations.* The places where operations are performed.
- *Storages and queues.* Places where entities accumulate until they are ready to be processed further.
- *Resources.* Personnel, forktrucks, etc. used to enable processing.

When defining a model using a manufacturing-oriented simulation package, a modeler specifies the processing sequence of entities through workstations and queues and what operation times are required for entities at each workstation. Once a model that accurately captures the processing logic of the system is built, it is ready to run.

10.2.2 Simulation Events

When a model is run, it translates the processing logic into the events that are to occur as time passes in the simulation. An event might be the arrival of a part, the completion of an operation, the failure of a machine, etc. Because a simulation runs by processing individual events as they occur over time, it is referred to as discrete-event simulation.

Simulation events are of two types: *scheduled* and *conditional*. A scheduled event is one whose time of occurrence can be determined beforehand and can therefore be scheduled in advance. Assume, for example, that an operation has just begun and has a completion time that is normally distributed with a mean of 5 min and a standard deviation of 1.2 min. At the start of the operation a sample time is drawn from this distribution, say 4.2 min, and an activity completion event is scheduled for that time into the future. Scheduled events are inserted chronologically into an event calendar to await the time of their occurrence.

Conditional events are events that are triggered when some condition is met or when a command is given. Their time of occurrence cannot be known in advance so they can't be scheduled. An example of a conditional event might be the capturing of a resource which is predicated on the resource being available. Another example would be an order waiting for all of the individual items making up the order to be assembled. In these situations, the event time cannot be known beforehand so the pending event is simply placed on a waiting list until the condition can be satisfied.

Discrete-event simulation works by scheduling any known completion times in an event calendar in chronological order. These events are processed one at a time and, after each scheduled event is processed, any conditional events that are now satisfied are processed. Events, whether scheduled or conditional, are processed by executing certain logic associated with that event. For example, when a resource completes a task, the state and statistical variables for the resource are updated, the graphical animation is updated, and the input waiting list for the resource is examined to see what activity to respond to next. Any new events resulting from the processing of the current event are inserted into either the event calendar or other appropriate waiting list. A logic diagram depicting what goes on when a simulation is run is shown in Fig. 10.3.

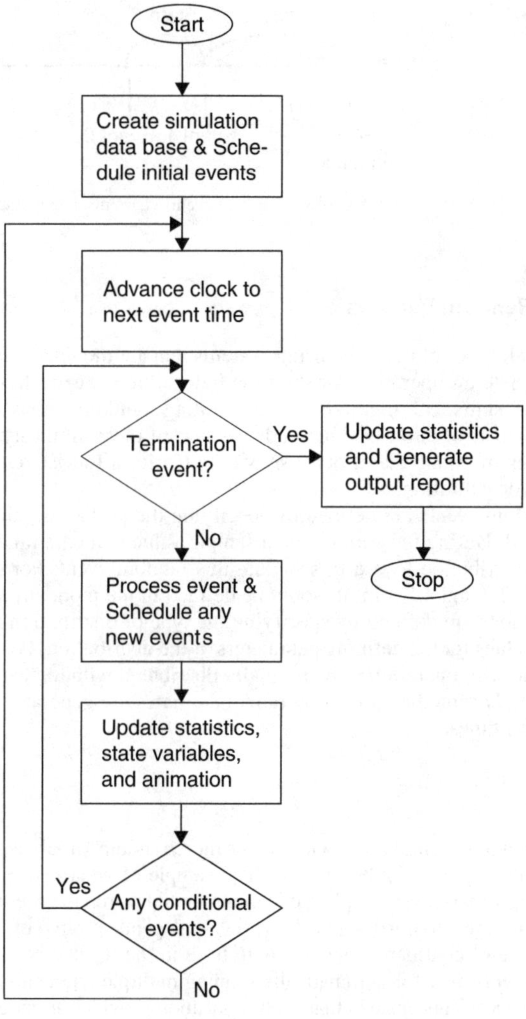

FIGURE 10.3 Logic diagram of how discrete-event simulation works.

10.2.3 State and Statistical Variables

State variables represent the current condition or status of a model element at any given point in time in the simulation. A state variable might be the number of items in a queue, or whether a machine is busy or idle. State variables in discrete-event simulation change only when some event occurs. Each time a state variable changes, the state of the model is changed since the model state is essentially the collective value of all the state variables in the model. (See Fig. 10.4).

During a simulation, statistics are gathered using statistical variables on how long model elements were in given states and how many different types of events occurred. These statistical variables are then used to report on the model performance at the end of the simulation. Typical output statistics include resource utilization, queue lengths, throughput, flow times, etc.

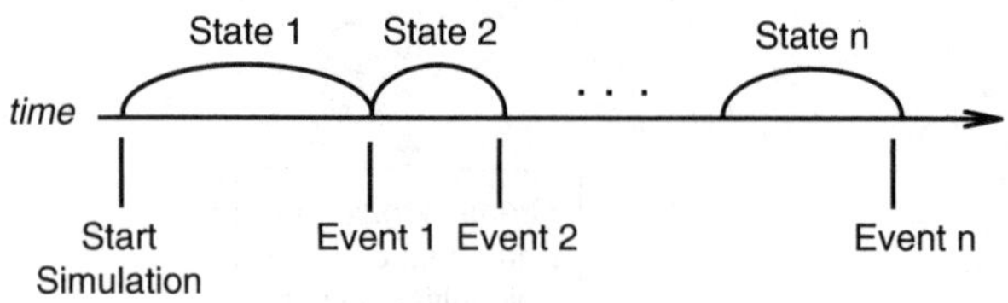

FIGURE 10.4 Discrete events cause discrete changes in states.

10.2.4 Generating Random Variates

Nearly all types of manufacturing systems that are modeled have random behavior such as the time to complete an operation, or the time before the next machine failure. Discrete-event simulation employs statistical methods for generating random behavior. These methods are sometimes referred to as Monte Carlo methods because of their similarity to the probabilistic outcomes found in games of chance, and because Monte Carlo, a tourist resort in Monaco, was such a popular center for gambling.

Random events are defined by specifying the probability distribution from which the events are generated. During the simulation, a sample value (called a random variate) is drawn from the probability distribution to use to schedule this random event. For example, if an operation time varies between 2.2 and 4.5 min, it would be defined in the model as a probability distribution. Probability distributions are defined by specifying the type of distribution (normal, exponential, etc.) and specifying values for the defining parameters of the distribution. For example, we might describe the time for a check-in operation to be normally distributed with a mean of 5.2 min and a standard deviation of 1 min. During the simulation random variates are generated from this distribution for successive operation times.

10.2.5 Replications

When running a simulation with one or more random variables, it is important to realize that the output results represent only one statistical sample of what could have happened. Like any experiment involving variability, multiple replications should be run in order to get an expected result. Usually anywhere from five to thirty replications (i.e., independent runs) of the simulation are made depending on the degree of confidence one wants to have in the results. Nearly all simulation software provides a replication facility for automatically running multiple replications, each with a different random number sequence. This ensures that each replication provides an independent observation of model performance. Averages and variances across the replications are automatically calculated to provide statistical estimates of model performance. Confidence intervals are also provided that indicate the range within which the true performance mean is likely to fall.

10.3 SIMULATION APPLICATIONS

Simulation began to be used in commercial applications in the 1960s. Initial models were usually programmed in Fortran and often consisted of thousands of lines of code. Not only was model building an arduous task, but extensive debugging was required before models ran correctly. Models frequently took upwards of a year or more to build and debug so that, unfortunately, useful results were not obtained until after a decision and monetary commitment had already been made. Lengthy simulations were run in batch mode on expensive mainframe computers where CPU time was at a premium. Long development cycles prohibited major changes from being made once a model was built.

It has only been in the last couple of decades that simulation has gained popularity as a decision-making tool in manufacturing industries. For many companies, simulation has become a standard practice when a new facility is being planned or a process change is being evaluated. It is fast becoming to systems planners what spreadsheet software has become to financial planners.

The surge in popularity of computer simulation can be attributed to the following:

- Increased awareness and understanding of simulation technology
- Increased availability, capability, and ease-of-use of simulation software
- Increased computer memory and processing speeds, especially of PCs
- Declining computer hardware and software costs

Simulation is no longer considered to be a method of "last resort," nor is it a technique that is reserved only for simulation experts. The availability of easy-to-use simulation software and the ubiquity of powerful desktop computers have not only made simulation more accessible, but also more appealing to planners and managers who tend to avoid any kind of solution that appears too complicated. A solution tool is not of much use if it is more complicated than the problem that it is intended to solve. With simple data entry tables and automatic output reporting and graphing, simulation is becoming much easier to use and the reluctance to use it is disappearing.

Not all system problems that could be solved with the aid of simulation should be solved using simulation. It is important to select the right tool for the task. For some problems, simulation may be overkill—like using a shotgun to kill a fly. Simulation has certain limitations of which one should be aware before making a decision to apply it to a given situation. It is not a panacea for all system-related problems and should be used only if the shoe fits. As a general guideline, simulation is appropriate if the following criteria hold true:

- An operational (logical or quantitative) decision is being made.
- The process being analyzed is well defined and repetitive.
- Activities and events are highly interdependent and variable.
- The cost impact of the decision is greater than the cost of doing the simulation.
- The cost to experiment on the actual system is greater than the cost to do a simulation.

The primary use of simulation continues to be in the area of manufacturing. Manufacturing systems, which include warehousing and distribution systems, tend to have clearly defined relationships and formalized procedures that are well suited to simulation modeling. They are also the systems that stand to benefit the most from such an analysis tool since capital investments are so high and changes are so disruptive. As a decision-support tool, simulation has been used to help plan and make improvements in many areas of both manufacturing and service industries (Fig. 10.5). Typical applications of simulation include

- Work-flow planning
- Capacity planning
- Cycle time reduction
- Staff and resource planning
- Work prioritization
- Bottleneck analysis
- Quality improvement
- Cost reduction
- Inventory reduction
- Throughput analysis
- Productivity improvement
- Layout analysis
- Line balancing
- Batch size optimization
- Production scheduling
- Resource scheduling
- Maintenance scheduling
- Control system design

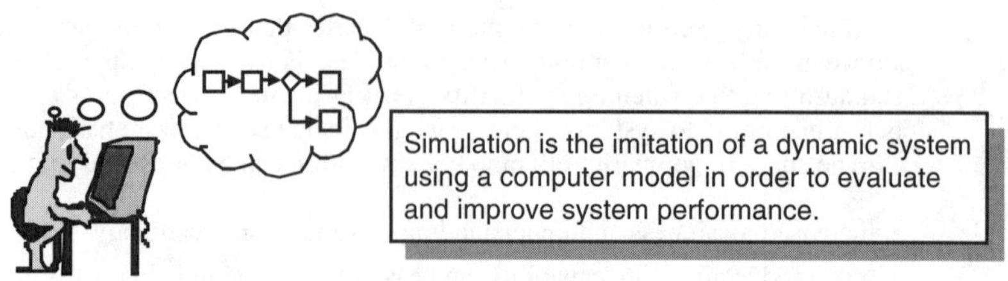

FIGURE 10. 5 Simulation is imitation.

10.4 CONDUCTING A SIMULATION STUDY

Simulation is much more than building and running a model of the process. Successful simulation projects are well planned and coordinated. While there are no strict rules on how to conduct a simulation project, the following steps are generally recommended:

Step 1: Define Objective, Scope, and Requirements. Define the purpose of the simulation project and what the scope of the project will be. Requirements need to be determined in terms of resources, time, and budget for carrying out the project.

Step 2: Collect and Analyze System Data. Identify, gather, and analyze the data defining the system to be modeled. This step results in a conceptual model and a data document that all can agree upon.

Step 3: Build the Model. Develop a simulation model of the system.

Step 4: Verify and Validate the Model. Debug the model and make sure it is a credible representation of the real system.

Step 5: Conduct Experiments. Run the simulation for each of the scenarios to be evaluated and analyze the results.

Step 6: Present the Results. Present the findings and make recommendations so that an informed decision can be made.

Each step need not be completed in its entirety before moving to the next step. The procedure for doing a simulation is an iterative one in which activities are refined and sometimes redefined with each iteration. The decision to push toward further refinement should be dictated by the objectives and constraints of the study as well as by sensitivity analysis which determines whether additional refinement will yield meaningful results. Even after the results are presented, there are often requests to conduct additional experiments. Figure 10.6 illustrates this iterative process.

In order to effectively execute these steps, it should be obvious that a variety of skills are necessary. To reap the greatest benefits from simulation, a certain degree of knowledge and skill in the following areas is recommended:

- Project management
- Communication
- Systems engineering
- Statistical analysis and design of experiments

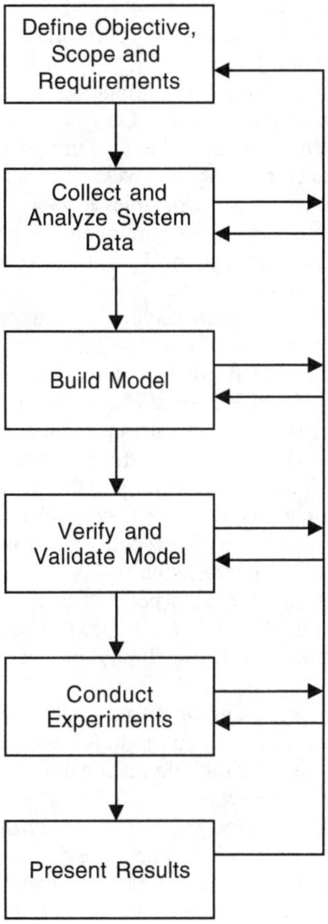

FIGURE 10.6 Iterative steps of a simulation project.

- Modeling principles and concepts
- Basic programming and computer skills
- Training on one or more simulation products
- Familiarity with the system being investigated

Modelers should be aware of their own inabilities in dealing with the modeling and statistical issues associated with simulation. Such awareness, however, should not prevent one from using simulation within the realm of one's expertise. Simulation can be beneficially used without being a statistical expert. Rough-cut modeling to gain fundamental insights, for example, can be achieved with only a rudimentary understanding of statistical issues. Simulation follows the 80–20 rule, where 80 percent of the benefit can be obtained from knowing only 20 percent of the science involved (just make sure you know the right 20 percent). It isn't until more precise analysis is required that additional statistical training and knowledge of design of experiments are needed.

If short on time, talent, resources, or interest, the decision maker need not despair. There are plenty of consultants who are professionally trained and experienced and can provide simulation services. A competitive bid will help get the best price, but one should be sure that the individual assigned to the project has good credentials. If the use of simulation is only occasional, relying on a consultant may be the preferred approach.

10.5 ECONOMIC JUSTIFICATION OF SIMULATION

Cost is always an important issue when considering the use of any software tool, and simulation is no exception. Simulation should not be used if the cost exceeds the expected benefits. This means that both the costs and the benefits should be carefully assessed. The use of simulation is often prematurely dismissed due to the failure to recognize the potential benefits and savings it can produce. Much of the reluctance in using simulation stems from the mistaken notion that simulation is costly and very time consuming. This perception is shortsighted and ignores the fact that in the long run simulation usually saves much more time and cost than it consumes. It is true that the initial investment, including training and start-up costs, may be between \$10,000 and \$30,000 (simulation products themselves generally range between \$1 and \$20,000). However, this cost is often recovered after the first one or two projects. The ongoing expense of using simulation for individual projects is estimated to be between 1 and 3 percent of the total project cost (Glenney and Mackulak, 1985). With respect to the time commitment involved in doing simulation, much of the effort that goes into building the model is in arriving at a clear definition of how the system operates, which needs to be done anyway. With the advanced modeling tools that are now available, the actual model development and running of simulations take only a small fraction (often less than 5 percent) of the overall system design time.

Savings from simulation are realized by identifying and eliminating problems and inefficiencies that would have gone unnoticed until system implementation. Cost is also reduced by eliminating overdesign and removing excessive safety factors that are added when performance projections are uncertain. By identifying and eliminating unnecessary capital investments, and discovering and correcting operating inefficiencies, it is not uncommon for companies to report hundreds of thousands of dollars in savings on a single project through the use of simulation. The return on investment (ROI) for simulation often exceeds 1,000 percent, with payback periods frequently being only a few months or the time it takes to complete a simulation project.

One of the difficulties in developing an economic justification for simulation is the fact that one can't know for certain how much savings will be realized until the simulation is actually used. Most applications in which simulation has been used have resulted in savings that, had they been guaranteed in advance, would have looked very good in an ROI or payback analysis.

One way to assess the economic benefit of simulation in advance is to assess the risk of making poor design and operational decisions. One need only ask what the potential cost would be if a misjudgment in systems planning was to occur. Suppose, for example, that a decision is made to add another machine to solve a capacity problem in a production or service system. The question should be asked: What are the cost and probability associated with this being the wrong decision? If the cost associated with a wrong decision is $100,000 and the decision maker is only 70 percent confident that the decision being made is correct, then there is a 30 percent chance of incurring a cost of $100,000. This results in a probable cost of $30,000 (0.3 × $100,000). Using this approach, many decision makers recognize that they can't afford not to use simulation because the risk associated with making the wrong decision is too high.

Tying the benefits of simulation to management and organizational goals also provides justification for its use. For example, a company committed to continuous improvement or, more specifically, to lead time or cost reduction can be sold on simulation if it can be shown to be historically effective in these areas. Simulation has gained the reputation as a best practice for helping companies achieve organizational goals. Companies that profess to be serious about performance improvement will invest in simulation if they believe it can help them achieve their goals.

The real savings from simulation come from allowing designers to make mistakes and work out design errors on the model rather than on the actual system. The concept of reducing costs through working out problems in the design phase rather than after a system has been implemented is best illustrated by the rule of tens. This principle states that the cost to correct a problem increases by a factor of 10 for every design stage through which it passes without being detected (see Fig. 10.7).

Many examples can be cited to show how simulation has been used to avoid making costly errors in the start-up of a new system. One example of how simulation prevented an unnecessary expenditure occurred when a Fortune 500 company was designing a facility for producing and storing subassemblies and needed to determine the number of containers required for holding the subassemblies. It was initially felt that 3,000 containers were needed until a simulation study showed that throughput did not improve significantly when the number of containers was increased from 2,250 to 3,000. By purchasing 2,500 containers instead of 3,000, a savings of $528,375 was expected in the first year, with annual savings thereafter of over $200,000 due to the savings in floor space and storage resulting from having 750 fewer containers (Law and McComas, 1988).

Even if dramatic savings are not realized each time a model is built, simulation at least inspires confidence that a particular system design is capable of meeting required performance objectives and thus minimizes the risk often associated with new start-ups. The economic benefits associated with instilling confidence were evidenced when an entrepreneur, who was attempting to secure bank financing to start a blanket factory, used a simulation model to show the feasibility of the proposed factory. Based on the processing times and equipment lists supplied by industry experts, the model showed that the output projections in the business plan were well within the capability of the proposed facility. Although unfamiliar with the blanket business, bank officials felt more secure in agreeing to support the venture (Bateman et al., 1997).

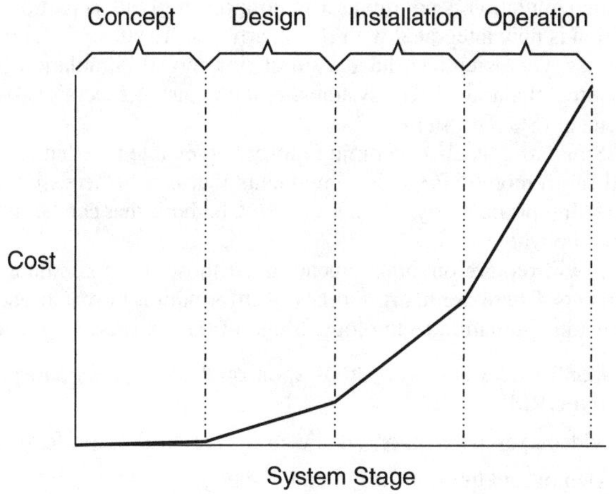

FIGURE 10.7 Cost of making changes at each stage of system development.

Often, simulation can help achieve improved productivity by exposing ways of making better use of existing assets. By looking at a system holistically, long-standing problems such as bottlenecks, redundancies, and inefficiencies that previously went unnoticed start to become more apparent and can be eliminated. "The trick is to find waste, or muda," advises Shingo, "after all, the most damaging kind of waste is the waste we don't recognize" (Shingo, 1992). Consider the following actual examples where simulation helped uncover and eliminate wasteful practices:

- GE Nuclear Energy was seeking ways to improve productivity without investing large amounts of capital. Through the use of simulation, they were able to increase the output of highly specialized reactor parts by 80 percent. The cycle time required for production of each part was reduced by an average of 50 percent. These results were obtained by running a series of models, each one solving production problems highlighted by the previous model (Bateman et al., 1997).
- A large manufacturing company with stamping plants located throughout the world produced stamped aluminum and brass parts on order according to customer specifications. Each plant had from 20 to 50 stamping presses that were utilized anywhere from 20 to 85 percent. A simulation study was conducted to experiment with possible ways of increasing capacity utilization. As a result of the study, machine utilization improved from an average of 37 to 60 percent (Hancock, Dissen, and Merten, 1977).

In each of these examples, significant productivity improvements were realized without the need for making major investments. The improvements came through finding ways to operate more efficiently and utilize existing resources more effectively. These capacity improvement opportunities were brought to light through the use of simulation.

10.6 FUTURE AND SOURCES OF INFORMATION ON SIMULATION

Simulation is a rapidly growing technology. While the basic science and theory remain the same, new and better software is continually being developed to make simulation more powerful and easier to use. New developments in the use of simulation are likely to be in the area of integrated applications where simulation is not run as a stand-alone tool but as part of an overall solution. For example, simulation is now integrated with flowcharting software so that, by the time you create a flowchart of the process, you essentially have a simulation model. Simulation is also being integrated into enterprise resource planning (ERP) systems, manufacturing execution systems (MES) and supply-chain management (SCM) systems.

Simulation is also becoming more Web enabled so that models can be shared across the internet and run remotely. Prebuilt components that can be accessed over the internet can greatly increase modeling productivity. Models can now be built that can be used for training purposes and accessed from anywhere.

It will require ongoing education for those using simulation to stay abreast of these new developments. There are many sources of information to which one can turn to learn the latest developments in simulation technology. Some of the sources that are available include

- Conferences and workshops sponsored by vendors and professional societies (e.g., SME, IIE, INFORMS)
- Videotapes, publications, and web sites of vendors, professional societies, and academic institutions
- Demos and tutorials provided by vendors
- Trade shows and conferences such as the Winter Simulation Conference
- Articles published in trade journals such as IIE Solutions, APICS Magazine, International Journal of Modeling and Simulation, and the like.

10.7 SUMMARY

Businesses today face the challenge of quickly designing and implementing complex production systems that are capable of meeting growing demands for quality, delivery, affordability, and service. With recent advances in computing and software technology, simulation tools are now available to help meet this challenge.

Simulation is a powerful technology that is being used with increasing frequency to improve system performance by providing a way to make better design and management decisions. When used properly, simulation can reduce the risks associated with starting up a new operation or making improvements to existing operations.

Because simulation accounts for interdependencies and variability, it provides insights that cannot be obtained any other way. Where important system decisions are being made of an operational nature, simulation is an invaluable decision making tool. Its usefulness increases as variability and interdependency increase and the importance of the decision becomes greater.

Lastly, simulation actually makes designing systems exciting! Not only can a designer try out new design concepts to see what works best, but the visualization makes it take on a realism that is like watching an actual system in operation. Through simulation, decision makers can play what-if games with a new system or modified process before it actually gets implemented. This engaging process stimulates creative thinking and results in good design decisions.

REFERENCES

Bateman, R. E., R. O. Bowden, T. J. Gogg, C. R. Harrell, and J. R. A. Mott, *System Improvement Using Simulation,* PROMODEL Corp., Orem, Utah, 1997.

Glenney, Neil E., and Gerald T. Mackulak, "Modeling and Simulation Provide Key to CIM Implementation Philosophy," *Industrial Engineering*, May 1985, p. 16.

Hancock, W., R. Dissen, and A. Merten, "An Example of Simulation to Improve Plant Productivity," *AIIE Transactions,* March 1977, pp. 2–10.

Law, A. M., and M. G. McComas, "How Simulation Pays Off," *Manufacturing Engineering,* February 1988, pp. 37–39.

Oxford American Dictionary, Oxford University Press, New York, 1980. [Eugene Enrich et al., comp.]

Schriber, T. J., "The Nature and Role of Simulation in the Design of Manufacturing Systems," *Simulation in CIM and Artificial Intelligence Techniques,* J. Retti, and K. E. Wichmann, eds., Society for Computer Simulation, San Diego, CA, 1987, pp. 5–8.

Shannon, R. E., "Introduction to the Art and Science of Simulation," *Proceedings of the 1998 Winter Simulation Conference,* D. J. Medeiros, E. F. Watson, J. S. Carson, and M. S. Manivannan, eds., Institute of Electrical and Electronics Engineers, Piscataway, NJ, 1998, pp. 7–14.

Shingo, S., *The Shingo Production Management System—Improving Process Functions,* A. P. Dillon, trans., Productivity Press, Cambridge, MA, 1992.

CHAPTER 11
INDUSTRIAL AUTOMATION TECHNOLOGIES

Andreas Somogyi
Rockwell Automation
Mayfield Heights, Ohio

11.1 INTRODUCTION TO INDUSTRIAL AUTOMATION

Industrial automation is a vast and diverse discipline that encompasses machinery, electronics, software, and information systems working together toward a common set of goals—increased production, improved quality, lower costs, and maximum flexibility.

But it's not easy. Increased productivity can lead to lapses in quality. Keeping costs down can lower productivity. Improving quality and repeatability often impacts flexibility. It's the ultimate balance of these four goals—productivity, quality, cost, and flexibility—that allows a company to use automated manufacturing as a strategic competitive advantage in a global marketplace.

This ultimate balance (a.k.a. manufacturing "nirvana") is difficult to achieve. However, in this case, the journey is more important than the destination. Companies worldwide have achieved billions of dollars in quality and productivity improvements by automating their manufacturing processes effectively. A myriad of technical advances—faster computers, more reliable software, better networks, smarter devices, more advanced materials, and new enterprise solutions—all contribute to manufacturing systems that are more powerful and agile than ever before.

In short, automated manufacturing brings a whole host of advantages to the enterprise; some are incremental improvements, while others are necessary for survival.

All things considered, it's not the manufacturer who demands automation. Instead, it's the manufacturer's customer, and even the customer's customer, who have forced most of the changes in how products are currently made. Consumer preferences—for better products, more variety, lower costs, and "when I want it" convenience—have driven the need for today's industrial automation. Following are some results of successful automation:

- *Consistency.* Consumers want the same experience every time they buy a product, whether it's purchased in Arizona, Argentina, Austria, or Australia.
- *Reliability.* Today's ultra-efficient factories can't afford a minute of unplanned downtime, with an idle factory costing thousands of dollars per day in lost revenues.
- *Lower costs.* Especially in mature markets where product differentiation is limited, minor variations in cost can cause a customer to switch brands. Making the product as cost-effective as possible without sacrificing quality is critical for overall profitability and financial health.

FIGURE 11.1 From alarm clocks to cereals to cars, automation is responsible for the products that people use and rely on every day.

- *Flexibility.* The ability to quickly change a production line on the fly (from one flavor to another, one size to another, one model to another, and the like) is critical at a time when companies strive to reduce their finished goods inventories and respond quickly to customer demands.

What many people don't realize is just how prevalent industrial automation is in our daily lives. Almost everything we come in contact with has been impacted in some way by automation (Fig. 11.1).

- It was used to manufacture the alarm clock that woke you up.
- It provided the energy and water pressure for your shower.
- It helped produce the cereal and bread for your breakfast.
- It produced the gas—and the car—that got you to school.
- It controlled the elevator that helped you get to your floor.
- It tracked the overnight express package that's waiting for you at home.
- It helped manufacture the phone, computer, copier, and fax machine you use.
- It controlled the rides and special effects at the amusement park you visited over the weekend.

And that's just scratching the surface.

11.1.1 What Is Industrial Automation?

As hard as it is to imagine, electronics and computers haven't been around forever, and neither has automation equipment. The earliest "automated" systems consisted of an operator turning a switch on, which would supply power to an output—typically a motor. At some point, the operator would turn the switch off, reversing the effect and removing power. These were the light-switch days of automation.

Manufacturers soon advanced to relay panels, which featured a series of switches that could be activated to bring power to a number of outputs. Relay panels functioned like switches, but allowed for more complex and precise control of operations with multiple outputs. However, banks of relay panels generated a significant amount of heat, were difficult to wire and upgrade, were prone to failure, and occupied a lot of space. These deficiencies led to the invention of the programmable controller—an electronic device that essentially replaced banks of relays—now used in several forms in millions of today's automated operations. In parallel, single-loop and analog controllers were replaced by the distributed control systems (DCSs) used in the majority of contemporary process control applications.

These new solid-state devices offered greater reliability, required less maintenance, and had a longer life than their mechanical counterparts. The programming languages that control the behavior of programmable controls and distributed control systems could be modified without the need to disconnect or reroute a single wire. This resulted in considerable cost savings due to reduced commissioning time and wiring expense, as well as greater flexibility in installation and troubleshooting.

At the dawn of programmable controllers and DCSs, plant-floor production was isolated from the rest of the enterprise—operating autonomously and out of sight from the rest of the company. Those days are almost over as companies realize that to excel they must tap into, analyze, and exploit information located on the plant floor.

Whether the challenge is faster time-to-market, improved process yield, nonstop operations, or a tighter supply chain, getting the right data at the right time is essential. To achieve this, many enterprises turn to contemporary automation controls and networking architectures. Computer-based controls for manufacturing machinery, material-handling systems, and related equipment cost effectively generate a wealth of information about productivity, product design, quality, and delivery.

Today, automation is more important than ever as companies strive to fine-tune their processes and capture revenue and loyalty from consumers. This chapter will break down the major categories of hardware and software that drive industrial automation; define the various layers of automation; detail how to plan, implement, integrate, and maintain a system; and look at what technologies and practices impact manufacturers.

11.2 HARDWARE AND SOFTWARE FOR THE PLANT FLOOR

11.2.1 Control Logic

Programmable Controllers. Plant engineers and technicians developed the first programmable controller, introduced in 1970, in response to a demand for a solid-state system that had the flexibility of a computer, yet was easier to program and maintain. These early programmable controllers took up less space than the relays, counters, timers, and other control components they replaced, and offered much greater flexibility in terms of their reprogramming capability. The initial programming language, based on the ladder diagrams and electrical symbols commonly used by electricians, was key to industry acceptance of the programmable controller.

There are two major types of programmable controllers—fixed and modular. Fixed programmable controllers come as self-contained units with a processor, power supply, and a predetermined number of discrete and/or analog inputs and outputs (I/O). A fixed programmable controller may have separate, interconnected components for expansion, and is small, inexpensive, and simple to install. However, modular controllers are more flexible, offering options for I/O capacity, processor memory size, input voltage, and communication type.

Originally, programmable controllers were used in control applications where I/O was digital. They were ideal for applications that were more sequential and discrete than continuous in nature. Over time, suppliers added analog and process control capabilities making the programmable controller a viable solution for batch and process applications as well.

It wasn't until microcontrollers were introduced that programmable controllers could economically meet the demands of smaller machines—equipment that once relied exclusively on relays and single board computers (SBCs). Microcontrollers are generally designed to handle 10 to 32 I/O in a cost-efficient package making them a viable and efficient replacement. In addition, this low-cost, fixed-I/O option has opened the door for many small-machine original equipment manufacturers (OEMs) to apply automated control in places where it wasn't feasible in the past. For instance, manufacturers can use a microprogrammable controller to power lottery ticket counting machines, elevators, vending machines, and even traffic lights.

FIGURE 11.2 Today, programmable controllers and PCs come in different sizes and scales of functionality to meet users' evolving needs.

When a technology like the programmable controller has been on the market for more than 25 years, the natural question that arises is, "What will replace it?" However, the same question was asked about relays 25 years ago, and they can still be found on the plant floor. So a more appropriate question for industrial automation may be, "What else is needed?"

There has been some push to promote *soft control* as an heir to the programmable controller. In essence, soft control is the act of replacing traditional controllers with software that allows users to perform programmable controller functions on a personal computer (PC). Simply put, it's a programmable controller in PC clothing. Soft control is an important development for individuals who have control applications with a high degree of information processing content (Fig. 11.2).

Soft control is, however, only part of a larger trend—that of PC-based control. PC-based control is the concept of applying control functions normally embedded in hardware or software to the control platforms. This encompasses not just the control engine, but all aspects of the system, including programming, operator interface, operating systems, communication application programming interfaces (APIs), networking, and I/O. PC-based control has been adopted by some manufacturers, but most continue to rely on the more rugged programmable controller. This is especially true as new-generation programmable controllers are incorporating features of PC-based control yet maintaining their roots in reliability and ruggedness.

Distributed Control Systems. Distributed control systems (DCSs) are a product of the process control industry. The DCS was developed in the mid 1970s as a replacement for the single-loop digital and analog controllers as well as central computer systems. A DCS typically consists of unit controllers, which can handle multiple loops, multiplexer units to handle a large amount of I/O, operator and engineering interface workstations, a historian, foreign device gateways, and an advanced control function in a system "box" or computer. All these are fully integrated and usually connected via a communications network.

DCS suppliers have traditionally taken the approach of melding technology and application expertise to solve a specific problem. Even the programming method, called function block programming, allows a developer to program the system by mimicking the actual process and data flow. DCSs allow for reliable communication and control within a process; the DCS takes a hierarchical approach to control with the majority of the intelligence housed in a centralized computer.

A good analogy for the DCS is the mainframe computer and desktop computers. Not long ago, it was unheard of for companies to base their corporate computing on anything other than a mainframe computer. But with the explosive growth in PC hardware and software, many companies now use a network of powerful desktop computers to run their information systems. This architecture gives them more power in a flexible, user-friendly network environment and at a fraction of the cost of mainframes. Likewise, DCS systems are more distributed than in the past.

DCSs are generally used in applications where the proportion of analog to digital I/O is higher than a 60/40 ratio and the control functions are more sophisticated. DCSs are ideal for industries where the process is continuous, has a high analog content and throughput, and is distributed across a large geographical region. It is also well suited for applications where down time is very expensive (e.g., pulp and paper, refining and chemical production).

While programmable controllers, DCSs, and PCs each have unique strengths, there is often no easy way to select one controller over another. For example, a DCS is the model solution when the application is highly focused such as load shedding. But if a company had a load-shedding application in years past and 50 ancillary tanks that feed into the process, it had to make a decision between a DCS or programmable controller (which can manage the tanks more effectively).

That's no longer a problem because suppliers have introduced a concept called "hybrid controllers." These allow a user to either have a programmable controller, DCS, or both in one control unit. This hybrid system allows a large amount of flexibility and guarantees tremendous cost savings compared to two separate solutions.

11.2.2 Input/Output

Generally speaking, I/O systems act as an interface between devices—such as a sensor or operator interface—and a controller. They are not the wires that run between devices and the controller, but are the places where these wires connect (Fig. 11.3).

The birth of industrial I/O came in the 1960s when manufacturers concerned about reusability and cost of relay panels looked for an alternative. From the start, programmable controllers and I/O racks were integrated as a single package. By the mid-to-late 1970s, panels containing I/O but no processor began to populate the plant floor. The idea was to locate racks of I/O closer to the process but remote from the controller. This was accomplished with a cabling system or network.

In some applications—material handling, for example—each segment of a line can require 9 or 10 points of I/O. On extensive material handling lines, wiring 10 or fewer points from a number of locations back to panels isn't cost effective. Companies realized that it would be much cheaper to locate small "blocks" of I/O as near as possible to the actuators and sensors. If these small blocks could house cost-effective communication adapters and power supplies, only one communication cable would have to be run back to the processor—not the 20 to 30 wires typically associated with 10 I/O points.

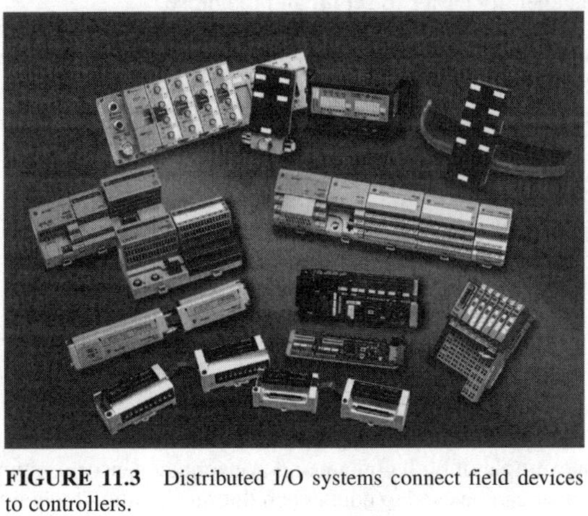

FIGURE 11.3 Distributed I/O systems connect field devices to controllers.

OEMs and end users also need greater flexibility than what's offered by small, fixed blocks of I/O. With flexible I/O, modules of varying types can be "snapped" into a standard mounting rail to tailor the combination of I/O to best suit the application.

Following is an overview of common I/O terminology:

- *Inputs (Sensors).* Field devices that act as information gatherers for the controller. Input devices include items such as push buttons, limit switches, and sensors.
- *Outputs (Actuators).* Field devices used to carry out the control instructions for the programmable controller. Output devices include items such as motor starters, indicator lights, valves, lamps, and alarms.
- *I/O Module.* In a programmable controller, an I/O Module interfaces directly through I/O circuits to field devices for the machine or process.
- *I/O Racks.* A place where I/O modules are located on the controller.
- *I/O Terminal.* Located on a module, block, or controller, an I/O terminal provides a wire connection point for an I/O circuit.
- *Distributed I/O Systems.* Standalone interfaces that connect the field devices to the controller.

11.2.3 Sensors

A sensor is a device for detecting and signaling a changing condition. Often this is simply the presence or absence of an object or material (discrete sensing). It can also be a measurable quantity like a change in distance, size, or color (analog sensing). This information, or the sensor's output, is the basis for the monitoring and control of a manufacturing process. There are two basic types of sensors: contact and noncontact.

Contact sensors are electromechanical devices that detect change through direct physical contact with the target object. Encoders and limit switches are contact sensors. Encoders convert machine motion into signals and data. Limit switches are used when the target object will not be damaged by physical contact.

Contact Sensors

- Offer simple and reliable operation
- Can handle more current and better tolerate power line disturbances
- Are generally easier to set up and diagnose

Noncontact sensors are solid-state electronic devices that create an energy field or beam and react to a disturbance in that field. Photoelectric, inductive, capacitive, and ultrasonic sensors are noncontact technologies. Since the switching components are not electromechanical and there is no physical contact between the sensor and target, the potential for wear is eliminated. However, noncontact sensors are not as easy to set up as contact sensors in some cases.

Noncontact Sensors

- No physical contact is required between target and sensor
- No moving parts to jam, wear, or break (therefore less maintenance)
- Can generally operate faster
- Greater application flexibility

An example of both contact and noncontact sensor use would be found on a painting line. A contact sensor can be used to count each door as it enters the painting area to determine how many doors have been sent to the area. As the doors are sent to the curing area, a noncontact sensor counts how many have left the painting area and how many have moved on to the curing area. The change to a noncontact sensor is made so that there is no contact with, and no possibility of disturbing, the newly painted surface.

11.2.4 Power Control and Actuation

Power control and actuation affects all aspects of manufacturing. While many in industrial automation view power control as simply turning motors off and on or monitoring powered components, those who properly apply the science of power control discover immediate increases in uptime, decreases in energy costs, and improvements in product quality.

Power control and actuation involves devices like electromechanical and solid-state soft starters; standard, medium-voltage, and high-performance servo drives; and motors and gears. These products help all moveable parts of an automated environment operate more efficiently, which in turn increases productivity, energy conservation, and profits.

Wherever there's movement in plants and facilities, there's a motor. The Department of Energy reports 63 percent of all energy consumed in industrial automation powers motors. Solid-state AC drives—which act as brains by regulating electrical frequencies powering the motors—help motors operate more efficiently and have an immediate, measurable impact on a company's bottom line. When applications require less than 100 percent speed, variable frequency drives for both low- and medium-voltage applications can help eliminate valves, increase pump seal life, decrease power surge during start-up, and contribute to more flexible operation.

Many motor applications, such as conveyors and mixers, require gear reduction to multiply torque and reduce speed. Gearing is a common method of speed reduction and torque multiplication. A gear motor effectively consumes a certain percentage of power when driving a given load.

Picking the right gear type allows cost-efficient, higher-speed reductions. Applications that require long, near-continuous periods of operation and/or those with high-energy costs are very good candidates for analysis. Proper installation of equipment and alignment of mechanical transmission equipment will reduce energy losses and extend equipment life.

Integrated intelligence in solid-state power control devices gives users access to critical operating information, which is the key to unlocking the plant floor's full potential. Users in all industries are seeking solutions that merge software, hardware, and communication technologies to deliver plant-floor benefits that go beyond energy savings such as improved process control and diagnostics and increased reliability. Power control products today are increasingly intelligent compared to their mechanical ancestors and that intelligence is networked to central controllers for data mining that delivers uptime rewards.

11.2.5 Human-Machine Interface

Even the simplest controller needs some sort of operator interface device—whether it's a simple pushbutton panel or a highly sophisticated software package running on a PC. There is a wide range of choices in between, and each can be evaluated based on the degree of responsibility/risk the user is willing to take on, as well as the capability and training of the operator.

From the time the first push button was developed decades ago, human-machine interface (HMI) applications have become an integral fixture in manufacturing environments. HMIs allow users to directly control the motion and operating modes of a machine or small groups of machines. Having an HMI system that increases uptime by streamlining maintenance and troubleshooting tasks is crucial to optimizing production processes.

The first HMI applications consisted of a combination of push buttons, lights, selector switches, and other simple control devices that started and stopped a machine and communicated the machine's performance status. They were a means to enable control. The interfaces were rudimentary, installed because the designer's overriding goal was to make the control circuitry as small as possible. Even

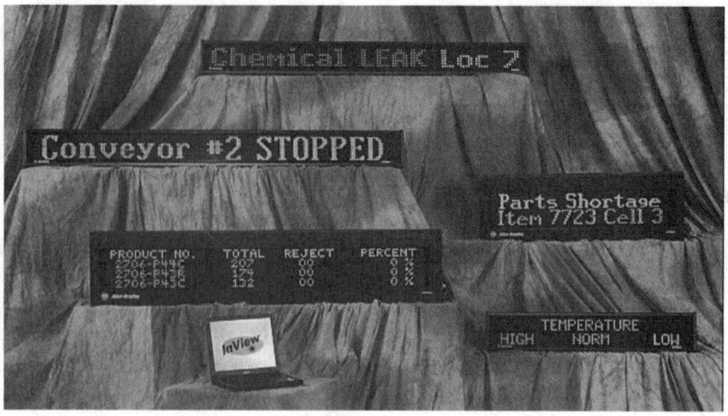

FIGURE 11.4 Message displays provide real-time information such as system alarms, component availability, and production information to plant-floor workers.

though troubleshooting aids were almost nonexistent, the HMI automation and controls made the systems less prone to problems and easier to troubleshoot than previous systems.

The first major upgrade in HMI applications came as an outgrowth of the programmable control system. By simply wiring additional devices to the programmable controller, the HMI could not only communicate that a machine had stopped, but also indicate why the stoppage occurred. Users could access the programming terminal and find the fault bit that would lead them to the problem source. Controllers could even be programmed to make an automated maintenance call when a machine stopped working—a big time-saver at large factory campuses.

At the same time, area-wide HMI displays—used to communicate system status and production counts within a portion of the plant—were becoming a common feature on the plant floor. The introduction of the numeric display—and later, the alphanumeric display—gave maintenance personnel information about the exact fault present in the machine, significantly reducing the time needed to diagnose a problem (Fig. 11.4).

The introduction of graphic display terminals was the next significant step in HMI hardware. These terminals could combine the functionality of push buttons, lights, numeric displays, and message displays in a single, reprogrammable package. They were easy to install and wire as there was only one device to mount and the only connection was through a single small cable. Changes could be easily made without requiring any additional installation or wiring to be done. The operator could press a function key on the side of the display or a touch screen directly on the display itself.

Functionality was added to these terminals allowing much more sophisticated control that could be optimized for many different types of applications.

Initially, these terminals used cathode ray tube (CRT) displays, which were large and heavy. Flat-panel displays evolved out of the laptop computer industry and found their way into these industrial graphic terminals. These displays allowed for much smaller terminals to be developed allowing graphic terminals to be used in very low-cost machines. As the flat-panel displays continue to improve and decrease in cost, they have almost completely taken over the operator terminal market with displays up to 20 in (diagonal).

The use of HMI software running on a PC has continued to grow substantially over the past 10 years. This software allows for the functionality of a graphic terminal, while also providing much more sophisticated control, data storage, and the like through the ever-growing power of PCs. Industrial computers with more rugged specifications are available to operate in an industrial environment. These computers continue to become more rugged while definite purpose operator terminals continue to become more sophisticated. The line between them becomes more blurred every year.

Distributed HMI Structures. Early HMI stations permitted both viewing and control of the machine operation or manufacturing processes but were not networked with other HMI stations. HMI evolved into a single central computer networked to multiple programmable controllers and operator interface. In this design, the "intelligence" rests within the central computer that performs all the HMI services including program execution. Recently, however, technological advancements have allowed companies to move from stand-alone HMI to a distributed model where HMI servers are networked together communicating with multiple remote client stations to provide an unprecedented distribution of HMI information.

As software continues to evolve, an innovative twist on the single server/multiple client architecture has surfaced. Companies are now able to implement multiple servers with multiple clients—adding an entirely new dimension to distributed HMI. The future of industry will increasingly see servers joined through a multilayered, multi-functioned distributed enterprise-wide solution in which a variety of applications servers—such as HMI, programmable controllers, single loop controllers, and drive systems—are networked together with "application generic clients" to exchange information.

The transformation to multiple servers and clients will eliminate the one risk associated with traditional distributed HMI—a single point of reliability. In a traditional single server/multiple client environment, all of the programming and control is loaded onto just one high-end computer.

But the built-in redundancy of the multiple server/client model means the failure of one server has a minimal impact on the overall system. With either single or multiple servers, if a client goes down, users can get information through other plant-floor clients and the process continues to operate.

The future of HMI shows great promise. However, decisions on what technology to use for HMI applications today are usually driven by cost and reliability, which can often exclude the devices and software with the highest functionality. As the latest HMI products are proven—either by a small installed base or through lab testing—system designers will be able to provide customers with even more efficient control systems that can meet future productivity demands.

11.2.6 Industrial Networks

Industrial automation systems, by their very definition, require interconnections and information sharing. There are three principal elements that a user needs from the networks that hold their automation systems together—control, configure, and collect. *Control* provides the ability to read input data from sensors and other field instruments, execute some form of logic, and then distribute the output commands to actuator devices.

Solutions for providing the control element might be very centralized, highly distributed, or somewhere in between the two. Centralized control typically entails a large-scale programmable or soft controller that retains most—if not all—of the control-system logic. All the other devices are then connected to the controller via hardwiring or a network. In a highly distributed control philosophy, portions of the overall logic program are distributed to multiple devices. This usually results in faster performance because there is more than one controller doing all the work.

The ability to *collect* data, which is the second element, allows the user to display or analyze information. This could involve trending, making mathematical calculations, using a database, or a host of other activities.

To collect data in a centralized control scheme, the user is usually dependent on the controller since much of the data reside there. The more centralized the control philosophy, the more likely it is that the user will retrieve data from the controller. Distributed architectures, on the other hand, provide more flexibility in collecting information independently from each device.

A mechanism for *configuring* devices enables the user to give a "personality" to devices, such as programmable controllers, operator interfaces, motion controllers, and sensors. This mechanism is typically required during system design and start-up. However, the user may need to modify device configurations during operation if, for instance, they change from recipe A to recipe B or change from one model of car to another. These modifications could entail a plant engineer editing the parameters of one or more devices on the manufacturing line, such as increasing a sensor's viewing distance.

FIGURE 11.5 Before—The control panel uses analog, hardwired I/O, which requires a significant investment in wiring and conduit.

FIGURE 11.6 After— The panel now features a digital communications network (DeviceNet, in this case) that reduces wiring and increases flexibility.

Configuration of devices can be done in two ways. Either the users are required to go to each device with a notebook computer or some other tool, or they may have a network that allows them to connect to the architecture at a single point and upload/download configuration files from/to each of the devices.

Digital Communication. Since the information going from the device to the controller has become much more detailed, a new means of communication—beyond the traditional analog standard—has become necessary. Today's field devices can transmit the process signal—as well as other process and device data—digitally. Ultimately, the use of digital communication enables the user to distribute control, which significantly reduces life-cycle costs.

First, adopting a distributed control model has the obvious benefit of reduced wiring. Each wiring run is shortened as the control element or the I/O point moves closer and closer to the field sensor or actuator. Also, digital communication provides the ability to connect more than one device to a single wire. This saves significant cost in hardware and labor during installation. And in turn, the reduced wiring decreases the time it takes to identify and fix failures between the I/O and the device (Figs. 11.5 and 11.6).

Digital communication also helps to distribute control logic further and further from a central controller. The migration path might involve using several smaller controllers, then microcontrollers, and eventually embedding control inside the field sensor or actuator and linking them with a digital network. By doing this, significant cost savings can be achieved during the design, installation, production, and maintenance of a process.

As devices become smarter, it is easy to see the value of incorporating digital networks within the plant. However, a single type of network can't do it all. The differences between how tasks are handled within a plant clearly indicate the need for more than one network. For example, a cost/manufacturing accountant might want to compile an annual production report. At the same time, a photoelectric sensor on a machine might want to notify the machine operator that it is misaligned (which could cause the machine to shut down). Each task requires communication over a network, but has different requirements in terms of urgency and data sizes.

The accountant requires a network with the capacity to transfer large amounts of data. But at the same time, it is acceptable if these data are delivered in minutes. The plant-floor sensor, on the other hand, requires a network that transfers significantly smaller data sizes at a significantly faster rate (within seconds or milliseconds). The use of more than one network in a manufacturing environment identifies the need for a way to easily share data across the different platforms.

Because of the different tasks in most control systems, there are typically three basic network levels: information, control and device. At the *information* level, large amounts of data are sent nondeterministically for functions such as system-wide data collection and reports. (EtherNet/IP is typically used at this level.) At the control level—where networks like ControlNet and EtherNet/IP reside—programmable controllers and PCs control I/O racks and I/O devices such as variable speed drives and dedicated HMI. Time-critical interlocking between controllers and guaranteed I/O update rates are extremely important at this level. At the device level there are two types of networks. The first type primarily handles communication to and from discrete devices (e.g., DeviceNet). The other handles communication to and from process devices (e.g., Foundation Fieldbus).

Producer/Consumer Communication. In addition to choosing the right networks, it is important to note that many networks don't allow the user to control, configure, and collect data simultaneously. One network may offer one of these services while another may offer two. And then there are other networks that offer all three services, but not simultaneously. This is why many users have identified the need for a common communications model like producer/consumer, which provides a degree of consistency regardless of the network being used.

Producer/consumer allows devices in a control system to initiate and respond when they have the need. Older communication models, such as source/destination, have a designated master in a system that controls when devices in the system can communicate. With producer/consumer, the users still have the option to do source/destination, but they can take advantage of other hierarchies, such as peer-to-peer communication, as well (Fig. 11.7). It also offers the advantage of numerous I/O exchange options such as change-of-state, which is a device's ability to send data only when there's a change in what it detects. Devices can also report data on a cyclic basis, at a user-configured frequency. This means that one device can be programmed to communicate every half-second, while another device may be set to communicate every 50 ms.

Producer/consumer also allows for devices to communicate information one-to-one, one-to-several, or on a broadcast basis. So in essence, devices are equipped with trigger mechanisms of when to send data, in addition to providing a broad range of audience choices. Rather than polling each device one-at-a-time (and trying to repeat that cycle as fast as possible), the entire system could be set for change-of-state, where the network would be completely quiet until events occur in the process. Since every message would report some type of change, the value of each data transmission increases.

Along with the producer/consumer model, a shared application layer is key to advanced communication and integration between networks. Having a common application-layer protocol across all industrial networks helps build a standard set of services for control, configuration, and data collection, and provides benefits such as media independence; fully defined device profiles; control

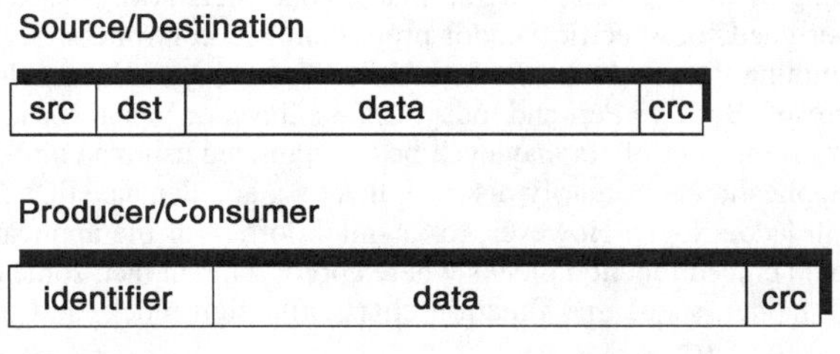

FIGURE 11.7 Producer/consumer communication—Instead of data identified as source to destination, it's simply identified with a unique number. As a result, multiple devices on a network can consume the same data at the same time from a single producer, resulting in efficient use of bandwidth.

services; multiple data exchange options; seamless, multi-hop routing; and unscheduled and scheduled communication.

11.2.7 Software—Proprietary/Open

Closely linked to the hardware/packaging choice is the operating system and software. The operating system, put simply, is software that defines how the information flows between the processor chip, the memory, and any peripheral devices. The vendor usually develops operating systems for programmable controllers and DCSs, optimizing a particular product offering and applications. In these cases, the operating system is embedded in firmware and is specific to that vendor only.

Today's programmable controller operating systems, like the hardware platform, are the result of more than 25 years of evolution to provide the determinism, industry-hardened design, repeatability, and reliability required on the plant floor. In the past, achieving these objectives meant choosing a vendor-specific operating system and choosing that vendor's entire control solution as well. Today, a real-time operating system (RTOS) helps eliminate the situation of being locked into any one vendor's controls.

Embedded RTOSs (e.g., VxWorks, QNX Neutrino, pSOS, etc.) were specifically developed for high-reliability applications such as those found in manufacturing or telecommunications industries. They have been successful in situations where the user is concerned about the reliability of the system, but not so concerned about the sacrifice flexibility and cost benefits associated with commercially available hardware. RTOSs typically come with a base set of programming tools specific to that vendor's offering, and the communication drivers to other third-party peripherals either have to be custom-written or purchased as add-ons.

Commercial-grade operating systems like Microsoft Windows 2000 have quickly come on the scene as viable choices for control on Intel-based hardware platforms. Earlier versions of Windows were deemed not to be robust enough for control applications. However, with the introduction of Windows NT, more control vendors are advocating this commercial operating system as a viable choice for users with information-intensive control applications. In addition, an industry standard operating system allows the user access to a wide range of development tools from multiple vendors, all working in a common environment.

11.2.8 Programming Devices

At one time, the "box" being programmed directly defined the programming methods. Relays require no "software" programming—the logic is hardwired. SBCs typically have no external programming by the end user. Programmable controllers always used ladder logic programming packages designed for specific vendor-programmable controllers. DCS systems used function block programming specific to the vendor, while PC users typically employed higher-level languages such as Microsoft Basic or Perl and today, C/C++, Java, or Visual Basic.

Now, some controllers may even be programmed using a combination of methods. For instance, if the application is primarily discrete, there's a good chance that the user could program the application in ladder logic. However, for a small portion of the application that is process-oriented, the user could embed function blocks where appropriate. In fact, some companies now offer a variety of editors (ladder, sequential function charts, function block, and the like.) that program open and dedicated platforms.

Most programming methodologies still mimic the assembly line of old: the system layout drawing is created, then the electrical design is mapped, then the application code is written. It's an extremely time-consuming, linear process. With programming costs consuming up to 80 percent of a control system's budget, manufacturers are looking to migrate from step-by-step programming to

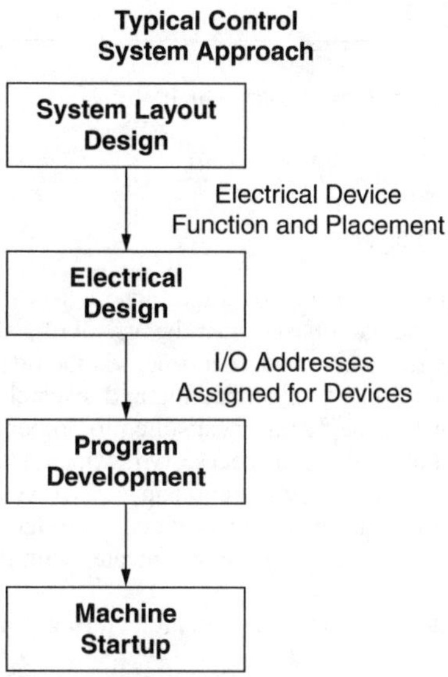

FIGURE 11.8 Due to software limitations, the traditional approach to control system programming has been very linear and prone to bottlenecks.

a more concurrent, multi-dimensional design environment.

Key to developing this new design environment is the identification of the current bottlenecks within the programming process. The most common include:

- Waiting to write code until the electrical design is complete
- Force-fitting an application into a fixed memory structure
- Maintaining knowledge of physical memory addresses to access controller operation data values
- Managing/translating descriptions and comments across multiple software products
- Tedious address reassignments when duplicating application code
- Debugging the system where multiple programmers mistakenly used the same memory address for different functions

To eliminate these bottlenecks, it's important to work with programming software that supports techniques like tag aliases, multiple-data scopes, built-in and user-defined structures, arrays, and application import/export capabilities. These techniques deliver a flexible environment that can significantly reduce design time and cut programming costs (Figs. 11.8 and 11.9).

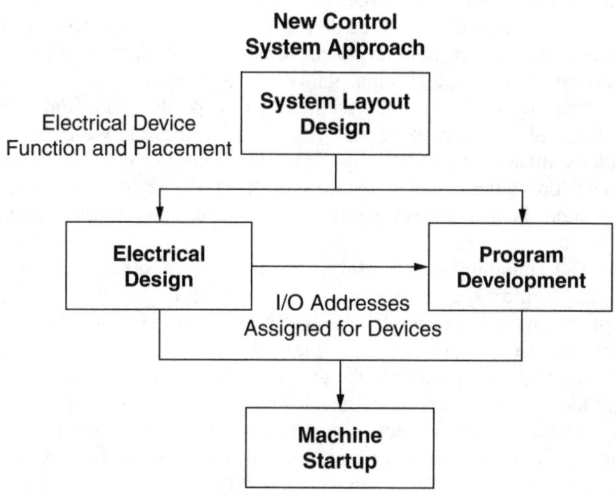

FIGURE 11.9 New programming software packages offer a flexible, concurrent environment that helps reduce design cycle times and cut costs.

11.3 FROM SENSORS TO THE BOARDROOM

Industrial automation is divided into three primary layers which are:

- The plant-floor automation layer
- The manufacturing execution system (MES) layer
- The enterprise resource planning (ERP) layer

The vision of modern manufacturing is to create an operation where these layers are wholly integrated. In this scenario, manufacturing is demand-based (i.e., the arrival of a single request causes a chain reaction). For example, a company receives a customer order via the Internet. The order enters the ERP layer and is transmitted to the MES layer for scheduling and dispatch to the plant floor. Or, in the case of a global enterprise, it is sent to the factory best suited to process the order.

On the plant floor, the manufacturing line receives the necessary raw materials as they are needed—the result of electronic requests made to suppliers by the company's ERP system. The control system automatically reconfigures the manufacturing line to produce a product that meets the given specifications. As the product is shipped, the plant floor communicates with the ERP/MES systems and an invoice is sent to the customer.

This seamless industrial environment is built on several manufacturing principles.

- Ever-changing customer demands require increased flexibility from manufacturers.
- Flexibility relies on the ability to exchange information among factories—planning, purchasing, production, sales, and marketing.
- Manufacturing success is measured in terms of meeting strategic business goals like reducing time to market, eliminating product defects, and building agile production chains.

11.3.1 The Plant-Floor Automation Layer

The plant-floor automation layer is a defined functionality area with equipment engaged to make a machine or process run properly. Typically, it includes sensors, bar-code scanners, switches, valves, motor starters, variable speed drives, programmable controllers, DCSs, I/O systems, human machine interfaces (HMIs), computerized numeric controllers (CNCs), robot controls, industrial networks, software products, and other plant-floor equipment.

The philosophies and strategies that shape the plant-floor automation layer have changed over time. And in many ways, they have come full circle with respect to a preferred architecture. With relays, manufacturers had a nearly one-to-one I/O ratio. However, control and automation requirements drove the industry toward more centralized programmable controllers and DCSs. This model for control had a distinct pyramid shape—with multiple functions run by a main computer or control system.

The pendulum has started to swing back the other way. As companies recognized the need to tightly control specific elements of the process, and as technology allowed for cost-effective distribution, engineers broke the control model into more logical, granular components. Where a large programmable controller once managed all functions in a manufacturing cell, a network of small (or even "micro") controllers currently reside. These systems are often linked to other types of controllers as well—motion control systems, PCs, single loop controllers, and the like.

Another recent development is that technologies prevalent in government and commercial sectors are now finding their way into industrial control systems at an incredibly fast rate. It took decades for automation to progress from the integrated circuit to the programmable controller. But Pentium chips—launched only a few years ago—are the standard for PC-based control systems. And Windows operating systems used in commercial applications are now powering hand-held operating stations on the plant floor. Likewise, many companies now use the controller area network (CAN)—originally developed for automobiles—to connect to and communicate with industrial devices. Ethernet, the undisputed networking champ for office applications, also has trickled down to the plant-floor layer.

As it stands, the plant-floor automation layer consists of two major components. One is the real-time aspect of a control system where equipment has to make decisions within milli- and micro-seconds. Picture a paper machine, for instance, producing 4,000 ft of paper a minute, 600 ton a day. Imagine how fast sensors, programmable controllers, HMIs, and networks must exchange data to control the behavior and outcome of the machine.

The second functional component is called *data acquisition* (DAQ), where information about the machine, the components used to drive the machine, and the environment is collected and passed to systems that execute based on this data. These systems can be on the plant floor or in any of the two other layers discussed later (MES, ERP). The main difference between real-time communication and DAQ is that equipment used for the latter does not operate under critical time constraints.

Communication networks are an integral part of the plant-floor automation layer. As discussed, a control system usually comprises programmable controllers, I/O, HMIs, and other hardware. All of these devices need to communicate with each other—functionality supplied by a network. Depending on the data relayed (e.g., a three-megabyte file versus a three-bit message), different network technologies are used. Regardless of the application, the trend in automation is definitely toward open network technologies. Examples of this are the CAN-based DeviceNet network and an industrialized version of Ethernet called EtherNet/IP.

One key to success at the plant-floor automation layer is the ability to access data from any network at any point in the system at any time. The idea is that even with multiple network technologies (e.g., CAN, Ethernet), the user should be able to route and bridge data across the entire plant and up to the other layers without additional programming. This type of architecture helps accomplish the three main tasks of industrial networks: controlling devices, configuring devices, and collecting data.

11.3.2 The ERP Layer

ERP systems are a collection of corporate-wide software solutions that drive a variety of business-related decisions in real time—order entry, manufacturing, financing, purchasing, warehousing, transportation, distribution, human resources, and others. In years past, companies used separate software packages for each application, which did not provide a single view of the company and required additional time and money to patch the unrelated programs together. Over the past few years, however, many companies have purchased ERP systems that are fully integrated.

ERP systems are the offspring of material requirement planning (MRP) systems. Starting in the mid 1970s, manufacturers around the world implemented some kind of MRP system to improve production efficiency. The next step in the evolution of this platform of applications was MRP II systems (the name evolved to manufacturing resource planning systems). MRP II systems required greater integration with the corporation business systems such as general ledger, accounts payable, and accounts receivable. These companies struggled with the integration efforts, which were compounded when you had a global company that had operations in different countries and currencies. The MRP and financial systems were not able to handle these challenges. So out of necessity a new solution, the ERP system, was born.

Within the last 10 years, ERP systems have matured, becoming a vital component of running a business. That's because ERP systems help companies manage the five Rs, which are critical to performance and financial survival:

- Produce the right product
- With the right quality
- In the right quantity
- At the right time
- At the right price

Key data elements collected from the plant floor and MES layers can be used by ERP execution software to manage and monitor the decision-making process. In addition, this software provides the status of open orders, product availability, and the location of goods throughout the enterprise.

As mentioned earlier, ERP systems help companies make critical business decisions. Here is a list of questions an ERP system could help analyze and answer:

- What is the demand or sales forecast for our product?
- What products do we need to produce to meet demand?
- How much do we need to produce versus how much are we producing?
- What raw materials are required for the products?
- How do we allocate production to our plant or plants?
- What is the target product quality?
- How much does it cost to make the product versus how much should it cost?
- How much product do we have in stock?
- Where is our product at any give point in time?
- What is the degree of customer satisfaction?
- Have the invoices been sent and payment received?
- What is the financial health of the company?

ERP systems are usually designed on a modular base. For each function, the company can choose to add an application-specific module that connects to the base and seamlessly merges into the entire system.

Sample ERP modules include:

- Financial accounting
- Controlling
- Asset management
- Human resources
- Materials management
- Warehouse management
- Quality management
- Production planning
- Sales and distribution

Evolving out of the manufacturing industry, ERP implies the use of packaged software rather than proprietary software written by or for one customer. ERP modules may be able to interface with an organization's own software with varying degrees of effort, and, depending on the software, ERP modules may be alterable via the vendor's configuration tools as well as proprietary or standard programming languages.

ERP Systems and SCADA. Supervisory control and data acquisition (SCADA) is a type of application that lets companies manage and monitor remote functions via communication links between master and remote stations. Common in the process control environment, a SCADA system collects data from sensors on the shop floor or in remote locations and sends it to a central computer for management and control (Fig. 11.10). ERP systems and MES modules then have access to leverage this information.

There are four primary components to a SCADA application—topology, transmission mode, link media, and protocol. TOPOLOGY is the geometric arrangement of nodes and links that make up a network. SCADA systems are built using one of the following topologies:

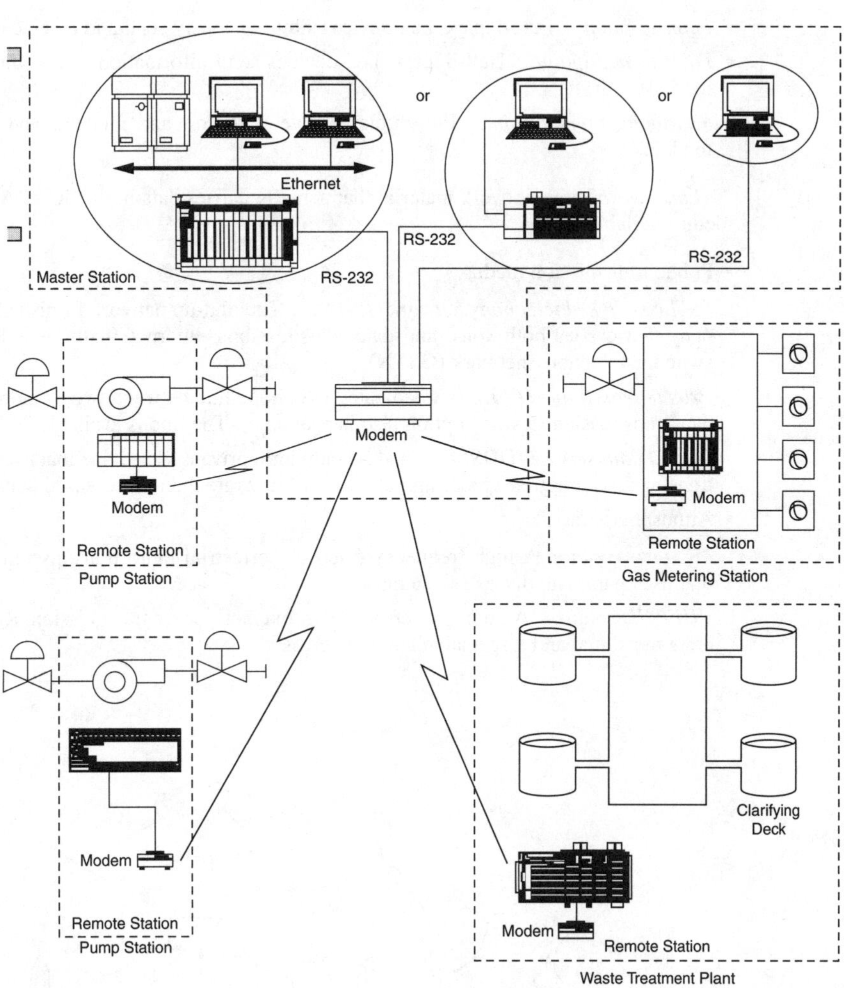

FIGURE 11.10 SCADA systems let engineers monitor and control various remote functions. They also collect valuable data that the ERP and MES layers can tap into.

- *Point-to-point.* Involves a connection between only two stations where either station can initiate communication, or one station can inquire and control the other. Generally, engineers use a two-wire transmission in this topology.
- *Point-to-multipoint.* Includes a link among three or more stations (a.k.a. multidrop). One station is designated as the arbitrator (or master), controlling communication from the remote stations. This is the main topology for SCADA applications. And it usually requires a four-wire transmission, one pair of wire to transmit and one pair to receive.
- *Multipoint-to-multipoint.* Features a link among three or more stations where there is no arbitrator and any station can initiate communication. Multipoint-to-multipoint is a special radio modem topology that provides a peer-to-peer network among stations.

Transmission mode is the way information is sent and received between devices on a network. For SCADA systems, the network topology generally determines the mode.

- *Point-to-point.* Full-duplex, i.e., devices simultaneously send and receive data over the link.
- *Point-to-multipoint.* Half-duplex, i.e., devices send information in one direction at a time over the link.
- *Multipoint-to-multipoint.* Full-duplex between station and modem, and half-duplex between modems.

Link media is the network material that actually carries data in the SCADA system. The types of media available are:

- Public transmission media

  Public switched telephony network (PSTN). The dial-up network furnished by a telephone company that carries both voice and data transmissions. Internationally, it is known as the general switched telephony network (GSTN).

  Private leased line (PLL). A dedicated telephone line between two or more locations for analog data transmission (a voice option also is available). The line is available 24 h a day.

  Digital data service (DDS). A wide-bandwidth, private leased line that uses digital techniques to transfer data at higher speeds and at a lower error rate than most leased networks.
- Atmospheric media

  Microwave radio. A high frequency (GHz), terrestrial radio transmission and reception media that uses parabolic dishes as antennas.

  VHF/UHF radio. A high frequency, electromagnetic wave transmission. Radio transmitters generate the signal and a special antenna receives it.

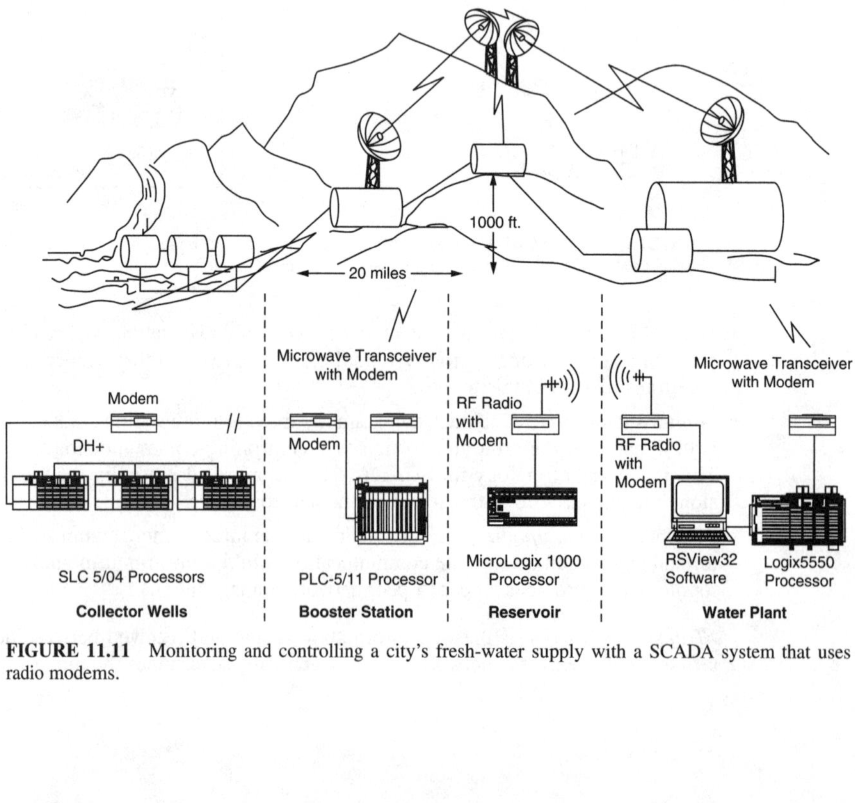

FIGURE 11.11 Monitoring and controlling a city's fresh-water supply with a SCADA system that uses radio modems.

Geosynchronous satellite. A high-frequency radio transmission used to route data between sites. The satellite's orbit is synchronous with the earth's orbit, and it receives signals from and sends signals to parabolic dish antennas.

- Power lines

  With special data communication equipment, companies can transmit and receive data over 120 V AC or 460 V AC power bars within a factory.

The trend in industrial automation is to use radio modems. As telephone voice traffic has shifted from landlines to the airwaves over the last decade, a similar transition is occurring in SCADA networks. In water/wastewater, oil and gas, and electric utility applications, where dedicated leased line connections once reigned supreme, radio modems are now flourishing (Fig. 11.11).

One of the reasons behind this evolution is the advent of spread-spectrum radio modem technology which allows multiple users to operate their radio modem networks in shared radio frequency bandwidth. The users can do so without any governmental licensing requirements and at transmission speeds that parallel the fastest communication rates of dedicated lines.

Another reason is that radio modem technology allows companies to take full control over the operation and maintenance of their SCADA media instead of relying on the telephone carriers to provide what can be less-than-reliable service.

The final element in a SCADA system is the PROTOCOL, which governs the format of data transmission between two or more stations including handshaking, error detection, and error recovery. A common SCADA protocol is DF1 half/full duplex, an asynchronous, byte-based protocol. Its popularity stems from benefits like remote data monitoring and online programming.

11.3.3 The MES Layer

According to the Manufacturing Execution System Association (MESA) International, a nonprofit organization comprised of companies that work in supply-chain, enterprise, product-lifecycle, production, and service environments:

> Manufacturing execution systems (MES) deliver information that enables the optimization of production activities from order launch to finished goods. Using current and accurate data, MES guides, initiates, responds to, and reports on plant activities as they occur. The resulting rapid response to changing conditions—coupled with a focus on reducing non value-added activities—drives effective plant operations and processes. MES improves the return on operational assets as well as on-time delivery, inventory turns, gross margin, and cash flow performance. MES provides mission-critical information about production activities across the enterprise and supply chain via bidirectional communications.

This definition describes the functionality of the MES layer. MES systems are able to provide facility-wide execution of work instructions and the information about critical production processes and product data. This information can be used for many decision-making purposes. For example, MES provides data from the facilities to feed historical databases, document maintenance requirements, track production performance, and the like. It is a plant-wide system to not only manage activities on production lines to achieve local goals, but also to manage global objectives. Key beneficiaries of these data are operators, supervisors, management, and others in the enterprise and supply chain.

The MES allows a real-time view of the current situation of the plant-floor production, providing key information to support supply chain management (SCM) and sales activities. However, this function in many plants is still handled by paper and manual systems. As this forces plant management to rely on the experience, consistency, and accuracy of humans, many recognize the value an MES can add to their operation. Manufacturers recognize that manual systems cannot keep up with the increased speed of end-user demands, which trigger changes in products, processes, and technologies. The responsiveness to customer demand requires higher flexibility of operations on the factory floor and the integration of the factory floor with the supply chain, which forces manufacturers to enable MES solutions in their plant to achieve this goal.

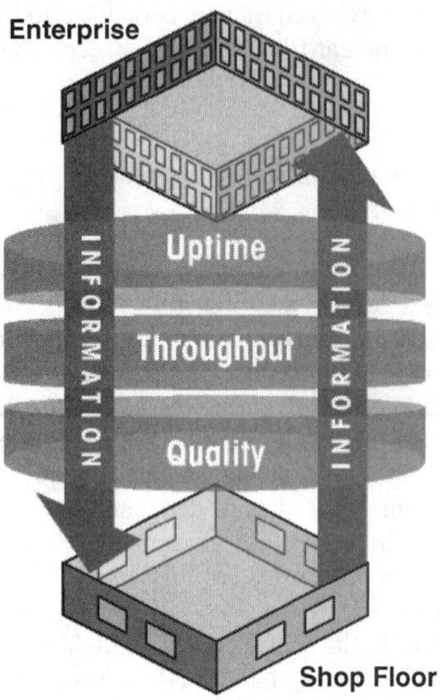

FIGURE 11.12 MES solutions provide an architecture that facilitates information sharing of production process and product data in a format that is usable by supervisors, operators, management, and others in the enterprise and supply chain.

MES modules that have been implanted include:

- Operation scheduling
- Resource allocation
- Document control
- Performance analysis
- Quality management
- Maintenance management
- Process management
- Product tracking
- Dispatching production units

Uptime, throughput, and quality—these are the driving factors for manufacturers' excellence (Fig. 11.12). Coupled with the pressures to meet government regulatory requirements and customer quality certifications, a manufacturer must find ways to quickly and cost-effectively improve production efficiency while reducing costs. Supply chain pressures demand that many manufacturers have an accurate view of where products and materials are at all times to effectively supply customers with product "just in time." The plant's status and capability to handle changes in production orders are additional key pieces of information for supply in today's markets. As a result, many manufacturing companies are now making the transition from paper and manual systems to computerized MES.

11.3.4 MES and ERP Standards

The industry has come a long way from a scenario of no standards; today there is a high degree of out-of-the box, standards-based functionality in most software packages. MES software vendors have always been (for the most part) industry specific. In batch, continuous, or discrete industries, the common production and business processes, terminology, and data models found their way into vendors' software products. But now there is a more coordinated offering of MES/ERP solution components both, from a functional as well as vertical industry standpoint. Product maturation has significantly tilted the balance to product configuration rather than customization. An indication of this maturation is the number of MES/ERP and data transport and storage model standards.

Helpful industry models and associations like the following now identify MES and ERP as required components in the corporate supply chain:

- Supply Chain Council
- Supply-Chain Operations Reference (SCOR)
- Manufacturing Execution System Association (MESA)
- S95 from the Instrumentation, Systems and Automation Society (ISA)
- Collaborative Manufacturing Execution from AMR Research

These frameworks were the first steps in addressing commonality among language and business processes, both of which are key aspects of MES interaction and synchronization with ERP business applications.

A common language also is emerging for corporate (vertical) and plant (horizontal) integration in the form of new standards from SCOR, MESA, and ISA, among others. That means cost effective business-to-business integration is on its way. In the last ten years industry standards have evolved from reference into object models and are now focused on the use of Web services as the method for integrations. All major ERP vendors and most MES vendors and developing Web services interface to facilitate integration between applications. These standards are gaining wide acceptance in the user community, perhaps because vendors are finally incorporating them into software products across both corporate and plant IT systems. This metamorphosis will simplify application, interface design, maintenance, and change management, thus reducing total cost of ownership (TCO).

11.3.5 Horizontal and Vertical Integration

Plant-wide integration has taken on meaning and context beyond its official boundaries, causing confusion. According to some ads, brochures, Web sites, and the like, all an engineer needs for a seamlessly integrated plant is to install a few high-end devices. But the truth is, plant-wide integration is a process, not an event. Integration requires a series of steps, the outcome of which is always unique to the user looking to update and integrate a facility. To properly prepare for plant-wide integration, start with the big picture and work down to hardware- and software-level detail.

It's important to define and underscore the business objectives and benefits of integration. Many companies have failed when they implemented projects that were focused on integrating different systems. This integration is successful only when it is focused on solving defined business problems and not on the technology involved.

In reality, there are two different levels of integration that fall under the plant-wide umbrella: horizontal and vertical.

Horizontal integration involves tying the floor of a manufacturing plant together through automation. In simple terms, it encompasses every step of the "making-stuff" process—from a rail car full of barley malt docking in receiving, to a truck full of lager kegs pulling away from shipping. The easiest way to describe horizontal integration, though, is to provide an example of a disjointed, fractured facility.

Sticking with the brewing theme, imagine that the process and packaging portions of a brewery use separate networks to transfer information and are driven by different, unconnected controllers. Basically, they are separate entities housed under the same roof. In this extremely common scenario, the left side doesn't know what the right is doing, and both suffer through an inordinate amount of "dead time" as a result. For instance, in a situation where a new order for additional product is being processed in the packaging area, the process area must be notified to ensure sufficient product is available and ready for transfer. If the communication is not done successfully, the packaging line may have to stop to wait for product and the customer may not receive its order in time.

Horizontal integration eliminates isolated cells of activity by merging the entire manufacturing process into a single coordinated system. Every corner of the plant is connected and can adjust and compensate to the changing business situation without considerable effort. That's not to say that the entire facility runs at optimal efficiency at all times, however. That's where vertical integration comes into play.

Vertical integration allows the transfer and execution of work instructions and the flow of information—from the simplest sensor on the plant floor to the company's Intranet and Extranet. This is accomplished via integration between the factory floor, MES, and ERP systems. The main goal of vertical integration is to reduce "friction" and transfer information in real time.

The best way to describe friction is with a modern, on-line example. A Web surfer is searching for information on vacation destinations. He locates a site of interest and wants to read more about visitor attractions in rural Montana. He spots the proper link, executes the standard click and … a registration screen pops up. The system requires user ID and password. This is friction. A roadblock, though small, has prevented instant access to key information.

The same happens on the plant floor. Many manufacturing operations lack a sound database structure, which—combined with the information it contains and applications that use it—is the only means to the right data, at the right time, in the right place. Meanwhile, operations that have sound databases are often segregated with several incompatible networks present, making filters and intermediaries to data abound.

Vertical integration removes these obstacles, providing real-time data to plant personnel and employees in other parts of the company. That means an operator with access to a PC can sit in an office and check the production status of any given line to make sure it is at peak productivity. On-the-spot access to this type of information provides an unlimited resource of knowledge. What areas of the plant are experiencing downtime? What line has the most output? Plus, the knowledge can be synthesized into process improvements: "We need to increase batch sizes to make sure the packaging lines are constantly running." "We need to make these adjustments to lines B and C so they will be as efficient as A."

The benefits of horizontal and vertical integration are obvious. The first advantage of an integrated plant is an increase in productivity. With a cohesive plant floor and the ability to gather information anywhere and at any time, engineers can drive out the inefficiencies that habitually plague production. The second benefit is the ability to manufacture more goods. If the entire plant is running efficiently, throughput will be amplified.

The need to be responsive to customer demand continues to lead manufacturers toward integrated solutions that reduce costs and, ultimately, create greater plant-wide productivity through a tightly coordinated system.

Integrating multiple control disciplines has other benefits as well. Design cycles are shortened, for example, speeding time-to-market for new goods. Software training and programming time also drop, and getting systems to work together is painless. Plus, an integrated architecture is synonymous with a flexible, scalable communications system. That means no additional programming is needed to integrate networks. And at the same time, networks are able to deliver an efficient means to exchange data for precise control, while supporting noncritical systems and device configuration at start-up and during run time.

The ability of a company to view plant information from anywhere in the world, and at any stage of production, completes an integrated architecture. A transparent view of the factory floor provides integration benefits like common user experience across the operator interface environment; configuration tools for open and embedded control applications; improved productivity with the ability to reuse technology throughout the plant; and overall reduced cycle costs related to training, upgrades, and maintenance.

From a practical standpoint, this kind of integration extends usability around the globe. Information entered into the system once can be accessed by individuals throughout the enterprise—from the machine operator or maintenance personnel on the factory floor to a manager viewing live production data via the Internet halfway around the world.

11.4 HOW TO IMPLEMENT AN INTEGRATED SYSTEM

In addition to hardware and software, a number of vital services and specific project steps are necessary for successful project implementation. Proper ordering and execution of the project steps, from specification, design, manufacture, and factory test, to start-up and maintenance provide for a system that meets the needs of the users. While emphasis on certain activities of project implementation may vary depending on project size, complexity, and requirements, all facets of the project must be successfully addressed.

Some integration companies provide a full complement of implementation services. The services discussed in this chapter are:

- Project management
- Generation of a functional specification
- System conceptualization and design

- Hardware/software engineering
- Assembly and system integration
- Factory acceptance test
- Documentation
- System training, commissioning, and startup

11.4.1 Project Management

A project team philosophy is key for a successful systems integrator. It takes a highly talented and experienced project manager to direct and coordinate projects to assure completion on time and within the established budget. Project managers are well versed in their industry and know the control and communication network technologies and application practices required to meet customer needs.

A project management team needs to work closely with the end user's project team to define, implement, and document a system that meets the needs of the user. Sharing of information through intense, interactive sessions results in the joint development of a system that fulfills the needs of the user community while remaining within budget and schedule constraints.

11.4.2 Generation of a Functional Specification

A critical phase in implementing a complex system is the establishment and documentation of the baseline system. This phase assures that the system delivered matches the needs and expectations of the end user. The project manager and the integrator's technical team need to assist the end user in establishing the baseline.

It is vital that input for the baseline system is solicited from all the end user's personnel who will be using the system. This includes personnel from operations, maintenance, management, quality, computer, and engineering departments. The baseline system is documented in a functional specification, which includes an interface specification, drawings, system acceptance procedures, and other appropriate documents. Once the baseline system is documented and agreed upon, system implementation begins.

A number of vital services and specific project steps are necessary for successful system project implementation. Proper ordering and execution of the project steps, from specification, design, manufacture, and factory test, to start-up and maintenance provide for a system that meets the needs of users. While emphasis on certain activities of project implementation may vary depending on project size, complexity, and requirements, all facets of the project must be successfully addressed.

11.4.3 System Conceptualization and Design

The technical team addresses the hardware design by considering the requirements defined in the system specification. Other factors considered in the selection include cost, complexity, reliability, expandability, operability, and maintainability. Then, application-specific factors such as heating, accessibility, and environmental requirements are considered.

In the system design phase, the hardware and software architectures are developed. Inter- and intra-cabinet wiring are defined and formally documented, and hardware is selected to populate the architecture. To assist in the hardware design, the speed and flexibility of computerized design and drawing systems are utilized. These systems utilize a standard parts database in addition to custom parts and libraries.

As with all review processes, the objective is to discover design errors at the earliest possible moment so that corrective actions can be taken. The end user may participate in the reviews during the design stages of the development process. Along with the various design reviews, other quality measures are continually applied to the development process to ensure the production of a well-defined, consistent, and reliable system.

Designs include, for example, standard or custom console packages, electrical panel layouts, wiring diagrams, enclosures, operator panels, network drawings, and keyboard overlays.

11.4.4 Hardware/Software Engineering

Once hardware is selected, the bills of material are finalized. Final drawings are released by drafting to the manufacturing floor to begin system implementation.

Any software that may be required as part of the overall system is designed to meet the user needs and requirements defined in the system specification. The design consists of user-configurable subsystems composed of modules performing standardized functions. This approach guarantees the user a highly flexible, user friendly, maintainable system.

The software design is accomplished by employing a consistent, well-defined development methodology. First, the functional specification is transformed into a system design. After the design and specification are in agreement, coding begins. Advanced software development techniques (e.g., rapid prototyping, iterative, or spiral methods) may also be deployed at this stage to accelerate the development cycle while maintaining project control.

11.4.5 Assembly and System Integration

Upon release to manufacturing, the equipment is accumulated and system assembly is initiated. Reviews are held with production control and assembly representatives to assess progress. In the event, problem areas are identified, action plans are formulated, and action items are assigned for scheduled maintenance. Where schedule erosion is apparent, recovery plans are formulated and performance is measured against these plans until the schedule is restored.

To assure compliance with the specified performance, the technical team vigorously tests the integrated system. If deficiencies are identified, hardware and software modifications are implemented to achieve the specified performance.

11.4.6 Factory Acceptance Tests

Where provided by contract, formal, witnessed, in-plant acceptance testing is conducted in the presence of the end user. These tests are performed in accordance with the approved acceptance test procedures to completely demonstrate and verify the compliance of the system performance with respect to specification.

11.4.7 System Level Documentation

The generation, distribution, and maintenance of system documentation are an important part of the success of any industrial automation system. All persons involved with the system must have current and complete documentation. The documentation must satisfy the requirements of all the engineering, installation, production, operations, and maintenance functions.

Here is an example of a typical set of documentation provided with a system:

- *Mechanical drawings.* Showing overall dimensions of each cabinet/enclosure along with a general panel layout identifying all components that collectively comprise a packaged subsystem/enclosure. A mechanical drawing package is generally provided on a per-enclosure basis.
- *Electrical drawings.* Of all packaged enclosures showing component interconnection circuits, termination points of external peripherals/field devices, and all enclosure-to-enclosure interconnections.

- *Standard product information.* Including complete sets of standard product data sheets on all major supplied components.
- *Application software documentation.* Documenting all program points used and a complete listing of all programs with a brief narrative explaining the function of each program module.
- *User's manual.* Including the procedures used for all operator interfaces and report generation. Start-up and shutdown procedures are addressed.
- *Acceptance test specification.* As defined in the system specifications.
- *Recommended spare parts.* Listing all major system components employed and recommended spare quantities based on anticipated failure rates.

11.4.8 System Training, Commissioning, and Startup

A smooth transition from the integrator to the end user requires adequate training prior to commissioning. The training program consists of two facets: standard-training courses on major system elements, and specialized instructions addressing specific systems. The technical team examines the scope of the system and provides a list of recommended coursework to be pursued prior to system delivery.

It is essential to have qualified technical personnel involved in the design and programming of the proposed system to assist in the installation, startup, and commissioning of the system.

11.5 OPERATIONS, MAINTENANCE, AND SAFETY

11.5.1 Operations

Today's manufacturing environment is dramatically changing with increasing pressure to reduce production costs while improving product quality and delivery times. In today's global economy, a company can no longer simply manufacture products for storage in inventory to meet the demands of their customers; instead, they must develop a flexible system where production output can be quickly adjusted to meet the demands of fluctuating market conditions. Investments in ERP, SCM, and trading exchanges have enabled companies to closely link the manufacturing plant with its suppliers and customers. This environment is commonly referred to as e-manufacturing.

To be successful, today's manufacturing companies must harness and leverage information from their operating systems. This is where initiatives like lean manufacturing drive out excess, achieving nonstop operations for maximum efficiency and throughput of production, and where techniques like Six Sigma reduce variability in processes to ensure peak quality.

Capitalizing on this information requires manufacturers to develop:

- An in-depth analysis to understand the business issues facing the company and the operational data needed to solve those issues
- A plant information plan that defines the systems for linking the factory floor to business system, and the collection and reporting of production information

Manufacturers face two core production issues. The first is how to optimize the performance of their supply chain process, from their suppliers to their clients. The second is how to improve the performance of their plants, both in production efficiency and equipment efficiency.

Most manufacturing companies have implemented software programs and procedures to monitor and optimize their supply chain. These programs collect real-time factory-floor data on raw material usage, yield, scrap rate, and production output. By tracking this data, companies can reduce raw material consumption, work-in-progress, and finished-goods inventory. It also allows companies to track the current status of all production orders so they can meet the needs of their customers.

For many manufacturers, factory-floor data is manually entered into business systems, increasing the probability of incorrect and outdated information. Much of this data may be entered into the system 8 to 48 h after execution. Consequently, supply chain programs may use data that is neither timely nor accurate to optimize operations, which can hinder a company's goal of meeting customer demand. This is certainly an area where optimization is desperately needed.

To effectively use data from the factory floor to report on operations, analyze results, and interface to business systems, a company must perform a thorough analysis of its operations and systems and then develop an integrated operations strategy and practical implementation plan. Each initiated project within the plan must have a calculated return that justifies new investment and leverages prior investments made in factory-floor control, data acquisition, and systems.

An important goal of every operation is to improve performance by increasing output and yield while reducing operating costs. To meet these production objectives, companies make large capital investments in industrial automation equipment and technology. With shareholders and analysts looking at return on investment (ROI) as a key factor in evaluating a company's health and future performance, manufacturers must emphasize the importance of optimizing the return on their assets.

11.5.2 Maintenance

Efficient maintenance management of all company assets—like materials, processes, and employees—ensures nonstop operations and optimum asset productivity. Without a solid, efficient foundation, it is very difficult to withstand the rigors of this fast-paced environment where growth and profits are demanded simultaneously.

A lean workforce, tight profit margins, and increased competitive pressures have manufacturers seeking new ways of producing more goods at higher quality and lower costs. Many companies are turning to maintenance, repair, and operations (MRO) asset management and predictive maintenance as a core business strategy for boosting equipment performance and improving productivity.

In the process industry, downtime can quickly erode profitability at an alarming rate—upwards of $100,000 an hour in some applications. These companies recognize that equipment maintenance is quickly evolving beyond simple preventive activities into a proactive strategy of asset optimization. This means knowing and achieving the full potential of plant floor equipment and performing maintenance only when it is warranted and at a time that minimizes the impact on the overall operation. To achieve this, companies need to be able to gather and distribute data across the enterprise in real-time from all process systems.

The Way Things Were. Until recently, the majority of condition monitoring was performed on a walk-around or ad hoc basis. In the past, companies couldn't justify the cost or lacked the sophisticated technology needed to efficiently gather critical machine operating data. Typically, this high level of protection was reserved for a privileged few—those machines deemed most critical to production. The protection systems that companies leveraged were centrally located in a control room and operated independently from the control system. This required extensive wiring to these machines and used up valuable plant-floor real estate. Additionally, many of the systems were proprietary, so they did not easily integrate with existing operator interfaces or factory networks. Not only was this approach costly to implement and difficult to troubleshoot, but it also gave plant managers a limited view of overall equipment availability and performance (Fig. 11.13).

After years of capital equipment investments and plant optimization, many manufacturers aren't able to make major investments in new technology and are looking to supplement existing equipment and processes as a way to bolster their predictive maintenance efforts.

Today, new intelligent devices and standard communication networks are opening up access to manufacturing data from every corner of the plant. By leveraging existing networks to gather information, new distributed protection and condition-monitoring solutions are providing manufactures with never-before imagined opportunities to monitor and protect the health of their plant assets. This includes real-time monitoring of critical machinery as well as implementing corrective actions before a condition damages equipment.

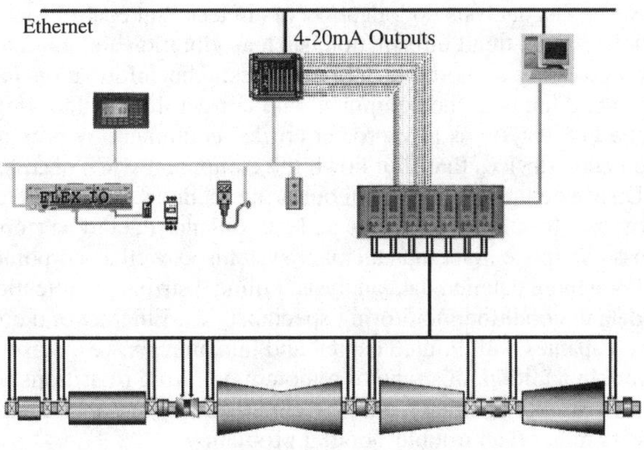

FIGURE 11.13 A traditional centralized rack solution.

Embracing the Future of Maintenance. Open communication is key to maximizing asset management technology. While condition monitoring equipment suppliers are now providing products that communicate using open protocols, this has not historically been the case for condition monitoring and asset management solutions. The development of industry standards by groups like OPC (OLE for process control) and MIMOSA (Machinery Information Management Open Systems Alliance) is giving MRO applications open access to condition monitoring—diagnostic and asset management information from intelligent instruments and control systems.

The Distributed Approach. Fueled by market demand, technological advancements have led to a new approach to condition monitoring—online distributed protection and monitoring. Building on the principles of distributed I/O and integrated control, distributed protection and monitoring systems replace large, centralized control panels with smaller control systems and put them closer to the process and machinery being monitored.

By using a facility's existing networking infrastructure, online distributed protection and monitoring requires significantly less wiring than traditional rack-based protection systems. Inherent in online distributed protection and monitoring systems is the scalability of the architecture. By using more modular components, manufacturers are able to connect more than one device to a wire and add machinery into the system as needed (Fig. 11.14).

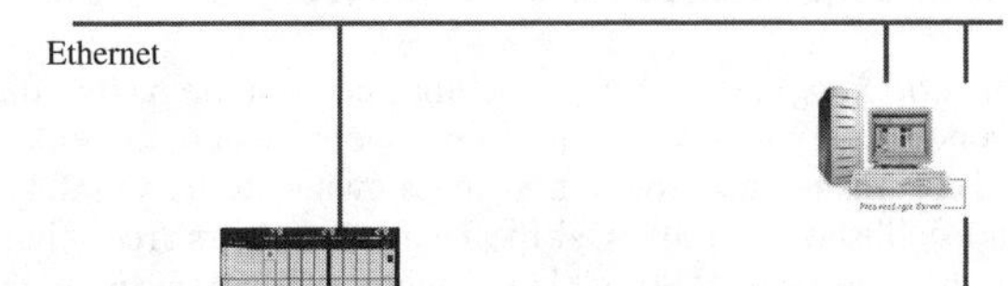

FIGURE 11.14 A distributed protection and monitoring architecture.

Since data analysis no longer occurs in a central control room, maintenance personnel can quickly view important trend information, such as vibration and lubrication analysis, directly from the operator's consoles or portable HMI devices. The information gathered allows operators to identify impending faults in the equipment and correct them before impacting production or compromising safety. These systems also protect critical equipment by providing alarm status data in real time to automation devices that shut down the equipment when necessary to prevent catastrophic damage.

Distributed protection and monitoring modules also can be connected with condition monitoring software. This allows all online and surveillance data to be stored in a common database and shared across enterprise asset management systems as well as corporate and global information networks.

For a more detailed data analysis, online distributed protection and monitoring systems can transfer data to condition monitoring specialists via Ethernet or a company's wide area network (WAN). For companies with limited capital and human resources, outsourcing this task offers a cost-effective option. In addition, this type of remote monitoring transitions on-site maintenance engineers from a reactive to a preventative mode—freeing them to focus their attention on optimizing the manufacturing process rather than troubleshooting problems.

The Future Is Now. Unplanned downtime need not cost companies millions of dollars each year. The technology exists today to cost-effectively embrace a proactive strategy of predictive maintenance and asset optimization. Progressive companies realize that capturing, analyzing, and effectively using machine condition information provides them with a strategic and competitive advantage, allowing them to maximize return on investment and making optimal maintenance a reality.

11.5.3 Safety

Maximizing profits and minimizing loss can be achieved by a number of methods, but there's one that most wouldn't expect—plant floor safety. Safety is, above all, about protecting personnel. Today's manufacturers, for the most part, view safety as an investment with a positive return in the sense that a safer workplace boosts employee morale; machine and process operators feel more comfortable with the equipment and are aware of the company's commitment to their safety. The result is increased productivity and savings attributed to a decrease in lost-time accidents, medical expenses, and possible litigation.

It took some time before manufacturers realized that safety measures weren't a hindrance to productivity and that safety was truly an investment with positive return. The acceptance of safety as a good business practice is evident as the number of workplace injuries continues to fall each year. But can that positive return be measured? In any business, there is increasing pressure to determine the most valuable programs, in financial terms, and areas where cuts can be made. In these cases, plant safety programs and safety professionals have historically been easy targets for cutbacks, simply because the true value of safety is not easily calculated. Hard data on the price tag of lost-time accidents is required to show that safety has economic value and is good business.

Safety and Progress. When machines entered the picture during the Industrial Revolution, the idea of worker safety was secondary to productivity and, more directly, money. Accidents were common, and there was no incentive for business owners to make safety a priority. A "laissez faire" system had been established that allowed the business owners free reign of their ventures without interference from the government. So while productivity was soaring higher than ever, unsafe machines and dismal working conditions were taking their toll on the workforce. In the 19th century, however, things took a turn for the better—edicts on acceptable working environments and safe machine practices began to emerge.

By the beginning of the 20th century, true machine safety products started to appear in the form of emergency stops. World War II saw the introduction of safety control relays that could provide electromechanical diagnostics through the use of interlocking contacts. But the most dramatic leap in machine safety started in the latter half of the century—and safety products haven't stopped evolving since.

In the 1960s fixed machine guards came to the fore as the primary method of protecting personnel from hazardous machinery. Driven by legislation, the installation of these cages and barriers basically prevented access to the machine. Fixed machine guarding (also known as "hard guarding") provides the most effective protection by not allowing anyone near the point of hazard but unfortunately it is not a feasible solution when the application requires routine access by an operator or maintenance personnel.

By the 1970s, movable guards with interlocking systems became the most prominent solution for applications requiring access to the machine. Hinged and sliding guard doors outfitted with safety interlock switches allow access to the machine but cut off machine power when the guard is open. Some interlocking systems also contain devices that will lock the guard closed until the machine is in a safe condition, a function known as *guard locking*. As the first step toward the integration of safety and machine control, the interlock solution allows for a modest degree of control while restricting access during unsafe stages of the machine's operation. In terms of the marriage between safety and productivity, the combination of movable guards and interlock switches is still the most reliable and cost-effective solution for many industrial applications. However, in processes requiring more frequent access to the machine, repeated opening and closing of guards is detrimental to cycle times—even a few seconds added to each machine cycle can severely hamper productivity when that machine operates at hundreds of cycles per day.

Presence sensing devices for safety applications made their way onto the plant floor in the 1980s with the introduction of photoelectric safety light curtains and pressure-sensitive floor mats and edges. Designed to isolate machine power and prevent unsafe machine motion when an operator is in the hazardous area surrounding a machine, safety sensors help provide protection without requiring the use of mechanical guards. They also are less susceptible than interlock switches to tampering by machine operators. The use of solid-state technology in sensors also provides a degree of diagnostics not previously possible in systems using relay control with electromechanical switches.

Fifty years' worth of safety advances culminated in the safety control domain of the 1990s—the integration of hard guarding, safety interlocks, and presence sensing devices into a safety system monitored and controlled by a dedicated safety controller and integrity monitoring. Trends show that this safety evolution will continue to move toward seamless control solutions involving electronic safety systems, high-level design tools, networking capabilities, and distributed safety implementation through embedded intelligence.

Global Safety Standards for a Global Market. Safety in automation is not new, but as global distribution of products becomes the norm, machinery manufacturers and end users are increasingly being forced to consider global machinery safety requirements when designing equipment. One of the most significant standards is the Machinery Directive which states that all machines marketed in the European Union must meet specific safety requirements. European law mandates that machine builders indicate compliance with this and all other applicable standards by placing CE—the abbreviation for "Conformité Européenne"—markings on their machinery. Though European in origin, this safety-related directive impacts OEMs, end users, and multinational corporations everywhere.

In the United States, companies work with many organizations promoting safety. Among them:

- Equipment purchasers, who use established regulations as well as publish their own internal requirements
- The Occupational Safety and Health Administration (OSHA)
- Industrial organizations like the National Fire Protection Association (NFPA), the Robotics Industries Association (RIA), and the Society of Automotive Engineers (SAE)
- The suppliers of safety products and solutions

One of the most prominent U.S. regulations is OSHA Part 1910 of 29 CFR (Title 29 of the Code of Federal Regulation), which addresses occupational safety and health standards. Contained within Subpart O are mandatory provisions for machine guarding based on machine type; OSHA 1910.217, for example, contains safety regulations pertaining to mechanical power presses. In terms of private sector voluntary standards (also known as *consensus standards*), the American National Standards Institute (ANSI) serves as an administrator and publisher, maintaining a collection of industrial safety standards including the ANSI B11 standards for machine safety.

With components sourced from around the world, the final destination and use of a product often remains unknown to its manufacturer. As a result, machine builders are looking to suppliers not only for safety products that meet global requirements and increase productivity, but also as a useful resource for an understanding of safety concepts and standards. And in an effort to more efficiently address customer concerns and stay abreast of the market, those suppliers have assumed an active role in the development of standards.

Safety Automation. Investing in solutions as simple as ergonomic palm buttons–designed to relieve operator strain and decrease repetitive motion injuries–helps manufacturers meet safety requirements while increasing production. In one example, a series of safety touch buttons was installed on an industrial seal line in which operators previously had to depress two-pound buttons during the entire 5-s cycle. Using standard buttons, these operators suffered neck and shoulder soreness during their shifts. After installing the safety touch buttons, employees no longer complained that the machine was causing discomfort. The buttons created better working conditions that have directly affected employee morale, decreased employee injuries, and led to a more productive plant (Fig. 11.15).

Another example of how advanced safety products can improve productivity involves light curtains—infrared light barriers that detect operator presence in hazardous areas. Typically, a safety interlock gate is used to help prevent machine motion when an operator enters the hazardous area. Even if it only takes 10 s to open and close that gate for each cycle, that time accumulates over the course of a 200-cycle day. If the traditional gates were replaced with light curtains, operators would simply break the infrared barrier when entering the hazardous area, and the operation would come to a safe stop. Over time, the light curtain investment would increase productivity and create a positive return.

In addition to the safety function, protective light curtains also may serve as the means of controlling the process. Known as *presence sensing device initiation* (PSDI), breakage of the light curtain's infrared beams can be used to initiate machine operation. Upon breakage of the beam, the machine stops to allow for part placement. After the operator removes his or her hands from the point of hazard, the machine process restarts.

Manufacturers' desire for continuous machinery operation without compromising safety has led to the merging of control and safety systems. The development of safety networks represents a major step forward in this evolution. Similar to its standard counterpart, a safety network is a fieldbus system that connects devices on the factory floor. It consists of a single trunk cable that allows for quick connection/disconnection of replacement devices, simple integration of new devices, easy configuration and communication between the devices, delivery of diagnostic data (as opposed to simple on/off status updates), and a wealth of other features to help workers maintain a safety system more efficiently. But unlike standard networks, which also provide this functionality but are designed to tolerate a certain number of errors, a safety network is designed to trap these errors and react with predetermined safe operation.

FIGURE 11.15 Ergonomic safety equipment increases safety and productivity.

This combined safety and control domain also will allow facility engineers to do routine maintenance or troubleshooting on one section while production continues on the rest of the line, safely reducing work stoppages and increasing flow rates. For example, in many plants, a robot weld cell with a perimeter guard will shut down entirely if an operator walks into the cell and breaches the protected area. Control systems using safety programmable controllers tested to Safety Integrity Level 3 (SIL 3)—the highest level defined by the IEC for microprocessor-based safety systems—can actually isolate a hazard without powering down an entire line. This permits the area undergoing maintenance to be run at a reduced, safe speed suitable for making running adjustments. The result is an easier-to-maintain manufacturing cell, and one that is quicker to restart.

In a downtime situation with a lock-out/tag-out operation, system operators may have to use five or six locks to safely shut down a line including electronic, pneumatic, and robotic systems. Shutting down the entire machine can be time consuming and inefficient. If a safety control system with diagnostic capabilities were installed, operators could shorten the lock-out/tag-out process, quickly troubleshoot the system, and get it running.

Previous generations of safety products were able to make only some of these things happen. But current safety products, and those of the future, can and will increase productivity from another perspective—not only are today's machine guarding products faster and safer, but their integration may actually boost productivity by enhancing machine control.

The increasing effect of standards and legislation—especially global—continues to drive the safety market. But even with those tough standards in place, today's safety systems have helped dispel the notion that safety measures are a burden. Ultimately, the good news for today's manufacturers is that safety products can now provide the best of both worlds: operator safety that meets global regulations and increased productivity.

11.6 CONCLUSION

For the past 75 years, industrial automation has been the linchpin of mass production, mass customization, and craft manufacturing environments. And nearly all the automation-related hardware and software introduced during this span were designed to help improve quality, increase productivity, or reduce cost.

The demand for custom products requires manufacturers to show a great deal of agility in order to adequately meet market needs. The companies that used to take months, even years, to move from design to prototype to final manufacturing may find themselves needing to merge design and manufacturing— eliminating interim prototype stages and paying closer attention to designing products based on their manufacturability.

In general, we're entering an incredible new era where manufacturing and business-level systems coexist to deliver a wider array of products than ever before. And advances in technology are giving us an incredible number of choices for controlling automation in this era. While there is no single solution that is right for every application, taking a systematic approach to planning and implementing a system will result in operational and maintenance savings throughout the life of the manufacturing process.

INFORMATION RESOURCES

Industrial Automation Research

Aberdeen Group, http://www.aberdeen.com/

AMR Research, http://www.amrresearch.com/

ARC Advisory Group, http://www.arcweb.com/

Forrester Research, http://www.forrester.com/

Gartner Research, http://www.gartner.com/

Venture Development Corp., http://www.vdc-corp.com/industrial/

Yankee Group, http://www.yankeegroup.com/

Industrial Automation Publications

A-B Journal magazine, http://www.abjournal.com/
Control and *Control Design* magazines, http://www.putmanmedia.com/
Control Engineering magazine, http://www.controleng.com/
Control Solutions International magazine, http://www.controlsolutionsmagazine.com/
IndustryWeek magazine, http://www.industryweek.com/
InTech magazine, http://www.isa.org/intech/
Maintenance Technology magazine, http://www.mt-online.com/
Managing Automation magazine, http://www.managingautomation.com/
MSI magazine, http://www.manufacturingsystems.com/
Start magazine, http://www.startmag.com/

Industrial Automation Directory

Information and services for manufacturing professionals, http://www.manufacturing.net/

Industrial Automation Organizations

ControlNet International, http://www.controlnet.org/
Fieldbus Foundation, http://www.fieldbus.org/
Instrumentation, Systems and Automation Society, http://www.isa.org/
Machinery Information Management Open Systems Alliance, http://www.mimosa.org/
Manufacturing Enterprise Solutions Organization, http://www.mesa.org/
OPC Foundation, http://www.opcfoundation.org/
Open DeviceNet Vendor Association, http://www.odva.org/
SERCOS North America, http://www.sercos.com/

Industrial Automation Resources

Collaborative manufacturing execution solutions, http://www.interwavetech.com/
Condition-based monitoring equipment, http://www.entek.com/
Factory management software, http://www.rockwellsoftware.com/
Industrial controls and engineered services, http://www.ab.com/
Linear motion systems and technology, http://www.anorad.com/
Mechanical power transmission products, http://www.dodge-pt.com/
Motors and drives, http://www.reliance.com/

CHAPTER 12
FLEXIBLE MANUFACTURING SYSTEMS

Paul Spink
Mori Seiki USA, Inc.
Irving, Texas

12.1 INTRODUCTION

Numerically controlled (NC) controls were adapted first to lathes followed by vertical machining centers and last to horizontal machining centers. This evolution began in the 1960s and gathered steam in the 1970s. It was during this period that horizontal spindle machines began the conversion to NC controls. There were a number of horizontal spindle machines manufactured, but the typical design consisted of a fixed column with a variety of table sizes in front of it. The headstock moved up and down on the side of the column and had a live spindle that extended out of the headstock, but the machine didn't have an automatic tool changer or pallet changer. Toward the second half of the 70s, tool changers started appearing on horizontal machines to improve the productivity of the machines. Then, as the control capability expanded with CNC in the early 80s, the builders started designing automatic pallet changers for these machines. It was in the late 70s that the fixed-spindle machining centers started appearing. These were horizontal machines with smaller tables since the tools had to reach the center of the table and tool rigidity was a major concern.

The adaptation of the horizontal machine to both the automatic tool changer and the automatic pallet changer was primary to the development of flexible manufacturing systems. Flexible manufacturing systems required the machines to have the capability to move parts on and off the machine plus the tool capacity to machine several different parts. After these options were available, it was a natural step to look for a means of improving the productivity of the equipment and that means was the flexible manufacturing system.

Some of the early system integration was done on larger horizontal machining centers. It was not unusual for a changeover to take one or two shifts, creating very low spindle utilization. Combining the ability to keep pallets stored on stands with setups completed and moving them to the machines when needed resulted in a tremendous output improvement immediately. During this period, there were machine tool builders who had also designed large vertical turning machines with tool changer and pallet changer capabilities. With these machines, there was now the ability to turn large parts and for vertical operations such as drilling, tapping, and counterboring holes and machining surfaces inside the part without moving the part to a vertical machining center. Then, by moving the pallet from the vertical turning machine to the horizontal, surfaces on five sides of a cube could be completed in one setup without operator interaction. This potential was very attractive to industrial segments such as the large valve industry, oil field equipment producers, turbines, pumps, and aerospace.

The natural design of the mechanism to move the pallets from machine to machine was a small rail vehicle. This vehicle ran on railroad type rails and was generally driven by a rack and pinion system. It was simple to make whatever length needed and the weight capacity of the vehicle was easily upgraded for heavier projects. With the requirement of rail being installed in a straight line, the term *linear* was given to the long straight type of system layout. Secondly, the storage of pallets along the track led to the use of the term *pallet pool.*

Systems generally required a cell controller to schedule parts to the machines, keep track of production, generate reports for the customer, and a number of other functions. Early cell controllers consisted of major computer installations utilizing mini computers with man-years of programming. Each system was unique and required the software to be developed and tested for each installation.

In the mid 1980s, the fixed spindle machining center started to appear designed with automatic tool changers and automatic pallet changers as standard equipment. This type of machine was more economical, faster, and more productive than the larger horizontals and the design made inclusion in a linear pallet pool system a natural. Unfortunately, because there was a good labor pool and machines were less expensive and capable of turning out more parts in a shorter time, interest in systems dwindled.

By the 90s, conditions had changed in the manufacturing industry. Machining center prices had increased over the years, production in nearly all segments of industry had improved so companies were looking for a way to boost productivity, and the skilled labor pool was not being replenished. Production was moving to smaller lot sizes. Flexible manufacturing was becoming the norm. All the new approaches were based on making a greater variety of parts each day or generating one part of each piece in an assembly to reduce inventory. But now the machinist was continually making setups that reduced the machine utilization and productivity. Management had to find ways to minimize the time needed for setups, increase machine utilization, improve production flexibility, and offset the lack of skilled personnel. The answer was *linear pallet pool systems.*

12.1.1 Theory of Flexible Manufacturing Systems

A flexible manufacturing system (Fig. 12.1) consists of at least one or more horizontal machining centers, an automatic guided vehicle (AGV) to move the pallets, sections of track to guide the movement of the AGV, pallet stands that hold pallets and parts used in the machines, and one or more setup stations located adjacent to the track to allow an operator to load material on the pallets.

Typically, there may be from one to eight machines, from six to 100 pallet stands, and from seven to 108 pallets in a flexible manufacturing system. Pallet stands can be single level or multilevel.

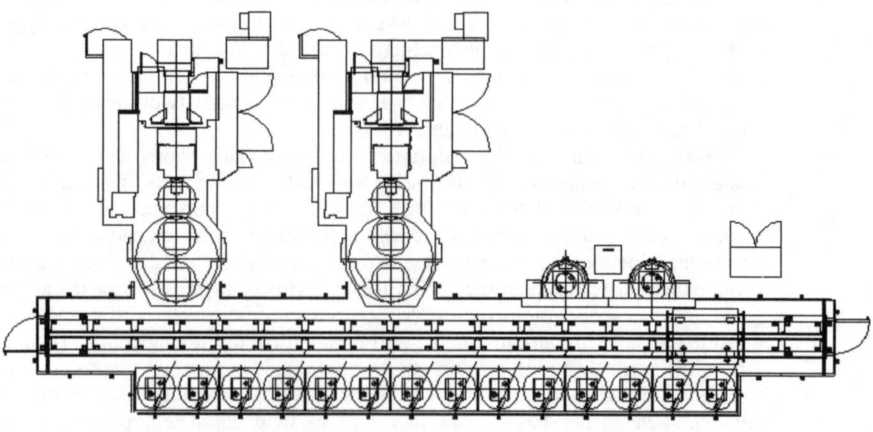

FIGURE 12.1 A flexible manufacturing system.

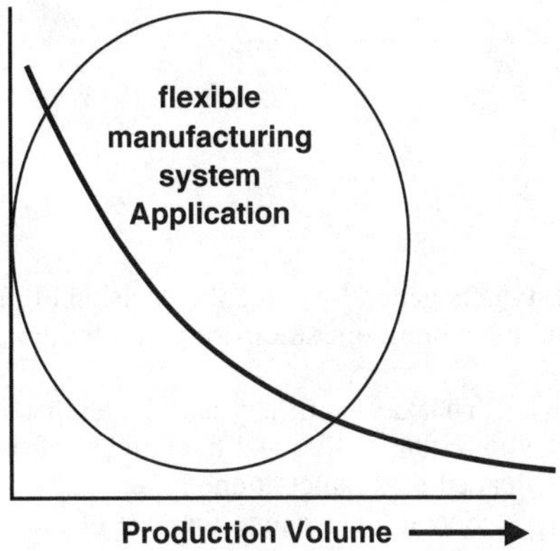

FIGURE 12.2 Set-up number vs. production.

The purpose of installing a flexible manufacturing system is to improve the machine utilization, improve the output of the machines, and boost the productivity of manufacturing by eliminating nonproductive idle time and optimizing use of manpower in producing parts. Machine utilization is increased by reducing and eliminating the time the machine sits as the machinist changes from one job to the next. Time is also saved while he or she is running a first article test piece at the start of each setup and by eliminating the dead time waiting for the test piece to be inspected and approved. This may take an hour or it may take a day depending on the part and the facility operation. With a number of jobs set up and held inside the system, changeover from job to job is eliminated. The fixtures and tools are sitting there ready for their turn to be moved into the machine. The time to change from one job to the next is nothing more than the 16 s to change a pallet. Profitability is greatly improved by both the increased production from the system and the ability to run one or more machines with personnel other than a highly trained machinist (Fig. 12.2).

What kind of operation could take advantage of the benefits offered by a flexible manufacturing system? The companies that would benefit generally have repetitive jobs.

The greatest savings are from operations that do small quantities and therefore have frequent setups. But high production installations that may change over once a week see savings by reducing the number of operators and eliminating the fatigue from walking between a number of machines continually for the entire day. These customers get the improvements because the fixtures and tools for their jobs have been setup and held inside the flexible manufacturing system until the job is to be run again. Most of the common cutting tools such as face mills, end mills, drills, and taps used on other jobs in the flexible manufacturing system are left in the machine tool magazine. Special tools may be removed from the magazine but are not broken down, so tool offsets remain viable. Now when the part is rerun, none of the holding devices, tools, or programs are changed and there is no need for the part to be inspected before production parts are machined. It is always recommended that the customer purchase the machine with as big a tool magazine as possible to keep all the tools in the machine all the time along with backup tools.

The very first time the job is run in the system, the part program and tools will have to be debugged and the part inspected and approved. Now, the machinist is finished at the system. Fixtures and tools are in the system and are not disturbed. By following this philosophy there's no setup needed when scheduling the job again. And because there are no setups being made, the machinist can spend his time working on other equipment that needs his skills. Loading of the system fixtures is left to a less skilled individual.

As orders arrive for the different jobs, production control sends a manufacturing schedule to the flexible manufacturing system cell controller. As jobs are completed, new jobs are automatically started. Following the commands of the cell controller, the AGV moves the pallet to the setup station—where material is loaded—and then to a machine for production. At the end of the machine cycle, the pallet is moved back to the setup station where the part is removed and new material loaded. It is common to design fixtures to hold several parts to extend the machining time. By installing the flexible manufacturing systems with 25, 30, or more pallets, the operational time of the system can be extended to run for long periods completely unattended or in a lights-out condition. During the day, the operator replaces the finished parts. When he leaves at the end of the day, the machines continue to cut parts. The AGV removes the finished pallets putting them on a pallet stand for storage and places a new pallet on the machine until all the pallets are machined or the operator arrives the next day. If all the jobs are completed during the night, the machines will shut down to save power until the operator shows up.

12.2 SYSTEM COMPONENTS

Let's discuss each of these system components.

12.2.1 Machines

First, let's decide on the machine size. Customers generally make the decision of the machine pallet size and axis travels based on the size and machining operations required by the largest part to be run on that machine (Fig. 12.3).

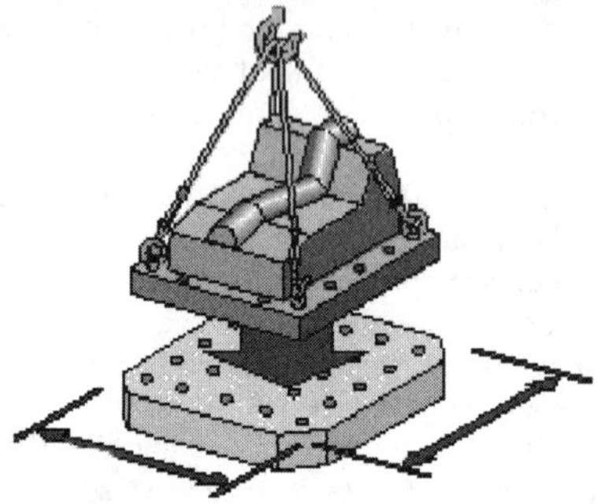

FIGURE 12.3 Pallet size.

That decision may also be tempered by considering the fixture design and the number of parts that could be loaded on a pallet at one time.

Second, determine the size of the tool magazine on the machine (Fig. 12.4).

The optimum size of tool magazine would hold all the tools needed to machine the parts to be run in the flexible manufacturing system *plus* backup tools (a second or third tool) for any tools that may wear out quickly. Purchasing a tool magazine large enough for all the parts to be machined allows any part to be run on any machine in the system without changing tools in the magazine. This prevents production delays, potential mistakes by the person changing the tools in a hurry, and it improves the machine productivity. If the material is abrasive, hard, tough, or subject to impurities that cause tools to wear out quickly, having more than one tool in the magazine enables the programmer to determine the useful life of the more critical tools and switch to the backup tool when the life of the first tool is approached. Again, this keeps the machine's parts running. Without the backup tools, when the tool reaches its useful life, the machine would stop and wait for someone to replace that tool, enter new tool offsets, and restart the machine. Depending on the system layout, the operator may see that the machine has stopped quickly or it may take a long time to become aware that tool life has been reached and the machine has stopped. Magazine capacity of 300 tools or more is available as standard equipment on most machines.

The cell controller can be programmed to route parts only to the machines that have the tools to produce that part. Several situations can warrant this capability, such as when the number of tools exceeds the magazine capacity or there are a large number of expensive or special tools. This method solves the tooling problem but may reduce the overall production of the system. Production will suffer when a part mix that requires routing of most of the parts to one machine is scheduled. At times like this, the other machines in the system will run intermittently or not at all. Operators can move duplicates of the needed tools to a second machine and the routing can be changed to utilize the other machines in the system. This procedure does take some time and has to be reversed when the production run is complete.

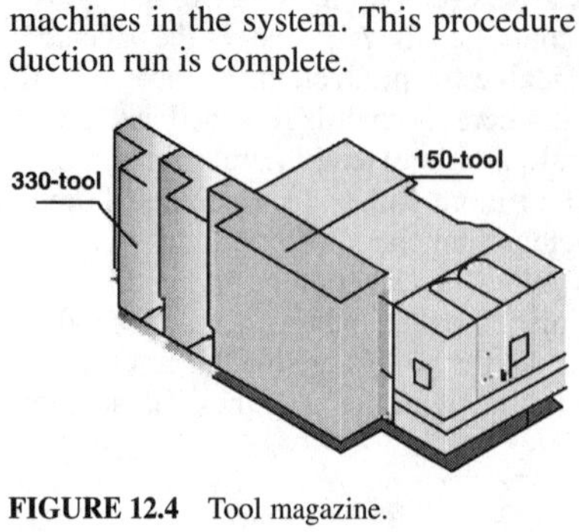

FIGURE 12.4 Tool magazine.

A major disadvantage can be caused by purchasing a machine or using an existing machine with a small tool magazine. A small magazine will allow only a limited number of parts to be run without changing the tool mix in the magazine and it will limit the backup tools needed for high-wear tools to run the system unattended. Using small tool magazines requires tools to be replaced in the magazine as jobs are changed. When tools mixes are changed, the operator has to enter the new tool offsets in the CNC control and here lies a possibility for errors. An operator could reverse a number in reading the tool offset or make a mistake on the keypad that would cause the tool to cut shallow and scrap the part or cause a crash with the part or fixture that can put the machine out of commission for days. To limit this activity, it is always recommended to purchase the largest magazine possible.

12.2.2 Fixtures

Another consideration that should be included in the evaluation of machine capacity is how the part will be processed through the system for the best machine efficiency and productivity. Most parts require machining on two or more faces and generally require two setups to complete the part.

It is common practice in manufacturing facilities to put a number of parts of the same operation on a pallet to extend the pallet cycle time and have the machine use the same tools to finish parts at several locations around the pallet (Fig. 12.5). Let's take a look at an example.

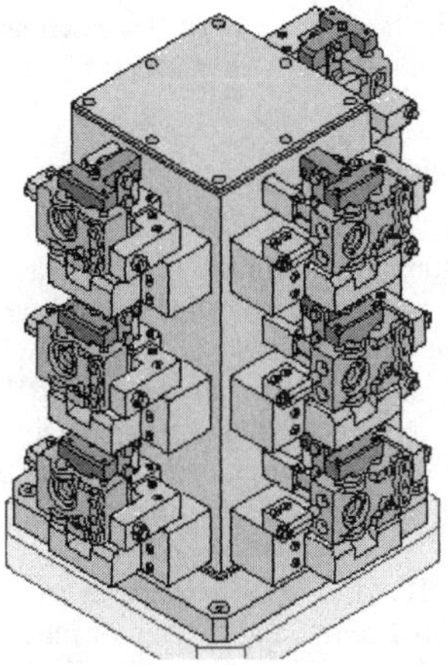

FIGURE 12.5 Part fixturing.

Assume there is a part to be generated that requires three operations to complete, and each of the fixtures designed holds 12 parts. When the job is scheduled to run, the fixture is moved to the setup station where the operator would load 12 parts for operation 10 in the fixture. That pallet is moved to the designated machine with the tools needed to machine operation 10 and the 12 parts are machined. The pallet is then returned to the setup station where the 12 parts are removed, placed on a holding table, and new material is loaded into the fixture. When the operation 20 fixture is moved to the setup station, the operator would remove the 12 finished parts from the fixture and place them on the holding table next to the 12 operation 10 parts. After the fixture is cleaned, the fixture would be reloaded with the 12 operation 10 parts completed earlier. When the operation 30 fixture arrives at the setup table, the operator would unload the 12 completed operation 30 parts from the fixture and place them in the completed parts location. The fixture is cleaned and reloaded with the 12 operation 20 parts completed earlier. The operator will always have some quantity of semifinished material on the holding table. The throughput time for the part (the time from the start of machining to the completion of the first part) will be fairly long because there is the machining time for 12 parts to be run through three fixtures plus the waiting time between operations 10 and 20 plus the waiting time between operations 20 and 30.

A different approach is to design the fixture with all the operations for the part on the same fixture (Fig. 12.6).

FIGURE 12.6 Design the fixture with all operations for the part on the same fixture.

If the part has three operations, design the fixture with three parts of operation 10, three parts of operation 20 and three parts of operation 30. This essentially eliminates in-process material and the storage of semifinished parts waiting for the next fixture to arrive. Each time the pallet returns to the setup station, the operator will remove the three finished operation 30 parts, clean the station, take off the three completed operation 20 parts, and move them to the operation 30 position. Now, clean the operation 20 location and take the three completed operation 10 parts on the fixture and move them to the operation 20 position. Clean the operation 10 location and load three new pieces of material in the operation 10 position. Each cycle of the pallet through a machine yields at least one finished part. In-process material consists of only the completed operation 10 and operation 20 parts on the pallet and the throughput time will be the cycle time to machine one part plus any queuing time in the system. This amounts to at least a 75 percent reduction in the throughput time for the first part over the earlier process.

By listing the parts to be machined on the flexible manufacturing system, the anticipated annual production of each, and the cycle time of each, simple mathematics will tell you how many hours will be required per year to generate the production of these parts. Next, divide this total time by the long-term efficiency expected from the flexible manufacturing system. Long-term efficiency can run from 75 percent to as high as 90 percent. It is not possible to run the system for an extended period at 100 percent efficiency. There will be some tooling problems, personnel problems, lack of material, broken tools, and required maintenance that will reduce the system operational time. In many facilities, there are labor contracts that limit the time the system can run and also reduce the operational time. Now, divide the adjusted total time needed for production by the number of hours (or minutes) run per year to get an estimate of the number of machines needed in the flexible manufacturing system.

Flexible manufacturing systems can vary from one machine to as high as eight machines. The arrangement of the machines will generally be determined by the space available in the plant, but it is common to position the machines in a line with the track in front of the machines.

12.2.3 Track

Track design varies from manufacturer to manufacturer based on the materials available, the speed of the AGV, cost, and a number of other factors. Track design can range from forged railroad rail to round bar stock with ball bushings to flat ground bar stock. One design has a single rail in the center of the AGV on the floor and a second rail directly above the first above the AGV (Fig. 12.7). This design is generally used in multilevel configurations. No matter the physical design of the track, the track is straight (linear). The track will generally utilize a system of cross ties that hold the rails parallel and enable the rail to be tied to the floor and leveled both across the rails and down the length of the rails.

Leveling is necessary to minimize the pitch and roll of the AGV as it moves from position to position and lets the builder control the elevation of a pallet relative to a stand or machine.

Track is generally made in sections to simplify manufacturing, shipping, and installation. Flexible manufacturing systems are sold in a multitude of configurations and sizes. By building the track in standard sizes, a larger system only requires that additional sections of track be installed. It also makes expansion of the flexible manufacturing system at some later date easier. Adding another machine would only require installing a few sections of track to allow the AGV to reach the machine. Shipping very long sections of track is expensive and difficult. With standard track sections that problem is eliminated. Handling the track during installation—to maneuver it into position and level it on the factory floor—is easier and faster with shorter standard track sections. Generally, the track will have the ends of the rail cut square to the rail or mitered to make adding additional sections simple.

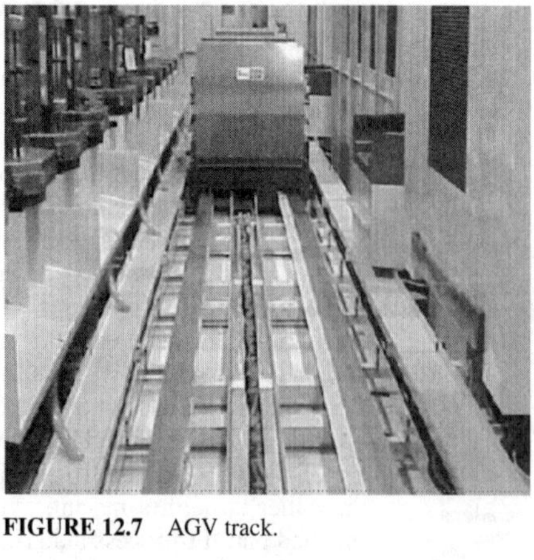

FIGURE 12.7 AGV track.

At the ends of the track, it is common to find heavy-duty safety bumpers capable of stopping the AGV if it were to move past the normal working area of the track. The bumpers are the last resort to stop the AGV. Software limits typically keep the AGV within travel limits, but usually there are also electrical limit switches below the track. These safety bumpers can be made from steel or rubber blocks. The track of a system normally would

FIGURE 12.8 Cable between AGV track.

extend beyond the machines, pallet stands, or setup stations to accommodate the ends of the AGV. Pallets are generally held in the center of the AGV vehicle and are moved perpendicular to the track centerline. When the AGV has stopped to transfer a pallet to the end module, the portion of the AGV that extends beyond the module has to be supported and contained.

Power to the AGV and feedback from the AGV are also part of the track design. Power is supplied in several ways: one, using open power strips along the side of the track with a slider mounted on the AGV (similar to an overhead crane installation); second, using cables running down the center of the track to the AGV; and third, using power cables hanging from rollers and an elevated cable at the side of the track connected to a mast on the AGV. Positioning instructions and feedback of the AGV position are also necessary to control the movement of pallets without crashes or damage. Instructions are given to the AGV and feedback is received from the AGVs using the power strips using an infrared beam between the AGV and a transceiver at the end of the track. Instructions and feedback on units using power cables are normally done using standard wire cables bundled with the power cables running down the center of the track or along the side the track (Fig. 12.8).

12.2.4 Automatic Guided Vehicle (AGV)

Automatic guided vehicles (AGVs) come in a variety of sizes, shapes, and configurations. The AGV is the heart of the pallet pool. Its basic function is to move pallets with fixtures and material around inside the pallet pool as directed by the cell controller. The most common AGV is a single level, four-wheeled vehicle driven by an electric servo motor that engages a gear rack running the length of the track. As the size of pallet pools has grown and the number of pallets contained in them has increased, space to store the pallets has become a problem. One solution is to use multiple level storage racks. Because it is necessary to service the pallets, the AGV configuration has also changed to allow it to reach two or three levels high. At first, this may not seem to be a difficult design concern; however, the AGV has to have the capability to lift a machine pallet with the full weight capacity of the machine on that pallet and place it on a rack 9 to 10 ft in the air and 3 to 4 ft to the side of the AGV centerline within less than 0.030 in. All this has to take place while moving this pallet at the programmed speed of the unit—both horizontally and vertically—repeatedly 24 h a day, 7 days a week for years without a failure.

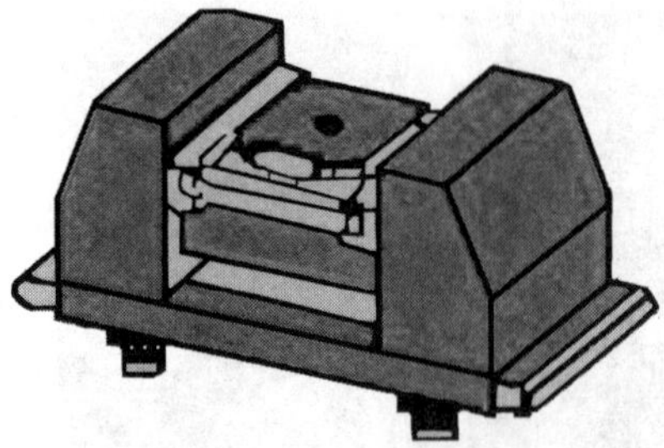

FIGURE 12.9 AGV.

Every manufacturer has a slightly different design for their AGV (Fig. 12.9).

First, the AGV has to have the capability to move down a track of some type with a pallet holding the maximum weight capacity of the machine and the maximum working part size of the machine at a given acceleration and maximum velocity. This AGV must be able to decelerate and stop at a position within some reasonable tolerance that will allow it to place a pallet on a machine table without damaging the locating mechanism. The AGV must have the ability to travel distances from 10 ft or less to 150 ft or more based on the design of the system; if the customer expands the system, the AGV should easily accommodate that modification. One of the designs used to drive the AGV is a rack-and-pinion system located under the AGV. The rack is bolted to the inside of the track and may have a protective cover to minimize contamination from chips, with the pinion on the AGV driven by a servo motor through a gear reduction. Expansion of the track only requires that the cables to the AGV be extended for the additional travel. Sections of the rack on the expansion track would match the current drive rack.

Second, the AGV must have a method of moving the pallet vertically. On a single level system, this vertical travel is only a few inches and is necessary to lift the pallet off the machine table locating-and-clamping mechanism and place the pallet on a stationary stand in a known position. On a multilevel system, this vertical travel can become significant, extending up to 10 ft or more. The lifting mechanism needs a positive drive such as a ball screw that can generate a positive feedback for vertical position, sustain heavy lifting forces, and work trouble free for extended periods. On single-level AGVs the lifting methods used vary greatly. There are straight vertical lifts with the ball screw on one side, a scissor linkage driven by a ball screw, and a rocking action that pivots in the center of the AGV and raises the end of the cross arms on either side of the AGV and others.

Third, the AGV must have the ability to move the pallet from side to side, perpendicular to the track. This enables the AGV to reach out and place the cross arms under the pallet on a machine, lift the pallet, and retract the cross arm with the pallet to the AGV. Then, at that same location or a different location, lift the pallet, extend the cross arm to the same or opposite side of the AGV, and place the pallet on another station. As before, the drive for this motion must be positive and allow feedback to the control so that it can place the pallet on the station accurately. The motion of the cross arms can be driven in a variety of ways. Two of the more common are using a sprocket and chain or a rack and pinion. The cross arms are spaced for the different pallet sizes to allow them to pass under the pallet and contact the outer surface to lift it. Generally, some type of locator is used on the cross arms that will locate in a mating hole in the bottom of the pallet and assure that the position of the pallet is maintained during the move.

When a pallet is removed from the machine, it is common for coolant and chips to have collected on the parts, on the fixture, and on the pallet. To prevent these coolants and chips from contaminating the system area causing slippery, dangerous conditions, the AGV is designed with a sheet metal pan under the lifting arms that extends out to the maximum size of the part that could be machined in the equipment. This pan is designed to collect any chips and coolant that drip from the pallet and direct them to one side of the AGV, where they run into a trough mounted on the front of the pallet stands. This system has proven to keep the installation clean and dry for extended periods.

Wheels on the AGV can vary from builder to builder. AGV's that run on forged rail generally use a solid steel wheel with a rim for guidance down the track. The more advanced design using a ground steel surface for the track can use polyurethane wheels with wipers to reduce noise and vibration. Wipers are used in front of these wheels to keep any contaminants from embedding themselves in the wheel surface and extend their life. Tracking or steering of the AGV to minimize any wear on the rack and pinion drive is accomplished by putting cam rollers on the outside edges of the track near each of the AGV wheels.

When the AGV is moving heavy pallets from one side of the track to the other, the AGV must have a method of counterbalancing this off-center load. Without some way of keeping the AGV solidly on the track, the AGV would rock or tip as the pallet is transferred to the station. On AGVs using forged rail, a roller is angled against the fillet on the underside of the forged rail at each wheel. When the pallet is transferred, the roller restricts the tipping of the AGV. On units with a flat track, the thickness of the rail is more controlled so a cam roller can be positioned on the underside of the track at each wheel to prevent any lifting during the pallet transfer.

Another design of AGV uses a single track on the floor and a single track at the top of the pallet stack. The AGV consists of two columns located between the upper and lower rails with the drives and mechanism for driving the AGV at the base of the column. A carriage moves up and down between the columns and carries the lifting arms that extend to either side to pick up the pallets. Expansion of the track is simple and designing the unit for multiple levels was the original focus. These systems were adapted for machine tool use from the warehouse palletizing industry.

12.2.5 Setup Station

Setup stations are the doorways to the working area of the system (Fig. 12.10).

This is where you remove or insert pallets or fixtures, load material, or remove finished parts. By using the setup station as the gateway, information can be loaded to the cell controller as the exchanges are taking place to keep the database updated. Setup stations can be located anywhere in the system that is convenient for the operator. Because there will be a substantial amount of both raw material and finished parts moving into and out of the system, it would be beneficial to have some storage space around the setup stations.

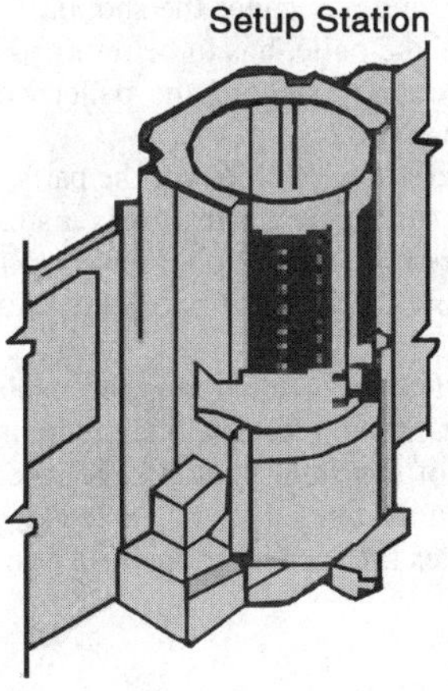

FIGURE 12.10 Part setup station.

It is common for a system to have more than one setup station. The number needed depends on a great variety of factors and can be determined after some detailed evaluation of the production requirements, fixturing, lot sizes, cycle times, and the work force. A system that consists of one machine and a few pallet stands would be a candidate for a single setup station. When the quantity of machines reaches two or more, it would be time to invest in a second setup station. By adding a second setup station the AGV can deliver a pallet to the second setup station when there is a pallet in the first setup station. When the operator completes the loading of the first pallet and the AGV has not removed it from the first setup station, the operator can be working on the second pallet in the other setup station. If he is working on modifying a fixture, he can still load parts to keep the machines operating. As the average cycle times get shorter or the loading time of the fixtures increases, more setup stations are required. What is the limit? That's a difficult question. Generally, it requires running a computer model to generate the data to help make that decision. In some cases, the decision may be based on experience. If in doubt, it is better to have more setup stations than too few. If it is found that the setup stations are the bottleneck of the system production, the flexible design of the pallet pool should allow additional stations to be added easily to eliminate the problem. System software can accommodate up to 10 setup stations.

Setup stations are designed to make the loading and unloading of parts as convenient as possible while protecting the operator from moving pallets, potential AGV mistaken commands, or debris from an operator on an adjacent setup station (Fig. 12.11).

Sliding doors and permanent guards accomplish this. There should be a door between the setup station and the AGV to prevent chips and coolant from being blown onto the track or AGV while cleaning fixtures and parts, and it will protect the operator from any accidental delivery of a pallet to the same setup station. There should be an interlocked door on the front of the setup station to protect an operator while the pallet is being delivered to or being taken from the setup station. The typical interlock is to have the front door locked when the door to the AGV is open, and the

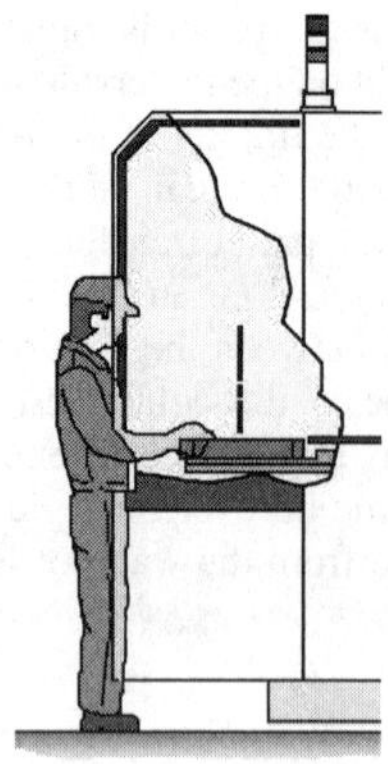

FIGURE 12.11 Setup station and operator ergonomics.

door to the AGV locked when the front door is open. They cannot be open at the same time. Permanent guarding around the setup station keeps the chips washed or blown off the parts and fixtures inside the setup station. Without the guarding, debris can contaminate the AGV track, injure people working around the setup station, make the work area around the setup station slippery and unsafe, and contribute to any number of other problems.

Inside the setup station, the stand that holds the pallet should have the ability to rotate freely. Generally, the pallet is placed in the setup station at the 0 degree position and it must be returned to that position before the AGV retrieves it from the setup station. A shot pin or some other method of restraint must hold the setup station stand at the 0°, 90°, 180°, and 270° positions. The shot pin enables the operator to work on the fixture without the stand rotating. In a typical pallet movement, the operator steps on a foot switch to release the shot pin, rotates the pallet to the next side, and reengages the shot pin by stepping on the foot switch again. At the end of the loading operation, the pallet has to be returned to the 0° position or the cycle start button will not actuate. This is necessary since if the pallet were delivered to a machine 180° out of position, a major machine crash could take place.

An option on the setup station is a *coolant gun* used to wash chips off the parts, fixtures, and pallets. The gun is similar to a garden hose pistol grip spray nozzle. It directs a stream of coolant to flush the chips into the base of the setup station where it goes through a screen to take out the chips and the coolant is recirculated. An air gun can also blow chips off the parts and fixtures into the setup station chip collection bin.

Also available on setup stations are built-in hydraulics for clamping and unclamping hydraulically-actuated fixtures. The hydraulic units are self-contained stand-alone pump and motor systems that have automatic connections on the bottom side of the pallet that engage when the pallet is set on the stand. The operator must actuate the clamp or unclamp valves when he is ready to change the parts. This type of system is requested by many higher production shops that want to minimize the loading time for a pallet.

12.2.6 Pallet Stands

Pallet stands are used to hold pallets in the system (Fig. 12.12).

The concept of a flexible manufacturing system is to take the recurring jobs in your facility, set up the fixtures that hold the jobs on pallets, indicate them in, load the tools, and debug the program. Now keep those pallets in the flexible manufacturing system for use whenever that part is needed. If the cutting tools for these jobs are kept in the machine tool magazine and are not taken apart after each use, the fixtures are not disassembled, and the program is not changed, then the part doesn't require first article inspection. Nothing regarding the manufacturing of the part has changed since the last time the part was cut a few days ago or a few weeks ago. When the manufacturing schedule calls for the part to be made, the part is entered in the cell controller schedule, the pallet arrives at the setup station where material is loaded, and the pallet is moved to a machine. Some time later an acceptable part is delivered to the setup station.

Based on the number of machines in the flexible manufacturing system, the number of different parts that are run in the system, and the volume of the parts run, the system is built with a number of pallet stands to hold the pallets. The number of stands in a system can vary from as few as six to as many as a 100 and can be made single level, two level, or even three levels high to conserve floor space.

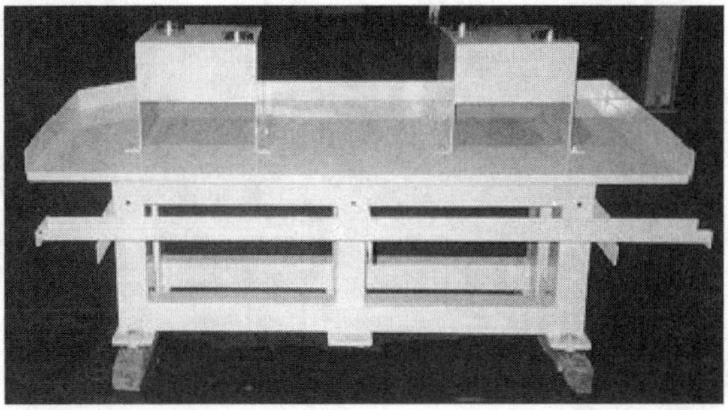

FIGURE 12.12 Pallet stand.

Each stand is made to hold the pallet weight plus the maximum weight capacity of the machine the pallet goes on. If the machine capacity is 2200 lb and the pallet weighs 500 lb, the stands would be designed to hold at least 2700 lb. The spacing of the stand in the system from side to side and the distance to the center of the track is based on the sum of two distances: the maximum part size that can be accommodated on the machines in the system and a small comfort zone. The comfort zone ensures that there are no collisions if two pallets with the maximum part sizes are on adjacent stands or one pallet is moved past the other on the AGV. For multiple-level stands, the vertical spacing of the stands has to accommodate the tallest part that the machines can handle plus the lifting distance of the AGV and a small comfort zone to prevent accidents.

At the base of the pallet stand, there should be a flange on each leg that enables the stand to be leveled relative to the AGV and track and allows anchoring to the floor to prevent movement and misalignment.

Under the pallet support pads at the top of the stand should be a sheet metal drip pan. The drip pan slopes toward the track and will catch any coolant or chips that fall from the parts, fixtures or pallets as they sit on the stand, and then drain that coolant into a collection trough mounted on the front of the stand. This trough is sloped from the center to each end where the coolant can be drained off periodically. On multilevel stands, a collection tube is mounted from the front of the drip pan down to the trough at the bottom level.

12.2.7 Guarding

A flexible manufacturing system is an automated flexible manufacturing system that has the potential to run completely unattended for extended periods of time in a lights-out situation. Even during the normal operation, there are one or two people at the setup stations loading material and there may be another person that replaces tooling as it wears. People unfamiliar with this type of operation may inadvertently move into areas of the system that can injure them. Therefore, it is absolutely mandatory that the system be completely surrounded by some type of guarding that will prevent anyone from coming in contact with the automation (Fig. 12.13).

Guarding of the machines has been done to prevent chips from being thrown from the work area and injuring someone, from coolant being splashed outside the machine table causing a slippery floor, from a tool change taking place when someone may be close to the exchange area, from movement of the machine parts, from pinch points and the like. Personnel safety is important to everyone. When a robot is installed in a facility, guarding is installed at the extremes of the robot's travels so that when it is doing its work, no one can be within that space. The robot has been programmed to take specific paths to accomplish the task at hand and those movements are generally very fast. The force of the moving robot arm could severely injure or kill someone if struck.

FIGURE 12.13 Equipment guarding.

The pallet pool AGV is nothing but a robot on wheels. It is confined to a track, but the pallets and the mass of the AGV are extremely high. At the speeds the AGV moves, if someone were accidentally on or near the track, he/she could be severely injured. The pallets and their workload can many times exceed 1000 lb and the work on the pallet can often extend beyond the actual pallet surface. A person can be standing next to the track and be struck by this workpiece. When the pallet is being transferred on or off the pallet stand, there is little noise generated by the AGV and a person standing next to the pallet stands can be hit by the moving pallet.

The end result of all this discussion is that a flexible manufacturing system is a dangerous place for a person during its operation and it must be guarded. Some companies take a very conservative approach to designing the guarding as you can see in the picture above. It is intended to protect personnel even if some catastrophic accident were to take place. Steel panels are heavy enough to prevent even a pallet from penetrating it. The panels are anchored to the floor with brackets, stabilized at the top with angle iron brackets running from one side to the other side, and they are bolted on the sides to the adjacent panels (Fig. 12.14).

Others take a passive approach and add simple welded wire panels around the periphery as an afterthought.

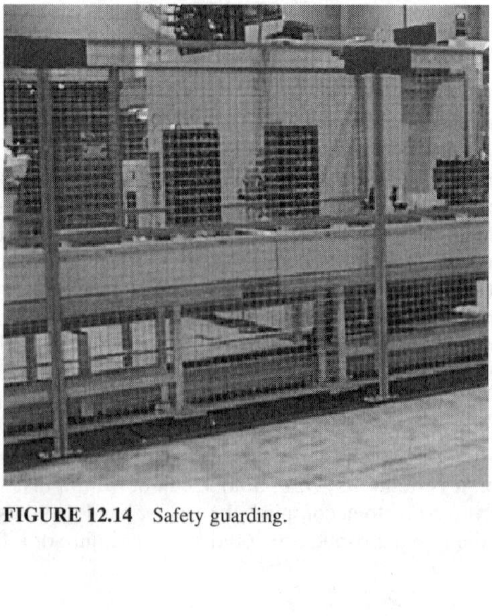

FIGURE 12.14 Safety guarding.

12.2.8 Cell Controller

A flexible manufacturing system is a sophisticated flexible manufacturing system that has the capability to machine several different parts in a prescribed manner without an operator at the machine to make adjustments and corrections. To accomplish the control of the system requires some type of master control system or cell controller.

Builders have generally settled on stand-alone PC based controllers to direct the operation of their systems running custom software designed and written by the builder or one of his exclusive suppliers. Most cell controllers will use standard database software and any number of other readily available programs to support the main cell controller program. There is little or no compatibility between the many different cell controller systems. Each cell controller is designed and built to interface with the builder's equipment only. There has been an effort by a few material handling companies to become the common thread between the builders. These material handling firms are trying to interface both the flexible manufacturing system equipment and the cell controller to a number of different builder's equipment. As they expand the number of builders using their equipment, it will be possible to incorporate different suppliers into the same flexible manufacturing system.

FIGURE 12.15 Cell controller.

Cell controllers are generally packaged as stand-alone units (Fig. 12.15).

The PC, monitor, interruptible power supply, and keyboard/mouse are housed in a NEMA 12 electrical enclosure that can be located wherever the user feels is appropriate. Since there are generally only one or two people working in the system, the cell controller is close to the setup stations. This location enables the operator loading parts to monitor the system operation, check the production schedule, and be alerted to any alarms that may occur as the system is running.

Cell controllers handle a number of functions:

- Show the movement and status of pallets, machines, setup stations and AGV in the flexible manufacturing system on a real-time basis (Fig. 12.16).

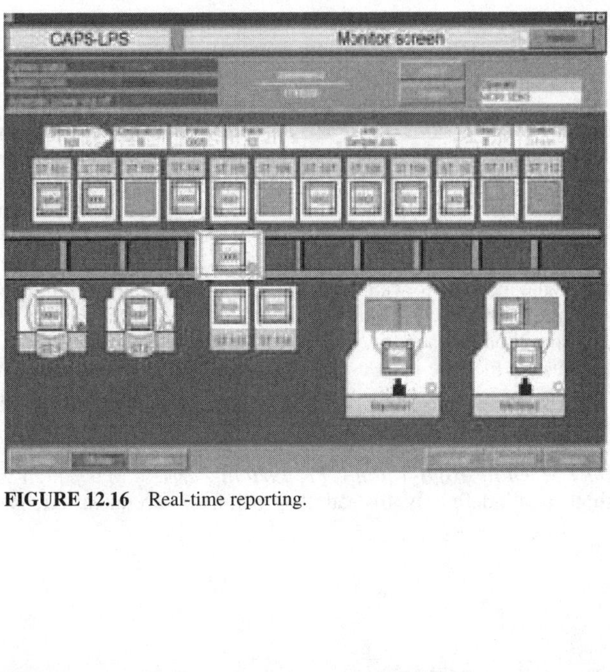

FIGURE 12.16 Real-time reporting.

- Store part programs for all the parts that can be run in the pallet pool and download them to the individual machine when the pallet is moved into the working area of the machine. By keeping the programs in a central location, any modification to the programs is easily accomplished since the program is only changed in one location. It was not uncommon at one time for the programs to reside in the machine control; however, when there were multiple machines and a program was changed, the programmer had to go to each machine and reload the modified program. Occasionally the programmer was distracted before he finished loading all the machines and parts were made using the old program.
- Store and execute a production schedule for a number of parts in the system. When a system contains from one to eight machining centers, from six to 100 pallets and from one to eight different parts per pallet, the production sequence of parts in the system can become extremely complex. It is common that the production schedule is a job-based system controlling the manufacturing of individual parts in the system. Job-based scheduling enables a pallet to have different parts on different sides of a tombstone fixture and those parts may have different production quantities. When the low production part quantities are completed, the cell controller will stop production of those parts. This type of scheduling requires the power of a PC to keep up with the needs of the system.
- Monitor the overall operation of the pallet pool operation for extended periods of time including tracking of pallet movement through any system station and generating a time log of all equipment.
- Generate reports on the operation, productivity, and utilization of the system.
- Simplify communication with the factory network using an Ethernet network connection.
- Interface to tool presetting equipment to download tool offsets and tool life information.
- Track cutting tool use, life, and availability so that only parts will be delivered to machines with the necessary tools and with the necessary tool life to machine those parts.
- Automatically notify an individual through email or a cell phone call if alarms occur during the operation of the system.
- Allows management to monitor the real-time operation and production of the system equipment through the Ethernet system.

12.3 BENEFITS OF A FLEXIBLE MANUFACTURING SYSTEM

Companies considering a flexible manufacturing system should look at their operation and evaluate the benefits a flexible manufacturing system can offer. Here are questions an owner or manager should ask.

Do I Machine a Number of Parts in Small Quantities? The flexible manufacturing system (FMS) system stores the fixtures and tools needed for frequently run jobs so that they can be called up whenever the production schedule requires. Because the fixture has not been broken down or changed, and tools have not been modified, first article inspection isn't necessary. If you do change the setup, use the FMS system for setting up the jobs at the setup station while the machines continue running current production. In a stand-alone machine situation, the tear down and setup of a new job generally causes the machine to wait for the operator to finish the setup. For new jobs introduced into the FMS system, the fixture can be set up while the machines run current production. To debug a program and checkout the tools needed, one of the machines is removed from automatic operation and the parts machined. Once the parts are complete and transferred back to the setup station where they can be removed for inspection, the machine is returned to automatic operation and continues to machine the parts in the production schedule.

Do I Machine High-Volume Production Parts? Many high volume manufacturers do not do a single part indefinitely. Instead, they will have several variations of the part or several similar parts.

Changeovers are done every few days or once a week. Rather than having the operator walk between 3, 4, 5 or even 6 machines loading and unloading parts from their fixtures eight hours a day, the fixtures and parts are delivered to the setup stations of the flexible manufacturing system where raw material and finished part bins are located. The operator doesn't become exhausted walking to and from machines carrying parts. Machines are kept loaded with material by the system AGV. The benefits here are generally ergonomic. Machining centers can't make parts as fast as a dial-type machine or transfer line, but the reliability of the machining center and FMS system easily offsets the difference in production. When one machining center stops, the remainder continue generating parts. Does that happen on the dial type or transfer lines? Certainly not! plus, on the FMS system any design changes or part changes can be made instantly using the cell controller to download the new program to the machine CNC controls. Dials and transfer lines were never intended to be flexible or versatile.

Do I Repeat Jobs Frequently? Every Week, Every Month? When jobs are rerun on a regular basis, leave the fixtures and tools in the system. When it is time to rerun the job, simply start the job in the cell controller schedule. Production is scheduled and the fixture is delivered to the setup station where you load new material. Using this procedure saves time setting up the part, and it eliminates the dead time for first article and last article inspection.

Do I Need to Deliver Parts on Short Notice? When the fixtures, tools, and programs are stored in the FMS system, all it takes to produce the part is material and a priority in the schedule. There is even an "express," or emergency condition, that puts your "hot" part at the top of the list for production and sends it to the next available machine.

Do I Make a Series of Parts Before I can Ship a Unit? When there is a series of parts that go into an assembly, the cell controller can be programmed to produce all the involved parts sequentially. Using this method, batch production of the parts is eliminated, in-process inventory is minimized, and assembly receives a continuous supply of material. If you have capacity on the system machines, a mix of standard production can still be run.

Do I Need to Supply Parts on a "Just-In-Time" Basis? Again—assuming you keep the parts, fixtures, and tools in the FMS system because you are running parts on a recurring schedule—you can produce your critically timed parts as you need them with the help of the cell controller.

Do I Have Frequent Setups That Delay Production? Setups, first article inspection, and final inspection can take many times longer than machining the parts. While the machinist is tackling these nonproductive operations, your machines are sitting idle. With an FMS system, the fixtures and tools remain untouched in the system. Since nothing has been changed on the job, you eliminate first article inspection and greatly improve machine utilization. By also using SPC inspections during production, you can track the overall accuracy of the parts. Studies have shown that not only will part production generally double from stand-alone machines to flexible manufacturing system operation, but profitability (machine output per labor hour of input) will increase by as much as five times.

Do I Have Great Operators but Low Spindle Utilization? Your machine operators work very diligently, but while they are setting up jobs and getting them inspected, the machine spindle doesn't do any cutting. It is not the fault of the operator that setups are complex or inspection takes a long time. The way to keep the machines productive is to eliminate the setups and first article inspection from the process of making the parts. With the flexible manufacturing system, repeating jobs stay untouched. Even if there are setups as part of the operation they are done on fixtures that are not being used at the time. While first article inspection is taking place, the machine is scheduled to continue working on other production parts until the new part has been accepted.

Do I Setup and Run Several Months of Production? Management understands production problems. In many cases, once they have invested the time to get fixtures set up, tools assembled and set, and the part machined, inspected and within tolerance, they don't want to stop production after a few hours or days of running. Instead, they run 3, 4, or 5 months of inventory to improve the machine's utilization. They willingly invest that money in parts sitting on the shelf so that they can minimize the changeover expense. As an alternative, they could invest in the FMS system, eliminate the setup and inspection time, run parts as needed, reduce any in-process material and most of the finished inventory, and apply the inventory savings directly to the system justification or the profit of the plant.

Do I Minimize In-Process Inventory? In many facilities, operation 10 is run on a machine, operation 20 on a second machine, operation 30 on a third, and so on. The entire lot of parts for operation 10 is machined, piled up on a skid, then moved to the next machine to have all the parts machined for operation 20, and so on. With an FMS system, you store the fixtures for all the operations in the FMS system. When operation 10 returns to the setup station to be unloaded, an unfinished part is loaded. Soon after, the fixture for operation 20 would arrive.

The finished operation 20 parts are removed and the finished operation 10 parts loaded. In-process inventory is reduced to the few parts on the fixtures in the FMS system. The other parts become finished inventory that can be assembled or shipped as they are completed. It is not necessary to wait for the entire lot to finish on the last machine.

An even more efficient operation is to have as many of the operations on one pallet as possible. When the pallet returns to the setup station, the operator removes the finished part, moves the subsequent operations up one station, and loads a piece of raw material. Now every cycle of the pallet generates at least one finished part, and in-process material are the few semifinished parts on the pallet.

Do I Maximize Your Manufacturing Flexibility? Horizontal machining centers are probably the most versatile machines in your shop. They are capable of machining a great variety of parts. The benefit of the FMS system is that it can store fixtures for different parts and have them available to machine parts needed by your own factory or your customers on an as needed basis and with fast turnaround.

Do I Want Reduced Manufacturing Costs? You can reduce manufacturing costs in a number of ways, some of which are reduction in labor or increased output per machine. You can keep FMS machines loaded with parts using fewer people than stand-alone machines. In many cases one man can easily keep 3 or 4 machines in an FMS system loaded with parts. Second, you can use personnel with a lower skill level to load parts in the FMS system. Once the part is running in the FMS system, parts only need to be loaded and unloaded, not setup and run from scratch each time. Third, machines in an FMS system run 20 to 50 percent more parts than stand-alone machines and so they produce more parts per day resulting in a lower cost per part. Reduced in-process inventory and finished inventory further lower the cost of manufacturing. Plus, reducing finished inventory reduces the taxes the company pays on finished inventory.

Do I Want Improved Ergonomics and Reduced Operator Fatigue? Because parts are loaded and unloaded in an FMS system at one or more setup stations, you can economically build platforms around the setup stations. This positions the material at proper ergonomic levels for the operators. With stand-alone machines, building platforms around all the machines can become very expensive since most ergonomic standards do not allow the operators to go up and down the stairs when moving from machine to machine.

Do I Want Simplified Material Movement? You can move material to and from the FMS much easier than to individual machines on the shop floor since the handling takes place at a central location rather than at each machine. Secondly, because the FMS performs all the part operations without the material moving from one location to another, you eliminate the movement of skids or bins of in-process material among machines.

Do I Have the Ability to Expand Easily? Buy the machines needed for production today and expand the FMS as production grows. With the modular design, you can add machines, setup stations, or pallet stands at will.

12.4 OPERATIONAL CONSIDERATIONS

12.4.1 Production and Productivity Improvements

Let's assume that you have looked at your operation and it is imperative that you improve the productivity of the facility. Skilled labor is becoming hard to find and even harder to keep. Costs of insurance and overhead are growing faster than your profits. Equipment in the factory is aging and it will be necessary to start replacing some of the machines to take advantage of the higher speeds, new cutting tool technology, and improved machine accuracy. Marketing has evaluated the marketplace and has initiated several new products that will be reaching production in the next 6 months to a year. The customers they have contacted want shorter delivery periods to minimize their inventory levels even though their order quantities may fluctuate from month to month during the year.

After some extensive evaluation of these requirements, it has been decided that running small lots on the vertical machining centers will reduce the spindle utilization to the range of 25 to 30 percent. Operators will spend most of their time setting up. Horizontal machining centers will improve the productivity because the load time for parts is eliminated with a pallet shuttle and some setup time can be done while the second pallet is being machined. Even with these changes, the best that we could realistically expect with the horizontal machining centers is around 40 to 45 percent spindle utilization. That's a 50 percent increase over the vertical machining centers. After further discussions with the machine tool people and visiting a few users, installing a flexible manufacturing system onto the horizontal machining centers would be expected to boost the horizontal machining center utilization up to the 85 percent level.

But, how can the additional expense of the flexible manufacturing system be justified?

First, if a flexible manufacturing system is installed, production output or the number of parts generated by the flexible manufacturing system will generate nearly *double* the number of parts that can be generated using stand-alone horizontal machining centers. That is a significant benefit by itself.

Second, using the benefits above, setup time is eliminated, first article inspection is eliminated, changeovers time is gone, parts can be manufactured to meet the customer's needs, and we can run the system with fewer operators instead of one machinist per machine.

What happens to the productivity or the production per man-hour of the machinist? Let's make a few assumptions. Assume the company works 16 h a day, 5 days a week and based on the anticipated production, it will take two horizontal machining centers to make the production. With smaller lot sizes and potential new customers the stand alone machines would be expected to do two setups per day on the average.

Here's what would happen:

- *Number of people needed.* Stand-alone machines would require one machinist per shift for each machine. Once the flexible manufacturing system has been set up, it can be run with one less experienced (and less expensive) person per shift.
- *Machine efficiency.* This was discussed earlier in this section.
- *Total machine hours/day available.* This is the time of 16 h worked per day multiplied by the number of machines. 16 h per day × 2 machines = 32 h/day. This is the same for either the stand-alone machines or the flexible manufacturing system.
- *Machine output or Hours run/day.* Output is determined by how much the machine runs per day. Since the stand-alone machines had an efficiency of 40 percent, the output is 32 h/day times 40 percent = 12.8 h/day. The flexible manufacturing system has an efficiency of 85 percent, so the output is 32 h/day times 85 percent = 27.2 h/day.

TABLE 12.1 Productivity of Stand-Alone System vs. Flexible Manufacturing System

	2 Stand-alone machines	2 Machines in linear pallet pool system
Number of people needed	4 machinists/day (1/mach/shift)	2 operators (1/shift) (Not necessarily machinists)
Machine operating efficiency	40%	85%
Total machine hours/day available	32	32
Machine output (h/day)	(32 × 40%) = 12.8 h/day	(32 × 85%) = 27.2 h/day
Labor input (man-hours/day)	(4 × 8) = 32 man-hours/day	(2 × 8) = 16 man-hours/day
Productivity-machine output/labor input (h/day)	12.8 h/day/32 man hours/day = 0.4	27.2 h/day/16 man hours/day = 1.7
Productivity comparison	100%	425% greater

- *Labor Input to run the machines (man-hours/day).* Stand-alone machines need one man per shift multiplied by the number of machines working 16 h per day. 4 men times 8 h/day = 32 man-hours/day. In comparison, the flexible manufacturing system uses 1 man per shift for a total of 2 men, each working 8 h/day = 16 man-hours/day.
- *Productivity of the machines.* It is the machine output per day (hours/day) from the chart above for the labor input that it takes to run the machines (hours/day). Stand-alone machine output is 12.8 h/day divided by the labor input of 32 man-hours/day = 0.4 units. The flexible manufacturing system output is 27.2 hours/day divided by the labor input of 16 man-hours/day = 1.7 units.

Comparing the productivity of the stand-alone machines to the flexible manufacturing system shows that the flexible manufacturing system generates 425 percent more product per hour of work than the same number of stand-alone machines (Table 12.1). More products are being generated at a lower cost!!

12.4.2 System Layout

Next, let's think about the configuration of the system. Should the system be located in the center of the shop or near one of the outside walls? Let's start with the machines. The machines need work room around them for regular maintenance and at some time in the future, for major maintenance. Tooling has to be loaded and unloaded in the tool magazine and chips must be removed from the rear of the machine along with the coolant. And, the machinists and tooling man need access to the machine control. Machines can't be too close together so that any doors that are opened will restrict access to the adjacent machine. If we install the machines close to the outside wall of the building, an aisle will have to be left behind the machines for forklift access. If they are installed in the interior of the building the rear of the machine should be close to an aisle but also permit forklift access.

Obviously, the track and the AGV are positioned in front of the machine.

Pallet stands can be positioned between the machines or on the opposite side of the track. If there are a large number of stands and a few machines, putting all the pallet stands on one side will make the system very long, but the machine side will not be utilized. Putting pallet stands on both sides will shorten the track. An alternative when there is a large number of pallet stands is to look at a multiple-level pallet stand arrangement to keep the layout compact (this requires a more expensive AGV capable of servicing multiple levels). When the manufacturer is generating the system layouts, it is best to give them the layout of the area in your plant where you would like the system to be installed. The drawing must show the column locations, existing foundation breaks, and any other obstructions the system must take into account. Many times machines or stands have to be relocated to avoid building supports.

Next is the location of the setup stations. The setup station is where the operator will spend his time, where material is loaded and unloaded from pallets, where partially finished parts will have to be stored until the needed pallet arrives to be loaded, and where fixtures may be built up or modified. Location of the setup station is critical to the operation of the system.

12.4.3 Material Movement

It is best if the setup stations and material storage are accessible by forklifts (Fig. 12.17).

When running a number of parts in the system, bins or skids of raw material and finished pieces for each of the parts being run must be relatively close to the setup station to eliminate long walks by the operator to access the material. This is a fatigue factor and it makes loading of fixtures very slow. With this in mind, locating the setup stations between machines would make the storage of material and parts close to the setup station difficult. Locating the setup stations adjacent to the machines on the same side of the track is better, but one side of the area is still taken by the machines. Positioning the setup stations on the opposite side of the track next to the pallet stands—as shown here—is probably the best location for access. Don't put this area adjacent to an aisle because the material cannot be placed close to the loading station without blocking the aisle.

FIGURE 12.17 Material storage and setup station accessibility.

For production of a limited number of heavy, larger parts being machined on a random basis, one method may be to consider putting each of the components on a gravity fed conveyor in front of the setup stations. This is one way of making the parts available. Look at the time it will take to use a hoist or overhead crane to move the finished parts from the setup station and replace it with an unmachined casting. If the operator is using a crane for an extended period, how will a second person unload and reload the adjacent setup station? Can one man load the necessary parts for the system? Should the setup stations be located under different overhead cranes? Should they be positioned so each setup station would have individual jib cranes that will not interfere with the other's operation? Do parts need to be loaded on either of the setup stations for versatility or will the production be balanced enough to allocate a list of parts to individual setup stations? Where do finished parts go? Do they move out of the area on a conveyor to inspection, cleaning, deburr, or other department? Do they get placed on skids to be taken to a different area of the plant? You get the idea.

When the flexible manufacturing system is in a high production facility, the location of the setup stations is even more critical. Here the cycle time of a pallet is generally short and the AGV is extremely busy. Load times of the pallet are fast, and there is hydraulic fixturing and automatic doors on the setup station to eliminate having the operator spend the time to open the doors. There will generally be two setup stations so that a pallet can be delivered or picked up during the loading of the adjacent fixture. Material movement through the setup station is rapid and consists of large quantities, so raw material must be located close to the setup station. In many cases, the parts are in bins that are raised and tilted for the convenience of the operator. Finished parts may be placed back into a second bin or placed on a conveyor to move to a secondary operation or assembly. Empty bins have to have a way of being moved out of the work area easily and without a lot of operator effort. They can be moved horizontally or vertically with some type of powered material handling. In some installations, the system is close to the receiving dock to enhance the supply of material.

In either case, the movement of material to and from the flexible manufacturing system should be seriously evaluated. If the system is setup in a difficult location in the facility, system utilization and productivity will suffer waiting for material to be loaded. Correcting the problem will be difficult and expensive.

12.4.4 Secondary Operations

Operators of the flexible manufacturing system have a condensed work period. When the pallet is in the setup station, it is necessary to unload, clean, and reload the parts to be machined. However, when cycle times of pallets are long, the operator will have time to handle other duties or secondary operations—parts deburring, inspection, drilling of a compound angle hole that would be difficult on the machining centers without another operation, marking the parts, or any of a hundred other things that may have to be done to the part at some time during its manufacturing. Operators may also handle replacement of worn tools—replacing tools that are not used for the current production—assemble replacement tools, or run a tool presetter. Duties may include moving skids of material or parts to make loading easier, sweeping up chips around the work area, or filling oil reservoirs on the machines.

When the productivity of the flexible manufacturing system is reviewed, these are small but beneficial operations that add to the justification of this equipment and utilize the operator's time.

12.5 TRENDS

12.5.1 Scheduling Software in the Cell Controller

When a facility with a flexible manufacturing system is running a mix of production parts to meet the needs of its customers, there are times when the utilization of the system may actually be much lower than expected. Some machines may run only occasionally or not at all for hours or days. But, the cell controller will move pallets to another machine or pick another part if the first part isn't ready to keep the machines cutting.

Yes, all that's true. The cell controller can do all these things *if* the parts have been programmed to be run on multiple machines, and the machines have the capability to run any of the parts. *But* many times parts may be programmed to run on a single machine because the tools needed are special and very expensive. The cost of purchasing duplicates for the other machines in the system would be prohibitive. Sizes of the machine tool magazines may limit the number of tools the machine can hold and the number of tools in a machine can only generate one or two different parts. Or, one machine may have a positioning table where the second has a full 4th-axis contouring table and that limits the machines some parts can be machined on.

However, production control departments generate schedules that are sent to the production floor based on orders received and inventory requirements. Very few actually run a simulation of the flexible manufacturing system to see what effect this mix of parts will have on the machine utilization or if the schedule can be met. When the schedule gets to the shop, the operator enters the jobs and the priority to get them done. The result is that the machines are not coming close to their potential. And in some cases, the parts being run are all to be done on only one machine. All the others are idle. If the volume of parts expected from the flexible manufacturing system is not met, the delivery schedule may not be met, assembly may not get the parts to stay busy, and the like. How is this problem overcome? Production control can enter the part mix into a simulation program that emulates the operation of the flexible manufacturing system and they can try a number of different production mixes until they get one that meets the production schedule and keeps the machines running. Not an easy task and it may take several hours a day to accomplish.

Programmers of cell controller software are developing software that will look at the production schedule and adjust it automatically. Production control will enter the due dates for each of the jobs, the run times of the part, and the sequence of operation in the system. The software will routinely look at the production of parts in the system and adjust the starting time of jobs based on the parts that have been completed and their scheduled completion dates and any downtime on the system's machines. If a current job is falling behind, it will boost the priority of the job to route it to the machines before the other jobs currently running. This software is expected to boost the long-term utilization of systems by at least 10 percent.

12.5.2 Improved Tooling Control

As manufacturing operations use less skilled people to run sophisticated machine-tool systems like a flexible manufacturing system, there are areas of the operation like the tool setting function that, if not done properly, can lead to accidents on the machines that can put them out of commission for days if not weeks.

When a system is running, tools such as drills, taps, mills, and boring tools wear out and require replacement. There are a number of tools that only require an insert to be replaced and no changes in the tool length are required for the tool to be used. Other tools such as drills and taps need to have the entire drill replaced with a sharp tool. In this case, the length of the tool must be measured and that tool length manually inserted into the CNC control. This length is important for drilling or tapping the holes to the correct depth. It is also critical that the tool length be correct so that the machine will position the tool above the part when it is moving into position.

There are a number of places mistakes can be made in this process and these can lead to a catastrophe. If the length is measured incorrectly, crash. If the length is measured correctly but written down incorrectly, crash. If the operator inadvertently reverses two number and inserts the wrong length in the CNC control, crash.

To eliminate the human element in setting tools, there is an option available for the system that will record the measurement from the tool presetter and send it to the cell controller through the Ethernet system where it is recorded in a database. When that tool is put in the machines tool magazine, the tool offset is downloaded to the CNC control. There is no reading or writing done by the people in the system. Even the ID number for the tool that is used in the database is read from an electronic chip imbedded in the tool holder so that even that number is not manually inserted.

In addition to recording the tool offset for the CNC control, the database will track the tool life for that tool. When the tool comes within some set amount of the programmed tool life, a message will alert the operator and the tool room that that tool will need to be replaced soon. It also enables the cell control computer to scan the machine controls and check the remaining tool life to be sure that there is enough tool life remaining in the machine to complete the next part. If tools exceed the programmed tool life, it will look for other machines with tool life available to machine the parts.

12.5.3 Integration of Secondary Operations

The design of the AGV enables it to place a pallet on a stand, machine, or setup station. But at the same time there is no reason the pallet could not be placed on the table of any other type of machine. This could include a welding table so that areas on the part could be welded as part of the operation and then the welded area machined to finish the part. Pallets can be loaded into wash machines where they are cleaned and dried before being delivered to an inspection CMM, a special boring machine, a press to insert bushings, or any special operation desired. These operations do require some interfacing to the cell controller. This permits the pallet to have a delivery destination and the cell controller to get a signal from the unit to know that the operation has been completed so that the AGV will pick the pallet up and deliver it to the next step in the operation.

Practicality is important here. Integrating both the CMM and the washing machines into a flexible manufacturing system is *not* recommended. When there are several machines in a system and the parts are to be washed and inspected as part of the sequence of operations, these two machines become a bottleneck. For example, washing cycles are generally moderately long at 6 to 10 min or more. Assuming that there are three machines in the system with cycle times at 20 min per pallet, there will be a pallet coming to the wash station every $6\frac{1}{2}$ min. Looking at the CMM, it is rare that all parts running are inspected on the CMM. Generally, there is one part out of 20, 50, or 100 parts that is checked. This is a random function that is extremely difficult to program in the cell controller. Second, inspecting a part seldom generates an absolute pass or fail response. Many times there are dimensions that are inspected that are extremely close to being acceptable and when reviewed are passed. With the CMM checking automatically, the cell controller and the operator would get the response of pass or fail. The parts would be taken off the fixture and put into the rejected bin. Unless the parts are serialized, there will be no way to find out if the part could be acceptable.

A better solution is to have the wash station and CMM adjacent to the system where the operator can put the parts through the washer and then load the necessary parts on the CMM for inspection. During the periods that the operator is not using the CMM, it can be used by others in the inspection department.

12.6 CONCLUSION

Flexible manufacturing systems can greatly improve the efficiency and profitability of any manufacturing facility. They are versatile, efficient, and extremely productive. There are some limitations to their use but a progressive company that meets the conditions can benefit significantly from their application.

BIBLIOGRAPHY

Niebel, B.W., Draper, A.B., and Wysk, R.A.: *Modern Manufacturing Process Engineering*, McGraw-Hill Publishing Company, New York, 1989.

Groover, M.P.: *Automation, Production System, and Computer-Integrated-Manufacturing*, 2d ed., Prentice Hall, Englewood Cliffs, NJ, 2000.

Change, T.C., and Wysk, R.A.: *Computer-Integrated-Manufacturing*, Prentice Hall, Englewood Cliffs, NJ, 1991.

Black, J.T.: *The Design of the Factory with a Future*, McGraw-Hill Publishing Company, New York, 1991.

CHAPTER 13
OPTIMIZATION AND DESIGN FOR SYSTEM RELIABILITY

Way Kuo
University of Tennessee
Knoxville, Tennessee

V. Rajendra Prasad
Texas A&M University
College Station, Texas

Chunghun Ha
Texas A&M University
College Station, Texas

13.1 INTRODUCTION

System reliability is an important factor to be considered in modern system design. The objective of this chapter is to provide a design guide for system engineers by providing an overview of the various types of reliability optimization problems and methodologies that can be used to solve the problems. The reliability optimization problems, which are frequently encountered, include redundancy allocation, reliability-redundancy allocation, cost minimization, and multiobjective optimization problems. All reliability optimization problems can be formulated as a standard form of mathematical programming problem. By employing the various techniques of mathematical programming, solutions to the various reliability optimization problems can be obtained efficiently.

13.1.1 Design and Reliability in Systems

Modern products have a short life cycle due to frequently changing customer demands and rapidly developing new technologies. In addition, increasingly sophisticated customer needs require more types and more complicated systems than before. However, short production times and complex systems result in loss of system quality and reliability, which is a contradiction of the customer needs. To meet these requirements simultaneously, comprehensive reliability analysis must be incorporated at the design stage. Various new technologies and tools have been developed to increase system reliability and simultaneously reduce production costs. The objective of this chapter is to review the whole stream of these technologies and tools and to assure their reliability in preventing possible loss due to the incorporation of complex system configurations. There are important relationships between reliability and concurrent engineering and system design.

The following definition of concurrent engineering is widely accepted: "Concurrent engineering is a systematic approach to the integrated, concurrent design of products and their related processes, including manufacture and support. This approach is intended to cause the developers, from the outset, to consider all elements of the product life cycle from concept through disposal, including quality, cost, schedule, and user requirements."[1]

The concurrent engineering approach faces uncertainty early and directly so as to successfully address the needs of rapid prototyping and design qualifications that guarantee high reliability. Also, through concurrent engineering we are able to evaluate the tradeoffs between reliability and cost.

System design usually includes performing preliminary system feasibility analysis during which one has to define alternative system configurations and technologies. It is also important to predict system reliability and define maintenance policies at the system design stage. Therefore, we must evaluate the technical and economic performance of each alternative solution in order to select the strategic criteria for determining the best performing alternative and then develop an implementation plan for installation.

13.1.2 System Reliability Function Induced by System Configuration

Reliability is the probability that a system performs its intended functions satisfactorily for a given period of time under specified operating conditions. Let T be the random variable that indicates the lifetime of a component. Since reliability is a probability measure, the reliability function of a component $R(t)$, with respect to a mission time, is a real valued function defined as follows:

$$R(t) = P(T > t) = \int_t^{\infty} f(x)dx \tag{13.1}$$

where $f(t)$ is a probability density function of the component distribution.

Computing the reliability of a component depends on the distribution of the component. However, deriving the system reliability function is not as simple as deriving the component reliability because a system is a collection of connected components. The system reliability function critically depends on the system structures. There are four basic system structures: *series, parallel, parallel-series,* and *series-parallel.* A series system is a system where each component is sequentially connected—the system fails if any component fails. A parallel system is connected in parallel and so the system works if any component works. A parallel-series system is a parallel system in which the subsystems are series systems and a series-parallel system is vice versa. The simplest and easiest way to describe the relation of each component in a system is a reliability block diagram in Fig. 13.1. The system reliability functions R_s of each configuration are defined as follows:

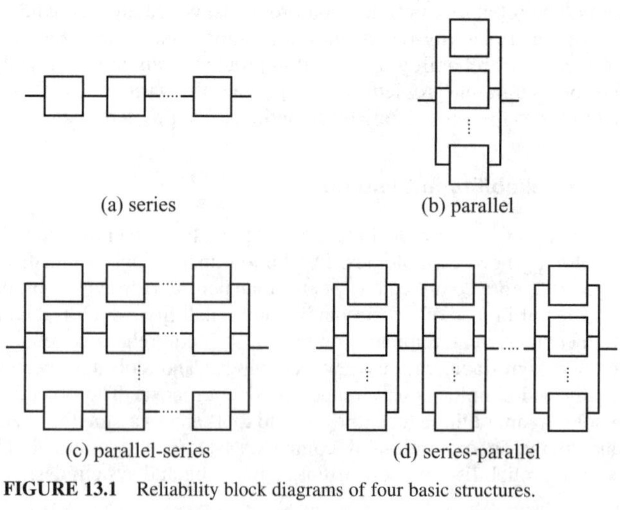

FIGURE 13.1 Reliability block diagrams of four basic structures.

- series system

$$R_s = \prod_{j=1}^{n} R_j$$

- parallel system

$$R_s = 1 - \prod_{j=1}^{n} (1 - R_j)$$

- parallel-series system

$$R_s = 1 - \prod_{i=1}^{k} \left(1 - \prod_{j=1}^{n_i} R_{ij} \right)$$

- series-parallel system

$$R_s = \prod_{i=1}^{k} \left[1 - \prod_{j=1}^{n_i} (1 - R_{ij}) \right]$$

where n = number of components in the system
k = number of subsystems
n_i = number of components in the ith subsystem
R_s = overall system reliability
R_j = component reliability for the jth component
R_{ij} = component reliability of the jth component in the ith subsystem
$\prod$ = product operator

If a system and its subsystems are connected to one of the basic structures, the system is called a *hierarchical series-parallel* (HSP) system. The system reliability function of HSP can easily be formulated using a parallel and series reduction technique.[2] An example of an HSP structure is depicted in Fig. 13.2(*a*). An interesting configuration is the *k*-out-of-*n* structure. A *k*-out-of-*n*:*G*(*F*) is a redundant system which works (fails) if at least *k* components work (fail) among a total of *n* components. In terms of the *k*-out-of-*n* structure, alternative descriptions of a series system and a parallel system are 1-out-of-*n*:*F* and 1-out-of-*n*:*G*, respectively. The reliability function of *k*-out-of-*n*:*G*, with the same component reliability *R*, can be computed as below:

$$R_s = \sum_{j=k}^{n} \binom{n}{j} R^j (1 - R)^{n-j}$$

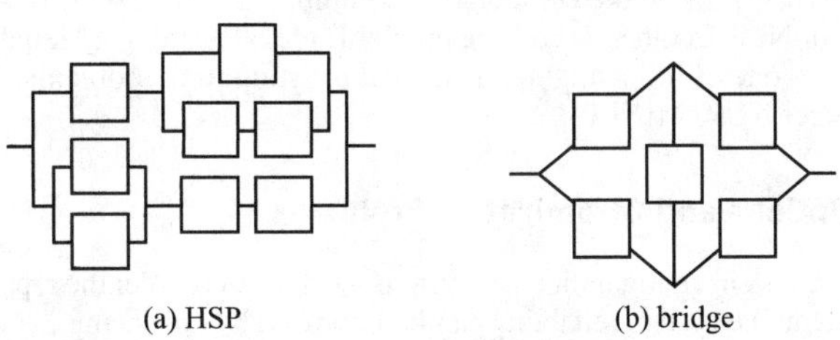

FIGURE 13.2 Nonbasic structures.

There are many variants of the *k*-out-of-*n* system including the consecutive-*k*-out-of-*n* and the multidimensional-consecutive-*k*-out-of-*n* system. To obtain further information on various *k*-out-of-*n* systems, refer to Ref. 1. If a system cannot be described as any of the above listed structures, the system is called a *complex* system. If a system is complex, it is very difficult to obtain the system reliability function. There are various methods for computing the system reliability function, e.g., pivotal decomposition, the inclusion-exclusion method, and the sum-of-disjoint-products method. A simple example of a complex structure is the bridge structure in Fig. 13.2(*b*). To obtain detailed information on the formal definitions of the above structures, other special types of structures, and various methods for computing the system reliability function, refer to Refs. 2 and 3.

Most practical systems are coherent systems. A system is coherent if it has an increasing structure function and all the components are relevant to the system reliability function. If a system is coherent, the system reliability function is increasing for each component. In this case, the optimal solution lies on the boundary (continuous case) or near the boundary (discrete case) of the feasible region. If the lifetime of $R(t)$ in Eq. (13.1) is fixed, the reliability function of a component does not depend on time and is thus considered scalar. Throughout this section, we consider only the reliability of the fixed time model.

13.1.3 Mathematical Programming

A mathematical programming problem is an optimization problem of the following form:

$$\mathbf{P}: \quad \text{maximize} \quad f(\mathbf{x}) \tag{13.2}$$

$$\begin{aligned} \text{subject to} \quad & g_i(\mathbf{x}) \le b_i, \quad && i = 1,\ldots,m \\ & h_i(\mathbf{x}) = c_i, \quad && i = 1,\ldots,l \\ & \mathbf{x} \in \Omega \end{aligned} \tag{13.3}$$

where $\mathbf{x} = (x_1,\ldots,x_n)$ and Ω is a subset of $\Re^n$. The function f is called the *objective function*; $g_i(\mathbf{x}) \le b_i$ is called the *inequality constraints* and $h_i(\mathbf{x}) = c_i$ is called the *equality constraints*; x_js for $j = 1,\ldots, n$ are called the *decision variables*. The objective of problem **P** can be replaced so as to minimize and the inequality constraints can be $g_i(\mathbf{x}) \ge b_i$ for some i according to the problem definition. A point $\mathbf{x}$ is a *feasible solution* if $\mathbf{x} \in \Omega$ and $\mathbf{x}$ satisfies all the constraints. A *feasible region* S that is a subset of Ω is defined as a set of all feasible points. A point $\mathbf{x}^*$ is called as an *optimum* or *optimal solution* if $\mathbf{x}^* \in S$ and $f(\mathbf{x}^*) \ge f(\mathbf{x})$ for any point $\mathbf{x} \in S$. All reliability optimization problems can be formulated as a form of **P**– the objective function can be the overall system reliability, total cost to attain a required reliability level, percentile life, and so on; the constraints are total cost, weight, volume, and so on; the decision variables are the reliability level and/or the redundancy level of the components.

If the objective function f and all constraint functions g_i and h_i are linear functions, we call the problem **P** a linear programming problem (LP); otherwise, a nonlinear programming problem (NLP). Since system reliability functions are nonlinear, all reliability optimization problems are classified into NLP. If a feasible region S is a subset of Z_+^n, which is a set of n-dimensional nonnegative integers, the problem is labeled as an integer programming problem (IP), and if some x_js are real and others are integers, then the problem is designated a mixed integer programming problem (MIP). From the computational complexity viewpoint, IP and MIP are not easier to solve than LP or NLP. In other words, the problems are NP-hard. Any reliability optimization problem is classified as one of following mathematical programming problems: NLP, integer NLP (INLP), or mixed integer NLP (MINLP).

13.1.4 Design Options and Optimization Problems

Let a system configuration be given. If we do not consider the repair and preventive maintenance for the system, the system reliability can be improved by enhancing the component reliability itself and/or by adding redundancy to some less reliable components or subsystems.[3] However, any effort at improving system reliability usually requires resources. Improvement of component reliability is another big research area. To understand enhancement of component reliability for modern semiconductor products, refer to Kuo and Kim[4] and Kuo and Kuo.[5] The problem deciding the reliability level of each component with some constraints—called the reliability allocation problems—was well developed in the 1960s and 70s as is documented in Tillman et al.[6] Recent developments are mainly concentrated on the redundancy allocation, reliability-redundancy allocation, cost minimization, and optimization of multiple objectives using heuristics and metaheuristics. More detailed design options and a description of various solution methodologies are well summarized by Kuo and Prasad [7] and Kuo et al.[3]

The diversity of objectives, decision variables, system structures, resource constraints and options for reliability improvement has led to the construction and analysis of numerous optimization problems. In practical situations, the following problems often appear:

Redundancy Allocation

$$\begin{aligned} \text{RA:}\quad & \text{maximize} \quad R_s = f(\mathbf{x}) \\ & \text{subject to} \quad g_i(\mathbf{x}) \le b_i, \qquad i = 1,\ldots,m \\ & \qquad\qquad \mathbf{l} \le \mathbf{x} \le \mathbf{u}, \qquad \mathbf{x} \in Z_+^n \end{aligned}$$

where $\mathbf{l} = (l_1, \ldots, l_n)$ and $\mathbf{u} = (u_1, \ldots, u_n)$ are the lower and upper bounds of $\mathbf{x}$, respectively. The objective of the redundancy allocation problem is to find an optimal allocation that maximizes system reliability where several resource constraints such as cost, weight, and/or volume are given. An allocation is a vector of the numbers of parallel or standby redundancies at the components. The problems in this category are pure INLP because the levels of redundancy of each component are nonnegative integers. Many developments in the reliability optimization have concentrated on this problem. We shall review the heuristic methods, metaheuristic algorithms, and exact methods used for solving problem **RA** in later sections.

Reliability-Redundancy Allocation

$$\begin{aligned} \text{RR:}\quad & \text{maximize} \quad R_s = f(\mathbf{r}, \mathbf{x}) \\ & \text{subject to} \quad g_i(\mathbf{r}, \mathbf{x}) \le b_i, \qquad i = 1,\ldots,m \\ & \qquad\qquad \mathbf{l} \le \mathbf{x} \le \mathbf{u}, \qquad \mathbf{x} \in Z_+^n, \mathbf{r} \in \Re^n \end{aligned}$$

where $\mathbf{r} = (r_1, \ldots, r_n)$ is a vector of the level of component reliabilities. The reliability of a system can be enhanced by either providing redundancy at the component level or increasing component reliabilities, or both. Redundancy and component reliability enhancement, however, lead to an increase in system cost. Thus, a tradeoff between these two options is necessary for budget-constrained reliability optimization. The problem of maximizing system reliability through redundancy and component reliability choices is called the reliability-redundancy allocation problem. This mixed-integer optimization problem represents a very difficult but realistic situation in reliability optimization. Mathematically, the problem can be formulated as RR. The problem RR is MINLP which means it is more difficult than a pure redundancy allocation problem.

In some situations, the component reliabilities $\mathbf{r}$, of problem RR, have discrete functions of integer variables instead of continuous values. The problem can be formulated so as to maximize $R_s = f(\mathbf{x}) = h(R_1(x_1), \ldots, R_k(x_k), x_{k+1}, \ldots, x_n)$ subject to $\mathbf{g}(\mathbf{x}) \le b$, $\mathbf{x} \in Z_+^n$ where $R_j(x_j)$ is the jth subsystem reliability function with respect to integer variable x_j, and h is a system reliability function with respect to the subsystem reliabilities. This problem is considered as a pure INLP.

Cost Minimization

$$\begin{aligned} \text{CM:}\quad & \text{maximize} \quad C_s = f(\mathbf{x}) = \sum_{j=1}^{n} c_j x_j \\ & \text{subject to} \quad g_i(\mathbf{x}) \le b_i, \qquad i = 1,\ldots,m \\ & \qquad\qquad \mathbf{l} \le \mathbf{x} \le \mathbf{u}, \qquad \mathbf{x} \in Z_+^n \end{aligned}$$

where C_s is the total cost of the system and c_j is the unit cost of the jth component. Like maximizing the system reliability, minimizing the total cost is a very important objective in reliability optimization. The problem can be formulated as the standard form of CM where the objective function is a total cost and one of constraints is the minimum required system reliability. In most cases, the total cost $C_s = \Sigma_{j=1}^{n} c_j x_j$ is a linear and separable function. However, since the system reliability function that is nonlinear is included in the constraints, this problem is classified as INLP.

Multiple Objectives Optimization

$$\begin{aligned} \text{MO:}\quad & \text{maximize} \quad z = [f_1(\mathbf{x}), f_2(\mathbf{x}), \ldots, f_s(\mathbf{x})] \\ & \text{subject to} \quad g_i(\mathbf{x}) \le b_i, \qquad i = 1, \ldots, m \\ & \qquad\qquad\quad \mathbf{l} \le \mathbf{x} \le \mathbf{u}, \qquad \mathbf{x} \in Z_+^n \end{aligned}$$

where $f_i(\mathbf{x})$ is the ith considered objective function for multiple objectives optimization and z is an objective vector of the values of those objective functions. In single objective optimization problems relating to system design, either the system reliability is maximized subject to limits on resource consumption or the consumption of one resource is minimized subject to the minimum requirement of system reliability and other resource constraints. While designing a reliability system, it is always desirable to simultaneously maximize system reliability and minimize resource consumption. When the limits on resource consumption are flexible or they cannot be determined properly and precisely, it is better to adopt a multiobjective approach to system design even though a single solution that is optimal with respect to each objective may not exist. A design engineer is often required to consider, in addition to the maximization of system reliability, other objectives such as minimization of cost, volume, and weight. It may not be easy to define the limits on each objective in order to deal with them in the form of constraints. In such situations, the designer faces the problem of optimizing all objectives simultaneously. This is typically seen in aircraft design. Suppose the designer is considering only the option of providing redundancy for optimization purposes. Mathematically, the problem can be expressed as an MO. A general approach for solving this multiobjective optimization problem is to find a set of nondominated feasible solutions and make interactive decisions based on this set.

The major focus of recent work has been on the development of heuristic methods and metaheuristic algorithms for the above four reliability optimization problems. The literature and relevant methodologies for solving various types of reliability optimization problems are summarized in Table 13.1. In the following pages, Sec. 13.2 contains the methods for optimal redundancy allocation whereas Sec. 13.3 describes the methods for reliability-redundancy allocation. Sections 13.4 and 13.5 describe work on cost minimization and multiobjective optimization in reliability systems, respectively. Section 13.6 concludes the reliability optimization models and methods.

Notations

R_s = system reliability
R_j = jth subsystem reliability
n = number of stages in the system
m = number of resources
x_j = number of components at stage j
l_j and u_j = lower and upper limit on x_j, respectively
r_j = component reliability at stage j
$\mathbf{x} = (x_1, \ldots, x_n)$
$\mathbf{r} = (r_1, \ldots, r_n)$
$g_{ij}(x_j)$ = amount of ith resource required to allocate x_j components at stage j

TABLE 13.1 Recent Developments for Reliability Optimization Problems

Problem	Methods	Description
RA	Exact	Surrogate constraints method[4] Near boundary of feasible region enumeration[36,38] Branch-and-bound[41] Three-level decomposition method[42] Lexicographic search method[39]
	Heuristics	One-neighborhood sensitivity factor[9–12] Two-neighborhood sensitivity factor[14] Sensitivity factor using minimal path set[13] Linear approximation[15] Boundary of feasible region search[15,16]
	Metaheuristics	Genetic algorithms[19,20,22–28] Simulated annealing[31,32] Tabu search[33]
RR	Heuristics	Stage sensitivity factor[45] Branch-and-bound with Lagrangian multiplier[47] Iterative solving nonlinear programming[46] Surrogate constraints dynamic programming[48] Renewal theory[51] Random search technique[50]
CM	Metaheuristics	Genetic algorithms[23,52–55]
MO	Heuristics	Decomposition and surrogate worth tradeoff[56] Goal programming[60] Sequential proxy optimization technique[57] Multicriteria optimization[58,59] Iterative dynamic programming[61]
	Metaheuristics	Genetic algorithms[62,63]

$g_i(\mathbf{x})$ = total amount of ith resource required for allocation $\mathbf{x}$

b_i = total amount of ith resource consumed

R_0 = minimum system reliability required

13.2 REDUNDANCY ALLOCATION

13.2.1 Heuristics

Iterative Heuristics Based on Sensitivity Factor. Almost all heuristics developed for solving the redundancy allocation problem RA have a common feature—in any iteration, a solution is obtained from the solution of the previous iteration by increasing one of the variables by a value of one with selection of the variable for the increment based on a sensitivity factor, where the sensitivity factor is a quantity of the impact of a component redundancy at each iteration. Nakagawa and Miyazaki[8] numerically compared the iterative heuristic methods of Nakagawa and Nakashima,[9] Kuo et al.,[10] Gopal et al.[11] and Sharma and Venkateswaran[12] for a redundancy allocation problem with nonlinear constraints. They carried out extensive numerical investigations and reported computational the time and relative solution errors for these methods.

Dinghua[13] has developed a heuristic method for the solution of the RA problem based on the approach of a unit increment at a time. It requires the determination of all minimal path sets of the reliability system. A set of components is called a path set if the system works when all components of the path set work. A minimal path set is a path set where proper subsets of the set are not a path set. In every iteration of this method, a stage is selected in two steps for a feasible redundancy increment. A minimal path set is selected in the first step on the basis of a sensitivity factor, and a stage is selected in the second step from the chosen path set using another sensitivity factor.

Kohda and Inoue[14] have developed a heuristic method in which the solutions of two successive iterations may differ on one or two variables. This method is applicable even when the constraints do not involve all nondecreasing functions. In each iteration one of the following improvements will hold: (i) a redundancy is added to the component which has the largest sensitivity factor; (ii) two redundancies are simultaneously increased on two distinct components, respectively; and (iii) one redundancy of the component which has the smallest sensitivity factor is discarded and a redundancy is added on the component which has the largest sensitivity factor. This heuristic is called a two-neighborhood heuristic in contrast to the one-neighborhood type of heuristic described above. During the process, cases (i) and (iii) use a single-stage sensitivity factor while case (ii) is based on a two-stage sensitivity factor.

Boundary Feasible Region Search. A search heuristic algorithm developed by Kim and Yum[15] for the redundancy allocation problem RA makes excursions to a bounded subset of infeasible solutions while improving a feasible solution. They have assumed that the system is coherent and the constraint functions are increasing for each decision variable. The algorithm starts with a feasible solution and improves it as much as possible by adding increments to the variables. Later it goes to another feasible solution, passing through a sequence of solutions in a predetermined infeasible region Ψ with a change (increase or decrease by value 1) in a single variable in each move. The resulting feasible solution is improved as much as possible through increments. The cycle is repeated until it reaches an infeasible solution outside Ψ.

Jianping[16] has recently developed a method called a bounded heuristic method for optimal redundancy allocation. This method also assumes that the constraint functions are increasing in each variable. A feasible solution is called a bound point if no feasible increment can be given to any variable. In each iteration, the method moves from one bound point to another through an increase of 1 in a selected variable and changes in some other variables. The method has some similarity with the method of Kohda and Inoue[14] in the sense that an addition and subtraction are simultaneously done at two stages in some iterations.

Linear Approximation. Hsieh[17] recently developed a linear approximation heuristic for redundancy allocation problem RA for a series-parallel system with multiple component choices. The method consists of two main stages—the approximation and improving stages. In the approximation stage, all integral decision variables are relaxed as real and the objective function is linearized by reformulation. Then the LP is solved and the solution is rounded off to its nearest integer solution. Then, a 0 to 1 knapsack problem with linear constraints is formulated for the residual resources and the integer LP is solved for improving the feasible integer solution. The main advantage of this method is that well developed LP techniques and solvers, e.g., CPLEX, LINGO, MINOS, and so on, can be used for solving the relaxed LP and 0 to 1 knapsack problems and thus the method can be easily applied to large-scale problems.

13.2.2 Metaheuristic Methods

In recent years, metaheuristics have been selected and successfully applied to handle a number of reliability optimization problems. In this subsection, however, emphasis is placed on solving the redundancy allocation problem. These metaheuristics, based more on artificial reasoning than classical mathematics-based optimization, include genetic algorithms (GA), simulated annealing (SA), and Tabu search (TS). Genetic algorithms seek to imitate the biological phenomenon of evolutionary production through the parent-children relationship. Simulated annealing is based on a physical process in metallurgy. Tabu search derives and exploits a collection of principles involved in intelligent problem solving.

Genetic Algorithms. A genetic algorithm (GA) is a probabilistic search method for solving optimization problems. Holland[18] made pioneering contributions to the development of genetic algorithms in the initial stages. There has been significant progress in the application of these methods during the 1980s and 1990s. The development of a genetic algorithm can be viewed as an adaptation of a probabilistic approach based on principles of natural evolution. The genetic algorithm approach can be effectively adopted for complex combinatorial problems. However, it gives only a heuristic solution. This approach was used in the 1990s by several researchers to solve reliability optimization problems. For a detailed description of applications of genetic algorithms to combinatorial problems, including reliability optimization problems, one may refer to Gen and Cheng.[19]

The general genetic algorithm has the following process: (i) represent the decision variables in the form of chromosomes, normally binary strings; (ii) generate a number of feasible solutions for the population; (iii) evaluate and select parents from the population; (iv) execute genetic operations such as crossover and mutation; (v) evaluate and select offspring for the next generation and include these offspring in the population; (vi) repeat the above procedures until reaching a satisfying termination condition. A major advantage of the GA is that it can be applied to very complicated problems because it does not require mathematical analysis and reformulation, e.g., derivatives of functions, system structures, and so on. On the other hand, the main weakness of the GA is that the basic process described above is sophisticated and hence the computation of GA is slower than one of iterative type of heuristics. This is the reason why many applications of GA select very complicated problems which are difficult to handle with iterative types of heuristics.

Redundancy Allocation With Several Failure Modes. Gen and Cheng[19] and Yokota et al.[20] have applied a GA to the problem of finding optimal redundancy allocation in a series system in which the components of each subsystem are subjected to two classes of failure modes: O and A. A subsystem fails when a class O failure mode occurs in at least one of the components or when a class A failure mode occurs in all components. The problem is originally considered by Tillman[21] using an implicit enumeration method. The objective function can be approximated as:

$$R_s \approx \prod_{j=1}^{n} \left\{ 1 - \sum_{u=1}^{h_j} \left[1 - (1 - q_{ju})^{x_j+1} \right] - \sum_{u=h_j+1}^{s_j} (q_{ju})^{x_j+1} \right\}$$

where x_{j+1} is the number of parallel components in the jth subsystem; q_{ju} is the probability that a component in subsystem j fails resulting in failure mode u; $1, \ldots, h_j$ modes belong to class O and $h_{j+1}, \ldots, s_j$ modes belong to class A.

MTBF Optimization With Multiple Choices. Painton and Campbell[22] have adopted the genetic algorithm approach to solve a reliability optimization problem related to the design of a personal computer (PC). The functional block diagram of a PC has a series-parallel configuration. There are three choices for each component: the first choice is the existing option and the other two are reliability increments with additional costs. The component failure rate for each choice of a component is random following a known triangular distribution. Due to the randomness in input, the system mean time between failure (MTBF) is also random. The problem considered by Painton and Campbell[22] is the maximization of the 5th percentile of the statistical distribution of MTBF over the choices of components subject to a budget constraint. The problem has both combinatorial and stochastic elements; the combinatorial element is the choice of components whereas the stochastic one is the randomness of input (component failure rates) and output (MTBF).

Redundancy Allocation With Multiple Choices. In the GA design—using the lower and upper bounds—the decision variables are converted into binary strings that are used in the chromosome representation of the solutions. A large penalty is included in the fitness of infeasible solutions. Coit and Smith[23] and[24] have developed GAs for a series-parallel system in which each subsystem is a k-out-of-n:G system. For each subsystem, there are multiple component choices available. If a problem is highly constrained, the optimal solution can be efficiently obtained through an infeasible region search. To increase the efficiency of GA search and provide a final feasible solution, Coit and Smith[23] applied a dynamic penalty function based on the squared constraint-violation determined by the relative degree of infeasibility.

Marsequerra and Zio[25] solved the redundancy allocation problem with the same configuration as Coit and Smith[23] using a GA. They selected the objective function as the net profit of the system operation for a given mission time which implicitly reflects the possible availability and reliability through the system downtime and accident costs, respectively. The net profit is computed by subtracting all the costs relative to the system implementation and operation, e.g., repair costs, system downtime costs, accident costs, and so on, from the service revenue.

Percentile of Distribution Optimization. Assuming that component reliabilities are random following known probability distributions, Coit and Smith[26] have developed a GA for problem RA with the objective replaced by maximization of a percentile of the statistical distribution of system reliability. The GA can also be used for maximization of a lower percentile of the distribution of system time-to-failure.[27,28] The main advantages of maximizing the percentile of time-to-failure are that there is no requirement of a specified mission time and risk avoidance property. In Ref. 10, Weibull distribution parameters, i.e., a shape and a scale parameters, are assumed as known and the objective function is evaluated using the Newton-Raphson search for both system reliability and system failure time. On the other hand, Coit and Smith[28] solved the same problem with more general assumption of Weibull distribution parameters, i.e., a known shape parameter and a distributed scale parameter. To solve the problem, they used GA and a Baysian approach which considers the uncertainty distribution as a prior distribution.

Simulated Annealing Method. The simulated annealing (SA) algorithm is a general method used to solve combinatorial optimization problems. It involves probabilistic transitions among the solutions of the problem. Unlike iterative improvement algorithms, which improve the objective value continuously, SA may encounter some adverse changes in objective value in the course of its progress. Such changes are intended to lead to a global optimal solution instead of a local one.

Annealing is a physical process in which a solid is heated up to a high temperature and then allowed to cool slowly and gradually. In this process, all the particles arrange themselves gradually in a low-energy ground state level. The ultimate energy level depends on the level of the high temperature and the rate of cooling. The annealing process can be described as a stochastic model as follows. At each temperature T, the solid undergoes a large number of random transitions among the states of different energy levels until it attains a thermal equilibrium in which the probability of the solid appearing in a state with energy level E is given by the *Boltzmann distribution*. As the temperature T decreases, equilibrium probabilities associated with the states of higher energy levels decrease. When the temperature approaches zero, only the states with lowest energy levels will have a nonzero probability. If the cooling is not sufficiently slow, thermal equilibrium will not be attained at any temperature and consequently the solid will finally have a metastable condition.

To simulate the random transitions among the states and the attainment of thermal equilibrium at a fixed temperature T, Metropolis et al.[29] developed a method in which a transition from one state to another occurs due to a random perturbation in the state. If the perturbation results in a reduction of energy level, transition to the new state is accepted. If, instead, the perturbation increases the energy level by $\Delta E(>0)$, then transition to the new state is accepted with a given probability governed by the Boltzmann distribution. This method is called the *Metropolis algorithm*. The criterion for acceptance of the transition is called the *Metropolis criterion*.

Based on a simulation of the annealing process, Kirkpatrick et al.[30] developed a simulated annealing algorithm for solving combinatorial optimization problems. Although SA gives satisfactory solutions for combinatorial optimization problems, its major disadvantage is the amount of computational effort involved. In order to improve the rate of convergence and reduce the computational time,

Cardoso et al.[31] introduced the nonequilibrium simulated annealing algorithm (NESA) by modifying the algorithms of Metropolis et al.[29] In NESA, there is no need to reach an equilibrium condition through a large number of transitions at any fixed temperature. The temperature is reduced as soon as an improved solution is obtained.

Improved Nonequilibrium Simulated Annealing. Ravi et al.[32] have recently improved NESA by incorporating a simplex-like heuristic in the method. They have applied this variant of NESA, denoted as I-NESA, to reliability optimization problems such as redundancy allocation RA and cost minimization CM. It consists of two phases: phase I uses a NESA and collects solutions obtained at regular intervals of the progress of the NESA and phase II starts with the set of solutions obtained in phase I and uses a simplex-like heuristic procedure to improve the best solution further.

Tabu Search Method. Tabu search (TS) is another metaheuristic that guides a heuristic method to expand its search beyond local optimality. It is an artificial intelligence technique which utilizes memory (information about the solutions visited up to that stage) at any stage to provide an efficient search for optimality. It is based on ideas proposed by Fred Glover. An excellent description of TS methodology can be found in Glover and Laguna.[33]

Tabu search for any complex optimization problem combines the merits of artificial intelligence with those of optimization procedures. TS allows the heuristic to cross boundaries of feasibility or local optimality which are major impediments in any local search procedure. The most prominent feature of Tabu search is the design and use of memory-based strategies for exploration in the neighborhood of a solution at every stage. TS ensures responsive exploration by imposing restrictions on the search at every stage based on memory structures. It is very useful for solving large complex optimization problems that are very difficult to solve by exact methods. To solve redundancy allocation problems we recommend that the Tabu search be used alone or in conjunction with the heuristics presented in Sec. 13.2.1 to improve the quality of the heuristics.

13.2.3 Exact Methods

The purpose of exact methods is to obtain an exact optimal solution to a problem. It is generally difficult to develop exact methods for reliability optimization problems which are equivalent to methods used for nonlinear programming problems. When dealing with redundancy allocation, the methods used become those of integer nonlinear programming problems. Such methods involve more computational effort and usually require larger computer memory. For these reasons, researchers in reliability optimization have placed more emphasis on heuristic approaches. However, development of good exact methods is always a challenge. Such methods are particularly advantageous when the problem is not large. Moreover, the exact solutions provided by such methods can be used to measure the performance of heuristic methods.

Surrogate Constraints Method. Nakagawa and Miyazaki[34] have adopted the surrogate constraints method to solve the reliability optimization problem RA when the objective function is also separable and there are exactly two constraints. For such problems, one can apply dynamic programming (DP) either using Lagrangian multipliers or defining the state space with respect to both the constraints. Of course, there is no guarantee that DP with Lagrangian multipliers will yield an exact optimal solution. With the surrogate constraints method, Nakagawa and Miyazaki[34] solved the surrogate problem. They have reported that the performance of their method is superior to DP with Lagrangian multipliers for the problem under consideration. They have also indicated that it is possible, although remotely, that this method may fail to yield an exact optimal solution.

Implicit Enumeration Methods. Misra[35] has proposed an exact algorithm for optimal redundancy allocation for problem RA based on a search near the boundary of the feasible region. This method was implemented later by Misra and Sharma,[36] Sharma et al.,[37] and Misra and Misra[38] for solving various redundancy allocation problems.

Prasad and Kuo[39] have recently developed a partial implicit enumeration method based on a lexicographic search with an upper bound on system reliability. During this process, a redundancy is added to the current position lexicographically until reaching the largest lexicographical allocation. If the current allocation is infeasible, the current position moves to the next position that is decided at the beginning of the algorithm, because increasing any redundancy in the allocation generates an infeasible allocation due to the fact that the system is coherent. The paper demonstrates that for both small and large problems the method is superior to conventional methods in terms of computing time. For more background on percentile system life optimization with the same methodologies see Ref. 40.

Branch-and-Bound Method for Multiple-Choices System. Recently Sung and Cho[41] applied the branch-and-bound method to the redundancy allocation problem RA. The system considered has a series-parallel configuration, and components of a subsystem can be selected from several different choices. A similar problem is also solved using GA by Coit and Smith.[22] The major approaches they used are several solution space reduction techniques and Lagrangian relaxation for finding sharper upper bounds.

Decomposition Methods for Large Systems. For large systems with a good modular structure, Li and Haimes[42] proposed a three-level decomposition method for reliability optimization subject to linear resource constraints. At level 1, a nonlinear programming problem is solved for each module. At level 2, the problem is transformed into a multiobjective optimization problem which is solved by the ε-constraint method of Chankong and Haimes.[43] This approach involves optimization at three levels. At level 3 (the highest level), the lower limits, ε_is on multiple objective functions, are chosen while Kuhn Tucker multipliers are chosen at level 2 for fixed ε_is. For fixed Kuhn-Tucker multipliers and fixed ε_is, a nonlinear programming problem is solved for each module of the system at level 1.

13.3 RELIABILITY–REDUNDANCY ALLOCATION

13.3.1 Iterative Heuristics Based on Stage Sensitivity Factor

For the reliability–redundancy allocation problem RR Tillman et al.[44] are among the first to solve the problem using a heuristic and search technique. Gopal et al.[45] have developed a heuristic method that starts with 0.5 as the component reliability at each stage of the system and increases the component reliability at one of the stages by a specified value δ in every iteration. The selection of a stage for improving a component's reliability is based on a stage sensitivity factor. For any particular choice of component reliabilities, an optimal redundancy allocation is derived by a heuristic method. Any heuristic redundancy allocation method can be used for this purpose. When such increments in component reliabilities do not yield any higher system reliability, the increment δ is reduced and the procedure is repeated with the new increment δ. This process is discontinued when δ falls below a specified limit ε.

Xu et al.[46] offered an iterative heuristic method for solving problem RR. In each iteration, a solution is derived from the previous solution in one of the following two ways: (i) one redundancy of a component is added and optimal component reliabilities are obtained corresponding to the new fixed allocation by solving a nonlinear programming problem; or (ii) one redundancy of a component is added and one redundancy is deleted from another component, then the optimal component reliabilities are obtained for the new fixed allocation by solving a nonlinear programming problem. Xu et al.[46] assume that the objective and constraint functions are differentiable and monotonic nondecreasing functions.

13.3.2 Branch-and-Bound Method With Lagrangian Multipliers

Kuo et al.[47] have presented a heuristic method for the reliability-redundancy allocation problem based on a branch-and-bound strategy and the Lagrangian multipliers method. The initial node is associated with the relaxed version of problem RR. The bound associated with a node is the optimal value of the relaxed version of problem RR with some integer variables fixed at integral values because the feasible region of the original problem is a subset of one of the relaxed problems. The method requires the assumption that all functions of the problem are differentiable. The relaxed problem, which is a nonlinear programming problem, is solved by the Lagrangian multipliers method. Kuo et al.[47] have demonstrated the method for a series system with five subsystems.

13.3.3 Surrogate Constraints Method

Hikita et al.[48] have developed a surrogate constraints method to solve problem RR. The method is for minimizing a quasi-convex function subject to convex constraints. In this method, a series of surrogate optimization problems are solved. In each surrogate problem, the objective is the same in problem RR but the single constraint is obtained by taking a convex linear combination of the m constraints. The surrogate constraint approach to problem RR is to find a convex linear combination that gives the least optimal objective value of the surrogate problem and to take the corresponding surrogate optimal solution as the required solution. Hikita et al.[48] use a dynamic programming approach to solve single-constraint surrogate problems. With this method, the requirement is that either the objective function f is separable or the surrogate problem can be formulated as a multistage decision-making problem. The surrogate constraint method is useful for special structures including parallel-series and series-parallel designs.

13.3.4 Software Reliability Optimization

Reliability-redundancy allocation problems arise in software reliability optimization also. The redundant components of software may result from programs developed by different groups of people for given specifications. The reliability of any software component can be enhanced by additional testing which requires various resources. Another feature of software systems is that the components are not necessarily completely independent. Chi and Kuo[49] have formulated mixed integer nonlinear programming problems for reliability-redundancy allocation in software systems with common-cause failures and systems involving both software and hardware.

13.3.5 Discrete Reliability-Redundancy Allocation

The discrete reliability-redundancy allocation problem is the same problem except that the component reliabilities of RR are discrete functions of the integer variables. Thus, this problem is included as pure INLP. Mohan and Shanker[50] adopted a random search technique for finding a global optimal solution for the problem of maximizing system reliability through the selection of only component reliabilities subject to cost constraints. Bai et al.[51] considered a k-out-of-n:G system with common-cause failures. The components are subjected not only to intrinsic failures but also to a common failure cause following independent exponential distributions. If there is no inspection, the system is restored upon failure to its initial condition through necessary component replacements. If there is inspection, failed components are replaced during the inspection. For both of the cases—with and without inspection—Bai et al.,[51] using renewal theory, derived an optimal n that minimizes the mean cost rate. They also demonstrated their procedure with numerical examples.

13.4 COST MINIMIZATION

13.4.1 Genetic Algorithm Using Penalty Function

Coit and Smith[23] have also considered the problem of minimizing total cost, subject to a minimum requirement on system reliability and other constraints such as weight. Their objective function involves a quadratic penalty function with the penalty depending on the extent of infeasibility. Later, Coit and Smith[52] introduced a robust adaptive penalty function to penalize the infeasible solutions.

This function is based on a *near-feasibility threshold* (NFT) for all constraints. The NFT-based penalty encourages the GA to explore the feasible and infeasible regions close to the boundary of the feasible region. They have also used a dynamic NFT in the penalty function which depends on the generation number. On the basis of extensive numerical investigation, they have reported that a GA with dynamic NFT in the penalty function is superior to GAs with several penalty strategies including a GA that considers only feasible solutions. Based on numerical experimentation, they have also reported that GAs give better results than the surrogate constraint method of Nakagawa and Miyazaki.[34]

13.4.2 Genetic Algorithm for Cost-Optimal Network

Genetic algorithms are also developed for cost-optimal network designs. Suppose a communication network has nodes 1, 2, . . . , n and a set of h_{ij} links are available to directly connect a pair of nodes i and j for $i = 1, \ldots, n$ and $j = 1, \ldots, n$ and $i \neq j$. The links have different reliabilities and costs and only one of the h_{ij} links is used if nodes i and j are to be directly connected. The network is in good condition as long as all nodes remain connected, that is, the operating links form a graph that contains a spanning tree. Let x_{ij} denote the index of the link used to connect the pair (i, j). If the pair (i, j) is not directly connected, then $x_{ij} = 0$.

Dengiz et al.[53] have designed a GA for cost-optimal network design when $h_{ij} = 1$ and the objective function is separable and linear. In this case, only one link is available to connect any particular pair of nodes. The evaluation of exact network reliability requires a lot of computational effort and also possibly requires a large computer memory. To avoid extensive computation, each network generated by the algorithm is first screened using a connectivity check for a spanning tree and a 2-connectivity measure. If the network passes the screening, then an upper bound on network reliability is computed and used in the calculation of the objective function (fitness of solution). For network designs for which the upper bound is at least the minimum required network reliability and total cost is the lowest, Monte Carlo simulations are used to estimate the reliability. The penalty for not meeting the minimum reliability requirement is proportional to $(R(\mathbf{x}) - R_0)^2$, where $R(\mathbf{x})$ is a network reliability function and R_0 is the minimum required network reliability. Deeter and Smith[54] have developed a GA for cost-optimal network design without any assumption on h_{ij}. Their penalty involves the difference between R_0 and $R(\mathbf{x})$—the population size and the generation number.

13.4.3 Genetic Algorithm for Multistate System

When components of a system have different performance levels according to the state of the contained components, the system can be considered a multistate system. Levitin[55] deals with the sum of investment cost minimization problems subject to the desired level of availability. The system considered is a series-parallel system consisting of several main producing subsystems (MPS), where MPSs are supplied different resources from resource generating subsystems (RGS). Each element of MPS that is connected in parallel consumes a fixed amount of resources and is selected from a list of available multistate components. The objective of the problem is to find an optimal system structure to minimize system cost. Levitin[55] solves the problem using a GA with a double-point crossover operation.

13.5 MULTIOBJECTIVE OPTIMIZATION

13.5.1 Dual Decomposition and Surrogate Worth Trade off Method

Sakawa[56] has adopted a large-scale multiobjective optimization method to deal with the problem of determining optimal levels of component reliabilities and redundancies in a large-scale system with respect to multiple objectives. He considers a large-scale series system with four objectives:

maximization of system reliability, minimization of cost, weight, and volume. In this approach he derives Pareto optimal solutions by optimizing composite objective functions which are obtained as linear combinations of the four objective functions. The Lagrangian function for each composite problem is decomposed into parts and optimized by applying both the dual decomposition method and the surrogate worth tradeoff method, treating redundancy levels as continuous variables. Later, the resulting redundancy levels are rounded off and the Lagrangian function is optimized with respect to component reliabilities by the dual decomposition method to obtain an approximate Pareto solution. Sakawa[57] has provided a theoretical framework for the sequential proxy optimization technique (SPOT) which is an interactive, multiobjective decision-making technique for selection among a set of Pareto optimal solutions. He has applied SPOT to optimize system reliability, cost, weight, volume, and product of weight and volume for series-parallel systems subject to some constraints.

13.5.2 Multicriteria Optimization

To solve multiobjective redundancy allocation problems in reliability systems, Misra and Sharma[58] have adopted an approach which involves the Misra integer programming algorithm[35] and a multicriteria optimization method based on the min-max concept for obtaining Pareto optimal solutions. Misra and Sharma[59] have also presented a similar approach for solving multiobjective reliability-redundancy allocation problems in reliability systems. Their methods take into account two objectives: the maximization of system reliability and the minimization of total cost subject to resource constraints.

13.5.3 Goal Programming

Dhingra[60] has adopted another multiobjective approach to maximize system reliability and minimize consumptions of resources: cost, weight, and volume. He uses the goal programming formulation and the goal attainment method to generate Pareto optimal solutions. For system designs in which the problem parameters and goals are not formulated precisely he suggests the multiobjective fuzzy optimization approach. He has demonstrated the multiobjective approach for a four-stage series system with constraints on cost, weight, and volume. Recently, Li[61] has considered iterative dynamic programming where multiobjective optimization is used as a separation strategy and the optimal solution is sought in a multilevel fashion. This method has been demonstrated in constrained reliability optimization.

13.5.4 Genetic Algorithms

Yang et al.[62] have applied a genetic algorithm to a multiobjective optimization problem for a nuclear power plant. The main difficulty in realistic application is defining the objective function. In the nuclear power plant, reliability, cost, and core damage frequency (CDF) must be considered simultaneously. Yang et al.[62] define the realistic objective function using value impact analysis (VIA) and fault tree analysis (FTA). The parameters for the GA are determined by performing sensitivity analysis.

Busacca et al.[63] used a multiobjective genetic algorithm to solve the multiobjective optimization problem, MO. Two typical approaches for solving MOs are the weighted summation of objectives into a single objective function and the consecutive imposition of objectives. On the other hand, the approach of Busacca et al.[63] is to consider every objective as a separate objective to be optimized where Pareto optimal solutions are obtained by multiobjective genetic algorithm. The Pareto solutions provide a complete spectrum of optimal solutions with respect to the objectives and thus help the designer to select the appropriate solution.

13.6 DISCUSSION

A major part of the work on reliability optimization is devoted to the development of heuristic methods and metaheuristic algorithms that are applied to redundancy allocation problems which can be extended to optimal reliability-redundancy allocation problems. It is interesting to note that these heuristics have been developed on the basis of a very distinct perspective. However, the extent to which they are superior to the previous methods is not clear. When developing heuristics, it is relevant to seek answers to two important questions: (i) under what conditions does the heuristic give an optimal solution? and (ii) what are the favorable conditions for the heuristic to give a satisfactory solution? The answers to these questions enhance the importance and applicability of the heuristic. We can understand the merit of a newly developed heuristic only when it is compared with existing ones for a large number of numerical problems. For the reliability-redundancy allocation problem, Xu et al.,[46] made a thorough comparison of a number of algorithms.

Genetic algorithms, treated as probabilistic heuristic methods, are metaheuristic methods which imitate the natural evolutionary process. They are very useful for solving complex discrete optimization problems and do not require sophisticated mathematical treatment. They can be easily designed and implemented on a computer for a wide spectrum of discrete problems. GAs have been designed for solving redundancy allocation problems in reliability systems. The chromosome definition and selection of the GA parameters provide a lot of flexibility when adopting the GA for a particular type of problem. However, there is some difficulty in determining appropriate values for the parameters and a penalty for infeasibility. If these values are not selected properly, a GA may rapidly converge to the local optimum or slowly converge to the global optimum. A larger population size and more generations enhance the solution quality while increasing the computational effort. Experiments are usually recommended to obtain appropriate GA parameters for solving a specific type of problem. An important advantage of a GA is its presentation of several good solutions (mostly optimal or near-optimal). The multiple solutions yielded by the GA method provide a great deal of flexibility in decision making for reliability design.

Simulated annealing is a global optimization technique that can be used for solving large-size combinatorial optimization problems. It may be noted that, unlike many discrete optimization methods, SA does not exploit any special structure that exists in the objective function or in the constraints. However, SA is relatively more effective when a problem is highly complex and does not have any special structure. The redundancy allocation problems in reliability systems are nonlinear integer programming problems of this type. Thus, SA can be quite useful in solving complex reliability optimization problems. Although several approaches are available in the literature for designing an SA, the design still requires ingenuity and sometimes considerable experimentation. A major disadvantage of SA is that it requires a large amount of computational effort. However, it has great potential for yielding an optimal or near-optimal solution.

Tabu search is very useful for solving large-scale complex optimization problems. The salient feature of this method is the utilization of memory (information about previous solutions) to guide the search beyond local optimality. There is no fixed sequence of operations in Tabu search and its implementation is problem-specific. Thus, Tabu search can be described as a metaheuristic rather than a method. A simple Tabu search method which uses only short term memory is quite easy to implement. Usually such methods yield good solutions when attributes, Tabu tenure and aspiration criteria are appropriately defined. A simple Tabu search can be implemented to solve redundancy allocation and reliability-redundancy allocation problems. One major disadvantage of Tabu search is the difficulty involved in defining effective memory structures and memory-based strategies which are problem-dependent. This task really requires good knowledge of the problem nature, ingenuity, and some numerical experimentation. A well designed Tabu search can offer excellent solutions in large scale system-reliability optimization.

To derive an exact optimal redundancy allocation in reliability systems, Misra[35] has presented a search method which has been used in several papers to solve a variety of reliability optimization problems including some multiobjective optimization problems. Very little other progress has been made on multiobjective optimization in reliability systems although such work could provide the system designer with an interactive environment. These problems belong to the class of nonlinear integer multiobjective optimization problems. A fuzzy optimization approach has also been adopted by Park[64] and Dhingra[60] to solve reliability optimization problems in a fuzzy environment.

ACKNOWLEDGMENTS

This section is based largely on an article that we obtained permission to use material from W. Kuo and V. R. Prasad, "An annotated overview of system reliability optimization," *IEEE Transactions on Reliability*, Vol. R-49 (2): 176–187, 2000. ©2000 IEEE

REFERENCES

1. R. I. Winner, J. P. Pennell, H. E. Bertrand, and M. M .G. Slusarezuk, *The Role of Concurrent Engineering in Weapon Systems Acquisition,* Institute for Defense Analysis, IDA Report R-338, Alexandria, VA.
2. W. Kuo, and M. Zuo, *Optimal Reliability Modeling: Principles and Applications,* John Wiley, New York, 2003.
3. W. Kuo, V. R. Prasad, F. A. Tillman, and C. L. Hwang, *Optimal Reliability Design: Fundamentals and Applications,* Cambridge University Press, Cambridge, UK, 2001.
4. W. Kuo, and T. Kim, "An Overview of Manufacturing Yield and Reliability Modeling for Semiconductor Products," *Proceedings of the IEEE,* Vol. 87(No. 8): 1329–1346, 1999.
5. W. Kuo, and Y. Kuo, "Facing the Headaches of ICs Early Failures: A State-of-the-Art Review of Burn-in Decisions, *Proceedings of the IEEE,* Vol. 71(No. 11): 1257–1266, 1983.
6. F. A. Tillman, C. L. Hwang, and Way Kuo, *Optimization of Systems Reliability,* Marcel Dekker, New York, 1980.
7. W. Kuo and V. R. Prasad, "An Annotated Overview of System Reliability Optimization," *IEEE Transactions on Reliability,* Vol. 49(No. 2): 176–187, 2000.
8. Y. Nakagawa and S. Miyazaki, "An Experimental Comparison of the Heuristic Methods for Solving Reliability Optimization Problems," *IEEE Transactions on Reliability,* Vol. R-30 (No. 2): 181–184, 1981.
9. Y. Nakagawa, and K. Nakashima, "A Heuristic Method for Determining Optimal Reliability Allocation, *IEEE Transactions on Reliability,* Vol. R-26(No. 3): 156–161, 1977.
10. W. Kuo, C. L. Hwang and F. A. Tillman, "A Note on Heuristic Methods in Optimal System Reliability, *IEEE Transactions on Reliability,* Vol. R-27(No. 5): 320–324, 1978.
11. K. Gopal, K. K. Aggarwal, and J. S. Gupta, "An Improved Algorithm for Reliability Optimization," *IEEE Transactions on Reliability,* Vol. R-27(No. 5): 325–328, 1978.
12. J. Sharma and K. V. Venkateswaran, "A Direct Method for Maximizing the System Reliability," *IEEE Transactions on Reliability,* Vol. R-20(No. 4): 256–259, 1971.
13. S. Dinghua, "A New Heuristic Algorithm for Constrained Redundancy-Optimization in Complex Systems," *IEEE Transactions on Reliability,* Vol.R-36(No. 5): 621–623, 1987.
14. T. Kohda and K. Inoue, "A Reliability Optimization Method for Complex Systems With the Criterion of Local Optimality," *IEEE Transactions on Reliability,* Vol. R-31(No. 1): 109–111, 1982.
15. J. H. Kim and B. J. Yum, "A heuristic method for solving redundancy optimization problems in complex systems," *IEEE Transactions on Reliability,* Vol. R-42(No. 4): 572–578, 1993.
16. L. Jianping, "A Bound Heuristic Algorithm for Solving Reliability Redundancy Optimization," *Microelectronics and Reliability,* Vol. 3(No. 5): 335–339, 1996.
17. Y. Hsieh, "A Linear Approximation for Redundancy Reliability Problems With Multiple Component Choices," *Computers and Industrial Engineering,* Vol. 44: 91–103, 2003.
18. J. H. Holland, *Adaptation in Natural and Artificial Systems.* University of Michigan Press, Ann Arbor, 1975.
19. M. Gen and R. Cheng, *Genetic Algorithms and Engineering Design,* John Wiley and Sons, New York, 1997.

20. T. Yokota, M. Gen, and K. Ida, "System Reliability of Optimization Problems With Several Failure Modes by Genetic Algorithm," *Japanese Journal of Fuzzy Theory and Systems,* Vol. 7(No. 1): 117–135, 1995.
21. F. A. Tillman, "Optimization by Integer Programming of Constrained Reliability Problems With Several Modes of Failure," *IEEE Transactions on Reliability,* Vol. R-18(No. 2): 47–53, 1969.
22. L. Painton and J. Campbell, "Genetic Algorithms in Optimization of System Reliability," *IEEE Transactions on Reliability,* Vol. 44: 172–178, 1995.
23. D. W. Coit and A. E. Smith, "Reliability Optimization of Series-Parallel Systems Using a Genetic Algorithm," *IEEE Transactions on Reliability,* Vol. 45(No. 2): 254–260, June 1996.
24. D. W. Coit and A. E. Smith, "Solving the Redundancy Allocation Problem Using a Combined Neural Network/ Genetic Algorithm Approach," *Computers and Operations Research,* Vol. 23(No. 6): 515–526, June 1996.
25. M. Marsequerra and E. Zio, "System Design Optimization by Genetic Algorithms," *Proc. Annual Reliability and Maintainability Symposium,* Vol. 72: 59–74, 2000.
26. D. W. Coit and A. E. Smith, "Considering Risk Profiles in Design Optimization for Series-Parallel Systems," *Proceedings of the 1997 Annual Reliability and Maintainability Symposium,* Philadelphia, PA, January 1997.
27. D. W. Coit and A. E. Smith, "Design Optimization to Maximize a Lower Percentile of the System Time-to-Failure Distribution," *IEEE Transactions on Reliability,* Vol. 47(No. 1): 79–87, 1998.
28. D. W. Coit and A. E. Smith, "Genetic Algorithm to Maximize a Lower-Bound for System Time-to-Failure With Uncertain Component Weibull Parameters," *Computers and Industrial Engineering,* Vol. 41: 423–440, 2002.
29. N. Metropolis, A. W. Rosenbluth, and M. N. Rosenbluth, "Equation of State Calculations by Fast Computing Machines," *J. Chemical Physics,* Vol. 21: 10–16, 1953.
30. S. Kirkpatrick, C. D. Gelatt Jr., and M. P. Vecchi, "Optimization by Simulated Annealing," *Science,* Vol. 220: 671–680, 1983.
31. M. F. Cardoso, R. L. Salcedo, and S. F. de Azevedo, "Nonequilibrium Simulated Annealing: A Faster Approach to Combinatorial Minimization," *Industrial Eng'g Chemical Research,* Vol. 33:1908–1918, 1994.
32. V. Ravi, B. Murty, and P. Reddy, "Nonequilibrium Simulated Annealing Algorithm Applied Reliability Optimization of Complex Systems," *IEEE Transactions on Reliability,* Vol. 46(No. 2): 233–239, 1997.
33. F. Glover and M. Laguna, *Tabu Search.* Kluwer Academic Publishers, Boston, MA, 1997.
34. Y. Nakagawa and S. Miyazaki, "Surrogate Constraints Algorithm for Reliability Optimization Problem With Two Constraints," *IEEE Transactions on Reliability,* Vol. R-30: 175–180, 1980.
35. K. B. Misra, "An Algorithm to Solve Integer Programming Problems: An Efficient Tool for Reliability Design," *Microelectronics and Reliability,* Vol. 31: 285–294, 1991.
36. K. B. Misra and U. Sharma, "An Efficient Algorithm to Solve Integer Programming Problems Arising in System Reliability Design," *IEEE Transactions on Reliability,* Vol. 40(No. 1): 81–91, 1991.
37. U. Sharma, K. B. Misra, and A. K. Bhattacharjee, "Application of an Efficient Search Technique for Optimal Design of Computer Communication Network," *Microelectronics and Reliability*, Vol. 31: 337–341, 1991.
38. K. Misra and V. Misra, "A Procedure for Solving General Integer Programming Problems," *Microelectronics and Reliability,* Vol. 34(No. 1): 157–163, 1994.
39. V. R. Prasad and W. Kuo, "Reliability Optimization of Coherent Systems," *IEEE Transactions on Reliability,* Vol. 49(3): 323–330, 2000.
40. V. R. Prasad, W. Kuo, and K. O. Kim, "Maximization of Percentile of System Life Through Component Redundancy Allocation," *IIE Transactions,* Vol. 33(No. 12): 1071–1079, 2001.
41. C. S. Sung and Y. K. Cho, "Reliability Optimization of a Series System With Multiple-Choices and Budget Constraints," *European Journal of Operational Research,* Vol. 127: 159–171, 2000.
42. D. Li and Y. Y. Haimes, "A Decomposition Method for Optimization of Large System Reliability," *IEEE Transactions on Reliability,* Vol. 41: 183–188, 1992.
43. V. Chankong and Y. Y. Haimes, *Multiobjective Decision Making: Theory and Methodology,* Elsevier, New York, 1983.
44. F. A. Tillman, C. L. Hwang, and W. Kuo, "Determining Component Reliability and Redundancy for Optimum System Reliability," *IEEE Transactions on Reliability,* Vol. R-26(No. 3): 162–165, 1977.
45. K. Gopal, K. K. Aggarwal, and J. S. Gupta, "A New Method for Solving Reliability Optimization Problem," *IEEE Transactions on Reliability,* Vol. R-29: 36–38, 1980.

46. Z. Xu, W. Kuo, and H. Lin, "Optimization Limits in Improving System Reliability," *IEEE Transactions on Reliability,* Vol. 39(No. 1): 51–60, 1990.

47. W. Kuo, H. Lin, Z. Xu, and W. Zhang, "Reliability Optimization With the Lagrange Multiplier and Branch-and-Bound Technique," *IEEE Transactions on Reliability,* Vol. R-36: 624–630, 1987.

48. M. Hikita, Y. Nakagawa, K. Nakashima, and H. Narihisa, "Reliability Optimization of Systems by a Surrogate-Constraints Algorithm," *IEEE Transactions on Reliability,* Vol. 41(No. 3): 473–480, 1992.

49. D. H. Chi and W. Kuo, "Optimal Design for Software Reliability and Development Cost," *IEEE Journal on Selected Areas in Communications,* Vol. 8(No. 2): 276–281, 1990.

50. C. Mohan and K. Shanker, "Reliability Optimization of Complex Systems Using Random Search Technique," *Microelectronics and Reliability,* Vol. 28(No. 4): 513–518, 1988.

51. D. S. Bai, W. Y. Yun, and S. W. Cheng, "Redundancy Optimization of K-out-of-N:G Systems With Common-Cause Failures," *IEEE Transactions on Reliability,* Vol. 40: 56–59, 1991.

52. D. W. Coit and A. Smith, "Penalty Guided Genetic Search for Reliability Design Optimization," *Computers and Industrial Engineering,* Vol. 30(No. 4):895–904, September 1996.

53. B. Dengiz, F. Altiparmak, and A. E. Smith, "Efficient Optimization of All-Terminal Reliable Networks Using an Evolutionary Approach," *IEEE Transactions on Reliability,* Vol. 46(No. 1): 18–26, 1997.

54. D. L. Deeter and A. E. Smith, "Economic Design of Reliable Network," IIE Transactions, Vol. 30: 1161–1174, 1998.

55. G. Levitin, "Redundancy Optimization for MultiState System With Fixed Resource-Requirements and Unreliable Sources," *IEEE Transactions on Reliability,* Vol. 50(No. 1): 52–59, 2001.

56. M. Sakawa, "Optimal Reliability-Design of a Series-Parallel System by a Large-Scale Multiobjective Optimization Method," *IEEE Transactions on Reliability,* Vol. R-30: 173–174, 1982.

57. M. Sakawa, "Interactive Multiobjective Optimization by Sequential Proxy Optimization Technique (Spot)," *IEEE Transactions on Reliability,* Vol. R-31: 461–464, 1982.

58. K. B. Misra and U. Sharma, "An Efficient Approach for Multiple Criteria Redundancy Optimization Problems," *Microelectronics and Reliability,* Vol. 31: 303–321, 1991.

59. K. B. Misra and U. Sharma, "Multicriteria Optimization for Combined Reliability and Redundancy Allocation in Systems Employing Mixed Redundancies," *Microelectronics and Reliability,* Vol. 31: 323–335, 1991.

60. A. K. Dhingra, "Optimal Apportionment of Reliability and Redundancy in Series Systems Under Multiple Objectives," *IEEE Transactions on Reliability,* Vol. 41: 576–582, 1992.

61. D. Li, "Interactive Parametric Dynamic Programming and its Application in Reliability Optimization," *J. Mathematical Analysis and Applications,* Vol. 191: 589–607, 1995.

62. J. E. Yang, M. J. Hwang, T. Y. Sung, and Y. Jin, "Application of Genetic Algorithm for Reliability Allocation in Nuclear Power Plants," *Reliability Engineering and System Safety,* Vol. 65: 229–238, 1999.

63. P. G. Busacca, M. Marsequerra, and E. Zio, "Multiobjective Optimization by Genetic Algorithms: Application to Safety Systems," *Reliability Engineering and System Safety,* Vol. 72: 59–74, 2001.

64. K. S. Park, "Fuzzy Apportionment of System Reliability," *IEEE Transactions on Reliability,* Vol. R-36: 129–132, 1987.

CHAPTER 14
ADAPTIVE CONTROL

Jerry G. Scherer
GE Fanuc Product Development
Charlottesville, Virginia

14.1 INTRODUCTION

Adaptive control is a method for performing constant load machining by adjusting the axis path feedrate in response to load variations monitored at the spindle drive. The system usually comprises a spindle drive which can output an analog (0 to 10 V) representation of the load at the drive, a controller which calculates a path feedrate based on the difference between the target load and the load represented by the spindle drive, and a motion control which can accept path feedrate changes through an external input.

By maintaining a constant load at the tool, machining can be optimized to achieve the best volumetric removal rate for the process. In this manner the time to machine the part will be reduced, increasing the throughput of the process. Because the feedrate is adjusted during the machining process to achieve the desired load, the surface finish will change during the machining process. In general, adaptive control is used during the rouging and semiroughing machining processes where surface is not an issue.

Adaptive control can either be performed as an application within the motion controller (i.e., CNC) or by using an external processing module. In many cases the external processing module is the preferred type of controller as it is viewed as a "bolt-on" option that can be retrofit into existing applications. The external processing module is also not "embedded" into a system, making it applicable to more than just one brand of motion controller.

Although most adaptive controllers today provide broken and worn tool detection, it should be cautioned that this capability might not be suitable for many applications. The capability within the adaptive control is usually, at best, rudimentary detection based only on load. Specialized tool monitors have been developed to capture information from several different sensors and develop a "signature" of the tool to determine if it has become worn or broken. Machining processes that utilize unattended operation need specialized tool monitoring to avoid unnecessary scrapped parts and possible damage to the machine. It cannot be expected that rudimentary monitoring performed by an adaptive control module can replace a sophisticated tool monitoring system costing several times the price of the adaptive control unit.

14.2 PRINCIPLE AND TECHNOLOGY

Adaptive control machining is based on the premise that during the machining process the tool load can be approximated to the spindle drive load with a bias ($T_1 \cong S_1 + b$). The tool load can also be approximated as inversely proportional to the axis path feedrate ($T_1 \cong 1/F_p$). Since the tool load can be

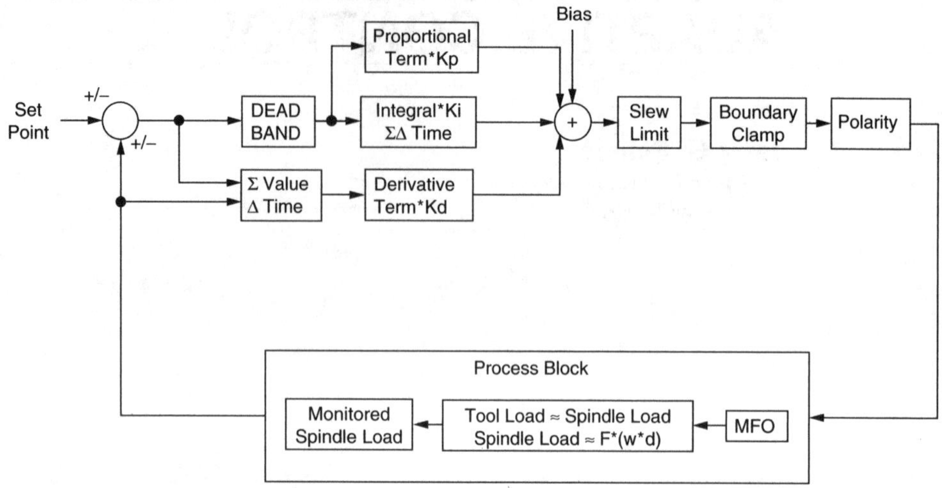

FIGURE 14.1 Independent term PID control loop.

equated to both the spindle drive load and the inverse of the axis path feedrate, it can be seen that changing one has an effect on the other ($T_1 = S_1 + b = 1/F_p \Rightarrow \Delta S_1 = 1/\Delta F_p$).

These factors allow the machining process to be controlled with a negative feedback, closed loop control algorithm. Different types of closed loop algorithms have been used in adaptive controllers. One example is the neural-type controller, which bases its output on "learned" patterns. The most used algorithm to date still remains the PID (Proportional/Differential/Integral) algorithm, which was chosen because it gave the best response to step input changes within a wide variety of process parameters and associated delays (Fig. 14.1).

14.3 TYPES OF CONTROL

14.3.1 Independent Term PID

One type of PID algorithm is the independent term PID, which operates by monitoring two input variables and outputting a correction variable such that the output correction will drive the process to make the input variables equal to each other. The two input variables are typically called the *set point* (SP) and the *process variable* (PV) and the output correction term is called the *control variable* (CV). Since the difference between the two input variables is of interest to the PID algorithm and it is a negative feedback control loop, this quantity is given its own name—*error term* (ε = SP – PV). In the independent term PID algorithm, the error term (ε) is observed and the corrective output term, control variable (CV), is calculated using the following equation:

$$\mathrm{CV} = (K_p \times \varepsilon) + (K_i \times \Sigma\varepsilon\,\delta t) + (K_d \times \delta\varepsilon/\delta t) \tag{14.1}$$

where K_p = proportional gain
K_i = integral gain
K_d = differential gain

It can be seen from Eq. (14.1) that the output control variable is made up of three terms. The first term is the proportional component of the corrective output. It is calculated by multiplying the error term by a constant known as the *proportional gain*. This term is proportional to the difference between the set point and the process variable. Its job is to apply a correction based simply on the difference between the set point (SP) and the monitored process variable (PV). It should be noted that by using only the proportional component, at steady state, there would be a constant difference between the desired set point and the process variable.

The second term is the integral component of the corrective output (CV). By multiplying the accumulated error by the accumulated time and then multiplying this quantity by a constant known as the *integral gain* the integral component is calculated. This term of the PID equation applies a correction based on the accumulated error over time ($\Sigma\varepsilon\,\delta t$). This term will drive steady state errors to zero, over time. The third term is the differential component of the corrective output (CV). It is calculated by multiplying the change in error between the prior sampling of the monitored control variable (PV) and the current sample and dividing it by the change in time between the samples and then multiplying it by a constant known as the *differential gain*. This term applies a correction based on the rate of change of the error term ($\delta\varepsilon/\delta t$), which will attempt to correct the output when changes in the set point (SP) or the process occur.

14.3.2 Dead Band Control

The PID control algorithm is a powerful method to calculate corrections to a closed loop process, but it can be overly sensitive to low-level noise or transients in the control loop. To overcome this problem, *dead band* control is usually added to the error term calculation to act as a filter. The function of the dead band control is to suppress small changes in the error term, which can be magnified by the PID calculation leading to unstable operation. Setting the error term to zero, when the error term is below a threshold value, provides the dead band control calculation. It can be expressed by the following pseudocode:

$$\begin{array}{c}\text{If } (\varepsilon < \text{dead band}) \\ \varepsilon = 0\end{array} \tag{14.2}$$

14.3.3 Slew Rate Control

Another issue with the PID control algorithm is that it is not sensitive to changes in inertial response. The machining process is traditionally accomplished through the removal of material and thus, changing the mass of the closed loop system results in a change in machining forces. If the control system is tuned using very low inertial mass, the control may become unstable when the inertial mass is significantly increased. This could lead to corrective solutions which could saturate (or overwhelm) the path axes moving the workpiece. To address this condition, a slew control algorithm has been added to the calculation. The slew control only allows the corrective output to change by a maximum amount. If the output of the PID calculation exceeds the prior solution by the slew limit, the new solution is clamped at the sum of the prior solution and the slew limit. Since the rate of change is clamped, the solution will be clamped to a fixed amount, which will limit the forces due to inertial response of the machine. The slew control can be expresses by the following pseudocode:

$$\begin{array}{c}\text{If } (\Delta\text{CV} < \text{slew limit}) \\ \text{CV} = \text{CV}_{\text{last}} + \text{slew limit}\end{array} \tag{14.3}$$

Thus it can be seen that ΔCV is acceleration, since CV is a path axis feedrate (velocity) command that is changing over time and from Newton's Law (F = ma) the following equation can be expressed:

$$F = m\Delta\text{CV} \tag{14.4}$$

From Eq. (14.4) we can see that by limiting the rate at which CV can change, the inertial response and resultant forces can be limited. This can provide higher gains in the PID control loop and protection against forces that could result from large changes in the mass being processed.

14.3.4 Maximum/Minimum Limit Control

It can be seen that the PID calculation could result in solutions that exceed the axis path feedrate of the machine or a negative feedrate (erroneous condition). To overcome this condition, a minimum/maximum clamp algorithm can be added. If the corrective output either exceeds the maximum limit or drops below the lower limit, the output will be clamped at that limit. The minimum/maximum control can be expressed by the following pseudocode:

```
If (CV > maximum clamp)
    CV = maximum clamp
If (CV < minimum clamp)                (14.5)
    CV = minimum clamp
```

14.3.5 Activation

Since adaptive control will adjust the path axis feedrate to obtain the desired tool load, we must control when it becomes active. Different designs have been used to control the activation of the adaptive controller. Some designs have been based on activating the adaptive control after a preset delay time has expired. Although this is one of the earliest and simplest methods, the user must have prior knowledge of the programmed feed rates and distance from the part. But even with this information, a change in the feedrate override by the operator or dimensional changes in the part could have disastrous results.

Another design for adaptive controller activation has been based on geometric information of the part. As this type of control is very difficult to accomplish in the external type controller, this type of activation has been mostly accomplished in the *embedded* type controller (running within the motion controller). Even this type of activation, although simpler than the previous method, has its own difficulties. Activation based on geometric information necessitates the previous knowledge of cutter path. If the cutter path, work offsets, or geometric part tolerances change, the adaptive control might not activate at the proper time.

To resolve these issues, some newer designs "learn" basic information about the process and attempt to control based on "experience." This design needs to "learn" a process before it can provide consistent results. If the design is used in an environment that will produce many of one part, it can provide satisfactory results. But if the production quantity is small or the process to be learned is not already optimized, satisfactory results may not be obtained.

A new concept has been added in the latest controller design, the *Demand Switch* algorithm. The demand switch automatically activates and deactivates the adaptive control system based on the load being monitored. This allows the user to either manually or programmatically "enable" the adaptive control feature. However, the design will activate only as needed. The activation or deactivation of the adaptive control is based on the monitored load exceeding or dropping, respectively, below a preset *arming limit*. Thus, if the user has enabled the adaptive control, the controller will actively send corrective solutions to the motion controller once the monitored load has exceeded the arming limit preset by the user. The adaptive controller will continue to calculate and send corrective solutions to the motion controller until the monitored load drops below the arming limit plus an offset.

The arming limit offset is adjustable and allows the controller to incorporate a *hysteresis* in the activation/deactivation condition. Hysteresis overcomes a possible condition where unstable or discontinuous operation could result when the monitored load is equal to the arming limit (i.e., control chattering between on and off states).

The demand switch has not been available on prior versions of adaptive control systems as they were also used for broken tool detection. To provide broken tool detection the adaptive control unit monitors when the monitored load drops below a threshold value. This is because when a tool breaks there is usually no engagement between the tool and the workpiece, resulting in no tool load. By removing the broken tool (no load) detector, the adaptive control can detect when the tool is engaged with the workpiece and when it is not. This makes the adaptive control an on-demand control system, simplifying operation.

One issue that has been difficult to overcome with prior versions of adaptive controllers is the *interrupted cut*. When the machining process comes to an area where there is no material (i.e., a hole in the part) the adaptive controller will increase the axis path feedrate to attempt to increase the load. The adaptive controller will ultimately request the maximum path feedrate to attempt to increase the load. The result is the tool proceeding through the *interrupt* at maximum speed and engaging the tool with catastrophic results. The demand switch design resolves this issue because activation is based on load and not temporal or positional information. When the adaptive control is enabled, but not yet activated, the corrective output (CV) is held to the programmed axis path feedrate of the motion controller. When the monitored process variable (PV) exceeds the arming limit, the corrective output CV is based on the PID control. When the monitored process variable drops below the arming limit plus an offset, the corrective output (CV) is again held to the programmed axis path feedrate of the motion controller. The demand switch algorithm can be expressed with the following pseudocode:

$$\begin{aligned} &\text{If } (PV_{scale} > \text{arming limit}) \\ &\quad CV_{out} = CV_{scale} \\ &\text{Else if } (PV_{scale} < \text{arming limit} + \text{offset}) \\ &\quad CV_{out} = \text{programmed feedrate} \end{aligned} \tag{14.6}$$

The earliest adaptive controllers were analog computers (arrangements of cascaded analog amplifiers) that were able to perform the most rudimentary control. With the advent of microprocessor technology, control algorithms became more sophisticated and resolve many of the issues found in early applications. The power to perform the adaptive control calculations can be provided by almost any microprocessor, digital signal processor (DSP) or microcontroller. It is the goal of the newest designs to simplify operation and provide processing information that can be used by other programs to further optimize the process and increase production throughput.

14.4 APPLICATION

As mentioned previously, the goal of adaptive control is to optimize the process and increase production throughput. This can be accomplished by a combination of both, decreasing the time to produce a part and decreasing the part rejection rate. In the past this has been accomplished by a skilled operator monitoring the process and maintaining both speed and quality during production. With the demands on business today to produce more and at a lower cost, many companies cannot afford to put a skilled operator at every machine on the production floor. It has now become the norm to have a skilled operator monitoring several machines at one time and augment with lower-skill operators.

Adaptive control attempts to bring the knowledge of the skilled operator to the machine control. As a skilled operator would make changes to the feedrate based on his or her sensory information, the adaptive controller will do the same. The skilled operators will use their sight, smell, and hearing to detect the load at the tool. Adaptive control has a more direct method to sense tool load—as described in the theory section—by monitoring the spindle load. As the operator would slow the process down when the chips demonstrated high-temperature discoloration or the sound of the process changed, adaptive control will also change the process speed to maintain consistent operation.

14.4.1 Installation

Although most installations of the adaptive controller will be different, there is some commonality among different vendors. First, we can categorize the level of integration into the motion controller's (i.e., CNC) tool management system as *standalone*, *semi-integrated* and *Fully Integrated.* Each category of integration necessitates an added level of work to implement the adaptive control scheme, but in addition provides a friendlier user interface; thereby simplifying its operation.

Standalone. The stand-alone configuration necessitates the least amount of integration between the motion controller and the adaptive control module. In the stand-alone configuration, the controller's user interface is provided through hardware inputs on the controller. Activation and the setting of internal information within the controller are performed through mechanical switches and possibly some connections to the motion controller's machine interface controller.

In this configuration, the adaptive controller has no integration with the tool management system within the motion controller. All control information is set by the user through the hardware inputs to the controller. Likewise, activation of the adaptive controller is performed through the assertion of a hardware input. Although this may seem like a cumbersome method to operate the adaptive controller, most installations of this type don't require much operator process intervention. The stand-alone configuration is the best example of a bolt-on (retrofit) application Fig.14.2). It requires the minimum interaction between the motion and the adaptive controllers, thereby requiring the minimum amount of time to install. Although the user interface is provided through mechanical switches, most applications will not necessitate process changes in normal operation.

Semi-Integrated. The semi-integrated configuration provides additional capabilities compared to the stand-alone configuration, but also requires additional interfacing to the motion controller's machine interface. The controller's interface is provided through connections to the motion controller's machine interface. In this manner, the motion controller can programmatically change activation and the setting of internal information within the controller.

In this configuration, the user sets and changes information within the adaptive controller through programmed requests of the motion controller. This necessitates some type of communication between the adaptive and motion-control units. The actual method of establishing communications

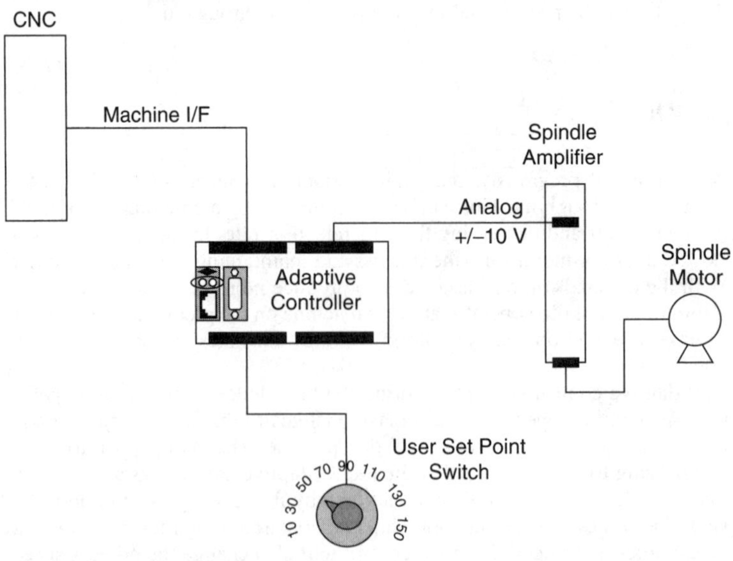

FIGURE 14.2 Standalone configuration.

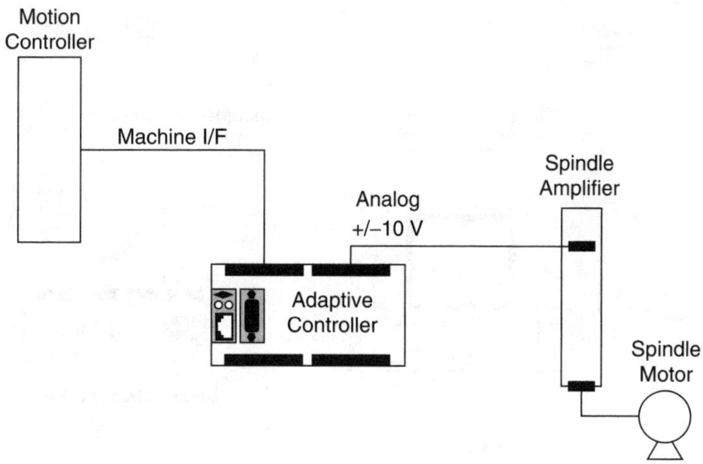

FIGURE 14.3 Semi-integrated configuration.

might be quite different between adaptive controllers, but the same information needs to be communicated. Most communications of this type are provided through generic (nonproprietary) methods such as serial or parallel connections. Some adaptive controllers also provide proprietary communication methods which can greatly reduce the time to interface but restrict that communication method to a specific motion controller.

Activation and changes to the internal settings of the adaptive controller are usually performed by issuing programmable G-codes (machine mode request) and M-codes (miscellaneous request) to the motion controller. The integrator of the adaptive controller will provide the specific programmable codes. These programmable codes are then available for manual data input (MDI) and normal part program operation.

The semi-integrated configuration is another example of a bolt-on (retrofit) application, but it requires the additional interaction between the motion and the adaptive controllers (Fig. 14.3). This translates to additional cost for installation but provides a simpler user interface. This type of configuration is ideally suited to processes that require changes to the adaptive control settings for optimized part throughput.

Fully Integrated. The fully integrated configuration provides the maximum capability to the user by directly interfacing with the motion controller's machine interface but at the cost of requiring the maximum amount of machine interface work. This type of configuration is usually performed by the machine tool builder (MTB) at the factory and is not generally suited to retrofit applications.

This configuration is usually integrated to the MTB's tool management system. In this manner, requesting different tools can change adaptive control settings. This does not mean that there can only be one set of adaptive control settings per tool. In modern tool management system, each tool can be redundant (more than one entry) in the tool table. This allows each tool to have more than one set of characteristics. In the case of adaptive control, redundant tool information is provided to allow different adaptive control settings for different operations. The redundancy is resolved during the request for a tool. Each tool has a number, but also a unique "member" identifier within a specific tool number. This allows the user to request a specific set of tool characteristics for each tool, based on the process. Like the semi-integrated configuration, activation and changes to the internal settings of the adaptive controller are usually performed by issuing programmable G-codes (machine mode request) and M-codes (miscellaneous request) to the motion controller. The integrator of the adaptive controller will provide the specific programmable codes. These programmable codes are then available for manual data input (MDI) and normal part program operation.

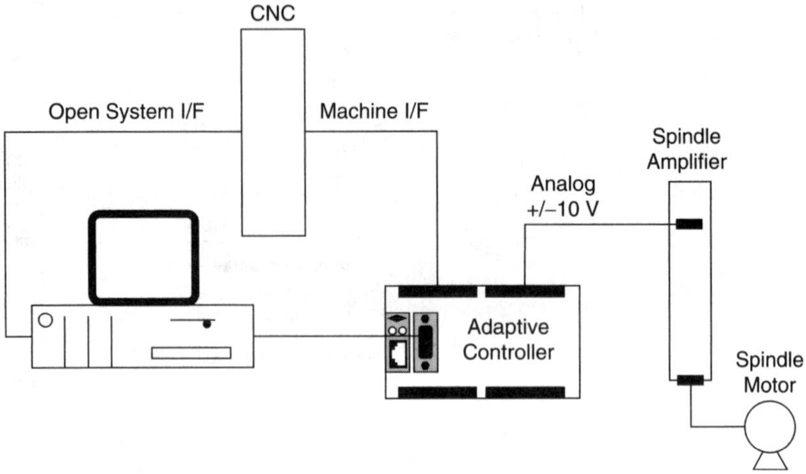

FIGURE 14.4 Fully integrated configuration.

The fully integrated configuration is not an example of a bolt-on (retrofit) application but some MTB's might provide a package for retrofitting to specific machine models (Fig. 14.4). Again, this translates to an additional cost for installation but provides the simplest user interface. This type of configuration is ideally suited to processes that require changes to the adaptive control settings for optimized part throughput.

14.5 SETUP

The setup of the adaptive controller can be broken down into three areas: hardware, application software and the user interface. Out of these three areas, the maximum time will be spent in setting up the user interface.

14.5.1 Hardware

To setup the hardware, it is understood that the adaptive and motion controllers are mounted and wired as per the connection information provided by the vendors. Specific information on the wiring of the units varies amongst vendors and therefore is beyond the scope of this book. But care should be taken to heed all the requirements specified by the vendor. The vendor has taken great care to ensure that their product conforms to the demanding requirements of the production environment. Failure to meet these requirements can yield unsatisfactory results or premature failure of the equipment.

Feedrate Override (FOV). In general, the adaptive controller provides path axis feedrate change requests through the motion controller's FOV feature. This can be accomplished by either wiring directly into the existing FOV control on the machine operator's panel or through the machine interface. Although it is easier to wire directly to the existing FOV control, this also provides the minimal capability available to the integrator and operator.

Monitored Load. The adaptive controller also needs to monitor some load with which it will attempt to maintain at a constant level. This connection is usually provided through an analog (0 to 10 V) signal output on the machine. In the case of a milling or turning machine, this will be an output on the spindle drive system. Since this is an analog signal, care should be taken to provide the "cleanest" signal possible to the adaptive control unit. This is accomplished by using standard noise suppression techniques (i.e., short length, shielded cable, proper grounding). Failure to do so will result in poor signal to noise ratio (SNR) which will limit the capability of the adaptive control unit.

Noise. The term SNR is a common term in electronics which refers to the amount of signal present compared to the noise. For example, if the adaptive controller is expected to track the input within 1 percent of the maximum load, this would mean 10 V (max scale) × 0.01 = 0.10 V. Now if the noise measured is 100 mV (0.10 V), then the SNR = 0.10/0.10 = 1.0. The lower the SNR, the less signal the controller has to work with and it means the controller will not only react to the signal present but also to the noise. The worst case SNR should be greater than 10.0, but the larger the SNR, the more signal the controller has to work with.

Think of it this way, have you ever tried to talk with someone else in a crowded room? Most of us have and it is difficult to hear and understand clearly what the person is saying. What is your method to contend with this situation? Normally it would be to raise your voice to improve the intelligibility of the conversation. This would be an example of increasing the signal strength while maintaining the noise level. Another method would be to retire to a quiet room where you could close the door and effectively filter out the noise. This is an example of reducing the noise the while maintaining the signal strength.

In the adaptive control case, you cannot normally increase the signal strength as it is fixed by the hardware involved. But you can either set the minimum signal strength with which you will operate, maybe increasing it by a factor of two or you could filter the noise. Active filtering is another discussion which is beyond the scope of this book, but filtering has its drawbacks. First, filtering implies decreasing the strength of the incoming signal. Second, it implies possible delays or "shifting" of the data compared to the original signal.

Grounding. The best method of handling any analog signals in the production environment is to follow proper grounding techniques (Fig. 14.5). Be sure that all grounds converge at one point

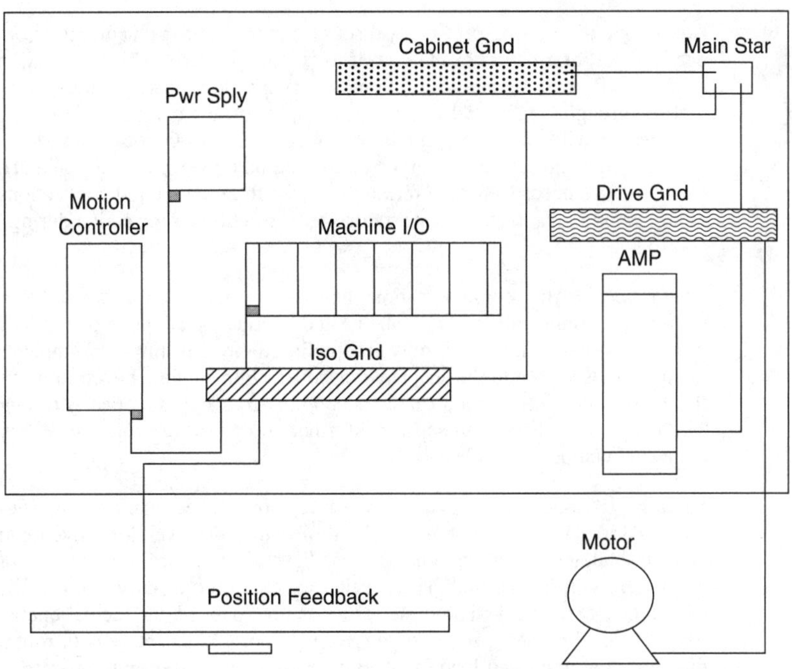

FIGURE 14.5 Grounding scheme.

(referred to as the star point) that is connected to the ground grid of the factory. Be sure that ground conductors are of sufficient size to carry the current, but also provide the best impedance match (AC resistance) among the ground conductors in the circuit. Avoid *ground loops*—do not connect ground conductors or shields to more than one ground point.

By minimizing the source of the noise, shielding the signal from noise, and providing the best signal strength, the adaptive control can provide the best operation possible. Techniques that involve active filtering should be avoided unless the noise frequencies are high and no other methods can provide adequate protection. By improving the SNR to the adaptive control unit, the controller will have the maximum signal to base its corrective adjustments on. Care taken in proper grounding techniques can be the difference between satisfactory and unsatisfactory operation.

14.5.2 Application Software

The application software is the software that is loaded by the vendor into the adaptive control unit. This software will generally have its internal registers initialized at the factory. In some cases the software might need initialization again during installation. In such situations the vendor will make provisions to reinitialize the unit in the field. This is usually accomplished through a hardware pushbutton or a utility command that can be issued by a maintenance person. In either case it is not a normal operation to reinitialize the adaptive controller; this should only be attempted by qualified personnel.

14.5.3 User Interface Software

The user interface software is the software that runs within the motion controller or an external personal computer. Like the application software it needs information to be initialized and loaded for normal operation. Each manufacturer of adaptive controllers has its own user interface software. It is beyond the scope of this book to go into the specific needs of your user interface, but there is some commonality among these interfaces (Fig. 14.6).

Communications. As previously discussed in the section on configurations, the adaptive controller needs to get information from the user to operate properly. This information might include set point (target load), *warning* and *alarm* limits for operation. The stand-alone configuration gets this information through mechanical switches and inputs on the controller. For the integrated solutions, this information will come through the motion controller as G-codes and M-codes. Thus, the integrated solutions must provide a communications method between the motion and adaptive controllers. The user interface needs to have information about the method and connections that provide these necessary communications. The integrator needs to set this information during the setup of the adaptive controller.

Commands. After communications have been established between the motion and adaptive controllers, the commands that will change and operate the adaptive controller need to be known by the motion controller. This is generally accomplished through simple assignment of G-codes and M-codes to commands known to the adaptive controller. In this manner everything from activation to altering the control loop information can be accomplished programmatically through the motion controller. The integrator will need to set this information up before programmatic operation on the adaptive controller can be accomplished.

Scaling. Although most motion controllers provide feedrate override (FOV) capability, the actual number input into this feature can be different among vendors. Even spindle manufacturers that provide analog output of spindle load can have different scaling factors. For these reasons, adaptive controllers will have a method to scale the monitored input and corrective output. The integrator should take care to calculate these values as commanded and actual operation might be different if not performed properly. A common error is to not correct for peak or root mean square (RMS) output from the monitored load input. Before completing the setup process ensure that the command and monitored load units agree.

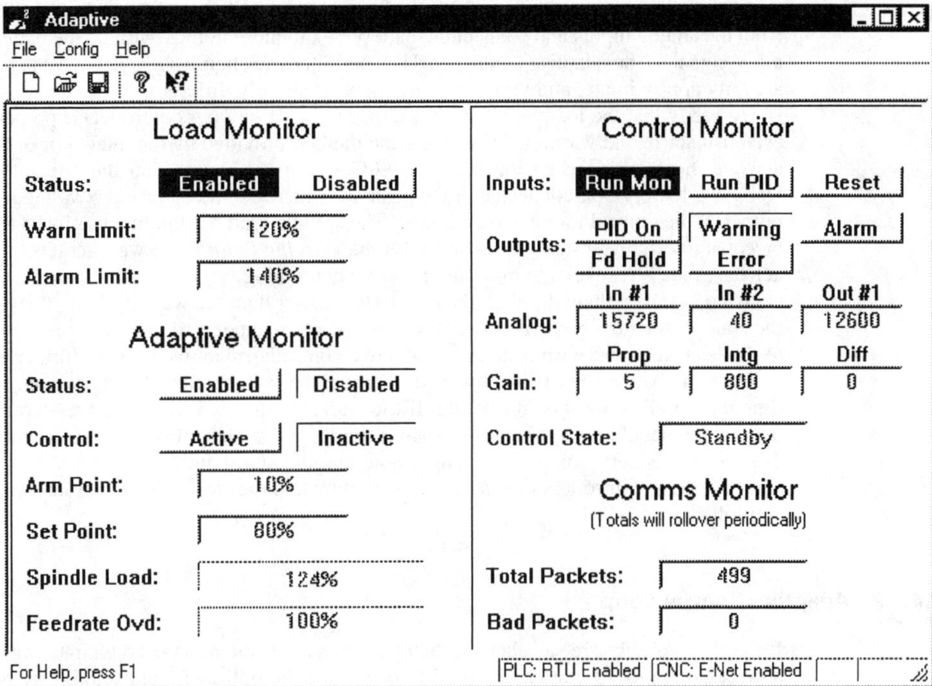

FIGURE 14.6 User interface screen.

Although we have not gone into all the setup operations that go into installing the adaptive controller, it should be known that this would be the most time consuming task in configuring your particular software. Careful planning and documentation of the setup process can minimize the additional machine setup time. It should also be noted that providing a common user interface among different machines will aid the user in best operating practices on each machine. Although the user interface software does not need to be the same, commands should be kept constant, if possible, across all machines using adaptive control on the production floor.

14.6 TUNING

The tuning of the PID gains within the adaptive controller can be approached using several different methods. Please feel free to use the method you feel most comfortable with. There have been many articles and research performed on the tuning of PID control loops. These can be broken down into analytical and practical approaches. The analytical approaches are based on the ability to instrument and determine closed loop response. These methods work well but can be confusing to persons not versed in control theory. The following is a general approach that can be used if the adaptive controller manufacturer does not provide instructions.

14.6.1 Load Monitoring (Optional)

Before starting the tuning process the load input should be tested. In this manner, not only can the spindle load calibration be checked, but also the load monitor interface prior to cutting. The first step should be to use a battery box to simulate the load input to the adaptive controller. Remove the wiring from the analog inputs and connect the battery box to these points. Turn on the battery box and apply a 5 V signal to the adaptive controller. The user interface should show about 50 percent, these values are only approximate, and your controller may be slightly different.

To verify that the load monitor feature (if this exists on your controller) is properly implemented, establish alarm and warning levels using the method provided by the integrator of this feature. Then activate the feature (using the M-Code, G-Code, or the switch that the integrator has provided). Using the battery box connected in the prior step, increase the battery box voltage until the warning output of the control monitor is activated. The spindle load read in the user interface display should be equal to or greater than the value set for the *warning limit*. If the warning level indicator does not activate, please recheck the installation before continuing.

Bringing the voltage level on the battery box lower than the warning limit, the warning output on the control monitor should deactivate. If this operation appears to work correctly, please continue on to the next step otherwise contact the adaptive controller manufacturer for further assistance.

With the battery box still connected, increase the voltage until the *alarm limit* is exceeded. This time more indicators should activate. If the warning limit is lower than the alarm limit (in normal operation it should always be), the warning, alarm, and possible FdHold outputs should all be active. Turn off the battery box and you might note that the alarm and FdHold outputs are still active. On some adaptive controllers this is normal operation and they will not deactivate until the *reset* on the controller is input.

14.6.2 Adaptive Control Loop

If the above checks are satisfactory, then you should be able to proceed with tuning. The following is a practical approach that can be used to determine the optimum gain settings by performing actual cutting and observing the stability of the loop control using the user interface screen. The first step is to select a material and tool that will allow up to 75 percent continuous spindle load to be achieved during cutting. For most machines this will mean some type of mild steel and the appropriate cutter.

The tuning will be performed at several load levels, but the tuning process will be the same. Begin by calculating a volumetric feed (depth and width of cut) that will—at a fixed feedrate—give an approximate 25 percent load on the spindle. If you are not sure of the proper feedrate and volume of cut, try different values of depth and use the feedrate override to determine a feed and volume that will give you approximately a 25 percent load in the spindle.

Next, we want to set the set point of the adaptive load control feature to 25 percent of the target load (use the method provided by the integrator for setting this target set point load). Make at least one cutting pass across the part to make sure that the spindle and feedrate are correct. Next, enable the load control feature using the method provided by the integrator (i.e., G126, if that is the method used). The user interface should indicate that the feature is enabled. Once the adaptive control feature is active, the feedrate control will try to adjust the path feedrate to attain the target set point load set by the user. Since we are just beginning, the load control feature will be very sluggish and might not attain the desired target set point. Do not despair, this is normal for an undertuned condition.

We want to first adjust the *proportional gain* of the controller to maintain stable operation and provide some margin of stability. To accomplish this task, go to the PID SETUP screen and increase the PROPORTIONAL GAIN (steps of 1 to 10 percent are used initially). Again make a cut at the part and watch the spindle load. If the load is steady, the loop is stable and you can increase it further. If the load changes abruptly from a low load to a large load alternating back and forth, the loop is unstable. If an unstable loop operation exists, stop the cut (i.e., feedhold) and decrease the PROPORTIONAL GAIN by approximately 10 percent and try again.

Continue adjusting the PROPORTIONAL GAIN and cutting the part until you believe you have found the maximum stable setting (i.e., no extreme load oscillations) and then decrease the gain setting by approximately 10 percent. This should provide stable operation of the proportional section of the PID loop control with a 10 percent margin of safety.

The next step is to adjust the INTEGRAL GAIN. Proceed as in the tuning of the proportional gain, but this time, adjust the integral gain. You will usually find that the integral gain can be adjusted in steps of 1 to 10 percent. Again you want to find the maximum setting that will provide stable operation and then back off the setting by approximately 10 percent.

The final step is to adjust the DIFFERENTIAL–PID GAIN. Again proceed as in the tuning of the proportional gain, but this time, adjust the differential gain. The differential gain is not normally used because its sensitivity is very high. Therefore you might find that even one count of gain will cause unstable operation. Please set to 0 in most cases.

Note The gain setting values are based on the spindle drive it is applied to. These values might be much larger than the initial values. Do not feel these settings are range limited. Please use the procedure described above or by the manufacturer. Once the loops have been tuned at 25 percent load, test loop stability at 50 percent, 75 percent, and 100 percent and readjust if necessary. You will find that after tuning the PID loops, stable accurate control will be attained by the adaptive control feature. In some case the user might find that different gear ranges might degrade the tuning and retuning of the loops should be performed. Please be sure to write down these gain settings as you might want to use them in the future. This should complete the tuning of load control feature. The steps for tuning are as follows:

- Check the load monitor feature

  Disconnect spindle load output from the adaptive controller.

  Connect the battery box to the inputs on the adaptive controller.

  Adjust battery box to around 5 V, should read 50 percent maximum spindle load on the display.

  Set the warning limit to 80 percent and the alarm limit to 120 percent.

  Adjust battery box until the warning indicator is displayed on the load monitor.

  Confirm that the spindle load reading is equal to or greater than the warn limit.

  Adjust battery box until the alarm indicator is displayed on the control monitor.

  Confirm that the FdHold indicator is also displayed on the control monitor. (optional)

  Turn off battery box, the alarm and the FdHold indicator should still be active. (optional)

  Assert the reset input on the VersaMax controller. (optional)

  Confirm that "alarm" and "FdHold" indicators are now deactivated. (optional)

- Tune the load adaptive feature

  Select tool and material for test cutting.

  Determine volumetric feed that will produce 25 percent spindle load.

  Make test cut on part to verify 25 percent spindle load occurs.

  Activate load adaptive feature.

  Verify that load control status is "ENABLED" on adaptive monitor display.

  Set a "SET POINT" target value of 25 percent.

  Start first test cut "off of the part."

  Verify that the tool depth, spindle speed, and path feedrate is correct.

  Observe spindle load display for large oscillation in load (unstable operation).

  If load is unstable, stop, cut, and decrease proportional gain (go to step j).

  If load is stable, at the end of cut increase proportional gain and repeat test (go to step j).

  Repeat steps until maximum stable gain value is attained.

  Decrease the value obtained in step m by 10 percent.

  Repeat steps h through n adjusting integral gain.

  If necessary, repeat steps h though n adjusting differential gain.

Note Step "o" will not normally need to be performed. In most adaptive control applications, no differential gain is necessary. Typically leave this setting at 0.

14.7 OPERATION

The most common difficulty in the operation of adaptive control is the determination of the set point load of the controller. There are two common methods to determine this setting. The first method is to monitor and learn an existing process. The second method is to calculate it based on the tooling and material under process. Each has its advantages and disadvantages.

14.7.1 Learned Set Point

The term "learning" is ambiguous and misleading in adaptive control. I prefer the term "analyzing." This might appear to be a matter of semantics but "learning" suggests that the process is being taught to the controller. In this line of thinking, the controller will "learn" the process so that it can replicate it over and over again. By "analyzing" the process, the controller captures the statistical information about the process with which it will attempt to optimize. In either case, data are acquisitioned by the controller and analyzed to determine the existing processing capabilities.

Learning is used when the process under control is understood from a processing point of view and not necessarily from a tooling and materials point of view. This in general means a process that is repeated over and over again. In this case, the adaptive controller will be given lots of information about the process with which it can analyze and optionally optimize. Some controllers have the capability to continue analysis during active control of the process and attempt further optimization automatically.

The disadvantage of the learning method is that it needs to perform processing to analyze. This is not always possible in a production environment. Some shops produce very limited quantities of a particular product. In this case, the learning method is not a satisfactory solution as much time is consumed in analyzing the part. The time consumed in the "learning" of the process might not be offset by the time gained in optimization by the adaptive controller.

14.7.2 Calculated Set Point

The calculated set point method is accomplished by using information about the tooling and material to calculate a set point. As previously noted, adaptive control is accomplished by attempting to maintain a constant load at the tool. This in turn is accomplished by monitoring the spindle load and adjusting the feedrate. The premise for this type of control is that the tool load can be controlled by changing the volumetric removal rate of the material under processing.

Tooling manufacturers have information available for most types of materials, providing the expected tool-life and horsepower requirements necessary for a given volumetric removal rate. Thus, knowing the tooling to be used, the material, the path geometry, and the spindle horsepower available, we should be able to set the load set point based on the tooling manufacturer information. An example will demonstrate this method for calculating the set point for the adaptive control.

Example

Given

Spindle: 20 HP – 2000 RPM, Efficiency = 70 percent
Tooling: 3/4 in × 11/2 in 4-flute End Mill

Material: Steel 1060 – BH 250
Path: Width = 1/2 in, Depth = 1/4 in, Feedrate = 100 in/min

$$\text{HPs} = \frac{Q \times P}{E}$$

where HPs = horsepower required spindle
Q = volumetric removal rate
P = unit horsepower factor
E = spindle efficiency

Base Hd on the tooling manufacturers information, unit horsepower for this tool and material:

$$P = 0.75 \qquad Q = \text{Fa} \times W \times D$$

$$= 100 \times 0.5 \times 0.25$$

$$= 12.5 \text{ in}^3/\text{min}$$

$$\text{HPs} = \frac{12.5 \times 0.75}{0.70} = 13.5 \text{ HP}$$

Set point:

$$\text{SP} = \frac{\text{HPs}}{\text{HPM}} \times 100$$

where SP = set point (based on percent maximum horsepower)
HPs = horsepower required at spindle
HPm = maximum horsepower at spindle

$$\text{SP} = \frac{13.5}{20} \times 100$$

$$= 67\%$$

As can be seen from the previous example, it is not difficult to calculate a set point based on tooling and material information. The only tooling information used in the previous example was for the unit horsepower rating for the tool in the given material. It should be noted that you will need to confirm the maximum rating for the tool based on the type of cutting to be performed. In general, tooling will have a higher horsepower figure during end-milling versus side-milling. Most tooling manufacturers provide the horsepower based on side-milling only as it is the limiting case. In these cases use the more limiting figure, even for end-milling, as it will provide an additional level of protection.

14.7.3 Range Versus Resolution

Adaptive control operates by modifying the axis path feedrate through the motion controllers FOV feature. The FOV provides these changes as percentages of the axis path-commanded feedrate. Though in most motion controllers the FOV's resolution is in increments of 1 percent, some motion controllers provide resolution down to 0.1 percent increments. Therefore the commanded feedrate and the increment of the FOV sets the resolution of the adaptive controller's axis path feedrate changes.

Some users of adaptive controllers have thought that all they have to do is command the maximum axis path feedrate and the controller will do the rest. In some cases this may be acceptable but in most it will not. If the commanded feedrate is too high, then the percentage change requested by the adaptive controller might also be too high. In the extreme case, the minimum feedrate override might still command a feedrate too high for the set point load to be accomplished. An example might demonstrate this case.

In our example let's assume we are trying to control a process such that the target load is 20 percent of the maximum spindle load. If the geometry of the cut is such that we need an axis path feedrate of 5.0 in/min and the programmed feedrate is 600 in/min, what would happen in this example?

Well, in the case of a standard motion controller, the minimum increment of the FOV is 1 percent. Thus, in our example the minimum feedrate available from the FOV is 600×0.01, which is equal to 6.0 in/min. Since this is larger than the axis path feedrate to maintain the 20 percent target load, the feedrate will drop to zero. When the axis path feedrate drops to zero, the tool will drop to 0 load (cut-free condition). In some controllers this is an illegal condition and will stop the machine process. In others the adaptive control will increase feed and start the cycle again. This operation will continue until the cut is complete or the operator intervenes.

One way to correct the condition is to decrease the commanded feedrate. If the maximum axis path feedrate to hold the target load is 150 in/min, why command 600 in/min? In our example above, by decreasing the commanded feedrate by a factor of 4, it will also decrease the minimum adaptive commanded feedrate by a factor of 4. This would give us a minimum adaptive feedrate of 1.5 in/min—much better than the case where we could not even get below the necessary minimum feedrate.

Even in the revised example, 1.5 in/min changes may result in an operation that is unacceptable. The increments of the path feedrate change may appear to "step" like in nature (discontinuous) and result in large demands on axis motion. This type of discontinuity can also excite resonance within the machine structure causing unacceptable anomalies and surface finish in the part. The best rule-of-thumb is to command the appropriate axis velocity to maintain the set point load (put it in the ballpark). With this, you can let the adaptive controller "drive" the machine for optimal processing.

14.7.4 Control Loop Constraints

We need to understand a little control theory, to get the most out of the adaptive controller. As mentioned in the section about adaptive control theory, the controller operates in a negative feedback closed control loop. This control loop is performed by measuring the load at the spindle and calculating the error between the measured load and the target load. A corrective output is calculated by the adaptive control algorithm and adjusts the motion controllers' FOV feature. The axis path feedrate that results from the corrective output changes the volumetric removal rate of the process. By changing the volumetric removal rate, the load required by the spindle will change. This control loop is performed over and over again, as long as the adaptive control feature is active.

Position Loop Gain The correction made through the FOV feature of the motion controller is generally applied during interpolation of the axis move. The motion controllers' interpolator then sends the command on to each axes servo control loop. This is the point we need to understand—the adaptive control loop commands the motion controllers servo control loop. But why is this important? In basic control theory we refer to control loops that command other loops as "outer" and "inner" control loops (Fig. 14.7). The outer loop is the commanding loop, the inner loop is the receiving loop. It can be shown that there is a maximum command rate at which the outer loop can command the inner loop and maintain stabile operation.

The theory involved is outside the scope of this book. However, in general, the maximum response rate of the outer loop is equal to one-third the inner loop's response rate. In terms of servo control, the response rate is equal to the inverse of the position loop gain (radians/sec). To calculate the minimum response rate for the adaptive controller we multiply the servo loop response rate times 3.). Thus, for a position loop gain of 16.67 rad/s, the servo response rate would be 0.06 s and the adaptive response rate would be 0.180 s.

Feed-Forward Control. To increase the response rate of the adaptive control loop we must increase the response rate of the servo control loop to maintain stabile operation. Features such as *feed-forward* control in the servo loop can further increase the response rates of the control loops by anticipating

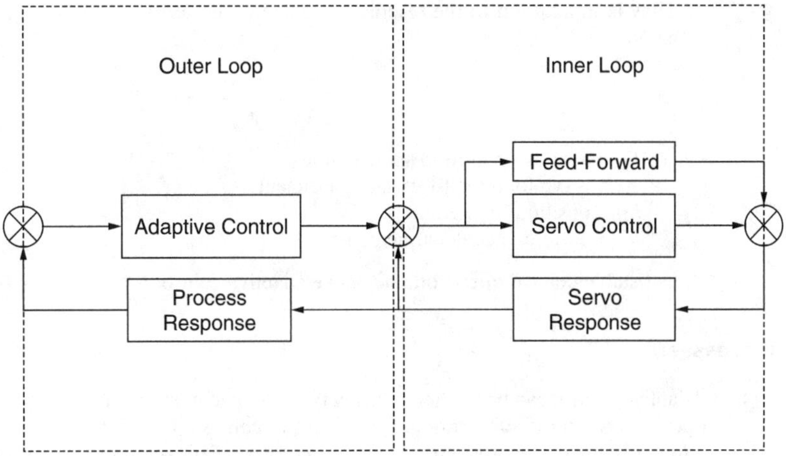

FIGURE 14.7 Inner and outer control loops.

the error due to closed loop control and "feeding it forward" in the servo control loop. The effect of using feed-forward control can be seen by the following equation:

$$Ke = \frac{K_P}{1-\alpha}$$

where Ke = effective position loop gain
Kp = position loop gain
α = feed-forward coefficient (0 – 1.0)

It can be seen that as the feed-forward coefficient increases, the effective position loop gain increase and therefore also the servo loop update rate. This then also allows the adaptive control response rate to increase. Thus it can be seen that the use of feed-forward control in the servo loop can help improve the response of the adaptive controller.

Acc/Dec Control. Another constraint that must be considered is the use of buffered acc/dec control in the motion controller. Most motion controllers now provide some means of controlling the rate at which the axis path feedrate changes. One of the earliest forms of acc/dec control was the use of the position loop gain to control the rate at which the axis path feedrate changed. Newer controls provide linear and acc/dec control which provide much smoother response. The issue is how the acc/dec control is performed.

One of the easiest methods to control acc/dec is to perform what is called *buffered acc/dec* (sometimes also referred to as acc/dec after the interpolator). In this type of acc/dec control all velocity changes occur over a fixed period of time. To make a velocity change all velocity commands are broken up into an integral number of segments and the requested velocity change is broken equally up into each. This does cause the desired acc/dec control but also delays the commanded velocity changes by the acc/dec control's fixed period of time.

As mentioned earlier, the adaptive control feedrate changes are performed through the FOV feature. The FOV feature then changes the motion controller's interpolator output, changing the commanded speed to the servo drives. Since buffered acc/dec occurs after interpolation, the delay caused by the acc/dec control will also delay the adaptive control changes. So to minimize any delays from the adaptive control unit, the buffered acc/dec time must also be minimized. The effect of this type of delay is in addition to the response rate of the servo drive system. This results in the following equation:

$$T_{ar} >= \frac{1-a}{K_p} \times 3 + \mathrm{T}_a$$

where T_{ar} = adaptive control response time
α = servo loop feed-forward coefficient
Kp = position loop gain
T_a = buffered acc/dec time constant

For stable control operation, the above adaptive control loop response time cannot be lower.

14.7.5 Processes

Adaptive control can be applied to many types of machining operations. Most people think of milling operation when the subject of adaptive control comes up. Both milling and turning operations have a long history of adaptive control. More recently, adaptive control has shown up on electronic discharge machines (EDM), drilling, boring, and even broaching machines.

The difficulty with adaptive control operation is the ability to maintain sensitivity over a broad range of input. Some people have thought that adaptive control cannot provide adequate control during cutting of an inclined surface. It is thought that the small change in thrust load is greatly offset by the side or bending load on the tool. This would be true if the load being measured by the adaptive control was the side load, but remember that the load being measured is the spindle load. The spindle load is a result of increased torque upon the tool because of the volumetric removal rate of the process. As the depth of the cut increases—due to the cutting of the inclined surface—the volume of material increases. This increases the torque demand upon the spindle drive and it is reflected through the analog load output of the drive system.

Don't think that all processes behave the same. Milling and turning operations are not the same—although the cutting physics are very similar, the machine construction and even the tooling can be very different. Materials, removal rates, and tooling can dictate the use of coolants which protect and aid the cutting processes. Make sure to purchase an adaptive control unit that has been developed for your type of processing. Failure to do so can increase the time it will take for you to obtain satisfactory results.

14.8 FINANCIALS

The decision to use adaptive control was based on increasing your part throughput while reducing your rejection rate. It simply does not make sense to be able to produce 30 *widgets* per day if they are all rejected. Similarly it does not make sense to produce less widgets per day if the rejection rate does not decrease. It comes down to the bottom-line—the more "acceptable" parts I can produce per unit time, the more units I can sell. Any new feature added to a process has to provide a return on investment (ROI) that outweighs putting our company funds into some other investment.

ROI attempts to quantify benefits (or profits) in terms of the expenditure. For production environment this will generally equate to weighing total benefits to total costs. It should be noted that there are incommensurable items that you will be unable to put a price tag on. For example, if your business produces a product for sale that is also used in the production of the product (i.e., Robots building Robots), what is the cost of using your own product? Sometimes true costs are lost on the person not versed in your production facilities. Be sure that all the pertinent information is used in the calculation of your ROI.

The adaptive control weighs the cost of the control unit versus profit enhancements using the control. The cost of the adaptive control will not require as much capital as that invested on the existing machinery. Adaptive control normally enhances only the roughing and semiroughing processes in your production environment. So please take this into account when evaluating your ROI.

A typical ROI calculation will involve the following:

Given

- Shift information

  Parts per shift

  Shifts per day

  Hour per day

  Hourly burden rate

  Production days per year

  Processing time per shift
- Adaptive information

  Adaptive productivity

  Adaptive installed costs

Calculate

- Production information

  Parts per day = parts per shift × shifts per day

  Cutting time per day = processing time per shift × shifts per day
- Production costs

  Cost per day = cutting time per day × hourly burden rate

  Cost per part = cost per day/parts per day

  Cost per year = cost per day × production days per year
- Production savings

  Annual cost savings = cost per year × adaptive productivity

  Daily cost savings = cost per day × adaptive productivity

  Part cost savings = cost per part × adaptive productivity
- Production capacity

  Additional daily capacity = hours per day × adaptive productivity

  Additional annual capacity = production days per year × adaptive Productivity
- Return on investment

  ROI = (annual cost savings − adaptive installed costs)/adaptive installed costs

  Payback (days) = adaptive installed costs/daily cost savings

  Payback (parts) = adaptive installed costs/part cost savings

Example

Given

- Shift information

  Parts per shift = 6.0 parts

  Shifts per day = 1.0 shifts

  Hour per day = 8.0 h

  Hourly burden rate = $80.00

Production days per year = 225 days
Processing time per shift = 6.0 h

- Adaptive information

  Adaptive productivity = 25 percent increase
  Adaptive installed costs = $10,000

Calculate

- Production information

  Parts per day = 6.0 × 1.0 = 6.0
  Cutting time per day = 6.0 × 1.0 = 6.0

- Production costs

  Cost per day = 6.0 × $80.00 = $480.00
  Cost per part = $480.00/6.0 = $80.00
  Cost per year = $480.00 × 225 = $108,000

- Production savings

  Annual cost savings = $108,000 × 0.25 = $27,000
  Daily cost savings = $480.00 × 0.25 = $120
  Part cost savings = $80.00 × 0.25 = $20.00

- Production capacity

  Additional daily capacity = 8.0 × 0.25 = 2.0 h
  Additional annual capacity = 225 × 0.25 = 56 days

- Return on investment

  ROI = ($27,000 - $10,000)/$10,000 = 1.70%
  Payback (days) = $10,000/$120 = 84 days
  Payback (parts) = $10,000/$20.00 = 500 parts

The above example demonstrates that the investment of using adaptive control will pay for itself within 84 days of operation and yield a 1.7 percent ROI. In a year, the investment will provide a theoretical profit of $13,680 ([production days per year – payback] × daily cost savings). This would tend to state that not only does the product pay for itself in much less than a year, it would also offset the cost of adding an additional unit to another machine. Adaptive control increases production capacity without any addition in floor space requirements.

I believe that the above example is a conservative estimate for adaptive control. Even more attractive ROI calculations have been run by manufacturers based on higher adaptive productivity and higher burden rates.

14.9 FUTURE AND CONCLUSIONS

Adaptive control has a long history leading back to the early 1960s. It has had many significant developments but they have occurred in bursts over the years. It has been like the technology has never been able to keep up with the requirements of adaptive control, at least not until now.

14.9.1 Technology

Advances in microprocessor and microcontroller technology have greatly reduced the number of components necessary to produce an adaptive control unit. What once took cabinets to house can now be placed in the palm of your hand. The restrictions in volume and space appear to be the only factors of power dissipation and human interaction.

The biggest advancements have been in the ease of operating the adaptive control units. With the advent of *open system* technology, integration with the motion controller has been greatly improved. The use of *neural-type* learning algorithms will become more preferred to the existing PID-style control algorithms. Neural-type algorithms have necessitated the computing power and memory of very large units. Advances have included not only the reduction is size and space of the computing units, but also reduction and simplifications in the neural-type algorithms.

User interaction has greatly improved with the newer style interfaces. The users are presented with graphical information, enabling them to integrate larger amounts of information in a shorter period of time. This not only improves the speed, but also the safety of machine operation. With newer 3-D virtual technology (available off the shelf) new developments are being made to integrate this intriguing technology. Simply by looking at a machine, all of its operating parameters are presented in a HUD (heads-up-display) for your easy viewing. The adaptive control might even warn you that something is wrong by "tapping you on the hand" or asking you to take a look at something.

The idea of adaptive control is to provide a knowledgeable assistant that can not only request help when needed but also take appropriate action if necessary. As sensor technology continues to advance, the adaptive control will also improve in its abilities. With the addition of sound sensors monitoring machine sounds such as chattering, the adaptive control can either take control to avoid the condition or suggest a change that the operator can make to avoid the condition. The cost of knowledgeable operators has increased while the availability has gone down. Adaptive control will aid this situation by providing the necessary assistance, no matter what the level of the operator. Integration has only just begun for the adaptive control unit. Newer user interfaces integrate directly into the machine's tool management system. In this manner the operator has to only input information into one area within the controller. In earlier nonintegrated systems, the adaptive information was separate from the tooling information. The operator had difficulty maintaining duplicate areas within the control. Data acquisition has become a new tool provided by the adaptive control system. Through data acquisition, the process can be analyzed at any time to aid in the development of more optimized processing capabilities. The data are also being incorporated into maintenance systems that can request maintenance, before it is absolutely needed. Further integration will see big changes not only in the adaptive control unit, but also the motion controller. Envision you are the operator of several machines. What a burden this must be—the machines run themselves but you must maintain the operation. With 3-D virtual technology, you will be able to view the process even through obstacles such as smoke or flood coolant. You will be able to feel if the tool is getting warm by just reaching out and touching the tool in virtual space. Your assistant gently taps you and asks for assistance with a problem on another machine. This is not science fiction. Developments are taking place today to enable this type of technology in the near future.

As processing power continues to increase with costs either staying or declining, advancements that were once thought of as fantasy are coming to fruition. Technology is allowing the innovation and creativity of engineers and designers to become reality. Need will drive the developments of tomorrow. Be sure the manufacturer of your adaptive control unit knows your needs; not for just today but also for the future.

CHAPTER 15
OPERATIONS RESEARCH IN MANUFACTURING

V. Jorge Leon
Texas A&M University
College Station, Texas

15.1 INTRODUCTION—WHAT IS OPERATIONS RESEARCH?

Operations research (OR) is a discipline based on applied mathematics for quantitative system analysis, optimization, and decision making. OR applications have benefited tremendously from advances in computers and information technologies. Developments in these fields have helped even very complicated analysis to be now conducted on a laptop or desktop computer. OR is general and has been successfully applied in manufacturing and service industries, government, and the military. Manufacturing examples of these successes include the improvement of car body production, optimal planning of maintenance operations, and the development of policies for supply chain coordination.

15.1.2 How Can It Help the Modern Manufacturing Engineer?

The modern manufacturing professional who is familiar with OR tools gains significant advantages by making data-driven decisions and a deeper understanding of the problem at hand. Often OR models lead to the formalization of intuition and expert knowledge—explaining why giving priority to produce the highest profit product on a bottleneck station may not be a good idea, or producing the most urgent job first is not necessarily the best option. OR helps the decision maker find not only a solution that works, but the one that works best. For instance, it guides the decision maker to form flexible cells and corresponding part families to minimize material handling and setup costs. With OR tools one can also assess the past and expected performance of a system. Finally, a family of OR tools is specifically designed to formulate decision problems in a variety of scenarios. In summary, OR tools can help the manufacturing engineering professional to:

- Better understand system properties and behavior
- Quantify expected system performance
- Prescribe optimal systems
- Make rational data-driven decisions

15.2 OPERATIONS RESEARCH TECHNIQUES

This section briefly describes a subset of OR techniques that have been successfully applied in manufacturing endeavors. Readers interested in a more complete yet introductory treatment of OR techniques can consult Hillier and Lieberman (2001) or Taha (2003). The techniques are classified based on whether they are suitable for system evaluation, system prescription and optimization, or general decision making—in all cases the results obtained from the OR analysis constitute the basis of quantitative information for decision making.

15.3 SYSTEM EVALUATION

System evaluation entails the quantification of past and future system performance. Sound system evaluation methods must explicitly account for the inherent variability associated with system behavior and errors associated with the data used in the calculations. The mathematical formalization of concepts of variability and expected behavior can be traced back to the seventeenth and eighteenth centuries where one can find the work of notable thinkers as B. Pascal, A. de Moivre, T. Bayes, C. F. Gauss, and A. Legendre, among others. Examples of system evaluation include predicting customer demands, determining work-in-process levels and throughput, or estimating the expected life of products. The main OR techniques for system evaluation are forecasting, queuing theory, simulation, and reliability theory.

15.3.1 Forecasting

Most decisions in the manufacturing business are directly influenced by the expected customer demand. Forecasting theory deals with the problem of predicting future demand based on historical data. These methods are also known as statistical forecasting. In practice, the decision maker typically modifies the numerical predictions to account for expert judgment and business conditions and information not captured by the general mathematical model. A suggested forecasting environment and main information flows are illustrated in Fig. 15.1.

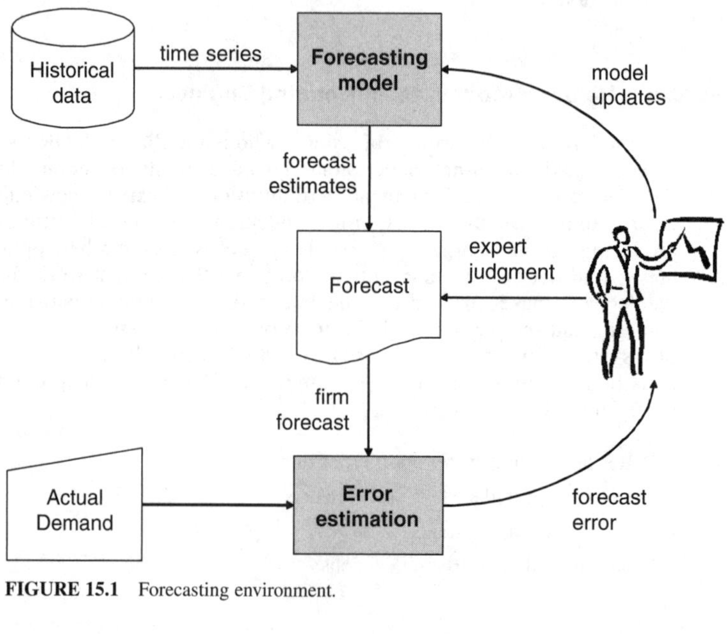

FIGURE 15.1 Forecasting environment.

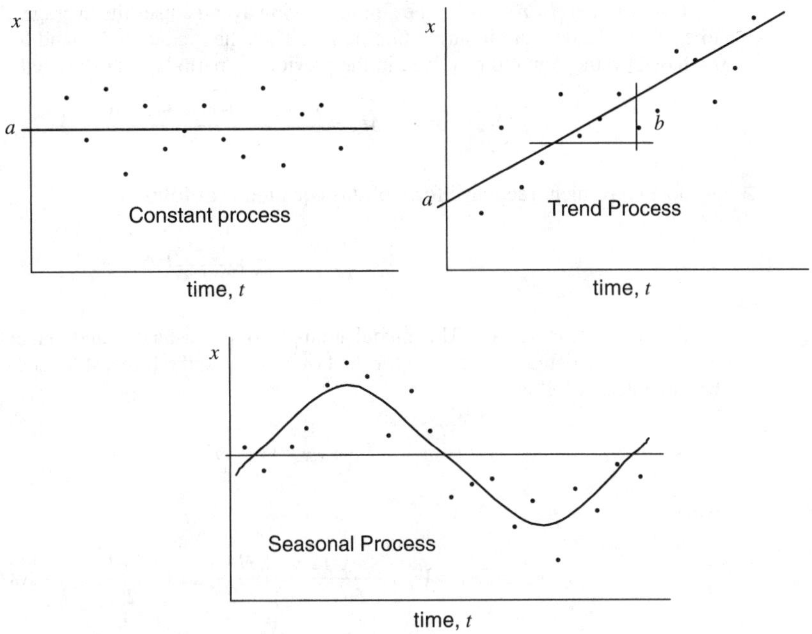

FIGURE 15.2 Time series patterns.

Three types of demand patterns can be forecasted using the models in this section, namely constant process, trend process, and seasonal process as illustrated in Fig. 15.2. The decision maker must plot the historical data and determine the appropriate demand pattern before applying any model. Popular OR models for single-item, short-term forecasting include simple moving average, simple exponential smoothing, and the Winters exponential smoothing procedure for seasonal demand.

Notation. The basic notation used in this section is as follows:

t = index denoting a time period

T = current time or time period at which the forecast decision is being made

x_t = demand level (units) in time period t

a = a constant representing a demand level (units)

b = a constant representing a demand trend (units/time)

c_t = a seasonal coefficient for time period t

It is convenient to assume that the actual demand levels are known up to the end of period $t = T$ (i.e., historical data up to period T), and that the demands for periods after T are predicted values. Typically the values of the constants a, b, and c_t are not known and must be estimated from the historical data. The estimated values, as opposed to actual values, are referred to using the "hat" notation $\hat{x}_t$, $\hat{a}$, $\hat{b}$, and $\hat{c}_t$. For instance, $\hat{a}$ reads *a-hat* and denotes the estimated or forecasted constant demand level.

Moving Average Methods. Moving average models use the last N demand levels to forecast future demand, as a new observation becomes available, the oldest one is dropped and the estimates are recalculated. Here "constant-" and "linear trend-" models are presented.

Constant Level Process. The simple moving average uses the average of the previous N periods to estimate the demand in any future period. Thus the predicted demand for any period after period t—based on the demand observed in the previous N periods—is estimated as follows:

$$\hat{x}_{T,N} = M_T = \frac{x_T + x_{T-1} + x_{T-2} + \cdots + x_{T-N+1}}{N}$$

A more convenient recursive form of this equation is as follows:

$$\hat{x}_{T,N} = \hat{x}_{T-1,N} + \frac{x_T - x_{T-N}}{N}$$

Linear Trend Process. This model attempts to forecast a demand that exhibits a linear trend pattern. Given the demand level data for the last N periods, the forecast level τ time periods after T, can be estimated as follows:

$$\hat{x}_{T+\tau,N} = \hat{a}_T + \hat{b}_T(T+\tau)$$

where

$$\hat{b}_T = W_T = W_{T-1} + \frac{12}{N(N^2-1)}\left[\frac{N-1}{2}x_T + \frac{N+1}{2}x_{T-N} - NM_{T-1}\right]$$

$$\hat{a}_T = M_T - \hat{b}_T\left(T - \frac{N-1}{2}\right)$$

M_T for linear trend process is the same as defined for the constant level process. Typical number of periods considered to calculate moving averages range within 3 to 12 (Silver and Peterson, 1985).

Exponential Smoothing. Exponential smoothing methods are popular forecasting methods because of their accuracy and computational efficiency when compared to the moving average methods. The basic idea is to make predictions by giving more weight to recent data and (exponentially) less weight to older observations.

Constant Level Process (Single Exponential Smoothing). Given the forecast for the previous period ($\hat{x}_{T-1}$), the latest observation (x_T), and a *smoothing constant* (α), the demand for *any* period after period t is estimated as follows:

$$\hat{x}_T = \alpha x_T + (1-\alpha)\hat{x}_{T-1}$$

An equivalent expression can be conveniently derived by rearranging the above formula in terms of the forecast error $e_T = (x_T - \hat{x}_{T-1})$ as follows:

$$\hat{x}_T = \hat{x}_{T-1} + \alpha e_T$$

Notice that in single exponential smoothing only the demand level at the current time period and the previous forecast need to be stored, and that the historical information is captured in the previous forecast value.

Trend Process (Double Exponential Smoothing). The forecast level at τ time periods after T, can be estimated as follows:

$$\hat{x}_{T+\tau} = \hat{a}_T + \tau\hat{b}_T$$

where

$$\hat{a}_T = [1-(1-\alpha)^2]x_T + (1-\alpha)^2(\hat{a}_{T-1} + \hat{b}_{T-1})$$

$$\hat{b}_T = \left[\frac{\alpha^2}{1-(1-\alpha)^2}\right](\hat{a}_T - \hat{a}_{T-1}) + \left[1 - \frac{\alpha^2}{1-(1-\alpha)^2}\right]\hat{b}_{T-1}$$

The use of regular unweighted linear regression is recommended to initialize a and b. Let 1, 2, 3,…, n_o be the available historical demand observations. The initial a_o and b_o values are calculated as follows:

$$\hat{b}_o = \frac{\sum_1^{n_o} t x_t - \frac{n_o+1}{2}\sum_{t=1}^{n_o} x_t}{\sum_{t=1}^{n_o} t^2 - (\sum_{t=1}^{n_o} t)^2/n_o}$$

$$\hat{a}_o = \frac{\sum_{t=1}^{n_o} x_t}{n_o} - \frac{\hat{b}_o(n_o+1)}{2}$$

Selection of Smoothing Constants. Smaller values of the smoothing constant α tend to give less importance to recent observations and more importance to historical data. Conversely, larger values of α tend to give more weight to recent information. Therefore smaller values of α are preferred in situations where the demand is stable and larger values of α should be used when the demand is erratic. Johnson and Montgomery (1974) recommend that the value of α is chosen between 0.1 and 0.3.

Seasonal Processes. Winters' method is described here for forecasting under processes exhibiting seasonal behavior. In addition to a level and a trend, this model incorporates a seasonal coefficient. It is assumed that the season has a period P. The forecast level at τ time periods after T can be estimated as follows:

$$\hat{x}_{T+\tau} = (\hat{a}_T + \tau\hat{b}_T)\hat{c}_{T+\tau}$$

where

$$\hat{a}_T = \alpha_s\left(\frac{x_t}{\hat{c}_{T+\tau-P}}\right) + (1-\alpha_s)(\hat{a}_{T-1} + \hat{b}_{T-1})$$

$$\hat{b}_T = \beta_s(\hat{a}_T - \hat{a}_{T-1}) + (1-\beta_s)\hat{b}_{T-1}$$

$$\hat{c}_{T+\tau} = \gamma_s\left(\frac{x_T}{\hat{a}_T}\right) + (1-\gamma_s)\hat{c}_{T+\tau-P}$$

The seasonal index $\hat{c}_{T+\tau-P}$ is the estimate available from the previous period.

Selection of Smoothing Constants. The seasonal smoothing constants α_s, β_s, and γ_s must be selected between 0 and 1. Silver and Petersen (1985) suggest that the initial values for α_s and β_s can be obtained in terms of the smoothing constant α used in the previous models as follows:

$$\alpha_s = 1-(1-\alpha)^2 \qquad \text{and} \qquad \beta_s = \frac{\alpha^2}{\alpha_s}$$

Moreover, for stability purposes the value of α must be such that $\alpha_s \gg \beta_s$. Experimentation is recommended to appropriately select the values of α_s, β_s, and γ_s.

Forecasting Error Estimation. The methods presented earlier in this section only give an expected value of the forecast for some time period in the future. In order to quantify the accuracy of the prediction it is useful to estimate the standard deviation associated with the forecast errors (recall that $e_t = x_t - \hat{x}_{t-1}$).

A common assumption is to consider the forecast errors distributed normally with mean zero and standard deviation σ_e. Given n past periods and corresponding forecast errors $e_1, e_2, \ldots,$ and e_n, the standard deviation of forecast errors is estimated as follows:

$$\sigma_e \approx s_e = \sqrt{\frac{\sum_{t=1}^{n}(e_t - \bar{e})^2}{n-1}}$$

An alternative method to estimate σ_e is using the *mean absolute deviation* (MAD) as follows:

$$\sigma_e \approx 1.25(\text{MAD}) = 1.25\left(\frac{\sum_{t=1}^{n} |e_t|}{n}\right)$$

Some practitioners prefer to use MAD because of its practical meaning; i.e., it is the average of the absolute value of the forecast errors.

Application Example—Exponential Smoothing. Consider the demand for a given product family summarized in Fig. 15.3. For illustration purposes, assume that data from January to June are known before starting to apply forecasting. The values after June compare the actual demand and the 1-month look-ahead forecast. The data suggests that a trend model may be appropriate to forecast future demand. First, the initial values $a_o = 522.35$ and $b_o = 37.33$ are calculated using the data from January to June. Assuming a smoothing constant of $\alpha = 0.15$, and given the demand for the month, the forecasts for the next month are obtained. The forecast plot in Fig. 15.1 is the result of applying the model in July, August, September, and so on. The standard deviation of the forecast error can be estimated using the MAD method: MAD = [|722 – 673| + |704–731| + |759 – 767| + |780 – 808| + |793 – 843| + |856 – 871|]/6 = 29.47; the standard deviation $\sigma_e \approx (1.25)(29.47) = 36.84$.

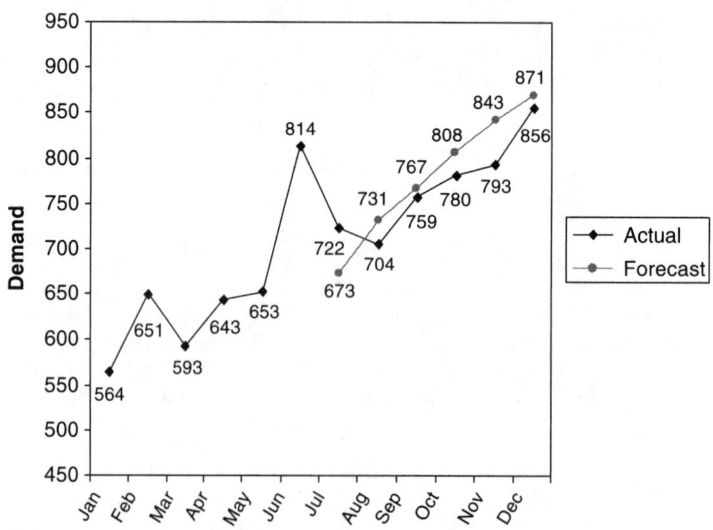

FIGURE 15.3 Actual and forecasted demand for the example.

15.3.2 Queuing

Queuing theory studies the performance of systems characterized by entities (e.g., customers, products) that must be processed by servers (e.g., bank tellers, machining centers), and waiting lines (or queues) of entities that form due to busy servers. In queuing theory the variability of systems is considered explicitly. Applications of queuing theory to manufacturing systems include the determination of important performance metrics such as lead-times and work-in-process, the specification of the buffer space needed between two work-centers, and the number of machines needed, among many other applications.

The elemental queuing system model consists of an *input source*, *queues*, *queue discipline*, and *service mechanism*. The input source (or population) can be *finite* or *infinite* and the pattern by which entities arrive to the system is specified by an *interarrival time*. The queues can also be *finite* or *infinite* depending on their capacity to hold entities. Queue discipline refers to the *priority rules* used to select what entity in the queue to select for service. Finally, the service mechanism is characterized by the *number of servers*, the *service time,* and *server arrangement* (i.e., parallel or serial servers). For instance, some basic queuing models assume that there is an infinite population of entities that arrive to the system according to a Poisson process, the queue capacity is infinite, the queue discipline is first-in-first-out (FIFO), there are a given number of parallel servers, and exponentially distributed service times. Figure 15.4 illustrates an elemental queuing system.

Definitions and Basic Relationships. Basic queuing concepts and the notation are summarized as follows:

s = number of parallel servers in the service facility

λ = mean arrival rate (expected number of arrivals per unit time)

μ = mean service rate (expected number of entities served per unit time)

$\rho = \lambda/s\mu$, utilization factor for the service facility

L = expected number of entities in the queuing system

L_q = expected number of entities in queue

W = expected waiting time in the system (for each entity)

W_q = expected waiting time in queue

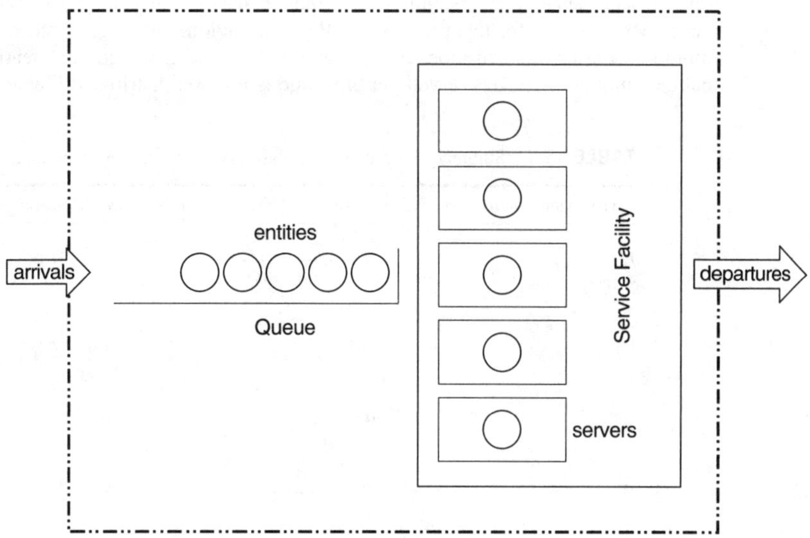

FIGURE 15.4 Elemental queuing model.

All these concepts have useful interpretation in the manufacturing context if entities represent products and servers represent machines. For instance, in the long run λ can be viewed as the demand rate, $1/\mu$ the mean processing time, L the mean work-in-process, and W the mean manufacturing lead-time.

Thus it is convenient to view L and W as system performance metrics. Queuing theory yields the following fundamental steady-state relationships among these performance metrics:

$$L = \lambda W \text{ (Little's Law)}$$

$$L_q = \lambda W_q$$

$$W = W_q + \frac{1}{\mu}$$

These fundamental relationships are very useful because they can be used to determine the performance metrics if any of them is calculated or known.

Also for stability it is important that $\rho < 1$.

Given that the interarrival and service times are random variables, the queuing models will depend on the underlying probability distributions. Covering all known cases is out of the scope of this manual. This section presents two types of queuing models. The first case assumes that the system exhibits Poisson arrivals and exponential service times. The second case assumes general distributions. Interested readers are referred to Buzacott and Shantikumar (1993) for an extensive treatment of queuing models for manufacturing systems.

Constant Arrival Rate and Service Rate—Poisson Arrivals and Exponential Service Times. This model assumes that the number of arrivals per unit time is distributed according to a Poisson distribution. This is characteristic of systems where the arrival of entities occurs in a totally random fashion. An important characteristic of Poisson arrivals is that the mean number of arrivals in a period of a given length is constant.

Exponential service times refer to service times that are distributed according to an exponential probability distribution. This system is characterized by totally random service times where the next service time is not influenced by the duration of the previous service (i.e., memoryless property), and the service times tend to be small but can take large values occasionally.

The exponential and Poisson distributions are related. Consider a process where the interarrival time of occurrences is exponentially distributed. It is possible to prove that the number of occurrences per unit time for this process is a Poisson random variable. In other words, Poisson arrivals imply that the interarrival times are exponentially distributed and exponential service times imply that the number of entities served per unit time is Poisson distributed (Table 15.1).

TABLE 15.1 Summary Formulas for Poisson Arrivals and Exponential Service Times

Performance metric	Single-server model ($s = 1$)	Multiple-server model ($s > 1$)
L	$L = \frac{\rho}{1-\rho} = \frac{\lambda}{\mu-\lambda}$	$L = \lambda\left(W_q + \frac{1}{\mu}\right) = L_q + \frac{\lambda}{\mu}$
L_q	$L_q = \frac{\lambda^2}{\mu(\mu-\lambda)}$	$L_q = \frac{P_o(\lambda/\mu)^s \rho}{s!(1-\rho)^2}$
W	$W = \frac{1}{\mu-\lambda}$	$W = W_q + \frac{1}{\mu}$
W_q	$W_q = \frac{\lambda}{\mu(\mu-\lambda)}$	$W_q = \frac{L_q}{\lambda}$

For the multiple-server model, P_0 is the probability that there are zero entities in the system. The expressions for single- and multiple-server models have been adapted from Hillier and Lieberman (2001).

Constant Arrival Rate and Service Rate—General Distributions. Often, in manufacturing, the arrival and service rates are better understood than what has been assumed in the previous section. Rather than assuming exponentially distributed time between events, the models in this subsection use the mean and standard deviation of the interarrival and service times. Emphasis is given to results that can be used to relate manufacturing lead-times, work-in-process, throughput, and utilization as a function of the system variability.

The following additional notation will be used in this section:

λ^{-1} = mean interarrival time
σ_a = interarrival time standard deviation
μ^{-1} = mean service time (without breakdowns)
σ_s = service time standard deviation (without breakdowns)
$V_a = \sigma_a/\lambda^{-1}$, variability ratio of interarrival times
$V_s = \sigma_s/\mu^{-1}$, variability ratio of service times
f = mean-time-to-fail, or mean time between equipment breakdowns
r = mean-time-to-repair, mean time to repair breakdown equipment
$A = f/(f + r)$, availability ratio (fraction uptime)

The performance metrics assuming a single server are calculated as follows:

$$\rho = \frac{\mu^{-1}/A}{\lambda^{-1}}$$

$$W_q = \frac{1}{2}\left(\frac{\rho}{1-\rho}\right)\left[V_a^2 + V_s^2 + \left(\frac{2rA(1-A)}{\mu^{-1}}\right)\right]\left(\frac{\mu^{-1}}{A}\right)$$

$$W = W_q + \frac{\mu^{-1}}{A}$$

$$L = \lambda W$$

The above equations apply for a single server, single step system. It is relatively straight forward to extend the analysis to a serial line configuration using the following linking equation to determine the arrival variability ratio for the next stage $V_{a,next}$ based on the parameters of the current stage.

$$V_{a,next}^2 = \rho^2\left(V_s^2 + \frac{2rA(1-A)}{\mu^{-1}}\right) + (1-\rho^2)V_a^2$$

A more comprehensive treatment of these types of general queuing models can be found in Hopp and Spearman (1996) or Suri (1998).

Application Example. Consider a machining work center that can nominally produce an average 100 units/h with a buffer space limited to a maximum of 25 units. If the buffer reaches its maximum, the previous stage stops production until buffer space is available. The problem observed is that average demand rates of 75 units/h cannot be achieved even though, at least nominally, the utilization appears to be 0.75 (or 75 percent). Further investigation reveals that the work center breaks down every 3 h, and it takes about 0.5 h to fix the problem on the average. The utilization can be updated by considering the availability factor, $A = 3/(3 + 0.5) = 0.86$; or utilization $= 0.75/0.86 = 0.87$. Notice that accounting for machine reliability yields a higher utilization but still less than 1.0, so it does not explain why the demand rate cannot be reached. Thus it is necessary to use queuing models to explicitly consider the variability in the system. Let's assume that the variability ratio for the interarrival times and service times are 0.6 and 0.3, respectively. The average time parts stay waiting to be serviced on the machine is:

$$W_q = \frac{1}{2}\left(\frac{0.87}{1-0.87}\right)\left[0.6^2 + 0.3^2 + \frac{2(0.5)(0.86)(1-0.86)}{(1/100)}\right]\left(\frac{(1/100)}{0.86}\right) = 0.49 \text{ h}$$

The average time in the system is $W = 0.49 + (1/100)/0.86 = 0.50$ h and the average wip level needed to produce 75 units at any point in time is $L = (75)(0.50) = 37.5$ units. Clearly, this exceeds the available buffer space and explains why the desired production levels are not achieved.

15.3.3 Other Performance Evaluation Techniques

Two additional performance evaluation tools in OR are *simulation* and *reliability*. These techniques are covered in detail in other chapters in this manual.

15.4 SYSTEM PRESCRIPTION AND OPTIMIZATION

An important class of OR techniques is aimed at prescribing the best (optimal) way of achieving a given goal or set of goals. Examples are the determination of the production plan that will minimize costs, the product mix that maximizes profit given the available capacity, and the best route to follow to minimize traveled distances. This section gives an introductory description of one important OR optimization technique; namely, mathematical programming.

OR employs mathematics to model the real situation and prescribes efficient solution methodologies to obtain the desired results. Often the mathematical model uncovers structural properties of the problem that become part of the decision maker's deep knowledge. The main elements of the mathematical model are *decision variables, objective functions,* and *constraints*. The mathematical properties of these elements typically determine the appropriate OR technique to utilize.

15.4.1 Linear Programming

Because of its broad applicability, the most popular type of mathematical programming is *linear programming* (LP). LP considers a single objective with multiple constraints where the objective and the constraints are linear functions of real decision variables. The following notation will be used in this section:

x_i = real valued decision variable, $i = 1,\ldots,N$

c_i = per unit objective function coefficient associated with decision variable i

a_{ij} = per unit constraint coefficient associated with decision variable i in constraint j, $j = 1,\ldots,M$

b_j = bound (i.e., right hand side) associated with constraint j

Z = objective function

The general form of a linear programming model is as follows:

$$\text{Maximize (or Minimize)} \quad Z = \sum_{i=1}^{N} c_i x_i$$

$$\text{Subject to:} \quad \sum_{i=1}^{N} a_{ij} x_i \ (\leq, \text{or} =, \text{or} \geq) b_j, \qquad j = 1,\ldots,M$$

x_i is a real variable.

TABLE 15.2 Input Data for LP Example

<table>
<tr><td colspan="2" rowspan="2">Consumption rate, a(i, j)</td><td colspan="2">Equipment, j</td><td rowspan="2">Unit profit, c(i)</td><td rowspan="2">Potential demand</td><td rowspan="2">Decision variable: Production level, x(i)</td><td rowspan="2">Profit</td></tr>
<tr><td>Lathe, L</td><td>Grinder, G</td></tr>
<tr><td rowspan="2">Product, i</td><td>A</td><td>9</td><td>0</td><td>12</td><td>6</td><td>6.00</td><td>72.0</td></tr>
<tr><td>B</td><td>16</td><td>10</td><td>16</td><td>10</td><td>10.00</td><td>160.0</td></tr>
<tr><td colspan="2">Available capacity, b(j)</td><td>140</td><td>110</td><td></td><td colspan="2">Total profit</td><td>232.0</td></tr>
<tr><td colspan="2">Potential cap. req.</td><td>214</td><td>100</td><td></td><td></td><td></td><td></td></tr>
<tr><td colspan="2">Actual cap. req.</td><td>214</td><td>100</td><td></td><td></td><td></td><td></td></tr>
</table>

Many decision problems facing the manufacturing professional can be modeled as a linear program. A simple scenario will be used to illustrate this technique. Consider a situation where a manufacturer has good market potential but the sales potential exceeds the available manufacturing capacity—LP can be used to determine the best quantities of each product to produce such that the profit is maximized and the available capacity and market potential are not exceeded. The input data and information for our example is summarized in Table 15.2.

An initial calculation assuming that all the potential market is exploited results in a profit of 232. However, this potential cannot be achieved because the available capacity at the Lathe work-center is exceeded (214 > 140). The decision here is to determine what should be the production levels that will maximize profit. The LP model for this problem can be expressed as follows:

Maximize profit: $Z = 12x_A + 16x_B$

Subject to:

Capacity constraint for the lathe: $9x_A + 16x_B \le 140$

Capacity constraint for the grinder: $0x_A + 10x_B \le 110$

Market constraint for product A: $x_A \le 6$

Market constraint for product B: $x_B \le 10$

Nonnegativity constraint: $x_A, x_B \ge 0$

LPs can be solved efficiently for formulations with thousands of variables and constraints using commercially available software. For smaller problems LP solvers are included in common office applications. Table 15.3 shows the solution obtained using MS Excel's *solver* tool. The *optimal* solution is to produce 6 units of product A and 5.38 units of product B, or $x_A = 6$ and $x_B = 5.38$, respectively. This yields a maximum profit of 158. This solution is called optimal because no other production mix can result in higher profit. In this particular instance—against most common sense solutions—the optimal strategy is not to produce the product that has higher per-unit profit, or higher demand potential.

An additional advantage of LPs is that the solution includes other useful information for decision making. In particular, *slack variables, shadow prices,* and *sensitivity analysis.*

Slack variables in the optimal solution provide information of how binding each constraint is. For instance, in our example the slack variables associated with the lathe's capacity and product A's

TABLE 15.3 Maximum Profit (Optimal) Solution for the LP Example

Consumption rate, $a(i,j)$		Equipment, j		Unit profit, $c(i)$	Potential demand	Decision variable: Production level, $x(i)$	Profit
		Lathe, L	Grinder, G				
Product, i	A	9	0	12	6	6.00	72.0
	B	16	10	16	10	5.38	86.0
Available capacity, $b(j)$		140	110			Total profit	158.0
Potential cap. req.		214	100				
Actual cap. req.		140	53.75				

demand have a value of zero indicating that these constrains are binding—i.e., they are restricting the possibility of more profit. On the other hand, the slack variables associated with the capacity of the grinder and product B's demand are 56.25 and 4.625, respectively, and they are not restricting the profit in the current solution.

Shadow prices in the optimal solution represent the amount of increase in objective function attainable if the corresponding constraint bound is increased by one unit. The shadow prices associated with the Lathe and Grinder capacity constraints are 1 and 0, respectively; this tells us that if we had the choice of increasing capacity, it would be most favorable to increase the capacity of the lathes. Similarly, the shadow price associated with the demand of product A is larger than that of product B; i.e., having the choice, it is better to increase the market potential of product A.

Sensitivity analysis provides the range of values of each model parameter such that the optimal solution will not change.

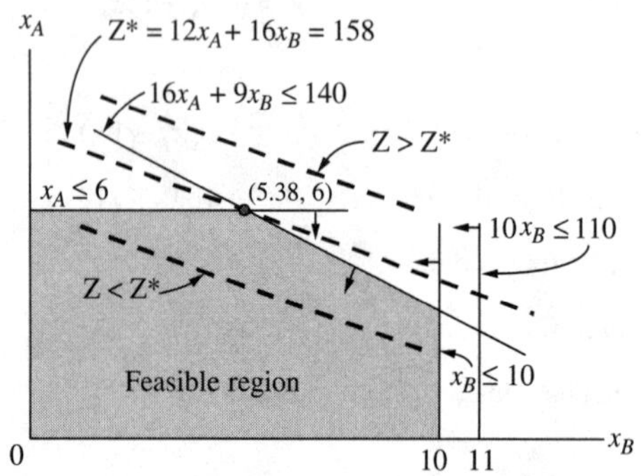

FIGURE 15.5 Graphical interpretation of the sample LP problem.

A graphical interpretation of the LP example is possible because it only deals with two decision variables as illustrated in Fig. 15.5. A solution that satisfies all the constraints is called a *feasible solution*. The region containing all the feasible solutions is called *feasible region*. The objective function for three values of profit is shown in dashed lines. Of notice is that the optimal solution is the intersection of the objective function with an *extreme point* or vertex of the feasible region.

Other classical LP problems include the *assignment* problem, the *transportation* problem, the *transshipment* problem, *production planning* problems, and many others (Hillier and Lieberman, 2001).

15.4.2 Other Mathematical Programming and Optimization Methods

Other mathematical programming approaches in OR include *integer programming, multiobjective programming, dynamic programming,* and *nonlinear programming*.

Integer programming (IP) has the same structure as an LP but the decision variables are restricted to take integer values. In *mixed-integer programming* (MIP) both continuous and integer decision variables are required. A characteristic of IP and MIP models is that, unlike LPs, they are typically very difficult to solve, requiring specialized software and significantly more computer resources. Interested readers are referred to Nemhauser and Wosley (1988).

Multiobjective programming (MOP) is used in situations where the goodness of a solution cannot be expressed with only one objective. MOP provides a variety of methods to handle such situations. It is often possible—through some mathematical manipulation—to convert the MOP into an LP. A well-known methodology to solve MOP is *Goal programming* (GP).

Nonlinear programming (NLP) is the set of techniques that can be applied when the objective or constraints in the mathematical program are nonlinear.

Dynamic programming (DP) techniques can optimally solve problems that can be decomposed in a sequence of stages and have an objective function that is additive over the stages. DP uses *recursion* to solve the problem from stage to stage. The recursion prescribes an algorithm that is repeatedly applied from stage to stage, from an initial state to the final solution. Both forward and backward recursions are possible. Interested readers are referred to Dreyfus and Law (1977).

Graphs and network models represent decision problems using networks consisting of nodes and interconnecting edges. Classical problems with this characteristic include *shortest path problems, critical path problems, production planning problems, maximal-flow problems*, and many others. Interested readers are referred to Evans and Minieka (1992) for a comprehensive treatment of network models.

15.5 DECISION MAKING

Although the techniques described earlier are aimed at aiding the decision maker, decision making in OR refers to techniques that explicitly consider the alternatives at hand and their comparison based on quantitative, qualitative, and subjective data. In this section two decision methodologies are covered that consider decision making with deterministic and probabilistic data. The contents of this section are based on Taha (2003) where interested readers can find more details and other techniques.

15.5.1 Deterministic Decision Making

In deterministic decision making there is no uncertainty associated with the data used to evaluate the different alternatives. The *analytical hierarchy process* (AHP) (Saaty, 1994) is a deterministic decision-making approach that allows the incorporation of subjective judgment into the decision process. The decision maker will quantify his or her subjective preferences, feelings, and biases into numerical *comparison weights* that are used to rank the decision alternatives. Another advantage of AHP is that the consistency of the decision maker's judgment is also quantified as part of the analysis.

The basic AHP model consists of the *alternatives* that are to be ranked, *comparison criteria* that will be used to rank the alternatives, and a *decision*. Figure 15.6 shows a single level decision hierarchy with c criteria and m alternatives. By inserting additional levels of criteria, the same model can be applied recursively to form multiple-level hierarchies. For clarity, the following discussion applies to a single level hierarchy model.

The objective of the procedure is to obtain *rankings* R_j for each alternative $j = 1,\ldots, m$ that represents a ranking of the alternatives based on the importance that the decision maker has attributed to each criterion.

The *comparison matrix*, $A = [a_{rs}]$, is a square matrix that contains the decision maker's preferences between pairs of criteria (or alternatives). AHP uses a discrete scale from one to nine where $a_{rs} = 1$ represents no preference between criteria, $a_{rs} = 5$ means that the row criterion r is strongly more important than the column criterion s, and $a_{rs} = 9$ means that criterion r is extremely more important than criterion s. For consistency, $a_{rr} = 1$ (i.e., comparison against itself), and $a_{rs} = 1/a_{sr}$.

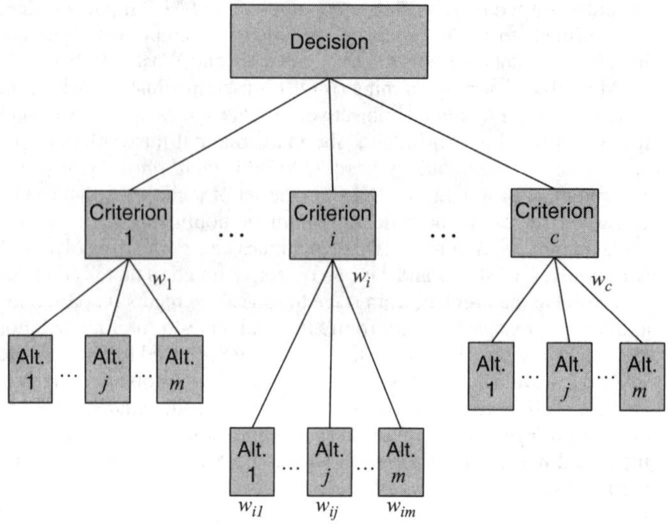

FIGURE 15.6 Single level AHP models.

The *normalized comparison matrix*, $N = [n_{rs}]$, normalizes the preferences in matrix A such that the column sums add to 1.0. This is obtained by dividing each entry in A by its corresponding column sum. If A is a $q \times q$ matrix, then the elements of N are:

$$n_{rs} = \frac{a_{rs}}{\sum_{k=1}^{q} a_{ks}}$$

The *weight* w_r associated with criterion r is the row average calculated from matrix N, or

$$w_r = \frac{\sum_{s=1}^{q} n_{rs}}{q}$$

The AHP calculations to determine the rankings R_j are:

Step 1. Form a $c \times c$ comparison matrix, A, for the criteria.

Step 2. Form $m \times m$ comparison matrices for the alternatives with respect to each criterion, or A_i, for $i = 1,\ldots, c$.

Step 3. Normalize the comparison matrices obtained in steps 1 and 2. Denote these normalized matrices N and N_i, $i = 1,\ldots, c$.

Step 4. Determine the weights for criteria and alternatives. Denote these weights w_i and w_{ij} for $i = 1,\ldots, c$ and $j = 1,\ldots, m$.

Step 5. Determine the rankings for each alternative $R_j = \sum_{i=1}^{c} w_i w_{ij}$, $j = 1,\ldots, m$.

Step 6. Select the alternative with highest ranking.

The *consistency* of the comparison matrix A is a measure of how coherent was the decision maker in specifying the pairwise comparisons. For a $q \times q$ comparison matrix A, the *consistency ratio* CR is calculated as follows:

$$CR = \frac{q(q_{\max} - q)}{1.98(q-1)(q-2)}$$

where

$$q_{\max} = \sum_{s=1}^{q}\left(\sum_{r=1}^{q} a_{sr} w_r\right)$$

Comparison matrices with values of $CR < 0.1$ have acceptable consistency, 2×2 matrices are always perfectly consistent, and matrices with $q_{max} = q$ are also perfectly consistent.

15.5.2 Probabilistic Decision Making

In probabilistic decision making there are probability distributions associated with the payoffs attainable through the alternatives. A common objective of the decision is to select the alternative that will yield the best expected value.

Decision trees are a convenient representation of probabilistic decision problems. The elements of a decision tree are *decision* nodes (Y), *alternative* branches, *chance* nodes (⊄), probabilistic *state* branches, and *payoff* leaves (see Fig. 15.7).

Associated with probabilistic state *j* there is a probability (p_j) that the system is in that state, and associated with the payoff leave for alternative *i* if the world is in state *j* is a payoff a_{ij}. The expected value of the payoff associated with alternative *i* can be calculated as follows:

$$EV_i = \sum_{j=1}^{n} p_j a_{ij}$$

The decision maker will tend to select the alternative with the best expected value.

The basic model presented here can be extended to include *posterior Bayes*' probabilities such that the result of experimentation can be included in the decision process. Decisions involving nonmonetary or decision maker's preferences can be dealt with *utility functions*.

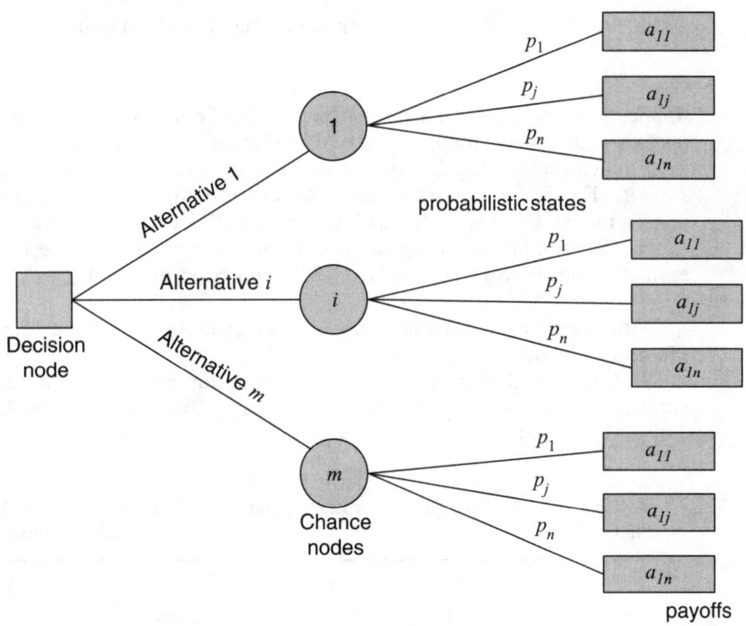

FIGURE 15.7 Probabilistic decision tree.

15.5.3 Decision Making Under Uncertainty

Decision making under uncertainty is similar to probabilistic decision making because in both cases the payoffs are associated to random states of the system. However, in decision making under uncertainty the state probabilities are unknown. Letting a_{ij} be the payoff obtained via alternative i, given that the system is in state j, the following criteria have been developed to make decisions without explicit knowledge of these probabilities:

The *Laplace* criterion assumes that all states are equally likely to occur and selects the alternative with the maximum average payoff. The selected alternative i^*:

$$i^* = \arg\max_i \left(\frac{1}{n} \sum_{j=1}^{n} a_{ij} \right)$$

The *maximin* criterion takes the most conservative attitude selecting the best out of the worst cases. The selected alternative i^*:

$$i^* = \arg\max_i \left(\min_j a_{ij} \right)$$

The *Savage regret* criterion is less conservative than the maxmin criterion via the following transformation of the payoff matrix:

$$r_{ij} = \max_k (a_{kj}) - a_{ij}$$

The method then applies the maximin criterion based on the transformed payoff matrix.

The *Hurwicz* criterion allows the decision maker to take from the most conservative to the most optimistic positions. This is accomplished using an *index of optimism* α that ranges from zero (most conservative) to one (most optimistic). The selected alternative i^*:

$$i^* = \arg\max_i \left\{ \alpha \max_j a_{ij} + (1-\alpha) \min_j a_{ij} \right\}$$

Application Example—Decision Making Under Certainty Using AHP. Consider the problem of opening a facility in a foreign country. The alternatives are open in country A, open in country B, or keep the current facility C. For this example the criteria for decision are *labor* cost and region *stability*. The decision maker expresses his or her preferences among criteria in the following comparison matrix A (Table 15.4*a*), and its corresponding normalized matrix (Table 15.4*b*):

The decision maker must generate comparison matrices among each alternative with respect to each criterion. With respect to labor the comparison matrix A_{labor} (Table 15.5*a*) and normalized matrix (Table 15.5*b*) are:

With respect to stability the comparison matrix $A_{stability}$ (Table 15.6*a*) and normalized matrix (Table 15.6*b*) are:

Next, the weights associated with each criterion and alternative are the row averages of the normalized matrices: $w_{labor} = 0.8$, $w_{stability} = 0.2$, $w_{labor,A} = 0.59$, $w_{labor,B} = 0.33$, $w_{labor,C} = 0.08$, $w_{stability,A} = 0.07$, $w_{stability,B} = 0.21$, and $w_{stability,C} = 0.72$.

TABLE 15.4a Comparison Matrix (Labor) and (Stability)

a_{rs}	Labor	Stability
Labor	1	4
Stability	1/4	1

ABLE 15.4b Normalized Comparison Matrix (Labor) and (Stability)

n_{rs}	Labor	Stability
Labor	0.8	0.8
Stability	0.2	0.2

TABLE 15.5a Comparison Matrix With Respect to Labor

n_{rs}	Country A	Country B	Local facility
Country A	0.61	0.63	0.54
Country B	0.30	0.31	0.38
Local facility	0.09	0.06	0.08

TABLE 15.5b Normalized Comparison Matrix With Respect to Labor

n_{rs}	Country A	Country B	Local facility
Country A	1	2	7
Country B	1/2	1	5
Local Facility	1/7	1/5	1

The rankings for each alternative are calculated as follows:

$$R_{country\,A} = w_{labor}\,w_{labor,A} + w_{stability}\,w_{stability,A} = (0.8)(0.59) + (0.2)(0.07) = 0.49$$

$$R_{country\,B} = w_{labor}\,w_{labor,B} + w_{stability}\,w_{stability,B} = (0.8)(0.33) + (0.2)(0.21) = 0.31$$

$$R_{current} = w_{labor}\,w_{labor,C} + w_{stability}\,w_{stability,C} = (0.8)(0.08) + (0.2)(0.72) = 0.21$$

AHP suggests building a plant in country A because it is the alternative with largest ranking.

The decision maker may desire to quantify how consistent each comparison matrix is. The 2×2 criteria comparison matrix is perfectly consistent. The consistency ratio CR is calculated for the 3×3 matrices as follows:

$$\text{Labor:} \quad q_{max} = 0.59(1 + 1/2 + 1/7) + 0.33(2 + 1 + 1/5) + 0.08(7 + 5 + 1) = 3.07$$

$$\text{Consistency ratio:} \quad CR_{labor} = \frac{3(3.07-3)}{1.98(3-1)(3-2)} = 0.05$$

$$\text{Stability:} \quad q_{max} = 0.07(1 + 5 + 8) + 0.21(1/5 + 1 + 6) + 0.72(1/8 + 1/6 + 1) = 3.42$$

$$CR_{stability} = \frac{3(3.42-3)}{1.98(3-1)(3-2)} = 0.32$$

Hence, the comparison matrix with respect to labor A_{labor} has acceptable consistency (i.e., $CR_{labor} < 0.1$). However, the comparison matrix with respect to stability is inconsistent (i.e., $CR_{stability} > 0.1$). So the decision maker must try to reassess the rating given in matrix $A_{stability}$.

15.5.4 Other Decision Making Methods

Other well known decision-making models include *game theory,* and *Markov decision processes*.

Game theory models the decision problem as a game among adversaries with conflicting payoff structures. The result of game theory analysis is often expressed as a set of strategies. Each strategy describes the decision maker's payoff and the effect on the opponents.

TABLE 15.6a Comparison Matrix With Respect to Stability

a_{rs}	Country A	Country B	Current facility
Country A	1	1/5	1/8
Country B	5	1	1/6
Current facility	8	6	1

TABLE 15.6b Normalized Comparison Matrix With Respect to Stability

n_{rs}	Country A	Country B	Current facility
Country A	0.07	0.03	0.10
Country B	0.36	0.14	0.13
Current facility	0.57	0.83	0.77

Markov decision processes can be viewed as a generalization of probabilistic decision making where the decision model considers that the system can be described by n states, with known transition probabilities between any two states and corresponding payoff matrix associated with these transitions.

Additional details on the models mentioned here can be found in the provided references.

15.6 FUTURE TRENDS

The application of OR is undergoing an explosive period due to advances in information and computer technologies. The internet, fast inexpensive computers, and user friendly software allow decision makers with different specialties to use OR techniques that could be applied only by specialists in the recent past.

Even the immense computer power available today is not sufficient to solve some difficult decision and optimization problems in reasonable time. Recent advances in mathematical programming theory are allowing practitioners to tackle these difficult problems. The widespread use of web-based applications and enormous amount of data that can be accessed ubiquitously, together with emerging data mining techniques, allow the extraction of useful information and new knowledge for competitive advantage. Finally, the connectivity among geographically and organizationally dispersed systems and decision makers will benefit from recent developments in distributed decision making and collaboration methodologies. Future trends indicate that OR will be at the heart of decision making software applications.

15.7 CONCLUDING REMARKS

This chapter has succinctly presented operations research and some of its most widespread applications. The material is presented at an introductory level such that the reader gains an appreciation for the type of analysis and problems that OR can be applied to. Due to space considerations some important models were only briefly described and the reader directed to appropriate references. Readers interested in OR practice and professional community will find the website of the Institute for Operations Research and Management Science (INFORMS, 2003) informative.

REFERENCES

Buzacott, J.A., and J.G. Shantikumar, 1993. *Stochastic Models of Manufacturing Systems,* Prentice Hall, New Jersey.

Dreyfus, S., and A. Law, 1977. *The Art and Theory of Dynamic Programming,* Academic Press, Florida.

Evans, J.R., and E. Minieka, 1992. *Optimization Algorithms for Networks and Graphs,* 2d ed., Marcel Dekker, New York.

Johnson L.A., and D.C. Montgomery, 1974. *Operations Research in Production Planning, Scheduling, and Inventory Control,* Wiley, New York.

Hillier, S.H., and G.J. Lieberman, 2001. *Introduction to Operations Research,* 7th ed., McGraw-Hill, New York.

Hopp, W.J., and M.L. Spearman, 1996. *Factory Physics,* Irwin, Illinois.

Nemhauser, G., and L. Wosley, 1988. *Integer and Combinatorial Optimization,* Wiley, New York.

Saaty, T.L., 1994. *Fundamentals of Decision Making,* RWS Publications, Pennsylvania.

Silver E.A, and R. Peterson, 1985. *Decision Systems for Inventory Management and Production Planning,* 2nd ed., Wiley, New York.

Suri, R., 1998. *Quick Response Manufacturing,* Productivity Press, Oregon.

Taha, H.A., 2003. *Operations Research: An Introduction,* 7th ed., Pearsons Education, Prentice Hall, New Jersey.

CHAPTER 16
TOOL MANAGEMENT SYSTEMS

Goetz Marczinski
CIMSOURCE Software Company
Ann Arbor, Michigan

ABSTRACT

This chapter describes the role of tool management systems (TMS) in a flexible manufacturing environment. The hardware and software components that make up a TMS are explained and how these increase the shop-floor productivity. The four step process of planning and implementation of a TMS is laid out, followed by practical advice about how to operate a TMS. Case studies are cited in the chapter to show that a TMS can yield substantial cost reductions and productivity increases. Future trends concerning the support of digital manufacturing environments with 3D CAD models of cutting tools conclude this chapter.

16.1 INTRODUCTION

Flexible manufacturing systems (FMS) obtain their flexibility, to a large extent, through CNC machines, which are capable of machining different parts in a single setting.[1] Computer controlled tool exchange mechanisms allow a large variety of different operations at one single machine. Up to 100 or more different metal cutting tools need to be stored locally at the machine's tooling system. With the physical tool, a set of information to identify and localize the tool as well as to feed the CNC control with performance data has to be made available at the machine. Because the part mix in the FMS may change and the tools may wear down or break, the local storage needs to be supplied with new tools. For both purposes, the supply of cutting tools and the respective information, an efficient working tool management system (TMS) is needed. The system should be designed in such a way that the CNC machines of an FMS do not need to stop machining because the required tool is not available.

Tool management in this context is a method to materialize the promises of new manufacturing technologies. What good are ever higher speeds and feeds or reduced chip-to-chip cycle times from the FMS, if a lack of tools causes machine downtimes or wrongly assigned tools yield rework or scrap? Further to that, professional tool management considers the total cost of cutting tools along the supply chain. Apart from the purchase costs of the tool, the supply chain includes the processing cost incurred by tool search, supplier selection and procurement, tool assembly and presetting, delivery, dismantling, refurbishment, or scrapping. Thus tool management is a service function in each manufacturing operation, geared to yield the following results:

- Increased availability of cutting tools
- Minimized stock level and variety of tools
- Minimized administrative effort in the tool supply chain

As the overall equipment efficiency (up time × speed × scrap rate) of the FMS is significantly, but not solely, driven by the availability of the tools, tool management has to be an integral part of every company's production strategy. Complementary building blocks, like total productive maintenance (TPM) or full service commodity supply need to be combined as integral parts of a competitive manufacturing strategy.

16.2 DEFINITION OF A TOOL MANAGEMENT SYSTEM (TMS)

Viewed from the perspective of an FMS, a tool management system is in the first place a systematic approach to and a set of business rules applied for tool changes. Software comes into play after the rules are set. It is currently the most widely used dynamic approach developed over time to avoid the pitfalls of the initial static approach.[1]

Within the static approach the tool changes occur in intervals. From the set of production orders, the loading procedure generates a specific part mix to be machined on the FMS and assigns workloads to the CNC machines. This in turn necessitates setup of the local tool magazines of the CNC machines. The planning and control system automatically informs the personnel what kinds of tools are required at which machine. Then these tools are assembled, preset, and usually placed manually into the local tool magazine. When the production of the given part mix is finished, the next mix is introduced, normally requiring retooling of the CNCs. This simple and robust approach was most widely used in the early days of FMS. However, it has the disadvantage of requiring additional set up time for retooling. Further to that, it has some pitfalls concerning sudden tool breakages. The usual way to work around the problems is to increase the stock level of the local tool storage such that many different tools are quickly available and thus reduce the necessity of retooling between each part mix change. The ever-increasing size of the tool magazine of the CNCs supports that view. Furthermore, redundant tools can be kept to bolster against tool breakage. Another strategy is to balance the FMS for predictable tool life of the weakest tool. In any case, this static approach to tool management drives the stock level of redundant tools and in the case of "predictable tool life" increases the number of tool changes. Further to that, extensive decentralized tool storage areas bear the threat of proliferation, which means that the variety of cutting tools will be higher than required by the operations. That is because the supervisors in charge of each FMS will use what they consider the right tool, with little information sharing to other departments. More often than not each employee's tool box or work station grows to a personal tool crib, which means that an overabundance of unneeded tools is available.

The more promising, but also more challenging, approach is the dynamic approach.[1] Within such a system, the tool components are pooled centrally in the tool crib and assembled and delivered by manufacturing supply orders. Tool changes occur while the CNC is running a different job. This allows a continuous part mix change. However, a dynamic tool management system is not restricted to closing the loop from the tool crib to the FMS and back, but also to the tool suppliers for procurement and replenishment of the tool inventories. Clear cut functional interfaces allow for the possibility of outsourcing the complete supply cycle.

A tool management system supports all activities necessary to control the supply and discharge of CNC machines with cutting tools (Fig. 16.1). The supply cycle includes the selection of the needed tool components as required by process engineering, the configuration and presetting of tool assemblies, the delivery to the FMS, and the stocking of the local CNC magazine. The discharge cycle includes the recollection of worn out tools to the tool crib, the dismantling of the tool assemblies, the

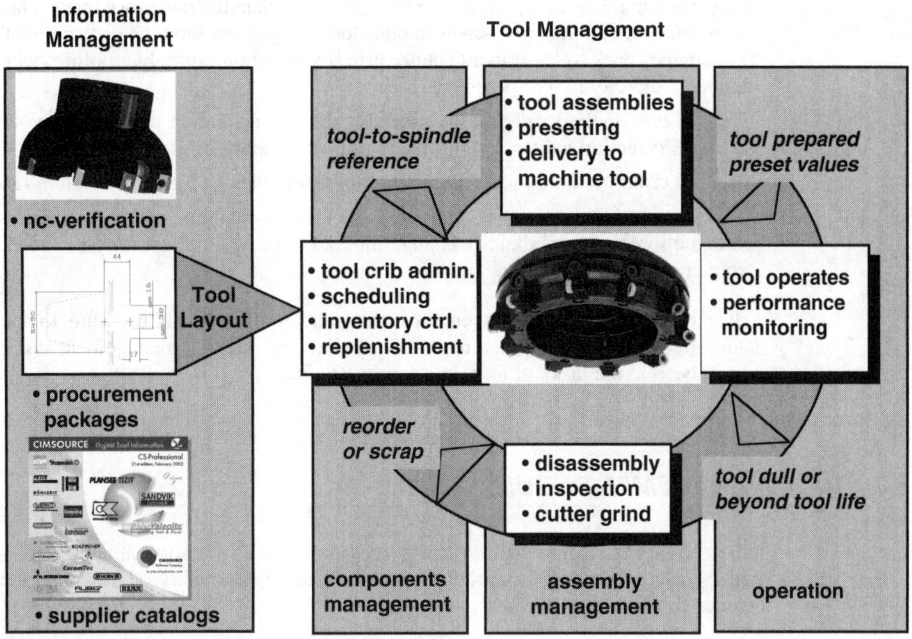

FIGURE 16.1 Tool management activities.

inspection of the tool components, and the assignment of tool components to regrinding, to refurbishment, or to scrap.

All activities to register, to stock, to retrieve, and to hand out cutting tools from systematically organized storage areas are covered by the term tool administration. To execute tool management as a business function means to collect all information relevant to the tool supply and discharge cycles, to continuously monitor it, and to make decisions according to the operations business goals.

A substantial part of tool management is information management. Before any physical tool is assembled, preset, and sent to the CNC a lot of information has to be exchanged between the respective departments, be it on paper or electronically. That's because the engineering department needs to pull together all relevant information for a machining process and consolidate this into a single tool assembly. But who sells tool assemblies? So purchasing must disintegrate the tool assembly to create procurement packages.

The tool layout is the central document of tool information; it helps process engineering to communicate tooling requirements for a certain job to the shop-floor. A tool layout captures the tool information in the language of engineering—drawings, bills of material, and parameter lists. A single tool layout refers to a single tool assembly for a certain operation performed with a specific spindle on a specific CNC. The layout documents all related components of the tool assembly, including spare parts. For example an average of 30 to 50 tool assemblies are assigned to a CNC in engine manufacture, with each assembly including some 50 to 150 components. The tool assembly also holds performance data (speeds and feeds) for the specific operation.

From the tool layout, crib personnel (for presetting), machine operators, and purchasing personnel (or a full service supplier) pick relevant information for their own needs and, in most cases, add information which is not available in digital form. Information has to be obtained and keyed in at various stages of the tool management cycle, including each of the following:

- Procurement generates tool packages. These are bills of material for all components, spare parts included, which are used to generate purchase orders.

- Tool crib administration assigns storage IDs to populate the inventory management system, adding the distinction of perishable versus durable tool components, and the distinction of returnable tooling (which goes back to the crib and cutter grind) versus nonreturnable tooling (which is assigned and delivered directly to a cost center).
- The tool crib operators physically assemble the tool according to the tool layout and perform presetting. Correction values are digitized on the tool or are directly sent to the CNC control.
- The tool crib operators inspect returned tool assemblies and generate failure reports.
- On-site cost reduction teams improve cycle times—changing speed and feed rates or calling for alternate tooling—and thereby change the tool specification, creating the need for a new release from engineering.

It is the job of the tool management system, and now we do talk software, to manage all this information. Because of the required responsiveness, a dynamic tool management system needs computer support beyond the tasks of tool data administration.

16.3 TOOL MANAGEMENT EQUIPMENT

Subject to tool management are the physical tools and information about the tools, which are managed by a combination of hardware and software components. Hardware for tool management includes (Fig. 16.2):

- Storage equipment, including mobile supply racks
- Identification system (e.g., bar code readers or radio frequency identification (RFID))
- Presetter and calibration devices

Software components include

- Tool administration software (inventory management and tool tracking)
- Tool database (master data)
- Work flow management of tool documents and engineering reference

The term *tool* addresses both tool components and tool assemblies. Further to that it is helpful to distinguish perishable from durable tooling. Perishables are tools which wear out during use—drills, inserts, taps, and the like. These tools drive the dynamics of the tool flow because they need to be replaced in direct relation to the machining volume. Durables are not consumed by use, like tool holders, collets and the like. Another important distinction is whether a tool is returnable or assigned to a machine or operator. Returnables are returned to the tool crib, refurbished, and reused. Nonreturnables are mostly specials which stay at a distinct machine, like CNC specific holders and collets. Each tool assembly on a CNC is in fact a combination of perishable and durable tooling which requires a preparatory step of assembly and calibration before the tool assembly is sent to the CNC.

Tool components are stored in respective cabinets. Apart from simple steel cabinets for local storage in smaller jobs shops or less automated manufacturing environments, automated horizontal or vertical "paternoster-type" cabinets are applied. These cabinets allow for automatically managing inventory levels and issuing tool components only by an unambiguous identification number. However, access to these cabinets is restricted mostly to the crib personnel, since the tool ID, which needs to be keyed in to issue the tool, comes with the tool BOM. Higher levels of automation apply barcode systems, but are still mostly run centrally from the tool crib where tool assemblies are built according to the respective tool layouts.

FIGURE 16.2 Hardware components of a TMS. (*Courtesy of Zoller, Inc.*)

For perishable tooling, automated dispensers are available to efficiently supply consumables like inserts and drills on the shop floor. These kinds of cabinets resemble vending machines for snacks and soft drinks. An operator would identify himself or herself at the machine and key in the required tooling. The cabinet will release only the required type and amount of tooling and charge it to the respective account. Usually a full service supplier will replenish the cabinet by minimum stock level.

The tool assemblies are brought to the CNCs on mobile supply racks. These racks may be stocked according to a fixed schedule and a person from the crib tours the plants regularly to restock the local storage areas at the machines. Or the racks are configured according to the job forecast for specific CNCs. In this context the tool racks are also used for local storage of redundant tooling. Finally the magazines of the CNCs themselves need to be considered as tool storage equipment because they are an important and integral part of a tool management system.

A pivotal prerequisite for the successful operation of the storage systems is an identification system, comprising both an unambiguous nomenclature for the identification of tool components and tool assemblies as well as the equipment needed to read it. For the time being tool components are mostly identified by their product code (e.g., CNMG for an insert), which means in this case that all inserts are considered equal, no matter what the state of usage. In cases such as advanced cutter grind operations this is not enough because the decision regarding whether or not another regrind cycle is feasible needs information on how often an individual tool has already been regrinded. That is why companies offering regrind or coating services use bar coding of tool items individually; only then are they able to control the number of regrinds and return the tool to the right customer.

On the shop floor the individual identification is restricted to tool assemblies. Advanced tooling systems use rewritable computer chips to carry both the calibration information and the ID of the tool assembly.[2] This is an important input for the CNC control. If the information record includes tool life, the CNC is able to issue a tool supply order since the actual usage is measured against the theoretical maximum.

Presetters are another important piece of tool management equipment. Consider a high precision CNC which needs precise information about the location and geometry of the cutting edge of a tool assembly. After the tool is assembled, reference points need to be identified, such as the overall length and correction values for the CNC control. All this information is measured individually using a presetter.[3] In smaller job shops the presetter might be solely in the crib. In large scale manufacturing environments each FMS might have a presetter to increase flexibility because as perishable tooling (indexable inserts) are replaced, the tools need to be calibrated. In some cases the presetter is directly linked to the CNC for information interchange; in others the information is conveyed via computer chip.

Talking about tool management systems, most people associate them with a piece of software, although more items are involved in tool management than just the software. Consider where you get the information to physically manage the flow of tools to and from the FMS and who needs it, and you'll find out what you need in terms of software. Three major modules are necessary (Fig. 16.3).

Most visible to the user is the tool administration software. It monitors stock levels, receives tool supply orders from the shop floor, and issues replenishment orders to suppliers. It also monitors the rework and repair cycle including the cutter grind area and (external) coating operations. Interfaces to physical storage areas (receiving and inspection, tool crib, local storages at the CNCs) are vital in this respect. Differential inventory control in this context means that the software compares the stock level "to-be" with the "as-is" status. This should include the comparison of actual tooling in a CNC with the tooling requirements for the next jobs. For that purpose some commercial software packages apply direct links to the DNC to receive the tooling requirements, and compare the local tool magazine of the CNC to identify the tooling differentials. For replenishment purposes, EDI capability or an open interface to the respective ERP system should be available where purchase orders are actually issued.

To support the assembly of tools a preset management system should be available, offering a direct link to presetters of different makes, the DNC, and work scheduling system if possible. The link to the work scheduling system supports another important feature called *kitting*.[4] Kitting means

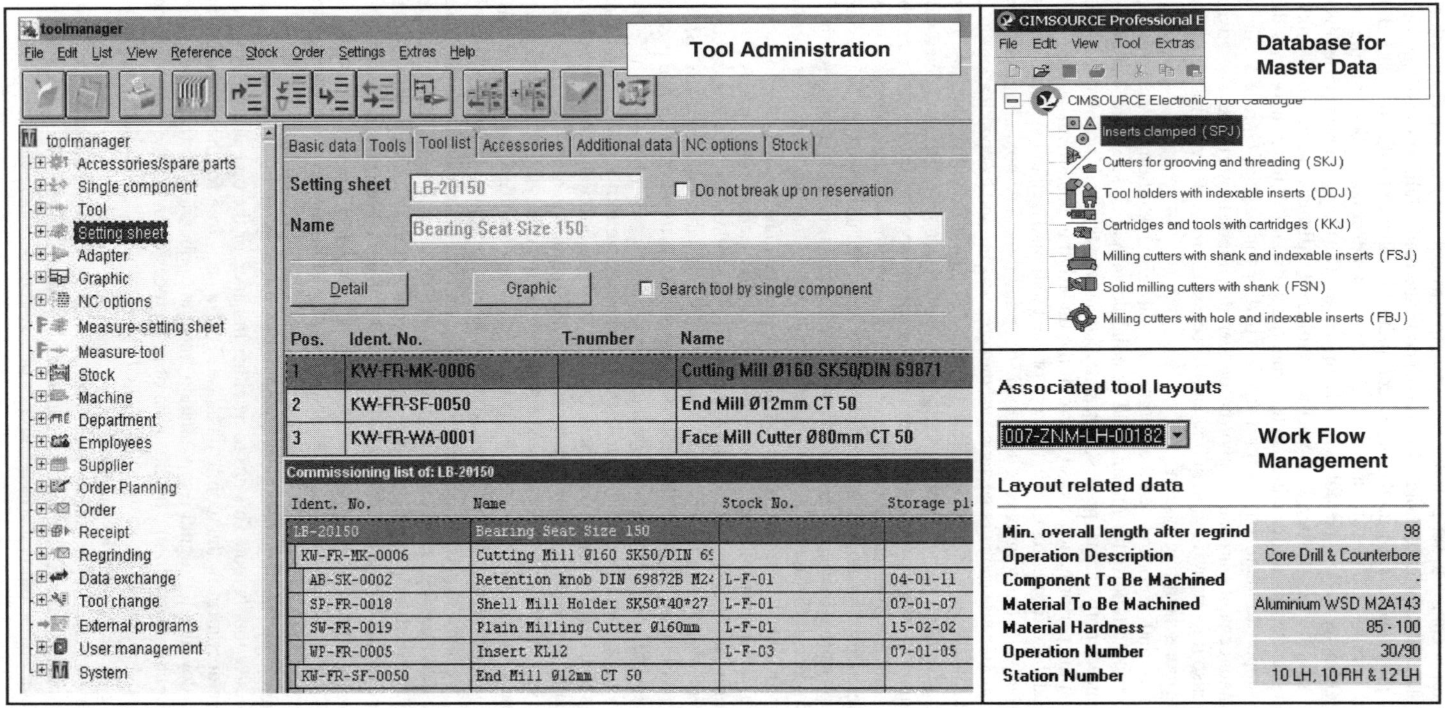

FIGURE 16.3 Tool management software. (*Courtesy of Zoller, Inc., and Cimsource, Inc.*)

to configure a complete set of tooling (several tool assemblies) for a specific job as specified in the process plan. The ability to hold all settings and usage data might be important for operations where safety is paramount, such as the aerospace industry. Further to the standard BOM functionality, advanced administration software offers where-used information both on the component-to-tool assembly level as well as the tool assembly-to-process (kit) level. In some cases even the where-used link to a specific part (process-to-part) is established. All these where-used links are important if advanced search capabilities are expected from the software. Without these links the software would offer no support to identify tools for a specific part or for a specific process. Also an audit trail for quality management purposes cannot be delivered without these links. The advanced where-used links are also important if tool requirements forecasts are to be derived from information drawn from the shop-floor scheduling system.

Analysis and reporting features complete the administration module of the software. This includes costing reports for internal and external departments, scrap and failure reports, as well as performance reports (actual process parameters and settings) for critical tools. These features depend heavily on the performance and structure of the underlying master database.

The database for master data is the core of each tool management system. It is the integration hub for all other software components described so far and thus must hold all information needed for any application. Because of the large variety of tool components and its combinations, relational structures and parametric systems are most widely used. Master data include the geometric (explicit) features of the tool, application ranges for speeds and feeds, as well as application guidelines. Cost information of each component, storage areas, and where-used and BOM information for tool assemblies are included. Further, administrative data (supply sources, replenishment times, and minimum inventory levels) and cross-references to spare parts and accessories for each tool are important. Some companies also consider graphics (photo images, DXF drawings, CAD models) important items for the master database.

It is important for an efficient tool management system that the database is able to handle different views on the data. The underlying classification scheme must be adaptable to the user's perspective, which might range from process planners to buyers to machine operators. A process planner might search for a tool starting from the application, which would only be feasible because the respective parameters would be in the database. A buyer might be satisfied with the ANSI code or a supplier ID. The machine operator would only be interested in the tool assembly and possibly which inserts fit as he or she needs to refurbish a dull tool.

Further, the link to electronic catalogs of the respective tool suppliers is crucial for the efficient operation of the system since only then could the software be automatically populated with data. This is especially important if NC-programming and NC-simulation are to be supported by the tool management system, because otherwise the extreme difficulty of keying in the respective data might hold people back from using those systems at all. An important feature of any tool database therefore is the support of standard data formats. To save time and effort, major tool manufacturers have agreed on a common standard to specify cutting tools. This standard, called StandardOpenBase, helps to easily populate different application systems with the tool data.[5] It is also highly recommended that this standard be used as a so-called master classification from which the different application views could be derived.[6] Another advantage is that a lot of interfaces to commercial software packages (NC-programming, NC-simulation, presetting, and the like.) are available off the shelf.

Finally tool management software includes some kind of work flow management because most operations still rely on paper documents. This includes the tool layouts for the operators. Especially in semiautomated manufacturing environments, each tool assembly is delivered to the machine with the respective tool layout. This has to be either in paper, or the operator needs a viewer to check the respective document in the tooling database. The same is true for the NC-programmer, who would prefer to download a digital representation into the application system. Even more important is a work flow component for the qualification of new processes and/or the communication with suppliers for special tooling. Advanced tool management systems keep track of the release status of tools including the respective documents.

As an engineering reference, some systems offer the documentation of test results and the set up of a knowledge base for application guidelines. Advanced TMS even provide a database of workpiece materials including a cross-reference of different national standards.

16.4 PRODUCTIVITY INCREASES

The previous explanation shows that professional tool management requires quite an investment. This investment needs to be justified by savings, which will occur in three productivity levels:

1. Tool spend (reduced purchasing volume)
2. Reduced downtimes (due to stock-outs of the FMS)
3. Efficient execution of tool management activities

The first reason for the amount spent for cutting tools to go down is because the tool management system will yield a lower tool usage. That is because constantly monitored inventory locations will lead to a personalized accountability of the tool use. Operators will only get what the requirements calculation offers for the job, and unused tooling will be returned to the crib or stored for further use. The tool management system will "recall" by difference calculation that a certain operation is still stocked. Further to that, the reporting tools of the software will indicate machine or part material problems before excessive tool usage takes place.

The second reason for decreasing cost of tooling is that the TMS will help to identify obsolete tooling, eliminate redundant tooling, and drive the use of standard tooling. This helps to focus both the supplier base and the variety of tools in use, yielding scale economies. These effects, summoned up with the term deproliferation, are achieved by the classification scheme and the respective search functions of the TMS. Different views could be cross-referenced, like supplier catalogues with internal product codes. Consider the case of an automotive supplier, using the CS-Enterprise system of a leading software supplier for its powertrain operations. CS-Enterprise uses an international description standard as a master classification from which different application views could be cross-referenced. This enables a direct link to the ToolsUnited master server of the software supplier, which holds the information on standard tooling of major tool manufacturers. If a tooling engineer has a machining problem, he or she will specify the application to search through the internal database. If no match is found, the same request goes out to the ToolsUnited server in search of standard tools to solve the problem. If there still is no match, only then can a new tool be created. This is a dramatic departure from the seemingly unbounded creativity that led to the high level of tool proliferation in the first place!

Deproliferation is also driven by purchasing. Because tools can be identified by application, the buyer gains engineering intelligence that was previously unavailable. Redundant tooling becomes obvious. As a result, purchasing can blind these tools from the database and wait for someone to complain. When no one complains (which is typically the case), the tools can be removed from the database. Indeed, in a typical manufacturing facility, nearly 10 percent of all listed items can be deleted in this way within the first four months of the system's life. Other cases reveal that up to 28 percent of tool items are either obsolete (not used for the past 24 months) or redundant (one of several tool items for the same application).

The up-time of the FMS will increase due to improved prekitting and tool requirement planning. Manufacturing facilities without a well-established TMS more often than not show the painful irony that even though $50,000 worth of cutting tools may be tied up per manufacturing line or flexible cell, stock-outs are still a recurring problem. As the TMS is linked to the scheduling system and able to read tooling requirements from the process plans, the tool supply cycle from the crib via assembly and presetting is tuned to actual load situation of each TMS.

Another impact for increased machine productivity is improved manufacturing processes because advanced TMS enable best practice benchmarking throughout the plant and actual speeds and feed rates can be fed back into the system from the shop floor. Dedicated tooling strategies could be applied. Consider the case of a capital equipment manufacturer trying to balance the CNCs of an FMS by using "predictable tool life." Only since TMS has been delivering technology parameters and application guidelines, have the necessary reference values for this endeavor been available. For further productivity increases, the feedback loop from the CNCs to the FMS is used.

Up-time of the FMS will also be positively influenced as the TMS helps to enforce quality procedures. The unambiguous identification of tooling and the restricted access to the crib and to automated storage systems, helps to prevent individuals from picking the wrong tool for a process. If machining problems lead to the withdrawal of an engineering release, the TMS helps to locate all the respective tool items and to physically collect them. Regrinding and recoating cycles no longer depend on personal judgement, but are automatically triggered according to preset technology parameters.

Finally the TMS reduces time and effort to conduct tool management activities. The most obvious help comes through the intelligent search functionality and the easy review of legacy case studies. The BOM and where-used references help to cut done time requirements to generate tool assemblies and for kitting. Including communication with tool suppliers and machine tool companies, this work alone can consume up to 50 percent of the typical tooling group's time. In the tooling group of the powertrain plant cited above, 25 percent of the shop-floor activities were tool searches, mainly to establish cross-references from supplier IDs to the company's IDs. The TMS helps to refocus the tooling group on engineering effort to use new technology.

Procurement activities become much more productive as automatic replenishment procedures for perishables are introduced. Further to that the identification of spare parts and accessories for each tool is easy because the relevant information is directly tied to the respective tool. In most cases the TMS would also provide for direct links to electronic catalogues of the tool suppliers, so that e-procurement is supported. Retaining the manufacturer's technical tool representation and machining expertise becomes feasible. With the respective interfaces, the TMS is able to distribute the tool information to other downstream systems (CAM, NC-simulation) which otherwise would have to be populated manually. At a rate of about 2 min per item, this translates to about 500 h for the average of 15,000 tool items per transmission plant. For an engine plant, the figure is approximately 200 h.

16.5 PLANNING AND IMPLEMENTATION

As the decision to introduce a tool management system is made, five steps are necessary to plan and implement it.

16.5.1 Baselining

Baselining means the assessment of the current situation. The starting point is the quantification of key characteristics. These include:

- Number of tool components and of the respective tool assemblies
- Dollar value of tool purchases per year, dollar value of inventory, items in stock
- Number of tool suppliers and number of purchase orders (including average order size)
- Number of NC programs and respective tool specs
- Number of automated storage spaces (paternoster in the crib, tool magazines of the CNC, and the like.)
- Number of tool supply orders (from the shop floor to the crib)
- Number of CNC machines (classified in 1, 2, or 3 shift operation)
- Number of tool changes per machine and shift

The quantification of the status quo is complemented by a review of the key business processes concerned with tool management:

- Adding intelligence to the tool description and assignment of unique number codes for tool assemblies and association with the job number for which these are intended
- Supplier selection and tool purchase, reordering of tools and restocking the tool crib (including inspection at receiving)

- Issuing of tools to the shop floor (who generates tool supply orders?) including presetting
- Returning tools from the shop floor, dismantling, and inspection
- Decision for regrind of refurbishment, scrapping of tools
- Sending out tools for rework (grinding, coating, and the like)

The business process review also reveals insights about who is involved in tool management activities. With those three key inputs—key characteristics, business processes, and manpower—the baselining phase delivers everything you'll need for an activity-based analysis.

16.5.2 Development of a Target Scenario

The goal to implement a tool management system has to follow a clearly defined business case. Soft goals like "improved transparency" or "easier operation" do not justify the investment in time and money. Therefore the results of the baselining phase have to be evaluated, e.g., by benchmarks which are available both from independent consultants as well as from vendors of tool management software.

But benchmarking is not enough. A sound concept could only be derived from a "green field" planning of the future tool management scenario. The green field planning could be compared to an industrial engineering approach for the relevant activities. It uses the key characteristics of the baselining phase as resource drivers and benchmarking figures to relate those to actual resource consumption. This leads to an ambitious target scenario, as no constraints from the actual setting in the respective companies are considered yet. For example, benchmarking figures might indicate that tool assembly and presetting for the given type of manufacturing could be done with 30 percent less effort than the baselining figures show. But most likely the baseline setting is constrained by realities which need to be taken into account. The experience shows that no matter what the specific scenario is, recurring constraints include

- No commonality in tool descriptions
- Documentation divided between different systems
- Graphics and tool data not in digital format
- No digital information about tool performance
- No information about which tools are actually in use

And to compound these obstacles, collaboration with outside suppliers is difficult because in general:

- There is no electronic supplier integration beyond electronic data interchange links with major suppliers.
- No communication standards are available for collaboration with engineering partners such as cutting tool and machine tool companies and less advanced presetters, lack of database support, or simply the fact that no clearly defined tool management rules are in place.

That is why the green field setting needs to be compared systematically to the "as is situation." The gap needs to be analyzed item by item to avoid jumping to conclusions too early, which would be more often than not "go get a piece of tool management software." But the analysis will reveal that it is not a software problem in the first place. It is a problem of complexity, to a large extent self-induced complexity through redundant tooling in unassigned inventories. Properly done, the gap analysis will demand prerequisites before any software comes into play. These prerequisites have to be specified in a KPI (key performance indicator) driven action plan and will most likely include:

- Identification and elimination of obsolete tooling
- Deproliferation of tool items (elimination of redundant or obsolete tooling)
- Elimination of "unassigned inventories" and declaration of defined storage areas

Beyond the prerequisites the development of a target scenario results in the specification of the future tool management system. The concept includes:

- The rules of the future tool management practice, including eventual organizational changes
- The hardware requirements (including tool storage equipment)
- The software requirements, including an answer to the question of data sources (how do we populate the system?)

The next step then is to mirror the hardware requirements and the software specification to the software market.

16.5.3. Selection of Software

This review of the tool management equipment makes it clear that tool management software is primarily an integration platform for the different hardware and software systems needed to efficiently manage the tooling requirements. The core of any such system is a database with application interfaces around it. This insight helps to classify the market of commercial software for these applications, which originated from either one of the following bases:

- Knowledge bases from tool suppliers have in some cases been expanded to a complete tool management package. For example, Sandvik's AutoTas system was developed that way.
- Coming from the hardware side, companies selling presetters expanded their system to control the full scope of tool management. Zoller's Tool Manager is an example.
- Tool databases of NC-programming or CAM-systems have been expanded for that purpose. The Resource Manager of EDS-Unigraphics is an example.

Software packages of the first two come as a single system whereas the third kind is mostly a module of comprehensive engineering software packages. In addition, many software companies doing business in the manufacturing industry offer tool management software. In effect, these companies develop a proprietary system to meet each customer's needs.

As the specification is mirrored against the features of prospective TMS packages, the key question is what database lies behind the system and what data structures are supported. If there is one general advice to be given, it is that "integration beats functionality." Rather trade the latest software feature for well-established interfaces. Make sure that whatever software you choose, data storage should clearly be separated from data processing. In commercial packages there should be a "base" module, including the database system, and application modules that could be customized to your specific needs. And as no system runs without data, make sure the system has features to populate it with commercial data. Electronic catalogues of the tool manufacturers are available for that purpose. Advanced TMS-packages rely on a combined multisupplier database.

The experience shows that smaller job shops with just a few CNCs will have the NC-programming or CAM-system as the dominating software package. Most of the tooling will be stored at the machines anyway, so an add-on for general tool reference will be enough. Small-scale professional packages which could easily be integrated into most of the popular CAM packages are recommended for that purpose.

Larger job shops might turn to the hardware supplier, e.g., the presetter for a solution. Again the main package is concerned with tool assemblies, and an open solution to reference tool components is recommended. Consider the case of the hardware-driven Tool Manager of Zoller, Inc. which runs in conjunction with a multisupplier database of Cimsource. The software is split in the cycle of tool assemblies, which go off the presetters to the machines and back, and the components which are stored in the crib. The TMS just holds the data of tools which are actually in use and mainly focuses on the control of the shop-floor activities. The multisupplier database is the reference for optional tooling and performance data. It is also linked to the ERP system to generate purchase orders. The components database also provides CAD information which is needed for tool layouts.

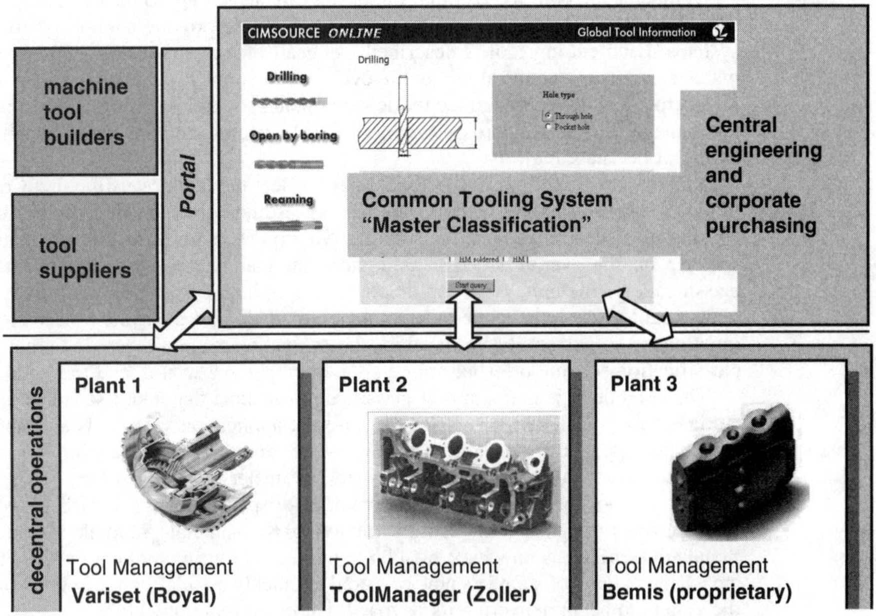

FIGURE 16.4 Concept of a TMS for a multiplant production system. (*Courtesy of Cimsource, Inc.*)

Multisite operations could be optimized using a combination of a centralized tool database and decentralized TMS-packages at the plant level (Fig. 16.4). At the core of this software is a relational database interfaced to shop-floor tool management and shop-floor requisitioning. For process engineering, a direct link to the drawing management system is available. A browser-based interactive user interface could be adapted to each user group's priorities. Import profiles could be designed so they could be tuned to different suppliers' content, allowing the database to be populated automatically—whether from the supplier directly or via a general master server. This latter source stores data covering the product ranges of various tool suppliers, saving users the time and expense of gathering this information on their own.

In cases where the proprietary solution is preferred instead of a commercial piece of software, still insist that a proper database and a widely accepted data structure are used. Some software companies offer, for example, their own database system to other software companies to build customized application packages around it.

16.5.4 Implementation

Implementation of a TMS follows the same route as the implementation of any manufacturing system. And like all projects where software is involved, the road to success lies in organizational and structural preparations. Streamlined business processes, a deproliferated range of tools, and an unambiguous tool classification are paramount for the overall success of the system. If there is a commonality of successfully implemented TMS, then it is:

1. The complete separation of content from applications.
2. The establishment of a master classification scheme that could serve as a central reference for different views of the system seen by different users.

Number 1 refers to the fact that proprietary databases are to be avoided in all cases. Instead, the TMS should refer to a relational database which would provide parameters for different application systems. Each cutting tool is described using all of the parameters necessary to meet the various management applications along its life cycle. Each application then accesses only the relevant parameters to populate its predefined models, templates, or tables. This technique is common practice in the management of standard parts within CAD systems, and any management information system (MIS) works the same way.

Number 2 is important as the TMS has to reflect the corporate structure. Consider again the case of the automotive supplier with its multisite powertrain operations of centralized engineering and purchasing and decentralized plant operations. No group should be forced to take on another group's perspective on the overall process, a requirement that relates most significantly to the underlying classification structures. If an engine plant uses an item called *mini-drill,* for example, then another plant might refer to the same product as *micro-drill,* and the system needs to allow for both designations to be correct. From the start, the project team should conclude that the push for a single classification scheme offering only a single perspective would lead to a fatal level of resistance.

The good news is that a master classification scheme that addresses the structure of the world of cutting tools as well as parametrics for different application systems is available. For the metal cutting industry, a consortium of tool manufacturers and their customers has worked to define a master classification scheme that includes requisite parameters for standard tools. The result is an industry standard for tool descriptions which also works for specials. This standard, called StandardOpenBase, is promoted through a joint venture involving Kennametal, Sandvik, CeraTizit, and Cimsource. StandardOpenBase is now used by a large number of cutting tool companies. That means that there are a large number of suppliers now prepared to quickly populate a newly implemented TMS, allowing the system to begin delivering its payback that much more quickly.

16.6 OPERATION AND ORGANIZATIONAL ISSUES

Usually the main users of the TMS will be the crib personnel and the tooling group. It is their responsibility to ensure the constant supply of the FMS with the right tools at the right time. In some cases, the tool supply might be outsourced to a commodity supplier. However, the responsibility stays the same. Shop-floor personnel have to act in a very disciplined way in order to use the TMS properly. That includes, above all, that crib personnel and machine operators record the issues and returns of the tool they are using. If they fail to do that, then the inventory counts are wrong and any justification of TMS involving inventory turns, reduced stock-outs, and the like is gone.

The same is true for quality of the tool master data, as the old saying about "garbage in garbage out" is especially true for a TMS with its thousands of components and tool assemblies to control. It is important to name individuals to be held accountable for each item of the master data. This should be somebody from process engineering or the tooling group who is involved in the technical specification of the tools to be used. This individual is also accountable that the tool databases of any engineering system (NC programming, NC simulation) are aligned with the TMS, with the TMS being the leading system. Master data changes should only occur in the TMS and be transferred to the other systems. Technical tool specs include regrind cycles and coating specs. Logistic parameters, like replenishment time and minimum stock levels per item have to be maintained by the crib personnel or an individual of the tooling group. If a commodity materials supplier (CMS) is in charge, this task is its responsibility.

The administrative characteristics of the master data, like prices, discounts and delivery specs should be maintained from purchasing as the main users. In most practical settings this will take place in the ERP-system which actually issues purchase orders for tools, so that the TMS needs a link to that system to receive the respective updates. Companies that operate e-procurement platforms will ask the suppliers to submit the respective data of their product range.

In the context of master data management a clear cut decision needs to be made as to what information is subject to the regular change mechanisms and release processes. Consider the fact that tool manufacturers constantly change their product range. New products are developed to replace others, but very seldom in a one-to-one relation. The TMS must be able to trace this back because the requirements planning will ask for a certain tool which may no longer be available. Sometimes the tools are improved but the product ID is not changed, so a manufacturing problem due to a certain tool is difficult to trace back as the manufacturer might have changed the internal spec of this tool incrementally.

Manufacturing management should also provide for the fact that the TMS will be the backbone for future tool supply. It has to be implemented and maintained in the same way as the ERP system of the company. This includes redundant servers for the software and the decoupling of the inventory controls of any automated storage equipment. If bad comes to worse, the storage devices have to be manually operated.

16.7 ECONOMY AND BENEFITS

Whether or not the introduction of the TMS will be a success is determined by tangible results which should occur in reduced spending for cutting tools and increased productivity of the FMS. More difficult to quantify is the commercial impact of efficient replenishment processes and higher turns of the tool inventory.

Experience shows that between 5 percent and 10 percent of the annual purchasing volume could be saved—the biggest chunk coming from deproliferation activities before any software is implemented. Savings in tool spending are also driven by the TMS because it helps that tools are returned after use and refurbishment is controlled. This rationale indicates that a "full size" TMS only makes sense in a larger manufacturing environment where at least $200,000 is spent on tooling annually. Smaller job shops should use public databases like CS-Pro which could be customized to a basic TMS functionality.

The productivity gain through machine up-time cannot be generally quantified, as the manufacturing conditions vary too much from operation to operation. However, if downtimes are analyzed systematically, the impact of missing or wrongly assigned tools as drivers for downtime could be qualified. Improved requirement planning, unambiguous tool identification and prekitting could bring down that fraction of the downtime close to zero. The commercial impact of that improvement depends on the machine cost per hour.

Several studies using activity-based costing revealed that the cost involved in the sourcing and replenishment process can exceed the purchase value of tools. This insight has led many manufacturers to switch to commodity suppliers, which in turn use TMS systems at least for the tool administration. This is also true for the inventories, so that these are no more in the focus of the manufacturer himself.

The conclusion is that the payback of a TMS ranges between 10 months and 2 years, depending on the size of the operation. This payback estimate excludes the purchase of large scale hardware components like vertical tool cabinets. And it should be clear at this stage that the tool classification and the population of the system with tool data both could be show stoppers for a successful TMS implementation. First, entering data is a big job even if a well organized classification system is available. Second, the way data is captured might make an enormous difference. If your current classification system has more data elements than that of your TMS you will be forced to prioritize the information to be entered into the system. It is highly recommended to use industrywide accepted classification standards for both reasons.

16.8 FUTURE TRENDS AND CONCLUSION

It is to be expected that tool management will grow beyond shop-floor activities and more and more include engineering activities because many of the decisions affecting the difficulty of tool management have already been made before any physical tool comes into play. But there is only so much that can be done at the shop-floor level. To look to the shop floor for the solution to these problems

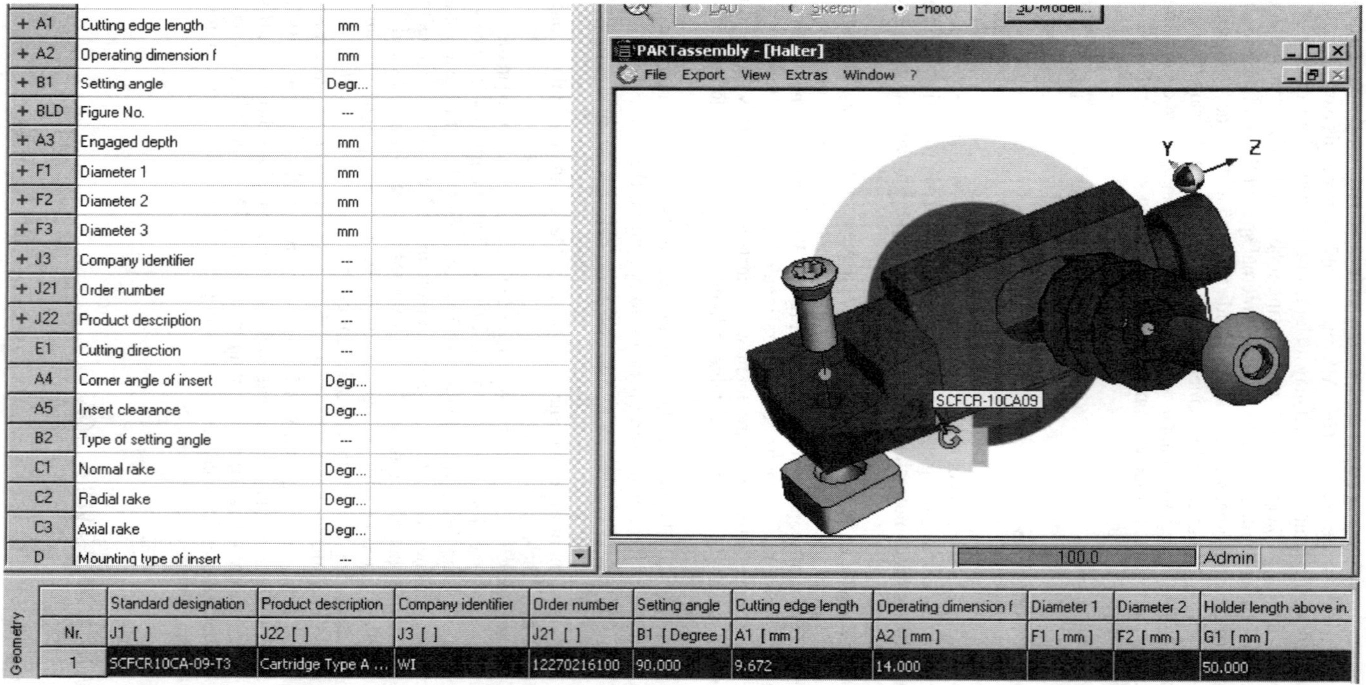

FIGURE 16.5 3D CAD models in the future TMS. (*Courtesy of Cimsource, Inc.*)

is to take for granted the tool as it was released from engineering, and treat tool management as merely a logistics and delivery challenge.

In fact, close to 70 percent of the difficulty in shop-floor tool management is created in engineering. Here the tool layout is fixed, and so is the supplier and the operation. Thus 70 percent of the life cycle cost of the tooling is fixed as well. The concept of *integrated tool management* allows for the fact that a cutting tool, over the course of its life time, changes from an engineering item to a logistics item.[7] During engineering, tooling is an area for process innovation. During the manufacturing cycle, tooling becomes a productivity driver for the machine tools.

Both of these separate realms can optimize their own flow of information. However, these areas typically don't view one another as customers. Engineering has no means to provide value-added service to manufacturing, and manufacturing has no means to channel its experience and requirements back to engineering. Providing the means—and therefore the integration—is the mission of integrated tool management.

Companies considering the introduction of a TMS should bear that future trend in mind, which mainly affects the decision for the database system and the master classification. Both should be powerful enough in scope that future requirements from engineering could be solved. Tests to generate 3D tool models from the relational database and submit those into any kind of commercial CAD system are already in progress (Fig. 16.5).

REFERENCES

1. Tetzlaff, A.W., "Evaluating the Effect of Tool Management on FMS Performance," *International Journal of Production Research,* Vol. 33, No. 4, 1995.
2. Tap, M., J.R. Hewitt, and S. Meeran, "An Active Tool-tracking System for Increased Productivity," *International Journal of Production Research,* Vol. 38, No. 16, 2000.
3. Albert, M., "A Shop Preset for Productivity," *Modern Machine Shop Magazine,* January 2000.
4. Plute, M., *Tool Management Strategies,* Hanser Gardner Publications, Cincinnati, OH, 1998.
5. Kettner, P., "Tool Base—The Electronic Tool Data Exchange," *CIRP STC C Meeting Procedures,* Paris, 1995.
6. Marczinski, G., and M. Mueller, "Cooperative Tool Management—An Efficient Division of Tasks Between Tool Suppliers and Tool Users Based on Standardized Tool Data," *VDI Reports No. 1399,* Duesseldorf, 1998.
7. Marczinski, G., "Integrated Tool Management—Bridging the Gap Between Engineering and Shop-floor Activities," *Modern Machine Shop Magazine,* November 2002.

CHAPTER 17
GROUP TECHNOLOGY FUNDAMENTALS AND MANUFACTURING APPLICATIONS

Ali K. Kamrani
Rapid Prototyping Laboratory
University of Houston
Houston, Texas

17.1 INTRODUCTION

Grouping objects (i.e., components, parts, or systems) into families based on the object features has been done using *group technology* (GT) approaches.[1,2,3,4] Similar components can be grouped into design families, and modifying an existing component design from the same family can create new designs. The philosophy of group technology is an important concept in the design of advanced integrated manufacturing systems. Group technology classifies and codes parts by assigning them to different part families based on their similarities in shape and/or processing sequence. The method of grouping that is considered the most powerful and reliable is *classification and coding*. In this method, each part is inspected individually by means of its design and processing features. A well-designed classification and coding system may result in several benefits for the manufacturing plant. These benefits may include:

- It facilitates the formation of the part families.
- It allows for quick retrieval of designs, drawings, and process plans.
- Design duplication is minimized.
- It provides reliable workpiece statistics.
- It aids production planning and scheduling procedures.
- It improves cost estimation and facilitates cost.

Classification is defined as a process of grouping parts into families, based on some set of principles. This approach is further categorized into the *visual method* (ocular) and *coding procedure*. Grouping based on the ocular method is a process of identifying part families, visually inspecting parts, and assigning them to families and the production cells to which they belong. This approach is limited to parts with large physical geometries, and it is not an optimal approach because it lacks accuracy and sophistication. This approach becomes inefficient as the number of parts increases. The coding method of grouping is considered to be the most powerful and reliable method. In this

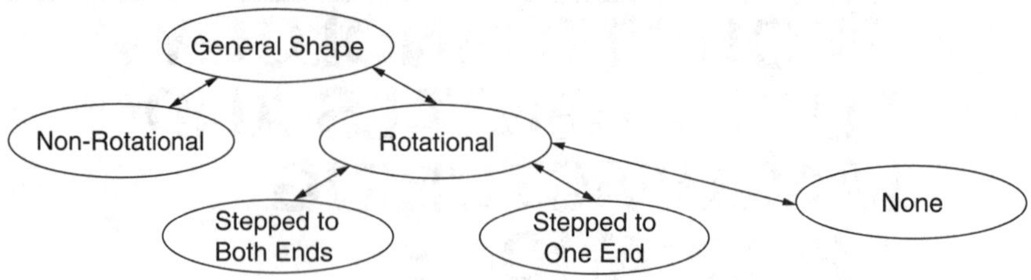

FIGURE 17.1 Monocode structure.

method, each part is inspected individually by means of its design and processing features. Coding can be defined as a process of tagging parts with a set of symbols that reflect the part's characteristics. A part's code can consist of a numerical, alphabetical, or alphanumerical string. Three types of coding structures exist.

17.1.1 Hierarchical (Monocode) Structure

In this structure, each digit is a further expansion of the meaning of a digit, dependent on the meaning of the previous digit in the code's string. The advantage of this method is the amount of information that the code can represent in a relatively small number of digits. However, a coding system based on this structure is complicated and very difficult to implement. Figure 17.1 illustrates the general structure of this method of coding.

17.1.2 Chain (Attribute or Polycode) Structure

In this structure, the meaning of each digit is independent of any other digit within the code string. In this approach, each attribute of a part is tagged with a specific position in the code. This structure is simple to implement, but a large number of digits may be required to represent the characteristics of a part. Figure 17.2 illustrates the general layout of a code based on this structure.

17.1.3 Hybrid Structure

Most of the coding systems available are implemented using the hybrid structure. A hybrid coding system is a combination of both the monocode and the polycode structures, taking advantage of the best characteristics of the two structures described earlier.

Digits	Feature	Code				
		1	2	3	4	...
1	External Shape	Symmetrical	Non-Symmetrical	Contour	Complex	
2	Number of Holes	0	3 to 5	6 to 9	More than 10	...
3	Holes' Diameter	<0.5	0.5 < hd < 1	1 < hd < 3	3 < hd < 8	...
...	...	...	...	...	...	...

FIGURE 17.2 Polycode structure.

Clustering is another technique used for grouping similar objects. This method requires the calculation of a clustering factor known as a *similarity* or a *dissimilarity coefficient* by assigning a clustering criterion as an objective to optimize system performance. Similarity and dissimilarity coefficients are calculated values that represent the relationships between parts. Most research has been based on the fact that these coefficients range from 0 to 1. This indicates that dissimilarity = 1.0—similarity or vice versa. Two methods widely used are hierarchical clustering and nonhierarchical clustering. The hierarchical method results in a graph known as a dendrogram which illustrates the data grouped into smaller clusters based on the value of similarities/dissimilarities. The nonhierarchical method uses partitioning clustering algorithms to search for a division of a set of objects into a number K of clusters in such a way that the elements of the same cluster are close to each other and the different clusters are well separated. Because the K clusters are generated simultaneously, the resulting classification is nonhierarchical. The number of clusters K can be either given or determined by the algorithm. When K is unknown, the algorithm can be repeated for several values of K. In this way it is possible to evaluate the different clusterings and sometimes select an optimal one.[2,5]

17.2 IMPLEMENTATION TECHNIQUES

17.2.1 Sorting Techniques

Sorting techniques begin by constructing an incidence matrix that describes the relationship between parts and machines. The part-machine incidence matrix a_{ij} consists of 0, 1 entries. An entry of 1 indicates that machine i is used to process part j while an entry of 0 indicates that machine i is not used to process part j. After constructing the matrix, the techniques will attempt to identify groups (clusters) existing in the matrix by rearranging the machines and parts into a more structured form. Most sorting techniques are based on the idea of arranging the matrix entries into a block diagonal form.

Rank Order Clustering (ROC) Algorithm. The rank order clustering algorithm was developed to group the entries of a part-machine incidence matrix around the diagonal by repeatedly arranging the columns and rows according to the sum of binary values.[6] The ROC was designed to solve the clustering problem of matrices with binary entries. The idea is to cluster the positive entries (1s) into groups and place them along the diagonal of the matrix. The part-machine incidence matrix is read as a binary word. Then the algorithm converts these binary words for each row (column) into decimal equivalents. After that, the algorithm rearranges the rows (columns) in order of descending values. This step is repeated until there is no change. Following is a step-by-step example of applying ROC for clustering.[2]

Step 1. Assign binary weights to rows.

		Part Number 1	2	3	4	5	6
Binary Weights		2^5	2^4	2^3	2^2	2^1	2^0
Machine Number	1			1	1		1
	2	1	1			1	
	3			1	1		1
	4	1	1			1	
	5	1	1				

Step 2. Calculate decimal equivalent.

Machine Number	Part 1	Part 2	Part 3	Part 4	Part 5	Part 6	Decimal Equivalent
1			1	1		1	13
2	1	1			1		50
3			1	1		1	13
4	1	1			1		50
5	1	1					48

Step 3. Sort the decimal equivalent in descending order.

Machine Number	Part 1	Part 2	Part 3	Part 4	Part 5	Part 6	Decimal Equivalent
2	1	1			1		50
4	1	1			1		50
5	1	1					48
1			1	1		1	13
3			1	1		1	13

Step 4. Assign binary weights and calculate decimal equivalent for columns.

Machine Number	Part 1	Part 2	Part 3	Part 4	Part 5	Part 6	Binary Weights
2	1	1			1		2^4
4	1	1			1		2^3
5	1	1					2^2
1			1	1		1	2^1
3			1	1		1	2^0
Decimal Equivalent	28	28	3	3	24	3	

Step 5. Sort the decimal equivalent in descending order.

Machine Number	Part 1	Part 2	Part 5	Part 3	Part 4	Part 6
2	1	1	1			
4	1	1	1			
5	1	1				
1				1	1	1
3				1	1	1
Decimal Equivalent	28	28	24	3	3	3

Step 6. If no separate groups are visible, repeat the previous steps; otherwise stop.

Machine Number	Part 1	Part 2	Part 5	Part 3	Part 4	Part 6
2	1	1	1			
4	1	1	1			
5	1	1				
1				1	1	1
3				1	1	1

The ROC provides a simple analytical technique that can be easily computerized. The ROC has fast convergence and relatively low computation time depending on the matrix size. The fact that ROC uses binary values will impose restrictions on the size of the matrix that can be solved.

Cluster Identification Algorithm (CIA). The cluster identification algorithm was designed to identify disconnected blocks if they exist.[7] If there are no disconnected blocks (the matrix is not mutually separable), the entire matrix will form one block. The algorithm begins by masking all columns that have an entry of 1 in any row (selected randomly), then all rows that have an entry of 1 in these columns are masked. This process is repeated until all intersecting columns and rows are masked. These masked columns and rows are placed in a block and removed from the matrix. This will continue until all the entries in the matrix are assigned in blocks and all separable blocks are identified.

Step 1. Consider the machine-part matrix shown below.[2] Select any row i of the incidence matrix and draw a horizontal line h_i through it. Row 3 of the matrix is selected randomly and a horizontal line h_3 is drawn.

Step 2. For each entry of 1 crossed by the horizontal line h_i draw a vertical line v_j. Three vertical lines v_2, v_6, and v_7 are drawn.

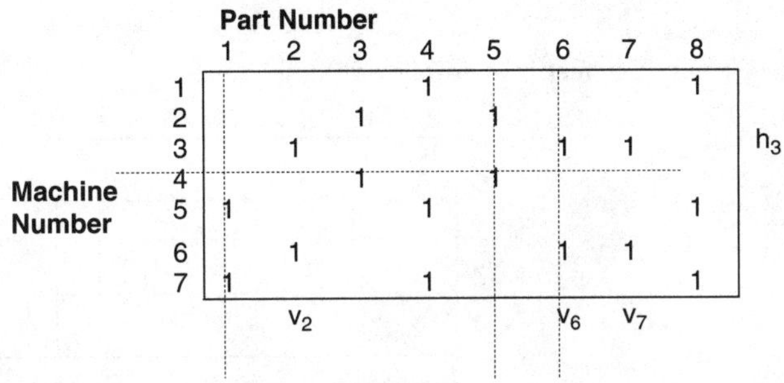

Step 3. For each entry of 1 crossed once by a vertical line v_j draw a horizontal line. A horizontal line h_6 is drawn through all the crossed-once entries of the matrix:

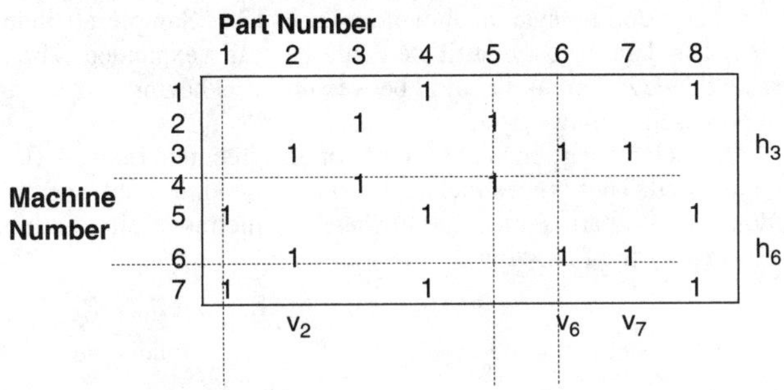

Step 4. The preceding steps are repeated until no more crossed-once entries are left. All the crossed-twice entries are grouped in a block and removed from the matrix. Parts 2, 6, 7 and machines 3, 6 are grouped in one block.

Machine Number \ Part Number	2	6	7	1	3	4	5	8
3	1	1	1					
6	1	1	1					
1						1		1
2					1		1	
4					1		1	
5				1		1		1
7				1		1		1

The grouped parts and machines are removed from the matrix.

	2	6	7
3	1	1	1
6	1	1	1

Step 5. The above procedure is repeated for the remaining matrix entries until all entries are grouped.

	1	3	4	5	8
1			1		1
2		1		1	
4		1		1	
5	1		1		1
7	1		1		1

The resulting matrix is as follows:

	2	6	7	8	1	4	3	5
3	1	1	1					
6	1	1	1					
5				1	1	1		
1				1		1		
7				1	1	1		
2							1	1
4							1	1

Mathematical Programming-Based Technique. A distance measure d_{ij} is considered in most mathematical programming models used in group technology. The distance measure is a real-valued symmetric function based on the three fundamental axioms.[2,3] These are reflexivity ($d_{ii} = 0$), symmetry ($d_{ij} = d_{ji}$), and triangle inequality ($d_{ij} \le d_{is} + d_{sj}$) where s is any part other than parts i and j. Consider the generic coding system shown in Table 17.1. Sample attributes of this coding system are also listed. The data for each attribute could be easily expanded. Three strategies are available for measuring the degree of association between objects having mixed variables types.[2] This is the case for the proposed coding system.

One such strategy involves the use of disagreement indices. Generally, four types of variables can be identified. These types include binary, nominal, ordinal, and continuous. The linear disagreement index between parts i and j for attribute k, which is of either a binary or a nominal variable type, is measured by the following:

$$d_{ijk} = \begin{cases} 1, & \text{if } R_{ik} \neq R_{jk} \\ 0, & \text{otherwise} \end{cases}$$

where d_{ijk} = disagreement index between parts i and j for attribute k
R_{ik} = rank of part i for attribute k
R_{jk} = rank of part j for attribute k

The linear disagreement index for an ordinal variable is determined by the following equation:

$$d_{ijk} = |R_{ik} - R_{jk}|/(m-1)$$

where m = number of classes for attribute k
$m - 1$ = maximum rank difference between parts i and j

The linear disagreement index for a continuous variable is determined by the following equation:

$$d_{ijk} = |R_{ik} - R_{jk}|/X_k$$

where X_k = range of values for the variable.

TABLE 17.1 Sample Generic Coding Structure

Design attributes	
Attribute 1: General shape of the part 1) Rotational (CA1-1) R-Bar (CA1-2) R-Tube (CA1-3) R-Hexagonal bar etc. 2) Nonrotational (CA1-4) NR-Plate (CA1-5) NR-Square bar (CA1-6) NR-Sheet plate (CA-7) NR-Rectangular bar etc. **Attribute 2: Material** (CA2-1) Aluminum alloys (CA2-2) Copper-zinc alloys (CA2-3) Steels (CA24) Cast irons (CA2-5) Plastics etc. **Attribute 3: Maximum diameter** (CA3-1) D <= 0.75 in (CA3-2) 0.75 < D <= 1.50 in (CA3-3) 1.50 < D <= 4.00 in	(CA3-4) D > 4.00 in (CA3-5) N/A (Nonrotational Part) etc. **Attribute 4: Overall length** (CA4-1) L <= 6 in (CA4-2) 6 < L <= 18 in (CA4-3) 18 < L <= 60 in (CA4-4) L > 60 in etc. **Attribute 5: Diameter of inside hole** (CA5-1) d <= 0.5 in (CA5-2) 0.5 < d <= 1.0 in (CA5-3) 1.0 < d <= 5.0 in (CA5-4) d > 5.0 in (CA5-5) No Hole etc. **Attribute 6: Product type** (CA6-1) Commercial/service (CA6-2) Electrical (CA6-3) Industrial/mechanical (CA6-4) Military (CA6-5) Special (CA6-7) Other
Manufacturing attributes	
Attribute 7: Number of processing steps **Attribute 8: Processing type and sequence** (CA8-1) Turning (CA8-2) Drilling (CA8-3) Reaming (CA8-4) Boring (CA8-5) Tapping (CA8-6) Milling (CA8-7) Grinding (CA8-8) Broaching (CA8-9) Cutting etc. **Attribute 9: Minimum number of machines required for processing** **Attribute 10: Processing machine type** (CA10-D CNC turning (CA10-2) CNC Drilling/tapping (CA10-3) Vertical/horizontal CNC Milling (CA10-4) External/internal grinding (CA10-5) Broaching (CA10-6) Band/circular sawing etc. **Attribute 11: Number of tool types** **Attribute 12: Tool type and sequence** (CA12-1) Insert (CA12-2) Twist drill (CA12-3) Adjustable reamer (CA12-4) Adjustable boring bar (CA12-6) Milling cutter (CA12-7) Grinding wheel (CA12-8) Broach (CA12-9) Band/circular saw blade etc. **Attribute 13: Number of fixture/jig types**	**Attribute 14: Fixture/jig Type** (CA14-1) Special fixture (CA14-2) Multipurpose fixture (CA14-3) Adjustable fixture (CA14-4) Rotational adjustable fixture (CA14-5) Nonrotational adjustable fixture (CA14-6) Special jig (CA14-7) Multipurpose jig (CA14-8) Adjustable jig etc. **Attribute 15: Number of end operations** **Attribute 16: End operation type and sequence** (CA16-1) Clean (CA16-2) Polish (CA16-3) Buff (CA16-4) Coat (CA16-5) Paint (CA16-6) Assemble (CA16-7) Packaging (CA16-8) Inspect etc. **Attribute 17: Number of devices required for end operations** **Attribute 18: Devices used for end operations** (CA18-1) Dip tank (CA18-2) Disk grinder (CA18-3) Process robot (CA18-4) Material handling robot (CA18-5) Painting robot (CA18-6) Assembly robot (CA18-7) Packaging machine (CA18-8) Vision system etc.

The linear disagreement index for attributes 1, 2, and 6 is calculated using a binary/nominal equation form, and the linear disagreement index for attributes 3, 4, and 5 is calculated using an ordinal equation form. The linear disagreement index for the process and end operation sequence is calculated using equation:

$$d_{ij} = 1 - \sum_{o} (q_{io} \times q_{jo}) \Big/ \sum_{o} (q_{io} + q_{jo} - q_{io} \times q_{jo})$$

where

$$q_{io} = \begin{cases} 1, & \text{if part } i \text{ requires operation } o \\ 0, & \text{otherwise} \end{cases}$$

The linear disagreement index for tools and fixtures is calculated by:

$$d_{ij} = (NT_i + NT_j - 2NT_{ij})/(NT_i + NT_j)$$

where NT_i = number of tools required by part i
NT_{ij} = number of tools common to both parts i and j

The linear disagreement index for process and end operation machines is calculated using the Hamming metric as follows:

$$d_{ij} = \sum_{m} \delta(X_{im}, X_{jm})$$

where

$$X_{im} = \begin{cases} 1, & \text{if part } i \text{ uses machine } m \\ 0, & \text{otherwise} \end{cases}$$

and

$$\delta(X_{im}, X_{jm}) = \begin{cases} 1, & \text{if } X_{im} \neq X_{jm} \\ 0, & \text{otherwise} \end{cases}$$

After the evaluation of these parameters, the analyst can assign weights to represent his or her subjective evaluation of variables and group parts based on their assigned priorities. The weights can be categorized in any ordinal form, 0 (low degree of importance) to 1 (high degree of importance). Finally, the weighted dissimilarity measure (DIS_{ij}) between parts i and j can be determined by the following:[2]

$$DIS_{ij} = \sum_{k} (w_k \times d_{ijk}) \Big/ \sum_{k} w_k$$

where w_k = weight assigned to attribute k
d_{ijk} = disagreement index between parts i and j for attribute k
DIS_{ij} = weighted dissimilarity coefficient between parts i and j

For example, codes are assigned to two sample parts as shown in Table 17.2. The disagreement measure between these two parts is then calculated using the proposed formulation.

TABLE 17.2 Sample Parts Used for Dissimilarity Analysis

Coding system attribute	Values: Part I	Values: Part II	Values: Dissimilarity index, d_{ijk}
Design			
General shape	NR-square bar: 5	NR-square bar: 5	0 (Match)
Material type	Steel: 3	Steel: 3	0 (Match)
Maximum diameter	N/A: 5	N/A: 5	0 (Match)
Overall length	18 < L ≤ 60 in: 3	L > 60 in: 4	0.33
Inside hole diameter	d ≤ 0.5 in: 1	1.0 < d ≤ 5.0 in: 3	0.5
Product type	Commercial: 1	Special: 5	1 (Mismatch)
Manufacturing			
Processing sequence	Drilling, tapping: 25	Drilling, reaming: 23	0.67
Processing machine	CNC drilling/tapping: 2	CNC drilling/tapping: 2	1
Tool types	Twist drill, tap: 25	Twist drill, reamer: 23	0.5
Fixture type	Adjustable fixture: 3	Nonrotating adjustable: 5	1
E. operation sequence	Clean, paint, and pack: 157	Clean and pack: 17	0.67
E. operation devices	Dip tank, process robot, and handling robot: 134	Dip tank and handling robot: 14	0.67

The dissimilarity of these two parts is 0.528 for all attributes considered critical. The p-medium mathematical technique is used to solve the grouping and part family formation problem. The formulation is as follows:
Model:

$$\text{Minimize} \sum_i \sum_j DIS_{ij} \times x_{ij} \qquad \forall i, j = 1, 2, 3, \ldots, p$$

subject to:

$$\sum x_{ij} = 1, \qquad \forall i, j$$
$$\sum x_{ii} = K, \qquad \forall i$$
$$xij \leq xjj \qquad \forall i, j$$

The coefficients for the mathematical model are as follows:

p = number of parts
K = required number of part families
DIS_{ij} = dissimilarity measure between parts i and j ($DIS_{ij} = DIS_{ji}$)
x_{ij} = 1, if part i belongs to group j and 0 otherwise

The first constraint assures that each part belongs to only one family. The required number of part families is set in the second constraint. Parts are assigned to part families only if that part family has already been created. This is the last and a conditional constraint to the problem. The model is a 0 to 1 integer linear program. The model is solved using a branch-and-bound technique. The branch-and-bound technique is a partial enumeration method in which the set of solutions to a problem is examined by dividing this set into smaller subsets. It can be shown mathematically that some of these subsets do not contain the optimal solution. One important way in which a subset can be examined is to determine a bound. A bound of a subset is a value that is less than or equal to the value of all solutions contained in the subset. By comparing the bound of a subset with an already-found solution,

TABLE 17.3 Disagreement Measures

Part	1	2	3	4	5	6	7	8	9	10	11	12	13	14	15
1	0.00	1.00	0.60	0.98	0.60	1.00	0.68	0.86	0.73	0.96	0.85	0.88	0.89	1.00	0.74
2	1.00	0.00	0.90	0.45	1.00	0.63	0.94	0.66	0.92	0.78	0.51	0.65	0.71	0.36	0.88
3	0.60	0.90	0.00	0.74	0.62	1.00	0.65	0.69	0.84	0.92	1.00	0.80	0.85	0.93	0.77
4	0.98	0.45	0.74	0.00	0.84	0.60	0.94	0.51	1.00	0.76	0.43	0.55	0.70	0.44	1.00
5	0.60	1.00	0.62	0.84	0.00	0.97	0.76	0.71	0.76	0.77	1.00	0.88	0.89	1.00	0.82
6	1.00	0.63	1.00	0.60	0.97	0.00	1.00	0.59	1.00	0.67	0.63	0.78	0.71	0.51	1.00
7	0.68	0.94	0.65	0.94	0.76	1.00	0.00	0.82	0.75	0.94	0.88	0.76	0.99	0.97	0.56
8	0.86	0.66	0.69	0.51	0.71	0.59	0.82	0.00	1.00	0.75	0.75	0.65	0.87	0.63	0.78
9	0.73	0.92	0.84	1.00	0.76	1.00	0.75	1.00	0.00	0.68	1.00	0.65	0.91	1.00	0.49
10	0.96	0.78	0.92	0.76	0.77	0.67	0.94	0.75	0.68	0.00	0.92	0.72	0.81	0.75	0.59
11	0.85	0.51	1.00	0.43	1.00	0.63	0.88	0.75	1.00	0.92	0.00	0.73	0.74	0.54	1.00
12	0.88	0.65	0.80	0.55	0.88	0.78	0.76	0.65	0.65	0.72	0.73	0.00	0.55	0.77	0.66
13	0.89	0.71	0.85	0.70	0.89	0.71	0.99	0.87	0.91	0.81	0.74	0.55	0.00	0.86	0.97
14	1.00	0.36	0.93	0.44	1.00	0.51	0.97	0.63	1.00	0.75	0.54	0.77	0.86	0.00	0.93
15	0.74	0.88	0.77	1.00	0.82	1.00	0.56	0.78	0.49	0.59	1.00	0.66	0.97	0.93	0.00

it is sometimes possible to eliminate the subset from further consideration. Table 17.3 lists the result of the dissimilarity analysis for 15 components. Using the proposed formulation and by setting p (required number of part families) to be four, parts and their associated families are as *G1*(1,3,5), *G2*(2,4,6,8,11,14), *G3*(7,9,10,15), and *G4*(12,13). The dissimilarity values are also used to set up the dendrogram of this example, shown in Fig. 17.3. For the threshold assignment of 0.60, the four families are as follows: *G1*(2,4,6,8,11,14), *G2*(7,9,10,15), *G3*(1,3,5), and *G4*(12,13). This solution is the same as proposed by the optimization model.

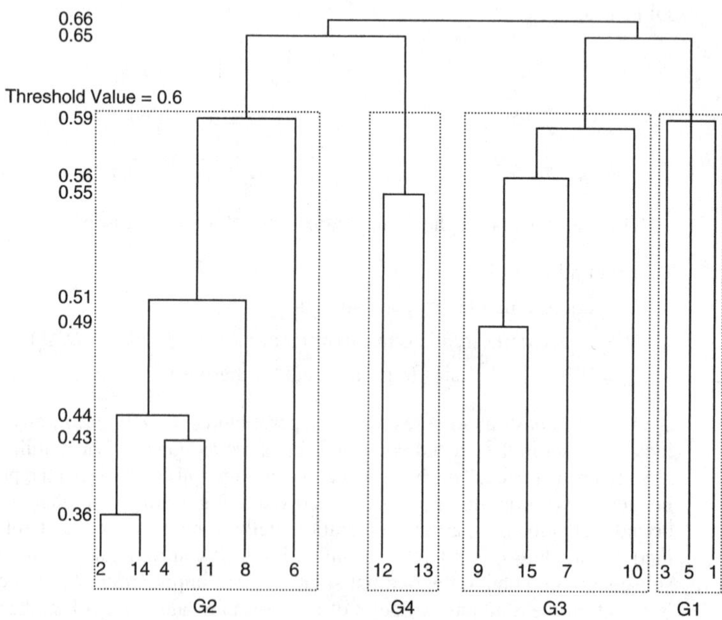

FIGURE 17.3 The dendrogram constructed for sample parts.

17.3 APPLICATIONS OF GROUP TECHNOLOGY IN MANUFACTURING

17.3.1 Cellular Manufacturing Systems

Cellular manufacturing (CM) utilizes the philosophy of group technology. Cellular manufacturing, also known as group production, is a manufacturing process that produces families of parts within a single line or cell of machines operated by a machinist or by robots that operate only within the line or cell. Some of the benefits of the cellular manufacturing system are:[8,9,10,11,12]

- Elimination or reduction of setup time
- Greater manufacturing flexibility
- Reduction of work-in-process and less inventory (just-in-time)
- Less floor space around the high-investment machines
- Product cost (goods, labor, etc.)

Manufacturing cells are further categorized into two groups—*manned cells* and *unmanned cells*. In manned cells, operators have been trained to operate more than one machine, each of which is conventional and programmable. Unmanned cells are those composed of programmable, automated machines and robots. These unmanned cells may be considered as part of a trend toward flexible manufacturing systems and computer-integrated manufacturing systems. Table 17.4 summarizes and compares some of the characteristics of traditional manufacturing systems and cellular manufacturing systems. Comparing the traditional manufacturing system to cellular manufacturing indicates that the use of a cellular manufacturing philosophy can improve the overall productivity of a manufacturing firm.

17.3.2 Process Planning and Computer-Aided Process Planning Systems

Process planning is simply the set of activities associated with the manufacturing of a part, presented in a detailed and structured format.[2,13] Process planning is the link between design and manufacturing. With the interpretation being made on the product design data, material and process activities are selected and sequenced. Further resources such as machine, tools, and fixtures are also selected. Time for processes is evaluated and documented.[14,15,16] The process planning task is experience-based and manual. As a result, modern manufacturing industries are facing a series of problems (e.g., the departure of the skilled labor force that can produce the part). Manual development of process plans is subjective and may result in the development of plans that are feasible but not optimal. Computer technology has had a great impact on the role of design and manufacturing engineers during the development and implementation

TABLE 17.4 Characteristics of Cellular vs. Traditional Manufacturing Systems

Cellular production	Job shop production	Flow shop production
Flexible and modular machines	Flexible and modular machines	Special-purpose machines
Programmable machines	General-purpose machines	Single-function machines
Mechanized/manual handling	Mechanized/manual handling	Transfer lines
Single setup	Long/variable setup	Long/variable setup
Small/medium lot size	Small/medium lot size	Large lot size
Low inventory	Large inventory	Large inventory
100% Quality assurance	Random sampling	Random sampling

TABLE 17.5 Overview of V-CAPP and G-CAPP Characteristics

Variant process planning	Generative process planning
Group technology concept	Consistent process plan, GT supported
Partially automated procedures	Fully automated procedures
Experienced planner required	AI applications
Low cost	High cost
Simple to develop	Complex architecture
Computer-assisted	Computer-aided

life cycle of products. *Computer-aided process planning* (CAPP) is one such tool that has freed the process planner from making many routine decisions. The CAPP system reduces process planning time and manufacturing cost and enables overall improvement of productivity. CAPP is based on two major approaches: variant and generative.[17] The variant approach uses the computer as a design-supporting tool for information retrieval and text editing. Order-specific process plans are stored in the process plan database and a standard plan for a similar part is retrieved and modified. Still, the skill and the experience of the process planner play a major role in this approach. The generative approach develops a new process plan for every new product designed. It takes advantage of information-management technologies and capabilities. Company-specific knowledge can be structured in a standardized process planning procedure. Table 17.5 lists the characteristics of each technique.

Development of a CAPP system is a difficult task. A list of technical objectives that should be considered is given below:[17]

- Capture the manufacturing engineering knowledge and make it available to the process planner.
- Provide editing capabilities for changes required to the developed process plan.
- Develop optimum and consistent process plans for part families.
- Provide an engineering database of manufacturing knowledge, its feasibility, and its availability to the process planner.
- Provide information and data management capabilities in text and graphics.
- CAD integration capabilities.
- Provide alternatives for cost analysis and attempt to select the most economical process plan by realization of all the cost associated with the production of the parts.

Figure 17.4 illustrates an overview of a generic CAPP system's characteristics.

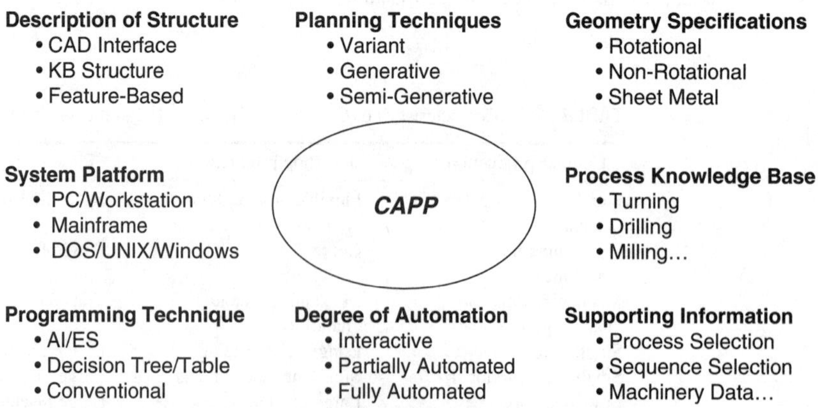

FIGURE 17.4 CAPP characteristics.

17.4 CONCLUSION

Group technology is based on the philosophy that many problems are similar and therefore by grouping these similar problems, a single possible solution could be determined. Design and manufacturing problems are both impacted by the application of the group technology concept. GT impacts design by promoting design standardization and impacts manufacturing by providing means for near net shape analysis and process planning. Group technology is considered to be a major element for integrated design and manufacturing environment.

REFERENCES

1. Arn, E. A., *Group Technology: An Integrated Planning and Implementation Concept for Small and Medium Batch Production*, Springer-Verlag, New York, 1975.
2. Kamrani, A., and S. M. Salhieh, *Design for Modularity,* 2d ed., Kluwer Academic Publishing, Norwell, Massachusetts, 2002.
3. Singh, N., and D. Rajamani, *Cellular Manufacturing Systems: Design, Planning, and Control*, Chapman & Hall, New York, 1996.
4. Snead, C. S., *Group Technology: Foundation for Competitive Manufacturing,* Van Nostrand Reinhold, New York, 1989.
5. Kusiak, A., A. Vannelli, and K. R. Kumar, "Clustering Analysis: Models and Algorithms," *Control and Cybernetics,* Vol. 15, No. 2, pp. 139–154, 1986.
6. Kusiak, A., "Intelligent Manufacturing Systems," *International Series in Industrial and Systems Engineering,* Prentice Hall, Englewood Cliffs, New Jersey, 1990.
7. Kusiak, A., *Intelligent Design and Manufacturing,* John Wiley & Sons, New York, 1992.
8. Black, J. T., "Cellular Manufacturing Systems Reduce Setup-Time, Make Small Lot Production Economical," *Industrial Engineering,* November 1983, pp. 36–48.
9. Dumalien, W. J., and W. P. Santernm, "Cellular Manufacturing Becomes Philosophy of Management at Components Facility," *Industrial Engineering,* November 1983, pp. 72–76.
10. Gettelman, K., "GT and MRP: Productivity," *Modern Machine Shop,* Vol. 51, No. 9, pp. 90–97, 1979.
11. McManus, J., "Some Lessons from a Decade of Group Technology," *Production Engineering,* Vol. 59, No. 11, pp. 40–42, 1980.
12. Zisk, B. I., "Flexibility is Key to Automated Material Transfer for Manufacturing Cells," *Industrial Engineering,* November 1983, pp. 58–64.
13. Zhang, H. C., and L. Alting, *Computerized Manufacturing Process Planning Systems,* Chapman & Hall, New York, 1994.
14. Bedworth, D. D., M. R. Henderson, and P. M. Wolfe, *Computer-Integrated Design and Manufacturing,* McGraw-Hill, New York, 1991.
15. Kamrani, A. K., "GD&T Classification Impact on Integrated Product and Process Design," *Proceedings of the 4th Industrial Engineering Research Conference,* Nashville, Tennessee, 1995.
16. Kamrani, A. K., and P. R. Sferro, "Critical Issues in Design and Evaluation of Computer-Aided Process Planning Systems," *Proceedings of the 17th International Conference on C&IE,* Miami, Florida, 1995.
17. Requicha, A. A. G., and J. Vandenbrande, "Automated Systems for Process Planning and Part Programming in *Artificial Intelligence: Implications for CIM*, ed. A. Kusiak, IFS Publications, Kempston, U. K., 1988, pp. 301–326.
18. McCormick, W. T., P. J. Schweitzer, and T. W. White, "Problem Decomposition and Data Reorganization by Cluster Technique," *Operations Research,* Vol. 20, No. 5, pp. 993–1009, 1982.

P · A · R · T · 3

HEAT TREATING, HOT WORKING, AND METALFORMING

CHAPTER 18
HEAT TREATMENT

Daniel H. Herring
The Herring Group, Inc.
Elmhurst, Illinois

18.1 PRINCIPLES OF HEAT TREATMENT

Heat treating is the controlled application of time and temperature to produce a predictable change in the internal structure of a material. Since the heat treatment process deals with changes to the internal structure of the material, it is often difficult to see the effects of heat treating. Often these changes can only be determined by physical, mechanical, or metallurgical testing. However, we can use the model of material science (Fig. 18.1) to help us to unlock the mystery of heat treatment by representing the technology as a series of interlocking rings underscoring the interdependence of each element in the model.

The model can be interpreted two ways as indicated by the arrows. An engineering-based methodology (down arrow) uses this model by considering the needs of the end user: that is, the consideration of the performance demands of a specific product which requires the design engineer to select a material having certain mechanical, physical, and metallurgical properties. These properties can only be developed in the selected material by producing a specific microstructure that defines a particular heat treatment process or series of processes. The science-based methodology (up arrow) begins with material selection that requires a specific process or series of processes to be performed to produce a predicted microstructure in the material that determines the mechanical, physical, and metallurgical properties which ultimately define the end use performance capability of the product.

What is clear from either use of this model is that the manufacture of any product depends to a great extent on heat treating, a core competency. Its contribution is vitally important for cost control, durability, and reliability. Heat treating represents a significant portion of the total manufacturing cost. If not properly understood and controlled, it can have a significant impact on all aspects of the manufacturing process.

For heat treatment to thrive in the twenty-first century it must be the most price- and performance-competitive technology. The areas of emphasis today are

- Lowering part cost reduced energy consumption
- Elimination of manufacturing steps
- Minimizing part distortion
- Use of environmentally friendly processes
- Integration into manufacturing

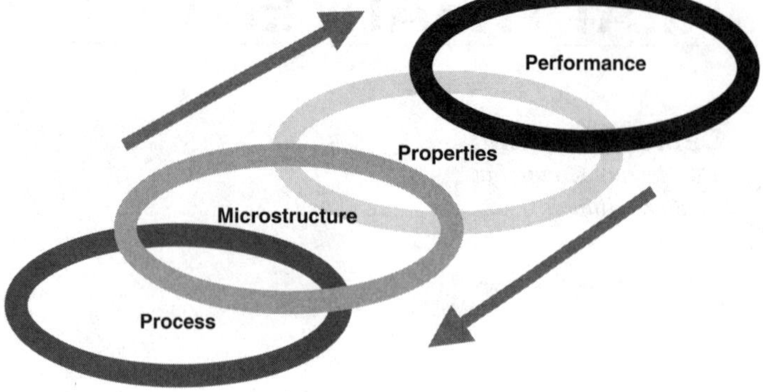

FIGURE 18.1 Model of material science.[1]

18.1.1 Why Do We Heat-Treat?

Heat treatment is a unique step in the manufacturing process because, in a single word, it is *manipulative*. In other words, by changing the type of heat treatment or the steps involved in the process, we can create different end results. One type of treatment will produce parts that are strong, such as automotive connecting rods or aircraft landing gear. By altering the treatment, parts can be made soft for use in such applications as automotive gaskets or rivets. Still other treatments will make parts tough enough to withstand sudden impact, such as railroad springs or safety shoe toe caps. Corrosion resistance can be enhanced in applications as diverse as aerospace turbine blades and kitchen sinks. Still other characteristics can be produced such as high hardness for bearing applications, or improvement in fatigue life and pitting resistance in gears. Virtually all materials can have their properties enhanced by heat treatment.

The heat treating process involves heating and cooling a material in such a way as to obtain desired specific conditions or properties. Some of the reasons for heat treating include the following:

- Remove stresses, such as those developed in processing a part.
- Refine the grain structure of the steel used in a part.
- Add wear resistance to the surface of a part by increasing its hardness; and, at the same time, increase its resistance to impacts by maintaining a soft, ductile core.
- Beef up the properties of an economical grade of steel, making it possible to replace a more expensive steel and reduce material costs in a given application.
- Increase toughness by providing a combination of high tensile strength and good ductility to enhance impact strength.
- Improve cutting properties.
- Upgrade electrical properties.
- Change or modify magnetic properties.

Our goal, however, is to avoid or eliminate this step in the manufacturing process. No one wants to heat treat; they must do so in order to balance the cost of producing a product with the performance demands of the end use application. We add this step to our manufacturing processes because quite often we can take an inexpensive raw material and, by heat treating, give it properties (such as hardness or wear resistance) of a more expensive material at a fraction of the cost of that more expensive material.

18.1.2 Heat Treatment Terminology

In the study of any specialized subject, a proper understanding of the terminology, as it applies to that specific topic, is of critical importance. No matter what the field of endeavor, an accurate knowledge of the vocabulary is essential to understanding the subject. In the field of heat treatment, considerable emphasis is placed on the proper use, meaning, and understanding of certain words or phrases. A brief summary of the terminology is presented below. For a more comprehensive list the reader is referred to Ref. 2.

Aging. Hardening by aging, typically after rapid cooling or cold working. Aging is a change in properties that generally occurs slowly at room temperature and more rapidly at higher temper atures. Other common names and related subjects are *age hardening, artificial aging, interrupted aging, overaging, precipitation hardening, precipitation heat treatment, progressive aging, quench aging,* and *strain aging.*

Alloy. A metallic compound of one or several materials, one of which is typically a metal.

Annealing. Heating to and holding at a suitable temperature and then cooling at a suitable rate, for such purposes as reducing hardness, improving machinability, facilitating cold working, producing a desired microstructure (for subsequent operations), or obtaining desired mechanical, physical, or other properties. Other common names and related subjects are *black annealing, intermediate annealing, isothermal annealing, malleabilizing, process annealing, quench annealing, recrystallization annealing,* and *spheroidizing.*

Brazing. Joining metals by flowing a thin capillary thickness layer, of nonferrous filler metal into the space between them. Bonding results from the intimate contact produced by dissolution of a small amount of base metal into the molten filler metal, without fusion of the base metal. The term *brazing* is used when the temperature exceeds some arbitrary value, typically 470°C (880°F).

Carbon Restoration. Heating and holding a metal above the *critical temperature* (Ac_1) in contact with a suitable carbonaceous atmosphere which may be solid, liquid, or gas for the purpose of restoring carbon by adding it to the surface. The amount of carbon added is usually sufficient to return the material to the original carbon content prior to a heat treat operation that reduced this carbon level.

Carbonitriding. The introduction of carbon and nitrogen into a solid ferrous alloy by holding above the critical temperature (Ac_1) in an atmosphere that contains suitable gases, such as hydrocarbons, carbon monoxide, and ammonia. The carbonitrided alloy is usually quench hardened. Other common names and related subjects are *case hardening, dry cyaniding, gas cyaniding, nicarbing,* and *nitrocarburizing* (obsolete).

Carburizing. The introduction of carbon into a solid ferrous alloy by holding above the critical temperature (Ac_1) in contact with a suitable carbonaceous atmosphere which may be solid, liquid, or gas. The carburized alloy is usually quench hardened. Other common names and related subjects are *case hardening, gas carburizing,* and *cyaniding.*

Cryogenic Treatment. The cooling of a material to a temperature low enough to promote the transformation of phases. Other common names and related subjects are *deep freezing, cold treatment,* and *subzero treatment.*

Ferrous. A metal of, related to, or containing iron.

Hardening. An increase in the hardness of a material by suitable treatment usually involving heating and cooling. Other common names and related subjects are *neutral hardening, quench hardening, flame hardening, induction hardening, laser hardening,* and *surface hardening.*

Homogenizing. Holding at high temperature to eliminate or decrease chemical segregation by diffusion. The resultant composition is uniform with coring and concentration gradients eliminated.

Materials Science. The science and technology of materials. Metals and nonmetals are included in this field of study.

Metals. (1) An opaque lustrous element that is a good conductor of heat and electricity and when polished a good reflector of light. Most metals are malleable and ductile and, in general, heavier than other elemental substances. (2) Metals are distinguished from nonmetals by their atomic binding and electron availability. Metallic atoms tend to lose electrons from the outer shells. (3) An elemental structure whose hydroxide is alkaline. (4) An alloy.

Metallurgy. The science and technology of metals. Metals are classified as *ferrous* and *nonferrous,* and divided into classifications of *wrought* and *powder* metal. *Process* (chemical) *metallurgy* deals with extraction and refinement of ores while *physical metallurgy* deals with the physical and mechanical properties of metals as they are affected by composition, mechanical working, and heat treatment.

Microstructure. The structure of polished and etched metals as revealed by a microscope at a magnification greater than 10 times.

Nitriding. Introducing nitrogen into a solid ferrous alloy by holding below the critical temperature (Ac_1) in contact with a suitable nitrogenous material which may be solid, liquid, or gas. Quench hardening is not required to produce a hard case. Other common names and related subjects are *case hardening*.

Nonferrous. A metal not containing, including, or related to iron.

Nonmetals. A chemical element that lacks typical metallic properties and is able to form anions, acidic oxides, acids and stable compounds with hydrogen.

Normalizing. Heating a ferrous alloy to a suitable temperature above the transformation range (and typically above the suitable hardening temperature) and then cooling in air to a temperature substantially below the transformation range.

Powder metallurgy. The art of producing metal powders and of utilizing metal powders for the production of massive materials and shaped objects.

Sintering. The bonding of adjacent powder particle surfaces in a mass of metal powders or a compact, by heating. Also, a shaped body composed of metal powders and produced by sintering with or without prior compaction. Other common names and related subjects are *cold/hot, isostatic pressing, liquid phase sintering,* and *metal injection molding*.

Solution Treating. Heating an alloy to a suitable temperature, holding at that temperature long enough to allow one or more of the constituents to enter into solid solution, and then cooling rapidly enough to hold the constituents in solution. The alloy is left in a supersaturated, unstable state, and may subsequently exhibit quench aging. (The aging process typically follows solution treating.)

Steel. Any alloy based on iron typically containing manganese and carbon.

Stress Relief. Heating to a suitable temperature, holding long enough to reduce residual stresses, and then cooling slowly enough to minimize the development of new residual stresses. Other common names and related subjects are *stress relieving*.

Tempering. Reheating a quench hardened or normalized ferrous alloy to a temperature below the transformation temperature and then cooling at any rate desired. This generally results in a reduction in strength and increase in ductility properties of the material. Other common names and related subjects are *draw, drawing,* and *temper*.

Wrought Metallurgy. The art of producing metals by conventional steel making practices, either in blast furnaces or electric arc furnaces. Iron oxide reduction in the former and recycling of scrap steel in the latter are two main sources of raw materials.

18.1.3 Alloy Designations

Ferrous Alloys. Irons and steels make up the vast majority of all materials used in the creation of manufactured goods. Table 18.1 provides a general categorization of the various classes of steel. With few exceptions, steel compositions are established by the American Iron and Steel Institute (AISI), and the Society for Automotive Engineers (SAE). Both use a four numerical series to designate standard carbon and alloy steels, specified to chemical compositional ranges. In several instances, five-digit designations are used. The general classifications are shown in Table 18.2.

The reader is directed to Ref. 5, pp. 15–20[5] for details regarding exact chemical composition of the grades listed.

Nonferrous Alloys. The principles that govern heat treatment of metals and alloys are applicable to both ferrous and nonferrous alloys. However, in practice, there are enough sufficient differences that it makes sense to classify these materials differently. The peculiarities of the alloys of each class with regard to their response to heat treatment require careful understanding of each (see Sec. 18.1.5). The common heat treatable alloys can be classified as shown in Table 18.3. The reader is directed to Ref. 5 for details regarding exact chemical composition of the grades listed.

TABLE 18.1 General Classification of Irons and Steels

Classification	Alloying level	Key uses
Iron	None	Magnets
Cast iron	High carbon levels	Large structures, castings, automobile engine blocks
Carbon steel, plain	≤0.8% carbon	General purpose materials requiring moderate strength levels
Carbon steel, low alloy	≤5% alloy additions + some carbon	Materials requiring relatively high strength levels
Silicon steel	≤8% silicon	Electrical transformers, motor laminations
Magnetic steel	≤50% nickel, many other compositions	Magnetic components
Stainless steel	≥12% chromium	Corrosion resistant material and/or applications requiring resistance to softening at moderate temperatures
Tool steel	30–50% alloying elements; >0.4% carbon, including carbide forming elements	Wear resistance, tooling, cutting, drilling, and machining applications
High-heat resistant alloy steel	≤25% alloying elements; high chromium and aluminum contents	Applications requiring resistance to softening at elevated temperatures; jet engines, furnace components, heating elements
Specialty alloy steel	≤50% alloying elements	Special applications, including electronics, glass seals, filters

***Source*:** From German.[3]

TABLE 18.2 Types and Approximate Percentages of Identifying Elements in Standard Carbon and Alloy Steels

Series designation	Description
Carbon steels	
10XX	Nonresulfurized, 1.00 manganese maximum
11XX	Resulfurized
12XX	Rephosphorized and resulfurized
15XX	Nonresulfurized, over 1.00 manganese maximum
Alloy steels	
13XX	1.75 manganese
40XX	0.20 or 0.25 molybdenum or 0.25 molybdenum and 0.042 sulfur
41XX	0.50, 0.80, or 0.95 chromium and 0.12, 0.20, or 0.30 molybdenum
43XX	1.83 nickel, 0.50 to 0.80 chromium, and 0.25 molybdenum
46XX	0.85 or 1.83 nickel and 0.20 or 0.25 molybdenum
47XX	1.05 nickel, 0.45 chromium, 0.20 or 0.35 molybdenum
48XX	3.50 nickel and 0.25 molybdenum
51XX	0.80, 0.88, 0.93, 0.95, or 1.00 chromium
51XXX	1.03 chromium
52XXX	1.45 chromium
61XX	0.60 or 0.95 chromium and 0.13 or 0.15 vanadium minimum
86XX	0.55 nickel, 0.50 chromium, and 0.20 molybdenum
87XX	0.55 nickel, 0.50 chromium, and 0.25 molybdenum
88XX	0.55 nickel, 0.50 chromium, and 0.35 molybdenum
92XX	2.00 silicon or 1.40 silicon and 0.70 chromium
50BXX (a)	0.28 or 0.50 chromium
51BXX (a)	0.80 chromium
81BXX (a)	0.30 nickel, 0.45 chromium, and 0.12 molybdenum
94BXX (a)	0.45 nickel, 0.40 chromium, and 0.12 molybdenum

Source: From Unterwieser.[4]

TABLE 18.3 General Classification of Common Nonferrous Alloys

Classification	Grades	Remarks
Aluminum and aluminum alloys	1060, 1100, 1350, 2014, 2017, 2024, 2036, 2117, 2124, 2219, 3003, 3004, 3105, 5005, 5050, 5052, 5056, 5083, 5086, 5154, 5182, 5254, 5454, 5456, 5457, 5652, 6005, 6009, 6010, 6061, 6063, 6066, 7001, 7005, 7049, 7050, 7075, 7079, 7178, 7475	Wrought alloys
Aluminum and aluminum alloys	204, 206, 208, 238, 242, 295, 296, 308, 319, 332, 336, 339, 354, 355, 356, 357, 359, 360, 380, 383, 384, 390, 413, 443, 514, 518, 520, 535, 712, 713, 771, 850	Cast alloys
Aluminum and aluminum alloys	601AB, 201AB, 602AB, 202AB, MD-22, MD-24, MD-69, MD-76	Powder metal alloys
Aluminum and aluminum alloys	Weldalite 049, 2090, 2091, 8090, CP 276	Light (aluminum–lithium) alloys
Copper and copper alloys	C10100–C10800, C11000, C11300–C11600, C12000, C12200, C14500, C14700,	Copper
Copper and copper alloys	C15500, C16200, C17000, C17500, C19200, C19400	Copper alloys
Copper and copper alloys	C21000, C22000, C22600, C23000, C24000, C26000, C26800, C27000, C27400, C28000, C33000, C33500, C33200, C34200, C35300, C35600, C36000, C36500, C36600, C36700, C36800, C37000, C37700	Brass
Copper and copper alloys	C60600, C60800, C61000, C61300, C61400, C62800, C62300, C62500, C61900, C63000, C63200, C64200, C63800, C65100, C65500, C66700, C67000, C67400, C67500, C68700, C68800	Bronze (phosphor bronze, aluminum bronze, silicon bronze, manganese bronze)
Copper and copper alloys	C70600, C71000, C71500	Copper nickel
Copper and copper alloys	C74500, C75200, C76400, C75700, C77000, C78200	Nickel silver
Copper and copper alloys	C81400, C81500, C81800, C82000, C82200, C82500, C82800, C95300–C95800	Cast alloys
Magnesium and magnesium alloys	AM100A, AZ63A, AZ81A, AZ91C, AZ92A, EZ33A, EQ21A, HK31A, HZ32A, QE22A, QH21A, WE43A, WE54A, ZC63A, ZE41A, ZE63A, ZH62A, ZK51A, ZK61A	Casting alloys
Magnesium and magnesium alloys	AZ80A, HM21A, HM31A, ZC71A, ZK60A	Wrought alloys
Nickel and nickel alloys	Nickel 200, 201; Monel 400; R-405; K-500	
Superalloys	A286; Discaloy; N155; Incoloy 903, 907, 909, 925	Iron based alloys
Superalloys	Astroloy; Custom Age 625 PLUS; Inconel 901, 625, 706, 718, 725; X-750; Nimonic 80A, Nimonic 90, Rene 41, Udimet 500, 700, Waspalloy	Nickel based alloys
Superalloys	S816	Cobalt based alloys
Titanium and titanium alloys	Ti-8Al-1Mo-1V, Ti-2.5Cu, Ti-6Al-2Sn-4Zr-2Mo, Ti-6Al-5Zr-0.5Mo-0.2Si, Ti-5.5Al-3.5Sn-3Zr-1Nb-0.3Mo-0.3Si, Ti-5.8Al-3.5Zr-0.7Nb-0.5Mo-0.3Si	Alpha or near-alpha alloys
Titanium and titanium alloys	T-6Al-4V, T-6Al-6V-2Sn, Ti-6Al-2Sn-4Zr-6Mo, Ti-4Al-4Mo-2Sn-0.5Si, Ti-4Al-4Mo-4Sn-0.5Si, Ti-5Al-2Sn-2Zr-4Mo-4Cr, Ti-6Al-2Sn-2Zr-2Mo-2Cr-2.5Si	Alpha–beta alloys
Titanium and titanium alloys	Ti-13V-11Cr-4Al, Ti-11.5Mo-6Zr-4.5Sn, Ti-3Al-8V-6Cr-4Mo-4Zr, Ti-10V-2Fe, Ti-15V-3Al-3CR-3Sn	Beta or near-beta alloys
Zinc and zinc alloys	No. 2, 3, 5, 7; ZA-8; ZA-12; ZA-27	Zamak alloys

Source: From Chandler.[5]

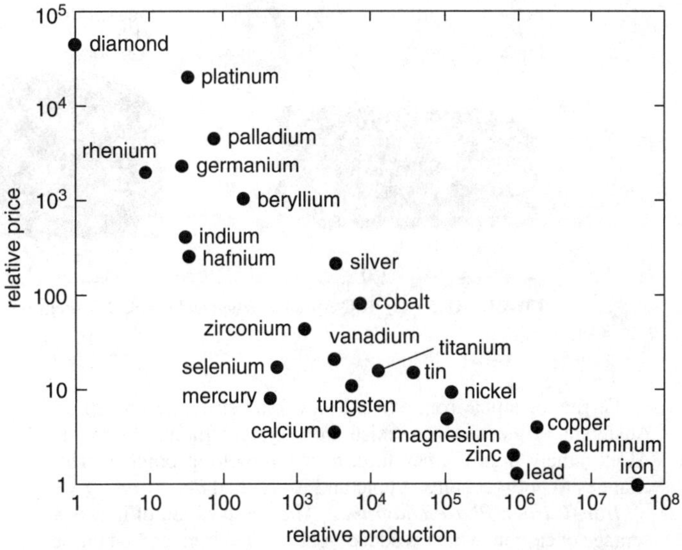

FIGURE 18.2 Relative cost of materials.[3]

18.1.4 Ferrous Metallurgy

Iron and Steel. One of the properties of iron that makes it very appealing as an engineering material is that it is extremely plentiful and inexpensive (Fig. 18.2). However, *pure iron is not very strong,* so by itself it would not be a good material from which to manufacture metal parts. This fact is a surprise to many people. Actually, pure iron is not even as strong as most plastics.

Adding certain chemical elements called alloying elements to iron, such as manganese and carbon, produces steel. Steel is not only stronger than iron but can also have its properties dramatically changed by heat treating, thus improving its usefulness as a material to manufacture the items we use everyday. Table 18.4 shows the dramatic strength improvement that takes place by heat treating.

Internal (Crystal) Structure. Iron, the basic building block of steel, has a very unique property. It is allotropic. That is, it changes its internal (crystal) structure as it is heated. This is an internal change in the material which occurs due to a rearrangement of the atoms. Iron changes, as the temperature increases, into

- alpha iron (symbol: α-iron) or ferrite
- gamma iron (symbol: γ-iron) or austenite
- delta iron (symbol: δ-iron)

TABLE 18.4 Strength Improvement by Heat Treating

Property	Condition	Yield Strength* MPa (ksi)	Tensile Strength* MPa (ksi)	Elongation (%)
Pure iron	Untreated	200 (29)	310 (45)	26
Steel†	Untreated	379 (55)	689 (100)	12
Steel†	Heat treat and slow (air) cool	428 (62)	773 (112)	10
Steel†	Heat treat and fast (water) quench	1380 (220)	1380 (220)	1

*Yield strength and ultimate tensile strength are measures of how strong the steel is after the heat treatment process shown.
†Eutectoid steel (0.77% C).

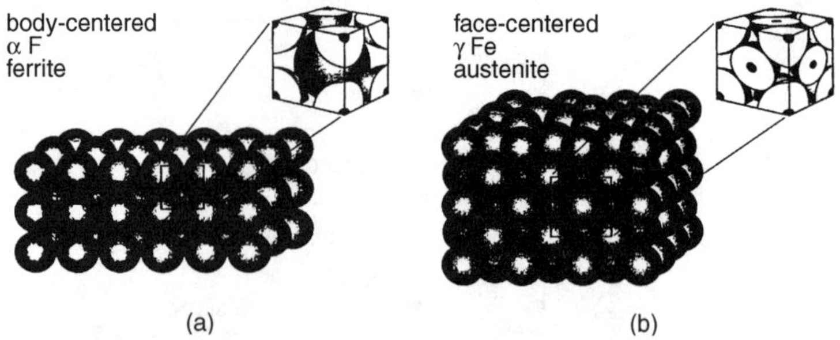

FIGURE 18.3 (a) BCC crystal structure and (b) FCC crystal structure.[3]

Ferrite, or alpha iron, exists in a crystal structure known as *body-centered* cubic (Fig. 18.3(*a*)). Austenite, or gamma iron, exists in a crystal structure known as *face-centered* cubic (Fig. 18.3(*b*)). At extremely high temperature, near the melting point of iron, the crystal structure of iron again changes to body-centered cubic and it is called *delta iron*.

Iron–Carbon Phase Diagram. The composition of iron is also a function of the amount, or percentage of carbon, which is added. A simplified iron–carbon phase diagram (Fig. 18.4) provides a "road map" and indicates where the internal changes will occur and in what form, or state, you will find the internal structure of the metal at any given time. This diagram plots these changes in crystal structure as a function of temperature (on the vertical axis) and percentage carbon (on the horizontal axis).

From this diagram we find that something special occurs at a carbon level around 0.77% C and at a temperature of 1333°F (723°C). You will see lines on the diagram come together here at a point, called the *eutectoid* point. These lines define as a function of carbon percentage, the critical temperature (Ac_1) of a material. Table 18.5 shows the beginning (Ac_1) and *complete transformation temperature* (Ac_3) of several steels to reach the austenite range.[6]

Why is it important to know the critical temperature of steel? In order to quench harden the material it must be heated above its critical temperature and held to stabilize in temperature prior to quenching (rapidly cooling). For example, in order for carbon to be absorbed into the surface of the steel in processes like carburizing or carbonitriding, the steel must be heated above its critical temperature and into the austenite range. Other case hardening processes, such as nitriding, take place below the critical (Ac_1) temperature.

Microstructure. The *iron–carbon phase diagram* (Fig. 18.4) tells us that by heating and/or cooling a steel component heat treatment can produce a change to the internal structure, or microstructure, with the resultant change in physical, mechanical, or metallurgical properties. This is accomplished either by heating to a temperature range where a phase or combination of phases are stable and/or heating or cooling between temperature ranges where the different phases are stable. The heat treating of steel involves the formation of structures called *pearlite, proeutectoid ferrite* and *cementite, martensite,* and *bainite*.

On slow cooling, the eutectoid (0.77% C) transformation of steels produces a microstructure termed *pearlite*. Pearlite is made up of ferrite and cementite (Fig. 18.5) in an alternating or lamellae pattern. The amounts of cementite and ferrite formed can be determined by calculation. A number of factors influence the rate of pearlite formation since carbon atom rearrangement must take place in order for the transformation from austenite to occur to low carbon ferrite and high carbon cementite.

In most steels, that is, those not at the eutectoid composition, the transformation from austenite begins well above the Ac_1 temperature. The ferrite and cementite that form by a mechanism other than the eutectoid transformation are called *proeutectoid* phases. In *hypoeutectoid* steels, those that contain less than the eutectoid carbon content, ferrite forms below the Ac_3 (Fig. 18.6). In hypereutectoid steels, those that contain more than the eutectoid carbon content, cementite forms below the Acm, the boundary between the austenite and cementite fields (Fig. 18.7). Proeutectoid ferrite and

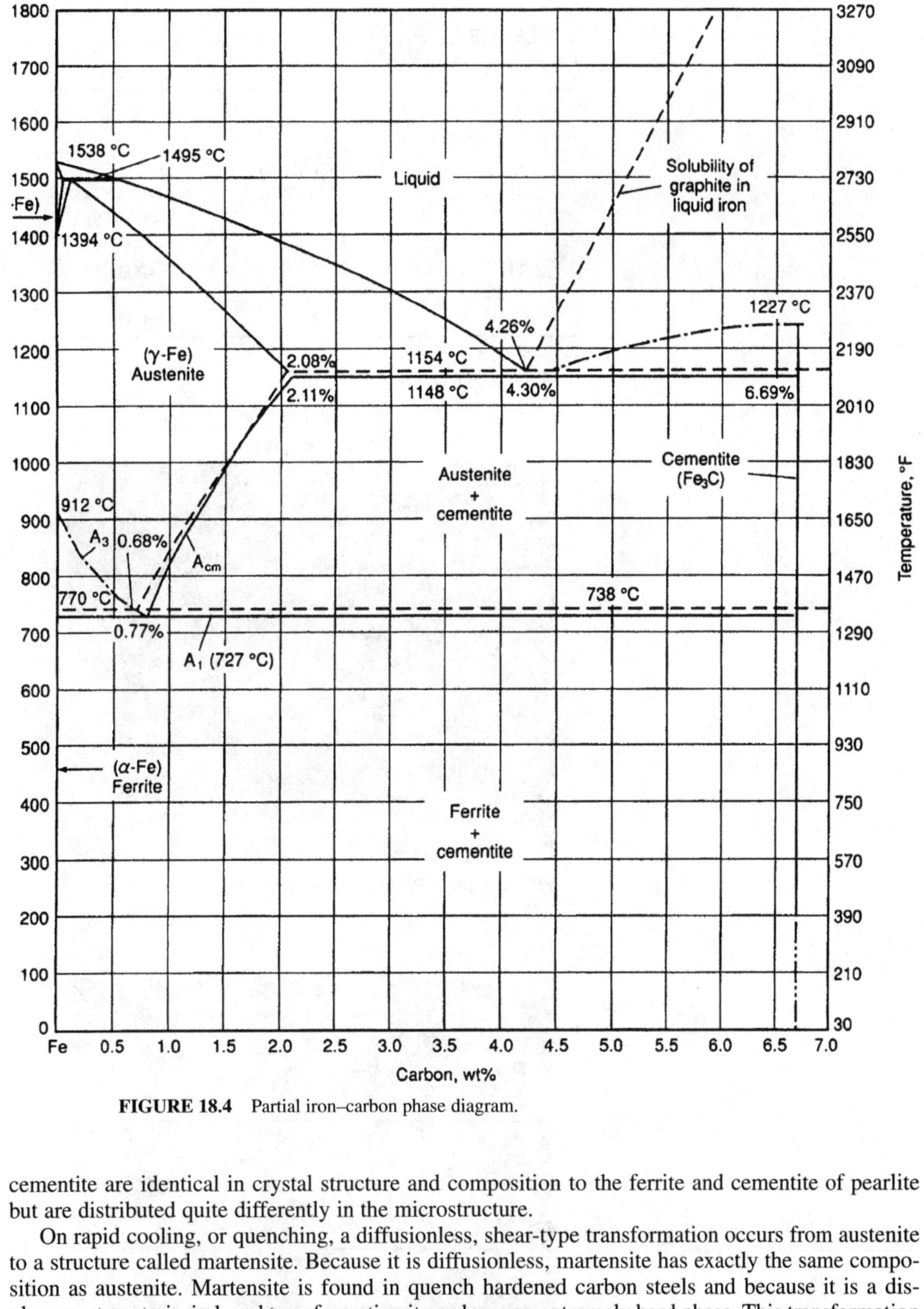

FIGURE 18.4 Partial iron–carbon phase diagram.

cementite are identical in crystal structure and composition to the ferrite and cementite of pearlite but are distributed quite differently in the microstructure.

On rapid cooling, or quenching, a diffusionless, shear-type transformation occurs from austenite to a structure called martensite. Because it is diffusionless, martensite has exactly the same composition as austenite. Martensite is found in quench hardened carbon steels and because it is a displacement or strain-induced transformation it produces an extremely hard phase. This transformation can also occur in many nonferrous systems as well.

Martensite is a unique phase that has its own crystal structure (body-centered tetragonal) and composition and is separated by well-defined interfaces from the other phases. However, it is

TABLE 18.5 Selected Critical Temperatures of Typical Carburizing Steels

Carburizing steel (AISI designation)	Ac_1, °F (°C)*	Ac_3, °F (°C)*
1018	1370 (743)	1565 (852)
3310	1335 (723)	1440 (782)
4118	1385 (750)	1500 (815)
4320	1355 (735)	1485 (808)
5120	1380 (749)	1525 (829)
8620	1350 (732)	1525 (829)
9310	1315 (713)	1490 (810)

*On heating.
Source: From Tinken.[6]

FIGURE 18.5 Pearlite microstructure.[17]

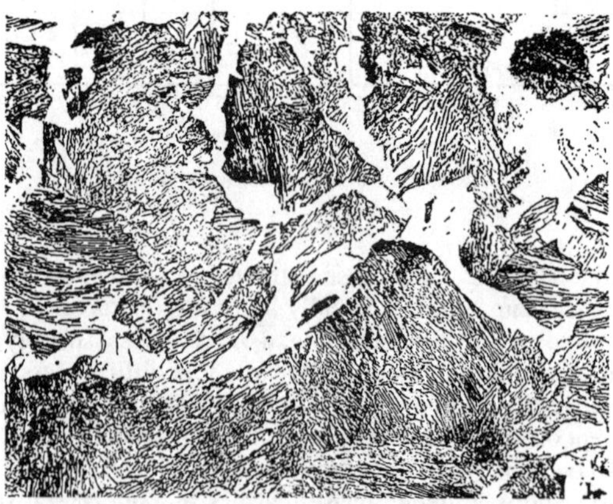

FIGURE 18.6 Proeutectoid ferrite.[17]

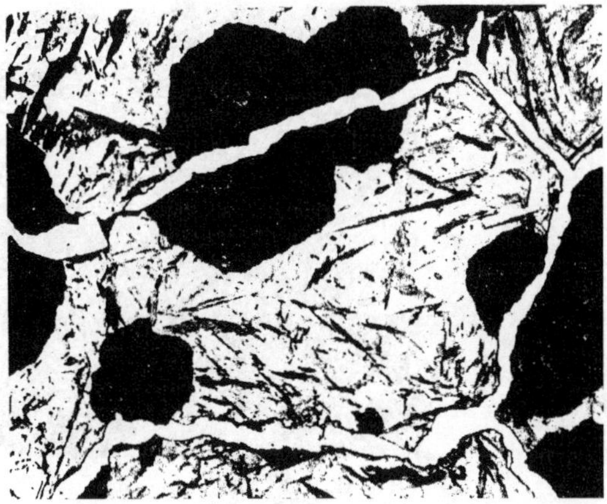

FIGURE 18.7 Proeutectoid cementite.[17]

metastable since diffusion has only been suppressed. If martensite is heated, for example in a service application, to a temperature where the carbon atoms have mobility, the martensite is replaced by a mixture of ferrite and cementite with a resultant volume change occurring in the component. To avoid this effect, and since the conversion takes place continuously with decreasing temperature, martensite can be stabilized by lowering the transformation temperature, often below zero, by cryogenic (deep freeze) treatment.

There are two types of martensite in iron–carbon systems, lath martensite (Fig. 18.8) and plate martensite (Fig. 18.9). Lath martensite forms in low and medium carbon steels (up to approximately 0.6% C) while plate martensite forms in high carbon steels (above 1.0% C). A mixture of lath and plate martensite can be found in carbon range between 0.6% and 1.0%. The terms *lath* and *plate* refer to the three-dimensional shapes of individual martensite crystals.

Finally, a phase called *bainite* forms under continuous cooling or isothermal transformation conditions intermediate to those of pearlite and martensite. Similar to pearlite, bainite is a mixture of phases of ferrite and cementite and is diffusion controlled. However, unlike pearlite, the ferrite and cementite are present in nonlamellar arrays whose characteristics are dependent on alloy composition

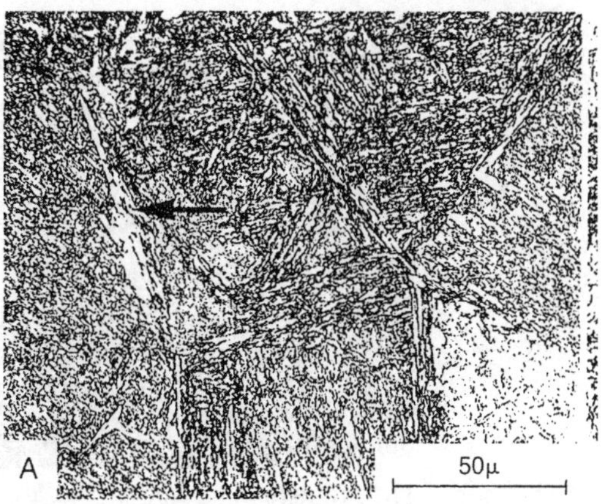

FIGURE 18.8 Lath martensite.[17]

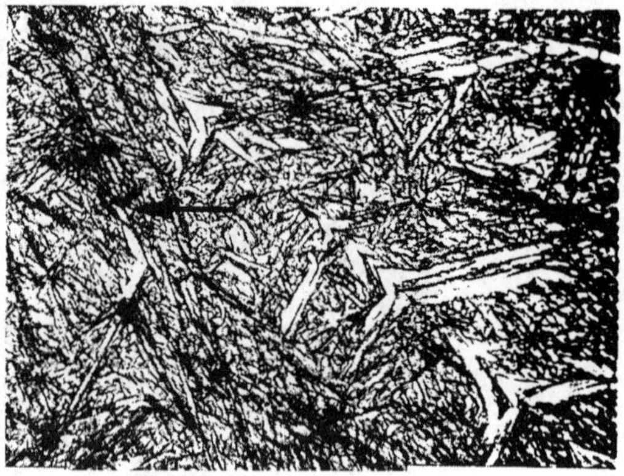

FIGURE 18.9 Plate martensite.[17]

and transformation temperature. Similar to martensite, the ferrite may be in the form of laths or plates containing a dislocation structure that involves shear as well as diffusion.

There are two major forms of bainite: upper bainite (Fig. 18.10) that forms in the temperature range just below that of the pearlite transformation, and lower bainite (Fig. 18.11) that occurs closer to the martensite transformation range.

Alloying Elements. The presence of alloying elements in steel is designed to enhance their physical and mechanical properties so that they will perform the end-use application for which they are intended. To understand why they are used, we will briefly look at what benefits can be gained from each.

Carbon is a key alloying element in steel, and the general effects that result from the addition of carbon into steel are as follows:

- Raising the ultimate strength
- Lowering the ductility and toughness
- Reducing the shock resistance
- Enhancing the abrasive (or wear) resistance

FIGURE 18.10 Upper bainite.[17]

FIGURE 18.11 Lower Bainite

- Increasing the resistance to machinability
- Lowering the hardening and quench temperature
- Increasing the depth of hardening
- Improving uniformity of hardness (especially in the quenching of large sections)
- Intensifying the fineness of fracture
- Lessening the heat conductivity
- Diminishing the electrical or magnetic conductivity

As you can see, carbon causes quite a number of changes to occur. Many other elements can be added to steel including manganese, silicon, nickel, chromium, molybdenum, vanadium, boron, tungsten, cobalt, and copper. Each element acting alone, or in combination with one another, produce different effects or intensify the effects of another addition.

Alloying elements used in steel can be classified into three general groups:

Group 1: Elements that completely dissolve into the iron.

Group 2: Elements that combine with carbon to form a compound called a carbide.

Group 3: Elements that partially dissolve into the iron and partly form a carbide.

Elements in Group 1 that enter into solid solution with iron usually enhance strength and toughness of steel. Elements in Group 2 that combine with carbon to form carbides increase the hardness and ultimate tensile strength of the steel. Elements in Group 3 that are partially dissolved into the iron and form carbides have the tendency to enhance both toughness as well as raise ultimate tensile strength.

Manganese is present in steel and is essential to steel production, not only in melting but also in rolling and other processing operations.

Nickel increases strength and toughness. It is a ferrite strengthener. Nickel steels are easily heat treated since nickel effectively lowers the critical cooling rate necessary to produce hardening on quenching. Carburized steels with high nickel contents tend to have large amounts of retained austenite which can lower the wear resistance of the case. Retained austenite, in certain applications such as gears and bearings, is useful typically in percentages between 10 and 15 percent. Nickel also intensifies the effects of chromium and other alloying elements. In combination with chromium, nickel produces alloy steels with higher elastic ratios, greater hardenability, higher impact, and fatigue resistance than is possible with carbon steels.

Chromium is essentially a hardening element since it is a carbide former similar to molybdenum and tungsten. Two of the most important properties of steels containing chromium are wear resistance and cutting ability. The eutectoid carbon level of steel is lowered by increased amounts of chromium. Chromium goes into solution slowly, so sufficient time at temperature is required in medium carbon grade materials.

Molybdenum raises the ultimate strength, hardness, and toughness by virtue of its solution in the iron matrix and its formation of a carbide. It is particularly valuable because it promotes the retention of high hardness at elevated service-use temperatures. Molybdenum steels in the quenched condition have significant resistance to tempering, requiring higher tempering temperatures.

Vanadium improves the strength and ductility, especially elasticity and shock resistance due to its formation of a carbide and its solution in the iron matrix. Vanadium steels show a much finer grain structure. In addition, vanadium gives other alloying effects of importance, namely increased hardenability where it is in solution in the austenite prior to quenching, a secondary hardening effect upon tempering, and increased hardness at elevated temperature.

Boron is usually added to improve hardenability. Boron is very effective when used with low carbon alloy steels but its effect is reduced as the carbon increases. Boron additions are usually in relatively small amounts and within these limits enhance the hardenability of other alloys, and for this reason boron is called an alloy "intensifier."

Hardening and Hardenability.[7] Hardness is simply the measure of the resistance of a material to an applied force, and involves the use of an indenter of fixed geometry under static load. The ability of the material to resist plastic deformation depends on the carbon content and microstructure of the steel. Therefore, the same steel can exhibit different hardness values depending on its microstructure, which is influenced by the cooling (transformation) rate.

Hardenability on the other hand is used to describe the heat treatment response of steels using either hardness or microstructure, both of which are interrelated. Hardenability is a material property, independent of cooling rate and dependent on chemical composition and grain size.

When evaluated by hardness testing, hardenability is defined as the capacity of the material under a given set of heat treatment conditions to harden "in depth." In other words, hardenability is concerned with the "depth of hardening" or the hardness profile obtained, not the ability to achieve a particular hardness value. When evaluated by microstructural techniques, hardenability is defined as the capacity of the steel to transform partially or completely from austenite to some percentage of martensite at a given depth when cooled under known conditions.[8] The reason why it is important to measure the hardenability of a steel is to make sure that we are making the right material choice for a specific engineering application.

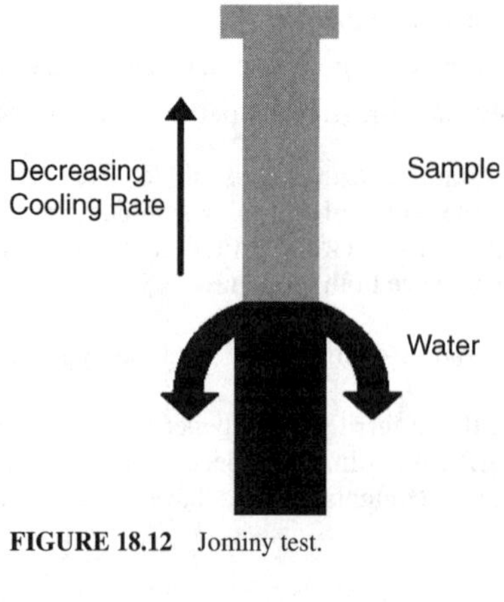

FIGURE 18.12 Jominy test.

The Jominy end quench test was developed in the 1930s along with a number of other testing methods as a cost and time effective way to determine the hardenability of steel. These tests were developed as alternatives to the creation of *continuous cooling transformation* (CCT) *diagrams*. The test sample is a cylinder with a length of 102 mm (4") and a diameter of 25 mm (1"). The steel sample is normalized to eliminate differences in microstructure due to previous forging and then austenitized. The test sample is quickly transferred to the test fixture, which quenches the steel as it sprays a controlled flow of water onto one end of the sample (Fig. 18.12). The cooling rate

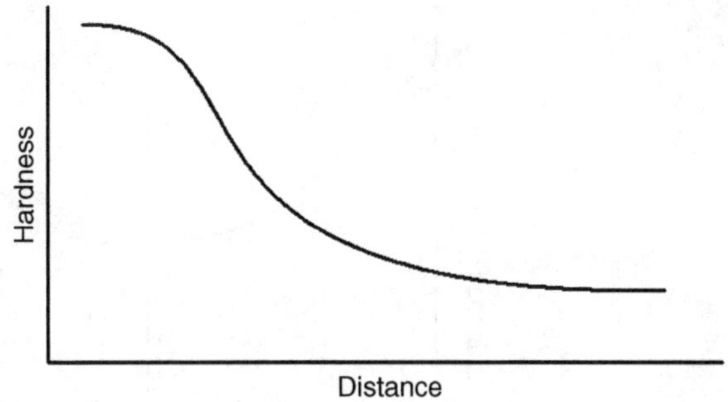

FIGURE 18.13 Jominy hardness profile.

varies along the length of the sample from very rapid at the quenched end where the water strikes the specimen, to slower rates that are equivalent to air cooling at the other end.[9]

The round specimen is then ground flat along its length on opposite sides to a depth of at least 0.38 mm (0.015") to remove decarburized material. Care should be taken that the grinding does not heat the sample, as this can cause tempering, which can soften the microstructure. The hardness is then measured at intervals from the quenched end. The interval is typically 1.5 mm (0.059") for alloy steels and 0.75 mm (0.030") for carbon steels, beginning as close as possible to the quenched end. The hardness decreases with distance from the quenched end (Fig. 18.13). High hardness occurs where high volume fractions of martensite develop. Lower hardness indicates transformation to bainite or ferrite/pearlite microstructures.

A common misconception associated with the Jominy test is that the hardness values achieved can be used directly for actual parts being oil quenched. This is not true. The results of the test allow us to compare steels to determine their equivalent hardenability. In other words, the Jominy curves can be used to predict the expected hardness distribution obtained on hardening steels of different dimensions in various cooling media.[10]

One criticism of the Jominy test is that it is not discriminating when applied to steel of low hardenability. For such steels, the S-A-C test is considered more reliable. The test sample is also a cylinder but this time with a length of 140 mm (5.5") and a diameter of 25.4 mm (1"). After normalizing and austenitizing above the Ac_3 (the temperature at which that transformation of ferrite to austenite is completed during heating) the specimen is quenched by submerging in a water bath. After quenching, a cylinder 25.4 mm (1") long is cut from the test specimen and the end faces are ground very carefully to remove any tempering effects induced by the cutting operation. Rockwell C hardness measurements are then made at four positions on the original cylinder face and the average hardness provides the surface, or S-value. Rockwell testing is carried out along the cross-section of the specimen from the surface to the center and provides the type of hardness profile (Fig. 18.14). The total area under the curve provides the area, or A-value in units of "Rockwell-inch" and the hardness at the center gives the C-value.[8] For example, a steel value reported as 65-51-39 indicates a surface hardness of 65 HRC, an area value of 51 Rockwell-inches and a core hardness of 39 HRC. The advantage of the S-A-C test lies in the fact that the three numbers give a good visual image of the hardness distribution curve. The surface hardness is influenced by the carbon content, and is important in placing the hardness distribution curve; the area under the curve provides both knowledge of the extent of hardening as well as a comparative index of various steels.

Other testing methods include the Jominy-Boegehold end quench test for measuring the hardenability of the carburized portion of carburized and hardened steels, the Cone Test of Post, Greene, and Fenstermacher for shallow hardening steels having hardenabilities in the range exhibited by carbon tool steels (typically 1.10% C), and the Shepherd P-F and P-V tests for shallow hardening steels. In the P-F test, P stands for penetration of hardening (hardenability) and F stands for fracture grain size. P-V indicates penetration of hardening on a V-shaped test piece.[11]

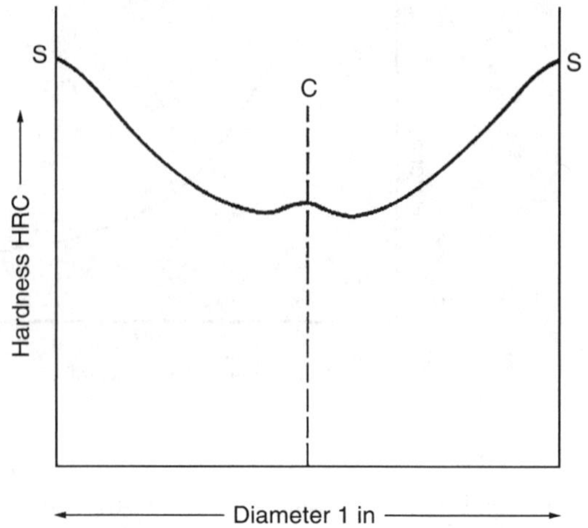

FIGURE 18.14 Schematic representation of the S-A-C hardenability curve.

More recently, Liscic and Filetin, Wunning, Tensi, and others have developed processes for evaluating quenching severity (intensity). The Liscic technique is based on a method for determining the heat flux (the rate of heat transfer across a given surface) at the surface of a part during quenching using a test probe. In addition to the heat flux data obtained from the probe, a second specimen with the same dimensions as the probe and a Jominy specimen of the same steel are quenched under identical conditions. The samples are evaluated and a database is established for various quenching conditions and for determining equivalent Jominy distances.[12]

Although the Jominy end quench test is relatively easy to perform, the procedure must be followed exactly to obtain good results. Perhaps the single most important aspect of the test is consistency. Each test must be done under identical conditions: austenitizing temperature; transfer time (from oven to quench); handling method; and water temperature, flow, and pressure. Water is an excellent and inexpensive quenchant, but its heat removal characteristics are highly variable, being dependent on temperature and the relative velocity between the water and the specimen.

There is a mixture of opinions on the best way to measure the hardness profile (Rockwell, Brinell, or Vickers). Choice of measurement depends to a great extent on the specimen dimensions and hardness. Rockwell "C" is typically used if the hardness is greater than 20 HRC. Brinell is used when the hardness is below 20 HRC. If a microhardness profile is required, Vickers is chosen. Standard hardness testing rules for minimum distance between readings and specimen thickness must be strictly followed.

Stainless Steels. A large family of heat and corrosion resistant alloys has been developed to satisfy many of the requirements for resistance to these effects not only on the surface of the component but throughout its cross section. These materials are known as stainless steels and can be classified in four basic groups:

1. Austenitic grades
2. Ferritic grades
3. Martensitic grades
4. Precipitation hardening grades

Specific requirements for corrosion-resistant or heat-resistant applications have lead to the development of approximately 50 grades that are considered standard. In addition there are approximately 100 or more nonstandard grade developed for specialized applications. The various standard types for each grade. are shown in Table 18.6.[4] The reader is directed to Ref. 4 for details regarding exact chemical composition of the grades.

TABLE 18.6 Types of Stainless Steels by Grade

Designation	Type number	Remarks
Austenitic grades	201, 202, 205, 301, 302, 302B, 303, 303Se, 304, 304L, 304N, 305, 308, 309, 309S, 310, 310S, 314, 316, 316F, 316N, 31, 317L, 321, 329, 330, 347, 348, 384	Grade 310 can be used for elevated temperature service to 1200°F (650°C)
Ferritic grades	405, 409, 429, 430, 430F, 430FSe, 434, 436, 442, 446	—
Martensitic grades	403, 410, 414, 416, 416Se, 420, 420F, 422, 431, 440A, 440B, 440C	—
Precipitation hardening grades	630, 631, 632, 633, 634, 660	630 is a martensitic type; 660 is an austenitic type; and the rest are semi-austenitic types
Casting grades	CA-6NM, CA-15, CA-40, CB-30, CB-7Cu, CC-50, CD-4M Cu, CE-30, CF-3, CF-8, CF-20, CF-3M, CF-8M, CF-8C, CF-16F, CG-8M, CH-20, CK-20, CN-7M	As a rule, there is little if any difference in corrosion resistance or response to heat treatment between the wrought and cast versions of the same alloy

Source: From Unterwieser.[4]

The austenitic grades are used widely in corrosive environments. They do not change their crystal structure upon heating and as such do not respond to conventional quench hardening techniques. The only heat treatments for these grades are full annealing by rapidly cooling from elevated temperature, stress relieving, and surface hardening by nitriding or nitrocarburizing.

The ferritic grades rank substantially higher in corrosion resistance than the martensitic grades but substantially lower than most of the austenitic grades. Like the austenitic grades they cannot be hardened by heating and quenching. Annealing to a lower hardness after cold working is the only heat treatment applied to ferritic grades, except for nitriding or nitrocarburizing in some instances.

The martensitic grades are capable of changing their crystal structure on heating and cooling. They can be quench hardened to a fully martensitic structure much the same as alloy steels. In addition to quench hardening, these materials respond well to surface hardening by nitriding. Since the chromium content is generally lower than the austenitic grades, in general, the corrosion resistance is far lower than the austenitic grades and somewhat lower than the ferritic grades.

The precipitation hardening grades vary considerably in corrosion resistance, but in general approach the values obtained by austenitic grades. Hardening techniques for the precipitation hardening grades are like those used for nonferrous metals. The general approach is to solution treat by heating to an elevated temperature, cool rapidly, and then age harden by heating to an intermediate temperature.

The element chromium is the key to corrosion resistance. For a metal to be considered a stainless steel, it must contain a minimum of 11.5 percent chromium. Other alloying elements and their effects are shown in Table 18.7.

Tool Steels. Tool steels vary in composition from plain carbon steels containing iron and up to 1.2% carbon with no significant amounts of other alloying elements to very highly alloyed grades in which the total alloy content approaches 50%. They are classified as shown in Table 18.8 and described below:

- The air hardening, medium alloy cold work tool steels (A series) cover a wide range of carbon and alloy contents. All have high hardenability and exhibit a high degree of dimensional stability in heat treatment. Depending on the grade selected, one can achieve high hardness, shock resistance, abrasion resistance, wear resistance, and toughness.

TABLE 18.7 Alloy Addition and Their Influence in Stainless Steels

AISI type	Changes in analyses from basic type
Austenitic, chromium nickel	
301	Cr and Ni lower for more work hardening
302	Basic type, 18% Cr + 8% Ni
302B	Si higher for more scaling resistance
303	P and S added for easier machining
303Se	Se added to improve machinability
304	C lower to avoid carbide precipitation
304L	C lower for welding application
305	Ni higher for less work hardening
308	Cr and Ni higher with C low for more corrosion and scaling resistance
309	Cr and Ni still higher for more corrosion and scaling resistance
309CT	Cb and Ta added to avoid carbide precipitation
309S	C lower to avoid carbide precipitation
310	Cr and Ni highest to increase scaling resistance
314	Si higher to increase scaling resistance
316	Mo added for more corrosion resistance
316L	C lower for welding application
317	Mo higher for more corrosion resistance and greater strength at high temperatures
318	Cb and Ta added to avoid carbide precipitation
321	Ti added to avoid carbide precipitation
347	Cb and Ta added to avoid carbide precipitation
347Se	Se added to improve machinability
348	Similar to 347, but low tantalum content (0.10)
384	Ni higher than 305 for severe cold heading
385	Similar to 384, but lower Cr and Ni
Austenitic, chromium nickel manganese	
201	Cr and Ni lower for more work hardening
202	Basic type, 18% Cr + 5% Ni + 8% Mn
204	C lower to avoid carbide precipitation
204L	C lower for welding application
Martensitic, straight chromium	
403	12% Cr adjusted for special mechanical properties
410	Basic type, 12% Cr
414	Ni added to increase corrosion resistance and mechanical properties
416	P and S added for easier machining
416Se	Se added to improve machinability
418Spec	W added to improve high-temperature properties
420	C higher for cutting purposes
420F	P and S added for easier machining
431	Cr higher and Ni added for better resistance and properties
440A	C higher for cutting applications
440B	C higher for cutting applications
440C	C still higher for wear resistance
440Se	Se added for easier machining
Ferritic, straight chromium	
405	Al added to 12% Cr to prevent hardening
430	Basic type, 17% Cr
430F	P and S added for easier machining
430Ti	Titanium stabilized
442	Cr higher to increase scaling resistance
446	Cr much higher for improved scaling resistance

Source: From Unterweiser.[4]

TABLE 18.8 Classification of Tool Steels

Tool steel designation	Identifying symbol	Common grades
Cold work steels (air hardening, medium alloy)	A	A2, A3, A4, A6, A7, A8, A9, A10
Cold work steels (high carbon, high chromium)	D	D2, D3, D4, D5, D7
Hot work steels	H	H10, H11, H12, H13, H14, H19, H21, H22, H23, H24, H25, H26, H42
Low alloy special purpose	L	L2, L6
High speed steels (molybdenum)	M	M1, M2, M3 Class 1, M3 Class 2, M4, M7, M10, M30, M33, M34, M35, M36, M42, M43, M44, M46, M47, M48, M50
Cold work steels (oil hardening)	O	O1, O2, O6, O7
Mold steels	P	P2, P3, P4, P5, P6, P20, P21
Shock resisting	S	S1, S2, S5, S6, S7
High speed steels (tungsten)	T	T1, T2, T4, T4, T6, T8, T15
Water hardening	W	W1, W2, W5

Source: From Unterwieser.[4]

- The high carbon, high chromium cold work tool steels (D series) have extremely high resistance to abrasive wear due to their carbon and vanadium contents. Hot forming and shearing operations rely on these grades for optimum performance.
- Hot work (H series) steels are divided into three groups: chromium, tungsten, and molybdenum. Tooling applications (tools and dies) in the automotive and aerospace industry are commonly made from this group of steels.
- The low alloy, special purpose tool steels (L series) span a wide range of alloy contents and mechanical properties. Die components and machinery are typical application uses.
- The high speed steels are divided into three groups: (1) those where molybdenum is the principle alloying element (M series); (2) those where tungsten is the principle alloying element (T series); and (3) those that are capable of being more highly alloyed for applications requiring unusually high hardness.
- The oil hardening, cold work tool steels (O series) have high hardenability and can be quench-hardened in oil.
- The mold steels (P series) are typically supplied for injection and compressive molds for plastics as well as zinc die casting molds and holder blocks. Certain grades can be supplied prehardened while other grades require case hardening after machining.
- Shock resistant tool steels (S series) have widely varying alloy contents resulting in a wide variation in hardenability within the grade. All grades are intended for applications requiring extreme toughness such as punches, shear knives, and chisels.
- The water hardening tool steels (W series) are among the least expensive tool steels. They are shallow hardening often requiring water quenching. These steels may be used for a variety of tools but do have limitations.

18.1.5 Nonferrous Metallurgy

One way to distinguish nonferrous alloys is to recognize that the eutectoid transformations, which play a prominent role in ferrous metallurgy, are seldom encountered. Thus many of the principles associated with these transformations and with martensite formation are not of primary importance. Nonferrous alloys rely on diffusion related mechanisms and thus there is both a time and a temperature related relationship involved with these structures and the changes that will occur during heat treatment.[3]

Nonferrous materials include

- Aluminum and aluminum alloys
- Copper and copper alloys (including beryllium copper)
- Lead and lead alloys
- Magnesium and magnesium alloys
- Nickel and nickel alloys (including beryllium nickel)
- Refractory metals and alloys
- Superalloys
- Tin and tin alloys
- Titanium and titanium alloys
- Zinc and zinc alloys
- Heat treatments common to all of these materials include annealing after cold working
- Homogenization of castings
- Precipitation hardening
- Development of two phase structures

Diffusion Processes. The diffusion process is involved in nearly all heat treatment processes for nonferrous alloys. Atoms in a lattice structure are not static but are vibrating around an atom position, normally at a lattice site. Therefore, diffusion can simply be thought of as the movement of these atoms within the material. This movement can be via the following mechanisms:

- Vacancy diffusion
- Chemical diffusion
- Interstitial diffusion
- Grain boundary diffusion

Diffusion primarily occurs by vacancy diffusion. If a vacancy exists in a lattice, then the movement of atoms to a new position requires much less energy. Common diffusion mechanisms are thought to be when two atoms move simultaneously to exchange positions and when four atoms move cooperatively to rotate simultaneously and move to new positions. The rate of diffusion increases exponentially when temperature is applied.

Chemical diffusion occurs when two metals or alloys are placed in contact. Atoms will begin to migrate across the contacting interface. This often occurs by vacancy diffusion. Interstitial diffusion occurs if a solute atom is sufficiently small to locate between the larger solvent atoms. This occurs because atoms jump from one interstitial site to another. Finally grain-boundary diffusion occurs along the core of dislocations and on free surfaces.

The rate of structural changes that occur is therefore controlled by the rate at which atoms change position within the lattice structure.

Nonferrous materials are generally cold or hot worked. Cold working is done when plastic deformation causes strengthening or hardening to occur. Cold working causes slip under plastic deformation and the material work or strain hardens. The result is an increase in hardness, yield strength, and tensile strength; and lowers ductility. It also increases electrical resistivity. In shaping of metals and alloys by cold working, there is a limit to the amount of plastic deformation attainable without failure by fracture. Proper heat treatment prior to reaching this limit restores the metal or alloy to a structural condition similar to that prior to deformation and then additional cold working can be conducted. This type of treatment is called annealing.

Another process is homogenization, an important heat treatment of nonferrous castings. Homogenizing involves raising the temperature of the metal high enough and holding at temperature long enough to eliminate or decrease significantly chemical segregation, or coring, as well as nonequilibrium second phase particles.

In designing for strength, alloys are developed in which the structure consists of particles dispersed throughout the matrix to impede dislocation movement. The finer the dispersion, the stronger the material. Such a dispersion can be obtained by choosing an alloy that at elevated temperature is single phase but which on cooling will precipitate another phase in the matrix. A heat treatment is then developed to give the desired distribution of the precipitate in the matrix. If hardening occurs from this structure then the process is called precipitation hardening or age hardening. A prerequisite to precipitation hardening is the ability to heat the alloy to a temperature range where all of the solute is dissolved, so that a single phase structure is obtained. This is called solution heat treating. Rapid cooling or quenching prevents the precipitate from forming, and the resultant room temperature structure is supersaturated with respect to the solute, and hence unstable.

Two-Phase Structures. In some nonferrous alloys the desired structure consists of a mixture of two phases of comparable quantities. Titanium-based alloys (Fig. 18.15) and copper-zinc alloys of high zinc content are examples. These are unlike the two-phase structure in precipitation-hardening alloys where the precipitate is in the minority. The microstructure and amount of each are varied by controlling the highest temperature used and the cooling rate from that temperature. The resultant microstructures can be complex, and the required treatments differ considerably for different systems.

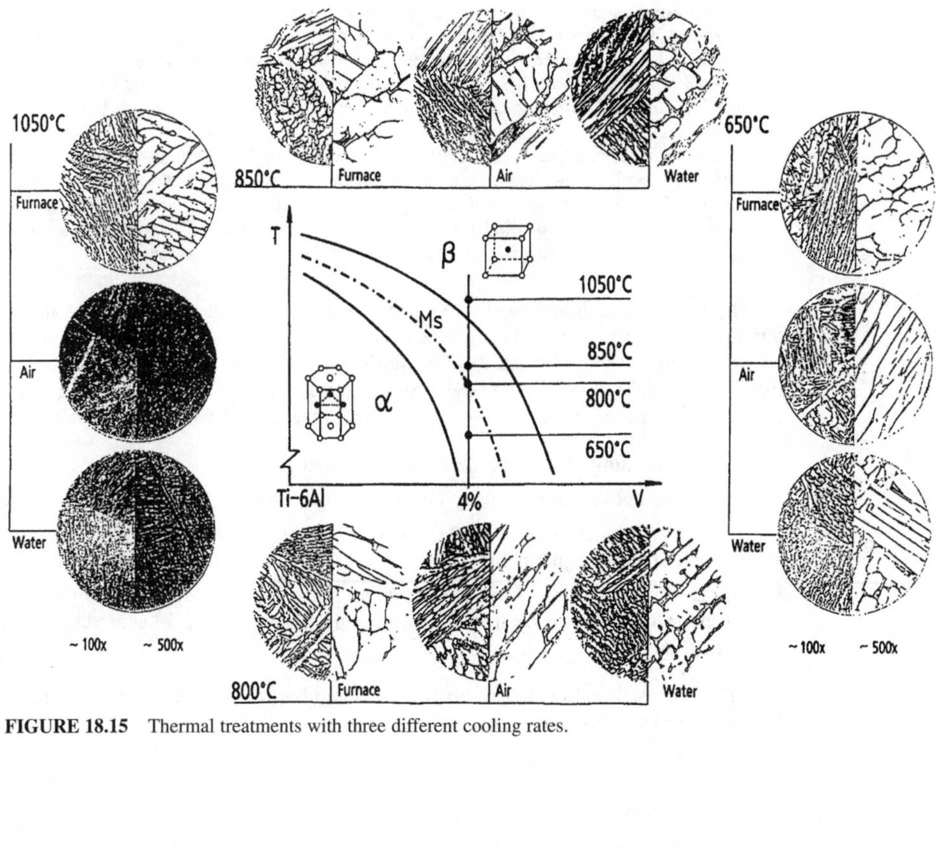

FIGURE 18.15 Thermal treatments with three different cooling rates.

18.2 FERROUS HEAT TREATMENT

A number of heat treatment processes are used on ferrous materials. These are covered in the sections below.

18.2.1 Annealing

Annealing is a process that consists of heating to and holding at a suitable temperature followed by cooling at an appropriate rate, primarily intended to soften the part and improve its machinability. Supercritical or full annealing involves heating a part above the *upper critical temperature* (Ac_3), that is the temperature at which austenite begins to transform to ferrite during cooling, and then slowly cooling in the furnace to around 315°C (600°F). Intercritical annealing involves heating the part to a temperature above the *final transformation temperature* (Ac_1), the temperature at which austenite begins to form during heating. Subcritical annealing heats the part to just below the Ac_1 point followed by a slow cooling in the furnace. The rate of softening increases rapidly as the annealing temperature approaches the Ac_1 point.

As noted by Payson,[13] the following guidelines have been established for the annealing process.

Rule 1. The more homogeneous the structure of the austenitized steel, the more completely lamellar will be the structure of the annealed steel. Conversely, the more heterogeneous the structure, the more nearly spheroidal will be the annealed carbide structure.

Rule 2. The softest condition in the steel is usually developed by austenitizing at a temperature less than 100°F (55°C) above Ac_1, and transforming at a temperature less than 100°F (55°C) below Ac_1.

Rule 3. Due to the long soak times that may be required for complete transformation at temperatures below 100°F (55°C) below Ac_1, allow most of the transformation to take place at the higher temperature, where a soft product is formed; and finish the transformation at a lower temperature, where the time required for completion of transformation is short.

Rule 4. After the steel has been austenitized, cool to the transformation temperature as rapidly as feasible in order to minimize the total duration of the annealing operation.

Rule 5. After the steel has been completely transformed, at a temperature that produces the desired microstructure and hardness, cool to room temperature as rapidly as feasible to decrease further the total time of annealing.

Rule 6. To ensure the maximum of lamellar pearlite in the structures of annealed 0.70–0.90% C tool steels and other low alloy and medium alloy carbon steels, preheat for several hours at a temperature about 50°F (28°C) below the *lower critical temperature* (Ac_1) before austenitizing and transforming as usual.

Rule 7. To obtain minimum hardness in annealed hypereutectoid alloy tool steels, heat to austenitizing temperature for a long time (about 10 to 15 hours) then transform as usual.

These rules are applied most effectively when the critical temperature and transformation characteristics of the steel have been established and when transformation by isothermal treatment is feasible.

Spheroidizing is performed to improve the cold forming of a material as well as to improve the machinability of hypereutectoid steels and tool steels. Spheroidizing consists of heating and cooling to produce a spheroidal or globular form of carbide in steel. Spheroidizing methods frequently used are

- Prolonged holding at a temperature just below Ac_1
- Heating and cooling alternatively between temperatures that are just above and just below Ac_1
- Heating to a temperature above Ac_1 or Ac_3, and then cooling very slowly in the furnace or holding at a temperature just below Ac_1
- Cooling at a suitable rate from the minimum temperature at which all carbide is dissolved, to prevent the reformation of a carbide network, and then reheating in accordance with one of the first two methods shown above

Different combinations of microstructures and hardness are important for machining. Optimum microstructures are shown in Table 18.9.

TABLE 18.9 Optimum Microstructures for Machining

Carbon (%)	Optimum microstructure
0.06–0.20	As rolled
0.20–0.30	Under 3" (75 mm), normalized, over 3" (75 mm) as rolled
0.30–0.40	Annealed to produce coarse pearlite, minimum ferrite
0.40–0.60	Annealed to produce coarse lamellar pearlite to coarse spheroidized carbides
0.60–1.00	Annealed to produce 100% spheroidized carbides, coarse to fine.

Source: From Metals.[12]

As the hardness of steel increases during cold working, ductility decreases and additional cold reduction becomes difficult. In these instances, the material must be annealed to restore its ductility. Such annealing between processing steps is simply referred to as process annealing, and consists of any appropriate annealing treatment. Subcritical annealing is often the least costly, and this process is most often associated with the term "process annealing."

Process annealing usually consists of heating to a temperature below Ac_1, soaking an appropriate amount of time, and then cooling, usually in air. In most instances, heating to a temperature from 20°F (10°C) to 40°F (20°C) below Ac_1 produces the best combination of microstructure, hardness, and mechanical properties.

18.2.2 Brazing

Brazing is a joining process between materials of the same or dissimilar composition achieved by heating to a suitable temperature where a "filler" metal with a melting point above 450°C (840°F) and below the melting point of the "base" metal is distributed between closely fitted surfaces by capillary action.[14] It will be discussed in details in a separate section.

18.2.3 Carbon Restoration

Carbon restoration, or recarburization, involves the replacement of carbon lost in the surface layer from a previous heat treatment process by carburizing this layer to return it to the original carbon level (see Carburizing). The process of restoring carbon can be applied to steels of low, medium, or high carbon content. Three types of atmospheres are generally used for carbon restoration:

1. Nitrogen based, enriched with hydrocarbon gas (natural or propane)
2. Exothermic, dried and purified
3. Endothermic with controlled dew point

Of these choices, endothermic gas is most widely used. Control of the dew point is best achieved by additions of air, nitrogen, methane, or hydrocarbon gas at the furnace rather than adjusting the air/gas ratio at the generator.

Successful restoration depends on the following.

- Type of furnace (batch or continuous)
- Atmosphere tightness of the furnace
- Surface condition of the material being carbon restored
- Distribution of the load
- Flow rate and degree of circulation of the furnace atmosphere

- Annealing or normalizing temperature
- Type and effect of quench
- Composition of the furnace atmosphere
- Control of the dew point
- Composition of the steel

It is difficult to control the carbon potential of the atmosphere. A surface carbon level slightly below that of the base material is usually much less harmful than over carburizing.

18.2.4 Case Hardening

Case hardening is used to produce a hard, wear resistant case, or surface layer, on top of a ductile, shock resistant interior, or core. A variety of techniques are used and the most common of these, carbonitriding, carburizing, nitrocarburizing, and nitriding are explained below.

Carbonitriding. Carbonitriding is a modified carburizing process, not a form of nitriding. This modification consists of introducing ammonia into the carburizing atmosphere in order to add nitrogen into the carburized case as it is being produced (Fig. 18.1). Typically, carbonitriding is done at a lower temperature than carburizing, between 1300°F (700°C) and 1650°F (900°F) and for a shorter time than carburizing. Along with the fact that nitrogen inhibits the diffusion of carbon this combination of factors result, in general, in a shallower case than is typical for carburized parts, typically between 0.003 in (0.075 mm) and 0.030 in (0.75 mm). It is important to note that a common contributor to nonuniform case depths is to begin ammonia additions before the load is stabilized at temperature (this is a common mistake in furnaces that begin gas additions after the furnace has recovered set-point rather than introduce a time delay for the load to reach temperature). Also, when the ammonia addition is halted, desorption of nitrogen will begin to occur.

The temperature range in which carbonitriding is performed is necessary, since at higher austenitizing temperatures the thermal decomposition of ammonia is too rapid, and limits nitrogen availability; while at lower temperature, a brittle structure is formed and operating furnaces below 1400°F (760°C) can be a safety concern.

The nitrogen in carbonitrided steel enhances hardenability and makes it possible to form martensite in plain carbon and low alloy steels that initially have low hardenability. Examples of these steels include 1018, 12L14, and 1117. The nitrides formed contribute to the high surface hardness. Nitrogen, like carbon, manganese, or nickel, is an austenite stabilizer so retained austenite is a concern after quenching. Controlling the ammonia percentage will reduce the amount of retained austenite and should be done if hardness or wear resistance are reduced. Another consequence of high nitrogen percentages is the formation of voids or porosity. In general it is recommended that the nitrogen content at the surface be limited to 0.40% maximum.

Several other points are worth mentioning. The presence of nitrogen in the carbonitrided case increases the resistance to softening on tempering, just like some alloying elements do. The higher the nitrogen content, the higher the resistance to softening. Higher tempering temperatures, up to 440°F (230°C), are often used on carbonitrided parts. The resistance to tempering manifests itself in wear properties, carbonitrided gears often exhibit better wear resistance than carburized gears. Many shallow case depth thin section parts of unalloyed steel, such as die cutting punches, can be used without tempering.

Carburizing. Gas atmosphere carburizing is the most common type of case hardening method performed today. It is most often done in the temperature range of 1550–1750°F (840–955°C). Temperature extremes as low as 1450°F (790°C) and as high as 2000°F (1090°C) have been used. Carburizing case depths can vary over a broad range, 0.005–0.325 in (0.13–8.25 mm) being typical.

The carburizing process begins when steel or another ferrous alloy of relatively low carbon content is heated into the austenite range. A change (called a solid state transformation) occurs in the crystal structure creating a high solubility for carbon. The steel is then brought in contact with an environment rich in carbon (that is, of high carbon potential) so that absorption of carbon takes place into the surface of the metal. The necessary carbon may be imparted in either a solid ("pack" carburizing), a liquid ("salt bath" carburizing) or gaseous ("gas atmosphere" or "low pressure vacuum/plasma" carburizing) medium.

Carbon is a relatively small atom and the amount or percentage of carbon absorbed increases as a function of increasing temperature. The depth of carbon penetration increases with increasing time at a fixed temperature.

When prolonged carburizing times are used for medium and deep case depths, a high carbon potential produces a high surface carbon content. If the part were hardened directly from carburizing temperature excessive amounts of retained austenite and free carbides would result, both of which have adverse effects on the properties of the case. In order to reduce the high surface carbon content, a diffusion cycle is usually utilized at the end of the carburizing cycle in which the carbon potential (and often the furnace temperature) is reduced. The carbon content of the case is allowed time to diffuse (or move inward) creating a slightly lower surface carbon, more suitable to produce good properties in the case on hardening. A comparison of carburizing methods is shown in Table 18.10 below.

If the parts being processed need uniform absorption of carbon into the surface, i.e., good case uniformity, you would want to select lower carburizing temperatures. If however increased productivity is the principle concern, then higher temperatures could be used. Recall that we cautioned that as the temperature increases, especially above 1800°F (980°C), changes to the internal microstructure (and resultant properties) must be carefully considered. Today there is a trend toward higher carburizing temperatures to reduce cycle times.

In gas carburizing, parts are heated to austenitizing temperature in an atmosphere that contains approximately 40 percent hydrogen, 40 percent nitrogen, and 20 percent carbon monoxide. Small percentages of carbon dioxide (up to 1.5 percent), water vapor (up to 1 percent), and methane (up to 0.5 percent) along with trace amounts of oxygen are also present.

This "neutral" atmosphere or "carrier gas" is neither carburizing (adding carbon) nor decarburizing (removing carbon) to the surface of the steel. The carrier gas is supplied to the furnace in sufficient volume to build up a positive pressure in the chamber and to prevent the infiltration of air, preventing oxidation from taking place on the surface of the part as it is heated to elevated temperature. Oxidation would inhibit the penetration of carbon into the surface of the steel. Although there are a number ways to produce this neutral atmosphere the two most common are a generated atmosphere produced by an endothermic gas generator (commonly called an *Endo generator*) or a synthetic atmosphere, produced from a mixture of nitrogen and methyl alcohol (methanol), a so-called "endo equivalent" atmosphere.

A common variation of the carburizing process is to introduce ammonia during the latter portion of the cycle, typically in the last $^1/_4$–$^1/_2$ hour before the load is quenched. Any loss of hardenability that might occur due to internal (or intergranular) oxidation is partially compensated by the nitrogen absorption.

In order to perform the carburizing or carbonitriding process we add an enriching gas to the carrier gas atmosphere and in the case of carbonitriding, enriching gas plus ammonia. The enriching gas is usually either natural gas which is about 90–95% methane or propane. A good rule of thumb is that the amount of enriching gas used is approximately 10–15% of the carrier gas flow and if ammonia is used, this percentage varies from about 2% to 12%. With automatic control systems these percentages will vary over time.

It is extremely important that the parts be at temperature before the furnace atmosphere is changed from "neutral" to "carburizing." Otherwise the surface of the parts will pick up free carbon in the form of soot and the amount of carbon absorption into the surface of the steel will be limited. Large variations in case depth uniformity result and damage to the furnace can occur with excessive sooting.

Other variables affect the amount of carbon transferred to the parts, including the degree of atmosphere circulation and the type and amount of alloy content of the parts and the materials of furnace construction.

Carburization of a metal surface is both a function of the rate of carbon absorption and the diffusion of the carbon away from the surface and into the metal. Once a high concentration of carbon

TABLE 18.10 Common Carburizing Methods

	Process		
	Atmosphere	Vacuum	Plasma
Gases	Endothermic nitrogen/ methanol Direct feed *plus* natural gas	Acetylene or propane or ethylene or methane	Methane or methane/ argon or propane/ hydrogen or methane/ argon/hydrogen
Enriching gas consumption	50–150 cfh	10–15 cfh	3–4 cfh
Gas disposal	Burnoff	Nitrogen dilution of pump effluent	Nitrogen dilution of pump effluent
Pressure range	Atmospheric (760 torr)	Subatmospheric (0.3–200 torr)	Subatmospheric (0.75–15 torr)
Carbon source	Gas dissociation (thermodynamic)	Gas dissociation (thermodynamic)	Activated by plasma
Case depth	Shallow, medium, and heavy (>0.100")	Very shallow, shallow, and, medium	Very shallow, shallow, and, medium
Case depth uniformity	From ±0.005" to ±0.010"	From ±0.002" to ±0.005"	From ±0.002" to ±0.005"
Carburizing speed	Function of carbon potential in the atmosphere and diffusion laws	Shallow and medium case depth quicker than atmosphere	Shallow case depths quicker than vacuum or atmosphere
Partial carburizing	Copper plating or stop-off paints	Copper plating or stop-off paints	Physical masking (covers or shields)
Process control	Oxygen probe or infrared or dew point	None	Plasma parameters
Loading (typical)	Random (bulk loading) or fixtured	Fixtured/orientated	Fixtured/orientated
Part geometry	Wide range excluding blind holes and contact areas (masked locations)	Wide range including blind holes	Wide range including blind holes and very complicated shapes
Internal oxidation	IGO (0.0003"–0.0005" typical)	None	None
Quenching Method	Oil, polymer, salt	Oil, high pressure gas (≤20 bar)	High pressure gas (≤20 bar)

Source: From Herring.[15]

has been developed on the surface of the metal during what is commonly referred to as the "boost stage," we normally introduce a "diffuse stage" where solid state diffusion is allowed to occur over time. This step produces a change in the carbon concentration (or carbon gradient) between the carbon rich surface and the interior core of the metal, resulting in a reduction of the overall carbon concentration at the surface while increasing the depth of carbon absorption.

Vacuum or plasma carburizing furnaces do not use a "carrier" gas atmosphere, but instead they use vacuum pumps to remove the atmosphere out of the chamber before the process begins. So for carburizing to take place in these vacuum furnaces, we need only provide a small and controlled amount of hydrocarbon gas additions. In carbonitriding, ammonia is also added.

Both vacuum carburizing and plasma (also called "ion") carburizing are low pressure processes and will see increased popularity and usage over time due to their environmentally friendly operating conditions (no heat given off to the room, minimal usage of gas, no fumes, smoke or flames). It has taken a long time for these processes to be properly controlled (the problem was soot which caused extremely short life of internal components of the furnace and required high maintenance). In addition, the cost of vacuum furnaces is much higher than conventional gas carburizing equipment. However, the sooting problems have been solved and the technology is poised to make a major leap forward.

Vacuum carburizing is a modified gas carburizing process in which the carburizing is done at low pressures far below atmospheric pressure. The typical pressure range for low pressure vacuum carburizing is under 20 torr, often as low as 3–4 torr. Some manufacturers recommend pressures less than 1 torr. This may increase cycle times.

The advantage of this method is that the metal surface remains very clean and the vacuum makes it easier to transfer carbon to the surface of the steel. Acetylene, ethylene, and propane have been found to be the best gases to use for low pressure vacuum carburizing. Cyclohexane, a liquid, is also used. In some situations methane is used. The carbon is free to penetrate into the surface of the steel, while the hydrogen byproduct is removed from the system by the vacuum pumps.

The breakdown of hydrocarbons in vacuum or plasma carburizing is not an equilibrium reaction as it is in gas carburizing. What this means is that the surface of the steel is very rapidly raised to the saturation level of carbon in austenite. This step in the carburizing process is, just as in gas carburizing called the "boost" stage. To achieve a proper surface carbon level, the boost stage must be followed by a "diffuse" stage in which carbon is allowed to diffuse away from the surface of the steel without adding any more carbon. Evacuating all of the hydrocarbon gas from the vacuum chamber and not allowing any more to go in does this. By repeating these boost and diffuse steps, any desired carbon profile and case depth can be achieved.

Plasma can also be used for carburizing. It is primarily used when selective carburizing is desired and when it is possible to use a physical masking technique to accomplish this (essentially, a plate or cap is used to cover the area that is not to be carburized). If this is not possible, other methods of selective carburizing should be used (see below) and plasma carburizing is not used. Two methods of plasma carburizing are commonly used today. Each uses a different hydrocarbon gas as the source of carbon. One process involves the use of methane, and the other propane, or propane diluted with hydrogen gas.

Given the same process parameters (time, temperature, gas pressure, and plasma conditions), the carbon transfer characteristics of these two gases differ. The carbon transfer with propane is so high that the carbon content of the part quickly reaches the limit of saturation of carbon in austenite. A series of short "boost" and "diffuse" cycles are used to develop the proper case depth and surface carbon levels. In contrast, plasma carburizing with methane is typically done in a single boost and diffuse cycle. The boost portion is about one-third of the total cycle time.

The plasma carburizing process takes place at temperatures between 1650°F (850°C) and 2000°F (1090°C). The use of low vacuum levels in the 0.75–15 Torr (1–20 mbar) range and in the presence of an electric field not only permits the use of methane or propane as the source of carbon but ensures that effective carburization can occur at very low levels of gas consumption, typically less than 3–4 cfh (0.08–0.10 m^3/h).

In plasma, or ion carburizing, the parts are heated to austenitizing temperature both by the gas plasma created and by auxiliary heaters in the furnace. An imposed voltage from a dc power supply ionizes gas introduced into the furnace. The purpose of ionization is to accelerate carbon transfer by creating ions of carbon generated from the introduction of a hydrocarbon gas such as methane or propane. The electrical attraction of the positively charged carbon atoms to the negative parts also promotes uniformity. These two factors result in extremely uniform carbon profiles in complex geometry (such as holes and gear teeth).

In many instances only a portion of the part needs to be carburized. For example, a shaft might be hardened only at areas where the bearings are located, leaving the balance of the shaft in the tougher, unhardened condition. In order to case harden locally, or selectively, several procedures are available:

- Carburize only the areas where a hardened case is desired.
- Remove the case from the areas that are desired to be soft, either before or after hardening.
- Carburize all over, but harden only selective areas.

Restricting the carburizing action to some selective area may be accomplished in several ways, one is to copper plate the area that is not to be carburized. The plating is usually done electrolytically, although a less satisfactory job can be done by dipping the steel into copper sulfate.

Carburizing can be accomplished in ways other than by gas atmosphere and vacuum techniques. These other methods are now less common for a variety of reasons, mainly environmental. Below presented is a brief summary of each method:

- *Pack carburizing* is a process that is seldom used today. Carbon monoxide from a solid compound decomposes at the metal surface into nascent carbon and carbon dioxide. The nascent carbon is then absorbed into the metal The carbon dioxide resulting from this decomposition immediately reacts with carbonaceous material present in the solid carburizing compound to produce fresh carbon monoxide. Carburizing continues as long as enough carbon is present to react with the excess of carbon monoxide.
- *Liquid carburizing* is a process that is most often carried out in molten salt bath equipment. Environmental concerns make this method unattractive for most heat treaters. Many liquid carburizing baths *contain* cyanide, which introduces both carbon and nitrogen into the case. A new type of bath uses a special grade of carbon rather than cyanide. This bath produces a case that contains only carbon as a hardening agent.

Light case and *deep case* are arbitrary terms that have been associated with liquid carburizing in baths containing cyanide; there is necessarily some overlapping in bath composition for the two types of case. In general the two types are distinguished more by operating temperature than by bath composition; the terms low temperature and high temperature are preferred.

Cases produced with high temperature liquid carburizing baths consist essentially of carbon dissolved in iron. Sufficient nascent nitrogen is available to produce a superficial nitride containing surface layer that aids in resisting wear and also resists tempering as the part is reheated.

- *Drip carburizing* is the introduction of a liquid, typically an alcohol such as benzene, toluene, or methanol into the furnace, typically impinging on a "hot target" which aids in the dissociation and forms an atmosphere high in carbon.

Nitrocarburizing. Nitrocarburizing is a modification of the nitriding process, not a form of carburizing. This modification consists of the simultaneous introduction of nitrogen and carbon into the steel, typically in its ferritic condition, i.e., below the temperature at which austenite begins to form during heating. Nitrocarburizing is typically performed in the temperature range of 1025–1110°F (550–600°C) in atmospheres of 50% endothermic gas + 50% ammonia, or 60% nitrogen + 35% ammonia + 5% carbon dioxide. Other atmospheres that vary the composition such as 40% endothermic gas + 50% ammonia + 10% air are also used. The presence of oxygen in the atmosphere activates the kinetics of nitrogen transfer. The thickness of the "white" or "compound" layer is a function of gas composition and gas volume (flow). Nitrocarburizing can be used to produce a 58 HRC minimum hardness, with this value increasing dependent on the base material. White layer depths range from 0.0013 to 0.056 mm (0.00005–0.0022 in) with diffusion zones from 0.03 to 0.80 mm (0.0013–0.032 in) being typical.

A complex sequence is involved in the formation of a nitrocarburized case. Of importance here is that normally a very thin layer of single-phase epsilon (ε) carbonitride is formed between 840°F (450°C) and 1095°F (590°C), and this compound layer has associated with it an underlying diffusion zone containing iron (and alloy) nitrides and absorbed nitrogen. The white layer has excellent wear and antiscuffing properties and is produced with minimum distortion. The diffusion zone, provided it is substantial enough, improves fatigue properties such as endurance limit, especially in carbon and low alloy steels. Some of the increased hardness of the case is due to a diffusion zone beneath the compound layer, especially in the more highly alloyed steels with strong nitride formers.

It is not uncommon to observe porosity of the compound layer due to the presence of a carburizing reaction at the steel surface influencing the nitriding kinetics and therefore the degree and type of porosity at the surface of the epsilon (ε) layer. Three different types of layers can be produced: no porosity; sponge porosity; or columnar porosity. Some applications require deep nonporous epsilon (ε) layers. Others applications where, for example, optimum corrosion resistance is needed benefit from the presence of sponge porosity. Still others benefit from columnar porosity where oil retention can enhance wear resistance.

Nitrocarburizing is often followed by an oxidizing treatment to both enhance corrosion resistance and surface appearance.

A lower temperature variant of carbonitriding is austenitic nitrocarburizing. This process takes place in the temperature range of 1250–1425°F (675–775°C) and can be controlled to produce a surface compound layer of epsilon (ε) carbonitride with a subsurface of bainite and/or martensite produced on quenching and resulting in a good support structure for the hard surface. The resultant microstructure is particularly useful in intermediate stress point contact resistance applications such as helical gears. Examples of gear steels that are commonly nitrocarburized include AISI 1018, 1141, 12L14, 4140, 4150, 5160, 8620, and certain tool steels.

Nitriding can also be performed using plasma, or ion methods. Adding small amounts of methane (CH_4) or carbon dioxide (CO_2) to the nitrogen-hydrogen mixture will produce a carbon containing eta (ε) phase ($Fe_2C_xN_y$, $Fe_3C_xN_y$). This process is called plasma nitrocarburizing and is typically used for unalloyed steels and cast irons.

Nitriding. Nitriding is typically done using ammonia with or without dilution of the atmosphere with dissociated ammonia or nitrogen/hydrogen in the temperature range of 925–1050°F (500–565°C). Single stage and two stage (Floe process) nitriding processes are used. In the single stage process a temperature range of 925–975°F (500–525°C) is used within an atmosphere of anhydrous ammonia. The dissociation rate (of ammonia to nitrogen and hydrogen) ranges from 15 to 30%. The process produces a brittle, nitrogen rich layer, known as the "white layer," at the surface. The two-stage process has the advantage of reducing the thickness of this white layer. The first stage is, except for time, a duplication of the single stage process. In the second stage, however, the addition of a dilutant gas (dissociated ammonia or nitrogen) increases the percent dissociation to around 65–85%. The temperature is raised to 1025–1075°F (550–575°C), and the result is the reduction of the depth of the white layer.

Case depth and case hardness properties vary not only with the duration and type of nitriding being performed but also with steel composition, prior structure, and core hardness. Typically case depths are between 0.008 in (0.20 mm) and 0.025 in (0.65 mm) and take from 10 to 80 hours to produce.

Nitriding is another surface treatment process that has as its objective increasing surface hardness. One of the appeals of this process is that rapid quenching is not required, hence dimensional changes are kept to a minimum. It is not suitable for all applications; one of its limitations being that the extremely high surface hardness case produced has a more brittle nature than, say, that produced by the carburizing process. Principle among the benefits of nitriding are these valuable properties unique to this case hardening process:

- Exceptionally high surface hardness
- Resistance to wear and antigalling properties (good in poor lubrication conditions)
- A minimum of distortion and deformation (compared to carburizing/hardening)
- Resistance to tempering (resistant to softening effect of heat, up to nitriding temperature at which conventional steels would soften)
- Stability of the nitrided case
- A beneficial effect on fatigue properties (improve fatigue life)
- Reduction in notch sensitivity
- A marked resistance to corrosion in several common media (except for nitrided stainless steels)
- Small volumetric changes (some growth does occur)
- To obtain a surface that is resistant to the softening effect of heat at temperatures up to the nitriding temperature

To insure the best nitriding results, the following precautions and recommendations should be followed:

- The steel should be hardened and tempered prior to nitriding so as to possess a uniform structure. The tempering temperature must be sufficiently high to guarantee structural stability at the nitriding temperature; the minimum tempering temperature being 50°F (10°C) higher than the maximum temperature to be used for nitriding.
- Before nitriding, the steel must be free from decarburization and precleaning is mandatory; residue on the parts will result in spotty cases.
- If freedom from distortion is of paramount importance, the internal stresses produced by machining or heat treating should be removed before nitriding by heating to a temperature of 1000°F (538°C) and 1300°F (705°C).
- Since some growth takes place on nitriding, this should either be allowed for in the final machining or grinding operation prior to nitriding or removed by lapping or by careful grinding. However, the removal of only a slight amount of the nitride case is permissible.
- If maximum resistance to corrosion is desired, the parts should be used with white layer intact (as processed).
- Nitrided steels of the nitralloy type should not be used where resistance to the corrosion of mineral acids is encountered or where resistance to sharp abrasive particles at high velocities is required, as in sand nozzles.
- If straightening is required after nitriding, it should be done hot if possible in the temperature range of 1200°F (650°C). Cold straightening techniques should be carefully reviewed.
- If maximum hardness and maximum resistance to impact are desired, and the question of maximum corrosion resistance is not of vital importance, the removal of 0.001–0.002 in (0.025–0.050 mm) of the nitrided case is desirable. The amount to be removed depends on the original case depth. This operation will remove the most brittle surface layer.
- If nitrided articles exhibit a shiny gray surface after their removal from the furnace, the results should be viewed with suspicion. Invariably, the case will be shallow and below hardness. The articles should have a matte gray appearance, although a slight discoloration does not indicate faulty nitriding. The opening of the furnace at too high a temperature or the presence of air leakage on cooling will account for the slight discoloration.

Nitriding can also be performed using plasma or ion methods. The reactive gas species in ion nitriding is nitrogen gas. Under ordinary conditions, nitrogen does not react with steel below approximately 1800°F (980°C). However, under plasma conditions, i.e., at pressures in the range of 1–10 torr (1–13 mbar) and in the presence of an electric field of 300–1200 V, nitrogen becomes reactive.

The nitrogen transfer is caused by the attraction of positively charged nitrogen ions to the workload held at a negative potential. This reaction takes place in a plasma "glow" discharge near the surface of the parts.

Ion nitrided components typically have a "case" that consist of two layers; an exterior, or compound zone ("white" layer) and a diffusion zone lying beneath it.

The current necessary to develop the required case depth is a product of the total area under glow and the value of the current density required to maintain it in the (abnormal) discharge range. A "rule of thumb" is 1 mA/cm^2 of workload area. The variables that control the rate of nitrogen transfer are temperature, current density (that is, plasma power), and gas dilution.

Nitrogen transfer increases with an increase in nitriding temperature. Ion nitriding is typically done between 750°F (400°C) and 1300°F (700°C). The depth of the diffusion layer is strongly dependent on both temperature and time. The current density has an impact on the thickness of the compound layer.

Ion nitriding is typically done in an atmosphere of 75% hydrogen and 25% nitrogen. The rate of nitrogen transfer is changed by diluting the gas with hydrogen. As the amount of hydrogen in the nitrogen–hydrogen mixture increases, the thickness of the compound zone is affected. At 98 percent hydrogen, no compound layer is produced. As hydrogen additions fall below 75 percent little effect on nitrogen transfer occurs. With increasing dilution, the composition and thickness of the compound layer changes. This compound layer consists of iron nitrides that develop in the outer region of the diffusion layer after saturation with nitrogen. Two iron nitrides are possible, the nitrogen-poor gamma prime (γ) phase (Fe_4N) and the nitrogen-rich eta (ε) phase (Fe_2N, Fe_3N).

18.2.5 Cryogenic Treatment

Cryogenic treatments are used to minimize retained austenite in a variety of heat treated components and enhance other heat treat operation such as stress relieving of castings and machined parts.

Common practice is for parts to be subzero cooled to a temperature not warmer than –120°F (–84°C) and held at this temperature or lower for a minimum of 1 hour per inch (25.4 mm) of cross sectional area but not less than 30 minutes, and allowed to warm back to room temperature in air on removal from the deep freeze. An alternative practice, so called "deep cooling" involves reducing the part temperature to approximately –310°F (–190°C).

To minimize retained austenite most parts are cryogenically treated immediately after hardening and before tempering. The choice of deep freezing or tempering often depends on the grade of steel and design of the part since cracking is a concern, especially in parts with stress risers, such as sharp corners or abrupt changes in section size.

Successful cryogenic treatment simply depends on reaching the minimum low temperature and holding at that temperature for a sufficient length of time to allow the part to reach a uniform temperature. Transformation then occurs. Materials of different compositions and different configurations can be cryogenically cooled together, irrespective of material chemistry or austenitizing temperature.

18.2.6 Hardening

Applied Energy. Various methods of hardening by use of applied energy are used in the manufacture of gears including flame hardening, laser surface hardening and induction.

Flame. Common flame hardening techniques involve either spinning or progressive heating of components. In the progressive heating method, the flames gradually heat the part in front of the flame head, and sometimes this effect must be compensated for by gradually increasing the speed of travel or by precooling. A wide range of materials can be hardened by this technique including plain carbon steels, carburizing grades, cast irons, and certain stainless grades.

The principle operating variables are rate of travel of the flame head or work; flame velocity and oxygen-fuel ratios; distance from the inner flame cone or gas burner to the work surface; and the type, volume, and angle of quench. The success of many flame hardening operations for small production runs is dependent on the skill of the operators. A principle advantage of this technique is that parts of extremely large size can be heat treated.

Induction. Induction hardening is a commonly used heat treatment technique for a multitude of reasons including the ability to process one part at a time, ease of integration into manufacturing and the need to only heat that portion of the part where the treatment will take place. Table 18.11 shows typical products that are induction heated.

Induction heating is a process which uses alternating current to heat the surface of a part. The area is then quenched resulting in an increase in hardness in the heated area. It is typically accomplished in a relatively short period of time. The type of steel, its prior microstructure and the desired part performance characteristics determine the required hardness profile and resulting strength and residual stress distribution. For example, external spur and helical gears, bevel and worm gears, and racks and sprockets are commonly induction hardened from materials, such as AISI 1050, 4140, 4150, and 8650.

The hardness pattern produced by induction heating is a function of the type and shape of inductor used as well as the heat mode. One technique for induction hardening is the use of a coil encircling the part. An inductor which is circumferential will harden either a stationary or moving part.

TABLE 18.11 Typical Induction Heating Products and Applications

Preheating prior to metalworking	Heat treating	Welding	Melting
Forging	Surface hardening, tempering	Seam welding	Air melting of steels
Gears	Gears	Oil-country tubular	Ingots
Shafts	Shafts	products	Billets
Hand tools	Valves	Refrigeration tubing	Castings
Ordnance	Machine tools	Line pipe	
	Hand tools		Vacuum induction melting
Extrusion			Ingots
Structural members	Through hardening, tempering		Billets
Shafts	Structural members		Castings
	Spring steel		"Clean" steels
Heading	Chain links		Nickel-base superalloys
Bolts	Tubular goods		Titanium alloys
Other fasteners			
	Annealing		
Rolling	Aluminum strip		
Slab	Steel strip		
Sheet (can, appliance, and automotive industries)			

Source: From Metals.[12]

Hardness patterns that are more like those found in a carburized case can also be developed. This type of induction hardening is called *contour hardening* and is produced via tooth-by-tooth or gap-by-gap techniques, by applying either a single-shot or scanning mode. Pattern uniformity is very sensitive to coil positioning.

Laser. Laser surface hardening is used to enhance the mechanical properties of highly stressed machine parts such as gears. The use of lasers for surface treatments is relatively limited mainly due to the high cost of large industrial lasers required for most metal-working operations. Adding to the expense is the fact that lasers are not very efficient from an energy standpoint. Medium carbon, alloy steels, and cast irons (gray, malleable, ductile) are good candidates for this technology.

Through Hardening. Through or direct hardening refers to heat treatment methods which do not produce a case. It is important to note that hardness uniformity should not be assumed throughout the part if the outside cools faster than the inside, a hardness gradient will develop. The final hardness of a material is dependent on the amount of carbon in the steel. The depth of hardness depends on the hardenability of the steel.

Hardening is achieved by heating the material into the austenitic range, typically 1500–1600°F (815–875°C), followed by rapid quenching and tempering. Successful hardening usually means achieving the required microstructure, hardness, strength, or toughness while minimizing residual stress, distortion, and avoiding cracking.

Quenching refers to the process of rapidly cooling a part from austenitizing temperature, or solution heat treatment temperature in the case of nonferrous materials. The selection of a quenchant medium depends on the following properties:

- Hardenability of the particular alloy
- Section thickness (ruling section) and shape
- Cooling rates required to achieve the desired microstructure

The most common quenchants are liquids and gases. Typical liquid quenchants include water (including brine), oils, aqueous polymers, and molten salt. Typical gas quenchants include air, nitrogen, argon, hydrogen, helium, and mixtures.

The rate of heat extraction by a quenching medium and the way it is used, substantially effects quenchant performance. Variations in quenching practices have resulted in the assignment of specific names to some quenching techniques:

- Direct quenching
- Time quenching
- Selective quenching
- Spray quenching
- Fog quenching
- Interrupted quenching

18.2.7 Normalizing

Normalizing is a process that involves heating the part above the upper critical temperature and then air cooling outside the furnace to relieve residual stresses and for dimensional stability (Fig. 18.16).

Normalizing is often considered from both a thermal and microstructural standpoint. In the thermal sense, normalizing is austenitizing followed by cooling in still or slightly agitated air or nitrogen. In a microstructural sense, the areas of the microstructure that contain about 0.8% carbon are pearlitic; while the areas of low carbon are ferritic. A normalized part is very machinable but harder than an annealed part. Good normalizing practice requires:

- The part be uniformly heated to a temperature high enough to cause complete transformation to austenite.
- The part remain at temperature long enough to achieve uniform temperature throughout the section size.
- The part be allowed to cool uniformly in still air to room temperature.

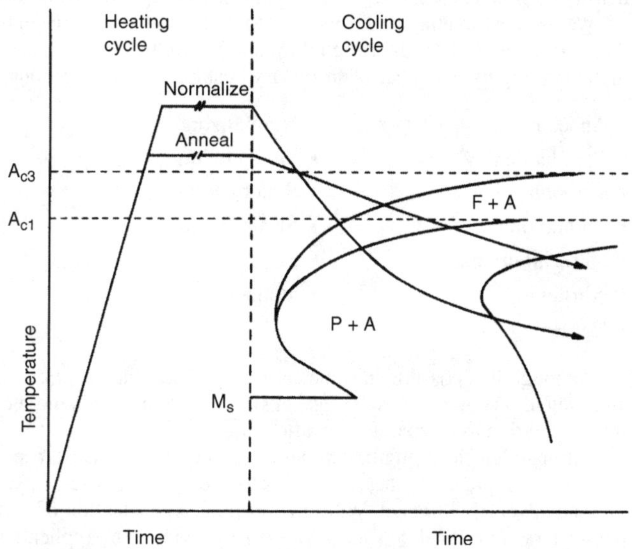

FIGURE 18.16 Schematic representation of time–temperature cycle for normalizing.[17]

The air cooling associated with normalizing produces a range of cooling rates dependent on section size. Uniform cooling requires that there be no area where the cooling has been either restricted or accelerated. Restriction of the cooling rate will produce properties more similar to those achieved in an annealing cycle while too rapid a cooling rate will partially quench the part.

Normalizing will produce a uniform structure of ferrite and pearlite and may soften, harden, or stress relieve the part depending on the condition of the material before normalizing. Normalizing may be used to

- Refine grain and homogenize the microstructure so as to improve the materials response to hardening
- Improve machinability
- Modify and refine cast dendridic structures
- Change mechanical properties

Normalizing may be applied to both carbon and alloy steels in both the cast and wrought forms. Depending on the mechanical properties required, normalizing may be substituted for hardening when the size or shape of the part is such that liquid quenching might result in cracking, distortion, or excessive dimensional change. Thus parts with sharp changes in section or with complex shape may be normalized and tempered provided the properties obtained are acceptable.

18.2.8 Sintering

The heat treatment of powder metal (P/M) parts requires an understanding of the differences between a "powder metal" and a conventional "wrought metal" part.[16]

The density of pure iron is 7.87 g/cc. This is referred to as *theoretical density* in powder metallurgy. Porosity, the amount of void space in a P/M part, means that it is difficult to achieve full density. Thus the density of a P/M component is often expressed as a ratio to theoretical density.

Various consolidation and/or sintering techniques can be performed on a *green* part to achieve high density including metal injection molding, hot isostatic pressing, P/M forging, and liquid-phase sintering. However, conventional "press and sinter" is the most widely used method and results in a typical density range of 6.8–7.2 g/cc. By comparison, a wrought (0.40%) carbon steel has a density of 7.84 g/cc.

We use heat treating after sintering to help us attain the desired final part strength and hardness. Approximately 60% of the current P/M steels have their physical and mechanical properties enhanced after sintering by a number of secondary heat treatment operations including the following:

- Annealing
- Case hardening
- Carbonitriding
- Carburizing
- Nitrocarburizing
- Nitriding
- Hardening
- Normalizing
- Precipitation hardening
- Sinter hardening
- Steam treating
- Stress relief
- Tempering

Although it is possible in some cases to perform the heat treating step as part of the cooling from sintering temperature, as is the case in sinter hardening, it is more common to perform the heat treatment operation as a separate step after sintering.

Ultimate tensile strength, hardness, wear, corrosion resistance, and compressive strength can, in general, be improved by heat treating; while other properties, such as impact resistance and ductility may be adversely affected. With this in mind, the selection of material chemistry and part process parameters are critical considerations for the successful application of heat treatment techniques to P/M parts.

The P/M factor is a term used to describe a multitude of variables which influence the heat treatment of ferrous P/M parts. The most critical of these parameters include the following:

- Part density
- Material composition
- Quenching or cooling method
- Process factors
- Equipment induced variables

Factors such as the type of base iron or steel powder, as well as the amount and type of alloy additions and sintering parameters are unique to the P/M industry. When planning or executing a secondary heat treatment operation, the most important variables to consider are density, microstructure, carbon and alloy content, and the heat treatment cycle.

The quench media and the hardenability of the material have a significant influence on as quenched properties. Oil quenching, though less severe than water or brine, is preferred due to improved distortion control and minimized cracking. Control of oil temperature assures load to load consistency and the use of a *fast oil* (9–11 second) is preferred because of improved heat transfer characteristics. Since P/M parts can absorb up to 3% oil (by weight) subsequent cleaning operations can be difficult. Incomplete cleaning leads to the creation of a great deal of smoke during tempering and potential safety concerns with respect to breathing these fumes and a concern about fire due to the presence of large amounts of oil in the tempering furnace and/or ventilation ducts.

Quenching in water, brine, or polymer as an alternative to oil can improve the rate of heat transfer but in many cases accelerates part corrosion due to residual fluid trapped near the surface. For this same reason, salt quenching can also create problems.

Temperature is one of the process variables that must be taken into consideration in secondary heat treatment operations. In certain applications, such as hardening and case hardening, it must be high enough to fully austenitize the material so that it will quench to a martensitic structure. Oil quenching, e.g., may require a higher austenitizing temperature to achieve a structure similar to water or brine. It is also important to note that some secondary operations, such as tempering or steam treating, do not raise the parts to austenitizing temperature. However, the uniform distribution and dissipation of heat is a major factor in the consistency of the end product.

Time is another process variable which influences secondary heat treatment. Soak times up to 50% longer than wrought materials are typical. This is due to lower thermal conductivity of the porous P/M material.

18.2.9 Special Heat Treatments

Austempering. In the austempering process, the steel is austenitized, quenched in molten salt and held at a temperature above which martensite will form so as to promote the transformation to bainite (Fig. 18.17). No tempering is required. Stresses are minimized due to the nature of the process, as the entire part reaches a uniform temperature and the transformation to bainite takes place almost in an isothermal condition. Austempering is another hardening treatment designed to reduce distortion and cracking in higher carbon steels. Toughness is greatly improved (relative to martensite) at the same hardness level.

Martempering. Martempering or interrupted quenching, is a hardening technique that consists of quenching to a temperature above where martensite will begin to form, holding for a sufficient time for the temperature to become uniform and then air cooling through the martensite transformation range to room temperature (Fig. 18.18). Tempering is performed as required. A steel suitable for martempering must have sufficient hardenability to compensate for both the reduced cooling rate and the fact that no other transformation product other than martensite is formed.

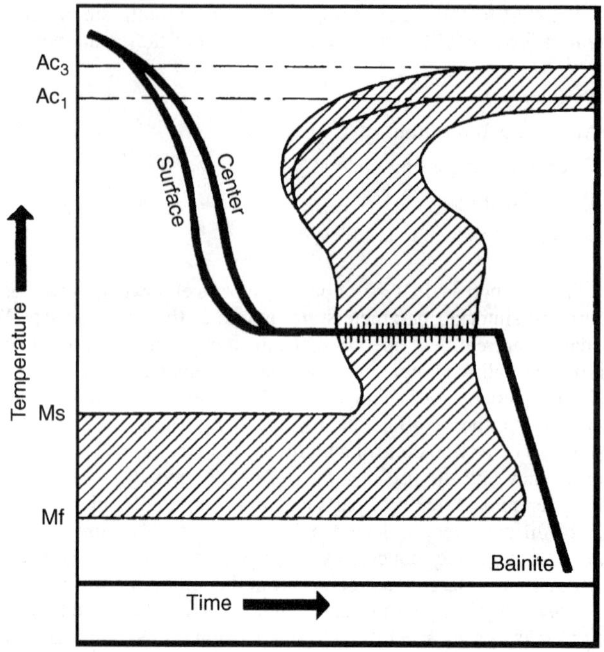

FIGURE 18.17 Austempering cycle.[17]

The equalization of temperature throughout the part prior to martensite formation ensures that the transformation stresses across the part will be minimal. The tendency for cracking and distortion are dramatically reduced.

Stress Relief. Stress relief involves heating to a temperature below the lower transformation temperature, as in tempering, holding long enough to reduce residual stress and cooling slowly enough, usually in air, to minimize the development of new residual stresses. Stress relief heat treating is used to relieve internal stresses locked in parts as a consequence of the various manufacturing steps.

A number of thermal and mechanical processes produce residual stresses which left unchecked may cause cracking, distortion, or premature failure in service below the design life of the component. One of the major sources of thermal stress is the (differential) cooling of the component from austenitizing temperature. Machining, cold working, drawing, extrusion, forging, welding, and heading operations are major sources of manufacturing induced stress.

The objective of stress relieving is not to produce major changes in mechanical properties by recrystallization as do the subcritical annealing treatments (see Sec.18.2.1). Rather, the stress relief is accomplished by recovery mechanisms that precede recrystallization. This allows the relief of residual stresses without significantly changing mechanical properties.

Tempering. Virtually all steels that are quench hardened or normalized are subjected to tempering, a process that takes place below the lower transformation temperature of the material primarily to increase ductility and toughness, but also to increase grain size. Tempering is also used to reduce hardness developed during welding and relieve stresses induced by forming and machining.

Tempering temperature, time at temperature, cooling rate from tempering temperature, and steel chemistry are the variables associated with tempering that affect the mechanical properties

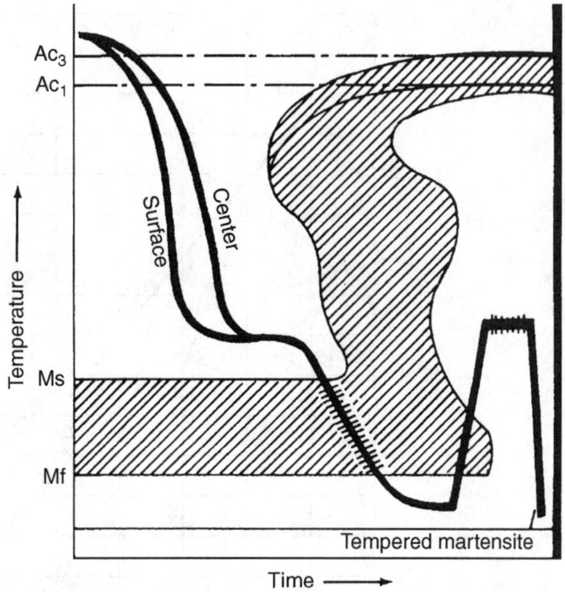

FIGURE 18.18 Martempering cycle.[17]

and microstructure of the finished part (Fig. 18.19). Changes to the microstructure by tempering typically decrease hardness, and strength (tensile and yield) while increasing ductility and toughness.

Tempering is said to occur in three distinct stages in overlapping temperature ranges as shown in Table 18.12. The precipitation of finely dispersed alloy carbides responsible for secondary hardening (in highly alloyed steels) is sometimes referred to as the fourth stage of tempering. The microstructure of steel quenched to a form martensite is highly unstable being in a strain induced state. The change of martensite during tempering into a mixture of cementite and ferrite typically results in a decrease in volume as a function of increasing tempering temperature.

Tempering time as well as temperature is an important consideration especially for the diffusion of carbon and alloying elements necessary for the formation of carbides.

Another factor that can affect the properties of steel is the cooling rate from tempering temperature. Although tensile properties are not affected, toughness (as measured by impact testing) can be decreased if the steel is cooled too slowly through a temperature range from 705°F (375°C) to 1065°F (575°C), especially in steels that contain carbide forming elements. Elongation and reduction in area may also be affected. This phenomenon is known as "temper embrittlement." Temper embrittlement will also occur if carbon and low alloy steels are held for an extended period of time in these temperature ranges. De-embrittlement can be performed by reheating to 1065°F (575°C), holding a short period of time and cooling rapidly. Several other forms of embrittlement are blue brittleness, tempered martensite embrittlement, and hydrogen embrittlement.

Temper color (or coloration) of steel is the result of a thin, tightly adhering oxide that forms when steel is heated and exposed to air (or an oxidizing atmosphere) for a short period of time. The color and thickness of the oxide layer varies with both tempering time and temperature. Table 18.13 below lists the various temper colors. The colors produced are typically not uniform, because of surface condition and fluctuation of temperature. Different steel chemistry also results in a variation in color.

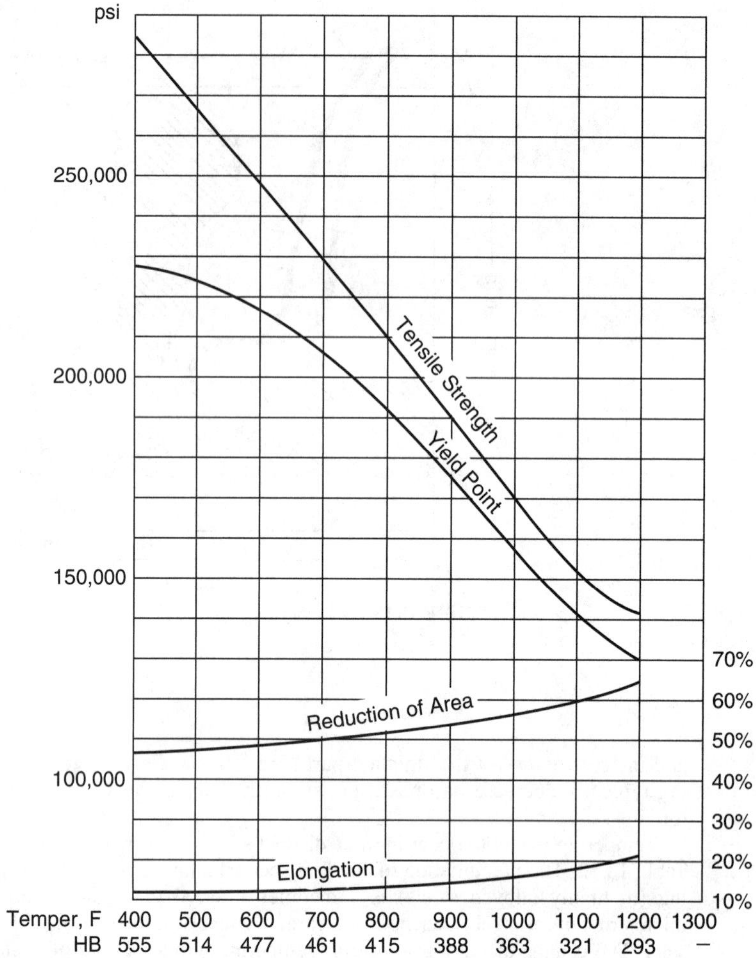

FIGURE 18.19 Changes in mechanical properties as a function of tempering temperature.

TABLE 18.12 Tempering Stages

Stage	Temperature range	Characteristic
One	210–480°F (100–250°C)	The formation of a transition carbide and the lowering of the carbon content of the martensite to approximately 0.25% C.
Two	390–570°F (200–300°C)	The transformation of retained austenite to ferrite and cementite (Fe_3C).
Three	480–660°F (250–350°C)	The replacement of transition carbide and low temperature martensite by cementite and ferrite.

Source: From Krauss.[17]

TABLE 18.13 Coloration of Steel

Temperature °F (°C)	Temper color
400 (205)	Faint straw yellow
425 (220)	Light straw yellow
440 (225)	Straw yellow
475 (245)	Deep straw yellow/light brown
500 (260)	Orange/brown
520 (270)	Bronze/dark brown
525 (275)	Light purple
545 (285)	Purple/peacock
560 (295)	Dark blue
590 (310)	Full blue
620 (325)	Grey
660 (350)	Grey-purple
705 (375)	Grey-blue
750 (400)	Dull grey
>750 (>400)	Black*

*Oxide adherence issues occur above 750°F (400°C), with surfaces appearing first as a velvet textured oxide, progressing to loose, flaky scale.

Source: From Herring.[18]

18.3 NONFERROUS HEAT TREATMENT

18.3.1 Aluminum and Aluminum Alloys

Heat treating processes for aluminum and aluminum alloys involve solution treating, quenching, age hardening, annealing, stress relieving, and tempering. Heat treating is employed primarily to increase strength and hardness in certain wrought and cast alloys. A three-step process is typical:

1. Solution heat treatment
2. Quenching
3. Age hardening

The purpose of solution heat treatment is the dissolution of the maximum amount of soluble elements in the alloy into solid solution. The process consists of heating and holding the alloy at a temperature sufficiently high and for a long enough period of time to achieve a nearly homogenous solid solution in which all phases have dissolved. Care must be taken to avoid overheating or underheating. In the case of overheating, eutectic melting can occur with a corresponding degradation of properties such as tensile strength, ductility, and fracture toughness. If underheated, solution is incomplete, and strength values lower than normal can be expected. In certain cases, extreme property loss can occur. In general, a temperature variation of ±10°F (±6°C) from control setpoint is allowable but certain alloys require much tighter tolerances. The time at temperature is a function of the section thickness of the material and may vary from several minutes to many hours. The time required to heat a load to the treatment temperature also increases with section thickness and loading arrangement and thus the total cycle time must take into consideration these factors.

Rapid and uninterrupted quenching in water or polymer is, in most instances, required to avoid precipitation detrimental to mechanical properties and corrosion resistance. The solid solution formed by solution heat treatment must be cooled rapidly enough to produce a supersaturated solution at room temperature which provides the optimal condition for precipitation hardening.

Hardening is achieved either at room temperature (natural aging) or with a precipitation heat treatment (artificial aging) cycle. The same general rules used in solution heat treatment (temperature uniformity, time at temperature) apply for precipitation hardening.

In general, the principles and procedures for heat treating wrought and cast alloys are similar. For cast alloys, however, soak times and quenching media are often different due to the section sizes involved. Soak times are longer and quenchants such as boiling water are used to reduce quenching stresses in complex shapes.

Not all aluminum alloys are heat treatable. The strength of the nonheat treatable alloys is primarily due to cold working.

Annealing can be used for both heat treatable and nonheat treatable alloys to increase ductility with a slight reduction in strength. There are several types of annealing treatments dependent to a large part on the alloy type, initial structure, and temper condition. In annealing it is important to ensure that the proper temperature is reached in all portions of the load. The maximum annealing temperature is also important in achieving good results. Full annealing (temper designation "O") produces the softest, most ductile, and most versatile condition. For both heat treatable and nonheat treatable aluminum alloys, reduction or elimination of the strengthening effects of cold working is accomplished by heating at temperature from about 500°F (260°C) to 825°F (440°C) . The rate of softening is strongly temperature dependent; the time required can vary from a few hours at low temperature to a few seconds at high temperature. Partial annealing (or recovery annealing) is done on nonheat treatable wrought alloys to obtain intermediate mechanical properties.

Stress relief annealing can be used to remove the effects of strain hardening in cold worked alloys. No appreciable holding time is required after the parts have reached temperature. Stress relief annealing of castings provides maximum stability for service applications where elevated temperatures are involved.

Tempering can be performed on heat treatable aluminum alloys to provide the best combination of strength, ductility, and toughness. These may be classified as follows:

- Designation "O," annealed
- Designation "W," solution heat treated
- Designation "T," heat treated to provide stable tempers other than "O"
- Designation "F," fabricated
- Designation "H," strain hardened

Table 18.14. summerises the various "T" temper conditions.

TABLE 18.14 "T" Temper Designations

Designation	Process
T1	Cooled from an elevated temperature shaping process and naturally aged
T2	Cooled from an elevated temperature shaping process, cold worked, and naturally aged
T3	Solution treated, cold worked, and naturally aged
T4	Solution treated and naturally aged
T5	Cooled from an elevated temperature shaping process and artificially aged
T6	Solution heat treated and artificially aged
T7	Solution heat treated and stabilized
T8	Solution heat treated, cold worked, and artificially aged
T9	Solution heat treated, artificially aged, and cold worked
T10	Cooled from an elevated temperature shaping process, cold worked, and artificially aged
Tx51	Stress relieved by stretching
Tx510	Stress relieved by stretching with no subsequent straightening
Tx511	Stress relieved by stretching with minor subsequent straightening
Tx52	Stress relieved by compressing
Tx54	Stress relieved by combining stretching and compressing
T42	Solution heat treated from the "O" or "F" temper and naturally aged
T62	Solution heat treated from the "O" or "F" temper and artificially aged

Source: From Unterwieser.[4]

18.3.2 Copper and Copper Alloys

Heat treating processes for copper and copper alloys involve homogenizing, annealing, stress relieving, solution treating, age (precipitation) hardening, quench hardening, and tempering.

Homogenizing involves raising the temperature of the metal high enough and holding at temperature long enough to eliminate or decrease significantly chemical segregation, or coring, in an alloy which required hot or cold working. Alloys of copper, such as tin bronzes, copper nickel alloys, and silicon bronzes are all subject to segregation. Homogenizing is primarily applied to copper alloys to improve hot and cold ductility. Temperatures for homogenization are dependent on the alloy analysis but generally are around 1400°F (760°C).

Annealing softens and increases the ductility and/or toughness of copper and copper alloys. Cold working of these alloys will usually cause work hardening. The alloy hardness then becomes dependent on the degree and amount of cold work. If the work hardening is not reduced by annealing, the alloy could fail in service. The temperatures for the annealing of pure electrolytic copper varies between 480°F (250°C) and 1250°F (675°C). The following factors are to be considered when annealing copper and its alloys:

- Amount of cold working
- Temperature
- Time

The reasons for these considerations are as follows:

- Impurities will affect the characteristics of the metal. The impurities are usually deliberately added to make the metal flow more freely on melting or to raise the softening temperature or to control the grain size. However, all have an effect on the annealing temperature.
- The increase of cold working prior to annealing will lower the recrystallizaton temperature.
- Time and temperature are both related and integrated in annealing. They are usually selected on the basis of specification or productivity, or economic considerations. The higher in temperature, the greater the grain growth, as with steels. If a high temperature is selected, then the time at temperature needs to be reduced.

Some of the copper and copper alloys can have their tensile strength and impact values improved by solution heat treating and age (precipitation) hardening. Solution treating temperatures are selected primarily because of the analysis of the material and achieved by raising the alloy to a high temperature, usually around 1400–1500°F (760–815°C), and quenching in a suitable solution of water based polymer quenchant or molten salt. After solution treatment, these alloys are age hardened by a precipitation treatment that involves reheating to temperatures between 350–480°F (175°C–250°C). The higher the temperature, the faster the precipitation hardening will occur.

18.3.3 Magnesium and Magnesium Alloys

Heat treating processes for magnesium and magnesium alloys involve annealing, stress relieving, solution treating, quenching, and age hardening. Most magnesium alloys are heat-treated to improve mechanical properties (strength, toughness, shock resistance) or as a means of conditioning prior to a specific fabrication technique. The type of heat treatment is dependent on the alloy composition, form (wrought or cast), and on the anticipated service condition.

Annealing is usually performed at temperature between 550°F (290°C) and 850°F (455°C) to reduce strain hardening or tempering. Since most forming operations on magnesium are done at elevated temperatures, the need for fully annealed wrought material is less than for other materials.

Stress relieving is performed on magnesium and magnesium alloys after shaping, forming, straightening, and welding operations that result in residual stresses from cold or hot working. The necessity to avoid distortion (warpage) in magnesium casting prior to machining and to prevent stress-corrosion

TABLE 18.15 Heat Treatments for Magnesium and Magnesium Alloys

Designation	Process	Alloys
F	As fabricated	—
O	Annealed and recrystallized (wrought only)	—
H	Strain hardened (wrought only)	—
H1	Strain hardened only	—
H2	Strain hardened and partially annealed	—
H3	Strain hardened and stabilized	—
W	Solution heat treated, unstable temper	—
T	Heat treat to produce stable tempers other than "F," "O," or "H"	—
T2	Annealed (cast only)	—
T3	Solution heat treated and cold worked	—
T4	Solution heat-treated	AM100A, AZ63A, AZ81A, AZ91c, AZ92A, ZK61A
T5	Artificially aged only	ZC71A, AZ80A, HM21A, HM31A, ZK60A, AM100A, AZ63A, EZ33A, HZ32A, ZE41A, ZH62A, ZK51A
T6	Solution heat-treated and artificial age	ZC71A, AM100A, AZ63A, AZ91C, AZ93A, EQ21A, HK31A, QE22A, QH21A, WE43A, WE54A, ZC63A, ZE63A, ZK61A
T7	Solution heat-treated and stabilized	—
T8	Solution heat-treated, cold worked, and artificially aged	HN31A
T9	Solution heat-treated, artificially aged and cold worked.	—
T10	Artificially aged and cold worked	—

Source: From Chandler.[5]

cracking in certain welded alloys makes it mandatory that the casting be stress relieved. Heat Treatments commonly used for wrought and cast alloys are shown in Table 18.15 below.

18.3.4 Nickel and Nickel Alloys

Heat treating processes for nickel and nickel alloys involve annealing and stress relieving. In many ways nickel and nickel alloys are easier to heat treat than many of the iron-based alloys that depend on complex microstructural changes. Nickel is an austenite former and as such no allotropic phase transformations occur.

Fabricating operations such as deep drawing, rolling, bending, or spinning requires softening before cold working can be continued. Annealing, that results in recrystallization, is used to produce complete softening of the material. Stress relieving can also be done after the fabrication operations. Table 18.16 summarizes the heat treatments involved.

18.3.5 Superalloys

Heat treating processes for superalloys involve stress relieving, annealing, solution heat treatment, quenching, and age hardening. Superalloys, whose service applications are among the most severe, can be divided into three distinct classifications of heat resistant alloys:

1. Nickel based
2. Iron–nickel based
3. Cobalt–nickel based

TABLE 18.16 Heat Treatments for Nickel and Nickel Alloys

	Soft annealing							
	Continuous				Batch			
	Temperature		Time	Cooling	Temperature		Time	Cooling
Material	°C	°F	(min)	method†	°C	°F	(h)	method
Nickel 200	815–925	1500–1700	0.5–5	AC or WQ	705–760	1300–1400	2–6	AC
Nickel 201	760–870	1400–1600	0.5–5	AC or WQ	705–760	1300–1400	2–6	AC
Monel 400	870–980	1600–1800	0.5–15	AC or WQ	760–815	1400–1500	1–3	AC
Monel R-405	870–980	1600–1800	0.5–15	AC or WQ	760–815	1400–1500	1–3	AC
Monel K-500	870–1040	1600–1900	0.5–20	WQ	870–1040	1600–1900	1–3	WQ
	Stress relieving				Stress equalizing			
	Temperature		Time	Cooling	Temperature		Time	Cooling
Material	°C	°F	(min)	method	°C	°F	(h)	method
Nickel 200	480–705	900–1300	0.5–120	AC	260–480	500–900	1–2	AC
Nickel 201	480–705	900–1300	0.5–120	AC	260–480	500–900	1–2	AC
Monel 400	540–650	1000–1200	0.5–120	AC	230–315	450–600	1–3	AC
Monel R-405	—	—	—	—	—	—	—	—
Monel K-500	—	—	—	—	—	—	—	—

†AC indicates aircooled; WQ indicates water quenched.
Source: From Chandler.[5]

Stress relieving of superalloys usually involves a compromise between maximum reduction of residual stresses and effects deleterious to high temperature performance properties and/or corrosion resistance. Time and temperature cycles vary considerably depending on the metallurgical characteristics of the alloy and on the type of and magnitude of the residual stresses developed in prior fabrication techniques.

Annealing is mainly used to reduce hardness and increase ductility to facilitate fabrication operations. Complete recrystallization, or full annealing, is done on most nonhardening wrought alloys. For most age hardenable alloys, the annealing cycle is the same as the solution heat treatment cycle but is done for resolution of second phases.

Solution heat treatment is intended to dissolve second phases to produce maximum corrosion resistance or to prepare for age hardening. For wrought superalloys a lower solution temperature is used for optimum short time tensile properties at elevated temperature, improved fatigue resistance via finer grain size, or improved resistance to notch rupture sensitivity. A higher solution temperature is specified for optimum creep-rupture properties. More grain coarsening and dissolution of carbides will take place at the higher temperatures.

As previously mentioned, the purpose of quenching after solution treating is to maintain, at room temperature, the supersaturated solution obtained at solution temperatures. Quenching also permits a finer age hardening precipitate size. Oil and water quenching are used as well as forms of air and inert gas cooling.

Factors that influence the selection and number of aging steps and aging temperatures include

- Type and number of precipitating phases
- Anticipated service temperature
- Precipitate size
- The combination of strength and ductility desired

Table 18.17 provides typical solution heat treatment and age hardening cycles for common wrought superalloys.

TABLE 18.17 Solution Treatment and Age Hardening Cycles for Wrought Superalloys

	Solution treating				Aging			
	Temperature				Temperature			
Alloy	°C	°F	Time (h)	Cooling procedure	°C	°F	Time (h)	Cooling procedure
Iron-base alloys								
A-286	980	1800	1	Oil quench	720	1325	16	Air cool
Discaloy	1010	1850	2	Oil quench	730	1350	20	Air cool
					650	1200	20	Air cool
N-155	1165–1190	2125–2175	1	Water quench	815	1500	4	Air cool
Incoloy 903	845	1550	1	Water quench	720	1325	8	Furnace cool
					620	1150	8	Air cool
Incoloy 907	980	1800	1	Air cool	775	1425	12	Furnace cool
					620	1150	8	Air cool
Incoloy 909	980	1800	1	Air cool	720	1325	8	Furnace cool
					620	1150	8	Air cool
Incoloy 925	1010	1850	1	Air cool	730*	1350*	8	Furnace cool
					620	1150	8	Air cool
Nickel-base alloys								
Astroloy	1175	2150	4	Air cool	845	1550	24	Air cool
	1080	1975	4	Air cool	760	1400	16	Air cool
Custom Age 625 PLUS	1038	1900	1	Air cool	720	1325	8	Furnael cool
					620	1150	8	Air cool
Inconel 901	1095	2000	2	Water quench	790	1450	2	Air cool
					720	1325	24	Air cool
Inconel 625	1150	2100	2	†	—	—	—	—
Inconel 706	925–1010	1700–1850	—	—	845	1550	3	Air cool
					720	1325	8	Furnace cool
					620	1150	8	Air cool
Inconel 706(c)	980	1800	1	Air cool	730	1350	8	Furnace cool
					620	1150	8	Air cool
Inconel 718	980	1800	1	Air cool	720	1325	8	Furnace cool
					620	1150	8	Air cool
Inconel 725	1040	1900	1	Air cool	730*	1350	8	Furnace cool
					620	1150	8	Air cool
Inconel X-750	1150	2100	2	Air cool	845	1550	24	Air cool
					705	1300	20	Air cool
Nimonic 80A	1080	1975	8	Air cool	705	1300	16	Air cool
Nimonic 90	1080	1975	8	Air cool	705	1300	16	Air cool
René 41	1065	1950	1/2	Air cool	760	1400	16	Air cool
Udimet 500	1080	1975	4	Air cool	845	1550	24	Air cool
					760	1400	16	Air cool
Udimet 700	1175	2150	4	Air cool	845	1550	24	Air cool
	1080	1975	4	Air cool	760	1400	16	Air cool
Waspaloy	1080	1975	4	Air cool	845	1550	24	Air cool
					760	1400	16	Air cool
Cobalt-base alloys								
S816	1175	2150	1	*	760	1400	12	Air cool

* To provide adequate quenching after solution treatment, cool below 540°C (1000°F) rapidly enough to prevent precipitation. Air cool thin sheet, oil or water quench heavier sections not subject to cracking.

18.3.6 Titanium and Titanium Alloys

Heat treating processes for titanium and titanium alloys involve stress relieving, annealing, solution heat treatment, quenching, and age hardening. Titanium and titanium alloys are heat-treated for the following purposes:

- To reduce residual stress developed during fabrication
- To produce an acceptable combination of properties (ductility, machinability, and dimensional stability)
- To increase strength
- To optimize special properties such as fracture toughness, fatigue strength, and high temperature creep strength

The response of these alloys to heat treatment depends on the composition.

Titanium and titanium alloys can be stress relieved without adversely affecting strength or ductility. Separate stress relief operations can be omitted when the manufacturing sequence can be adjusted to employ annealing or hardening as the stress relieving process. Table 18.18 presents combinations of time and temperature that are used for stress relieving.

Many titanium alloys are placed into service in the annealed condition. Since improvement in one or more properties is generally obtained at the expense of some other property, the annealing cycle should be selected according to the objective of the treatment. Common annealing processes include the following:

- Mill annealing
- Recrystallization annealing
- Beta annealing

Mill annealing is a general purpose heat treatment given to all mill products. It is not a full anneal, often leaving traces of cold or hot working in the microstructure. *Recrystallization annealing* is used

TABLE 18.18 Selected Stress Relief Treatments for Titanium and Titanium Alloys

Alloy	°C	°F	Time (h)
Commercially pure Ti (all grades)	480–595	900–1100	1/4–4
Alpha or near-alpha titanium alloys			
Ti-5Al-2.5Sn	540–650	1000–1200	1/4–4
Ti-8Al-1Mo-1V	595–705	1100–1300	1/4–4
Ti-6Al-2Sn-4Zr-2Mo	595–705	1100–1300	1/4–4
Ti-6Al-2Cb-1Ta-0.8Mo	595–650	1100–1200	1/4–2
Ti-0.3Mo-0.8Ni (Ti Code 12)	480–595	900–1100	1/4–4
Alpha–beta titanium alloys			
Ti-6Al-4V	480–650	900–1200	1–4
Ti-6Al-6V-2Sn (Cu + Fe)	480–650	900–1200	1–4
Ti-3Al-2.5V	540–650	1000–1200	1/2–2
Ti-6Al-2Sn-4Zr-6Mo	695–705	1100–1300	1/4–4
Ti-5Al-2Sn-4Mo-2Zr-4Cr (Ti-17)	480–650	900–1200	1–4
Ti-7Al-4Mo	480–705	900–1300	1–8
Ti-6Al-2Sn-2Zr-2Mo-2Cr-0.25Si	480–650	900–1200	1–4
Ti-8Mn	480–595	900–1100	1/4–2
Beta or near-beta titanium alloys			
Ti-13V-11Cr-3Al	705–730	1300–1350	1/12–1/4
Ti-11.5Mo-6Zr-4.5Sn (Beta III)	720–730	1325–1350	1/12–1/4
Ti-3Al-8V-6Cr-4Zr-4Mo (Beta C)	705–760	1300–1400	1/6–1/2
Ti-10V-2Fe-3Al	675–705	1250–1300	1/2–2
Ti-15V-3Al-3Cr-3Sn	790–815	1450–1500	1/12–1/4

Parts can be cooled from stress relief by either air cooling or slow cooling.
Source: From Chandler.[5]

TABLE 18.19 Selected Annealing Data for Titanium and Titanium Alloys

Alloy	Temperature °C	Temperature °F	Time (h)	Cooling method
Commercially pure Ti (all grades)	650–760	1200–1400	1/10–2	Air
Alpha or near-alpha titanium alloys				
Ti-5Al-2.4Sn	720–845	1325–1550	1/6–4	Air
Ti-8Al-1Mo-1V	790*	1450*	1–8	Air or furnace
Ti-6Al-2Sn-4Zr-2Mo	900†	1650†	1/2–1	Air
Ti-6Al-2Cb-1Ta-0.8Mo	790–900	1450–1650	1–4	Air
Alpha-beta titanium alloys				
Ti-6Al-4V	705–790	1300–1450	1–4	Air or furnace
Ti-6Al-6V-2Sn (Cu + Fe)	705–815	1300–1500	3/4–4	Air or furnace
Ti-3Al-2.5V	650–760	1200–1400	1/2–2	Air
Ti-6Al-2Sn-4Zr-6Mo	‡	‡	—	—
Ti-5Al-2Sn-4Mo-2Zr-4Cr (Ti-17)	‡	‡	—	—
Ti-7Al-4Mo	705–790	1300–1450	1–8	Air
Ti-6Al-2Sn-2Zr-2Mo-2Cr-0.25Si	705–815	1300–1500	1–2	Air
Ti–8Mn	650–760	1200–1400	1/2–1	§
Beta or near-beta titanium alloys				
Ti-13V-11Cr-3Al	705–790	1300–1450	1/6–1	Air or water
Ti-11.5Mo-6Zr-4.5Sn (Beta III)	690–761	1275–1400	1/6–1	Air or water
Ti-3Al-8V-6Cr-4Zr-4Mo (Beta C)	790–815	1450–1500	1/4–1	Air or water
Ti-10V-2Fe-3Al	‡	‡	—	—
Ti-15V-3Al-3Cr-3Sn	790–815	1450–1500	1/12–1/4	Air

* For sheet and plate, follow by 1/4 h at 790°C (1450°F), then air cool.
† For sheet, follow by 1/4 h at 790°C (1450°C), then air cool (plus 2 h at 595°C or 1100°F, then air cool, in certain applications). For plate, follow by 8 h at 595°C (1100°F), then air cool.
‡ Not normally supplied or used in annealed condition.
§ Furnace or slow cool to 540°C (1000°F), then air cool.
Source: From Chandler.[5]

to improve toughness by heating the alloy into the upper end of the alpha–beta range, holding, and then very slow cooling. *Beta annealing* also uses a slow cool from temperatures above the beta transformation. Table 18.19 provides information on annealing treatments.

A wide range of strength levels can be obtained in alpha–beta or beta alloys by solution treatment and age hardening. The reason for this lies in the instability of the high temperature beta phase at lower temperatures. Table 18.20 provides typical solution heat treatment and age hardening cycles for common titanium alloys.

18.3.7 Zinc and Zinc Alloys

Heat treating processes for zinc and zinc alloys involve stress relieving and thermal stabilization.

Stresses can result from the rapid chilling rate inherent in zinc die castings resulting in minor property and dimensional changes over time, particularly if the castings are quenched from the die rather than air cooled. Stress relieving these materials can significantly improve their service life.

Similarly, stabilization helps tensile strength and hardness of certain die castings adversely affected by changes in wall thickness and by significant aging that takes place with time at room temperature. Creep strength is also reduced.

TABLE 18.20 Solution Treatment and Age Hardening Cycles for Titanium and Titanium Alloys

Alloy	Solution temperature °C	Solution temperature °F	Solution Time (h)	Cooling method
Alpha or near-alpha alloys				
Ti-8Al-1Mo-1V	980–1010*	1800–1850*	1	Oil or water
Ti-6Al-2Sn-4Zr-2Mo	955–980	1750–1800	1	Air
Alpha–beta alloys				
Ti-6Al-4V	955–970†,‡	1750–1775† ‡	1	Water
	955–970	1750–1775	1	Water
Ti-6Al-6V-2Sn (Cu + Fe)	885–910	1625–1675	1	Water
Ti-6Al-2Sn-4Zr-6Mo	845–890	1550–1650	1	Air
Ti-5Al-2Sn-2Zr-2Zr-4Mo-4Cr	845–870	1550–1600	1	Air
Ti-6Al-2Sn-2Zr-2Mo-2Cr-0.25Si	870–925	1600–1700	1	Water
Beta or near-beta alloys				
Ti-13V-11Cr-3Al	775–800	1425–1475	1/4–1	Air or water
Ti-11.5Mo-6Zr-4.5Sn (Beta III)	690–790	1275–1450	1/8–1	Air or water
Ti-3Al-8V-6Cr-4Mo-4Zr (Beta C)	815–925	1500–1700	1	Water
Ti-10V-2Fe-3Al	760–780	1400–1435	1	Water
Ti-15V-3Al-3Cr-3Sn	790–815	1450–1500	1/4	Air

Alloy	Aging temperature °C	Aging temperature °F	Aging time (h)
Alpha or near-alpha alloys			
Ti-8Al-1Mo-1V	565–595	1050–1100	—
Ti-6Al-2Sn-4Zr-2Mo	595	1100	8
Alpha—beta alloys			
Ti-6Al-4V	480–595	900–1100	4–8
	705–760	1300–1400	2–4
Ti-6Al-6V-2Sn (Cu + Fe)	480–595	900–1100	4–8
Ti-6Al-2Sn-4Zr-6Mo	580–605	1075–1125	4–8
Ti-5Al-2Sn-2Zr-2Zr-4Mo-4Cr	580–605	1075–1125	4–8
Ti-6Al-2Sn-2Zr-2Mo-2Cr-0.25Si	480–595	900–1100	4–8
Beta or near-beta alloys			
Ti-13V-11Cr-3Al	425–480	800–900	4–100
Ti-11.5Mo-6Zr-4.5Sn (Beta III)	480–595	900–1100	8–32
Ti-3Al-8V-6Cr-4Mo-4Zr (Beta C)	455–540	850–1000	8–24
Ti-10V-2Fe-3Al	495–525	925–975	8
Ti-15V-3Al-3Cr-3Sn	510–595	950–1100	8–24

* For certain products, use solution temperature of 890°C (1650°F) for 1 h, then air cool or faster.
† For thin plate or sheet, solution temperature can be used down to 890°C (1650°F) for 6 to 30 min, then water quench.
‡ This treatment is used to develop maximum tensile properties in this alloy.
Source: From Chandler.[5]

18.4 HEAT TREATING EQUIPMENT

Heat treating furnaces can be divided into two main types: batch and continuous.

The fundamental difference between these two styles is not in the materials of construction, although there are differences due to inherent design requirements. Instead, the key difference lies in how workloads are positioned in the units and how they interact with the atmosphere within the furnaces. The primary sources of energy to heat the equipment are (natural) gas and electricity. Alternative energy sources, such as oil and other hydrocarbon fuels are also used.

TABLE 18.21 Classification of Furnaces

Criteria	Distinguishing feature	Remarks
Method of heating	Combustion of fuel	Gas (natural, other hydrocarbon, manufactured, tank) or oil (tar)
—	Electricity	Electrical resistance (metallic, ceramic, other), electric arc (melting), electrical induction (heat treating, melting)
Method of handling charge	Batch	Work remains stationary
—	Continuous	Work moves continuously within the equipment
—	Intermittent	Work moves periodically
Internal atmosphere	Air	—
—	Other	Generated, synthetic, elemental, mix
Exposure of charge to atmosphere	Open	Exposed charge, single heat transfer
—	Closed	Muffle design (isolated charge, double heat transfer)
Type of hearth	Stationary	Slab, skid, rails
—	Moveable	Belt, car, roller, rotating table, screw, shaker
Liquid bath	Salt	—
—	Other	Molten lead and fluidized bed

Heat treating furnace equipment can further be divided into furnaces and ovens. Today, oven construction can be used in temperature applications up to 1400°F (760°C), although 1000°F (538°C) is a traditional upper limit. Oven technology utilizes convection heating, i.e., the circulation of air, products of combustion, or an inert gas, as the primary means to heat a workload to temperature. Oven construction also varies considerably from furnace construction. Furnaces can be classified in a number of ways as summarized in Table 18.21.

Batch units tend to involve large, heavy workloads processed for long periods of time. In a batch unit the work charge is typically stationary, so that interaction with changes in the furnace atmosphere is performed in near equilibrium conditions. Batch furnace types include

- Bell furnaces
- Box furnaces
- Car bottom furnaces
- Elevating hearth furnaces
- Fluidized bed furnaces
- Gantry furnaces
- Mechanized box furnaces (also called "sealed quench" or "integral quench" or "in-out" furnaces)
- Pit furnaces
- Salt pot furnaces
- Split or Wraparound furnaces
- Tip-up furnaces
- Vacuum furnaces

Of all the batch furnace types, integral quench furnaces are the most common.

In a continuous unit the workload is moving in some manner and the environment surrounding the workload changes dramatically as a function of the position of the work charge. Continuous furnace types include

- Cast link belt furnaces
- Humpback furnaces
- Mesh belt furnaces
- Rotary drum (rotary retort) furnaces
- Rotary hearth furnaces
- Shaker hearth furnaces

- Monorail furnaces
- Pusher furnaces
- Roller hearth furnaces
- Vacuum furnaces
- Walking beam furnaces

Of all the continuous furnace types, pusher furnaces are the most common.

In addition, there are a number of special purpose furnaces including the following:

- Continuous slab and billet heating furnaces
- Electron beam surface treatment equipment
- Induction heating systems
- Laser heat treating equipment
- Quartz tube furnaces
- Resistance heating systems
- Rotating finger furnaces
- Screw conveyor furnaces

Table 18.22 summarizes the different processes that can be used in each type of equipment.

18.4.1 Atmosphere Furnaces

Atmosphere furnaces are characterized by their use of a protective atmosphere to surround the work-load during heating and cooling. The most common furnace atmosphere, however, is air. Often times, nothing more is needed. When an air atmosphere is used, such as in a low temperature tempering operation, the final condition of the material's surface (or skin) is not considered important.

Furnace atmospheres play a vital role in the success of the heat treating process. It is important to understand why we use them and what the best atmosphere for a specific application is. There are many different types of atmospheres being used, and it is important to understand how a particular atmosphere is chosen as well as its advantages and disadvantages and to learn how to control them safely.

The purpose of a furnace atmosphere varies with the desired end result of the heat treating process. The atmospheres used in the heat treating industry have one of two common purposes:

1. To protect the material being processed from surface reactions, i.e., to be *chemically inert (or protective)*
2. To allow the surface of the material being processed to change, i.e., to be *chemically active (or reactive)*

The types of atmospheres used in heat treating furnaces are summarized in Table 18.23 below. Some atmospheres on the list such as argon and helium are often associated with vacuum furnaces and are used at partial pressure (pressure below atmospheric pressure). Others, such as sulfur dioxide are used for very special applications.

Generated atmospheres produce combinations of gases of specific composition prepared on site by use of *gas generators* that are designed for this purpose. The "feed stock" (what goes in, i.e., the hydrocarbon fuel gas used in combination with air to create the atmosphere) is typically natural gas or propane. During operation, the volume of protective atmosphere required for safe use in a particular heat treating furnace depends to a great extent on the following:

- Type and size of furnace
- Presence or absence of doors and/or curtains
- Environment (especially drafts)
- Size, loading, orientation, and nature of the work being processed
- Metallurgical process involved

TABLE 18.22 Common Applications of Heat Treating Furnaces

Furnace style	Application use
Bell	Aging, bluing, hardening, nitriding, solution heat treatment, stress relieving, tempering
Box	Aging, annealing, carburizing, hardening, malleabilizing, normalizing, solution heat treatment, stress relieving, tempering
Car bottom	Annealing, carburizing, hardening, homogenizing, malleabilizing, normalizing, spheroidizing, stress relieving, tempering
Cloverleaf	Annealing, carbon restoration, carbonitriding, carburizing, hardening, normalizing, tempering
Continuous slab	Carburizing, homogenizing, solution heat treatment
Conveyor	Austempering, annealing, brazing, carbon restoration, carbonitriding, carburizing, hardening, homogenizing, spheroidizing, tempering
Electron beam	Hardening (surface)
Elevator hearth	Aging, annealing, hardening, malleabilizing, solution heat treatment, stress relieving, tempering
Fluidized bed	Carbonitriding, carburizing, hardening, nitriding, nitrocarburizing, steam treating, tempering
Humpback	Annealing, brazing, hardening, stress relieving, sintering
Induction	Hardening, tempering
Integral quench	Austenitizing, annealing, carbon restoration, carbonitriding, carburizing, hardening, nitrocarburizing, normalizing, stress relieving, tempering
Ion	Carbonitriding, carburizing, nitriding, nitrocarburizing
Laser	Annealing
Monorail	Annealing, hardening, normalizing, stress relieving, tempering
Pit	Annealing, bluing, carbon restoration, carbonitriding, carburizing, hardening, homogenizing, nitrocarburizing, nitriding, normalizing, solution heat treatment, steam treating, stress relieving, tempering
Pusher	Annealing, carbon restoration, carbonitriding, carburizing, hardening, malleabilizing, metallizing, nitrocarburizing, normalizing, solution heat treatment, sintering, spheroidizing, stress relieving, tempering
Quartz tube	Hardening, sintering
Resistance heating	Aging, annealing, carbonitriding, hardening, normalizing, stress relieving
Roller hearth	Bluing, carbon restoration, carbonitriding, carburizing, hardening, malleabilizing, normalizing, solution heat treatment, spheroidizing, stress relieving, tempering
Rotating finger	Annealing, hardening, normalizing, stress relieving, tempering
Rotary hearth	Annealing, austempering, carbon restoration, carbonitriding, carburizing, hardening, tempering
Salt bath	Austempering, carbonitriding, carburizing, hardening, malleabilizing, martempering, nitrocarburizing, normalizing, tempering
Screw conveyor	Annealing, hardening, stress relieving, tempering
Shaker hearth	Annealing, carbonitriding, carburizing, hardening, normalizing, stress relieving, tempering
Split	Annealing, stress relieving
Tip-up	Annealing, hardening, malleabilizing, normalizing, spheroidizing, stress relieving, tempering
Vacuum	Annealing, brazing, carbon deposition, carbonitriding, carburizing, degassing, hardening, nitrocarburizing, normalizing, solution heat treatment, sintering, stress relieving, tempering
Walking beam	Annealing, hardening, normalizing, sintering, stress relieving, tempering

Source: From Greenberg.[19]

TABLE 18.23 Common Types of Furnace Atmospheres

Type	Chemical symbol	Remarks
Air	$N_2 + O_2$	Air is approximately 79% nitrogen and 21% oxygen
Argon	Ar	Argon is considered an inert gas
Carbon dioxide	CO_2	—
Carbon monoxide	CO	—
Custom blends	—	Alcohols, combinations of N_2 and other gases
Generated atmospheres	—	Endothermic, exothermic, dissociated ammonia
Helium	He	Helium is considered an inert gas
Hydrocarbons	CH_4, C_3H_8, C_4H_{10}	Methane (CH_4), Propane (C_3H_8), Butane (C_4H_{10})
Hydrogen	H_2	—
Nitrogen	N_2	—
Oxygen	O_2	—
Products of combustion	—	A mixture of a hydrocarbon fuel gas and air whose composition is dependent on the air/gas ratio
Steam	H_2O	Water vapor
Sulfur dioxide	SO_2	—
Synthetic atmospheres	—	Nitrogen/methanol
Vacuum	—	Vacuum is the absence of an atmosphere

In all cases, the manufacturer's recommendations should be followed since they have taken these factors into account during the design of the equipment. Remember that to purge air out of a furnace prior to introduction of a combustible furnace atmosphere requires a minimum of five volume changes of the chamber. This is to ensure that the oxygen content of the chamber is below 1% prior to the introduction of the atmosphere.

Generated atmospheres are classified according to the relative amounts of the individual gases produced. Table 18.24 below provides a list of these classifications according to the American Gas Association (AGA). The gases are divided into six main classes.

Exothermic reactions are heat producing while endothermic reactions require heat to promote the reaction. The composition of the atmospheres produced can be changed in a number of ways. Varying the gas/air ratio, or using a different feed stock (e.g., natural gas or propane) will cause the chemistry of the gas to change.

TABLE 18.24 Classification of Gases

Class	Base type	Description
100	Exothermic	An atmosphere created from the products of partial or complete combustion of an air–gas mixture in a water cooled combustion chamber.
200	Nitrogen	A prepared atmosphere using an exothermic base with a large percentage of the carbon dioxide and water vapor removed.
300	Endothermic	An atmosphere created by partial reaction of an air–gas mixture in an externally heated, *catalyst* filled chamber.
400	Charcoal	Uncommon today. Formed by passing air over a bead of incandescent charcoal.
500	Exothermic–Endothermic	An atmosphere created by complete combustion of a mixture of gas and air, removing a large percentage of the water vapor, and reforming most of the carbon dioxide to carbon monoxide by reaction with fuel gas in an externally heated catalyst reactor.
600	Ammonia	Any atmosphere created using ammonia as the primary constituent, including *nascent* (raw) ammonia, dissociated ammonia, or partially or completely combusted ammonia with a large percentage of the water vapor removed.

18.4.2 Ovens

Ovens may be designed for intermittent loading, that is, one batch at a time, or for a continuous flow of work using some form of conveyance through the unit. Oven equipment sizes vary dramatically, from small bench top units in laboratory environments to huge industrial systems with thousands of cubic feet (cubic meters) of capacity. Ovens operate with air atmospheres but may be designed to contain special atmospheres, such as nitrogen or argon, or incorporate special construction, such as adaptations for retorts, that allow the use of special atmospheres for the processing of very specialized applications.

The source of heat may be derived from combustion of fuel or electricity. Heat is transferred to the work primarily by natural gravity or forced convection, or by radiant sources if the temperature is high enough. As stated earlier, oven construction can be used in temperature applications up to 1400°F (760°C), although temperature ratings of 1250°F (675°C) or 1000°F (538°C) are more common.

Selection of the type of oven involves the careful consideration of several variables:

- Quantity of material to be processed
- Uniformity in size and shape of the product
- Lot size
- Temperature tolerances
- Effluent evolution, if any

Batch systems may be classified as follows:

- Bell
- Bench top
- Cabinet
- Truck
- Walk-in

Continuous systems include the following components:

- Belt
- Drag chains
- Monorail
- Pusher
- Roller hearth
- Rotary drum (or retort)
- Screw
- Walking beam

There are several design criteria for oven construction including the following:

- Operating temperature
- Heating method
- Thermal expansion of materials
- Atmospheres
- Airflow patterns

The range of operating temperature is one of the main determinants of oven construction. Typically, all ovens are constructed of a double wall of sheet metal with insulation and reinforcing members sandwiched between the sheets. The insulation may be glass fiber, mineral wool, or lightweight fiber material. The sheet metal lining for ovens may be of low carbon steel, galvanized steel, zinc-gripped steel, aluminized steel, or stainless steel depending on the temperature requirement.

Several distinct changes occur in oven construction as the temperature increases. Problems with expansion and interior (heat and atmosphere) sealing become much more significant at higher temperatures. For example, an oven designed for operation at 400°F (205°C) will have mineral wool insulation, 4" (100 mm) thick. By contrast, for a 700°F operating temperature, a thickness of 7° (175 mm) is required. Thermal expansion in large ovens is generally compensated for by the use of telescoping panel joints in the walls, ceiling, and floor. Door construction must incorporate similar expansion joints.

The type and quantity of airflow is important. For example, ovens designed for handling explosive volatiles such as paint drying or solvent extraction have special considerations including large air flow volumes to dilute the volatile, explosion relief hatches, purge cycles, powered exhausters, airflow safety switches, and fresh air dampers. Several different patterns of air flow can be used depending on the workload configuration. These are

- Horizontal
- Vertical
- Combination uniflow

The method of heating an oven often depends not only on the availability of a particular fuel but also on the process itself. Many processes cannot tolerate products of combustion from direct fired systems, so indirect (radiant tube) firing or alternate energy sources need to be considered. In addition, some means of heat transfer, such as microwave heating, are severely limited in the type of product that can be processed. Ovens are commonly heated by fuel (natural gas or other hydrocarbons), steam, or electricity. Infrared heating and microwave (radio frequency) can also be used.

18.4.3 Vacuum Furnaces

Vacuum furnaces can be classified, according to the mode of loading, into horizontal and vertical furnaces and can be batch or continuous (multi-chamber) designs. A large number of configurations exist that are described in detail in the literature.[1]

Heat treatment in vacuum furnaces is characterized by special conditions with regard to the design of the furnaces as well as the control of temperature and vacuum level during the heat treatment. The design of the furnaces generally depends on the size of the load, the pressure and temperature to be attained, and the medium to be used in cooling the load. The main parts of a vacuum furnace include

- Vessel
- Pumping system
- Hot zone
- Cooling system

Vacuum furnace vessels can be grouped into so-called hot wall and cold wall designs. A typical hot wall furnace has a retort that is commonly metallic or ceramic dependent on the temperature. The heating system is usually located outside of the retort and consists of resistive heating elements or an induction coil. Limitations of this retort-type furnace are the restricted dimensions of the heating zones and the restricted temperature range of the metallic retort, usually limited to 2000°F (1100°C). With cold wall furnaces, the vacuum vessel is cooled with a cooling medium (usually water) and is kept near ambient temperature during high temperature operations.

In comparison to the hot wall furnace, the features of the cold-wall furnace are as follows:

- Lower heat losses and less heat load released to the surroundings
- Faster heating and cooling performance
- Greater temperature uniformity control

Higher operating temperature ranges with 2400°F (1315°C) standard and 3000°F (1650°C) or higher practical. A disadvantage over the retort design is the greater absorption of gases and water vapors on the cooled furnace walls and in the insulation after opening of the furnace. The cold wall vacuum furnace has become the dominant design for high temperature furnaces since the late 1960s. The construction of the pumping system depends on the following factors:

- Volume of the vessel
- Surface area of the vessel and the type of furnace internals
- Outgassing of the workload and related fixturing
- Time required for evacuation down to the final pressure

It is important to note that the pumping system must maintain the process vacuum level without being overwhelmed by the outgassing of the workload. Pumping systems are usually divided into two subsystems, pumps for rough vacuum (micron range) and pumps for high vacuum (submicron range). For certain applications a single pumping system can handle the entire range and cycle. The pumps themselves are usually classified in two general categories, mechanical pumps and diffusion pumps. There are other specialized types of vacuum pumps for use in achieving higher vacuum ranges such as ejectors, ion pumps, cryo pumps, turbo-molecular pumps, and chemical getter pumps.

For the insulation of the heating chamber, or hot zone, the following designs and materials are in common use:

- All metallic (radiation shields)
- Combination radiation shields and other (ceramic) insulating material
- Multiple-layer (sandwich) insulation
- All graphite (board, fiber, carbon–carbon composite)

Radiation shields are manufactured from following metals and alloys

- Tungsten or tantalum having a maximum operating temperature of 4350°F (2400°C)
- Molybdenum having a maximum operating temperature of 3100°F (1700°C)
- Stainless steel or nickel alloys having a maximum operating temperature of 2100°F (1150°C)

Most metallic designs consist of a combination of materials, e.g., three molybdenum shields backed by two stainless steel shields would be typical for 2400°F (1150°C) operation. Radiation shields adsorb only small amounts of gases and water vapors during opening of the furnace. They are, however, expensive to purchase and maintain and often require greater pumping capacity to remove any moisture trapped between the shields. Compared with other types of insulation, their heat losses are high and become higher with loss of emissivity (reflectivity) due to the gradual contamination of the shields.

Sandwich insulation is composed of one or more radiation shields typically with ceramic wool insulation between them. Combinations of graphite fiber sheets and ceramic insulation wool are also used. These versions are cheaper to buy and maintain, but absorb higher levels of water vapor and gases (due to the very large surface area of the insulation wool). Their heat losses are considerably lower than those of radiation shields.

Graphite fiber insulation designs cost somewhat more than sandwich insulation designs. However, as their heat losses are lower, a smaller thickness is sufficient. In these designs, the absorption of gases and water vapor is considerably reduced. Furthermore, the heating costs are lower, and the lifetime of this type of insulation is much longer. The maximum operating temperature is around 3630°F (2000°C). The lifetime depends strongly on the purity of the graphite. In some applications such as brazing, a sacrificial layer is used to protect the insulation beneath. For most heat treatments in vacuum furnaces, graphite insulation is used.

In general, the heating elements for heating systems in vacuum furnaces are made from one of the following materials:

- Nickel/Chromium alloys that can be used up to 2100°F (1150°C). Above 1475°F (800°C), there is a risk of evaporation of chromium.
- Silicon carbide with a maximum operating temperature of 2200°F (1200°C). There is a risk of evaporation of silicon at high temperatures and low vacuum levels.
- Molybdenum with a maximum operating temperature of 3100°F (1700°C). Molybdenum becomes brittle at high temperature and is sensitive to changes in emissivity brought about by exposure to oxygen or water vapor.
- Graphite can be used up to 3630°F (2000°C). Graphite is sensitive to exposure to oxygen or water vapor resulting in reduction in material thickness due to the formation of carbon monoxide (CO) that will be evacuated by the pumps. The strength of graphite increases with temperature.
- Tantalum has a maximum operating temperature of 4350°F (2400°C). Tantalum, like molybdenum, becomes brittle at high temperatures and is sensitive to changes in emissivity brought about by exposure to oxygen or water vapor.

Uniformity of temperature is of great importance to heat treatment results. The construction of the heating system should be such that temperature uniformity in the load during heating is optimal; it should be better than ±10°F (5°C) after temperature equalization. This is realized with single or multiple temperature control zones and a continuously adjustable supply of heating power for each zone.

In the lower temperature range, below 1550°F (850°C), the radiant heat transfer is low and can be increased by convection-assisted heating. For this purpose, after evacuation the furnace is backfilled with an inert gas up to an operating pressure of 1–2 bar, and a built-in convection fan circulates the gas around the heating elements and the load. In this way, the time to heat different loads, especially those with large cross section parts to moderate temperatures, typically under 1600°F (870°C), can be reduced by as much as 30–40%. At the same time the temperature uniformity during convection-assisted heating is much better, resulting in less distortion of the heat-treated part.

The following media (listed in order of increasing intensity of heat transfer) are used for the cooling of components in vacuum furnaces:

- Vacuum
- Sub-atmospheric cooling with a static or agitated inert gas (typically Ar or N_2)
- Pressurization (up to 20 bar or more) cooling with a highly agitated, recirculated gas (Ar, N_2, He, H_2, or mixtures of these gases)
- Oil, still or agitated

After heating in vacuum, the bright surface of the components must be maintained during the cooling. Today, sufficiently clean gases are available for cooling in gas. Permissible levels of impurities amount to approximately 2 ppm of oxygen and 5–10 ppm of water by volume. Normally nitrogen is used as a cooling medium because it is inexpensive and relatively safe.

With multichamber furnaces, such as a vacuum furnace with an integral oil quench, an additional cooling medium, namely oil, is also available. These oils are specially formulated (evaporation-resistant) for vacuum operation.

One variation worth noting are the plasma or ion furnaces. Plasma furnaces exist in all styles, horizontal in single or multiple-chamber configurations, as well as vertical designs such as bell furnaces and bottom loaders. The basic differences between these designs and conventional vacuum furnaces are the electrical isolation of the load from the furnace vessel via load support isolators; the plasma current feed-through; the high-voltage generator, which creates the plasma; and the gas dosage and distribution system. Plasma furnaces also utilize conventional vacuum furnace chamber and pumping systems.

Depending on the specific application, they are either low temperature furnaces of 1400°F (750°C) for plasma (ion) nitriding, or high temperature furnaces up to 2400°F (1100°C) for plasma (ion) carburizing. Low temperature furnaces for plasma nitriding are constructed as cold-wall or hot-wall furnaces. High temperature furnaces are usually cold-wall furnaces with water-cooled double walls. They can be equipped either with a high-pressure gas quench system or an integrated oil quench tank.

The generator needed to create a plasma glow discharge inside a plasma furnace has to be a high-voltage dc generator (up to 1000 V). Currently there are two types of generators in use; one type has continuous-current outputs and the other has pulsed-current output.

REFERENCES

1. Lohrmann M., and D. H. Herring, "Heat Treating Challenges in the 21st Century," *Heat Treating Progress,* June/July 2001.
2. *Properties and Selection of Metals* (Definitions of Metals and Metalworking), Vol. 1, *Metals Handbook,* 8th ed., ASM International, Materials Park, OH, 1961, pp. 1–41.
3. German, Randall M., *Powder Metallurgy of Iron and Steel,* Wiley, New York, 1998.
4. Unterwieser, Paul M., Howard E. Boyer, and James J. Kubbs, eds., *Heat Treaters Guide: Standard Practices and Procedures for Steel,* ASM International, Materials Park, OH, 1982.
5. Chandler, Harry, ed., *Heat Treater's Guide: Practices and Procedures for Nonferrous Alloys,* 2d ed., ASM International, Materials Park, OH, 1996.
6. *Practical Data for Metallurgists,* 14th ed., The Timken Company.
7. Herring D. H., "Jominy Testing: The Practical Side," Industrial *Heating,* October 2001.
8. Llewellyn D. T., and R. C. Hudd, *Steels: Metallurgy and Applications,* 3d ed., Butterworth Heinemann, Oxford, 1998.
9. Marrow, James, "Understanding the Jominy End Quench Test," *Industrial Heating* —, September 2001.
10. Thelning, Karl-Erik, *Steel and Its Heat Treatment,* Butterworths, London, 1975.
11. Grossmann, M. A., and E. C. Bain, *Principles of Heat Treatment,* 5th ed., American Society for Metals, Materials Park, OH, 1964, pp. 112–118.
12. *Heat Treating*, Vol. 4, *Metals Handbook,* ASM International, Materials Park, OH, 1991.
13. Payson P., "The Annealing of Steel," (series) *Iron Age,* June/July 1943.
14. Herring, D. H., *Fundamentals of Brazing,* Materials Engineering Institute, ASM International, Materials Park, OH, 2002.
15. Herring D. H., "Plasma Assisted Surface Treatments," *Heat Treating Progress,* March 2002.
16. Herring, D. H., and P. Hansen, "Heat Treating of Ferrous P/M Parts," *Advanced Materials and Processes,* April 1998.
17. Krauss G., *Steels: Heat Treatment and Processing Principles,* ASM International, Materials Park, OH, 1990.
18. Herring D. H., "Surface Oxidation Effects," *Heat Treating Progress,* December 2001.
19. Greenberg, J. H., *Industrial Thermal Processing Equipment Handbook,* ASM International, Materials Park, OH, 1994.

FURTHER READING

Bradley, Elihu F., ed., *Superalloys: A Technical Guide,* ASM International, Materials Park, OH, 1988.

Chandler, Harry, ed., *Heat Treater's Guide: Practices and Procedures for Irons and Steels,* 2d. ed., ASM International, Materials Park, OH, 1995.

Donachie, Jr., Matthew J., ed., *Titanium: A Technical Guide,* ASM International, Materials Park, OH, 1988.

Edenhofer, B., J. Boumann, and D. Herring "Vacuum Heat Treatment," Chap. 7 in *Steel Heat Treatment Handbook,* 2d ed., Marcel Dekker, New York, 2003.

Haga, L., *Practical Heat Treating,* Metal Treating Institute.

Haga, L., *Principles of Heat Treating,* Metal Treating Institute.

Heat Treating, Cleaning, and Finishing, Vol. 2, *Metals Handbook,* 8th ed., ASM International, Materials Park, OH, 1964.

Herring, D. H., "An Update on Low Pressure Carburizing Techniques and Experiences," 18th *Annual Heat Treating Conference Proceedings,* ASM International, Materials Park, OH.

Herring, D. H., "Comparing Carbonitriding and Nitrocarburizing," *Heat Treating Progress,* April/May 2002.

Melgaard, Hans L., *Ovens and Furnaces,* Metals Engineering Institute, ASM International, Materials Park, OH, 1977.

Otto, F., and D. H. Herring "Gear Heat Treatment," *Heat Treating Progress,* June and July/August 2002.

Schwartz, M. Mel, *Brazing,* ASM International, Materials Park, OH, 1987.

Totten, George E., and Maurice, A. H. Howes, *Steel Heat Treatment Handbook,* Marcel Dekker, New York, 1997.

CHAPTER 19
METALCASTING PROCESSES

Ian M. Kay
Vice President Education
Cast Metals Institute, Inc.
American Foundry Society, Inc.
Des Plaines, Illinois

19.1 INTRODUCTION

Although history does not record when or where the first metal casting was made or who produced it, artifacts and Biblical references indicate this probably occurred in ancient Mesopotamia about 5000–6000 years ago. The oldest casting in existence is believed to be a copper frog cast in Mesopotamia about 3200 B.C., and its complexity indicates that the artisans of that time had already developed considerable expertise in the art and science of metalcasting. Early historical information on metalcasting is difficult to find for two major reasons. First, the production of metal castings predates writing, thus there is little written record of the early events. Second, few metal artifacts exist since these were quite precious and the metal could be easily remelted and recast into other useful objects. You might say that metalcasters were the first to practice recycling.

Man learned to improve the properties of castings through alloying. Bronze (copper–tin) castings began to appear about 3000 B.C. The first production of iron castings is attributed to the Chinese about 1000 B.C. Cast crucible steel was made in India about 500 B.C., but the process was lost and not rediscovered until the late 1800s. Sir Humphrey Davy of Great Britain discovered the existence of aluminum in 1808, and, by 1884, the total U.S. production of aluminum was a mere 125 pounds per year. In 2003, the U.S. will produce 9.3 million tons of gray and ductile iron castings, 1.1 million tons of steel castings, 2.3 million tons of aluminum castings, and 314,000 tons of copper-base castings.

Newly developed casting processes have revolutionized the casting industry allowing metalcasting technology to advance further in the last 50 years than in the previous 5500 years. This astounding progress—achieving better cast surfaces, closer dimension tolerances, and faster turnaround times—has been primarily due to the ingenuity of the many developers of new processes as well as the application of modern scientific instrumentation to probe the science of sand technology, metallurgy, and metal solidification.

The metalcasting process consists of pouring molten metal into a mold containing a cavity of the desired shape of the casting, and allowing the metal to solidify. There are many processes used to make metal castings. What differentiates one process from another is the material from which the mold is made, the type of pattern used (permanent or expendable), and the amount of pressure, or force (positive or negative), that is used to fill the mold cavity with molten metal. Conventional green sand, chemically bonded, V-process and plaster molds utilize permanent patterns, but the mold is used only once. Permanent mold and diecasting dies are machined in metal or graphite, and the die can be used repeatedly. Investment casting and the lost foam process involve an expendable pattern as well as an expendable mold.

19.2 METALCASTING PROCESSES

19.2.1 Sand Casting Processes

Fundamentally, a mold is produced by shaping a suitable refractory material to form a cavity of desired shape, such that molten metal can be poured into this cavity. The mold cavity needs to retain its shape until the metal has solidified and the casting is removed from the mold. This sounds easy enough to accomplish, but depending on the choice of metal, certain definite characteristics are demanded of the mold. When granular refractory materials, such as silica, olivine, chromite, or zircon sands are used, the mold must be

- Strong enough in its construction to sustain the weight of the liquid metal
- Constructed to permit any gases formed within the mold or mold cavity to escape into the air
- Resistant to the erosive action of molten metal during pouring and the high heat of the metal until the casting is solid
- Collapsible enough to permit the metal to contract without undue restraint during solidification
- Able to cleanly strip away from the casting after the casting has sufficiently cooled
- Economical, since large amounts of refractory material are used

Green Sand Molding. The most common method used to make metal castings is the green sand process. In this process a granular, refractory sand is coated with a mixture of bentonite clay, water, and, in some cases, other additives.

When the sand mixture is compacted against the pattern, the clay/water "glue" binds the refractory grains in place. Thus, when the pattern is removed, the mold cavity retains the shape of the pattern surfaces (Fig. 19.1).

The granular refractory most often used to produce green sand molds is silica sand. Silica is used most often because of its abundance and availability in the United States and its cost effectiveness when compared to other materials. More costly refractory materials, such as zircon, chromite, olivine, mullite, and carbon sand are used for special applications.

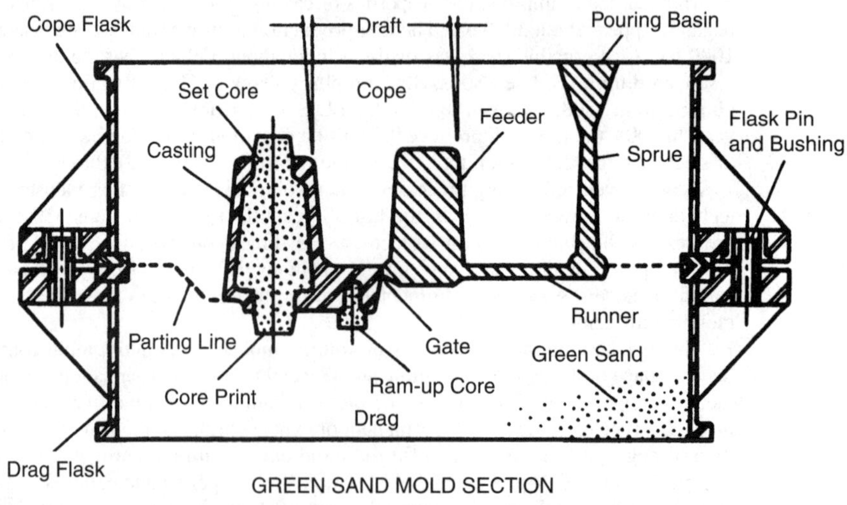

FIGURE 19.1 This schematic shows the main features of a green sand mold, the most common molding process used to make metal castings.

Because the surface of the metal casting is in immediate contact with the sand mold, the quality of the casting (especially its surface) will reflect the quality of the molding sand. For this reason, the goal of every foundry is consistent and close control of the sand mixture. Most foundries use extensive sand testing procedures and automatic sand preparation systems to aid in attaining this goal.

Following the mixing or mulling, the green sand mixture is ready for molding. There are various methods used to compact the molding sand around the pattern. Method selection is dependent upon the desired mold rigidity, which also determines its ability to hold casting dimensions. The material used to build the pattern is determined, in part, by the compaction process employed.

In hand molding, a molder physically compacts the sand around the pattern or uses pneumatic hand tools to compact the sand into a mold. Wood or plastic patterns can be used with this method of compaction. In some cases the pattern is loose; that is, it is not attached to a pattern board or plate. This type of molding is usually done for one-of-a-kind or larger castings. If the pattern is mounted to a pattern board or plate, the process can be sped up.

A mechanical force on the molding sand will produce better compaction than hand molding. The mechanical force can be induced by slinging, jolting, squeezing or, a more recent innovation, by impact/impulse. A sand slinger uses centrifugal force to throw the green sand against the pattern. The operator manipulates the slinger over the patter, which is in the flask, and builds up layers of molding sand. The sand slinger is used primarily for large castings, which are made in a flask or a pit.

Machine molding may be of several types: jolt or squeeze, jolt and squeeze, or impact/impulse. With these types, a molding machine is used to make the molds. The casting size is dependent largely on the size of the molding machine. In the case of smaller castings, multiple casting can be made in one mold.

The term *high-density* molding refers to sand molds that are compacted with pressure equal to or greater than 100 psi. These pressures are achieved consistently on automatic molding machines. These machines may use a combination of jolting and squeezing, or rely solely on a squeezing action for compaction. More recently, a form of compaction called *impact/impulse molding* has been developed and is used in some foundries. It is said to offer better uniformity of compaction and density in green sand molds.

In impact/impulse molding, the molding sand is filled into the flask over the pattern surface under gravity. Once filled with the green sand mixture, an impact/impulse pressure of 40–75 psi is exerted against the sand, which in turn further compacts the sand. Due to the permeability of the molding sand, this pressure is distributed throughout and develops a more uniformly compacted dense mold.

In addition to the machines already discussed. Some high-pressure molding machines are capable of making molds that are parted vertically. These machines are highly automated and compact the sand by squeezing. Normally, these machines are used for high-production runs, but can be used for short runs if necessary. The high sand density of molds made on this machine produces castings with excellent surface finish and good dimensional control.

The following points should be taken into account when considering the green sand molding process: For many metal applications, green sand processes are the most cost-effective of all metal forming operations. These processes readily lend themselves to automated systems for high-volume work as well as short runs and prototype work.

In the case of hand molding, slinging, manual jolt, or squeeze molding, wood or plastic pattern materials can be used. High-pressure, high-density molding methods almost always require metal pattern equipment.

High-pressure, high-density molding normally produces a well-compacted mold, which yields better surface finishes, casting dimensions, and tolerances.

The properties of green sand are adjustable within a wide range, which makes it possible to use this process with all types of green sand molding equipment and for a majority of alloys poured. Titanium and manganese-steel castings cannot be produced in silica sand molds.

Chemically Bonded Molding Systems. This category of sand casting molding processes is widely used throughout the foundry industry because of the economics and improved productivity each offers. Each process uses a unique chemical binder and catalyst to cure and harden the mold. Some processes require heat to facilitate the curing mechanism, while others do not.

Chemically bonded systems are currently being used to produce cores and molds. Cores are used to create internal cast surfaces within the casting. These processes are well suited for large castings as well as smaller castings where intricate design and better dimensional accuracy are required.

Shell Process. Shortly after World War II, a process requiring heat to cure a mold or core composed of sand grains coated with a thermosetting resin was brought to this country. This molding and coremaking method was originally called the Croning Process, but today it is better known as the shell process.

Typically, silica sand is used in the shell process, but zircon or other sands can be used. The sand is precoated with a phenolic novalac resin containing a hexamethylenetetramine catalyst. The resin-coated sand is dumped, blown, or shot into a metal corebox or over a metal pattern that has been heated to 450–650°F. Shell molds are made in halves that are glued or clamped together before pouring. Cores, on the other hand, can be made whole, or, in the case of complicated coring applications, can be made of multiple pieces glued together.

The term *shell* came from the fact that the cores are typically hollow, or, in the case of the shell mold, have a relatively thin mold wall. The thickness of the mold or core wall is dependent on the temperature and amount of molten metal that will surround the core or must be contained by the mold. The shell thickness is determined by the length of time the resin-coated sand is in contact with the heated corebox or pattern. At the proper time, the corebox or pattern is inverted and vibrated, allowing the coated sand not affected by the heat to drain out of the center of the core or away from the mold. This leaves a shell of bonded sand, typically $^1/_4$- $^3/_8$-in thick, adhering to the corebox or pattern. Ejector pins on the corebox or pattern are then used to strip the core or mold.

Shell cores and molds can be used together, or in conjunction with other molding and coremaking processes. Shell molds and cores are widely used throughout the foundry industry, and despite the required energy and metal tooling costs, their use continues to grow. Benefits of the shell process include:

- An excellent core or mold surface, resulting in good casting finish
- Good dimensional accuracy in the casting because of mold rigidity
- Storage for indefinite periods of time, which improves just-in-time delivery
- High volume production
- Selection of refractory material other than silica for specialty applications
- A savings in materials usage through the use of hollow cores and thin shell molds

Nobake or Airset Systems. In order to improve productivity and eliminate the need for heat to cure mold and core binders, foundry binder manufacturers developed a series of resin systems referred to as "nobake" or "airset" binders. As with the shell process, nobake binders can be used to produce both molds and cores.

In this system, silica sand, or another refractory material, is mixed with one or two liquid resin components and a liquid catalyst component. As soon as the resin(s) and catalyst combine, a chemical reaction begins to take place which hardens (cures) the binder. The curing time can be lengthened or shortened based on the amount of catalyst used and the temperature of the refractory sand.

Nobake resins, catalyst, and sand can be mixed in a batch mixer or in a high-production continuous mixing operation. The sand mixture from a high-intensity continuous mixer is generally deposited directly into a corebox or into a flask over a pattern. Although these sand mixtures have good flowability, some form of compaction, usually vibration, is used to provide densification of the sand in the mold or core. After a period of time, the core or mold has cured sufficiently to allow stripping from the corebox or pattern without distortion. The cores or mold are then allowed to sit and thoroughly cure. After curing, the cores can accept a refractory wash or coating, which provides a better surface finish on the casting and protects the sand in the mold from the heat and erosive action of the molten metal as it enters the mold cavity.

One advantage of these nobake processes can be seen when making intricate core assemblies. In many applications, because of the complexity of the internal passageways of a casting, highly configured core assemblies require much time consuming handwork. Today, if the size of the assembly permits, "take-down" coreboxes can be used, which allow these core assemblies to be produced as one integral piece.

Nobake systems do not always lend themselves to high production because of the time necessary to thoroughly cure the mold or core. But in many applications, castings of all sizes and complexities can and are being made using these processes. In the case of complex casting shapes, the mold can be assembled with cores to shape the outside surfaces and the internal passageways of the casting at the same time.

Along with the advantages already mentioned, these nobake processes provide other positive features, such as wood, and in some cases, plastic patterns and coreboxes can be used. Due to the rigidity of the mold, good casting dimensional tolerances are readily achievable. Casting finishes are very good. Most of the systems allow easy shakeout, the separation of the casting from the mold after solidification is complete; cores and molds can be stored indefinitely.

Gas-Catalyzed or Coldbox Systems. One drawback of the nobake systems is that the sand mixture, once prepared, needs to be used quickly. This is because the sand mixture contains both the resin and catalyst components and when mixed together a chemical reaction occurs which hardens the sand mass. Any leftover sand mixture from making a mold or core is already cured, and so the material is not fit for producing another mold or core. Resin manufacturers developed a family of binders where the catalyst is not added to the sand mixture, just the sand and resin component. As a result, this mixture will not cure until it is brought in contact with a catalyst agent. In the coldbox process, catalysts in the form of a gas or vapor are used to accomplish this. The sand-resin mixture is blown into a corebox to compact the sand, and then a catalytic gas or vapor is permeated through the sand mixture where the catalyst reacts with the resin component hardening the sand mixture almost instantly. Any sand mixture that has not come in contact with the catalyst is still capable of being cured, so many small cores can be produced from a large batch of mixed sand. Since mixing is not required for each core, production rates are very high as cores can be produced in a few seconds. While best suited for small core production, coldbox processes can be used to produce large cores and molds for some casting applications.

Coldbox systems are available in both organic and inorganic processes. The inorganic process utilizes a sodium silicate resin, which is cured using CO_2 gas. This process is environmentally friendly, since few fumes are produced on pouring. However, the inorganic bond does not degrade when the molten metal comes in contact with the mold or core, so shakeout can be difficult. This is especially true with higher pouring temperatures, so the process is often limited to the production of aluminum castings. Another factor to consider is that the air contains CO_2, so the sand mixture must be protected from lengthy exposure to the air, or the sand will begin to cure and be difficult to compact.

Several organic processes exist, including phenolic urethane/amine vapor, furan/SO_2 and acrylic/SO_2. These organic processes overcome most of the difficulties described for the inorganic process. In general, the organic processes offer the following benefits:

- Very good dimensional accuracy of the cores since the cores are cured in the corebox without the use of heat
- Excellent surface finish of the casting
- Excellent for high production runs since production cycles are very short
- Excellent shelf life of the cores and molds is obtained

Counter Gravity Pouring Processes. Two casting processes that use a vacuum to assist the filling of the mold cavity in chemically bonded molds have been developed. These are patented processes, and their use is limited to those foundries holding the patent or license to use the process.

The first uses a silicon carbide tube, which is attached to a shell or nobake mold. The mold and one end of the tube are held in a sealed metal container. The container with the mold in it is placed over a furnace that contains molten metal. The end of the silicon carbide tube not attached to the mold is submerged into the molten metal bath. A vacuum is drawn in the container, and the molten metal is drawn into the mold. The vacuum is maintained until the casting has solidified, after which it is released, and the remaining liquid in the tube drains back into the furnace.

The second, and most recently developed process, uses the same vacuum-assist, mold-filling operation. The main difference is that, instead of using the silicon carbide tube, a portion of the mold is submerged into the bath of liquid metal. This process is called the CLAS process.

With the CLAS process, the mold cross-sectional area is a circle. The two halves of the mold are glued together, and once again, the shell molding or nobake molding process is used. The drag half of the mold has openings that are connected with the casting cavities in the mold. The completed mold is then "screwed" into a round metal chamber. The metal chamber has internal threads, which, as the chamber rotates, cut into the circumference of the mold, usually just below the parting line of the mold.

The chamber and mold are then swung over and lowered into the furnace holding the liquid metal. The container and mold are lowered until a portion of the drag half of the mold is submerged into the bath of liquid metal. A vacuum is then drawn in the container and the liquid metal is pulled up through the openings into the mold cavities. When the castings have solidified, the vacuum is released and the mold withdrawn from the bath.

The container is then rotated in reverse, and the mold is "unthreaded" from the container. A normal shakeout then takes place. A variation of the process that uses a ceramic shell mold also was developed.

While this process is best used for long production runs with small-to-medium-sized castings, CLAS offers the following benefits:

- Flow rate into the mold cavity is accurately controlled by the amount of vacuum drawn, improving overall casting soundness.
- Only clean metal is drawn into the mold cavities, reducing the potential for inclusions in the castings.
- Microstructure of the castings is said to be better, thus imparting better mechanical properties.
- Castings have good surface finishes.
- Dimensional tolerances are good.

Unbonded Sand Processes. Unlike the sand casting processes that use various binders to hold the sand grains together, two unique processes use unbonded sand as the molding media. These include the lost foam process and the V-process.

Lost Foam Casting or Expendable Pattern Casting. In this process, the pattern is made of *expendable polystyrene* (EPS) beads. For high-production runs, the patterns can be made in a die by injecting EPS beads into a die and bonding them together using a heat source, usually steam. For shorter runs, pattern shapes are cut from sheets of EPS using conventional woodworking equipment, and then assembled with glue. In either case, internal passageways in the casting, if needed, are not formed by conventional cores but are part of the mold itself.

The pattern is coated with a refractory coating, which covers both the external and internal surfaces of the pattern. With the gating and risering system attached to the pattern, the assembly is suspended in a one-piece flask. The flask is then placed onto a compaction or vibrating table. As the dry, unbonded sand is poured into the flask and pattern, the compaction and vibratory forces cause the sand to flow and become rigid. The sand flows around the pattern and into the internal passageways of the pattern.

After compaction, the mold is moved to the pouring area. As the molten metal is poured into the mold, it replaces the EPS pattern, which vaporizes. The sand stays in place due to its rigidity. After the casting has solidified, the mold is moved to the shakeout area where the unbonded sand is dumped out of the flask leaving the casting with an attached gating system. The sand forming the internal passageways in the casting drains out at the same time. In the case of large castings, the coated pattern is first covered with a facing of a chemically bonded sand. The facing sand is then backed up with a weaker chemically bonded sand or green sand.

The lost foam process offers the following advantages:

- No size limitations for castings
- Improved surface finish of metal castings due to the pattern's refractory coating

- No fins around coreprints or parting lines
- In most cases, separate cores are not needed
- Excellent dimensional tolerances

V-Process. The primary differences between the V-process and conventional sand casting are that the V-process uses a thin plastic film heated to its deformation point and then vacuum-formed over a pattern on a hollow carrier plate. Like the lost foam process, the V-process uses dry, free-flowing, unbonded sand to fill the special flask set over the film-coated pattern. The sand contains no water or organic binders and is kept under a vacuum during the molding process.

Because permeability of the sand is not a concern as in green sand, finer sand can be used to achieve improved casting surfaces. Slight vibration quickly compacts the fine grain sand to its maximum bulk density. The flask is then covered with a second sheet of plastic film. The vacuum is drawn on the flask, and the sand between the two plastic sheets becomes rigid. Releasing the vacuum originally applied to the pattern permits easy stripping.

The other half of the mold is fashioned in the same manner. The cope and drag are then assembled forming a plastic-lined mold cavity. Sand hardness is maintained by holding the vacuum within the mold halves at 300–600-mm Hg. As molten metal is poured into the mold, the plastic film melts and is replaced immediately by the metal. After the metal solidifies and cools, the vacuum is released and the sand falls away.

The V-process offers the following benefits:

- Smooth surface finish
- Excellent dimensional accuracy
- Zero draft
- Thin-wall capabilities
- Excellent reproduction of details
- Low tooling costs
- Unlimited pattern life because only plastic contacts the pattern—there is no sand to cause wear, reduce surface finish, or to open up the tolerances
- "User-friendly" patterns—easy revisions to patterns, no metal tooling, good for prototypes
- Fast turnarounds and short lead times

19.2.2 Permanent Mold Casting

At least five families of molding and casting processes can be categorized as "permanent mold" processes. These include diecasting, permanent mold casting, squeeze casting, graphite mold, and centrifugal casting. Unlike sand casting processes in which a mold is destroyed after pouring to remove the casting, permanent mold casting uses the mold repeatedly.

Diecasting. Diecasting is used to produce small-to-medium-sized castings at high-production rates. The metal molds are coated with a mold surface coating and preheated before being filled with molten metal. A premeasured amount of liquid metal is forced under extreme pressure from a shot chamber into the permanent mold or die. Castings of varying weights and sizes can be produced. Nearly all die castings are produced in nonferrous alloys with limited amounts of cast iron and steel castings produced in special applications.

Die castings and the diecasting process (Fig. 19.2) are suitable for a wide variety of applications in which high part volumes are needed. Benefits include

- Excellent mechanical properties and surface finish
- Dimensional tolerances of 0.005 to 0.010 in
- Recommended machining allowances of 0.010 to 0.030 in
- Thin section castings

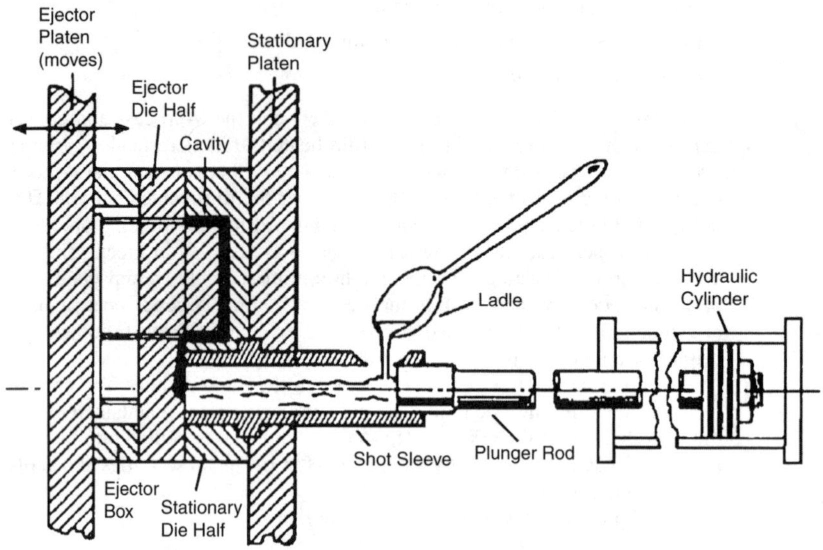

FIGURE 19.2 This schematic shows a cold-chambered diecasting machine, in which metal is ladled into shot sleeves and forced under extreme pressure into the permanent mold or die.

Permanent Molding (Gravity Diecasting). Another form of permanent molding is where the molten metal is poured into the mold, either directly or by tilting the mold into a vertical position. In this process, the mold is made in two halves, male and female, from cast iron or steel. If cores are to be used, they can be metal inserts, which operate mechanically in the mold, or sand cores, which are placed in the molds before closing. If sand cores are used, the process is called "semipermanent molding."

The mold halves are preheated and the internal surfaces are then coated with a refractory. If static pouring is to be used, the molds are closed and set into the vertical position for pouring; thus, the parting line is in the vertical position. In the case of tilt pouring, the mold is closed and placed in the horizontal position at which point molten metal is poured into a cup(s) attached to the mold. The mold is then tilted to the vertical position, allowing the molten metal to flow out of the cup(s) into the mold cavity.

Various Permanent Mold Techniques. Gravity pour, tilt pour, and semipermanent molding offer a variety of advantages for a variety of metalforming applications. Benefits include the following:

- A casting with superior mechanical properties is produced because the metalmold acts as a chill.
- The castings are uniform in shape and have excellent dimensional tolerances because the molds are made of metal.
- Excellent surface finishes are obtainable.
- The process lends itself to high-production runs.
- Sections of the mold can be selectively insulated or cooled, which helps control the solidification and improves overall casting properties.

Low-Pressure Permanent Molding. In this process, a low pressure is used to force the liquid metal into the mold rather than gravity pouring. The amount of pressure, from 3 to 15 psi, is dependent on the casting configuration and the quality of the casting desired. When pressure is used to fill the mold cavity, this pressure also is used to feed shrinkage in the casting. Metal can be fed directly into the casting or through a gating system. When internal passageways are required, they can be made by

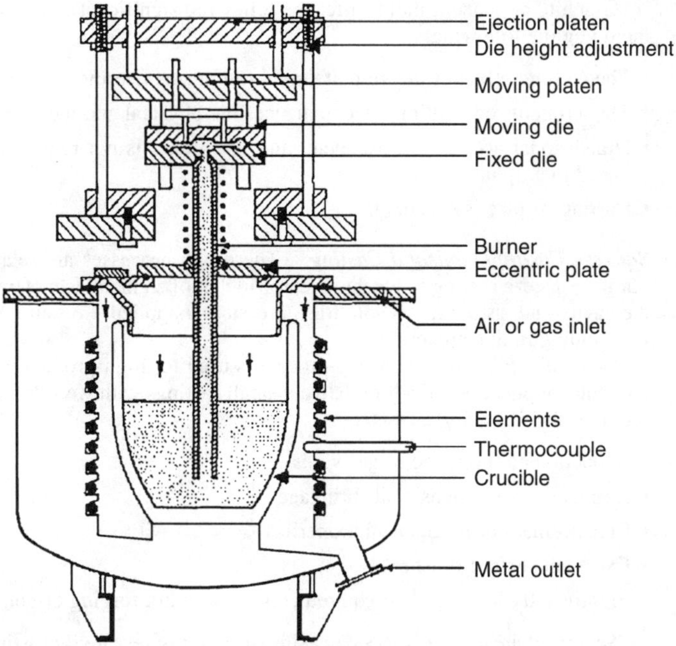

FIGURE 19.3 This schematic illustrates the principal components of a low-pressure permanent mold machine. "Low pressure" means that the liquid is forced into the mold rather than poured.

either mechanically actuated metal inserts or sand cores. A low-pressure permanent mold machine is depicted in Fig. 19.3.

Nearly all of the low-pressure permanent molding (LPPM) castings produced are made of aluminum, other light alloys, and, to a lesser extent, some copper-base alloys. Because it is a highly controllable process, LPPM offers the following advantages:

- When liquid metal is fed directly into the casting, excellent yields are realized.
- Odd casting configurations and tooling points for machining can be placed in areas where gates and risers normally would be placed.
- When liquid metal is fed directly into the casting, it reduces the need for additional handwork.
- The solidification rate in various sections of the casting can be controlled through selective heating or cooling of the mold sections, thus offering excellent casting properties.
- Surface finish of castings is from good to excellent.

It should be mentioned that the processes using metal permanent molds are used primarily for casting aluminum and magnesium alloys. However, copper-base alloys also are cast and poured statically in permanent molds. Some small, thin-section steel castings also are cast.

Graphite Mold Casting. Another form of permanent molding uses molds constructed of graphite. This process is used mostly for specialized types of castings, such as railroad car wheels, and is usually coupled with a special pouring operation, such as pressure pouring. In addition, the geometry of the casting must be such that solidification shrinkage moves away from the graphite mold to prevent hot tearing of the casting and damage to the mold. Graphite molds have been used effectively with the family of zinc-aluminum alloys in certain applications.

Graphite permanent molds offer the following significant advantages in specialized applications including the following:

- The chilling effect of the graphite mold minimizes risering.
- This pronounced chilling effect enhances the physical and mechanical properties of the casting.
- Dimensional accuracy is excellent, and machining is not required on many of the castings produced in graphite molds.
- Casting surface is excellent.

Squeeze Casting/Semisolid Casting. These two processes are relatively new to the metalcasting family. Squeeze casting is a technique in which molten metal is metered into a metal permanent mold die cavity and, as the metal solidifies, pressure is applied. Pressure of 8000 psi or more is required for molding in this process.

Generally, the squeeze casting process is used for high-production runs in aluminum alloys. If a particular application calls for relatively small castings with well-defined geometries, squeeze casting provides the following advantages:

- Reduction of shrinkage or gas voids
- Elimination of dimensional shrinkage
- Enhancement of mechanical properties
- Excellent surface finishes
- Significantly less metal is required compared to hot forging or conventional casting.

Semisolid metalcasting (SSM) is similar to high-pressure diecasting in that metal is injected into a reusable steel die under pressure. However, rather than using liquid metal, SSM uses metal that is about 40 percent liquid and 60 percent solid (Fig. 19.4). Currently, aluminum is the major alloy used with this process, and the major users are automakers.

Centrifugal Casting. Generally speaking, the centrifugal casting process can be categorized as a permanent mold method of casting, though variations of the process use other materials. It has been used for many years as an economical method for producing cylinders and tubes.

With centrifugal casting, a permanent metal mold revolves at very high speeds in a horizontal, vertical, or inclined position as the molten metal is being poured. Centrifugal castings can be made in almost any required length, thickness, and diameter. Because the mold forms only the outside surface and length, castings of many different wall thicknesses can be produced from the same size mold. The centrifugal force of this process keeps the casting hollow, eliminating the need for cores.

FIGURE 19.4 An aluminum billet is heated to the consistency of ice cream for use in the semisolid process.

Horizontal centrifugal casting machines are used for the production of pipe and tubing up to 40 ft long. The length and outside diameter are fixed by the mold cavity dimensions while the inside diameter is determined by the amount of molten metal poured into the mold.

Castings other than cylinders and tubes also can be produced in vertical casting machines. Castings such as controllable pitch propeller hubs, for example, are made using this variation of the centrifugal casting process.

Molds for centrifugal castings generally are divided into three classifications. One is a *permanent mold* made of steel, iron, or graphite. This type of mold usually is coated on the inside surface with a thin refractory coating to increase mold life. The mold is preheated before the coating application to dry the coating and improve its adherence to the mold surface.

A second type of mold is a *rammed mold*. It consists of a metal flask, usually steel, lined with a layer of refractory molding mix, which has been rammed into place. The lining is coated with a refractory wash and then baked until dry and hard.

A third type of mold is the *spun* or *centrifugally cast mold*. It consists of a metal flask into which a predetermined weight of refractory material in slurry form is poured. The flask is rotated rapidly and the refractory material is centrifuged onto the wall of the flask. The flask rotation is then stopped, and the liquid part of the slurry is drained off. This leaves a mold with a refractory coating, which is then baked until dry prior to use.

Molten metal is then poured into a rotating mold where it is accelerated to mold speed. Centrifugal force causes the metal to spread over and cover the mold surface. Continued pouring of the molten metal increases the thickness to the intended cast dimensions. Rotational speeds vary but sometimes reach more than 150 times the force of gravity on the outside surface of the castings.

Once the metal has been distributed over the mold surface, solidification begins immediately. Most of the heat in the molten metal is extracted through the mold. This induces progressive solidification. During solidification, the liquid head of metal feeds the solid–liquid interface as it progresses toward the bore. This, combined with the centrifugal pressure being applied, results in a sound, dense structure across the wall with impurities generally being confined near the inside surface. The inside layer of metal can be removed by boring if an internal machined surface is required.

For specialized engineered shapes, centrifugal casting offers the following distinct benefits:

- Any alloy common to static pouring can be produced centrifugally.
- Mechanical properties of centrifugal castings are excellent.
- Cleaner, denser metal is found on the outside of the casting, and the impurities are on the inside surface, where they can be bored out.

19.2.3 Ceramic, Plaster, and Special Casting Processes

This family of casting processes is unique in that alternative materials are used as molding media, most noticeably ceramic and plaster. These processes offer a high degree of precision in regard to dimensions as well as excellent surface finishes.

Investment Casting. The investment casting process was one of the first processes used to make metal castings. The process has been described as the lost wax process, precision casting, and investment casting. The latter name generally has been accepted to distinguish the present industrial process from artistic, medical, and jewelry applications.

The basic steps of the investment casting process are as follows:

1. Production of heat-disposable wax or plastic patterns
2. Assembly of these patterns onto a gating system
3. "Investing," or covering the pattern assembly with ceramic to produce a monolithic mold
4. Melting the pattern assembly to leave a precise mold cavity
5. Firing the ceramic mold to remove the last traces of the pattern material while developing the high-temperature bond and preheating the mold ready for casting
6. Pouring
7. Knockout, cutoff, and finishing

The patterns are produced in dies. For the most part, the patterns are made of wax, however, there are patterns that are made of plastic or polystyrene. In all cases, the patterns are made in injection molding machines. Because the tooling cost for the wax patterns is high, investment casting is normally used when high volumes are required.

When cores are required, they are made of soluble wax or ceramic materials. In the case of soluble wax cores, they are removed from the pattern before the pattern is "invested." Ceramic cores, on the other hand, stay in the mold during the entire casting process and are removed during the casting cleaning process.

There are two types of molding processes: the solid mold process and the ceramic shell process. The more common of these processes is the ceramic shell.

The ceramic shell is built around a pattern/gating assembly by repeatedly dipping the "tree," as it is called, into a thin refractory slurry (Fig. 19.5). After dipping, a refractory aggregate, such as silica, zircon, or aluminum silicate sand, is rained over the wet slurry coating. After each dipping and stuccoing is completed, the assembly is allowed to thoroughly dry before the next coating is applied. Thus, a shell is built up around the assembly. The required thickness of this shell is dependent on the size of the castings and temperature of the metal to be poured.

FIGURE 19.5 Investment casting, or the lost wax process, offers a high degree of precision in regard to dimensions as well as excellent surface finishes. This photograph shows covering the pattern assembly with a ceramic slurry to produce a monolithic mold.

After the ceramic shell is completed, the entire assembly is placed into an autoclave to melt and remove a majority of the wax. The shell is then heated to about 1800°F (982°C) to burn out any residual wax and to develop a high-temperature bond in the shell. The shell molds can then be stored for future use or molten metal can be poured into them immediately. If the shell molds are stored, they have to be preheated before molten metal is cast.

The vast majority of investment castings weigh less than 5 lb, but there is a distinct trend to produce larger castings in the 10–30-lb range. Castings weighing up to 800 lb have been poured in this process. Some of the advantages of investment casting are

- Excellent surface finishes
- Tight dimensional tolerances
- Lends itself to the production of titanium castings as well as the other superalloys
- Machining can be reduced or completely eliminated.

Ceramic Molding. Another method of molding in which a ceramic material is used is simply referred to as *ceramic molding*. This process and its offshoots also are known as the Shaw Process, the Unicast Process, the Osborn-Shaw Process, and the Ceramicast Process.

Generally, these processes employ a mixture of graded refractory fillers (in some cases, hydrolyzed ethyl silicate and a liquid catalyst), which are blended to a slurry consistency. Various refractory materials can be used as filler material. The slurry is then poured over a pattern that has been placed in a container.

First, a gel is formed in a pattern, and it is stripped from the mold. Once the ceramic slurry gels, the pattern is stripped from the mold. The mold is then heated to a high temperature until it becomes rigid. After the molds cool, molten metal is then poured into them, with or without preheating.

The ceramic molding processes have proven effective with smaller size castings in short- and medium-volume runs. At the same time, these processes offer several advantages including the following:

- Excellent surface finish
- Good casting dimensional tolerances
- Adaptability to intricate castings

Plaster Molding. Plaster molding is used to produce castings of the lower melting temperature metals, such as the aluminum alloys. The four generally recognized plaster molding processes are as follows:

- Conventional plaster molding
- Matchplate-pattern plaster mold casting
- Antioch process
- Foamed-plaster process

Plaster, a slurry containing gypsum, is poured into a flask that contains the pattern. After the slurry has set, the pattern and flask are removed, and the drying cycle to remove the moisture from the mold begins. After the mold has been allowed to cool, the cores and mold are assembled. After assembly, molds are preheated before pouring. Because these molds have very low permeability, vacuum-assist or pressure is usually required during pouring.

The plaster mold processes are especially suited for short run and prototype work with the lower temperature alloys, particularly aluminum. In addition to these benefits, plaster molding offers the following advantages:

- Castings have especially smooth surfaces, and intricate designs and details are readily obtainable in plaster molds.
- Dimensional accuracy of castings is good.
- Because of the mold material and vacuum-assist, thinner wall castings can be produced.
- Slow cooling of plaster molds minimizes warpage and promotes uniformity of structure and mechanical properties in the casting.

19.2.4 Rheocasting and Thixomolding

In 1976, a new metalforming process called Rheocasting was developed at the Massachusetts Institute of Technology. This process uses a phenomenon called thixotropy that involves the vigorous agitation of a semisolid metal to produce a highly fluid, diecastable alloy. Advantages of the process are reportedly longer die and chamber life, finer grained castings with fewer defects and greater economy (less loss) of metal fed to the diecasting machine. The use of nonmetallic materials to produce composites also was developed. Today, the principles of this process are used in the thixomolding process.

The major difference between Rheocasting and thixomolding is that, in Rheocasting, the metal alloy completely liquifies and then cools. In thixomolding, the metal alloy is heated only to a *mushy* state between the liquidus and solidus temperatures.

Some advantages of thixomolding include the following:

- Cast parts have less porosity than conventionally diecast parts.
- Lower porosity levels allow parts to be heat treated to improve mechanical properties.

- There is improved material flow in thin-wall sections.
- Component warpage is greatly reduced after the part is removed from the mold.

19.3 CASTING ECONOMICS

Due to the many different types of foundries, the variety of metals cast, and the wide range of casting processes employed, it is difficult to offer a generic formula for estimating the cost of castings. However, general guidelines can be established to assist the buyer or designer in determining factors that impact the cost of castings.

Metal type plays a significant role in the cost of a casting. But generally speaking, regardless of the metal type, casting cost is directly linked to process and production requirements. Table 19.1 compares various characteristics offered by a variety of molding processes. Each process offers various capabilities and constraints, with certain processes better suited to specific casting designs and applications. Even within a given process, different methods may exist that lend themselves to greater efficiencies at different production levels. Also, processes utilized for tight dimensional tolerances, thin sections, and casting complexity tend to be more costly, thus casting design and application have a substantial impact on cost. Designs requiring tight tolerances and minimal draft may require extensive coring or special molding processes and will also cost more.

Other major factors contributing to the cost of the casting are casting end use, tooling costs, production volume, and foundry production capabilities.

TABLE 19.1 A Comparison of Various Molding and Casting Processes

	Green Sand	Precision Molding			Chemically Bonded
	Sand Casting	Permanent Molding	Die Casting	Ceramic and Investment Casting	Shell, CO_2 Nobake
Typical dimensional tolerances, in	±.010 in ±.030 in	±.010 in ±.050 in	±.001 in ±.015 in	±.010 in ±.020 in	±.005 in ±.015 in
Relative cost—large run	Low	Low	Lowest	Highest	Moderately high
Relative cost—small run	Lowest	High	Highest	Moderate	Moderately high
Permissible weight of casting	Unlimited	100 lb	75 lb	Oz-100 lb	Shell Oz-250 lb CO_2 and nobake 0.5 lb-Tons
Thinnest section castable	1/10 in	1/8 in	1/32 in	1/16 in	1/10 in
Relative surface finish	Fair to good	Good	Best	Excellent	Shell good CO_2 fair
Ease of casting complex design	Fair to good	Fair	Good	Best	Good
Ease of changing design in production	Best	Poor	Poorest	Fair	Fair
Range of alloys	Unlimited	Aluminum base and copper base preferable	Aluminum base preferable	Unlimited	Unlimited

19.4 ENVIRONMENTAL AND SAFETY CONTROL

The need for environmental protection and safety are major concerns in the metal casting industry. Breathing air that contains hazardous contaminants can cause health problems. Water that contains toxic pollutants can contaminate surface and ground water. Waste sand and other solid materials can adversely affect the environment when disposed of in landfills and may need to be cleaned or treated before disposal, or beneficially recycled or reused.

The use of specialized equipment and procedures to insure the protection of workers and the public is a necessary and mandatory part of foundry operations. Larger foundries often have an entire department devoted to overseeing environmental health and safety programs and government regulations.

There are four major areas of environmental and safety control in the metalcasting industry:

- Emissions
- Water pollution
- Solid waste
- Safety

Emission Control. Equipment needed to clean the air in the foundry is generally located where the molds and cores are made, where metal is melted, poured, cooled, and shaken out; and where castings are cleaned by blasting, grinding, and chipping.

Particulate matter is controlled by fabric filter baghouses or wet scrubbers. Gaseous pollutants VOCs, including carbon monoxide and organic compounds are controlled by thermal oxidation or chemically scrubbing.

Water Pollution Control. The sources of discharged water in foundry operations vary significantly. Major sources of polluted and nonpolluted water are wet scrubbers (mentioned under emission control), and cooling water used in the melting and slag quenching areas. The treatment and discharge of these waters is regulated by various government agencies.

Foundry water treatment systems can consist of a number of different options, depending on the discharge water quality from in-plant processes. Water may contain large amounts of suspended solids, heavy metals, or oil and grease. Treatment will depend on the nature of the pollutants and may include the use of settling ponds or clarifiers, chemical treatment for heavy metals, pH adjustment, activated carbon beds, and other options to meet specific permit limit conditions.

Solid Waste Management. During the metalcasting process, many of the materials used for molding, coremaking, and the final casting are no longer usable in their present state. These solid materials have to be dealt with by disposal or by recycling and reuse. Most metallic waste can be remelted. Some of the sand from molds and cores can be treated and added to other sand. Everything else needs to be disposed of in the proper manner and according to government regulations.

Today, the metalcasting industry has found increasing opportunities to recycle by-products of the manufacturing process, such as molding and core sands. This sand is the largest single waste stream generated by foundries. Individual foundries have found the following beneficial reuse options:

Construction fill/road subbase

Flowable fill

Grouts and mortars

Potting and specialty soils

Cement manufacture

Precast concrete products

Highway barriers
Pipe bedding
Asphalt
Cemetery vaults
Brick pavers
Landfill daily cover

Safety and Health Programs. Safety in the foundry is paramount to all other activities within the facility. When safety is given major priority, production and quality will be achieved to expected levels. No job is so urgent that it cannot be done in a safe and timely manner. The metalcasting process presents many unsafe situations to employees: molten metal, moving conveyor chains and pulley drives, confined spaces, hazardous atmospheres, harmful noise, heat stress, electrical exposure; ergonomic hazards associated with melting and pouring of metals, shakeout, degating, grinding, machining; and material handling of castings. To protect employees and to eliminate identified hazards, foundries utilize comprehensive safety and health programs based on management commitment and employee involvement, worksite analysis, hazard prevention and control, and safety and health training.

BIBLIOGRAPHY

Engineered Casting Solutions, vol. 4, no. 3, pp. 22–31, American Foundry Society, 2002.
Kotzin, E.L., *Metalcasting and Molding Processes*, American Foundry Society, 1981.
Schleg, F.P., *Technology of Metalcasting*, American Foundry Society, 2003.

CHAPTER 20
POWDER METALLURGY

Chaman Lall
Metal Powder Products Company
Westfield, Indiana

20.1 INTRODUCTION

The *powder metallurgy* (P/M) technology is capable of producing large quantities of complex, discrete, components and is, therefore, suited for many industrial markets, including automotive, trucks, off-the-road vehicles, power tools, lawn and garden, hardware, medical, marine, aerospace, computers, fluid power pumps, and locks, to mention just a few examples.

While the technology is an ancient art-form, with artifacts created by the P/M process dating back to AD 400 (iron pillar of Delhi, India)[1] and earlier, this fabrication technology did not become an accepted industrial manufacturing process until the twentieth century. The development of successful procedures to process refractory metals such as tungsten, molybdenum, and tantalum pioneered the commercialization of the P/M technology. The Coolidge process[2] for the manufacture of tungsten wire, used in incandescent light bulbs, is essentially the same process in use today, more than 100 years later. In essence, the process consists of compacting very fine W powder, presintering at 1200°C (2190°F), and fully sintering at 3000°C (5430°F) by use of electrical current applied directly to the compacts. The next step was the real secret in that mechanical working to produce fine filaments must begin near 2000°C (3630°F), as this is where the metal has usable ductility. Furthermore, as the degree of deformation increases, the ductility level also increases, permitting the process temperature to be reduced slowly to fairly low levels. These material processing developments in the early part of the century were an essential ingredient for the growth of the electrical, electronics, and aerospace industries that was to follow.

Another significant development was the process for making cemented carbides, which provided materials with excellent wear characteristics for forming dies and for cutting tool tips. In this process, very fine tungsten carbide particles are partially alloyed or "cemented" together with cobalt; the result is a composite with extremely hard carbide particles in a background of cobalt metal that provides good ductility and toughness. The tungsten carbide particle morphology may be modified to make rounded particles for wear applications, such as forming dies or tooling, or angular shapes for sharp cutting tips in machining operations. It is hard to imagine the industrial progresses of the twentieth century without this class of materials to aid the traditional forming processes such as machining, grinding, rolling, and drawing.

Bearings and filters are more widely known applications of P/M in which metal powders are bonded together by sintering, while still retaining an interconnecting, porous structure. The size and distribution of the metal powder is carefully controlled to produce the desired degree of porosity for the intended application. Filters usually have more open porosity to enable the fluid to flow thorough the pores, while "self-lubricating bearings" have sufficient porosity to soak up the lubricating oil during the impregnation process. When used as a bearing, the initial frictional resistance between the shaft and the P/M part causes heating, which expands the oil. The expanding oil seeps to the surface and generates a lubricating film, which decreases the frictional resistance and, thereby, the bearing temperature—causing the oil to contract. Capillary action pulls the oil back into the pores. This self-regulating process enables the correct amount of lubrication to be presented to the functional interface. Bearings can be made in both ferrous as well as nonferrous alloys, sometimes incorporating a solid lubricant (graphite) within the metal matrix.

Structural applications using ferrous parts made by P/M advanced rapidly in the last half of the century, driven by the needs of the automobile industry for mass produced components. Powder metallurgy offered the opportunity to produce high volumes of components such as gears, hubs, and cams, at an inexpensive price. Beyond ferrous parts for structural applications, P/M also offered the same values in manufacturing soft magnetic components such as pole pieces, and conductive components such as brushes for motor applications. As materials and processes have developed, the P/M technology has grown to include more difficult components such as hot forged connecting rods (Fig. 20.1).

At the beginning of the new century, more than half a billion metric tons of metal powder were being shipped by the P/M industry in North America, on an annual basis.[4] The majority (70%) of the parts produced by the P/M industry are consumed by the automotive industry. Examples of such products include components for stainless steel exhaust systems (flanges, sensor bosses), engines (connecting rods, main bearing caps, crankshaft and camshaft sprockets, valve seats and guides, cam caps, and oil and water pump gears), brakes (antilock brake systems, tone rings, and sensor pole pieces), manual and automatic transmissions (planetary carriers, pinion gears, clutch races, synchronizer rings, and hubs), interior (rear view mirror mounts, latches, and locks), and chassis systems (steering column locking pawls, shock absorber pistons, and rod guides). An estimated 40 pounds of P/M parts are contained in a typical car manufactured in North America, a number that continues to grow each year, even though total steel usage is trending downward.

The nonautomotive markets are much smaller[5] and include recreation, hand tools and hobby (16%), industrial motor control and hydraulics (3%), household appliances (3%), hardware (1.5%), business machines (1.5%), while the remaining markets served make up the balance of about 5%. Nonautomotive applications include lock components, lawn and garden gears and bushings, fluid power vanes, and wear plates.

The powder metallurgy technology includes the very precise process of powder production, followed by forming into shapes, and sintering to gain the functional properties of the material being processed. Powder production is an integral part of the industry as it determines the quality and consistency of the components that the parts making sector of the industry is capable of producing.

FIGURE 20.1 Powder forged automotive components, connecting rods on the right.[3]

In many cases the powder producers have lead the way in R&D programs to advance the technology and the materials that this fabrication method can offer.

Various techniques can be used to produce powder,[6,7] depending upon the oxidation resistance of the metal or alloy, the desired chemistry, particle size distribution, and morphology. An estimated 95% of the global production of commercial metal powders is made by atomization, a process which refers to the breakup of liquid or molten metal into fine droplets. The process begins with heating of the scrap metal and/or virgin metal to form an alloy melt. This molten liquid is refined and its chemistry controlled to exacting standards before being passed through an orifice. Water or gas is used to break up or *atomize* this stream of molten metal. In the case of inert gas atomization, the powder particles are spherical and usually have smooth surfaces. The powders are sufficiently clean that they just need to be classified into different sizes and shipped to the customer.

In the case of water atomization, the powders become oxidized during the initial process and therefore need to be treated further. The slurry of water and metal powder is pumped from the atomization chamber and dried in several stages. The resulting product is ball-milled and then decarburized and deoxidized under a controlled atmosphere. Further granulation and annealing may be needed to customize the powder particle size distribution for the P/M industry. The higher energy and cooling rate of the water jets creates very irregular shapes and rough surfaces, which enhance the strength of the as-molded compacts. In contrast, the smooth surfaces of the gas atomized powders offer no such interlocking mechanism, making them unsuitable for the conventional cold pressing P/M technology.

The following subsections provide an overview of the P/M process, but detailed descriptions of the powder metallurgy technology are contained in a very comprehensive handbook and a number of excellent texts.[7–12] The focus of this section is on processes that convert the metal powders to structural components with defined shapes and functions. Uses of these materials in the powder form (e.g., food additives, rocket propellants, paint pigments, and plastic fillers) are beyond the scope of this brief review.

20.2 POWDER METALLURGY PROCESSES

Figure 20.2 depicts the traditional or conventional P/M process. This process begins with the preparation of a specific blend or mixture of metal powders together with a lubricant and other additives. The molding grade metal powders are usually in the range of 10–200 μm, while the additives (e.g., Cu, Ni, C) may be on the low end of that range, the lubricant is necessary to permit ejection of the part from the die. This mixture is molded at room temperature in mechanical or hydraulic compacting presses, with most of the tooling motion in the vertical direction. For iron-based alloy systems, compacting pressures near 552 MPa (80,000 psi). The resulting "green" compact is sintered in furnaces set at the appropriate thermal cycle and atmosphere for the material being processed. Sintering is often performed near 1120°C (2050°F), while higher temperatures are preferred for soft magnetic components and high performance materials.[13,14] While nitrogen and dissociated ammonia was a popular mixed atmosphere for sintering steels in the twentieth century, the trend now is toward use of nitrogen–hydrogen mixtures. Higher levels of hydrogen, including pure hydrogen, are used for those metal systems that contain particularly stable oxides or where there is a need to minimize the detrimental effects of nitrogen (e.g., soft magnetic materials and stainless steels).[13,15,16]

Optional postsinter, or "secondary," operations may be added to complete the processing of the part. Examples of such operations are repressing (to densify the part or qualify dimensions), machining, grinding, heat treatment, resin impregnation, and plating or coating. The normal cutting, drilling, and grinding procedures may need to be altered slightly from the wrought metal processing parameters, because of the intermittent cutting action due to the porosity of the P/M product. Similarly, the process parameters for the heat treatment of P/M steel components need to accommodate the fact that the porosity allows the carburizing gas to enter the body of the part, causing rapid through-hardening of the P/M part. For this reason, case hardening is relatively difficult in P/M parts, unless surface densification techniques are used or the overall density is greater than about 7.2 g/cc, when the porosity starts to become nonconnected.

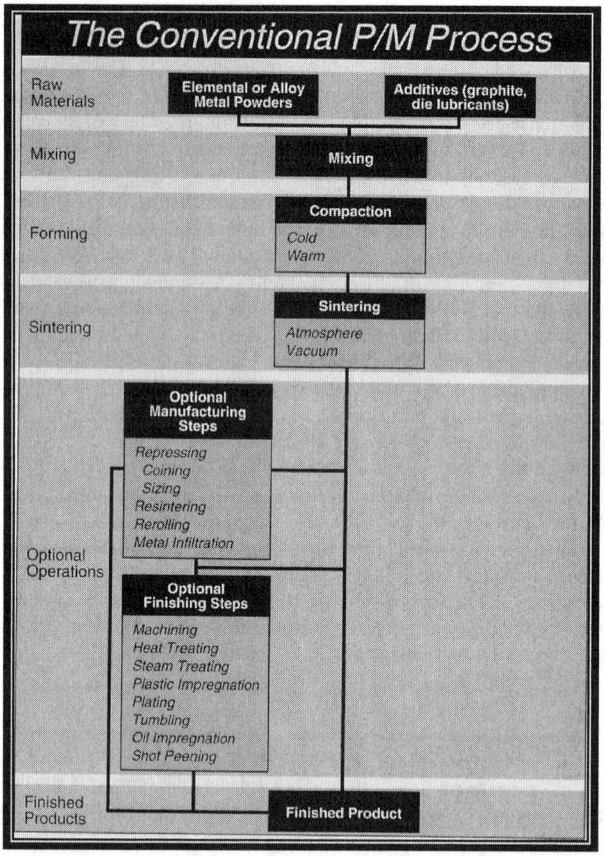

FIGURE 20.2 Flow diagram for conventional P/M process.[3] *(Courtesy of Metal Powder Industries Federation, Princeton, NJ.)*

The presence of interconnected porosity is also the reason why plastic or resin impregnation is used for P/M parts intended for use in pressure applications or if the subsequent operation is plating or coating. One common resin impregnation process involves the immersion of the P/M parts in a liquid resin system inside a vacuum chamber. As a vacuum is pulled, the air bubbles out of the pores and is replaced by the liquid monomer. The anaerobic resin begins to convert from a liquid monomer to a solid polymer, aided by catalysts in the system. Once the parts are brought back to atmospheric pressure, they are washed and dried, ready for the next operation. Resin impregnation can improve machinability by reducing the intermittent cutting phenomena, as well as preventing cutting fluids from entering the part. In the same manner, during the plating or coating operations, the solid resin/plastic prevents the chemicals from the process tanks entering the pores. If the plastic/resin impregnation step were not performed, chemicals would be carried from one tank to the next, very quickly compromising the delicate chemical balance that must be maintained in the tanks for a plating or coating to be applied successfully.

The sealing ability of the plastic enables parts to be used in hydraulic pressure applications up to about 65 Mpa (10,000 psi). Repeated resin impregnation of the same parts, or the application of pressure once the liquid monomer has entered the part will ensure that the pores are properly filled. Most anaerobic resin systems are designed for continuous use at temperatures up to 205°C (400°F). For applications above this temperature, copper infiltration is often used. This process involves placement of the infiltration material on the green compact (or presintered part), and passing both into the

FIGURE 20.3 Compressor valve plate body. Molded as three separate pieces, which are brazed together and sintered in one pass through the furnace. (*Courtesy of Metal Powder Products, Westfield, IN.*)

FIGURE 20.4 Copper infiltrated and furnace brazed three-piece planetary carrier for automotive transmission. (*Courtesy of Metal Powder Industries Federation, Princeton, NJ.*)

sintering furnace. As the copper-based material melts, it seeps into the pores of the steel structure by capillary action. Some degree of diffusion and alloying will occur between the copper and the steel, by the time the composite passes through the furnace and solidifies. Beyond the value of sealing pores, copper infiltration also offers advantages by increasing the strength of the base steel by alloying, as well as "blunting" the sharp angular porosity, which can act as initiation sites for cracks. In addition, copper infiltrated parts can be heat-treated, but plastic impregnated parts cannot, since the plastic degrades rapidly at the heat treatment temperatures.

A variety of techniques can be used to join P/M parts to each other or to wrought parts. These include brazing, welding, and mechanical methods, such as swaging. P/M parts can also be joined together during the sintering process, by use of copper infiltration or special brazing pastes. Care must be taken with P/M parts because of the inherent porosity, which will suck up the traditional low viscosity braze materials. Special brazing pastes designed for P/M have higher viscosities and solidify rapidly at the application temperatures. Figure 20.3 is an example of three parts that have been joined together in the sintering furnace, while Fig. 20.4 is an example of a furnace brazed product for a demanding automotive application. Figure 20.5 is an example of two stainless steel parts that have been welded together, replacing an investment cast part with a significant cost reduction.

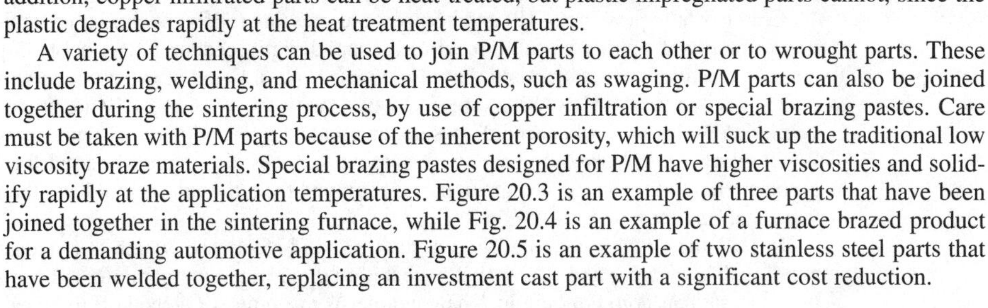

(a) (b)

FIGURE 20.5 Welded P/M parts. (*a*) Two-piece welded stainless steel latch bolt. (*b*) Two-piece welded stainless steel lock set retractor. (*Courtesy of Metal Powder Products, Westfield, IN.*)

The conventional cold-press-and-sinter P/M process accounts for more than 85% by volume of current production by the P/M technology. The balance is produced by the more involved and expensive processes called P/M forging, metal injection molding, cold and hot isostatic pressing. The P/M forging process, also referred to as powder forging, or P/F, begins with the compaction of a preform by conventional means.[10] This preform is heated along with the tools, to an elevated temperature and compacted to full density. Since most of the internal porosity is eliminated during this process, the resultant product has properties very close to wrought components. The additional benefit is that the P/F process can produce much finer-grained microstructures and accompanying improvements in mechanical properties.

Metal or ceramic powders can be combined with a carefully designed binder system and molded in conventional plastic injection molding machines.[11,12] Molding pressures are typically less than 20 MPa (3000 psi). The process is referred to by the general terminology Powder Injection Molding (PIM), or specifically, Metal Injection Molding (MIM) and Ceramic Injection Molding (CIM). The majority of the successful systems use much finer powders (usually, 20 μm or less) than conventional P/M as these provide the high surface-to-volume ratios that provide the energy to drive densification during the sintering process. Once the parts are molded, the binder system no longer serves any useful function and is removed by a variety of extraction techniques. Water-based binder systems can be removed by immersing in an aqueous bath, while organic-based binder systems are processed in an organic solvent. Another group of binder systems (consisting mostly of waxes) are thermally removed at low temperatures. Wall thicknesses of MIM and CIM parts are usually limited to about 6 mm (0.25 in) in order to keep the binder extraction times from being too excessive, although new binder systems are being developed to overcome this limitation. Once the binder has been extracted, the parts are sintered in vacuum or atmosphere furnaces at temperatures appropriate for the material being processed. Many ferrous-based alloy systems are sintered in the range 1100–1300°C (2000–2400°F), while ceramic systems such as alumina require 1500°C (2700°F). The dimensional change during sintering can be fairly high (15–20%) as the fine powders consolidate to densities in the range of 95–98% of theoretical. Full density is rarely reached since some of the fine porosity gets closed off and trapped inside the body of the part.

Cold isostatic pressing (CIP) and hot isostatic pressing (HIP) are similar technologies that permit the formation of a part with full three dimensional design freedom. CIP consists of pouring the powder into a rubber or flexible mold that is immersed into a liquid at room temperature, which is then pressurized up to 414 MPa (60,000 psi). Green densities achieved by CIP are much higher than convention P/M since the rubber mold does not create the high frictional resistance that the walls of hard P/M tooling do. Spherical powders with smooth surfaces also cannot be used for CIP, because the green strengths are so low. The green part is removed from the rubber mold and then sintered in conventional furnaces.

In HIP, the powder is poured into a glass or solid steel case, which is evacuated and sealed by welding. The powder and container are heated to an elevated temperature and pressurized under an inert gas such as argon or helium. The container is machined or etched off to reveal a product that is fully dense, which may be machined or ground further into the desired shape. Both CIP and HIP can produce large parts with uniform density. These processes are used to process difficult materials, such as tool steels, titanium, nickel-base superalloys, and refractory materials, such as molybdenum and tungsten.

While the various processes described thus far in this section all begin with powder, the final product and the cost of that end product can be significantly different. The conventional press and sinter is the least costly process, which can produce parts routinely from about 5 g to 2000 g, although parts weighing up to 16,000 g have been reported. Part sizes can range from a few mm in diameter to over 20 cm, using conventional P/M. The size is limited simply by the capacity of the compacting press, which must apply a stress of at least 414 MPa (60,000 psi) for steel powders during the molding stage. Furthermore, the configuration of the product is limited in design flexibility by the fact that undercuts are very difficult, although not impossible, to mold in; the vertical motion of the molding tools (punches and core rods) tend to fracture any tooling member that may be used as a retractable slider to form that undercut. As a consequence, undercuts and cross holes are normally machined-in by including a secondary operation, after the product has been sintered.

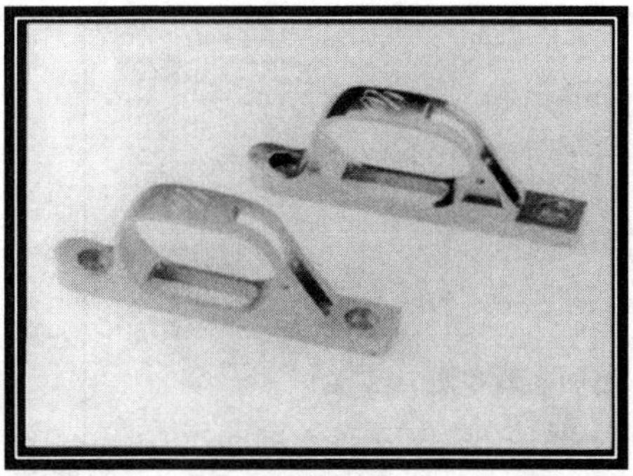

FIGURE 20.6 Metal injection molded rifle trigger guard weighing 70 g (0.15 lb). Made from 4640 low alloy steel, with an ultimate tensile stress of 1160 MPa (240,000 psi) and a yield stress of 1480 MPa (215,000 psi) at 7.55 g/cc). *(Courtesy of Metal Powder Industries Federation, Princeton, NJ.)*

In contrast, metal injection molding, and cold and hot isostatic pressing are capable of total 3-D design flexibility since almost any shape can be molded in. This ability to produce the final part design, or "near net shape" capability, makes these manufacturing processes particularly valuable for materials that are difficult to machine or deform. Since the cost of the starting powders is inherently higher than for conventional molding grade powders, plus the fact these advanced processes require longer and more involved steps, the cost of components produced by these technologies can be significantly higher.

CIP and HIP are often used to produce tool steels and components for demanding applications such as aerospace and military applications. MIM is being used successfully for producing components for the orthodontic field and medical fields, as well as the firearms industry (Fig. 20.6).

20.3 PART DESIGN CONSIDERATIONS

One of the most important concepts to grasp when designing parts for conventional uniaxial compaction is that one can produce a prism of any shape in the die; other tooling members are used to provide additional shapes or features. The periphery of that prism is the profile that is cut into the die, which can be a regular shape, such as a square, rectangle, circle, or any shape that can be cut by a wire electrical discharge machine. Holes also have this latitude of design freedom. To form a shallow (1–2 mm) feature in the faces of that prism, that detail can be formed in the top and bottom tool punches. For deeper features, or "steps," separate punches will be needed. Another technique is to add a "shelf" in the die. Figure 20.7 illustrates these concepts for some parts made with multiple tooling punches entering from both the top and bottom of the die. A solid core rod is used to make the holes.

As in any design for a part that is to be put under mechanical stress, sharp corners and angles should be avoided when designing a component. The P/M tool designer also has the same objective when designing the tooling that will mold the part, since that tooling is highly stressed in use. This is why edge breaks and generous radii are encouraged, wherever possible, so that tool life can be maximized.

As mentioned earlier, undercuts and cross holes are not practical for routine molding by die compaction, and must be machined in after sintering. Configurations such as splines and pinion gears are relatively easy to mold by the P/M technology, since these profiles run parallel to the pressing axis. Helical gears are more difficult to mold, but are produced routinely by use of special bearing packs that synchronize the rotation of the tooling punches with the vertical motion of those tools, so as not to break the teeth. Gears of helix angles up to about 25° can be made without too much difficulty.

The dimensional control of P/M parts is similar to machined products, but not as well as ground parts. As-compacted parts have very tight tolerances orthogonal to the pressing direction, as these are controlled by solid die walls. Even though there is a slight elastic expansion as the part is released from the die, and this can vary for different materials, die dimensions can be held to better than 0.05 mm (0.002 in) in the green state. Dimensional control in the pressing direction is not as tight, since this is dictated by the amount of powder in the die cavity, the apparent density of the powder, and the degree of control of tool motion by the press. Because there are finite differences between "shots" of powder, green parts can be held to about 0.1 mm (0.004 in) in the pressing direction.

FIGURE 20.7 Examples of P/M parts illustrating internal and external diametral detail, as well as the steps or "levels" that can be formed by use of top and bottom punches. (*Courtesy of Metal Powder Industries Federation, Princeton, NJ.*)

Once the parts are molded, they are sintered at elevated temperatures which introduce additional dimensional variations. The degree of variability is dependent on the alloy system being processed and the specific sintering parameters used. In general, P/M parts can be held to 0.25 mm (0.01 in) along the pressing axis and about 0.125 mm (0.005 in) normal to this axis. Note that additional thermal processes, e.g., heat treatment will tend to degrade dimensional control.

20.4 MATERIALS AND PROPERTIES

The ability to process almost any material combinations in economical quantities provides tremendous flexibility for the P/M technology to tailor-make the metal powder blend for the specific application under consideration. While this flexibility is an advantage over other processes where large tonnage runs are commonplace, the P/M industry has developed material standards for structural parts[17] and properties that can be the basis for considerations in new applications and input data for *finite elemental analysis* (FEA) design packages. Figure 20.8 shows an example of the kind of data that is supplied in the materials standards generated by the P/M industry. *Ultimate tensile strengths* (UTS) for these selected low alloy steels range from about 200 MPa (28,000 psi) to 600 MPa (86,000 psi) in the as-sintered condition. Once heat treated, UTS values can reach almost 1300 MPa (187,000 psi). The specific properties, be they mechanical, wear, fatigue, corrosion resistance, electrical, or magnetic, will be determined by the specific chemistry and the processing steps employed. The role of porosity in conventional P/M parts should be considered when evaluating P/M parts for a specific intended function.

Low Alloy Steel

P/M Material Properties **(SI Units)**

Material Designation Code	MINIMUM VALUES (A): Minimum Strength (A) (E) Yield	Minimum Strength (A) (E) Ultimate	TYPICAL VALUES (B) — TENSILE PROPERTIES: Ultimate Strength	Yield Strength (0.2%)	Elongation (in 25.4 mm)	ELASTIC CONSTANTS: Young's Modulus	Poisson's Ratio	Unnotched Charpy Impact Energy	Transverse Rupture Strength	Compressive Yield Strength (0.1%)	HARDNESS: Macro (apparent)	HARDNESS: Micro (converted)	Fatigue Limit 90% Survival	Density
	MPa	MPa	MPa	MPa	%	GPa		J	MPa	MPa	Rockwell	Rockwell	MPa	g/cm³
FL-4205 -35	240		360	290	1.0	130	0.27	8	690	290	60HRB		140	6.80
-40	280		400	320	1.0	140	0.27	12	790	320	66	N/A	190	6.95
-45	310		460	360	1.5	150	0.27	16	860	360	70		220	7.10
-50	340		500	400	2.0	160	0.28	23	1030	390	75		280	7.30
FL-4205 -80HT		550	620		<0.5	115	0.25	7	930	550	28HRC	60HRC	210	6.60
-100HT		690	760	(D)	<0.5	130	0.26	9	1100	760	32	60	260	6.80
-120HT		830	900		<0.5	140	0.26	11	1280	970	36	60	300	7.00
-140HT		970	1030		<0.5	155	0.27	16	1480	1170	39	60	340	7.20
FL-4405 -35	240		360	290	1.0	120	0.25	8	690	270	60HRB		140	6.70
-40	280		400	320	1.0	135	0.27	15	860	310	67	N/A	190	6.90
-45	310		460	360	1.5	150	0.27	22	970	360	73		220	7.10
-50	340		500	400	2.0	160	0.28	30	1140	390	80		280	7.30
FL-4405 -100HT		690	760		<1.0	120	0.25	7	1100	930	24HRC	60HRC	230	6.70
-125HT		860	930	(D)	<1.0	135	0.27	9	1380	1070	29	60	290	6.90
-150HT		1030	1100		<1.0	150	0.27	12	1590	1210	34	60	330	7.10
-175HT		1210	1280		<1.0	160	0.28	19	1930	1340	38	60	400	7.30
FL-4605 -35	240		360	290	1.0	125	0.27	8	690	290	60HRB		140	6.75
-40	280		400	320	1.0	140	0.27	15	830	310	65	N/A	190	6.95
-45	310		460	360	1.5	150	0.28	22	970	360	71		220	7.15
-50	340		500	400	2.0	165	0.28	30	1140	390	77		280	7.35
FL-4605 -80HT		550	590		<0.5	110	0.24	6	900	630	24HRC	60HRC	200	6.55
-100HT		690	760	(D)	<0.5	125	0.25	8	1140	790	29	60	260	6.75
-120HT		830	900		<0.5	140	0.26	11	1340	960	34	60	320	6.95
-140HT		970	1070		<0.5	155	0.27	16	1590	1170	39	60	370	7.20
FLN-4205-40	280		400	320	1.0	115	0.25	8	720	310	64HRB		140	6.60
-45	310		460	360	1.0	130	0.27	11	860	340	70	N/A	190	6.80
-50	340		500	400	1.5	145	0.27	18	1030	390	77		220	7.05
-55	380		600	430	2.0	160	0.28	30	1210	410	83		280	7.30
FLN-4205 -80HT		550	620		<1.0	115	0.25	7	900	860	24HRC	60HRC	190	6.60
-105HT		720	790	(D)	<1.0	130	0.27	9	1170	1000	30	60	250	6.80
-140HT		970	1030		<1.0	145	0.27	12	1590	1170	36	60	320	7.05
-175HT		1210	1280		1.0	160	0.28	19	2000	1380	42	60	400	7.30

Standard 35, Materials Standards for P/M Structural Parts
1997 Edition

FIGURE 20.8 Typical detail of properties illustrated in the P/M industry material standards. (*Courtesy of Metal Powder Industries Federation, Princeton, NJ.*)

Pure iron as well as many of its alloys can be fabricated by the P/M industry for both structural and magnetic applications. Other popular soft magnetic materials beyond pure Fe include the Fe–P, Fe–Si, and Fe–Ni alloy systems. Both soft and hard magnets can be produced by the powder metal route. Soft magnetic materials need to be as clean as possible, while hard magnets are made with as many inclusions and defects as possible, so that these can act as barriers to domain wall movement.[13] Cermets and the rare-earth magnets (Nd–Fe–B), e.g., are all processed with minor modifications to the P/M technology. Both dry pressing in rigid tooling and injection molding are used to form the hard magnet shapes.

Stainless steels may also be used for soft magnetic applications as well as for structural applications where corrosion resistance is important. Magnetic properties of the ferritic (400 series) stainless steels are used in *antilock brake systems* (ABS), magnetic clutches, and limit switches. The nonmagnetic austenitic (300 series) stainless steels are generally more corrosion resistant and, therefore, are used in watches, computer hardware, threaded fasteners, filters, water and gas meters, shower heads, sprinkler system nozzles, etc. Additional automotive examples include the rear view mirror mounts, brake components, and windshield wiper pinions. In more recent applications, flanges and sensor bosses for automotive exhaust systems were found to exhibit better oxidative corrosion resistance to thermal cycling when small additions of niobium were made to the ferritic stainless steels.

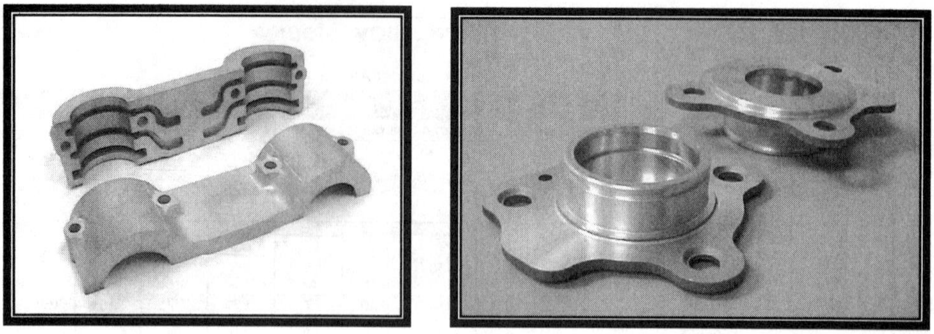

(a) (b)

FIGURE 20.9 Examples of P/M aluminum structural parts. (*a*) Cam cap for dual overhead cam engine, overall length about 17 cm. (*b*) Hub guard for hand tool application, hub diameter about 4 cm. (*Courtesy of Metal Powder Products, Westfield, IN.*)

Material standards for P/M self-lubricating bearings, P/F steel parts and metal injection molded parts have also been developed. These can be used to provide guidance on the properties of products to be expected from these technologies.

Nonferrous materials, e.g., brass and bronze are often used to make components for bearings, locks, shock absorber, and counterweights. Pure copper is used in electrical applications as well as for thermal management. Usage of P/M aluminum alloys took off at the end of the century when a major automotive company switched to P/M from die-cast aluminum as a process to make camshaft caps for a dual overhead cam engine.[18] The P/M structural aluminum was chosen over a die cast part as it offered improved dimensional consistency and mechanical properties, as well as the elimination of any secondary machining. Beyond the automotive industry, P/M structural aluminum components are being used in the hand tool (Fig. 20.9), business machine and marine markets.

As mentioned above, the more exotic materials such as tool steels, superalloys, refractory metals, and titanium are processed by CIP and HIP, as well as newer, specialized processes.

20.5 COMPARISON TO COMPETING METALWORKING TECHNOLOGIES

The P/M technology has the ability to produce inexpensive components that compete with alternate fabrication methods, such as machining from billets/bar stock, screw machining, and castings.

In general, the screw machined product made on multiple spindle machines, requiring little or no additional handling will be less expensive than P/M. However, when additional operations are needed on the screw machined part, the price of the P/M part becomes more competitive. Recall that the shape of screw machined parts is limited to configurations that are symmetrical to the axis of rotation.

The clear advantage in terms of cost for P/M occurs when the alternate machining process (CNC, lathe turning, milling, drilling, etc. generates excessive scrap, since most (>95%) of the initial raw material in P/M ends up in the sintered product. Much of the challenge in designing a part for P/M is to ensure that as many of the features as possible that are desired in the final product are produced during molding. Any operation that is added to the process after sintering will of course add cost, and must be evaluated from the perspective of part function. Even with additional secondary operations, such as machining, the P/M process may still have a clear advantage, as it is able to produce the rest of the configuration in a consistent and reliable manner.

P/M competes very well with die-cast parts, from the perspective of material availability, but the degree of design freedom is much greater with the casting approach. Die cast parts are usually less expensive and can produce intricate shapes comparable to injection molded parts. They can also be much larger and have tighter dimensional control than P/M parts. However, die castings are limited to lower melting point materials (aluminum, magnesium, and zinc) and their inherent limited properties. Additives are also added to the melt to aid fluid flow, which may degrade properties. Most castings will contain some degree of shrinkage porosity and minor amounts of trapped inclusions from the melt surface.

Castings made in sand molds may also contain small amounts of ceramic material from the mold or the furnace and ladle materials used to handle the melt. Sand or "foundry" castings offer similar materials to P/M with lower tool and material costs, but higher manufacturing costs. Therefore, castings may be preferred for large parts at low production volumes, while P/M is more competitive for higher quantities. Powder metallurgy may also offer advantages if there is significant surface machining or finishing that needs to be performed on the casting.

Stamping or sheet metal forming processes using one die are often less expensive than P/M, but the low material utilization of these processes may swing the advantage to P/M for certain components. Powder metallurgy also becomes more economical, if several successive dies are needed to form the desired part, because the stamping, tooling, and manufacturing costs are higher. When the thickness of the sheet exceeds about 3.2 mm (0.125 in), conventional stamping production becomes more difficult. Specialized techniques such as fine-flow blanking can be used effectively up to about 6.4 mm (0.25 in), if the slight edge curling is acceptable. As long as the material usage and dimensional flatness is not a concern, sheet metal forming is the preferred method of manufacturing over P/M. This is certainly the case when producing large instrument covers and panels.

20.6 CONCLUSION

This short introduction to the P/M fabrication technology was intended to bring to the forefront for the engineering community, a parts-making methodology that many are not exposed to. The technology has the potential to provide a method of producing almost any part that can fit in the palm of ones hand. It is highly-suited for mass-production quantities, since each part is so consistent. As there is some capital investment in the molding tools that must be amortized over the price of the parts, the P/M technology is not normally suited for cheap parts in quantities of less than ten thousand. At the same time it should be understood that the cost of P/M tooling is considerably less than that for cold heading, injection molding and die casting.

Brief descriptions of the process steps were provided with a view to providing the pros and cons of the P/M process when compared to competing fabrication technologies. Detailed technical discussions of the idiosyncrasies of each of the process steps were not provided, even though a thorough understanding by the P/M engineers is essential to make the technology so successful. The referenced publications and web sites can be used as a source of such processing details as well additional information about the P/M industry.

These technical publications also provide descriptions of many process variations and novel technological advancements that were not covered in this section. For example, research work on advanced techniques, such as nanocrystalline materials, rapid solidification, plasma processing, spray deposition, and metal matrix composites portend a future of exciting possibilities for the parts-making capabilities of the P/M technology.[19]

REFERENCES

1. "New Delhi; India's Mirror," *National Geographic,* Vol. 167, No 4, April 1985.
2. U.S. Patent 963, 872, 1910.
3. *Powder Metallurgy Design Manual,* 3d ed., Metal Powder Industries Federation, Princeton, NJ, 1998.
4. D. White, "State-of-the-North American P/M Industry—2001," *Advances in Powder Metallurgy & Particulate Materials1,* Metal Powder Industries Federation, Princeton, NJ, 2001.
5. D. White, "State-of-the-North American P/M Industry—2002, *Advances in Powder Metallurgy & Particulate Materials,* Metal Powder Industries Federation, Princeton, NJ, 2002.
6. A. Lawley, *Atomization; the Production of Metal Powders,* Metal Powder Industries Federation, Princeton, NJ, 1992.
7. ASM Handbook,Vol. 7, *Powder Metal Technologies and Applications,* ASM International, Materials Park, OH, 1998.
8. F. V. Lenel, *Powder Metallurgy—Principles and Applications,* Metal Powder Industries Federation, Princeton, NJ, 1980.
9. R. M. German, *Powder Metallurgy Science,* 2d ed., Metal Powder Industries Federation, Princeton, NJ, 1994.
10. H. A. Kuhn and B. L. Ferguson, *Powder Forging,* Metal Powder Industries Federation, Princeton, NJ, 1995.
11. R. M. German, *Powder Injection Molding of Metals and Ceramics,* Metal Powder Industries Federation, Princeton, NJ, 1990.
12. R. M German and A. Bose, *Injection Molding of Metals and Ceramics,* Metal Powder Industries Federation, Princeton, NJ, 1997.
13. C. Lall, *Soft Magnetism: Fundamentals for Powder Metallurgy and Metal Injection Molding,* Metal Powder Industries Federation, Princeton, NJ, 1992.
14. W. D. Badger and H. J. Sanderow, *Power Transmission Components, Advances in High Performance Powder Metallurgy Applications,* Metal Powder Industries Federation, Princeton, NJ, 2001.
15. E. Klar and P. K. Samal, "Powder Metallurgy Stainless Steels," ASM Handbook,Vol. 7, *Powder Metal Technologies and Applications,* ASM International, Materials Park, OH, 1998, pp. 774–785.
16. C. Lall, "Principles and Applications of High Temperature Sintering," *Reviews in Particulate Materials,* Vol. 1, 1993, pp. 75–107.
17. "Materials Standards for P/M Structural Parts—2000," *MPIF Standard 35,* Metal Powder Industries Federation, Princeton, NJ, 2000.
18. C. Lall and W. Heath, "P/M Aluminum Structural Parts—Manufacturing and Metallurgical Fundamentals", *Int. J. Powder Metall.,* Vol. 36, No. 6. pp. 45–50, 2000.
19. F. H. Froes, "Advances in Powder Metallurgy Applications," ASM Handbook, Vol. 7, *Powder Metal Technologies and Applications*, ASM International, Materials Park, OH, 1998, pp. 16–22.

INFORMATION RESOURCES

Organizations

ASM international:www.asminternational.org

European Powder Metal Association: www.epma.com

Metal Powder Industries Federation (MPIF): www.mpif.org

Parts Producers

ASCO Sintering: www.ascosintering.com

Burgess-Norton Manufacturing: www.burgessnorton.com

Chicago Powder Metal Products: www.cpmp.net

GKN Sinter Metals: www.gknsintermetals.com
Keystone Powdered Metal: www.keystonepm.com
Metal Powder Products: www.metalpowder.com
Pacific Sintered Metal: www.pacificsintered.com
Stackpole Limited: www.stackpole.on.ca

Powder Suppliers

Ametek: www.ametekmetals.com
Hoeganaes: www.hoeganaes.com
North American Hoganas: www.northamericanhoganas.com
Quebec Metal Powders: www.qmp-powders.com

CHAPTER 21
WELDING, FABRICATION, AND ARC CUTTING

Duane K. Miller
Manager, Engineering Services
The Lincoln Electric Company
Cleveland, Ohio

21.1 INTRODUCTION

21.1.1 The Role of Welding, Brazing, and Thermal Cutting in Industry

Welding is an important and complex manufacturing process used in a wide variety of industries. It is used to join metals and some nonmetals, eliminating fasteners such as bolts and rivets. Stampings, castings, forgings, rolled plate and sheet, bar stock, rolled shapes, and other material forms can be joined by welding to create a one piece assembly. Within limits, dissimilar materials can be joined by welding, such as stainless steel to carbon steel.

Within the general category of welding is included brazing and soldering. These forms of joining provide the manufacturing engineer with more options as these processes permit joining of materials that would not normally be welded with other processes. Also in the general field of welding are thermal cutting processes that allow materials to be severed.

These topics are extensively covered in this chapter. Additionally, the manufacturing engineer should be aware of several related issues that are not normally within the scope of his/her responsibility, including material selection, connection design, and nondestructive testing. References are included so that the reader can investigate issues of interest that are beyond the scope of coverage for this chapter.

21.1.2 Relevance to the Manufacturing Engineer

The manufacturing engineer's role as it relates to welding may be quite varied, depending on the size of company by which he/she is employed, how functions and responsibilities are determined, and other factors. The basic assumptions of this chapter are that a product design engineer is responsible for the selection of material types, thickness, joint designs, calculation of stresses in the part, and stresses across the joint, and has determined the weld type and size. The basic responsibilities of the manufacturing engineer are to determine the means and methods by which a product can be produced at the most economical rate, with the required quality level, and in a safe manner for all involved with the manufacturing operations. There is obviously interaction between the product design engineer and the manufacturing engineer, and understanding the constraints imposed upon the other party can help achieve the goals of low cost, quality, and safety. For example, a seemingly minor material change can have significant welding implications. Welds that have been specified in design to be larger than needed will automatically increase production costs. Ongoing interaction between the product design engineer and manufacturing engineer can help optimize operations.

21.2 FUNDAMENTAL PRINCIPLES OF FUSION

Welding is a process of joining materials. In order for materials to join, the atoms of each material must be brought close enough such that the atoms of each material share electrons and form an atomic bond. Fusion, or welding, can be achieved when two conditions are met: atomic closeness and atomic cleanliness. The atoms must be close enough to each for the atomic bonds to form. On an atomic level, even apparently flat surfaces have peaks and valleys that are many angstroms in height, creating significant atomic gaps between the atoms of the two pieces of material. Atomic cleanliness is the second requirement. Again, this is rarely achieved because of the great tendency of metal to oxidize. This oxide neutralizes the atomic forces that would encourage fusion.

For welding to occur, heat and/or pressure are introduced to achieve atomic closeness and atomic cleanliness. Heat is used to achieve atomic closeness in many ways. The most common method is the use of an electric arc. The intense heat of an electric arc achieves atomic closeness by locally melting the two metals, and upon cooling and solidification of the metal, an atomic bond is formed. Chemical reactions such as oxygen together with a fuel gas, as well as resistance heating of materials either by electricity or friction, are other methods in which heat can be used to achieve atomic closeness. In terms of fusion related processes, little or no pressure is required to achieve atomic bonding, since local melting and solidification of the material will achieve atomic bonding. However, in solid-state joining processes, mechanical pressure must be exerted on the materials to achieve atomic bonding. The mechanical pressure forces the atoms of the two materials together, enabling atomic closeness to occur.

Atomic cleanliness is the second requirement for fusion. Surface oxides, as well as other contaminants, such as oil, grease, moisture, paint, various platings, and other foreign material, inhibit atomic cleanliness. For welding to be successful, the material must either be free of surface contaminants that would prevent proper bonding, or the surface material must be removed. Removal is accomplished by mechanical means, or chemical reactions. The mechanical methods may be as simple as wire brushing, shot blasting or scraping the part. Alternately, the cleaning may be part of the welding process itself, as in friction welding where one part is rubbed against another, and surface contaminates are mechanically removed. Particularly in arc welding, surface contaminants are removed by chemical reactions and melting the surfaces on which the foreign material resides. Specially formulated fluxes are used in soldering, brazing, and arc welding to make certain the base metal surfaces are atomically clean so that bonding can occur.

After the surfaces to be joined are atomically close, and atomically clean, they must be protected from further contamination until the metallurgical bond has been formed. The time period involved is generally short, in the range of seconds to a few minutes. During the time when materials are hot and molten, they must be protected from the atmosphere, which contains nitrogen and oxygen, both materials that can contaminate molten and hot metals. To accomplish this purpose, inert shielding gases, fluxes, and slags are used. Gases displace the atmosphere, precluding contact of the hot metal with nitrogen and oxygen. Fluxes and slags function as mechanical barriers, inhibiting the ability of the atmosphere from coming into contact with the hot metal.

All welding processes do approximately the same thing: permit atomic closeness and atomic cleanliness, and protect hot metal from contamination until the welding operations are complete. The means by which these functions are performed distinguishes one welding process from another. Some involve the application of only heat, while others apply only pressure. Some involve both. The processes may involve operations whereby the surfaces are mechanically or chemically cleaned; other processes require clean material to begin with, in order for welding to be successful. Recognizing how each of these functions is being achieved will help in the selection of the process best suited for a given application.

21.3 PROCESS SELECTION

More than 40 different welding processes are used to join metals, and while not all are covered in this chapter, the most popular commercial processes are described. Proper process selection will depend upon the materials to be joined, the service requirements of the assembly and the fabrication costs. Welding processes can be divided into two main categories: fusion-related processes and solid-state joining of materials.

21.3.1 Types of Processes

A fusion process is one that melts the juncture of two metal parts, and upon cooling and solidification, a metallurgical bond is formed between the parts. Fusion-related processes include: *shielded metal arc welding* (SMAW), *gas metal arc welding* (GMAW), *oxyacetylene welding* (OAW), *gas tungsten arc welding* (GTAW), and many others. Solid-state welding creates metallurgical bonds between two materials without melting in the juncture of the two materials. Solid-state welding processes include, but are not limited to, friction welding, ultrasonic welding, and brazing.

21.3.2 Modes of Application

Welding processes are also categorized by their mode of application. The four basic modes of application are manual, semiautomatic, mechanized/automatic, and robotic. Manual welding processes are those that require the operator to manipulate the weld pool while manually adding filler metal if needed. Manual welding processes include SMAW, GTAW, and OAW. Manual welding processes typically require the greatest operator skill. A semi-automatic process is one that still requires the operator to manipulate the weld pool, yet the filler metal is mechanically fed into the weld pool. The most popular semiautomatic welding process is GMAW. A mechanized/automatic welding process, or hard automation, is a machine-controlled process that mechanically moves the arc along the weld joint, or moves the workpiece relative to a stationary arc. The semiautomatic processes can all be automated; furthermore, some processes can only be applied in the automatic mode. Robotic welding, or flexible automation, is an application in which a robot typically welds around a fixtured assembly. The high precision and repeatability of robotic welds are typically used to optimize welding speeds on high production parts requiring multiple welds. Semiautomatic processes can be adapted to the robotic mode of application, although some processes lend themselves to adaptation more easily than others. In arc welding, GMAW and GTAW are commonly used for robotic applications.

Arc Welding. Arc welding is one of several fusion processes for joining metal. By the generation of intense heat, the juncture of two metal pieces is melted and mixed—directly or, more often, with an intermediate molten filler metal. Upon cooling and solidification, the resulting welded joint metallurgically bonds the former separate pieces into a continuous structural assembly, typically called a weldment.

In arc welding, the intense heat needed to melt metal is produced by an electric arc (Fig. 21.1). The arc forms between the workpieces and an electrode, i.e., either manually or mechanically moved along the joint; conversely, the work may be moved under a stationary electrode. The electrode generally is a specially prepared rod or wire that not only conducts electric current and sustains the arc, but also melts and supplies filler metal to the joint; this constitutes a consumable electrode. Carbon or tungsten electrodes may be used, in which case the electrode serves only to conduct electric current and to sustain the arc between the tip and workpiece. The electrode is not consumed; with these electrodes, any filler metal required is supplied by rod or wire introduced into the region of the arc and melted there. Filler metal applied separately, rather than via a consumable electrode, does not carry electric current.

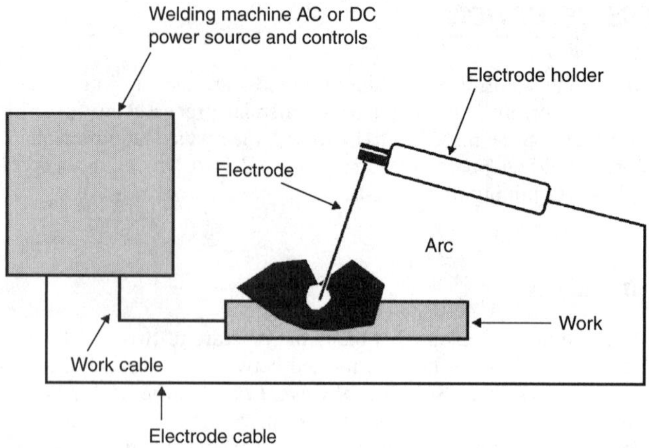

FIGURE 21.1 Typical welding circuit.

21.3.3 Shielded Metal Arc Welding

The SMAW process (Fig. 21.2), commonly known as stick welding, or manual welding, is a popular and widespread welding process. It is versatile, relatively simple to do, and very flexible in how it is applied. SMAW is used in the shop and in the field for fabrication, erection, maintenance, and repairs. Because of the relative inefficiency of the process, it is seldom used for production.

SMAW may utilize either direct current (dc) or alternating current (ac). Generally, dc is used for smaller electrodes, usually less than $^3/_{16}$ in diameter. Larger electrodes utilize alternating current to eliminate undesirable *arc blow* conditions. ("Arc blow" is the deflection of an arc from its normal path because of magnetic forces.)

Electrodes used with ac must be designed specifically to operate in this mode, in which the current flow changes direction 120 times/s with 60-Hz power. All ac electrodes will operate acceptably on direct current. The opposite is not always true.

Flexibility is perhaps SMAW's strongest suit. With one power supply, a wide range of materials can be welded by simply changing the electrode to be used, and the power source output setting. The same machine can, e.g., weld material from $^1/_8$ in thick to several inches in thickness, as well as weld steel, stainless steel, and cast iron. For these reasons, SMAW remains very popular for maintenance and repair applications.

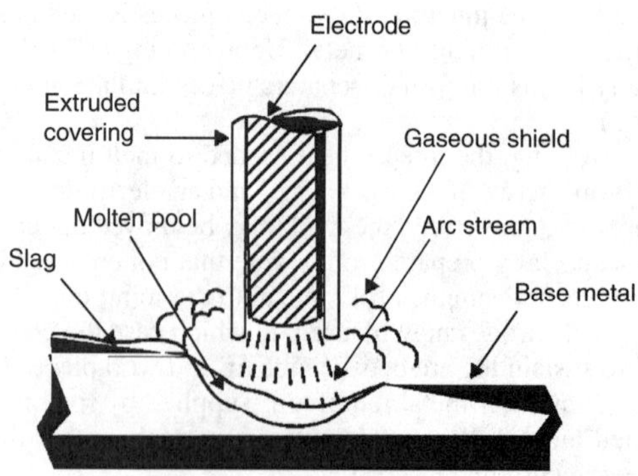

FIGURE 21.2 SMAW process.

The greatest shortcoming of SMAW is its relative inefficiency. The electrodes are used in lengths of 9–18 in, and approximately 2 in of each electrode is unusable. Once the electrode is consumed, the welder must stop welding, remove the stub, insert another electrode, and start welding again. These work interruptions are inherent to the process, and reduce production rates accordingly.

21.3.4 Gas Metal Arc Welding

GMAW is a process that employs a solid or metal cored electrode that is fed by a wire feeder into a gun and cable assembly, and through a tube called a contact tip. At the contact tip, the electrical current is passed into the electrode. A short length of electrode extends beyond the contact tip, known as the electrode extension. An arc is formed at the end of the electrode, and the arc region is shielded by a gas that displaces the atmosphere. The arc melts the base metal as well as the electrode (see Figs. 21.3 and 21.4).

GMAW has several distinct advantages over SMAW described above. The electrode is dispensed from a coil or reel, ranging in size from a pound to several hundred pounds, with 20–60 lb packages being typical. The welding operations no longer need to be interrupted to replace each 9–18 in of electrode. Welding is typically permitted to continue until the weld is complete. The process does not involve flux and slag as does SMAW. When the welding is complete, the weld is typically in a "finished" condition. There is no slag to remove, and only minimal levels of smoke are deposited on the part being welded. With the proper machine settings, semi-automatic GMAW requires less skill than does manual SMAW.

GMAW may be referred to as metal inert gas, solid wire and gas, and miniwire or microwire welding. The shielding gas may be carbon dioxide or blends of argon with CO_2 or oxygen, or both. GMAW is usually applied in one of four ways: short arc transfer, globular transfer, spray arc transfer, or pulsed arc transfer.

Short arc transfer is ideal for welding thin-gage materials. In this mode, a small electrode, usually of 0.035–0.045-in diameter, is fed at a moderate wire feed speed (WFS) at relatively low voltages. The electrode contacts the workpiece, resulting in a short circuit. The arc is actually quenched at this point, and very high current will flow through the electrode, causing it to heat and melt. A small amount of filler metal is transferred across the arc and is added to the weld puddle. The cycle will repeat itself when the electrode short-circuits to the work again. This occurs between 60 and 200 times per second, creating a characteristic buzz. This mode of transfer is ideal for sheet metal, but often results in fusion problems if applied to thicker sections, where cold lap or cold casting

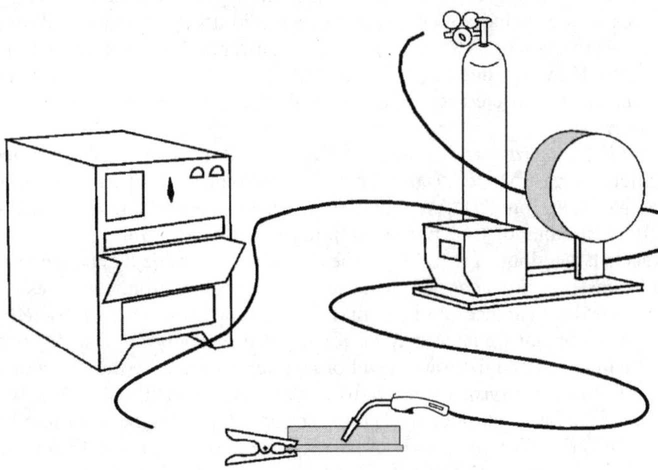

FIGURE 21.3 GMAW and FCAW equipment.

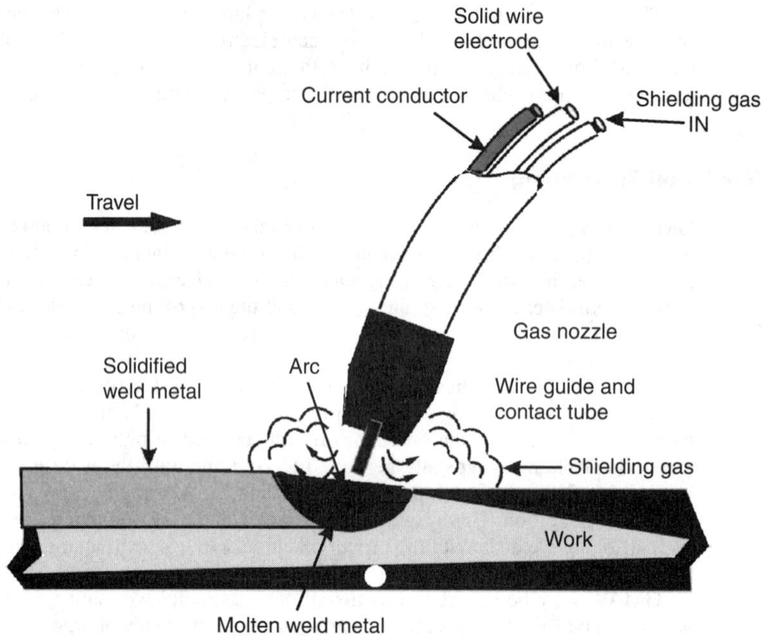

FIGURE 21.4 GMAW welding process.

can result, and the deposited metal may not fuse to the base metal. This is unacceptable since the welded connection will have virtually no strength. Caution must be exercised, particularly if the short arc transfer mode is applied to sections over $^1/_8$ in thick.

Spray arc transfer is characterized by high WFSs at relatively high voltages. A fine spray of molten filler metal, all smaller in diameter than the electrode, is ejected from the electrode toward the work. Unlike with short arc transfer, the arc in spray transfer is maintained continuously. High-quality welds with particularly good appearance are obtained. The shielding gas used in spray arc transfer is composed of at least 80 percent argon, with the balance either CO_2 or O_2. Typical mixtures would include $^{90}/_{10}$ argon/CO_2, and $^{90}/_5$ argon/O_2. Relatively high arc voltages are used with spray arc transfer. Gas metal spray arc transfer welds generally have excellent appearance and good fusion. However, due to the intensity of the arc and gravitational pull on the weld puddle, spray arc transfer is restricted to applications in the flat and horizontal positions, and to relatively thick materials.

Globular transfer is a mode of GMAW that results when high concentrations of CO_2 are used for shielding gas. Moderate to high WFSs are typical. Carbon dioxide is not an inert gas, rather it is active. Therefore, GMAW that uses CO_2 may be referred to as MAG, for *metal active gas*. With high concentrations of CO_2 in the shielding gas, the arc no longer behaves in a spray-like fashion, but ejects large globs of metal from the end of the electrode, hence the term globular transfer. This mode of transfer, while generally resulting in deep penetration, generates relatively high levels of spatter, and weld appearance can be poor. Like the spray mode, it is restricted to the flat and horizontal positions. Globular transfer may be preferred over spray arc transfer because of the low cost of CO_2 shielding gas and the lower level of heat experienced by the operator.

Pulsed arc transfer is a relatively new development in GMAW. In this mode, a background current is applied continuously to the electrode. A pulsing peak current is applied at a rate proportional to the WFS. With this mode of transfer, the power supply delivers a pulse of current, which ideally ejects a single droplet of metal from the electrode. The power supply then returns to a lower background current to maintain the arc. This occurs between 100 and 400 times/s. One advantage of pulsed arc transfer is that it can be used out-of-position. For flat and horizontal work, it will not be as fast as spray arc transfer. However, when it is used out-of-position, it is free of the problems associated with the gas metal arc short-circuiting mode. Weld appearance is good, and quality can be excellent. The disadvantages of pulsed arc transfer are that the equipment is slightly more complex and is relatively costly.

Metal cored electrodes comprise another new development in GMAW. This process is similar to *flux cored arc welding* (FCAW, discussed below in 21.3.5) in that the electrode is tubular, but the core material does not contain slag-forming ingredients. Rather, a variety of metallic powders are contained in the core, resulting in exceptional alloy control. The resulting weld is slag-free, as are other forms of GMAW.

The use of metal cored electrodes offers many fabrication advantages. Compared to spray arc transfer, metal cored electrodes require less amperage to obtain the same deposition rates. They are better able to handle mill scale and other surface contaminants. When used out-of-position, they offer greater resistance to the cold lapping phenomenon, common with short arc transfer. Finally, metal cored electrodes permit the use of amperages higher than may be practical with solid electrodes, resulting in higher metal deposition rates.

21.3.5 Flux Cored Arc Welding

In FCAW, the arc is maintained between a continuous tubular metal electrode and the weld pool. The tubular electrode is filled with flux and a combination of materials that may include metallic powders. FCAW may be done automatically or semiautomatically. FCAW has become the workhorse in fabrication shops practicing semiautomatic welding. Production welds that are short, change direction, are difficult to access, must be done out-of-position (i.e., vertical or overhead), or are part of a short production run generally will be made with semiautomatic FCAW.

FCAW utilizes equipment that is very much like that used with GMAW (Fig. 21.3), in many situations the equipment can be used interchangeably. FCAW electrodes are generally larger in diameter than GMAW electrodes, and thus the wire feeder may require different drive rolls. The amperage levels used with FCAW are usually greater than for GMAW, so power source output levels may be higher for FCAW.

Like GMAW, FCAW uses an electrode fed from a large coil or reel that permits uninterrupted welding. Unlike GMAW, FCAW leaves behind a slag layer that covers the completed weld, as is the case with SMAW. This layer must be removed before welding over the previous weld bead. Thus, there are some extra cleaning activities associated with FCAW. This slag, however, is the result of fluxes that are contained within the electrode. These fluxes render FCAW more tolerant of surfaces contaminated with substances such as scale and rust, than is GMAW.

Within the category of FCAW, there are two specific subsets: *self-shielded flux cored arc welding* (FCAW-ss) (Fig. 21.5) and *gas-shielded flux cored arc welding* (FCAW-g) (Fig. 21.6). Self-shielded flux cored electrodes require no external shielding gas. The entire shielding system results

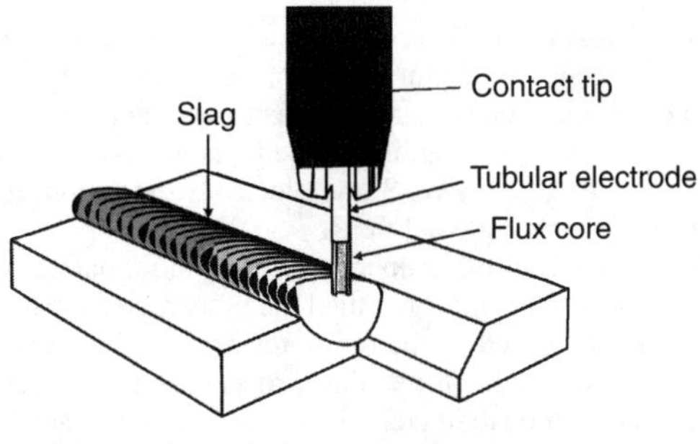

FIGURE 21.5 Self-shielded FCAW.

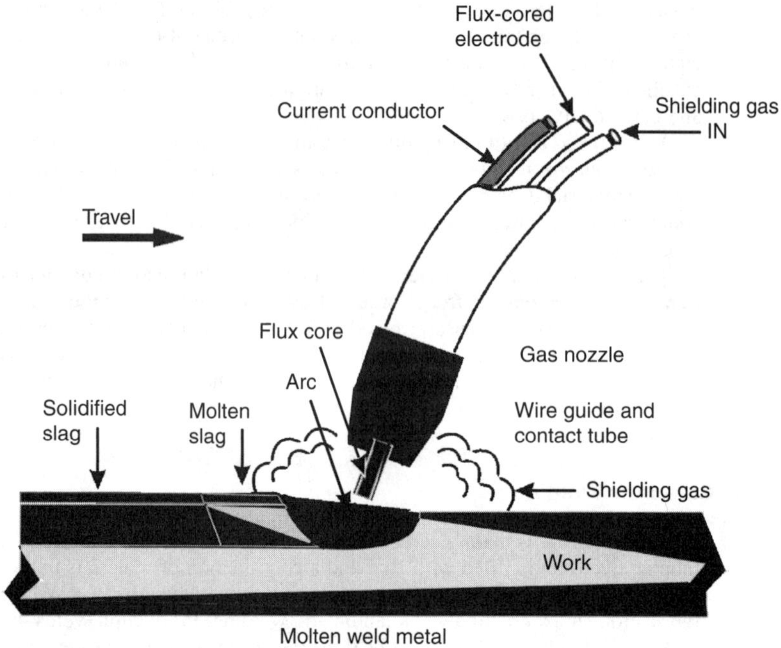

FIGURE 21.6 Gas-shielded FCAW.

from the flux ingredients contained in the tubular electrode. The gas-shielded variety of flux cored electrode utilizes, in addition to the flux core, an externally supplied shielding gas. Often, CO_2 is used, although other mixtures may be employed.

Self-shielded flux cored electrodes are ideal for field welding operations because no externally supplied shielding gas is required and the process may be used in high winds without adversely affecting the mechanical properties or quality of the weld metal deposited. With any gas-shielded processes, wind shields must be erected to preclude wind interference with the gas shield. A momentarily opened door in a shop, or a fan used to cool a welder in a hot building, can disrupt the gas shielding. Self-shielded FCAW may offer advantages in such situations.

Gas-shielded flux cored electrodes tend to be more versatile than self-shielded flux cored electrodes and, in general, provide better arc action. Operator acceptance is usually higher. The gas shield must be protected from winds and drafts, but this is not difficult for most manufacturing shops. Gas shielded FCAW electrodes are generally available for more steel alloys and strength levels than for the self shield variation.

FCAW Equipment and Procedures. Like all wire-fed welding processes, FCAW requires a power source, wire feeder, and gun and cable assembly (Fig. 21.3). The power supply is a dc source, although either electrode positive or electrode negative polarity may be used. The four primary variables used to determine welding procedures are voltage, wire feed speed, electrode extension, and travel speed. For a given WFS and electrode extension, a specified amperage will be delivered to maintain stable welding conditions.

As WFS is increased, amperage will be automatically increased by the equipment. On some equipment, the WFS control is called the amperage control, which, despite its name, is just a rheostat that regulates the speed of the dc motor driving the electrode through the gun. The most accurate way, however, to establish welding procedures is to refer to the WFS, since electrode extension, polarity, and electrode diameter will also affect amperage. For a fixed WFS, a shorter electrical stick-out will result in higher amperages. If procedures are set based on the WFS, the resulting amperage verifies that proper electrode extensions are being used. If amperage is used to set welding procedures, an inaccurate electrode extension may go undetected.

Advantages and Limitations of FCAW. FCAW offers two distinct advantages over SMAW. First, the electrode is continuous and eliminates the built-in starts and stops that are inevitable with SMAW using stick electrodes. An economic advantage accrues from the increased operating factor; in addition the reduced number of arc starts and stops largely eliminates potential sources of weld discontinuities. Second, increased amperages can be used with FCAW. With SMAW, there is a practical limit to the amount of current that can be used. The covered electrodes are 9–18 in long, and if the current is too high, electric resistance heating within the unused length of electrode will become so great that the coating ingredients may overheat and break down. With continuous flux cored electrodes, the tubular electrode is passed through a contact tip, where electric current is transferred to the electrode. The short distance from the contact tip to the end of the electrode, known as electrode extension or "stickout," inhibits heat buildup due to electric resistance. This electrode extension distance is typically 1 in for flux cored electrodes, although it may sometimes be as much as 2 or 3 in.

Smaller-diameter flux cored electrodes are suitable for all-position welding. Larger electrodes, using higher electric currents, usually are restricted to use in the flat and horizontal positions. Although the equipment required for FCAW is more expensive and more complicated than that for SMAW, most fabricators find FCAW much more economical than SMAW.

21.3.6 Gas Tungsten Arc Welding

The GTAW process (Fig. 21.7), colloquially called TIG welding, uses a nonconsumable tungsten electrode. An arc is established between the tungsten electrode and the workpiece, resulting in heating of the base metal. If required, a filler metal is used. The weld area is shielded with an inert gas, usually argon or helium. GTAW is ideally suited to weld nonferrous materials such as stainless steel and aluminum, and is very effective for joining thin sections.

Highly skilled welders are required for GTAW, but the resulting weld quality can be excellent. The process is often used to weld exotic materials, such as titanium. Critical repair welds, as well as root passes in pressure piping, are typical applications. The completed GTAW weld, when properly

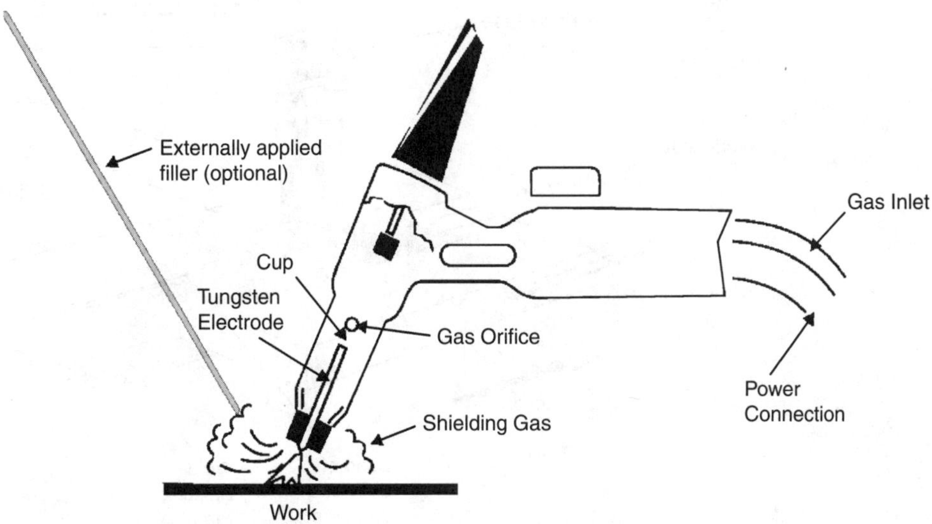

FIGURE 21.7 Gas tungsten arc welding.

made, has excellent appearance, and contains no slag that requires removal. GTAW is inherently slow, and thus, often used only in situations where no other process is viable. It is typically used in the manual mode, although automated and robotic applications are possible.

Tungsten electrodes can become contaminated when touched to the workpiece during the act of initiating an arc. To preclude this possibility, different starting circuits may be employed to eliminate the need to touch the electrode to the work. One such approach is to employ a high frequency voltage that is capable of jumping a gap between the electrode and the work. While very effective at this function, the same high frequency power can interfere with other electrical equipment. Proper installation of systems equipped with such starting circuits is essential.

21.3.7 Submerged Arc Welding

Submerged arc welding (SAW) differs from other arc welding processes in that a blanket of fusible granular flux is used to shield the arc and molten metal (Fig. 21.8). The arc is struck between the workpiece and a bare-wire electrode, the tip of which is submerged in the flux. The arc is completely covered by the flux and it is not visible, thus the weld is made without the flash, spatter, and sparks that characterize the open-arc processes. The flux used develops very little smoke or visible fumes.

Typically, the process is operated fully automatically, although semiautomatic operation is possible. The electrode is fed mechanically to the welding gun, head, or heads. In semiautomatic welding, the welder moves the gun, usually equipped with a flux-feeding device, along the joint.

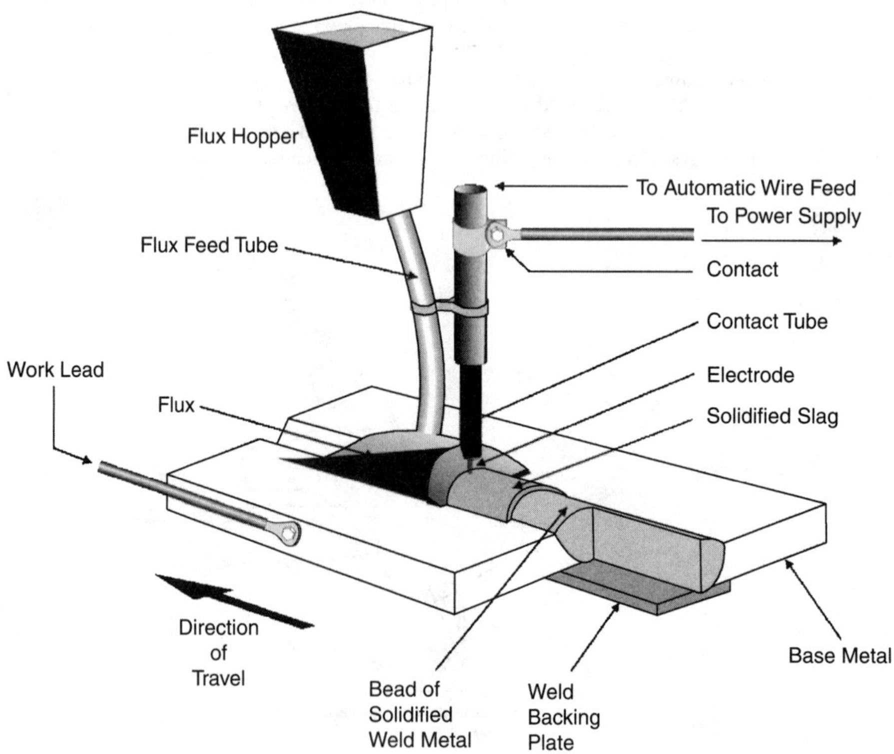

FIGURE 21.8 SAW process.

Flux may be fed by gravity flow from a small hopper atop the torch and then through a nozzle concentric with the electrode, or through a nozzle tube connected to an air-pressurized flux tank. Flux may also be applied in advance of the welding operation or ahead of the arc from a hopper run along the joint. Many fully automatic installations are equipped with a vacuum system to capture unfused flux left after welding; the captured, unused flux is recycled for use once more.

During welding, arc heat melts some of the flux along with the tip of the electrode. The electrode tip and the welding zone are always shielded by molten flux and a cover layer of unfused flux. The electrode is kept a short distance above the workpiece. As the electrode progresses along the joint, the lighter molten flux rises above the molten metal to form slag. The weld metal, having a higher melting (freezing) point, solidifies while the slag above it is still molten. The slag then freezes over the newly solidified weld metal, continuing to protect the metal from contamination while it is very hot and reactive with atmospheric oxygen and nitrogen. Upon cooling and removal of any unmelted flux, the slag is removed from the weld.

Advantages and Limitations of SAW. High currents can he used in SAW, and extremely high heat input can be developed. Because the current is applied to the electrode a short distance above the arc, relatively high amperages can be used on small-diameter electrodes. The resulting extremely high current densities on relatively small cross-section electrodes permit high rates of metal deposition.

The insulating flux blanket above the arc prevents rapid escape of heat and concentrates it in the welding zone. Not only are the electrode and base metal melted rapidly, but also fusion is deep into the base metal. Deep penetration allows the use of small welding grooves, thereby minimizing the amount of filler metal to be deposited and permitting fast welding speeds. Fast welding, in turn, minimizes the total heat input to the assembly and thus tends to limit problems of heat distortion. Even relatively thick joints can be welded in one pass with SAW.

SAW is versatile, in that can be applied in more ways than other arc welding processes. A single electrode may be used, as is done with other wire feed processes, but it is possible to use two or more electrodes in submerged arc welding. Two electrodes may be used in parallel, sometimes called twin arc welding, employing a single power source and one wire drive. In multiple-electrode SAW, up to five electrodes can be used, but most often, two or three arc sources are used with separate power supplies and wire drives. In this case, the lead electrode usually operates on direct current while the trailing electrodes operate on alternating current.

SAW is restricted to welding in the horizontal and flat position where gravity holds the granular flux in place. The lack of an open arc is both an advantage and limitation of the process. While the hidden arc makes it easy for the operator to oversee the general welding operations, it precludes observation of the weld pool and makes it more difficult to know for certain that weld placement is exact.

21.3.8 Electroslag and Electrogas Welding

Electroslag (ESW) and *electrogas* (EGW) *welding* (Figs. 21.9 and 21.10) are closely related processes that allow high deposition, automatic welding in the vertical plane. Properly applied, these processes offer tremendous savings over alternative, out-of-position methods and, in many cases, savings over flat-position welding. Although the two processes have similar applications and mechanical setup, there are fundamental differences in the arc characteristics.

Electroslag and electrogas are mechanically similar in that both typically utilize water cooled copper dams, or shoes, that are applied to either side of a square-edged butt joint. An electrode or multiple electrodes are fed into the joint. Usually, a starting sump is applied for the beginning of the weld. As the electrode is fed into the joint, a puddle is established that progresses vertically. The water-cooled copper dams chill the weld metal and prevent its escape from the joint. The weld is completed in one pass.

As an alternative to the copper dams that are removed once the weld is completed, it is possible to use steel dams that fuse to the metal being joined. In this case, the dams become part of the final product.

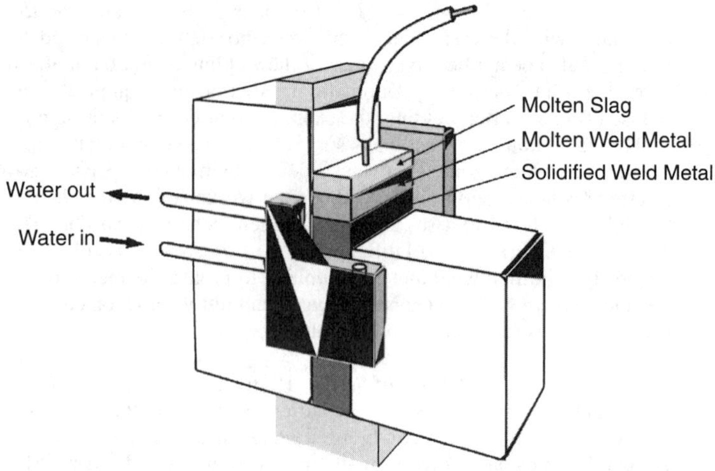

FIGURE 21.9 ESW process.

Both ESW and EGW are specialized processes and usually involve dedicated tooling and fixturing. They are best applied to thicker materials, usually over 1 in thick. Multiple electrodes can be used to join sections 4 in and thicker.

Electrogas welding is similar to GMAW: the electrode is fed into the puddle and shielded with gas. Electroslag welding is somewhat like SAW: the weld puddle is shielded with flux. Unlike SAW flux, ESW flux is electrically conductive. When the ESW weld begins, there is an arc under the flux blanket. Once the flux melts and forms the slag covering, the protective slag becomes electrically conductive, and electrical energy is transferred from the electrode directly into the slag. This hot slag then begins to melt the electrode, and the arc is extinguished. Thus, the ESW process is technically not an arc welding process, but a resistance welding process.

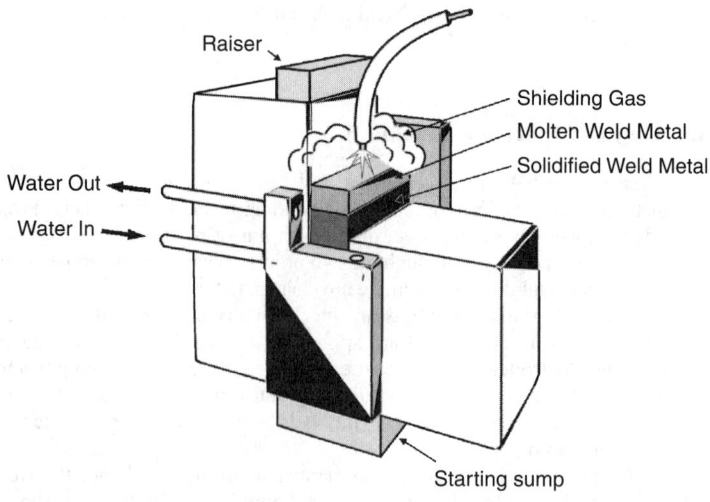

FIGURE 21.10 EGW process.

21.3.9 Plasma Arc Welding

Plasma arc welding (PAW) uses a constricted arc to generate very high, localized heating. PAW may utilize either a transferred or a nontransferred arc. In the transferred arc mode, the arc occurs between the electrode and the workpiece, much as in GTAW, the primary difference being the constriction afforded by the secondary gases and torch design. With the nontransferred arc mode, arcing is contained within the torch between a tungsten electrode and a surrounding nozzle.

The constricted arc results in higher localized arc energies than are experienced with GTAW, resulting in faster welding speeds. Applications for PAW are similar to those for GTAW. The only significant disadvantage of PAW is the equipment cost, which is higher than that for GTAW.

Most PAW is done with the transferred arc mode, although this mode utilizes a nontransferred arc for the first step of operation. An arc and plasma are initially established between the electrode and the nozzle. When the torch is properly located, a switching system will redirect the arc toward the workpiece. Since the arc and plasma are already established, the reliable transfer of the arc to the workpiece is easily accomplished. For this reason, PAW is often preferred for automated applications. PAW can be applied manually, automatically, or robotically.

21.4 RESISTANCE WELDING

In resistance welding, coalescence is produced by the heat obtained from the electric resistance of the workpiece to the flow of electric current in a circuit of which the workpiece is a part, and by the application of pressure. The specific processes include *resistance spot welding, resistance seam welding,* and *projection welding*. Figure 21.11 shows diagrammatic outlines of the processes.

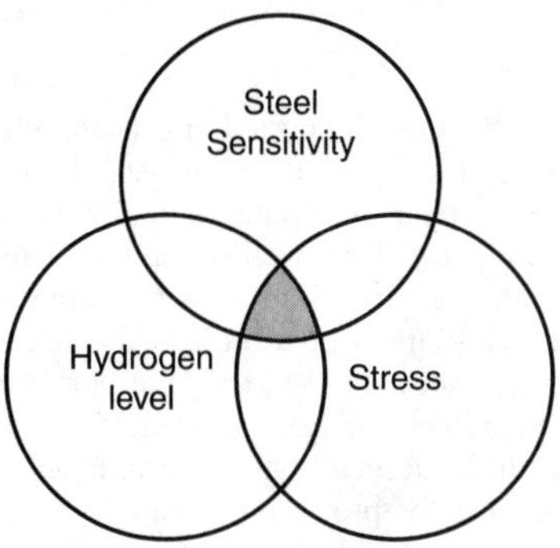

FIGURE 21.11 Resistance welding.

The resistance of the welding circuit should be at maximum at the interface of the parts to be joined, and the heat generated there must reach a value high enough to cause localized fusion under pressure. Pressure is applied through the electrodes that clamp the parts. During the welding cycle, material at the interface melts. Material surrounding the molten region is heated and expands. The expansion in turn generates increased pressure, encouraging fusion.

Electrodes are of copper alloyed with such metals as molybdenum and tungsten, with high electrical conductivity, good thermal conductivity, and sufficient mechanical strength to withstand the high pressures to which they are subjected. The electrodes are water-cooled. The resistance at the surfaces of contact between the work and the electrodes must be kept low. This may be accomplished by using smooth, clean work surfaces, and a high electrode pressure. For all forms of resistance welding, the parts to be joined must be relatively clean, as the process does not utilize any flux to reduce surface oxides.

In resistance spot welding, the parts are lapped and held in place under pressure. The size and shape of the electrodes control the size and shape of the welds, which are usually circular.

Resistance spot welding machines vary from small, manually operated units to large, elaborately instrumented units designed to produce high-quality welds, as on aircraft parts. Portable gun-type machines are available for use where the assemblies are too large to be transported to a fixed machine. Spot welding guns are often mounted on robots that allow for rapid repositioning of the guns. Spot welds may be made singularly or in multiples, with the latter generally made using special purpose machines. Spacing of electrodes is important to avoid excessive shunting of welding current.

The resistance seam welding process produces a series of spot welds made by circular or wheel type electrodes. The weld may be a series of closely spaced individual spot welds, overlapping spot welds, or a continuous weld nugget. The weld shape for individual welds is rectangular, and continuous welds are about 80 percent of the width of the roll electrode face.

A mash weld is a seam weld in which the finished weld is only slightly thicker than the sheets, and the lap disappears. It is limited to thicknesses of about 16 gage and an overlap of $1^1/_2$ times the sheet thickness. Operating the machine at reduced speed, with increased pressure and uninterrupted current, a strong quality weld may be secured that will be 10–25 percent thicker than the sheets. The process is applicable to mild steel but has limited use on stainless steel; it cannot be used on nonferrous metals. A modification of this technique employs a straight butt joint. This produces a slight depression at the weld, but the strength is satisfactory on some applications, e.g., on electric-welded pipe and tubing.

Cleanliness of sheets is of even more importance in seam welding than in spot welding. Best results are secured with cold-rolled steel, wiped clean of oil—the next best with pickled hot-rolled steel. Grinding or polishing is sometimes performed, but not sand- or shot-blasting.

In projection welding, the heat for welding is derived from the localization of resistance at predetermined points by means of projections, embossments, or the intersections of elements of the assembly. The projections may be made by stamping or machining. The process is essentially the same as spot welding, and the projections seem to concentrate the current and pressure. Welds may be made singularly or in multiple with somewhat less difficulty than is encountered in spot welding. When made in multiple, all welds may be made simultaneously. The advantages of projection welding are (1) the heat balance for difficult assemblies is readily secured, (2) the results are generally more uniform, (3) a closer spacing of welds is possible, and (4) electrode life is increased. Sometimes it is possible to make projection welds that could not be made by other means.

21.5 SOLID-STATE WELDING

Solid-state welding encompasses a group of processes in which joining is accomplished by bringing clean metal surfaces into intimate contact under certain specific conditions. No melting of the base metals occurs. Highly specialized, these processes tend to be automatically applied by dedicated machines made for specific applications. They are particularly well suited for joining materials that, typically for metallurgical reasons, cannot be melted and resolidified without major difficulties.

Friction welding involves rotating one part while the second part is fixed. Pressure is exerted between the two parts, resulting in friction that generates intense local heat. Surface oxides and contaminants are mechanically removed from the interface. The parts are heated, but not to the melting point of the metal. Rotation is stopped at the critical moment that bonding is achieved. Base metal properties across the joint show little change because the process is so rapid.

Friction welding applications typically involve round parts. Hydraulic cylinders, stems to valves, and drilling heads for well drilling are typical friction welding applications. While capital costs may be high, the per unit costs may be very low with this process.

Ultrasonic welding employs mechanical vibrations at ultrasonic frequencies plus pressure to create the intimate contact between faying surfaces needed to produce a weld. The welding tool is essentially a transducer that converts electric frequencies to ultra-high-frequency mechanical vibrations. By applying the tip of the tool, or anvil, to a small area on the external surface of two lapped parts, the vibrations and pressure are transmitted to the faying surfaces. Foils, thin-gage sheets, or fine wires can be spot- or seam-welded to each other or to heavier parts.

Ultrasonic is capable of welding metals to metals, plastics to plastics, and metals to nonmetals. It is restricted to relatively thin materials and is used to weld electronic components. It has been used to join encapsulating material for explosives where high temperatures would be unacceptable. It is often used in applications that might otherwise be joined by resistance welding, but where the size of the part, or the need to avoid heating the part, makes ultrasonic welding the preferred process.

21.6 OXYFUEL GAS WELDING

The heat for oxyfuel gas welding is supplied by burning a mixture of oxygen and a suitable combustible gas. The gases are mixed in a torch that controls the welding flame. The torch is moved along the juncture of two metals, and a filler metal, if used, is manually fed into the weld pool by the welder. Perhaps the biggest advantage of the oxyfuel gas welding process is the increased ability of the operator to control heat input into the base metal, compared to other welding processes. However, the oxyfuel gas welding process is relatively slow, is not easily automated, and is rarely used for production welding.

Acetylene is almost universally used as the combustible gas for oxyfuel welding because of its high flame temperature. This temperature, about 6,000°F, is so far above the melting point of all commercial metals that it provides a means for the rapid localized melting essential in welding. The oxyacetylene flame is also used in cutting ferrous metals.

The same torch used for oxyfuel welding can be used for brazing, soldering and thermal cutting. It can be used to heat parts as well. This flexibility makes oxyfuel welding sets popular tool for maintenance, repair, and light fabrication. The slow productivity of the process is a major restriction that has essentially eliminated its use for production welding applications. The thermal cutting capability of oxygen is reviewed in following section on Oxygen Cutting.

21.7 THERMAL CUTTING

21.7.1 Air Arc Gouging

The air carbon arc gouging system (Fig. 21.12) utilizes an electric arc to melt the base material; a high-velocity jet of compressed air subsequently blows the molten material away. The air carbon gouging torch looks much like a manual electrode holder, but it uses a carbon electrode instead of a metallic electrode. Current is conducted through the base material to heat it. A valve in the torch handle permits compressed air to flow through two air ports. As the air hits the molten material, a combination of oxidation and expulsion of metal takes place, leaving a smooth cavity behind. The air carbon arc gouging system is capable of removing metal at a much higher rate than it can be deposited by

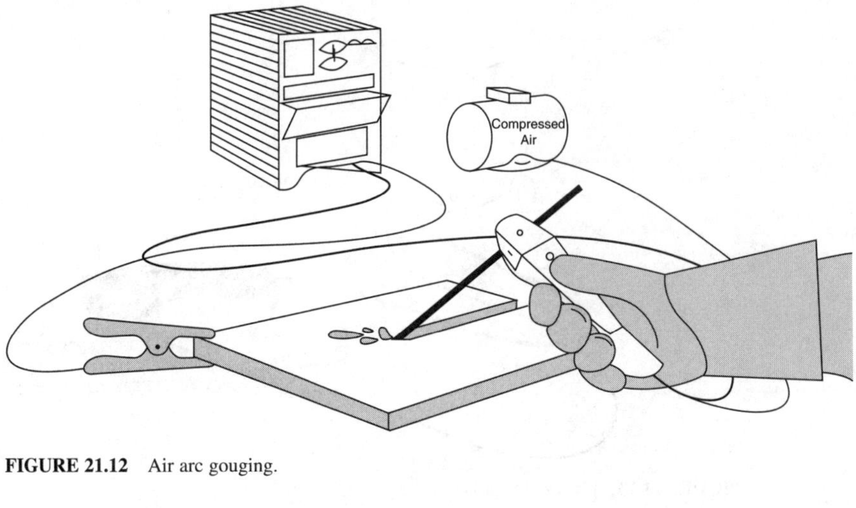

FIGURE 21.12 Air arc gouging.

most welding processes. It is a powerful tool used to remove metal at low cost. It is most often used in a production situation to remove defective material before repairing the part by welding.

21.7.2 Plasma Arc Cutting

The *plasma arc cutting* (PAC) process (Fig. 21.13) was developed initially to cut materials that do not permit the use of the oxyfuel process—stainless steel and aluminum. It was found, however, that plasma arc cutting offered economic advantages when applied to thinner sections of mild steel, especially those less than 1 in thick. Higher travel speed is possible with plasma arc cutting, and the volume of heated base material is reduced, minimizing metallurgical changes as well as reducing distortion.

PAC is a thermal and mechanical process. To utilize PAC, the material is heated until molten and expelled from the cut with a high-velocity stream of compressed gas. Unlike oxyfuel cutting (discussed below) the process does not rely on oxidation. Because high amounts of energy are introduced through the arc, PAC is capable of extremely high-speed cutting. The thermal energy generated during the oxidation process with oxyfuel cutting is not present in plasma; hence, for thicker sections, PAC is not economically justified. The use of PAC to cut thick sections usually is restricted to materials that do not oxidize readily with oxyfuel.

PAC may be used manually, or can be automated. Larger parts can be on large cutting tables. The tables may be filled with water, and the parts being cut submerged in the water. This keeps noise, smoke, and distortion to a minimum.

21.7.3 Oxygen Cutting

Oxyfuel Cutting (OFC) (Fig. 21.14) is used to cut steels and to prepare bevel and vee grooves. In this process, the metal is heated to its ignition temperature, or kindling point, by a series of preheat flames. After this temperature is attained, a high-velocity stream of pure oxygen is introduced, which causes oxidation or "burning" to occur. The force of the oxygen stream blows the oxides out of the joint, resulting in a clean cut. The oxidation process also generates additional thermal energy, which is radially conducted into the surrounding steel, increasing the temperature of the steel ahead of the cut. The next portion of the steel is raised to the kindling temperature, and the cut proceeds.

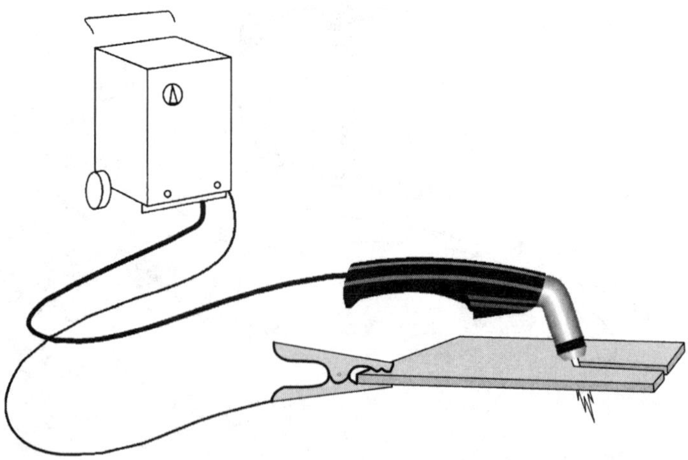

FIGURE 21.13 Plasma arc cutting.

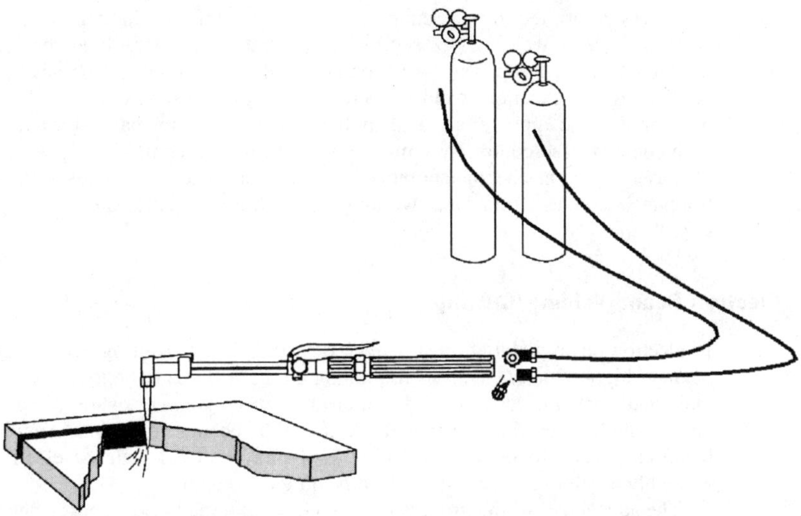

FIGURE 21.14 Oxyfuel cutting.

Carbon and low-alloy steels are easily cut by the oxyfuel process. Alloy steels can be cut, but with greater difficulty than mild steel. The level of difficulty is a function of the alloy content. When the alloy content reaches the levels found in stainless steels, oxyfuel cutting cannot be used unless the process is modified by injecting flux or iron-rich powders into the oxygen stream. Aluminum cannot be cut with the oxyfuel process. Oxyfuel cutting is commonly regarded as the most economical way to cut steel plates greater than $^1/_2$ in thick. A variety of fuel gases may be used for oxyfuel cutting, with the choice largely dependent on economics; they include natural gas, propane, acetylene, and a variety of proprietary gases offering unique advantages. Because of its role in the primary cutting stream, oxygen is always used as a second gas. In addition, some oxygen is mixed with the fuel gas in proportions designed to ensure proper combustion.

Oxygen cutting can be performed manually or automatically. Large cutting tables with automatically positioned torches can be used to cut multiple parts at one time. Steel of thicknesses over 12 in can be cut in this manner. The greatest limitation of the process is its inability to cut nonferrous materials.

21.8 HIGH ENERGY DENSITY WELDING AND CUTTING PROCESSES

21.8.1 Laser Welding/Cutting

In laser welding and/or cutting, a beam of coherent electromagnetic radiation at ultraviolet, visible, or infrared frequency is used to heat the parts. Because high-energy laser beams are coherent, they can be highly concentrated to provide the high-energy density needed for welding, cutting, and heat-treating metals.

As applied to welding, pulsed and continuously operating solid-state lasers and lasers that produce continuous-wave (cw) energy have been developed to the point that multikilowatt laser beam equipment based on CO_2 is capable of full-penetration, single-pass welding of steel of thicknesses of up to $^3/_4$ in.

Lasers do not require a vacuum in which to operate, so that they offer many of the advantages of electron beam welding (see section below) but at considerably lower equipment cost and higher production rates. Deep, narrow welds are produced at high speeds and low total heat input, thus duplicating the excellent weld properties and minimal heat effects obtained from electron beam welding in some applications. The application of lasers to metals—for cutting or welding—coupled with computerized control systems, allows their use for complex shapes and contours.

Laser cutting has become popular for processing sheet metal parts, eliminating expensive tooling for punch presses. Both laser welding and cutting are automated and require significant capital investments.

21.8.2 Electron Beam Welding/Cutting

In electron beam welding, coalescence of metals is achieved by heat generated by a concentrated beam of high-velocity electrons impinging on the surfaces to be joined. Electrons have a very small mass and carry a negative charge. An electron beam gun, consisting of an emitter, a bias electrode, and an anode, is used to create and accelerate the beam of electrons. Auxiliary components such as beam alignment, focus, and deflection coils may be used with the electron beam gun; the entire assembly is referred to as the electron beam gun column.

The advantages of the process arise from the extremely high energy density in the focused beam which produces deep, narrow welds at high speed, with minimum distortion and other deleterious heat effects. Major applications are with metals and alloys highly reactive to gases in the atmosphere or those volatilized from the base metal being welded.

EB welding generally takes place in a vacuum chamber, restricting the size of parts that can be welded to those that can fit within the chamber. Another disadvantage of the process lies in the necessity for providing precision parts and fixtures so that the beam can be precisely aligned with the joint to ensure complete fusion. Gapped joints are not normally welded because of fixture complexity and the extreme difficulty of manipulating filler metal into the tiny, rapidly moving, weld puddle under high vacuum. When no filler metal is employed, it is common to use the keyhole technique. Here, the electron beam makes a hole entirely through the base metal, which is subsequently filled with melted base metal as the beam leaves the area. Other disadvantages of the process arise from the cost, complexity, and skills required to operate and maintain the equipment, and the safety precautions necessary to protect operating personnel from the X-rays generated during the operation.

21.9 WELDING PROCEDURES

Within the welding industry, a *welding procedure specification* (WPS) is used to define the variables that are to be used to make a certain weld. The terms "welding procedure," or simply "procedure," are often used interchangeably. The discussion that follows is directly related to arc welding processes.

21.9.1 WPS Variables

The welding procedure is somewhat analogous to a cook's recipe. It outlines the steps required to make a quality weld under specific conditions. The "ingredients," or variables, consist—at a minimum—of the following:

Process (SMAW, FCAW, etc.)

Electrode specification (AWS A5.1, A5.20, etc.)

Electrode classification (E7018, E7028, etc.)

Electrode diameter ($^1/_8$ in, $^5/_{32}$ in, etc.)

Electrical characteristics (ac, dc+, dc–)

Base metal specification (A36, A572, Gr50, etc.)

Minimum preheat and interpass temperature

Welding current (amperage)

Arc voltage

Travel speed

Position of welding

Post weld heat treatment

Gas flow rate

Joint design details

The effects of the variables are somewhat dependent on the welding process being employed, but general trends apply to all the processes. It is important to distinguish the difference between constant current (CC) and constant voltage (CV) electrical systems. SMAW is always done with a CC system. Flux cored welding and GMAW generally are performed with CV systems. Submerged arc may utilize either. The summaries provided here will assume submerged arc welding that utilizes a CV system. (Note: the complexity of the CC system used with submerged arc is beyond the scope of this summary.)

Amperage is a measure of the amount of current flowing through the electrode and the work. It is a primary variable in determining heat input (discussed below). Generally, an increase in amperage means higher deposition rates, deeper penetration, and more admixture. The amperage flowing through an electrical circuit is the same, regardless of where it is measured. It may be measured with a tong meter or with the use of an electrical shunt. The role of amperage is best understood in the context of heat input and current density considerations.

Arc voltage is directly related to arc length. As the voltage increases, the arc length increases, as does the demand for arc shielding. In submerged arc welding, more flux will be melted with a higher arc voltage. Arc voltage also controls the width of the weld bead, with higher voltages generating wider beads. Arc voltage has a direct effect on the heat input computation.

The voltage in a welding circuit is not constant, but is composed of a series of voltage drops. Consider the following example: assume the power source delivers a total system voltage of 40 V. Between the power source and the welding head or gun, there is a voltage drop of perhaps 3 V associated with the input cable resistance. From the point of attachment of the work lead to the power source work terminal, there is an additional voltage drop of, say, 7 V. Subtracting the 3 V and the 7 V from the original 40, this leaves 30 V for the arc. This example illustrates how important it is to ensure that the voltages used for monitoring welding procedures properly recognize any losses in the welding circuit. The most accurate way to determine arc voltage is to measure the voltage drop between the contact tip and the work piece. If this measurement is made, system voltage losses will not affect the results.

Travel speed, measured in inches per minute, is the rate at which the electrode is moved relative to the joint. All other variables being equal, travel speed has an inverse effect on the size of the weld beads. As the travel speed increases, the weld size will decrease. Excessively low travel speeds may result in a reduction in penetration, as the weld puddle rolls ahead of the arc. Travel speed is a key variable used in computing heat input.

Electrode extension, also known as *stickout,* is the distance from the contact tip to the end of the electrode. It applies only to the wire fed processes. As the electrode extension is increased in a constant voltage system, the electrical resistance of the electrode increases, causing the electrode to be heated. This is known as *I^2R heating.* As the amount of heating increases, the arc energy required to melt the electrode decreases. Longer electrode extensions may be employed to gain higher deposition rates at a given amperage. When the electrode extension is increased without any change in wire feed speed (WFS), the amperage will decrease. This results in less penetration and less admixture. With the increase in electrical stickout, it is common to increase the machine voltage setting to compensate for the increased voltage drop across the electrode.

In constant voltage systems, it is possible to simultaneously increase the electrical stickout and WFS in a balanced manner so that the current remains constant. When this is done, higher deposition rates are attained.

Electrode Diameter. Larger electrodes can carry higher welding currents. For a fixed amperage, however, smaller electrodes result in higher deposition rates. This is because of the effect on current density discussed below.

Polarity is a definition of the direction of current flow. Positive polarity (or reverse) is achieved when the electrode lead is connected to the positive terminal of the dc power supply. The work lead would connect to the negative terminal. Negative polarity (or straight) occurs when the electrode is connected to the negative terminal. Alternating current is not a polarity, but a current type. With ac, the electrode is alternately positive and negative. Submerged arc is the only process that commonly uses both electrode positive and electrode negative polarity for the same filler metal, ac may also be used. For a fixed WFS, a submerged arc electrode will require more amperage on positive polarity than on negative. For a fixed amperage, it is possible to utilize higher deposition rates with negative polarity than with positive—ac exhibits a mix of both positive and negative polarity characteristics.

Heat input is proportional to the welding amperage, times the arc voltage, divided by the travel speed. Higher heat inputs relate to larger weld cross sectional areas, and larger heat affected zones, which may negatively affect mechanical properties in that region. Higher heat input generally results in slightly decreased yield and tensile strength in the weld metal, and generally lower notch toughness because of the interaction of bead size and heat input.

Current density is determined by dividing the welding amperage by the cross sectional area of the electrode. The current density is therefore proportional to I/d^2. As the current density increases, there will be an increase in deposition rates, as well as in penetration. The latter will increase the amount of admixture for a given joint. Notice that this may be accomplished by either increasing the amperage or decreasing the electrode size. Because the electrode diameter is a squared function, a small decrease in diameter may have a significant effect on deposition rates and plate penetration.

21.9.2 Purpose of the WPS

The purpose of a WPS is to communicate to the welder and inspectors the parameters under which welding is to be performed. Any information that must be conveyed to the welder should be included on the written procedure, whether required by a code or not. The welding procedure becomes the "recipe" for making a particular weld.

Welding procedures are not developed by welders, but by welding engineers or welding technicians. While welders must be able to read and follow a written welding procedure in order to be qualified, they need not understand how and why each variable was selected for the procedure.

Depending on the industry in which the welding is being done, a variety of codes and specifications may detail specific requirements for welding. For example, pressure vessels are often governed by codes written by the American Society of Mechanical Engineers (ASME), while bridges and buildings are fabricated in accordance with American Welding Society (AWS) codes. There are a host of other agencies that may control various aspects of welding. These standards often prescribe specific requirements for welding procedures, and the applicable specifications should be reviewed to make certain compliance with those standards is achieved.

Once a welding procedure is developed, its value is lost if the welding performed in production is not in conformance with the WPS. There are a variety of reasons why an individual welder may not follow a WPS. In piece-work, incentive shops, a welder may be inclined to use a procedure that features higher production rates, but at the sacrifice of quality. In other situations, a welder may use less productive procedures because the welding is easier. Adherence to the WPS is key for controlling both productivity and quality.

Supervisors and inspectors can be helpful in making certain approved welding procedures are used in production. Rather than monitoring only the completed welds, spot inspections can be employed to ensure that welding is being performed in accordance with the WPS.

21.10 BASIC METALLURGY FOR THE MANUFACTURING ENGINEER

The nature of the base metals must be understood and recognized prior to welding. Knowledge of the mechanical properties of the base metal will allow the designer to ensure that the weld metal deposited will have properties equal to those of the base metal. The chemical composition of the base metal will affect the selection of the filler metal and/or electrode, and well as the required preheat and interpass temperature (see 21.12). Finally, the metallurgical structure of the base metal as it comes to the welding operation (hot-worked, cold-worked, quenched, tempered, annealed, etc.) will affect the weldability of the metal and, if it is weldable, the degree to which the final properties are as dictated by design requirements. Welding specifications may address these matters, and base metal suppliers can provide additional data as to the weldability of the metal.

In some cases, the identity of the base metal is not known with certainty. The results of welding a metal without knowing what it is may prove disastrous. Identification may be aided by general characteristics, some of which may be self-evident: carbon steel (oxide coating) vs. stainless steel (unoxidized), brush-finished aluminum (lightweight) vs. brush-finished Monel metal (heavy), etc. Ultimately, it may become necessary to subject the unknown metal to chemical, mechanical, and other types of laboratory tests to ascertain its exact nature.

21.10.1 Steel

Most common steels are very easily welded. Weldability, which is defined as the ease with which a material can be welded, decreases as the carbon content in the steel increases. Increases in the level of alloys similarly decrease weldability. Carbon and alloys are added to steels to increase their strength, but these additions also usually make higher strength steels more difficult to weld.

Low-Carbon Steels (Carbon up to 0.30 percent). Steels in this class are readily welded by most arc and gas processes. Preheating is unnecessary unless parts are very heavy or welding is performed below 32°F. Torch-heating steel in the vicinity of welding to 70°F offsets low temperatures. More preheat may be required for heavier sections. GTAW is usable only on killed steels; rimmed steels produce porous, weak welds. Resistance welding is readily accomplished if carbon is below 0.20 percent; higher carbon requires heat-treatment to slow the cooling rate and avoid hardness. Brazing with BAg, BCu, and BCuZn filler metals is very successful.

Medium-Carbon Steels (Carbon from 0.30 to 0.45 percent). This class of steel may be welded by the arc, resistance, and gas processes. As the rapid cooling of the metal in the welded zone produces a harder structure, it is desirable to hold the maximum carbon as close to 0.30 percent as possible. These hard areas are proportionately more brittle and difficult to machine. The cooling rate may be diminished and hardness decreased by preheating the metal to be welded above 300°F and preferably to 500°F. The degree of preheating depends on the thickness of the section. Subsequent heating of the welded zone to 1100 to 1200°F will restore ductility and relieve thermal strains. Brazing may also be used, as noted for low-carbon steels above.

High-Carbon Steels (Carbon from 0.45 to 0.80 percent). These steels are rarely welded except in special cases. The tendency for the metal heated above the critical range to become brittle is more pronounced than with low- or medium-carbon steels. Thorough preheating of metal, in and near the welded zone, to a minimum of 500°F is essential. Subsequent annealing at 1350 to 1450°F is also desirable. Brazing is often used with these steels, and is combined with the heat treatment cycle.

Low-Alloy Steels. The weldability of low-alloy steels is dependent upon the analysis and the hardenability. Hardenability is a term used to define how a steel responds to rapid cooling. Alloys increase hardenability, and make the material more difficult to weld (i.e., lower weldability). Those exhibiting low hardenability are welded with relative ease, whereas those of high hardenability require preheating and postheating. Sections of 1/4 in or less may be welded with mild-steel filler metal and may provide joint strength approximating base metal strength by virtue of alloy pickup in the weld metal and weld reinforcement. Higher-strength alloys require filler metals with mechanical properties matching the base metal. Special alloys with creep-resistant or corrosion-resistant properties must be welded with filler metals of the same chemical analysis. Low-hydrogen-type electrodes (either mild- or alloy-steel analyses) permit the welding of alloy steels, minimizing the occurrence of under-head cracking.

Free Machining Steels. A special classification of steels deserves mention due to their popularity and their sensitivity to welding. Free machining steels are those of the SAE class 11XX and 12XX, where deliberate additions of sulfur and/or phosphorus are made. These elements are added to aid in machinability. Unfortunately, they make welding more difficult, and welds made on such materials are particularly prone to cracking in fabrication. While some measures can be taken to mitigate these effects, selection of alternate materials is generally the best method to overcome the cracking tendency.

21.10.2 Stainless Steel

Stainless steel is an iron-based alloy containing in excess of 11 percent chromium. A thin, dense surface film of chromium oxide which forms on stainless steel imparts superior corrosion resistance. Its protective nature inhibits scaling and prevents further oxidation, hence the designation "stainless." This oxide coating, however, can be deceptive: it makes welding somewhat more difficult.

There are five types of stainless steels, and depending on the amount and kind of alloying additions present, they range from fully austenitic to fully ferritic. Most stainless steels have good weldability and may be welded by many processes, including arc welding, resistance welding, electron and laser beam welding, and brazing. With any of these, the joint surfaces and any filler metal must be clean.

The coefficient of thermal expansion for the austenitic stainless steels is 50 percent greater than that of carbon steel; this must be taken into account to minimize distortion. The low thermal and electrical conductivity of austenitic stainless steel is generally helpful. Low welding heat is required because the heat is conducted more slowly from the joint, but low thermal conductivity results in a steeper thermal gradient and increases distortion. In resistance welding, lower current is used because electric resistivity is higher.

Ferritic Stainless Steels. Ferritic stainless steels contain 11.5–30 percent chromium (Cr), up to 0.20 percent carbon (C), and small amounts of ferrite stabilizers, such as aluminum (Al), niobium (Nb), titanium (Ti), and molybdenum (Mo). They are ferritic at all temperatures, do not transform to austenite, and are not hardenable by heat treatment. This group includes types 405, 409, 430, 442, and 446. To weld ferritic stainless steels, filler metals should match or exceed the Cr level of the base metal.

Martensitic Stainless Steels. Martensitic stainless steels contain 11.4 to 18 percent chromium (Cr), up to 1.2 percent carbon (C), and small amounts of manganese (Mn) and nickel (Ni). They will transform to austenite on heating and, therefore, can be hardened by formation of martensite on cooling. This group includes types 403, 410, 414, 416, 420, 422, 431, and 440. Welding on these stainless steels is difficult. Weld cracks may appear on cooled welds as a result of martensite formation. The Cr and C content of the filler metal should generally match these elements in the base metal. Preheating and interpass temperature in the 400 to 600°F range is recommended for welding most martensitic stainless steels. Steels with over 0.20 percent C often require a postweld heat treatment to avoid weld cracking.

Austenitic Stainless Steels. Austenitic stainless steels contain 16 to 26 percent chromium (Cr), 10 to 24 percent nickel (Ni) and manganese (Mn), up to 0.40 percent carbon (C), and small amounts of Mo, Ti, Nb, and tantalum (Ta). The balance between Cr and Ni + Mn is normally adjusted to provide a microstructure of 90 to 100 percent austenite. These alloys have good strength and high toughness over a wide temperature range, and they resist oxidation to over 1000°F. This group includes types 302, 304, 310, 316, 321, and 347. Filler metals for these alloys should generally match the base metal, but for most alloys should also provide a microstructure with some ferrite to avoid hot cracking. Austenitic stainless steels are commonly welded. Two problems are associated with welding austenitic stainless steels: sensitization of the weld-heat-affected zone and hot cracking of weld metal.

Sensitization is caused by chromium carbide precipitation at the austenitic grain boundaries in the heat-affected zone, when the base metal is heated to 800 to 1600°F. Chromium carbide precipitates remove chromium from solution in the vicinity of the grain boundaries, and this condition leads to intergranular corrosion. The problem can be alleviated by using low-carbon stainless steel base metal (types 302L, 316L, etc.) and low carbon filler metal. Alternately, there are stabilized stainless steel base metals and filler metals available which contain alloying elements that react preferentially with carbon, thereby not depleting the chromium content in solid solution and keeping it available for corrosion resistance. Type 321 contains Ti and type 347 contains Nb and Ta, all of which are stronger carbide formers than Cr.

Hot cracking is caused by low-melting-point metallic compounds of sulfur and phosphorus which penetrate grain boundaries. When present in the weld metal or heat-affected zone, these compounds will penetrate grain boundaries and cause cracks to appear as the weld cools and shrinkage stresses develop. Hot cracking can be prevented by adjusting the composition of the base metal and filler metal to obtain a microstructure with a small amount of ferrite in the austenite matrix. The ferrite provides ferrite-austenite boundaries which control the sulfur and phosphorus compounds and thereby prevent hot cracking.

Precipitation-Hardening Stainless Steels. Precipitation-hardening (PH) stainless steels contain alloying elements such as aluminum which permit hardening by a solution and aging heat treatment. There are three categories of PH stainless steels: martensitic, semiaustenitic, and austenitic. Martensitic PH stainless steels are hardened by quenching from the austenitizing temperature (around 1900°F) and then aging between 900 and 1150°F. Semiaustenitic PH stainless steels do not transform to martensite when cooled from the austenitizing temperature because the martensite transformation temperature is below room temperature. Austenitic PH stainless steels remain austenitic after quenching from the solution temperature, even after substantial amounts of cold work.

If maximum strength is required of connections made with martensitic PH and semiaustenitic PH stainless steels, matching, or nearly matching, filler metal should be used, and before welding, the work pieces should be in the annealed or solution-annealed condition. After welding, a complete solution heat treatment plus an aging treatment is preferred. If postweld solution treatment is not feasible, the components should be solution-treated before welding and then aged after welding. Thick sections of highly restrained parts are sometimes welded in the overaged condition. These require a full heat treatment after welding to attain maximum strength properties.

Austenitic PH stainless steels are the most difficult to weld because of hot cracking. Welding is preferably done with the parts in solution-treated condition, under minimum restraint and with minimum heat input. Filler metals of the Ni-Cr-Fe type, or of conventional austenitic stainless steel, are preferred.

Duplex Stainless Steels. Duplex stainless steels are the most recently developed type of stainless steel, and they have a microstructure of approximately equal amounts of ferrite and austenite. They have advantages over conventional austenitic and ferritic stainless steels in that they possess higher yield strength and greater stress corrosion cracking resistance. The duplex microstructure is attained in steels containing 21 to 25 percentchromium (Cr) and 5 to 7 percent nickel (Ni) by hot-working at 1832 to 1922°F, followed by water quenching. Weld metal of this composition will be mainly ferritic because the deposit will solidify as ferrite and will transform only partly to austenite without hot working or annealing. Since hot-working or annealing most weld deposits is not feasible, the metal composition filler is generally modified by adding Ni (to 8 to 10 percent), this results in increased amounts of austenite in the as-welded microstructure.

21.10.3 Cast Iron

Even though cast iron has a high carbon content and is a relatively brittle and rigid material, welding can be performed successfully if proper precautions are taken. The difficulties of welding cast iron, however, limit its use in production situations. Welding of cast iron is typically limited to repair conditions.

Optimum conditions for welding cast iron include the following: (1) A weld groove large enough to permit manipulation of the electrode or the welding torch and rod. The groove must be clean and free of oil, grease, and any foreign material. (2) Adequate preheat, depending on the welding process used, the type of cast iron, and the size and shape of the casting. Preheat temperature must be maintained throughout the welding operation. (3) Welding heat input sufficient for a good weld but not enough to superheat the weld metal, i.e., welding temperature should be kept as low as practicable. (4) Slow cooling after welding. Gray iron may be enclosed in insulation, lime, or vermiculite. Other irons may require postheat treatment immediately after welding to restore mechanical properties. ESt and ENiFe identify electrodes of steel and of a nickel-iron alloy. Many different welding processes have been used to weld cast iron, the most common being manual shielded metal-arc welding, gas welding, and braze welding.

21.10.4 Aluminum and Aluminum Alloys

Aluminum and aluminum alloys are routinely welded, but the welding conditions are different than those used for steel due to aluminum's inherent properties. The properties that distinguish the aluminum alloys from other metals determine which welding processes can be used and which particular procedures must be followed for best results. Among the welding processes that can be used, choice is further dictated by the requirements of the end product and by economic considerations.

Physical properties of aluminum alloys that most significantly affect all welding procedures include low melting-point range, approximately 900 to 1215°F, high thermal conductivity (about two to four times that of mild steel), high rate of thermal expansion (about twice that of mild steel), and high electrical conductivity (about three to five times that of mild steel). Interpreted in terms of welding, this means that, when compared with mild steel, much higher welding speeds are demanded, greater care must be exercised to avoid distortion, and for arc and resistance welding, much higher current densities are required.

Aluminum alloys are not quench-hardenable. However, weld cracking may result from excessive shrinkage stresses due to the high rate of thermal contraction. To offset this tendency, welding procedures, where possible, require a fast weld cycle and a narrow-weld zone, e.g., a highly concentrated heat source with deep penetration, moving at a high rate of speed. Shrinkage stresses can also be reduced by using a filler metal of lower melting point than the base metal. The filler metal ER4043 is often used for this purpose.

Welding procedures also call for the removal of the thin, tough, transparent film of aluminum oxide that forms on and protects the surface of these alloys. The oxide has a melting point of about 3700°F and can therefore exist as a solid in the molten weld. Removal may be by chemical reduction or by mechanical means such as machining, filing, rubbing with steel wool, or brushing with a stainless-steel wire brush.

Most often, aluminum is welded with GTAW or GMAW. GTAW usually uses alternating current, with argon as the shielding gas. The power supply must deliver high current with balanced wave characteristics, or deliver high-frequency current. With helium, weld penetration is deeper, and higher welding speeds are possible. Most welding, however, is done using argon because it allows for better control and permits the welder to see the weld pool more easily.

GMAW employs direct current, electrode positive in a shielding gas that may be argon, helium, or a mixture of the two. In this process, the welding arc is formed by the filler metal, which serves as the electrode. Since the filler metal is fed from a coil as it melts in the arc, some arc instability may arise. For this reason, the process does not have the same precision as the GTAW process for welding very thin gages. However, it is more economical for welding thicker sections because of its higher deposition rates.

21.10.5 Copper and Copper Alloys

In welding commercially pure copper, it is important to select the correct type. Electrolytic, or "tough-pitch," copper contains a small percentage of copper oxide, which at welding heat leads to oxide embrittlement. For welded assemblies it is recommended that deoxidized, or oxygen-free, copper be used and that welding rods, when needed, be of the same analysis. The preferred processes for welding copper are GTAW and GMAW; manual SMAW can also be used. It is also welded by oxyacetylene method and braze-welded; brazing with brazing filler metals conforming to BAg, BCuP, and RBCuZn-A classifications may also be employed. The high heat conductivity of copper requires special consideration in welding: generally higher welding heats are necessary together with concurrent supplementary heating. Copper alloys are extensively welded in industry. The specific procedures employed are dependent upon the analysis, and reference should be made to the AWS "Welding Handbook." Filler metals for welding copper and its alloys are covered in AWS specifications.

21.11 DESIGN OF WELDED CONNECTIONS

This material is presented to give the manufacturing engineer a general understanding of the issues that must be considered in the design of welded connections. Product design engineers typically determine joint type and generally specify weld type and the required throat dimension. Fabricators usually select the joint details.

21.11.1 Joint Types

When pieces of steel are brought together to form a joint, they will assume one of the five configurations presented in Fig. 21.15. Joint types are descriptions of the relative positions of the materials to be joined and do not imply a specific type of weld.

21.11.2 Weld Types

There are several major weld types, including the following: fillet, groove, plug, slot, and spot welds. Depending on the joint type, welding process type, and the loads that are transferred across the connection, a weld type (Fig. 21.16) will be selected for the application. The selection of the weld type is typically assigned to the product design engineer.

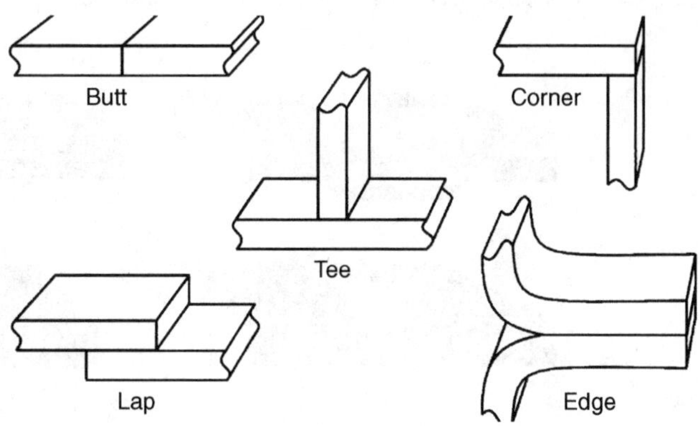

FIGURE 21.15 Joint types.

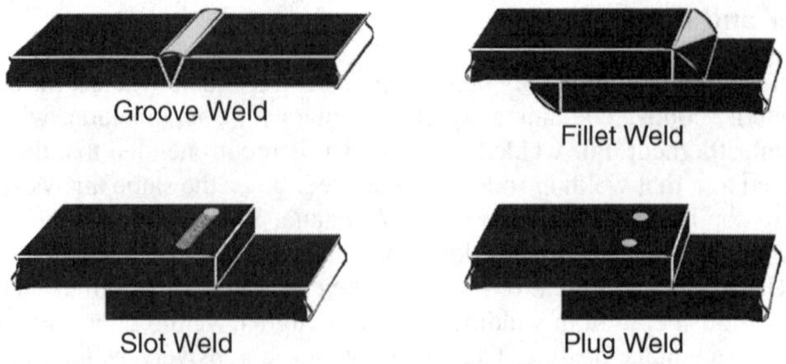

FIGURE 21.16 Weld types.

Fillet welds have a triangular-like cross section and are applied to the surface of the materials they join. By themselves, fillet welds do not fully fuse the cross-sectional areas of parts they join, although it is still possible to develop full-strength connections with fillet welds. The size of a fillet weld is usually determined by measuring the leg, even though the weld is designed by specifying the required throat. For equal-legged, flat-faced fillet welds applied to plates that are oriented 90° apart, the throat dimension is found by multiplying the leg size by 0.707.

Groove welds comprise two subcategories: *complete joint penetration* (CJP) and *partial joint penetration* (PJP) groove welds (Fig. 21.17). By definition, CJP groove welds have a throat dimension equal to the thickness of the material they join; a PJP groove weld is one with a throat dimension less than the thickness of the materials joined.

Plug, slot, and spot welds are applied to lap joints. Functionally, they are similar to bolted or riveted joints. They are used in situations where the connection is lightly loaded, or where the number of welds can be increased in order to increase connection strength. Spot welds are the most common welds made with the resistance welding process.

21.11.3 Weld Size and Strength

In terms of their application, welds fall into two broad categories: primary and secondary. Primary welds are critical welds that directly transfer the full applied load at the point at which they are located. These welds must develop the full strength of the members they join. Complete joint penetration groove welds are often used for these connections. Secondary welds are those that merely hold the parts together to form a built-up member. The forces on these welds are relatively low, and fillet welds are generally utilized in these connections.

Filler Metal Strength may be classified as matching, undermatching, or overmatching. Matching filler metal has the same, or slightly higher, minimum specified yield and tensile strength as the base

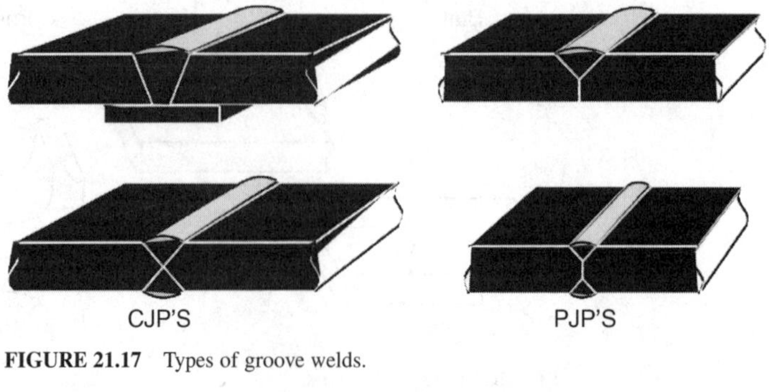

FIGURE 21.17 Types of groove welds.

metal. CJP groove welds in tension require the use of matching weld metal—otherwise, the strength of the welded connection may be lower than that of the base metal. Undermatching filler metal deposits welds of a strength lower than that of the base metal. Undermatching filler metal may be deposited in fillet welds and partial joint penetration groove welds as long as the designer specifies a throat size that will compensate for the reduction in weld metal strength. An overmatching filler metal deposits weld metal that is stronger than the base metal; this is undesirable unless, for practical reasons, lower-strength filler metal is unavailable for the application. When overmatching filler metal is used, if the weld is stressed to its maximum allowable level, the base metal can be overstressed, resulting in failure in the fusion zone. Designers must ensure that connection strength, including the fusion zone, meets the application requirements. Additionally, overmatching increases the residual stresses and the weld cracking potential.

Allowable Strength of Welds Under Steady Loads. A structure, or weldment, is only as strong as its weakest point, and *allowable* weld strengths are specified by AWS, the American Institute of Steel Construction (AISC), and various other professional organizations to ensure that a weld will deliver the mechanical properties of the members being joined. Allowable weld strengths are designated for various types of welds for steady and fatigue loads.

Complete joint penetration groove welds are considered full-strength welds, since they are capable of transferring the equivalent capacity of the members they join. In calculations, such welds are allowed the same stress as the plate, provided the proper strength level of weld metal is used (e.g., matching filler metal). In such CJP welds, the mechanical properties of the weld metal must at least match those of the base metal. If the plates joined are of different strengths, the weld metal strength must at least match the strength of the weaker plate.

21.11.4 Minimizing Distortion

Distortion occurs due to the nonuniform expansion and contraction of weld metal and adjacent base material during the heating and cooling cycles of the welding process (see also 21.12, Thermal Considerations). At elevated temperatures, hot, expanded weld and base metal occupies more physical space than it will at room temperatures. As the metal contracts, it induces strains that result in stresses being applied to the surrounding base materials. When the surrounding materials are free to move, distortion results. If they are not free to move, as in the case of heavily restrained materials, these strains can induce cracking stresses. When distortion or cracking do not occur, the shrinkage results in residual stresses within the part. Distortion, cracking, and residual stresses are all caused by the same phenomonen: nonuniform expansion and contraction of material during thermal welding cycles.

It should be emphasized that not only the weld metal, but also the surrounding base material is involved in this contraction process. For this reason, welding processes and procedures that introduce high amounts of energy into the surrounding base material will cause more distortion.

Stresses resulting from material shrinkage are inevitable in all welding that involves the application of heat. Distortion, however, can be minimized, compensated for, and predicted. Through efficient planning, design, and fabrication practices, distortion related problems can be effectively minimized.

Design Concepts to Minimize Distortion. The manufacturing engineer who is aware of the effects of distortion can design measures into the welded assemblies that will minimize the amount of distortion. These concepts include the following:

Minimize the amount of weld metal. Any reduction in the amount of weld metal will result in a decrease in the amount of distortion:

- Use the smallest acceptable weld size.
- Use intermittent welds where acceptable.
- Utilize double-sided joints vs. single-sided joints where applicable.
- Use groove details that require the minimum volume of weld metal per length.

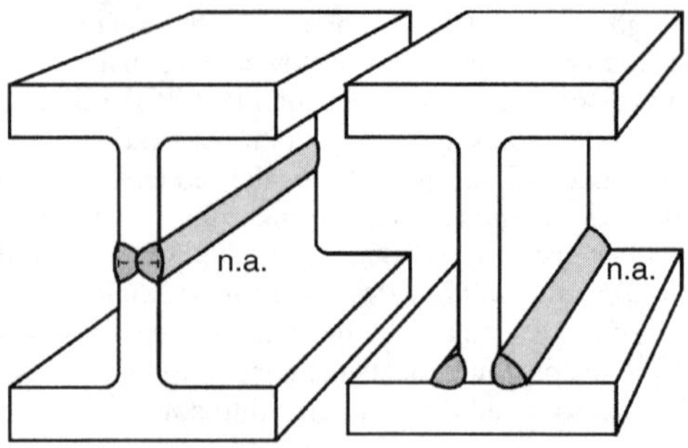

FIGURE 21.18 Neutral axis.

Place welds near the neutral axis. When welds are placed near the neutral axis, less longitudinal sweep or camber results. This is illustrated in Fig. 21.18.

Balance welds about the neutral axis. When welds are balanced around the neutral axis, the shrinkage force of one weld will be counteracted by the shrinkage force of another. This is illustrated in Fig. 21.19.

Production Practices That Minimize Distortion. Manufacturers can use techniques that will minimize distortion. These include the following:

Use as few weld passes as possible. Fewer passes are desirable in as much as they limit the number of heating and cooling cycles to which the joint will be subjected. The shrinkage stresses of each pass tend to accumulate, increasing the amount of distortion when many passes are used.

Avoid overwelding. Overwelding results in more distortion than is necessary. Holding weld sizes to print requirements will help avoid unnecessary distortion.

Obtain good fit-up. Poor fit-up, resulting in gaps and larger included angles for bevel preparations, means more weld metal is placed in the joint than is required, contributing to excessive distortion.

Use high productivity, low heat input welding procedures. Generally speaking, high productivity welding procedures (those using high amperages and high travel speeds) result in a lower net heat input than low productivity procedures. At first, high amperage procedures may seem to be high heat input procedures. However, for a given weld size, the high amperage procedures are high travel speed procedures. This will result in a decreased amount of heat affected zone and reduced distortion.

Use clamps, strongbacks, or fixtures to restrict the amount of distortion. Any tooling or restraints that avoid rotation of the part will reduce the amount of distortion experienced. In addition, fixturing may be used to draw heat away, particularly if copper chill bars and clamps are used in the vicinity of the joint. The arc should never impinge on copper as this could cause cracking.

Use a well planned welding sequence. A well planned welding sequence is often helpful in balancing the shrinkage forces against each other.

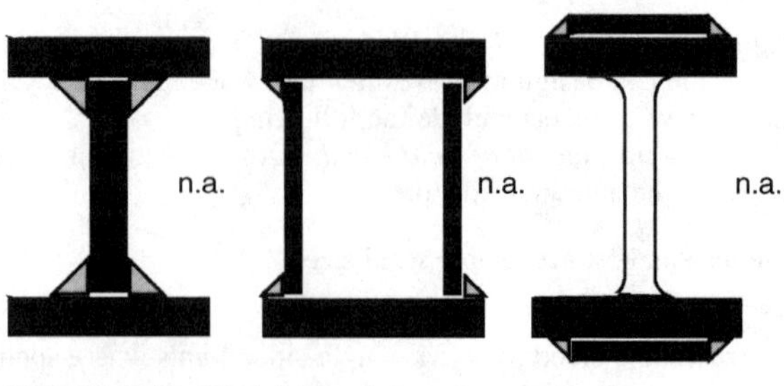

FIGURE 21.19 Welds balanced about the neutral axis.

Preset or precamber parts before welding. Parts to be joined may be preset or precambered before welding. When weld shrinkage causes distortion, the parts will be drawn back into proper alignment.

21.12 THERMAL CONSIDERATIONS

When joining or cutting materials, the metal adjacent to the weld metal or cutting area, known as the *heat affected zone* (HAZ), is subjected to extreme thermal conditions. In fusion welding, the HAZ experiences rapid thermal cycling from that of near melting to ambient temperature. Such thermal cycling in the heat affected zone affects various alloys in different ways. Steels, e.g., will typically experience an increase in hardness in the HAZ, while aluminum alloys will typically soften. Thus, the manufacturing engineer should consider the potential effects to the base metal when selecting a welding or cutting process.

21.12.1 Preheat

Preheat involves heating the base metal, either in its entirety or just the region surrounding the joint, to a specific desired temperature prior to welding. Heating may be continued during the welding process, but frequently the heat from welding is sufficient to maintain the desired temperature without continuing to supply an external source of heat. The four major reasons to utilize preheat are as follows:

1. It slows the cooling rate in the weld metal and base metal, producing a more ductile metallurgical structure with greater resistance to cracking.
2. The slower cooling rate allows hydrogen that may be present to diffuse out harmlessly, reducing the potential for cracking.
3. It reduces the shrinkage stresses in the weld and adjacent weld metal.
4. It raises some steels above the temperature at which brittle fracture would occur in fabrication.

When applicable codes or standards specify preheat, such specifications must, of course, be followed. When no codes apply to a given situation, the manufacturing engineer must determine whether or not preheat is needed, and what temperature will be required for a given base metal and section thickness. Preheat usually is not required on low carbon steels less than 1 in thick, but as chemistry, diffusible hydrogen level of the weld metal, restraint or section thickness increases, the need for preheat also is increased. Reference 1 provides guidelines for methods of determining proper amounts of preheat for typical steels used in building construction. Other sources address preheat requirements for steels with higher alloy contents.

21.12.2 Interpass Temperature

The temperature of the material in the weld area immediately before the second and each subsequent pass of a multiple-pass weld is called the interpass temperature. In fact, the minimum specified interpass temperature is often equal to the minimum specified preheat temperature, but this is not necessarily required. Both the yield and ultimate tensile strengths of the weld metal are functions of the interpass temperature. Especially on sensitive base metals, the interpass temperature must be high enough to prevent cracking, but low enough to allow for adequate mechanical properties. Temperature indicating crayons can be used to control interpass temperature. One such crayon is used to measure both the minimum specified preheat and the minimum specified interpass temperature, while a second, higher temperature crayon is used to measure the maximum specified interpass temperature (if required). The effects of the welding process, procedures, and sequence of welding must always be considered in order to maintain interpass temperature within the proper range. Finally, any specified interpass temperature should be maintained throughout the full thickness of the base metal and some reasonable distance away from the weld, usually approximately 1 in, unless codes or standards specify otherwise.

21.12.3 Postweld Heat Treatment

Any heat treatment applied after welding to improve the properties of a weldment is called *postweld heat treatment* (PWHT). In steel fabrication, the most commonly applied PWHT procedures are post heating and stress relieving.

The need for post heating assumes the potential for hydrogen induced cracking. Post heating is not necessary for most applications, but may be an essential requirement for specialized applications. Although various code and service requirements can dictate a variety of temperatures and hold times, a common level would be 450°F, maintained for 1 h/in of thickness.

Stress relief heat treatment is used to reduce the stresses that remain locked in a structure as a consequence of manufacturing processes. When the temperature of the weldment is elevated, residual stresses can relieve themselves. Upon return to room temperature, the remaining residual stresses will be significantly reduced. Stress relief is typically utilized for one of two reasons: to achieve specific metallurgical properties, or to produce a more dimensionally stable weldment.

Some materials, in the as-deposited state, are very strong but brittle. A stress relief treatment can lower the strength and restore ductility. This is frequently the case for steel alloys that will be placed into elevated temperature service conditions, such as pressure piping and pressure vessels.

The second major reason a weldment may receive stress relief is to provide for dimensional stability. When machining an as-welded weldment, the process of metal removal will naturally remove some material that holds residual stress. When the material is removed, the residual stresses must realign themselves. This can cause the weldment to change its geometric shape. A machinist may complain that the weldment is "walking," i.e., changing shape, as one cut follows another. Stress relief can minimize these tendencies.

When determining whether or not to postweld heat treat, the alloying system and previous heat treatment of the base metal must be considered. PWHT may adversely affect the properties of quenched and tempered alloy steels, for example, if the temperature exceeds the tempering temperature of the base metal. For specific heat treatment recommendations, the filler metal manufacturer or the steel producer should be consulted.

21.12.4 Annealing

Annealing, which consists of heating to and holding at a suitable temperature followed by cooling at a suitable rate, may be used to improve machinability (by making the material softer), facilitate cold work, improve mechanical or electrical properties, and/or stabilize dimensions (just like stress relief). In ferrous alloys, annealing typically is done above the upper critical temperature, but time–temperature cycles vary greatly both in maximum temperature achieved and in cooling rate applied, depending on composition, material condition, and desired results. In nonferrous alloys, annealing is designed to either remove part or all of the effects of cold working, or cause coalescence of precipitates from solid solution in relatively coarse form or both. When weldments are annealed, changes to the properties of the weld deposit can occur. The properties of the annealed weld metal must be determined.

21.12.5 Normalizing

Normalizing involves higher temperatures than postweld heat treatment, usually 1560–1740°F for ferritic steels. Cooling takes place in still air, resulting in a refinement of the metallurgical grain size which leads to increased yield strength and better fracture resistance. For some types of steel, better control over mechanical properties can be achieved by subsequent tempering at temperatures below 1275°F. Knowledge of the normalizing and tempering history of the parent material is critical when determining PWHT holding temperature ranges. When a steel has been normalized and subsequently tempered to achieve a given set of mechanical properties, the upper limit of the hold temperature should not exceed the tempering temperature, or mechanical properties may be adversely affected. Normalized weld deposits can have significantly changed properties as compared to the as-welded condition. Investigation into the suitability of the resultant properties is essential.

21.13 QUALITY

Welding is a process with many variables that should be monitored. When performing welding and thermal cutting operations on metals, the base materials are typically subjected to rapid heating and cooling in localized areas. Such thermal conditions can produce distortion and result in detrimental mechanical properties of the materials if not controlled. Contaminants on the material to be welded, such as grease, dirt, rust, and other debris should be cleaned and removed. If such impurities are not removed before welding, the strength of the welded assembly could be compromised. Nondestructive testing methods, such as radiographic testing and dye penetrant are often used to detect flaws, such as cracks, porosity, lack of fusion, and slag entrapment. Destructive testing methods, such as bend tests and tensile tests are often performed to determine if a weldment is of adequate strength and soundness.

21.13.1 Definition of Quality

The quality of a weld may be defined as its ability to perform in the given design environment. Criteria used to determine weld quality include the weld's appearance, mechanical properties, soundness, and chemical composition. If any of these criteria are not monitored, the integrity of the weld could be compromised.

21.13.2 Appearance

Visual inspection of the appearance of the weld should be the first area addressed to determine weld quality. However, it should be noted that although a weld passes visual inspection, it still may be inadequate for service as measured by other criteria due to internal porosity, underbead cracking, incomplete fusion or slag inclusions. The shape and size of the weld bead strongly influences the design strength of the weld. In fillet welds, it is typically desirable to have a flat-to-slightly convex face on the weld. However, if a weld is too concave, the operator risks increased difficulty of slag removal at the toes of the weld. A weld that is too convex may also cause stress risers at the toes of the weld, as well as greatly increase welding costs.

Undercut into the base metal is also a sign of improper welding procedures. Undercut typically results from a travel speed too fast for a given current. In other words, the arc removes more base metal than can be replaced by filler metal. Typical remedies to reduce undercut include decreasing current and/or travel speed.

21.13.3 Mechanical Properties

Yield/Tensile Strength. Grades of steel are typically designated based on their minimum yield strength. Associated with that yield strength is a corresponding tensile strength. In the case of welding filler materials, just the opposite is true: the primary reference is to the tensile strength. In terms of quality, it is important that the deposited weld metal exhibit the strength properties intended by the design. This starts with the use of the proper filler metal, and continues with compliance with the intended welding procedures. The WPS parameters that deviate from recommendations may result in welds with either lower, or higher, values of yield or tensile strength.

Notch toughness may be defined as the ability of a material to absorb energy in the presence of a flaw. Stated another way, notch toughness may be considered the property of the material that resists the onset of a brittle fracture. The notch toughness of deposited weld metal is particularly sensitive to changes in welding parameters. Adherence to the welding procedure is again important. It is essential that the WPS values selected for the application be capable of delivering weld metal with the required level of notch toughness. This often necessitates testing the procedure to ensure that the specific combination will achieve the desired results.

21.13.4 Soundness

Soundness refers to the relative quality of a weld in terms of freedom from weld discontinuities. Discontinuities are defined as irregularities in a weld. They may or may not be acceptable, depending on the application, location, size, or distribution of such discontinuities. Defects are discontinuities that are unacceptable. Weld soundness generally refers to freedom from excessive porosity, slag inclusions, and other discontinuities.

Porosity is a flaw that occurs as gas bubbles are trapped in solidified weld metal. Most porosity is not visible, and severe porosity can weaken the weld. To minimize porosity, keep the weld puddle molten for a longer period of time, so that gases may boil out before the metal freezes.

Spatter does not affect weld strength but does produce a poor appearance and increases cleaning costs.

Penetration refers to the depth the weld enters into the base metal. For full-strength welds, penetration to the bottom of the joint is required. To overcome lack of penetration, use higher currents and/or slower travel. Using smaller electrodes will also allow the operator to reach into deep, narrow grooves.

Inclusions occur when slag or other impurities are located in the weld metal. Slag inclusions occur in single- or multiple-pass welds. Slag inclusions result from improper welding procedure and may significantly weaken the weld. To reduce or eliminate slag inclusions, follow proper welding procedures and techniques for the given process.

Proper fusion exists when the weld bonds to both walls of the joint and forms a solid bead across the joint. Lack of fusion is often visible and must be avoided for a sound weld. Increasing the current and using a slower travel speed may decrease the potential for lack of fusion.

21.13.5 Weld Cracking During Fabrication

Whereas there are acceptable limits for slag inclusions and porosity in welds, cracks are never acceptable. Cracks in the vicinity of a weld indicate that one or more problems exist that must be addressed. A careful analysis of crack characteristics will make it possible to determine their cause and appropriate corrective measures.

For the purposes of this section, *cracking* will be distinguished from weld failure. Welds may fail due to overload, underdesign, or fatigue. The cracking discussed under this section has to do with solidification, cooling, and the stresses that develop due to weld shrinkage. Weld cracking occurs close to the time of fabrication. Hot cracks are those that occur at elevated temperatures and are usually solidification related. Cold cracks are those that occur after the weld metal has cooled to room temperature and may be hydrogen related. Neither is the result of service loads.

Most forms of cracking result from the shrinkage strains that occur as the weld metal cools. If the contraction is restricted, the strains will induce residual stresses that cause cracking. There are two opposing forces: the stresses induced by the shrinkage of the metal, and the surrounding rigidity of the base material. The shrinkage stresses increase as the amount of shrinking metal increases. Large weld sizes and deep penetrating welding procedures increase the shrinkage strains. The stresses induced by these strains will increase when higher strength filler metals and base materials are involved. When the stresses exceed the yield point, no further accumulation of stress is seen. The material simply yields. With a higher yield point, however, higher residual stresses will be present.

Opposing these residual stresses induced by shrinkage strains is the base metal rigidity. When thick, highly restrained plates are utilized, the strains imposed by the weld metal are largely self absorbed. That is, all shrinkage is restricted to the weld bead. When plates are relatively thin and flexible, however, the shrinking weld metal may pull the base material, redistributing the shrinkage strains. This causes distortion. Heavily restrained, normally ductile plates may be restricted from exhibiting their ductility, due to the very high triaxial (multidirectional) tensile strain that may develop in this region. The state of strain may cause the yield point to climb to a much higher value than the uniaxial yield point. This condition of strain causes the weld metal to experience even higher residual stresses which may exceed the uniaxial yield point.

Factors that increase the amount of strain on the metal will obviously increase cracking tendencies. However, particular attention must be paid to the restraint offered by the surrounding base material. Unfortunately, there is no simple way to describe restraint mathematically. However, experience has shown that when higher strength plate (greater than 50 ksi yield strength) in thicker sections (greater than $1^1/_2$ in) intersect from all three geometrical directions, high restraint will be experienced at the intersection of these plates.

Under conditions of high restraint, extra precautions must be utilized to overcome the cracking tendencies which are described in the following sections. It is essential to pay careful attention to welding sequence, preheat and interpass temperature, post weld heat treatment, joint design, welding procedures, and filler material. The judicious use of peening and in-process stress relieving may be necessary to fabricate highly restrained members.

Weld cracking may be characterized in one of the following ways: centerline cracking, heat affected zone cracking, and transverse cracking. Contained below is a summary of the factors that cause these types of cracks. The summary is directly applicable to arc welds made on steel. These factors are generally applicable to other materials and welding processes, but depending on the specific conditions involved, factors in addition to those discussed below may affect cracking tendencies in other applications. This summary is not intended to be applicable to all materials and welds made with all processes.

Centerline Cracking. Centerline cracking is characterized as a separation in the center of a given weld bead. If the weld bead happens to be in the center of the joint, as is always the case on a single pass weld, centerline cracks will be in the center of the joint. In the case of multiple pass welds, where several beads per layer may be applied, a centerline crack may not be in the geometric center of the joint, although it will always be in the center of the bead (see Fig. 21.20).

Centerline cracking is the result of one of the following three phenomena: segregation induced cracking, bead shape induced cracking, or surface profile induced cracking. Unfortunately, all three phenomena evidence themselves in the same type of crack, and it is often difficult to identify the cause. Moreover, experience has shown that often two or even all three of the phenomena will interact and contribute to the cracking problem. Understanding the fundamental mechanism of each of these types of centerline cracks will help in determining the corrective solutions.

Segregation Induced Cracking. Segregation induced cracking occurs when low melting point constituents in the admixture separate during the weld solidification process. Low melting point

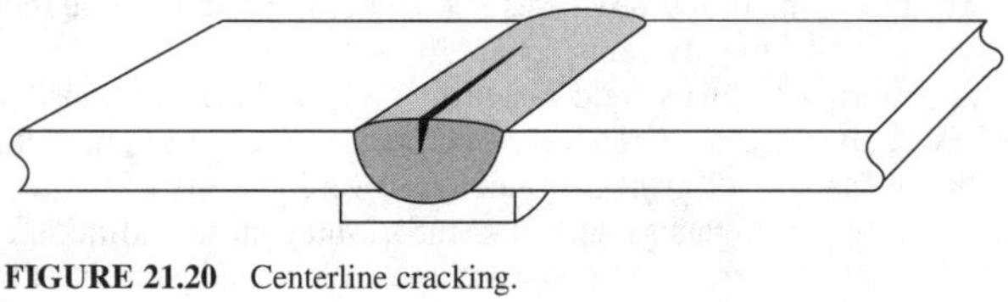

FIGURE 21.20 Centerline cracking.

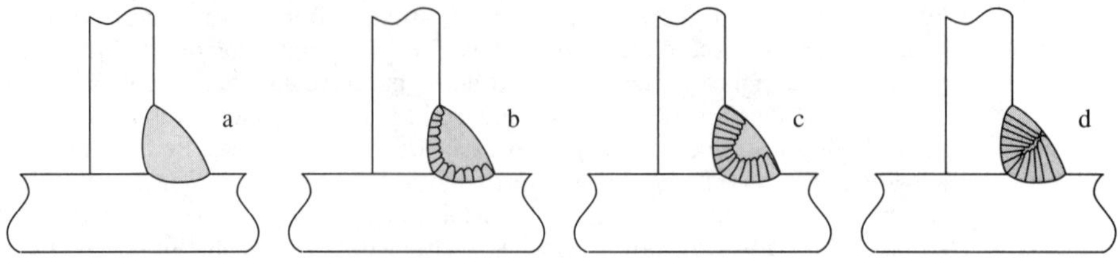

FIGURE 21.21 Segregation induced cracking.

components in the molten metal will be forced to the center of the joint during solidification, since they are the last to solidify and the weld tends to separate as the solidified metal contracts away from the center region containing the low melting point constituents.

When intermixed materials have a significantly different melting point than the basic iron-carbon weld metal, it is possible to have a liquid mixture in the center of the joint well after the majority of the weld has solidified. This is illustrated in Fig. 21.21. In Fig. 21.21(*a*), the weld nugget is entirely molten. In Fig. 21.21(*b*), solidification has begun. When materials solidify, segregation may occur. The result is a change in composition throughout the cross section of the solidified material. The grains of steel have begun to grow, perpendicular to the fusion interface. As this solidification proceeds, segregation occurs. In an iron-carbon system, the first materials to solidify are typically lower in carbon content. This is due to the fact that pure iron has a higher freezing point than iron-carbon alloys.

The low carbon layer that begins to form, forces higher levels of carbon into the center. Any low melting point constituents will also be forced into this liquid center. Part of the cross section is solidified, while a portion remains liquid. In Fig. 21.21(*c*), solidification has progressed further. In Fig. 21.21(*d*), solidification is nearly complete. Notice that the last portion of the bead to solidify is in the center of the cross section. This is the location that will contain any remaining components that have the lower freezing point.

If the steels contain higher than desirable levels of sulfur, phosphorus, lead, or copper, these elements will segregate into the center of the solidifying weld bead. Perhaps the most frequently encountered contaminant from steel is sulfur. In the presence of iron, the sulfur will combine to form iron sulfide (FeS). Iron sulfide has a melting point of approximately 2200°F. Steel, on the other hand, has a melting point of approximately 2800°F. As the grains grow, FeS are forced into the center of the joint. Well after all of this steel has solidified, the liquid iron sulfides with a melting point 600°F less than the steel will be contained in the center of the weld bead. As the steel cools, it contracts, pulling on the center of the weld bead which contains the liquid iron sulfide. As shown in Fig. 21.21(*d*), the weld bead will crack.

Phosphorus, lead, and copper will act in a similar manner. The primary difference with these elements is that they do not form compounds, but are present in their basic form. The commercial welding processes are all capable of tolerating low levels of these contaminants. However, when higher levels are experienced, the segregation occurs and may result in centerline cracking. Whereas these elements may come from the filler material, they are more commonly the result of base material compositions; therefore, they must be controlled in the base materials.

When centerline cracking induced by segregation is experienced, several solutions may be implemented. Since the contaminant usually comes from the base material, the first consideration is to limit the amount of contaminant pickup from the base material. This may be done by limiting the penetration of the welding process. In some cases, a joint redesign may be desirable. The extra penetration afforded by some of the processes is not necessary and can be reduced. This can be accomplished by using lower welding currents.

A buttering layer of weld material (see Fig. 21.22), deposited by a low energy process such as SMAW, may effectively reduce the amount of pickup of contaminant into the weld admixture.

In the case of sulfur, it is possible to overcome the harmful effects of iron sulfides by preferentially forming manganese sulfide (MnS). Manganese sulfide is created when manganese is present in sufficient quantities to counteract the sulfur. Manganese sulfide has a melting point of 2900°F.

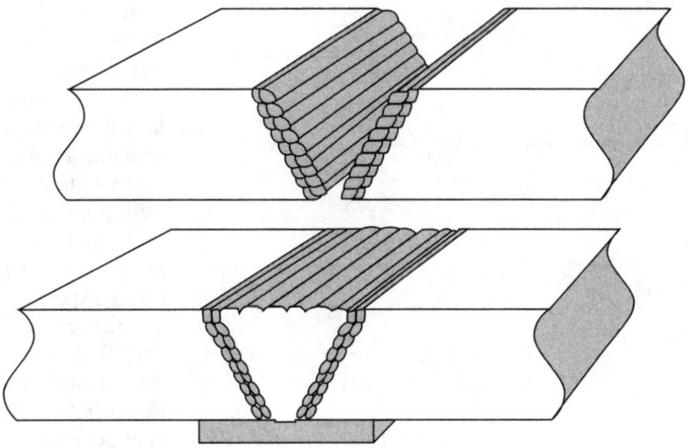

FIGURE 21.22 Buttering.

In this situation, before the weld metal begins to solidify, manganese sulfides are formed which do not segregate. Steel producers utilize this concept when higher levels of sulfur are encountered in the iron ore. In welding, it is possible to use filler materials with higher levels of manganese to overcome the formation of low melting point iron sulfide. Unfortunately, this concept cannot be applied to contaminants other than sulfur.

Bead Shape Induced Cracking. The second type of centerline cracking is known as bead shape induced cracking. This is illustrated in Fig. 21.23, and is associated with deep penetrating processes such as SAW and gas shielded FCAW. When a weld bead is of a shape where there is more depth than width to the weld cross section, the solidifying grains growing perpendicular to the steel surface intersect in the middle, but do not gain fusion across the joint. To correct for this condition, the individual weld beads must have at least as much width as depth. Recommendations vary from a 1:1 to a 1.4:1 width-to-depth ratio to remedy this condition. The total weld configuration, which may have many individual weld beads, can have an overall profile that constitutes more depth than width. If multiple passes are used in this situation, and each bead is wider than it is deep, a crack-free weld can be made.

Joint design affects the tendency toward centerline cracking induced by bead shape. The prequalified joints in welding specifications have taken this into consideration. Consider, for example, the three joints listed in Fig. 21.24. All three are prequalified for standard use in the AWS D1.1 Structural Welding Code-Steel. Known as a B-U2a, the combination of root opening and included angle is adjusted to ensure the bead shape of the root pass has an acceptable width-to-depth ratio. As the included angle is decreased, the tendency toward a narrow, deep bead increases. To compensate for this, a larger root opening is used. For PJP groove welds, the preferred configuration for a single bevel joint is to have a minimum of a 60° included angle when the submerged welding process is used.

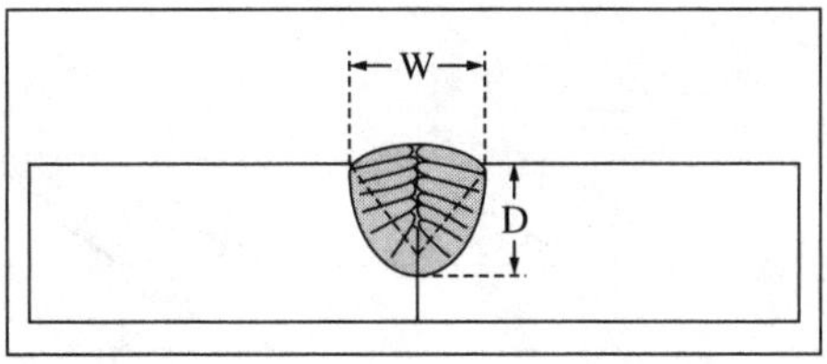

FIGURE 21.23 Bead shape induced cracking.

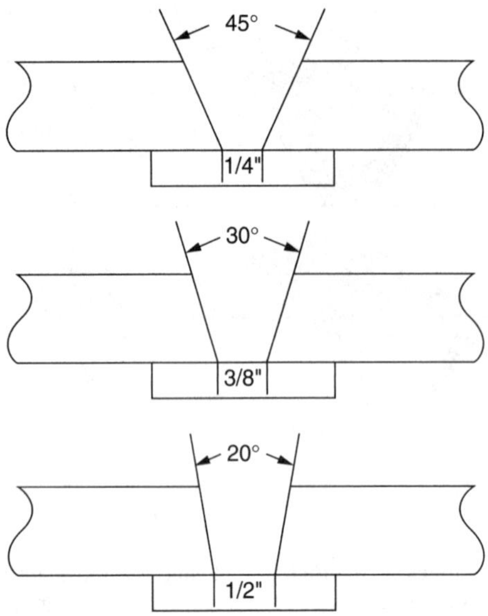

FIGURE 21.24 Acceptable root configurations.

With the deep penetration afforded by submerged arc, a 45° included angle could lead to unacceptable centerline cracking. Other processes may use smaller included angles, because they lack the penetration that would create this unacceptable relationship. However, these processes rarely obtain fusion down to the root under these tight joint configurations.

Centerline cracking due to bead shape may occur in either groove welds or fillet welds. It is rarely experienced in fillet welds when applied to 90° tee joints. However, when skewed joints are specified, and the acute angle side is less than 70°, centerline cracking may occur. This is particularly true when the weld process has significant penetrating capability.

When centerline cracking due to bead shape is experienced, the obvious solution is to change the width-to-depth relationship. This may involve a change in joint design. Since the depth is a function of penetration, it is advisable to reduce the amount of penetration. This can be accomplished by utilizing lower welding amperages and larger diameter electrodes. All of these approaches will reduce the current density and limit the amount of penetration. For submerged arc, using a direct current, negative polarity electrode will reduce penetration for a given WFS.

Surface Profile Induced Cracking. The final mechanism that generates centerline cracks is surface profile conditions. When concave weld surfaces are created, internal shrinkage stresses will place the weld metal on the surface into tension. Conversely, when convex weld surfaces are created, the internal shrinkage forces pull the surface into compression. Concave weld surfaces frequently are the result of high arc voltages. A slight decrease in arc voltage will cause the weld bead to return to a slightly convex profile and eliminate the cracking tendency. High travel speeds may also result in this configuration. A reduction in travel speed will increase the amount of fill and return the surface to a convex profile. Vertical down welding also has the tendency to generate these crack sensitive, concave surfaces. Vertical-up welding can remedy this situation by providing a more convex bead.

Heat Affected Zone Cracking. Heat affected zone cracking is characterized by separation that occurs immediately adjacent to the weld bead, as shown in Fig. 21.25. The cracking occurs in the base material, and although it is certainly related to the welding process, the crack does not occur in the filler material. This type of cracking is also known as *underbead cracking* or *delayed cracking*. The small, potentially brittle region surrounding the weld bead, called the heat affected zone or HAZ, results from the thermal cycle experienced by this region during welding. Initially, the energy of the welding process raises this zone to a temperature above the transformation temperature, but lower

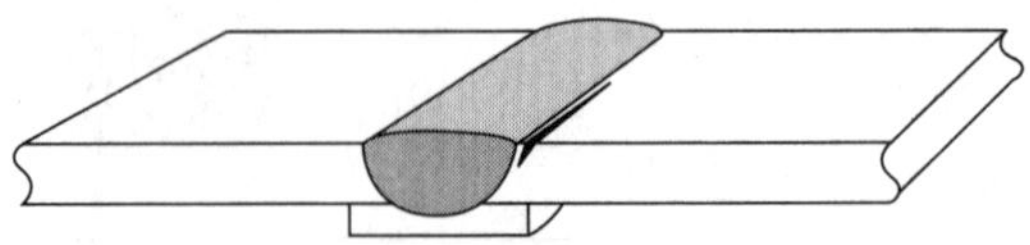

FIGURE 21.25 HAZ cracking.

than the melting point. After the steel has been raised to this temperature, its final properties will be dependent upon the cooling rate it experiences. The surrounding colder steel quickly cools the small region around the weld bead. It is possible for this zone to have hardnesses that approach the maximum possible for the steel. Along with this increase in hardness comes a decrease in ductility. This hardening is generally isolated to the base material (as opposed to the weld metal) due to the higher carbon and carbon equivalency values they possess. The probability of HAZ cracking increases as carbon and carbon equivalency levels are increased in the base materials. Since the filler material is not cracking, a change in filler material would generally not affect HAZ cracking.

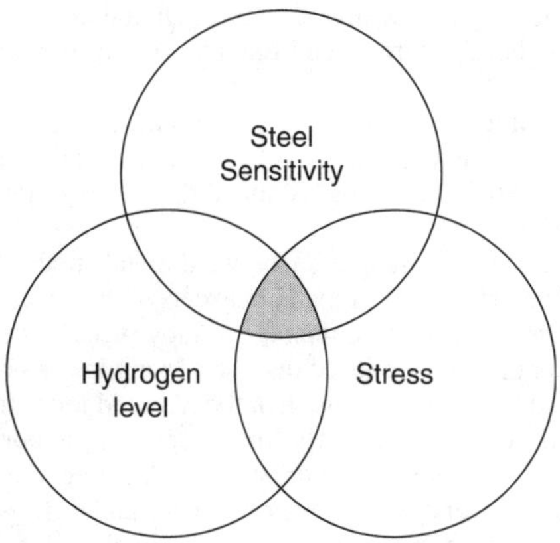

FIGURE 21.26 Hydrogen cracking variables.

Hydrogen is another factor that contributes to heat affected zone cracking. Moisture and organic compounds are the primary sources of hydrogen. They may be present on the steel, electrode, shielding materials, or even in the atmospheric humidity. Significant quantities of hydrogen can be dissolved in liquid weld material. As the weld metal begins to solidify, and particularly as it transforms from austenite to ferrite, the solubility for hydrogen decreases. Hydrogen moves within the weld metal at a rate exponentially proportional to the temperature. At low temperatures, the diffusion rates greatly decrease. Free hydrogen, moving at random through the solidified weld metal, tends to accumulate around discontinuities in the heat affected zone that is significantly lower in temperature than the weld material. As atomic hydrogen combines to form molecular hydrogen, a tremendous increase in internal pressure is experienced in this heat affected zone. A susceptible structure, the presence of hydrogen, and the residual stresses of welding will combine to cause cracking to occur. As is shown in Fig. 21.26, three elements are required for hydrogen assisted cracking to occur. If any one of the three is decreased, hydrogen cracking can be minimized.

If steels that are not sensitive to hydrogen cracking are used, even fairly high levels of hydrogen can be experienced before the onset of any problems. These insensitive steels include low carbon and low carbon equivalency steels. Generally speaking, these are also low strength steels. Welding processes with even high levels of diffusible hydrogen have been successfully used without generating cracks, but they may be used only on steels which are lower in yield strength and less hardenable. Increases in carbon, alloy, and strength will increase hydrogen cracking tendencies. To overcome these problems, hydrogen levels must be controlled more carefully, and increased preheat and interpass temperatures must be employed.

The diffusion of hydrogen through the weld metal requires time. Cracking may occur hours and even days after the time of fabrication. When steel is subject to this type of delayed cracking or welding, inspection periods are frequently delayed, typically by 48 h, to ensure that this cracking will not occur after inspection.

The most effective way to reduce the hardness of the HAZ is to preheat the steel before welding and control the heat input and interpass temperature during welding. The preheat will slow the rate of cooling experienced by the heat affected zone, and the microstructures that develop will be softer and more ductile. The hydrogen diffusion rate is increased and the level of hydrogen is effectively lowered as well. In addition, there is a slight decrease in the residual stresses when preheat is applied. Thus, preheat addresses all three variables that contribute to this heat affected zone cracking problem.

It is possible to confuse underbead cracking with lamellar tearing. Lamellar tearing is a subsurface crack in the base metal with a basic orientation parallel to the base metal surface. It is caused by shrinkage stresses acting on the through-thickness direction of the base metal. When strained by shrinking weld metal, the metal between small planar shaped nonmetallic inclusion can tear, creating a step-like pattern of cracking. A lamellar tear generally reveals the grainlike structure in the base material. It usually is highly oxidized and very coarse in appearance. On the other hand, heat affected zone cracking generally exhibits a smoother fracture resembling that of a brittle failure.

Transverse Cracking. Transverse cracking, also called cross cracking, is characterized as a crack within the weld metal perpendicular to the direction of travel. This is the least frequently encountered type of cracking, and is generally associated with weld metal that is very high in strength, significantly overmatching the base material.

When transverse cracking is encountered, the strength of the weld metal should be reviewed first. Emphasis is placed upon the weld metal because the filler metal may deposit lower strength, highly ductile metal under normal conditions. However, with the influence of alloy pickup, it is possible for the weld metal to exhibit extremely high strengths with reduced ductility. Using lower strength weld metal is an effective solution, but caution should be taken to ensure that the required joint strength is attained.

Hydrogen is a second factor in transverse cracking. Hydrogen cracking is normally associated with heat affected zone cracking. However, on lower carbon steels, HAZ cracking due to hydrogen may be replaced by hydrogen induced transverse cracking in the weld metal. Lower strength weld material, with a corresponding decrease in carbon equivalencies, effectively solves the problem.

Preheat may have to be applied to alleviate transverse cracking. The preheat will assist in diffusing the hydrogen. As preheat is applied, it will additionally expand the length of the weld joint, allowing the weld metal and the joint to contract simultaneously, and reducing the applied stress to the shrinking weld. This is particularly important when making circumferential welds. When the circumference of the materials being welded is expanded, the weld metal is free to contract along with the surrounding base material, reducing the transverse shrinkage stress. Finally, postweld heat treatments that involve holding the steel at 450°F will assist in diffusing any residual hydrogen.

21.13.6 Chemistry

For many welding applications, the chemistry of the weld is inconsequential; the joint is designed for strength only. However, significant consideration should be given to other alloy systems that may involve corrosive environments or high temperature service applications. The selection of the filler metal used to join a given material should take into consideration the service environment of the welded assembly. The variables of the welding process can also influence the deposit chemistry. To ensure that the chemical composition of the weld metal meets expectations, proper procedures for a given process should be developed and followed.

21.14 TESTING

Welds are inspected to ensure that they comply with the requirements of a given specification. Welds are acceptable when they conform to all the requirements in a given specification or code. Weld quality is directly tied to the code or specification under which the work is being performed.

Many welds contain discontinuities, which are defined as irregularities in the weld that may or may not be acceptable under a given specification. Defects are discontinuities that are rejectable by a given specification. Weld cracks should never be permitted and should be repaired with proper procedures. It is impossible to define a perfect weld. However, if it were possible, the cost associated with such a weld would make it prohibitive for most commercial applications. With today's modern nondestructive testing equipment, a host of discontinuities can be found by the skilled technician. These findings must be compared to specifications to determine if the discontinuities are acceptable.

Five major nondestructive methods are used to evaluate weld metal integrity. Each has unique advantages and limitations. Some discontinuities are revealed more readily with one method as compared to another. It is important for the manufacturing engineer to understand the capabilities and limitations of these inspection methods, particularly in the situations where interpretation of the results is questionable.

21.14.1 Visual Testing

Visual testing (VT) is by far the most powerful inspection method available. Because of its relative simplicity and lack of sophisticated equipment, some people discount its power. However, it is the only inspection method that can and should be applied during each step of the welding process, rather than after the weld has been made, it is the only method that can actually increase the quality of fabrication and reduce the generation of welding defects.

Most standards require that all welds be visually inspected. Visual inspection begins long before an arc is struck. Materials that are to be welded must be examined for quality, type, size, cleanliness, and freedom from defects. The pieces to be joined should be checked for straightness, flatness, and dimensions. Alignment and fitup of parts should be examined. Joint preparation should be verified. Procedural data should be reviewed, and production compliance assured. All of these factors precede any welding that will be performed.

During welding, visual testing includes verification that the procedures used are in compliance with the welding procedure specification. Upon completion of the weld bead, the individual weld passes are inspected for signs of porosity, slag inclusion, and any weld cracks. Bead size, shape, and sequences can be observed. Interpass temperatures can be verified before subsequent passes are applied. Visual inspection can ensure compliance with procedural requirements. Upon completing the weld, the size, appearance, bead profile, and surface quality can be inspected.

Visual testing should be performed by the weld inspector, as well as by the welder. Good lighting and eye sight is imperative. In most fabrication shops, some type of auxiliary lighting is required for effective visual inspection. Magnifying glasses, gauges, and workmanship samples all aid in visual testing.

21.14.2 Penetrant Testing

Penetrant testing (PT) involves the application of a liquid which by a capillary action is drawn into a surface discontinuity, such as a crack or porosity. When the excess, residual dye is removed from the surface, a developer is applied, which will absorb the penetrant that is contained within the discontinuity. This results in a stain in the developer showing that a discontinuity is present.

Penetrant testing is limited to revealing surface discontinuities. It has no ability to read subsurface discontinuities, but it is highly effective in accenting the surface discontinuities that may be overlooked or be too small to detect with visual inspection.

21.14.3 Magnetic Particle Testing

Magnetic particle testing (MT) utilizes the change in magnetic flux that occurs when a magnetic field is present in the vicinity of a discontinuity. This change in magnetic flux density will show up as a different pattern when magnetic powders are applied to the surface of a part. The process is effective in locating discontinuities that are on the surface and slightly subsurface. For structural applications, magnetic particle testing is more effective than dye penetrant inspection. Magnetic particle inspection can reveal cracks. It can also reveal slag inclusions and porosity, discontinuities which within prescribed limits may be acceptable. The distance between the discontinuities, as well as the size, will dictate their acceptance.

The magnetic field is created in the material to be inspected in one of two ways. Current is either directly passed through the material, or a magnetic field is induced through a coil on a yoke. With the first method, electrical current is passed through two prods that are placed in contact with the surface. When the prods are initially placed on the material, no current is applied. After intimate contact is assured, current is passed through. Small arcs may occur between the prods and the base material, resulting in an arc strike, which may create a localized brittle zone. It is important that prods be kept in good shape and that intimate contact with the work is maintained before the current is passed through the prods.

The second method of magnetic field generation is through induction. In what is known as the yoke method, an electrical coil is wrapped around a core with flexible ends. Electrical current is passed through the coil, creating a magnetic field in the core. When placed in contact with the part being inspected, the magnetic field is induced into the part. Since current is not passed into the part, the potential for arc strikes is eliminated. Along with this significant advantage, comes a disadvantage: the yoke method is not as sensitive to subsurface discontinuities as the prod method.

Cracks are most easily detected when they lie perpendicular to the magnetic field. With the prod method the magnetic field is generated perpendicular to the direction of current flow. For the yoke method, just the opposite is true. Magnetic particle inspection is most effective when the region is inspected twice: once with the field located parallel to, and once with the field perpendicular to the longitudinal weld axis.

While magnetic particle inspection can read some subsurface discontinuities, it is best viewed as an enhancer of visual inspection. Most codes prescribe that part of the fillet welds be inspected with this method. Another common use of MT is for the inspection of intermediate passes on large groove welds, particularly in those situations that are known to be crack sensitive.

21.14.4 Radiographic Testing

Radiographic testing (RT) uses X rays or gamma (γ) rays that are passed through the weld and expose a photographic film on the opposite side of the joint. X rays are produced by high voltage generators, while γ rays are produced by atomic disintegration of radioactive isotopes.

Whenever radiography is used, precautions must be taken to protect workers from exposure to excessive radiation. Safety measures dictated by the Occupational Safety and Health Administration (OSHA), the National Electrical Manufacturer's Association (NEMA), the Nuclear Regulatory Commission (NRC), the American Society of Nondestructive Testing (ASNT), and other agencies should be carefully followed when radiographic inspection is conducted. Radiographic testing relies on the ability of the material to pass some of the radiation through, while absorbing part of this energy within the material. Different materials have different absorption rates. Thin materials will absorb less radiation than thick materials. The higher the density of the material, the greater the absorption rate. As different levels of radiation are passed through the materials, portions of the film are exposed to a greater or lesser degree than the rest. When this film is developed, the resulting radiograph will bear the image of the cross section of the part. A radiograph is actually a negative. The darkest regions are those that were most exposed when the material being inspected absorbed the least amount of radiation. Thin parts will be darkest on the radiograph. Porosity will be revealed as small, dark, round circles. Slag is also generally dark, and will look similar to porosity, but will be irregular in its shape. Cracks appear as dark lines. Lack of fusion or underfill will show up as dark spots. Excessive reinforcement on the weld will result in a light region.

Radiographic testing is most effective for detecting volumetric discontinuities: slag and porosity. When cracks are oriented perpendicular to the direction of the radiation source, they may be missed with the RT method. Tight cracks that are parallel to the radiation path have also been overlooked with RT.

Radiographic testing has the advantage of generating a permanent record for future reference. With a "picture" to look at, many people are more confident that the interpretation of weld quality is meaningful. However, reading a radiograph and interpreting the results requires stringent training, so the veracity of a radiograph depends to a great degree on the skill of the technician.

21.14.5 Ultrasonic Testing

Ultrasonic inspection relies on the transmission of high frequency sound waves through materials. Solid, discontinuity—free materials will transmit the sound throughout a part in an uninterrupted fashion. A receiver "hears" the sound reflected off of the back surface of the part being inspected. If a discontinuity is contained between the transmitter and the back side of the part, an intermediate signal will be sent to the receiver indicating the presence of this discontinuity. The pulses are read on a screen. The magnitude of the signal received from the discontinuity indicates its size. The relationship of the signal with respect to the back wall will indicate its location. Ultrasonic inspection is sensitive enough to read discontinuities that are not relevant to the performance of the weld. It is a sophisticated device that is very effective in spotting even small discontinuities. *Ultrasonic testing* (UT) is most sensitive to planar discontinuities, i.e., cracks. Under some conditions, uniformly cylindrical or spherical discontinuities can be overlooked with UT.

The UT results must be interpreted by the UT technician. Depending on the geometry of the part being inspected, and the type of signal received, there may be significant differences in opinion among skilled technicians as to what is being detected. In contrast with other NDT methods that rely more on visual indications, UT results can be debated, often resulting in lack of confidence in the findings. More often than not, the debates surround whether an indication is cause for rejection or not. Experience is key to understanding what the signal means, and to thereby avoid unnecessary rejection or repair of otherwise acceptable discontinuities.

21.15 WELDING COSTS

Welding is a labor intensive technology. Electrodes, gases, and fluxes constitute a small portion of the total welding cost. Therefore, the prime focus of cost control will be reducing the amount of time required to make a weld.

The following example is given to illustrate the relative costs of materials and labor, as well as to assess the effects of proper process selection. The example to be considered is a simple $^5\!/_{16}$-in fillet weld. Two welding processes will be considered: SMAW and submerged arc welding using parallel electrodes. Either would generate high quality welded connections.

Example To calculate the cost per foot of a weld of this nature, an equation taking the following format is used:

$$\text{Cost per foot} = \frac{\text{Labor and Overhead Rate}}{5(\text{Travel Speed})(\text{O.F.})} + \frac{(\text{Weight of Weld/Foot})}{\text{Electrode Efficiency}}$$

In this equation, the two major components are labor and materials. Affecting the labor part of the equation are the labor and overhead rate, the travel speed, and the *operating factor* (OF). It is impossible to keep the arc initiated 100 percent of the time. To compensate for this, the term "operating factor" has been introduced. This represents the percentage of time in the day that the arc is actually lit.

For the material portion of the equation, data have been taken from actual welds and represent the amount of material required to make one foot of weld. A $^5\!/_{16}$ in fillet requires approximately 0.19 lbs of metal per foot. In the case of SMAW, part of the weight of the electrodes is consumed in the slag. Moreover, it is impossible to use the whole length of the electrode that has been purchased. To compensate for this, electrode efficiency is used to determine how many electrodes will be required. In the example, the SMAW weld required 0.32 lb of electrode per foot while the SAW weld required only 0.19 lbs. However, SAW also requires flux. In this particular case, 0.21 lb of flux were consumed per foot of weld.

	Procedures	
	SMAW	SAW*
Electrode classification	E7018	F7A2-EM12K
Electrode diameter	$^1/_4$ in	Two $^2/_{32}$ in.
Amperage	350925	
Voltage	NA28	
Travel Speed	7.5 ipm	50 ipm
Weight of weld per ft	0.19 lb/ft	0.19 lb/ft
Electrode required per ft	0.32 lbs/ft	0.19 lbs/ft
Electrode efficiency	60%	100%
Flux weight per foot	N.A.	0.21 lbs
Polarity	ac	dc+
Weld Size-Fillet	5/16 in	5/16 in

*(parallel electrodes)

To complete the computation, the following were assumed:

Labor and overhead rate	$25 per hour
SMAW operating factor	25%
SAW operating factor	30%
SMAW electrode cost	$0.70 per pound
SAW electrode cost	$0.60 per pound
SAW flux cost	$0.50 per pound

When inserted into the preceding formula, the following equations result:

For SMAW

$$\text{Cost per foot} = \frac{25}{5(7.5)(0.25)} + \frac{(0.19)(0.70)}{0.6}$$
$$= 2.66 + 0.22 = \$2.88/\text{ft}$$

For SMAW

$$\text{Cost per foot} = \frac{25}{5(50)(0.3)} + \frac{(0.19)(0.60)}{1.0} + \frac{(0.21)(0.50)}{1.0}$$
$$= 0.33 + 0.11 + 0.10 = \$0.54/\text{ft}$$

In the case of the SMAW weld, the cost per foot was $2.88. The same weld, made with the SAW and the parallel electrode technique, resulted in a cost of $0.54 per foot, or 81 percent less.

In the first example, 92 percent of the cost of the weld was in labor. In the second, only 61 percent of the cost of the weld was in labor. The cost of materials remained about the same. The conscientious manufacturing engineer should always be looking for new production methods that may result in lower fabrication costs.

21.16 SAFETY

Welding is safe when sufficient measures are taken to protect the welder from potential hazards. When these measures are overlooked or ignored, welders can be subjected to electric shock, overexposure to radiation, fumes and gases, and fire and explosion. Any of these can be fatal. Everyone associated with welding operations should be aware of the potential hazards and help ensure that safe practices are employed. Infractions must be reported to the appropriate responsible authority. Basic safe welding practices are outlined in Figs. 21.27, 21.28, and 21.29.

Operation	Electrode Size 1/32 in. (mm)	Arc Current (A)	Minimum Protective Shade	Suggested[1] Shade No. (Comfort)
Shielded metal arc welding	Less than 3 (2.5) 3–5 (2.5–4) 5–8 (4–6.4) More than 8 (6.4)	Less than 60 60–160 160–250 250–550	7 8 10 11	— 10 12 14
Gas metal arc welding and flux cored arc welding		Less than 60 60–160 160–250 250–500	7 10 10 10	— 11 12 14
Gas tungsten arc welding		Less than 50 50–150 150–500	8 8 10	10 12 14
Air carbon arc cutting	(Light) (Heavy)	Less than 500 500–1000	10 11	12 14
Plasma arc welding		Less than 20 20–100 100–400 400–800	6 8 10 11	6 to 8 10 12 14
Plasma arc cutting	(Light)[2] (Medium)[2] (Heavy)[2]	Less than 300 300–400 400–800	8 9 10	9 12 14
Torch brazing		—	—	3 or 4
Torch soldering		—	—	2
Carbon arc welding		—	—	14
	Plate thickness			
	in.	mm		
Gas welding Light Medium Heavy	 Under 1/8 1/8 to 1/2 Over 1/2	 Under 3.2 3.2 to 12.7 Over 12.7		 4 or 5 5 or 6 6 or 8
Oxygen cutting Light Medium Heavy	 Under 1 1 to 6 Over 6	 Under 25 25 to 150 Over 150		 3 or 4 4 or 5 5 or 6

(1) As a rule of thumb, start with a shade that is too dark to see the weld zone. Then go to a lighter shade which gives sufficient view of the weld zone without going below the minimum. In oxyfuel gas welding or cutting where the torch produces a high yellow light, it is desirable to use a filter lens that absorbs the yellow or sodium line in the visible light of the (spectrum) operation.

(2) These values apply where the actual arc is clearly seen. Experience has shown that lighter filters may be used when the arc is hidden by the workpiece.

Data from ANSI/ASC Z49.1-88

FIGURE 21.27 Guide for shade numbers.

WARNING — **ARC WELDING can be hazardous.**

PROTECT YOURSELF AND OTHERS FROM POSSIBLE SERIOUS INJURY OR DEATH. KEEP CHILDREN AWAY. PACEMAKER WEARERS SHOULD CONSULT WITH THEIR DOCTOR BEFORE OPERATING.

Read and understand the following safety highlights. For additional safety information it is strongly recommended that you purchase a copy of "Safety in Welding & Cutting - ANSI Standard Z49.1" from the American Welding Society, P.O. Box 351040, Miami, Florida 33135 or CSA Standard W117.2-1974. A **Free** copy of "Arc Welding Safety" booklet E205 is available from the Lincoln Electric Company, 22801 St. Clair Avenue, Cleveland, Ohio 44117-1199.

BE SURE THAT ALL INSTALLATION, OPERATION, MAINTENANCE, AND REPAIR PROCEDURES ARE PERFORMED ONLY BY QUALIFIED INDIVIDUALS.

ELECTRIC SHOCK can kill.

1.a. The electrode and work (or ground) circuits are electrically "hot" when the welder is on. Do not touch these "hot" parts with your bare skin or wet clothing. Wear dry, hole-free gloves to insulate hands.

1.b. Insulate yourself from work and ground using dry insulation. Make certain the insulation is large enough to cover your full area of physical contact with work and ground.

In addition to the normal safety precautions, if welding must be performed under electrically hazardous conditions (in damp locations or while wearing wet clothing; on metal structures such as floors, gratings or scaffolds; when in cramped positions such as sitting, kneeling or lying, if there is a high risk of unavoidable or accidental contact with the workpiece or ground) use the following equipment:

- **Semiautomatic DC Constant Voltage (Wire) Welder.**
- **DC Manual (Stick) Welder.**
- **AC Welder with Reduced Voltage Control.**

1.c. In semiautomatic or automatic wire welding, the electrode, electrode reel, welding head, nozzle or semiautomatic welding gun are also electrically "hot".

1.d. Always be sure the work cable makes a good electrical connection with the metal being welded. The connection should be as close as possible to the area being welded.

1.e. Ground the work or metal to be welded to a good electrical (earth) ground.

1.f. Maintain the electrode holder, work clamp, welding cable and welding machine in good, safe operating condition. Replace damaged insulation.

1.g. Never dip the electrode in water for cooling.

1.h. Never simultaneously touch electrically "hot" parts of electrode holders connected to two welders because voltage between the two can be the total of the open circuit voltage of both welders.

1.i. When working above floor level, use a safety belt to protect yourself from a fall should you get a shock.

1.j. Also see Items 4.c. and 6.

ARC RAYS can burn.

2.a. Use a shield with the proper filter and cover plates to protect your eyes from sparks and the rays of the arc when welding or observing open arc welding. Headshield and filter lens should conform to ANSI Z87. I standards.

2.b. Use suitable clothing made from durable flame-resistant material to protect your skin and that of your helpers from the arc rays.

2.c. Protect other nearby personnel with suitable non-flammable screening and/or warn them not to watch the arc nor expose themselves to the arc rays or to hot spatter or metal.

FUMES AND GASES can be dangerous.

3.a. Welding may produce fumes and gases hazardous to health. Avoid breathing these fumes and gases.When welding, keep your head out of the fume. Use enough ventilation and/or exhaust at the arc to keep fumes and gases away from the breathing zone. **When welding with electrodes which require special ventilation such as stainless or hard facing (see instructions on container or MSDS) or on galvanized, lead or cadmium plated steel and other metals which produce toxic fumes, keep exposure as low as possible and below Threshold Limit Values (TLV) using local exhaust or mechanical ventilation. In confined spaces or in some circumstances, outdoors, a respirator may be required.**

3.b. Do not weld in locations near chlorinated hydrocarbon vapors coming from degreasing, cleaning or spraying operations. The heat and rays of the arc can react with solvent vapors to form phosgene, a highly toxic gas, and other irritating products.

3.c. Shielding gases used for arc welding can displace air and cause injury or death. Always use enough ventilation, especially in confined areas, to insure breathing air is safe.

3.d. Read and understand the manufacturer's instructions for this equipment and the consumables to be used, including the material safety data sheet (MSDS) and follow your employer's safety practices. MSDS forms are available from your welding distributor or from the manufacturer.

3.e. Also see item 7b.

WELDING SPARKS can cause fire or explosion.

4.a. Remove fire hazards from the welding area. If this is not possible, cover them to prevent the welding sparks from starting a fire. Remember that welding sparks and hot materials from welding can easily go through small cracks and openings to adjacent areas. Avoid welding near hydraulic lines. Have a fire extinguisher readily available.

4.b. Where compressed gases are to be used at the job site, special precautions should be used to prevent hazardous situations. Refer to "Safety in Welding and Cutting" (ANSI Standard Z49.1) and the operating information for the equipment being used.

4.c. When not welding, make certain no part of the electrode circuit is touching the work or ground. Accidental contact can cause overheating and create a fire hazard.

4.d. Do not heat, cut or weld tanks, drums or containers until the proper steps have been taken to insure that such procedures will not cause flammable or toxic vapors from substances inside. They can cause an explosion even though they have been "cleaned." For information purchase "Recommended Safe Practices for the Preparation for Welding and Cutting of Containers and Piping That Have Held Hazardous Substances", AWS F4.1-80 from the American Welding Society (see address above).

4.e. Vent hollow castings or containers before heating, cutting or welding. They may explode.

FIGURE 21.28 Arc welding safety precautions

4.f. Sparks and spatter are thrown from the welding arc. Wear oil free protective garments such as leather gloves, heavy shirt, cuffless trousers, high shoes and a cap over your hair. Wear ear plugs when welding out of position or in confined places. Always wear safety glasses with side shields when in a welding area.

4.g. Connect the work cable to the work as close to the welding area as practical. Work cables connected to the building framework or other locations away from the welding area increase the possibility of the welding current passing through lifting chains, crane cables or other alternate circuits. This can create fire hazards or overheat lifting chains or cables until they fail.

4.h. Also see item 7c.

CYLINDER may explode if damaged.

5.a. Use only compressed gas cylinders containing the correct shielding gas for the process used and properly operating regulators designed for the gas and pressure used. All hoses, fittings, etc. should be suitable for the application and maintained in good condition.

5.b. Always keep cylinders in an upright position securely chained to an undercarriage or fixed support.

5.c. Cylinders should be located:
- Away from areas where they may be struck or subjected to physical damage.
- A safe distance from arc welding or cutting operations and any other source of heat, sparks, or flame.

5.d. Never allow the electrode, electrode holder or any other electrically "hot" parts to touch a cylinder.

5.e. Keep your head and face away from the cylinder valve outlet when opening the cylinder valve.

5.f. Valve protection caps should always be in place and hand tight except when the cylinder is in use or connected for use.

5.g. Read and follow the instructions on compressed gas cylinders, associated equipment, and CGA publication P-l, "Precautions for Safe Handling of Compressed Gases in Cylinders," available from the Compressed Gas Association 1235 Jefferson Davis Highway, Arlington, VA 22202.

FOR ELECTRICALLY powered equipment.

6.a. Turn off input power using the disconnect switch at the fuse box before working on the equipment.

6.b. Install equipment in accordance with the U.S. National Electrical Code, all local codes and the manufacturer's recommendations.

6.c. Ground the equipment in accordance with the U.S. National Electrical Code and the manufacturer's recommendations.

FOR ENGINE powered equipment.

7.a. Turn the engine off before troubleshooting and maintenance work unless the maintenance work requires it to be running.

7.b. Operate engines in open, well-ventilated areas or vent the engine exhaust fumes outdoors.

7.c. Do not add the fuel near an open flame welding arc or when the engine is running. Stop the engine and allow it to cool before refueling to prevent spilled fuel from vaporizing on contact with hot engine parts and igniting. Do not spill fuel when filling tank. If fuel is spilled, wipe it up and do not start engine until fumes have been eliminated.

7.d. Keep all equipment safety guards, covers and devices in position and in good repair. Keep hands, hair, clothing and tools away from V-belts, gears, fans and all other moving parts when starting, operating or repairing equipment.

7.e. In some cases it may be necessary to remove safety guards to perform required maintenance. Remove guards only when necessary and replace them when the maintenance requiring their removal is complete. Always use the greatest care when working near moving parts.

7.f. Do not put your hands near the engine fan. Do not attempt to override the governor or idler by pushing on the throttle control rods while the engine is running.

7.g. To prevent accidentally starting gasoline engines while turning the engine or welding generator during maintenance work, disconnect the spark plug wires, distributor cap or magneto wire as appropriate.

7.h. To avoid scalding, do not remove the radiator pressure cap when the engine is hot.

ELECTRIC AND MAGNETIC FIELDS may be dangerous

8.a. Electric current flowing through any conductor causes localized Electric and Magnetic Fields (EMF). Welding current creates EMF fields around welding cables and welding machines.

8.b. EMF fields may interfere with some pacemakers, and welders having a pacemaker should consult their physician before welding.

8.c. Exposure to EMF fields in welding may have other health effects which are now not known.

8d. All welders should use the following procedures in order to minimize exposure to EMF fields from the welding circuit:

8.d.1. Route the electrode and work cables together - Secure them with tape when possible.

8.d.2. Never coil the electrode lead around your body.

8.d.3. Do not place your body between the electrode and work cables. If the electrode cable is on your right side, the work cable should also be on your right side.

8.d.4. Connect the work cable to the workpiece as close as possible to the area being welded.

8.d.5. Do not work next to welding power source.

FIGURE 21.28 *(Continued).*

Hazard	Factors to Consider	Precaution Summary
Electric shock can kill 	• Wetness • Welder in or on workpiece • Confined space • Electrode holder and cable insulation	• Insulate welder from workpiece and ground using *dry* insulation. Rubber mat or dry wood. • Wear *dry, hole-free* gloves. (Change as necessary to keep dry.) • Do not touch electrically "hot" parts or electrode with bare skin or wet clothing. • If wet area and welder cannot be insulated from workpiece with dry insulation, use a semiautomatic, constant-voltage welder or stick welder with voltage reducing device. • Keep electrode holder and cable insulation in good condition. Do not use if insulation damaged or missing.
Fumes and gases can be dangerous 	• Confined area • Positioning of welder's head • Lack of general ventilation • Electrode types, i.e., manganese, chromium, etc. See MSDS • Base metal coatings, galvanize, paint	• Use ventilation or exhaust to keep air breathing zone clear, comfortable. • Use helmet and positioning of head to minimize fume in breathing zone. • Read warnings on electrode container and material safety data sheet (MSDS) for electrode. • Provide additional ventilation/exhaust where special ventilation requirements exist. • Use special care when welding in a confined area. • Do not weld unless ventilation is adequate.
Welding sparks can cause fire or explosion 	• Containers which have held combustibles • Flammable materials	• Do not weld on containers which have held combustible materials (unless strict AWS F4.1 procedures are followed). Check before welding. • Remove flammable materials from welding area or shield from sparks, heat. • Keep a fire watch in area during and after welding. • Keep a fire extinguisher in the welding area. • Wear fire retardant clothing and hat. Use earplugs when welding overhead.
Arc rays can burn eyes and skin 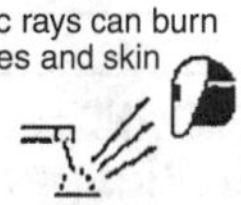	• Process: gas-shielded arc most severe	• Select a filter lens which is comfortable for you while welding. • Always use helmet when welding. • Provide non-flammable shielding to protect others. • Wear clothing which protects skin while welding.
Confined space 	• Metal enclosure • Wetness • Restricted entry • Heavier than air gas • Welder inside or on workpiece	• Carefully evaluate adequacy of ventilation especially where electrode requires special ventilation or where gas may displace breathing air. • If basic electric shock precautions cannot be followed to insulate welder from work and electrode, use semiautomatic, constant-voltage equipment with cold electrode or stick welder with voltage reducing device. • Provide welder helper and method of welder retrieval from outside enclosure.
General work area hazards 	• Cluttered area	• Keep cables, materials, tools neatly organized.
	• Indirect work (welding ground) connection	• Connect work cable as close as possible to area where welding is being performed. Do *not* allow alternate circuits through scaffold cables, hoist chains, ground leads.
	• Electrical equipment	• Use only double insulated or properly grounded equipment. • Always disconnect power to equipment before servicing.
	• Engine-driven equipment	• Use in only open, well ventilated areas. • Keep enclosure complete and guards in place. • See Lincoin service shop if guards are missing. • Refuel with engine off. • If using auxiliary power, OSHA may require GFI protection or assured grounding program (or isolated windings if less than 5KW).
	• Gas cylinders	• Never touch cylinder with the electrode. • Never lift a machine with cylinder attached. • Keep cylinder upright and chained to support.

FIGURE 21.29 Welding safety checklist.

Comprehensive information about welding safety is available from the American Welding Society, 550 NW LeJeune Road, Miami, FL 33126, www.aws.org.

REFERENCE

1. *Structural Welding Code—Steel,* Annex XI, AWS D1.1-2000.

CHAPTER 22
ROLLING PROCESS

Howard Greis
CEO and President
Kinefac Corporation
Worcester, Massachusetts

22.1 ROLLING PROCESS BACKGROUND

The rolling process described in this article is one whereby a surface of revolution, such as a thread, knurl, spline, gear, flange, or groove is formed in a deformable material by a flat or circular die configuration which is forced into a cylindrical blank while rotating in substantially uniform surface velocity with respect to it. This process can be traced back to ancient times when a potter, moving his hands backward and forward while pressing them together, roll formed a cylindrical piece of clay.

This type of rolling was first used as a metalworking process in the early 1800s and the first U.S. patent in 1831 described what is now known as a flat die rolling machine for producing wood screws. In 1850 the first cylindrical die rolling machine was patented for rolling machine screws.

From its inception as a simple and fast process for rolling screw threads on fasteners, it continued to be used for almost a century, predominantly for the production of a wide range of small low precision screws, nuts, and bolts of all types.

In the early 1900s various cylindrical die rolling machines for gear rolling were patented, but none appear to have found their way into significant commercial use. In the early 1940s the demand for high precision bolts for engines combined with the existing limitations of the flat die process resulted in the development of precision cylindrical die rolling machines. The first modern two-cylindrical die machine was developed and patented in Germany, followed shortly thereafter by a three-die machine for precision aircraft bolts patented in the United States.

To meet the ever increasing demands for small fasteners in the automotive, appliance, and other high-volume production industries, planetary die rolling machines were developed which could operate at rates as much as five times that of a flat die machine.

Parallel to this, rolling attachments and heads designed to roll threads on lathes and screw machines made it possible to achieve rolled thread quality on turned parts without a secondary operation.

Finally, in the second half of the twentieth century the rolling process began to be used to produce a wide variety of other helical, annular, and axial forms for other applications. These rolling processes and their theory and practice, and the tools and equipment applied to them, are described in this chapter.

As rolling is still an underexploited process, it is hoped that this information will enable metalworking manufacturers to take further advantage of its unique characteristics and capabilities.

22.1.1 Rolled Product Characteristics

Rolled forms are generally designed into products for several purposes. The first and by far most extensive use is to create a screw thread or other helical form to fasten mechanical elements together.

Threads produced by rolling have excellent surface finish and therefore can be torqued more effectively. In addition, the excellent root surface finish makes rolled threads more fatigue resistant. Finally, the cold working of the surface increases the surface hardness, especially in work-hardenable materials, and therefore improves the fastener contact strength and shear strength.

Rolled helical forms developed as a byproduct of the thread rolling process are now widely used to produce axial motion. Their consistent smooth surface finish and profile make rolled actuator and lead screws provide lower friction characteristics than those made by cutting. The work-hardened surface also improves wear life when such screws are used with a sliding nut.

The third major use is creation of forms for the transmission of torque. For the direct transmission of torque between a shaft and an element which is pressed onto it, rolled diamond knurls are commonly used. Axial knurls are also used and are modified in some cases to produce penetration into the mounted element. Such penetrating knurl joints generally provide superior torque slip resistance to machined or smooth interference fits.

Rolled splines are rapidly becoming the primary means for sliding or fixed axial transmission of torque in automotive, appliance, and other power transmission and linkage applications. In these applications, the work-hardened surface, the smooth radius roots, and surface following grain flow provide superior torsional and surface fatigue resistance. Parallel shaft torque transmission, using rolled gears which have similar characteristics, is a developing area of rolling application but process design requirements and die life consideration have thus far limited its use to the forming of relatively shallow helical gears.

The rolling of fins on tubing has been the primary means of producing heat exchanger tubes for many years, and a variety of other textures and convolutions have been rolled on tubes to enhance their heat transfer characteristics.

Rolling is now used extensively for the improvement of surface finish on shafts, ball joints, and other rotating or sliding joints. In such applications, the very smooth surface with rounded asperity shapes provides for low surface friction, minimal stick-slip friction variation, and low mating part wear.

Thrufeed rolling of annular forms on screw machine bar stock is sometimes used to produce significant material savings. The thrufeed rolling of annular ribs on concrete reinforcing bar is another useful application. That capability led to development of the through feed roll forming and cutoff of various pins, bearing and valve ball blanks, projectiles, and seal rings. In all these applications one of the most valuable process characteristics is the very repeatable shape and volume of the cutoff elements.

22.1.2 Process Benefits

In addition to the improved product characteristics described above, the use of the rolling process provides the metalworking manufacturer with a number of other general process benefits.

The first is the high productivity of the rolling process. The speed of the rolling operation is generally limited only by the ability to apply forming energy to the part being rolled and the ability to control the parts position within the dies during the rolling process. In most thread or form rolling applications where the rolling is being performed on a cold formed or forged blank there is significant material savings.

Since there is a relatively limited relative motion between the rolling die and the workpiece, the dies do not fail due to abrasive wear, but mostly due to long cycle surface fatigue. Therefore, generally the tool cost per piece is significantly lower than the cost per piece of cutting tools for creating the same form. The long tool life also minimizes lost time due to tool change and the tool change labor costs.

The absence of gradual abrasive wear means that it is not necessary to inspect a rolled form as frequently as would be necessary to inspect the same machined form. This reduces lost machine time due to form inspection and reduces inspection labor cost.

Although in some cases the capital equipment cost of a rolling machine is higher than an equivalent capability metal cutting machine, because of the much higher productivity of the rolling machine, the capital cost based on the output of parts per hour is generally much lower. Since rolling is a chipless process, provisions for clearing the chips from the work area and transporting and disposing of them is eliminated. In addition to the above benefits, groove rolling, flange rolling, fillet rolling, roll straightening, and roll deknicking provide significant benefits that warrant further application.

In view of the many general benefits the rolling process offers, increased awareness of its specific capabilities and characteristics should provide the basis for its increasing use throughout the metalworking industries.

22.1.3 Data Origins

The origins of the data in this article fall into four categories. The first is a body of empirical information concerning the theory and application of the rolling process and its benefits derived from extensive study testing and analysis of the function and use of the rolling process by the author's company and its personnel. The majority of the data comes from thread and other helical form rolling experience.

The second category is a group of diagrams and their explanations developed by me and my associates for corporate or customer rolling courses and for contribution to various industry magazines and handbooks.

The third is a group of charts of blank diameter, rolling load, rolling power, material characteristics, die life, and other rolling attributes. These represent aggregations of data put into the public domain by various manufacturers of machinery, tools, rolling material, and die material for the purpose of selling or helping customers apply their products.

The final category is a group of observations, experiences, and opinions of the author developed during forty-two years of direct involvement in the operation of rolling machinery, the design and manufacturer of rolling machinery, tooling, and systems, and the application of the rolling process to the design of customers' products.

22.2 GENERAL CHARACTERISTICS OF THE ROLLING PROCESS

22.2.1 Metal Flow in Rolling

Since rolling is a material forming operation rather than a material removal operation, it can only be used to form materials which at room temperature or some elevated operating temperature have the ability to be permanently deformed by the application of an external force. Therefore, the rolling process is generally applied to metals in a ductile state and to a small group of deformable plastics. Since metals are by far the most common materials rolled and the only ones where there is a useful body of data, we will use them as the basis for all of our theoretical analysis and discussions.

When a load is applied to a metal blank, it can react in either of three ways, depending on the size and direction of the applied load. It can elastically deform, it can plasticly deform, or it can fail. In all forming processes the key to success is to apply the forming load to the blank at the correct location, at the correct rate, and in the correct amount and direction so that the blank is deformed to the desired shape without failure and with proportionately small elastic spring back. Since most metal flow occurs in shear, and most failure occurs in tension, it follows that the way to produce optimum metal forming conditions is to maximize the shear stress while minimizing tension forces which may cause surface or core failure.

To understand how this is achieved in a typical thread rolling situation it is first necessary to examine how the blank material behaves under a typical three dimensional load condition. Figure 22.1 illustrates a small element of the blank material being loaded by the rolling die rib. All of the forces can be resolved into three principal compression stresses. The planes of maximum shear stress in any

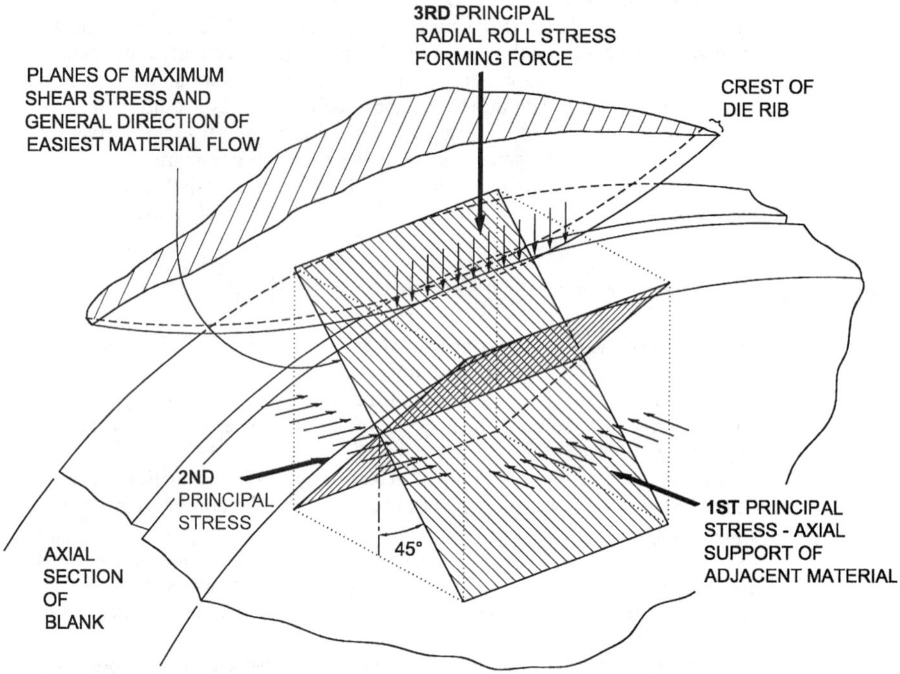

FIGURE 22.1 Flow stress-simplified triaxial stress model showing maximum shear stress under planes of penetrating die rib.

element of material will occur at 45° to the direction of application of the largest compression stress and the smallest compressive stress. The two planes of maximum shear stress intersect at a line in the direction of the second principle stress which is the middle stress in compression.

In the rolling process metal flow is accomplished by forcing the die rib during no slip rolling contact into the surface of the blank. This can be achieved with either a cylindrical or a flat die with the thread form along its surface. In a typical thread rolling operation as is shown in Fig. 22.1, the metal flow at and around the 30° flank angle of the forming rib approximates the 45° angle of the plane of easiest flow, the plane of maximum shear stress.

However, as a rolling die rib penetrates the blank, flow in other directions can occur as well. When this occurs, the material under the rib is loaded beyond its shear yield strength, and as a result there are three possible directions in which metal flow can occur. Figure 22.2 illustrated each of them in three section views of a forming rib of a cylindrical die penetrating into a blank.

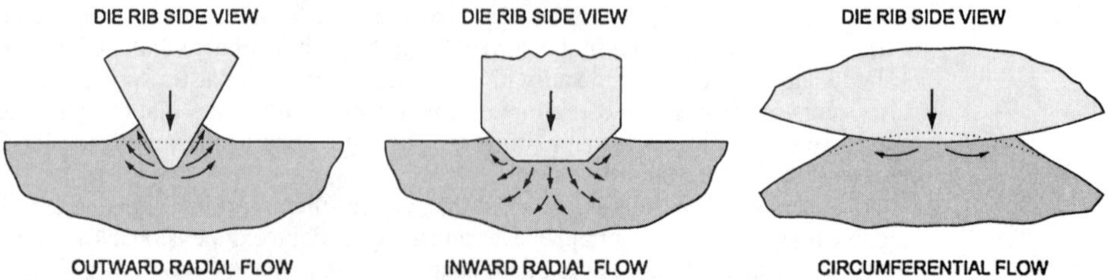

FIGURE 22.2 The three possible directions of metal flow in a rolling situation.

The first view is a cross section parallel to the axis of rotation of the blank showing the thread rolling die rib with a 30° flank angle and a rounded crest. This type of form tends to produce mostly outward radial metal flow which is used to create thread forms, involute teeth, flanges and similar shapes that protrude beyond the original surface of the blank.

The second section view, also taken parallel to the axis of rotation shows a die rib with a 45° flank angle, a broad flat crest, and smaller corner radii. This type of die rib produces mostly inward axial flow which can be used to create necked down local areas on a blank, or as a means for progressively rolling down a long section of blank. In the former, the dies use successively broadened die ribs to induce tensile stress which supplements the inward radial forming force produced by the penetrating die crest.

The third section view taken perpendicular to the axis of rotation shows possible circumferential flow. Some of this tends to occur in all die penetration situations. It increases when the die design is creating inward axial flow and axial elongation of the blank is constrained. Circumferential flow is to be avoided since it creates a wave of blank surface material in front of the die. When such a material wave circulates around the blank during die penetration, it can produce subsurface metal failure and material flaking under the die.

To achieve either of the first two useful types of flow, the rolling situation must be designed to apply sufficient compressive load radially to initiate plastic deformation. Depending on whether or not one wants to achieve inward axial or outward radial flow, the dies and the rolling system must develop the correct amounts of compressive and tensile stresses. In either event, the best rolling situation results when the axial stress is in compression at a level less than the circumferential compressive stress or is in tension. When that is achieved, the easiest directions of flow are more closely in line with the planes of maximum shear and approximately 45° to the axis of rotation and optimum metal flow results.

Because the mathematics of three-dimensional stress analysis in rolling helical and axial forms is exceedingly complex, it is difficult to apply the foregoing information quantitatively. However, a generalized understanding of the rolling metal flow phenomena may aid in successfully understanding and applying the rolling process.

22.2.2 Trochoidal Action

In addition to the effect of rib or tooth shape, the action on a rolling die as it penetrates into the workpiece to create a form has a number of other attributes which determine the outcome of the rolling process. The next most important of these is its trochoidal action. Figure 22.3 shows the shape of the path of each element of the forming die as it approaches the blank, performs the forming operation, and departs from it. Although it shows this action as it occurs with a flat die, it is substantially the same with a cylindrical die since a flat die is a cylindrical die with an infinite radius.

The die and the blank are mutually tangent and the tangency point represents the *no slip* contact point between them. The no slip point may vary slightly during the rolling process as the formed surface on the blank moves inwardly or outwardly with respect to the penetrating die. However, this does not significantly change the key fact. At the final point of penetration the penetrating point of the die is moving substantially radially with respect to the blank and at virtually zero radial velocity.

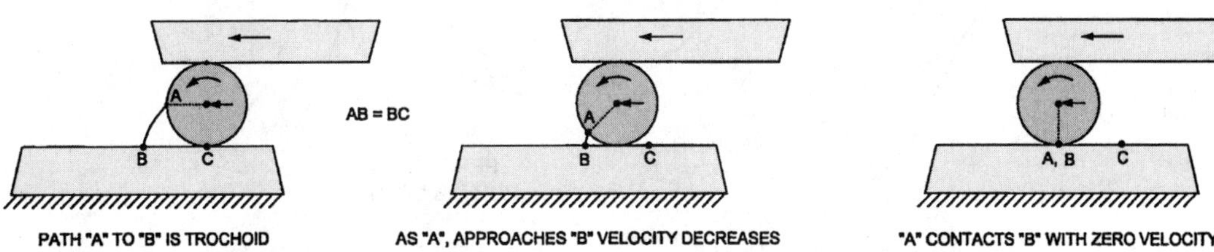

FIGURE 22.3 Trochoidal rolling action with no slip.

If the form being rolled is shallow relative to the part diameter there is very little sliding motion between the die and the blank. When the form is deep or has a high lead angle, the amount of sliding which occurs above and below the contact line increases. However, it is rarely significant enough to effect die and process design.

Because the penetrating action of the die rib is essentially radial, when the rolled form is annular or has a low lead angle, the flow patterns and displacement volumes can be effectively analyzed by use of two dimensional evaluation in a plane through the center of rotation of the blank.

For axial and high lead helical forms, the penetrating action of the die rib is still radial relative to the blank. However, it is accompanied by a sweeping action in the plane perpendicular to the axis of rotation of the blank similar to a progressively deepening gear mesh. Therefore, in this case the displacement and flow are best analyzed in the plane perpendicular to the axis of rotation of the blank.

Finally, since the relative radial velocity of the die rib approaches zero as it begins its penetration into the blank and there is very low sliding velocity of the die rib relative to the surface being formed, the rolling process variables and outcome are generally independent of blank rotational speed. Therefore, as will be discussed later in more detail, the most critical factor in the rolling process is the radial penetration per die contact and its relationship to blank and die diameter.

22.2.3 Intermittent Forcing Action

When viewed in operation, rolling appears to be a continuous process. However, examining the actual metal forming on any single area of the blank surface shows that it occurs intermittently but at very high frequency. Figure 22.4 shows an end view of a part being rolled between two cylindrical dies. The area of the part which is in the plastic deformation zone being worked by the right hand die would, after being deformed, move out from under the right die in a counter clockwise direction and be free from any radial deformation forces for almost 180° of its rotational travel. It would then move into the plastic deformation zone under the left hand die and be deformed further. Thereafter it would move out from the left die deformation zone and continue clockwise until it again contacted the right hand die. With a high die penetration per die contact of 0.001 in when rolling a 1/4"–20 TPI thread,

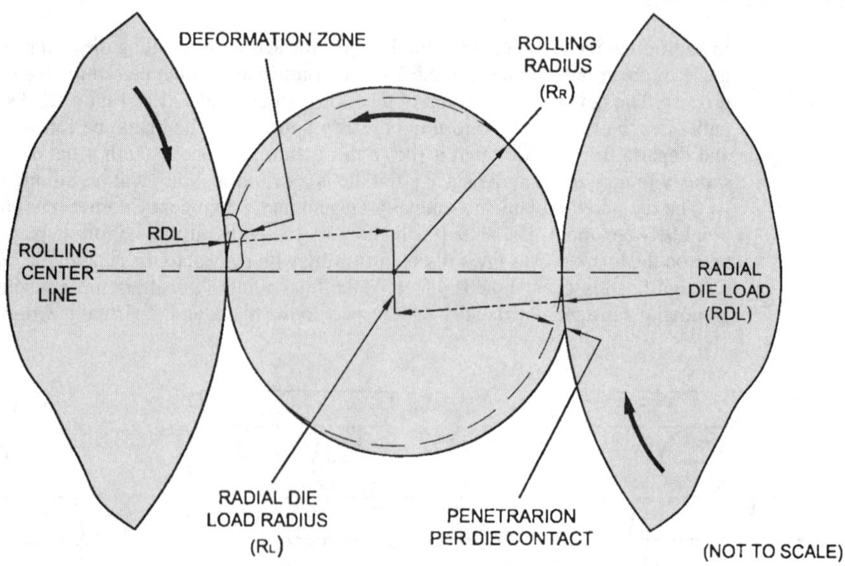

FIGURE 22.4 Cross section showing two cylindrical die system penetrating blank during annular form rolling.

any circumferential section of the part surface is undergoing deformation for 3 percent of each 360° of blank rotation.

Because the deforming force is essentially radial, only a small area of the circumference is worked at any one time, and the actual velocity of penetration approaches zero but the number of forming contacts is high. The closest traditional metal working analogy to rolling is crank press open die forging carried out at an ultrahigh cyclical rate. Using the forging cycle analogy, we can examine the throughfeed rolling of continuous 3/8–16 threaded bar in a two-cylindrical die machine. This can easily be done at 30 ft/min with 6-in diameter annular dies rotating at 360 rpm. Since each die has 10 progressively deeper, penetrating ribs, each point on the thread being formed is subjected to 20 separate forming contacts per blank revolution. With a blank rotating at 6300 rpm, the forging rate is about 126,000 contacts per minute, each of which is 0.001 in deep. Therefore, the equivalent forging rate is about 7200 contacts per minute.

22.2.4 Penetration Nonround

During the penetration phase of the rolling process, at each contact the die must penetrate further into the blank being formed. In order to provide the radial die force necessary to produce this penetration, the blank being formed is generally supported or held in equilibrium about its centerline between opposing dies. Such a configuration is shown in Fig. 22.4 for a two-cylindrical die rolling machine. To simplify the example, an annular form is being rolled and the penetration note is exaggerated.

As the blank passes under the ribs of the die, the die penetration reduces the diameter (and radius) under the forming rib. The amount of the reduction of the radius is the penetration per die contact.

The section of the blank between the two deformation zones has been progressively reduced by the contact with the previous die. As a result, the effective cross section of the part will remain in a nonround condition until the die penetration ends and a dwell section of the die or the infeed rolling cycle holds the die ribs in a fixed radial position and corrects the out of roundness of the rolled part. If the penetration stops abruptly without any dwell, the part will retain most of this out of round condition.

The annular analogy is complicated, as lead is added to the rolled format, but it holds true from a rotating torque versus forming resistance torque viewpoint for low lead angle forms. As the lead angle increases above about 20°, other blank-to-die contact conditions greatly complicate the ability to directly apply this analogy.

22.2.5 Friction Effect

For the rolling process to proceed in a rolling machine, the dies must drive the blank while at the same time penetrating into it to create the rolled form. The torque diagram for a simplified rolling situation is shown in Fig. 22.4. The couple of circumferential friction forces between the penetrating dies and the blank must be sufficient to overcome the resisting couple of the opposing radial die loads required to achieve the die penetration of that particular die contact. That is

$$2 \cdot \mathrm{RDL} \cdot \mathrm{M} \cdot \mathrm{R_r} > 2 \cdot \mathrm{RDL} \cdot \mathrm{R_2}$$

Simplifying the equation, it becomes clear that for successful rolling, if there is only friction driving the rolling, the product of the rolling coefficient of friction and the *rolling radius* (R_R) on the blank must exceed the effective radial die *load radius* (R_L):

$$MR_R > R_L$$

If the penetration rate per contact is increased, effective die R_L increases. If the increased R_L exceeds the product of the *circumferential friction* (M) and the R_R, then the rolling action will stall. If the rolling process does not stall and the dies will be able to penetrate to full depth, then the typical rolling system will maintain the die in the correct final position for several revolutions so as to round up the rolled part. Thereafter, the forming surface on the die will move away from the formed part gradually until all of the part and rolling system spring is relieved.

As the lead on the rolled part increases, the friction driving action is supplemental by the axial component of penetrating tooth form. As the lead angle grows beyond 30°, it becomes a major source of blank rotation torque. As the lead angle exceeds 60°, it becomes the primary source. Obviously, for an axial form surface contact friction is not necessary for rolling to occur.

Unfortunately, reliable values for rolling coefficients of friction are difficult to determine and experience indicates that it varies widely with die and blank materials, die surface condition, and the rolling lubricant used.

The foregoing holds true in virtually all rolling machines where the blank is driven by contact with the dies. However, in rolling attachments used to perform rolling operations on turning machines and on some fixed blank rolling machines, the blank is driven and the die rotation results from the circumferential frictional and axial interference forces transmitted to the dies from the driven blank. In those cases, the inverse holds true and the circumferential die rotating force must be greater than the rotational friction torque created by the product of the radial die load, the die to shaft coefficient of friction, and the shaft radius, all divided by the radius of the die.

The depth of penetration that can be achieved during each die contact to blank is generally limited by the effective coefficient of friction between the die and the blank. Therefore, the number of work revolutions required and thus the speed of the rolling cycle is very sensitive to the conditions which improve or limit coefficient of friction between the blank and the dies and the dies and their supporting shafts.

In special situations where it is not practical to achieve the desired penetration rate per contact by using the circumferential friction force between the die and the blank or vice versa, special rolling systems are built in, in which it is possible to drive the blank and the die in phased rotation with respect to one another.

In addition to limiting the penetration per die contact, during die penetration the effective coefficient of friction between the blank and the die is particularly critical during the start of the rolling process. In virtually all rolling machines, the initial contact between the dies and the blank must immediately accelerate the whole mass of the part around its axis of rotation. This can require a very high angular acceleration rate. For instance, if a 5-in diameter die rotating at 200 rpm is rolling a 1/4-in blank, it must accelerate the blank almost instantly to 4000 rpm for there to be no slip between the die and the blank. The initial circumferential friction force must also overcome any resisting frictional torque which is produced by the tooling which introduces the part into the dies and holds it in position for the rolling to begin.

To create the necessary frictional torque to produce such high rotational acceleration upon initial contact between the dies and blank with flat dies, single revolution cylindrical dies, and thrufeed cylindrical dies, the starting areas are frequently made rough by various mechanical means such as sand blasting, cross nicking or carbide particle deposition. However, no such friction supplementation is possible for infeed cylindrical die systems rolling annular and low lead helical forms.

If there is insufficient frictional torque to rotate the blank as the dies move relative to it, then the blank will not begin to rotate and the moving dies will destroy the surface of the blank and may cause welding between the blank and the dies.

Once the blank begins to rotate in a no slip relationship with the dies then the effect of friction between the dies and the part creates a force which restricts the radial flow of the material along the die surfaces. The material has no further space available for outward radial flow. If the rolling system continues to produce die penetration, then it produces circumferential flow. This results in a rapid increase in the rotation resisting torque which, if maintained, can possibly cause a stalling of the blank and adds to the rotational resistance torque as the rolled form fills the dies.

22.2.6 Constant Volume Process

During the rolling process the material in the blank is relocated by successive contacts between the blank and the dies. Since inward axial metal flow is not frequently used, and since no metal is cut

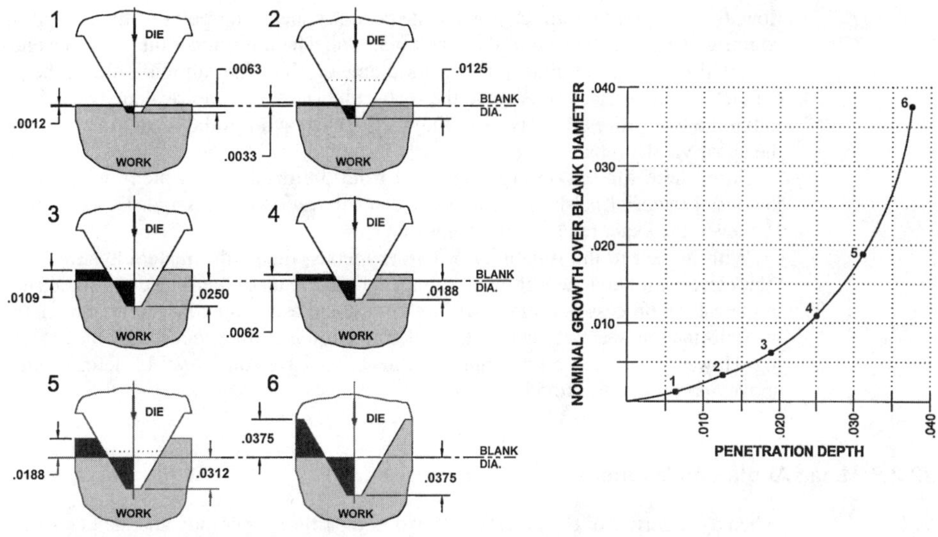

FIGURE 22.5 Thread rolling radial growth versus penetration in constant volume model.

away, for most rolling applications the volume of the blank has a major effect on the final size and form of the part being produced. This is illustrated by Fig. 22.5 showing an axial section view of the progressive formation of a screw thread on a blank which is not allowed to stretch during rolling.

As the thread form on the die penetrates into the blank, inward axial flow is constrained by the series of adjacent thread crests simultaneously penetrating into the O.D. of the blank. The rolling system controls the penetration rate of the die so that no significant circumferential flow occurs. The metal flow is outward radially, and as the die ribs begin to penetrate the blank, the radially growth is significantly smaller than the penetration depth. As the penetration reaches its midpoint, the radial growth rate and penetration rate equalizes. Then as the die continues to penetrate toward the center of the blank and the root approaches its final diameter, the free flowing outside diameter of the part grows much more rapidly than the root declines. This continues until the die becomes full. Then, if the rolling system attempts to cause the die to penetrate further, there is no available escape path either inward axially or outwardly radially.

Since it is not possible to further reduce the diameter of the part, if the die continues to be forced into the blank, it will create a circumferential wave of material which will thereafter cause surface failure and flakes. The diagram shown above illustrates the situation if surface flow is minimized so all of the displaced material flow is outward radially at the same rate, and its volume is not compensated for its change in radial location. Nevertheless, the chart below provides a good indication of the relationship between penetration and radial growth.

22.2.7 Work in Plastic Zone

As noted earlier, a rolling process creates the part form by the progressives deformation of the surface of a cylindrical blank. Each time as the forming surface of the die contacts an area of the blank it must first bring that area of contact to the elastic limit of the blank material. Then as a penetration continues, it moves that area and the adjacent area around it beyond the elastic limit into plastic zone. Most of this flow results from movement of the material in shear. However, due to the contact friction between the die surface, the blank surface, and the support of the adjacent material resisting flow, the unit pressure directly under the central penetrating point of the die will greatly exceed the compressive yield strength of the blank material. The multiplication factor depends on the material, the relationship of die diameter to work diameter, the penetrating die form, the penetration per die contact, the die surface finish, and the surface lubrication. Empirical analysis of die contact area and radial die load from a variety of rolling tests has shown it to be from three to five times higher than the shear yield strength.

From these limited observations, and from the manner in which the radial flow changes with increased material hardness, it appears that this ratio is to a large degree a function of hardness. However, the exact relationship is unclear.

With respect to the rollability of harder metals, the harder materials have less distance between their elastic limit and their ultimate strength. The amount of material deformation that can be achieved in the rolling process is limited by the size of the plastic zone. When the material work hardens during deformation, the materials range of deformation is further reduced by the amount of work hardening. However, virtually all wrought steel materials have some level of deformability unless hardened above approximately Rc 55.

22.2.8 Lead Angle and Diameter Matching

When a cylindrical or flat die or dies perform the rolling operation and the blank makes more than one revolution in contact with the dies, then the track laid down by one die must be correctly engaged by the form on the next die that contacts it. This matching of the form on the dies in phased relationship to each other and to the blank is similar to the conjugate relationship which occurs in gear meshes. This can be seen in Fig. 22.6, which illustrates a radial feed thread rolling setup where, the die axes are parallel to the blank axis and there is only a single lead on the thread. For systems where the pitch diameter, which is the no slip rolling diameter for a standard thread, is 1 in and there are four leads on the die, then the pitch diameter of the die must be 4 in. When the rolling die axes are parallel to the axis of rotation of the blank and no axial feed is designed, then the following relationship must be maintained:

$$\frac{\text{Die rolling diameter}}{\text{Number of leads or teeth on die}} = \frac{\text{Part rolling diameter}}{\text{Number of leads or teeth on part}}$$

If there were four leads on the thread and it was 1 in. in diameter and the die had a pitch diameter of $4^1/_2$ in, then there would need to be 18 leads on the die. This same relationship holds true for splines or other axial forms.

It should be noted that for three die radial feed rolling the maximum die diameter to part diameter ratio is five. When the die diameters are more than that, they clash with one another before contacting the blank.

Looking at the same parallel axis radial feed rolling setup in the plane tangent to the dies and blank illustrated in Fig. 22.6, it can be seen that the lead angle of the thread form on the die must be equal to the lead angle of the thread form on the part, if they are to mesh correctly. Since the dies and the blank are rolling together, then the hand of the lead angle of the die must be the opposite to

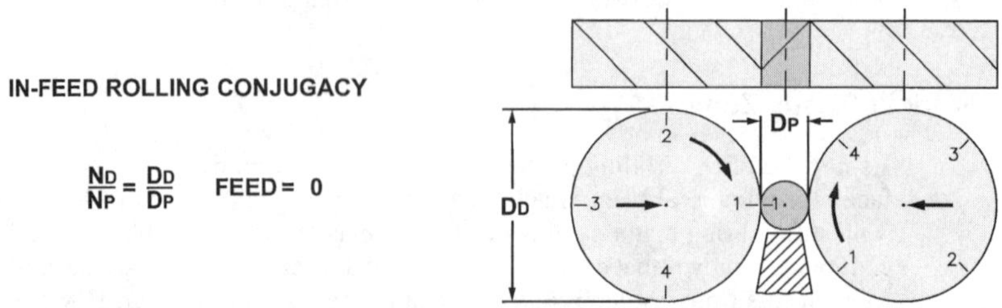

FIGURE 22.6 Parallel axis radial feed conjugacy.

the hand of the lead angle of the part. Therefore, in the radial feed mode of rolling, a left hand die must be used to form a right hand thread and vice versa.

22.2.9 Axial Feeding

When the die and blank rolling axes are parallel, there is a mismatch of the die to blank diameter ratio. There is also a mismatch of the lead angles. Therefore, either of two things may occur. If the resulting mismatch is small, up to about 10 percent of the pitch, then the blank will screw forward or backward with respect to the rotating dies, creating an axial feed. If the mismatch significantly exceeds that, the result, in addition to the axial feeding, would be distorted or damaged tooth forms.

Therefore, to use a parallel axis cylindrical die rolling system to produce axial feeding one can use controlled mismatch to create threads or other low lead helical forms that are somewhat longer than the die face width. However, where longer continuous thread or other low lead angle helical forms are needed, then it is necessary to produce axial feed by skewing the die with respect to the axis of rotation of the blank.

In this skewed axis, two-cylindrical die rolling system as shown in Fig. 22.7, it is also necessary to match the angle—the rib on the die makes—relative to the axis of rotation of the blank to the lead

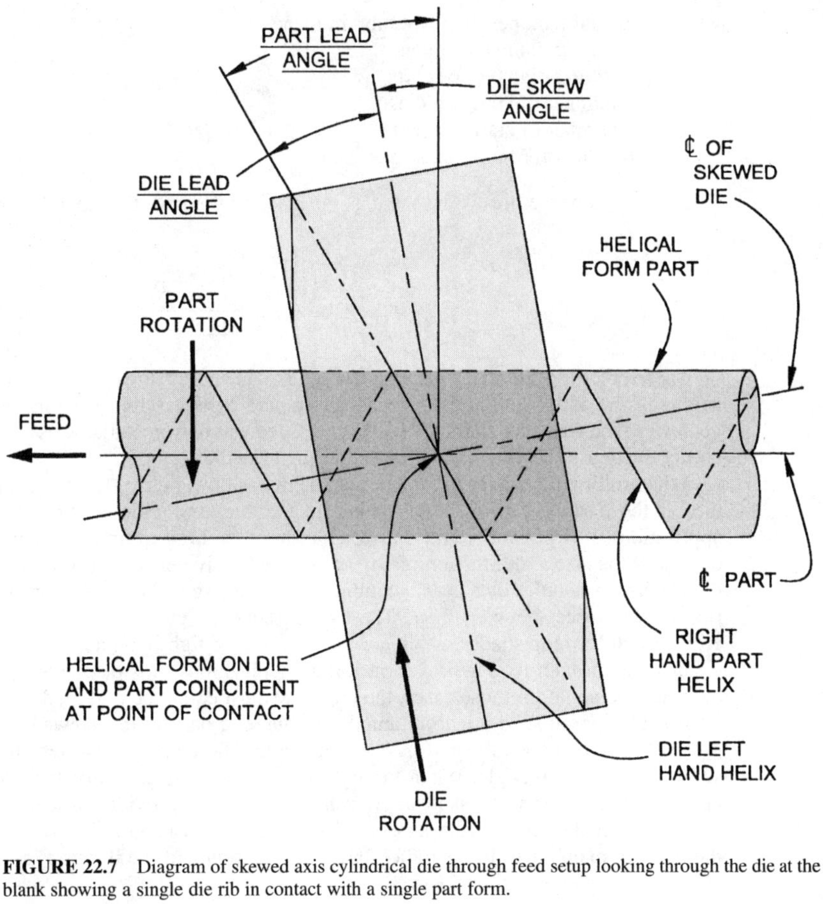

FIGURE 22.7 Diagram of skewed axis cylindrical die through feed setup looking through the die at the blank showing a single die rib in contact with a single part form.

TABLE 22.1 Hand Part and Hand Die Relationship

Hand part	Hand die	Relationship
RH	LH	Die Lead + Skew = Part Lead
RH	RR	Die Lead – Skew = Part Lead
LH	RH	Die Lead + Skew = Part Lead
LH	LH	Die Lead – Skew = Part Lead

angle of the form being rolled at their point of contact. Therefore, depending on the hand of the die and the hand of the rolled form, the following relationship holds for the lead angle of the die, the skew angle of the die and the lead angle of the part (see Table 22.1).

When the die hand is opposite the part hand, then the axial feed of the part is defined by the following relationship:

$$F = \frac{\left[\frac{D_D}{D_P}\right] N_P - N_D}{\text{TPI}}$$

where F = axial feed per die revolution in inches
D_D = rolling diameter of dies
D_p = rolling diameter of part
N_D = number of starts on dies
N_P = number of starts on part
TPI = helical forms (threads) per inch

If the hand of the dies is the same as the hand on the part, the equation changes to

$$F = \frac{\left[\frac{D_D}{DP}\right] N_P + N_D}{\text{TPI}}$$

Therefore, to speed up the through feed rate of an application the hand of the die is made the same as that of the part. The die diameter is determined by the rolling machine size and the skew by its geometry. The feed rate is limited by the die speed and power available as well as the rolling systems ability to control the part as it is passed through the dies.

When rolling threads or other low lead angle shallow, helical forms, it is possible to use annular dies. In those cases, the die lead angle is zero and the skew angle must be set at the part lead. In such applications, it is possible to have a variable shape forming rib in the starting relief area of the dies to control the size and direction of any seam which may develop during rolling.

The use of annular dies is also common in fin rolling where it is necessary to have very thin die ribs and the spaces between them. These dies frequently consist of a series of highly polished discs with special forming shapes which are locked together tightly on a hub.

In some through feed helical form rolling applications, it is desirable to use annular dies which have no lead angle. In those cases, the die skew must equal the part lead.

It is also possible to roll form annular forms on a part by the skewed axis through feed rolling process. In those cases, the part has no lead angle. Therefore, the die lead must be equal and opposite to the die skew angle. When through feed rolling splines, all of the above relationships hold. However, the use of a low skew angle generally does not provide sufficient axial feed force to create typical involute forms without excessive starting relief lengths. To roll such splines or serrations the dies are maintained in a parallel axis configuration and the through feeding action is produced by applying an external axial force.

When through feed roll finishing or straightening smooth bars, the dies are also smooth and the through feed action is solely related to the axial component of the surface of the die. Therefore, the feed rate per die revolution is equal to the die circumference times the sine of the skew angle.

22.2.10 Rolling Torque

Because the resistance to deformation of materials varies widely and the penetration area is defined by the shape of the rib, the penetration per die contact is extremely difficult to measure and since the effective resistance to flow of the adjacent material is not definable, there are no reliable means of predicting the rolling torque of a specific rolling application. There are, however, some general statements which are useful in understanding the rolling torque phenomena. *All other conditions being equal or constant:*

1. Torque is independent of blank rotation speed.
2. Torque increases generally proportional to initial hardness in low work hardening materials.
3. Torque increases as the rolled form depth increases.
4. Torque during penetration increases generally proportional to die penetration per contact.
5. Torque is relatively independent of rolling lubricant viscosity except in very deep forms.
6. Rolling machine input torque requirements vary widely depending on the power train and spindle bearing friction losses and work support and work handling friction losses.

22.2.11 Scalability

As can be seen in the foregoing geometric characteristics of the rolling process, most follow simple relationships and are therefore linearly scalable. If one observes a picture of a typical rolling machine and does not have any available size reference, it is not possible to know how large the machine is. Even the smallest machine of a given type, will have similar proportions to the largest.

The two main areas where nonlinearity occurs is in those relationships where friction and material formability are involved. Since there is no slip between the work piece and the blank at the rolling diameter, for the most part static friction and rolling friction are similar. However, as the forms being rolled get deeper and their flank angles lower, the radial sliding of the formed material outwardly along the flank of the die occurs at a different radius from the true rolling diameter. This produces a disproportionate increase in the effective torsional friction loss.

With respect to material hardness and formability, they are generally related but the work hardening effects and material characteristics prevent direct use of hardness as a formability indicator.

22.2.12 Operating Temperature

Most rolling is done without heating the blank prior to rolling it, and initially all of the data provided herein is based on tests made and experience gained with the blank starting at room temperature.

During the rolling operation, the energy applied to the blank to create the plastic deformation results in significant heating of the formed material. Where a coolant and lubricant fluid is used, much of the heat is carried away by it. Some of it is transferred to the dies through contact with the work. The balance of it is carried away in the rolled part.

When a large amount of rolling deformation is performed on a small diameter blank, the increase in blank temperature can create significant temporary elongation during the rolling operation. In those cases when the part cools down, the rolled form may be shorter than the die form.

This situation is more prevalent during through feed types of rolling, and in those cases the die pitch must be elongated to compensate for this.

In some rolling applications such as the production of aircraft fasteners from high hardness materials, blank heating to provide a larger plastic zone is used to reduce the required rolling force and to improve die life. In those cases, the heating is kept below the tempering temperature of the blank material to prevent loss of part strength.

Finally, in the single revolution roll forming of large automotive shaft blanks and the through feed annular roll forming and cutoff of grinding mill balls, the material is heated to some level of red heat prior to rolling. This creates the very plastic condition needed for major cross section area reduction on parts up to 3 in diameter.

22.3 ROLLING SYSTEM GEOMETRICS AND CHARACTERISTICS

To effectively apply the extensive application potential of the rolling process, the use of a variety of rolling systems is required. However, all of them must have three basic elements. The first is a means of creating the rolling contact between the die and the blank. The second is the means to create a controlled penetration of the die into the blank. And the third is a die support structure which places, adjusts, and then maintains the dies in the correct relative position of the dies to the blank and to one another.

22.3.1 Rolling Motion Sources

There are two common means of creating the rolling torque:

1. Drive the die which through friction drives the blank.
2. Drive the blank which through friction drives the die.

In some special situations if the friction between the dies and the blank are not sufficient to maintain no slip rolling contact between the die and the blank, both the die and the blank may be driven.

22.3.2 Die Penetration Sources

There are five commonly used means of creating rolling die penetration force:

1. *Radial feed.* By applying a radial force to driven cylindrical dies with parallel axes to move the dies into the blank.
2. *Parallel feed.* By using the driven parallel relative motion of two flat dies, by using the constant center distance rotation of a driven cylindrical die relative to a concave fixed die, by using the constant center distance rotation of two cylindrical dies on parallel axes. In each of these configurations, the dies have forming surfaces, teeth, threads, or ribs which progressively rise out of them.
3. *Through feed.* By using the axial, rotating, through feed movement of the blank along the centerline between two and three driven cylindrical dies, on skewed or parallel axis at fixed center distances which have forming surfaces, teeth, threads, or ribs which progressively rise out of them.
4. *Tangential feed.* By applying a radial force on two parallel axis, free wheeling cylindrical dies phased and on a fixed center distance to move them tangentially with respect to the rotating blank.

5. *Forced through feed.* By applying a force to the freewheeling blank axially along the centerline between two or three driven cylindrical dies on fixed parallel axes which have straight or low helix angle teeth which progressively rise out of the dies.

To achieve the desired rolled shape to the required tolerance, length, and location, it is frequently necessary to sequentially or simultaneously combine radial and through feed die penetration action in one rolling machine, and in one rolling cycle.

22.3.3 Die Support and Adjusting Structure

All rolling systems have substantially the same basic relative die and blank positions, motions, and adjustment. Figure 22.8 shows those involved in a generalized rolling system which uses two-cylindrical dies.

This diagram and the nomenclature can also apply to a flat die system. Flat dies are essentially cylindrical dies of infinite radius, and one die has the equivalent of a fixed centerline and the other has an equivalent centerline with its axis moving parallel to the face of the fixed die. During the rolling operation, the centerline of the blank will move parallel to the face of the fixed die. Similar analogies exist with all of the other rolling geometries.

22.3.4 Common Rolling Systems and Rolled Parts

Figure 22.9 shows in diagrammatic form the rolling systems in regular use along with a listing of the characteristics of each. Many cylindrical die machines provide combined system capabilities such as radial and through feed capability and can be adapted for other die or blank driving arrangements.

Figure 22.10 shows some of the capabilities of these rolling systems. Each part shown has one or more rolled forms on it which are integral to its ultimate function.

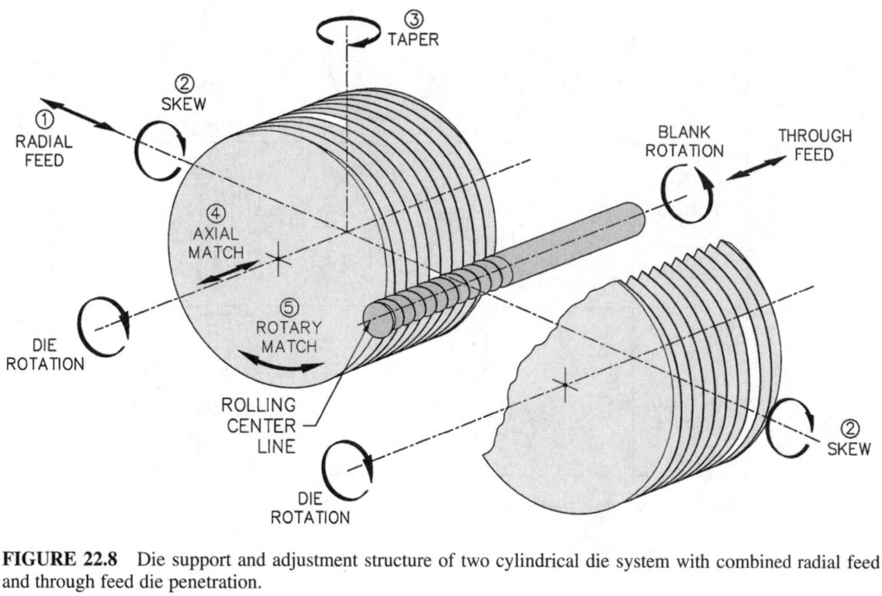

FIGURE 22.8 Die support and adjustment structure of two cylindrical die system with combined radial feed and through feed die penetration.

ROLLING SYSTEM GENERIC NAME	ROLLING SYSTEM DIAGRAM REAR END VIEW / LEFT SIDE VIEW	WORK DRIVEN	DIES DRIVEN	DIE AXIS FIXED	DIE PENETRATION RADIAL FEED	DIE PENETRATION PARALLEL FEED	DIE PENETRATION THROUGH FEED	DIE AXIS PARALLEL	DIE AXIS SKEWED
a FLAT DIE MACHINE	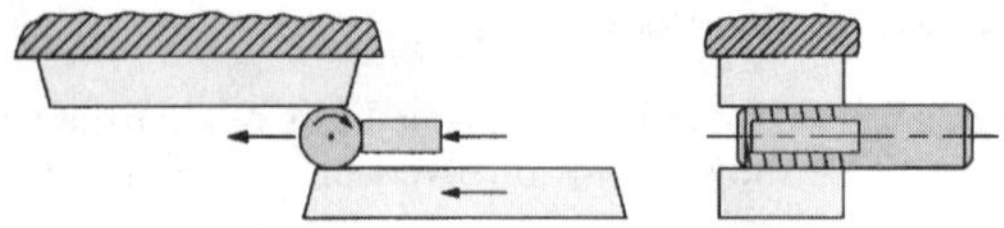		●			●		●	
b PLANETARY MACHINE	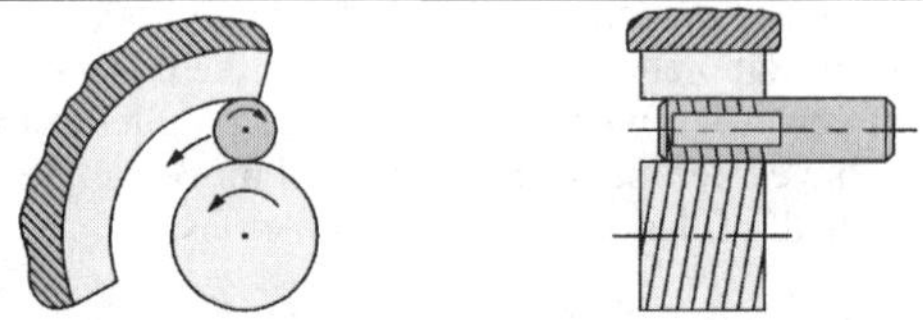		●	●		●		●	
c RACK MACHINE	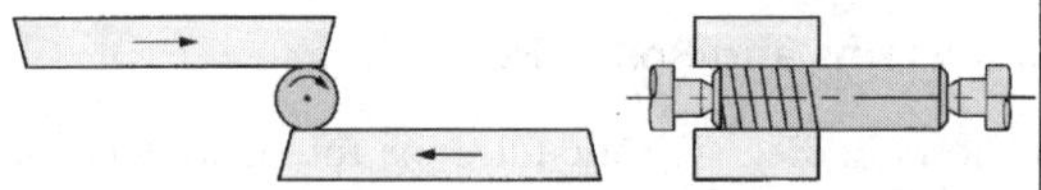		*4					●	
d TWO DIE IN-FEED MACHINE	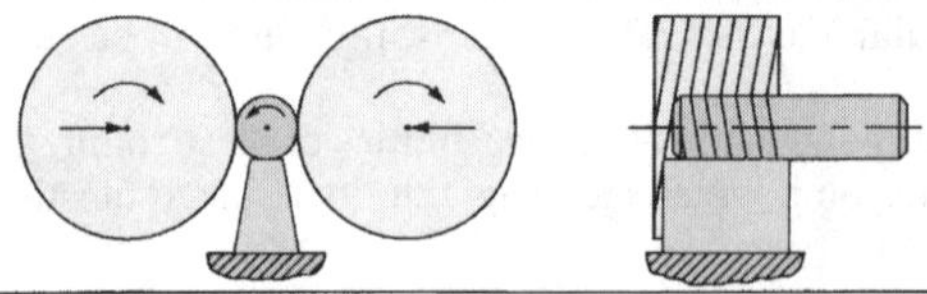		●		●			●	
e TWO DIE THROUGH FEED MACHINE	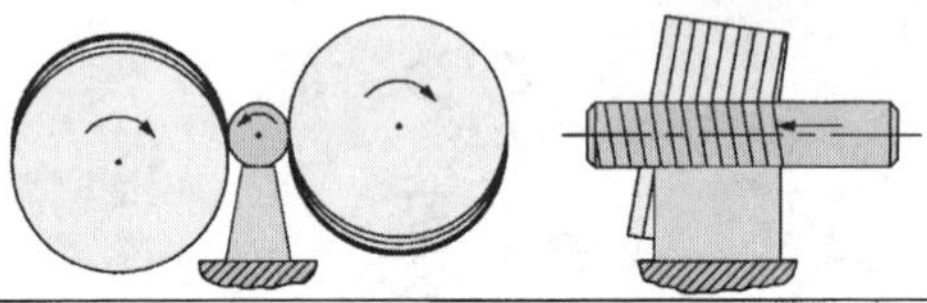		●	●			●		●
f TWO DIE SCROLL FEED MACHINE	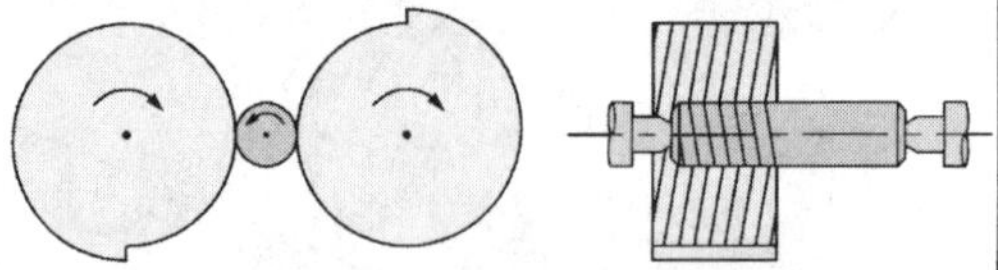		●	●	●			●	
g THREE DIE IN-FEED MACHINE	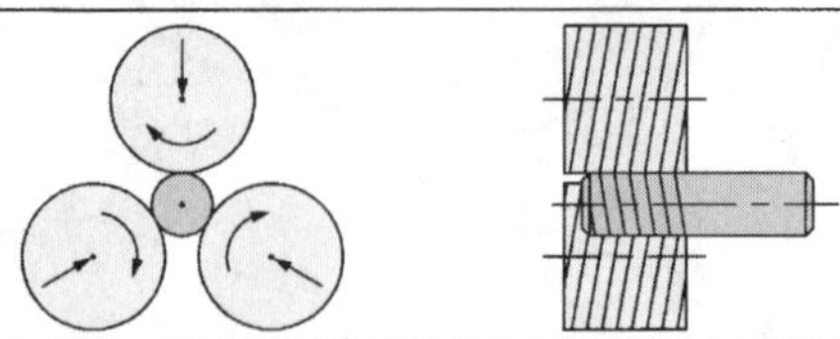		●		●			●	

FIGURE 22.9 Basic rolling system geometries

h THREE DIE TROUGH FEED MACHINE	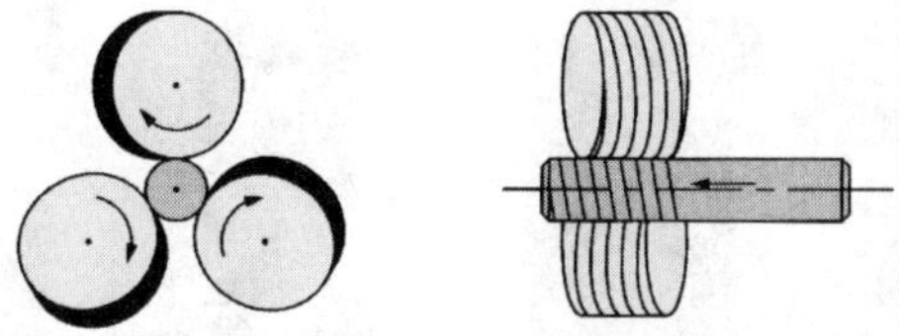		●				●		●
i THREE DIE FORCED TROUGH FEED MACHINE	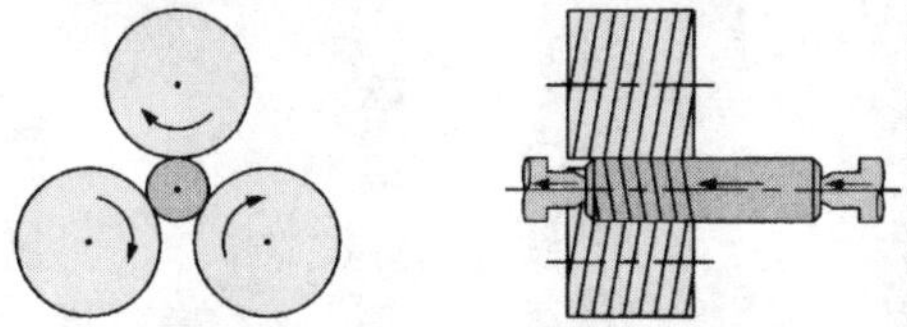		●				*3	●	●
j CONVERGENT AXIS MACHINE	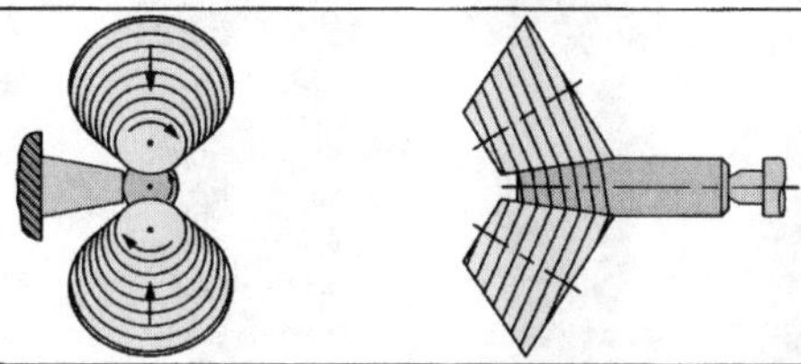		●		●			*1	
k BUMP ROLL ATTACHMENT		●			●			●	
l TANGENTIAL FEED ATTACHMENT	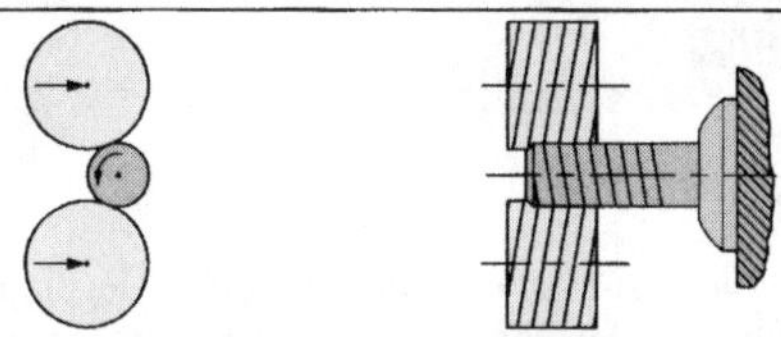	●			*2			●	
m RADIAL FEED ATTACHMENT	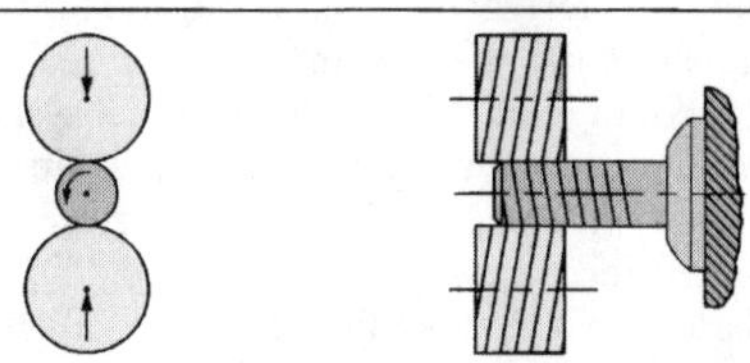	●	●						
n THREE DIE END-FEED HEAD	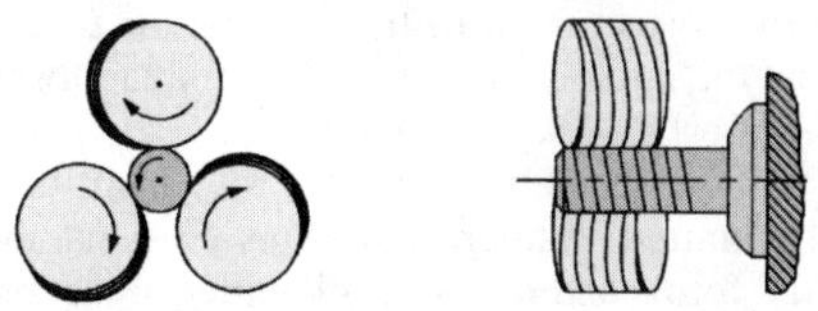	●							

*1 VERTICALLY PARALLEL AND HORIZONTALLY CONVERGENT
*2 RADIAL FEED BY TANGENTIAL MOVEMENT
*3 THROUGH FEED CREATED BY AXIAL FORCE
*4 OPTIONAL ONE DIE WITH CENTER LINE FIXED

FIGURE 22.9 (*Continued*).

FIGURE 22.10 Common rolled parts.

22.4 ROLLING EQUIPMENT

22.4.1 Flat Die Machines

As noted earlier, the first type of rolling machine developed was the *flat die* rolling machine. These machines were initially designed to roll threads on turned or headed blanks and thus to improve productivity by eliminating the need for much slower lead screw or die threading. That continues to be the primary use of rolling and the flat die machine system evolved for that purpose.

In the early machining era when the flat die machine was being developed, planers were the common method of removing material from a flat surface on a part. The planer used an axially moving work table which carried the work piece past a cutting tool which was indexed across the work. In applying a similar concept to threading, a horizontally oriented die moved axially in front of a horizontally fixed die each of which had a series of mating threads on its face. A blank was introduced between them at the correct point of time so that the threads formed by one die correctly engaged with the threads by the other die as the next die thread forms contacted the blank.

The flat dies were tilted with respect to one another to create the penetration, and after the moving die passed beyond the fixed die the rolled part fell out.

Because of the limited number of work revolutions, the absence of dwell in the dies, the low precision of the die forms and the poor rollability of the available material, the early rolled threads tended to be out of round and with rough finish. They had a reputation of low quality and continued to be considered inferior to cut threads until well into the twentieth century.

The basic machine design was simple and remains so. The reciprocating moving die is driven from a fly wheel by a crank and connecting arm system with the fly wheel providing sufficient angular momentum to start and sustain the rolling as the dies penetrate into the blank. A cam system directly connected to the crank shaft of the machine actuates a feeder blade which is timed so as to force the blank into the die gap at the correct point in the cycle to achieve match and to continue to push the blank briefly to insure the start of the friction driven rolling action.

This same basic concept shown in Fig. 22.9 continues to be used in virtually all modern flat die machines. The slide systems have been improved to allow high speed, high load reciprocation with good reliability and the drive systems have been upgraded to the point where it is not uncommon to operate small flat die machines at up to 600–800 parts per minute.

Initially, the dies were rough rectangular blocks of steel on which a planer had been used to cut the thread forms. They then were surface hardened sufficiently to penetrate the softer blank to be rolled without deformation.

The dies are now ground with precise forms having a controlled penetrating area, a dwell area, and a finish relief to provide precise control of the roll-forming action. The dies are designed with a correct lead and, therefore, in a precise well-maintained machine that can be mounted into the die pockets without taper or skew adjustment. If they are prematched, only a limited amount of match timing adjustment is required. Therefore, the primary process adjustment is for size. This is accomplished in a variety of ways by backup screws, draw bolts, or wedge systems.

The size capability of a flat die machine is limited predominantly by the length and width of the die that can be accommodated in the machine, its die drive capability, and the stiffness of its frame and slide system. Approximately seven work revolutions are generally necessary to roll a thread and more are desirable if the thread is hard or deep. The diameter that can normally be rolled on a flat die machine can be estimated by dividing the length of the moving die by seven times π. Therefore, if a flat die machine has an 8-in die, it can roll threads up to approximately $^3/_8$ in. in diameter. It should be noted that although there is no minimum diameter for a rolling machine, if a standard die is used in a machine, it is generally preferable not to roll a part below $^1/_3$ of the maximum capability since it will result in using excessive work revolutions to form the part.

Flat die machines are generally used on threads $^3/_4$ in. in diameter and below, but there are some very large machines capable of rolling threads up to $1^1/_2$ in. Flat die machines are also frequently used to roll knurls. As noted above, the cyclical speeds of up to 800 per minute are possible on the smaller machines with the larger machines running up to approximately 60 per minute.

Initially, when the flat die machines rolling screws and other fasteners were hand-fed, the rolling axis was vertical and the die travel horizontal which simplified hand-loading. The larger machines are still frequently hand loaded and have a vertical axis. For rolling the ends of long threads on the ends of long bars, machines with horizontal axis have been built, but are no longer common.

It should be noted that the ejection of the completed parts from flat die machines is automatic, as the moving die passes beyond the fixed die the part falls from the die gap. In cases where the parts are heavy and the material is soft it is sometimes necessary to provide means to prevent the threads from being knicked by the fall.

The use of flat die machines to produce wood screws and sheet metal screws from pointed blanks led to the development of rolling die designs which would permit the sharp pointing and cutoff of the blank ends. Recently, these progressive forming techniques have been applied to the production of small shafts and pins for automotive, connector, and other high volume uses.

Today most flat die machine are inclined to simplify automatic loading of the headed blanks into the feeding position from a gravity-driven input track which in turn is fed from a vibratory feed bowl or a mechanically operated blade type hopper.

22.4.2 Planetary Machines

Since the flat die rolling machines described above could roll only one part during each die cycle, and the time to return the moving die to the starting position was wasted, the only way to increase productivity was to increase the speed of the reciprocating die. This posed significant machine vibration problems. As a result, the planetary rolling machine was developed. As shown in Fig. 22.9 a centrally located round die rotates continuously. Opposing it is a concave die which is a segment of a circle. As the center die rotates and the threads on the round die come into match with those on the segment die, the blank is introduced into the die gap by a feed blade.

As it rolls through the die gap, the form on the segment die is located so it gradually comes closer to the form on the rotary die so that the die form progressively penetrates into the blank. By the time it reaches the end of the die gap, typically seven or more work revolutions, the thread or form is rolled full and the part is ejected from the dies.

As in the flat die machines, the timing of the introduction of the part into the dies and the dynamic match is accomplished by a cam which is mounted directly to the die shaft. The die is located in a supporting mechanism which allows for size and penetration rate adjustment. Both the round die and the concave segment die are of constant radius and, therefore, must be controlled precisely to provide for correct penetration and size with minimum size adjustment.

The longer the segment die, the more parts can be rolled at any one time. The maximum number is equal to the segment length divided by the circumference of the blank diameter. However, it is sometimes limited by the stiffness of the die support system, the amount of torque available at the die shaft and the ability to precisely feed blanks into the load position from the input track at very high speed. It is not uncommon to have four or five parts rolling simultaneously. Therefore, with a center die speed of 300 rpm, it is possible to produce 1500 parts per minute.

The majority of the planetary die applications are on machine screw type threads or nonpointed, self-tapping sheet metal screws up to $^3/_8$ in in diameter. They are also widely used for the rolling of shallow spiral grooves on grip nails. However, they are rarely used to produce more complex forms because of the difficulty of creating such die forms on the concave fixed die.

To simplify the loading and unloading of parts at these high production rates, the rolling axes of planetary die machines, like automatic feed flat die machines, are inclined and are equipped with similar bulk feeders. Some versions with horizontal axes are available to roll the ends of long bars. In addition, double-headed, horizontal-axis machines are built for the simultaneous rolling of threads on both ends of double-end studs. These horizontal machines are most commonly used for parts from $^1/_4$ to $^3/_4$ in. in diameter.

22.4.3 Rack Machines

The ability to roll straight knurls on flat die machines showed the capability of that general type of rolling geometry to produce small involute forms, such as 45° straight knurls very effectively on diameters up to $^3/_8$ in. As the use of serrations and splines, which are essentially coarser knurls, grew in the automotive industry, there was a need to reduce the cost of these forms which were then being produced predominantly by hobbing. Even the largest flat die machines with die lengths of up to 16 in did not provide adequate work revolutions to produce these deeper forms and it was difficult to accurately control the axis of rotation of the moving part perpendicular to the die travel which in turn resulted in lead error in the rolled form.

As a result, an American manufacturer produced the first rack rolling machines in the 1950s. To solve the lead control problem, the part being rolled was held on fixed centers and both of the dies were traversed perpendicular to the part axis in phase relationship with one another. Each of the dies was located on a stiff slide, the two slides being one above the other in a very large rigid frame. The slides were interconnected by a rack and pinion system with low back lash and each slide was actuated by a hydraulic cylinder connected to a common hydraulic power supply unit.

Diametrial size is adjusted by spacers or an adjustable wedge system to individually bring each of the dies to the correct radial position with respect to the center line of the part. The match of the dies is achieved by axially adjusting one of the dies with respect to the other.

These large rack machines for spline rolling have the dies traversing in planes parallel to the floor and typical die lengths for these machines were 24, 48, and 72 in. As a result, the basic machines combined with the separate hydraulic power unit occupied a large amount of floor space.

The dies are essentially gear racks in which the teeth form a conjugate mesh with respect to the part being rolled. The blank prior to rolling is about the same diameter as the pitch diameter of the final form. To achieve correct step off on the initial blank, the addendum of the die in the starting area is cut away almost down to the pitch diameter. Therefore, it has a tooth spacing equivalent to the correct circular tooth spacing on the blank at initial contact. As the die progresses, the die addendum gradually increases until the die reaches full depth. This die penetration area normally runs for five to eight work revolutions. After the die teeth have fully penetrated, a dwell area of from one to four work revolutions assures that the rolled form is round. After the dwell, the die teeth drop away slightly to relieve the spring in the machine.

After the rolling cycle is complete, the rolled part is withdrawn from the die gap area and the dies are returned to their starting position.

The rack machines currently perform the great majority of the spline rolling currently being done throughout the world. When correctly set up with good dies and rolling blanks held to 0.0005 in, "over wires" pitch diameter tolerances as low as 0.0015 in are being consistently achieved on 45° and 37 $^1/_2$° splines up to about 1 $^1/_4$ in diameter.

Because the dies are heavy and difficult to handle, mount, and adjust, setup is slow. In addition, handling of the parts in and out of the centers which support the part during rolling take additional time. Typical production rates are from four to seven parts per minute.

Since rack machines are mostly used for high-volume production of heavy parts, they are generally automated. To accomplish this, some types of conveyor or walking beam transfer is generally located parallel to the travel of the dies in front of the machine. A lifting mechanism places the part between the centers. One of the centers is actuated to grip the part and the combined center and part arrangement is axially transferred to the rolling position between the dies. The dies are actuated and at the completion of their stroke the center and part system is axially returned to the conveyor area where it is unloaded. During the unload, the dies return to their original position and the cycle begins again.

In some cases where two adjacent or closely positioned forms are required on a smaller diameter part, then two short dies are arranged so that they operate sequentially. Where there are two forms, one on each end of a part, two full length sets of dies are located parallel to each other. The form on one end is rolled during the forward stroke, the part is axially shifted and the form on the other end is rolled during the return stroke. This technique is very efficient from an operational point of view, but the interrelated matching of the two sets of dies makes setup slow and difficult.

In Japan, rack-type rolling machines are being used for rolling deep worms and other forms which require varying tooth forms during the penetration of the die into the part. Since most of these forms are smaller in diameter, to save floor space the dies have been oriented vertically rather than horizontally. In addition, in these machines and in some of the spline rollers, electromechanical servo drives have replaced the hydraulic cylinders to actuate the slides. By this means, the matching of the dies has been simplified.

Finally, in Europe versions of the rack-type machine with extremely long dies up to 72 in are being used to roll, reduce, and form large-diameter stepped shaft blanks up to about 4 in. in diameter with lengths up to 24 in. This *rotary forging operation* performed hot is attempting to compete with extrusion and swaging as a means of preforming shaft blanks to provide major material and machining savings. Its primary advantage over conventional forging and extrusion is its ability to produce almost square corners in the blanks.

22.4.4 Two-Die Infeed Machines

Although the concept of two-cylindrical die parallel axis rolling machines with radial feed goes back to the 1800s, they were not built as a standard machine tool until several European companies recognized their advantages in the early 1930s. Flat and Planetary die machines could provide only limited numbers of work revolutions, the blank centerline had to move a large distance during the rolling cycle, and the blank had to be introduced at exactly the right moment to provide proper match between the two dies. Therefore, they were not well adapted to producing large diameter threads or deep forms. So as rolled threads became more acceptable for structural parts and critical threads, the cylindrical die concept appeared to open up new rolling applications. A machine with two-cylindrical dies geared together in constant match with their axes parallel, one of which is forced radially toward the other while the work is held between them, made it possible to roll larger diameters of work or deeper forms and harder materials, all of which require more work revolutions than can be

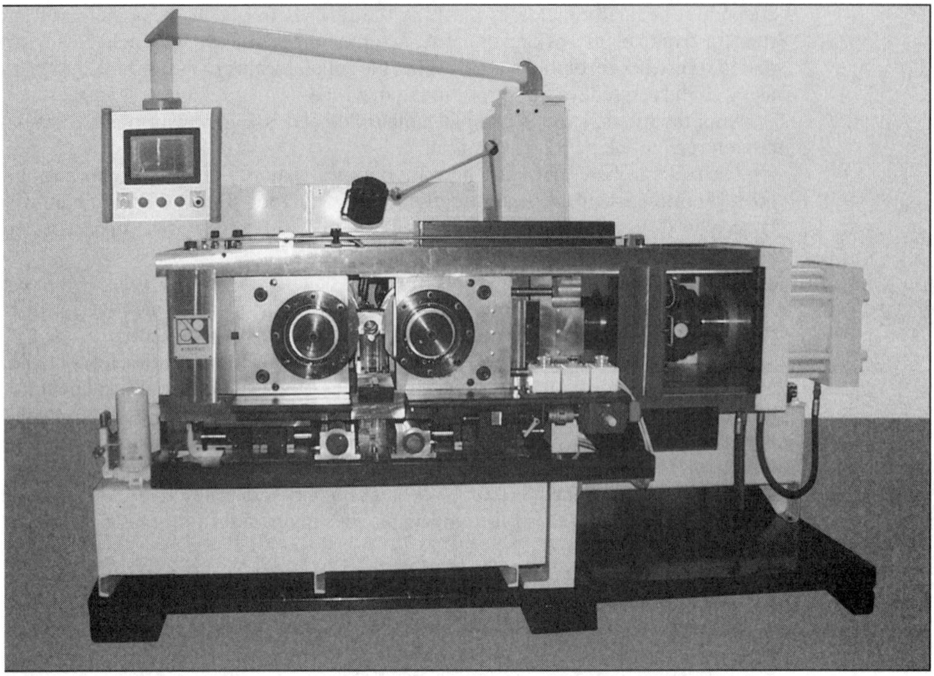

FIGURE 22.11 Two-die infeed rolling machine with 660,000 pound radial die load capability.

obtained on flat die or planetary rolling systems. This two-die configuration also lends itself to the rolling of threads on the ends of long shafts. With these advantages, it began to be widely used for infeed rolling of automotive and appliance parts and other similar high volume components. In almost all two-die machines the die axes are horizontal which simplifies the handling of shafts and similar long parts. The Infeed actuation of the moving die is most frequently done hydraulically. However, where a cyclical rate above 30–40 per minute or a precisely repeatable penetration pattern is required, a mechanically driven cam system can be used for die actuation.

To further simplify the handling many newer two-die rolling machines have both dies actuated toward a horizontally fixed work centerline. In most such double acting die machines, the work is supported between the dies, generally slightly below center on a blade or roller which is mounted between the dies. Where the part is stiff and the form deep then some type of external bushing or center system may be used to position the blank.

Most of the cylindrical two die systems now in use have die diameters between 3 and 8 in and with these machines it is normal to roll forms ranging from about $^1/_4$ to about 3 in. in diameter. There have been a few two-die machines built which use dies up to 12 in. in diameter that are capable of infeed rolling parts up to 6 in. in diameter. The largest of these machines which is shown in Fig. 22.11 is capable of applying a radial die load of 660,000 lb. It is used to roll threads on gas turbine rotor tie bolts which are approximately 3 in. in diameter up to 8 in long in a high strength alloy steel at Rc 47 hardness.

Two-cylindrical die machines with horizontal rolling axes are the most versatile. They are used extensively for secondary operations in the production of high-volume shaft-like parts.

A wide range of automatic feeding systems are available for this type of rolling machine. Where there are diameters larger than the rolled diameter, they generally move the part axially into the rolling position support it there and then return it to the front where it is unloaded. Where the body diameter of the part is about the same or smaller than the threaded area, the completed part can be ejected out the back end of the machine while the next part is being loaded into the front. Where these machines are integrated into systems, the incoming part is frequently brought to the preload point by a conveyor or walking beam structure. For separate second operation rolling machines, a bulk feeder or hopper is used with an escapement and feed track to bring the part to the preload position. For cell operation in place of a hopper there is generally a short input magazine and a simple unloading device.

Recently, to handle small headed parts with rolled areas from $^{1}/_{8}$ to $^{3}/_{8}$ in diameter, a vertical axis two-die rolling machine has been introduced.

22.4.5 Two-Die Through Feed Machines

To produce threads which were longer than the die length on two-cylindrical-die radial feed machines, it was common to build a small amount of lead mismatch into the dies so that the part would feed axially at a relatively slow rate. When the required rolled thread length was achieved, a switch or timer would open the dies. When this method was used to produce long lengths of threaded bar, the feed rate was too slow and the use of mismatched dies decreased the die life. As a result, the capability to skew the cylindrical dies was added to conventional two-die radial feed machines to increase the through feed rate and improve die performance. This capability was generally provided by the connection of each spindle to its respective drive gear in a gearbox at the rear of the machine through a double universal joint system which provided an ability to skew the dies upward or downward to angles of up to 10°.

The typical gearbox consisted of two independent worm gear systems connected to a common worm drive shaft. The worm gears provided both a speed reduction and a torque increase. The worm gears were separated sufficiently so that long continuous threaded bars could be fed from the front, threaded, and passed out the rear of the machine along the rolling centerline. Worm gear systems are limited in speed, relatively inefficient and, as a result, tend to run quite hot. To produce the significantly higher spindle speeds, in some cases up to 600 rpm, necessary for low cost production of continuous threaded rods, two die through feed machines with spur gearboxes and a hollow central intermediate gear were introduced. This type of gearbox provided high efficiency and long life with low backlash and is now commonly used on all types of two and three cylindrical die machines.

The two-die through feed rolling system is used predominately for production of continuous threaded studs, set screws, jack screws, and lead screws. The full threaded bar is generally fed into the rolling machine from bar feed units similar to those used on automatic screw machines. Long studs are frequently fed from magazine-type feeders, and set screws and short studs are most often fed from vibratory bowls. In the latter, case, the dies must be designed to pull the part into the penetration area. Otherwise, a separate feed wheel or belt system is required to supplement the feed force from the bowl. Recently, incline axis machines have been introduced to provide simple gravity feed into the dies.

Typical feed rates for $^{3}/_{8}$- to $^{1}/_{2}$-in threaded rod range from 10 ft per minute to 300 ft per minute depending on the machine power, spindle system, and die design. The higher through feed rates require good input blank control, close support of the finished bar so that completed threads are not damaged, and extensive rolling lubrication and cooling. It should be pointed out that to produce $^{3}/_{8}$-16 threaded bar at 3 ft/s in a two-die machine with 8 in dies skewed at 10°, it will require that the dies rotate at 500 rpm and, as a result, the threaded bar will spin at 11,600 rpm. Finally, it should be pointed out that there is virtually no rolling length or minimum diameter limitation for horizontal cylindrical die through feed rolling machines. Currently 100-ft-long continuous threaded rod and lead screws down to 0.008 in diameter have been rolled.

22.4.6 Scroll Feed Machines

In an attempt to increase automated cylindrical die production rates above 20–30 parts per minute, the scroll, cam, or incremental configuration shown in Fig. 22.9 evolved. By building a variable radius die, cylindrical dies with progressively increasing radii to provide the penetration action, a constant radius dwell, and a relief radius, the need for radial actuation of the spindles was eliminated. Therefore, the speed of operation could be greatly increased. The feeding of the blank was accomplished by a cage system in which a sleeve with a series of pockets surrounded one of the dies. As the cage is indexed the pockets are used to bring the blank into the rolling position, support it while it is being rolled, and then discharge the rolled part. The scroll became widely used in Europe and Japan to roll bolts from approximately $^1/_4$ in up to as large as 1 in at production rates of 50–60 per minute. However, the higher cost of the profiled cylindrical threading dies and the complexity of the cage feed and its loading have limited its use in the United States. Until recently, it has been used here primarily for double-end stud rolling, double-form rolling, and similar high-volume special applications where the job warrants the use of a more complex die and feed mechanism.

Recently, as the use of splines on shafts below $^1/_2$ in diameter has become more common, the two-die scroll feed system has provided a cost effective means of this production. In those single-die revolution applications, the shaft is supported on centers or in a bushing during rolling. For high volume applications, the shafts are automatically loaded into the bushing or center system. By this method, production rates of 15–20 parts per minute are possible.

To provide the variable radius dies, special hobbing or grinding machines are required and the rolling machines are of similar design to other two-die machines. However, the die diameters must be larger since the blank must be loaded, rolled, and unloaded in one-die revolution. Typically a minimum of eight work revolutions is desirable for rolling and 90° of the die surface is needed for loading and unloading. Therefore to roll a $^3/_4$ in diameter thread or spline, an 8 in diameter die is used.

In addition, the rolling machine must have a die drive which can start and stop repeatably in less than 90° of a die rotation.

22.4.7 Three-Die Infeed Machines

During World War II the need for high-precision, high-quality, high-strength aircraft cylinder head studs prompted aircraft engine builders to encourage the development of an American-built cylindrical die rolling machine. As a result, the three-die infeed machine came into being. These machines featured three smaller dies located with their axes vertical at 120° intervals around the periphery of the work to be rolled as shown in Fig. 22.12.

The rolling axis was vertical to simplify manual feeding of headed bolt blanks, and the infeed actuation cycling was continuous so the blank feeding and unloading was rhythmic like the manual operation of a flat die machine.

This three-die system has several advantages. Because the work is trapped in all directions by the dies, it does not require any work rest. Therefore, there was nothing in contact with the crest of the threads while they were being formed. As a result, the crests of the threads are free of any scuffing or deformation. In addition, the extra point of contact makes the three-die system better able to roll threads on hollow spark plug bodies and hydraulic fittings without using an internal mandrel for support.

In the early machines, the dies were moved on an arc radially inward, toward the centerline of the blank by the action of mechanical linkage and cam system. This complex system has been replaced in newer machines by direct radially moving dies which are hydraulically actuated. In some versions each die is independently actuated by a separate hydraulic cylinder and the infeed rate is balanced by a common hydraulic supply line. However, for precise penetration uniformity and rate control, an actuation system using a hydraulically actuated fully enveloping camming ring is used. The newer die drive system is essentially the same as that found in the spur gear drives used on some two-die machines but with a third spindle added.

Three-die machines are ideally suited for rolling precision fasteners and hollow parts from $^1/_2$ up to 3 in. in diameter. They are not practical for smaller work because the minimum die diameter that can be used on any size is only about five times the root diameter of the part being rolled since the dies clash with one another when they exceed this ratio. Above 3 in the overall machine structure becomes very large and inconvenient for vertical rolling axis orientation.

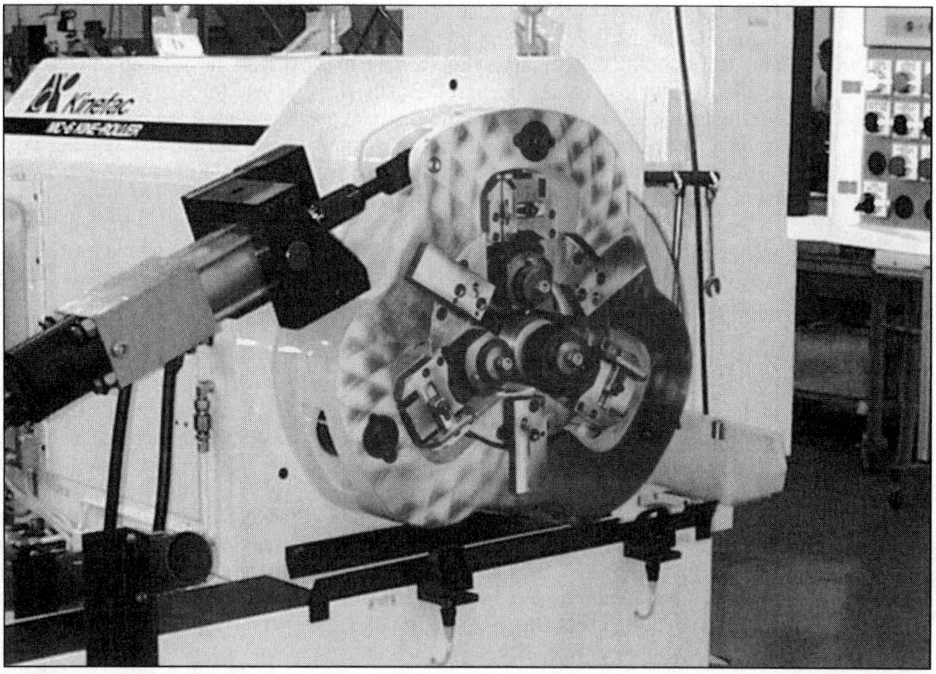

FIGURE 22.12 MC-6 infeed rolling machine with 30,000 pound radial die load capability.

This characteristic coupled with the added cost of a more complex mechanism to provide for die rotary match and the infeed actuation required by the three-die system have in recent years negated its earlier advantages in the rolling of solid parts. Therefore, the cylindrical three-die infeed system is now used mostly for spark plug shells, tube fittings, pipe plugs, and similar medium sized hollow work.

22.4.8 Three-Die Skewed Axis Machines

The development of three-cylindrical-die skewed axis through feed rolling evolved parallel to the adoption of two-die rolling machines. As a result, skewed axis capability was added to the three-die parallel axis machines as shown in Fig. 22.9. At the same time, the die axes were turned from a vertical to a horizontal position to facilitate the through feed rolling of long bars. With these changes, the cylindrical-three-die skewed axis system quickly gained acceptance for the production of high strength studs and threaded rod from $^1/_2$ to 2 in. in diameter.

Its three-point contact was also well suited to the rolling of hollow parts and finned tubing. However, because of their die diameter limitations, spindle bearing size constraints and their complexity and cost of structure, three die skewed axis through feed rolling is now used mostly in such specialized applications.

22.4.9 Forced Through Feed Machines

Initially rolling shallow low helix (high lead) angle or axial forms by the skewed axis system used the frictional component of the radial die load to produce the axial through feed force. This rolling friction force did not provide an adequate axial force component to produce the required metal forming in the starting relief on the dies. As a result, to roll deeper axial involute forms, such as splines and serrations it is necessary to replace the skewed axis die feeding force with an external force applied axially to the blank being rolled. This through feed force is generally applied through the use of a hydraulically actuated ball bearing center operating against the outboard end of the part. By controlling the hydraulic flow, the desired through feed rate through the parallel axis dies is achieved. Typically, this feed rate ranges from 0.030 in per part revolution to as high as 0.090 in, depending on the ratio of the die diameter to the work diameter and the length of the penetration area of the die.

Initially, when rolling splines by this system, the phasing of the dies with respect to one another was accomplished by the central gearbox of the machine from which the phased die drive torque was transmitted to each die independently through a drive shaft with two universals and a sliding center member which allowed for radial die position adjustment. The level of backlash in this arrangement was too great to produce satisfactory spacing error for precision splines. To remedy this condition, a phasing plug which is an exact analog to the spline to be rolled is now used. It is inserted between the dies and rotates in a tight mesh with them. It is connected by a face driver or some other means to the blank as it being forced into the rotating dies. By this means, when the blank enters the penetrating area of the dies, it is in exact conjugant relationship with them. Therefore, as the teeth begin to form, they are in the correct spacing relationship with respect to one another and thereafter continue to be formed with good spacing accuracy.

The radial stiffness of most three-die machines is adequate for radial feed and through feed thread rolling. However, when these systems are used for forced through feed spline rolling, it is frequently necessary to the preload the spindles in order to control the final pitch diameter of the spline. To stiffen these three-die machines when they are used for precision spline rolling, a system using a preload ring is interspersed between the three spindles. This protects the phase plug and makes it possible to hold the spline over wires size to precise tolerances. Such a system under well controlled rolling and blank conditions can produce splines as coarse as $^{20}/_{40}$ DP with spacing error less than 0.001 in and over wires pitch diameters of similar precision.

This three-die process is particularly useful when it is necessary to form splines on hollow blanks. The conventional rack process operates on the full length of the spline tooth continuously during the rolling process. Because the blank is subjected to two very high radial forces acting from diametrally opposite directions, it tends to collapse or go out of round. The forced through feed three-die process forms the spline progressively along the rolling axis with no more than a 0.500 in. forming area on each die in contact with the blank at any time. This low forming load is applied to the tubular blank from three directions, greatly minimizing the collapsing effect. This makes it possible for splines be rolled on tubes with walls as thin as $^{1}/_{5}$ the outside diameter, providing that the tooth form is less than about $^{1}/_{3}$ of the wall thickness. It should be noted that these limits are not exact and are also effected by the form, flank angle, blank material, die starting relief length, and depth of allowable crest seam. In cases where the I.D. of the hollow blank is closely held and the wall thickness is uniform, a mandrel can be used to allow even thinner wall blanks to be spline rolled by the forced through feed method.

Currently, this process is limited to splines up to approximately $2^{1}/_{2}$ in. in diameter with the tooth forms up to 16/32 DP and with flank angles as low as 30°.

22.4.10 Convergent Axis Machines

Conventional parallel axis rolling machines cannot effectively roll conical thread forms or shallow helical involute gears with cone angles of about 10° or more because the die to blank diameter ratio can only meet the integer requirement at one axial point on the die face. The balance of the form on either side of the match point becomes excessively out of match and, therefore, prevents full conjugate rolling action across the complete die face. This results in distorted forms and poor die life.

To solve this problem, convergent axis rolling machines are used. These machines are structurally the same as parallel axis machines except that the dies are aligned with their axes such that they converge with the rolling centerline of the blank at a point where the pitch cones of the dies and blank meet.

This alignment results in all points along the face of the die being in match with all points along the face of the blank.

Because the die axes converge, when the die contacts the blank it tends to move axially away from the convergence point. Therefore, in this rolling system it is necessary to hold the part axially in the dies. In addition, it is necessary to maintain the blank on the plain of the die centerlines, with the result that this type of rolling is frequently done on centers or in a bushing. In addition, the spindles of these machines must be able to support the resulting axial component of the radial die load.

22.4.11 Bump-Rolling Machines and Attachments

In many cases where a simple rolled form is needed, it is desirable to combine the rolling process with the blank shaping process which frequently takes place on a lathe or an automatic screw machine. If the form to be rolled is relatively shallow and the diameter of the blank is sufficiently large so that it will not bend away under the rolling load, a single rolling die can be mounted in a simple attachment on the cross slide of the turning machine and forced into the work to produce the desired form as is shown in Fig. 22.9. A typical two roll knurling attachment is a good example of a bump rolling configuration situation. In bump rolling the rotation of the dies is produced by contact with the driven work piece. The infeed force if produced directly by the slide and the turning machine spindle takes the full rolling force. Because of these characteristics, bump rolling is best done on rugged turning machines.

Based on early experience in bump rolling knurls in turning machines, and in an effort to simplify and reduce the cost of the machinery necessary to roll deeper involute forms, single cylindrical die rolling machines have been built. In those machines, the die may be driven in phase with the blank and it is fed radially by some form of hydraulic actuation. This technique is only practical where there is a bore in the blank or a protruding shaft large enough to support the roll forming load. Because of these limitations, the use of this configuration is generally limited to fine pitch involute form rolling or coarse pitch gear roll finishing.

22.4.12 Tangential Feed Attachments

For thread rolling on a turning machine, where the form is too deep to permit bump rolling or the work too thin or too long to support the rolling load, a tangential feed attachment is used. In this type of attachment as shown in Fig. 22.13, two dies are mounted in a rigid frame with their axes parallel to the axis of rotation of the work. The position of the two dies with respect to one another is such that when they are fed into the work they contact it tangentially. As the feed continues, the dies penetrate until the plane of the centerline between them moves to a point where it meets the axis of rotation of the work.

By use of this configuration, the bulk of the rolling force is taken by the attachment frame and only a small part of it is taken by the spindle, the slide, and the part. As in all attachments, the dies are rotated by their contact with the driven work-piece. When used to roll annular forms, the dies are free to rotate independently. When rolling helical or axial forms with this type of attachment the dies are geared together to provide correct match of one with respect to the other. This type of attachment has received wide acceptance in the last 40 years for the rolling of threads and knurls and screw machine parts near the collet or behind the shoulders. In fact it may be, by numbers, the most widely used rolling device.

22.4.13 Radial Feed Rolling Attachments

Where it is necessary to roll a thread or other helical form on a turning machine, and the part or spindle cannot provide adequate radial support, or the slide cannot provide the necessary radial die load,

FIGURE 22.13 Tangential feed attachment on turning machine.

it is necessary to provide a rolling attachment which independently produces that necessary radial die load. The radial feed attachment accomplishes that by mounting two parallel axis phased and free rotating dies in a scissor like frame which is provided with a wedge or linkage system, generally pneumatically actuated, to close the dies onto the blank.

The attachment is moved by the turning machine slide to the point where the blank and die axes are in a common plane, and then the scissors system is actuated to produce the die penetration into the blank. Since both dies penetrate with balanced radial force, and their centers are in plane with the blank, the spindle and slide encounter no radial load.

The dies are rotationally phased by a gear train and produce good results. Because it further expands larger turning machine capabilities, it is finding growing use.

22.4.14 Endfeed Rolling Head

To put a thread on the outer end of a part which is being turned in a lathe or a screw machine, it has been common practice for many years to use a die head in which there are three tools called chasers that cut the thread when the die head is axially fed on to the work. With the growing acceptance of the fact that a rolled thread had certain superior physical characteristics and better finish, it became desirable in these situations to replace cutting with rolling. Therefore, the end feed die head evolved with a configuration almost identical to that of the thread chasing head. In it there are normally three dies mounted on skewed axes and axially matched in such a way that when the blank is rotated and the head is fed axially, it will produce a rolled thread from a normal blank. This configuration as shown in Fig. 22.9 is used with dies ranging from about $^1/_2$ in diameter to over 2 in. After the dies reach the end of their axial travel, they are automatically opened by a cam system and the attachment is withdrawn from the work. The dies are annular and have progressive penetration built into them. As a result, they cannot form threads close to shoulders.

In addition to their use in machines where the work piece is rotated, end feeding heads can also be applied by rotating the head and holding the work piece from rotating. This can be done in a low speed drill press, a machining center, or on a special threading machine which is made to use this device.

22.4.15 Delta Feed Rolling Machine

Another approach to speed the production rate of the cylindrical-two-die parallel axis rolling machine was conceived a number of years ago but never achieved significant industrial usage except in Japan. It is the differential diameter die or delta feed system. If one replaces the two equal diameter dies with one large and one small die and then introduces a blank to be rolled into the dies above center, the larger diameter of the downward moving die will draw the blank downward into the decreasing die gap and therefore produce the penetration of the form into the blank without having had to actuate the die axes toward one another. Although well suited to the high speed rolling of annular form, its out of match die geometry makes the production of helical forms, such as threads, susceptible to match errors and other helix problems.

22.4.16 Internal–External Rolling Machines

This group of machines represents a logical extension of the rolling principle to the forming of hollow work. Although there are several standard types being built currently, none are constructed in significant numbers. They are used for such things as light bulb sockets, bearing retainers, tube fittings and other similar parts. Generally, they consist of a single die in the center of the work which is moved radially to form the work between it and an external developing die. The actuation is often hydraulic and in some applications, each die has an independent power source.

22.5 OPERATIONAL USES OF ROLLING

22.5.1 Primary Operations

The ability to roll helical, axial, and annular forms on bar provides the opportunity to preform bar prior to its subsequent processing in automatic lathes, screw machines, or other automatic bar to part processing systems. The most common of these applications is the rolling of full length threaded bar for subsequent drilling, broaching, and cutting off of set-screws. Another common product made from threaded bar is high strength studs for a variety of fastening applications. In those cases, the subsequent operations are cutting off, chamfering, and end-marking. In all of these applications, it is necessary to deburr the chamfered area of the thread—either during the turning operation or separately on individual parts. Worms, Acme lead screw, and jack screws are other helical parts—often made from preformed bar stock.

Knurls, splines, worms, and shallow-pinion stock are axial forms frequently rolled in bar length to save material and decrease subsequent processing time. This approach is used in the production of dental tool handles, plastic molding inserts, hardware products, small gear reducers, and similar high volume applications.

The rolling of annular preforms on bar for subsequent part production in screw machines and automatic lathes also has significant potential for material and processing cost savings, if the part designer is aware of the potentialities and limitations of the thrufeed annular form rolling process. However, the lack of process capability data available to part designers as well as the high cost of the initial tool development has limited its applicability to very high volume small flanged parts.

When any form is rolled on a bar for subsequent processing by a metal cutting system, it is necessary to take into consideration the effect of the clamping or collet forces on the rolled O.D. of the bar. Forms with broad flat crest are used where possible. However for threads, worms or actuator screws which have a sharper crest form, it is generally necessary to provide an area above the active flanks which can be deformed during the collet gripping, without effecting the operating surface of the flanks.

In all of these cases the preformed bar is generally between $^1/_4$ and 1 in. in diameter. Depending upon the length of the form and the diameter of the work, the bar production thrufeed rate ranges from about 100 to 400 inches per minute.

The use of thrufeed annular form rolling, carried to the point where the roll formed element is cut off, has found use for the very high speed (up to 4000/min) primary production of small simple shaped parts, such as projectiles, bearing ball and roller blanks, valve ball blanks, contact pins, and seal rings. Most of the solid blanks have been made from heat treatable steels $^3/_8$ in. in diameter and smaller. Because the roll cutoff end must be conical or have a small cylindrical protuberance, the process application has been limited to those applications where the end condition is not a consideration.

The annular through feed roll forming and cutoff process, performed hot, has been used for over 50 years for the production of chemical grinding mill balls up to 3 in. in diameter. Finally, large single-die revolution cylindrical die systems which have been developing during the past 40 years have reached the level of performance necessary for selective application to hot forming shaft blanks for automotive transmission and steering shafts in much the same manner as very large rack machines.

22.5.2 Secondary Operations

The most common secondary application is the thread rolling of headed blanks for small fasteners. These are produced in very large volumes in flat die and planetary machines using conventional rolling technology. Threading of turned or headed blanks for structural applications is the second most common area of secondary rolling application. Here the size ranges from as small as $^3/_8$ in up to 10 in. in diameter. Secondary thread rolling operations are performed on the full range of shaft-like parts for industrial and consumer products. A very wide variety of thread forms and materials are used in these fastening applications. For high strength fasteners, the rolling is frequently performed on prehardened blanks. In addition, the fillets between the body and head of such bolts are frequently rolled.

The next major secondary rolling operation of helical thread forms is for worms on actuator screws, speed reducer shafts, and similar shafts used to increase torque or convert torque direction. This operation is most commonly used for high volume automotive and appliance worms, and medium volume power transmission devices.

Another major secondary rolling operation is knurling. It is most commonly used for creating press fits between shafts and mating parts. These knurls range from conventional diamond knurls to special all addendum straight knurls which are designed specifically for the mounting of laminations, slip rings, worms, gears, and other rotating parts onto motor or other power transmitting shafts. Secondary rolling of diamond knurls is also commonly used to create joints between injection molded parts and turned shaft or hub inserts. Most of this type of knurling is on shafts ranging from $^1/_8$ to 1 in. in diameter. However, it is practical to roll knurls up to 10 in. in diameter or more if the rolling machine has the required diametral capacity.

Initially rack type rolling systems were the primary method for rolling automotive splines and still represent the majority of spline rolling systems for diameters from $^1/_2$ to $1^1/_2$ in. The advent of cylindrical die forced through feed spline rolling has enabled rolling to further displace hobbing or broaching as the more cost effective secondary operation for producing splines of medium to high volume torque transmitting shafts. For the secondary rolling of involute serrations and splines of

$^{24}/_{48}$–$^{12}/_{24}$ diametral pitch on shafts from $^1/_2$ to $1^1/_2$ in. in pitch diameter with good spacing and pitch diameter tolerances, both forced through feed and rack rolling machines perform equally effectively. For $^1/_2$ in and below, scroll feed machines are more cost effective, and from $1^1/_2$ to 3 in diameter forced through feed rolling machines appear to have significant advantages.

A wide range of annular forms for bearing raceways, plastic press fits, stop flanges, snap ring grooves, and similar applications can be produced by infeed rolling. These are being performed as secondary operations where it is not possible or practical to produce them in the original turning operation due to a lack of machine stations, inadequate part support, or other primary process limitations.

It should be noted that in virtually all of these applications the constant volume rule holds and there is only minimal stretch during the rolling operation. Therefore, a key element in the application design is the ability to determine, and, where possible, to control, the amount and direction of the outward radial flow.

All of the above secondary operations require 40 or less work revolutions. Therefore, depending on the part diameter, die diameter and machine speed, such secondary rolling operations generally take between $^1/_2$ and 6 s. This does not include the loading and unloading of the rolling machines which are discussed in other areas. As a result, they form an important cost effective method of performing secondary operation.

22.5.3 Supplemental Rolling Operations

As described earlier, the forming of knurls in single or multispindle automatic lathes with a knurl-rolling tool is a very old application of the rolling process used to supplement turning machine capability. That experience led to the development of the single cylindrical die for the rolling of shallow threads on parts while they are being produced by the turning machines. To produce deeper and longer rolled threads of superior quality as a part of the primary turning operation, it was necessary to develop additional types of radial penetration rolling attachments and through feed axial rolling heads. These rolling units are now used extensively to eliminate the need for secondary thread rolling operations, mostly on shaftlike parts from $^1/_4$ to 3 in. in diameter. Units capable of rolling threads up to 9 in. in diameter are available, but their use is very limited.

22.5.4 Roll-finishing and sizing

In the production of turned parts, the ability to obtain surface finishes of 16 μin and below is limited by the axial tool feed action of the cutting process. The rolling process with its trochoidal, no slip action when applied with precisely controlled high force creates excellent surface finishes on nonheat-treated surfaces which have been turned by axial feed. On these turned surfaces with untorn finishes of 32 μin or better, rolling can create surface finishes as low as 2 μin on most surfaces of revolution. These include journal surfaces on shafts, spherical surfaces on ball studs and tie rods, ball and roller bearing raceways, and similar sliding or rolling contact surfaces.

Operational experience with assemblies using roll finished surfaces in plastic bearings indicates that a rolled surface provides superior wear life to that of ground surface of the same μin finish. The empirical explanation is that the radially deformed asperities have a significantly smoother texture than that produced by the abrasive action of the grinding process. Because of this characteristic, roll finishing is finding growing application in permanently lubricated automotive and appliance components.

In most roll finishing applications, there is a small diametral change of up to 0.0005 in. This change results predominantly from the deformation of the surface asperities. Therefore, it is not practical to use roll finishing as a sizing operation. However, there is a two-step sizing process where an annular preform, which is rolled in the first step, is then roll sized to tolerances of as low as 0.0003 in by a precisely controlled roll finishing operation.

Since most roll finishing operations generally require less than 10 work revolutions, they can be performed in a few seconds.

22.5.5 Roll Straightening

For small diameter bent bars, heat-treated shafts, or other long cylindrical blanks, it is possible to improve their straightness by skewed axis through feed rolling on two-cylindrical die rolling machines. This relatively old process uses concave and convex dies rotating opposite one another. They overbend the part to be straightened as the through feed rolling action passes the rotating part axially through the dies. The deflection created is sufficient to raise the stress in the skin and significantly below the surface of the part above the elastic limit of the material. As it spirals through the die gap while supported on a blade, the rolling increases the stress beyond the level existing in the initial bend of the part and then gradually relieves the stress level in a spiral pattern. This leaves a substantially symmetrical residual stress around the cross section of the shaft and improves its straightness. However, there is generally some small residual stress unbalance near the neutral axis of the part which cannot be removed by the level of overbending from the rolling action and, therefore, it is not possible to produce a perfectly straight part by this method.

22.5.6 Roll De-Nicking

Roll threaded parts or other similar ribbed forms which have been dented or nicked at their crests during handling or heat-treating can be repaired by rerolling. Since in most cases the nick or dent is a depression which is surrounded by the outward radially displaced material, it is generally possible to return the displaced material into its original position by a second rolling operation. This can usually be done in the same machine which originally rolled the parts or in some simpler rolling system. The dies are designed to primarily contact the outer surfaces to the original rolled form while maintaining conjugate rolling action with the part.

22.5.7 Fin Rolling

The rolling of fins on tubing to increase its heat transfer capability is an old process. The fins are helical and normally produced by three skewed annular die assemblies. In some systems the die assemblies are used in a three-cylindrical-die through feed machine and the dies drive the tube. In other systems, particularly for rolling small diameter fin tube, the free wheeling die assemblies are mounted in a three-cylindrical-die skewed axis head which is rotationally driven while the tube is fed axially through the system while being rotationally constrained. In both cases a support mandrel is often used.

22.6 ROLLABLE FORMS

The rolling process is capable of producing helical, annular and axial forms. The configuration of these forms is generally defined by three basic elements: the flank slope, the lead angle and the form depth. Each of these elements can be varied widely, but all are closely interrelated. However, in all cases the form must be capable of conjugate rolling action with the forming die. In general, the flank slope and form depth are limited by the ability of the die tooth to enter the blank, penetrate, create the rolled form and depart from the part being rolled without distorting the rolled form and without overstressing the forming die due to bending loads. Figure 22.14 shows the geometric characteristics of the more common rolled forms.

In addition to these relatively standard forms, a wide range fins, flanges, grooves, and other functional forms and surfaces can be rolled, provided that the material is capable of sustaining the required deformation without failure.

Function	Product	Tooth Form	Flank Angle	Flank Shape	No. of Teeth Min. - Max.		T.P.I. or D.P. Min. - Max.		
Fastener	Machine Screws		30°	Straight	-	1	4	80	T.P.I.
	Wood Screws		30°	Straight	-	1	8	32	T.P.I.
	Self Tapping Screws		30°	Straight	-	1	12	48	T.P.I.
Linear Actuation	Acme Screws		14 1/2°- 30°	Straight	1	5	4	16	T.P.I.
	Lead Screws		10° - 30°	Straight	1	7	4	32	T.P.I.
	Ball Screws		45°	Gothic	1	2	2	10	T.P.I.
Torque Connection	Knurls		45°	Involute	20	240	64	160	D.P.
	Grip Knurls		45°	Involute	20	90	24	60	T.P.I.
	Serrations		45°	Involute	20	48	8/16	32/64	D.P.
	Splines		30° - 37 1/2°	Involute	17	48	12/24	32/64	D.P.
Torque Transmission	Worms		10° - 20°	Involute	1	4	3.5	16	T.P.I.
	Gears		20° Stub (20° Helix)	Involute	12	48	12	32	D.P.
Fluid Connection	Pipe Threads		30°	Straight	-	1	8	27	T.P.I.

FIGURE 22.14 Common rolled forms from 1/4 inches to 1 1/2 inches outside diameter.

22.6.1 Tooth Form Shapes

The initial development of the rolling process was directed toward the high speed production of threads on fasteners. As a result, the bulk of the early process development work dealt predominantly with machine screw threads and the bulk of the early rolling experience came from rolling helical forms which had a 30° flank angle and a single lead on flat die machines. As the use of screws for jacks and other actuation purposes increased, the rolling process was applied to Acme screws which had 14$^1/_2$° flank angles to maximize the ratio of useful actuation force to wasted radial tooth load. The majority of the Acme actuation screws are single start, but as plastic nuts are being used for stepping motor and other motion devices, multistart high lead angle lead screws are becoming more common.

Flank angle tolerances on 60° thread forms for fasteners are generally ±$^1/_2$°. As the straight flanked low helix forms are used for actuation and to carry moving load, they are frequently reduced to as low as ±$^1/_4$°.

With the advent of antifriction ball nuts, the associated ball screws required a curved flank to mate with the balls in the nut. Since the ball nuts operated in both directions and preload was desired, the mating helical form had to contact the balls in the same way as in an angular contact ball bearing. To do this, the *gothic arc* flank form was developed. These forms consist of two symmetrical arcs meeting at a pointed root and are designed so that the flank contacts the nut ball at the correct pressure angle. It is generally about 45° and the mating arc form is generally slightly larger than the ball. These gothic arc forms are held to very precise tolerances which are designed to produce specified areas of ball contact at the desired pressure angle. Typical rolled ball screws use balls ranging from approximately $^1/_8$ in diameter to $^1/_2$ in diameter.

The rolling of knurls for gripping or press fitting purposes began using simple milled dies which were cut using the corner of a 90° cutter to cut the teeth. As a result, the 45° flank angle became the standard and remains so today. However, the knurls now have many starts and high lead angles. As the lead angle increases to as much as 60°, the rolling action creates a slightly involute shape on the flank. Generally, the pitch of these knurls ranges from 128 up to approximately 24 TPI (teeth per inch).

Early knurl designs used a full depth symmetrical form with a sharp crest and root. When rolling straight knurls on shafts with these sharp crested forms the shallow metal flow produced by the sharp penetrating die teeth created excessive crest seams. When these knurls were used to produce press fits into mating gears or motor laminations, the crests failed, causing poor joints.

To correct this, special all addendum knurl forms were developed. In these grip knurls, which are generally between 20 and 60 TPI, the tooth root is flat and is as much as 2$^1/_2$ times the width of the base of the tooth. As a result, during the rolling, the associated broad die crest creates deep metal flow, readily filling the shallow adjacent tooth forms without any seam.

In addition, because these knurls can be rolled very full without the tendency to produce circumferential flow and flake, the O.D. tolerance after rolling is about the same as the blank before rolling. Therefore with precision dies and blanks, it is possible to hold the knurl O.D. to tolerances as low as 0.001 in.

The use of very course pitch straight knurls (splines) for axial torque connections opened up new applications for the rolling process in the automotive, appliance, and other high volume machinery applications. Here again, to maximize their effective torque transmission capability and to minimize the wasted radial forces in the joint, the flank angles have been reduced as far as possible while still maintaining form rollability. Currently 37$^1/_2$° and 30° are becoming more common.

To provide optimal performance, the rolled involute forms can be held to tight tolerances. Profile errors as low as 0.0005 in, generally negative, to provide a bulged tooth are common. Tooth spacing errors of below 0.001 in can be held with good blank and rolling conditions, depending on the rolling system.

With respect to pitch diameter tolerance capability, the pressure angle has a significant effect. As it is decreased, variations in tooth spacing and tooth thickness cause increasing change in the "over wires" measurement of that dimension. All other conditions being equal, the change in over wires measurement due to spacing and tooth thickness is inversely proportional to the tangent of the pressure angle.

Another important consideration in rolling splines and pinions is the requirement that all of the involute forms of the flanks must be outside of the base circle of the generated involute. This characteristic has the effect of limiting the minimum number of teeth on the rolled part. The lower the pressure angle on a given blank and the lower the dedendum of the teeth, the higher the minimum number of teeth that can be rolled without causing the die to undercut the dedendum area of the flanks.

It should be noted that this limitation is mitigated by the addition of a helix angle. For that reason helical forms do not encounter this limitation.

For right angle worm gear torque transmission, rolled worms operating with plastic worm wheels have achieved widespread use. Windshield wiper drives and power window lift drives started with 20° flank angles and as the need for size reduction and efficiency increased, they have been reduced to as low as 10°.

In these and similar applications, to balance the effective bending strength of the steel worm teeth to the plastic worm gear teeth and to make more compact drives special tooth forms have been developed. The worm teeth have been made thinner by as much as 30 percent and the worm gear teeth increased a commensurate amount. In addition, the ratio of tooth depth to shaft diameter has been increased to as much as 25 percent, depending on the number of leads in the worm.

Rolled worm pitch diameter tolerances of ±0.0005 in are possible, but as the tooth depth to pitch diameter ratio increases, bending of the worm at its juncture to the shaft overrides the pitch diameter error, especially on small diameter single lead worms. To minimize this bending which can be as much as 0.007 in TIR on $^3/_8$ in shafts, special die end runouts and rolling tooling can be used, but if runouts of 0.0015-in T.I.R. or below are required, a subsequent straightening operation is required.

Finally, for parallel axis helical gear torque transmission, conventional 20° stub tooth gear forms are being rolled, but generally not to full depth. They are used mostly for low precision applications. However, high precision automotive transmission planet gears have been roll finished for many years. In those applications, a limited amount of surface metal flow occurs and the pitch diameter is not controlled by the rolling, only the form and surface finish. The form tolerance depends almost entirely on the rolling die and blank design and precision and can be as low as 0.0003 in.

All of the foregoing tooth shapes are generally produced with symmetrical, i.e., balanced, forms. However, in a number of special cases, nonsymmetrical forms are produced to handle unidirectional loads. For threads, the buttress form with flank angles as low as 5° on the loaded flank and 45° on the unloaded flank is common. In addition, for locking worms, such as those used on truck air brake slack adjusters, a smaller level of nonsymmetry is used.

In all of the above applications, flank angle limitations are also closely related to the depth of the form and the flank shape. For a given pitch, as the depth increases, high flank angle forms quickly become pointed. For that reason, involute splines have a standard diametral pitch but are truncated in depth by 50 percent. So, a $^{20}/_{40}$ pitch spline will have its tooth spacing the equivalent of a 20 diametral pitch gear but the tooth depth of a 40 diametral pitch gear.

For annular forms, where there is radial growth such as a flange, the radial growth and depth of the form are primarily determined by the material rollability and the ability of the dies to gather that material and create radial flow while preventing axial flow and minimizing the tendency for circumferential flow. For most such applications, it is generally not practical to obtain radial flow equal to the die rib penetration.

Finally, the depth of the rolled form and its shape are greatly affected by the rollability of the material, the pattern of die penetration, the die design, the die surface finish, and the rolling system used. Figure 22.15 shows an extreme rolling application, the actual cross section of a single lead heat transfer fin rolled in a soft aluminum casing on a steel tube using a three-die through feed process.

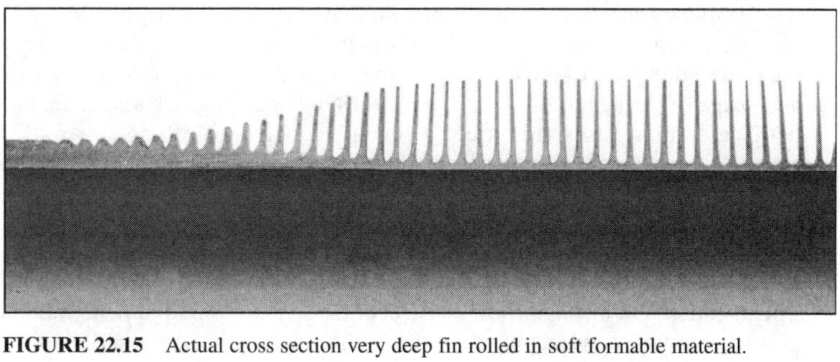

FIGURE 22.15 Actual cross section very deep fin rolled in soft formable material.

22.6.2 Lead and Pitch

Standard fastener screw threads have a single lead with lead angles ranging from 2° to 5°, which provides the torque self-locking characteristics needed in a screw fastener. Most screw threads are held in place by a nut or tapped hole which has at most two diameters of engagement, therefore, standard screw specifications do not have any lead angle or lead accuracy and no specific tolerances for them are specified in the standards.

When helical forms are used for actuation purposes, then lead can become an important characteristic. Most rolled actuation screws below $^1/_4$ in. in diameter have a single lead and lead angles of 10° or less. When rolled by the radial feed method where the full rolled area is within the width of the die, very precise lead repeatability can be achieved. With blank material having consistent yield point and blank held to diameter tolerances of 0.001 in, it is possible to get leads consistent within 0.0007 inches per inch. Therefore, if the die lead is compensated, generally by trial and error for axial spring back, then 0.0002 inches per inch is practical.

When the screw or other helical form is longer than the available die face in the rolling machine, the through feed method must be used for rolling the form. In that case, the control of lead is more difficult because the skew of the dies, the penetration and dwell of the dies and their setup, in addition to the material variability, all affect lead produced.

Since the die is only acting on a small section of the part as the part passes through the dies, then any error that occurs during that period remains in the part. Therefore, through-feed lead error is cumulative.

The lead of the part, plus or minus the designed skew setting for the setup, plus any compensation for axial stretch or spring back, is built into the die. If all is estimated correctly, the setup is right, the lead angle is low and the depth of the form is shallow relative to the overall diameter of the part, the lead will be satisfactory and repeatable. For normal single lead Acme type screw actuation applications, a lead tolerance of ±0.001 inches per inch can be readily achieved without die compensation.

As the lead angle of the part increases, which occurs when the form is deeper relative to the O.D. or starts are added to get a longer axial motion per screw rotation, then all the variables come into play. In these cases, adjusting the skew of the machine can have some limited effect. In addition, taper adjustment of the machine can be used to compensate for stretch effects, but in many situations, a trial and error modification in the die load must be made. If the die diameter is not changed, then this will require commensurate adjustment of the die tooth and space thickness.

For precision lead screws or ball screws which must be hardened after rolling, an additional lead compensation must be made. In spite of this additional variable, some manufacturers are achieving lead accuracies of ±0.0001 inches per inch from the theoretical lead of the screws after heat-treating.

For axial forms, the control of lead encounters different variables, depending on the rolling system used. For systems which use radial die penetration, the lead is determined primarily by the lead of the die, the machine setup and the precision of the alignment of the blank as it is introduced into the rolling machine. The latter is naturally affected by the manner in which the blank is machined and supported. Assuming no blank run-out between the supporting and, therefore, measurement location, lead errors of below 0.0002 inches per inch from theoretical are practical.

For splines and other axial forms rolled by the forced through feed method, there is an additional variable which may affect lead. This is the very small spirally tendency that may be induced by the penetrating form of the die. Depending on the depth and pressure angle of the form and the length and shape of the penetration area of the dies, this can produce a lead change of up to 0.0005 inches per inch. In most cases, this effect is negligible, but when rolling larger and longer splines, it is necessary to have a precision die skew adjustment in the rolling machine to compensate.

22.6.3 Form Depth

The total depth of a rolled form consists of the amount of radial penetration of the die into the blank plus the amount of radial growth of the displaced material. As described in section 2F, in most rolled forms, this is essentially a constant volume process in which the displaced material generally flows symmetrically about the penetrating rib. Therefore, the depth of form that can be produced is a function of the conditions which support and limit the local radial flow. These variables include the flank angle, tooth thickness to height ratio, blank material rollability, die penetration rib shape, die penetration per contact, available die penetration contacts, die surface condition, process lubrication and, in some situations, the ratio of rolled O.D. to root diameter.

Because of this wide range of interacting variables, it is not practical to define quantitatable relationships or numerical limits to the depth of form that it is possible to roll. Once again, there are some general statements which are useful in describing means for maximizing the depth of a rolled form:

- Maximize die rib crest corner radii.
- Minimize die crest flat length.
- Select materials with highest percent elongation.
- Avoid materials which work harden rapidly.
- Provide best possible surface finish on crests and flanks of die form.
- Avoid circumferential surface discontinuities on die flanks.
- Maximize axial constraint on blank by die design or some external means.
- Provide flood lubrication of rolling process with high film strength lubricant.
- Maximize penetration per die contact.

22.6.4 Root and Crest Forms

In general, rolled forms require radii at the root and crest for a variety of reasons. Since the root of the die forms the crest of the part, crest radii are added to the part for two purposes. In the case of fine pitch thread forms 24 TPI and smaller, the radii are necessary to enable the root of the die to be ground without grinding wheel breakdown. They are also necessary to move any rolling seam away from the flanks. In most cases, the smallest die root radius that can be easily ground is 0.002 in.

A root radius on the part which allows a crest radius on the die is used predominately to improve metal flow around the crest of the die as it enters the blank. This smooth metal flow around the crest of the die generally results in improved part root surface finish. The combination of the improved surface finish and the stress mitigating effect of the root corner radii also results in improved fatigue resistance for the rolled form.

When rolling forms which require a broad root flat, the addition of corner radii is generally necessary to prevent flaking of the material which would result from turbulent metal flow which occurs when the penetrating die ribs have sharp corners.

For radial feed rolling applications where a very wide root space is required on a rolled form, such as a deep worm and the space width on the part must be maintained to the bottom of the flanks, it is sometimes necessary to point the crest of the die. This prevents trapping of the displaced material under the wide penetrating die crest. This trapping action, if not eliminated, can cause major flake generation on some of the less rollable materials.

In through feed rolling of broad root flat forms, painting of the die is also used to centralize any crest seam which may develop. In those cases, the point of the die crests in the penetrating area is side shifted toward the dwell area in order to balance the metal-flow up each side of the die flank.

22.6.5 Surface Characteristics

In virtually all conventional helical and annular form rolling applications such as those described in Fig. 22.14, a rolled surface of the flanks of 8 μin or better is readily achievable on parts of ductile metals. However, the crest and root of a rolled form are subject to aspects of the rolling process which frequently degrade the surface finish in those areas.

During the forming of the crest of a rolled form, the material being formed generally moves up both sides of the converging die flanks more rapidly than the material in the middle. As a result, as it fills in the root of the die, a crest seam may be formed. In many applications, if the part is rolled full, the seam has been closed by the material flowing up the flanks, meeting in the crest flat. If die penetration stops just as the part is rolled full, the seam will disappear and the crest will have a surface finish similar to that on the flanks, provided the part is rolled by systems which do not use a support blade.

If the part is not rolled full and the radial growth does not reach the root of the die, there may be sharp edges on the crest. Generally, the best rolling conditions from the standpoint of die life and part quality are somewhere between the two, rolling just full enough to leave a thin trace seam in the middle of the part crest.

It should be pointed out that in virtually all applications a thin trace seam in the crest of most rolled forms has no negative effect on the forms' function or service life. In static threaded fasteners, it does change their load carrying capability. For fasteners undergoing cyclical loading, the main source of fatigue failure is the thread root. For actuation screws or worms, the crest seam would not effect the wear capability of the flanks or the torsional strength of the central shaft area.

One of the very few applications where a trace crest seam might cause some problem is where the O.D. of the rolled form serves as a bearing or centralizing function for the rolled form. In those cases, rolling full may be necessary. Unfortunately, in many applications, the negative cosmetic effect of a crest seam causes the user to require the form to be rolled full.

The surface conditions found in the root of rolled forms varies very widely depending mostly on the shape of the penetrating die rib, the material characteristics and the level of overrolling that may arise. As noted above, dies with penetrating ribs, which have good corner radii and are pointed, tend to produce very smooth roots on conventional rolled forms and, even with moderate overrolling, do not tend to cause root finish degradation.

In difficult rolling applications, all of the factors which tend to increase the rolled form depth capability also help to eliminate root surface problems.

22.6.6 Subsurface Characteristics

As a rolling die penetrates radially into the blank, the displaced metal can flow in any of three directions, but in low lead angle helical and annular form rolling applications the metal flow is either outward radial or inward axial. In either case, the flow then is essentially in a plane passing through the axis of rotation of the blank.

Therefore, the shape of the tooth space which generally corresponds to the shape of the penetrating rib on the die is the primary factor in the resulting subsurface condition and grain flow pattern that results from the rolling operation. As with other product features, it is also to a smaller degree affected by the penetration per die contact, the material characteristics, and the ratio of form depth to blank O.D. Since there are no specific quantitative measures of grain flow, one can only provide general observations about the effects of the common rolled forms. First, grain lines follow the surface near the flanks of the form. If the flank angles are low, they follow the surface further toward the center of the teeth. As the surface flow continues toward the crest of the form, the grain lines converge to meet the radial flow in the center of the tooth and curve around the corners of the crest, ending at the seam area.

With typical thread forms, the grain flow in the root follows the shape of the penetrating die crest and this crest form following pattern attenuates to the point where it is hardly discernable at a distance equal to about one half a thread depth below the rest of the form.

With broader root flat forms the grain flow pattern grows deeper and the center tooth pattern goes higher. If the form is unbalanced with a wider space and narrower tooth, this increases the depth and height of the grain flow pattern. On the other hand, increasing the flank angle has the reverse effect.

For axial forms almost all of the metal flow occurs in the plane perpendicular to the axis of rotation of the blank. In addition, the direction of the metal flow is *not* symmetrical about the penetrating die tooth. Since most axial forms are involute, they behave like a gear set in which the die is the driver and the blank is the driven gear. Since the tip of the driving flank tooth of the die contacts the O.D. of the blank first, as it drives the blank along its line of action and penetrates it, that flank tooth penetrates with an inward radial motion. The resulting metal flow along that flank is similar to the penetration of a thread rolling die. But as the die tooth rotates conjugately with respect to the blank, the trailing flank of the die forms the other flank of the part with an outward wiping motion.

This action, although different than that of helical or annular form rolling, produces substantially the same subsurface grain flow patterns as those forms, except at the crest. There the wiping action of the trailing flank of the die adds to the rate of flow of metal up that flank and thereby tends to shift the point where the two flank flows meet toward the driven flank of the die. Therefore, if a seam forms it may in some cases encroach on that flank.

In all cases, the grain direction is generally parallel to the flank surface of the tooth form and increases the abrasive wear resistance of these surfaces. In addition, the grain pattern which parallels the root form increases the bending fatigue resistance of the teeth. These improvements in subsurface strength are further enhanced by any work hardening which results from the induced strain in the material.

22.7 ROLLING MATERIALS

Virtually all commonly used metals are sufficiently ductile so that they can be roll formed while at room temperature, however, some of the harder alloys must be rolled at elevated temperatures to achieve useful deformation. In evaluating a material for an individual rolling application four questions must be asked:

a. Can the material be formed sufficiently so that the rolling operation can produce the shape required?
b. What is the rolling force required to achieve the desired deformation?
c. How will the material behave during the rolling deformation?
d. What will the material characteristics be after the rolling operation?

By evaluating the materials ductility and related physical characteristics, it is possible to get useful qualitative answers but not specific quantitative answers. The yield point, ultimate strength, modulii of both shear and tension, percent elongation, percent reduction in area, hardness, and work hardening rate along with the actual rolling process variables—all effect the answers to the above questions. Each characteristic has some significance, but depending on the question some provide more information than others.

22.7.1 Material Formability

A generally accepted understanding of the metal forming phenomena is that metal flow occurs along slip planes whereby layers of grains of the material under load slide with respect to one another. As they slip, the grains transfer their molecular bonds progressively to succeeding adjacent areas without breaking away. This flow can continue in a slip plane until some distortion of the grain pattern or foreign element in it, a dislocation, impedes the smooth slip. At that point any further movement in that slip plane is blocked. As more and more of the slip planes in the material are used up by the deforming process, it requires progressively more force to continue to deform the material. This increase in the required deforming force caused by the previous deformation is referred to as strain hardening or work hardening.

When the material is deformed beyond the ability of its grain structure to accommodate further movement of the grains with respect to one another without breaking the intermolecular bonds, it fails. A material's formability is the degree to which such deformation can be made before failure.

This characteristic is generally defined as ductility. It has traditionally been evaluated by pulling a standard tensile specimen to the point of failure, measuring the amount elongation which occurred before failure, and converting that amount to a percentage of the standard specimen length of 2 in. The result is called "percent elongation." It should be noted that during this test, as the specimen elongates, it necks down by flowing in shear. This continues until no more shear movement can be sustained. Then it finally fails with a "cup cone" break in which the cone angle generally approximates 45°.

The test for percent elongation uses simple unidirectional tension loading, but rolling is a complex three-dimensional compressive deformation process. This limits its use as a quantitative indicator of rollability. However, it can help to answer to the first question, can the material be rolled to the desired form without the failure. Experience indicates that with any conventional thread rolling process, material with a 12 percent elongation or more can be readily formed. With an optimized penetration rate, good lubrication, and blank heating is practical to roll relatively fine threads in material at Rc 47 has only a 4 percent elongation. For simple roll finishing, it is possible to create a significantly improved surface on a ground cylindrical part with a hardness up to Rc 52 and about 2 percent elongation. At the other end of the formability range, it is possible to roll deep fins as shown in Fig. 22.15 on aluminum or copper tube with a percent elongation of about 75. Clearly the higher the percent elongation, the deeper the form that can be rolled. Therefore, for the purpose of comparing and selecting materials for rolling applications, percent elongation is the best rollability indicator available.

Materials with a high percent elongation do not fail easily in shear, therefore, roll well. Materials which fail in shear easily, machine well, but do not roll well, therefore, materials which machine easily generally will roll poorly and vice versa. Steels to which lead, sulfur, and boron have been added to improve machinability are not generally as capable of being rolled into deep forms. They tend to flow less well, particularly around sharp die corners, and they are particularly prone to flaking when overrolled.

22.7.2 Resistance to Rolling Deformation

As noted above, the rolling process creates forms and surfaces by the plastic deformation of the blank material. As the individual elements of the blank material are subjected to compressive force from the dies they deform without failure by flow in shear. Therefore the resistance to rolling deformation is primarily a function of the shear yield strength of the material. Since this physical characteristic is not readily available for most of the materials which are used for rolled parts and it cannot easily be measured, the material hardness is the next best practical indicator.

The measure of a material's hardness is normally determined by the depth of penetration produced by a penetrating point or ball forced into the material under a controlled load. That penetrating action is similar enough to the deformation that occurs in the rolling process, so that the materials hardness is closely related to its resistance to rolling deformation. In a rolling operation the actual radial force applied by the dies to overcome this resistance and produce a fully rolled form is called the *radial die load.* Although it is primarily a function of the blank's initial hardness, it is significantly affected by the work hardening tendency of a material. This causes the hardness in the area being deformed to increase as the die penetration progresses. The radial die load for any rolling system is also a function of the ratio of die diameter to blank diameter, the shape and fullness of the rolled form, and to some degree the coefficient of friction between the die and the work piece, of the grain structure of the material and its previous cold-work history. Nevertheless, the measured hardness material is the best easily available starting point in determining how much radial die load is required to roll a specific form in any rolling system.

With this range of material and process variables affecting the radial die load required to roll any given part, it is not even possible to provide an effective estimating means to cover the wide range of rolling applications and rolling systems. However, it is possible to develop a generalized guide by selecting a set of process variables from which it is possible extrapolate the maximum amount of useful radial die load data and then make a controlled test using those variables. The two cylindrical

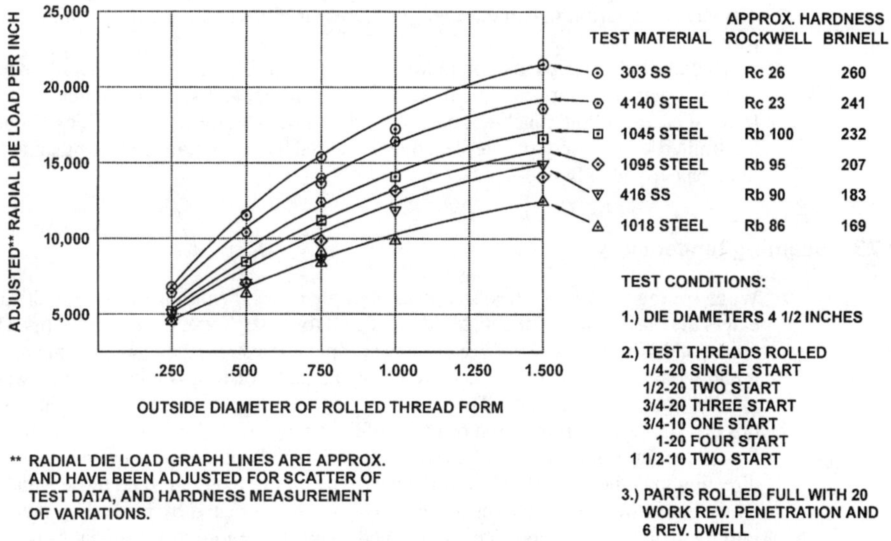

FIGURE 22.16 Thread rolling die load estimation chart.

die system provides the most controllable means of applying radial die load. Medium pitch thread forms are the most common rolled form, and steels are by far the most common rolled materials. The 18 work revolution penetration cycle in a two cylindrical die system with the die diameter to work diameter ratio of from 4:1 to 10:1 provides a set of mid-range process conditions. Therefore, those elements were the basis for tests, which produced the data shown in Fig. 22.16.

To use this *radial die load estimation chart* to estimate *radial die loads* (RDL) for other applications and systems the following conditions should be understood.

a. The test threads were rolled on 1-in long blanks. For radial-, parallel-, or scroll-feeding parts longer than 1 in, the approximate RDL can be obtained by multiplying the chart reading by the actual rolled length. For longer parts rolled by the through feed method, use the combined length of the penetrating area of the die and the dwell as the multiplier.

b. The rolled forms were standard 10 and 20 TPI, 60° threads. The load for thread forms from 32 TPI to 4 TPI are substantially the same for a given outside diameter. As the rolled form increases in depth, decreases in flank angle and has broader root flats, the RDL may increase as much as 10–15 percent. If the flank angle decreased, the RDL will increase slightly. However, for this type of form, there is not any controlled test data.

c. The test threads were rolled to 100 percent fullness without any overrolling. Any reduction in fullness will result in a directly proportionate decrease in RDL.

d. These RDL results are based on average die diameter to part diameter ratio of about 5:1. For cylindrical, flat, and planetary die systems with higher ratios, the RDL could be as much as 10 percent higher.

e. Other process, machine, or material factors can effect the actual radial die load encountered in a rolling application, therefore the chart data is for informative purposes only and the determination of actual radial die load for an application requires specific tests.

Where the rolling system being used has a limitation on the applicable radial die load, it is possible to reduce that requirement to a small degree by such steps as increasing the work revolutions, decreasing the level of fullness of the rolled form or reducing the die diameter. However, when a rolling machine is operating at its maximum radial die load capability, minor changes in blank diameter and hardness can cause significant variability in the fullness and diameter of the rolled form.

Radial die load capacity limitations for various rolling systems generally depend upon machine stiffness, actuation capacity, die drive, force or torque, and slide or spindle load carrying capability. In most cases, rolling machines cannot apply all the maximum capabilities at once and applications are limited by a variety of interdependent factors. Therefore, in evaluating an application, radial die load capacity is only one key factor.

22.7.3 Seaming Tendency

When rolling threads, splines, and other multiple ribbed forms, as the crest of the penetrating die form enters the blank it forces the material in its path to move. Some material is trapped in front of it, which creates a deforming pressure wave. As the rib penetrates into the blank, the metal-flow divides around it and the displaced material, has three possible paths of flow—it can flow outward radially, inward-axially, or circumferentially. The geometry of the rolling situation and the material flow characteristics determine what proportion of metal will flow in each of these directions. In rolling any common form experience indicates that the softer materials will tend to flow more easily in the outward radial direction than they will flow inward axially. The harder materials, and those which tend to work harden during rolling, tend to transmit the pressure wave created by the rolling die contact more deeply into the material, and therefore are inclined to produce more inward-axial flow.

When rolling forms in softer, more ductile materials, the mostly outward radial flow tends to occur directly under and near to the crests of the penetrating die. It travels more quickly out along the flanks of the die while the area in between the two penetrating ribs flows outward more slowly. As the die space begins to fill, the two waves of metal which are advancing outwardly along the surface of the die, meet at the root of the die before the material in the center of the wave arrives. This leaves a hollow pocket just below the crest of the rolled form. As the die penetrates slightly deeper, this pocket then fills in. However, the material cannot reweld itself, so a discontinuity or seam remains near the crest of the thread form.

Hard materials and those which work harden significantly during rolling tend to produce more inward axial flow; however, in rolling threads and other helical and annular forms, the axial flow component is restrained by the adjacent penetrating die teeth. This causes the deforming pressure wave to go much deeper into the blank. The increased subsurface flow and the die to blank friction cause the material in the center of the thread being formed to move outward at the same rate as that on the flanks. As a result, no seam and possibly some bulge is formed at the crest.

In order to provide means of categorizing the seam formation tendency off materials, one can put two levels of seaming tendency in between the extremes described above. Cross sections of seam formation in these four categories are illustrated in Fig. 22.17.

1. No tendency to form seam
2. Limited tendency to form seam
3. Moderate tendency to form seam
4. Strong tendency to form seam

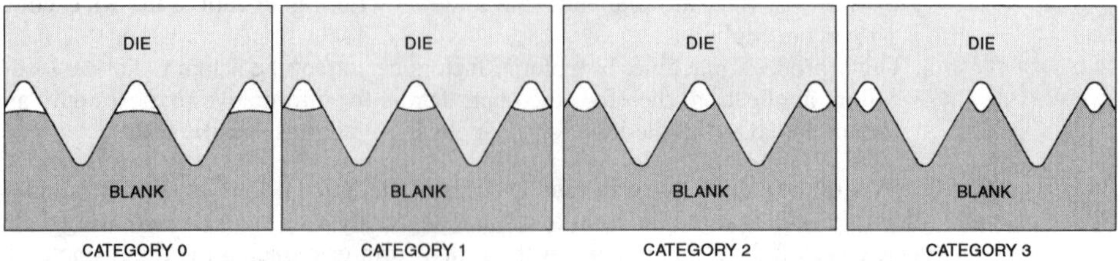

FIGURE 22.17 Seam formation threads of various materials.

These generally represent the four levels of seam found in threads and similar helical and annular forms rolled by the radial feed method.

The soft forms of most nonferrous materials such as copper and aluminum fall into category 3, as they are alloyed or hardened they fall into category 2. The plain low carbon steels with up to about 20 percent elongation in typical "as drawn" condition also fall into category 2. As they are more heavily drawn, heat-treated, or alloys added, they fall into category 1 along with the austenitic stainless steels. Most of the standard high quality fastener materials, such as 4140 also fall into category 1. As those materials are hardened to Rc 27–30 they move toward category 0.

The austenitic 300 series of stainless steels fall between 0 and 1 depending on how much cold work they have had prior to rolling. Most of the aerospace fastener materials when rolled hard at room temperature do not produce seam and are in category 0.

There are also some metals among the red alloys of copper which for some reason tend to fall into category 1 and there are other very soft materials which fall in category 2. Care should be taken to check out materials for their seaming tendency particularly before rolling very deep forms.

Factors other than material can have a major effect on seam formation. Anything which drives the deforming pressure wave deeper will reduce the seam formation. This includes increasing the rolling penetration per die contact or adding broader die crests, provided that they have adequate corner radii.

Another related condition is crest seam which is not central to the tooth form. When through feed rolling threads, worms and similar forms with cylindrical dies, a series of die ribs penetrate progressively deeper into the blank as it moves axially through the dies. Since the outward radial flow of material tends to divide evenly around the crest of each penetrating die rib and each subsequent rib is the deeper than the previous one, any seam tends to be displaced away from the deeper rib, back toward the entry of the die. If the form is deep with limited corner radii, the penetration per die contact is high, and the materials seaming tendency is 2 or 3, the seam will probably shift sufficiently to distort the flanks of the thread. In those cases, it is necessary to repoint the penetrating crest of the die off center toward the front to balance the material displaced into the thread crests and thereby centralize the seam.

The foregoing applies to the rolling of annular and most helical forms with a lead angle up to about 30°. When rolling splines and other axial or low helix high lead angle involute forms an additional motion is superimposed on the basic radial penetration. That is the conjugate gear rolling motion. During this involute rolling action, the die and the blank engage as mating gears interacting on one another. Both flanks of the die and part are in opposed contact with each other as the teeth are being formed. As a result, there are two forming contact lines of action during the die to part mesh.

The drive side mesh is the same as in any spur gear set with the first contact starting near the crest of the driving die tooth and blank O.D. As it forms the part tooth space, the contact proceeds downward along the driven flank of the part as shown in Fig. 22.18. The second contact line of action is on what is called the coast side in normal gear meshes. This line of action starts near the root of the part flank as the tooth space is being formed, and proceeds outward along this "coast" flank. This nonsymmetrical metal flow around the die teeth when superimposed on the basic seam forming tendency that occurs in the general rolling action tends to shift whatever crest seam is produced toward the driven flank of tooth being formed.

Once again the depth and shape of the seam and the degree to which it is shifted is a function of the rolling system used, the tooth form being rolled the die design, the penetration per contact and the part material.

22.7.4 Flaking tendency

Under most rolling process conditions the only discontinuity in the rolled surface is the seam at the crest of the rolled form, and the rolled form surface is smooth, however, under some conditions, the root of the form and the adjacent areas of the flanks may become rough and flaky. This condition may be material related. Materials to which elements have been added to improve machinability, tend to form less well particularly around sharp die quarters, and are particularly prone to flaking in

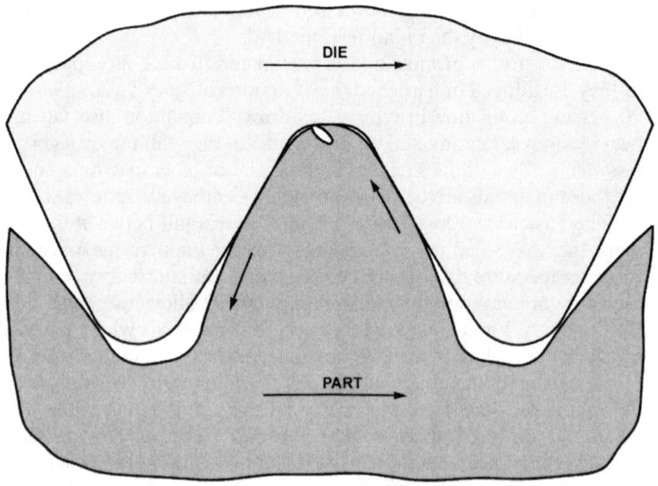

FIGURE 22.18 Seam formation rolling an involute axial on category 2 material.

those areas when overrolled. Materials which work harden quickly and have had excessive cold work prior to rolling are also prone to flake when being rolled to deep forms.

In evaluating an application where flaking occurs, the broadness of the root form, the corner radii of the die crest, the penetration rate per contact and the number of work revolutions in the dwell should also be examined.

22.7.5 Work Hardening

During any rolling operation, the rolled material flows in shear, uses up its slip planes and becomes progressively harder. The extent of this work hardening effect is a function of the amount of local deformation (strain rate) of the blank material and its work hardening characteristics. For deep forms in high work hardening material, the deformed material can increase significantly in hardness. For high alloy steels and some stainless steels, the hardness can increase as much as Rc 12.

The increase is highest at the rolled surface and decreases proportionate to the level of local material flow. The depth of hardness change is generally proportional to the maximum increase. For conventional steel, the work hardening rate is somewhat proportional to the amount of carbon and is also affected by the amount of cold working the material had after the last annealing and prior to the rolling operation. As nickel, chrome, and vanadium are added to create higher strength alloys, the work hardening rate increase significantly. It is highest for the austenitic stainless steels.

For fine pitch threads or knurls rolled in medium carbon steel the work hardening is slow and does not extend significantly below the thread form. As the form deepens and the flank angles go below 30°, the depth of work hardening increases and the whole cross section of the rolled form may be harder than the original material. In addition, since the penetrating dies trap an area of the original material in front of them, they can drive the deformation significantly below the root of the rolled form. In some cases, where these forms have broad crests the work hardening affect can extend below the root of the rolled form by as much as 20–30 percent of the form depth.

In general, the net strain rate the material has encountered at any point is the primary determinant of the work hardening and it is not affected by the sequence of the forming actions. In the rolling process, therefore, the work hardening effect is independent of the number of work revolutions which was used to create the form.

22.7.6 Core Failure

During the rolling operation, as the dies penetrate, the blank cross section is not round as noted in section 2. Since the penetrating forces of the die on the nonround work apply a couple to it, they also create a significant shear force acting through the center of the section being rolled. If the root of the rolled form is small proportionate to the depth of the form being created and the penetration per contact is high, the shear stress rapidly increases. As the dies rotate the part, this stress rotates around the center of the root cross section and by short cycle fatigue action, the material in the center begins to fail. As the failure progress, a small core opens up in the center which weakens the rolled part. When this occurs, the root of the rolled part may break up and expand in a nonround way. However, in some cases, this failure may not be apparent from the surface and the only way of detecting it is to measure to diameter the rolled root of the part.

This condition most commonly occurs when rolling two-start worms or deep annular forms with broad root flat in a two-die machine. In this situation, if the crests of the opposing dies penetrate directly opposite each other, and if the part has a root diameter which is less than $^1/_2$ of the O.D., the core will probably not be solid. Reduction of the penetration per contact and/or pointing the die crests can inhibit this possibility.

22.7.7 Fatigue Strength

The rolling process generally improves the bending and tensile fatigue resistance of the rolled surface. It accomplishes this in four ways. First, it increases the hardness and therefore the tensile strength of the rolled surface. The rolling flow action arranges the grain flow lines of the material so that they follow the shape of the rolled form. Rolling creates very smooth corners and root surfaces. Finally, it leaves the material near the surface with a residual compressive stress.

The higher strength obviously increases the point at which cyclical loading begins to initiate failure. The smooth surface tends to eliminate stress raising points from which fatigue cracks can originate. The grain flow lines parallel to the surface tend to inhibit the propagation of the fatigue cracks, and the residual compressive stress superimposed on the applied structural stress reduces the net fatigue stress the rolled surface undergoes during the load cycle.

Because of their improved fatigue strength, rolled threads are almost universally used in dynamic load applications. To obtain this improvement, the rolling must be done after heat-treating. Compared with threads which are cut and then heat-treated to the same hardness as the prerolled blank, the fatigue strength is 20–50 percent better. Published tests also show that for a given loading condition, use of rolled threads can produce similar improvements in the number of cycles to failure. However, it should be emphasized that only by tests under actual loading conditions can any such improvements be validated.

Rolling can also be used to improve the bending fatigue resistance at the inside corners of bolt heads and stepped shafts. There is no quantitative information on the level of improvement. It is dependent on the material and the amount and character the material displacement made by the rolling operation. At present, aircraft bolts and other critical high strength fasteners have rolled fillets.

From a material viewpoint, the greater the work hardening tendency of the material, the more the improvement in fatigue strength due to rolling. Also, the higher the blank hardness when the rolling is performed, the greater the improvement is in fatigue life. However, all of the above is limited by the point at which rolling process itself may bring the rolled surface to incipient failure.

22.8 ROLLING BLANK REQUIREMENTS AND RELATED EFFECTS

22.8.1 Diameter, Solid Blanks

In rolling full forms on virtually all types of helical, axial, or annular forms on solid blanks, where there is no significant stretch, the constant volume rule holds. After rolling the form full with no open crest, the final form dimension will follow the original blank dimension quite closely. Therefore, in such cases, the blank diameter prior to rolling must only be held to a tolerance which is slightly tighter than the finished form diameters. The small difference is generally necessary due to die form tolerance variations. For example, when rolling typical small, high precision threads with a given set of dies in a well set-up machine, it is possible to hold the pitch diameter and O.D. to total tolerances of as low as 0.0005 in with a blank O.D. tolerance of 0.0003 in, once the correct blank diameter (i.e., volume) is established.

When rolling a thread or similar form in which the die form will not be full at the end of its penetration and all of the metal flow is radial, the form diameter (i.e., pitch diameter of a thread) and the outside diameter move in opposite directions. Therefore, if it is not necessary to roll the form full and the outside diameter is allowed a significantly greater tolerance, then the blank may not require as precise a tolerance as that required for the form diameter. However, in all cases, it is also necessary to put in a factor for the diametral repeatability of whatever rolling system is being used.

For example, when rolling a standard 60° class 2a thread, the O.D. grows at more than twice the rate that the pitch diameter is reduced by the penetrating die. Therefore, if the maximum blank is designed to produce the maximum allowable O.D. with the minimum allowable P.D., then, since the O.D. has a total tolerance range of about $2^1/_2$ times that of the pitch diameter, if there were no other error sources, the blank diameter could be the selected maximum diameter minus up to 0.004 in. However, to minimize the possibility of producing bad parts due to die, machine and set-up errors, the typical blank diameter tolerance range for a 2a class thread would be 0.002 in.

In some rolling situations, more precise control of the blank diameter is necessary to ensure correct step-off from the die upon an initial blank contact. This is especially true in rolling high lead angle worms and lead screws where blank diameter variation can result in index error and lead drunkenness. Excessive blank diameter variation also creates the tendency of the part to feed axially during radial feed rolling.

When rolling axial or low helix angle helical forms by the parallel, scroll, or radial feed, blank diameter variations have a direct effect on tooth-to-tooth spacing errors. The amount of the spacing error directly attributable to the variation in blank diameter is a function of the flank angle, number of teeth, pitch diameter, and the die to blank match situation. Based on a review of a variety of such spline rolling applications below 1-in diameter with pitch $^{24}/_{48}$ or less, it is generally necessary to hold the developed blank diameter to a tolerance of 0.001 in to hold the maximum tooth-to-tooth spacing error below 0.0025 in.

For forced through feed spline rolling where the blank is driven in phase with the dies at initial contact, variations in blank diameter have less effect on tooth spacing or pitch diameter, but have a major effect on the O.D.

An additional factor which affects blank diameter determination is any stretch of the part which might occur during through feed thread or worm rolling. This is not easily predictable, but should generally be considered anytime the depth of form being rolled exceeds $^1/_5$ of the final diameter. In such cases, the type of material, the thread form, through feed rate, and the rolling die starting relief are the primary determinants of stretch. Materials which have a good rollability, generally stretch less because the flow tends to be more toward the surface of the blank. Forms with broad flat roots stretch more.

To counteract the tendency to stretch during through feed rolling, the starting relief of the die can be lengthened and the thread form pointed to prevent backward escape of blank material as it is pulled into the die. Even with the dies designed to prevent stretch, acme-threaded bars rolled with high form depth to O.D. ratios still stretch to the degree that the blank O.D. must be enlarged to obtain the specified P.D. and O.D.

22.8.2 Diameter and Wall Requirements for Hollow Parts

Rolling hollow parts has become a more common requirement as the need to remove weight from vehicles and other mobile or portable equipment becomes more important. In all such rolling applications the rolling blank must be able to support the radial die load without collapsing or it must be supported to prevent collapse. The tendency to collapse can be resisted by either of three means—increasing the wall thickness, adding to the number of dies contacting the outside diameter, or supporting the inside diameter. If the wall thickness cannot be increased, then more dies in contact can help. A three-die system decreases the deflection produced by the rolling load by about 80 percent. This has about the same effect as increasing the wall thickness by as much as 50 percent. However, if that is not possible, a two-die system must be used. The last alternative providing internal support is the most difficult to use effectively.

In any event, specifying blank diameter size and tolerance when rolling hollow parts presents a very complicated problem. The various considerations are the ratio of the wall thickness to the outside diameter of the blank, the ratio of the depth of form to the wall thickness, the blank material, the type of machine or attachment being used, and, finally, whether an internal support mandrel is being used under the rolling area.

Rolling a hollow part without any internal support requires that the rolled area wall thickness, with whatever support is available from the adjacent area, is sufficient to support the radial die load without permanent deflection or collapse. Since most threads on hollow parts are located near the ends of the parts, the adjacent area available for support becomes a significant factor in determining the ability to roll the thread. To establish the correct blank diameter, the degree of fullness required must also be taken into consideration since a small amount of over rolling tends to collapse the wall of the hollow parts.

In some cases, if a full thread is required all the way to the end, then a tapered blank O.D. may be required. In general, the blank diameter for solid threads is a good starting point for thread on hollow blanks when the thread length is $3/4$ or less of the thread diameter and the wall thickness is $1/5$ of the O.D.

For two-cylindrical die rolling systems, through feed rolling a thread on a steel tube—generally requires a ratio of wall thickness to outside diameter of approximately 0.3 depending on the material and depth of form. A three-die system greatly improves the stiffness of the system and allows the ratio to be reduced to be about 0.2 in.

To allow the rolling of thinner wall parts a mandrel is sometimes used. However, this introduces major internal diametral control issues. For effective internal support the diameter of the support mandrel should be such that when the blank is deflected inwardly during rolling, the deflection does not produce permanent radial deformation. This generally requires tight control of the I.D. and good concentricity of the bore to the blank O.D.

Use of a mandrel can produce another reverse result. As the mandrel rotates with the blank to support the area directly under the die, the blank is subjected to a rolling mill type of situation under which it tends to elongate circumferentially and increase in diameter. As a result, the final form diameter may be enlarged as well as the O.D., and the final I.D. will vary significantly depending on the length of the rolling dwell.

22.8.3 Roundness

The roundness required in a solid preroll blank is also related to whether or not the part will be rolled to a full form in the die. If it is, and there is typical spring in the rolling system, the rolled form will have an out of roundness substantially the same as the original blank. If the rolling system is very stiff and there is excessive blank out of roundness, there will be localized over rolling where the blank is high and it may cause localized flake and surface degradation.

If the part is to be rolled open in a stiff rolling system and two point measurements of out of roundness of the blank are all within the specified blank roundness tolerance, then when it is rolled, the pitch diameter will exhibit only a slight increase in out of roundness. However, there will be proportionately greater variations in fullness of the rolled form around the periphery.

It should be noted that all of the above information is based on two point measurement which is measured diametrally. For all rolling systems, except three-die machines or attachments, both the rolling process and the diametral measurement do not react to three point blank out of roundness. In fact, if the blank has significant three point out of roundness but is of constant diameter (isodiametral), then the rolled form could be perfect when measured by a micrometer but not fit into the "go" ring gage.

This characteristic can cause problems when rolling precision threads on centerless ground blanks which tend to have this type of roundness error. On the other hand, it is useful in roll forming fluteless taps and similar special isodiametral forms.

In cases of extreme blank out of roundness—such as are encountered in thread rolling hot rolled bar stock—an additional problem arises. Starting of the blank in either through feed or infeed rolling machines can be inhibited. This can cause deformed starting threads and, in extreme cases, may cause the blank rotation to stall with resulting damage to the dies.

22.8.4 Surface Finish

In general, the output quality of the form rolling process is not greatly affected by the surface finish of the input blank, provided that it has good continuity. In most rolling situations where the dies are ground, a blank with a turned surface finish of 125 μin or better will result in a rolled surface finish of 8 μin or better. However, any unfilled crest seam area will generally show a trace of the original blank surface finish.

When using the rolling process for roll finishing the blank surface condition becomes critical. Typically, the objective is to convert a turned surface into a very smooth mirror like surface with a surface finish of 4 μin or better. In those cases, the prerolled surface must be continuously turned or shaved without tears to a surface finish of approximately 32 μin. If turned, the operation should generally be performed at an axial feed rate of 0.006 in per revolution or less with a tool nose radius of $^1/_{32}$ in or larger.

22.8.5 Surface Continuity

Since rolling is a metal forming process, any blank surface diameter continuity breaks up the normal metal flow. The most common discontinuity is the existence of axial surface seams in the blank. When such a surface seam passes under the die, it tends to extrude a trailing flake from the seam which will significantly effect the rolled surface finish and to some degree the fatigue strength of the part. Even very tight surface seams which would not show up during a machining operation will become visible during rolling. For that reason, when rolling worms or splines which transmit torque during rotation or sliding, it is frequently necessary to remove the skin from any bar prior to rolling such a form directly on it.

22.8.6 Chamfers

Blanks are end-chamfered to enhance the start of through feed rolling applications, as well as to eliminate end cupping where the thread runs on and off the end of the part or bar. Chamfering also minimizes chipping failure in the dies. The chamfer depth is generally carried to a point slightly below the root diameter of the form to be rolled. As a rule—the lower the chamfer angle, the better. The depth and angle depend on the requirements of the thread and the material being rolled. Typically, a 30° chamfer angle with respect to the blank centerline is preferred. After rolling, this produces an effective crest chamfer of about 45° on most standard thread forms.

22.8.7 Undercuts

When rolling forms adjacent to a shoulder or head, undercuts are often specified to provide room for a full thread depth breakout. For infeed, tangential feed and radial feed applications, the breakouts should be at least $1^1/_2$ pitches long. In addition, the blank should be back chamfered to the same degree that the front end is.

22.8.8 Cleaning

For good surface finishes as well as the improvement of die life, it is imperative that blanks be cleaned prior to the rolling process. Tramp abrasives, embedded chips, and heat-treating oxides all have a deleterious effect on the process and the dies.

In addition, headed and drawn blanks which retain a significant amount of residual coating from the previous process may also require cleaning if the subsequent rolling operation requires a high penetration per contact. In the starting area, the residual lubricant tends to cause part stalling. This can cause material pickup on the dies and bad parts.

22.9 DIE AND TOOL WEAR

22.9.1 Die Crest Failure

Depending on the type of rolling process, a rolling die forming surface undergoes a variety of stresses depending on its location. In all rolling systems, where the rolling die has a protruding tooth or thread form that contacts the blank during the process, then every area of the crest of the die undergoes a repetitive tension, compression, and tension cycle. This stress pattern results from the radial force of the rolling contact between the die crest and the work as shown in Fig. 22.19.

Depending on the ratio of die diameter to part blank diameter, the hardness of the blank material at the point of contact, the crest shape, and the surface characteristic of the die, the compression stress on the die crest in the center of the area of the die to blank contact can reach up to five or six times the yield point of the material being formed. In rolling helical or annular forms, as the area of contact moves along the die crest, it undergoes the high tensile stress before and after the area of contact undergoes the high compressive stress. After many cycles of this repeated reversal from tension to compression, very small radial tensile fatigue cracks appear. As the circumferential crest area between the cracks makes subsequent repeated contacts with the blank, the resulting subsurface shear stress creates circumferential cracks which join at the radial cracks.

This causes the gradual spalling away of the die crests, which is the primary cause of most helical and annular die failures. A similar die tooth crest failure pattern occurs when rolling axial forms. However, it occurs tooth by tooth as each die tooth nears the center of the line of action and the die tooth crest is in direct compression while the area behind it is in tension. The effect of this compression tension cycle is also increased by a simultaneous tooth tip bending cycle.

As die crest failure begins, a typical thread, worm, or spline rolling die will continue to perform quite effectively with only minor die crest height loss. However, the root of the rolled form will take on a mottled appearance. Under static loading conditions, this root roughness will have no significant effect on the thread tensile strength or the spline tooth bending strength. However, it will somewhat degrade the fatigue strength since it provides stress points at which fatigue cracks may begin. Therefore, depending on the thread or spline application, the functional die life may vary widely, since the operations decision to stop using a set of dies is frequently based on the appearance of the rolled product and not the products' numerical tolerances.

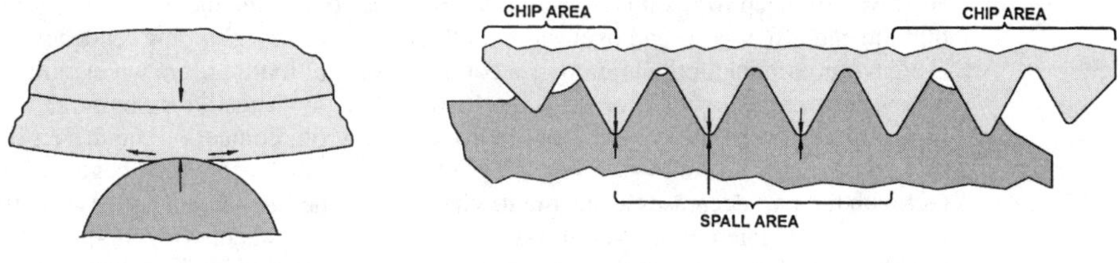

FIGURE 22.19 Die failure diagram

TABLE 22.2 Radial Feed Material Hardness and Die Life Relationship

Approximate part hardness–Rc	Approximate tensile strength 1000-PSI	Approximate crest fatigue life based on crest fatigue failure—parts thread rolled
5	82	480,000
10	90	240,000
15	100	120,000
20	110	60,000
25	123	30,000
30	138	15,000
35	153	7,500
40	180	3,500

Since crest fatigue failure is the predominant cause of form rolling die failures and the hardness of the rolled part has a disproportionately large affect on such die failure, it is important to try to establish an understanding of that phenomenon. Unfortunately, it is difficult to establish a precise relationship of die life to part hardness because there are so many other process variables which have significant affect on it. The following chart is based on an aggregation of published thread rolling die life estimates for typical radial and parallel feed, small diameter (below 3/4 in), and thread rolling setups. It provides a general indication of how rapidly rolling die life decreases due to die crest failure with an increase in the hardness of the part being rolled. If the blank hardness is increased by 5 points on the Rockwell C scale (or its equivalent in tensile strength) the die life is cut in half (see Table 22.2).

Although this data is presented to provide a general idea of the die life that commonly occurs in rolling standard 60° threads in a normal radial feed or parallel feed rolling machine, with average dies, setup and blank conditions, it is not uncommon to encounter process variable situations where the actual crest fatigue die life is cut in half or doubled. In general, crest spalling is slow to materialize and even after it begins, it progresses slowly. To put this in perspective, if 3/4"–10 TPI screw thread is rolled by a two-die radial-feed rolling machine with a 6-start die using 24 work revolutions and the dies produce 300,000 parts, then every area of the crest of the die ribs would have undergone approximately 1,200,000 tensile, compressive, tensile stress cycles before failure.

22.9.2 Die Tooth Chipping

Axial chipping of helical die teeth and circumferential chipping of axial die teeth is the second most common source of die failure. It results from bending loads produced on the flanks of the die by the workpiece moving past the die teeth. In thread rolling dies, this load is caused by the uneven side loads on the die teeth as they contact the start and/or end of the threaded area of the part as shown in Fig. 22.18. This type of chipping failure which can occur in all types of threadrolling dies generally starts toward the crest of the inner flank and ends near the pitch diameter on the outer flank. However, on deep forms it can extend down to the root of the die. Chamfering the end of the die so that the end effect is spread over more teeth can reduce this chipping tendency.

Circumferential tooth chipping is a common form of tooth failure when radial feed rolling splines or low helix angle involute forms. It results from the unbalanced forming force which creates a bending couple between the crest of the die drive flank tooth contact on the driven flank of the part and that on the flank also shown in Fig. 22.9. This force unbalance which begins at initial die crest contact with the part decreases as the die tooth comes to the line of centers between the die and part, and then increases again as the die tooth is about to leave contact with the part. This causes a failure which starts near the tip of the coast flank of the die and ends at or below the pitch diameter on the drive flank.

In forced through feed rolling of splines and other axial forms, the crest crumbling mode of failure is more common. On axial form forced through feed dies, extending the length of the penetration area distributes the forming load over a longer area of the teeth and can often eliminate this type of die failure.

22.9.3 Die End Breakout

When through feed rolling a form on straight cylindrical blanks or bars in cylindrical die machines, the part being rolled enters one end of the die and exits from the other end. As it enters the dies, it is gradually deflected radially away from the center of rotation of the bar by an amount equal to the radial die load divided by the spring constant of the rolling machine system. The range of this machine spring depends on the size of the rolling load and the stiffness of the rolling machine. When the machine stiffness is low and the load is high, this outward deflection can be as much as 0.030 in. In those applications, radial reliefs are added to the back ends of the dies to relieve this spring gradually. In deep forms on harder materials, if this relief is not adequate, radial and circumferential fatigue cracks begin to develop in the last few roots of the form. After a number of bars a major section of the back face of the die may then chip out. To compensate for this machine spring effect, blanks are fed through end to end where possible. This prevents the rolling dies from springing back in again at the end of each part or bar.

22.9.4 Surface Abrasion

Because of the trochoidal nature of the rolling contact when rolling shallow forms, the relative slip between die flanks and the part flanks is very small. Therefore, on hard dies with smooth flanks rolling clean blanks, abrasive wear is virtually never seen. Through feed rolling coarse threads on hot rolled bar stock which has a heavy residual oxide coating is one of the rare situations where abrasive wear of the die teeth may be encountered.

In roll finishing ball forms and other deep surfaces with high form diameter to blank diameter ratios, the sliding action is greatly increased. In those cases, if there is not adequate process lubricant or any tramp abrasive enters the rolling area, then the die surface will show gradual finish deterioration. The dies may then become unusable for further roll finishing purposes even though the die did not undergo any physical destruction or form shape change.

22.9.5 Die Material Effects

The degree to which die crest failure and die chipping occur is greatly affected by the selection of the die material and its heat-treating. Initially, flat dies were made of the same steel as ball bearings since ball bearings exhibit a similar spalling type failure to that which occurs on rolling die crests. As the rolling die requirements became more severe, various types of tool steels have become the most common die materials. M1, M2, and D2 tool steels or their close derivatives are currently used for the majority of rolling die applications. They are generally heat treated to hardness in the range from Rc-57 to Rc-62, depending on the expected type of die failure. For applications susceptible to chipping, die hardness is at the low end, but for most machine screw and similar rolling applications on low and medium hardness blanks, the higher end of the range is used.

When rolling harder parts, it is desirable to maximize the differential hardness between the rolling blank and the dies. Therefore, for rolling blanks heat-treated to Rc 32 and above, die materials which can be heat-treated to above Rc-62 and still maintain their toughness, are often used.

Because most helical and annular rolling dies are ground to their finished form, the use of such materials does not present a manufacturing problem. However, when rolling splines and other axial forms, the dies generally must be manufactured by hobbing or other metal cutting techniques. In those cases, D2 or other machinable, air-hardening tool steels are used.

22.9.6 Support Blade Wear

When radial or through feed rolling threads and other forms on a two-cylindrical die machine, it is necessary to support the blank prior to introduction and during rolling on or slightly below the die centerline. Where the thread or form is short and there is a cylindrical area adjacent to this thread which is precise, concentric, and with a length-to-diameter ratio of $1\frac{1}{2}$ or more, then the blank can be supported in a bushing directly in front of the dies. However, for longer infeed rolled forms and for the through feed rolling of threaded set screws, studs or bar, it is generally necessary to support the blank on the surface being rolled. This is generally done by a work support blade which is normally surfaced with carbide and lapped to a fine surface finish. If the rolling is done close to the centerline of the dies, the rubbing load is quite low and, if the process is lubricated, the wear will be quite slow and of a purely abrasive nature. In those cases, only relapping is required. However, for some blank materials which tend to weld to carbide, and for high speed through feed thread rolling, the abrasive wear leads to pickup on the blade and major surface erosion. In such cases, to provide the necessary wear resistance, a cubic-boron nitride or polycrystalline diamond surface is added to the blade. This can increase the blade life as much as ten times that of a carbide blade. These provide extremely long wear.

22.10 PROCESS CONTROL AND GAGING

22.10.1 CNC Radial Die Actuation

For rolling machines which require radial actuation of the dies into the blank to perform the rolling operation, the most common method is to hydraulically actuate the slide upon which one of the spindles is mounted. The actuation cylinder, piston and rod, which are connected to a moving spindle system are actuated under hydraulic pressure and move the die radially to penetrate the blank until a calibrated dial on an external piston rod at the opposite end of the cylinder bottoms out against the cylinder head on machine frame.

Another method used in hydraulically actuated radial feed die systems is to adjust the length of the piston rod connecting two opposing radially moving spindle systems. The hydraulic pressure actuates the cylinder so as to bring the dies toward one another to penetrate the blank as it remains on the rolling centerline. The stroke of the piston is manually adjusted so as to determine the depth of die penetration.

With conventional hydraulic actuation on these rolling machines, the radial penetration rate of the dies is controlled by a manually adjustable hydraulic flow control valve. However, since penetration per die contact is a function of both the die speed and radial penetration rate, to adjust that critical variable, one must adjust the die speed as well. For most ordinary rolling setups, the manual adjustment of size, radial penetration rate, and die rotation speed are quite adequate and cost effective.

However, when a rolling machine is changed over frequently, or is integrated into a multimachine part production system, it may become desirable to make all of these adjustments by CNC control. With that capability, many application programs can be readily stored for easy operator installation.

Electromechanical servo drives with ball screws, as noted in Fig. 22.20, are the most common type of CNC radial die actuation. CNC control, using a hydraulic servo system, has not provided comparable die position repeatability under varying rolling loads.

Generally, feed back from resolvers integral to the servomotors provided die position to the control system. Recently, the use of glass scales providing feed back of the position of the actuated spindle system housing has decreased the effect of spindle, machine frame, and ball screw spring; but still does not fully compensate for the effects of radial die load variation on machine accuracy.

Because of the higher cost of these electromechanical systems, as hydraulic servo systems with internal cylinder position sensing and improved servo valves have become less sensitive to variable rolling loads, they are being used more often to provide lower cost but acceptable CNC rolling size control systems.

FIGURE 22.20 Mc-4 Kine-Roller ® small radial feed rolling machine with ball screw CNC actuation.

22.10.2 True Die Position Size Control

Conventional CNC size control adjustment does not always result in an equal change in the size of the rolled part. When subjected to high variable rolling force on the dies, the deflection of the spindle system, the machine structure, and the actuation system can attenuate the affect of the CNC size adjustment. Therefore, for precision rolling applications with high, variable radial die loads which can result from blank hardness or blank diameter variability, there is in some applications, a significant difference between the actual operating surface location of the rolling die and the CNC input position. The only way to eliminate this difference is to sense true die position continuously during the rolling cycle. Systems are available to do this which sense the location of a cylindrical surface on the dies which is precisely concentric to the roll forming surface. This location information is fed back to a servo actuated radial die positioning system. Based on this information, by controlling the actual final dwell position of the dies, actual diametrial control of the rolling process and rolling size accuracy can be achieved under such high varying radial die loads.

This type of direct die positioning servo size control system requires special dies and the mounting of the die position probing system in the limited area above the die rolling area. It does not lend itself to quick changeover and is expensive and, therefore, not in common use.

22.10.3 CNC Rotational Die Matching

For the external adjustment of rotary match on cylindrical die rolling machines, mechanical or electromechanical rotary phasing units can be interposed between the spindle drive gear box output shafts and the spindles. When equipped with angular die position feed back of the dies with respect to one another, and servo actuation, this match adjustment can be CNC controlled. However, this type of CNC matching is normally done outside of the rolling cycle.

For dynamic rotary matching of the rolling dies within a single die rotation, some rolling machines are equipped with independent servo drives on each spindle and continuous rotational position feedback on each spindle. With CNC motion control, such machines can dynamically change die phasing during die penetration. This capability is sometimes useful when infeed rolling of axial involute forms with multirevolution cylindrical dies. However, it is not cost effective in most conventional cylindrical die rolling systems.

22.10.4 Radial Die Load Monitoring

Radial die load monitoring systems using strain gage load measuring techniques are readily available for application on almost all types of rolling machines. Since most machine structures are quite complex and the loading of the structure will vary dynamically during the rolling cycle, the ability to calibrate the strain gage read out directly to the actual radial die load is very difficult. In addition, the strain gage readout will vary depending on the relationship of its position to that of the dies in the die mounting area, the area of the dies in contact during the rolling cycle, and the structural location of the strain gage or gages. The measured radial die load accuracy will also depend on the response time of the measurement system.

During a typical rolling cycle, the radial die load begins gradually and peaks when the rolled form reaches 100 percent fullness. Then, if the die position dwells and releases, the measured load will quickly decrease and end. The peak measurement will be a good indication of the maximum radial die load. However, if there is even a small amount of overrolling, when the dies try to penetrate beyond the position required to create a full form, they will try to produce a wave of circumferential metal flow and the measured radial die load will rise very rapidly. Therefore, rolling load monitoring systems should include a load versus time readout to provide meaningful information.

To convert this unscaled data into accurate radial die load readings, such radial die load monitoring would require the system to be calibrated with a high force, precisely controllable load applying device in a very confined space. As a result, these systems are most often used for comparative rather than quantitative purposes.

22.10.5 Cylindrical Die Torque Monitoring

Measuring the amperage drawn by the spindle drive motors of cylindrical die rolling machines can provide a general indication of the actual rolling torque applied to the part.

Since the motor amperage input to torque output relationship is very nonlinear at low output, and gear and bearing losses also vary widely, it is not practical to get precise quantitative rolling torque data from this measurement. In addition, the response time and the overshooting tendency of conventional amperage meters further limit their usefulness.

There are some applications, particularly through feed rolling long bars with deep forms in cylindrical die machines, where measured rolling torque variations can provide an indication of process difficulty such as die failure, blade wear or over rolling due to system heatup. However, for radial feed rolling applications with their relatively short cycles, die torque variations are not generally useful as specific problem indicators.

22.10.6 Rolling Signature Monitoring

The ability to observe the radial die load versus time and torque versus time relationships during a radial infeed, scroll feed, or parallel feed rolling cycle above can be valuable from a process control viewpoint by creating a rolling cycle signature. Computerized monitoring systems which store and display those signature relationships can be used to detect such conditions as incorrect setup, hard or oversized rolling blanks, die crest failures, blade failure or pickup, incomplete rolling cycles, and any conditions where the rolling cycle signature of an abnormal rolling cycle on a bad part may differ significantly from that of a normal rolling cycle or a good part. Rolling signature monitoring systems can also be used to program die or tooling change and machine maintenance.

A number of systems which provide such rolling signature monitoring are available and have proven cost effective in applications where there are good strain gage mounting locations and the normal radial die load versus time pattern is stable.

22.10.7 Postprocess Feed Back Size Control

In continuous through feed rolling of thread or worm bar, or highly repetitive infeed and scroll feed rolling applications, machine heat up may cause gradual changes in the finished pitch diameter of the rolled parts. By the introduction of automatic pitch diameter measurement of the finished bars or parts, and the transmittal of any diametrial change to the size control system of a CNC cylindrical die rolling machine, it is possible to compensate for heatup of the rolling system. However, the time required for measuring of the form diameter and the costs of such systems limit their use.

22.11 PROCESS ECONOMIC AND QUALITY BENEFITS

22.11.1 General Advantages

Each of the above rolling systems has applications for which it is better suited than the others, and each has its limitations. Considering both, rolling, where it can produce the form required, offers some very significant advantages over competing processes. The most important of these advantages are

1. High speed
2. Material savings
3. Form repeatability
4. Superior physical characteristics
5. Improved surface conditions
6. East of setup
7. Low capital equipment cost
8. Low operating cost
9. Less waste material
10. System integration capability

22.11.2 High Speed

Unlike cutting operations, where the actual material removal rate is limited by the systems ability to remove heat from the interface between the cutting tool and the work, rolling has no such negative limitations. Experience to date indicates that the main factor that limits the speed with which the metal may be deformed is the rolling machine's ability to apply controlled forming energy. Because of this, rolling cycles rarely exceed 4 or 5 s and sometimes are as low as 100th of a second. The rolling system descriptions cite typical production rates with various types of rolling applications. In almost every case the speed is from 2 to 20 times faster than the comparable metal cutting process speed.

Achieving these production rates, requires blank or raw material which is rollable and can be automatically fed. When these two conditions are met, rolling becomes one of the fastest metalworking processes available to the manufacturing engineer.

22.11.3 Material Savings

Since rolling is a chipless process, it can provide rather large material savings when it replaces a cutting process. In rolling threads, worms, and serrations where the metal flow is generally outward and radial, the saving is quite obvious since the blank used to produce the form starts out considerably smaller than the final outside diameter (D_o). For such forms which are generally symmetrical above and below the pitch diameter (D_P), the blank diameter is close to the pitch diameter. Therefore, approximate material savings can be estimated by the following simple formula:

$$\%\ \text{Material Saving} = \frac{D_o^2 - D_p^2}{D_c^2} \times 100$$

For instance, by rolling a 3/8"–16 TPI thread rather than cutting, about a 20 percent material savings can be achieved. It should also be noted that the simple low carbon steels which roll well are from 5 to 20 percent less expensive than the steels which have sulfur and various other elements added to them to make them more easily machinable.

In through feed parts forming of ball blanks, bearing rollers, pins, and similar parts where the metal flow is inward and axial the finished part has about the same diameter as the original blank. In those cases, the material savings are a function of the metal's stretch. The latter can only be calculated by comparing the actual volume of the finished part to a cylinder of the same outside diameter and overall length from which one would have machined the part.

22.11.4 Improved Physical Characteristics

Because rolling is essentially a forging process, parts produced by rolling generally exhibit improved physical properties over parts produced by cutting directly from wrought material. These physical improvements fall into three categories; improved tensile strength, improved fatigue characteristics (endurance limit) and improved skin hardness. The improved tensile strength, which comes from rolling a part, is primarily the result of its increase in hardness. Therefore, threads and other forms rolled on low carbon steel will not exhibit any significant increase in static tensile strength. On the other hand those rolled in stainless steels, nickel steels, and other work hardening materials will exhibit improved static tensile stress which is a direct function of the integrated value of the increased cross section hardness which resulted for the cold work of rolling.

Since the increase in hardness due to rolling is a function of the shape of the form being rolled, the penetration rate, the previous cold work history of the material, as well as the chemical composition of the material, the exact amount of improvement in static tensile strength is difficult to predict. Therefore, it is not generally advisable to change peak static load safety factors when using rolled parts.

However, significant increases in fatigue life can be produced by use of rolling. Because a roll formed surface is work hardened, is normally very smooth and has flow lines following the shape of the form rolled, it has a tendency to resist the nucleation of fatigue cracks. In addition, because of the residual compression stresses in rolled surfaces, the magnitude of the stress reversals at the surface under certain load conditions is reduced. Because of this, on cyclical stressed fasteners, thread rolling has in many cases doubled fatigue life of high carbon steel bolts and, in work hardening or heat treated alloy materials, rolling has improved the fatigue life as much as 3 times.

Because fatigue life improvement is also a function of the shape of the rolled form, the penetration rate, material composition, and condition and a number of other variables, it is not possible to predict the improvement by calculation. However, representative examples described earlier give some specific idea of the general improvements achieved by rolling threads rather than cutting them. In general, this improvement occurs mostly in cases where the rolling takes place after heat treatment. However, in the bearing industry there have recently been some published cases where preforming of inner raceways by rolling prior to heat treating and finish grinding has produced some improvements of 15–30 percent in bearing life.

With a large body of qualitative data and a significant number of specific quantitative examples to further verify it, one can certainly be sure that worthwhile improvements in fatigue life can be achieved by correctly applying the rolling process. However, before making any changes in part design or loading to take advantage of the process, actual fatigue tests should be made.

22.11.5 Superior Surface Conditions

The basic nature of the rolling process is such that when properly applied, the resulting rolled surfaces normally have surface finishes ranging from 8 to 24 μin RMS. This is far superior to the surfaces that one could achieve with virtually any other conventional metal cutting process and equal to that produced by the better grinding and lapping techniques. In addition, the asperity shapes on a rolled surface of 24 μin are more rounded and have a lower surface coefficient than surfaces of the same microinch measurement on a finish produced by grinding. As a result, although the rolled surfaces may not always have the brightness of color which is produced by cutting they will be considerably smoother.

Because rubbing wear results primarily from asperity welding and pullout, the rounded over asperities produced by rolling tend to be more resistant to this class of wear. In addition, these rounded over asperities tend to produce a lower coefficience of friction when the rolled surface is in sliding contact with a mating part. This in turn results in decreased wear levels on the mating parts. In fact, the change from a hobbed worm to a rolled worm running with a molded nylon gear has frequently produced improvements in the wear life of the plastic gear of from 50 to 200 percent.

The decrease in the coefficient of friction on rolled threads provides another secondary advantage. A given amount of input torque on a rolled bolt can produce a considerably higher level of bolt preload then the same torque applied to a cut thread bolt. Finally, the very smooth surfaces produced by rolling with their associated skin hardness are more nick and corrosion resistant than surfaces produced by cutting and grinding techniques.

22.11.6 Process Repeatability

When the rolling process is correctly applied, its ability to repeatably reproduce the exact mating form on the surface of the rolled part is outstanding. Because of this characteristic where the external form is the critical design feature such as in threads or worms, the process is invaluable. Examination of a standard thread form on the first thread rolled by a new set of dies on a piece of conventional low carbon threaded rod and that of the 300,000th thread will show no measurable difference in form.

As the form gets deeper in proportion to the O.D., uniformity of material may affect the axial spring back of the rolled part and thereby affect lead, but even then the repeatability of the individual form will still be virtually unaffected. Naturally, the diameter of the form and its fullness may vary if the blank diameter is not controlled.

The repeatability characteristic is particularly useful in through feed roll forming and cutoff applications, since it permits the precise control of part length, except for breakoff area. Therefore, this characteristic makes it possible to produce various types of simple annular form blanks in soft materials at high speed with very repeatable volume and weight. Finally, when this form repeatability is coupled with good diametral control of the blanks to be rolled, the result is high precision parts at low cost. These range from class 3A to 5A threads to ball blanks which are round within 0.002 in.

22.11.7 East of Setup

Because the rolling process normally involves a single rapid pass of a blank through a set of dies, and these dies contain the precise form which is to be created by the process, the number of variables which the operator must control in the setup are quite limited. Therefore, in a well designed rolling machine, where the calibrated adjustments can be quickly and precisely made, the setup time required to change over to a different proven out job is generally less than one half hour. If an automatic feeder is involved, there may be an additional hour or two required. Even with this additional time, setup of a typical rolling system is generally easier and less time consuming than the setup of comparable form cutting machines, such as lead screw thread cutters, thread millers, hobbers, or thread grinders.

Once set up, a rolling die will last for many thousands of parts before it needs to be changed, whereas cutting tools rarely last into the thousands before they must be replaced or reshaped. Therefore, tool change frequency for rolling is far less than for comparable cutting processes.

22.11.8 Lower Capital Costs

The usual thread rolling machine consists of a stress support frame and base unit, die mounting units, a die drive system, a work support or feed device and a series of controls to coordinate these components. Since these components are designed primarily to produce the correct placement, interrelationship, and drive of two or three dies and support or feed the blank during rolling, the total amount of mechanism in most rolling machines is less than in comparable lathes, hobbers, automatic screw machines, etc. Furthermore, their overall productivity rates are frequently many times greater than the same size metal cutting machines. Because of this, even though rolling machines are built in much smaller quantities than the aforementioned cutting machines, the cost per unit of production per minute of a rolling machine is generally far lower.

22.11.9 Lower Operating Cost

The high speed of rolling also creates a proportionate decrease in the floor space requirement per unit of production per minute. Thus, where rolling can be used to do a job, the total space cost required will generally be considerably less for a given rate of production than the cost of competing chip making systems. Studies have also shown that the electric power costs of rolling in kilowatt hours per unit of production are lower than competing metal cutting process. Partly, this is due to the speed of the rolling process, and it is also due to the fact that metal cutting processes use a significant portion of the input power in overcoming the chip friction against the cutting tool.

22.11.10 No Chip Removal Costs

The absence of chips provides several major benefits of the rolling process. First, it eliminates the need for chip disposal. Second, if heavy coolant lubricants must be used in the comparable metal cutting process, those chips must be cleaned and the collected coolant/lubricant reused to decrease both the new coolant costs as well as the hazardous waste removal costs. Rolling also uses process lubricant coolants but much less per unit of production and the material is only carried out by the rolled part and is therefore, more easily blown off and returned to the rolling system. Finally, the absence of chips eliminates the need to find speeds, feeds, and other means that a metal cutting system must employ to prevent chips from disturbing the ongoing operation of the process.

22.11.11 System Integration Ability

Cylindrical die rolling machines are generally configured to simplify automatic loading and unloading. They are, therefore, well suited to integration into multiprocess automated systems. This characteristic makes rolling especially well suited to the high speed automated production of high volume shaft like parts which can be produced from cold formed blank. The system shown in Fig. 22.21 shows an integrated small motor shaft system which extruded the blank diameter, rolled the worm,

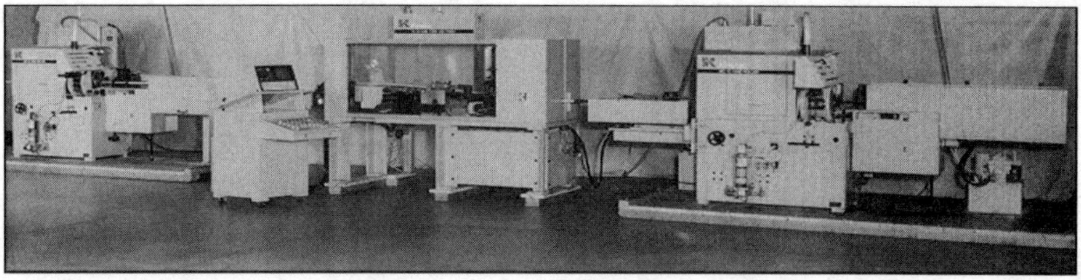

FIGURE 22.21 Integrated small motor shaft rolling and gaging system.

turned a support bearing diameter concentric to the worm, and then rolled the commutator and lamination grip knurls automatically and approximately 8 parts per minute.

It is typical of systems which are used to produce automotive alternator, starter, wiper, window lift, and seat adjuster shafts as well as a variety of automotive steering components and appliance shafts.

22.12 FUTURE DIRECTIONS

To meet the continuing need for manufacturing cost reduction, rolling is finding broader applications in five areas.

1. Near net shape blanks
2. Torque transmission elements
3. Combined metal cutting and rolling operations
4. Ultra high speed production of simple discreet parts or blanks
5. High speed roll sizing

In addition, the wide range of relatively unexploited rolling systems, supported by the necessary application engineering and die development, can provide a vast reservoir of special metal forming capabilities which can provide significant quality improvements, product cost reductions, and ecological benefits.

22.12.1 Near Net Shape Blanks

The need to reduce material costs, and minimize subsequent machining operations has in the past been answered by various types of casting, forging, and extrusion processes. Recently, the availability of rack and scroll type rolling systems with greatly increased die lengths, and with new die designs and materials, it is becoming more practical to roll multidiameter shaft blanks with relatively square corners and closer diametral tolerances. Such machines will find growing application for automotive and truck shafts up to 2 in. in diameter.

22.12.2 Torque Transmission Elements

The increased complexity of automotive power trains, steering and other actuated units and the use of higher speed components in appliances and other outdoor power equipment has greatly increased the number of areas where splines are used. The new forced through feed spline rolling machinery with automatic die match and CNC spline positioning and size control provides a cost effective means of producing the required high precision splines for this growing demand. It also allows the splines to be rolled on hollow shafts, which meets the rapidly increasing need for component weight reduction.

22.12.3 Combined Metal Cutting and Rolling Operations

The trend toward multitask machine tools has been exploited for many years through the use of thread and knurl rolling attachments on screw machines. As the capabilities of various other types of rolling attachments grow, one can expect a variety of new spline rolling, roll sizing, roll finishing, and flanging attachments to be used to further minimize the need for secondary operations. This will require the upgrading of the related turning machines to handle increased radial and axial rolling loads.

22.12.4 Ultrahigh-Speed Production of Simple Discreet Parts or Blanks

For almost 50 years the annular through feed forming and cutoff process has been used to create ore grinding mill balls and other large tumbling media. The same process has a long history of creating small spray can and cosmetic pump valve balls at production rates of over 2000 per minute with minimum grinding allowances. In addition, it has been used effectively for the production of small caliber bullet cores and projectile rotating bands. The success of these limited applications point to great potential for very high speed rolling applications where the product can be designed to accommodate the processes unusual cutoff characteristics.

22.12.5 High-Speed Roll Sizing

The use of rolling machines and attachments to roll finish a variety of high volume of automotive, appliance, and electronic parts has proven the speed and reliability of that process. However, it has not been practical to achieve sizing improvements at the same time. The development of ultrastiff die systems and of size control by the feed back of true die position, which can be precisely controlled, opens up new roll sizing opportunities. This new capability, combined with the use of annular preformed surfaces on the parts, now makes it practical to size bearing mounts and other precision diameters to total tolerance ranges of 0.0003 in at production rates of as high as 30 per minute.

22.12.6 Microform Rolling

As the miniaturization of electronic and medical products continues, the need to produce small, connecting devices and actuations, there is the need for rolling machinery that can create helical and annular forms as small as 0.010 in. in diameter.

CHAPTER 23
PRESSWORKING

Dennis Berry
SKD Automotive Group
Troy, Michigan

23.1 INTRODUCTION

Pressworking is one of the basic processes of modern high-volume manufacturing with many applications in the automotive, aerospace, and consumer products industries. Although processes ranging from coining to drop forging fall under the basic definition of pressworking, this discussion will be limited to operations in which relatively thin sheets of material are cut and/or formed between male and female dies in a mechanically or hydraulically powered press.

In very broad terms, pressworking is divided into two classes of operation—blanking and forming. In *blanking,* parts are physically removed from the feed stock, either in preparation for subsequent operations, or as finished components in and of themselves. Examples include literally thousands of products, such as washers, brackets, motor laminations, subassembly components, and the like.

Forming involves shaping sheet material into a three-dimensional object by forming it between mating male and female dies. Automobile body panels are a primary example, although an almost limitless variety of manufactured goods are produced from formed components. Blanked parts are very frequently also formed in subsequent press operations, and formed parts are also blanked after forming in some instances.

There are two basic press architectures in common use: the open-frame, or "C," type and the straight-sided type. Open-frame presses are further divided into *open back inclinable* (OBI) and *open back stationary* (OBS) types, and straight-side presses into those using solid columns and those using tie-bars. Each type of press, and each design within that type, has advantages and disadvantages which make it best suited to a particular range of applications.

Open-frame presses, for example, are not as rigid as straight-side presses and the resulting deflection tends to accelerate die wear and limit precision. Straight-side presses, on the other hand, tend to be more expensive for a given capacity and frequently offer more difficult access for die installation and maintenance. There is, however, considerable overlap in the capabilities of the two designs, and the choice is often dictated by factors other than press performance—capital budgets and availability within the enterprise's existing inventory being prominent among them.

Regardless of the type of press being used, material is typically fed either as precut blanks of various shapes, as individual sheets, or as a continuous strip from a coil. The forming operation itself may be performed in a single stroke of the press, progressively through multiple strokes of the same press, or through multiple synchronized presses connected by various types of automation.

Dies may be simple or complex, and often include secondary operations, such as welding, tapping, stud insertion, clinch nut attachment, and the like. Dies may be machined from solid blocks of steel, or cast over foam models of the finished part depending on the application and volume of parts to be produced.

Regardless of how they are produced, dies fall into several categories, including line dies which perform a single function, progressive dies which perform sequential shaping operations on a part as it is indexed through them while attached to the original feed strip, and transfer dies which perform multiple operations on a part as it is moved from die to die within the press. Die and tool design are complex subjects which are covered in a section of their own below.

23.2 COMMON PRESSWORKING PROCESSES

23.2.1 Blanking

Blanking is the process of producing a shaped part from a solid sheet of material using various combinations of punches and dies or shears and slitting tools. In open blanking smaller shapes are simply cut with shears or rotary cutters from larger sheets or strips of material. In closed blanking, a part is cut from the interior of a feed strip or sheet, washers providing a typical example although much more complex shapes easily can be produced with this process (see Fig. 23.1).

Closed blanking typically uses a punch and die arrangement in which the punch contacts the sheet producing elastic deformation and finally shearing as the force exceeds the material's shear strength. The remaining material in the original stock tends to spring back and grip the punch after the blanked part is removed, necessitating the use of a stripper to keep it in place as the punch is withdrawn.

Blanking punches need to be slightly smaller than the dimension of the finished part; the exact amount of undersizing depends on the thickness and strength of the material being punched. This is related to the "shear and tear" phenomenon, in which the punch shears through the workpiece material for roughly one-third of its thickness, and the material then fractures or tears away leaving a rough edge. Where a precision hole is required, it is frequently necessary to perform two operations, one with a "roughing" punch, and a second with a "shaving" punch to produce a smooth, precisely sized hole.

Calculating the force required for a shearing operation is done by multiplying the perimeter of the blanked shape by the thickness and tensile strength of the material being blanked. A number of techniques are available for reducing the amount of force required for blanking operations, most of which involve altering the punch to provide some sort of progressive engagement of the workpiece. Blanking punches are commonly beveled, pointed, or produced with a hollow "V" face to control cutting forces. Any treatment of the punch face should provide balanced cutting forces for best results.

Fine or *precision* blanking is a specialized blanking process in which the workpiece material is clamped around its entire perimeter before being processed. This creates a pure shear condition in which there is no tearing of the material, producing a smooth-edged, precisely sized part with no fracturing. Fine blanked parts can frequently be used as-is, with little or no subsequent processing required.

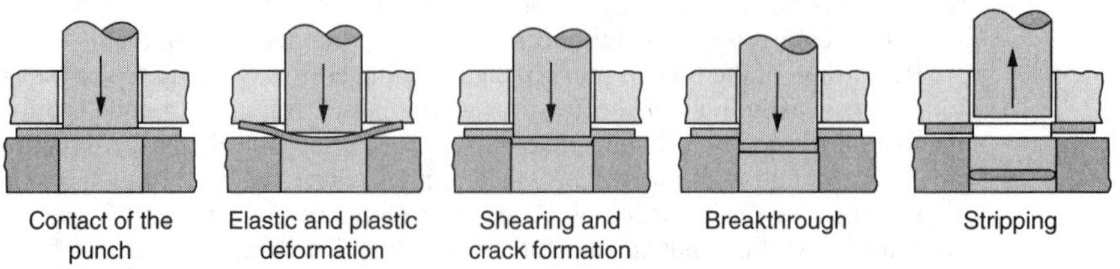

FIGURE 23.1 Blanking.

23.2.2 Forming

Forming is the process of shaping a workpiece between male and female dies to produce a three-dimensional part that is still attached to the parent blank material. There is no shearing or cutting involved in a forming operation; rather the material is bent, stretched, and/or flowed into the final configuration. Formed parts are frequently blanked after forming to produce the required final shape.

A variety of different forming techniques are commonly used based on tension, compression, bending, shear loading, and various combinations of these operations. Perhaps the most common of these operations is *drawing,* in which the workpiece is subjected to both tension and compression while being drawn into a die cavity by a die punch. While the term is applied with varying degrees of precision to many different forming processes, a true drawing operation has the following characteristics:

- The workpiece material is restrained by various devices generically called blankholders.
- The workpiece material is drawn into the die cavity by a punch.
- The workpiece material undergoes both tension and compression during the operation.

Drawing is very commonly used to produce cup-shaped parts, although very complex profiles are possible, automotive oil pans being a good example. Many variations of the basic process are in use, including those in which the punch is fixed and the die cavity moves, and those with multi-part punches designed to produce multidiameter parts, flanged parts, or parts with annular rings on their sidewalls.

Depending on the application, the blankholder may simply clamp the workpiece material to the die surface to prevent wrinkling during the forming operation, or it may exert a programmed amount of force on the workpiece to control its behavior during the operation. In the latter case, the work-piece material may be allowed to flow into the cavity during part of the stroke and be firmly restrained during other parts of the stroke. Such an arrangement permits the amount of stretching and metal-flow created by the forming process to be precisely controlled.

Hydroforming is also a variation of drawing in which hydraulic pressure is used to force the workpiece material into the die cavity to produce very complex shapes. Other commonly used drawing processes include

- Single and multiple stroke deep drawing
- Flanging
- Wrinkle bulging
- Stretch forming
- Embossing
- Roll and swivel bending

Crash forming is another widely used operation which involves bending a blank over or in a die with a single stroke of a punch. Many brackets and similar "U"-shaped parts are crash formed (see Fig. 23.2). The process is particularly useful for parts which do not require extreme dimensional precision.

Restriking is a technique used to overcome the tendency of sheet metal parts to spring back after being formed due to residual stresses. Essentially, the part is formed twice using the same tools, or struck a second time while constrained in the die. In either case, the restrike operation is intended to improve the geometry of the finished part by overcoming spring back.

Piercing is another major category of press operations. As the name implies, piercing involves the use of a punch and die to create a hole in a workpiece. It may be performed as a primary operation, or as a secondary operation during forming or blanking. Secondary piercing operations are often performed with cam-operated accessories which are actuated by the press ram. The major difference between piercing and blanking is that the piercing operation produces a feature on the finished workpiece, while in blanking with a punch, the cut-out form is the workpiece for subsequent operations.

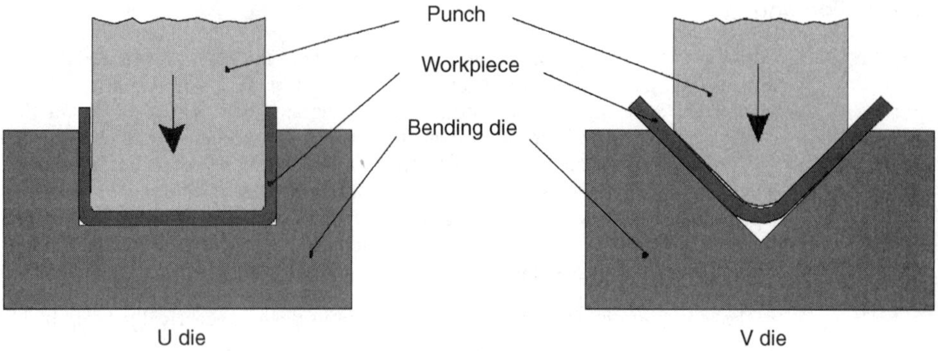

FIGURE 23.2 Die bending.

Hemming and *flanging* are processes for joining two or more sheet metal parts, inner and outer automotive door panels being a good example. In flanging an offset area is produced around the periphery of a workpiece which mates to a similar flanged offset in a mating part.

The mating flanged sections are then bent back upon themselves in the hemming operation to produce an assembly with a smooth, rounded edge in which the two parts are mechanically joined. Hemming and flanging are widely used in the automotive industry to manufacture *closures,* which include doors, hoods, deck lids, and similar components.

Secondary in-die operations while not strictly part of the forming process, are frequently incorporated into forming operations. These can include welding of various kinds, tapping, stud insertion, clinch-nut insertion, and a broad range of other operations.

Opportunities to incorporate secondary operations are limited only by the physical constraints of the press and tooling, and the imagination of the process designer. The ability to perform multiple operations in a single stroke of the press is one of the major economic advantages of metalforming as a production process.

23.3 TOOLING FUNDAMENTALS

Common tooling terms include

Blankholder. A device used to restrain a blank during the forming process. The blankholder prevents wrinkling and controls the amount of stretching during a forming operation.

Bolster. A thick plate attached to the press bed to which the die is mounted. Bolsters typically have precisely spaced holes or T-slots to facilitate die mounting.

Die. Used to describe both the complete set of tooling used for a forming operation, and the female portion of the punch/die set.

Die pad. A moveable component of a female die which functions to eject a part. Die pads may be actuated by springs, hydraulics, or mechanical means.

Die set. An upper and lower plate which are able to move on guides and bushings to keep them in alignment as the press strokes and to which the punch and die are attached. Die sets are available in a wide range of sizes and configurations.

Die shoes. The upper and lower plates of a die set, to which the actual punch and die are attached. The die shoes are attached to the bolster and ram face of the press.

Die space. The space available in a press for mounting dies. Die space includes both the vertical distance between the ram and bolster, and the available mounting space on the bolster plate itself.

Die spring. A heavy coil spring used to actuate various moving components of a die set.

Draw bead. A projection on the workpiece clamping portion of a die or blankholder which serves to help restrain the movement of the sheet during the forming operation.

Draw cushion. A hydraulically or pneumatically powered blankholding device used in deep drawing operations to prevent winkling and also as a workpiece ejector.

Ejector. a pneumatic, hydraulic, or mechanical device to remove a workpiece from a die after forming.

Guide pins/bushings. Precision ground pins and bushings that are attached to die shoes to guide them during the forming operation. Pins and bushings must be fully engaged prior to any forming to ensure proper alignment.

Heel block/plate. A block or plate usually attached to the lower die which engages and aligns the upper die before the guide pins enter the guide bushings. The heel plate compensates for any press misalignment and minimizes deflection of punches, cams, or other die components.

Knockout/shedder/liftout. A spring-loaded pin used to separate a part from the tool after forming. Shedder pins are often inserted into punches or dies to remove parts that tend to stick to the oil film coating them.

Line die. One of a series of dies used to produce a finished part by manually moving the workpiece between them. Multiple line dies may be installed in a single press, or in a series of presses.

Nitrogen die cylinder. A cylinder charged with nitrogen gas used in place of springs or die cushions where high initial pressure is required during the forming operation. Unlike springs or pads, nitrogen die cylinders provide high pressure from the moment of contact, making them very useful for certain drawing and forming operations.

Pad. A generic term for any component or die feature that provides pressure to hold the workpiece during the forming or drawing operation.

Pilot. A component used in transfer and progressive dies to ensure proper workpiece positioning. The pilot mates with a locating hole in the workpiece or blank to locate it.

Pin plate. A plate used to protect the working surface of a cushion or lower slide from wear caused by pressure pins.

Pressure pin. A hardened pin attached to a moving die component that transfers force to the pressure plate.

Pressure plate. A plate mounted under the bolster and supported by hydraulic or pneumatic cylinders. The pressure pins bearing against the pressure plate provide uniform pressure over the entire press stroke.

Progression. The uniform fixed distance between stations in a progressive die. The pilot (which sees) ensures that the stock is indexed accurately according to the individual die's progression.

Progressive die. A die with multiple stations arranged in a linear geometry. Each station performs a specific operation on the workpiece which remains attached to the parent strip until the final blanking or parting operation is performed. The distance between the individual dies is the progression (which sees) and a pilot is used to assure accurate indexing.

Punch. The male member of the punch/die combination. Depending on the application, the punch may be attached to the bolster or the slide.

Slide or ram. The moving element of a press, typically located at the top.

Slug. The scrap material produced when a hole is punched.

Stripper. A die component that surrounds a punch to literally strip away the workpiece material that tends to spring back and cling to the punch after a hole is pierced. Strippers may be either fixed in place or moveable depending on application requirements.

Transfer die. A die with multiple stations arranged in a nonlinear geometry. Workpieces are transferred between stations by various means including robots, pick-and-place mechanisms, or manually.

Vent. A small hole in a punch or die that allows air to enter or escape to prevent the formation of air pockets or vacuums which might interfere with proper die operation.

23.3.1 Press Tooling Components

The foundation of most production dies is a standard die set, which is available in a broad range of sizes and configurations. Major differences in die sets include size, number, and location of guide pins and bushings, plate thickness, and mounting arrangements to attach the die set to the bolster and slide.

Die sets are also rated according to the precision with which the upper and lower plates are guided, which determines how accurately the punch and die will mate in operation. Obviously, higher precision in the die set translates into higher cost, and it is not generally advisable to specify a more precise unit that is actually required by the application.

Guide pins and bushings are available in two basic types, matched or ball. Matched bushings are, in effect, journal bearings which slide along the pin and depend on a close fit for correct alignment. Ball bushings depend on the rolling contact between hardened steel balls and the precision ground surface of the pin for their alignment. The choice is influenced by many factors including press speed, required life of the die set before bushing replacement, maintenance requirements, and cost.

Springs and nitrogen cylinders are used in dies to either produce motion, or to provide a counter force during some portion of the forming or drawing operation. Metal springs are relatively inexpensive, but they require careful sizing and design to ensure proper operation and maximum life.

Nitrogen cylinders, while more costly than springs, can be more precisely controlled and have the additional advantage of supplying full force over the entire stroke. They can also be connected via a manifold to provide uniform pressure at multiple locations which can help to keep forces balanced in large dies.

Resilient elastomeric springs and bumpers are also available which provide cost and performance advantages in certain applications.

Pads and strippers are incorporated into the structure of dies and punches wherever possible to reduce complexity and cost. Their function is to control workpiece movements, and to strip workpiece materials away from punches. Careful design frequently permits a stripper or pad to perform more than one function, for example both holding a workpiece and stripping it, or locating a blank and controlling drawing force.

Many other standard components are commonly used in dies, including knockouts, stops, kicker and shedder pins, pushers, guides, heel plates, wear plates, pilots, retainers, punches, and buttons. Each of these items is, in turn, available in a variety of different sizes and designs to suit various application requirements.

Dies and die components are often given various surface treatments to improve physical properties such as hardness and wear resistance. Among those most commonly used are nitriding, vapor or plasma coating with materials such as titanium nitride (TiN), and hard chromium plating. Surface coatings may be applied to an entire punch or die, or only to selected areas depending on the specific application of the tool.

23.3.2 Die Materials

Historically, dies have been produced by highly skilled machinists who cut the desired profiles from homogeneous blocks of tool steel using various machines and hand operations. This is a costly and time-consuming process, but the resulting dies have excellent properties including wear resistance, impact/shock resistance, and repairability.

TABLE 23.1 Properties of Homogeneous and Cast Tool Steels Commonly Used in Die Manufacture

	Homogeneous tool steels							
	Oil/water hardening		Air hardening				Cast tool steels	
	W2	0-6	S-7	A-2	D-2	M-2	A2363	S7
Alloy content	Low	Med	Med	Med/High	High	High	Med/High	Med
Wear resistance	Fair/Poor	Fair	Fair/Poor	Good	Very Good	Best	Good	Fair/poor
Toughness	Good/ Fair	Fair/Poor	Best	Fair	Poor	Poor	Poor	Fair/good
Machinability	Good	Good	Fair	Fair/Poor	Poor	Poor	Fair/Poor	Fair
Distortion in heat treat	High	Med	Low	Low	Low	Low	Low	Low
Resistance to decarburization	Good	Fair	Poor	Poor	Poor	Poor	Poor	Poor
Flame hardenability	Good	N/A	Good	Fair	Poor	N/A	Fair	Good
Depth of hardness	Shallow	Med	Deep	Deep	Deep	Deep	Deep	Deep
Weldability	Good	Med	Fair/Poor	Fair/Poor	Poor	Very Poor	Poor	Poor
Cost index (compared to D2)	0.35	0.87	0.79	0.75	1.0	1.4	1.1	1.1

More recently, the technology has been developed to produce cast dies from a range of materials offering adequate performance for many applications. In this process, a mold is made around an accurately machined plastic foam replica of the part which vaporizes when the hot metal is poured into the mold. The resulting die requires relatively little machining to finish and is, therefore, substantially less expensive to produce.

The casting process is best applied to dies that are relatively large and not subject to high stress, as it provides the optimum cost benefits under these conditions. It must be noted, however, that cast dies are extremely difficult to repair if damaged, and require substantially more maintenance than tool steel dies. They also offer less wear resistance and less shock/impact resistance, making them more suitable for lower volume applications.

Table 23.1 gives the main properties of homogeneous and cast tool steels commonly used in die manufacture.

In addition to dies, various other components are made from a variety of materials which may be either cast or machined from homogeneous stock. Materials for these components include all of the above tool steels and other materials ranging from gray, pearlitic, nodular, and alloy cast iron to hot and cold rolled low-carbon steel and various alloy steels.

23.3.3 Line Dies

Line dies are the simplest method for accomplishing multiple operations on a blank. Each line die is independent of the others, and in many cases each die is installed in a separate press. The workpieces are manually moved from die to die until all operations have been performed (see Fig. 23.3).

Despite their simplicity, line dies are capable of performing very complex and precise operations on a workpiece. A part may be blanked, drawn, formed, pierced, trimmed, flanged, and hemmed in a set of line dies just as precisely as the same series of operations could be performed in a much more complex transfer or progressive die. The difference is one of production rate and overall cost per part.

The decision to use line dies rather than more complex transfer or progressive dies is based on several factors: part size, production volume, and equipment availability being among the most critical. For very large parts, line dies may be the only choice since there is not likely to be enough room in the die set, or on the bolster, for more than a single die. They also tend to be more economical than transfer or progressive dies, making them a good choice for lower volume production. Finally, they may permit the economical use of smaller presses, particularly where such equipment is

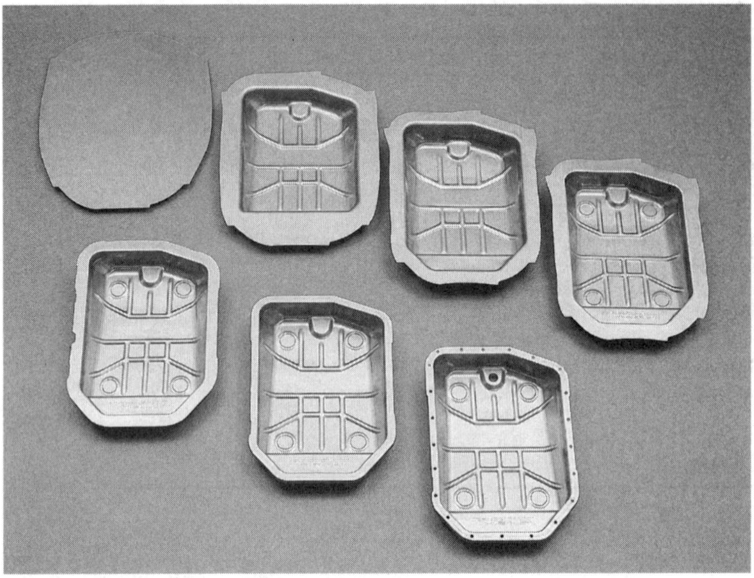

FIGURE 23.3 Line dies—production steps for the manufacture of an oil pan.

already in inventory, since the tonnage requirement for each operation can be matched closely to press capabilities.

23.3.4 Transfer Dies

A transfer die performs multiple operations on a preblanked workpiece as it moves from die to die either within a single press, or between multiple presses. Between-die part handling is automatic in a transfer die production process, and may be accomplished by robotics, pick-and-place automation, or a variety of "hard" automation.

Because the workpiece is preblanked, transfer dies can perform operations that are not possible with progressive dies, such as working on the full perimeter of the part, or forming features that require tilting or inverting the workpiece. It is also somewhat simpler to incorporate secondary operations like clinch nut insertion, or threading into a transfer die operation. Transfer die operations also tend to be more economical of material since there is no requirement for a carrier strip which becomes scrap.

On the other hand, transfer dies tend to be more costly than progressive dies. Transfer dies are best suited to parts that do not nest well, or require a precise orientation of the grain structure of the parent metal. Because of the cost, transfer dies are not generally suitable for low-volume applications.

23.3.5 Progressive Dies

A progressive die performs multiple operations on a workpiece which remains attached to a carrier strip of parent metal until all operations have been performed. The workpiece moves from die to die in a uniform progression which is typically regulated by a pilot of some type to ensure accurate alignment. Progressive dies are always used in a single press, somewhat limiting the maximum part size that can be processed with this method (see Fig. 23.4).

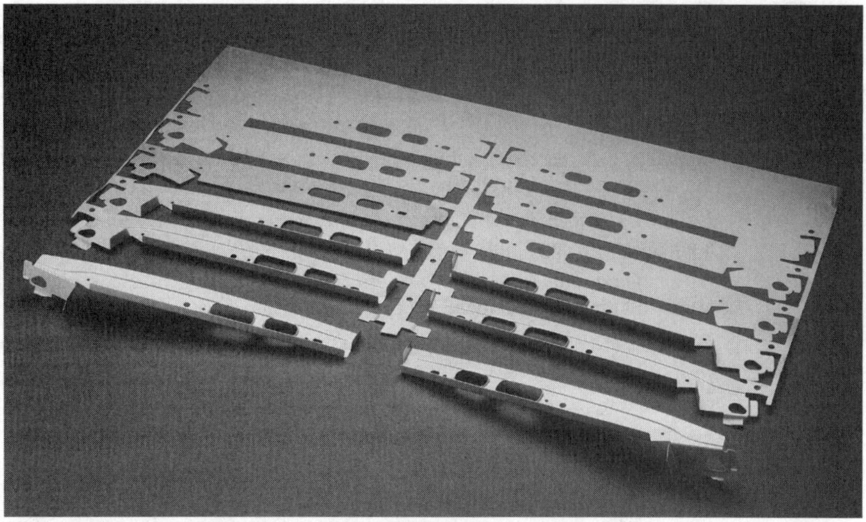

FIGURE 23.4 Progressive dies—reinforcing part of a car produced in a strip.

For parts that nest well, progressive dies can provide very low per-part cost because of their ability to operate at high press speeds and relatively short strokes. In some cases, highly nestable parts can eliminate the need for a carrier strip entirely, creating even greater production economies.

Progressive dies are not well suited for most secondary operations, and require high-precision carrier strip indexing systems which add to cost and maintenance requirements. They also can be difficult to use in operations requiring substantial metal-flow during forming. In most cases the carrier strips are good only for scrap.

The decision to use either a transfer or a progressive die can be a very complex one. Factors to be considered include die cost, part size and complexity, secondary operation requirements, production requirements, the number of parts to be produced, and equipment availability.

23.4 PRESS FUNDAMENTALS

Common press terms include

Adjustable bed/knee. A bed attached to an open-frame press that can be moved up and down with a jackscrew. The term is also used to describe a moveable bed installed on some straight-side presses.

Bed. The stationary base of a press to which the bolster, or sometimes the lower die, is attached.

Bolster. A thick plate attached to the press bed to which the die is mounted. Bolsters typically have precisely spaced holes or T-slots to facilitate die mounting.

Capacity. The force a press is rated to deliver at a specific distance above the bottom of the slide stroke.

Closed/shut height. The distance from the face of the slide to the top of the bolster when the slide is fully down and the bolster, if adjustable, is fully up. The maximum amount of space available for the die set and any auxiliary components within the press. Also called "daylight."

Clutch. A device to couple the flywheel to the crankshaft in a mechanical press.

Crown. The topmost part of a straight-side press structure. The crown usually contains the drive mechanism on a mechanical press, or the cylinder or cylinders on a hydraulic press.

Die space. The space available in a press for mounting dies. Die space includes both the vertical distance between the ram and bolster, and the available mounting space on the bolster plate itself.

Flywheel. A massive rotating wheel used to store kinetic energy. When the clutch is engaged the kinetic energy from the flywheel is transmitted to the crankshaft.

Frame. The main structure of a press. The frame may be a monolithic casting, a multipart casting, a weldment, or any combination of these.

Gibs. Guides that maintain the position of a moving machine element. Gibs are normally attached to the vertical members of a straight-side press.

Platen. The slide or ram of a hydraulic press; the moving member of such a press.

Press. A machine with a stationary element, the bed, and a moving element, the slide or ram, reciprocating at right angles to it designed to apply force to workpiece materials placed between the bed and ram. When used in conjunction with dies, a press is capable of forming metal and other materials into very complex, three-dimensional shapes.

Ram/slide. The moving component of a press.

Stroke. The distance a slide moves between full-up and full-down positions. Also one complete movement of the slide from full-up to full-up, used as a measure of press speed expressed as strokes per minute.

Throat depth/gap. The distance between the frame and the centerline of the slide in an open-frame press.

Tie rod. Steel rods with threaded ends used to prestress the vertical frame members of a straight-side press, or to prevent deflection in an open-frame press.

23.4.1 Press Architectures

Presses are generally defined by the basic architecture of the machine, the means used to generate force, and the amount of force available.

The basic function of a press frame is to absorb the forces generated during the pressing operation and maintain the precise alignment of the dies. The frame also serves as a mounting for the drive system and various peripheral devices necessary to support production. While most presses are built to operate in a vertical orientation, horizontal models are also available for special purpose applications.

There are two basic frame architectures, open frame and straight side. Within these basic architectures are a wide variety of subtypes and variations that have been developed to meet specific process and production requirements. Each type has its own advantages and limitations, which will be examined briefly below.

23.4.2 Open Frame Presses

Open frame presses, also known as *gap frame, C-frame,* or *open front* presses, consist of a frame shaped roughly like the letter "C," with a bed at the bottom and a guide and support structure for a moving ram at the top. The frame may be a large casting, or a steel fabrication that is either welded or riveted. The bed, in turn, may be an integral part of the frame, or it may be moveable to adjust the distance between the ram and bed.

One of the most common open frame press designs is the OBI press in which the frame is mounted on a pivot in the base which permits it to be tilted off vertical to facilitate stock handling or scrap removal. Other common types are the OBS, the adjustable bed stationary (ABS), and a variety of knee-type units with various systems of table adjustment and support.

The major advantages of an open frame press design are economy of construction and unhindered access to the die area. Inclinable models and those with moveable beds or tables also offer a great deal of versatility, making them particularly useful for short run production or job shop applications.

Open frame presses are available with force capacities ranging from a ton (8.9 kN) for small bench-type units, up to about 450 tons (4000 kN). The limiting factor on the size of open frame presses is the lack of stiffness inherent in their design. In operation the frames tend to deflect in both a linear and an angular fashion as they are loaded. The angular deflection component is particularly undesirable because it tends to cause misalignment between punches and dies which leads to rapid wear, loss of precision, and even tool breakage.

Various means are available to counteract deflection in open frame presses including the installation of prestressed tie rods spanning the front opening to connect the upper and lower ends of the frame, and the use of tie rods inside tubular spacers to prevent misalignment caused by the tie rod prestress. The drawback to these methods is that they largely negate the die access advantage which is one of the most important benefits of the open frame design. In general, when an open frame press is not sufficiently stiff for a particular application the best course is to either move the work to a larger press, or move it to a straight-side press.

One additional limitation of the open frame design is the fact that such presses are generally limited in practice to the use of single dies. This is a result of several factors including the lack of stiffness and the typically small force capacity and die area of open frame presses.

23.4.3 Straight-Side Presses

A straight-side press consists of a bed and a crown which are separated by upright structures at each end of the bed. The bolster is attached to the bed, and a sliding mechanism moves up and down on gibs and ways attached to the vertical members. The drive is typically mounted on the crown in a straight-side press.

Straight-side presses may be monolithic castings, or fabrications of castings or weldments held together with tie rods, welds, or mechanical keys and fasteners. Larger presses tend to be fabricated because of the difficulty of transporting large castings or weldments from the press builder to the user's location.

Straight-side presses have two major advantages over openframe designs. First, they can be very large. Mechanical straight-side presses with capacities up to 6000 tons (53.376 mN) have been built, although the upper limit for standard presses tends to be about 2000 tons (17.8 mN). Straight-side hydraulic presses have been built with capacities up to 50,000 tons (445 mN), but these machines are generally used for specialized forging applications rather than traditional forming and drawing operations.

The second advantage of straight-side presses is that they tend to deflect in a much more linear fashion under load than an open-frame press. They also tend to deflect much less for a given load. Taken together, these two characteristics of a straight-side press translate into greater precision and longer tool life because linear deflection does not cause punch and die misalignment in the way angular deflection does.

Linear deflection is a result of the balanced geometry of the straight-side design, and the fact that the slide can be guided at all four corners during its entire stroke. As long as the press loading is applied symmetrically a straight-side press will deflect symmetrically with virtually no effect on punch and die alignment.

The slide, which is normally a box-shaped weldment, may be connected to the press drive system at a single point, or at multiple points. Typically, in mechanical presses, this connection is via one or more connecting rods that are driven by a crankshaft in the crown.

Other systems include gear drives, and a variety of linkages designed to produce controlled slide motion. Bottom-drive presses are also available. Hydraulic presses use hydraulic cylinders to supply the required force, and may also be single or multiple-point designs.

Many small straight-side presses have a single point connection between the drive and the slide. Thus, any resistance not centered directly below the point of connection will tend to tilt the slide and cause misalignment.

Presses using two or more connections are called multipoint presses, and they provide substantially more ability to compensate for uneven loading of the slide since the load is spread among the connecting points. Such presses are normally of larger size than single-point units and, therefore, more costly. Multipoint connection is recommended for progressive and single-press transfer die operations, although they can be accommodated on single-point presses if carefully designed.

Some presses are designed with multipart slides which are actuated by different connections to the crankshaft. These are most commonly of the box-in-a-box type with a center slide surrounded by a hollow rectangular secondary slide. Multiple slide presses are designated at double or triple action, depending on the number of slides present.

The columns or vertical members of a straight-side press may be monolithic castings, mechanically fastened multipart castings, or weldments. Often tie bars attached to the base and crown are used to compress the vertical members and provide uniform, adjustable resistance to vertical deflection.

The gibs and ways are designed to prevent tipping of the slide, so the resultant misalignment largely is a function of the precision with which they are fit, and the length of contact. Any misalignment, of course, will result in wear to the components involved and a loss of precision in the operation being performed. Since the fit between gibs and ways is never perfect, special care must be taken in designing dies for a single-point press to ensure uniform loading.

The gib and way configuration also contributes to slide stability, and there are many variations. Square, "V," box, and 45° gibs are all used, as well as various roller systems. Both six- and eight-point-of-contact gibs are used, with the eight-point system being preferred for larger presses with high operating forces and on tie bar type presses. Six-point gibs are more commonly used on solid frame presses.

23.4.4 Press Drives

Nearly all mechanical presses utilize an electric motor-driven flywheel to store the energy necessary for the forming or drawing operation. The flywheel is connected to the slide either directly, or via single or multiple gears.

In direct systems, the flywheel is connected to the crankshaft, which may run either from side to side or from front to back depending on the machine design. Direct drive presses are generally used for lighter operations and applications requiring very high speed. Because of the limited amount of energy available, they are generally best applied to operations where the maximum force is needed near the end of the stroke.

Single or multiple gear reductions between the flywheel and the slide provide substantially more usable energy, but at the expense of operating speed. Single-reduction presses normally operate at 150 strokes per minute or less.

Multiple gear reduction systems are very commonly used on large presses with massive slides, and for forming and drawing of heavy gauge workpieces. Multiple-reduction presses normally operate at 30 strokes per minute and less.

Gear reduction presses may use a single set of gears on one end of the main shaft, a so-called single-end drive, or a set of gears at each of the shaft in a double or twin-end drive. Double or twin-end drives are often used in very large, or very long, narrow presses to reduce the amount of torsional deflection in the shaft by applying energy simultaneously at each end. Quadruple-drives consisting of two double-drive systems are also used in very large presses.

Regardless of the drive type, the flywheel is connected to the drive or crank shaft with a clutch and brake of some type. Clutches are divided into full-revolution and part-revolution types, and into positive engagement and friction types.

A full-revolution clutch cannot be disengaged until the crankshaft has made a full revolution. A part-revolution clutch is one that can be disengaged at any point in the cycle regardless of how far the crankshaft has moved. Part-revolution clutches are much safer than full-revolution systems and are the preferred system in all but a few very specialized applications.

Positive engagement clutches provide a mechanical connection between the driven and driving components while engaged, typically through the use of multiple jaws, keys, or pins. Friction-type clutches utilize friction materials pressed into contact with springs or hydraulic or pneumatic cylinders. Friction clutches are further subdivided into wet, running in oil, or dry types.

Eddy current clutches and brakes are often used on very large presses. These systems make use of electromagnetic phenomena to generate clutching and braking force without any physical contact between the driven and driving elements. Eddy current clutches and brakes are generally part of an integrated adjustable speed drive system.

The brake is used to stop the moving components of the system at the end of each stroke. This is especially important in presses operating in single-stroke mode. Brakes are normally mechanically actuated and designed to fail-safe.

The drive system is most commonly located at the top of the press, in the crown. However, bottom-drive presses are also available. Their major advantage is that the drive mechanism is located in a pit, which minimizes the amount of overhead space required for the press installation. Bottom-drive systems use linkages to pull the slide down, whereas top-drive systems push the slide down, making them somewhat simpler mechanically.

23.4.5 Hydraulic Presses

In a hydraulic press the force is supplied by one or more hydraulic cylinders which replace the motor drive, flywheel, clutch/brake, gearing, and linkages found in mechanical presses (see Fig. 23.5). All of the common press architectures—straight-side, open-frame, OBI, OBS, etc.—are available in hydraulic versions, as well as a number of special designs including horizontal presses.

Hydraulic presses offer several advantages over mechanical presses. They develop full force over the entire stroke, for example, and are able to dwell at full-stroke for extended periods of time, which can be important in some types of forming applications. It is also relatively easy to control the amount of force generated in a hydraulic press independent of ram position, which can be useful in cases where stock thickness is not uniform, and in operations like coining and assembly.

Because they are mechanically simple and essentially self-lubricating, hydraulic presses tend to be more reliable than mechanical presses. Perhaps the greatest advantage of hydraulic presses, however, is the ability to produce very large amounts of force in a relatively compact machine. Hydraulic presses have been built with capacities as large as 50,000 tons (445 mN), far in excess of the practical limits for a mechanical press.

Until recently, the main disadvantage of a hydraulic press has been speed. Advances in hydraulic valve technology have helped close the gap between hydraulic and mechanical presses, but the speed advantage still lies with the mechanical unit, particularly at the larger end of the spectrum, and probably will for the foreseeable future. However, in some applications, small short-stroke hydraulic presses are now competitive with high-speed mechanical presses.

23.4.6 Other Press Types

There are several other types of presses in common use, most of which are special purpose adaptations of standard designs. Perhaps the most common is the high-speed press, which is typically a straight-side press optimized for very rapid stroking. These are most often used in high-volume production of relatively small parts using progressive dies.

Another very common type is the transfer press, which uses a series of individual dies to perform multiple operations on a workpiece that is moved from die to die by an automatic transfer device. Transfer presses range from very small units used to make small metal parts, to large systems used to produce automotive body components. Transfer presses are typically used in mid- to high-volume applications.

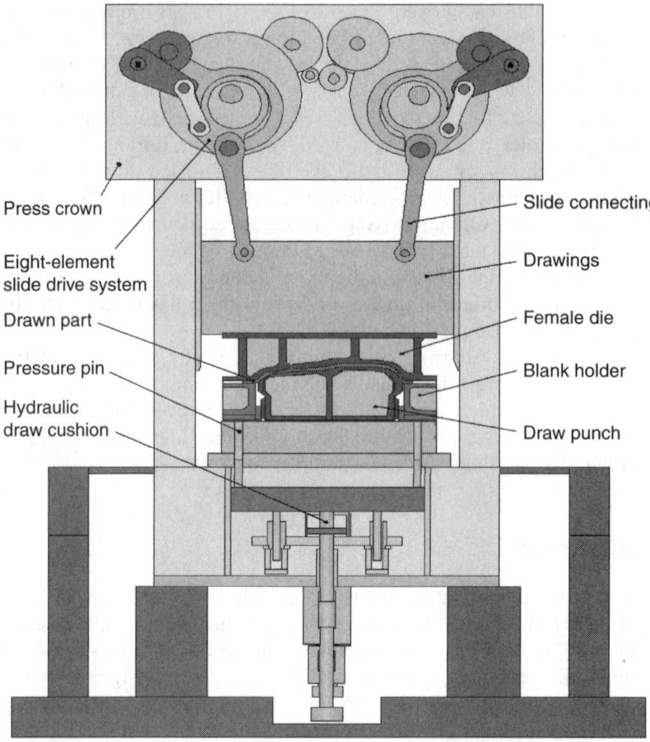

FIGURE 23.5 Single action mechanical press with draw action.

23.4.7 Press Accessories

In most production applications, the press is supported by several devices designed to handle stock preparation, feeding, and removal. These include

- Straighteners, which remove the curvature from coiled stock
- Feeders, which feed stock into the press at a controlled rate
- Coil handlers
- Stacker/destacker units, which handle and feed sheet-type blanks

Another important class of press accessories are systems designed to lubricate and/or coat the workpiece, most typically with materials designed to minimize die wear and/or prevent rust.

23.5 COMMON MATERIALS FOR PRESSWORKING

Common terms applied to materials for pressworking include

Bending stress. A result of the nonuniform distribution of tensile and compressive forces in the inside and outside radii of a bend.

Circle grid. A regular pattern of small circles marked on a sheet metal blank as an aid to analysis. By observing the deformation of the circle grid an engineer can visually verify stretch and metal flow in the forming process.

Creep. Plastic deformation which occurs over time in metal subject to stresses below its yield strength.

Deformation limit. In deep drawing, the point at which the force required to deform the workpiece flange exceeds the tensile strength of the material in the part wall.

Drawing. Any process in which a punch is used to deform the workpiece by drawing it into a die cavity. Depths less than half of the part radius are characterized as shallow draws. Those greater than half the part radius are characterized as deep draws.

Ductility. The ability of a material to deform permanently before fracturing when subjected to tensile stress.

Elastic limit. The maximum stress that does not induce permanent deformation in a metal.

Elongation. In tensile testing the amount of permanent stretch in the area of the fracture. Expressed as a percentage of the original length, e.g., 20 percent in 3 in (76.2 mm).

Hardness. The resistance of a metal to indentation.

Modulus of elasticity. The ratio of stress to strain. In compression the elastic modulus is called "Young's Modulus."

Shear strength. The maximum stress required to fracture a metal when the load is applied parallel to the plane of stress.

Springback. The tendency of formed metal to partially return to its pre-formed shape as the forming force is removed.

Tensile strength. The maximum tensile stress required to break a metal by a gradual, uniformly applied load. Also called "Ultimate Strength."

Torsional strength. The maximum torsional stress required to break a metal.

Ultimate compressive strength. The compressive stress required to fracture a brittle material.

Yield point. The stress at which certain steels deform appreciably with no increase in load. Not all steels exhibit this property which is primarily seen in low- and medium-carbon alloys.

Yield strength. The stress required to deform a ductile material.

23.5.1 Mild (Low-Carbon) Steel

The combination of formability, weldability, strength, and relatively low cost make these steels among the most commonly used for formed products in the automotive and other high-volume industries. Yield strengths for mild steels are generally in the 25–35 ksi range (172–241 MPa). Typical mild steels include SAE 1006 and 1008, which are highly ductile and easily formed, and SAE 1010 and 1012, which are somewhat less ductile, but considerably stronger. These steels are offered in sheet and coil form in a wide range of thickness and widths, and are easily blanked, sheared, and slit.

23.5.2 High-Strength Steel

These stronger materials offer an opportunity for weight reduction in many components because their higher strength makes it possible to use a thinner sheet to achieve the same mechanical or structural properties as compared to mild steel. Yield strengths for high-strength steels are generally in the 35–80 ksi range (241–552 MPa). While their increased strength makes them somewhat less formable than mild steels, they are still able to be processed efficiently on standard production systems and can be welded and painted with little difficulty.

In addition to strength, high-strength steels also have superior toughness, fatigue resistance, and dent resistance. This latter property is a reason high-strength steels are an increasingly popular choice for automotive body panels and similar applications. Like mild steels, high-strength steels are available as sheets and coils in a wide range of thickness and widths which can be blanked, sheared, and slit.

23.5.3 High-Strength, Low-Alloy Steel

These materials utilize alloy materials including silicon, chromium, molybdenum, copper, and nickel at very low levels; and microalloying materials including columbium, vanadium, titanium, and zirconium in various combinations to produce a low-carbon steel with relatively high strength and good formability, weldability, and toughness. In effect, *high-strength, low-alloy* (HSLA) steels provide the best of both worlds, and come close to achieving this goal in many uses.

In practice, HSLA steels tend to be similar to mild steels in forming and drawing properties, but have less elongation tolerance, and are considerably more difficult to use in deep drawing operations. They also exhibit more springback than mild steels. HSLA steels are available as sheets and coils in a wide range of thickness and widths, all of which can be blanked, sheared, and slit.

23.5.4 Ultrahigh Strength Steel

These very strong materials are intended for applications in which strength is the major requirements, and are only moderately formable and weldable. Yield strengths for ultrahigh strength steels are in the 85–200 ksi range (586–1,379 MPa). Special component engineering and die design practices are required to efficiently process ultrahigh strength steels as the results of direct substitution are seldom satisfactory.

A recent development in ultrahigh strength steels are the so-called "bake hardenable" grades which achieve the ultimate physical properties after passing through a paint oven with a soaking temperature in the 350°F (175°C) for 20–30 min. These steels are considerably more formable than regular ultrahigh strength steels prior to baking, and develop comparable strength and dent resistant properties after baking.

23.5.5 Coated Steels and Nonferrous Materials

The automotive industry consumes large quantities of single- and double-side galvanized sheet steel for use in body manufacture. These are essentially processed according to the properties of the underlying steel.

Aluminum is also gaining favor as an automotive metal due largely to its much lower density than steel which contributes to weight reduction. Aluminum is quite ductile and readily formable, but requires different tool designs and different material handling, coating, and lubrication processes.

23.6 SAFETY CONSIDERATIONS FOR PRESSWORKING

Common terms applied to pressworking safety include

Antirepeat. A control system component intended to ensure that the press does not perform more than one stroke even in the case of clutch or brake failure or other component malfunction.

Brake monitor. A control system component intended to detect and warn of any diminution in brake performance.

Guard. A physical barrier that prevents the entry of any part of a worker's body into a dangerous area of the press or other equipment.

Light curtain. A device which senses the presence of an object, such as a part of a worker's body, between a sending and receiving element positioned on either side of a pinch point or other dangerous area. In practice, a properly functioning light curtain takes the place of a mechanical guard with the advantage of offering unobstructed vision for operators and unobstructed access for workpiece feeding.

Pinch point. Any point on a machine at which it is possible for any part of a worker's body to be caught between moving parts. Pinch points must be suitably guarded.

Point of operation. The area of the press in which the workpiece is being acted upon by the dies.

Repeat. An unintended stroke of the press following immediately upon an intended stroke. Also called "Doubling."

Single stroke. A complete movement of the slide from full-up to full-up.

Stop. An operator-activated control intended to immediately stop the slide motion. Often called an "Emergency Stop" although it is normally used for routine operational purposes.

The presses used in metalforming are extremely powerful machines which must be used appropriately and guarded properly to ensure worker safety. Operator protection is largely a question of proper procedures and proper guarding, both of which must be continuously monitored and reinforced. Mechanical guards, lock-outs, interlocks, light curtains, and other safety equipment must never be intentionally disabled for any reason, and the proper operation of such devices must be monitored constantly to ensure safety.

Presses and dies are also costly items of capital equipment which are not easily replaced if damaged, and so must be operated and controlled properly to prevent damage. This is accomplished via the machine control system, and in a worst case situation by overload prevention devices. The amount of energy available in a typical press is fully sufficient to damage or destroy the machine if uncontrolled.

The machine control system must be designed to prevent common failures, such as repeats, and to provide warning of potential failures as indicated by performance changes in brakes and clutches. It must also monitor the operation of auxiliary devices such as part transfers to detect double-feeds or other part-handling failures.

Many press controls incorporate tonnage monitors to record the actual amount of force required to perform a given operation. While this information is useful for quality and maintenance purposes, it does not prevent catastrophic damage in the case of a failure and should not be confused with devices designed to prevent damage to the press in case of overload.

Of these devices, the hydraulic overload is one of the most common. This device consists of one or more permanently pressurized hydraulic cylinders located in the connection between the columns or tie-rods and the crown in a straight-side press. In an overload situation the stress causes the pressure in the cylinder to rise until it reaches a preset limit at which time the fluid is released via a valve or a blow-out plug. A variation of this system is also offered for bottom-drive presses. Similar hydraulic supports are sometimes used between the base and the bolster to achieve the same result.

Other methods of overload protection include mechanical shear washers, stretch links, and various electrical systems based on strain gauges and similar devices. Hydropneumatic overload protection systems are also available.

23.7 TECHNOLOGY TRENDS AND DEVELOPMENTS

Most of the recent advances made in pressworking relate to the use of computer technology in the design of presses, tooling, and processes. With increasing computer and software capabilities much of what used to be "art" has been reduced to predictable, repeatable "science" in both part and tool engineering.

Today's software permits engineers to design a tool as a three-dimensional solid, based on mathematical information from the CAD model developed by the original designer. This capability eliminates many of the error-prone processes formerly used in die development and greatly speeds the tool production process (see Figs. 23.6 and 23.7).

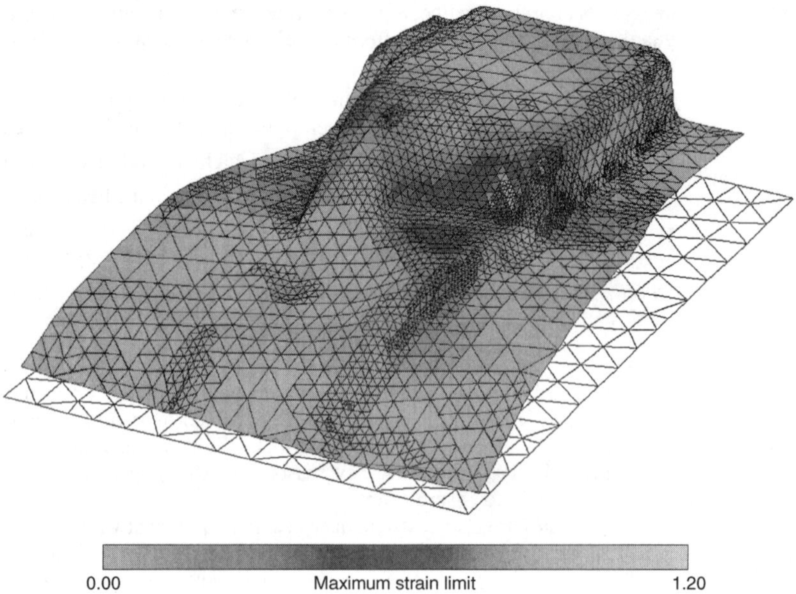

FIGURE 23.6 Stress distribution by means of FEA.

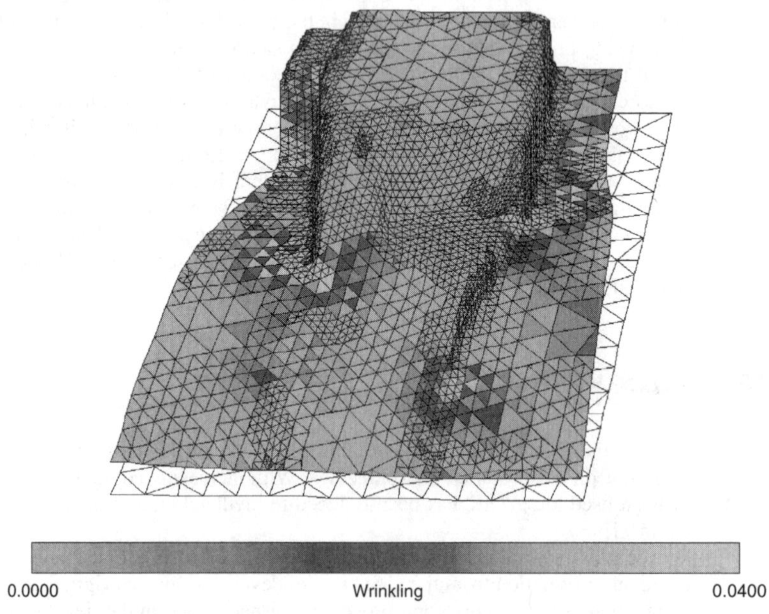

FIGURE 23.7 Possible wrinkle formation by means of FEA.

Draw simulation software is available to accurately model the behavior of various materials during the drawing process to help determine the manufacturability of a given product design. Blank development using nesting software helps to minimize scrap. And *finite element analysis* (FEA) is used to evaluate the mechanical behavior of the tooling under the forces encountered during production.

Once the tooling is designed, other software permits the engineer to test its performance, even to the point of realistically simulating its behavior in the real-world production environment. Analytical and simulation software can help eliminate tooling problems during the design stage, and streamline the entire production process to achieve maximum efficiency long before any metal is cut or formed.

In a modern die shop, the same data is used to drive the computer-controlled machine tools that cut the metal to form the dies. Thus, the loop is closed from design concept to production via the use of consistent computer data to control the entire process.

Many of the same tools are used by press designers to create more efficient and reliable machines which are also smaller, lighter, and less costly than their predecessors. The ability to analyze the reaction of various machine elements to the stresses induced during production has led to significant changes in the way presses are designed and built.

Computers are also being applied increasingly as elements of advanced press control systems. In this application they not only introduce intelligence to the control scheme, but also the ability to collect operating data and perform sophisticated analyses on it to help identify potential failures before they occur.

Pressworking may be among the oldest and most fundamental of industrial processes, but it is also one of the most sophisticated users of cutting edge computer, material, and control technologies in today's industrial arena. Its position at the heart of high-volume precision manufacturing appears to be assured as far into the future as anyone can see.

CHAPTER 24
STRAIGHTENING FUNDAMENTALS

Ronald Schildge
President, Eitel Presses, Inc.
Orwigsburg, Pennsylvania

24.1 INTRODUCTION

Becoming a world-class manufacturer of components for the metal working industry requires an ever increasing focus on improving quality and controlling the manufacturing process. Six Sigma and ever higher CpK requirements demand a manufacturing process that reduces waste and increases the statistical process controls and traceability of parts thoughout the system. Straightening can provide that process control and improve the quality of the parts by automating that function. This section will focus on these improvements both in the part and in the pre- and post-processing of that part.

24.2 CAUSES OF DISTORTION

The need to straighten parts results from distortions caused by the manufacturing processes specific to these parts. They can include processes such as

- Forming processes, such as extrusion or upsetting. Parts such as axle shafts and pinions which are formed in this manner distort due to the extreme forces placed on the part. Worn or misaligned tooling can further exacerbate the problem.
- Cut-to-length operations can result in distortions at the ends of parts if cut off tooling wears, material quality varies, or if fixturing devices fail.
- Material handling or improper storage of parts can lead to distortion.
- Heat treatment is a significant cause of distortion in parts. This is especially true if the part quenching process is not well maintained. The reason that parts distort in heat treatment is the differential cooling rates for different cross sections of the workpiece.

Typical parts that require straightening due to these factors include

- Transmission shafts and drivetrain components such as pinions
- Axle shafts
- Camshafts and crankshafts

- Steering components, such as steering racks and steering pinions
- Pumpshafts and compressor shafts
- Electric motors and armature shafts

24.3 JUSTIFICATIONS FOR USING A STRAIGHTENING OPERATION

To compensate for this distortion, the manufacturer can either use starting material with sufficient excess stock that it can be removed to meet part process tolerances or he or she can choose to straighten the part. The advantage of straightening is clear:

- You save material costs by buying starting material that is closer to the near net shape of the part.
- You reduce the amount of grinding or turning required by straightening to a closer tolerance. Straightening is always less expensive than a material removal process as it requires no abrasives or cutting tools and it does not require coolant. The straightening process is also faster than a metal removal process and will increase the production throughput. The cost of the equipment is also less as one straightener can replace the need for multiple grinders to meet the required throughput.
- The quality of heat-treated parts improve considerably due to more uniform case depth hardness. If a part is not straightened before grinding, it will have more stock removed on the high side than the low side resulting in a shallow case depth on one side of the part.

Given these facts, it is clear that straightening can result in a better part and it is a more economical and productive process than existing material removal processes. It also stands to reason that the closer tolerance you can achieve in straightening will result in even greater cost savings and even better part quality. The obstacle to this in the past was that straightening was a manual process and the manufacturer was dependent on the operator to determine with manual gauging whether the part was within tolerance or not. As a result, acceptable tolerances were typically in the range of about 0.1 mm TIR (total indicator runout, or the total difference measured between the high and low point of the workpiece in one full rotation of the workpiece on its linear axis). Straightening times were also a function of the skill of the operator and could fluctuate greatly.

24.4 THE STRAIGHTENING PROCESS

The straightening process begins with the determination of what constitutes a good part and how this can be measured. Straightness is a linear measurement that determines the deviation from a theoretical centerline of the workpiece measured from one end of the part to the other. Since this poses difficulties in fixturing and measuring the part in a production process, straightening measurements are determined by measuring TIR at critical surfaces along the linear axis of the workpiece. Total indicated runout is measured by placing a transducer under the part at that critical surface and rotating the part 360°. This results in a sinus curve depicting the high and low point of the measurement (Fig. 24.1).

Knowing the high and low point of the deflection at each straightening point enables the control to determine the theoretical centerline of the workpiece. The centerline is equal to exactly one half of the TIR.

An automatic straightening machine uses a servo-driven center tool or roller device for rotating the part 360°. The servo drive has a pulse counter that takes about 200 measurements for each full revolution of the part and records the measurement data from the transducer at each of these points. With the high-speed PC controls available on the market, most machines can measure and store this data for up to seven different straightening points along the workpiece in as short a time as 0.5 s.

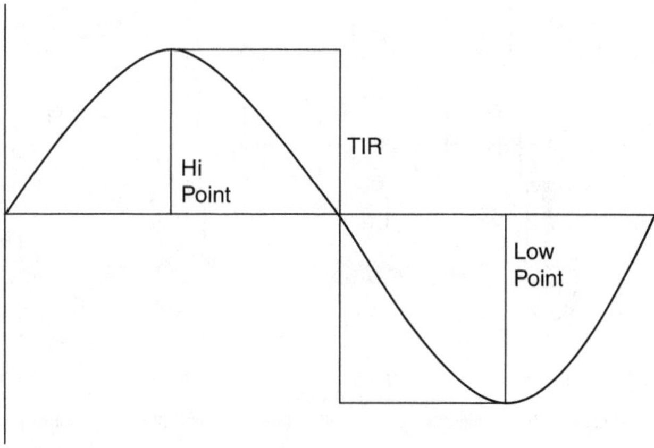

FIGURE 24.1 Measuring deviation from straightness using TIR.

It is critical when determining the straightening locations to also consider the base datum for these measurements. The choice of straightening locations and the base datum are made based on the functionality of the part as well as the manufacturing processes employed. Some of these considerations are as follows:

- The straightening points should include bearing surfaces so as to remove the possibility of vibration in the parts' ultimate application.
- The straightening points should include surface areas that come into contact with opposing parts in the final assembly. For example, matching gear sets should provide for measuring and straightening within the pitch diameter of the gear.
- Base datum for measuring TIR should either be the OD of the part or the centers of the part. If all future manufacturing processes are done between centers, the straightening should be done relative to the centers. If the part is to be ground in a centerless grinder after straightening, the base datum should be the OD of the part.

At this point of the automatic straightening cycle, the part has been transferred into the straightening station of the machine, clamped between centers or on rollers, rotated 360°, and measurements have been taken at all the straightening points relative to the base datum. For each straightening point, the machine control has recorded the TIR along with the angular displacement of the high and low point of the deflection. The straightening process can now follow either a predetermined straightening sequence or, as is more common, start straightening at the point with the greatest deflection (Fig. 24.2).

In the above example, a camshaft has been clamped between two male centers, and TIR measurements have been taken at surfaces Z1, Z2, and Z3 relative to the base datum at X and Y. Assuming that the deflection is greater at Z2 than at Z1 or Z3, the machine would start at Z2. The process is as follows:

1. The servo-driven center rotates the part so that the high point at Z2 is located directly underneath punch 2. As the acceptable tolerance has been set for this straightening point, a value equal to one half of the final TIR is determined to be the straightening target. For example:
   - Initial TIR is 0.100 mm
   - Required straightening tolerance is 0.030 mm
   - Target tolerance is 0.015 mm or $^1/_2$ of 0.030 mm TIR. In reality, the target is set slightly below $^1/_2$ of the acceptable TIR, so that on the final revolution the part is well within the required TIR tolerance. In this case the target would be 0.013 or 0.014 mm.

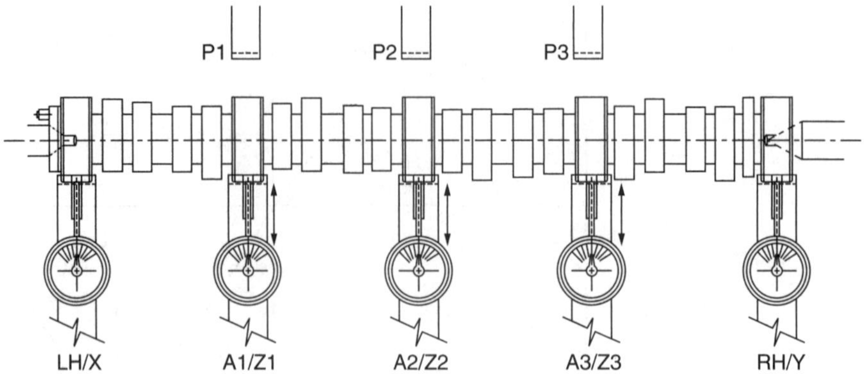

FIGURE 24.2 Measuring and tooling layout for straightening a camshaft.

2. The straightening ram advances a length that is calculated based on the distance between the ram starting point, the part surface, and the measured deflection. Most straightening systems on the market automatically adjust the stroke based on the measured deflection, so that the part can be straightened with the fewest possible strokes.
3. The ram holds the part briefly at the maximum bending moment, then retracts to a position just above the part. The transducer then records its present position relative to the target tolerance. If the target of 0.013 mm has been reached, the part will be rotated one time to confirm that the TIR is within the allowable tolerance.
4. If the part has not reached its target tolerance, the stroke will be adjusted once again by the remaining deflection and the ram will stroke again. This will continue as necessary until the part is within tolerance at Z2.
5. Once Z2 is within tolerance, the same process is repeated at Z1 and Z3 or at as many straightening points as required until the part is within tolerance.
6. After the last straightening point is within tolerance, the part is rotated once again and all surfaces are measured to confirm that the part is within tolerance over its entire length.
7. If the part is within tolerance it is picked up and transported through the machine to the unload conveyor. If the part could not be straightened within a preset time limit or if the part was determined to be cracked, the part is rejected and will be separated into a reject unload station.

24.5 ADDITIONAL FEATURES AVAILABLE IN THE STRAIGHTENING PROCESS

The previous section explained the process of measuring the straightness of the part and the process by which the part is flex straightened to the required tolerance. In addition to this process, various other methods can be used to improve the quality of the part and to meet the required tolerance within an acceptable production rate. A brief description of these methods follows:

- For through-hardened workpieces that cannot be flex straightened due to the danger of part breakage, peen straightening can be used. In this process, the part is positioned with its low point under the ram and the ram strikes the part with a quick blow at the required straightening point. This process leaves a mark on the workpiece but it results in the release of stress at that point and the part growing in the direction from which it is hit. Since the ram does not bend the part, it does not cross the tensile threshold—thus not breaking the part. This process is suitable for brittle parts such as cast iron camshafts where the peen force is applied to the surface areas between the bearing and lobe surfaces of the camshaft. It is not suitable for high tolerance straightening and the best possible TIRs are in the range of 0.10 mm.

- For parts in the green before heat treatment, roller straightening can be used for straightening and stress relieving. This is often used for extruded axle shafts that are roll straightened after extrusion but before cutoff and centering operations. Roll straightening involves clamping the part in a chuck and rotating it while bending it under rollers to a certain deflection. By controlling the length of stroke, the speed of rotation, and the hold down time under pressure, parts can be straightened to tolerances between 0.5 and 1.0 mm TIR.
- Crack detection can be incorporated into the straightening process using devices such as acoustic emission, eddy current and ultrasonic crack detectors. These can be installed in the straightening station or in the unload conveyor to provide for 100 percent inspection. Parts that are cracked will be rejected and a separate counter will keep track of that number independently of any other rejected parts.
- Measurements of the center runout relative to the OD of the part can be taken and parts can be rejected if the center runout exceeds an allowable amount.
- Using an algorithm known as *fast fourier transform* (FFT) parts with rough surfaces can be measured and a theoretical centerline can be determined. This measurement then is a true measurement of the deflection of the part independent of errors in the geometry of the part or surface condition of the part. This is necessary for straightening parts, such as
  - Tubing that might be out of round
  - Hardened parts with heat treat scale
  - Hexagonal or square shafts
  - Gears that have form error greater than the allowable straightening tolerance
- Using master gears attached to the measuring transducers, the pitch diameter of gear surfaces can be measured. This ensures that the runout at the meshing point of matching gear sets are within tolerance. By using the FFT described above, one can also measure the difference between the TIR of the part on the pitch surface with the filter on or off. This results in measuring the form error of the part independent of the deflection of the part. Parts whose form error relative to deflection is greater than an allowable tolerance can than be rejected.
- Most automatic straightening presses available now offer PC controls that provide for connection via serial link or better still by Ethernet to a factory information system. This provides for real time data tracking of the manufacturing process. All incoming measurements, cycle times, reject rates and types, and final measurements can be transmitted to a factory information system to be used to analyze and improve the process.

24.6 SELECTING THE PROPER EQUIPMENT

Traditionally, straightening has been done utilizing hydraulic presses due to the infinitely adjustable stroke length and the ability to adjust pressure necessary to overcome the resistance of the workpiece. Lately, there have been advances in mechanical drives that provide for easy adjustment of the stroke length. These electromechanical presses offer the following advantages to the traditional hydraulic drives as follows:

- Smaller footprint because no hydraulic power units are required
- Less energy consumption
- Better environmental considerations
- Lower maintenance requirements

Hydraulic presses though still have the advantage in applications requiring longer stroke lengths, such as parts with a high initial runout and/or a high degree of elasticity. Apart from the decision as to whether to choose a mechanical or hydraulic drive, a more important consideration is the degree of automation desired. This decision should be made based on the following considerations:

- Are parts going to be processed in large lot sizes or in small batches?
- Do the parts to be straightened fit into family groups that allow for automatic changeover?
- Will the straightening equipment be installed in a production line or in a manufacturing cell?
- How close do you need to straighten?
- What are the financial resources available for investment?

There are presses available on the market for manual straightening, semiautomatic straightening, and fully automatic straightening. A brief analysis of their competitive advantages is as follows:

Manual	Semiautomatic	Automatic
Pros		
Inexpensive	Automated straightening sequence—100% inspection	Fastest cycle times
Easy changeover	Low maintenance	Fits into automatic production lines
Easy to operate	Easy changeover Ideal for cells	Small footprint
Cons		
Accuracy depends on operator	Not as fast as a fully automatic machine	Most expensive More involved tool changeover for different family of parts
Slower cycle time	Part travels as opposed to straightening tooling	

Due to the many offerings available on the market, it is suggested that a full investigation be completed before selecting the proper equipment for your application. If possible, straightening tests are advisable to determine actual production rates based on your particular part characteristics.

INFORMATION RESOURCES

www.amtonline.org
www.eitelpresses.com
www.hess-eng.com
www.galdabini.it
www.dunkes.de

CHAPTER 25
BRAZING

Steve Marek
*Lucas-Milhaupt, Inc.**
Cudahy, Wisconsin

25.1 INTRODUCTION

Brazing is defined by the American Welding Society (AWS) as a "group of welding processes which produce a coalescence of materials by heating them to a suitable temperature and by using a filler metal having a liquidus temperature above 840°F (450°C) and below the solidus of the base materials. The filler metal is distributed between the closely fitted surfaces of the joint by capillary attraction."[1]

Breaking this definition down into several parts, the group of welding processes described are the methods in which heat may be applied to the base materials/joint. These heating methods are commonly fuel gas torches, induction, resistance, and furnaces. Other methods of heating such as dip, laser, electron beam, infrared, etc. are also viable heating methods, which may be employed in supplying the heat necessary. The "production of a coalescence of materials" is the joining or combining of the base materials. This joining or union of the base materials is accomplished when a metallic bond takes place between the base metal and filler metal. Base materials are commonly the ferrous and nonferrous metal groups, such as copper, brass, steel, stainless steel, etc.; as well as nonmetals, such as carbides, cermets, ceramics, and diamonds.

This joining process takes place only when the base materials are heated to a suitable temperature. This temperature is based on the temperature of the filler metal used in the brazing process. This temperature is typically 50°F to 100°F above the liquidus of the filler metal. Note it is very important that the base materials reach the temperature required for the filler metal to melt, or metallic bonds will not form.

These filler metals have a liquidus temperature above 840°F (450°C) but below that of the solidus of the base materials being joined. Various organizations stipulate chemical compositions for the different types of filler metals available. The filler metal is distributed between the surfaces of the base materials by capillary attraction. This capillary attraction will be affected by the cleanliness of the base materials surface. Precleaning and the flux and/or atmosphere used during the brazing process will enhance the capillary attraction and bonding by creating a clean surface free of oils, grease, dirt, and oxides.

*A Handy-Harman Company

25.2 WHY BRAZE

Brazing is a versatile method of joining metals and in some cases nonmetals.[2] One the most important advantage of brazing is the ability to join of dissimilar materials. The joining of materials with very different melting characteristics such as brass and steel preclude joining with methods such as welding. Very unique combinations of base materials can be joined such as diamonds to carbide.

Brazed joints are strong. Properly designed brazed joints can obtain strengths equal to or greater than the base metals joined. Brazed joints are ductile and can withstand considerable shock and vibration. The joining of thin to thick cross sections is also more easily accomplished using a brazing technique. Brazed joints are generally easy and rapid to make. Skills necessary to accomplish brazing are readily acquired. Braze joints seldom require any post finishing, such as grinding, or filing besides post braze cleaning. The fillets that are created are smooth and continuous. Brazing can be performed at relatively low temperatures, thus reducing the possibilities of warping, overheating, or melting the base materials. Brazing is economical. Cost per joint compare favorably with other metal joining methods. Brazing is also adaptable to both hand and automated processes. This flexibility enables one to match production techniques with production requirements.

25.3 BASE MATERIALS

The list of commonly brazed engineering materials includes ferrous and nonferrous metals and their respective alloys, glass sealing alloys, metallized ceramics, nonmetallized ceramics, graphite, and diamond.

In general the ability to braze a particular material is based on the ability to create a clean surface free of oxides that can be wet by the filler metal. Elements that do not oxidize readily, such as gold, silver, platinum, palladium, copper, iron, nickel, cobalt, and many of their alloy systems are easily brazed. Elements from the refractory group—tungsten, molybdenum, and tantalum—present more challenges. The reactive group of elements in the periodic chart form more aggressive oxide layers. These elements such as aluminum, chromium, beryllium, titanium, zirconium, and their respective alloys can be more difficult to braze, typically requiring special cleaning procedures, special fluxes, and/or stringent atmosphere requirements. Even when such elements as titanium or aluminum are present in small quantities of 0.5 percent or greater as in certain stainless steel or nickel-based alloys, brazing becomes more difficult. In these situations, where the brazing process cannot provide an oxide-free surface, the surface will need to be plated with either copper or nickel to facilitate wetting and bonding.

As with all heat related processes, the temperature at which the filler metal melts may affect the metallurgical properties of the base metal. In many cases the cold-worked or heat-treated base material may anneal at brazing temperatures. If this is unacceptable, consideration must be given to material changes or heat treating and brazing simultaneously. Other considerations with various base metals may be metallurgical reactions, including interalloying, carbide precipitation or sensitization, stress cracking, and liquid metal, hydrogen, sulfur, or phosphorous embrittlement.

With the nonmetals such as carbon, graphite, alumina, etc. that are difficult to wet, brazing can be accomplished on a metallized surface deposited on the substrate, such as copper or nickel. An alternative to the metallized surface is to braze using active braze filler metals. These filler metals usually contain titanium, zirconium, vanadium, or chromium as the reactive element responsible for the bonding of the nonmetallic surface.

25.4 FILLER METALS

Filler metals as defined by the AWS have a temperature above 840°F but below the solidus of the base materials. Various organizations such as American Welding Society (AWS), Society of

Automotive Engineers (SAE), American Society of Mechanical Engineers (ASME), and the federal government and military have written specifications to standardize filler metal compositions. In addition to the standard compositions available, proprietary filler metals are available by vendors. One of the most common specifications used is AWS A5.8—"Specification for Filler Metals for Brazing and Braze Welding."

Filler metals must be chosen for each application. When reviewing the application, the following criteria must be reviewed to choose the appropriate/cost effective filler metal. First the filler metal must be compatible with the base material. The filler metal must wet the base metal. By definition the filler metal must flow out and cover the base material surface with a thin film of liquid. If wetting does not occur, the filler metal will bead/ball up on the surface. In addition the interaction of the filler metal with the base material should not form detrimental metallurgical compounds. Table 25.1 gives suggested filler metal–base material combinations. This table uses the AWS designation for classifying filler metals in accordance with the AWS specification A 5.8. This designation begins with the letter "B" for braze filler, followed by the chemical symbol for the major alloying element(s) in the filler metal. If a specific composition is being called out, the element will be followed by a numeric value corresponding to a specific composition within that classification, i.e., BCu-1, BAg-24, and BCuP-5. If the filler metal has been designed to be used in vacuum applications the letter "V" will immediately follow the "B," i.e., BVAg-8. If the letter "B" is preceded by the letter "R," it indicates the filler metal can be used in rod form for braze welding as well as brazing. Braze welding is a joining method in which braze filler metals are used, but the filler metal is deposited at the joint surface similar to welding. Capillary is not necessary for the joint to be made.

The filler metals in the AWS specification are categorized into seven groups, aluminum, cobalt, copper, gold, magnesium, nickel, and silver. Examples of filler metals by group and classification are shown in Table 25.2. This table represents only a sample of the more common filler metals used, their temperature range, composition, and general comments.

In addition to the filler metal being compatible with the base materials, the heating method, service requirement, required braze temperature, joint design, cosmetic requirements, and safety issues should be considered when determining the appropriate filler metal for a specific application.

Filler metals can be presented to the joint in various forms. The most common form is wire in both straight length and coil form for hand feeding. Filler metals can also come in strip or ribbon form. In both wire and strip various preforms can be wound or blanked. These preforms allow for controlled volume of filler metal being presented to the joint at time of brazing. Automation can also be used to place these preforms into position prior to brazing. In some brazing methods such as furnace brazing the filler metal cannot be added by hand and therefore must be preplaced using a preform. A third form of the filler metal is powder. These powders can be used as is or mixed in proprietary binder systems creating dispensable paste. These paste products can be made with or without flux depending on the filler metal and heating method chosen. Paste products like preforms control the volume of alloy, and are easily dispensed in automated systems.

25.5 FUNDAMENTALS OF BRAZING

In order to make a braze joint the following six categories should be addressed in each brazing operation.

25.5.1 Joint Design

What is your primary consideration in a braze joint? Are you looking for high strength, corrosion resistance, electrical conductivity, ductility, or hermeticity? These requirements should be considered because different joint designs will dramatically affect the performance characteristics of the joint. Here are some general rules when designing a joint for brazing.

TABLE 25.1 Base Metal-Filler Metal Combinations

	Al & Al alloys	Mg & Mg alloys	Cu & Cu alloys	Carbon & low alloy steels	Cast iron	Stainless steel	Ni & Ni alloys	Ti & Ti alloys	Ba, Zr, & alloys (reactive metals)	W, Mo, Ta, Ch & alloys (refractory metals)	Tool steels
Al & Al Alloys	BAlSi										
Mg & Mg alloys	X	BMg									
Cu & Cu alloys	X	X	BAg, BAu, BCuP, RBCuZn	BNi							
Carbon & low alloy steels	BAlSi	X	BAg, BAu, RBCuZn, BNi	BAg, BAu, BCu, RBCuZn, BNi							
Cast iron	X	X	BAg, BAu, RBCuZn, BNi	BAg, RBCuZn, BNi	BAg, RBCuZn, BNi						
Stainless steel	BAlSi	X	BAg, BAu	BAg, BAu, BCu, BNi	BAg, BAu, BCu, BNi	BAg, BAu, BCu, BNi					
Ni & Ni alloys	X	X	BAg, BAu, RBCuZn, BNi	BAg, BAu, BCu, RBCuZn, BNi	BAg, BCu, RBCuZn	BAg, BAu, BCu, BNi	BAg, BAu, BCu, BNi				
Ti & Ti alloys	BAlSi	X	BAg	BAg	BAg	BAg	BAg	Y			
Be, Zi & alloys (reactive metals)	X BAlSi (Be)	X	BAg	BAg, BNi*	BAg, BNi*	BAg, BNi*	BAg, BNi*	Y	Y		
W, Mo, Ta, Cb & alloys (refractory metals)	X	X	BAg, BNi	BAg, BCu, BNi*	BAg, BCu, BNi*	BAg, BCu, BNi*	BAg, BCu, BNi*	Y	Y	Y	
Tool steels	X	X	BAg, BAu, RBCuZn, BNi	BAg, BAu, BCu, RBCuZn, BNi	BAg, BAu, RBCuZn, BNi	BAg, BAu, BCu, BNi	BAg, BAu, BCu, RBCuZn, BNi	X	X	X	BAg, BAu, BCu, RBCuZn, BNi

Note: Refer to AWS Specification A5.8 for information on the specific compositions within each classification.
X—Not recommended; however, special techniques may be practicable for certain dissimilar metal combinations.
Y—Generalizations on these combinations cannot be made. Refer to the Brazing Handbook for usable filler metals.
*—Special brazing filler metals are available and are used successfully for specific metal combinations.

Filler Metals:

BAlSi—Aluminum	BCuP—Copper phosphorus
BAg—Silver base	RBCuZn—Copper zinc
BAu—Gold base	BMg—Magnesium base
BCu—Copper	BNi—Nickel base

TABLE 25.2 Common Braze Filler Metals in Accordance with AWS A5.8

Filler metal type aluminum	AWS classification	Solidus		Liquidus		Alloy composition			Comments
		F	C	F	C	Al	Cu	Si	
Aluminum filler metals	BAISi-3	970	521	1085	585	86	4	10	Aluminum filler metals are used to braze aluminum base metals. Typical base metals joined are the 1100, 3000, and 6000 series aluminum alloys. Aluminum brazing requires tighter process parameters than most brazing processes due to the close relationship between the melting point of the filler metals and base metals.
	BAISi-4	1070	576	1080	582	88		12	

Filler metal type copper	AWS classification	Solidus		Liquidus		Alloy composition	Comments
		F	C	F	C	Cu	
Copper filler metals	BCu-1	1981	1083	1981	1083	99.9 Min.	Copper filler metals are primarily used in furnace brazing ferrous base materials such as steels and stainless steels. Note BCu-1 is produced only in wire and strip form. BCu-1a is the powder form of BCu-1.
	BCu-1a	1981	1083	1981	1083	99.0 Min.	

Filler metal type copper	AWS classification	Solidus		Liquidus		Alloy composition						Comments
		F	C	F	C	Cu	Zn	Sn	Fe	Mn	Ni	
Copper /zinc filler metals	RBCuZn-C	1590	866	1630	888	58	40	0.95	0.75	0.25		The copper/zinc based filler metals are used in joining steels, copper, copper alloys, nickel, nickel alloys, and stainless steels. Heating methods typically used are torch and induction with flux. Due to the high zinc content of these filler metals, they are rarely used in furnace brazing process.
	RBCuZn-D	1690	921	1715	935	48	42				10	

Filler metal type copper	AWS classification	Solidus		Liquidus		Alloy composition			Comments
		F	C	F	C	Ag	Cu	P	
Copper/phosphorus filler metals	BCuP-2	1310	710	1460	793		92.7	7.3	The copper/phosphorus filler metals are used to braze copper and copper alloys. The can also be used to braze electrical contacts containing cadmium oxide or molybdenum. These filler metals are considered self-fluxing on copper base metals.When used to braze various copper alloys (i.e., brass) a mineral type flux is recommended. Do not use these filler metals on ferrous materials or nickel bearing material in excess of 10% nickel as brittle phosphides will be formed at the braze interface.
	BCuP-3	1190	643	1495	813	5	89	6	
	BCuP-4	1190	643	1325	718	6	86.7	7.3	
	BCuP-5	1190	643	1475	802	15	80	5	

TABLE 25.2 Common Braze Filler Metals in Accordance with AWS A5.8 (*Continued*)

Filler metal type silver	AWS classification	Solidus F	Solidus C	Liquidus F	Liquidus C	Ag	Cu	Zn	Cd	Sn	Ni	Comments
Cadmium bearing silver filler metals	BAg-1	1125	607	1145	618	45	15	16	24			Silver based filler metals can be used to braze a variety of base materials. In general all ferrous and non-ferrous base materials can be joined. Note that the temperature range of the liquidus temperatures of the silver based filler metals (1145°F to 1761°F) preclude them from being used to join aluminum or magnesium. This large temperature range for the silver group provides for a selection of filler metals to be utilized in brazing at the lowest temperature, or brazing at a temperature in which heat treated properties may be obtained in the base materials. Silver based filler metals can be used by all heating methods; however, when choosing a filler metal to be used in an atmosphere or vacuum process, the content of the filler metal should not contain cadmium or zinc. Cadmium and zinc can volitalize from the filler metal contaminating the work and/or furnace. Silver filler metals that contain cadmium as a principal constituent require care to avoid exposure to cadmium fumes. Filler metals which contain 1% to 5% nickel are found to be efffective in wetting carbide materials. They will also inhibit or prevent interface corrosion on stainless steels.
	BAg-2	1125	607	1295	701	35	26	21	18			
	BAg-3	1170	632	1270	687	50	15.5	15.5	16		3	
Cadmium free silver filler metals	BAg-4	1240	670	1435	779	40	30	28			2	
	BAg-5	1225	662	1370	743	45	30	25				
	BAg-7	1145	618	1205	651	56	22	17		5		
	BAg-13	1325	718	1575	856	54	40	5			1	
	BAg-13a	1420	770	1640	892	56	42				2	
	BAg-21	1275	690	1475	801	63	28.5			6	2.5	
	BAg-24	1220	659	1305	707	50	20	28			2	
	BAg-28	1200	648	1310	709	40	30	28		2		
	BAg-34	1200	648	1330	720	38	32	28		2		
	BAg-36	1195	643	1251	677	45	27	25		3		
	BVAg-8Gr2	1435	779	1435	779	72	28					
	BVAg18Gr2	1115	601	1325	718	60	30			10		
	BVAg29Gr2	1155	623	1305	707	61.5	24			14.5		

Filler metal type gold	AWS classification	Solidus F	Solidus C	Liquidus F	Liquidus C	Ag	Au	Cu	Pd	Ni	Comments
Gold filler metals	BAu-1	1815	991	1860	1016		37.5	62.5			Gold based filler metals are used to join steels, stainless steel, nickel based alloys, where ductility and resistance to oxidation or corrosion is required. Gold filler metals readily wet most base materials including super alloys and are especially good for brazing thin sections due to their low interaction with most base materials. Most gold based filler metals are rated for continuous service up to 800°F. Gold filler metals are typically brazed in either a protective atmosphere or vacuum process.
	BAu-3	1785	974	1885	1029		35	62		3	
	BVAu-4	1740	949	1740	949		82			18	
	BAu-6	1845	1007	1915	1046		70		8	22	
	BVAu-8	2192	1200	2264	1240		92		8		

Filler metal type nickel	AWS classification	Solidus		Liquidus		Alloy composition						Comments
		F	C	F	C	Ni	Cr	Si	B	Fe	P	
Nickel filler metals	BNi-1	1790	977	1900	1038	73.1	14	4	3.1	4.5		Nickel based filler metals are used to braze ferrous and non-ferrous, high temperature base materials, such as stainless steels and nickel based alloys. These filler metals are generally used for their strength, high temperature properties, and resistance to corrosion. Some of these filler metals can be used in continous service up to 1800°F (980°C), and 2200°F (1205°C) for short periods of time. Nickel base filler metals melt in the range of 1610°F (C) to 2200°F (C) but can be used at higher temperatures when the melting depressant elements in the filler metal such as silicon and boron are diffused from the filler metal into the base maetal altering the composition of the joint.
	BNi-2	1780	971	1830	999	82.3	7	4.5	3.1	3		
	BNi-6	1610	877	1610	877	89					11	
	BNi-7	1630	888	1630	888	75.9	14				10.1	

For more information on the filler metal types above, or BCo, BMg, BPd refer to the American Welding Societies "Specification for Filler Metals for Brazing and Braze Welding."

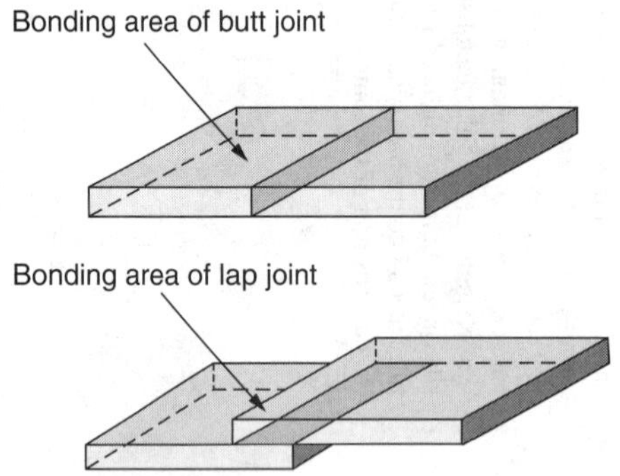

FIGURE 25.1 Bonding areas showing butt and lap joint.

The basic joint designs start with either a butt or lap joint (Fig. 25.1). The butt joint is simplest to prepare, but has limited tensile strength due to the cross sectional area available to be bonded. The lap joint increases the bonding area, and changes the stress from tensile to shear forces to produce a strong joint. Actual results will depend on the length of the developed overlap.

Joint strength in the lap joint is a function of joint length. Properly designed, the strength of the braze joint can equal or exceed that of the base metal. Generally the lap area should be at least three times the thickness of the thinner joint member. This rule will vary with higher strength materials. Simplistic calculations can be made to verify required overlaps. Two modifications to the butt and lap joints are the modified-butt joint and the scarf joint. Both types of joints require more preparation.

Clearances between the two surfaces must be maintained to allow capillary to occur. The strongest joints are typically made when the joint clearance is maintained between 0.002" and 0.005" for a mineral flux brazed joint. For joints brazed in a protective atmosphere or vacuum clearances are typically 0.000"–0.002". This clearance must be maintained at brazing temperature. When brazing two materials of similar *coefficients of thermal expansion* (C.T.E.), the room temperature clearance should provide ample clearance. If the two materials are vastly different in their C.T.E. adjustments must be made to either increase or decrease room temperature clearance such that at brazing temperature proper clearance is achieved.

Design joints to be self-venting. This will help reduce entrapment of flux, air, or gases. Design joints to carry load forces by shear or tensile forces. Never design a joint to be loaded in a peel mode. In addition prevent concentration of stress from weakening the joint. If necessary impart flexibility to a heavy or stiff section, or add strength to a weaker member. Surface finish for braze joint are typically 30 to 80 μin. If either electrical or corrosion resistance are important, joint clearance should be maintained to the minimum. This provides less material to minimize resistivity, or exposure of filler metal to the corrosive environment.

25.5.2 Precleaning[2]

Capillary action will work properly only when the surfaces of the metal are clean. If they are contaminated, i.e., coated with oil, grease, rust, scale, or just plain dirt, those contaminates must be removed. If they remain, they will form a barrier between the base materials surface and the braze filler metal. An oily base material will repel the flux, leaving bare spots that oxidize under heat, resulting in voids. Oils and grease will carbonize upon heating, forming a film over which the filler metal will not flow. Rust and scale (oxides) will also inhibit brazing filler metal from flowing and bonding.

Cleaning the base materials is seldom a complicated task. The sequence in which cleaning is done, however, is very important. Oils, grease, and dirt should first be removed. These materials if not removed first may inhibit the actions of the cleaning process, which follow to remove oxides (rust, scale). Pickling solutions meant to remove surface scale or oxides were not designed as effective degreasers. In addition if mechanical means such as emery cloths, wire brush, or grit blasting are used to remove oxides, these methods themselves may become contaminated, or do nothing more than spread the contaminate around.

Therefore start the precleaning of all base materials by removing oils, grease, and dirt. This can be done in most cases by degreasing solvents, vapor degreasing, alkaline, or aqueous cleaning. It is advisable to identify what types of material the base material have come in contact with during their processing to enable one to choose the proper methods of cleaning.

Again if the base materials have an oxide film, it can be removed either chemically or mechanically. For most metals an acid pickling treatment is available. There are some base materials such as aluminum, which may require a caustic solution to achieve oxide removal. Most importantly again check that the chemicals used are compatible with the base material, and that parts have been rinsed thoroughly so that no traces of chemical solutions remain in crevices or blind holes. Mechanical methods of removing oxides are emery cloth, sand paper, grinding wheels, files, or metallic shot blast. After mechanical cleaning it is advisable to rinse the parts free of any remaining particulates. Once cleaned it is recommended that brazing takes place within 48 h. This minimizes any chance for recontamination.

25.5.3 Proper Flux/Atmosphere

Once the parts to be brazed have been cleaned, they must remain clean throughout the brazing process in order to promote wetting and bonding of the filler metals to the base materials. Of primary concern is the formation of oxides on both the filler metal and base material upon heating. There are two basic ways to protect parts while brazing. The first is by using a mineral based type flux. The second method is by using a protective atmosphere around the part.

Mineral fluxes can provide for a clean surface in two ways. First the flux will inhibit oxygen from reaching the surface so that the formation of oxides is minimized. Second the flux can actively reduce oxides, or remove oxides. Note the fluxes are meant to protect the base materials from forming oxides, and removing residual oxides, they are not formulated to remove dirt, grease, or oil. Mineral fluxes are usually composed of various salts of borates, fluorides, or chlorides. They can be supplied in various forms, such as pastes, slurries, liquids, and dispensable paste. The most common is the paste form. Application of flux is done either by dipping or brushing. If production quantities warrant it, the flux can be dispensed at the joint through pache or positive displacement type equipment. Besides the various forms, fluxes like the filler metals have various compositions. Selection criteria for the fluxes include base material, temperature range, heating method, concentration of fluxes, etc. As with the filler metals, a specification by AWS exists for the different variations of fluxes. The specification is AWS A5.31-92—"Specification for Fluxes for Brazing and Braze Welding."

Brazing parts can also be accomplished in a protective atmosphere, which can be either a gas or vacuum. This protective atmosphere surrounds the parts to be brazed during the heating and cooling cycle. The primary function of the atmosphere is to prevent the formation of oxides during brazing. Some of the atmospheres available for brazing are argon, helium, hydrogen, dissociated ammonia, nitrogen, combinations of hydrogen and nitrogen or argon, combusted fuel gases (exothermic, endothermic generated atmosphere), and vacuum. Having a protective atmosphere is not sufficient however to protect most base materials. In order to effectively protect various base materials the oxygen content of the atmosphere must be controlled. Measuring the dew point of the atmosphere controls this. The dew point indicates the oxygen content within the protective furnace atmosphere. Information such as a metal–metal oxide chart can be used to approximate at what dew point various oxides will be reduced. As base materials vary in composition so will the atmosphere requirements to successfully accomplish brazing. Many sources are available to determine what type and quality of atmosphere is required for a particular base material. Protective atmospheres and vacuum atmospheres are typically associated with furnace brazing and induction brazing processes. Care must be exercised with all atmospheres as they may be explosive or create a suffocation hazard.

25.5.4 Brazing Fixtures

Once the parts have been prepared and are ready for brazing, the assembly must be held to assure proper alignment during the brazing process. The assemblies are either self-fixtured (held in position without external aids.), or by use of external fixturing. When possible the self-fixturing is the preferred method. When practical the use of gravity to hold parts in an assembly is the simplest. Other methods of self fixturing parts are press fits, use of straight knurls in press fitting, swaging, staking, locating bosses pierced joint, folds and interlocks, tack welds, and pins or screws. It is very important to remember that the fixturing method holds the alignment of the part as well as maintains the proper gap for the filler metal/flux combination. Therefore the use of a press fit in combination with a mineral flux and standard silver braze filler metal would be a poor choice. Self-fixtured parts are very often used when furnace brazing is the heating method.

External fixtures are commonly used when the shape and weight of the part dictate additional support. When designing external type fixtures several key points should be remembered. The fixture should be designed such that the framework is thin yet ridged. The design should be kept as open as possible to allow for access to the part with the heating method above and below the intended joint area. The fixture must be designed to allow for the expansion and contraction of the base metal upon heating and cooling. If pressure must be applied to the parts during brazing, movable weights, weighted pins or levers, cams, or stainless/inconel springs can be used to apply pressure, yet still allow for the base metal to expand. The part is best held in position by point or line contact. This minimizes the fixture acting as a heat sink, unless it is the intention of a fixture design to do so. Materials often used for fixtures include, steel, stainless steel, inconel, and titanium. These are most often used in flame brazed assembles. Ceramic and carbon materials are common in induction and furnace-brazed assemblies.

25.5.5 Heating Methods

Several methods are available to apply the necessary heat to obtain the temperature within the parts to melt and flow the filler metal.

Torch brazing is one of the most common and versatile methods of heating. Heat is applied to the parts broadly by the combustion of fuel gas and either oxygen or air. Common fuels used are acetylene, propane, natural gas, and hydrogen. Torch brazing is usually done in combination with a mineral flux. Flames are typically adjusted to either a reducing or neutral combustion ratio so as not to oxidize the parts during brazing. Torch brazing is very common to the manual hand braze operation where filler metal is hand fed. When torch brazing systems are automated for increased production the filler metal is either preplaced or automatically fed.

Induction brazing is a very rapid and localized form of heating. The heat necessary to braze parts is generated by the resistance of the base material to a flow of current induced into the part. This current is induced by a primary coil carrying an alternating current placed near or around the part. Brazing usually requires the use of a mineral flux and preplaced filler metal. Induction coils can be placed inside a protective atmosphere if brazing without flux is a requirement.

Furnace brazing is commonly used for large volumes of self-fixtured parts, or multiply joint assemblies. Furnace brazing relies on heat being transferred to the part due to convection by the air or protective atmosphere, or by radiation as in a vacuum furnace. The heat is typically generated by fuel gas or electrical resistance elements. Mineral flux can be used with preplaced filler metal, however, the advantages of using a protective atmosphere eliminates flux related discontinuities, and post-cleaning. The types of furnaces available are batch, retort/bell, continuous belt, and vacuum.

Resistance brazing like induction brazing generates the heat necessary for brazing by the base material resistance to flow of a current (i.e., I^2R losses). The electrical current is provided to the part by a pair of electrodes that make physical contact with the part. The electrodes are typical made of a high resistance material such as carbon, graphite, tungsten, or molybdenum. These materials provide heat to the parts as well, thus not relying on the base materials resistance to the flow of current solely. Brazing usually requires the use of a mineral flux and preplaced filler metals.

Dip brazing is accomplished by either dipping the part to be brazed in a bath of molten filler metal or dipping the part with preplaced filler metal into a bath of molten flux. Like furnace brazing, temperatures can be controlled accurately. Dip brazing using a molten flux bath is predominately used for aluminum brazing.

Others methods found in brazing are laser, electron beam, and infrared.

25.5.6 Postcleaning

Postcleaning of the brazed assembly is usually necessary only if a mineral flux has been used during the brazing process. The glassy material, which remains after brazing, should be removed to avoid contamination of other parts, reduce or eliminate corrosion, improve inspectability, and improve appearance. Fluxes can generally be removed by submersing the assembly or parts in hot water (150°F or hotter). This is due to the chemical nature of the fluxes being salts and water-soluble. In addition parts may be pressure washed or steam cleaned. Chemical solutions as well as mechanical methods (ultrasonic tanks, brushing, grit blasting) are available if hot water cleaning is not sufficient to clean flux residue.

Parts that have been brazed in protective atmospheres or vacuum should come out as clean as they went into the brazing operation. If parts emerge from these atmospheres oxidized, problems with the furnaces may be indicated.

25.6 BRAZING DISCONTINUITIES

Braze discontinuities commonly identified in braze joints are lack of fill, voids, porosity, flux inclusions, cracks, distortion, unsatisfactory surface appearance, interrupted or noncontinuous fillets, and base metal erosion.

25.7 INSPECTION METHODS

Inspection methods common to brazing include both destructive and nondestructive testing techniques. Nondestructive techniques include visual inspection that is the most widely used form. Verifying that the filler metal has flowed into the joint can easily be accomplished if both sides of the joint are visually accessible. Other nondestructive forms of inspection are proof testing, pressure testing, vacuum testing, helium testing, radiographic, ultrasonic, dye penetrant, and thermal techniques. Destructive techniques include mechanical testing of joints in tensile, shear, fatigue, or torsion modes. Two simple destructive tests to evaluate braze joint are metallographic sections and peel testing, which provide a view of the internal discontinuities.

25.7.1 Safety

In brazing, as is the case in many manufacturing processes, potential safety hazards exist. In general eye/face and protective clothing should be worn. Ventilation should be provided to extract fumes or gases emanating from base metals, base metal coatings, filler metal constituents such as zinc or cadmium, and fluorides from fluxes. Ventilation fans or exhaust hoods are recommended. Make sure base materials are clean. Any unknown contaminate on the surface could add to hazardous fumes. Review constituents in base metals, filler metals, and fluxes. Material safety data sheets should be consulted for all materials being used to identify potential hazardous elements or compounds. In addition safe operating procedures should be established for the brazing processes and equipment, such as compressed gas cylinders for torch brazing. For additional information on safety, consult the American National Standard Z49.1, "Safety in Welding and Cutting" and the "Brazing Handbook" published by the AWS, Occupational Safety and Health Administration (OSHA) regulations, the Compressed Gas Association (CGA) and National Fire Protection Agency (NFPA).

REFERENCES

1. AWS committee on Brazing and Soldering, *Brazing Handbook,* 4th ed., American Welding Society, Miami, FL, 1991.
2. *The Brazing Book,* Lucas-Milhaupt/Handy & Harman, Cudahy, WI, 2000; www.handyharmanpmfg.com

FURTHER READING

Brazing, Mel. M. Schwartz, ASM International, Metals Park, Ohio.

Metals Handbook, Vol. 6, ASM International, Metals Park, Ohio.

Giles Humpston, David M. Jacobson, *Principles of Soldering and Brazing*, 1993 ASM International, Metals Park, Ohio.

Recommended Practices for Design, Manufacture, and Inspection of Critical Brazed Components, AWS C3.3-80, American Welding Society, Miami, FL.

Recommended Practices for Ultrasonic Inspection of Brazed Joints, AWS C3.8-90, American Welding Society, Miami, FL.

Safety in Welding, Cutting, and Allied Processes, ANSI Z49.1:1999, American Welding Society, Miami, FL.

Specification for Filler Metals for Brazing and Braze Welding, AWS A5.8-92, American Welding Society, Miami, FL.

Specification for Fluxes for Brazing and Braze Welding, AWS A5.31-99, American Welding Society, Miami, FL

Specification for Furnace Brazing, AWS C3.6-99, American Welding Society, Miami, FL.

Specification for Induction Brazing, AWS C3.5-99, American Welding Society, Miami, FL.

Specification for Resistance Brazing, AWS C3.9-99, American Welding Society, Miami, FL.

Specification for Torch Brazing, AWS C3.4-99, American Welding Society, Miami, FL.

Standard for Brazing Procedure and Performance Qualification, AWS B2.2-92, American Welding Society, Miami, FL.

Welding Handbook, 8th ed., Vol. 2, American Welding Society, Miami, FL.

CHAPTER 26
TUBE BENDING

Eric Stange
Tools for Bending, Inc.
Denver, Colorado

26.1 PRINCIPLES OF TUBE BENDING

There are several methods of bending tube or extruded shapes. However, the economic productivity of a bending facility depends not only on the most effective method, but also on the use of proper tooling and proven techniques. Of course, the operator is a factor, but the right equipment and tooling minimize the degree of craftsmanship and expertise required.

Two principles apply to all three primary methods—compression (Fig. 26.1), press (Fig. 26.2), and rotary bending (Fig. 26.3).

First, the material on the inside of the bend must compress. Second, the material on the outside of the neutral axis must stretch (Fig. 26.4). A fourth method, crush bending, uses press bending to achieve bends.

26.1.1 Bend Die Functions

When the ratio of the tube diameter to wall thickness is small enough, the tube can be bent on a relatively small radius (CLR = 4 × tube O.D.) without excessive flattening or wrinkling of the bend. The outside of a bend tends to pull toward the center line flattening the tube. A conventionally grooved bend die supports the tube along the center line and the inherent strength of the round or square tube help prevent flattening (see Fig. 26.5).

26.1.2 Compression Bending

There are three basic steps to compression bending:

1. The work piece is clamped to a bend die (or radius block).
2. The wipe shoe (or slide block) is brought into contact with the work piece.

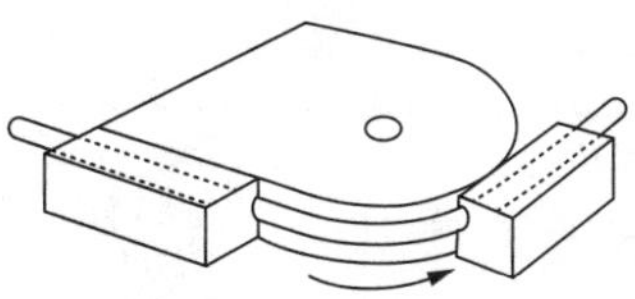

FIGURE 26.1 Compression bending.

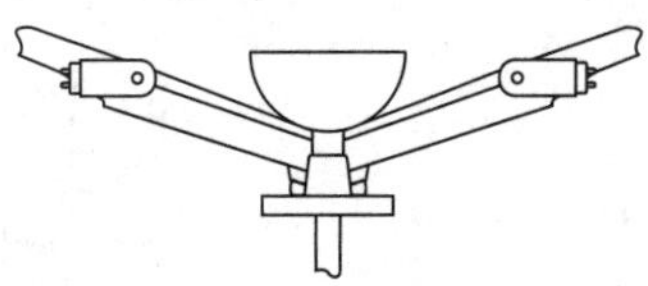

FIGURE 26.2 Press bending.

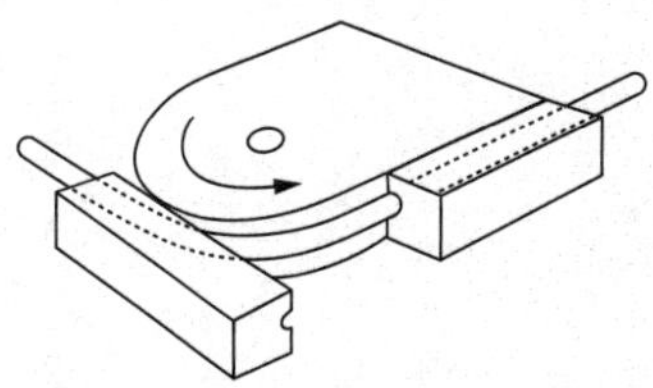

FIGURE 26.3 Rotary draw bending.

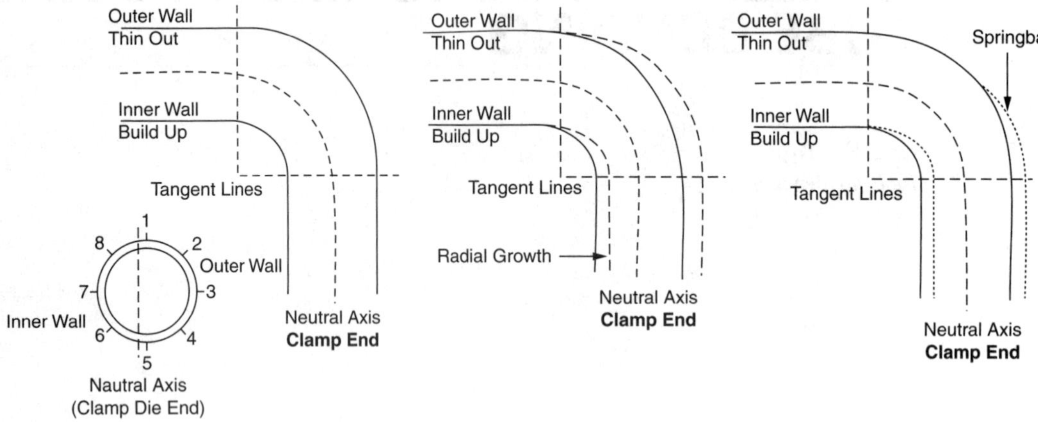

FIGURE 26.4 Reaction of the tube to bending.

3. As the wipe shoe rotates around the static bend die, it bends the work piece to the radius of the bend die.

Depending on the tube and bend specifications, compression bending can range from a simple to complex procedure. It is relatively simple when the bend radius is generous (e.g., 4 × O.D.) and the wall-to-tube ratio is low.

26.1.3 Press Bending

This method utilizes three steps:

1. A ram die with the desired radius of bend is fitted to the press arm.
2. The ram die forces the tubing down against the opposing two wing dies.
3. The wing dies, with resisting pressure, pivot up, and force the tubing to bend around the ram.

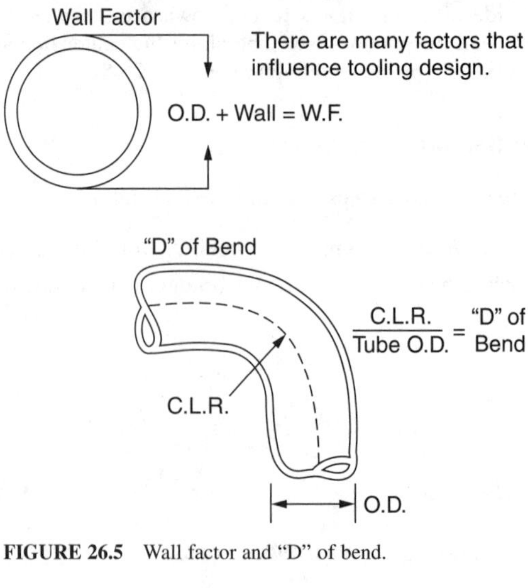

FIGURE 26.5 Wall factor and "D" of bend.

Because of its high rate of bending, press bending probably bends more miles of tubing than any other method. However, considerable distortion can occur since the tubing is not internally supported. For example, the tube may flatten on the outside of the bend and wrinkle or hump on the inside of the bend.

26.1.4 Rotary Draw Bending

This is probably the most common, versatile, and precise bending method (Fig. 26.6). It consistently produces high-quality bends, even with tight radii and thin wall tubes. Only three tools are required for bending heavy-walled tube to a generous radius (see Tables 26.1 and 26.2):

1. The work piece is locked to the bend die by the clamp die.
2. As the bend rotates, the follower type pressure die advances with the tube.
3. As the wall of the tube becomes thinner and/or the radius of bend is reduced, a mandrel and/or wiper are required.

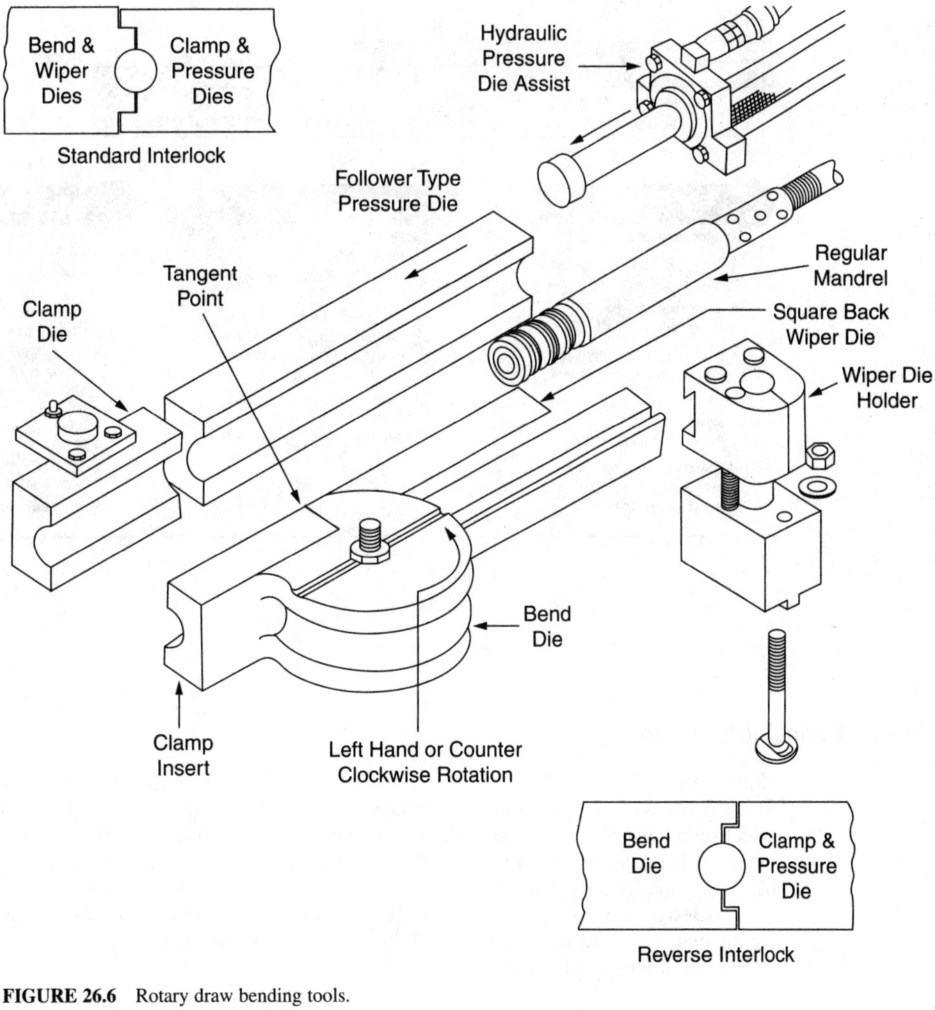

FIGURE 26.6 Rotary draw bending tools.

TABLE 26.1 Rotary Draw Bending—Design and Set-Up of Tooling

Typical Example:
2.0" O.D. × .065 Wall on 4" Centerline
Wall Factor 30 - 2 × "D" of Bend

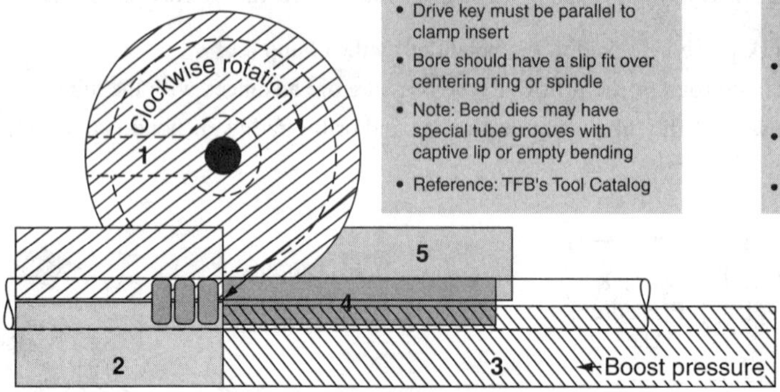

1. Bend die

- Hardened tool steel alloy steel, heat-treated and nitrided
- Clamp insert is secured with cap screws and dowel pins
- Drive key must be parallel to clamp insert
- Bore should have a slip fit over centering ring or spindle
- Note: Bend dies may have special tube grooves with captive lip or empty bending
- Reference: TFB's Tool Catalog

2. Clamp die

- Hardened tool steel alloy steel, heat-treated and nitrided
- Preferable length is $3\frac{1}{2}$ × tube O.D.
- Tube groove is grit blasted or may be serrated if less than preferred length
- With tube held in bend die, advance clamp die and adjust for vertical alignment
- Adjust for parallel contact with entire length of clamp
- Adjust for pressure

3. Pressure die

- Alloy steel and nitrided
- Tube groove must be parallel to back of die
- If follower type pressure die is used, length equals 180° + 2 O.D.
- If a boosted system is used, groove should be grit blasted
- With tube damped to bend die, advance pressure die and adjust for vertical alignment
- Start with minimum pressure and increase as required in small increments

4. Mandrel

- Type of mandrel and number of balls indicated by Tooling Selection Guide which is on back of this wall chart.
- Aluminum, bronze, chrome or Kro-lon mandrels for ferrous tubing.Only chrome madrels non-ferrous
- Gain best results with most mandrels when shank projects a small amount past tangent (bend & try)
- Lube L.D. of each tube

5. Wiper die

- The Tooling Selection Guide (on back of this wall chart) indicates when a wiper may be required
- Push tube over properly located mandrel and bring damp and pressure dies up to bending position
- Slide wiper along tube as far as possible into bend die then secure to holder
- Unclamp pressure and clamp dies, tip of wiper should be "very close" to tangent
- Adjust for rake and vertical alignment
- Lube each tube and the wiper

26.1.5 Springback Control

"Springback" describes the tendency of metal which has been formed to return to its original shape. There is excessive springback when a mandrel is not used, and this should be a consideration when selecting a bend die. Springback causes the tube to unbend from 2 to 10 percent depending on the radius of bend, and this can increase the radius of the tube after bending. The smaller the radius of bend, the smaller the springback.

The design and manufacture of tools is influenced by several factors. Wall factor and "D" of bend are the two most critical considerations followed by desired production rate, tubing shape and material, and required quality of bends.

TABLE 26.2 Rotary Draw Bending—Corrections for Poorly Bent Tubes

After the initial tooling set-up has been made. Study the bent part ot determine what tools to adjust to make a better bend. Keep in mind the basic bending principle of stretching the material on the outside radius of bend and compressing the material on the inside of bend. Make only one adjustment for each trial bend unless the second adjustment is very obviously needed. Avoid the tendency to first increase pressure die force rather than adjust the wiper die or mandrel location. Start with a clean deburred and lubed tube with the elongation properties sufficient to produce the bend.

Note: There are certainly other corrections that could be made for the following problems. These illustrations are a few examples of how to "read" a bend and improve the tooling set-up

1. Problem

Hump at end of bend.

Correction

1) Adjust mandrel slightly back from tangent until hump is barely visible. This is also a good system to find the best location for the mandrel.

2) Increase force on pressure die assist.

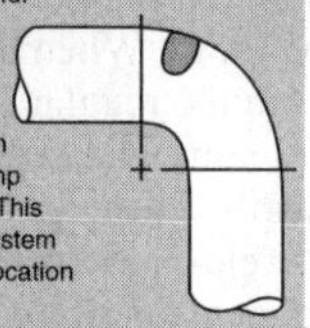

Clamp end

2. Problem

Tool marks on centerline of bend.

Correction

1) Re-adjust vertical alignment of clamp under pressure die.

2) Undersized tube groove in bend die.

3) Tooling not purchased from TFH.

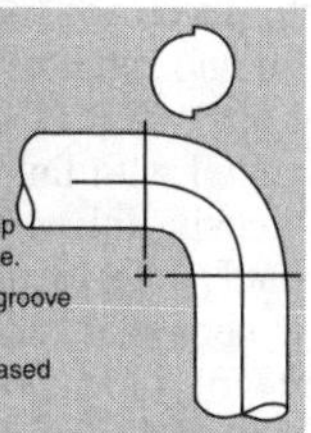

Clamp end

3. Problem

Tool marks on centerline of bend in clamp and pressure die area.

Correction

1) Reduce pressure and clamp die source.

2) Oversized tube or undersized tube groove from bad tooling source.

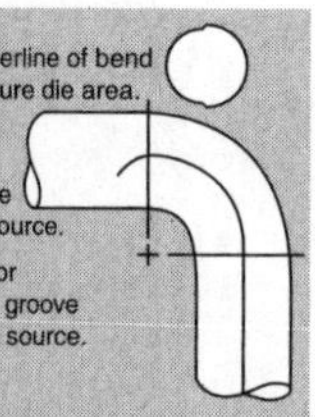

Clamp end

4. Problem

Wrinkling throughout bend even extending into wiper die area.

Correction

1) Advance wiper die closer to tangent.

2) Decrease rake of wiper die.

3) Recut worn wiper by TFB.

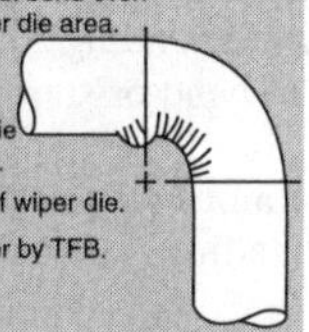

Clamp end

5. Problem

Bad mark at start of bend over bend for 90°.

Correction

1) Removable clamping portion of bend die not matched properly to round part of bend die.

2) Clamping portion of bend die not parallel to the key way.

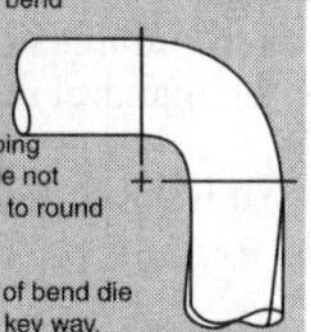

Clamp end

6. Problem

Wrinkling occurring for only a portion of the bend (45° out of 90°).

Correction

1) Bend die out of round. Bad centering ring or counter bore.

2) Taper in pressure die (from bottom of tube groove to back of die).

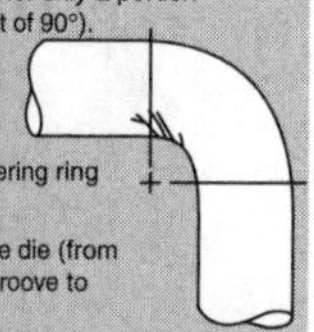

Clamp end

7. Problem

Wrinkles throughout bend area with wiper and mandrel in known proper position.

Correction

1) Check for under-sized mandrel.

2) Increase pressure die force only after checking wiper fit and mandrel location.

3) Reduce force on pressure die advance.

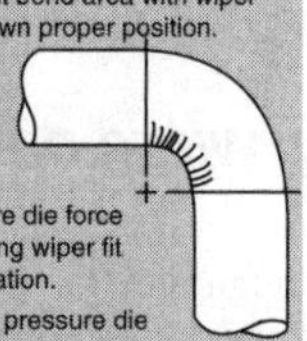

Clamp end

8. Problem

Excessive collapse with or without wrinkling throughout entire bend.

Correction

1) Advance mandrel toward tangency. untill slight hump occurs (? mandrels must project some what past tangency).

2) Need more balls on mandrel.

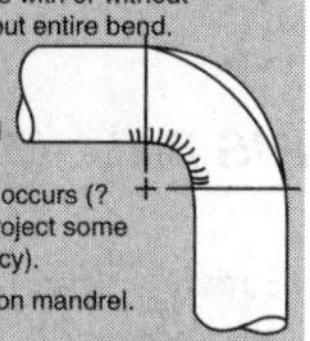

Clamp end

9. Problem

Mandrel ball humps

Correction

1) Too much drag on tube; back off pressure die force—increase wiper die rake

2) May require close pitch mandrel ball assembly.

3) Tubing material too soft.

4) Increase force on pressure die assist.

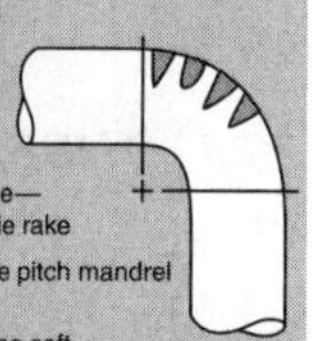

Clamp end

10. Problem

Excessive collapse after tubing is pulled off mandrel balls.

Correction

1) Check for too much drag on tube back off pressure die force. Increase rake on wiper die etc.

2) Increase mandrel to support change from a plug to a one ball; a 3 ball instead of a 2 ball mandrel, etc.

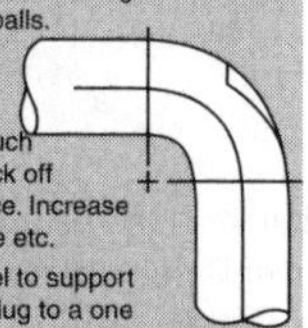

Clamp end

11. Problem

Deep scratches throughout the bend and in wiper die area.

Correction

1) Increase rake.

2) Check for undersized mandrel.

3) Increase pressure die force only after checking wiper fit and mandrel location.

4) Reduce force on pressure die advance.

5) Use more and/or a better lube.

6) Recut tube groove at TFB.

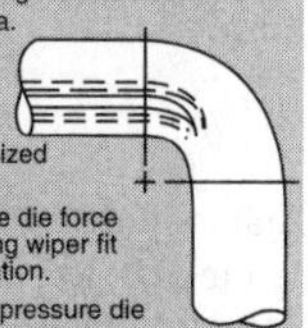

Clamp end

12. Problem

Heavy wrinkles through bend area only and linear scratches in grip area indicating clamp slippage

Correction

1) Reduce pressure die force.

2) Check location of mandrel and wiper die (and lube).

3) Increase pressure on clamp die.

4) Use serrated or carbide spray in tube groove of clamp die.

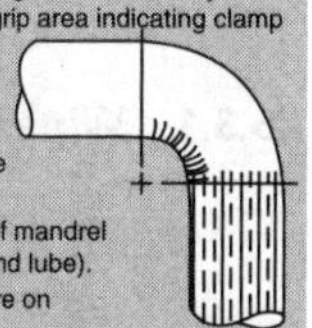

Clamp end

26.2 TYPES OF MANDRELS

The mandrel of choice has been the universal flexing steel-link mandrel in various forms including regular, close-pitch, ultraclose pitch. Single-plane, flexing, and brute mandrels are still being used.

Universal flexing mandrels rotate much like your wrist. Single plane-of-flex mandrels bend like your finger.

As the wall factor increases and D-of-Bend decreases, closer ball support to the tube is improved by reducing the size and pitch of the link. For example, a regular size/pitch link will work with a 1.500" O.D. on a 1.500" CLR (1 × D Bend) × 0.065" wall. When the wall is reduced to 0.042", go to a close-pitch mandrel with links down one size from the regular size.

Brute linkage or chain-link construction is ideal for nonround bending such as square, rectangular ("E" easy and "H" hard plane) extrusions, and rolled shapes. There are unique and special considerations for mandrels used in nonround bending applications—weld-flash height and consistent location, corner radius, material integrity and elongation, temper, dimensional consistency, and distance between plane of bend changes and surface finish.

The pressure die should be adjusted for a moderate pressure against the tube. The pressure die has three purposes. (1) It holds the tube against the bend die during bending. (2) It also keeps the mandrel from bending. (3) Finally, the pressure die maintains a straight tube behind starting tangent of bends (the portion of tubing still on the mandrel after bending).

The location of the mandrel relative to the point of bend or starting tangent affects the degree of springback.

The mandrel must be brought forward (toward clamp) when the radius is increased. However, there is no simple formula for the exact mandrel setting, so it should be determined with test bends.

When the tube breaks repeatedly, the material might be too hard for the application. Hard material lacks elongation properties and does not stretch sufficiently. Working with recently fully annealed material can help preclude this possibility. Breakage can also occur when the material is set too far forward or the tube slips minutely in the clamp die.

26.3 TUBE BENDING USING BALL MANDRELS AND WIPER DIES

These two tools are reviewed together because, although they have different functions, they generally perform in conjunction with one another. Ball mandrels and wiper dies are used with the previously discussed tools (bend, clamp, and pressure dies) when the ratio of tube diameter-to-wall thickness exceeds a certain value and bent on a given radius (Table 26.3).

The wiper die is used to prevent wrinkles. The ball mandrel performs essentially like the plug mandrel with the balls keeping the tube from collapsing after it leaves the mandrel shank.

26.3.1 Wiper Dies

Wiper dies are available in the conventional square back configurations. It is important to stress that the tip of the wiper die should be 0.005" to 0.010" thick depending on the size and material of the wiper die. The tip should never extend pass tangent, but it should be set as close as possible. The CLR machined surfaces should be a given percentage larger than the root diameter of the bend die. This accommodates for rake and some adjustment for wear.

26.3.2 Bending Thin Wall Tubing

When making tight bends or bending thin wall tubing containing the material during compression becomes increasingly difficult. The pressure is so intense the material is squeezed back past tangent

TABLE 26.3 Tooling Selection Guide

$$\text{"D" of Bend} = \frac{\text{centerline radius}}{\text{tube outside diameter}}$$

(2.0 C.L.R. ÷ 1.0" O.D. = 2 × D)

$$\text{Wall factor} = \frac{\text{tube outside diameter}}{\text{wall of tube}}$$

(2.0" O.D. ÷ .032 = 62.5 W.F.)

"D" of Bend	1 × D		1.50 × D		2 × D		2.50 × D		3 × D		3.50 × D	
Degree of bend	90°	180°	90°	180°	90°	180°	90°	180°	90°	180°	90°	180°
10 Ferrous	P	P	P	P								
Non-Ferrous	P	P	P	P	P	P						
★ 20 Ferrous	RP-1	RP-1	RP-1	RP-1	RP-1	RP-1	P	P				
Non-Ferrous	RP-1	RP-2	RP-2	RP-2	RP-2	RP-2	RP-1	RP-1	P	P		
30 Ferrous	RP-2	RP-2	RP-2	RP-2	RP-2	RP-2	RP-1	RP-1	P	P		
Non-Ferrous	RP-3	RP-3	RP-3	RP-3	RP-3	RP-3	RP-2	RP-2	RP-1	RP-1	P	P
40 Ferrous	RP-3	RP-3	RP-3	RP-3	RP-2	RP-2	RP-2	RP-2	RP-1	RP-1	P	P
Non-Ferrous	CP-4	CP-4	CP-4	CP-4	RP-3	RP-3	RP-3	RP-3	RP-3	RP-3	RP-2	RP-2
50 Ferrous	CP-4	CP-4	CP-3	CP-3	RP-3	RP-3	RP-2	RP-2	RP-2	RP-2	P	P
Non-Ferrous	CP-4	CP-4	CP-4	CP-4	CP-4	CP-4	RP-3	RP-3	RP-3	RP-3	RP-2	RP-2
60 Ferrous	CP-4	CP-4	CP-4	CP-4	CP-4	CP-4	RP-3	RP-3	RP-3	RP-3	RP-1	RP-1
Non-Ferrous	CP-5	CP-5	CP-4	CP-4	CP-4	CP-4	CP-4	CP-4	RP-3	RP-3	RP-1	RP-1
70 Ferrous	CP-5	CP-6	CP-6	CP-5	CP-4	CP-5	CP-4	CP-4	RP-3	RP-4	RP-1	RP-1
Non-Ferrous	UCP-6	UCP-6	UCP-6	UCP-6	CP-4	CP-4	CP-4	CP-4	CP-4	CP-4	RP-2	RP-2
80 Ferrous	CP-5	CP-5	CP-5	CP-6	CP-4	CP-5	CP-4	CP-4	RP-3	RP-3	RP-1	RP-1
Non-Ferrous	UCP-6	UCP-8	UCP-6	UCP-8	UCP-5	UCP-6	CP-4	CP-5	CP-4	CP-4	RP-3	RP-3
90 Ferrous	UCP-6	UCP-8	UCP-5	UCP-5	CP-4	CP-4	CP-4	CP-4	CP-4	CP-4	RP-3	RP-3
Non-Ferrous	UCP-8	UCP-10	UCP-8	UCP-10	UCP-6	UCP-6	UCP-6	UCP-6	CP-4	CP-4	RP-3	RP-3
100 Ferrous	UCP-6	UCP-8	UCP-6	UCP-6	UCP-5	UCP-5	UCP-5	UCP-5	UCP-5	UCP-5	CP-4	CP-4
Non-Ferrous	UCP-8	UCP-8	UCP-8	UCP-8	UCP-8	UCP-8	UCP-6	UCP-6	UCP-5	UCP-5	CP-5	CP-5
125 Ferrous			UCP-6	UCP-6	UCP-6	UCP-6	UCP-5	UCP-5	UCP-5	UCP-5	CP-4	CP-4
Non-Ferrous					UCP-6	UCP-6	UCP-6	UCP-6	UCP-6	UCP-6	CP-4	CP-4
150 Ferrous	CAUTION:				UCP-8	UCP-8	UCP-6	UCP-6	UCP-6	UCP-6	CP-5	CP-5
Non-Ferrous							UCP-8	UCP-8	UCP-6	UCP-6	UCP-6	UCP-6
175 Ferrous			BETTER CALL:				UCP-6	UCP-6	UCP-6	UCP-6	CP-6	CP-6
Non-Ferrous									UCP-8	UCP-8	UCP-8	UCP-8
200 Ferrous			TOOLS FOR BENDING, INC.						UCP-6	UCP-6	CP-6	CP-6
Non-Ferrous												

KEY
P-Plug or Empty-Bending
RP-Regular Pitch
CP-Close Pitch
UCP-Ultra Close Pitch
No. indicates suggested number of balls

Note
1. The Empty-Bending system (without a mandrel or wiper die) is recommended for applications above the dotted line.
2. A wiper die is recommended for applications below the dotted line.
3. "H" style brute, chain link mandrel in regular pitch, close pitch, and ultra-close pitch.
4. All mandrels are available with tube holes and grooves and finished in chrome, Kro-Lon, AMPCO bronze.

where is not supported by the bend die and wrinkles. This area must be supported so the material will compress rather than wrinkle, and this is the primary purpose of the wiper die.

Bending thin wall tubing has become more prevalent in recent years, and tight-radius bends of center line radius equaling the tube outside diameter (1 × D) have accompanied thin wall bending. To compound the problem, new alloys have been developed that are extremely difficult to bend, and the EPA has restricted the use of many good lubricants.

In bending square or rectangular tubing, material builds up on the inside of bend and binds the tube in the bend die preventing easy removal. There are several ways to eliminate this. In leaf construction, the bend die captures one or both plates on the top and bottom of the pressure die; but this does not provide a high quality bend. A better approach is to capture three sides of the square tube in the bend die. After the bend is completed, the top plate is lifted by a manual or hydraulic actuator.

26.4 EXAMPLE CASE STUDY

26.4.1 Application

- 2.0"O.D. × .035" wall on a 2" centerline radius bend
- Tubing material is 6061-T4 aluminum, one bend 90°, 4" long legs
- Tooling to fit "Conrac" No. 72 with pressure die advance system
- Total parts 2000 pieces. Aircraft Quality
- Factor: 60 – 1 × D

26.4.2 Recommendation

- *Bend die.* Type 3 (Fig. 26.7), one piece construction with a partial platform for rigidity. Reverse interlocking for ease of set-up and quality bend. Hardened 58-60Rc, 6" long clamp. Radius portion to have 0.060 lip or 1.060 deep tube groove to minimize possible tool marks.
- *Clamp die.* Light grit blast in tube groove for improved grip. Interlocked to bend die for ease of set-up and minimize clamp marks.
- *Pressure die.* Interlocked to bend die for ease of set-up. Negative lip—preventing pressure die from hitting the bend die. Tube groove with light grit blast to enhance benefit of pressure die advance.
- *Wiper die.* 4130 alloy material preheated 28–32 Rc. Interlocked to pressure die.
- *Mandrel.* Close pitch series to prevent wrinkles. Four balls for additional support. Hardened tool steel with hard chrome surface to minimize drag. (Kro-Lon surface is not used for soft or nonferrous tubing.)
- *Tooling set-up.* Much more attention is required to properly position the wiper die and mandrel. The bender is fitted with a pressure die advance to increase pressure applied through the tube against the wiper and bend without the normal drag which can stretch the wall and rupture. To conserve material and expedite production, the work piece will be bent 90° on each end, clamping twice in the center, when parted making two parts.
- *First bend.* Excessive collapse of over 5 percent of O.D. occurred. Wrinkles of 0.040 high appeared only in wiper die area.
- *Correction.* Mandrel advanced 0.070. The blunt end of the wiper die is located closer to the tube reducing rake. Obviously, to achieve a successful bend for this application, several more adjustment would have been made. It is prudent to make only one adjustment at a time.

26.4.3 Guidelines for Thin Wall Bending

The tubing should be a firm slip fit on the mandrel and clearance should not exceed 10 percent to 15 percent of wall thickness. This same clearance also applies to the pieces of outside tools.

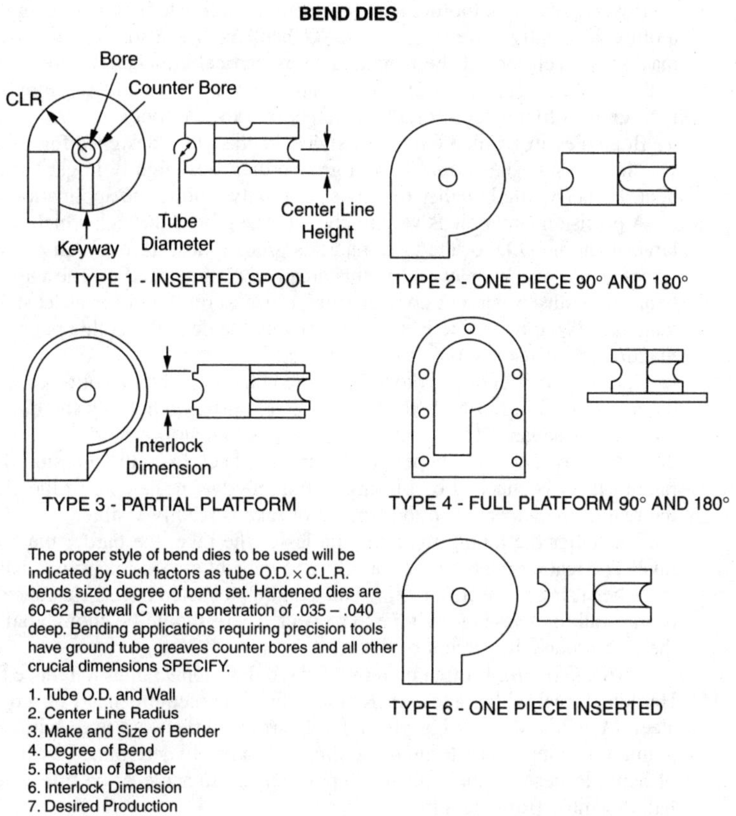

FIGURE 26.7 Bend dies.

Few tube bending machines are capable of bending thin wall, 1 × D tubing. Even machines designed for this special bending must be in excellent condition and be large enough to assure tooling rigidity. The mandrel rod should be as large as possible to eliminate its stretching. Wiper dies and their holders must be solid. Clamp and pressure die slides and tool holders must be tight.

A full complement of controls is essential for bending thin wall tubing. The machine must be capable of retracting and advancing the mandrel with the clamp and pressure dies closed. A direct acting hydraulically actuated pressure die is desirable because it provides consistent pressure on the tube regardless of wall variation.

A pressure die advance should also be available. This counteracts the drag of the pressure die, mandrel, and wiper die, and pushes the tube into the bending area which prevents excessive wall thin-out.

Without a pressure die advance, the normally expected thinning is about three-quarters of the elongation of the outer wall. Therefore, a 2-in tube bent to a 3-in centerline radius will thin about 25 percent.

26.4.4 Lubrication

A controlled amount of lubricant can be applied to the mandrel and inside the tube. The lubricant must cover the entire inside of the tube. Wiper dies, and especially mandrels, can be machined to permit autolubrication.

Reverse interlock tooling may represent the ultimate in tube bending tooling. Complete interlock tooling originally developed for CNC benders has also proven advantageous for conventional machines. Each tool of the matching set is vertically locked in alignment. The clamp die is locked to the bend die. The wiper die located and locked to the bend die. Finally, the pressure die is locked in alignment to the bend for a completely interlocked tool set.

Bend dies are available in many styles. Each style is designed for different bending requirements.

The pressure die should have a groove diameter slightly larger than the O.D. of the tube to be bent. Properly fitted quality tooling should only require the application of containing pressure.

A precision wiper die is very important. The groove through which the tube slides must be slightly larger than the O.D. of the tube, and this groove must have a high polish lubricated with a thin oil. Excessive or overly heavy oil in this area can cause wrinkles. The wiper die must fit radially to the bend die with 85 percent contact from 12:00 to 6:00 and for at least 15 to 20 degrees back from tangency. If the wiper die is not supported by the bend die at this point, it will spring away from the mandrel and cause the tube to wrinkle.

The proper fit of wiper die to bend die is facilitated by a solid bar or thick walled tube the exact diameter of tubing to be bent. While the set-up bar is held by the clamp die and pressure die, the wiper die is advanced to the most forward position and secured to the wiper die holder. To minimize drag, the flat end of the wiper can be brought back from the pressure die or "rake." The amount of rake or taper is checked by placing a straight edge in the core of the clamp groove so it extends to the rear of the wiper. Then the amount of rake is readily visible.

The softer the tubing material, the lesser the rake; the harder the tubing material, the more the rake. The feather edge must be as close to tangent as possible, obviously never past tangent.

When using a universal flexing ball mandrel, it should have a clearance of approximately 10 percent of the wall thickness of the tube to be bent. There should be enough balls on the mandrel to support the tube around 40 percent of the bend.

AMPCO bronze is often preferred for stainless applications to reduce friction and prevent marking. Hardened steel with chrome or Kro-Lon finish is recommended for commercial bending of carbon steel. Mandrels with a high polish hard chrome surface are used with nonferrous materials such as aluminum, copper, etc. Mandrel settings are partially determined by the tubing materials and radius of bend. Project the mandrel shank past tangent to achieve the full benefit of the shank protect the ball assembly from breaking.

26.5 CONCLUSION

This chapter attempts to separate facts and modern good practices from misconceptions and antiquated methods. Admittedly, there are and will continue to be isolated instances where deviations from these recommendations will be required. New techniques and extensions of systems that have been discussed here will continue to be developed.

P • A • R • T • 4

METALWORKING, MOLDMAKING, AND MACHINE DESIGN

CHAPTER 27
METAL CUTTING AND TURNING THEORY

Gary Baldwin
Director, Kennametal University
Latrobe, Pennsylvania

27.1 MECHANICS OF METAL CUTTING

Metal cutting is a process in which a wedge-shaped cutting tool engages the workpiece to remove a layer of material in the form of a chip.

As the cutting tool engages the workpiece, the material directly ahead of the tool 1 is deformed (Fig. 27.1). The deformed material then seeks to relieve its stressed condition by flowing into the space above the tool as the tool advances. Workpiece section 1 is displaced by a small amount relative to section 2 along specific planes by a mechanism called plastic deformation.

When the tool point reaches the next segment, the previously slipped segment moves up further along the tool face as part of the chip. The portions of the chip numbered 1 to 6 originally occupied the similarly numbered positions of the workpieces. As the tool advances, segments 7, 8, 9, and so on, which are now part of the workpieces, will become part of the chip. Evidence of shear can be seen in the flow lines on the inner surface of the chip. The outer surface is usually smooth due to the burnishing effect of the tool.

Metal deforms by shear in a narrow zone extending from the cutting edge to the work surface. The zone of shear is known as the *shear plane*. The angle formed by the shear plane and the direction of the tool travel is called the *shear angle* (Fig. 27.2).

27.1.1 Types of Chips

Two types of chips may be produced in the metal cutting process. The type of chip produced depends upon workpiece material, tool geometry, and operating conditions.

- *Discontinuous chips.* They consist of individual segments, which are produced by fracture of the metal ahead of the cutting edge. This type of chip is most commonly produced when machining brittle materials, especially cast irons, which are not ductile enough to undergo plastic deformation (Fig. 27.3).
- *Continuous chips.* They are produced when machining ductile materials like steels and aluminums. They are formed by continuous deformation of the metal without fracture ahead of the tool. If the rake surface of the insert is flat, the chip may flow in a continuous ribbon. Usually, a chip groove is needed to control this type of chip (Fig. 27.4).

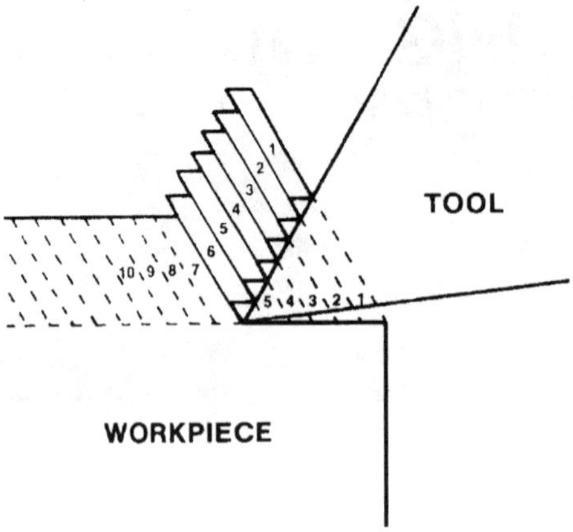

FIGURE 27.1 Deformed material.

Regardless of the type of chip formed, compressive deformation will cause it to be thicker and shorter than the layer of workpiece material removed. The work required to deform this material usually accounts for the largest portion of forces and power involved in a metal removal operation. For a layer of work material of given dimensions, the thicker the chip, the greater the force required to produce it. The ratio of chip thickness, to the undeformed chip thickness (effective feed rate) is often called the *chip thickness ratio*. The lower the chip thickness ratio, the lower the force and heat, and the higher the efficiency of the operation. The chip thickness ratio can never reach 1.0—if no deformation takes place then no chip will be formed. Chip thickness ratios of approximately 1.5 are common. The following formula will assist in calculating chip thickness.

$$\frac{t_2}{t_1} = \frac{\cos(\phi - \sigma)}{\sin \phi}$$

where t_1 = undeformed chip thickness
t_2 = chip thickness after cutting
θ = shear angle
σ = true rake angle

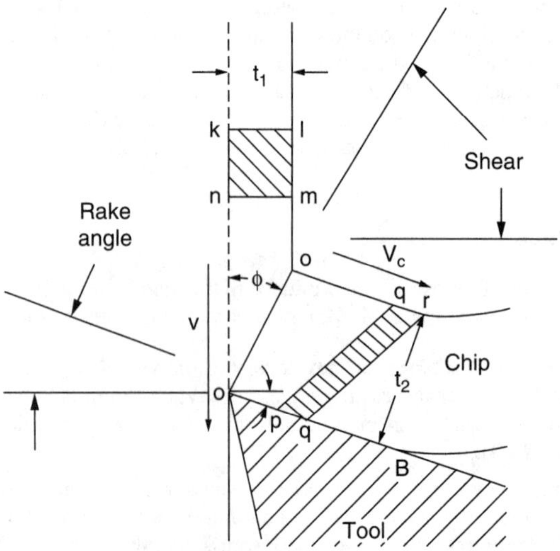

FIGURE 27.2 Shear angle.

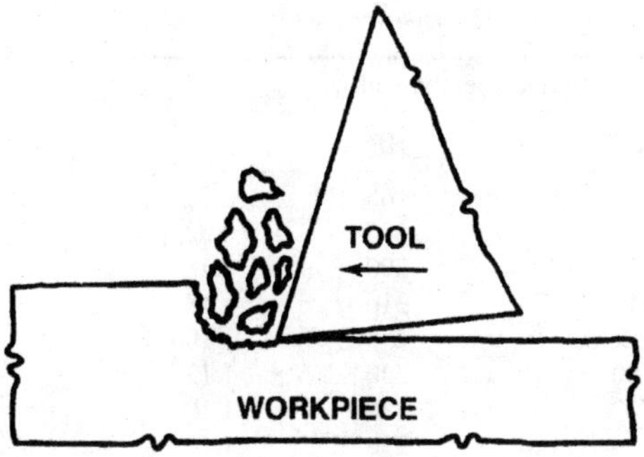

FIGURE 27.3 Discontinuous chips.

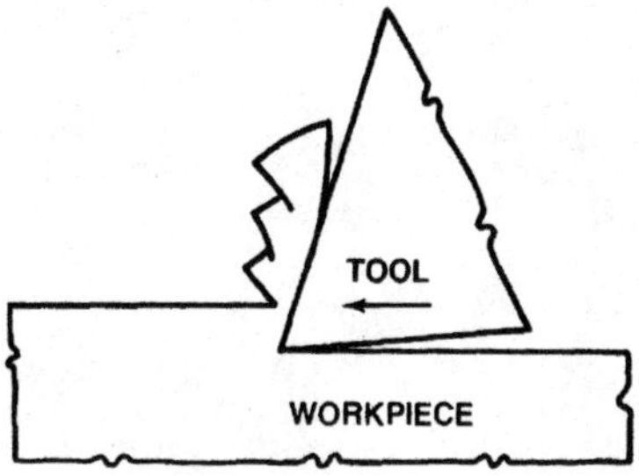

FIGURE 27.4 Continuous chips.

27.1.1 Power

One method of estimating power consumption in a turning or boring operation is based on the metal removal rate. Metal removal rate can be determined by the following formula:

$$Q = 12 \times V_t \times F_r \times d$$

where Q = metal removal rate (cubic inches per minute)
V_t = cutting speed (surface feet per minute)
F_r = feed rate (inches per revolution)
D = depth of cut (inches)

The *unit horsepower factor* (P) is the approximate power required at the spindle to remove 1 in^3/min of a certain material. Unit horsepower factors for common materials are given in the following chart (Tables 27.1 and 27.2). The approximate horsepower to maintain a given rate of metal removal can be determined by the following formula:

$$\mathrm{HP}_s = Q \times P$$

where HP_s = horsepower required at the spindle
Q = metal removal rate
P = unit horsepower factor

In practice, P is also dependent on the cutting speed, undeformed chip thickness, true rake angle, and tool wear land. If the formula shown above is used to calculate horsepower requirements, the result should be increased by approximately 50 percent to allow for the possible effects of these other factors. If a more accurate estimate is needed, the effect of these other factors can be included in the basic formula as described below (Fig. 27.5).

When machining most materials, as cutting speed increases up to a certain critical value the unit horsepower factor is reduced. This critical value varies according to the type of material machined. Once it is reached, further increases in cutting speed will not significantly affect unit horsepower.

As the undeformed chip thickness is increased, horsepower required per unit of metal removal is reduced. Increasing the undeformed chip thickness by increasing the feed rate will increase horsepower consumption. But, the increase in horsepower consumption will be proportionately smaller than the increase in metal removal rate. This is because extra power is required to deform the metal in the chip that passes over the cutting tool. As chip thickness is increased, this power becomes smaller in comparison to the total power required.

The undeformed chip thickness depends upon the *feed per revolution* (IPR) and the *lead angle* of the toolholder. In a single point operation with no lead angle the undeformed chip thickness will

TABLE 27.1 Unit Horsepower Factor

High temperature alloys		
Material	HB	P
A 286	165	0.82
A 286	285	0.93
CHROMOLOY	200	0.78
CHROMOLOY	310	1.18
HASTELLOY-B	230	1.10
INCO 700	330	1.12
INCO 702	230	1.10
M-252	230	1.10
M-252	310	1.20
TI-150A	340	0.65
U-500	375	1.10
4340	200	0.78
4340	340	0.93

Nonferrous metals and alloys	
Brass	P
Hard	83
Medium	50
Soft	33
Free machining	25
Bronze	
Hard	83
Medium	50
Soft	33
Copper	
Pure	90
Aluminum	
Cast	25
Hard (rolled)	33
Monel	
Rolled	1
Zinc alloy	
Die cast	25

equal the feed per revolution. The effect of a lead angle for a given feed per insert is to reduce the undeformed chip thickness. When a lead angle is used, the undeformed chip thickness (Fig. 27.6) can be determined by the following formula:

$$t = F_r \times \cos c$$

where t = undeformed chip thickness (inches)
F_r = feed rate (inches per revolution)
c = lead angle (degrees)

The element of tool geometry with the greatest effect on unit horsepower consumption is the *true rake angle* (*compound rake angle*). True rake angle is the angle formed on the top of the toolholder and the rake face of the insert, measured in a plane perpendicular to the cutting edge. As the true rake

TABLE 27.2 Ferrous Metals and Alloys

		Brinnell hardness number					
		150–175	176–200	201–250	251–300	301–350	351–400
ANSI	1010–1025	0.58	0.67				
	1030–1055	0.58	0.67	0.80	0.96	—	—
	1060–1095	—	—	0.75	0.88	1.00	—
	1112–1120	0.5	—	—	—	—	—
	1314–1340	0.42	0.46	0.50	—	—	—
	1330–1350	—	0.67	0.75	0.92	1.10	—
	2015–2115	0.67	—	—	—	—	—
	2315–2335	0.54	0.58	0.62	0.75	0.92	1.00
	2340–2350	—	0.50	0.58	0.70	0.83	—
	2512–2515	0.5	0.58	0.67	0.80	0.92	—
	3115–3130	0.5	0.58	0.70	0.83	1.00	1.00
	3160–3450	—	0.50	0.62	0.75	0.87	1.00
	4130–4345	—	0.46	0.58	0.70	0.83	1.00
	4615–4820	0.46	0.50	0.58	0.70	0.83	0.87
	5120–5150	0.46	0.50	0.62	0.75	0.87	1.00
	52100	—	0.58	0.67	0.83	1.00	—
	6115–6140	0.46	0.54	0.67	0.83	1.00	—
	6145–6195	—	0.70	0.83	1.00	1.20	1.30
	PLAIN CAST IRON	0.3	0.33	0.42	0.50	—	—
	ALLOY CAST IRON	0.3	0.42	0.54	—	—	—
	MALLEABLE IRON	0.42			—	—	
	CAST STEEL	0.62	0.67	0.80	—	—	

angle is increased (made more positive), cutting forces are reduced, horsepower consumption is reduced, and tool life is generally improved. On the other hand, the insert is placed in a weaker cutting position and the number of available cutting edges may reduce as a result of the increase in the true rake angle. True rake angle for any toolholder can be determined with the following formula:

$$\delta = \tan -1(\tan a \sin c + \tan r \cos c)$$

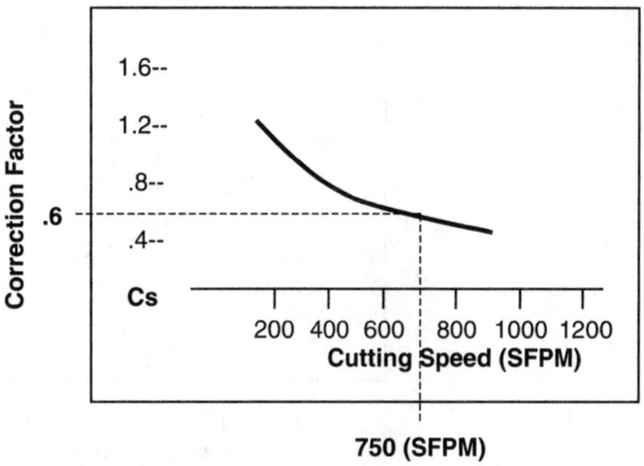

FIGURE 27.5 Cutting speed correction factor.

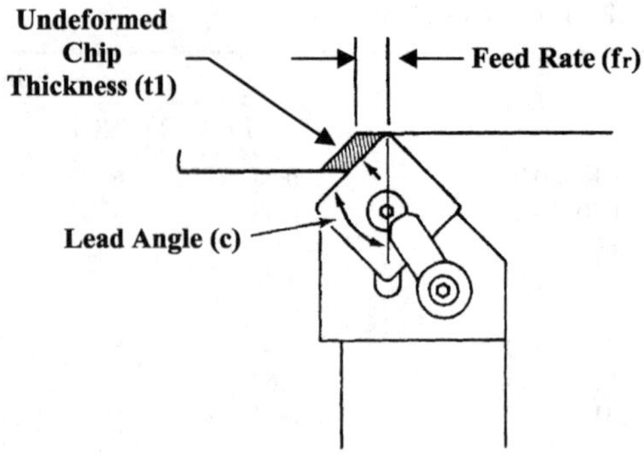

FIGURE 27.6 Undeformed chip thickness.

where δ = true rake angle
a = back rake angle
c = lead angle
r = side rake angle

The effect of true rake angle on unit horsepower is explained by the chip thickness ratio formula. This formula shows that when the rake angle is increased (becomes more positive), the shear angle increases and the ratio of chip thickness after cutting to the undeformed chip thickness is reduced. With thinner chips, less deformation occurs and less horsepower is required to remove a chip of a given thickness (Fig. 27.7).

Dull tools require more power to make a given cut. The full width of the wear land contacts the machined surface. So, as the wear land increases, power consumption increases. In typical operations, using a factor of 1.25 in the horsepower consumption formula can compensate for the effect of tool wear.

Unit horsepower factors allow the calculation of horsepower required at the spindle. They do not take into account the power required to overcome friction and inertia within the machine. The efficiency of a machine depends largely on its construction, type of bearings, number of belts or gears driving the spindle, carriage, or table, and other moving parts. The following chart provides typical *efficiency values* (*E*) for common machines used for turning and boring. The efficiency value is equal to the percentage of motor horsepower available at the spindle (Table 27.3).

$$HP_m = \frac{HP_s}{E}$$

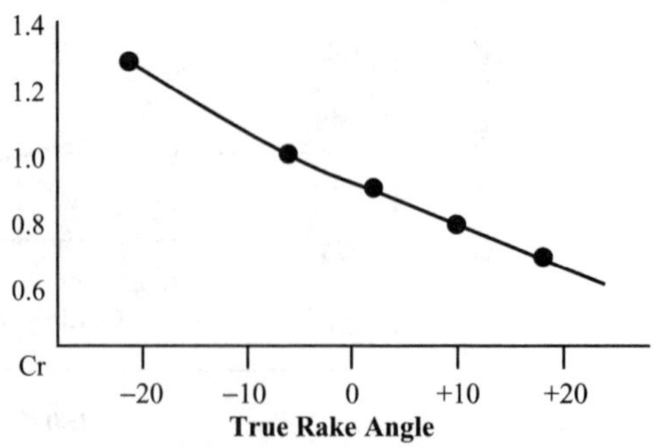

FIGURE 27.7 Rake angle correction factor (*Cr*).

TABLE 27.3 Spindle Efficiency

	Spindle efficiency (*E*)
Direct spindle drive	90%
One belt drive	85%
Two belt drive	70%
Geared head	70%

where HP_m = horsepower required at the motor
HP_s = horsepower required at the spindle
E = spindle efficiency

The correction factors are given by:

Cs = cutting speed correction factor

Ct = chip thickness correction factor

Cr = rake angle correction factor

1.25 = tool wear correction factor

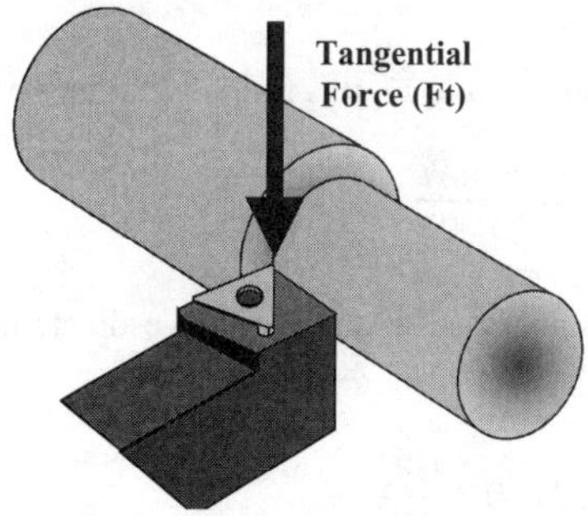

FIGURE 27.8 Tangential force.

When metal cutting occurs, three-force components act on the cutting tool:

Tangential force (*Ft*) acts in a direction tangent to the revolving workpiece and represents the resistance to the rotation of the workpiece (Fig. 27.8). Tangential forces are normally the highest of the three force components and account for nearly 99 percent of the total power required by the operation.

Longitudinal force (*Fi*) acts in a direction parallel to the axis of the work and represents resistance to the longitudinal feed of the tool (Fig. 27.9). Longitudinal force is approximately 50 percent as great as tangential force. Feed velocity is normally low when compared to the velocity of the rotating workpiece and accounts for only about 1 percent of total power required.

Radial force (*Fr*) acts in a radial direction from the centerline of the workpiece (Fig. 27.10). Increases in lead angle, or nose radius result in increased radial cutting forces. The radial force is the

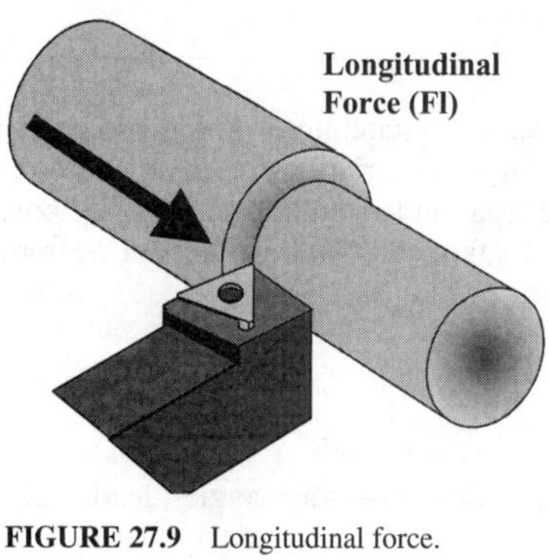

FIGURE 27.9 Longitudinal force.

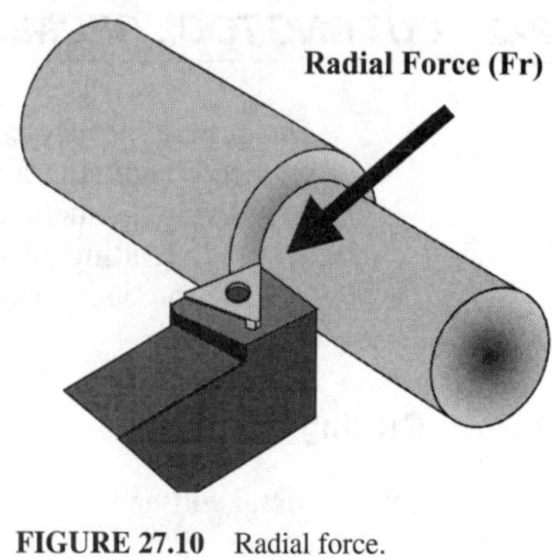

FIGURE 27.10 Radial force.

smallest of the three force components, representing approximately 50 percent of the longitudinal force, or 1/2 of 1 percent of the total cutting forces.

The total force acting on the cutting tool is the resultant of these three force components, and often denoted by *FR*. The numerical value of *FR* can be determined by the following formula:

$$FR = Ft^2 + Fl^2 + Fr^2$$

where FR = resultant force on the cutting tool
Ft = tangential force
Fl = longitudinal force
Fr = radial force

A fixed relationship exists between horsepower consumed at the spindle and cutting force. It is demonstrated by the following formula:

$$\mathrm{HP_s} = \frac{Ft \times Vt}{33{,}000} + \frac{Fl \times Vl}{33{,}000} + \frac{Fr \times Vr}{33{,}000}$$

where $\mathrm{HP_s}$ = horsepower required at the spindle
Vt = tangential force
Vl = longitudinal force
Vr = radial force

Since Vl and Vr are usually quite small in relation to Vt, this formula can be simplified to:

$$\mathrm{HP_s} = \frac{Ft \times Vt}{33{,}000}$$

Then, by solving for Ft, the following formula can be developed to estimate tangential cutting force:

$$Ft = 33{,}000 \times \frac{\mathrm{HP_s}}{Vt}$$

where Ft = tangential force
$\mathrm{HP_s}$ = horsepower at the spindle
Vt = cutting speed (SFPM)

27.2 CUTTING TOOL GEOMETRY

The general principles of machining require an understanding of how tools cut. Metal cutting is a science comprising a few components, but with wide variations of these components. To successfully apply the metal cutting principles requires an understanding of (1) how tools cut (geometry), (2) grade (cutting edge materials), (3) how tools fail, and (4) the effects of operating conditions on tool life, productivity, and cost of workpieces.

27.2.1 Cutting Geometry

Metal cutting geometry consists of three primary elements—rake angles, lead angles, and clearances angles.

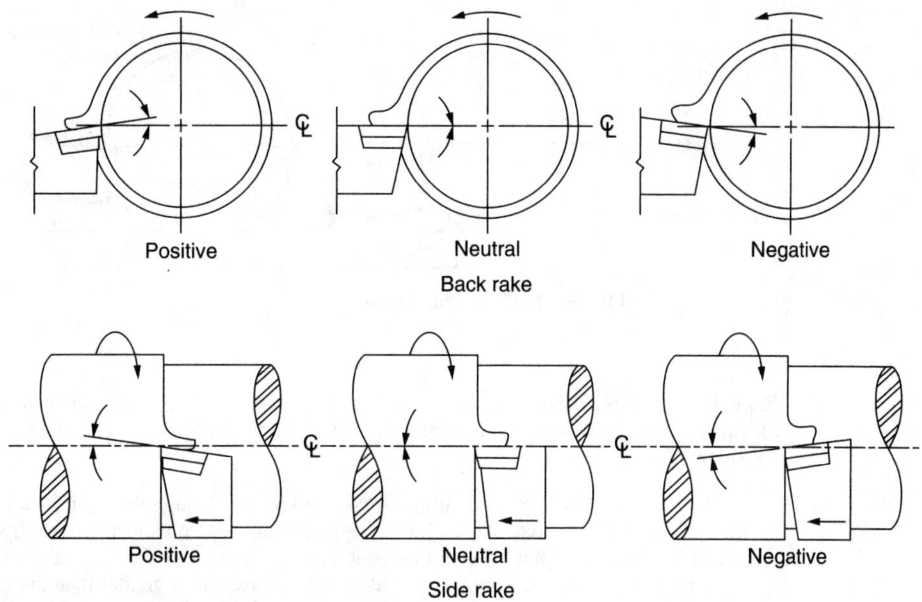

FIGURE 27.11 Back rake.

Rake Angles. A metal cutting tool is said to have rake when the face of the tool is inclined to either increase or decrease the keenness of the edge. The magnitude of the rake is measured by two angles called the *side rake angle* and *back rake angle*.

- *Side rake.* Side rake is measured perpendicular to the primary cutting edge of the tool (the cutting edge controlled by the lead angle). Side rake is the inclination of a line that is perpendicular to and on top of the cutting edge.
- *Back rake.* Back rake is measured parallel to the primary cutting edge of the tool (90 degrees from the side rake). Back rake is the inclination of a line that is parallel to and on top of the cutting edge (Fig. 27.11).

If the face of the tool did not incline but was parallel to the base, there would be no rake—the rake angles would be zero.

- *Positive rake.* If the inclination of the tool face is such as to make the cutting edge keener or more acute than when the rake angle is zero, the rake angle is defined as positive.
- *Neutral rake.* If the tool face is parallel to the tool base, there is no rake. The rake angle is defined as neutral.
- *Negative rake.* If the inclination of the tool face makes the cutting edge less keen or more blunt than when the rake angle is zero, the rake angle is defined as negative.

Dependent Rakes. Dependent rakes are rake angles that are applied based on the lead angle of the tool (dependent on lead). Both the side and the back rakes are based from the lead angle of the tool. Side rake is always measured perpendicular to the primary cutting edge while the back rake is measured parallel to the primary cutting edge. Dependent rakes follow the lead angle, changing position as the lead angle changes.

Independent Rakes. Independent rakes are rakes that are based on the tool axis and are independent of the tool lead angle. Side rake (axial) is measured parallel to the tool axis and back rake (radial) is measured perpendicular to the tool axis, regardless of the lead angle of the tool.

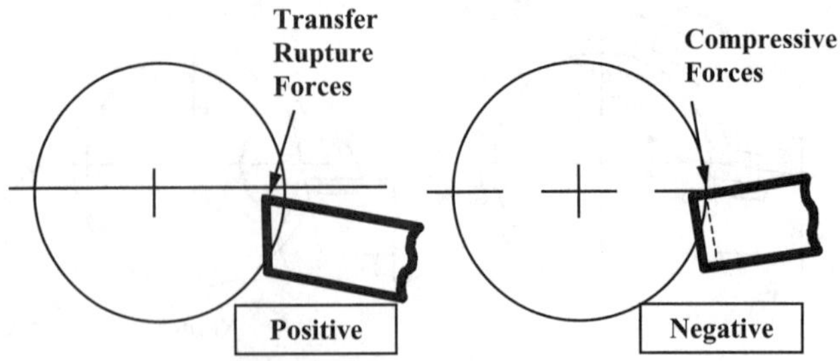

FIGURE 27.12 Cutting forces.

Rake Angle Usage. There are distinct advantages to each rake, which aid in the selection process for particular applications. Rake Angles impact the strength of the cutting edge and overall power consumed during the cut.

Strength of the Cutting Edge. Cutting forces tend to run through the cutting tool at right angles to the rake surface. Positive rake angles place the cutting edge under transverse rupture forces while negative rake angles place the cutting edge under compressive forces. The compressive strength of some cutting tool materials may be as much as three times greater than the transverse rupture strength.

Cutting Forces. Cutting forces change as the rake angles change from positive to negative. In mild steel, cutting forces change by approximately 1 percent per degree of rake change (Fig. 27.12).

Single point toolholders, except neutral handed, use rakes that are dependent on the lead angle while boring bars, and neutral handed tool holders use rakes that are independent of the lead angle. The workpiece material will determine if positive or negative rakes are to be used.

- Independent rakes will generally be used on tooling with internal applications (e.g., boring bars) or on OD tooling designed to be neutral handed. The advantage gained is the ease of determining the required rake angle to clear a given bore. This is due to the fact that the radial rake is applied perpendicular to the tool axis and is not related to the cutting edge. The axis of the internal tool and the axis of the workpiece are parallel under this condition.
- Dependent rakes will generally be used on tooling with external applications where there is no requirement to clear a minimum bore. The application of the rakes along and perpendicular to the cutting edge provides greater control of the cutting surface that is presented to the workpiece. The use of dependent rake orientation when using only one rake will permit the entire cutting edge to be parallel with the base of the tool.

Lead Angle (Side Cutting Edge Angle, Bevel Angle). It is defined as the angle formed between the cutting edge and the workpiece (Fig. 27.13). The direction of radial cutting forces is determined by the lead angle of the cutting tool. As the lead angle increases, the forces become more radial. Cutting forces tend to project off the cutting edge at right angles to the lead angle. In turning operations, at a low lead angle (0°) the forces are projected into the axis of the workpiece, while at a high lead angle (45°) the forces are projected across the radius of the workpiece. Lead angles do not impact total cutting forces, only the direction of the cutting force.

Lead angles control the chip thickness. As the lead angle increases the chip tends to become thinner and longer. As the lead angle decreases the chip tends to become thicker and shorter. Neither the volume of the chip nor the power consumed change with changes in lead angle.

It is important to note, that the amount of resultant (measured) cutting forces changes very little with changes in lead angle.

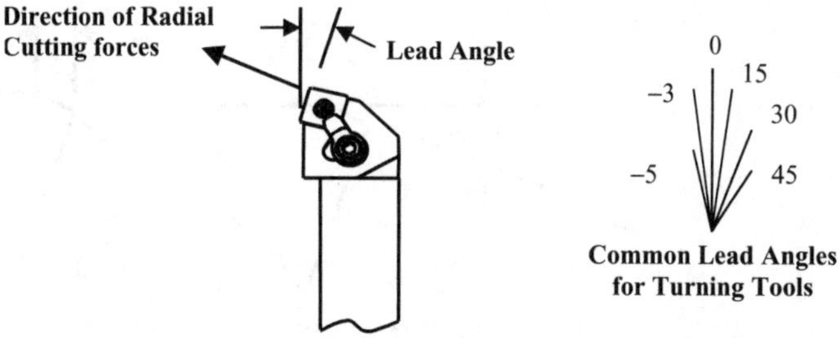

FIGURE 27.13 Lead angle.

Clearances Angles. They are workpiece material dependant and allow the tool to cut without rubbing on the flank surface of the cutting tool. Softer workpiece materials may require greater clearance angles than harder workpiece materials making the same cut. Primary clearance angles of approximately 5° are common on most cutting tools. This is adequate for most steels and cast irons but may be inadequate for aluminums and softer materials. Clearance angles may be 20° or greater for cutting tools designed for cutting certain soft workpieces materials.

27.2.2 Edge Preparation

The term *edge preparation,* as applied to a cutting tool, is a modification to both rake and clearance surfaces. Edge preparation is applied to the cutting edge of a tool for three primary reasons

1. To strengthen the cutting edge and reduce the tendency for the cutting edge to fail by chipping, notching, and fracture
2. To remove the minute burrs created during the grinding process
3. To prepare the cutting edge for coating by the *chemical vapor deposition* (CVD) process

This discussion will concentrate on strengthening the cutting edge and the resulting effect on the metal cutting process.

Edge preparation generally falls into three categories—sharp edge, honed, and T-landed cutting edges (Fig. 27.14).

Sharp Edge. The cutting edge on a carbide or ceramic cutting tool is never "sharp" when compared to an HSS cutting edge. The flash generated during the pressing and centering operations leaves irregularities that reduce the keenness of the edge. When the carbide or ceramic cutting tool is

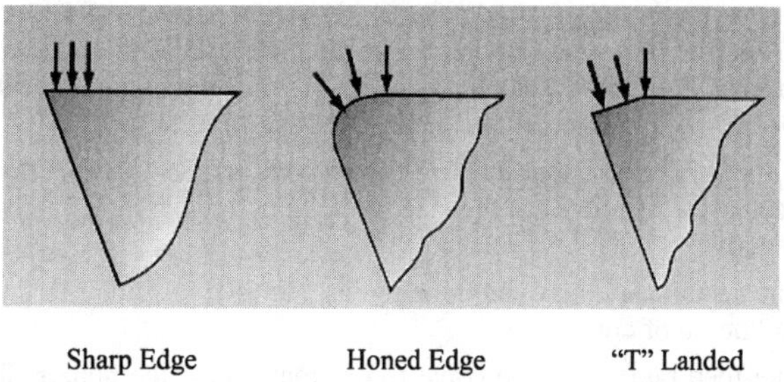

FIGURE 27.14 Types of edge preparation.

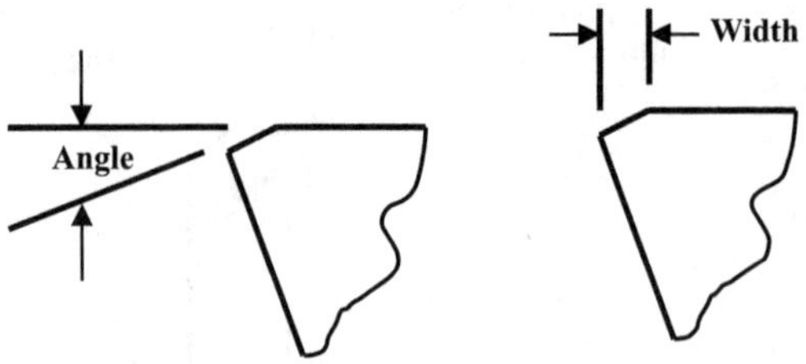

FIGURE 27.15 Angles.

ground, slight burrs are created that again reduce the keenness of the edge. Honing, lapping, or polishing of the rake and flank surfaces is necessary to gain the optimum edge keenness of the cutting edge.

T-Lands. T-lands are chamfers ground onto the cutting edge, which produce a change in the effective rake surface of the cutting tool. These chamfers, which make the rake surface more negative, are designed with a specific width and angle.

- Angles
  - The angle of the T-land is designed to provide the desired increase in edge strength. Increasing the chamfer angle directs the cutting forces through a thicker section of the cutting edge, placing the cutting edge more into compression.
  - Cutting forces increase as the chamfer angle increases (Fig. 27.15).
- Width
  - T-lands designed with a width greater than the intended feed rate change the total rake surface. This provides the maximum strength advantage but increases power consumption
  - T-lands designed with a width less that the intended feed form a compound rake surface. This limits the increase in power consumption while maintaining adequate edge strength.

The optimum T-land is one with the smallest angle and width that eliminates mechanical failure, (chipping, notching, and fracture). Angles greater than necessary to eliminate mechanical failure decrease useable tool life by *prewearing the insert.*

Increasing the angle of the T-land increases impact resistance. The smallest angle that eliminates mechanical failure is the optimum T-land angle (Fig. 27.16).

Increasing edge prep—including increasing the angle of T-lands—will decrease useable tool life if abrasive wear is the failure mechanism.

Hone Radius. The honing process can produce a radius, or angle and radius on the cutting edge. This *hone radius* strengthens the cutting edge by directing the cutting forces through a thicker portion of the cutting tool. The size of the hone radius is designed to be fed dependently.

- If the intended feed rate is greater than the size of the hone radius, a compound rake surface is formed. The hone radius forms a varying rake directly at the cutting edge with the actual rake surface forming the remainder of the effective rake surface.
- If the feed is less that the size of the hone radius, the hone forms the rake surface. As feed rate decreases, compared to the size of the hone radius, the effective rake becomes increasingly more negative.
- The relationship between hone radius and feed is similar to the relationship between *nose radius* and depth of cut.
- The feed rate should be equal to or greater than the hone radius on the cutting edge. i.e., if the cutting edge has a hone radius of .003 in the feed should be .003 in IPR/IPT or greater.

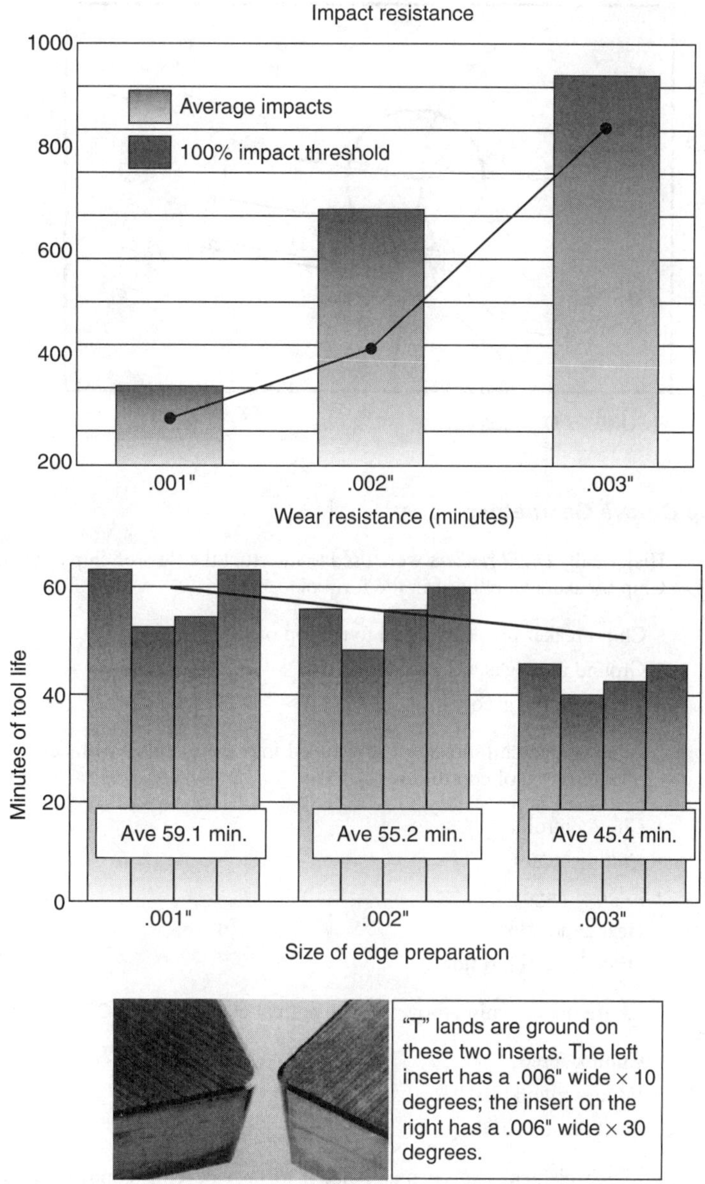

FIGURE 27.16 Edge preparation.

Clearance. The hone radius reduces the clearance directly under the cutting edge. In soft ductile materials this can create *built up edge* (BUE) on the flank of the tool. On work hardening workpiece materials this reduction of clearance can result in chipping of the cutting edge. This chipping is a result of heat generated from the hardened workpiece material rubbing the flank of the cutting tool. This excess heat causes thermal expansion of the cutting edge and results in thermal chipping of the rake surface (Fig. 27.17).

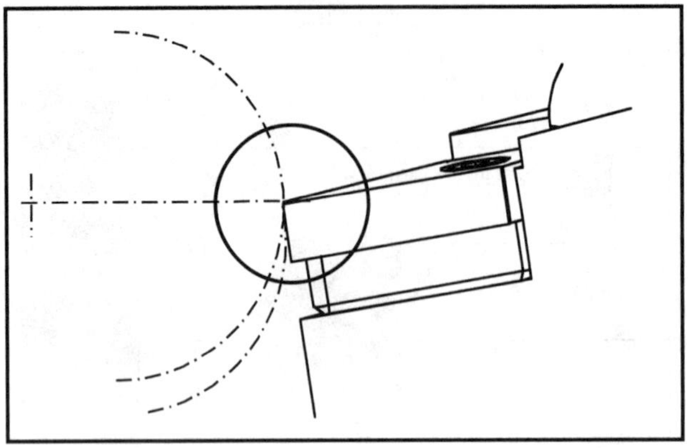

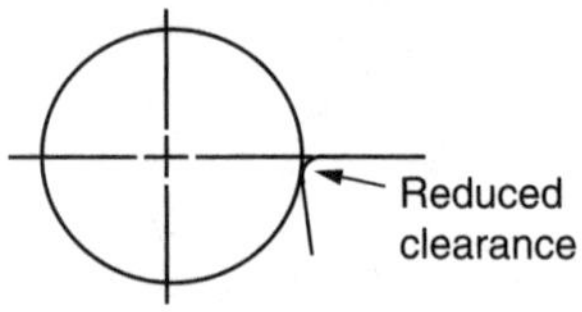

FIGURE 27.17 Clearance.

27.2.3 Chip Groove Geometries

Historically *chip breakers* were used to interrupt the flow of chips causing them to break (Fig. 27.18). Chip breakers were used in the form of

- Chip breaker plates clamped to the top of the insert
- Ground in ledges and grooves
- Traditional G grooves

The simple chip breaker has evolved into topographic surfaces that alter the entire rake surface of the cutting tool controlling:

- Chip control
- Cutting forces
- Edge strength
- Heat generation
- Direction of chip flow

A traditional chip groove has six critical elements (Fig. 27.19). Each affects

- Cutting force
- Edge strength
- Feed range

Each element can be manipulated to provide chip control, optimum cutting force, and edge strength for particular applications.

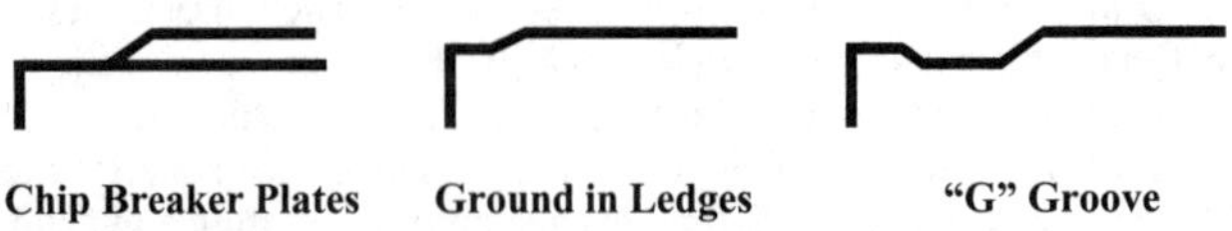

FIGURE 27.18 Chip groove geometries.

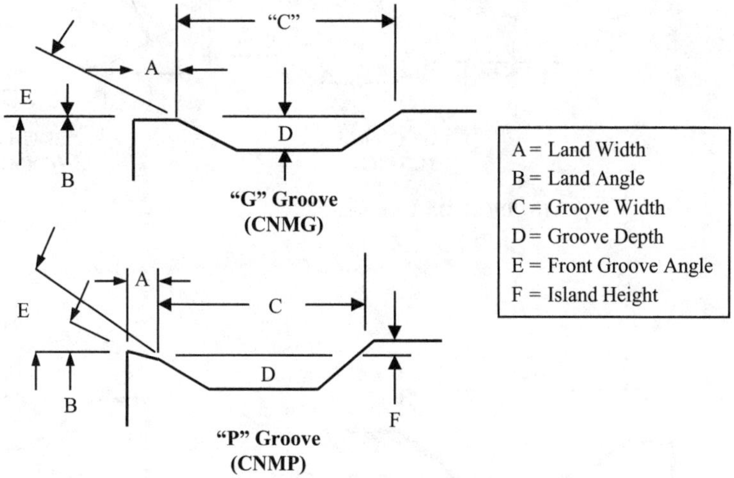

FIGURE 27.19 Traditional chip groove geometry.

Land width controls the point where chip control begins. This dimension corresponds to the "W" dimension used with mechanical chip breakers. A traditional industry standard $^1/_2$ in I.C. turning insert had a land width of between .010 in and .012 in (Fig. 27.20).

Land angle is a continuation of the rake surface and controls cutting forces and insert edge strength (Fig. 27.21).

Groove width (*C*) provides an interruption to normal chip flow, providing a controlling point for feed range. Excess feed rate will produce a "hairpin" chip which increases cutting forces (Fig. 27.22).

Groove depth effects cutting forces and chip flow characteristics of the insert. A deeper groove decreases cutting forces, weakens the insert, and produces a tighter chip and better chip control. A shallower groove increases cutting forces, strengthens the insert, and produces a loose chip and less chip control (Fig. 27.23).

Front groove angle. A steeper angle provides greater force reduction and better chip control but weakens the insert. A shallower angle reduces chip control while increasing forces and strengthening the insert (Fig. 27.24).

Island height. The height of the island is maintained above the cutting edge to provide greater interruption in chip flow while providing a resting surface for the insert that does not allow the cutting edge to be damaged (Fig. 27.25).

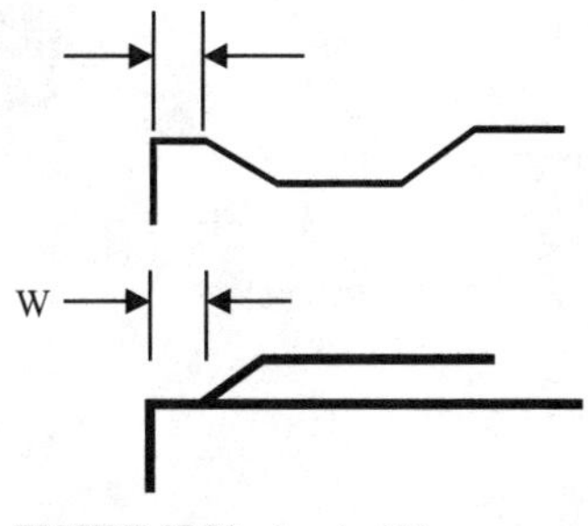

Typical Land Width
1/4" I.C. = .005"
3/8" I.C. = .007"
1/2" I.C. = .012"
3/4" I.C. = .018"

FIGURE 27.20 Land width.

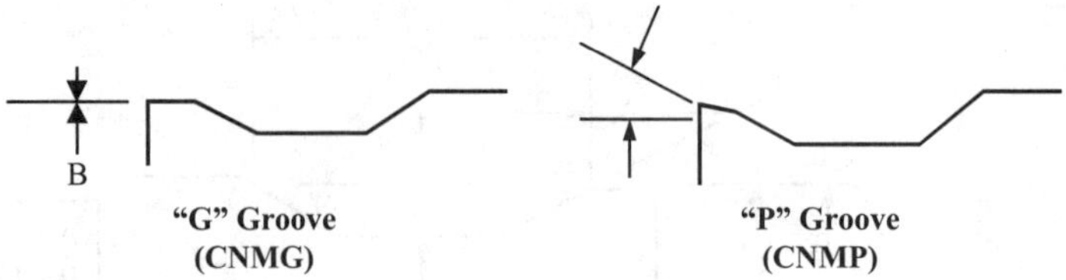

FIGURE 27.21 Land angle.

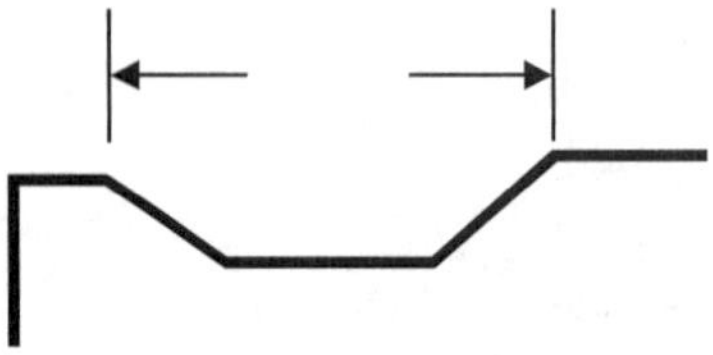

FIGURE 27.22 Groove width.

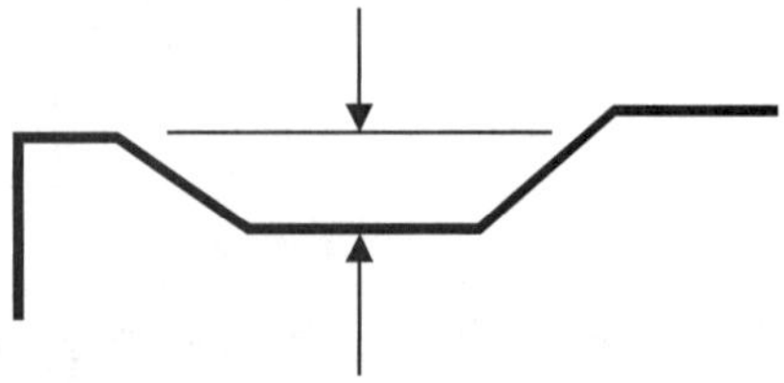

FIGURE 27.23 Groove depth.

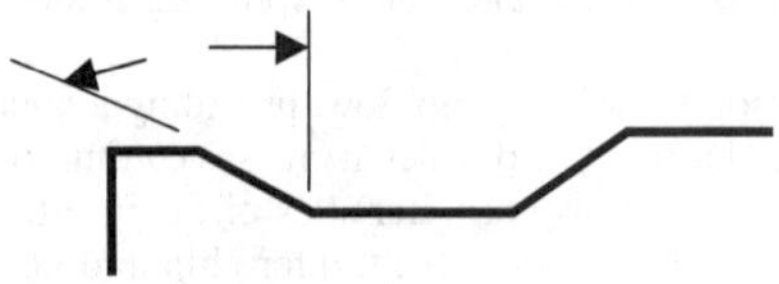

FIGURE 27.24 Groove angle.

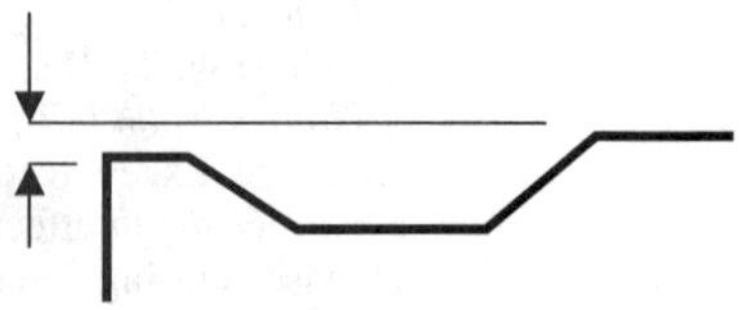

FIGURE 27.25 Island height.

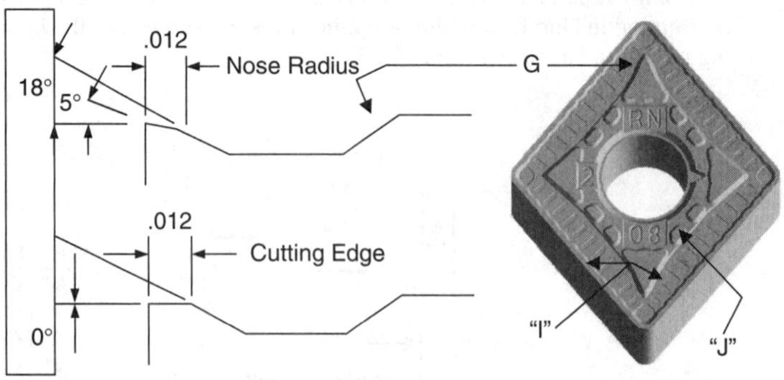

FIGURE 27.26 Nose.

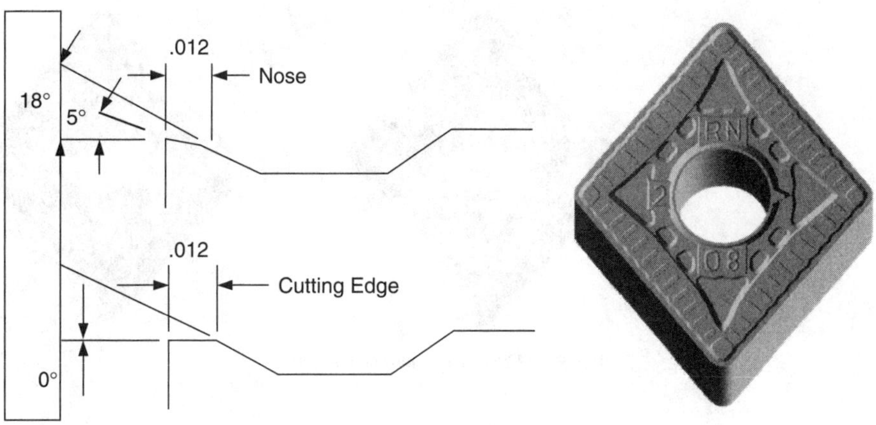

FIGURE 27.27 Nose radius geometry.

The latest chip groove designs have expanded the capabilities of cutting tools to control chips, control cutting forces, impact the strength of the cutting edge, control surface contact, and resultant heat and cutting forces, while deflecting the chips away from the finished surface of the workpieces (Fig. 27.26).

Angled back walls serve to deflect the chip flow away from the finished surface of the work-pieces.

Nose radius geometry. Many chip groove designs have different geometry on the nose radius than on the cutting edge of the insert. This allows an insert to serve as both a finishing insert at low depth of cut and reduced feed rates, and be effective at general purpose machining (Fig. 27.27).

Scalloped edges (I) located on the front wall of the groove, the floor of the groove, and on the island serve to suspend the chip. This reduces surface contact between the chip and the insert reducing heat and cutting forces. This allows greater productivity and increased tool life (Fig. 27.28).

Spheroids and bumps (J) serve to both impede chip flow providing chip control while reducing surface contact reducing heat and cutting forces (Fig. 27.29).

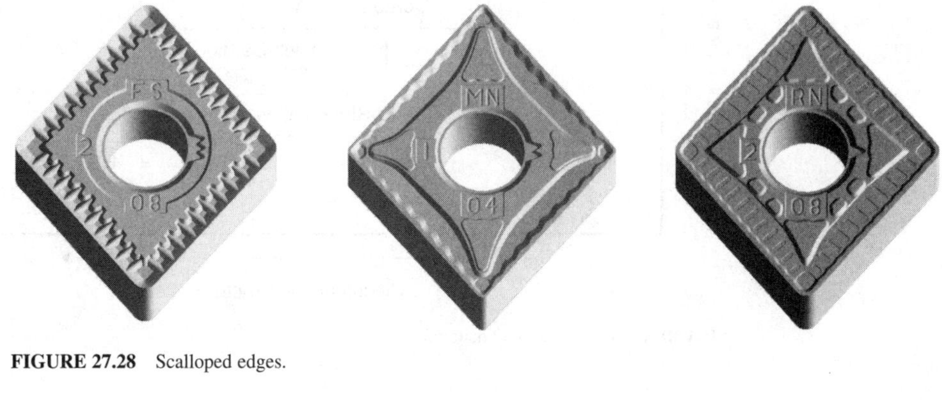

FIGURE 27.28 Scalloped edges.

FIGURE 27.29 Spheroids and bumps.

27.3 CUTTING TOOL MATERIALS

Many types of cutting tool materials, ranging from high-speed steel to ceramics and diamonds, are used as cutting tools in today's metalworking industry. It is important to be aware that differences and similarities exist among cutting tool materials (Fig. 27.30). All cutting tool materials can be compared using three variables;

- Resistance to heat (hot hardness)
- Resistance to abrasion (hardness)
- Resistance to fracture (toughness)

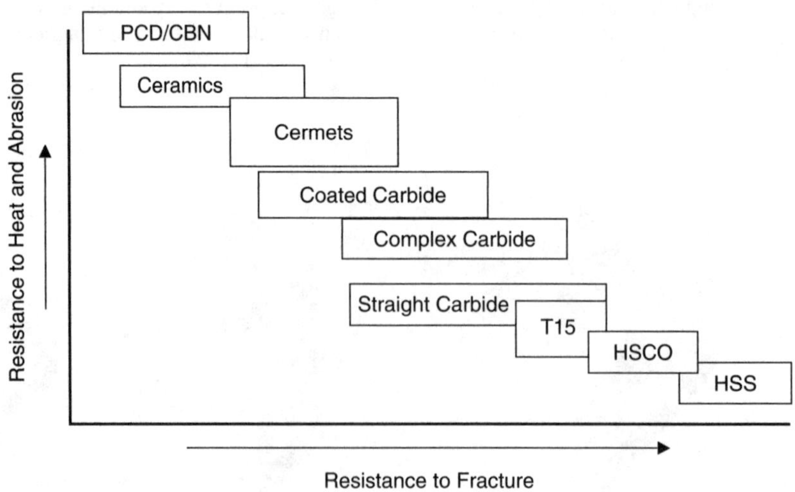

FIGURE 27.30 Cutting tool materials.

27.3.1 Material Selection

Factors affecting the selection of a cutting tool material for a specific application include:

- Hardness and condition of the workpiece material
- Operations to be performed-optimum tool selection may reduce the number of operations required
- Amount of stock to be removed
- Accuracy and finish requirements
- Type, capability, and condition of the machine tool to be used
- Rigidity of the tool and workpiece
- Production requirements influencing the speeds and feeds selected
- Operating conditions such as cutting forces and temperatures
- Tool cost per part machined, including initial tool cost, grinding cost, tool life, frequency of regrinding or replacement, and labor cost—the most economical tool is not necessarily the one providing the longest life, or the one having the lowest initial cost.

While highly desirable, no single cutting tool material is available to meet the needs of all machining applications. This is because of the wide range of conditions and requirements encountered. Each tool material has its own combination of properties making it the best for a specific operation.

27.3.2 High Speed Steels

Since the beginning of the twentieth century, high-speed steels (HSSs) have been an essential class of cutting tool materials used by the metalworking industry. HSSs are high-alloy steels designed to cut other materials efficiently at high speeds, despite the extreme heat generated at the cutting edges of the tools.

Classification of HSSs. Because of the wide variety of tool steels available, the American Iron and Steel Institute (AISI) has classified HSSs according to their chemical compositions. All types, whether molybdenum or tungsten, contain about 4 percent chromium; the carbon and vanadium contents vary. As a general rule, when the vanadium content is increased, the carbon content is usually increased (Table 27.4).

Molybdenum types of HSSs are identified with the prefix "M"; the tungsten types, with the prefix "T". Molybdenum types M1 through M10 (except M6) contain no cobalt, but most contain some tungsten. The cobalt-bearing—molybdenum-tungsten—premium types are generally classified in the M30 and M40 series. Super HSSs normally range from M40 upward. They are capable of being heat treated to high hardnesses.

The tungsten type T1 does not contain molybdenum or cobalt. Cobalt-bearing tungsten types range from T4 through T15 and contain various amounts of cobalt.

TABLE 27.4 Classification of HSSs

	Carbon (C)	Tungsten (W)	Molybdenum (Mo)	Chromium (Cr)	Vanadium (V)	Cobalt (Co)
M2	0.85	6.00	5.00	4.00	2.00	—
M7	1.00	1.75	8.00	4.00	2.00	—
M42	1.10	1.50	9.50	3.75	1.15	8.00
T1	0.75	18.00	—	4.00	1.00	—
T15	1.50	12.00	—	4.00	5.00	5.00

Advantages of HSS Tools. For good cutting tool performance, a material must resist deformation and wear. It must also possess a certain degree of toughness—the ability to absorb shock without catastrophic failure—while maintaining a high hardness at cutting edge temperatures. Also, the material must have the ability to be readily and economically brought to a final desired shape.

HSSs are capable of being heat treated to high hardnesses within the range of R_c63–68. In fact, the M40 series of HSSs is normally capable of being hardened to R_c70, but a maximum of R_c68 is recommended to avoid brittleness. HSSs are also capable of maintaining a high hardness at cutting temperatures. This hot hardness property of HSSs is related to their composition and to a secondary hardening reaction, which is the precipitation of fine alloy carbides during the tempering operation.

HSSs also possess a high level of wear resistance due to the high hardness of their tempered martensite matrix and the extremely hard refractory carbides distributed within this martensitic structure. The hardness of molybdenum-rich carbide M_6C is approximately R_c75 while the hardness of vanadium-rich carbide MC is about R_c84. Therefore, increasing the amount of MC increases the wear resistance of HSS. Although the higher vanadium HSSs (with up to 5 percent vanadium) are more wear resistant, they are more difficult to machine or grind.

HSS tools possess an adequate degree of impact toughness and are more capable of taking the shock loading of interrupted cuts than carbide tools. Toughness in HSSs can be increased by adjusting the chemistry to a lower carbon level or by hardening at an austenitizing temperature lower than that usually recommended for the steel, thereby providing a finer grain size. Tempering at a temperature range between 1100–1200°F (593649°C) will also increase the toughness of HSS. When toughness increases, however, hardness and wear resistance decrease (Fig. 27.31).

When HSSs are in the annealed state they can be fabricated, hot worked, machined, ground, and the like, to produce the cutting tool shape.

Limitations of HSS's. A possible problem with the use of HSSs can result from the tendency of the carbide to agglomerate in the centers of large ingots. This can be minimized by remelting or by adequate hot working. However, if the agglomeration is not minimized , physical properties can be reduced and grinding becomes more difficult. Improved properties and grindability are important advantages of powdered metal HSSs.

Another limitation of HSSs is that the hardness of these materials falls off rapidly when machining temperatures exceed about 1000–1100° F (538–593°C). This requires the use of lower cutting speeds than those used with carbides, ceramics, and certain other cutting tool materials.

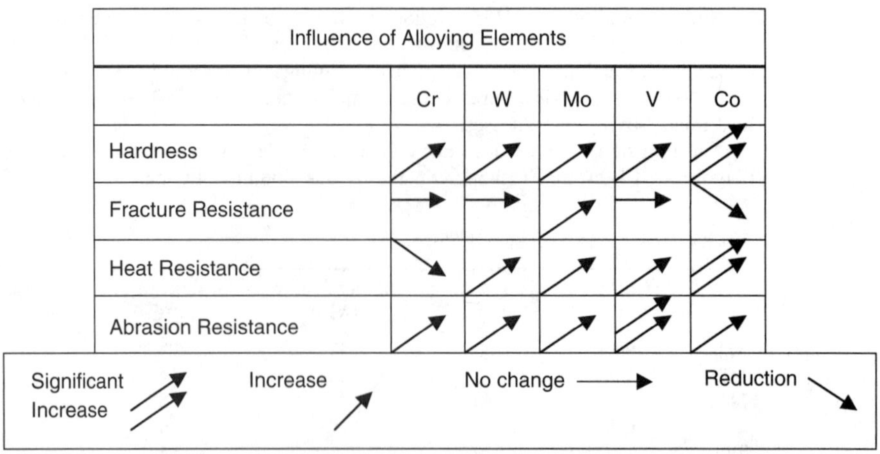

FIGURE 27.31 Influence of alloying elements.

Applications of HSS Tools. Despite the increased use of carbides and other cutting tool materials, HSSs are still employed extensively. Most drills, reamers, taps, thread chasers, end mills, and gear cutting tools are made from HSSs. They are also widely used for complex tool shapes such as form tools and parting (cutoff) tools for which sharp cutting edges are required. Most broaches are made from HSSs.

HSS tools are usually preferred for operations performed at low cutting speeds and on older, less rigid, low-horsepower machine tools. Reasons for the continued high usage of HSS tools include their relatively low-cost and easy fabrication, toughness, and versatility (they are suitable for virtually all types of cutting tools).

27.3.3 Powdered Metal High-Speed Tool Steels

High-speed tool steels made by powder metallurgy processes generally have a uniform structure with fine carbide particles and no segregation. Powder metal HSSs provide many advantages, and tools made from these materials are being increasingly applied.

Material Advantages. While HSSs made by the powder metal process are generally slightly higher in cost, tool manufacturing and performance benefits may rapidly outweigh this premium. In many cases, however, tools made from these materials are lower in cost because of reduced material, labor, and machining costs, compared to those made from wrought materials. Near-net shapes produced often require only a minimum of grinding, and the more complex the tool, the more savings possible.

Another important advantage is that the powder metal process permits more design flexibility. This is because complex tools shapes can be produced economically. Also, the method may allow the use of better-grade, higher-alloy steels that would be uneconomical to employ for tools with conventional production methods.

Applications. Milling cutters are becoming a major application for powder metal HSS (PM) tool steels. Metal removal rates can generally be increased through higher cutting speed and/or feed rate. In general, the feed per cutter tooth is increased for roughing operations, and the cutting speed is boosted for finishing.

27.3.4 Cast Cobalt-Based Alloys

Proprietary cutting tool materials are available as cast from cobalt-chromium-tungsten alloys. Molten metal is cast in chill molds made from graphite. Rapid cooling results in a fine grained, hard surface of complex carbides with a tough core.

Advantages. Tools cast from cobalt-based alloys are sometimes referred to as the intermediate tools for applications requiring properties between those of high-speed steel tools and carbide tools. They have proven effective for machining operations that are considered too fast for high-speed steel tools and too slow for carbide tools.

Cutting tools cast from cobalt-based alloys are particularly suited for machines with multiple tooling setups in which spindle speeds are restricted

Cast cobalt-based alloys cutting tools have greater fracture resistance than carbide and greater hot hardness than other high speed steels. Their high transverse rupture strength permits making interrupted cuts often not possible with carbide tools. Also, the high strength and low coefficient of friction of these tools make them ideal for slow speed, high-pressure operations such as cutoff and grooving.

27.3.5 Cemented Tungsten Carbides

Cemented carbides include a broad family of hard metals produced by powder metal techniques. Most carbide grades are made up of tungsten carbide with a cobalt binder.

Advantages of Cemented Carbides. High hardness at both room and high temperatures makes cemented carbides particularly well suited for metalcutting. The hardness of even the softest carbide used for machining is significantly higher than the hardest tool steel. Hot hardness—the capacity of WC-Co to maintain a high hardness at elevated temperatures—permits the use of higher cutting speeds. Critical loss of hardness does not occur until the cobalt binder has reached a temperature high enough to allow plastic deformation.

Cemented carbides are also characterized by high compressive strength. The compressive strength is influenced by Cobalt content, increasing as the cobalt content is increased to about 4 to 6 percent, then decreasing with additional amounts of cobalt.

Cemented carbides are classified into two categories;

- *Straight Grades.* They comprise tungsten carbide (WC) with a cobalt binder (Co) and are best suited for workpieces materials normally associated with abrasion as the primary failure mode, i.e., cast iron, nonferrous, and nonmetals.
- *Complex Grades.* They comprise tungsten carbide, titanium carbide (TiC), tantalum carbide (TaC) and often niobium carbide (NbC) with a cobalt (Co) binder. Complex grades of cemented carbide are best suited for "long chip" materials such as most steels (Fig. 27.32).
  - Titanium carbide provides a resistance to cratering and built-up edge. Hot hardness is improved with the addition of TiC. TiC reduces the transverse rupture, compressive, and impact strengths of the carbide.
  - Tantalum carbide provides a resistance to thermal deformation. TaC has lower hardness than TiC at room temperature but greater hot hardness at higher temperatures. The coefficient of thermal expansion for TaC more closely matches that for WC-Co, resulting in better resistance to thermal shock.

Carbide Grade Design. The cutting-tool grades of cemented carbides are divided into two groups depending on their primary application. If the carbide is intended for use on cast iron, which is a nonductile material, it is graded as a *straight carbide grade*. If it is to be used to cut steel, a ductile material, it is graded as a *complex carbide grade*. Cast-iron carbides must be more resistant to abrasive wear. Steel carbides require more resistance to cratering and heat. The tool-wear characteristics of various metals are different, thereby requiring different tool properties. The high abrasiveness of cast iron causes mainly edge wear to the tool. The long chips of steel, which flows across the tool at normally high cutting speeds, causes cratering and heat deformation to the tool.

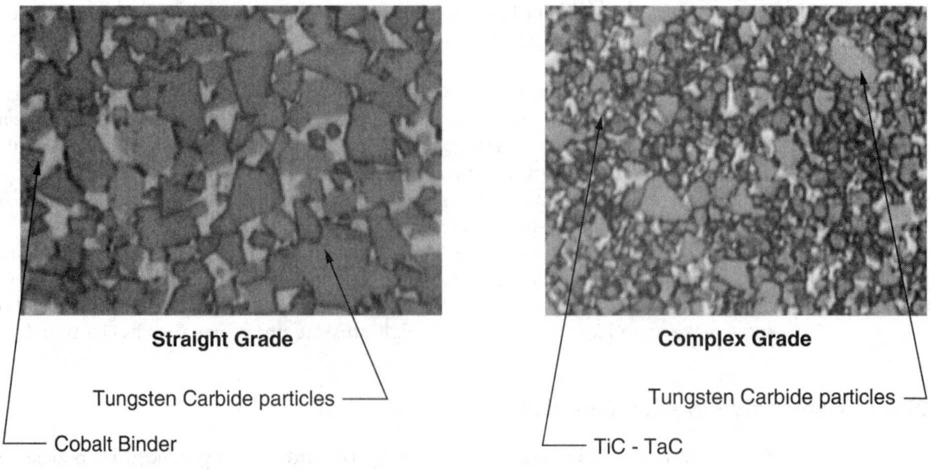

FIGURE 27.32 Categories of carbride.

It is important to choose the correct carbide grade for each application. Several factors make one carbide grade different from another and therefore more suitable for a specific application. The carbide grades may appear to be similar, but the difference between the right and wrong carbide for the job can mean the difference between success and failure.

Tungsten carbide is the primary ingredient of the carbide tool and is often used when machining materials such as cast iron. Tungsten carbide is extremely hard and offers the excellent resistance to abrasive wear.

Large amounts of tungsten carbide are present in all of the grades in the two cutting groups, and cobalt is normally used as the binder. The more common alloying additions to the basic tungsten carbide/cobalt material are TaC and TiC.

Some of these alloys may be present in cast-iron grades of cutting tools, but they are primarily added to steel grades. Tungsten carbide is abrasive resistant and is effective with the abrasive nature of cast iron.

The addition of the alloying materials such as tantalum carbide and titanium carbide offers many benefits:

- The most significant benefit of TiC is a reduction in the tendency of the tool to fail by cratering.
- The most significant contribution of TaC is that it increases the hot hardness of the tool, which in turn reduces thermal deformation.

Varying the amount of cobalt binder in the tool material largely affects both the cast-iron and steel grades in three ways.

- Cobalt is far more sensitive to heat than the carbide around it.
- Cobalt is also more sensitive to abrasion and chip welding. The more cobalt present, the softer the tool, making it more sensitive to thermal deformation, abrasive wear, chip welding, and leaching, which results in cratering.
- Cobalt is stronger than carbide. Therefore, more cobalt improves the tool strength and resistance to shock. The strength of a carbide tool is expressed in terms of *transverse rupture strength* (TRS).

Classification Systems. In the C-classification method grades C-1 through C-4 are for cast iron and grades C-5 through C-8 for steel. The higher the C-number in each group, the harder the grade; the lower the C-number, the stronger the grade. The harder grades are used for finish-cut applications; the stronger grades are used for rough-cut applications.

Many manufacturers produce and distribute charts showing a comparison of their carbide grades with those of other manufacturers. These are not equivalency charts even though they may imply that one manufacturer's carbide is equivalent to that of another manufacturer. Each manufacturer knows his carbide best, and only the manufacturer of that specific carbide can accurately place that carbide on the C-chart.

The ISO classification is based on application and is becoming more prevalent today. The ISO system separates carbide grade by workpiece material and indicates the wear and strength characteristics; i.e., P-20, M-20, K-20. The letter indicates the material (P = steels, M = stainless steels, K = cast iron), and the number indicates relative wear resistance (05 would be the most wear resistant, while 50 would be the most fracture resistant).

Many manufacturers, especially those outside the United States, do not use the C-classification system for carbides. The placement of these carbides on a C-chart by a competing company is based upon similarity of application and is at best an educated guess. Tests have shown a marked difference in performance among carbide grades that manufacturers using the C-classification system have listed in the same category.

27.3.6 Coated Carbides

Carbide inserts coated with wear-resistant compounds for increased performance and longer tool life represent the fastest growing segment of the cutting tool materials spectrum. The use of coated carbide inserts has permitted increases in machining rates up to five or more times over machining rates possible with uncoated carbide tools. Many consider coated carbide tools the most significant advance in cutting tool materials since the development of WC tooling.

The first coated insert consisted of a thin TiC layer on a conventional WC substrate. Since then, various single and multiple coatings of carbides and nitrides of titanium, hafnium, and zirconium and coatings of oxides of aluminum and zirconium, as well as improved substrates better suited for coating, have been developed to increase the range of applications for coated carbide inserts.

Coating. Coating falls into two categories;

- *Chemical vapor deposition* (CVD). The CVD process is the most common coating process for carbide cutting tools. It produces a significant heat shield, providing increased speed capability. The CVD process cannot be applied to a sharp cutting edge.
  - TiC, TiCN, TiN, Al_2O_3
  - Generally multilayer
  - Deposition temperature 900 to 1000°C
  - Thickness 5 to 20 μm
- *Physical vapor deposition* (PVD). The PVD process is a "line of sight" process suggesting that the coating will grow at a different thickness at different places within the reactor. The PVD process can coat a sharp edge (Fig. 27.33).
  - TiN, TiCN, TiAlN, ZrN, CrN, TiB_2
  - Deposition temperature 300 to 600°C
  - Thickness 2 to 8 μm
  - Line of sight process— requires tool fixture rotation

The capability for increased productivity is the most important advantage of using coated carbide inserts. With no loss of tool life they can be operated at higher cutting speeds than uncoated inserts. Longer tool life can be obtained when the tools are operated at the same speed. Higher speed operation, rather than increased tool life, is generally recommended for improved productivity and reduced costs. The feed rate used is generally a function of the insert geometry, not of the coating.

Increased versatility of coated carbide inserts is another major benefit. Fewer grades are required to cover a broader range of machining applications because the available grades generally overlap several of the C classifications for uncoated carbide tools. This simplifies the selection process and reduces inventory requirements. Most producers of coated carbide inserts offer three grades: one for machining cast iron and nonferrous metals and two for cutting steels. Some, however, offer more grades.

PVD Coating

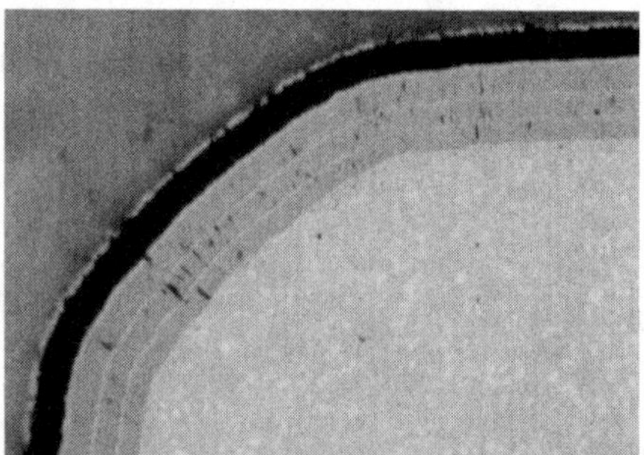

CVD Coating
Multi-Layered

FIGURE 27.33 Coating categories.

Limitations. CVD coated carbide inserts are not suitable for all applications. For example, they are generally not suitable for light finishing cuts including precision boring and turning of thin-walled workpieces—two operations which usually require sharp cutting edges for satisfactory results.

Coated carbide inserts are slightly higher in cost. However, a cost analysis should be made because the higher cutting speeds possible often increase productivity enough to more than offset their cost premium.

27.3.7 Ceramics

Ceramic or aluminum-oxide cutting tools were first proposed for machining operations in Germany as early as 1905—21 years before the introduction of cemented carbides in Germany, in 1926. Patents on ceramic tools were issued in England in 1912 and in Germany in 1913. Initial work on ceramic tools began in the United States as early as 1935, but it was not until 1945 that they were considered seriously for use in machining. Ceramic cutting tool inserts became commercially available in the United States during the 1950s.

Initially, these cemented-oxide, nonmetallic tools produced inconsistent and unsatisfactory results. This was partially because of the nonuniformity and weakness of the tools, but primarily because of lack of understanding and misapplication by the users. Ceramic tools were often used on older machines with inadequate rigidity and power.

Since then, many improvements have been made in the mechanical properties of ceramic tools as the result of better control of microstructure (primarily in grain size refinement) and density, improved processing, the use of additives, the development of composite materials, and better grinding and edge preparation methods. Tools made from these materials are now stronger, more uniform, and higher in quality. Consequently, resurgence of interest in their application has arisen.

Types of Ceramic Tools. Two basic types of ceramic cutting tools are available:

1. Plain ceramics, which are highly pure (99 percent or more) and contain only minor amounts of secondary oxides. One producer of ceramic cutting tools, however, offers two grades with a higher amount of a secondary oxide-zirconium oxide. One grade contains less than 10 percent and the other less than 20 percent of zirconium oxide. Cutting tool inserts made from plain ceramics are often produced by cold pressing fine alumina powder under high pressure, followed by sintering at high temperature, which bonds the particles together. The product, white in color, is then ground to finished dimensions with diamond wheels. Another processing method—hot pressing—simultaneously combines high-pressure compacting and high-temperature sintering in a single operation to produce inserts that are light gray in color. Hot isostatic pressing, which simplifies the production of chip breaker geometries, is also used.
2. Composite ceramics, sometimes incorrectly called cermets, are A1203-based materials containing 15–30 percent or more titanium carbide (TiC) and/or other alloying ingredients. Cutting tool inserts made from these materials are hot pressed or hot isostatically pressed and are black in color.

Ceramic Compositions

- Sialons (Si_3N_4)
- Black ceramics (Al_2O_3-TiC)
- White ceramics (Al_2O_3-ZrO_2)
- Whisker ceramic (Al_2O_3-SiCw)
- Coated Si_3N_4 (Al_2O_3/TiCN coatings)
- Coated black ceramic (TiCN coating)

Advantages. A major advantage of using ceramic cutting tools is increased productivity for many applications. Ceramic cutting tools are operated at higher cutting speeds than tungsten carbide tools.

In many applications, this results in increased metal removal rates. Favorable properties of ceramic tools that promote these benefits include good hot hardness, low coefficient of friction, high wear resistance, chemical inertness, and low coefficient of thermal conductivity. (Most of the heat generated during cutting is carried away in the chips, resulting in less heat buildup in the workpiece, insert, and toolholder.)

Another important advantage is that improved-quality parts can often be produced because of better size control resulting from less tool wear. In addition, smoother surface finishes aid size control. Also, ceramic tools are capable of machining many hard metals, often eliminating the need for subsequent grinding. Machining of hardened steel rolls used in rolling mills is an important application.

Limitations. Despite the many improvements in physical properties and uniformity of ceramic tools, careful application is required because ceramic tools are more brittle than carbides. Mechanical shock must be minimized, and thermal shock must be avoided. However, stronger grades now available plus the use of proper tool and holder geometry help minimize the effects of lower strength and ductility.

While ceramic tools exhibit chemical inertness when used to cut most metals, they tend to develop built-up edges, thereby increasing the wear rate when machining refractory metals such as titanium and other reactive alloys, and certain aluminum alloys. Tools made from ceramic materials are being used successfully for interrupted cuts of light-to-medium severity, but they are usually not recommended for heavy interrupted cutting.

Another possible limitation of using ceramic tools is that thicker inserts, sometimes required to compensate for the lower transverse rupture strength of the tools, may not be interchangeable in toolholders used for carbide inserts. Some milling cutters and other toolholders are available, however, that permit interchangeability.

Applications. Ceramic cutting tools are used successfully for high speed machining of cast irons and steels, particularly those requiring a continuous cutting action. They are generally good replacements for carbide tools that wear rapidly, but not for applications in which carbide tools break. Face milling of steel and iron castings is being done successfully, but heavy interrupted cutting is not recommended. Also, while ceramic cutting tools are useful for machining abrasive materials and most chemically reactive materials, they are not suitable, as previously mentioned, for cutting refractory metals such as titanium and reactive metal alloys and certain aluminum alloys.

27.3.8 Single-Crystal Diamonds

Increased use of both single-crystal and polycrystalline diamond cutting tools is primarily due to the greater demand for increased precision and smoother finishes in modern manufacturing, the proliferation of lighter weight materials in today's products, and the need to reduce downtime for tool changing and adjustments to increase productivity. More widespread knowledge of the proper use of these tools and the availability of improved machine tools with greater rigidity, higher speeds, and finer feeds have also contributed to increased usage.

Diamond is the cubic crystalline form of carbon that is produced in various sizes under high heat and pressure. Natural, mined single-crystal stones of the industrial type used for cutting tools are cut (sawed, cleaved, or lapped) to produce the cutting-edge geometry required for the application.

Advantages. Diamond is the hardest known natural substance. Its indentation hardness is about five times that of carbide. Extreme hardness and abrasion resistance can result in single-crystal diamond tools retaining their cutting edges virtually unchanged throughout most of their useful lives. High thermal conductivity and low compressibility and thermal expansion provide dimensional stability, thus assuring the maintenance of close tolerances and the production of smooth finishes.

Although single-crystal diamond tools are much more expensive than those made from other materials, the cost per piece machined is often lower with proper application. Savings result from reduced downtime and scrap, and in most cases, the elimination of subsequent finishing operations. Because of the diamond's chemical inertness, low coefficient of friction, and smoothness, chips do not adhere to its surface or form built-up edges when nonferrous and nonmetallic materials are machined.

Limitations. Selection of industrial single-crystal diamonds is critical. They should be of fine quality, free of cracks or inclusions in the cutting area. Also, skillful orientation is required in the tools for maximum wear. The stone must be mounted so that the tool approaches the workpiece along one of its hard planes–not parallel to soft cleavage planes (which are parallel to the octahedral plane)—or the tool will start to flake and chip at the edge. Orienting the diamond in the soft direction will cause premature wear and possibly flaking or chipping.

Tools with a low impact resistance require careful handling and protection against shock. Such tools should only be used on rigid machines in good condition. Rigid means for holding the tool and workpiece are also essential, and balancing or damping of the workpiece and its driver are often required, especially for turning. Three-jaw chucks are generally not recommended because they cannot be dynamically balanced. If required, they should be provided with dampers. Damping of boring bars is also recommended.

Single-crystal diamond tools are not suitable for cutting ferrous metals, particularly alloys having high tensile strengths, because the high cutting forces required may break the tool. The diamond tends to react chemically with such materials, and it will graphitize at temperatures between 1450 and 1800°F (788 and 982°C). Single-crystal diamond tools are also not recommended for interrupted cutting of hard materials or for the removal of scale from rough surfaces.

Applications. Single-crystal diamond cutting tools are generally most efficient when used to machine:

- Nonferrous metals such as aluminum, babbitt, brass, copper, bronze, and other bearing materials.
- Precious metals such as gold, silver, and platinum.
- Nonmetallic and abrasive materials including hard rubber, phenolic or other plastics or resins, cellulose acetate, compressed graphite and carbon, composites, some carbides and ceramics, fiberglass, and a variety of epoxies and fiberglass-filled resins.

Diamond crystals can be lapped to a fine cutting edge that can produce surface finishes as smooth as 11A in (0.025 pm) or less. For this reason, single-crystal diamond tools are often used for high-precision machining operations in which a smooth, reflective surface is required. The need for subsequent grinding, polishing, or lapping of workpieces is generally eliminated. One plant is using these tools on a specially built machine tool to produce an optical finish on copper-plated aluminum alloy mirrors.

Other parts machined with single-crystal diamond tools include computer memory discs, printing gravure and photocopy rolls, plastic lenses, lens mounts, guidance system components, ordnance parts, workpieces for which the cost of lapping and polishing can be eliminated, and parts with shapes, or made from materials, that do not lend themselves to lapping or polishing.

27.3.9 Polycrystalline Diamond Cutting Tools

Polycrystalline diamond blanks, introduced in the United States in about 1973, consist of fine diamond crystals that are bonded together under high pressure and temperature. Both natural and synthetic diamond crystals can be sintered in this way, and cutting tool blanks and inserts are currently being produced from both types of crystals.

Various shapes are compacted for cutting tool purposes, and some are made integral with a tungsten or tungsten carbide substrate. Polycrystalline diamond cutting tools are generally recommended only for machining nonferrous metals and nonmetallic materials and not for cutting ferrous metals.

Advantages. An important advantage of polycrystalline diamond cutting tools is that the crystals are randomly oriented so that the agglomerate does not have the cleavage planes found in single crystal diamond cutting tools. As a result, hardness and abrasion resistance are uniformly high in all directions. Hardness is about four times that of carbide and nearly equals that of single-crystal natural diamond. When polycrystalline diamond blanks are bonded to a tungsten or tungsten carbide substrate, cutting tools are produced that are not only high in hardness and abrasion resistance but also greater in strength and shock resistance.

Polycrystalline diamond cutting tools often cost less than single-crystal diamond tools, depending on their design and application, and they have proven superior for most machining applications. They generally show more uniformity, often allowing production results to be predicted more accurately. The compacts are also tougher than single-crystal diamonds and provide increased versatility, permitting the production of a wider variety of cutting tools with more desirable shapes. While smoother surface finishes can be produced with single-crystal diamond tools, polycrystalline diamond tools are competitive in this respect for some applications.

In comparison with carbide cutting tools, cutting tools made from polycrystalline diamond can provide much longer tool life, better size control, improved finishes, increased productivity, reduced scrap and rework, and lower tool cost per machined part for certain applications. The capability of using higher cutting speeds and feeds plus the reduction in downtime by eliminating many tool changes and adjustments can result in substantial increases in productivity.

Limitations. One limitation to the use of polycrystalline diamond tools, which also applies to single-crystal diamond tools, is that they are not generally suitable for machining ferrous metals such as steel and cast iron. Diamonds—both natural and synthetic—are carbon which reacts chemically with ferrous metals at high cutting temperatures and with other materials that are tough and have relatively high tensile strengths that can generate high pressures and induce chipping.

The high cost of polycrystalline and single-crystal diamond tools limits their application to operations in which the specific advantages of the tools are necessary. Such applications include the machining of abrasive materials that results in short life with other tool materials and the high-volume production of close-tolerance parts that require good finishes.

Applications. Tools made from polycrystalline diamond are most suitable for cutting very abrasive nonmetallic materials such as carbon, presintered ceramics, fiberglass and its composites, graphite, reinforced plastics, and hard rubber; nonferrous metals such as aluminum alloys (particularly those containing silicon), copper, brass, bronze, lead, zinc, and their alloys; and presintered carbides and sintered tungsten carbides having a cobalt content above 6 percent.

They are being increasingly applied because more nonferrous metals, plastics, and composites are now being used to reduce product weights. Increased demand for parts with closer tolerances and smoother finishes, and the availability of improved machines with higher speeds, finer feeds, and greater rigidity have also boosted the use of these tools.

Polycrystalline diamond tools have proven to be superior to natural, single-crystal diamonds for applications in which chipping of the cutting edge rather than wear has caused tool failure. They can better withstand the higher pressures and impact forces of increased speeds, feeds, and depths of cut and are suitable for many interrupted cut applications such as face milling. Sharpness of their cutting edges, however, is limited, and natural, single-crystal diamonds are still preferable for operations in which very smooth surface finishes are required.

Applications exhibiting excessive edge wear with the use of carbide cutting tools generally are good candidates for polycrystalline diamond tools. Other applications include operations where materials build up on the cutting edge resulting in burrs, operations with smeared finishes, and operations that produce out-of-tolerance parts. For certain applications, polycrystalline diamond tools outlast carbide tools by 50:1 or more.

27.3.10 Cubic Boron Nitride

Cubic boron nitride (CBN), a form of boron nitride (BN), is a super abrasive crystal that is second in hardness and abrasion resistance only to diamond. CBN is produced in a high-pressure/

high-temperature process, similar to that used to make synthetic diamonds. CBN crystals are used most commonly in super abrasive wheels for precision grinding of steels and super alloys. The crystals are also compacted to produce polycrystalline cutting tools.

Advantages. For machining operations, cutting tools compacted from CBN crystals offer the advantage of greater heat resistance than diamond tools. Another important advantage of CBN tools over those made from diamonds is their high level of chemical inertness. This provides greater resistance to oxidation and chemical attack by many workpiece materials machined at high cutting temperatures, including ferrous metals. Compacted CBN tools are suitable, unlike diamond tools, for the high speed machining of tool and alloy steels with hardnesses to R_c70, steel forgings and Ni-hard or chilled cast irons with hardnesses from R_c45–68, surface-hardened parts, and nickel or cobalt-based super alloys. They have also been used successfully for machining powdered metals, plastics, and graphite.

The high wear resistance of cutting tools made from compacted CBN has resulted in increased productivity because of the higher cutting speeds that may be utilized and/or the longer tool life possible. Also, in many cases, productivity is substantially improved because the need for grinding is eliminated. The relatively high cost of compacted CBN tools as well as diamond tools has, however, limited their use to applications such as difficult-to-machine materials, for which they can be economically justified on a cost-per-piece production basis.

Applications. Applications of cutting tools made from compacted CBN crystals include turning, facing, boring, and milling of various hard materials. Many of the applications eliminate the need for previously required grinding or minimize the amount of grinding needed. With the proper cutting conditions, the same surface finish is often produced as with grinding.

Many successful applications involve interrupted cutting, including the milling of hard ferrous metals. Because of their brittleness, however, CBN cutting tools are not generally recommended for heavy interrupted cutting.

Metal removal rates up to 20 times those of carbide cutting tools have been reported in machining super alloys.

27.4 FAILURE ANALYSIS

The forces and heat that are generated by the machining process inevitably cause cutting tooling to fail. Tool life is limited by a variety of failure mechanisms and those most commonly encountered are discussed below.

Cutting tools rarely fail by one mechanism alone. Normally several failure mechanisms are at work simultaneously whenever metal cutting occurs. Failure analysis is concerned with controlling all of the failure mechanisms so that tool life is limited only by abrasive wear.

Abrasive wear is viewed as the only acceptable failure mechanism because other failure mechanisms yield shorter and less predictable tool life. Recognizing the various failure mechanisms is essential if corrective action is to be taken. Control actions are considered effective when tool life becomes limited solely by abrasive wear.

There are eight identifiable failure mechanisms that fall into three categories;

1. Abrasive wear
2. Built-up edge
   - Rake surface
   - Flank surface
3. Thermal/mechanical cracking/chipping
4. Cratering
5. Thermal deformation

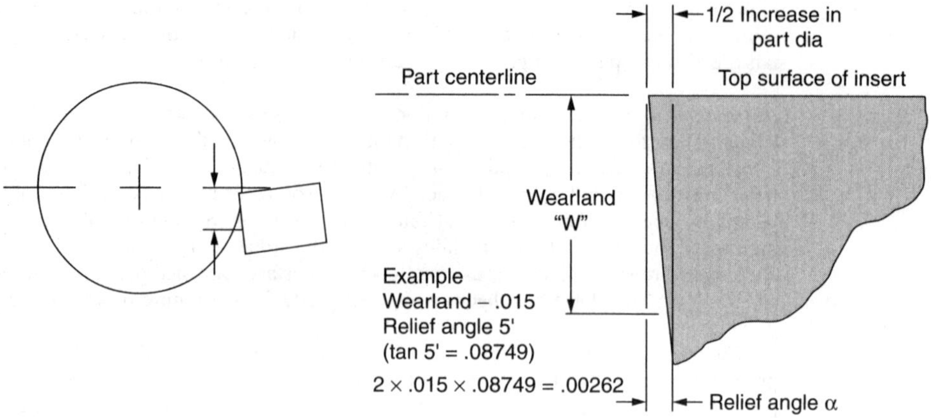

FIGURE 27.34 Abrasive wear.

6. Chipping
- Mechanical
- Thermal expansion

7. Notching

8. Fracture

Each failure mechanism will be discussed in detail along with control actions designed to inhibit that particular failure mechanism. It is essential that an accurate diagnosis of the failure mechanism be made. Misdiagnosis and application of the wrong control actions can result in worsening of the situation. The most effective way to accurately diagnose failure is to observe and record the gradual development of the symptoms.

27.4.1 Abrasive Wear (Abrasion)

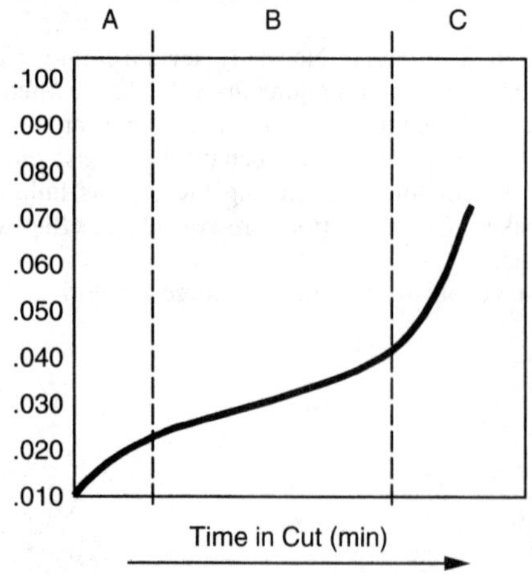

FIGURE 27.35 Wear curve.

Abrasive wear occurs as a result of the interaction between the workpiece and the cutting edge. This interaction results in the abrading away of relief on the flank of the tool. This loss of relief is referred to as a *wear land* (Fig. 27.34).

The amount of interaction that can be tolerated is primarily a function of the workpiece tolerance (both dimensional and surface finish), the rigidity of the machine tool, the set-up, and the strength of both the workpiece and the cutting edge.

The width of the wear land is determined by the amount of contact between the cutting edge and the workpiece.

Typically, wear curves caused by normal abrasive wear will exhibit an S-shaped configuration. The S is composed of three distinct zones which occur in the generation of the flank wear land (Fig. 27.35).

Zone A is commonly referred to as the break-in period and it exhibits a rapid wear land generation.

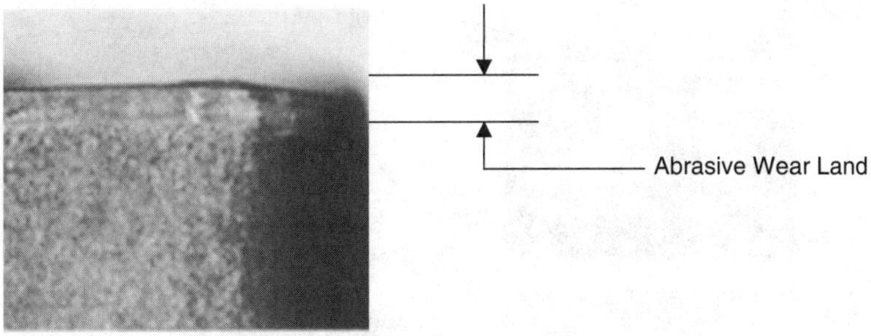

FIGURE 27.36 Abrasive wear land.

This occurs simply because the cutting edge is sharp and the removal of small quantities of tool material quickly generates a measurable wear land.

Zone B, which consumes the majority of the time in cut, constitutes a straight-line upward trend. The consistency of zone B is the key to predictable tool life.

Zone C occurs when the wear land width increases sufficiently to cause high amounts of heat and pressure which, in turn, will cause mechanical or thermal mechanical failure.

The total life span of the cutting edge, when abrasive wear is the failure mechanism, spans zones A and B. The insert should be indexed toward the end of zone B. This practice will generally reduce the incidence of fracture (Fig. 27.36).

27.4.2 Heat Related Failure Modes

Cratering (Chemical Wear). The chemical load affects crater (diffusion) wear during the cutting process. The chemical properties of the tool-material and the affinity of the tool-material to the workpiece material determine the development of the crater wear mechanism. Hardness of the tool-material does not have much affect on the process. The metallurgical relationship between the materials determines the amount of crater wear. Some cutting tool materials are inert against most workpiece materials while others have a high affinity (Fig. 27.37).

Tungsten carbide and steel have an affinity to each other, leading to the development of the crater wear mechanism. This results in the formation of a crater on the rake face of the cutting edge.

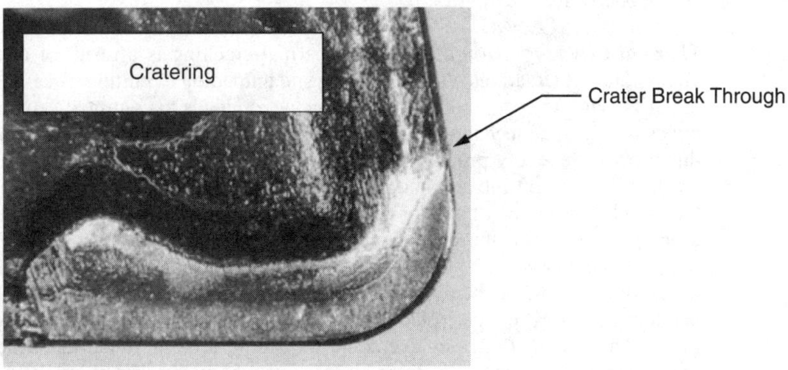

FIGURE 27.37 Chemical wear.

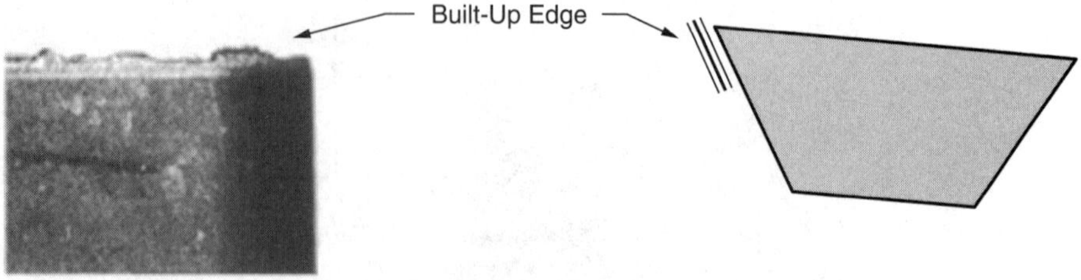

FIGURE 27.38 Built-up edge.

The mechanism is very temperature-dependent, making it greatest at high cutting speeds. Atomic interchange takes place with a two-way transfer of ferrite from the steel into the tool. Carbon also diffuses into the chip.

Cratering is identified by the presence of a concave wear pattern on the rake surface of the cutting edge. If unchecked, the crater will continue to grow until a breakthrough occurs at the cutting edge.

Build-Up Edge (Adhesion). It occurs mainly at low machining temperatures on the chip face of the tool. It can take place with long chipping and short-chipping workpiece materials—steel and aluminum. This mechanism often leads to the formation of a built-up edge between the chip and edge. It is a dynamic structure with successive layers from the chip being welded and hardened, becoming part of the edge. It is common for the build-up edge to shear off and then to reform. Some cutting materials and certain workpiece materials, such as very ductile steel, are more prone to this pressure welding than others. When higher cutting temperatures are reached, the conditions for this phenomenon are, to a large extent, removed (Fig. 27.38).

At certain temperature ranges, affinity between tool and workpiece material and the load from cutting forces combine to create the adhesion wear mechanism. When machining work-hardening materials, such as austenitic stainless steel, this wear mechanism can lead to rapid build-up at the depth of cut line resulting in notching as the failure mode.

Increased surface speeds, proper application of coolant, and tool coatings are effective control actions for built-up edge.

Build-up edge also occurs on the flank of the cutting tool, below the cutting edge. This is associated with the cutting of very soft materials such as soft aluminum or copper. Flank build-up is a result of inadequate clearance between the cutting edge and the workpieces resulting from material spring back after shearing.

Thermal Cracking (Fatigue wear). Thermal cracking is a result of thermo mechanical actions. Temperature fluctuations plus the loading and unloading of cutting forces lead to cracking and breaking of the cutting edge. Intermittent cutting action leads to continual generation of heating and cooling as well as the mechanical shocks generated from cutting edge engagement. Cracks created from this process generally propagate in a direction perpendicular to the cutting edge. Growth of these cracks tends to start inboard and grow toward the cutting edge. This failure mechanism stems from the inability of the cutting edge material to withstand extreme thermal gradients during the cutting process. Some tool materials are more sensitive to the fatigue mechanism. Carbide and ceramics are relatively poor conductors of heat. During metal cutting the heat generated is concentrated at or near the cutting edge while the remainder of the insert remains relatively cool. The expansion due to temperature increases that take place in the interfacial zone is greater than that of the remainder of the insert. The resultant stresses overcome the strength of the material, which results in cracks. The cracks that are produced isolate small areas of tool material, making them vulnerable to dislodging by the forces of the cut (Fig. 27.39).

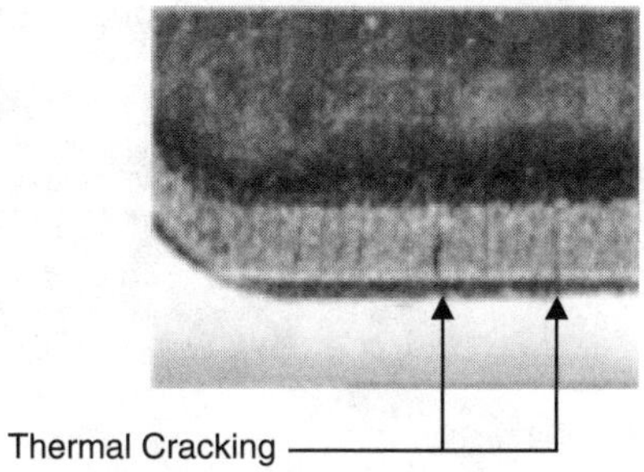

FIGURE 27.39 Thermal cracking.

Thermal Deformation (Plastic Deformation). It takes place as a result of a combination of high temperatures and high pressures on the cutting edge. Excess speed and hard or tough workpiece materials combine to create enough heat to alter the hot hardness of the cutting edge. As the cutting edge loses its hot hardness the forces created by the feed rate cause the cutting edge to deform (Fig. 27.40).

The amount of thermal deformation is in direct proportion to the depth of cut and feed rate. Deformation is a common failure mode in the finish machining of alloy steels.

27.4.3 Mechanical Failure Modes

Chipping (Mechanical). Mechanical chipping occurs when small particles of the cutting edge are broken away rather than being abraded away in abrasive wear. This happens when the mechanical load exceeds the strength of the cutting edge. Mechanical chipping is common in operations having variable shock loads, such as interrupted cuts. Chipping causes the cutting edge to be ragged altering both the rake face and the flank clearance. This ragged edge is less inefficient, causing forces and temperature to increase, resulting in significantly reduced tool life. Mechanical chipping is best identified by observing the size of the chip on both the rake surface and the flank surface. The forces are normally exerted down onto the rake surface producing a smaller chip on the rake surface and a larger chip on the flank surface.

Mechanical chipping is often the result of an unstable setup, i.e., a toolholder or boring bar extended to far past the ideal length/diamcter ratio, unsupported workpieces, and the like (Fig. 27.41).

Chipping (Thermal Expansion). Chipping occurs when the workpieces/cutting edge interface does not have adequate clearance to facilitate an effective cut.

- This may be the result of misapplication of a cutting tool with inadequate clearance for the workpieces material being cut.

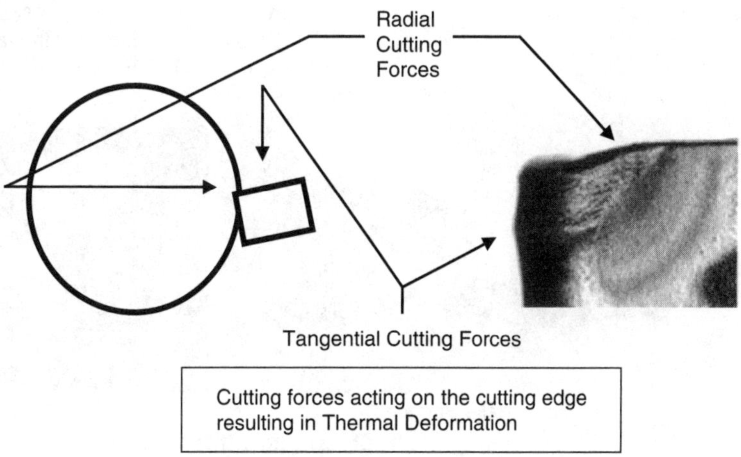

FIGURE 27.40 Thermal deformation.

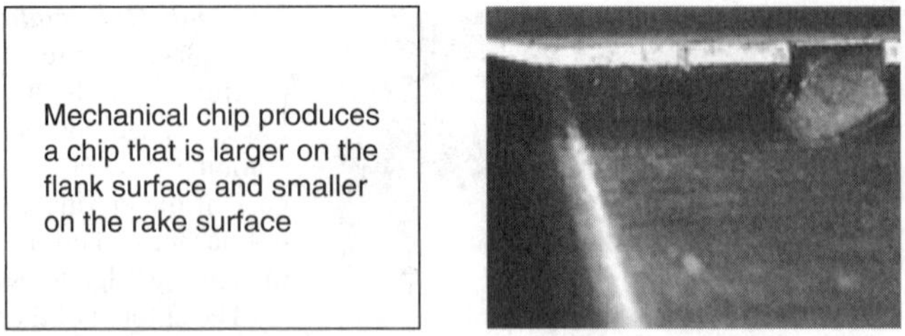

FIGURE 27.41 Mechanical chipping.

• This may be the result of an edge prep (hone) that is significantly greater than the feed rate (IPT/ IPR). For example, a cutting edge with a .005 in hone and a feed rate of .002 in IPR would produce a burnishing effect, causing heat to build up and causing the rake surface to explode into a chip.

The identifying characteristic of chipping by thermal expansion is a small chip on the flank surface and a larger, layered, chip on the rake surface. These chips appear to be flaking of the carbide or coating on the rake surface but are the result of thermal expansion of the cutting edge (Fig. 27.42).

Cutting Edge Notching. The failure mechanism called *notching* is a severe notch-shaped abrasive wear pattern that is localized in the area where the rough stock OD. contacts the cutting edge (depth of cut line). Both the flank and rake surfaces of the insert are affected by this failure mechanism.

Workpiece scale formed on the stock during casting, forging, or heat treating is primarily composed of a variety of oxides. This scale material is usually very hard and when machined produces

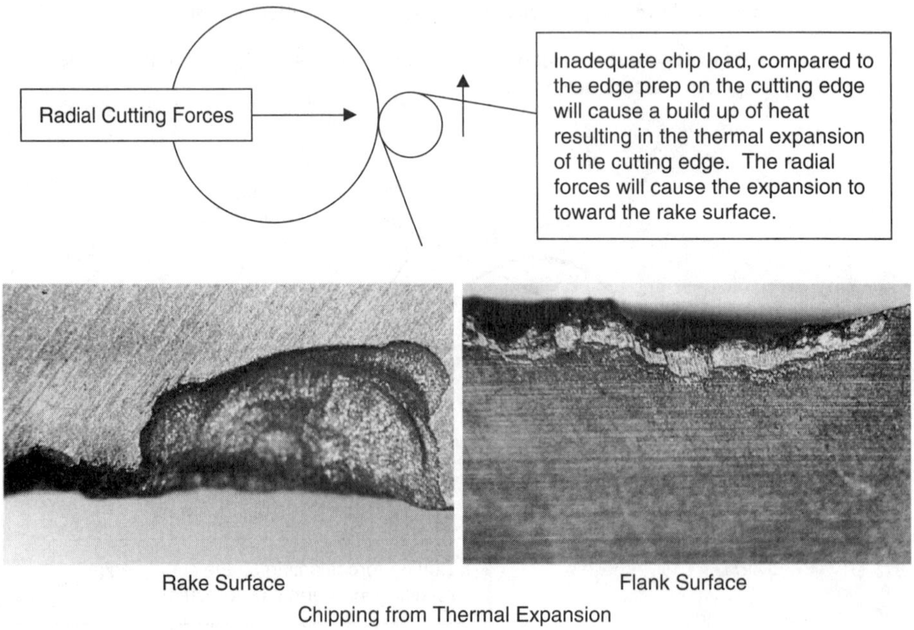

FIGURE 27.42 Chipping.

accelerated abrasive wear on the insert and, because it is caused by the workpiece OD., the wear is concentrated at the depth of cut line.

Typically, workpiece materials that are easily work hardened will notch the insert at the depth of cut line. High temperature/high strength alloys are good examples of notch producing workpiece materials.

Insert Fracture. When the edge strength of an insert is exceeded by the forces of the cutting process the inevitable result is the catastrophic failure called *fracture*. Excessive flank wear land development, shock loading due to interrupted cutting, improper grade selection or improper insert size selection are the most frequently encountered causes of insert fracture. Insert fracture is an intolerable failure mechanism that demands an immediate effective control action.

27.5 OPERATING CONDITIONS

In metal cutting, one of the most important aspects is the process of establishing operating conditions (depth of cut, feed rate, and surface speed). Operating conditions control tool life, productivity, and the cost of the part being machined. When operating conditions are changed to increase the metal removal rate, tool life normally decreases. When operating conditions are altered to reduce the metal removal rate, tool life normally increases. Metal removal rate (MRR) is normally measured in cubic inches per minute removed (in^3/min) and dictates both productivity and power consumption (HP).

27.5.1 Depth of Cut

Depth of cut is defined as the radial engagement for lathe tools and drills, and axial engagement for milling cutters (Fig. 27.43).

27.5.2 Feed Rate

Feed rate is defined as the axial movement for lathe tools and drills, measured in inches per revolution (IPR) and inches per tooth (IPT) for milling cutters. Please note that the chip thickness changes throughout the arch of the cut for milling. The centerline along the axis of movement is the only place where the chip load matches the calculated feed rate (Fig. 27.44).

27.5.3 Surface Speed

Speed in metalcutting will be defined by the amount of metal passing the cutting edge in a given amount of time (Fig. 27.45). The most common measurements are *surface feet per minute* (SFM)

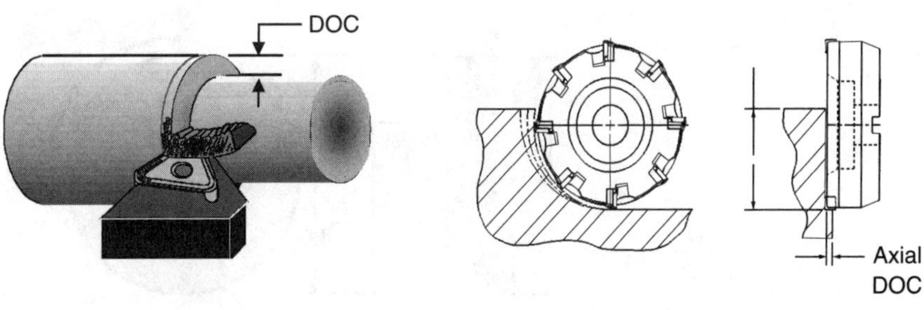

FIGURE 27.43 Depth of cut.

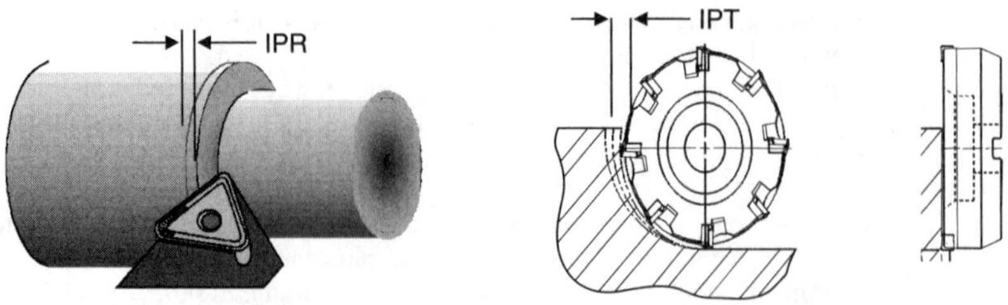

FIGURE 27.44 Feed rate.

and *surface meters per minute* (MPM). This is a relationship between the diameter of the moving part and the rpm.

27.5.4 Effects of Feed, Speed and Depth of Cut on the Metal Removal Rate

When depth of cut, feed rate, or surface speed is increased, the metal removal rate correspondingly increases. Reduce any one of these operating conditions and the metal removal rate will decrease. Changes in operating conditions are directly proportional to the metal removal rate, i.e., change feed, speed, or depth of cut by 10 percent and the metal removal rate in cubic inches per minute (in^3/min) will change by 10 percent. In all cases when one variable is changed the other two must be maintained.

- In lathe operations feed is measured in inches per revolution (IPR) connecting it to the speed (RPM).
- In milling operations feed and speed are not connected. When changes are made to speed, the feed rate, in inches per minute (IPM), must be changed in order to maintain feed in inches per tooth (IPT).

27.5.5 Effects of Metal Removal Rate on Tool Life

When the metal removal rate is increased, the friction and resultant heat generated at the cutting edge also increase causing a decrease in tool life. Assuming abrasive wear is the predominate failure mechanism, reducing the metal removal rate will produce an increase in tool life. However, changes in the three operating conditions do not impact tool life equally. Changes in depth of cut, feed rate, and surface speed each affect tool life differently. These differences, establish the process for establishing economically justifiable operating conditions.

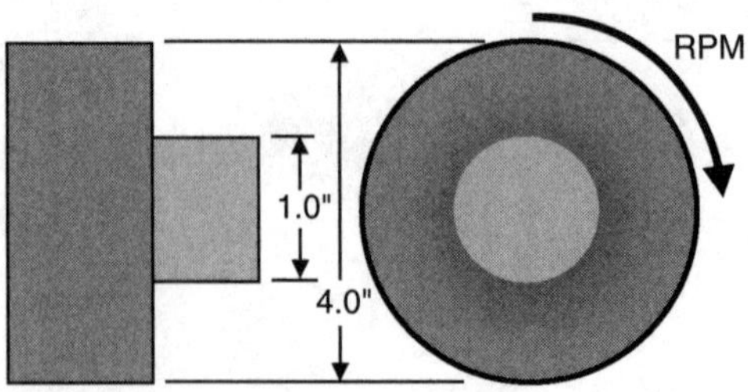

FIGURE 27.45 Surface speed.

Observation and Specification of Tool Life. The life of a cutting tool may be specified in various ways:

1. Machine time—elapsed time of operation of machine tool
2. Actual cutting time—elapsed time during which tools were actually cutting
3. Volume of metal removed
4. Number of pieces machined

The actual figure given for tool life in any machining operation or cutting test depends not only on the method used for specifying tool life, but also on the criteria used for judging tool failure. These criteria vary with the type of operation, the tool material used, and other factors. Some of the more common criteria for judging tool failure are:

1. Complete failure—tool completely unable to cut
2. Preliminary failure—appearance on the finished surface or on the shoulder of a narrow, highly burnished band, indicating rubbing on the flank of the tool
3. Flank failure—occurrence of a certain size of wear area on the tool flank (Usually based on either a certain width of wear mark or a certain volume of metal worn away.)
4. Finish failure—ccurrence of a sudden, pronounced change in finish on the work surface in the direction of either improvement or deterioration
5. Size failure-occurrence of a change in dimension(s) of the finished part by a certain amount (for instance, an increase in the diameter of a turned piece—of a specific amount—based on the diameter originally obtained with the sharp tool)
6. Cutting-force (or power) failure—increase of the cutting force (tangential force), or the power consumption, by a certain amount
7. Thrust-force failure—increase of the thrust on the tool by a certain amount; indicative of end wear
8. Feeding-force failure—increase in the force needed to feed the tool by a certain amount, indicative of flank wear

27.5.6 Tool Life vs. Depth of Cut

Depth of cut has less affect on resultant tool life than does feed rate or surface speed. As depth of cut increases tool life will decrease consistently until a depth of approximately 10 times the feed rate is achieved. Once the depth of cut reaches a level equal to 10 times the feed rate (0.050 in DOC with a feed rate of 0.005 in IPR), further increases have a decreasing affect on tool life. Tool life models developed to measure changes in tool life as depth of cut increases show significant changes in tool life below the 10× point and nearly no change in tool life above the 10× point. This change in tool life characteristics is a result of increasing chip thickness. As chip thickness increases, so does its ability to absorb heat generated in the cut (Fig. 27.46).

27.5.7 Tool Life vs. Feed Rate

Tool life models developed to measure changes in tool life as feed rate (IPR/IPT) increases show a near straight line relationship between changes in feed rate and changes in tool life. This relationship illustrates that feed rate changes have a greater effect on tool life than does depth of cut. In mild steel this relationship is nearly 1:1, suggesting that a 10 percent increase in feed rate (IPR) will result in nearly a10 percent reduction in measured tool life. The actual amount of change will vary depending upon the work piece material.

In terms of cost per cubic inch of metal removed, feed rate increases are more costly than depth of cut increases (Fig. 27.47).

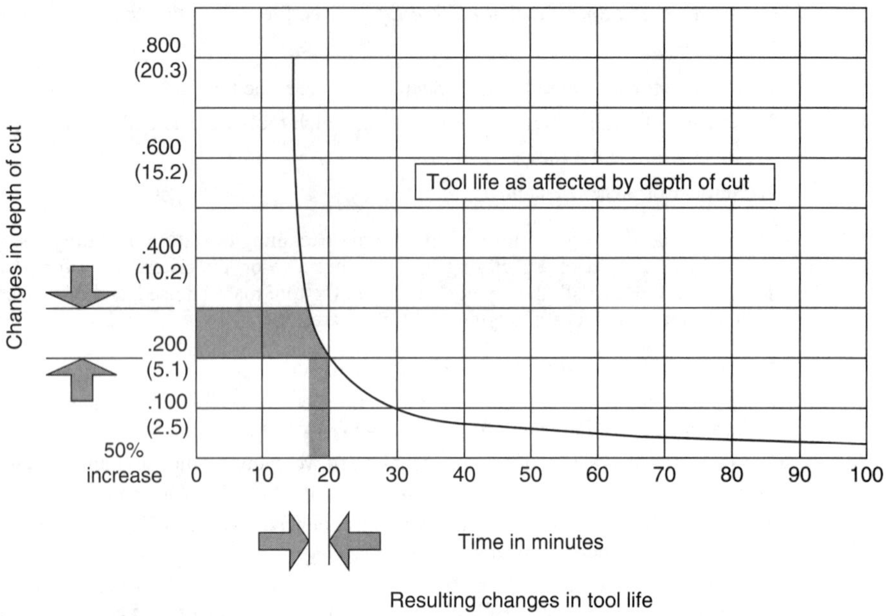

FIGURE 27.46 Tool life vs. DOC.

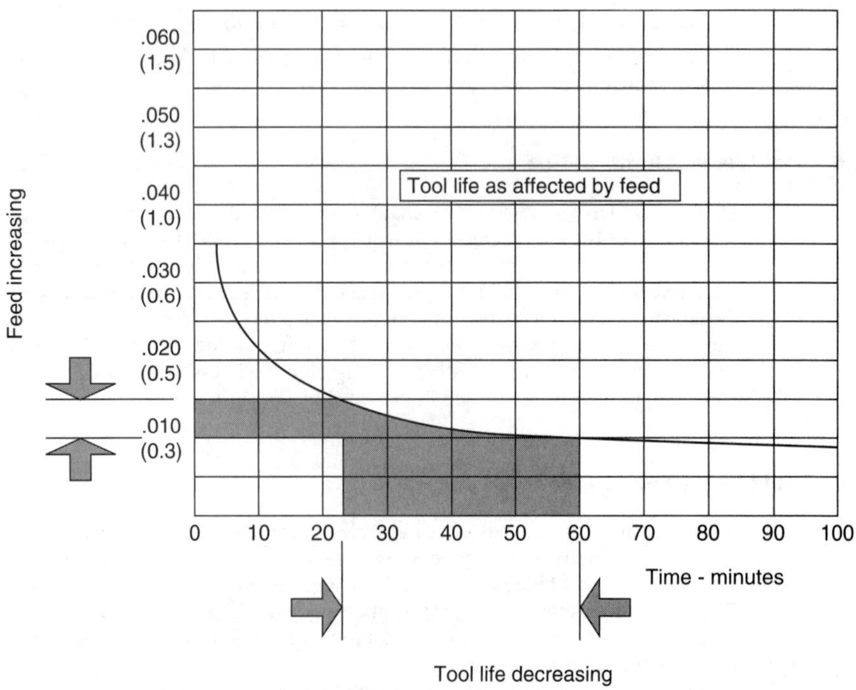

FIGURE 27.47 Tool life vs. feed rate.

27.5.8 Tool Life vs. Surface Speed

Tool life models developed to measure changes in tool life as cutting speed (SFM) increases show a near straight line relationship between changes in cutting speed and changes in tool life. This relationship illustrates that surface speed changes have a greater effect on tool life than does feed rate (IPR) or depth of cut (DOC). In mild steel this relationship is nearly 1:2, suggesting that a 10 percent increase in cutting speed (SFM) will result in nearly a 20 percent reduction in measured tool life. The actual amount of change will vary depending upon the work piece material (Fig. 27.48).

Cutting speed (SFM) has the greatest effect on tool life of the three basic operating conditions.

Tool life is less affected by changes in depth of cut and feed rate than by changes in surface speed. Increasing feed rate, like depth of cut, is judged a cost-effective action and should be maximized in order to achieve the least expensive cost per cubic inch of metal removed.

27.5.9 Rule of Thumb for Establishing Operating Conditions

1. Select the heaviest depth of cut possible (maximize DOC).
2. Select the highest feed rate possible (maximize feed rate).
3. Select a surface speed, which produces tool life that falls within the desired range based on desired productivity and /or cost of part machined (optimize cutting speed).

27.5.10 Limitations to Maximizing Depth of Cut

1. Amount of material to be removed
2. Horsepower available on the machine tool

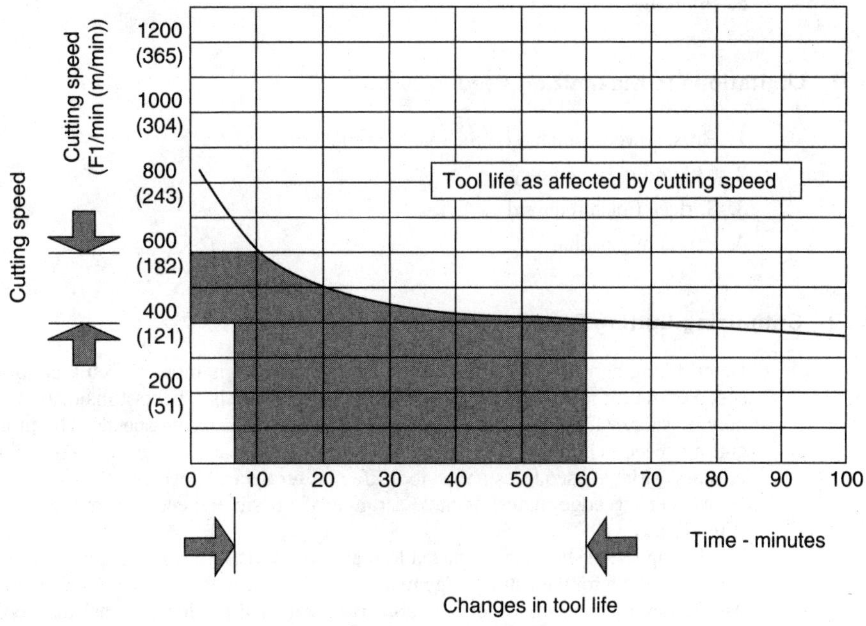

FIGURE 27.48 Tool life vs. surface speed.

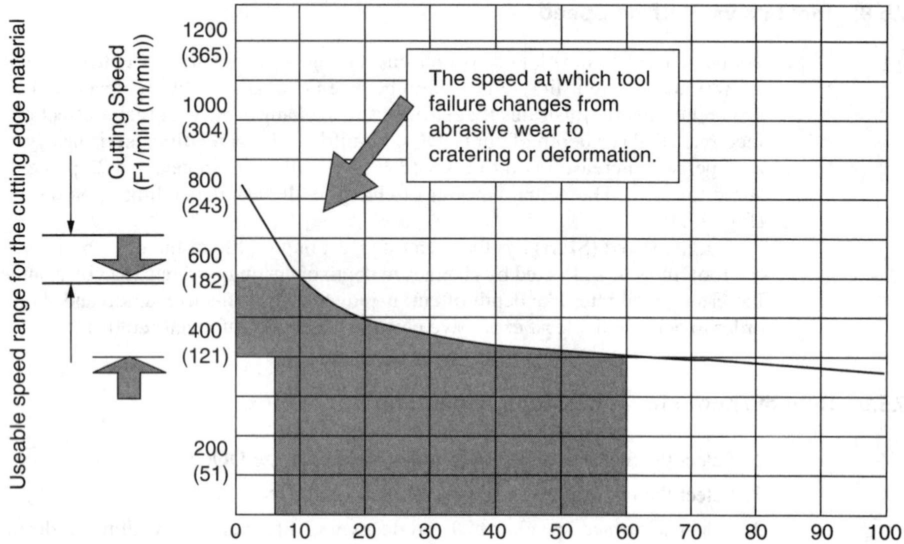

FIGURE 27.49 Optimizing cutting speed.

3. **Cutting edge**
   a. Cutting edge material
   b. Insert size & thickness
4. Workpiece configuration
5. Fixturing

27.5.11 Limitations to Maximizing Feed Rate

1. Horsepower available on the machine tool
2. Chip groove geometry
3. Surface finish required
4. Part configuration

27.5.12 Optimizing Cutting Speed

Determining an economically justifiable surface speed is more difficult as no single "best" cutting speed exists for most workpiece or cutting edge materials. The vast majority of workpiece materials may be successfully machined within a broad range of cutting speeds. The problem of establishing cutting speed is a question of desired tool life rather than "proper machining." Cutting speed is the primary variable used to establish tool life and production levels.

All cutting edge materials have a range of possible speeds for any given workpiece material (Fig. 27.49).

Cutting speed should be adjusted to maintain *abrasive wear* as the primary failure mode. Cutting speeds too high for the cutting edge material will result in failure by *cratering* or *thermal deformation.* Any failure mechanism, other than abrasive wear, will produce inconsistent tool performance and a resultant reduction in both tool life and productivity.

CHAPTER 28
HOLE MAKING

Thomas O. Floyd
Carboloy, Inc.
Warren, Michigan

28.1 DRILLING

In most instances drilling is the most cost-effective and efficient process for producing holes in solid metal workpieces. A drill is an end cutting tool with one or more cutting edges. The rotation of the drill relative to the workpiece, combined with axial feed, causes the edges to cut a cylindrical hole in the workpiece. Since drilling occurs in the workpiece interior, the chips formed and the heat generated must be removed. A twist drill has one or more flutes to evacuate chips and to allow coolant to reach the cutting edges.

Two methods of drilling are:

- Rotating applications—the drill rotates and the workpiece is held stationary, as on a mill.
- Nonrotating applications—the drill is stationary and the workpiece rotates, as on a lathe.

Drills are long relative to their diameters, therefore rigidity and deflection are major concerns.

A drill's resistance to bending is called *flexural rigidity*. Flexural rigidity is proportional to the drill diameter raised to the fourth power. Consider two drills of the same length, one is $^1/_4$ in diameter and the other is $^1/_2$ in diameter. The $^1/_4$ in drill has only one-sixteenth the rigidity of a $^1/_2$ in drill.

Deflection is proportional to the drill overhang raised to the third power. Deeper holes require longer drill overhangs, increasing the forces that cause deflection.

Because drill rigidity and deflection are influenced by length and diameter, holes are classified as either short or long based on the ratio of the hole length to the hole diameter, called the *L/D ratio*.

Short holes are usually drilled in a single pass. Holes up to 1.2 in diameter with *L/D* ratios of up to approximately five to one are considered short. Larger diameter holes with depths up to 2.5 diameters are also considered short holes. (These are general guidelines for HSS twist drills. Carbide drills are covered in a later section.)

Trepanning is often used to produce large diameter short holes. In trepanning, a ring is cut into the workpiece around a solid cylinder or core which is removed. Less workpiece material is reduced to chips, making it possible to drill large diameter holes on smaller, horsepower-limited machines.

Deep hole drilling is a more demanding operation than short hole drilling. Chips and heat are more difficult to remove, plus the cutting forces at the tip of a long tool make drilling a straight hole difficult. Often deep holes are pecked. When using a conventional drill, the drill is periodically withdrawn from the hole, clearing chips and allowing the workpiece material and the drill tip to cool.

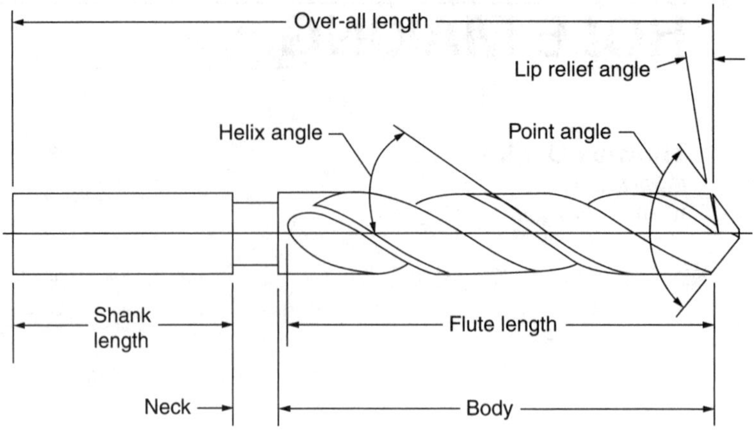

FIGURE 28.1 Drill-1: Twist drill.

Various hole configurations are encountered in machining operations. They include:

- Through holes
- Blind holes
- Interrupted holes (holes which intersect other holes or internal features)
- Holes with nonsquare entry or exit surfaces

Operating conditions (feed rate and cutting speed) often must be reduced when these configurations are encountered.

Drilling is generally considered a roughing operation. When hole specifications exceed the capabilities of the drill and machine tool, the hole must be finish-machined. Common hole finishing operations include reaming, boring, roller burnishing, and honing.

28.1.1 Twist Drills

Twist drills are drills with helical flutes. A standard twist drill is shown in Fig. 28.1 with key features labeled. The drill can be divided into three main parts—shank, flutes, and drill point.

The shank allows the drill to be held and driven. Straight shanks are used in collets and tapered shanks are mounted directly into spindles or spindle sleeves.

Chips are formed and heat is generated by the drill point. Flutes evacuate chips and allow coolant to reach the drill point. The helix angle of the flutes is dependent on the workpiece material. For steel and cast iron, a standard helix angle of 35° to 40° is used.

The workpiece material is cut by the drill point which comprises a chisel edge with two cutting lips or edges. The point angle on standard twist drills is 118° with a lip relief, or clearance angle of between 7° and 20°.

Drills with helix angles of 15° to 20° are called *low helix angle* or *slow-spiral* drills. They break chips into small segments and are capable of evacuating large volumes of chips. Low helix angle drills withstand higher torque forces than standard helix angle drills because of their greater rigidity. They are suited for brass, plastics, and other soft materials.

Drills with helix angles of 35° to 40° are called *high helix angle* or *fast-spiral* drills. The high helix angle and wide flutes provide good chip clearance when drilling relatively deep holes in materials with low tensile strength such as aluminum.

28.1.2 General Drill Classifications

Twist drills are available in a wide variety of types and sizes. Within the universe of twist drills are large capability overlaps—there are likely several, or even many, drills capable of producing a given hole.

Drill type connotes the configuration or geometry of the drill—point geometry, number, and geometry of flutes; web thickness, diameter, and length.

Drills are commonly classified based on diameter, web size, shank type, and length.

Drill diameters are classified by four methods:

- Fractional series—$^1/_{64}$ in and larger, in $^1/_{64}$ in increments
- Millimeter series—0.15 mm and larger
- Numerical series—number 97 (0.0059 in) to number 1 (0.228 in)
- Lettered series—A (0.234 in) to Z (0.413 in)

The web size of a twist drill determines whether it is a general purpose or heavy-duty drill. General purpose drills with standard two-flute designs are inexpensive, versatile, and available in a range of sizes. General purpose drills are commonly employed in high production applications on cast iron, steel, and nonferrous metals.

Heavy duty twist drills have a thicker web than general purpose drills. The heavier web results in greater rigidity and torsional strength. These drills are used in tougher applications such as drilling steel forgings, hard castings, and high hardness ferrous alloys.

Twist drills can also be classified by shank type. Twist drills with straight shanks are used in collets and are available in three series, based on drill length:

- Screw machine length (short length)
- Jobber-length drills (medium length)
- Taper length (long length)

Some machines require ANSI tapered shanks which are mounted directly into the machine spindle or into the spindle sleeve. Tapered shank twist drills are available in a wide variety of drill point and flute geometries, web thicknesses, diameters, and lengths.

28.1.3 Common Types of Twist Drills

Conventional high speed steel (HSS) twist drills are the most common drills currently used. Conventional twist drills have a point angle of 118°, two helical flutes, and either a straight or tapered shaft. Generally speaking, conventional twist drills are available in three size ranges:

- Microdrills range from 0.0059 to 0.125 in diameter.
- Standard sizes are from 0.125 to 1.5 in diameter.
- Large sizes are from 1.5 to 6.0 in diameter.

Conventional HSS twist drills offer several benefits. Twist drills:

- Are very versatile
- Are available in a wide range of sizes
- Can drill most materials
- Can be used in high production applications
- Have a low initial cost
- Can be resharpened to extend tool life

HSS twist drills have some limitations. HSS twist drills:

- Must be run at lower feed rates and cutting speeds than carbide drills and therefore productivity is not as great.
- Must be resharpened accurately or tool life and part quality may suffer.
- Are primarily a roughing tool. Holes frequently need a finishing operation.

A spade drill comprising a replaceable blade on a holder produces shallow holes, normally from 1.0 to 6.0 in diameter, but some range up to 15 in diameter. The removable blades, made from HSS or carbide, are held in place by a screw. Solid carbide spade drills are available in smaller sizes. Spade drills are suited for drilling large diameter, deep holes. Blade replacement is relatively inexpensive plus blades can be replaced while the drill is on the machine, eliminating resetting time. In addition, spade drills

- Are a low-cost alternative to twist drills in many applications
- Have a heavier cross section than twist drills (spade drills resist end thrust and torque better than twist drills, resulting in less vibration, chipping, and blade breakage.)
- Can be used on lathes (stationary tool) or mills (stationary workpiece)
- Are capable of close diameter, concentricity, and radii tolerances.
- Are available with multiple diameters for cutting chamfers.

Blades are available in HSS and carbide. Solid carbide blades are capable of higher penetration rates and longer tool life. A rigid setup is crucial, however. Solid carbide blades work well on low-carbon steels and low-alloy steels, hard or abrasive materials, and some soft materials (but not aluminum or magnesium). HSS blades are used on machines with RPM limitations and on very difficult applications.

Spade drills have some limitations.

- Spade drills require high torque and thrust forces to assure good chip evacuation.
- Spade drills should be used on rigid machines and setups.
- Spade drills must be run on machines with cutting speed and feed rate control.
- Cutting speed can vary from 50 to 400 SFPM depending on the workpiece material and whether the blades are carbide or HSS.
- Entering or exiting nonflat surfaces, or drilling fragile workpieces can cause problems because of the thrust and torque forces.
- The chisel edge of a carbide spade drill is susceptible to crushing and premature tool wear. To maximize tool life spade drills should be run at high cutting speeds and low feed rates.

Gun drills are used for producing very deep holes with tight tolerances—hole accuracy approaches that of reamed holes. The single cutting face of a gun drill is offset sharpened to form two cutting lips which break chips into short segments for easier evacuation. Gun drills have a single, straight, V-shaped flute and generally have an internal hole for delivering high pressure coolant to the cutting edge. Unbalanced forces resulting from the single cutting edge are often counterbalanced by carbide wear pads. Wear pads keep the drill centered.

Gun drills offer several benefits:

- Gun drills produce holes with high *L/D* ratios at close tolerances.
- If the setup is sufficiently rigid, finish reaming may not be required.
- A gun drill will not drift from centerline more than 0.0005 in after 2 in of penetration, if started properly.
- Carbide tips can be removed and reground.

When using gun drills

- The machine and setup must be rigid.
- The use of wear pads will maximize hole accuracy (straightness and roundness).
- Cutting fluid must be used at the cut and between the wear pads and workpiece material.
- Gun drills must be run at faster cutting speeds and lower feed rates than twist drills.
- The accuracy of deep holes may need to be enhanced by reaming or broaching.

28.1.4 Physics of Drilling

How does a drill cut and what common modifications are made to drill geometry to maximize drill performance?

As the drill rotates, the cutting speed at any point along the lip is described by this formula:

$$\text{Cutting speed (SFPM)} = \frac{\pi \times D \times \text{RPM}}{12}$$

where D is the drill diameter.

Though cutting speed is measured at the periphery of the drill, it varies from a maximum at the periphery to zero at the axis of the drill. At the chisel edge—where the cutting speed is zero and the axial rake is highly negative—the workpiece material is extruded until it reaches the lip, where it can be cut.

Drill "walk" at the start of a hole is caused by the high thrust forces which result from the extrusion at the chisel edge.

The point angle of a drill is analogous to the lead angle in turning and milling. In turning and milling, increasing the lead angle spreads the cutting forces over a longer section of the cutting edge (see Fig. 28.2). In drilling, this is accomplished by decreasing the point angle. The following two

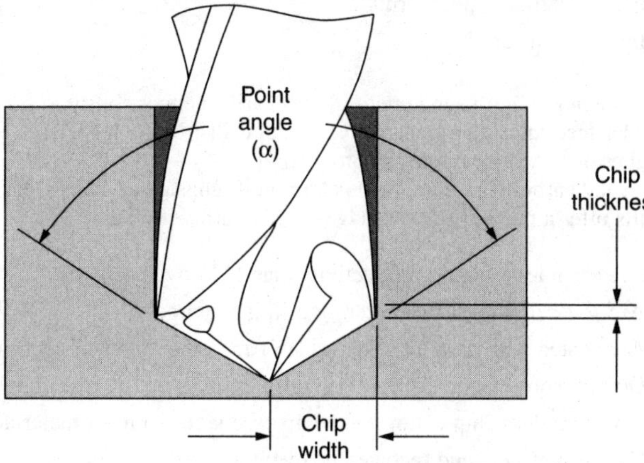

$$\text{chip width} = \frac{D}{2 \times \sin\left(\frac{\alpha}{2}\right)}$$

$$\text{chip thickness} = S_n \times \sin\left(\frac{\alpha}{2}\right)$$

Where:

α = point angle
D = drill diameter, in.
S_n = feed rate, IPR

FIGURE 28.2 Physics of drill.

formulas describe chip thickness and width as functions of the point angle:

$$\text{Chip thickness} = \text{feed per lip} \times \sin(1/2 \times \text{point angle})$$
$$\text{Chip width} = \text{drill diameter}/[2 \times \sin(1/2 \times \text{point angle})]$$

As the drill point angle decreases

- Chips become wider and thinner
- Cutting edge lengthens (increases tool life in materials that produce discontinuous chips, like cast iron)
- Axial forces decrease
- Radial and thrust forces increase

To maximize drill performance in as many applications as possible, drill manufacturers offer a wide selection of drill point modifications. A small sampling includes:

- Conventional single point drill—118° point angle
- Double-angle points
- Reduced-rake points
- Split points

28.1.5 Carbide Drills

Drills that exclusively employ carbide cutting edges (see Fig. 28.3) include:

- Solid carbide drills
- Brazed carbide tipped drills
- Indexable insert drills

The most significant benefit of these drills is their ability to cut at much faster cutting speeds and higher feed rates than most conventional drills. More aggressive operating conditions translate into higher metal removal rates or productivity!

Solid carbide drills are made entirely of tungsten carbide—shank, body, and point. Solid carbide drills offer a number of advantages. Solid carbide drills:

- Are capable of higher productivity than HSS twist drills
- Are self-centering—no centering drill is required.
- Are coated with titanium nitride (TiN) and other types of coatings for increased tool life
- Do not require peck cycles when drilling holes up to three diameters deep
- Give excellent chip control and chip evacuation in most materials
- Can be reground and recoated to keep tooling costs low

Solid carbide drills have some limitations:

- While many solid carbide drills have an internal coolant supply, smaller drills do not—they require the application of external coolant.
- The machine must have sufficient rigidity and horsepower to withstand the cutting forces at the higher cutting speeds and feed rates.
- They are designed primarily for rotating applications.

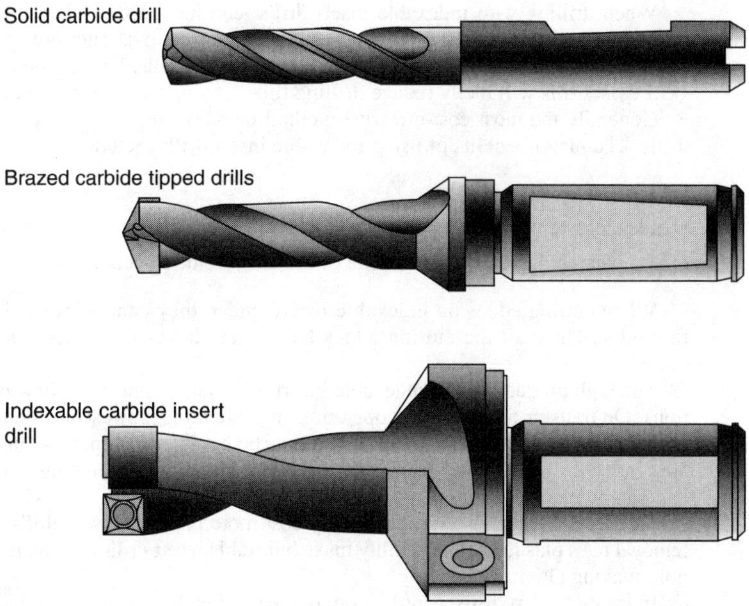

FIGURE 28.3 Carbide drills.

- Runout should not exceed 0.0015 in. in a rotating spindle and in nonrotating applications the distance between the drill point centerline and the workpiece centerline should not exceed 0.0008 in.
- Sufficient coolant pressure is essential to cool the cutting edges and for chip evacuation. Smaller diameter drills require coolant pressure of at least 500 psi.
- Solid carbide drills are not available in diameters greater than 0.75 in because of cost.

A drill with tungsten carbide inserts brazed onto a steel body is called a brazed carbide tipped drill. The advantages of brazed tipped carbide drills include:

- High productivity
- Brazed inserts can be resharpened and recoated to keep tooling costs low
- Slightly lower initial cost than an indexable insert drill
- Self-centering geometry
- Excellent chip control
- Optimized flute width and helix angle for high degree of stability and good chip evacuation
- Low cutting forces

In some applications, brazed carbide tipped drills can produce holes with high surface finish, diameter tolerance, and positioning accuracy without a finishing operation.

A significant advance in drilling technology occurred in the early 1970s—the development of indexable carbide insert drills. Indexable tungsten carbide inserts held onto steel drill bodies make it possible to produce short holes (up to five diameters in depth) faster and more economically than any other hole cutting process in most applications.

Indexable insert drills generally have two inserts, two helical or straight flutes, and a straight shank. Larger diameter indexable insert drills may employ three, four, or more inserts. The flute helix angle varies by the drill size to maximize the bending moment opposing the cutting forces, increasing stiffness. The stiffer drill is better able to resist deflection and minimize chatter.

When drilling with indexable insert drills, cutting speeds and feed rates approaching those of turning and milling can be used. The resultant metal removal rates are approximately equal to those of brazed tip drills and less than that of solid carbide drills. Using indexable insert drills in place of HSS twist drills will likely reduce drilling time—by up to 90 percent in some applications.

Generally the most cost-effective method of producing holes is to use indexable carbide insert drills. The major benefits of using indexable insert drills include:

- High metal removal rates
- Indexable, replaceable inserts
- No resharpening like solid carbide or brazed carbide tipped drills

When cutting edges on indexable inserts wear, they can be indexed to engage an unused cutting edge. Once all the cutting edges have been depleted, inserts can be replaced with unused ones.

The high productivity of indexable insert drills has implications beyond machining time calculations. On transfer lines, drilling operations are often the slowest steps and therefore pace the entire transfer line—other tools wait for drilling to be completed. Since the metal removal rate of indexable insert drills often approaches that of turning and milling, the productivity of an entire transfer line can be increased.

Indexable carbide insert drills initially cost more than HSS twist drills. However, the higher metal removal rates plus insert indexability make indexable insert drills a far more productive and economical hole-making alternative.

Indexable insert drills afford a high degree of flexibility:

- Inserts—select from several grade and chip control options to maximize performance on a specific application.
- Inserts can often be indexed while on the machine, minimizing resetting time.
- Machine-type flexibility—indexable insert drills can be used in rotating applications (mills and machining centers), or in nonrotating applications (lathes).
- They can be used to enlarge existing holes.
- Centering drills are not required—indexable insert drills are designed to be self-centering.

Despite the attractive benefits of indexable insert drills, they do have some limitations related to hole specifications and the available machine.

Limitations of indexable insert drills related to the hole include:

- They are available in diameters from 0.500 to 2.375 in and larger.
- They are considered roughing tools.
- The maximum *L/D* ratio available is 5 to 1 (most are 3 to 1).

Machine-related considerations include:

- *Horsepower.* To run at the cutting speeds required to achieve high metal removal rates the machine must have adequate horsepower and RPM.
- *Rigidity.* Indexable insert drills require a very rigid machine and fixturing.
- *Flow rates and pressure.* Coolant must reach the cutting edges at sufficient flow rates and pressure.
- *Coolant pressure.* Coolant pressure requirements increase as drill diameter decreases and drill length increases.

One important safety consideration when drilling a through hole on a lathe (workpiece rotates): a slug is produced at the drill exit which breaks free and flies outward at high speed. Safety guards must be in place!

Inserts on indexable drills are described as being the *center insert* or the *periphery insert*. Square inserts afford maximum strength and have more cutting edges per insert than other shapes. Some indexable drills feature square inserts in both the center and the periphery positions. This is the strongest configuration and is used for short hole drilling at the highest metal removal rates.

Some indexable drills use other insert shapes in one or both positions, usually to improve balance and minimize deflection when drilling deeper holes. Generally, substituting an insert shape with less strength than a square reduces the maximum feed rate of the drill as well as the number of cutting edges per insert.

Because indexable insert drills are capable of higher cutting speeds and feed rates than other drills, they produce more heat. Therefore, heat removal is critical. Effective heat removal is dependent on:

- *Effective chip control at the cut*. Chips must be broken and evacuated.
- *Coolant delivery*. Coolant must reach the cutting edges at sufficient flow rate and pressure.
- *Hole depth*. Chip removal and coolant delivery are both influenced by hole depth. Deeper holes make both more difficult.

Most irons and steels can be drilled with indexable insert drills. Softer and more ductile materials pose some challenges:

- Many ductile materials like copper alloys and aluminum can be drilled but chip evacuation may be difficult and will have to be monitored.
- Using neutral or negative rake inserts to cut gummy materials may produce thicker chips which can hang up in the drill flutes. Proper chip control and evacuation is critical.
- Indexable insert drills are not capable of drilling in soft materials like plastics, rubber, and copper

28.1.6 Selecting a Drill

The objective when selecting a drill is to select a drill capable of producing a hole that meets specifications while keeping the cost per hole at a minimum.

Several factors influence the selection of a drill for a given application. They are:

- Hole geometry and specifications
- Workpiece material
- Machine and setup
- Costs

What is the geometry of the hole?

- Diameter?
- Length?
- *L/D* ratio?
- Blind or through?
- Interrupted cut?
- Oblique entry or exit angle?

What tolerances are required on key hole dimensions?

- Diameter?
- Straightness?
- Location accuracy?
- Surface finish?

Are finishing operations required to bring the hole dimensions into the specified tolerance range?

- Reaming?
- Boring?
- Roller burnishing?
- Honing?

Are finishing operations required to add a feature to the hole?

- Countersinking?
- Counterboring?
- Spotfacing?
- Can these be accomplished by the same tool that produces the hole?

Judging the capabilities of drills based on handbooks and the manufacturers' literature can only be considered an approximation because the capability of a given drill in a specific application is dependent on the workpiece material properties, rigidity of setup, horsepower of the machine, and hole geometry (*L/D* ratio). These factors, in addition to the economics of the operation, must be considered during the selection process. For reference, several drill types can be ranked based on accuracy capability:

- Most accurate—gun drills and solid carbide drills.
- Medium accurate—brazed carbide tipped drills and HSS twist drills.
- Primarily roughing tools—indexable carbide insert drills and spade drills.

Remember that drilling is generally considered a roughing operation but some drills (solid carbide and gun drills) produce tighter tolerance holes that may not require finishing operations.

Two rules of thumb:

- For the highest accuracy, drill the hole undersized and then ream or bore to the finish specification.
- To maximize rigidity, select the shortest drill capable of producing the desired hole.

Three important properties of the workpiece material are:

- Hardness
- Tensile strength
- Work hardening tendency

As the hardness and tensile strength of the workpiece material increases, it becomes more important to select a strong drill design and to have a rigid setup.

Also, work hardening properties should be considered. Ideally, a workpiece that work hardens should be drilled in one continuous stroke. If the hole is pecked, the drill is forced to cut the work-hardened zone at the hole bottom with each peck. Tool wear will accelerate and tool life will likely be shortened.

What machine tool is available for the application in question?

- Does the tool rotate or is it stationary?
- Does the machine have sufficient RPM?
- Does the machine have adequate horsepower to achieve the metal removal rate required?
- Do the machine and setup have sufficient rigidity for the length and diameter of the drill selected?
- Does the machine require a drill with a tapered or straight shank?
- Does the machine have sufficient coolant capacity and pressure?

The total machining cost to produce a hole is the sum of the following cost elements:

Tool cost per piece
Machining cost per piece
Nonproductive cost per piece
Tool change cost per piece
Total machining cost per piece

Though general purpose HSS twist drills are inexpensive to purchase and are capable of producing a large variety of holes, they are not necessarily the lowest cost hole-producing tool to operate.

28.1.7 Considering Carbide Drills

Carbide drills should be considered whenever possible because of their high metal removal rates and productivity. Before making the decision, however, all the factors pertinent to the application must be considered:

- Hole geometry and specifications
- Workpiece material
- Machine and setup
- Costs

Hole accuracy and size will likely be the determining factors when deciding which type of carbide drill to use (see Fig. 28.4).

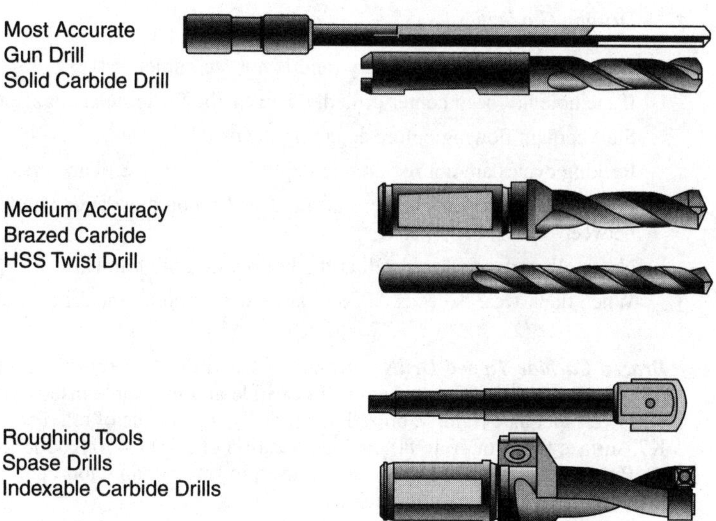

FIGURE 28.4 Carbide drills.

The three types of carbide drills (solid carbide, brazed insert, and indexable insert) vary from each other in accuracy capability, productivity, and sizes available. Each type is divided into families based on:

- Diameter ranges
- Length to diameter ratio (*L*/*D*)
- Internal versus external coolant supply

Solid Carbide Drills. The most accurate drilled holes are produced using solid carbide drills. Made from a general-purpose micro-grain carbide grade and coated with PVD titanium nitride (PVD TiN), solid carbide drills can be run at cutting speeds much greater than HSS twist drills.

Solid carbide drills are capable of the highest penetration rates for a given drill diameter. For example, a 0.500 in diameter solid carbide drill is capable of feed rates (inches per minute, IPM) approximately three to five times that of brazed carbide tipped or indexable carbide insert drills of the same diameter.

Solid carbide drills are available in several families—all manufactured to the same tolerance specifications which are a function of drill diameter. Solid carbide drills can produce surface finishes of 80 μin RMS, if the machine and setup have adequate rigidity.

Solid carbide drills can be used in rotating and nonrotating applications in steel, stainless steel, cast iron, and aluminum.

To assure maximum accuracy when using solid carbide drills:

- Run out in a rotating spindle should not exceed 0.0015 in
- In nonrotating applications the distance between the drill point centerline and the workpiece centerline should not exceed 0.0008 in
- Use an end mill holder, collet chuck, or milling chuck for maximum performance and tool life
- EPB end mill holders and ER collet chucks improve drill performance because of their tight manufactured tolerances
- Solid carbide drills can be reground and recoated when flank wear is approximately 0.008 in at its widest point

Drilling Guidelines.

Center drilling is not necessary (and is not recommended).

If the hole has been center drilled, decrease the feed rate at engagement by 50 percent.

Start coolant flowing before engaging the drill.

Pecking cycles are not required at depths of less than 3 diameters.

For deeper holes, use slower cutting speed recommendations and increase coolant pressure to between 500 and 800 psi.

Never allow the drill to dwell at the bottom of a blind hole.

When flank wear exceeds 0.008 in at its widest point, the drill should be resharpened.

Brazed Carbide Tipped Drills. Brazed insert drills represent the middle range of accuracy capability and productivity relative to solid carbide and indexable insert drills.

Given adequate rigidity, brazed insert drills are capable of tolerances as good as hole diameter of K7, surface finish of 40 to 80 μin, and location of ±0.0004 to 0.0008 in.

Brazed carbide tipped drills can be used in rotating and nonrotating applications in steel, stainless steel, cast iron, and aluminum.

To assure maximum accuracy when using brazed insert drills the runout in a rotating spindle should not exceed 0.0015 in. In non-rotating applications the distance between the drill point centerline and the workpiece centerline should not exceed 0.0008 in.

Brazed insert drills can be reground and recoated when flank wear is approximately 0.008 in at its widest point.

Drilling Guidelines

- Center drilling is not necessary.
- If the hole has been center drilled, decrease the feed rate at engagement by 50 percent.
- When using drills with high *L/D* ratios, the feed rate must be reduced until the drill is fully engaged.
- Brazed insert drills should be reground and recoated when flank wear exceeds 0.008 in at its widest point.

Indexable Carbide Insert Drills. Indexable insert drills are capable of the highest cutting speeds and metal removal rates of any of the carbide drill types (see Fig. 28.5). Therefore, indexable insert drills offer the largest potential productivity gain.

Indexable, replaceable inserts afford a high degree of flexibility:

- Multiple cutting edges on each insert—often inserts can be indexed on the machine.
- Several grades and geometries are available—fine tune a drill for maximum performance on a given application.

Indexable carbide insert drills can be used in rotating and nonrotating applications to produce holes in steel, stainless steel, cast iron, aluminum, and high-temperature alloys.

The various families of indexable carbide insert drills are differentiated by features such as:

- Diameter range
- Drilling depth
- Whether through-the-tool coolant holes are available or not
- Insert geometry

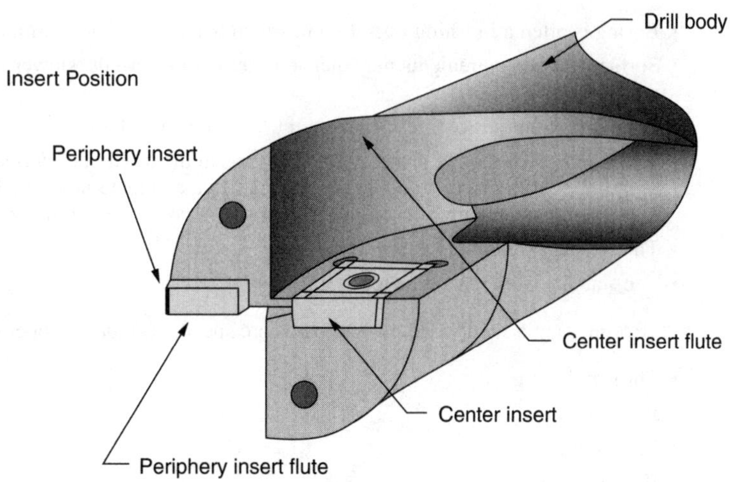

FIGURE 28.5 Indexable carbide.

Some indexable carbide insert drill configurations offer superior balance which allows them to drill offset from the workpiece centerline—fewer drill diameters are required to produce a range of hole diameters. However, there are tradeoffs for this flexibility:

- Cutting forces increase when an offset is introduced.

The inserts used to maximize balance may not be as strong as other insert shapes.
Drilling Guidelines.

- When flank wear exceeds 0.008 to 0.012 in at its widest point, inserts should be indexed.
- Cutting speed recommendations are based on tool life for the periphery insert of between 20 and 30 min.

Indexable carbide insert drills offer some adjustability of hole diameter. Refer to adjustability tables in the manufacturer's literature.

- The diameter of the drilled hole can be adjusted by moving the machine slide.
- When using a rotating drill, adjustments are made via adjustable holders.

Note that an adjustable holder produces tighter tolerances when using a roughing tool. By setting an indexable insert drill in a presetter it should be possible to drill a hole within ±0.001 in.

28.2 BORING

Boring is an internal turning operation used to enlarge drilled or cast holes or other circular contours. The tool path is controlled relative to the centerline of the rotating workpiece, allowing close dimensional tolerances and surface finishes. Boring is typically used to improve the hole location, straightness, dimensional accuracy, and surface finish.

Generally, boring operations are expected to be able to hold ±0.001 in location and as good as a 32 μin surface finish, although greater tolerances can be had with extra care.

The same metal cutting theory is used to determine insert, toolholder size, geometry, and operating conditions and is also used for both OD turning and ID boring operations.

However, ID boring is constrained by one or more factors that are likely to limit the metal removal rate.

- Boring is often a finishing operation in which the *depth of cut* is limited.
- Surface finish requirements may dictate faster cutting speeds, slower feed rates, and smaller nose radius.
- Chip control in the confines of a bore must be considered.
- The tolerance of the cut is affected by how much the toolholder deflects, which in turn is a function of the cutting forces, and the length, diameter, and material of the boring bar. The bar must be long enough to perform the operation yet its cross section is limited by the ID bore.
- The size and weight of the workpiece.
- The stability of the machine tool and clamping device.

Boring can be subdivided into several more specific ID turning operations

- Through boring
- Blind boring
- ID profiling
- ID chamfering

- ID grooving
- ID threading
- Reaming

Through boring is cutting completely through the workpiece—the hole penetrates two outer surfaces of the workpiece.

Blind boring is when the hole does not extend completely through the workpiece—only one outer surface of the workpiece is penetrated.

In *ID profiling* the tool travels in a combination of paths to machine a specific profile. Some examples would be:

- Diagonally toward or away from the workpiece center to produce an angular or conical contour
- Parallel to the axis of the rotating workpiece, producing a cylindrical bore
- On a curved path, toward or away from the workpiece center to produce curved contours

ID chamfering is the breaking of the edge of a corner where stress can build.

ID grooves for thread relief, O-rings, snap rings, and lubrication are machined using special grooving inserts and toolholders.

ID threading is used to make threads concentric with the workpiece centerline and can be done on diameters that are not practical to tap because of their size. ID threading is similar to OD threading but it is constrained by depth of cut, surface finish requirements, chip control, and toolholder deflection.

Reamers are used to enlarge drilled or cast holes to final size specifications with a high degree of dimensional accuracy and excellent surface finishes. Typically an H6 tolerance (±0.0003 to 0.0006 in). Reamers cannot improve the existing hole location because they follow the existing hole.

28.3 MACHINING FUNDAMENTALS

28.3.1 Cutting Forces

The cutting force acting on an insert is the cumulative effect of three component forces (Fig. 28.6)—

Tangential force
Radial force
Axial force

Each of the three force components acts on the insert in its own direction and the magnitude of each is dependent on several factors.

Operating conditions
Tool geometry
Workpiece material

Tangential force acts on an insert along the tangent in the direction that the workpiece is rotating. The tangent is a line that intersects the bore (or circle) and is perpendicular to the radius at that point. Tangential force is typically the largest of the three force components and, if sufficiently large will deflect the boring bar.

The magnitude of the tangential force is determined by

1. The area of contact between the chip and the insert face, called the undeformed chip thickness (depth of cut times feed rate)

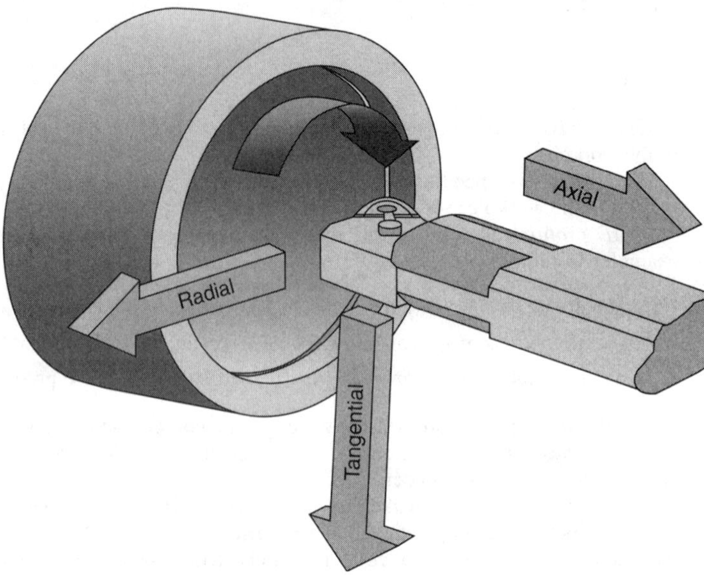

FIGURE 28.6 Cutting force acting on an insert.

2. The true rake angle
3. The chip forming and breaking tendencies of the workpiece material

The *radial force* component acts to push the insert inward along the radius of the bore. If great enough, it will cause the boring bar to deflect in the radial direction, reducing the depth of cut and negatively affecting the diametrical accuracy of the cut. As the chip thickness fluctuates, so does the magnitude of the radial force. This interaction may cause vibration. The factors that directly influence the magnitude of the radial force include the lead angle, nose radius, rake angle, depth of cut and the workpiece material.

The *axial force* component, sometimes called the *longitudinal force* acts opposite the direction of the toolholder feed, along the workpiece axis. The axial force is the least of the three force components and is directed into the strongest part of the setup making it the least concerned.

28.3.2 Rake Angle

The angle of inclination of the cutting surface of the insert in relation to the centerline of the workpiece is called the *rake angle* (Fig. 28.7). The true rake angle is a function of three angles—back rake, side rake, and lead angle.

In OD turning it is common to select negative rake tooling whenever the workpiece and machine tool allow the minimization of tooling cost (more cutting edges). However, negative rake tooling tends to increase forces and because of the nature of a negative rake insert, 90 degree sides, they also require more room in the tool or boring bar. This makes them less desirable for ID boring operations.

Positive rake tooling is typically used in ID boring operations for their lower cutting forces and for their ability to be used in smaller bars. There are two tradeoffs when using a positive rake insert.

1. The insert nose is unsupported and therefore weaker than on a negative rake insert.
2. There are fewer usable cutting edges.

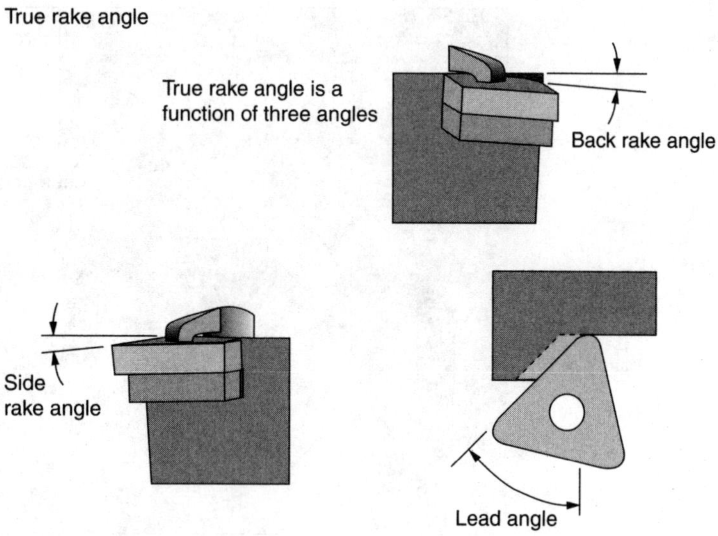

FIGURE 28.7 Rake angle.

28.3.3 Lead Angle

The angle formed by the leading edge of the insert and the projected side of the toolholder is called the *lead angle*.

The lead angle affects the relative magnitude of the radial and axial force components.

At a 0 degree lead angle, axial force is maximized and radial force is minimized.

As the lead angle is increased radial force increases and axial force decreases.

The larger the lead angle the thinner the chip thickness, figure six. This allows higher feed rates and increased productivity. However as the lead angle is increased, the radial force component is increased. As the radial force increases deflection of the boring bar will cause chatter and/or part tolerance problems. Because of this, most boring tools have lead angles of 0 to 15 degrees. In this range some benefits from lead angles (chip thinning) can be had while not overloading the radial forces.

28.3.4 Nose Radius and Depth of Cut

The relative magnitudes of the axial and radial force components are also affected by the relationship between the insert nose radius and the depth of cut.

Figure 28.8 shows the relationship between the radial and axel cutting forces as the depth of cut increases in relation to the nose radius. In these examples (at a lead angle of 0 degrees) the cutting forces around the nose radius are represented by a force arrow perpendicular to the cord of the arc and have both axial and radial components. The cutting force along the insert edge beyond the nose radius consists only of axial force.

When the nose radius is less than the depth of cut a beneficial tradeoff between axial and radial forces occurs, Radial forces decrease and axial forces increase.

This is a desirable tradeoff because radial forces can cause deflection and vibration, while axial forces are directed into the strength of the toolholder and clamping mechanism. The total resultant force comprises a large axial component and a relatively small radial component—the preferred relationship in boring.

When the depth of cut is greater than the nose radius, the radial force is determined by the lead angle—radial force increases as the lead angle increases.

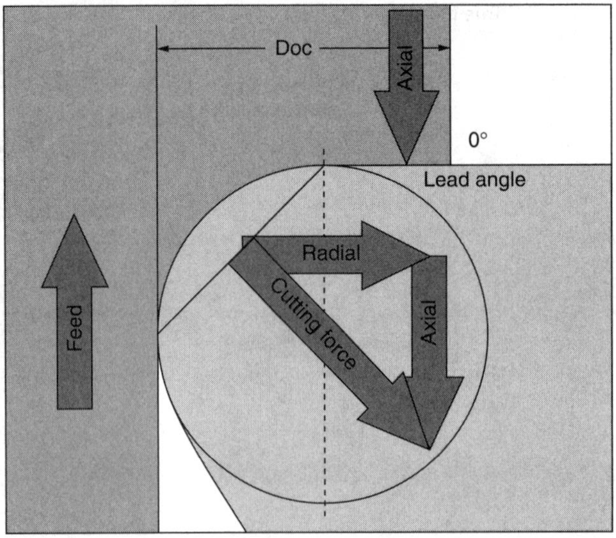

FIGURE 28.8 Cutting forces in relation to depth of cut.

When the depth of cut equals the nose radius—not the preferred practice in boring—the effects of the radial force become more significant.

Since the insert edge beyond the nose radius is no longer engaged in the cut, there are no axial forces along the insert edge—axial forces exist only at the insert nose radius. The forces are directed perpendicular to the cord that is formed as shown in Fig. 28.8.

As a rule, the depth of cut should always be larger than the nose radius of the insert.

28.3.5 Internal Clearance Angle

The internal clearance angle is the angle formed by the insert flank and the tangent at the point the insert contacts the workpiece ID. Insufficient clearance increases contact between the rotating workpiece and the insert flank, resulting in friction, increased flank wear, and abnormal horsepower consumption. The probability of vibration increases as well as the likelihood of an insert failure.

Tool geometry greatly influences the internal clearance angle; however the size of the bore diameter and the magnitude of the tangential force component also play roles. When a boring bar is deflected by the tangential force component the internal clearance angle decreases. The loss of clearance is more critical in small-diameter bores.

28.4 TOOLHOLDER DEFLECTION

For both OD and ID turning operations the toolholder deflection must be minimized because it compromises part quality and tool life. When selecting tooling and determining operating conditions, deflection is a greater concern in ID turning than in OD turning. OD turning operations often afford the flexibility of using very short toolholder overhang combined with a large toolholder cross section. However, in ID turning, the toolholder must extend into the ID a sufficient distance such that the cut can be completed, plus the boring bar diameter is constrained by the workpiece ID.

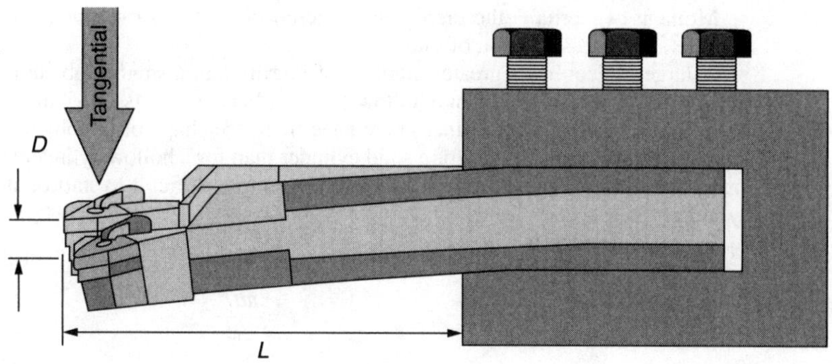

$$D = \frac{F \times L^3}{3 \times E \times I}$$

D = Deflection, in.
F = Force, lbs.
L = Length of overhang, in.
E = Modulus of elasticity, psi
I = Moment of inertia, in.^{4}

FIGURE 28.9 Deflection formula.

A general rule for all ID turning operations is: minimize deflection and improve stability by selecting a toolholder with the smallest length to diameter ratio (*L/D*) possible, such that it is capable of fitting inside the bore and extends far enough to perform the operation.

Proper management of these four factors is vital for minimizing deflection.

- The length of unsupported toolholder, or overhang.
- The cross section of the boring bar
- The material of the boring bar
- Cutting forces

Changes in the length of the overhang (*L*) result in drastic changes in the magnitude of the deflection because in the deflection formula, the length is cubed, Fig. 28.9.

In other words, deflection is proportional to the length of the overhang raised to the third power. Hence, shortening an overhang a small amount can significantly reduce deflection.

To further minimize deflection, a boring bar should be selected such that it has a strong shape or cross section that resists bending and the material is inherently stiff.

The bar cross section and composition determine how difficult a bar of fixed length is to bend. The relationship—called flexural rigidity—is part of the deflection equation.

Flexural rigidity ($E \times L$) appears in the denominator of the deflection formula. Making either E or L larger increases the flexural rigidity, thereby decreasing deflection.

The modulus of elasticity is a measure of how difficult it is to elastically deform a bar of a specific material. To deform elastically means to bend or stretch the bar such that when released it returns to its original size and shape. A material's modulus of elasticity is a function of the material and is independent of shape.

A bar made of a stiffer material is less likely to deflect. The quantitative relationship is—deflection is inversely proportional to the modulus of elasticity of the bar.

The most common boring bar materials are steels and tungsten carbide. The modulus of elasticity of tungsten carbide is approximately three times higher than that of steel. Therefore, a tungsten carbide boring bar will deflect only one-third as much as a steel bar in identical applications.

Moment of inertia is the inertia encountered in starting or stopping rotation. It is dependent on the size and the distribution of mass.

A larger object has a greater moment of inertia than a smaller object and a solid cylinder has a greater moment of inertia than a hollow cylinder because of its distribution of mass.

Equations for moment of inertia are specific to the shape of the object. Note that the moment of inertia is significantly greater for a solid cylinder than for a hollow cylinder (in fact different equations are used). It makes sense that a solid cylinder is more difficult to start or stop rotating that a hollow cylinder.

For a solid round boring bar

$$I = \frac{\pi d_1^{\ 4}}{64}$$

where I = moment of inertia
d_1= bar diameter
π = 3.1416

Deflection for boring bars with axial coolant passages can be determined by the formula:

$$I = \frac{\pi\left(d_1^4 - d_2^4\right)}{64}$$

where d_1 = diameter of the bar
d_2 = diameter of the coolant passage bore

The diameter term in the moment of inertia formula is raised to the fourth power. Therefore, a small increase in the boring bar diameter results in a large increase in the moment of inertia. Deflection is reduced by a proportionately large amount. Another way to state the relationship is: Deflection is inversely proportional to the boring bar diameter raised to the fourth power.

Of the three force components—tangential, axial, and radial—axial force typically has the least effect on deflection because it is directed back through the toolholder. Tangential and radial forces however, can have substantial impact on boring bar deflection and vibration.

Possible effects of tangential deflection are:

- Vibration
- Increased heat at the cutting edge and possible nose deformation
- Increased flank wear
- Reduced internal clearance angle which may cause flank wear in smaller diameter bores

Possible effects of radial deflection are:

- Varying undeformed chip thickness could cause vibration
- Increased heat at the cutting edge
- Loss of diametrical accuracy
- The depth of cut could be reduced, in turn reducing the axial force component and magnifying the effect of the radial force component.

If the magnitude of the tangential and radial force deflection is known, many machines can be adjusted to compensate.

For the tangential deflection, most likely a sensing instrument will be required near the cutting edge. The radial deflection is the difference between the desired inner diameter and the actual inner diameter.

To compensate for tangential deflection, position the cutting edge above the workpiece centerline a distance equal to the measured tangential deflection.

To compensate for radial deflection, increase the depth of cut by a distance equal to the measured radial deflection.

28.5 VIBRATION

Vibration is the periodic oscillation motion that occurs when an external force disturbs a body's equilibrium and a counteracting force tries to bring the body back into equilibrium.

Example

- High tangential force overcomes the properties of the bar (overhang, material, modulus of elasticity, and moment of inertia). The bar deflects.
- As the bar deflects, tangential force decreases.
- Once the tangential force has decreased sufficiently that the bar properties can counteract the deflection, deflection decreases.
- The process repeats.

Vibration is a common problem in boring and can be caused by setup and operating conditions, the workpiece, the machine, or other external sources. *Chatter* is a special case of vibration that occurs between the workpiece and the cutting tool.

When the boring bar deflects and vibration occurs, the distance the boring bar deflects in one direction is the amplitude of the vibration—it is the distance the bar travels from its state of equilibrium. The full range of movement of the boring bar between two unbalanced positions is called the *period of the vibration* and the time it takes to cover the distance is the *frequency*.

The bar diameter (moment of inertia) and the material (modulus of elasticity) determine the natural frequency of a boring bar—amplitude reaches a maximum at the natural frequency. The higher the natural frequency of a boring bar, the greater its dynamic stiffness, or its ability to resist vibration.

The amount of vibration and its increase or decrease are functions of:

- The cutting forces
- The rigidity of the machine, the insert clamping, and the toolholder clamping
- The stiffness of the boring bar
- The amount of overhang and the cross section of the boring bar (L/D)
- Interacting oscillations

Interacting oscillations are when two or more vibrating motions contact each other. The vibrations can interact in several ways:

- The forces combine to produce a vibration of a larger period
- The forces act opposite each other to dampen the overall period
- The forces combine to form a vibration of irregular period, with some larger and some smaller than the original period
- The forces combine to form resonance

Resonance occurs when both forces have the same natural frequency. It is potentially dangerous because the period does not stabilize. Rather it continues to increase until it is out of control. Resonance may cause premature and sudden failure of the insert, boring bar, workpiece, or machine.

A curious effect occurs when the frequency of the external force (e.g., the machine) is greater than the natural frequency of the boring bar. The amplitude of the boring bar vibration decreases and increasing the external amplitude reduces the amplitude of the boring bar vibration even more.

A tuned boring bar can be used to dampen vibration by counteracting oscillation.

A tuned bar contains a fluid that oscillates at a different frequency than the bar. Oscillation at the cutting edge is transferred into the tuned bar and is dampened by the fluid.

One type of boring bar uses an integral sensor that determines the natural frequency, then duplicates it 180 degrees out of phase.

28.5.1 Guidelines to Minimize Vibration

- Minimize *L/D*.
- Use tungsten carbide boring bar whenever economically possible.
- Optimize operating conditions and tool geometry to minimize tangential and radial forces.
- Check the rigidity of the clamping mechanisms.
- Make sure the machine is in good operating condition.
  - Balance rotating components
  - Check for bent or poorly positioned shafts
  - Check power transmission belts
  - Check for loose machine parts
  - Inspect hydraulic systems
- Use a plug dampener—a weight at the end of the workpiece opposite the cut which acts as a counterbalance to reduce vibration.
- Use an inertia disk dampener—varying diameter disks that move randomly against the boring bar to reduce vibration.
- Use a tuned boring bar.
- Use oil or cutting fluids delivered under pressure to the support bearing to reduce vibration by providing a film on the bearing.

28.6 CHIP CONTROL

Effective chip control is critical in ID machining operations, especially in deep hole applications. Poor chip control can cause high cutting forces, deflection, and vibration that could have detrimental effects on virtually all aspects of the machining operation—part quality, tool life, and machining cost—and cause damage to the workpiece, the tool, and the machine.

Effective chip control consists of breaking chips into the proper shape and size, then evacuating the chips from the ID.

Relatively short spiral chips, not the typical six or nine shape, but chips that are short enough to be managed tend to require less horsepower and smaller increases in cutting forces.

Very short tightly curled chips should be avoided because they require more horsepower to break with increased periodic forces.

Likewise long chips should be avoided because they can be trapped in the ID and recut. The workpiece and the insert could then be damaged.

The factors that affect chip control are the same factors that determine cutting forces—tooling geometry, operating conditions, and workpiece material.

Use boring bars with internal coolant supply whenever possible. The coolant stream flushes chips away from the cutting edge and out of the ID.

28.7 CLAMPING

A boring bar must be securely clamped in place—the bar will deflect if there is any movement in the clamped section.

Tangential and radial force components acting on the insert are opposed by equal forces at the outer edge of the clamping mechanism. The clamping surfaces must be hard (45 RC minimum) and smooth (32 μin R_a minimum) in order to withstand these forces. If the outer edge of the clamping surface deforms, deflection and possibly vibration will result.

Clamping recommendations:

- Use clamping mechanisms that completely encase the boring bar shank (rigid or flange mounted or divided clamping block).
- If possible, avoid clamping mechanisms in which tightening screws contact the boring bar shank.
- The clamped length of the boring bar shank should be three to four times the diameter of the shank.

28.8 GUIDELINES FOR SELECTING BORING BARS

- Select a boring bar capable of reaching into the bore and performing the operation
- Maximize the bar diameter
- Minimize the overhang
- The clamped length should be no more than three to four times the bar diameter
- Select a bar of sufficient stiffness based on the *L/D* ratio of the application
- Use a bar with internal coolant supply if possible

28.9 GUIDELINES FOR INSERTS

- Select the nose radius and feed rate to produce the desired surface finish. Ideally the nose radius should be less than the depth of cut.
- The lead angle should be between 0 and 15 degrees.
- Select a chip breaker according to the manufacture's guidelines for the application and workpiece material.
- Use positive true rake geometries if possible to minimize forces.
- Use a tougher carbide grade than that of the OD operation. This will help to withstand higher stresses from possible chip jamming, deflection, and vibration.
- Compensate for tangential and radial deflection if the machine allows.

28.10 REAMERS

Reamers are used to enlarge drilled or cast iron holes to final size specifications with a high degree of dimensional accuracy (H6 tolerance, ±0.0003 to 0.0006 in) and excellent surface finishes.

Reamers can be used on through holes, blind holes, and tapered holes.

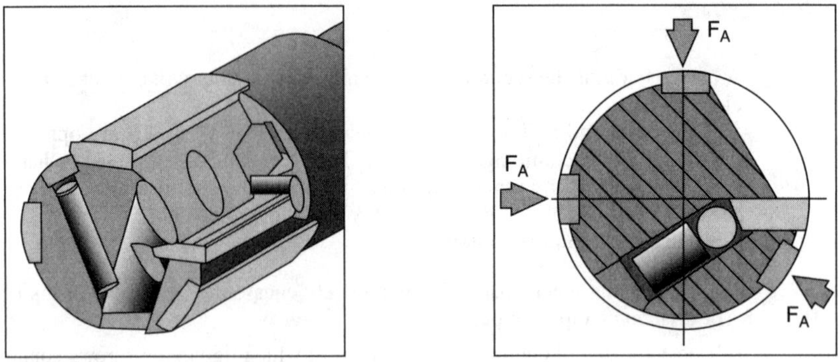

FIGURE 28.10 Indexable carbide blade reamers.

Reamers cannot improve existing hole location because they follow the existing hole—only boring can eliminate hole eccentricity.

There are many types of machine reamers available: *chucking reamers—straight and tapered shank* and *straight and helical flutes, tapered shank straight flute jobber reamers, rose reamers, shell reamers, morse taper reamers, helical taper pin reamers, helical flute tapered bridge reamers, carbide tipped (Brazed) reamers,* and *indexable carbide blade reamers.*

Indexable carbide blade reamers can be used on through and blind holes. An indexable carbide blade reamer typically has one carbide blade and three cermet wear pads (Fig. 28.10). The blade and pads counterbalance the radial forces of the single-point reamer.

The indexable blade is available in several grades to afford flexibility and efficient reaming across a spectrum of work materials.

The advantage of indexable carbide blade reamers includes: higher feed rates than conventional reamers, the efficiency of an indexable blade, high dimensional accuracy, and excellent surface finish capability.

CHAPTER 29
TAPPING

Mark Johnson
Tapmatic Corporation
Post Falls, Idaho

29.1 INTRODUCTION

Tapping is a process for producing internal threads. A tap is a cylindrical cutting or forming tool with threads on its outer surface that match the configuration of the internal threads it is designed to produce. The tap must be rotated and advanced into the hole an exact distance for each revolution. This distance is called the *pitch* of the tap. After the tap has advanced into the hole to the desired depth, its rotation must be stopped and reversed in order to remove the tap from the threaded hole.

A wide variety of machines may be used for tapping from manually controlled drill press and milling machines to CNC-controlled machining or turning centers. The type of machine being used and the application conditions determine the most suitable tap holding device for the job.

There are many factors that influence a tap's performance. Some of these include the workpiece material, fixturing of the part, hole size, depth of the hole, and type of cutting fluid being used. Selecting the correct tap for the specific conditions will make a big difference in your results. Manufacturers of taps today produce taps with special geometries for specific materials and make use of various surface treatments that allow them to run at higher speeds and produce more threaded holes before the tap must be replaced.

29.2 MACHINES USED FOR TAPPING AND TAP HOLDERS

A machine can be used for tapping if it has a rotating spindle and the ability to advance the tap into the hole, either automatically or manually. If you can drill on a machine, you can also tap by choosing the right tap holder or tapping attachment.

29.2.1 Drill Press or Conventional Mills

Drill presses and conventional milling machines are commonly used for tapping. The most efficient way to tap on these machines is by using a compact self-reversing tapping attachment. A self-reversing tapping attachment is mounted into the spindle of the machine. The tap is then held by this tapping attachment, and the operator of the machine manually feeds the tap into the hole while the machine spindle rotates. When the tap has reached the desired depth, the operator retracts the machine spindle, and this causes the tapping attachment to automatically reverse the tap's rotation. As the operator continues to retract the machine spindle, the tap is fed out of the hole. The drive spindle of the tapping attachment, which holds the tap, has the ability to float axially in tension and compression. This means that the operator does not have to perfectly match his or her feed of the machine spindle to

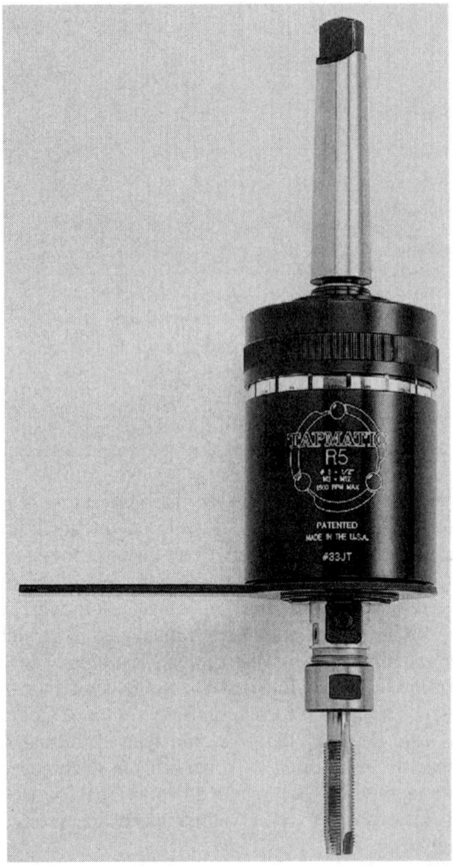

FIGURE 29.1 Self-reversing tapping attachments.

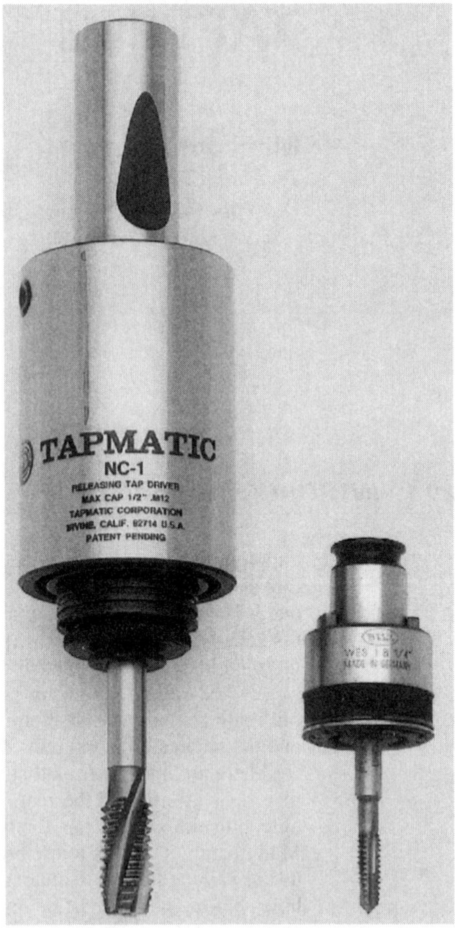

FIGURE 29.2 Nonreversing tap driver.

the pitch of the tap. Tapping attachments use a gear mechanism to drive the reversal and this requires a stop arm that must be restrained from rotating. A photograph of a typical tapping attachment installation on a drill press is shown in Fig. 29.1.

29.2.2 Conventional Lathes

Tapping on conventional lathes is also possible, but in this case a self-reversing tapping attachment cannot be used, since on a lathe, the workpiece rotates and the tap does not. A self-reversing tapping attachment must be driven for the automatic reversal to function. For a conventional lathe a tension compression tap driver can be used to tap holes on center. Since it is difficult for the operator to control the tapping depth manually, the best holders for this application include a release to neutral. To tap a hole, the workpiece is rotated and the operator feeds the tap into the hole until he or she reaches the machine stop.

A momentary dwell permits the tap to continue into the hole the self-feed distance of the tap holder. When the self-feed distance is reached, the drive of the tap holder releases and the tap begins to turn with the workpiece. At this point, the machine spindle holding the workpiece is stopped and reversed, and then the operator feeds the tap out of the hole. A typical nonreversing tap driver with release to neutral is shown in Fig. 29.2.

29.2.3 Machining Centers

Today most high production tapping is performed on CNC machines. Machining centers include canned cycles for tapping. There are two types of canned cycles. The older tapping cycle—still included on many machines—is employed with a tension/compression axial floating tap driver. A tapping speed is selected and the appropriate feed rate is specified in the program. The axial float of the tap driver compensates for the difference between the machine feed and the actual pitch of the tap. When the tap reaches the program depth, the machine spindle stops and reverses. Since the CNC control makes the movements of the machine consistent, a release to neutral may not be required.

The newest cycle for tapping on machining centers is called a synchronous or rigid tapping cycle. In this cycle the machine feed rate is synchronized to the revolutions of the spindle and it is possible to program the machine to match the pitch of the specific tap being used. Since the cycle is synchronized it is possible to tap with a solid holder that does not have axial float. It has been found, however, that thread quality and tap life are not ideal under these conditions. Since it is impossible for the machine to match the pitch of a given tap perfectly, there is an unavoidable deviation between the machine feed and the tap pitch.

Even a slight deviation between the machine synchronization and the pitch of the tap causes extra forces on the tap making it wear more quickly. It also produces a negative effect on thread quality.

A tap holder has now been developed with the ability to compensate for these slight deviations. It employs a precision machined flexure with a very high spring rate and only a precise, predictable amount of axial and radial compensation. Unlike a normal tension compression holder with a large amount of compensation and a relatively soft spring rate, the depth control for tapping remains very accurate with these new holders for rigid tapping. Improvements in tap life as great as 200 percent have been achieved by using these new holders in place of conventional tap drivers. Figure 29.3 is a photograph showing examples of the tap holders.

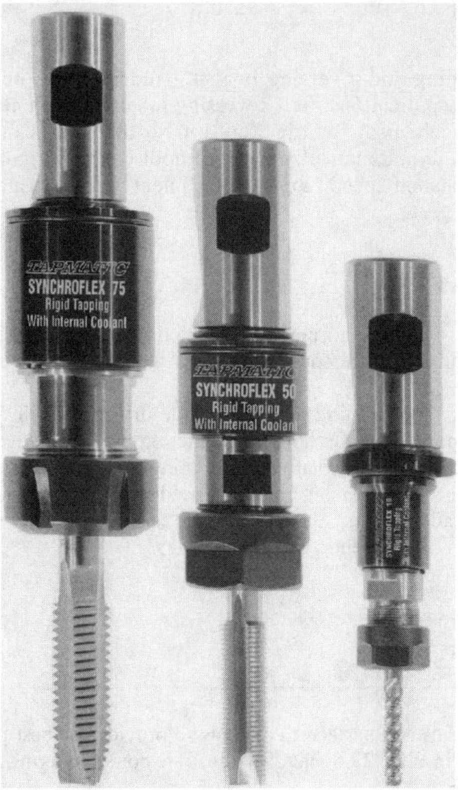

FIGURE 29.3 Tap holder's fixture.

A disadvantage of using the tapping cycle on a machining center is that it requires the machine spindle to reverse. It takes time for the machine spindle to stop and reverse rotation, and it must do this twice for each tapped hole. Once at the bottom of the hole to reverse the tap out and again to change back to forward rotation before entering the next hole. The mass of the machine spindle cannot make these changes in direction instantaneously, especially in the normally short feed distances associated with tapping. Taps perform best when running continually at the correct speed. The deceleration of the machine spindle required as the tap reaches depth has a negative effect on tap life. A self-reversing tapping attachment can be used on a machining center to eliminate these problems. Cycle time is faster—since the machine's forward rotation never stops, the machine spindle only has to feed the tap in and out of the hole. When the machine retracts, the tapping attachment instantly reverses the taps rotation. A constant speed is maintained during the tapping cycle and this allows the tap to continuously cut at the proper speed for the optimum tap life. Wear and tear to the

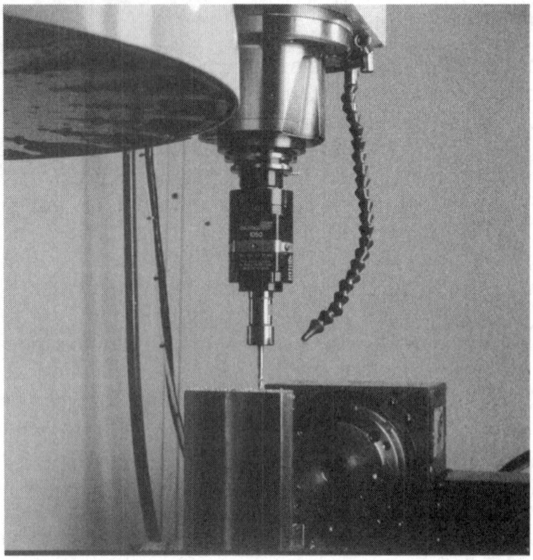

FIGURE 29.4 Self-reversing tapping attachment.

machine spindle caused by stopping and reversing is also avoided. These benefits are especially helpful in high production applications. Self-reversing tapping attachments for machining centers include a locking mechanism for the stop arm so that they can be loaded and unloaded from the machine spindle automatically during a tool change. Figure 29.4 shows a photograph of a self-reversing constant speed tapping attachment used on a machining center.

29.2.4 CNC Lathes and Mill-Turning Centers

Tapping on center on a CNC lathe can be performed much like on a machining center by using a tension compression tap holder. The only difference is that on a lathe the workpiece rotates instead of the tap.

CNC lathes with live tooling or mill-turning centers include driven tools in the turret of the machine. With driven tooling it is possible to tap holes off center on the face of the part, or even on the side of the part by stopping the workpiece rotation and turning on the tool's rotation. Since the tool is driven it is also possible to use self-reversing constant speed tapping attachments in this application. Tapping attachments are available with adaptations such as the commonly used VDI shank, shown in Fig. 29.5, to fit the turrets of different types of machines.

29.3 TAP NOMENCLATURE

Taps today are made with geometries and surface treatments to provide the best performance in a specific application. The drawings in Figs. 29.6 and 29.7 illustrate common terms used for describing the taps.

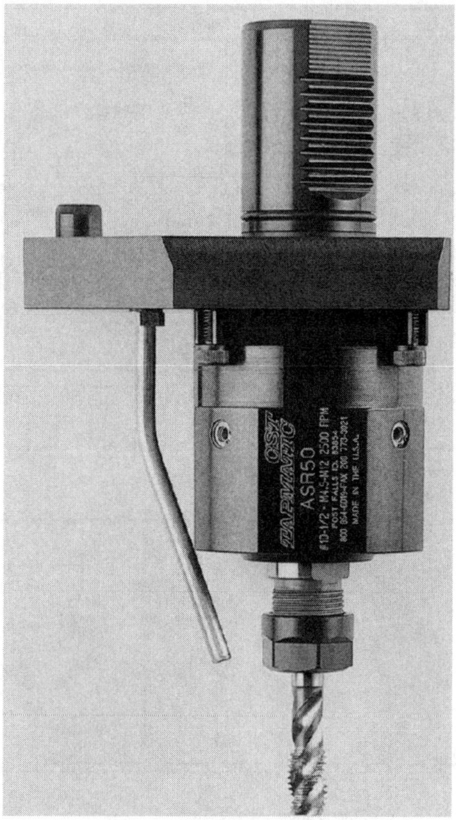

FIGURE 29.5 VDI shank.

29.4 INFLUENCE OF MATERIAL AND HOLE CONDITION

Two of the most important factors affecting tap selection are the material of the workpiece and conditions of the hole.

In general, harder materials are more difficult to tap than softer materials. An exception to this is certain soft materials that are gummy and form chips that readily adhere to the tap. The following drawings explain how certain features of the tap are selected based on material and hole conditions (Fig. 29.7).

An important function of the tap is the removal of chips from the hole. The flutes of a tap provide the cutting edges but also serve as a means for chip removal. When tapping through holes, spiral pointed taps are often used to push the chips forward and out of the hole. In blind holes, spiral fluted taps are used to pull the chips out of the hole. Blind holes are more difficult to tap due to problems of chip removal. Figure 29.8 shows some common geometries based on material and hole conditions.

29.5 EFFECTS OF HOLE SIZE

The size of the hole has a major impact on the tapping process, since the hole size determines the percentage of full thread or amount of material being removed. The higher the percentage of thread (i.e., the smaller the hole size), the more difficult it becomes to tap the hole (Fig. 29.9).

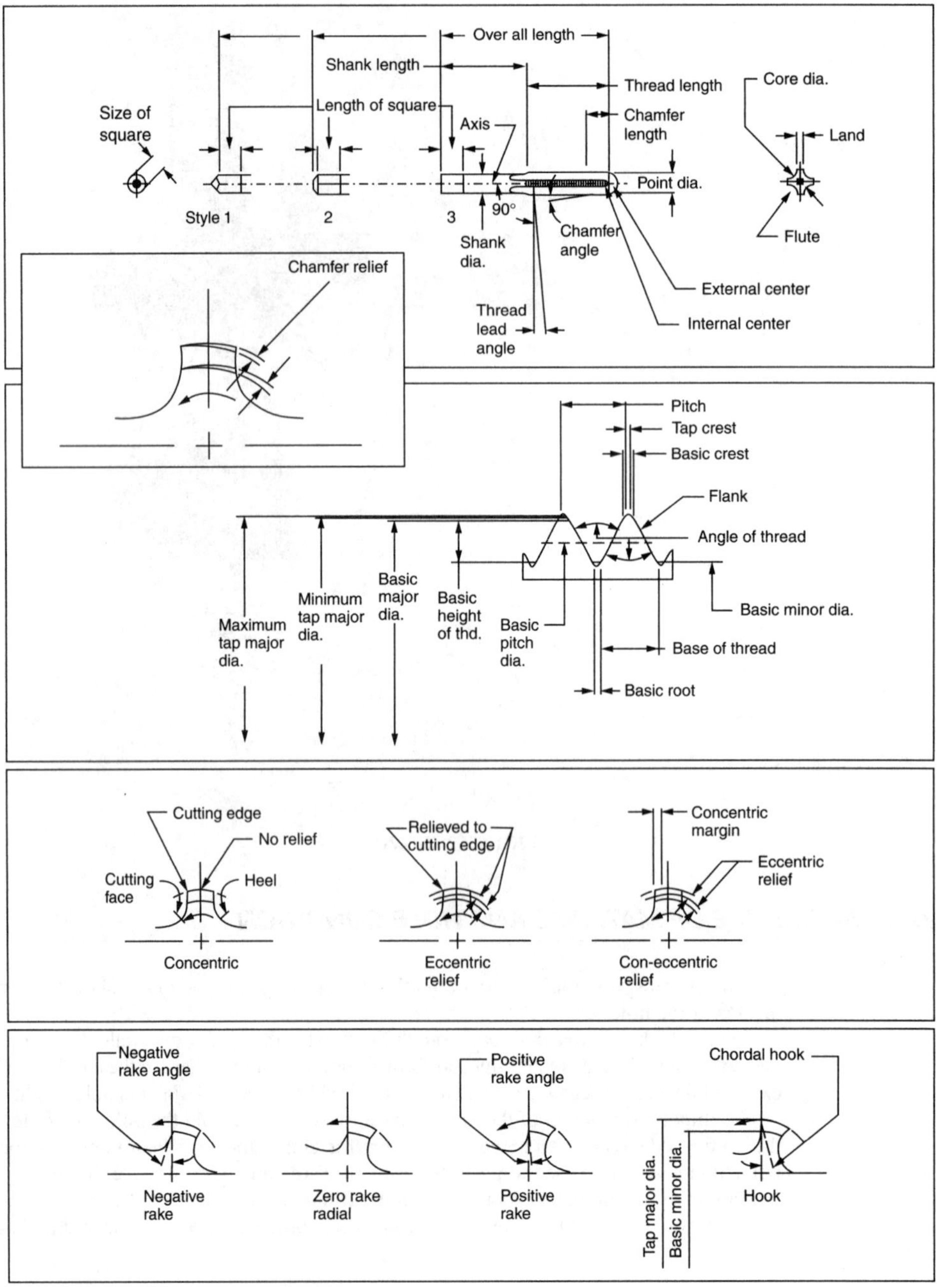

FIGURE 29.6 Illustration of tap terms.

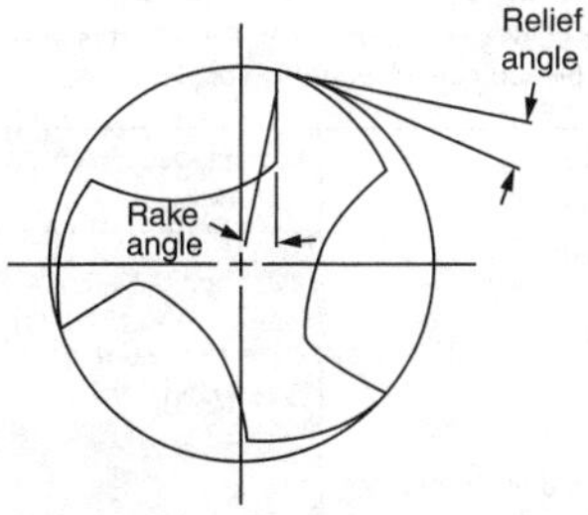

Rake angle
The best rake angle for a tap depends on the material. Materials that produce long chips normally require a tap with greater rake angle. Materials that produce short chips require a smaller rake angle. Difficult materials like titanium or inconnel require a compromise between greater rake angle for longer chips and smaller rake angle for more strength.

Relief angle in the lead of a tap
A small relief angle can be used in soft materials. Harder materials like stainless steel can be cut easier with a tap having a greater relief angle which reduces the friction. Tough materials like inconnel and nickel can be cut more easily with an even greater relief angle.

The relief angle is smaller on taps for blind holes than on taps for through holes so that the chip root can be sheared off when the tap reverses without breaking the taps cutting edge.

Chamfer length (lead)

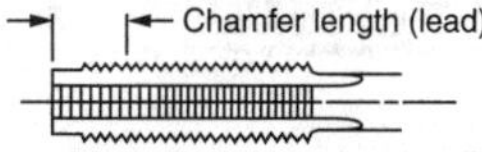

The actual cutting of the thread is done by the lead of the tap. When there are more threads in the chamfer length or lead the torque is reduced, producing the thread is much easier, and the life of the tap will be increased. In blind holes where there is not enough room to drill deep enough for a tap with a longer lead, taps with short leads are used. In some cases the lead of the tap is reduced to as little as 1.5 threads. This greatly increases torque and reduces tap life. Even when using taps with shortened lead it is still important to drill deep enough for adequate clearance. It is recommended to allow one thread length plus one mm beyond the lead of the tap as drill clearance.

Relief angle in the thread profile (pitch diameter relief)
The relief angle affects true to gage thread cutting, and also the free cutting ability and life of the tap. It has an effect on how the tap is guided when it enters the hole. If the relief angle is too great pitch guidance and self centering of the tap cannot be guaranteed especially in soft materials. In materials like stainless steel or bronze the relief angle should be larger to allow free cutting and to allow more lubrication to reach the cutting and friction surfaces. A bigger relief angle can allow higher tapping speed provided the tap is guided concentrically into the hole by the machine and tap holder.

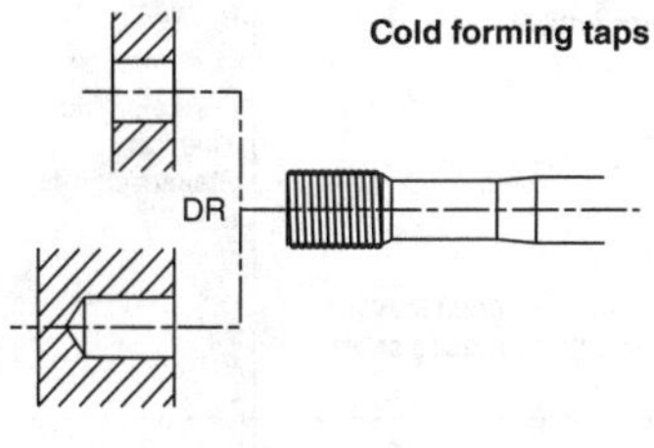

Cold forming taps

Workpiece materials	Recommended tap surface treatments
All ductile materials	

These taps form the thread rather than cut. Since no chips are produced they can be used in blind or through holes. Cold forming is possible in all ductile materials. Advantages include no waste in the form of chips, no mis-cutting of threads, no pitch deviation, higher strength, longer tool life, and higher speed. Please note that the core hole diameter must be larger than with a cutting tap. Good lubrication is important, more torque is required, and the minor diameter of the thread will appear rough due to forming process.

FIGURE 29.7 General tap recommendations for specific materials.

29.6 WORKPIECE FIXTURING

For the tap to cut properly, it must enter the hole concentrically. If the tap enters the hole at an angle or off center it can create oversize threads or cause the tap to be broken. For best results, the workpiece must be clamped securely so that it cannot rotate or lift, and so that the hole is lined up with the tap. In some cases, when a small amount of misalignment is unavoidable, a tap driver with radial float can be used to allow the tap to find the hole and center concentrically.

Tap manufacturers offer their own unique geometries for specific materials and applications. This chart is meant to provide general information. For a specific tap recommendation for your application, please consult your tap supplier.

Standard straight fluted tap with 6 to 8 threads chamfer length or lead.

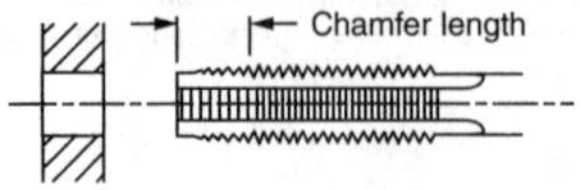

These taps do not transport the chips out of the hole. For this reason, they should not be used for deep hole tapping. They work best in shallow depth through holes and in materials that produce short chips.

Workpiece materials	Recommended tap surface treatments
Cast iron	Nitrided or TiN
Brass, short chipping	Nitrided
Cast aluminum	Nitrided
Short chip hard	Nitrided or TiN

Straight fluted taps with spiral point with 3.5 to 5 threads chamfer length or lead.

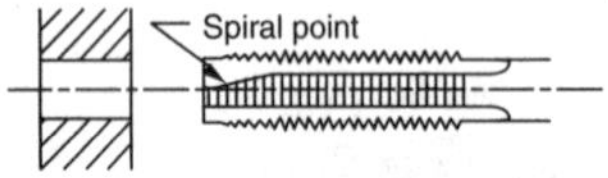

These taps push the chips forward. The chips are curled up to prevent clogging in the flutes. They are used for through holes.

Workpiece materials	Recommended tap surface treatments
Aluminum long chip	Bright, or Cr or TiN
Exotic alloys	Nitrided or TiN
Stainless steel	Nitrided or TiN
Steel	Bright or TiN or TiCN

Left hand spiral fluted tap with approximately 12 degrees spiral flutes with 3.5 to 5 threads chamfer length.

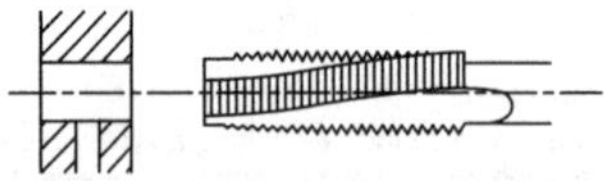

These taps are mostly used in thin walled parts or for holes interrupted by cross holes or longitudinal slots.

Workpiece materials	Recommended tap surface treatments
Titanium Special hole condition	Nitrided or TiN

Right hand spiral fluted tap with approximately 15 degree spiral flutes with 3.5 to 5 threads chamfer length.

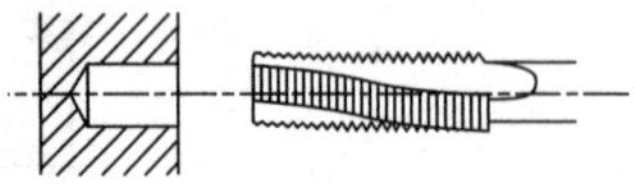

The spiral flutes transport chips back out of the hole. These taps are used in blind holes less than 1.5 times the tap diameter deep with materials that produce short chips.

Workpiece materials	Recommended tap surface treatments
Cast aluminum	Nitrided
Titanium	Nitrided or TiN
Stainless steel	Bright or TiN
Steel	Bright or TiN or TiCN

Right hand spiral fluted tap with 40 degrees to 50 degrees spiral flutes.

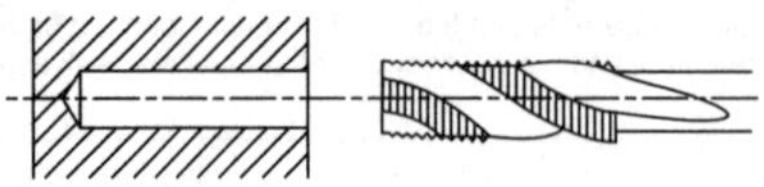

The greater helix angle provides good transport of chips back out of the hole. These taps are used only in blind holes in materials that produce long chips. They can also be used in deeper holes up to 3 times the tap diameter.

Workpiece materials	Recommended tap surface treatments
Aluminum long chip	Bright or Cr, or TiN
Stainless steel	Bright or TiN
Steel alloy Cr-NI	Bright or TiN or TiCN
Soft material	Bright

FIGURE 29.8 General tap recommendations.

Suggested percentage of full threads in tapped holes

It stands to reason that it takes more power to tap to a full depth of thread than it does to tap to a partial depth of thread. The higher the metal removal rate, the more torque required to produce the cut.

It would also stand to reason that the greater the depth of thread, the stronger the tapped hole. This is true, but only to a point. Beyond that point (usually about 75% of full thread) the strength of the hole does not increase, yet the torque required to tap the hole rises exponentially. Also, it becomes more difficult to hold size, and the likelihood of tap breakage increases. With this in mind, it does not make good tapping sense to generate threads deeper than the required strength of the thread dictates.

As a general rule, the tougher the material, the less the percentage of thread is needed to create a hole strong enough to do the job for which it was intended. In some harder materials such as stainless steel, Monel, and some heat-treated alloys, it is possible to tap to as little as 50% of full thread without sacrificing the usefulness of the tapped hole.

Workpiece material	Deep hole tapping	Average commercial work	Thin sheet stock or stampings
Hard or tough cast steel drop forgings Monel metal nickel steel stainless steel	55%–65%	60%–70%	–
Free-cutting aluminum brass bronze cast iron copper mild steel tool steel	60%–70%	65%–75%	75%–85%

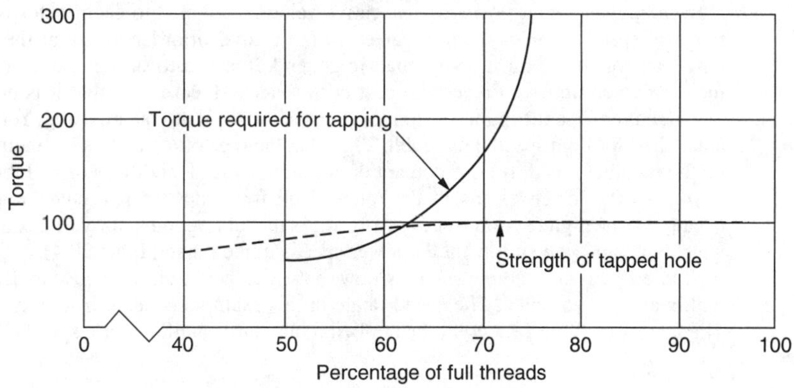

Cutting taps: formula for calculating percentage of thread

Inch size (Dimensions in inches)

$$\% \text{ of full thread} = \text{threads/in} \times \frac{\text{major diameter of tap minus drill diameter}}{.01299}$$

Metric size (Dimensions in mm)

$$\% \text{ of full thread} = \frac{76.980}{\text{Metric pitch}} \times (\text{basic major diameter (mm) minus drilled hole (mm)})$$

Metric taps

FIGURE 29.9 Tapping torque vs. thread strength.

29.7 TAP LUBRICATION

Lubrication of the cutting tool is more important in tapping than in most other machining operations because the cutting teeth of a tap are more easily damaged by heat, and it is easy for chips to clog the hole or threads since the cutting edges are completely surrounded by the workpiece material. A good extreme pressure lubricant makes a big difference in improving thread quality, finish, and tap life. Taps made with holes for internal coolant through the tap can also greatly improve performance, especially in blind holes as the coolant also helps to flush out the chips. See Table 29.1 for more recommendations for cold form tapping.

29.8 DETERMINING CORRECT TAPPING SPEEDS

Table 29.2 is a compilation of guidelines from tap manufacturers and other sources for cutting or cold-forming of threads in relations to workpiece material.

Cutting Speed for Tapping. Several factors, singly or in combination, can cause very great differences in the permissible tapping speed. The principal factors affecting the tapping speed are the pitch of the thread, the chamfer length on the tap, the percentage of full thread to be cut, the length of the hole to be tapped, the cutting fluid used, whether the threads are straight or tapered, the machine tool used to perform the operation, and the material to be tapped.*

Following are charts based on recommendations from several tap manufacturers. As you can see in the charts, a wide range of possible tapping speeds is given for each tap size and material. Listed in Table 29.2 are the factors influencing tap performance, which determine the best speed to use in a given application. A positive or negative value is assigned to each factor to help you narrow the range of speed recommendations given in the charts. In order to run at the maximum end of the range, all conditions must be optimal. In general, it is best to start at the lower end of the range and increase speed until you reach the best performance. Please note that it is best to consult your tap manufacturer for a recommendation for the specific tap that you are using. You can see from the separate chart for high speed taps (Table 29.3) that the type of tap and its geometry have a major impact on the possible speed. If your coolant does not contain EP additives or its lubrication quality is low, start from the lower speeds in the range. Roll form taps, in particular, require good lubrication because of the high friction forces involved. As the lubrication quality of a coolant is often unknown, we recommend you start from the lower speeds in the range (Table 29.4).

Table 29.2 is an example showing how to use the factors shown above to determine cutting speeds within a specified range. The speed range in this example is taken from the chart for standard taps. The factors in Table 29.2 could be applied to any tap manufacturer's speed chart.

*Erik Oberg, Franklin D. Jones, and Holbrook L. Horton, *Machinery's Handbook,* 23d ed., Industrial Press, New York, 1998.

TABLE 29.1 Machining Recommendations for Cold Form Tapping

Cold Forming Internal Threads With Taps

Internal threads can be produced by a cold forming or swaging process. The desired thread is formed in the metal under pressure and the grain fibers, as in good forging, follow the contour of the thread. These grain fibers are not cut away as in conventional tapping. The cold forming tap has neither flutes nor cutting edges and therefore it produces no chips and cannot create a chip problem. The resulting thread has a burnished surface.

Material Recommended

Care must be taken to minimize surface damage to the hole when tapping materials that are prone to work harden. This may be accomplished by using sharp drills, and correct speed and feeds. Surface damage may cause torque to increase to a point of stopping the machine or breaking the tap.

Cold forming taps have been recommended for threading ductile materials. Examples of material classes which have been tapped are:

- Low carbon steels
- Leaded steels
- Austenitic stainless steels
- Aluminum die casting alloys (low silicon)
- Wrought aluminum alloys (ductile)
- Zinc die casting alloys
- Copper and copper alloys (ductile brasses)

Cold Forming Tap Application Information

Tapping Action the Same

Except for changes in hole size, the application of cold forming taps differs in no way from conventional cutting taps.

Blind Hole Tapping Possible

Whenever possible, in blind holes, drill or core deep enough to permit the use of the plug style taps. These tools, with four threads of taper, will require less torque, will produce less burr upon entering the hole, and will give greater life.

Torque

One of the most important factors with roll form tapping is the torque required. Torque is influenced by the percentage of thread, workpiece material, depth of hole, tap lubrication, and tapping speed. Depending on these conditions, the torque required can vary from no additional torque to as much as four times more in comparison to cutting taps. Roll form taps have a very long tap life, but as they wear the torque increases and the torque required for the initial reversal of the tap becomes even higher. Since a roll form or fluteless tap has greater strength than a cutting tap the forces to break it may exceed the strength of the gearing or drive components in a compact tapping attachment or tap driver. This should be taken in to account when selecting the tap holder and when determining how frequently to replace the tap.

No Lead Screw Necessary

These taps work equally well when used in a standard tapping head, automatic screw machine, or lead screw tapper. It is unnecessary to have lead screw tapping equipment in order to run the cold forming tap because the tool will pick up its own lead upon entering the hole.

Standard Lubrication

In general it is best to use a good cutting oil or lubricant rather than a coolant for cold forming taps. Sulfur base and mineral oils, along with most of the lubricants recommended for use in cold extrusion or metal drawing, have proven best for this work.

Spindle Speeds

For most materials, spindle speeds may be doubled over those recommended for conventional cutting type taps. Generally, the tap extrudes with greater efficiency at high RPM's but it is also possible to run the tap at lower speeds with satisfactory results. The drilling speed may be used as a starting point for cold forming taps.

Counter Sinking or Chamfering Helpful

Because these taps displace metal, some metal will be displaced above the mouth of the hole during tapping. For this reason it is best to countersink or chamfer the hole prior to tapping, so that the extrusion will raise within the countersink and not interfere with the mating part.

Tapping Cored Holes Possible

Cored holes may be tapped with these taps provided the core pins are first changed to form the proper hole size. Because core pins have a draft or are slightly tapered the theoretical hole size should be at a point on the pin that is one-half the required length of engagement of the thread to be formed. In designing core pins for use with these taps, a chamfer should be included on the pin to accept the vertical extrusion.

Drill Size

With roll form tapping, material flows inward to create the minor diameter of the thread. For this reason, a different hole size is needed vs. a cutting tap. A theoretical hole size is determined for a desired percent of thread.

The formula for these theoretical hole size determinations is as follows:

Theoretical Hole Size

(core, punch, or drill size)

$$= \text{Basic Tap O.D.} - \frac{0.0068 \times \text{percent of Thread}}{\text{Threads per inch}}$$

Example: To determine the proper drill size to form 65 percent of thread with a 1/4-20 cold form tap.

Basic Tap O.D. = 1/4in or 0.250 in

Threads per Inch = 20

$$\text{drill size} = 0.250 - \frac{0.0068 \times \text{percent of Thread}}{\text{Threads per inch}}$$

$$\text{drill size} = 0.228$$

TABLE 29.2 Determining Speed Within Specified Range

Compilation of guidelines from tap manufacturers and other sources for cutting or cold-forming of threads in relation to workpiece material

Cutting speed for tapping: Several factors, singly or in combination, can cause very great differences in the permissible tapping speed. The principal factors affecting the tapping speed are the pitch of the thread, the chamfer length on the tap, the percentage of full thread to be cut, the length of the hole to be tapped, the cutting fluid used, whether the threads are straight or tapered, the machine tool used to perform the operation, and the material to be tapped. *From Machinery's Handbook 23rd edition.*

If your coolant does not contain EP additives or its lubrication quality is low, start from the lower speeds in the range. Roll form taps in particular require good lubrication because of the high friction forces involved. As the lubrication quality of a coolant is often unknown, we recommend you to start from the lower speeds in the range.

Check your tap manufacturer's tapping speed recommendations then use these guidelines for determining exact correct speed for your application.

These factors apply to everyone's tapping speed charts

−%	Ten factors requiring lower speeds		Ten factors permitting higher speeds	+%
−20	Poor lubrication	1	Good lubrication	+20
−15	High tensile strength of material	2	Low tensile strength of material	+15
−15	Large thread diameter	4	Small thread diameter	+15
−10	High alloy materials	3	Low alloy materials	+10
−10	Thread depth More than 1.5 × dia.	5	thread depth 1.5 × dia. or less	+10
−10	Thread pitch coarse	6	Thread pitch fine	+10
−5	Drill size More than 65% of thread	7	Drill size 65% or less thread	+5
−5	Tap lead less than 3.5 threads	8	Tap lead more than 3.5 threads	+5
−5	Blind holes	9	Through holes	+5
−5	Free running spindle inaccurate pitch control hydraulic/air feed	10	Synchronous spindle lead screw CNC control	+5

Below is an example showing how to use the factors above to determine cutting speeds within a specified range. The speed range in this example is taken from the chart for standard taps.

Example: Tap size: 1/4 in -28 coated, **Material:** Aluminum die cast, **From chart:** 688-1375 RPM, **RPM spread** = 687

Minus factors:		Plus factors:	
High tensile strength	−15	Coolant with good EP	+20
Thread depth 3 × dia.	−10	Small thread diameter	+15
Drill size = 75 percent Thd.	−5	Pitch fine	+10
Blind hole	−5	Lead 3.5 threads	+5
Total	−35	CNC machine	+5
		Total	+55

Apply the factors against the RPM spread of 687

+0.55 × 687 = 378 added to minimum RPM 688 = **1066 new minimum RPM**

−0.35 × 687 = 240 subtracted from maximum RPM 1375 = **1135 new maximum RPM**

Common sense rule: Begin with minimum RPM and work up to optimum efficiency and tap life.

TABLE 29.3 High Speed Taps—Speed Recommendation.

Tap size		Low carbon steel, medium carbon steel	High carbon steel, high strength steel, tool steel	Stainless 303, 304, 316	Stainless 17-4 annealed	Aluminum alloys	Aluminum die cast	Magnesium	Copper	Cast iron
		Surface feet per minute								
		165–200	25–100	30–80	20–40	65–200	65–100	100–130	100–130	130–165
		RPM range based on SFM theoretical RPM range actually possible*								
	0	10505–12733 6000	1592–6366 1592–6000	1910–5093	1273–2546	4138–12733 4138–6000	4138–6366 4138–6000	6366–8276 6000	6366–8276 6000	8276–10505 6000
	1	8634–10465 6000	1308–5233	1570–4186	1047–2093	3401–10485 3401–6000	3401–5233	5233–6808 5233–6000	5233–6808 5233–6000	6808–8634 6000
M2	2	7329–8884 5000	1110–4442	1333–3554	888–1777	2887–8884 2887–5000	2887–4442	4442–5774 4442–5000	4442–5774 4442–5000	5774–7329 5000
	3	6367–7717 5000	964–3858	1157–3096	772–1543	2508–7717 2508–5000	2508–3858	3858–5015 3858–5000	3858–5015 3858–5000	5015–6367 5000
M3	4	5628–6821 5000	853–3411	1023–2728	682–1364	2217–6821 2217–5000	2217–3411	3411–4434	3411–4434	4434–5628 4434–5000
	5	5042–6122 4000	764–3056	917–2445	611–1222	1986–6122 1986–4000	1986–3056	3056–3973	3056–3973	3973–5042 3973–4000
	6	4567–5536 4000	691–2764	829–2211	553–1106	1799–5536 1799–4000	1799–2764	2764–3592	2764–3592	3592–4567 3592–4000
M4	8	3843–4659 3843–4000	583–2330	699–1864	466–932	1514–4659 1514–4000	1514–2330	2330–3029	2330–3029	3029–3843
	10	3317–4021 3317–4000	502–2009	603–1607	402–804	1307–4021 1307–4000	1307–2009	2009–2612	2009–2612	2612–3317
M5	12	2918–3537	442–1769	531–1415	354–707	1150–3537	1150–1769	1769–2300	1769–2300	2300–2918
M6 M7	1/4	2521–3056	382–1528	458–1222	306–611	993–3056	993–1528	1528–1986	1528–1986	1986–2521

TABLE 29.3 High Speed Taps—Speed Recommendation (*Continued*)

Tap Size		Low carbon steel, medium carbon steel	High carbon steel, high strength steel, tool steel	Stainless 303, 304, 316	Stainless 17-4 annealed	Aluminum alloys	Aluminum die cast	Magnesium	Copper	Cast iron
		Surface feet per minute								
		165–200	25–100	30–80	20–40	65–200	65–100	100–130	100–130	130–165
		RPM range based on SFM theoretical RPM range actually possible*								
M8	5/16	2017–2449	306–1222	367–978	245–489	796–2449	796–1222	1222–1589	1222–1589	1589–2017
M9	3/8	1681–2037	255–1019	306–815	204–407	662–2037	662–1019	1019–1324	1019–1324	1324–1681
M10	7/16	1441–1748	219–873	262–698	175–349	568–1748	568–873	873–1135	873–1135	1135–1441
M12	1/2	1261–1528	191–764	229–611	153–306	497–1528	497–764	764–993	764–993	993–1261
M14	9/16	1121–1359	172–687	206–550	137–275	442–1359	442–687	687–893	687–893	893–1121
M16	5/8	1008–1222	153–611	183–489	122–244	397–1222	397–611	611–794	611–794	794–1008
M18 M20	3/4	840–1019	128–509	153–407	102–203	331–1019	331–509	509–662	509–662	662–840
M22	7/8	720–873	109–437	131–350	87–175	284–873	284–437	437–568	437–568	568–720
M24 M25	1	630–764	96–382	115–306	76–153	248–764	248–382	382–497	382–497	497–630

****Note:*** For certain smaller size taps it is not possible to reach sfn recommendations due to limits of machines and tap holders.

TABLE 29.4 Roll Form Taps—Speed Recommendations

Tap size		Low carbon steel, medium carbon steel	High carbon steel, high strength steel tool steel	Stainless 303, 304, 316	Titanium alloys	Aluminum alloys	Aluminum die cast
		Surface feet per minute uncoated tap coated tap					
		30–50 65–100	25–65	20–25 25–35	25–40	35–50 50–65	35–65
		RPM range uncoated RPM range coated					
	0	1910–3183 4138–6000	1592–4138	1273–1592 1592–2228	1592–2546	2228–3183 3183–4138	2228–4138
	1	1570–2617 3401–5233	1308–3401	1047–1308 1308–1831	1308–2093	1831–2617 2617–3401	1831–3401
M2	2	1333–2221 2887–4442	1110–2887	888–1110 1110–1555	1110–1777	1555–2221 2221–2887	1555–2887
	3	1157–1929 2508–3858	964–2508	772–964 964–1351	964–1543	1351–1929 1929–2508	1351–2508
M3	4	1023–1705 2217–3411	853–2217	682–853 853–1194	853–1364	1194–1705 1705–2217	1194–2217
	5	917–1528 1986–3056	764–1986	611–764 764–1070	764–1222	1070–1528 1528–1986	1070–1986
	6	829–1382 1799–2764	691–1799	553–691 691–969	691–1106	969–1382 1382–1799	969–1799
M4	8	699–1165 1514–2330	583–1514	466–583 583–815	583–932	815–1165 1165–1514	815–1514
	10	603–1005 1307–2009	502–1307	402–502 502–704	502–804	704–1005 1005–1307	704–1307
M5	12	531–884 1150–1769	442–1150	354–442 442–619	442–707	619–884 884–1150	619–1150
M6 M7	1/4	458–764 993–1528	382–993	306–382 382–535	382–611	535–764 764–993	535–993
M8	5/16	367–611 796–1222	306–796	245–306 306–429	306–489	429–611 611–796	429–796
M9	3/8	306–509 662–1019	255–662	204–255 255–357	255–407	357–509 509–662	357–662
M10	7/16	262–437 568–873	219–568	175–219 219–306	219–349	306–437 437–568	306–568
M12	1/2	229–382 497–764	191–497	153–191 191–267	191–306	267–382 382–497	267–497
M14	9/16	206–344 442–687	172–442	137–172 172–238	172–275	238–344 344–442	238–442
M16	5/8	183–306 397–611	153–397	122–153 153–214	153–244	214–306 306–397	214–397
M18 M20	3/4	153–255 331–509	128–331	102–128 128–178	128–203	178–255 255–331	178–331

TABLE 29.5 Standard Taps—Speed Recommendations

Tap size		Low carbon steel, medium carbon steel	High carbon steel, high strength steel, tool steel	High strength steel, tool steel hardened	Stainless 303, 304, 316	Stainless 410, 430, 17-4 hardened	Stainless 17-4 annealed	Titanium alloys	Nickel base alloys	Aluminum alloys	Aluminum die cast	Magnesium	Brass, bronze	Copper	Cast iron
		Surface feet per minute uncoated tap coated tap													
		25–50 50–80	6–30 10–35	6–12	12–35 20–50	12–15	12–15 12–25	3–15	10–15	50–65	40–65 45–90	45–100	30–65	50–60 65–100	35–50 50–65
		RPM range uncoated RPM range coated													
	0	1592–3183 3183–5093	382–1910 637–2228	382–764	764–2228 1273–3183	764–955	764–955 764–1592	191–955	637–955	3183–4138	2546–4138 2865–5730	2865–6000	1910–4138	3183–3820 4138–6000	2228–3183 3183–4138
	1	1308–2617 2617–4186	314–1570 523–1831	314–628	628–1831 1047–2617	628–785	628–785 628–1308	157–785	523–785	2617–3401	2093–3401 2355–4710	2355–5233	1570–3401	2617–3140 3401–5233	1831–2617 2617–3401
M2	2	1110–2221 2221–3554	267–1333 444–1555	267–533	533–1555 888–2221	533–666	533–666 533–1110	133–666	444–666	2221–2887	1777–2887 1999–3999	1999–4442	1333–2887	2221–2665 2887–4442	1555–2221 2221–2887
	3	964–1929 1929–3086	231–1157 386–1351	231–463	463–1351 772–1929	463–579	463–579 463–964	116–579	386–579	1929–2508	1543–2508 1736–3472	1736–3858	1157–2508	1929–2315 2508–3858	1351–1929 1929–2508
M3	4	853–1705 1705–2728	205–1023 341–1194	205–409	409–1194 682–1705	409–512	409–512 409–853	102–512	341–512	1705–2217	1364–2217 1535–3069	1535–3411	1023–2217	1705–2046 2217–3411	1194–1705 1705–2217
	5	764–1528 1528–2445	183–917 306–1070	183–367	367–1070 611–1528	367–458	367–458 367–764	92–458	306–458	1528–1986	1222–1986 1375–2750	1375–3056	917–1986	1528–1833 1986–3056	1070–1528 1528–1986
	6	691–1382 1382–2211	166–829 277–969	166–332	332–969 553–1382	332–415	332–415 332–691	83–415	277–415	1382–1799	1106–1799 1246–2487	1246–2764	829–1799	1382–1658 1799–2764	969–1382 1382–1799
M4	8	583–1165 1165–1664	140–699 233–815	140–280	280–815 466–1165	280–349	280–349 280–583	70–349	233–349	1165–1514	932–1514 1048–2097	1048–2330	699–1514	1165–1398 1514–2330	815–1165 1165–1514
	10	502–1005 1005–1607	121–603 201–704	121–241	241–704 402–1005	241–302	241–302 241–502	60–302	201–302	1005–1307	804–1307 905–1808	905–2009	603–1307	1005–1205 1307–2009	704–1005 1005–1307
M5	12	442–884 884–1415	106–531 177–619	106–212	212–619 354–884	212–265	212–265 212–442	53–265	177–265	884–1150	707–1150 796–1592	796–1769	531–1150	884–1061 1150–1769	619–884 884–1150

TABLE 29.5 Standard Taps—Speed Recommendations (*Continued*)

Tap size		Low carbon steel, medium carbon steel	High carbon steel, high strength steel, tool steel	High strength steel, tool steel hardened	Stainless 303, 304, 316	Stainless 410, 430, 17-4 hardened	Stainless 17-4 annealed	Titanium alloys	Nickel base alloys	Aluminum alloys	Aluminum die cast	Magnesium	Brass, bronze	Copper	Cast iron
		Surface feet per minute uncoated tap coated tap													
		25–50 50–80	6–30 10–35	6–12	12–35 20–50	12–15	12–15 12–25	3–15	10–15	50–65	40–65 45–90	45–100	30–65	50–60 65–100	35–50 50–65
		RPM range uncoated RPM range coated													
M6 M7	1/4	382–764 764–1222	92–458 153–535	92–183	183–535 306–764	183–229	183–229 183–382	46–229	153–229	764–993	611–993 688–1375	688–1528	458–993	764–917 993–1528	535–764 764–993
M8	5/16	306–611 611–978	73–367 122–429	73–147	147–429 245–611	147–184	147–184 147–306	37–184	122–184	611–796	489–796 551–1100	551–1222	367–796	611–733 796–1222	429–611 611–796
M9	3/8	255–509 509–815	61–306 102–357	61–122	122–357 204–509	122–153	122–153 122–255	31–153	102–153	509–662	407–662 458–917	458–1019	306–662	509–611 662–1019	357–509 509–662
M10	7/16	219–437 437–698	52–262 87–306	52–105	105–306 175–437	105–131	105–131 105–219	26–131	87–131	437–568	349–568 393–786	393–873	262–568	437–524 568–873	306–437 437–568
M12	1/2	191–382 382–611	46–229 76–267	46–92	92–267 153–382	92–115	92–115 92–191	23–115	76–115	382–497	306–497 344–688	344–764	229–497	382–458 497–764	267–382 382–497
M14	9/16	172–344 344–550	41–206 68–238	41–82	82–238 137–344	82–102	82–102 82–172	20–102	68–102	344–442	275–442 306–619	306–687	206–442	344–412 442–687	238–344 344–442
M16	5/8	153–306 306–489	37–183 61–214	37–73	73–214 122–306	73–92	73–92 73–153	18–92	61–92	306–397	244–397 275–550	275–611	183–397	306–367 397–611	214–306 306–397
M18 M20	3/4	128–255 255–407	31–153 51–178	31–61	61–178 102–255	61–76	61–76 61–128	15–76	51–76	255–331	203–331 229–458	229–509	153–331	255–306 331–509	178–255 255–331
M22 M24	7/8	109–218 218–350	26–131 44–153	26–52	52–153 87–218	52–65	52–65 52–109	13–65	44–65	218–284	175–284 196–392	196–437	131–284	218–262 284–437	153–218 218–284
M25	1	96–191 191–306	23–115 38–134	23–46	46–134 76–191	46–57	46–57 46–96	11–57	38–57	191–248	153–248 172–344	172–382	115–248	191–230 248–382	134–191 191–248

CHAPTER 30
BROACHING

Arthur F. Lubiarz
Manager, Product Development/Machines
Nachi America, Inc.
Macomb, Michigan

30.1 HISTORY OF BROACHING

Broaching was originally a rudimentary process used by blacksmiths back in the late 1700s. (see Fig. 30.1.) A tool called a drift was piloted in steel forging bores and driven through with a hammer. Successively larger drifts were pounded through until the desired size and configuration were achieved. In 1873, Anson P. Stevenson developed what some considered the first broaching machine (see Fig. 30.2). This was a rack and pinion hand powered type press which he used to cut keyways in various pulleys and gears.

Broaching as we know it today is a machining process which removes metal by either pushing or pulling a multiple-toothed tool (broach) through or along the surface of a part. It is extremely accurate, fast, and generally produces excellent surface finishes. Almost unlimited shapes can be produced on a broach surface as long as there is no obstruction in the way of the broach as it is pushed or pulled along its axis. Commonly broached parts would be metals, plastics, wood, and other nonferrous materials.

Broaching should be considered as the preferred process when there are large production requirements, or complex shapes which frequently cannot be produced economically by any other means.

FIGURE 30.1 The first broaches or drifts.

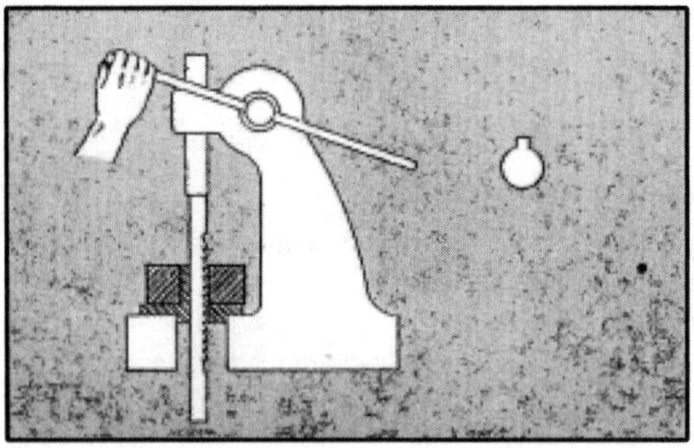

FIGURE 30.2 The first hand broaching machines.

30.1.1 Broach Terminology

Broach terminologies are defined in Fig. 30.3.

- *Back-off (B/O) angle.* The relief angle back of the cutting edge of the broach tooth.
- *Back radius.* The radius on the back of the tooth in the chip space.
- *Broach.* A metal cutting tool or bar or slab shape, equipped with a series of cutting teeth.
- *Burnishing button.* A broach tooth without a cutting edge. A series of buttons is sometimes placed after the cutting teeth of the broach, to produce a smooth surface by material compression.
- *Chipbreaker.* Notches in the teeth of broaches, which divide the width of chips, facilitating their removal. On round broaches, they prevent the formation of a solid ring in the chip gullet.
- *Chip per tooth.* The depth of cut which determines chip thickness.
- *Chip space.* The space between broach teeth, which accommodates chips during the cut. Sometimes called the *chip gullet,* it includes the face angle, face angle radius, and back radius.
- *External broach.* A broach, which cuts on the external surface of the workpiece.
- *Face angle.* The angle of the cutting edge of a broach tooth. It is sometimes called the *hook angle*.
- *Face angle radius.* The radius just below the cutting edge that blends into the back of the tooth radius.
- *Finishing teeth.* The teeth at the end of the broach arranged at a constant size for finishing the surface.
- *Follower dia.* That part of the broach which rests in the follower support bushing and which may be used as a retriever on the return stroke.
- *Front pilot.* The guiding portion of a broach (usually internal) which serves as a safety check to prevent overload of the first roughing tooth.
- *Gullet.* The name sometimes applied to the *chip space*.
- *Hook angle.* The name sometimes applied to the *face angle* of the tooth.
- *Internal broach.* A broach which is pulled or pushed through a hole in the work piece to bring the hole to the desired size or shape.
- *Land.* The thickness of the top of the broach tooth.
- *Land, straight.* A land having no back-off angle and used for finishing teeth to retain broach size after a series of sharpenings.

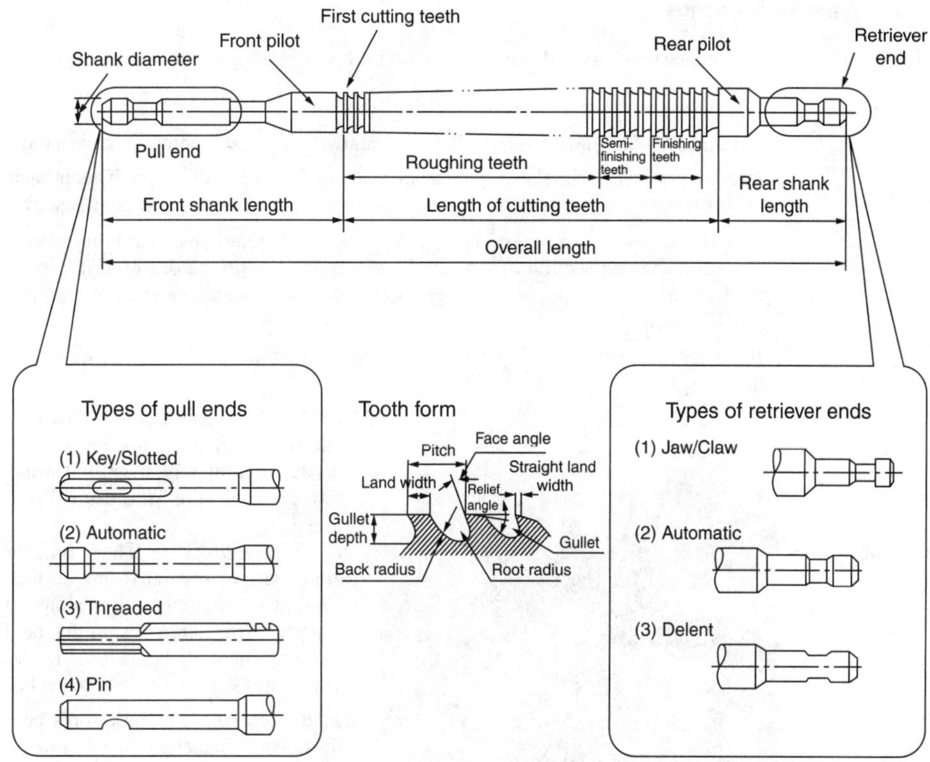

FIGURE 30.3 Broach terminology.

- *Overall length.* The total length of the broach.
- *Pitch.* The measurement from the cutting edge of one tooth to the corresponding point on the next tooth.
- *Pull broach.* A broach that is pulled through, or over the face of the workpiece.
- *Pulled end.* That end of the broach at which it is coupled to the puller of the broaching machine.
- *Push broach.* A broach which is pushed through or over the surface of the work piece.
- *Roughing teeth.* The teeth which take the first cuts in any broaching operation. Generally they take heavier cuts than semifinishing teeth.
- *Round broach.* A broach of circular section.
- *Semifinishing teeth.* Broach teeth, just ahead of the finishing teeth, which take the semifinishing cut.
- *Shank length.* The portion of broach in front of teeth which is the pull end.
- *Shear angle.* The angle between the cutting edge of a shear tooth and a line perpendicular to the broach axis or the line of travel on surface broaches.
- *Spiral gullet.* Also referred to as the *helical gullet,* chip space which wraps around the broach tool spirally like a thread which applies a shear cutting principle to internal broaches.
- *Surface broach.* An external broach which cuts a flat or contoured surface.
- *Tooth depth.* The height of tooth or broach gullet from root to cutting edge.

30.1.2 Types of Broaches

There are various types of broach tools for almost any production part. Some of the broach types are as follows:

- *Keyway broach.* Used to cut a keyslot on any type of part, internal, external or surface.
- *Internal broach.* Finishes the internal hole configuration of a part. This broach can be designed to finish round holes, such as pinions, flattened rounds, oval, polygonal, square, and hexagon.

A. Spline broach. An internal finishing tool which generates a straight-sided or involute spline. A profile shaving section or tool can be incorporated into the design if necessary. This may be in the form of a solid one-piece tool, or as an assembly (two pieces) with a roughing body and a side shaving shell.

B. Helical broach shell. This type of broach tool is identical to the above spline broach but has a helical spline form.

C. Blind spline broach. This tool broaches external splines that cannot be generated by passing a broach completely over or through the part. This tool generates the form using a series of progressive dies.

D. Surface broach. These broaches normally generate flats or special forms on the surface of parts. Contours can range from simple flats to complex shapes (Xmas tree) as might be found in turbine wheels. These tools are usually mounted in fixtures which are typically called broach holders.

E. Broach rings. These tools produce a profile on the outside diameter or a part. Profiles normally would be (but not limited to) splines or cam forms. A broach ring assembly consists of individual wafers, each with a single compliment of cutting teeth. These wafers are stacked and assembled inside a holder. As a set, the rings generate the desired form. (Fig. 30.4)

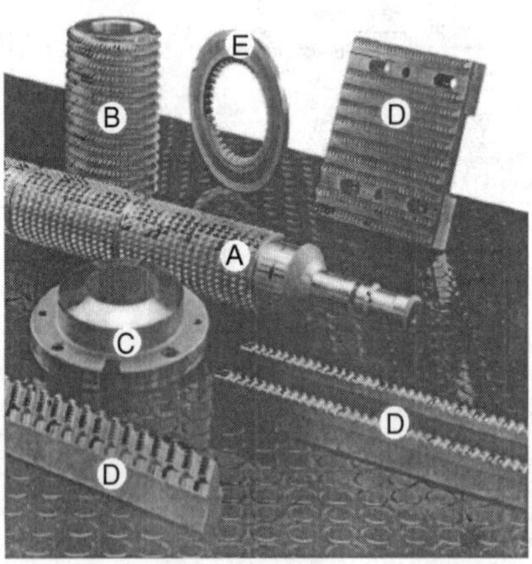

FIGURE 30.4 Broach sample A to E.

30.2 BROACHING PROCESS

30.2.1 Typical Broach Processes

Blind Spline. This process produces either internal or external splines which cannot be generated by passing a broach completely over or through the part. The form is generated using a series of progressive broach punches or rings which are usually mounted in a rotary table or in line transfer.

Chain Broach. This is mechanical form of surface broach where broach tools are mounted in a stationary holder. The parts are held in fixtures (carriers) which are pulled through the tools by a chain. Generally used where high production rates are required.

High Speed. On certain materials, broach ram speeds of up to 200 ft/min are used in this process. Advantages realized are better finish, reduced burr, increased production, and in some instances, better tool life.

Internal Broach. Finishes the internal hole configuration of a part. Machines are varied in this application such as, pull down, pull up, and table up. In the pull applications, tools move through the part. In the table up, tools are stationary, and parts are pushed up through the tool.

Pot Broach. These are used typically to broach external forms. The tools (sticks or rings) are mounted in a pot, and the parts are pushed through the pot.

Surface Broach. This tooling normally produces flats or forms on part surfaces. Broach rams on which the tooling is mounted are normally hydraulically driven and are either vertical or horizontal.

Vibra Broach. This is a reciprocating type of hard-gear broaching using coated tooling to resize parts which have distorted after heat treat.

30.2.2 State of the Art

In many cases, today's new broaching machines are of a new breed. Improvements in almost all aspects of the machines have been realized, with the most notable being in the electronics area. CNC controls with CRT touch screens are now the norm. Gone are the days of push buttons and relays. Electric drive improvements make the use of ball and roller screw simpler. CNC is used in applications such as helical broaching. Machines are designed ergonomically, and made more environment friendly. High efficiency motors and pumps make for reduced energy consumption. All in all, new technology has given an old process a new life.

30.3 APPLICATION

30.3.1 Economy and Flexibility of Broaching

Broaching is economical not only because single cuts can be made quickly and subsequent finishing operations omitted, but also because a number of cuts both external and internal can be made simultaneously, all in proper dimensional relationship with one another and the entire width of each surface machined in one pass. Mass production, high accuracy, and consistency result in maximum savings using the broaching process.

Broaching is flexible and has a wide rage of applications. Today it is finishing both external and internal surfaces of nearly any shape provided that all elements of the broach surface remain parallel with the axis of the broach. There also cannot be any obstructions in the plane of the broached surface, and wall thickness of the part must be such, as to adequately support the broaching thrust. Usually any material that can be machined by conventional methods such as milling, drilling, and shaping, can be broached. Table 30.1 illustrates sample applications; Figure 30.5 shows typical components.

30.3.2 How to Order Broaches

General Information to be Supplied with Inquiry

1. Type and size of the machine on which the broach is to be used.
2. Type and size of the puller and retriever to be used, if it is an internal application. Dimension to the first tooth, from the front edge of the tool.

TABLE 30.1 Sample Application Chart

Work material and hardness	Broach material	Machine speed	Coolant type
4340, Rc 28 max.	M-2	20–30 F.P.M.	Chlorinated water soluble oil
Aluminum	M-2	30–60 F.P.M.	Chlorinated water soluble oil
Cast Iron, Rc 28 max.	M-3	20–30 F.P.M.	Chlorinated water soluble oil
9260, Rc 30 max.	M-4	15–30 F.P.M.	Chlorinated water soluble oil
416 S.S., Rc 40 max.	T-15	8–15 F.P.M.	Chlorinated/sulphurized straight oil with EP additives

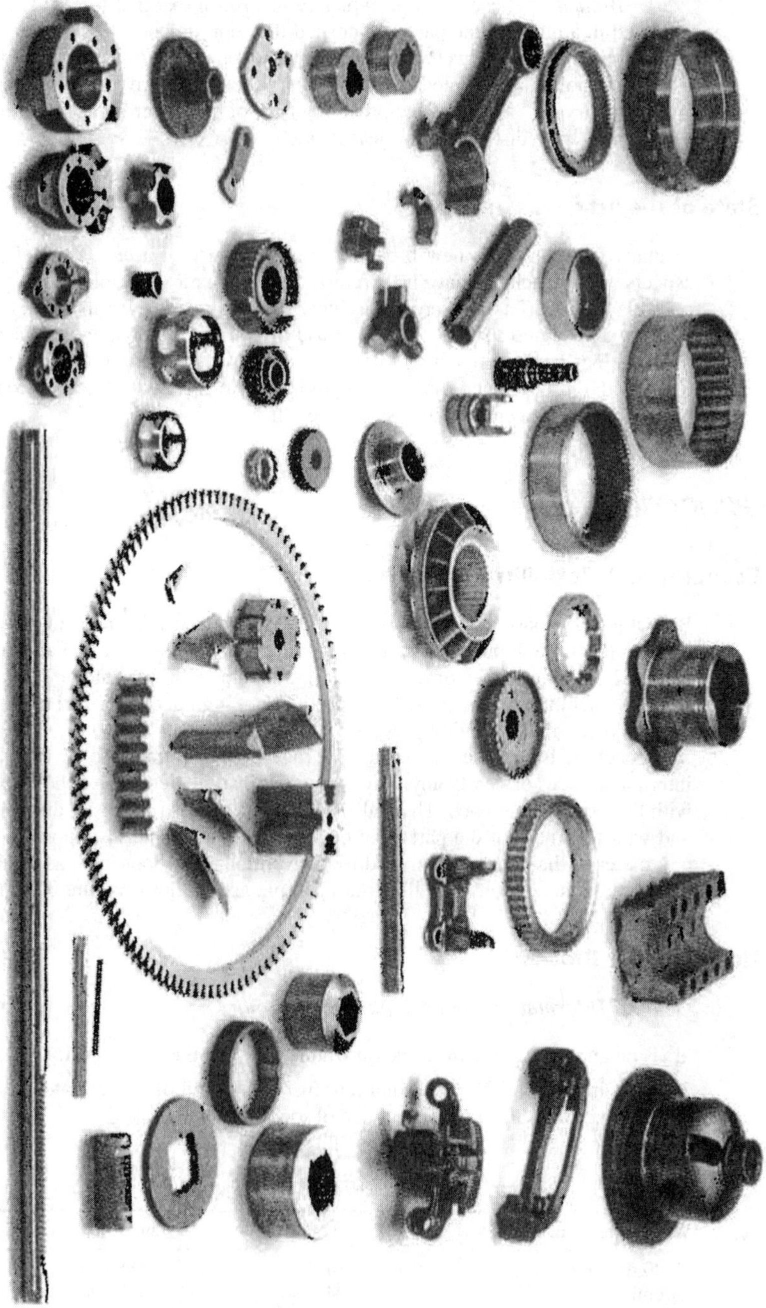

FIGURE 30.5 Components typically broached.

3. Condition of the part at the time of broaching.
4. Part print/process sheets and/or information on area to be broached, with dimensions and limits, prior machining data, and locating surfaces available.
5. Type of material to be broached, length of cut, hardness at time of broaching, and heat treatment after broaching if any.
6. Quality of finish required.

30.4 TROUBLESHOOT

30.4.1 Care of Broaches

Broaches are valuable precision tools and should be treated accordingly. They should be stored in individual compartments made of wood, or other soft material. Internal broaches, if stored horizontally, should contact the rack throughout its length to prevent any tendency to sag. When in transit between tool room and broaching machine, separation between tools should be adhered to as described above. Avoid shocks of any kind. Tools that are stored for any length of time should be coated with a rust preventative and carefully wrapped. Shipping tools by truck will require sturdy containers, and again, separation, support and the like. Containers should be marked "fragile do not drop" as a precaution for freight handlers.

30.4.2 Sharpenings

A broach needs sharpening when the cutting edges of the teeth show signs of being worn or if a definite land from wear begins to form (see Fig. 30.6). Such a condition is readily apparent when the teeth are examined through a magnifying glass. Dullness is also manifested by an increase in power needed to operate the broach, by the tool cutting small, and by broach surfaces becoming rough or torn. Wear on the cutting edges should never be permitted to become excessive before the broach is sharpened. It is suggested that users should have at least two standby tools/sets for each one in service. One kept in the tool room or stored near the machine ready for use while the other is being sharpened.

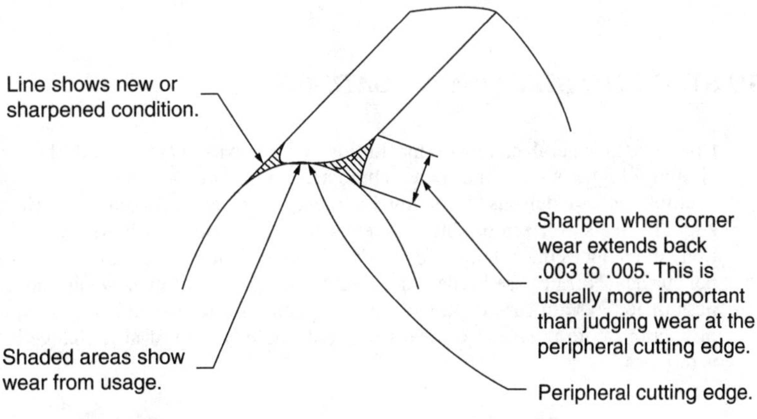

FIGURE 30.6 Signs that broach requires sharpening.

30.4.3 Solving Problems

Broached Surface Rough or Torn

1. Dull broach? Resharpen as required.
2. Is coolant the correct type? Are ratios correct?
3. Is material extremely soft? Check hook and B/O angles.
4. Is material extremely hard? Check part.
5. Does tool have excessive sharpening burr? Stone edges.
6. Straight land in finish teeth may be excessive. Re-B/O teeth.
7. Are chips packing in the tooth gullets? Check gullet depth, hook angle, coolant wash.
8. Has part material changed? Galling (material pick-up)? Check tooth surfaces, remove as required.

Chatter

1. Teeth have feather edge after sharpening. Stone edges.
2. Ram surging. Check the hydraulic system.
3 Surface applications. Check if the part is securely clamped.
4. Is the part rocking on thrust locators? Correct the locator or parts as required.
5. Is material extremely hard? Check part.

Broken Teeth

1. On surface applications, are parts oversize? Check stock.
2. Are chips packing in the tooth gullets? Check qullet depth, hook angle, coolant wash. Has part material changed?
3. Hard spots in part. Check part.
4. Teeth hitting fixture. Check for clearance.
5. On surface applications, are parts loose in fixture? Check clamps, clamping pressures and the like.
6. Galling (material pick-up) check tooth surfaces, remove as required.

30.5 HIGH-STRENGTH STEEL (HSS) COATINGS

Coatings, in general, can lower machining cost per part, increase tool life, improve part finishes, and allow for higher speeds and feeds. There are a multitude of coatings on the market. TIN Coat, as an example, is a widely used coating for machining low and medium carbon steel parts, and high-strength steels. Surface hardness exceeds Rc 80. It also adds lubricity to the surface reducing friction, and reduces the potential for galling. Typical coating thickness ranges from 0.00015 to 0.00030 per surface. Nitride case hardening , used for machining of most steels and irons, gives surface hardness of Rc 75 with case depths of 0.001 to 0.005 in. It also adds lubricity, lowering the coefficient of friction. With this form of heat treatment, there is little, if any, change in part dimensional characteristics.

CHAPTER 31
GRINDING

Mark J. Jackson
Tennessee Technological University
Cookeville, Tennessee

31.1 INTRODUCTION

More than twenty-five years of high-efficiency grinding have expanded the field of application for grinding from classical finish machining to high-efficiency machining. High-efficiency grinding offers excellent potential for good component quality combined with high productivity. One factor behind the innovative process has been the need to increase productivity for conventional finishing processes. In the course of process development it has become evident that high-efficiency grinding in combination with preliminary machining processes close to the finished contour enables the configuration of new process sequences with high performance capabilities. Using appropriate grinding machines and grinding tools, it is possible to expand the scope of grinding to high-performance machining of soft materials. Initially, a basic examination of process mechanisms is discussed that relates the configuration of grinding tools and the requirements of grinding soft materials.

There are three fields of technology that have become established for high-efficiency grinding. These are:

- High-efficiency grinding with cubic boron nitride (cBN) grinding wheels
- High-efficiency grinding with aluminum oxide (Al_2O_3) grinding wheels
- Grinding with aluminum oxide grinding wheels in conjunction with continuous dressing techniques (CD grinding)

Material removal rates resulting in a super proportional increase in productivity for component machining have been achieved for each of these fields of technology in industrial applications (Fig. 31.1). High equivalent chip thickness h_{eq} values between 0.5 and 10 μm are a characteristic feature of high-efficiency grinding. cBN high-efficiency grinding is employed for a large proportion of these applications. An essential characteristic of this technology is that the performance of cBN is utilized when high cutting speeds are employed.

Cubic boron nitride grinding tools for high-efficiency machining are subject to special requirements regarding resistance to fracture and wear. Good damping characteristics, high rigidity, and good thermal conductivity are also desirable. Such tools normally consist of a body of high mechanical strength and a comparably thin coating of abrasive attached to the body using a high-strength adhesive. The suitability of cubic boron nitride as an abrasive material for high-efficiency machining of ferrous materials is attributed to its extreme hardness and its thermal and chemical durability.

High cutting speeds are attainable, above all, with metal bonding systems. One method that uses such bonding systems is electroplating, where grinding wheels are produced with a single-layer

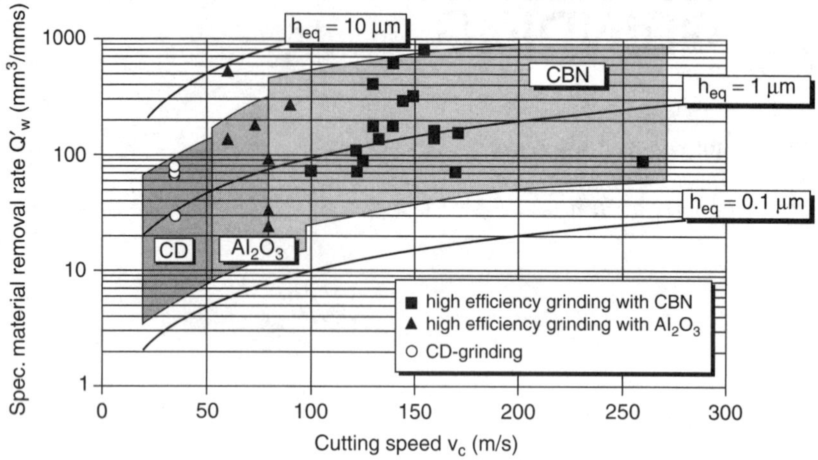

FIGURE 31.1 Main fields of application in high efficiency grinding.

coating of abrasive cBN grain material. The electro-deposited nickel bond displays outstanding grain retention properties. This provides a high-level grain projection and large chip spaces. Cutting speeds exceeding 280 ms^{-1} are possible, and the service life ends when the abrasive layer wears out. The high roughness of the cutting surfaces of electroplated cBN grinding wheels has disadvantageous effects. The high roughness is accountable to exposed grain tips that result from different grain shapes and grain diameters. Although electroplated cBN grinding wheels are not considered to be dressable in the conventional sense, the resultant workpiece surface roughness can nevertheless be influenced within narrow limits by means of a so-called touch-dressing process. This involves removing the peripheral grain tips from the abrasive coating by means of very small dressing infeed steps in the range of dressing depths of cut between 2 to 4 μm, thereby reducing the effective roughness of the grinding wheel.

Multilayer bonding systems for cBN grinding wheels include sintered metal bonds, resin bonds, and vitrified bonds. Multilayer metal bonds possess high bond hardness and wear resistance. Profiling and sharpening these tools is a complex process, however, on account of their high mechanical strength. Synthetic resin bonds permit a broad scope of adaptation for bonding characteristics. However, these tools also require a sharpening process after dressing. The potential for practical application of vitrified bonds has yet to be fully exploited. In conjunction with suitably designed bodies, new bond developments permit grinding wheel speeds of up to 200 ms^{-1}. In comparison with other types of bonds, vitrified bonds permit easy dressing while at the same time possess high levels of resistance to wear. In contrast to impermeable resin and metal bonds, the porosity of the vitrified grinding wheel can be adjusted over a broad range by varying the formulation and the manufacturing process. As the structure of vitrified bonded cBN grinding wheels results in a subsequently increased chip space after dressing, the sharpening process is simplified, or can be eliminated in numerous applications.

31.2 HIGH-EFFICIENCY GRINDING USING CONVENTIONAL ABRASIVE WHEELS

High-efficiency grinding practice using conventional aluminum oxide grinding wheels has been successfully applied to grinding external profiles between centers and in the centerless mode, grinding internal profiles, threaded profiles, flat profiles, guide tracks, spline shaft profiles, and gear tooth profiles. These operations require dressing the grinding wheel with a highly accurate and precise rotating wheel that is studded with diamond. Operations that are carried out using high performance conventional grinding wheels as a matter of routine are the grinding of:

- *Auto engine.* Crankshaft main and connecting-rod bearings, camshaft bearings, piston ring grooves, valve rocker guides, valve head and stems, head profile, grooves, and expansion bolts
- *Auto gearbox.* Gear wheel seats on shafts, pinion gears, splined shafts, clutch bearings, grooves in gear shafts, synchromesh rings, and oil-pump worm wheels
- *Auto chassis.* Steering knuckle, universal shafts and pivots, ball tracks, ball cages, screw threads, universal joints, bearing races, and cross pins
- *Auto steering.* Ball joint pivots, steering columns, steering worms, and servo steering pistons and valves
- *Aerospace industry.* Turbine blades, root and tip profiles, and fir-tree root profiles

31.2.1 Grinding Wheel Selection

The selection of grinding wheels for high performance grinding applications is focused on three basic grinding regimes, namely, rough, finish, and fine grinding. The grain size of the grinding wheel is critical in achieving a specified workpiece surface roughness. The grain size is specified by the mesh size of a screen through which the grain can just pass while being retained by the next smaller size. The general guidelines for high performance grinding require 40 to 60 mesh for rough grinding, 60 to 100 mesh for finish grinding, and 100 to 320 mesh grain for fine grinding. When selecting a particular grain size one must consider that large grains allow the user to remove material economically by making the material easier to machine by producing longer chips. However, finer grain sizes allow the user to achieve better surface roughness and achieve greater accuracy by producing shorter chip sizes with a greater number of sharp cutting points. Table 31.1 shows the relationship between abrasive grain size and workpiece surface roughness for aluminum oxide grains. Grinding wheel specifications are specific to a particular operation and are usually well documented by the manufacturers who supply them. The suggested operating conditions are also supplied by applications engineers who spend their time optimizing grinding specifications for a variety of tasks that are supported by case studies of similar operations. A list of web sites for those companies who supply grinding wheels and expertise is shown at the end of this chapter.

31.2.2 Grinding Machine Requirements for High-Efficiency Dressing

Diamond dressing wheels require relative motion between the dressing wheel and the grinding wheel. Dressing form wheels require relative motion generated by the path of the profile tool for the generation of grinding wheel form, and the relative movement in the peripheral direction. Therefore, there

TABLE 31.1 Relationship Between Abrasive Grain Size and Workpiece Surface Roughness

Surface roughness, R_a (μm)	Abrasive grain size (U.S. mesh size)
0.7 – 1.1	46
0.35 – 0.7	60
0.2 – 0.4	80
0.17 – 0.25	100
0.14 – 0.2	120
0.12 – 0.17	150
0.1 – 0.14	180
0.08 – 0.12	220

must be a separate drive for the dressing wheel. The specification of the drive depends on the following factors: grinding wheel specification and type, dressing roller type and specification, dressing feed, dressing speed, dressing direction, and the dressing speed ratio. A general guide for drive power is that 20 W/mm of grinding wheel contact is required for medium-to-hard vitrified aluminum oxide grinding wheels.

The grinding machine must accommodate the dressing drive unit so that the dressing wheel rotates at a constant speed between itself and the grinding wheel. This means that grinding machine manufacturers must coordinate the motion of the grinding wheel motor and the dressing wheel motor. For form profiling, the dressing wheel must also have the ability to control longitudinal feed motion in at least two axes. The static and dynamic rigidity of the dressing system has a major effect on the dressing system. Profile rollers are supported by roller bearings in order to absorb rather high normal forces. Slides and guides on grinding machines are classed as weak points and should not be used to mount dressing drive units. Therefore, dressing units should be firmly connected to the bed of the machine tool. Particular importance must be attached to geometrical run-out accuracy of the roller dresser and its accurate balancing. Tolerances of 2 μm are maintained with high accuracy profiles with radial and axial run-out tolerances not exceeding 2 μm. The diameter of the mandrel should be as large as possible to increase its stiffness; typically, roller dresser bores are in the range of 52 to 80 mm diameter. The class of fit between the bore and the mandrel must be H3/h2 with a 3 to 5 μm clearance. The characteristic vibrations inherent in roller dresser units are bending vibrations in the radial direction and torsional vibrations around the baseplate. Bending vibrations generate waves in the peripheral direction, while torsional vibrations generate axial waves and distortions in profile. The vibrations are caused by rotary imbalance and the dressing unit should be characterized in terms of resonance conditions. A separate cooling system should also be designed in order to prevent the dressing wheel from losing its profile accuracy due to thermal drift.

31.2.3 Diamond Dressing Wheels

The full range of diamond dressing wheels is shown in Fig. 31.2. There are five basic types of dressing wheels for conventional form dressing of conventional grinding wheels. The types described here are those supplied by Winter Diamond Tools (www.winter-diamantwerkz-saint-gobain.de). Figure 31.2 shows the five types of roller dressing wheels.

- *UZ type (reverse plated—random diamond distribution).* The diamond grains are randomly distributed at the diamond roller dresser surface. The diamond spacing is determined by the grain size used, and the close-packed diamond layers give a larger diamond content than a hand set diamond dressing roll. The manufacturing process is independent of profile shape. The process permits concave radii to be greater than 0.03 mm and convex radii to be greater than 0.1 mm. The geometrical and dimensional accuracy of these dressers is achieved by reworking the diamond layer.
- *US type (reverse plated—hand set diamond distribution).* Unlike the UZ design, the diamonds are hand set which means that certain profiles cannot be produced. However, the diamond spacing can be changed and profile accuracy can be changed by reworking the diamond layer. Convex and concave radii greater than 0.3 mm can be achieved.
- *TS type (reverse sintered—hand set diamond distribution).* Diamonds are hand set which means that certain profiles cannot be produced. However, the diamond spacing can be changed and profile accuracy can be changed by reworking the diamond layer. Concave radii greater than 0.3 mm can be produced.
- *SG type (direct plated—random diamond distribution, single layer).* The diamonds grains are randomly distributed. Convex and concave radii greater than 0.5 mm are possible.
- *TN type (sintered—random diamond distribution, multilayer).* The diamond layer is built up in several layers providing a long life dressing wheel. The profile accuracy can be changed by reworking the diamond layer.

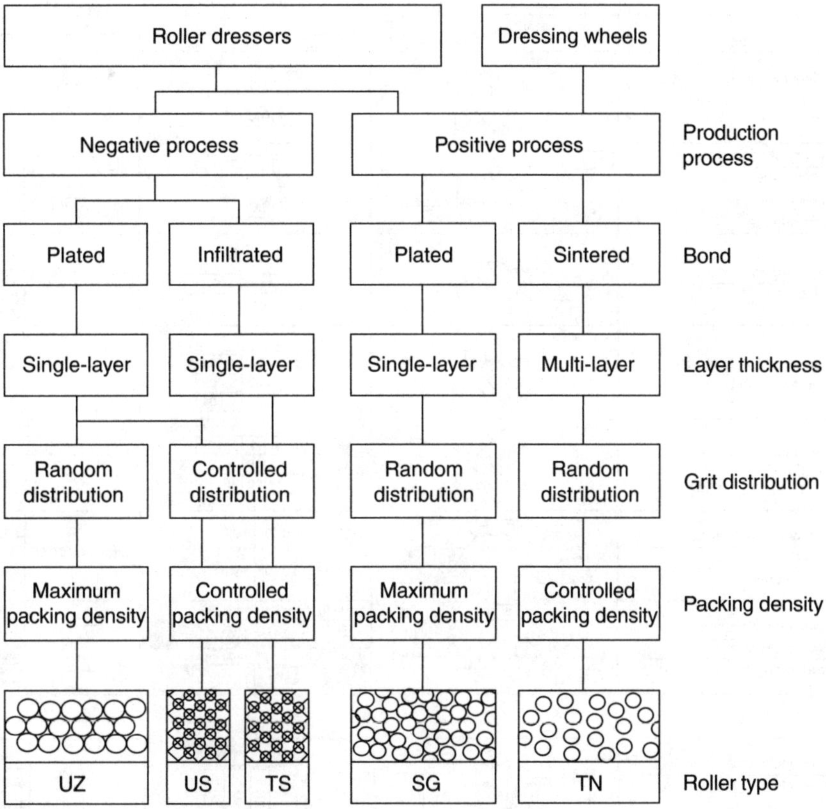

FIGURE 31.2 Types of dressing rolls and wheels.

The minimum tolerances that are attainable using diamond dressing rolls and wheels are shown in Fig. 31.3. The charts show tolerances of engineering interest for each type of dressing wheel.

31.2.4 Application of Diamond Dressing Wheels

The general guide and limits to the use of diamond dressing wheels is shown in Table 31.2. The table shows the relationship between application and the general specification of dressing wheels.

31.2.5 Modifications to the Grinding Process

Once the grinding wheel and dressing wheel have been specified for a particular grinding operation, adjustments can be made during the dressing operation that affects the surface roughness condition of the grinding wheel. The key factors that affect the grinding process during dressing are: dressing speed ratio V_r/V_s between dressing wheel and grinding wheel; dressing feed rate a_r per grinding wheel revolution; and the number of running-out or dwell revolutions of the dressing wheel n_a. Figure 31.4 shows the effect of speed ratio on the effective roughness of the grinding wheel.

The relationship shown is shown for different feed rates per grinding wheel revolution. It can be seen that the effective roughness is much greater in the unidirectional range than in the counter-direction. The number of dwell revolutions also influences the effective roughness by reducing the roughness as the number of dwell revolutions increases. Therefore, by changing the dressing conditions it

Dimensions and tolerances in mm

		Illustration	UZ	US	S	TS	T
Hole dimensional tolerance Cylindrical shape tolerance of hole	T_B ⌭	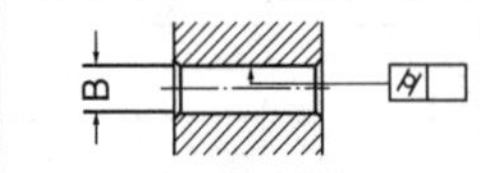	H3 ⌭ 0.003		H5 ⌭ 0.005	H3 ⌭ 0.003	H5 ⌭ 0.005
Angularity of contact face with respect to hole Parallelity tolerance of contact faces	⊥ //	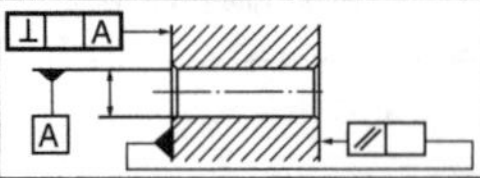	⊥ 0.002 A // 0.002		⊥ 0.005 A // 0.005	⊥ 0.002 A // 0.002	
True-running tolerance of profile	↗	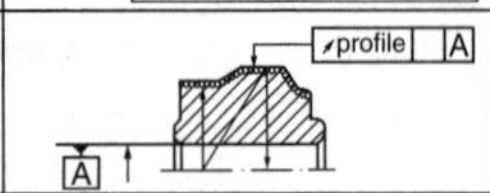	↗ profile 0.004 A		↗ profile 0.02 A	ground. ↗ profile 0.004 A unground. ↗ profile 0.02 A	↗ profile 0.02 A
Cylindrical shape tolerance referred to length L	⌭	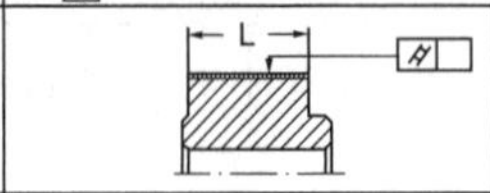	L / Tz ≦ 50 ⌭ 0.002 ≦ 80 ⌭ 0.003 ≦ 130 ⌭ 0.004		—	L / Tz ≦ 50 ⌭ 0.002 ≦ 80 ⌭ 0.003 ≦ 130 ⌭ 0.004	—
Angular tolerance referred to length L	T_α	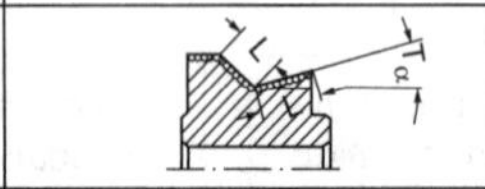	accuracy condition: tolerance essentially ±≧0,004 or ±≧1′ (see values below)				
Radii dimensional tolerance concave/convex referred to angle	T_R	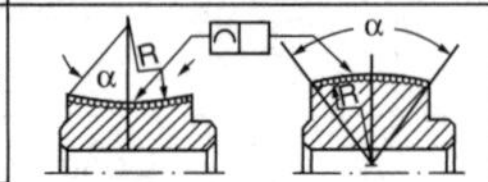	≦90° ±0.002 >90°–180° ±0.003	≦90° ±0.002 >90°–180° ±0.005	unground. ±0.05	≦90° ±0.002 >90°–180° ±0.005	— —
Linear shape tolerance concave/convex referred to angle	⌒	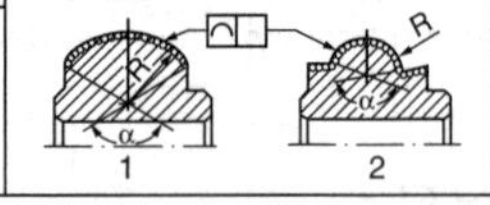	radii Fig. 1: ∢α≦ 90° ⌒ 0.002 >90º180° ⌒ 0.003 radii Fig. 2: ∢α≦ 90° ⌒ 0.004 >90º180° ⌒ 0.006	radii Fig. 1: ∢α≦ 90° ⌒ 0.006 >90º180° ⌒ 0.008 radii Fig. 2: ∢α≦ 90° ⌒ 0.006 >90º180° ⌒ 0.008	— —	radii Fig. 1: ∢α≦ 90° ⌒ 0.006 >90º180° ⌒ 0.008 radii Fig. 2: ∢α≦ 90° ⌒ 0.010 >90º180° ⌒ 0.016	— —

Angular tolerance referred to length L — accuracy condition: tolerance essentially ±≧0,004 or ±≧1′

type: leg length. ≦L	1	2	3	4	5	6	7	8	9	10	15	20	25	30	35	40	45	50	55	60
UZ,US,TS	10′	5′	3′30″	2′30″	2′	1′36″	1′24″	1′12″	1′		6	8	10	12	14	16	18	20	22	24
S,T	60′	38′	29′	23′	19′	15′	10′				60	80	100	120	140	160	180	200	220	240
	values in minutes										values in microns.									

		Illustration	UZ	US	S	TS	T
Symmetry tolerance α_1 to α_2 referred to leg length L	⌯	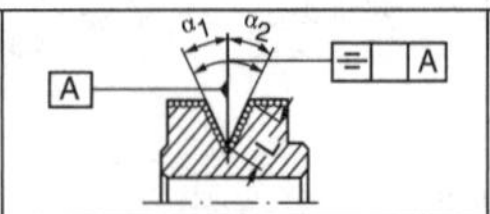	L / Tw ≦1 ⌯ 7′ A ≦5 ⌯ 4′ A >5 ⌯ 2′ A		—	L / Tw ≦1 ⌯ 10′ A ≦5 ⌯ 7′ A >5 ⌯ 3′ A	—
Rectangularity tolerance of diamond studded plane faces	⊥	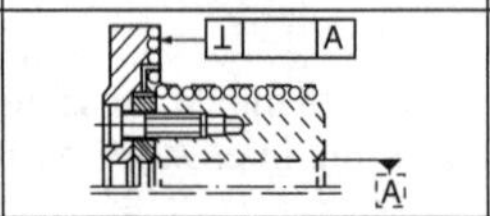	—	see TS	—	tolerance per 1mm face: face equal to workpiece 0.001 face not equal to workpiece 0.005	—
Step-Tolerance in difference between two associated diameters on different dressing rolls	T_s	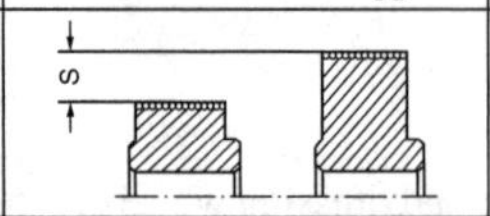	±0.002		±0.05	±0.002	±0.01
Step-Tolerance in difference between two associated diameters on one truing roll Linear dimensional tolerance of two associated faces	T_s T_L	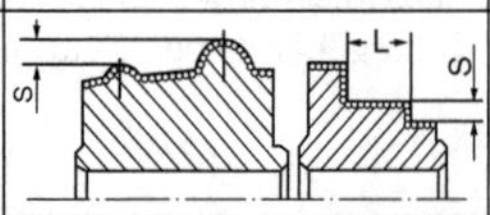	±0.002 ±0.005		±0.05	±0.002 ±0.005	—
Linear dimensional tolerance of two opposite faces	T_L	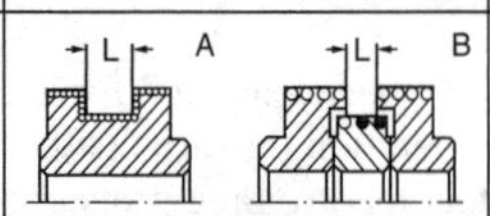	±0.02 A		—	±0.003 B	—
Dimensional tolerance on pitsch Cylindrical shape tolerance for profile	T_L ⌭	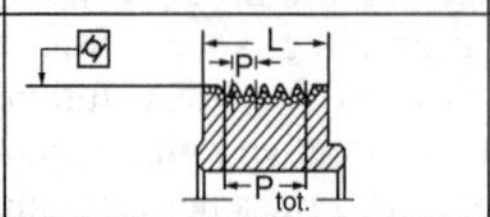	single pitch P:±0.002 P total for ≦16:±0.002 > 16: pro 10 mm = 0.00125 ⌭ profile 0.002 per 10 mm thread length	—	—	—	—
Straightness tolerance	—	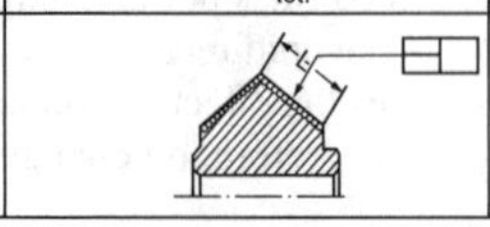	L / Tz ≦ 50 — 0.002 ≦ 80 — 0.003 ≦130 — 0.004		—	L / Tz ≦ 50 — 0.002 ≦ 80 — 0.003 ≦130 — 0.004	—

FIGURE 31.3 Minimum tolerances attainable with dressing wheels.

TABLE 31.2 General Specification and Application of Diamond Dressing Wheels

Market	Application	Wheel specification, grain size, hardness	Surface Roughness, R_a (μm)	Output	TS type, diamond size (mesh)	UZ type, diamond size (mesh)
Auto	Rough grinding	40–60, K–L	0.8 – 3.2 (hand set) 0.4 – 1.0 (random)	High	80–100 (hand set) 100–150 (random)	Not applicable
Auto	Transmission, gear box	60–80, J–M	0.2–1.6	High	200–250	200 – 250
Auto	Bearings, CV joints	80–120, J–M	0.2–1.6	High	200–300	200–300
Aero	Creep feed, continuous dressing	Porous structure	0.8–1.6	Low	Limited	250–300

is possible to rough and finish grind using the same diamond dressing wheel and the same grinding wheel. By controlling the speed of the dressing wheel, or by reversing its rotation, the effective roughness of the grinding wheel can be varied in the ratio of 1:2.

31.2.6 Selection of Grinding Process Parameters

The aim of every grinding process is to remove the grinding allowance in the shortest time possible while achieving the required accuracy and surface roughness on the workpiece. During the grinding operation the following phases are usually present:

- Roughing phase during which the grinding allowance removed is characterized by large grinding forces causing the workpiece to become deformed
- Finishing phase for improving surface roughness

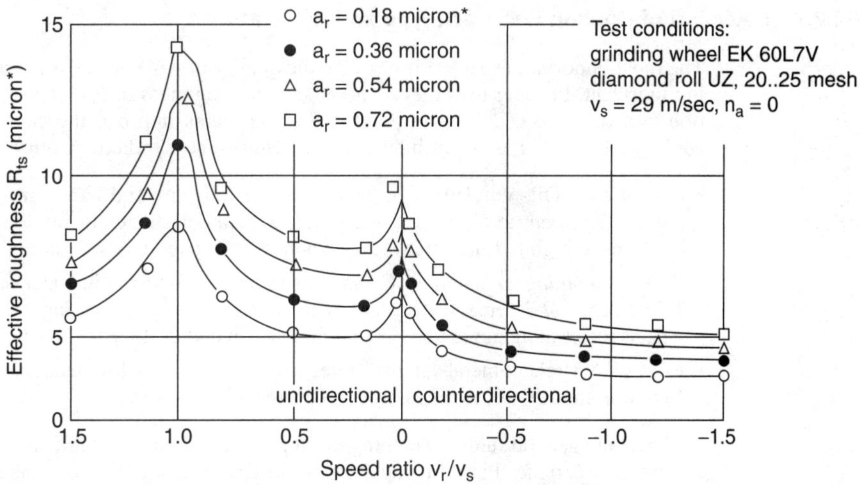

FIGURE 31.4 Effect of dressing speed ratio on the effective roughness of the grinding wheel.

- Spark-out phase during which time workpiece distortion is reduced to a point where form errors are insignificant

Continuous dressing coupled with one or more of the following phases has contributed to high efficiency grinding of precision components:

- Single-stage process with a sparking-out regime
- Two-stage process with a defined sparking out regime coupled with a slow finishing feed rate
- Three-stage and multistage processes with several speeds and reversing points
- Continuous matching of feed rate to the speed of deformation reduction of the workpiece

For many grinding processes using conventional vitrified grinding wheels, starting parameters are required that are then optimized. A typical selection of starting parameters is shown below:

- Metal removal rates for plunge grinding operations: when the diameter of the workpiece is greater than 20 mm then the following metal removal rates Q'_w (in mm^3/mm.s) are recommended: roughing—1 to 4, finishing—0.08 to 0.33. When the workpiece is less than 20 mm, roughing—0.6 to 2, finishing—0.05 to 0.17.
- Speed ratio q is the relationship between grinding wheel speed and workpiece speed. For thin-walled, heat-sensitive parts q should be in the range 105 to 140, for soft and hard steels, q should be 90 to 135, for high metal removal rates q should be 120 to 180, and for internal grinding q should be in the range 65 to 75.
- Overlap factors should be in the range 3 to 5 for wide grinding wheels and 1.5 to 3 for narrow grinding wheels.
- Feed rates should be between 0.002 and 0.006 mm per mm traverse during finish grinding.
- Number of sparking-out revolutions should be between 3 and 10 depending on the rigidity of the workpiece.

However, it is important that the user ensures that the specifications of grinding wheel and dressing wheel are correct for the grinding task prior to modifying the grinding performance. A list of web sites is available at the end of this chapter that details grinding wheel manufacturers who provide expertise with specific applications.

31.2.7 Selection of Cooling Lubricant Type and Application

The most important aspect of improving the quality of workpieces is the use of high-quality cooling lubricant. In order to achieve a good surface roughness of less than 2 μm, a paper-type filtration unit must be used. Air cushion deflection plates improve the cooling effect. The types of cooling lubricant in use for grinding include emulsions, synthetic cooling lubricants, and neat oils.

- *Emulsions.* Oils emulsified in water are generally mineral-based and are concentrated in the range 1.5 percent to 5 percent. In general, the "fattier" the emulsion the better the surface finish but leads to high normal forces and roundness is impaired. Also, susceptible to bacteria.
- *Synthetic cooling emulsions.* Chemical substances dissolved in water in concentrations between 1.5 percent and 3 percent. Resistant to bacteria and are good wetting agents. They allow grinding wheels to act more aggressively but tend to foam and destroy seals.
- *Neat oil.* Highest metal removal rates achievable with a low tendency to burn the workpiece. Neat oils are difficult to dispose of and present a fire hazard.

There are general rules concerning the application of cooling lubricants but the reader is advised to contact experienced grinding applications engineers from grinding wheel suppliers and lubricant suppliers. A list of suppliers is shown in the internet resource section of this chapter.

31.3 HIGH-EFFICIENCY GRINDING USING CBN GRINDING WHEELS

High-efficiency grinding practice using superabrasive cBN grinding wheels has been successfully applied to grinding external profiles between centers and in the centerless mode, grinding internal profiles, threaded profiles, flat profiles, guide tracks, spline shaft profiles, and gear tooth profiles. These operations require dressing the grinding wheel with a highly accurate and precise rotating wheel that is studded with diamond. Operations that are carried out using high performance conventional grinding wheels as a matter of routine are the grinding of:

- *Auto engine.* Crankshaft main and connecting-rod bearings, camshaft bearings, piston ring grooves, valve rocker guides, valve head and stems, head profile, grooves, and expansion bolts
- *Auto gearbox.* Gear wheel seats on shafts, pinion gears, splined shafts, clutch bearings, grooves in gear shafts, synchromesh rings, and oil-pump worm wheels
- *Auto chassis.* Steering knuckle, universal shafts and pivots, ball tracks, ball cages, screw threads, universal joints, bearing races, and cross pins
- *Auto steering.* Ball joint pivots, steering columns, steering worms, and servo steering pistons and valves
- *Aerospace industry.* Turbine blades, root and tip profiles, and fir-tree root profiles

31.3.1 Grinding Wheel Selection

The selection of the appropriate grade of vitrified cBN grinding wheel for high-speed grinding is more complicated than for aluminum oxide grinding wheels. Here, the cBN abrasive grain size is dependent on specific metal removal rate, surface roughness requirement, and the equivalent grinding wheel diameter. As a starting point, when specifying vitrified cBN wheels, Fig. 31.5 shows the relationship between cBN abrasive grain size, equivalent diameter, and specific metal removal rate for cylindrical grinding operations.

However, the choice of abrasive grain is also dependent on the surface roughness requirement and is restricted by the specific metal removal rate. Table 31.3 shows the relationship between cBN grain size and their maximum surface roughness and specific metal removal rates. The workpiece material has a significant influence on the type and volume of vitrified bond used in the grinding wheel. Table 31.4 shows the wheel grade required for a variety of workpiece materials that are based on cylindrical (crankshaft and camshaft) grinding operations.

Considering the materials shown in Table 31.4, chilled cast iron is not burn sensitive and has a high specific grinding energy owing to its high carbide content. Its hardness is approximately 50 HRc and the maximum surface roughness achieved on machined camshafts is 0.5 μm Ra. Therefore a standard structure bonding system is used that is usually between 23 and 27 percent volume of the wheel. The cBN grain content is usually 50 percent volume, and wheel speeds are usually up to 120 m/s.

Nodular cast iron is softer than chilled cast iron and is not burn sensitive. However, it does tend to load the grinding wheel. Camshaft lobes can have hardness values as low as 30 HRc and this tends to control wheel specification. High stiffness crankshafts and camshafts can tolerate a 50 volume percent abrasive structure containing 25 volume percent bond. High loading conditions and high contact reentry cam forms require a slightly softer wheel where the bonding system occupies 20 volume percent of the entire wheel structure. Low stiffness camshafts and crankshafts require lower cBN grain concentrations (37.5 volume percent) and a slightly higher bond volume (21 volume percent). Very low stiffness nodular iron components may even resort to grinding wheels containing higher strength bonding systems containing sharper cBN abrasive grains operating at 80 m/s. The stiffness of the component being ground has a significant effect on the workpiece/wheel speed ratio. Figure 31.6 demonstrates the relationship between this ratio and the stiffness of the component.

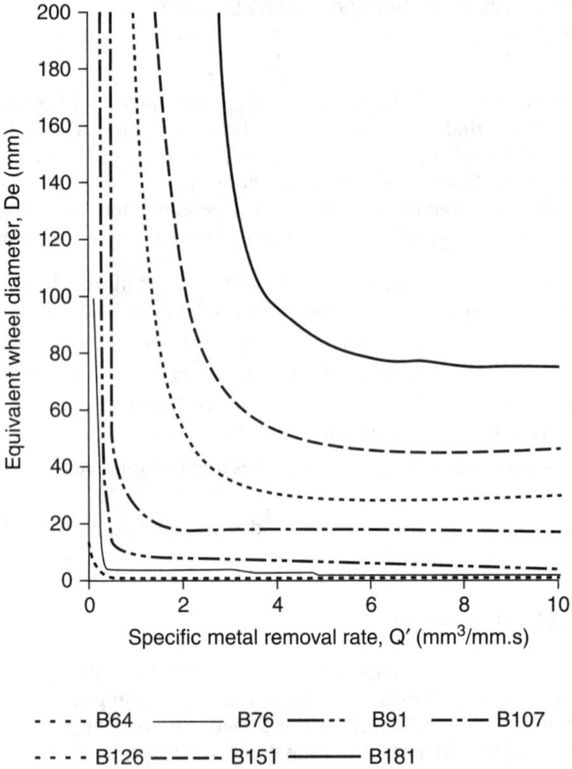

FIGURE 31.5 Chart for selecting cBN abrasive grit size as a function of the equivalent grinding wheel diameter D_e and the specific metal removal rate Q'_{w}.

Steels such as AISI 1050 can be ground in the hardened and the soft state. Hardened 1050 steels are in the range 68-62 HRc. They are burn sensitive and as such wheels speeds are limited to 60 m/s. The standard structure contains the standard bonding systems up to 23 volume percent. The abrasive grain volume contained at 37.5 volume percent. Lower power machine tools usually have grinding wheels where a part of the standard bonding system contains hollow glass spheres (up to 12 volume percent)

TABLE 31.3 CBN Abrasive Grain Selection Chart Based on Camshaft and Crankshaft Grinding Applications

cBN grain size	Surface roughness, R_a (μm)	Maximum specific metal removal rate, $Q'_{w\,max}$ (mm^3/mm.s)
B46	0.15 – 0.3	1
B54	0.25 – 0.4	3
B64	0.3 – 0.5	5
B76	0.35 – 0.55	10
B91	0.4 – 0.6	20
B107	0.5 – 0.7	30
B126	0.6 – 0.8	40
B151	0.7 – 0.9	50
B181	0.8 – 1	70

TABLE 31.4 Vitrified cBN Grinding Wheel Specification Chart and Associated Grinding Wheel Speeds Based on Camshaft and Crankshaft Grinding Applications

Workpiece material	Grinding wheel speed, v_s (m/s)	Vitrified cBN wheel specification	Application details
Chilled cast iron	120	B181R200VSS	High Q'_w (over 70 $mm^3/mm \cdot s$)
		B126P200VSS	Medium Q'_w (between 30 and 70 $mm^3/mm \cdot s$)
		B107N200VSS	Low Q'_w (up to 30 $mm^3/mm \cdot s$)
Nodular cast iron	80	B181P200VSS	Little or no wheel loading in previous grinding operations
		B181K200VSS	Wheel loading significant in previous grinding operations
		B181L150VSS	Low stiffness workpiece
		B181L150VDB	Very low stiffness workpiece
AISI 1050 steel (hardened)	80	B126N150VSS	Standard specification wheel for 1050 steel at 80 m/s
		B126N150VTR	For use on low power machine tools
AISI 1050 steel (soft condition)	120	B181K200VSS	Standard specification wheel
High speed tool steel	60	B107N150VSS	Standard specification wheel
Inconel (poor grindability)	50	B181T100VTR B181T125VTR B181B200VSS	Form dressing is usually required with all wheel specifications

exhibiting comparable grinding ratios to the standard structure system. These specifications also cover most powdered metal components based on AISI 1050 and AISI 52100 ball bearing steels. Softer steels are typically not burn sensitive but do tend to 'burr' when ground. Maximum wheel and work speeds are required in order to reduce equivalent chip thickness. High-pressure wheel scrubbers are required in order to prevent the grinding wheel from loading. Grinding wheel specification is based on an abrasive content in the region of 50 volume percent and a bonding content of 20 volume percent using the standard bonding system operating at 120 m/s.

Tool steels are very hard and grinding wheels should contain 23 volume percent standard bonding system and 37.5 volume percent cBN abrasive working at speeds of 60 m/s. Inconel materials are extremely burn sensitive, and are limited to wheel speeds of 50 m/s, and have large surface roughness requirements, typically 1 μm Ra. These grinding wheels contain porous glass sphere bonding systems with 29 volume percent bond, or 11 volume percent bond content using the standard bonding system.

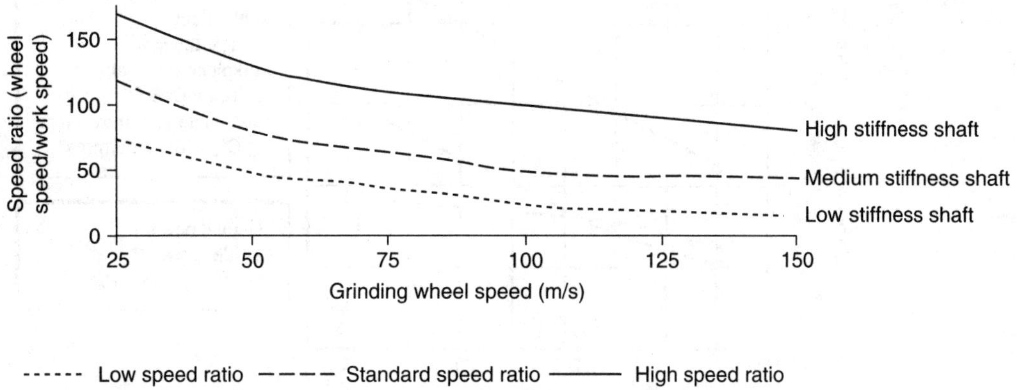

FIGURE 31.6 Work speed selection chart for camshaft and crankshaft grinding operations.

High-efficiency grinding with internal grinding wheels is limited by the bursting speed of the grinding wheel and the effects of coolant. The quill is an inherently weak part of the system and must be carefully controlled when using cBN in order to avoid the problems associated with changes in normal grinding forces. Quills should have a large diameter and a short length, and should be made from a stiff material such as Ferro-TiC, which has a relatively low density.

31.3.2 Grinding Machine Requirements for High-Efficiency Grinding

The advantages of high-speed cBN grinding can only be realized in an effective manner if the machine tool is adapted to operate at high cutting speeds. In order to attain very high cutting speeds, grinding wheel spindles and bearings are required to operate at speeds in the order of 20,000 rev min^{-1}. The grinding wheel/spindle/motor system must run with extreme accuracy and minimum vibration in order to minimize the level of dynamic process forces. Therefore, a high level of rigidity is required for the entire machine tool. Balancing of high-speed grinding wheels is also necessary at high operating speeds using dynamic balancing techniques. These techniques are required so that workpiece quality and increased tool life is preserved.

Another important consideration is the level of drive power required when increases in rotational speed become considerable. The required total output is composed of the cutting power P_c and the power loss P_l,

$$P_{total} = P_c + P_l \tag{31.1}$$

The cutting power is the product of the tangential grinding force and the cutting speed,

$$P_c = F'_t \cdot v_c \tag{31.2}$$

The power loss of the drive comprises the idle power of the spindle P_L and power losses caused by the coolant P_{KSS} and by spray cleaning of the grinding wheel P_{SSP} thus

$$P_l = P_L + P_{KSS} + P_{SSP} \tag{31.3}$$

The power measurements shown in Fig. 31.7 confirm the influence of the effect of cutting speed on the reduction of cutting power. However, idling power has increased quite significantly. The grinding power P_c increases by a relatively small amount when the cutting speed increases and all other grinding parameters remain constant.

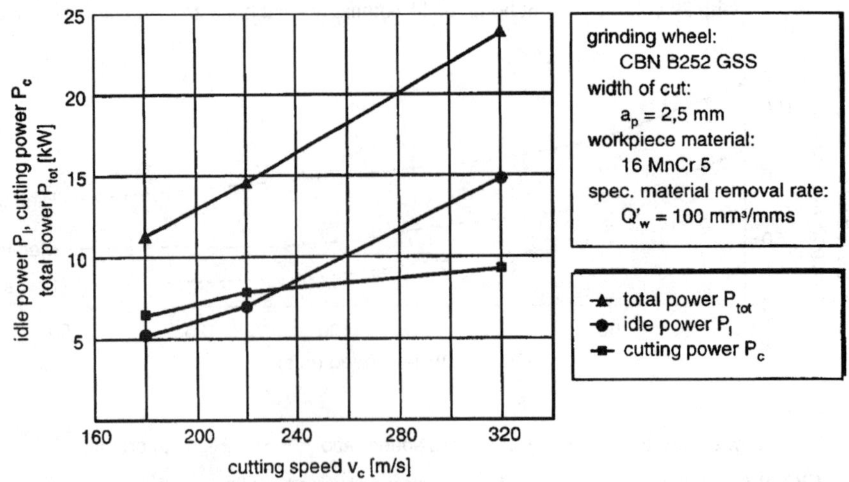

FIGURE 31.7 The effect of cutting speed on spindle power.

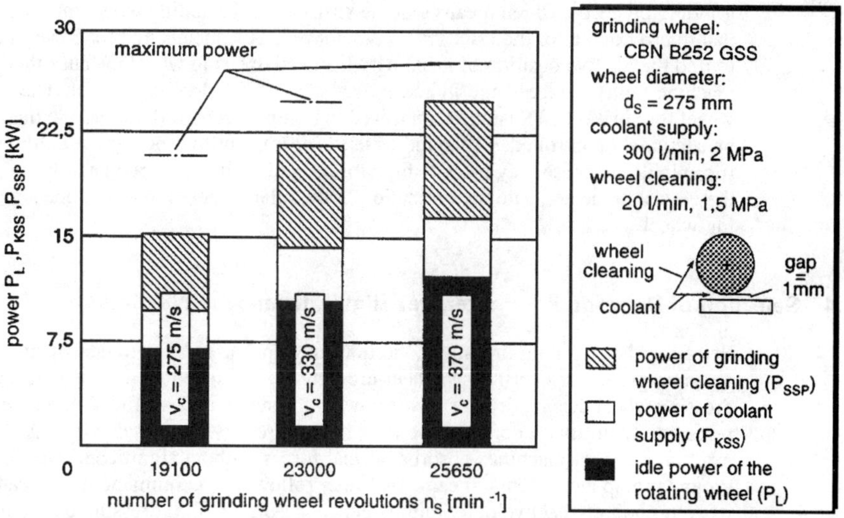

FIGURE 31.8 Levels of idling power, coolant power supply, and grinding wheel cleaning power in a machine tool using an electroplated cBN grinding wheel.

However, this means that the substantial power requirement that applies at maximum cutting speeds results from a strong increase in power due to rotation of the grinding wheel, the supply of coolant, and the cleaning of the wheel. The quantities and pressures of coolant supplied to the grinding wheel and the wheel cleaning process are the focus of attention by machine tool designers. This is shown in Fig. 31.8. The power losses associated with the rotation of the grinding wheel are supplemented by losses associated with coolant supply and wheel cleaning. The losses are dependent on machining parameters implying that machine settings and coolant supply need to be optimised for high-speed grinding.

In addition to the advantage of effectively reducing the power required for grinding, optimization of the coolant supply also offers ecological benefits as a result of reducing the quantities of coolant required. Various methods of coolant supply are available such as the free-flow nozzle that is conventionally used, the shoe nozzle that ensures "reduced quantity lubrication," and the mixture nozzle that ensures "minimum quantity lubrication." The common task is to ensure that an adequate supply of coolant is presented at the grinding wheel-workpiece interface. The systems differ substantially regarding their operation and the amount of energy required supplying the coolant.

A shoe nozzle, or supply through the grinding wheel, enables coolant to be directed into the workpiece-wheel contact zone. A substantial reduction in volumetric flow can be achieved in this way. In comparison to the shoe nozzle, supply through the grinding wheel requires more complex design and production processes for the grinding wheel and fixtures. An advantage of this supply system is that it is independent of a particular grinding process. Both systems involve a drastic reduction in supply pressures as the grinding wheel effects acceleration of the coolant. A more effective reduction in the quantity of the coolant results in minimal quantity coolant supply amounting to several milliliters of coolant per hour. As the cooling effect is reduced, dosing nozzles are used exclusively to lubricate the contact zone.

31.3.3 Dressing High Efficiency cBN Grinding Wheels

Dressing is the most important factor when achieving success with cBN grinding wheels. CBN abrasive grains have distinct cleavage characteristics that directly affect the performance of dressing tools. A dress depth of cut of 1 μm produces a microfractured grain, whilst a depth of cut of 2 to 3 μm produces a macrofractured grain. The latter effect produces a rough workpiece surface and lower cBN tool life. The active surface roughness of the grinding wheel generally increases as grinding proceeds, which means that the vitrified cBN grinding wheel must be dressed in such a way that micro fracture of the cBN grains is achieved. This means that touch sensors are needed in order to find the relative position of grinding wheel and dressing wheel because thermal movements in the machine tool far exceed the infeed movements of the dressing wheel. The design of the dressing wheel for vitrified cBN is based on traversing a single row of diamonds so that the overlap factor can be accurately controlled. The spindles used for transmitting power to such wheels are usually electric spindles because they provide high torque and do not generate heat during operation. Therefore, they assist in the accurate determination of the relative position of the dressing wheel and the grinding wheel.

31.3.4 Selection of Dressing Parameters for High Efficiency cBN Grinding

Selecting the optimum dressing condition may appear to be complex owing to the combination of abrasive grain sizes, grinding wheel-dressing wheel speed ratio, dressing depth of cut, diamond dressing wheel design, dresser motor power, dressing wheel stiffness, and other factors. It is not surprising to learn that a compromise may be required. For external grinding, the relative stiffness of external grinding machines absorbs normal forces without significant changes in the quality of the ground component. This means that the following recommendations can be made: dressing wheel/grinding wheel velocity ratio is between +0.2 to +0.5; dressing depth of cut per pass is 0.5 to 3 μm; total dressing depth is 3 to 10 μm; and the traverse rate is calculated by multiplying the dressing wheel r.p.m. by the grain size of cBN and multiplying by 0.3 to 1 depending on the initial condition of the dressing wheel. It is also right to specify the dressing wheel to be a single row or a double row of diamonds. For internal grinding, the grinding system is not so stiff, which means that any change in grinding power will result in significant changes in normal grinding forces and quill deflection. In order to minimize the changes in power, the following dressing parameters are recommended: dressing wheel/grinding wheel velocity ratio should be +0.8; dressing depth of cut per pass is 1 to 3 μm; total dressing depth is 1 to 3 μm; and the traverse rate is calculated by multiplying the dressing wheel r.p.m. by the grain size of cBN (this means that each cBN grain is dressed once). It is also right to specify the dressing wheel to be a single row of diamonds set on a disc.

31.3.5 Selection of Cooling Lubrication for High Efficiency cBN Grinding Wheels

The selection of the appropriate cooling lubrication for vitrified cBN is considered to be environmentally unfriendly. Although neat oil is the best solution, most applications use soluble oil with sulphur- and chlorine-based extreme pressure additives. Synthetic cooling lubricants have been used but lead to loading of the wheel and excessive wear of the vitrified bond. Using air scrapers and jets of cooling lubricants normal to the grinding wheel surface enhance grinding wheel life. Pressurized shoe nozzles have also been used to break the flow of air around the grinding wheel. In order to grind soft steel materials, or materials that load the grinding wheel, high pressure, low volume scrubbers are used to clean the grinding wheel. It is clear that the experience of application engineers is of vital importance when developing a vitrified cBN grinding solution to manufacturing process problems.

INFORMATION RESOURCES

Grinding Wheel Suppliers

Suppliers of cBN and diamond grinding and dressing tools, www.citcodiamond.com

Suppliers of cBN and diamond grinding and dressing tools, www.tyrolit.com

Suppliers of cBN and diamond grinding and dressing tools, www.noritake.com

Suppliers of cBN and diamond grinding and dressing tools, www.sgabrasives.com

Suppliers of cBN and diamond grinding and dressing tools, www.nortonabrasive.com

Suppliers of cBN and diamond grinding and dressing tools, www.wendtgroup.com

Suppliers of cBN and diamond grinding and dressing tools, www.rappold-winterthur.com/usa

Dressing Wheel Suppliers

Suppliers of cBN and diamond grinding and dressing tools, www.citcodiamond.com

Suppliers of cBN and diamond grinding and dressing tools, www.tyrolit.com

Suppliers of cBN and diamond grinding and dressing tools, www.noritake.com

Suppliers of cBN and diamond grinding and dressing tools, www.sgabrasives.com

Suppliers of cBN and diamond grinding and dressing tools, www.nortonabrasive.com

Suppliers of cBN and diamond grinding and dressing tools, www.wendtgroup.com

Suppliers of cBN and diamond grinding and dressing tools, www.rappold-winterthur.com/usa

Grinding Machine Tool Suppliers

Suppliers of camshaft and crankshaft grinding machine tools, www.landis-lund.co.uk

Suppliers of camshaft and crankshaft grinding machine tools, www.toyoda-kouki.co.jp

Suppliers of camshaft and crankshaft grinding machine tools, www.toyodausa.com

Suppliers of cylindrical grinders and other special purpose grinding machine tools, www.weldonmachinetool.com

Suppliers of cylindrical grinders and other special purpose grinding machine tools, www.landisgardner.com

Suppliers of cylindrical grinders and other special purpose grinding machine tools, www.unova.com

Suppliers of internal grinding machine tools, www.voumard.ch

Suppliers of internal grinding machine tools, www.bryantgrinder.com

For a complete list of grinding machine suppliers contact:

Index of grinding machine tool suppliers and machine tool specifications, www.techspec.com/emdtt/grinding/manufacturers

Case Studies

Case studies containing a variety of grinding applications using conventional and superabrasive materials, www.abrasives.net/en/solutions/casestudies.html

Case studies of cBN grinding wheel and diamond dressing wheel applications, www.winter-diamantwerkz-saint-gobain.de

Grinding case studies, www.weldonmachinetool.com/appFR.htm

Cooling Lubricant Suppliers

Suppliers of grinding oils/fluids and lubrication application specialists www.castrol.com

Suppliers of grinding oils/fluids and lubrication application specialists (formerly Master Chemicals) www.hays-siferd.com

Suppliers of grinding oils/fluids and lubrication application specialists, www.quakerchem.com

Suppliers of grinding oils/fluids and lubrication application specialists, www.mobil.com

Suppliers of grinding oils/fluids and lubrication application specialists, www.exxonmobil.com

Suppliers of grinding oils/fluids and lubrication application specialists, www.nocco.com

Suppliers of grinding oils/fluids and lubrication application specialists, www.marandproducts.com

Suppliers of grinding oils/fluids and lubrication application specialists, www.metalworkinglubricants.com

CHAPTER 32
METAL SAWING

David D. McCorry
Kaltenbach, Inc.
Columbus, Indiana

32.1 INTRODUCTION

The sawing of metal is one of the primary processes in the metalworking industry. Virtually every turned, milled, or otherwise machined product started life as a sawn billet. Sawing is distinct from other cut-off processes, such as abrasive and friction cutting, in that a chip is created. It is essentially a simple process, whereby a chip (or chips) is removed by each successive tooth on a tool which is rotated and fed against a stationary, clamped workpiece. The forces acting on the tool and material can be complex and are not as well researched and documented, for example, as the turning process. Traditionally, sawing has been looked upon as the last step in stores. Companies who have done the math realize that a far more productive approach is to view sawing as the first stage in manufacture. A quality modern metal saw is capable of fast, close tolerance cuts and good repeatability, and the sawing process itself is reliable and predictable. With certain caveats SPC can be successfully applied to the sawing process and in many applications judicious use of the sawing process can reduce downstream machining requirements, in some cases even render them unnecessary.

In the history of metal sawing three basic machine types have prevailed—the reciprocating hacksaw, the bandsaw, and the circular saw. The latter is sometimes referred to as "cold" sawing, although this is a misnomer since correct usage of the terms "hot" and "cold" sawing refers to the temperature of the metal as presented to the saw and not the sawing method used. For the purposes of this article we are concerned with the sawing of metal at ambient temperature.

32.2 THE HACK SAW

In hack sawing, a straight, relatively short, blade is tensioned in a bow, powered back and forth via an electric motor and a system of gears, and fed through a stationary, clamped workpiece either by gravity or with hydraulic assistance. The hacksaw therefore basically emulates the manual sawing action. Cutting is generally done on the "push stroke," i.e., away from the pivot point of the bow. In more sophisticated models, the bow is raised slightly (relieved) and speeded up on the return, or non-cutting, stroke, to enhance efficiency. By its very nature, however, hack sawing is inherently inefficient since cutting is intermittent. Also, mechanical restrictions make it impossible to run hacksaws at anything but pedestrian speed. The advantages of hack sawing are: low machine cost; easy setup and maintenance; very low running costs; high reliability and universal application—a quality hydraulic hacksaw can cut virtually anything, albeit very slowly. For this latter reason, in mainstream industrial applications hack sawing has all but disappeared in favor of band sawing or circular sawing.

32.3 THE BAND SAW

Band sawing uses a continuous band, welded to form a loop. The band sawing process is continuous. The band is tensioned between two pulleys—known as bandwheels—mounted on a bow (the nomenclature is derived from hack sawing). One of the band wheels is the "lay" or nondriven wheel. Generally, this wheel is arranged in an assembly which allows the band to be tensioned either mechanically or—more usually in modern machinery—via a hydraulic cylinder. The other wheel is driven by an electric motor and gearbox configuration. The band runs through a system of guides—usually roller bearings and preloaded carbide pads—to keep it running true through the material. Hitherto, mechanical variators arranged between motor and gearbox were employed to provide variable band speed. More recently, frequency-regulated motors feeding directly into the gearbox have become the norm (Fig 32.1).

Although a plethora of different designs exist, bandsaws split into two basic configurations—vertical and horizontal—based on the attitude of the band. Vertical machines are commonly used in tool room applications and in lighter structural applications. Heavy-duty billet sawing and the heavier structural applications favor the horizontal arrangement since this allows a closed, "box" type construction whereby the bow assembly runs in a gantry straddling the material and rigidly attached to the base of the machine. This setup allows maximum band tension and counteracts bow/blade deflection. Lighter-weight horizontal machines often employ a pivot-mounted bow which arcs in a similar fashion to the hack saw outlined above.

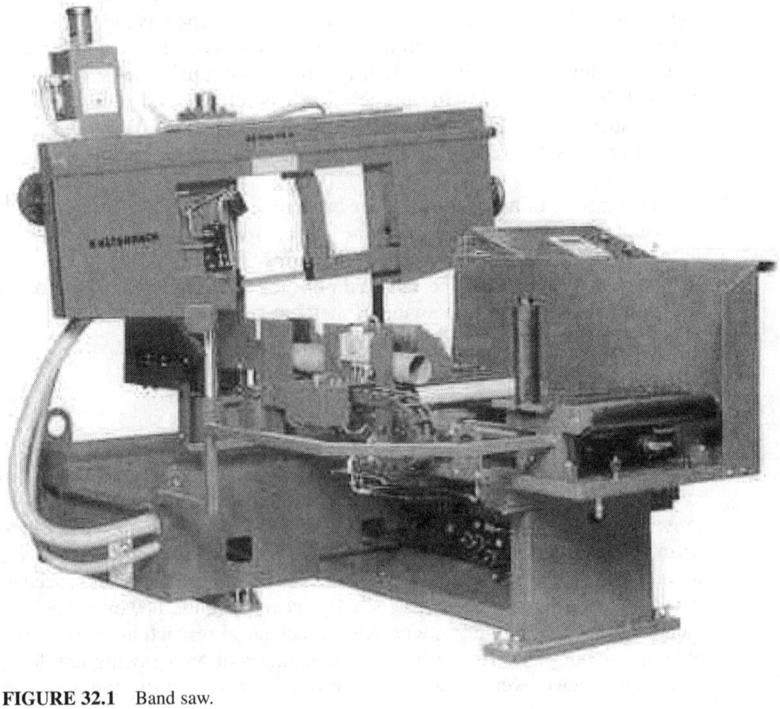

FIGURE 32.1 Band saw.

Band sawing offers the following advantages:

Relatively low machine cost

Relatively high efficiency over a wide range of applications

Relatively low operating costs

High reliability and relatively high durability

32.4 THE CIRCULAR SAW

Circular sawing differs fundamentally from both hacksawing and bandsawing, by virtue of the tool itself. As the nomenclature implies, the circular saw employs a round blade with teeth ground into the periphery. The blade is mounted on a central spindle and rotated via an electric motor and gearbox configuration (or via pulleys if higher peripheral speeds are required). Feed may be by gravity or via a hydraulic cylinder. In a few special cases machinery may also be fitted with a ball screw feed system—usually to provide exceptionally smooth movement at low feed rates. The rotating circular blade is fed against a clamped, stationary workpiece. The circular sawing process is continuous (Fig. 32.2).

Mechanically, the circular blade is capable of sustaining a far higher chip load than either hack sawing or band sawing. Also, only the blade itself rotates—a relatively low and well-balanced mass (as opposed to large bandwheel assemblies on a band saw or the geartrain/bow assembly on a hack saw). Circular saws are therefore capable of far higher peripheral speeds (the speed of the tooth through the cut) than any other sawing method. For this reason, circular saws are first choice in the sawing of nonferrous metals, particularly aluminum, where, in order to provide speed of cut and/or the desired surface finish, very high peripheral speeds are required. Thanks to its compact dimensions and the higher feed rates it allows, coupled with superior surface finish and consistently good squareness of cut, the circular saw offers advantages in many areas including high-tonnage structural steel fabricating, automotive parts manufacture, and any application where accuracy and/or speed and/or surface finish are prerequisites. Furthermore, the circular saw—particularly the "upstroking"

FIGURE 32.2 Circular saw.

or "rising" type (where the blade is fed from below through a slot in the machine table)—readily accepts jigs and fixtures and offers unparalleled scope for customization. "Specials" have been developed for many different applications such as exhaust manufacture (where tube in the bent condition is trimmed to finished length), splitting connecting rod big ends, and trimming "risers" off castings.

32.5 FERROUS AND NONFERROUS MATERIALS

Much confusion exists with regard to sawing fundamentally different metals—in particular steel and aluminum. The central issue here is peripheral blade speed. Much like any chip-making process, for any given grade of material there is a correct peripheral blade speed range, based on the shear strength of the workpiece material and the type of blade used (HSS, carbide, coatings, and the like). Heat is generated at the flow zone and, generally, the faster the peripheral speed the more heat will be generated. In the main, ferrous sawing machines use HSS blade technology and are designed to raise a fairly heavy chip at a moderate blade speed. Nonferrous machines (almost exclusively circular) use carbide blade technology designed to raise a light chip at extremely high blade speeds of up to 20,000 FPM (for comparison, mild steel would generally be cut in the 50 to 200 FPM range). For this reason, nonferrous sawing machines cannot be used to cut ferrous materials. However, ferrous machines can cut nonferrous materials—albeit with much longer cutting times, with poorer surface finish, and with wider tolerances. Bandsaws are able to cut ferrous and nonferrous materials but cannot compete with purpose-built nonferrous circular machinery in terms of speed, finish, or accuracy.

A simple rule for circular sawing is: a ferrous machine can be used to cut ferrous and non-ferrous materials, but on a nonferrous machine you may only cut nonferrous materials.

32.6 CHOOSING THE CORRECT SAWING METHOD

What follows, by necessity, is generalization and is based on ideal conditions. In the field a given job is often assigned to any available machinery physically capable of handling it—even if it is not ideally suited to that job. However, as a rough guideline the following rules may be applied:

For small materials of similar grades up to around 6 in, particularly if miter-cutting is required, a circular saw is often the better technology.

For larger materials and wherever a very wide variety of material grades is to be sawn by one machine, bandsaws have the advantage.

For very low production requirements on smaller materials a hack saw may be the correct choice.

In the sawing of aluminum extrusions, high-speed carbide circular machines are virtually mandatory.

Bandsaws, on the other hand, are practically the only choice today for the cutting of ultra-high-tensile superalloys, as well as large blocks of tool steel.

For small-batch, low-volume jobs choose a stand-alone semiautomatic machine with a good length measuring system—much of the total time on these applications is wasted in material handling and set-up (i.e., nonproductive time); for small-batch, high volume jobs (i.e., high set-up time and low production time, if done manually) choose a CNC machine with preprogramming facilities and the ability to store and download multiples of jobs. The CNC will help reduce set-up and nonproductive time, thus maximizing production throughput. For high-volume work in long batch runs (i.e., low set-up time and high production), choose an automatic machine—as opposed to CNC—and if necessary customize it for maximum efficiency. Remember, if you are producing 360 parts per hour and you shave 1 s off the cycle, you save 6 min (in which time you can produce another 40 parts).

32.7 KERF LOSS

Kerf loss is the amount of material removed by the blade as the cut is taken. It is generally smaller on bandsaws than on circular saws. However, because circular saws can usually cut to closer tolerances and always with better surface finish, it is possible to cut parts with less "oversize." Also, in many cases, subsequent machining can be reduced or rendered unnecessary. Finally, most modern automatic circular sawing machines have a shorter minimum end piece (the amount required for the automatic feed system to hold on to) than an automatic bandsawing machine, so some of the kerf loss can also be recouped here. Like any other form of machining, the entire process should to be taken into consideration to establish which sawing method gives better material utilization. Also, in many cases the kerf loss becomes irrelevant—for instance, if tolerance and/or surface finish requirements demand a circular saw.

32.8 ECONOMY

Many factors influence the economics of metal sawing—initial cost of machinery, operating costs of machinery including tooling and other consumables, man costs, material costs (waste and potential scrap), subsequent operations (perhaps they can be reduced, or even eradicated). Tolerances, surface finish, and other application-specific requirements should also be taken into account. Concerning the pure sawing process, however, in broad terms there is generally little difference in the cost per cut on a circular saw as compared with a bandsaw.

Note (for structural steel fabrication only): if one machine has to handle a wide range of structural steel section sizes—say from 1 in × 1 in tube up to 40 in beams and heavy columns—a bandsaw is the better choice. If the machine is to handle only larger sections and higher tonnage throughputs are required, or best possible surface finish is needed, a circular machine has the advantage. As a rule of thumb, when comparing modern, high quality bandsaws and modern, high quality circular saws, the circular is capable of nearly double the output, i.e., it can cut most sections twice as fast (particularly heavier beams and columns). Kerf loss is not normally an issue in structural fabrication. The choice between circular and bandsaw in structural fabrication is made almost entirely on the basis of tonnage requirements.

32.9 TROUBLESHOOTING

Given that any sawing machine is "merely pushing a moving blade through material" the vast majority of sawing problems have their roots either in the tool or in those parts of the machine which are directly related with the job of powering/guiding/feeding the tool through the material. Problems with blades can best be diagnosed with the help of one of the many useful guides published by the band and blade manufacturers. However a couple of commonly encountered problems are wrong blade pitch (number of teeth per inch of blade) and poor coolant/lubrication. The latter is easy to remedy by following the instructions supplied by the coolant/lubricant manufacturer and by regularly checking coolant mix ratio with the help of an inexpensive pocket refractometer. Blade pitch selection is a somewhat trickier issue, particularly when sawing sections or tube (intermittent cut and a wide variance of number of teeth in the cut). Also, this issue is compounded by the influences of machine design and machine condition—a good, new, heavy-duty machine will allow a much coarser pitch to be used on a given job than would an old, worn, lightweight machine. Again, blade manufacturers give excellent recommendations in respect of pitch selection in their catalogues, but any good machine supplier will be more than happy to discuss your specific application with you and offer advice on blade selection.

For all machine tools, a common adage is, look after the machine and it will look after you; keep it clean, make sure routine servicing is carried out according to the manufacturer's recommendations, and if there is a known problem, have it seen to sooner rather than later, otherwise it may cause further consequential damage. Use quality blades and consumables and run the machine within its design parameters. Don't cut any faster than you need to, e.g., if the next machining process takes 55 s, there is normally no point in sawing the part in 25 s.

Caveat—despite being a fundamental process in the metalworking industry since its earliest beginnings, metal sawing is generally not well understood. Notwithstanding the above guidelines, there is therefore no substitute for experience when choosing the correct sawing technology for a given application. Any reputable machine builder will be happy to discuss your sawing problems and answer any questions you may have.

32.10 FUTURE TRENDS

As with all machining processes, the sophistication of control systems is center stage in current developments. Machinery already exists, which is capable of automatically feeding, cutting (straight and miter cuts) sorting good pieces from scrap, and distributing good parts to predesignated bin positions—all from data downloaded from a remote workstation. The emphasis is on reducing manual inputs, both mechanical and electronic, and on further simplification of the programming process. Windows-based programs are becoming the norm and user-friendly operation via touchscreens with advanced graphics have elevated the humble sawing machine to the ranks of mainstream machine tools. Future developments may very well focus on further improving data transfer and enhancing the ability of machinery to automatically adapt to different material sizes and grades without operator intervention. There is also scope for further improvements in the prediction of tool life and/or failure. The introduction of higher technology even on lower-price machinery will further reduce maintenance and down-time and secure the position of the metal-cutting saw as one of the lowest cost machine tools on the shop floor. Trends in blade technology—currently the weaker link in the sawing process—are gradually moving towards higher peripheral speeds and greater accuracies whereby increased tooling costs may be offset by higher production and less waste.

FURTHER READING

Bates, Charles, "Band or Circular," *American Machinist,* April 2002.

Koepfer, Chris, "Making the First Cut," *Modern Machine Shop,* July 2000.

Lind, Eric, "Cutting to the Chase," *The Fabricator,* September 1999.

McCorry, David, "Cutting to the Chase," *The Tube & Pipe Journal,* September 2001.

McCorry, David, "What Is Your Factory Cut Out For?" *The Fabricator,* April 2001

TALK SAW magazines (case studies in metal sawing) published by Kaltenbach, Inc. 1-800-TALK-SAW.

CHAPTER 33
FLUIDS FOR METAL REMOVAL PROCESSES

Ann M. Ball
Milacron, Inc.
CIMCOOL Global Industrial Fluids
Cincinnati, Ohio

33.1 FLUIDS FOR METAL REMOVAL PROCESSES

The key factors in metal removal processes are the machine, the cutting tool or grinding wheel, the metal removal fluid, and the type of material being worked. Metalworking fluid is applied for its ability to improve process performance, productivity, tool life, energy consumption, and part quality.

33.1.1 What Is a Metalworking Fluid?

Metalworking fluids may be divided into four subclassifications—metal forming, metal removal, metal treating, and metal protecting fluids. In this discussion, the focal point will be on *metal removal fluids* which are the products developed for use in applications such as machining and grinding, where the material, typically metal, is removed to manufacture a part. It is important to note that metal removal fluids are often referred to interchangeably by other terms such as machining, grinding and cutting fluids, oils or coolant, or by the broadest term metalworking fluids. One technical definition of a *cutting fluid* is "A fluid flowed over the tool and work in metal cutting to reduce heat generated by friction, lubricate, prevent rust, and flush away chips."[1]

Metal removal fluids are generally categorized as one of four product types: (1) straight (neat) oils, (2) soluble oils, (3) semisynthetics, or (4) synthetics. The distinctive difference between each type is based mainly on two formulation features—the amount of petroleum oil in the concentrate and whether the concentrate is soluble in water.

Straight oil, as defined by Childers,[2] is petroleum or vegetable oil used without dilution. Straight oils are often compounded with additives to enhance their lubrication and rust inhibition properties. Straight oils are used "neat" as supplied to the end-user. Soluble oil, semisynthetic and synthetic metal removal fluids are water dilutable (miscible) fluids.

Soluble oil (or emulsifiable oil) fluid is a combination of oil, emulsifiers, and other performance additives that are supplied as a concentrate to the end user. A soluble oil concentrate generally contains 60 percent to 90 percent oil.[3] Soluble oils are diluted with water, typically at a ratio of one part concentrate to 20 parts water or 5 percent. When mixed with water they have an opaque, milky appearance. They generally are considered as general purpose fluids, since they often have the capability to be used with both ferrous and nonferrous materials in a variety of applications.

TABLE 33.1 Distinguishing Characteristics of Metal Removal Fluid Types

Product type	Heat control	Physical lubricity	Cleanliness	Fluid control level	Residue characteristics
Straight oil	Poor	Excellent	Poor	Low	Oily
Soluble oil	Good	Good	Fair	Moderate	Oily
Semisynthetic	Good to excellent	Fair to good	Good	Moderate	Oily to slightly oily
Synthetic	Excellent	Poor	Excellent	High	Slightly oily/tacky
Emulsifiable Synthetic	Good	Good	Good	High	Oily

Semisynthetic fluids have much lower oil content than soluble oils. The concentrate typically contains 2 percent to 30 percent[3] oil. When mixed with water, characteristically at a ratio of one part concentrate to 20 parts water or 5 percent, the blend will appear opaque to translucent. Foltz[3] notes that these fluids have also been referred to as chemical or preformed chemical emulsions since the concentrate contains water and the emulsion or dispersion of oil occurred during formulation, which contrasts soluble oil where emulsion does not form until diluted for use. These fluids usually have lubricity sufficient for applications in the moderate to heavy-duty range (i.e., centerless and creep feed grinding or turning and drilling). Their wetting and cooling properties are better than soluble oils which allow for faster speeds and feed rates.

Synthetic fluids contain no mineral oil. Most synthetic fluids have a transparent appearance when mixed with water. There are some synthetic fluids that are categorized as synthetic emulsions, which contain no mineral oil but appear as an opaque, milky emulsion when mixed with water. Synthetic fluids have the capability to work in applications ranging from light (i.e., double disk, surface, or milling) to heavy-duty (i.e., creep feed, threading, and drilling). Synthetic fluids generally are low foaming, clean, and have good cooling properties allowing for high speeds and feeds, high production rates, and good size control.

Fluids within and between each class will provide the user the choice of a broad range of performance characteristics and a variation of duty ranges, from light to heavy. Selecting the fluid for the specific process will usually require tradeoffs of certain characteristics for other, more critical, characteristics after a review of all the variables for the shop and application. The fluid types and their distinct characteristics are listed in Table 33.1 for purpose of comparison.

33.1.2 Functions of a Metal Removal Fluid

Metal removal fluids provide two primary benefits:

- *Cooling.* A tremendous amount of heat is produced in the metal removal process making it important to extract that heat away from the part and the wheel or tool. Dissipating heat from the workpiece eliminates temperature related damage to the part such as finish and part distortion. Removing heat from the cutting tool or grinding wheel extends their life and may allow for increased speeds.

  In metal removal applications, the metal removal fluid carries away most (96 percent) of the input energy with its contact to the workpiece, chips, and tool or grinding wheel. The input energy ends up in the fluid where it will be transferred to its surroundings by evaporation, convection, or in a forced manner, by a chiller.[4] Methods for cooling metalworking fluid are discussed in detail by Smits.[3]

- *Lubrication.* Fluids are formulated to provide lubrication that reduces friction at the interface of the wheel and the part. The modes of lubrication are described as being physical, boundary, or chemical.

  Physical lubrication in metal removal fluid is provided by a thin film of a lubricating component. Examples of these components may be a mineral oil or a nonionic surfactant.

Boundary lubrication occurs when a specially included component of the metal removal fluid attaches itself to the surface of the workpiece. Boundary lubricants are polar additives such as animal fats and natural esters.

Chemical lubrication occurs when a constituent of the fluid (i.e., sulphur, chlorine, phosphorous) reacts with a metallic element of the workpiece, resulting in improved tool life, better finishes, or both. These additives are known as extreme pressure (EP) additives.

In addition to the primary functions performed by a fluid, as previously described, there are other functions required from a fluid. These include providing corrosion protection for the workpiece and the machine, assisting in the removal of chips or swarf (build-up of fine metal and abrasive particles) at the wheel workpiece interface (grinding zone), transporting chips and swarf away from the machine tool, and lubricating the machine tool itself.

33.1.3 The Fluid Selection Process

When selecting a fluid, a total process review should be completed to achieve optimum performance. Process variables that should be considered include:

1. *Shop size.* Generally small shops that work with a variety of metals, in multiple applications will look for a general-purpose product to help minimize cost and number of products to maintain. Larger facilities with specific requirements will choose a product that is application and operation specific.
2. *Machine type.* Some machine designs, especially older models, may require that the fluid serve as a lubricating fluid for the machine itself. Some machine specifications may require only straight oil type fluids or only waterbased fluids to be compatible with machine components and materials. It is also important to consider if the machine is a stand alone with its own fluid sump or if many machines will be connected to one large central fluid system. Not all fluids are meant for both types of situations.
3. *Severity of operations.*[5] The severity of the operation will dictate the lubricity requirements of the fluid. Metalworking operations are categorized as light-duty (i.e., surface, double disc grinding, milling), moderate-duty (i.e., tool room, internal, centertype and centerless/shoe type grinding), heavy-duty (i.e., creep feed, thread/flute, form grinding, sawing, drilling, tapping), and extremely heavy duty (form and thread grinding, broaching). When determining the severity category, consideration should be made for the speeds, feeds, stock removal rates, finish requirements, and the metal types. The most critical operation to be performed and its severity will usually determine fluid choice. The belief that you need straight oil or a high oil-containing product to attain good grinding or machining results is not true with today's technology. For example, Yoon and Krueger[6] provide data exhibiting how the grinding process can be optimized, using a synthetic emulsion containing an EP lubricant along with an MSB grinding wheel.
4. *Materials.* Information on the workpiece material (i.e., cast iron, steel, aluminum, glass, plastic carbide, exotic alloy, and the like) to be used in the operation is a necessity when making a fluid selection. Often fluids are formulated for use with specific materials, i.e., glass, aluminum, and the like or the fluid does not have corrosion control compatible to the material in use.
5. *Water quality.* The main component of a water-based metal removal fluid mix is water, approximately 95 percent. This makes water quality a major factor in fluid performance. Poor water quality may promote corrosion, produce fluid instability, and create residue. The use of water treated by deionization (DI) or reverse osmosis (RO) is recommended if the water to be used is of questionable quality.
6. *Freedom from side effects.* The product should be chosen and used with health and safety in mind, (i.e., mild to the skin, minimizing cobalt leaching, low mist, and the like). It should not leave a residue or cause paint to peel off of the machine.

7. *Chemical restrictions.* Facilities, often, have restrictions on certain materials because of health and safety, environmental, or disposal issues. Often there are restrictions on, or allowable limits for materials due to interference with a process or end use (i.e., halogens in the aerospace industry), which may also influence the product choice. These should be identified before selecting a fluid.

33.2 APPLICATION OF METAL REMOVAL FLUIDS

Understanding the application of metal removal fluids to the grind zone or cutting area is an important aspect of the metal removal process. Grinding wheel and tooling life are greatly influenced by the way the cutting fluid is applied.[7] Generally, metal removal fluids are held in a fluid reservoir and pumped through the machine to a fluid nozzle which directs the fluid to the work zone. The fluid leaves the work zone, flowing back to the fluid reservoir where it should be filtered before being circulated back to the workpiece. The machine tool's electrical power in kilowatts is the basis for determining the systems fluid flow rate. The following helpful rule may be used to determine the applicable flow rate.

General purpose machining and grinding: Fluid flow rate (m^3/s) = machining power (kW)/120

High production machining and grinding: Fluid flow rate (m^3/s) = machining power (kW)/60 to 30

The fluid reservoir's capacity must allow sufficient retention time to settle fines and cool the fluid.[8] Suggested values for general-purpose machining and grinding operations would be:

Grinding: Tank volume = (flow/min) × 10

Machining cast iron and aluminum: Tank volume = (flow/min) × 10

Machining steel: Tank volume = (flow/min) × 8

Applications with high stock removal obey the same formulas for tank size since flow rate will be increased in relation to the machine horsepower.

Flooding coolant at low pressure to the work zone is considered the most efficient method for flushing chips while providing the best tool life and surface finish. Gun drilling and some reaming applications are exceptions where fluid is fed through the tool under high pressure. A *high-pressure* coolant application is beneficial in machining where chip packing is an issue or in grinding where the air currents generated by the grinding wheel need to be broken.[8]

Misting or *manual* coolant application methods are sometime used. Mist application of metal removal fluid is used, usually, where, due to part size or configuration, fluid could not be rechanneled back to a fluid reservoir. In this method, mist is generated by pumping fluid, from a small fluid receptacle, through a special nozzle where it mixes with air, creating a fine mist, which is then delivered to the cut zone. Manual application is not often recommended unless it is done in conjunction with a flood application system. An example of this would be the use of manually applied tapping compound in a threading operation where added friction-reducing materials are needed to provide the required tool life and part finish.

When applying fluid by any method it is important to reduce exposure to the metalworking fluid mist that is generated by providing adequate ventilation, utilizing mist collection equipment, machine guarding, and reducing fluid flow. Various types of nozzles may be used for fluid application; a description of *dribble, acceleration zone, fire hose, jet,* and *wrap around nozzle* types can be found in Smits.[4]

Smits[4] offers an informative discussion regarding how fluid flow rate, fluid speed entering the flow gap, fluid nozzle position, and grinding wheel contact with the work piece all influence the results of the grinding process.

33.3 CONTROL AND MANAGEMENT OF METAL REMOVAL FLUIDS

It is important to utilize shop and fluid management practices and controls to improve the workplace environment, extend fluid life, and enhance the fluid's overall performance. Control and management of metal removal fluids will be more effective and efficient if a program of *Best Practice Methods* is incorporated along with various types of control procedures and equipment. The use of a fluid management and control program is part of the recommendations made by Organization Resources Counselors in the "Management of the Metal Removal Environment"(http://www.aware-services.com/orc/metal_removal_fluids.htm) *and* by the Occupational Safety and Health Administration in the "METALWORKING FLUIDS: Safety and Health Best Practices Manual" (http://www.osha.gov./SLTC/metalworkingfluids/metalworkingfluids_manual.html), Management and control of the shop environment in conjunction with a quality fluid management program will help reduce health, safety, and environmental concerns in the shop.

33.3.1 Best Practices Management Program

The basis for a Best Practices program is the development and implementation of a written management plan. The plan, defining how the shop systems will be maintained, may be simple or complex, and should include the following key elements.

1. Goals and commitment of the program
   - Include references to managing fluids and other materials used in the process, improving product quality and the control and prevention of employee health and safety concerns.
2. Designation of program responsibility
   - A team or individual should be named to coordinate the management program.
3. Control of material additions to the fluid system
   - All system additions of any kind should be controlled and recorded by designated individuals.
4. A standard operating procedure should be written for fluid testing and maintenance
   - The SOP should include:

     How often samples should be collected and tested

     What action should be taken as a result of the test data

     Specific protocols for each test, and the like.
5. Data collection and tracking system
   - Data should include:

     System observations

     Lab analyses

     Material additions
   - Use the collected data to determine trends for improving the system and process management.
   - Examples of parameters to be tracked—(1) concentration, (2) pH, (3) water quality, (4) system stability, (5) additive additions; amount and timing, (6) biological test results (bacteria, mold counts), (dissolved oxygen-DO), (7) tramp oil, (8) biocide levels, and (9) corrosion data.
6. Employee participation and input
   - To have an effective fluid management program employees must be enlisted to aid in the constant observation and evaluation of the systems operation. Training should be provided to help employees develop an understanding of how the lab tests and results influence overall maintenance of the fluid system.

7. Training programs
   - Provide training to help employees develop an understanding of how the lab tests and results influence overall maintenance of the fluid system.
   - Training should include:

     Health and safety information

     The proper care of the fluid system for optimum performance

     An understanding of types and functions of metal removal fluids
8. Evaluation of fluid viability and subsequent disposal
   - All fluid systems eventually reach the end of their useful life; guidelines need to be written to help determine when the end of the fluid's life has been reached and how disposal will be addressed.

33.4 METAL REMOVAL FLUID CONTROL METHODS

Water-soluble metalworking fluids are each formulated to operate within a set of specified conditions. The operational ranges usually include parameters for concentration, pH, dirt levels, tramp oil, bacteria, and mold. Performance issues can develop if fluid conditions fall outside any of the specified operating parameters.

33.4.1 Concentration Measurement

Water-based metal removal fluids are typically formulated to operate in a concentration range of 3 to 6 percent, although concentrations up to 10 percent are not uncommon for many heavy-duty applications. Concentration is *the* most important variable to control. Concentration is not an absolute value but rather a determination of a value for an unknown mix based on values obtained from a known mix. There are certain inaccuracies, variables, and interferences in any method. This must be considered when evaluating the data.[5]

The fluid supplier should specify the desired procedure and method to be used to check and control the concentration of an individual fluid. The available techniques include, refractometer, chemical titration, and instrumental methods. For a detailed description of these methods please refer to Foltz.[5]

33.4.2 pH Measurement

The acidity or alkalinity level of a metal removal fluid is measured by pH determination. Metal removal fluids are typically formulated and buffered to operate in a pH range of 8.5 to 9.5.[5] Measuring pH is a quick indicator of fluid condition. A pH below 8.5 typically is an indicator of biological activity in the fluid, which can affect mix stability, corrosion, and bacterial control. When pH is greater than 9.5 the fluid has had an alkaline contamination or build-up which will affect the mildness of the fluid. The pH level may be measured by pH meter or pH paper. There are many types of pH meters available, both bench-top and handheld versions. The meters should always be standardized with buffer and the electrodes cleaned prior to checking the pH of any fluid mix to ensure that an accurate reading is observed. The pH level can also be measured by observing color change to pH paper after it has been dipped in the fluid mix. The recommended method for monitoring this control parameter is the pH meter.

33.4.3 Dirt Level

The dirt or total suspended solids (TSS) measurement usually indicates what the chip agglomeration and settling properties are of the fluid and/or the effectiveness of the filtration system. The types of dirt and TSS found in a fluid mix include metal chips and grinding wheel grit. Any quantity of recirculating dirt, small or large, can eventually lead to dirty machines, clogged coolant lines, poor part finish, and corrosion. There are many methods for determining a fluid's dirt load. We suggest discussing this with your fluid supplier. Typically, dirt volumes in excess of 500 ppm or 20 μm in size can lead to problems.[5]

33.4.4 Oil Level

Oil is found in most metal removal fluids either as a product ingredient or a contaminant. It is very useful to know the level of product oil and contaminant (tramp) oil present in a fluid mix. The level of product oil is an indicator of the mix concentration while tramp oil level indicates the amount of contamination in the mix.

Tramp oil is found in one of two forms, *free* or *emulsified.* Free oil is not emulsified or mixed into the product and floats on the top of the fluid mix. Free oil, usually, is easily removable from the product with the use of oil skimmer, wheel, or belt. Emulsified tramp (nonproduct) oil is chemically or mechanically emulsified into the product and is difficult to remove, even with a centrifuge or a coalescer. Tramp oil sources include way or gear lube leaks, hydraulic leaks, and/or leaks in the forced lubrication system often found on machines. Oil leakage should be kept to a minimum since high tramp oil levels will influence a product's effectiveness, reducing cleanliness, filterability, mildness, corrosion, and rancidity control. High tramp oil can also increase the level of metalworking fluid mist released into the plant air.

33.4.5 Bacteria and Mold Levels

Metalworking fluids do not exist in a sterile environment and can develop certain levels of organism growth.[9] Products are formulated in various ways to deal with biological growth—some contain bactericides and fungicides, others include ingredients that won't support organism growth, and then some are formulated to have low odor with organism growth. Monitoring and controlling microbial growth is important for maintaining and extending fluid life. The various methods used for monitoring microbial growth includes plate counts, bio-sticks, and dissolved oxygen.

33.4.6 Conductivity

Fluid condition may also be monitored by a parameter called *conductivity*. Conductivity is measured as a microSieman (μS) unit using a conductivity meter. In general, a 5 percent metalworking fluid mix in tap water will have 1500 μS conductivity. Mix concentration, increased water hardness, mix temperature, dissolved metals, and other contaminants all can change conductivity. Observing trends in the conductivity readings over time can help assess mix condition, and enhance problem solving of residue issues and mix instability.

33.4.7 Filtration of Metalworking Fluids

It is important to provide filtration at the fluid reservoir to remove wheel grit, metal fines, dirt, and other fluid contaminants that gather with use. Proper filtration of the fluid will extend fluid life, reduce plugged coolant lines, and improve part finish. Brandt[10] discusses the types of filtration (pressure, vacuum, gravity, media, and the like) available and how to determine the best filtration method for your application.

REFERENCES

1. D. Lapedes, ed., *McGraw-Hill Dictionary of Scientific and Technical Terms,* 2d ed., McGraw-Hill, New York, 1978, p. 396.
2. J. C. Childers, "The Chemistry of Metalworking Fluids," in *Metalworking Fluids,* J. Byers, ed., Marcel Dekker, New York, 1994, pp. 170–177.

3. G. Foltz, "Definitions of Metalworking Fluids," in *Waste Minimization and Wastewater Treatment of Metalworking Fluids,* R. M. Dick, ed., Independent Lubrication Manufacturers Association, Alexandria, VA, 1990, pp. 2–3.
4. C. A. Smits, "Performance of Metalworking Fluids in a Grinding System," in *Metalworking Fluids,* J. Byers, ed., Marcel Dekker, New York, 1994, pp. 100–132.
5. G. J. Foltz, "Management and Troubleshooting,"in *Metalworking Fluids,* J. Byers, ed., Marcel Dekker, New York, 1994, p. 307.
6. S. C. Yoon and M. Krueger, "Optimizing Grinding Performance by the Use of Sol-Gel Alumina Abrasive Wheels and A New Type of Aqueous Metalworking Fluid," 3(2), p. 287, *Machining Science and Technology,* 1999.
7. S. Krar and A. Check, "Cutting Fluids—Types and Applications," *Technology of Machine Tools,* 5th ed., Glencoe/McGraw-Hill, Ohio,1997, pp. 252–261.
8. G. Foltz and H. Noble, "Metal Removal: Fluid Selection and Application," in *Tribology Data Handbook,* E. R. Booser, ed., CRC Press LLC, New York, 1997, pp. 831–839.
9. E. O. Bennett, "The Biology of Metalworking Fluids," *Lubr. Eng* 227, July 1972.
10. R. H. Brandt, "Filtration Systems for Metalworking Fluids," in *Metalworking Fluids,* J. Byers, ed., Marcel Dekker, New York, 1994, pp. 273–303.

INFORMATION RESOURCES

The following resources are recommended to learn more about the technology, application, and maintenance of metal removal fluids.

"*Management of the Metal Removal Environment*" found at http://www.aware-services.com/orc/metal_removal_fluids.htm, Copyright © 1999 Organization Resources Counselors.

Metalworking Fluids, J. Byers, ed., Marcel Dekker, Inc., New York, (1994).

Waste Minimization and Wastewater Treatment of Metalworking Fluids, R. M. Dick, ed., Independent Lubrication Manufacturers Association, Alexandria, VA, (1990).

Cutting and Grinding Fluids: Selection and Application, J. D. Silliman, ed., Society of Manufacturing Engineers, Dearborn, Michigan, (1992).

Occupational Safety & Health Administration, "METALWORKING FLUIDS: Safety and Health Best Practices Manual" at http://www.osha.gov./SLTC/metalworkingfluids/metalworkingfluids_ manual.html, OSHA Directorate of Technical Support, (2001).

CHAPTER 34
LASER MATERIALS PROCESSING

Wenwu Zhang
General Electric Global Research Center
Schenectady, New York

Y. Lawrence Yao
Columbia University
New York, New York

34.1 OVERVIEW

LASER is the acronym of *light amplification by stimulated emission of radiation*. Although regarded as one of the nontraditional processes, *laser material processing* (LMP) is not in its infancy anymore. Einstein presented the theory of stimulated emission in 1917, and the first laser was invented in 1960. Many kinds of lasers have been developed in the past 43 years and an amazingly wide range of applications—such as laser surface treatment, laser machining, data storage and communication, measurement and sensing, laser assisted chemical reaction, laser nuclear fusion, isotope separation, medical operation, and military weapons—have been found for lasers. In fact, lasers have opened and continue to open more and more doors to exciting worlds for both scientific research and engineering.

Laser material processing is a very active area among the applications of lasers and covers many topics. Laser welding will be discussed in a separate chapter. In this chapter, laser machining will be discussed in detail while other topics will be briefly reviewed. Some recent developments, such as laser shock peening, laser forming, and laser surface treatment, will also be reviewed to offer the reader a relatively complete understanding of the frontiers of this important process. The successful application of laser material processing relies on proper choice of the laser system as well as on a good understanding of the physics behind the process.

34.2 UNDERSTANDING OF LASER ENERGY

34.2.1 Basic Principles of Lasers

Lasers are photon energy sources with unique properties. As illustrated in Fig. 34.1, a basic laser system includes the laser medium, the resonator optics, the pumping system, and the cooling system. The atomic energy level of the lasing medium decides the basic wavelength of the output beam, while nonlinear optics may be used to change the wavelength. For example, the basic optical

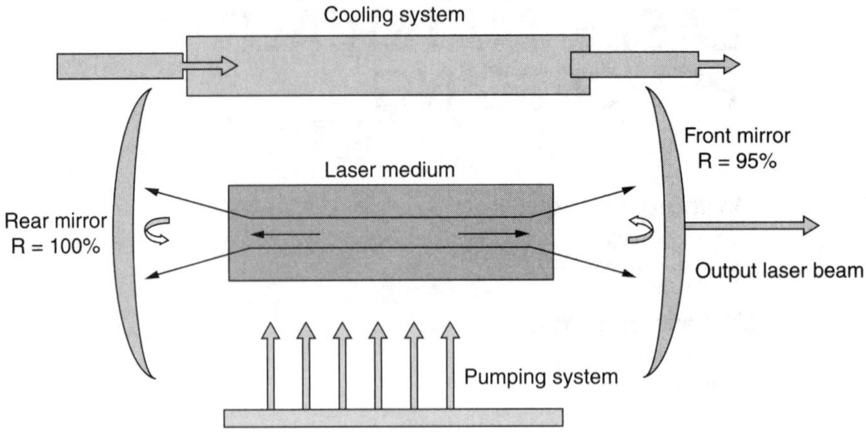

FIGURE 34.1 Illustration of a basic laser system.

frequency of the neodymium-doped yttrium aluminum garnet (Nd:YAG) laser at 1.06 μm wavelength may be doubled or tripled by inserting nonlinear crystals in the resonator cavity, getting the wavelengths of 532 nm and 355 nm. The lasing mediums, such as crystals or gas mixtures, are pumped by various methods such as arc light pumping or diode laser bar pumping. Population inversion occurs when the lasing medium is properly pumped, and photons are generated in the optical resonator due to stimulated emission. The design of the optical resonator filters the photon energy to a very narrow range, and only photons within this narrow range and along the optical axis of the resonator can be continuously amplified. The front mirror lets part of laser energy out as laser output. The output beam may pass through further optics to be adapted to specific applications such as polarizing, beam expansion and focusing, and beam scanning. The in-depth discussion of the principles of lasers can be found in Ref. 1, information on common industrial lasers can be found in Refs. 2 and 3, a web based tutorial on laser machining processes can be found in Ref. 4, and mounting literature on laser material processing can be found from many sources.

Understanding the physics in laser material interaction is important for understanding the capabilities and limitations of these processes. When a laser beam strikes on the target material, part of the energy is reflected, part of the energy is transmitted and part of it is absorbed. The absorbed energy may heat up or dissociate the target materials. From a microscopic point of view the laser energy is absorbed by free electrons first, the absorbed energy propagates through the electron subsystem, and then is transferred to the lattice ions. In this way laser energy is transferred to the ambient target material, as illustrated by Fig. 34.2. At high enough laser intensities the surface temperature of the

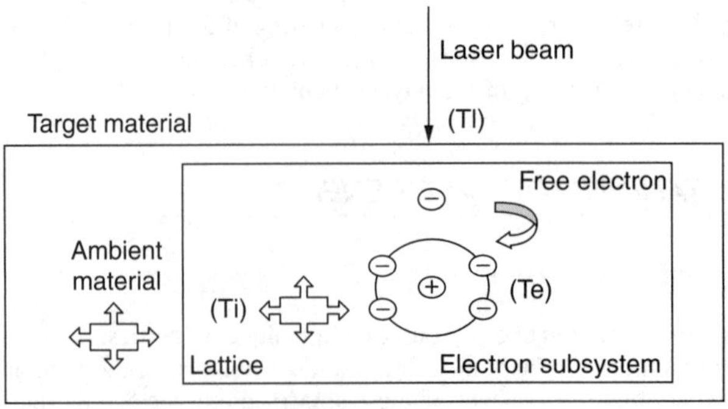

FIGURE 34.2 Laser energy absorption by target material.

target material quickly rises up beyond the melting and vaporization temperature, and at the same time heat is dissipated into the target through thermal conduction. Thus the target is melted and vaporized. At even higher intensities, the vaporized materials lose their electrons and become a cloud of ions and electrons, and in this way plasma is formed. Accompanying the thermal effects, strong shock waves can be generated due to the fast expansion of the vapor/plasma above the target.

Given the laser pulse duration, one can estimate the *depth of heat penetration*, which is the distance that heat can be transferred to during the laser pulse.

$$D = \text{sqrt}(4 \times alfa \times dT)$$

where D is the depth of heat penetration, *alfa* is the diffusivity of materials, and dT is the pulse duration. Laser energy transmission in target material is governed by Lambert's law:

$$I(z) = I_0 \times \exp(-a \times z)$$

where I is laser intensity, I_0 is the laser intensity at the top surface, z is the distance from the surface, and a is the absorption coefficient that is wavelength dependent. Metals are nontransparent to almost all laser wavelengths and a is about 100,000 cm^{-1}, which implies that within a depth of 0.1 μm, laser energy has decayed to 1/e of the energy at the surface. Many nonmetals such as glasses and liquids have very different a values. Laser-material interaction thus can be surface phenomena when the laser pulse duration is short and when the material has rich free electrons. Laser energy may also be absorbed over a much larger distance in nonmetals than in metals during its transmission.

When considering the laser power in material processing, the effective energy is the portion of energy actually absorbed by the target. A simple relation for surface absorption of laser energy is: $A = 1 - R - T$, where A is the surface absorptivity, R is reflection, and T is transmission. For opaque material, $T = 0$, then $A = 1 - R$.

It's important to understand that reflection and absorption are dependent on surface condition, wavelength, and temperature. For example, copper has an absorptivity of 2 percent for CO_2 lasers (Wavelength = 10.6 μm), but it has much higher absorptivity for UV lasers (about 60 percent). Absorption usually increases at elevated temperatures because there are more free electrons at higher temperatures.

34.2.2 Four-Attributes Analysis of the Laser Material Processing Systems

Laser material interaction can be very complex, involving melting, vaporization, plasma and shock wave formation, thermal conduction, and fluid dynamics. Modeling gives the in-depth understanding of the physics in the study of laser material processing processes. Many research centers are still working on this task and volumes of books and proceedings are devoted to it. We won't cover modeling in this chapter, but as a manager or process engineer, one can get a relatively complete picture of the laser material processing system following the *four-attributes analysis*—time, spatial, magnitude, and frequency.[4]

Time Attribute. Laser energy may be continuous (CW) or pulsed, and laser energy can be modulated or synchronized with motion. For CW lasers, the average laser power covers a wide range, from several watts to over tens of kilowatts, but their peak power may be lower than pulsed lasers. CW lasers may be modulated such as ramping up or ramping down the power, shaping the power, or synchronizing the on/off of the shutter with the motion control of the system. The common range of pulse duration is in the ms level, and the smallest pulse duration is normally larger than 1 μs. CW lasers can operate in pulsed mode with the shutter in open/close position. Despite these quasi-pulsed modes, the laser is still operating in CW mode inherently, in which lasing is still working in CW mode. No higher peak power than CW mode is expected normally. For a CW laser one should understand its capability of power modulations, focusing control, and energy-motion synchronization.

There are many types of pulsed lasers. The major purpose of pulsating the laser energy in laser material processing is to produce high peak laser power and to reduce thermal diffusion in processing.

Taking Q-switched solid-state lasers for example, lasing condition of the cavity is purposely degraded for some time to accumulate much higher levels of population inversion than continuous mode, and the accumulated energy is then released in a very short period—from several nanosecond (10^{-9} s) to less than 200 ns. Even shorter pulse durations can be achieved with other techniques as discussed in Ref. 1. Lasers with pulse duration less than 1 ps (10^{-12} s) are referred as *ultrashort pulsed lasers*. Pulsed lasers have wide range of pulse energies, from several nJ to over 100 J. These pulses can be repeated in certain frequencies called the *repetition rate*. For pulsed lasers, basic parameters are the pulse duration, pulse energy, and repetition rate. From these parameters, peak power and average power can be calculated. Similar to CW lasers, one should also understand the capability of power modulations, focusing control, and energy-motion synchronization for pulsed lasers. Peak laser intensity is the pulse energy divided by pulse duration and spot irradiation-area. Due to several orders of pulse duration difference, pulsed laser can achieve peak laser intensities $\gg 10^8$ W/cm^2, while CW lasers normally generate laser intensities $<10^8$ W/cm^2.

Spatial Attribute. Laser beam out of a cavity may have one or several modes, which are called *transverse electromagnetic mode* (TEM) . For laser material processing, we are concerned with the spatial distribution of the beam that affects the thermal field on the target. Laser intensity usually has a Gaussian beam distribution. For Gaussian beam with beam radius r and for a material with absorption $A = 1 - R$, where R is the reflectivity and $P(t)$ is the time dependent laser power, the spatial distribution of absorbed laser intensity on the target surface is:

$$I(x, y, t) = (1 - R)I_0(t) \exp(-(x^2 + y^2)/r^2)$$

Where $I_0(t) = 2P(t)/(\pi r^2)$, is the average laser intensity. Laser energy distribution may take other shapes, such as flat-hat shape, in which the laser intensity at the center is uniform. In general, the formula for laser energy transmitted to the material at depth z is:

$$I(x, y, z, t) = A \times I_0(t) \exp(-a \times z)SP(x, y)$$

where A = fraction of laser energy absorbed by the material at the surface
$I_0(t)$ = temporal distribution of laser intensity
a = absorption coefficient
SP = spatial distribution of laser intensity

Special optics can be used to change the beam shape and spatial distribution. For example, the beam can be changed from circular to square and uniform.

Laser beam radius is normally defined as the distance from the beam center within which 86.4 percent or $(1 - 1/e^2)$ of total energy is included. Beam radius at the focus is called the *focused spot size*. Frequently spot size variation with standoff distance (the distance from the focusing lens to the target) is needed. For lower intensities, laser energy profiler can be used to directly measure the intensity distribution. The laser beam size close to the focus is usually difficult to measure directly, especially for cases when the focused spot size is below tens of microns or when the laser power is high. One crude solution for high-power lasers is to measure the diameter of laser burnt holes in suitable thin sheet material. For a Gaussian beam, a more accurate solution is to combine experimental measurements with optical calculations. The spot size at large defocus can be measured either by the profiler or the knife-edge method. More than three measurements at different locations are measured to obtain (Z_n, D_n), $n = 1, 2, 3, \ldots$, where D_n is the beam size at location Z_n. The propagation of laser beams in air satisfies the following equation:

$$D_n^2 = D_0^2 + \left(\frac{4M^2\lambda}{\pi}\right)^2 \frac{(Z_n - Z_0)^2}{D_0^2}, \qquad n = 1, 2, 3, \ldots$$

where D_0 is the beam waist, Z_0 is the beam waist location, and M^2 is the beam quality parameter. Knowing (Z_n, D_n), D_0, Z_0, and M^2 can be determined. Then one can calculate the spot size at any location along the optical axis. Knowing M^2, one can also calculate the beam divergence and *depth*

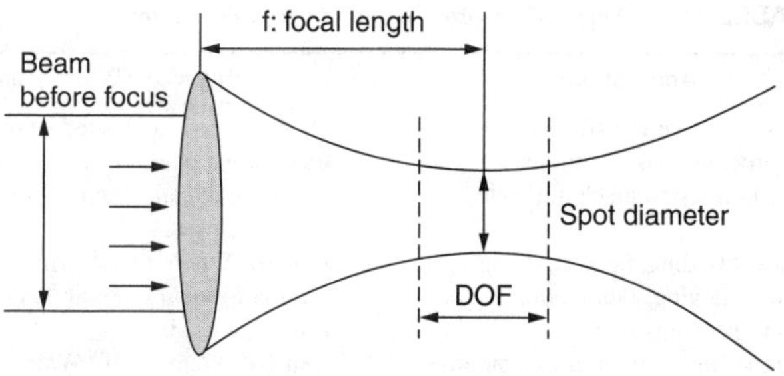

FIGURE 34.3 The DOF of laser light.

of focus (*DOF*). Depth of focus is the range of distance over which the spot size changed from the focused spot size by 5 percent. Figure. 34.3 illustrates the propagation, the beam waist, and the DOF of laser beam.

Laser intensity changes with defocus. Laser material processing is claimed noncontact because the highest intensity is at the focus while laser optics are some distance away from the target. It is not always convenient to change the focus in processing. The limited depth of focus limits laser machining to relatively thin materials (usually <15 mm).

In material processing, one can move the beam while keeping the part fixed, or move the part on a stage while keep the beam fixed, or move both of them. An XY or XYZ motorized stage is commonly used. Laser beams can be quickly scanned across specified locations by computer controlled reflection optics. This makes high-speed marking or drilling possible. The spatial resolution of laser material processing is influenced by the focused spot size. Shorter wavelength lasers are thus used for precision machining tasks.

Magnitude Attribute. Major magnitude parameters of laser energy are power (unit: watt), pulse energy (unit: Joule), and intensity (unit: W/m^2 or W/cm^2). The average power of laser is relatively low compared to other energy sources: over 1 kW is already regarded as high-power, and a pulsed laser normally has an average power of less than 100 W. The strength of laser energy is that it can have very high local energy intensity, and this intensity can be well controlled in time, space, and magnitude.

When the interaction between energy field and target is not continuous, energy intensity is usually the deciding factor. Depending on the laser type, laser pulse energy can be varied from below 10^{-9} J to far over 1 J, the spot size can be varied from sub-microns to over 10 mm, and pulse duration can be varied from several fs (1 femtosecond (fs) = 10^{-15} s) to over 1 s. For pulsed lasers, the laser intensity is equal to $E_0/(t_p \times pi \times R^2)$, where E_0 is pulse energy, t_p is pulse duration, and R is beam radius. For laser pulse energy of 0.1 J, if the pulse repetition rate can vary in the range from 1 Hz ~ 4 kHz, then the average power is 0.1 ~ 400 W. Let's vary the pulse length and the acting area and compute the peak intensity. With R = 0.5 μm, peak intensity of a 10 fs pulse is 10^{22} W/cm^2, the intensity of a 10^{-6} s pulse is10^7 W/cm^2 and the intensity of a 0.001 s pulse is only 10^4 W/cm^2. It is clear that laser intensity can be flexibly controlled to achieve a very wide range of laser intensities.

Depending on the absorbed laser intensity, different physical phenomena are involved. Applications at various laser intensities and deposition times are briefly shown in Table 34.1.

Many material properties such as thermal conductivity and reflectivity vary with material temperature and state, which are further decided by the magnitude of energy input. We tacitly assume that only one photon is absorbed by one electron at a specific time at normal laser intensities, but when the laser intensity is extremely high as in the case of ultrafast lasers (pulse duration <10^{-12} s), more than one photon can be absorbed by one electron simultaneously. This is termed as multiphoton absorption. Material optical property is then highly nonlinear and is very different from single photon absorption. Material can act as if it were irradiated by a frequency doubled or tripled laser source. In this meaning, we can say that extremely high magnitude of laser intensity can be equivalent to shorter wavelengths.

TABLE 34.1 Applications of Lasers in Material Processing

Applications	Intensity (W/cm^2) and laser material interaction
Laser surface transformation hardening, laser forming, laser assisted machining, etc.	$<10^5$ W/cm^2, target heated below melting temperature, phase transformation may occur that can harden the material, elevated temperature can soften the material. Pulse duration $>10^{-3}$ s, CW lasers are used.
Laser welding, laser cladding and alloying, rapid tooling, and laser machining	From 10^6 W/cm^2 to 10^8 W/cm^2, material melts, some vaporization and plasma formation possible. Pulse duration normally $>10^{-3}$ s. CW lasers are used.
Higher intensity laser machining—marking, grooving, drilling, and cutting	From 10^7 W/cm^2 to 10^9 W/cm^2, material melts and strong vaporization occurs, shock wave and plasma formation possible. Pulse duration normally $<10^{-3}$ s, 10^{-9} to 10^{-6} s pulse duration are common, while for micromachining even shorter pulses are used. CW lasers or pulsed lasers are used.
Laser shock processing, laser surface cleaning	Intensity $>10^9$ W/cm^2 and pulse duration $<10^{-7}$ s, very intense surface vaporization induces strong shock pressure toward the target.

Optical filters, polarizers, attenuators, and beam expanding and focusing systems can be used to modulate laser intensity and intensity spatial distribution so that one can match the laser output to a specific application without disturbing the internal laser source.

Frequency Attribute. The characteristic frequency of energy field is important because materials may respond very differently to energy fields at different frequencies. The characteristic frequency of laser is its EM oscillation frequency, and more frequently we use its equivalence—wavelength. The frequency decides the individual photon energy of the laser beam. Lasers usually have very narrow spectral width, while other energy sources may have very broad and complex spectral distributions.

The diffraction limited spot size is proportional to wavelength. For circular beams, the focal spot size is: $Dmin = 2.44f \times \lambda/D$, where f is the focus length, λ is wavelength, and D is the unfocused beam diameter. Thus for high-precision applications, shorter wavelength lasers are preferred. UV laser ablation of organic polymers can be very different in mechanism compared to infrared or visible laser ablation. The infrared and visible laser ablation is mainly photo-thermal degradation, while UV laser ablation may involve direct photo-chemical dissociation of the chemical bonds.

Materials show very different absorption properties at different wavelengths. Metals tend to have low absorption at far infrared (CO_2 laser 10.6 μm) while absorption increases with decreasing wavelength. Nonmetals such as ceramics and liquids have strong absorption at far infrared, much decreased absorption at visible wavelengths, and increased absorption at UV. At deep UV (some people call it *extreme UV*), almost any material has very strong absorption. That's why different materials may need to use lasers at different wavelengths for high energy coupling efficiency.

Keep in mind that absorption also depends on temperature, purity, and surface condition. Thin layers of black coating can be used to increase the energy coupling of CO_2 laser into metals. Defects or impurity in a transparent media may strongly absorb laser energy and thus create a local thermal point and finally break down the transparent condition. Also keep in mind that at high enough laser intensity, multiphoton absorption may occur, material reacts nonlinearly to the irradiation, and the beam acts as if its frequency is doubled or tripled. And once the surface temperature rises, absorption tends to increase, which forms a positive feedback. In this meaning, very high laser intensity may be regarded as wavelength-independent in material processing.

In general, the four attributes analysis can be applied to other energy forms. From here one can see the advantages and the limitations of a process and realize that many things are relative rather than absolute, such as the energy coupling efficiency and wavelength. Laser material processing can be very complex and modeling work is still actively going on around the world to better predict the process. Caution should be used when collecting the material properties from literature. In laser material processing, material properties are highly temperature-, wavelength-, geometry-, and intensity-dependent.

34.3 LASER SAFETY

Lasers used in material processing are usually high-power lasers that may inflict hazards to both the operator and the visitor. Strict safety rules must be followed to prevent such potential hazards. Once proper safety practices are followed, laser material processing is as safe as other material processing techniques.

The most common danger is the damage to the eye. Laser light, even at very low power level, can be much brighter than normal light sources. Laser light can be focused into smaller spot sizes by the lens structure of human eyes. Light in the range of 0.4 to 1.4 μm can be focused on the retina and cause damages, while light in the far infrared can cause thermal damage of the cornea. There are three major cases of eye damage. The first is the direct beam damage in which the eye is within the light path. Since the beam is collimated, this is extremely dangerous. This usually happens during laser alignment. The second case is the specula beam damage in which case light from reflective surfaces is reflected into the eye. The reflected light can still be collimated and is as dangerous as the direct beam. Mirrors, metal surfaces, or even a wristwatch, and the like can all be the potential reflective surfaces that cause specula beam damage. The third is the diffusely reflected beam. These beams are usually diverged and are less dangerous than the previous two cases. But for high-power lasers used in material processing, even the diffusely reflected beams can cause damage to the eye and skin.

Laser beams may do harm to skin in the form of skin burning. CW high-power lasers and pulsed lasers are especially dangerous for the skin, and even a short exposure in the beam can cause serious skin burning. Specula and stray beams are also dangerous in the case of high-power lasers. Skin absorption of laser energy is wavelength and intensity dependent. Far infrared and UV light are well absorbed while visible light has relatively higher reflection and transmission. For this reason, high-power CO_2 lasers are more dangerous than Nd:YAG lasers at the same power level.

There are other potential hazards associated with laser material processing. Some of these risks are electric shock from the laser power supply, possible explosion of the pumping arc light tube, leakage of the gases and liquids used in laser system, and possible toxic vapor or powder in material processing, and the like.

Due to the potential risks in laser material processing, installation of laser material processing system should be guided by the laser safety officer, only trained and qualified personnel should be allowed to operate lasers, and safety procedures must be followed in both laser operation and laser component disposition.

Some good practices are:

Never put your eyes in the beam path.

Wear coats and suitable safety goggles in laser processing.

Minimize the hazards of reflected light: try to contain the laser light.

Post warning signs and warning signal.

Restrict access, install interlock systems and flash light to prevent accidental intrusion into the dangerous working zone.

Try to have at least two people in the processing.

Have emergency treatment close by.

Routinely check the eye and skin health of the operator.

Report any accident immediately and treat it seriously.

Laser safety eyewear is applicable only to specified wavelengths, and is not assumed to apply to those out of that range. Even under the protection of the safety eyewear, one should never look into the laser beam directly. Laser safety eyewear is specified by the *optical density* (O.D.) numbers which are defined as O.D. = $\text{Log}_{10}(I_0/I_1)$, in which I_0 is the incident light intensity and I_1 is the transmitted light intensity. Thus the higher the O.D. number the higher the decay. An O.D. of 8 at 1.06 μm means 10^8 times decay of the incident light at 1.06 μm wavelength.

ANSI standard developed by the Z-136 Committee of America National Standard Institute is the most widely accepted laser safety standard.[13] *Maximum permissible exposure* (MPE) levels to laser light, laser safety classification, and definition of safety practices for each kind of laser are included. According to the ANSI standard, lasers are divided into four classes.

- *Class 1 laser.* Laser irradiation exposure is below the levels in which harmful effects will occur. Examples are CW He-Ne laser with power much less than 10 μW Class 1 laser can also be a high-power laser that is interlocked in such a manner that the user cannot access the beam.
- *Class 2 laser.* They are low-power visible lasers that do not have enough output power to injure a person accidentally, but may produce retinal injury when stared at for a 1000 s exposure. Examples are mW level He-Ne and Argon lasers.
- *Class 3 lasers.* Medium power lasers for which direct beam exposure can produce immediate hazard.
- *Class 4 lasers.* They are lasers that not only produce a hazardous direct or specularly reflected beam but also can produce a skin exposure hazard, a fire hazard, or produce a hazardous diffuse reflection.

Most lasers used in laser material processing fall in the class 4 lasers. Detailed safety definition and practices should refer the standard in Ref 13. The ANSI laser safety standard is voluntary. Individual states and employers have their mandatory regulations. There are also mandatory regulations from the Food and Drug Administration (FDA), the Occupational Safety and Health Administration (OSHA).

34.4 LASER MATERIAL PROCESSING SYSTEMS

A laser material processing system consists of the laser source, the beam delivery system, the motion and material handling system, and the process control system. Some systems may integrate the sensing unit to improve process quality. The individual subsystems are discussed below.

34.4.1 Common Lasers Used in Lmp

There are many kinds of lasers which cover a wide range of wavelengths, power levels, pulse durations, and beam quality. Lasers can be generally divided into gas lasers, liquid lasers, and solid state lasers. Gas lasers can be further divided into neutral gas, ion, metal vapor, and molecular lasers. Table 34.2 summarized the features of common lasers. The most widely used lasers in material processing are CO_2 lasers and Nd:YAG lasers. These lasers have a wide range of laser power. CO_2 lasers can have very high CW powers, up to tens of KW, while Nd:YAG laser can have powers up to several KW. Nd:YAG system usually comes with fiber coupling which makes it very flexible in processing. Diode lasers are in rapid development. They are used in pumping of other lasers, but material processing by direct diode laser beam is now practical with over KW diode lasers commercially available. Detailed discussion of lasers can be found in many of the references of this chapter. Specific lasers relating to a process will be further described in the relevant sections.

34.4.2 Beam Delivery and Material Handling Considerations

Laser beam out of the laser source is delivered to the target by beam delivery systems. The location of the energy deposition is determined by the relative motion between the laser head and the material. Beam delivery schemes are summarized in Table 34.3.

Note that in laser material processing, some assisting gases may be used to enhance machining, protect the optics, or prevent oxidation. The gas can be integrated with the laser head in various

TABLE 34.2 Common Industrial Lasers

Type	Wavelength (nm)	General feature
CO_2	10600	Power: wide range, from several watts to tens of KW Very wide applications in laser material processing, good absorption for nonmetals.
He-Ne	632.8	Low power, CW power 0.5 mW to 50 mW High beam quality Typical application: alignment, bar code reading, image and pattern recognition, and the like.
Ion lasers	Ar 514.5, 488 Kr 647.1 Xe 995-539.5	Low power, mW to several watts. Typical application: surgery, Raman spectroscopy, and holography
Metal vapor laser	Cu: 511, 578	Pulsed, can have short pulse and high peak power Typical application: surgery, laser micromachining
Excimer lasers	XeCl 308, XeF 351, KrCl 222, KrF 248, ArF 193, F_2 157	UV wavelength, beam shape is usually rectangular, pulse width from several ns to over 100 ns, pulse energy from 1 to 1000 mJ Typical application: semiconductor and other material machining
Ruby laser	694.3	First laser used for diamond drilling, can be Q-switched, pulse energy over 1 J, pulse duration in ns and ps; hole drilling and spot welding
Nd:YAG	1064, 532, 355	Power: wide range, from mW to KW, CW, and pulsed; commonly delivered by fibers Very wide applications in laser material processing
Nd:Glass	1064	Can have very high pulse energy (>100 J) and very short pulse duration (ps and fs). Applications: pulse welding, hole drilling, shock processing and the like
Diode laser	UV to IR	Can have high CW power output (>1 KW), beam has relatively large divergence. Can be coupled with fibers, very compact in size Typical application: signal processing, pumping, and direct material processing

TABLE 34.3 Beam Delivery

Beam delivery scheme	Description	Comments
Fixed beam	Laser beam is fixed while workpiece moves on motorized stages. Optics usually contained in metal tubes.	Simple to implement, laser experiences little external disturbance
Flying optics	Relevant motion between the laser head and the workpiece is realized by moving optics, such as an inclined mirror that moves with the processing head.	Beam quality may change at different locations. This change can be compensated by adaptive optical design.
Fiber or other flexible waveguide	Laser beam is coupled into the fiber or the flexible waveguide, such flexible structure can be further mounted on robot arm.	Nd:YAG lasers, some diode lasers have fiber coupling output, CO_2 lasers can use special waveguides such as hollow metal tube. Highly flexible in moving the laser sources in 3D space.
Coordinated scanning of laser beam	Galvanometer-driven mirrors reflect/deflect the laser beam onto the desired location on the target	Mirrors can have much less mass and very high scanning speed can be achieved. Commonly used in masking, scribing, and high-speed laser drilling.

TABLE 34.4 Motion and Material Handling Schemes

Scheme	Description	Comments
Fixed workpiece and moving laser	The whole laser moves relative to the target.	Applies to small mass lasers or when the workpiece is inconvenient to move. Diode lasers, low-power CO_2 lasers and the like can use this scheme. Small work floor requirement.
Fixed laser and moving workpiece	Laser and optics are fixed, while workpiece moves on XY stage, XYZ stage or 5-axis stations.	Applies to small to medium mass workpiece, speed is limited by the mass. This scheme is most popular. It has the advantage of little external disturbance to the laser. Larger work floor requirement.
Flying optics or moving flexible waveguide and fixed workpiece	Only part of the laser beam delivery system moves relative to the workpiece.	Due to the low mass of the flying optics, high speed and high flexibility possible. Small work floor requirement.

forms, for example, concentric gas jet with the laser output, or gas jet at an angle to the target surface. Gas jet may also be outside the laser head.

Beam delivery and material handling should be an integrated part in setting up the laser material processing system. Table 34.4 summarizes the considerations of motion and material handling system. Normally linear motors, polar robot, or gantry motion systems are used to move the workpiece.

34.4.3 Sensing and Process Control

High quality laser material processing relies on the optimal control of many parameters such as power, stand-off distance or spot size, energy deposition time, speed, scanning contour, path planning, gas pressure, and direction. Suitable sensing system is needed to control the important parameters such as spot size and surface temperature which cannot be directly defined by the laser controller.

Attention should also be paid to experimentally validate the settings on the controller. For example, the nominal power is the power directly out of the laser source, not the power out of the final optics. In reality, the customers usually build up their own optics to adapt the laser source to their specific applications. The beam out of the laser source is normally expanded, homogenized, polarized and so forth, and finally focused or defocused onto the target surface to achieve desired focus spot size or surface temperature.

A mechanical contact or a distance sensor can be used to control the distance from the lens to the target. An ideal focus control system should have high spatial resolution and can operate in real time. One potential technique to reach this aim is the on-axis monitoring system making use of the light reflected back from the workpiece. Machining quality can be improved when the laser energy is suitably modulated, for example, one can modulate the laser power in laser cutting to avoid the negative effects of the edge or control the taper in laser percussion drilling.

The stability of the laser energy should be considered in carrying out the control schemes. Lasers usually cannot change their power in real time because they need some time to stabilize when the settings are changed. A good solution is to modulate the power externally while keeping the laser power at a stable level. With automation of these external power modulators, laser power can be modulated in real time.

In summary, the complete consideration to build a laser material processing system should consider the laser source, the material to be processed, the optics to achieve desired energy level and energy deposition, the material handling system, and the control scheme among many other things such as precision, floor space, and cost. It is usually important to synchronize the laser settings with the motion control, i.e., make the energy and motion talk to each other. To make 2D or 3D motion paths, the motion can be manually programmed or can be generated from CAD tools. The laser supplier should be consulted in building up the accessories of the system, and the literature can be referred to save some effort for a successful process.

34.5 LASER MACHINING PROCESSES

Laser machining processes refers to material removal processes that use laser energy directly. In this section we will discuss the laser systems, basic mechanisms, and the process capability of typical laser machining processes. Laser material removal processes require higher laser intensities than that in laser welding and laser material processing. Complex physics is involved in laser machining. However, we won't cover the modeling of these processes, which have too much content to be fitted into this chapter. Readers interested in the modeling aspects are encouraged to refer to Chapter 3 of the LMP module in Ref. 4, and other references of this chapter. In general, laser machining processes are noncontact, flexible, and accurate machining processes applicable to a wide range of materials.

34.5.1 Laser Cutting

Lasers Used in Laser Cutting. The lasers used in laser cutting are mainly CO_2, Nd:YAG, and excimer lasers. Industrial lasers for cutting typically have power levels from 50 W to 5 KW, although higher powers are used to cut thick section parts. Because CO_2 lasers have higher average powers with cheaper cost-per-watt and they also have an early history of success in industrial laser cutting, today the majority of cutting operations are carried out by CO_2 lasers, especially for nonmetals which have better absorption at far infrared wavelength. Nd:YAG laser has shorter wavelength, smaller focused spot size, and is better absorbed by metals than CO_2 lasers. Multikilowatts Nd:YAG lasers are commercially available and they usually are delivered by fibers. All these factors lead to the increasing popularity of Nd:YAG lasers in industrial laser cutting, especially for metals. Q-switched Nd:YAG lasers are dominant in pulsed laser cutting. Excimer lasers have UV wavelengths that are strongly absorbed by both metals and nonmetals, the spatial resolution are higher than visible and infrared lasers, and thus they are mainly used for high-precision laser cutting, especially for polymers and semiconductors. Recently, conventional lasers using diode pumping and direct diode lasers are reducing their size and increasing their average power quickly, which may change the dominant role of bulky conventional lasers in industrial laser cutting. For example, 1 kW direct diode lasers at 808 nm wavelength with fiber coupler are now commercially available. Although suitable for laser welding and surface treatment, they can be used in laser cutting.

In laser micromachining, a much wider variety of lasers with short pulse durations and high pulse repetition rates are used, such as frequency doubled (Green 532 nm) and tripled (UV 355 nm) Nd:YAG laser, copper vapor lasers, ultrashort pulsed lasers, and excimer lasers. The shorter wavelength and shorter pulse duration helps increase spatial resolution and reduce the heat affected zone in laser cutting, the higher pulse repletion rate at smaller pulse energy makes it easier to get a smoother machined edge. But the average power of these systems is much lower than industrial lasers, typically the powers of lasers for micromachining are less than 50 W, although higher laser intensity may be reached by using smaller focused spot size. High-power industrial lasers are commonly used to cut through larger thickness parts with sufficient speed while micromachining lasers are used to generate small features with high precision.

The laser cutting system generally consists of the laser source, the beam delivery and focusing system, the material handling system, and the process monitoring and control system. Assisting gas is commonly used in laser cutting. Selection of the beam delivery and material handling scheme depends on the type of material to be cut, the thickness and mass of the part, and the affordable investment of the cutting system. The discussion in Section 34.4 applies to the laser cutting system and will not be repeated here.

Laser Cutting Mechanisms and Quality Issues. Almost any kinds of materials can be cut with a suitable laser. To achieve successful laser cutting, the material should have sufficient absorption to the incident laser energy and the part should be within certain thickness. This thickness depends on the material type, the laser, and the process parameters. Laser cutting is mainly a thermal process in which the material absorbs the focused laser energy and gets heated, melted, and vaporized. Deep UV laser machining of polymers may also involve the photon chemical dissociation process in which the chemical bonds of the material are directly separated by individual photons that have energy comparable with the molecular bonding energy. Industrial laser cutting is mainly a thermal material removal process. The laser energy can be CW or pulsed. Thick sections are mainly cut by high-power CW lasers. Pulsed laser cutting can reduce the heat affected zone and has better control of precision features such as sharp corners.

There are traditionally three laser cutting mechanisms—laser fusion cutting, laser oxygen cutting, and laser sublimation/vaporization cutting.

In laser fusion cutting, the material is melted by the laser beam, and either a gas jet is used to blow out the molten material or a vacuum device is used to suck away the molten material. A cutting front is formed at one end of the cutting kerf—the laser supplies the energy for melting and thermal diffusion while the gas jet provides the momentum to remove the molten material. To prevent oxidation, inert gases such as argon, nitrogen, or helium are normally used.

Laser oxygen cutting applies to reactive materials such as low carbon steel and titanium. In laser oxygen cutting, the laser is used to heat the material to the point where the exothermic reaction with oxygen will begin. The material is burnt through by the chemical reaction mainly. In this process the oxygen gas jet is used. This reduces the requirements on laser power. Under the same power level, higher cutting speed and thicker section cutting can be achieved using laser oxygen cutting than laser fusion cutting.

Laser sublimation/vaporization cutting generally applies to materials with low conductivity and low latent heat of vaporization, such as organic materials. Chemical reaction with oxygen may be uncontrollable for these materials. In laser micromachining, however, this mechanism applies to a much wider range of materials, including metals and ceramics. For this mechanism, no oxygen is used and the material is vaporized or sublimated by the laser energy only. This mechanism requires highest laser power and laser intensity among the three mechanisms. Protective gas jets are commonly used to protect the lens.

Quality issues in laser cutting include recast layer, dross or attachment, redeposition, taper, heat affected zone, wall roughness and striation, possible microcracks and the like. Laser energy creates a transient high temperature field in the target, a heat affected zone remains after the processing, and the resolidification of the molten material forms a recast layer. The kerf is usually not of the strictly rectangular shape, instead a taper normally exists from the top to the bottom. The molten material may attach to the bottom of the cutting kerf and may splash over the top surface resulting in attachment and redeposition. The wall surfaces usually show striations. The surface can be very rough if not well controlled.

With suitable control of the process parameters, however, high quality cutting can be achieved. Important process parameters in laser cutting are: laser power, laser spot size, stand-off distance, focus position, scanning speed, gas pressure, gas flow rate and direction, and gas composition. The quality of laser cutting depends on both the material and the laser.

Comparison With Other Cutting Processes. Laser cutting holds the largest market share (~38 percent) of all laser applications. It has gained wide acceptance in manufacturing due to the many advantages and benefits over other competing cutting methods. Table 34.5 compares the advantages and disadvantages of popular cutting technologies. Each technology has its niche, and the user should weigh their concerns carefully when facing the choice of these processes.

Process Capability of Laser Cutting. Organic materials such as paper, rubber, plastics, cloth, wood, and inorganic materials such as ceramics and glass have better absorption at 10.6 μm than at 1.06 μm. Thus CO_2 lasers are most commonly used for nonmetal material cutting, and a CW CO_2 laser with 100 W is adequate for many of the cutting tasks. Nonmetal materials are commonly cut directly by vaporization. Inert gas may be used to prevent scorching of organic materials in laser cutting.

TABLE 34.5 Comparison of Common Cutting Processes

Processes	Advantages	Disadvantages
Mechanical cutting—punching, sawing, turning, milling and the like.	Relatively low capital cost; high material removal rate; precision cutting front control due to direct mechanical contact; good cutting surface finish and excellent cutting kerf geometry. Matured technology, best fit for bulk material removal, wide range of precision achievable.	Have tool wear; need complex fixture due to large reacting force in cutting; cutting is material dependent, some materials are very difficult to cut or simply cannot be cut; too thin and too thick materials are difficult to cut due to too delicate or too bulky structure. Aspect ratio 1:1.
Water-jet cutting	Can cut a wide range of materials using the same system, including metals, ceramics, and organic materials; very little thermal damage; can cut thick sections; high material removal rate and good surface finish; no direct mechanical tool contact, easy fixturing in cutting.	High capital cost; have tool wear; spatial resolution limited by the focusing of the water jet, may show taper in the cross-section.
Wire electro-discharge machining	Negligible cutting force; good tolerance control and can cut complex geometry; excellent edge finish; can cut thick metals.	Applies only to conductive materials such as metals; have electrode wear; relatively slow cutting speed; have a larger heat affected zone than laser cutting. Aspect ratio 1:1.
Plasma arc cutting	High cutting rate; can cut complex geometry; cut thick materials well.	Poor tolerance control, large kerf and large heat affected zone; rough cutting edge; may need post processing.
Laser cutting	Noncontact cutting, no tool wear; small cutting kerf; versatile, almost any material can be cut; negligible cutting reaction force, easy fixturing, fast setup, and rapid design change; capability to cut complex geometry easily; high cutting speed for reasonable thickness materials; high cutting quality possible at suitable parameters; more flexible than other systems, especially with flexible beam delivery; cutting, drilling and welding can be done by one system; high spatial resolution possible; small heat-affected zone; low operating cost; very high reliability and repeatability; can be easily automated.	High capital cost; relatively slow material removal rate; difficult to cut thick sections; inherently a thermal material removal process, may have some quality issues such as taper, heat affected zone, and attachment.

Fixturing is easy for laser cutting—a vacuum chuck can be used to hold the material. Table 34.6 lists some cases of nonmetal laser cutting. These data are experimental data, they give the reader some idea of the capabilities of the process but not necessarily represent the optimal processing condition.

Higher average power is needed in laser cutting of metals compared to nonmetals. CO_2 lasers are commonly used for laser cutting of metals but high-power Nd:YAG lasers are increasingly widely used, especially when equipped with fiber laser energy coupling. Table 34.7 shows some experimental results of CO_2 laser cutting of metals. These experimental data do not necessarily represent optimal processing conditions, but they provide some general idea of the process capabilities.

TABLE 34.6 CO_2 Laser Cutting of Nonmetals*

Material	Thickness (in)	Laser power (W)	Cutting speed (in/min)	Gas assist	Reference
Soda lime glass	0.08	350	30	Air	19
Quartz	0.125	500	29		20
Glass	0.125	5000	180	Yes	21
Alumina ceramic	0.024	250	28	Air	22
Plywood	0.19	350	209	Air	19
Plywood	1	8000	60	None	23
Fiberglass epoxy composite	0.5	20,000	180	None	23
Acrylic plate	0.22	50	12	Nitrogen	24
Cloth	Single pass	350	2400	None	25

*See Ref. 5.

34.5.2 Laser Drilling

Lasers Used in Laser Drilling. Laser drilling is a process by which holes are formed by the removal of material through laser beam and material interaction. Laser drilling is one of the oldest applications of laser machining processes. The first ruby laser was demonstrated for laser drilling of diamonds. Nowadays, laser drilling has found successful applications in automobile, aerospace, electronic, medical, and consumer goods industries. A well-known example of laser drilling is the drilling of airfoil cooling holes of aircraft engines.

High-power CW lasers are difficult to focus to small spot size because of their poor beam quality.

Lasers used for drilling require higher laser intensities than in laser cutting. With finite pulse energy, high laser intensity can be achieved by tight focus and by short pulse duration. Normally, pulsed Nd:YAG lasers or pulsed CO_2 lasers are used. Similar to laser cutting, CO_2 lasers are better fit for nonmetals and Nd:YAG lasers are better suited for metals. The laser pulse duration is normally less than 1 ms. The average power of the laser may not be as high as that used in laser cutting, but the achievable laser intensity is higher than laser cutting due to shorter pulse duration and smaller spot size. Lasers can be used to drill very small holes with high accuracy and high repeatability. The diameters of holes range from several microns to about 1 mm. For extremely small diameter holes, tighter focus is needed and green or UV lasers, such as frequency doubled or tripled Q-switched Nd:YAG lasers, are used.

When the pulse duration is short and the pulse repetition rate is high, laser can drill when the part is moving. Thus very high drilling speed is possible. Laser drilling system may take all schemes discussed in Section 4 and will not be repeated here.

TABLE 34.7 Experimental Results of CO_2 Laser Cutting of Metals, Oxygen Assisted*

Metal	Thickness (in)	Power (W)	Cutting speed (in/min)	Reference
Titanium	0.67	240, O_2 assist	240	19
Stainless steel 410	0.11	250, O_2 assist	10	26
Rene 41	0.02	250, O_2 assist	80	26
Aluminum alloy	0.5	5,700	30	25
Steel 304	1.0	15,000	20	27
Titanium	0.25	3000	140	25
Titanium	2.0	3000	20	25
Rene 95	2.2	18,000	2.5	27

*See Ref. 5.

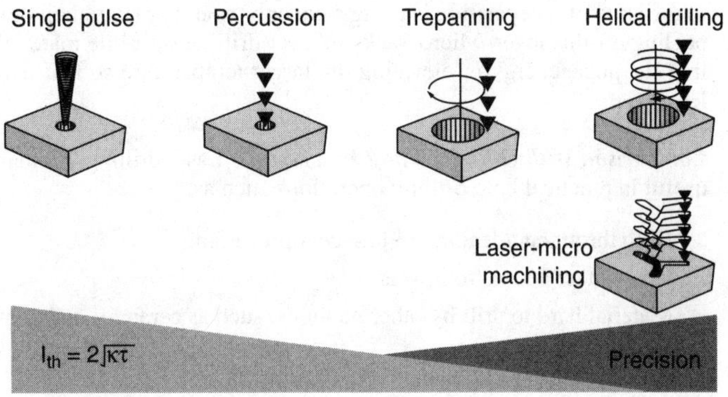

FIGURE 34.4 Various techniques in laser drilling.[12]

Laser Drilling Mechanisms and Quality Issues. In laser hole drilling, the high-intensity laser beam is focused on the target surface or slightly under the surface. The material is quickly heated over its vaporization temperature, and is removed (ablated) through direct vaporization or removed in bulk molten droplets. Figure 34.4 illustrates various drilling techniques—*single pulse drilling, percussion drilling, trepanning, and helical drilling*. When the target is thin relative to the available pulse energy, a single pulse can drill through the material. This is the case for thin film drilling, thin foil drilling, or thin plate hole drilling. Percussion drilling is widely used when one pulse cannot drill through the sample. In this case, consecutive laser pulses with pulse duration normally less than 1 ms are applied at the same location until the hole is drilled through. Percussion drilling is commonly used in cooling hole drilling of aircraft engines. Pulsed lasers can have high repetition rate. Thus using single pulse drilling or percussion drilling, thousands of small holes can be drilled in a short period compared to mechanical drilling and EDM drilling. But the diameters of holes are limited to the focused spot size, which should be small enough to gain high enough laser intensity.

Trepanning is the standard technique for drilling of larger holes, such as holes over 500 μm in diameter. It is essentially a percussion drilling process followed by a cutting procedure. Using this technique, noncircular geometry can be easily realized. The application of nanosecond pulses to trepanning can increase the quality of drilling.

All of the three techniques will generate an inherent taper along the thickness section, although under proper conditions this tapering issue is not serious. To decrease the taper, the helical drilling technique can be used. In this method, the material is gradually drilled through, not drilling at each location and followed by contour cutting. This method can be used to machine out a blind feature or drilling out a larger thickness target that is impossible for trepanning.

It's important to protect the focusing lens in laser drilling because the ablated material may contaminate the lens and cause damages. A shielding gas jet is commonly used to blow away the ablated material and a protective flat glass plate can be attached in front of the lens.

Quality issues in laser drilling include: taper; deviation from the circular or desired geometry; redeposition of ablated material around the hole; microcracks due to thermal stress, especially in drilling of brittle materials. Laser can drill holes with height to diameter ratios of up to 50. At low height to diameter ratios, tapering is not an issue but when the aspect ratio is high, taper can be a concern. Because material is removed dynamically in gas and liquid form, the geometry may show deviation from circular or desired geometry. With good beam quality, however, the geometry can be very close to circular, the wall normally shows roughness less than 5 μm, and the process can be very accurate and repeatable. Redeposition of ablated material is due to the fact that a large fraction of material is ablated in bulk liquid form instead of direct vaporization or sublimation. To decrease redeposition, shorter pulses, such as nanosecond or even picosecond and femtosecond pulses instead of microsecond pulses, can be used. But keep in mind that the average power of shorter pulse lasers may be lower and the drilling rate is usually lower than longer pulses. An alternative solution to the issue of redeposition is using a cover or coating material on top of the target, and after drilling, peeling off this layer. Microcracks in laser drilling of brittle materials can be alleviated by controlling the pulse energy or elevating the target temperature so that temperature gradient in drilling is less steep.

Comparison With Other Drilling Processes. Laser drilling has many advantages that make it very useful in practical hole drilling operations such as:

High throughput leading to low-cost processing

Noncontact and no tool wear

Material hard to drill by other methods, such as ceramics and gemstones can be drilled with high quality

Heat affected zone is small around the hole

Smaller holes can be drilled in thin materials

Capacity for a high degree of beam manipulation, including the ability to drill at shallow angles and to drill shaped holes

Highly accurate and consistent process quality

Can be easily automated

The same laser system can be used for multiple purposes such as cutting, drilling, and marking.

It is economical to drill relatively small holes that can be drilled through by lasers in a short period. Larger diameter holes can be drilled by mechanical method. Aspect ratio >25 is usually a challenge for laser drilling, drilling of thick sections can be very difficult due to multiple reflection and the limited depth of focus of the laser beam. Table 34.8 compares laser drilling with its major competing processes, namely mechanical drilling and EDM drilling.

TABLE 34.8 Comparison of Laser Drilling With Mechanical Drilling and EDM Drilling

Process	Advantages	Disadvantages
Mechanical drilling	Matured process for large and deep hole drilling; high material removal rate; low equipment cost; straight holes without taper; accurate control of diameter and depth. Applicable to wider range of materials than EDM but narrower range of materials than laser drilling. Typical aspect ratio 1.5:1.	Drill wear and breakage; low throughput and long setup time; limited range of materials; difficult to drill small holes and high aspect ratio holes; difficult for irregular holes.
Electrical discharge machining	Large depth and large diameter possible; no taper; low equipment cost; can drill complex holes. Mainly applicable to electrical conductive materials. Typical aspect ratio 20:1.	Limited range of materials; slow drilling rate; need to make tools for each type of hole, long setup time; high operating cost.
Laser drilling	High throughput; noncontact process, no drill wear or breakage; low operating cost; easy fixturing and easy automation; high speed for small hole drilling; high accuracy and high consistency in quality; easy manipulation of drilling location and angle; complex geometry possible; high quality and thick depth in drilling of many nonmetal materials. Applicable to a very wide range of materials. Typical aspect ratio 10:1.	Limited depth and not economical for large holes; hole taper and material redeposition for drilling of metals; high equipment cost.

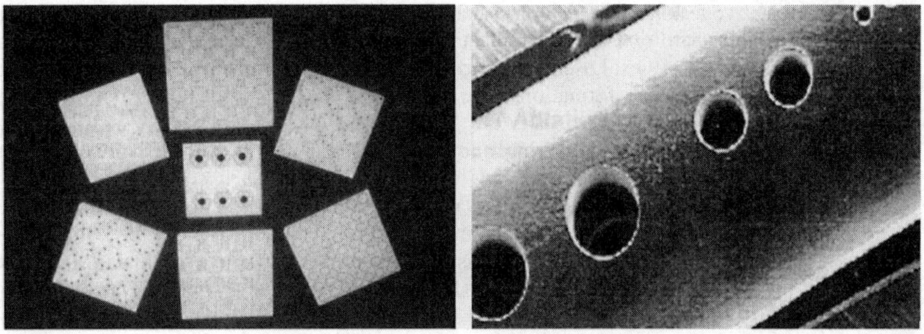

FIGURE 34.5 Left: Examples of patterns of laser-drilled holes in aluminia ceramics substrates (*Photograph courtesy of Electro Scientific Industries, Inc.*); Right: Cylindrical holes (25 μm, 100 μm, 200 μm) in catheter (*Illy, Elizabeth K, et al., 1997*).

Process Capability of Laser Drilling. Lasers can drill very small holes in thin targets with high speed. Many of the applications for laser hole drilling involve nonmetallic materials. A pulsed CO_2 laser with an average power of 100 W can effectively drill holes in many nonmetallic materials with high throughput. Laser drilling of nonmetallic materials tends to have higher drilling quality than metals because nonmetallic materials are normally less conductive and are easier to be vaporized. Laser drilling of metals may have the quality issues of taper, redeposition, and irregular geometry. Both CO_2 and Nd:YAG lasers are commonly used for drilling of metals. Nanosecond lasers or even shorter pulsed lasers are used to drill metals in order to alleviate the quality issues. Figure 34.5 shows examples of laser-drilled holes.

Holes from about 0.008 in (0.2 mm) to 0.035 in (0.875 mm) can be typically percussion drilled in material thickness of up to 1.00 in with standard high-power drilling lasers. The longest possible focal length should be chosen for materials thicker than 0.15 in. Smaller diameter holes can be drilled with green or UV lasers. Larger holes can be drilled by trepanning or helical drilling.

Lasers can drill special geometry holes easily. The laser beam can be programmed to contour out the specified geometry. Lasers are also good at drilling holes on slant surfaces, which can be difficult for mechanical methods. Lasers can be flexibly manipulated to drill holes on 3D surfaces or reflected to drill difficult-to-reach areas. The taper in laser drilling is normally within 2 degrees, and the edge finish normally varies within 5 μm. The aspect ratio in laser drilling can be over 20:1. The maximum depth of laser drilling for both CO_2 and Nd:YAG lasers is summarized in Table 34.9.

34.5.3 Laser Marking and Engraving

Lasers for Marking and Engraving. Laser marking is a thermal process that creates permanent contrasting marks in target materials by scanning or projecting intense laser energy onto the material. In some cases, the target is removed a shallow layer to make the marks, while in other cases, strong laser irradiation can create a color contrasting from nonirradiated area. Lasers are also used to

TABLE 34.9 Capabilities of Laser Drilling

Materials	CO_2 lasers	Nd:YAG lasers
Aluminum alloy	6.25 mm	25 mm
Mild steel	12.5 mm	25 mm
Plastics	25 mm	Not applicable
Organic composite	12.5 mm	Not applicable
Ceramics	2.5 mm	Not applicable

engrave features into materials such as wood or stone products. Laser marking holds around 20 percent market share of all laser applications and represents the largest number of installations among all laser applications. Lasers can mark almost any kind of material. Laser marking can be used for showing production information, imprinting complex logos, gemstone identification, engraving artistic features, and the like.

Lasers used for marking and engraving are mainly pulsed Nd:YAG lasers, CO_2 lasers, and excimer lasers.

In general, there are two fundamental marking schemes: one is marking through beam scanning or direct writing, and the other is marking through mask projection. In beam scanning or direct writing method, the focused laser beam is scanned across the target, and material is ablated as discrete dots or continuous curves. XY-tables, flying optics, and galvanometer systems are commonly used, and galvanometer systems turn out to be the most powerful. In the mask projection method, a mask with desired features is put into the laser beam path. Laser energy is thus modulated when it passes through the mask and a feature is created on the target. The mask can contact the target directly or can be away from the target and be projected onto the target by optics. The features in the mask projection method are usually produced with only one exposure. This mask projection method has been used in IT industry to produce very minute and complex features with the assistance of chemical etching. Beam scanning marking has more flexibility than mask projection marking while mask projection marking can be much faster than beam scanning marking.

Q-switched Nd:YAG lasers and excimer lasers are commonly used for beam scanning marking and CO_2 lasers operating in the range of 40 to 80 W are used to engrave features in wood and other nonmetallic materials. CO_2 TEA lasers and excimer lasers are widely used in mask projection laser marking.

Comparison With Competing Processes. Laser marking has proven to be very competitive with conventional marking processes such as printing, stamping, mechanical engraving, manual scribing, etching, and sand blasting. Beam scanning laser marking system is very flexible, it is usually highly automated, and can convert digital information into real features on any material immediately. Mask projection laser marking systems are very efficient. One can consider laser marking as a data driven manufacturing process. It's easy to integrate a laser marking system with the database, and the database has the same role as the tooling in conventional marking processes.

Compared to other marking systems, laser marking demonstrates high speed, good performance, and high flexibility, along with many other advantages, and the only downside seems to be the initial system cost. However, many practical examples show that the relatively higher initial investment in laser marking system can gain their payback in a short term. For example, an automobile and aerospace bearing manufacturer previously utilized acid-etch marking system to apply production information on the bearing. Turning to a fully automated laser marking system reduced the per piece cost by 97 percent, and the consumable and disposal materials were eliminated. In another case, a company needs to ensure close to 100 percent quality marking on the products, but failed to do so using the print marking method, which may have had problems of outdated information or poor quality of printing. Turning to laser marking, the quality is ensured and the marking information is directly driven by the production management database.

In summary, the advantages of laser marking include:

High speed and high throughput

Permanent and high quality features

Very small features easily marked

Noncontact, easy fixturing

Very low consumable costs, no chemistry, and no expendable tooling

Automated and highly flexible

Ability to mark wide range of materials

Digital based, easy maintenance

Reliable and repeatable process

FIGURE 34.6 Laser marking examples. (Left) A PC keyboard; (Middle) graphite electrode for EDM; and (Right) Laser marking of electronic components. (*Courtesy of ALLTEC GmbH Inc.*)

Environmental friendly, no disposal of inks, acids, or solvents

Low cost of operation.

Figure 34.6 shows some examples of laser marking.

34.6 REVIEW OF OTHER LASER MATERIAL PROCESSING APPLICATIONS

Laser energy is flexible, accurate, easy to control, and has a very wide range of freedom in spatial, temporal, magnitude, and frequency control. This unique energy source has found extraordinarily wide applications in material processing. In this section, we will review some important applications other than the more well-known processes described in previous sections.

34.6.1 Laser Forming

When a laser beam scans over the surface of the sheet metal and controls the surface temperature to be below the melting temperature of the target, laser heating can induce thermal plastic deformation of the sheet metal after cooling down without degrading the integrity of the material. Depending on target thickness, beam spot size and laser scanning speed, three forming mechanisms or a mixture of the mechanisms can occur. The three mechanisms are the *temperature gradient mechanism* (TGM), the *buckling mechanism* (BM), and the *upsetting mechanism* (UM).[14] Lasers used in laser forming are high-power CO_2 lasers, Nd:YAG lasers, and direct diode lasers.

Laser forming (LF) of sheet metal components and tubes requires no hard tooling and external forces and therefore is suited for dieless rapid prototyping and low-volume, high-variety production of sheet metal and tube components.[15] It has potential applications in aerospace, shipbuilding, automobile, and other industries. It can also be used for correcting and repairing sheet metal components such as prewelding "fit-up" and postwelding "tweaking." Laser tube bending involves no wall thinning, little ovality and annealing effects, which makes it easier to work on high work-hardening materials such as titanium and nickel super-alloys. LF offers the only promising dieless rapid prototyping (RP) method for sheet metal and tubes. Figure 34.7 shows pictures of laser-formed sheet metal and tubes. With strong government support and active research work, laser forming of complex 3D shape will be feasible in the near future.

34.6.2 Laser Surface Treating[5]

Lasers have been used to modify the properties of surfaces, especially the surfaces of metals. The surface is usually treated to have higher hardness and higher resistance of wear.

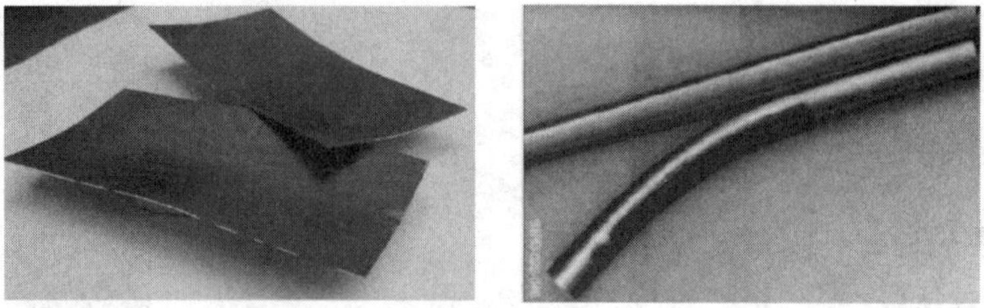

FIGURE 34.7 Laser forming of sheet metals and tubes. *(Courtesy of MRL of Columbia University and NAT Inc.)*

Laser Hardening. In laser hardening, a laser beam scanning across the metal surface can quickly heat up a thin top layer of the metal during laser irradiation, and after the irradiation it quickly cools down due to heat conduction into the bulk body. This is equivalent to the quenching process in conventional thermal treating. When favorable phase transformation occurs in this laser quenching process, such as in the case of carbon steels, the top surface hardness increases strikingly. Laser hardening involves no melting. Multikilowatt CO_2 lasers, Nd:YAG lasers, and diode lasers are commonly used. The hardened depth can be varied up to 1.5 mm and the surface hardness can be improved by more than 50 percent. Laser hardening can selectively harden the target, such as the cutting edges, guide tracks, grooves, interior surfaces, dot hardening at naps, and blind holes. The neighboring area can be uninfluenced during laser hardening. By suitable overlapping, a larger area can be treated.

Laser Glazing. In laser glazing, the laser beam scans over the surface to produce a thin melt layer while the interior of the workpiece remains cold. Resolidification occurs very rapidly once the laser beam passes by, thus the surface is quickly quenched. As a result, a surface with special microstructure is produced that may be useful for improved performance such as increased resistance to corrosion. The surface layer usually has finer grains and may even be amorphous. Laser glazing of cast iron and aluminum bronze has demonstrated much enhanced corrosion resistance.

Laser Alloying. In laser alloying, powders containing the alloying elements are spread over the workpiece surface or blown over to the target surface. By traversing the laser beam across the surface, the powder and the top surface layer of the workpiece melt and intermix. After resolidification, the workpiece has a top surface with alloying elements. Surface alloying can produce surfaces with desirable properties on relatively low cost substrates. For example, low carbon steel can be coated with a stainless steel surface by alloying nickel and chromium.

Laser Cladding. Laser cladding normally involves covering a relatively low performance material with a high-performance material in order to increase the resistance to wear and corrosion. In laser cladding, the overlay material is spread over the substrate or continuously fed to the target surface. Laser beam melts a thin surface layer and bonds with the overlay material metallurgically. The difference with laser alloying is that the overlay material doesn't intermix with substrate. Cladding allows the bulk of the part to be made with low cost material and coat it with a suitable material to gain desired properties. Good surface finish is achievable. Compared to conventional cladding processes, such as plasma spraying, flame spraying, and tungsten-inert gas welding, laser cladding has the advantage of low porosity, better uniformity, good dimensional control, and minimal dilution of the cladding alloy.

34.6.3 Laser Shock Processing or Laser Shock Peening (LSP)

High intensity (>GW/cm^2) laser ablation of materials generates plasma that has high temperature and high pressure. In open air, this pressure can be as high as sub GPa and the expansion of such high-pressure plasma imparts shock waves into the surrounding media. With the assistance of a fluid layer which confines the expansion of the plasma, 5 to 10 times stronger shock pressure can be induced. This multi-GPa shock pressure can be imparted into the target material and the target is thus laser shock peened. Laser shock processing can harden the metal surface and induce in-plane compressive residual stress distribution. The compressive residual stress refrains from crack propagation and greatly increases the fatigue life of treated parts. Compared to mechanical shot peening, LSP offers a deeper layer of compressive residual stress and is more flexible, especially for irregular shapes. It has been shown that LSP can improve fatigue life of aluminum alloy by over 30 times and increase its hardness by 80 percent.[16,17] Materials such as aluminum and aluminum alloys, iron and steel, copper, and nickel have been successfully treated. Laser shock processing has become the specified process to increase the fatigue lives of aircraft engine blades.

Conventional laser shock processing requires laser systems that can produce huge pulse energy (>50 J) with very short pulse duration (<50 ns), and Q-switched Nd:YAG lasers are commonly used. Such laser systems are expensive and the repetition rate is low (several shots per minute). Historically this has restricted the wider application of LSP in industry. This situation is improving with more and more cheaper high-power systems becoming commercially available. On the other hand, this technique can be extended to low pulse energy lasers with short pulse duration and tight focus. Two key requirements for a successful processing are the over GW/cm^2 laser intensity and short enough pulse duration (<50 ns). Microscale LSP using micron-sized laser beam has been developed and has been successfully applied to microcomponents. Microscale LSP has higher spatial resolution, is more flexible, and is low cost to implement. It is shown that the copper sample treated by a UV laser with 50 ns pulse can be increased by more than 300 percent.[18]

34.6.4 Other Applications

There are many other laser material processing applications in which difficult problems are solved by lasers, such as laser assisted machining of super-alloys and ceramics, laser assisted etching, laser surface cleaning, and laser coating removal. In laser assisted machining, laser is used to locally heat the work material prior to the cutting tool in an attempt to improve machinability of difficult-to-machine materials such as supper alloys and ceramics. It has been experimentally shown that laser assisted machining can extend the tool life, increase the removal rate, and also improve the surface quality of the machined surface. Etching rate is sensitive to temperature, thus laser beam can be used to enhance etching rate locally. This is in fact one way of direct writing. With the combination of laser heating and chemical etching, semiconductor devices can be etched 10 to 100 times faster using laser assisted chemical etching than that with conventional procedures. Laser induced shock wave can be used to clean very minute particles on a silicon wafer, and laser ablation has also been used to remove rust or peel off coatings. In these applications, only a very thin surface layer is affected. Lasers are indispensable energy sources in the majority of *rapid prototyping manufacturing* (RPM) and *rapid tooling* (RT) manufacturing systems. In RPM and RT, laser energy is used to cure the liquid material, melt solid material, or cut the contour of laminated material, and then manufacture complex 3D parts layer by layer. All these are possible because laser energy can be accurately controlled spatially and temporally by digital information.

34.7 CONCLUDING REMARKS

Laser material processing (LMP) processes have become indispensable engineering solutions in many cases. We have seen many dazzling applications of lasers. These processes are still in dynamic evolution due to the dynamic progress of laser sources. High initial capital cost is one of the major obstacles in choosing the laser material processing processes. This situation may change in the future. High-power lasers already have the same order of output power as mechanical systems (15 KW), and higher processing rates will be feasible with lower capital costs. Diode lasers offer great potential for increased power and lower costs. If the beam quality can be improved, diode lasers may change the world of material processing drastically. Extensive research work on LMP is going on throughout the world and the reader is encouraged to explore these processes by going to the references and browsing the world wide web. One has good reasons to expect seeing a constantly and fast improving world of laser material processing, such as higher machining rates, deeper holes, thicker section cutting, improved thermal coupling, and much improved quality.

REFERENCES

1. Svelto, O., *Principles of Lasers,* 4th ed., Plenum Press, New York, 1998.
2. Luxaon, J. T., and David E. Parker, *Industrial Lasers and Their Applications,* Prentice-Hall, Englewood Cliffs, New Jersey, 1985.
3. Schuocker, D., *High Power Lasers in Production Engineering,* Imperial College Press, London, 1999.
4. Zhang, W., Y. L. Yao, and Jin Cheng, Manufacturing Research Lab. of Columbia University, http://www.columbia.edu/cu/mechanical/mrl/ntm/index.html.
5. Ready, J. F., *Industrial Applications of Lasers,* 2d ed., Academic Press, San Diego, 1997.
6. Niku-Lari, A., and B. L. Mordike, *High Power Lasers,* Pergamon Press, Oxford, 1989.
7. Mordike, B. L., *Laser Treatment of Materials,* DGM Göttingen, Germany, 1987.
8. Chryssolouris, G., *Laser Machining,* Springer-Verlag, New York, 1991.
9. Steen, W. M., *Laser Material Processing,* 2d ed. Springer-Verlag, London, 1994.
10. Nikogosyan, D. N., *Properties of Optical and Laser-Related Materials, A Handbook,* Wiley, Chichester, England, New York, 1997.
11. Madelung, O., *Landolt-Börnstein Numerical Data and Functional Relationships in Science and Technology,* Springer-Verlag, London, 1990.
12. Dausinger, F., "Drilling of High Quality Micro Holes," *ICALEO'2000,* Section B, pp. 1–10, 2000.
13. Standard Z136.1, *Safe Use of Lasers,* Laser Institute of America, Orlando, FL, 1993.
14. Vollertsen, F., "Mechanism and Models for Laser Forming," *Laser Assisted Net Shape Engineering, Proceedings of the LANE'94,* Vol. 1, pp. 345–360, 1994.
15. Li, W., and Y. L. Yao, "Laser Bending of Tubes: Mechanism, Analysis, and Prediction," *ASME Trans. J. of Manufacturing Science and Engineering,* Vol. 123, No. 4, pp. 674–681, 2001.
16. Clauer, A. H., and J. H. Holbrook, "Effects of Laser Induced Shock Waves on Metals," in *Shock Waves and High Strain Phenomena in Metals-Concepts and Applications,* New York, Plenum, 1981, pp. 675–702.
17. Peyre, P., X. Scherpereel, L. Berthe, and R. Fabbro, "Current Trends in Laser Shock Processing," *Surface Engineering,* Vol. 14, No. 5, pp. 377–380, 1998.
18. Zhang, W., and Y. L. Yao, "Micro-Scale Laser Shock Processing of Metallic Components," *ASME Trans. J. of Manufacturing Science and Engineering,* Vol. 124, No. 2, pp. 369–378, May 2002.
19. Harry, J. E., and F. W. Lunau, *IEEE Trans. Ind. Appl.* IA-8, 418, 1972.
20. Feinberg, B., *Mfr. Eng. Development,* December 1974.
21. Chui, G. K., *Ceramic Bull.*, 54:515, 1975.
22. Longfellow, J., *Solid State Tech.,* August 1973, p. 45.
23. Locke, E. V., E. D. Hoag, and R. A. Hella, *IEEE J. Quantum Electron,* QE-8, 132, 1972.
24. Appelt, D., and A. Cunha, "Laser Technologies in Industry" (O. D. D. Soares, ed.), *SPIE Proc.,* Vol. 952, Part 2, SPIE, Bellingham, WA, 1989.
25. Wick, D. W., *Applications for Industrial Laser Cutter Systems,* SME Technical Paper MR75-491, 1975
26. Williamson, J. R., "Industrial Applications of High Power Laser Technology," *SPIE Proc.,* Vol. 86, SPIE, Palos Verdes Estates, CA, 1976.
27. Charschan, S. S., ed., *Guide to Laser Materials Processing,* Laser Institute of America, Orlando, FL, 1993.

CHAPTER 35
LASER WELDING

Leonard Migliore
Coherent Inc.
Santa Clara, California

35.1 MECHANISM

Light carries energy. This energy is typically converted into heat when it is absorbed by a material. Even metals, which are good reflectors of light, can be heated to their melting points with enough applied power, creating a weld. There are two types of generally available lasers that have enough power to melt metals—the carbon dioxide (CO_2) laser and the neodymium-doped yttrium aluminum garnet (Nd:YAG) laser. These lasers have somewhat different characteristics and tend to be used for different kinds of welding.

35.1.1 Keyhole Welding With CO_2 Lasers

CO_2 lasers operate in the mid-infrared at a wavelength of 10.6 μm. Most metals reflect more than 95 percent of light at this wavelength, so it would seem that CO_2 lasers would be very inefficient heat sources for welding. This is not the case because CO_2 lasers have so much power and can be so tightly focused that even the small percentage that is absorbed is enough to melt metal. An applied irradiance of at least 10^6 W/cm^2 is required to initiate this sequence, shown in Fig. 35.1. Once the metal is liquid, its absorption increases greatly (generally to about 50 percent) so it is quickly raised to the boiling point. The metal vapor creates a channel (called a keyhole) below the surface of the workpiece. The light then bounces off the walls of the hole many times and is also absorbed in the metallic vapor that fills the keyhole, resulting in extremely high absorption of the incoming power. Energy is transferred to the workpiece through the entire depth of the keyhole, which can be quite deep (at least 10 mm with a 12 kW laser). This phenomenon allows single-pass full-penetration welds to be made at high speed through metals up to 15 mm thick (electron beam welders, using the same phenomenon, are able to weld even deeper because they operate in vacuum and can deliver more power).

35.1.2 Conduction Welding With Nd:YAG Lasers

If a Nd:YAG laser beam with an irradiance of at least 10^4 W/cm^2 impinges on a metal surface, the absorbed energy is sufficient to raise the surface temperature to the melting point in milliseconds. Continued application of power causes the melt front to go deeper into the material, creating a roughly hemispherical fusion zone. The Nd:YAG laser, operating at 1.06 μm in the near infrared, is very effective in this mode. Many metals have appreciable absorption of 1 μm light, so surface heating is reasonably efficient. More importantly, the change in absorption between solid and liquid is not as great for Nd:YAG as it is for CO_2 lasers. This allows for a stable process.

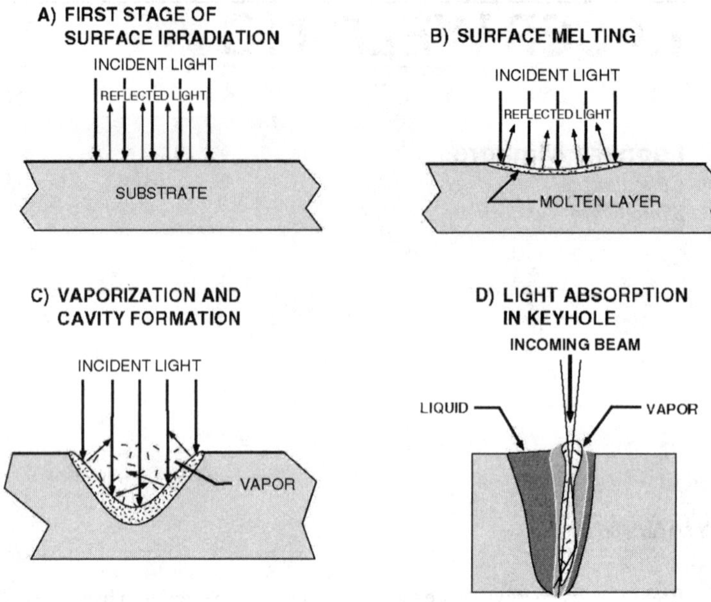

FIGURE 35.1 Sequence of events in keyhole welding.

35.2 IMPLEMENTATION OF LASER WELDING

35.2.1 Focusing

The output beams of high-power lasers range in diameter from 5 to 50 mm. Achieving suitable irradiance requires that the beams be focused to spots in the range of 100 to 800 μm. Lenses are used for focusing Nd:YAG lasers and small CO_2 lasers in configurations similar to those shown in Fig. 35.2. High powered CO_2 laser beams are focused with metal mirrors, as indicated in Fig. 35.3, since no sufficiently durable lens material exists.

The light from Nd:YAG lasers may be delivered through silica fibers (there are no practical fiber materials that work at the wavelength of CO_2 lasers). The working diameters of the fibers range from 200 to 600 μm. The laser beam is first focused into the fiber by a lens. The fiber can then transmit the beam with negligible losses for 100 meters or more. When the laser light exits the fiber it may be reimaged by another lens as shown in Fig. 35.4. Magnification is typically 1×, although 0.5× demagnification is sometimes used to achieve smaller spot sizes.

Because laser welding requires a high irradiance, it is important that the materials being welded are in the focal range of the lens. Depth of focus ranges from ±5 to ±0.2 mm depending on the exact configuration of the optical system. In general, smaller spots have a smaller depth of focus.

35.2.2 Gas Shielding in Laser Welding

While laser beam welding may be done in vacuum or in inert gas atmospheres, it is generally most convenient to operate in air. As with other fusion welding processes, laser beam welding in air requires a cover gas to protect the metal from oxidation. Welding systems that use lenses typically

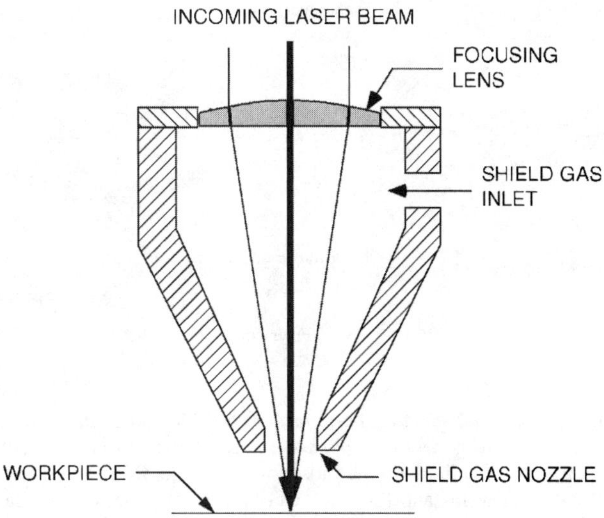

FIGURE 35.2 Schematic drawing of laser welding head using a focusing lens.

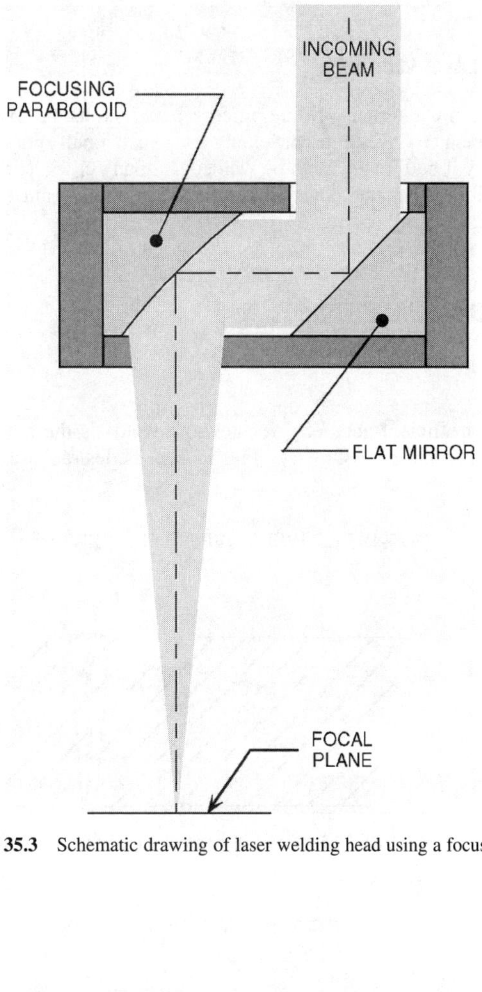

FIGURE 35.3 Schematic drawing of laser welding head using a focusing mirror.

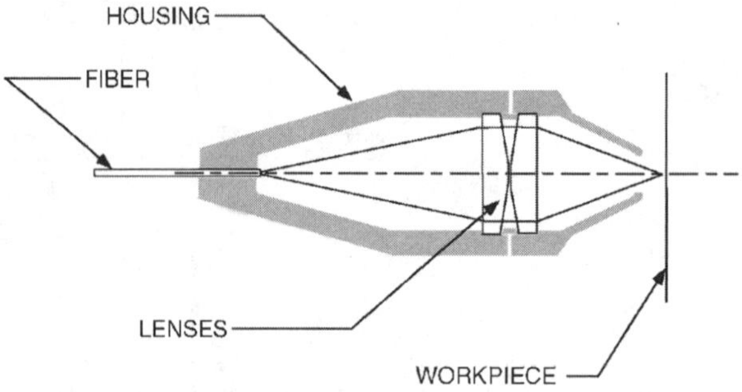

FIGURE 35.4 Fiber optic focusing head.

deliver the cover gas to a hood around the lens, allowing it to flow out through a coaxial nozzle. Mirror systems generally apply cover gas off-axis through a gas lens.

Argon and helium are the most common shield gases for laser welding. Argon is used for almost all Nd:YAG welds and for some CO_2 welds. It ionizes rather easily, which causes a problem with high-power CO_2 beams. Long-wave light is absorbed strongly by the argon plasma, reducing the power that gets to the metal. Helium, which does not ionize as easily as argon, is used for high-power CO_2 welds.

35.2.3 Fixturing for Laser Welding

In common with other welding processes, parts must be fixtured so that they are held securely in the proper geometry. Since laser welding uses such small spot sizes, the fixturing must be very accurate, especially if butt joints are to be welded. In many cases, parts are press-fit prior to welding to guarantee good fitup. Another constraint on fixturing is that it must keep the weld seam at the proper distance from the focusing optics.

35.3 LASER WELD GEOMETRIES

35.3.1 Butt Welds

The most efficient geometry for keyhole welds is the butt joint, where two edges of material are brought in contact as shown in Fig. 35.5. It is desirable to have the edges fit together with no gaps

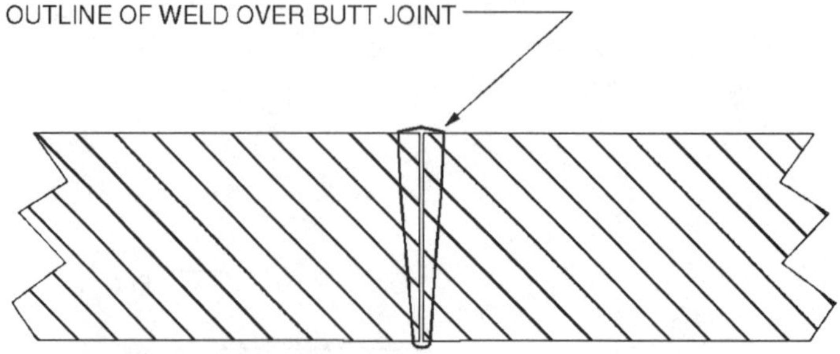

FIGURE 35.5 Cross section of butt joint with superimposed weld fusion zone.

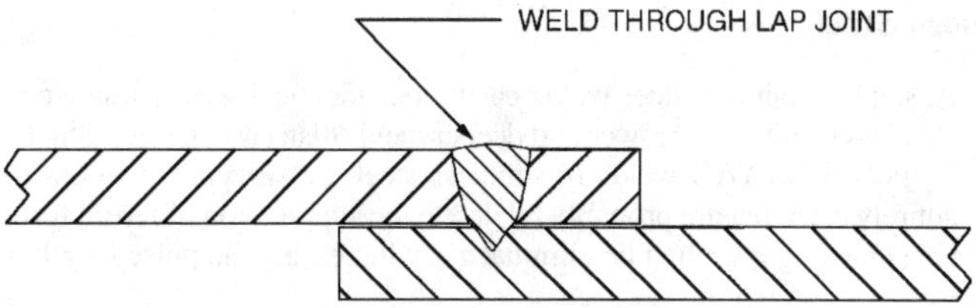

FIGURE 35.6 Cross section of welded lap joint.

because laser welding is fastest when no filler metal is used. The downside of a butt joint is that it requires very high accuracy in beam delivery—if the laser's focal spot is 200 μm, you have about $\pm$100 μm tolerance in its location with respect to the seam. This is not hard to achieve with circular parts such as transmission gears because their diameters can be well controlled but is much harder with sheet metal parts.

35.3.2 Lap Welds

The problem of aligning the focal spot to the seam is greatly reduced by adopting a lap joint (Fig. 35.6). Now, the only requirement is that the beam impinges on an area where the parts are in contact. The disadvantage of the lap joint is that most of the fusion zone is of no structural use; all that's holding the parts together is the weld cross-section at the seam.

35.4 CHARACTERISTICS OF METALS FOR LASER BEAM WELDING

While all the rules of welding metallurgy apply to laser welding, there are some areas where implementation is different. These arise from the optical properties of metals and from the high heating and cooling rates characteristic of high energy density processes. Also, the very high temperatures that exist in the keyhole vaporize volatile constituents. If the volatile materials are impurities, the weld nugget is refined. Materials such as zinc and magnesium escape from the melt pool, leaving the nugget depleted. Since laser welding is mostly autogenous, some normally weldable alloys cannot be laser welded.

35.4.1 Carbon Steel

Low carbon steels are readily laser welded. As with most welding processes, better grades of steel give better results. The fusion zone becomes very hard if the steel has more than 0.30 percent carbon, and cracking may occur. Preheating eliminates this by preventing the martensitic transformation responsible for the hardening. Because of the low heat input from the laser, unusually high preheat temperatures are required. It is not unusual to need a 500°C preheat for a satisfactory weld in 1040 steel. Postheating is typically of no use.

Many sheet steels are coated. In the automotive industry, galvanized sheet is being used to improve the corrosion resistance of car bodies. Zinc is a damaging impurity in steel welds because it has a low boiling point. If galvanized steel is lap welded, the zinc vapor exits through the keyhole, pushing steel ahead of it. The resulting welds have gross porosity. Attempts have been made to suppress this by coating the weld region with FeO (to form ZnO, a nonvolatile oxide) and by spacing the sheets apart during welding to allow the zinc vapor to exit. Both methods work but are difficult to implement in production.

35.4.2 Stainless Steel

Austenitic stainless steel welds easily. For identical weld parameters, the low thermal conductivity of stainless results in narrower and deeper welds than carbon steel. Shield gas is not required in some cases.

Pulsed Nd:YAG welds in stainless steel are susceptible to cracking. The reasons for this are not entirely clear, but are probably related to the suppression of ferrite formation from rapid cooling. In practice, cracking can often be eliminated by increasing the pulse length, which reduces the cooling rate.

35.4.3 Aluminum

Common aluminum alloys such as 6061, 5052, and 5086 are not very suitable for laser welding because they must be welded with alloy additions to prevent cracking. While it is possible to insert shims into butt joints and thus make satisfactory welds, the cost involved in preparing such joints makes it an uncommon procedure. The aluminum-lithium alloys currently being examined for aerospace use seem to benefit from rapid heating and cooling of the fusion zone, so that they may require laser welding.

35.4.4 Copper

Copper is highly reflective to infrared light. This, along with its high thermal conductivity, makes it difficult to weld with a laser. The most successful tool for welding copper is a pulsed Nd:YAG laser that can deliver a "spiked" pulse that has a very high initial power.

35.4.5 Titanium

As long as oxygen and nitrogen are excluded, titanium is an excellent material for laser welding. Since it is very easy to set up a sealed work area in a laser welder, the process should be common. In fact, there is very little laser welding of titanium being done. A possible reason is that most titanium is used in aerospace applications, and specifications that allow laser welding are uncommon. Electron beam welding is approved by many specifications, and, since it is done in a vacuum, contamination of the fusion zone is unlikely.

35.5 LASER WELDING EXAMPLES

Lasers are used for high-speed welding of steel with thicknesses ranging from 0.5 to 10 mm. While lasers can weld up to 25 mm under laboratory conditions, electron beam welders begin to be more cost effective beyond 10 mm. The automotive industry is a major user of laser welding because it requires large numbers of parts—the initial expense of a laser welding system is rapidly covered by its high productivity. Transmissions, sheet metal, cam followers, and numerous other assemblies are laser welded by the millions.

Nd:YAG lasers are used for precision spot welding, primarily in the electronics industry. Disc drive suspensions have been laser welded for decades. Multiblade razors are held together with laser spot welds.

35.6 LASER WELDING PARAMETERS

35.6.1 Power

Welding is melting metal. To melt more metal per unit time, you need more power. For a given average power level, there is a maximum weld penetration. At a given penetration, higher power allows greater travel speeds. Large (5 to 9 kW) CO_2 lasers have proven to be economical for a wide range of weld penetrations.

For precision welding, pulsed Nd:YAG lasers can deliver high peak power with low heat input. The power control possible with an Nd:YAG makes it easy to weld thermally sensitive electronic devices.

35.6.2 Mode

Laser output beams have a characteristic distribution of energy called a *mode*. Many lasers have a centrally-peaked mode which is useful for cutting. It is generally better to weld with a mode that has more power at the edges because this provides more tolerance for positioning the beam.

35.6.3 Travel Speed

One feature of laser welding is its ability to join metals at very high speeds. A common production application—the welding of aluminum window spacers—is done at a speed of 1 m/s. Speeds this high produce rough welds, but that is acceptable in this instance. Good quality welds can be made at speeds up to 250 mm/s before hydrodynamic effects cause the bead to become irregular.

35.7 PROCESS MONITORING

In production, it is desirable to get some feedback about the quality of the welds being made. While it is possible to measure current with arc, resistance, or electron beam welding, no similar quantity is available for laser welds. Several devices are being used that monitor weld plasma. These, along with pyrometers that check surface temperature, allow some indication of process quality but are not definitive. At this time, only destructive testing can guarantee the quality of laser welds. Fortunately, a properly set up laser welding process has such high reliability that the frequency of destructive tests can be held to a very low level.

CHAPTER 36

DIODE LASER FOR PLASTIC WELDING

Jerry Zybko
Leister Technologies, LLC
Itasca, Illinois

36.1 INTRODUCTION

Plastics are continuously being challenged to endure an increasing level of harsh environments. It is not uncommon for a plastic assembly to be required to perform flawlessly in temperatures between –40°F and 140°F. Additionally, as we enter the micro age, the new challenge in the plastics industry is to manufacture and assemble smaller and smaller parts. New polymers are being developed everyday to achieve these higher standards for strength and survivability. Polymer design is becoming more complex as mineral, glass, or other fillers are being used to meet the demands. As uses for plastics continue to grow, methods to assemble the products are being challenged. Standard joining methods, such as adhesives, fasteners, ultrasonic, or vibration welding may no longer suffice. Laser has emerged as an assembly method that allows more degrees of freedom for design engineers.

36.2 CO_2, Nd: YAG, AND DIODE LASERS

CO_2 laser was the first laser to develop a wide use in industry, though its primary use was with metals. The CO_2 laser has a high level of power (up to 50 kW) and a very long wavelength (10,600 nm) limiting its use for plastics. The yttrium, aluminum, garnet (Nd:YAG) laser offered much lower powers (up to 5 kW) and a much more useful wavelength (1064 nm). The lower power and shorter wavelength made the Nd:YAG a viable candidate for bonding plastics; however, the fixed wavelength of 1064 nm also poses limitations as some polymers work best in the 800–950 nm range. The diode laser has emerged as the premier source for plastic, due to its lower wavelength (800–980 nm) and lower power (from 0.06 to >1 kW).

Its history can be traced to infrared welding techniques using halogen and parabolic lamps from the 1970s.[1] Robert N. Hall developed the first semiconductor diode laser, while working for General Electric in 1962.[2] The first *high-powered diode laser* (HPDL) had an output wavelength of 840 nm, generated considerable heat, and was, therefore, only able to operate in the pulsed mode. Today, the materials that generally compose HPDL are a mixture of gallium (Ga), indium (In), and aluminum (Al) on one side and phosphorus (P), arsenic (As), and antimony (Sb) on the other side.[3] Today's diode lasers for welding of thermoplastics operate in a continuous mode between 800 and 980 nm, and high wattage diodes can incorporate a closed-looped water-cooling system for increased useful life.

Another desirable feature of the HPDL is the ability to easily adjust the power level. A standard HPDL system has the flexibility to adjust from 0.1 W to 150 W of power and beyond. In general, 5.0–20.0 W of power is sufficient to surpass the *glass transition* (Tg) of a standard thermoplastic material. Utilizing increased wattages, greater than 100 W, on materials that require only low wattages to create the melt will allow the material to reach the desired temperature more quickly, hence providing for higher throughput. Materials with higher Tg levels typically require higher wattages to reach the desired temperatures necessary to create a softening of the material.

36.3 LASER WELDING PLASTIC MATERIALS

The laser bonding method, called *through transmission infrared* (TTIr) welding, is accomplished by passing laser light through a top, laser-transparent plastic material and onto a bottom, laser-absorbent plastic material, where the laser is absorbed and transformed into heat (see Fig. 36.1).

The transmissive top layer need not be clear; it needs only the ability to allow the specific wavelength of light to pass through to the lower, absorbing layer. It can be translucent or any one of a myriad of colors, including black, accomplished by tinting the material with heavy concentrations of red or other pigments.

When heat is used to bond plastics, the challenge is to provide enough energy to bring the material to a workable state but not too much, which would result in burning or decomposition. For amorphous materials we refer to this as the glass transition (Tg). For crystalline polymers the workable state is defined as a melting transition. The second challenge is to develop plastic materials with different absorptive characteristics. When plastic is laser welded, one piece is stacked on top of the other piece and the laser light must pass through the top piece and stop at the surface of the base material, allowing the light to be absorbed and to create the temperature required to melt both materials.

The development of the diode laser provided the necessary energy and control required to address these challenges. This chapter reviews the capabilities of the diode laser as a means to create a bond. This chapter reviews the capabilities of the diode laser as a means to create a bond between two plastics and describes the most recent methods for presenting the laser light to the components being assembled.

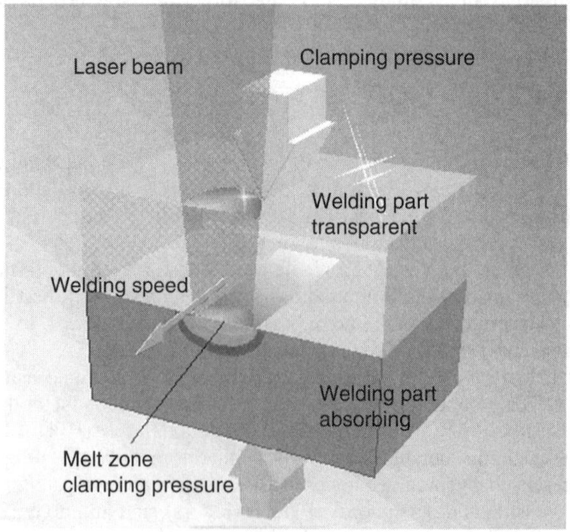

FIGURE 36.1 Transmission welding principle. *(Courtesy of LEISTER Technologies.)*

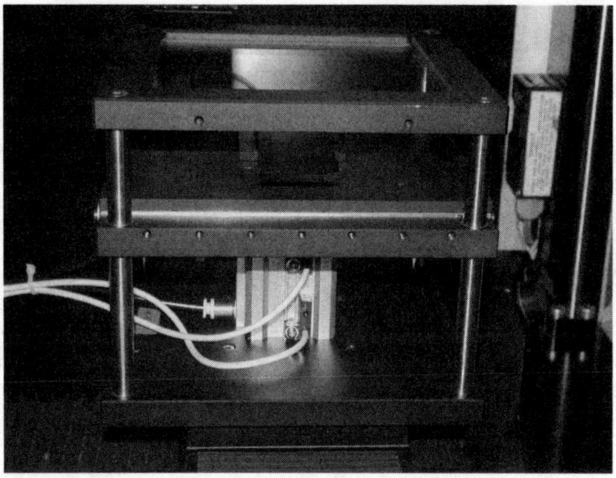

FIGURE 36.2 Clamping mechanism. (*Courtesy of LEISTER Technologies.*)

36.3.1 The Welding Process

The welding process begins by clamping or pressing the two components together in order to create intimate contact between the layers in the desired joining area. This process often incorporates a glass plate on which to press the assembly (see Fig. 36.2).

The joining area is prepared by incorporating a flat-to-flat joint design with no energy directors or texture. With the parts clamped together, the heat is conducted into both components, where it softens both parts, creating a hermetic bond. Most any thermoplastics can be joined with laser, including advanced materials, such as *liquid crystal polymer* (LCP), *polyetheretherketone* (PEEK), *polyimide* (PI), and *polyetherimides* (PEI). Less rigid material, such as *thermoplastic elastomer* (TPE) can also be welded because laser welding creates the required heat for welding by absorption of the laser energy by the material, and not by friction. A number of dissimilar material combinations are also possible, provided that the workable temperature ranges—between Tg and decomposition—are overlapping and there is chemical compatibility. Diode lasers have a shorter wavelength and have now reached output powers that allow them to be used to produce a controlled melt, or welding, of thermoplastics. Weld lines as narrow as 0.1 mm (0.004 in) have been achieved using diode laser welding systems.

36.3.2 Material Selection

The transmissive and absorptive properties of the material are important characteristics to consider when choosing polymers for laser welding. Let's begin with transmission—the ability to pass through space or material. In general, amorphous materials are much more transmissive than semicrystalline. A PMMA, for instance, would allow light to pass through with very little diffraction (bending) or scattering. Passing diode-generated light through 5 or 6 in of PMMA is not a problem. Semicrystallines, on the other hand, scatter the light more and have a higher level of diffraction. Therefore, the thickness of a semicrystalline material will have an effect as to how much light will pass through.

Figure 36.3 graphically displays an example of the transmissive properties of a semicrystalline material. Light can transmit at some wavelengths and not at others. Therefore, if one was working with a diode laser that operated at 808 nm wavelength, product "A" should be chosen for the application.

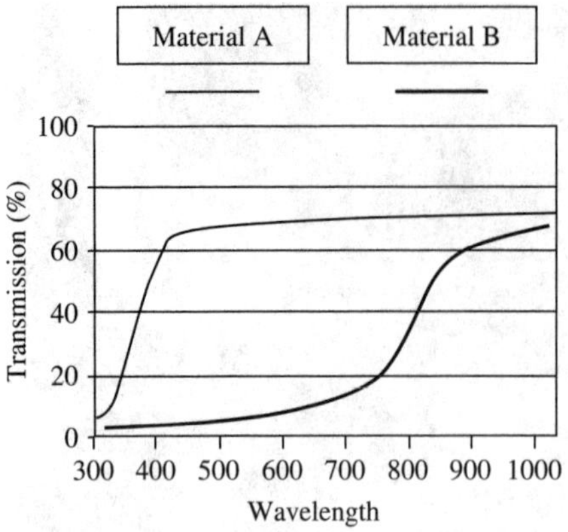

FIGURE 36.3 Absorption. (*Courtesy of LEISTER Technologies.*)

The absorptive portion of the assembly, or the bottom component, will be receiving the light after it passes through the top, transmissive layer. The desire is for the light to completely stop or absorb at the initial surface of the bottom layer. When the light is absorbed it creates heat. A simple method incorporated to increase the absorptive properties of the bottom material is to add carbon black to the polymer. The downside is that even small levels of carbon black (0.5–1.0 percent) result in a part that is black to the human eye. The major benefit is the absorptive properties of the carbon black. One must make sure not to add too much carbon black, as this will conduct the heat away from the interface of the two materials being bonded

Figure 36.4 illustrates the desired concepts of transmission and absorption. In the figure, assume that the laser light is moving from left to right across the graphic. The graphic is broken into three frames. The first frame illustrates the desire to allow as much light as possible to pass through the material. This would be the top material in the assembly. The third frame illustrates the desire to have all the light stop at the top surface of the bottom material. The middle frame illustrates neither a good top piece nor a good bottom piece. If the middle frame represented the top piece, not enough light would transmit through. Conversely, if it were the bottom piece, too much of the light would be absorbed below the surface, reducing the amount of heat that could be generated at the surface.

If dark colors are not functional or aesthetically functional, clear coatings are available which can be applied to the weld area of one of the parts prior to assembly, allowing for clear-to-clear assemblies. The solution can be applied to the parts days, weeks, or even months, in advance. The use of this type of absorption process is of particular necessity when downstream utilization of different wavelength of light or vision systems is incorporated to analyze the specimens. For example, the solution can be blended to absorb wavelengths greater than 810 nm and permit the 740-nm wavelength to pass through.

Consideration must also be taken when considering fillers used in the polymers. Glass is a common additive used to create more strength in the polymer. Unfortunately, glass melts at temperatures

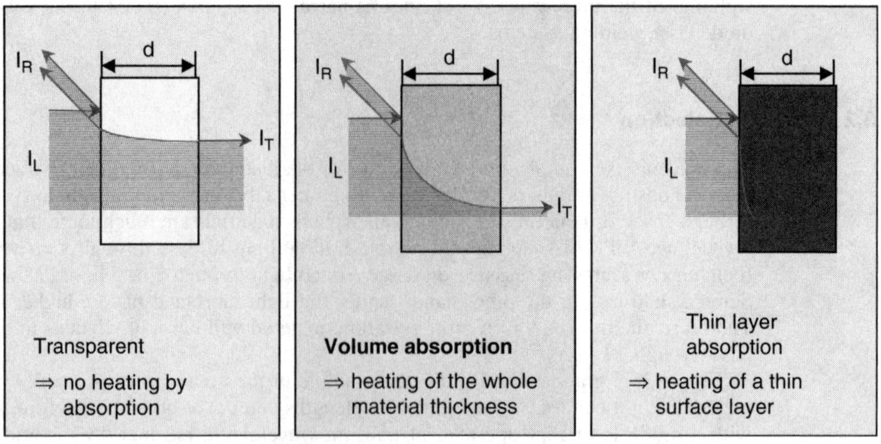

FIGURE 36.4 Transmission and absorption. (*Courtesy of LEISTER Technologies.*)

far greater than the melting range of plastics. The solid fibers of glass act to scatter the light, preventing it from penetrating through the top layer of plastic, and also limit the level of hermetic seal that can be accomplished. Other fillers, such as talc or titanlum dioxide, a common additive used to create white material, also affect the transmissive level of top layer.

36.4 METHODS OF BRINGING LASER TO THE PART

Once the materials have been chosen that create the desired transmittance, absorption, and resultant heat required to bond the substrates, the shaping of the laser light needs to be addressed. Unlike the light emitted from a standard Class II laser pointer, the light emitted from a diode laser, as with the light emitted from a standard flashlight, is divergent and immediately scatters. Typical HPDL assemblies are composed of diode bars that are 10-mm wide and 0.1-mm high. As the laser light emerges from the diode it diverges in both the perpendicular and parallel direction, with more divergence in the perpendicular (30–60°) than the parallel (25–35°) direction. Therefore, the first step in harnessing the light is a collimation of the beam followed by optical coupling, if necessary (see Fig. 36.5).

For some applications, such as a mask process, collimation provides the final form of light. Crisp line patterns approximately 1.0-mm wide × 60.0-mm long can be collimated and presented to the substrate with a balanced light density.

For other application methods, further beam shaping is required, typically beginning with optically coupling the light into a fiber optic cable whereby various lenses can be utilized downstream producing spots, lines, ring patterns, and the like. The following sections will describe these methods.

36.4.1 Spot (Contour) Welding

As discussed earlier, the first step in designing a laser assembly is the determination of laser transmissive and laser absorption combinations. The contour method is commonly used for this purpose, due to its focal concentration of light and pattern versatility.

This method utilizes fiber optic cable to bring the light from the diode bar to the work area. The fiber optic is terminated into a lens block where a special lens shapes the emitting light into a conical shape (see Fig. 36.6).

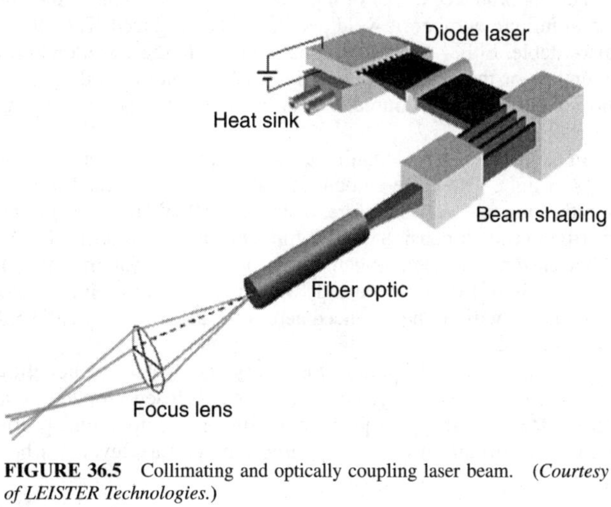

FIGURE 36.5 Collimating and optically coupling laser beam. (*Courtesy of LEISTER Technologies.*)

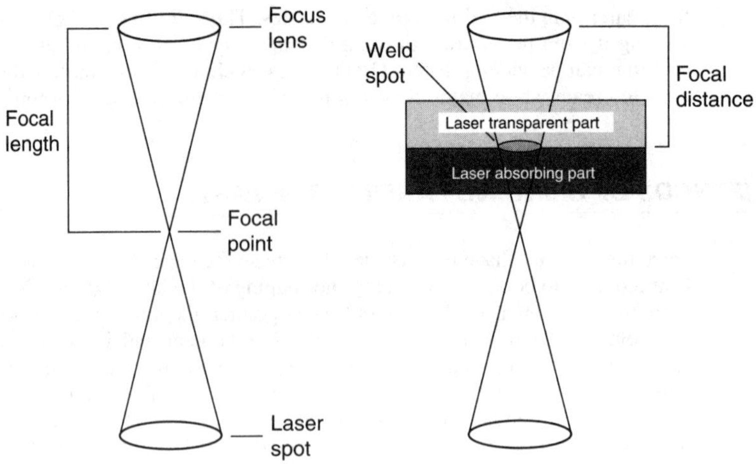

FIGURE 36.6 Fiber optic laser spot delivery (contour). (*Courtesy of LEISTER Technologies.*)

The distance between the focal lens and the interface between the two pieces being welded sets the width of the laser spot. Standard optic lenses, producing a conical distribution of light, can emit single spots ranging from 0.6- to 3.0-mm diameters, with the resultant diameter being controlled by the distance between the optic and work piece. The emerging light is Gaussian in distribution, i.e., the light is more concentrated in the center and is evenly distributed and dissipated as it moves outward from the center. As the spot size enlarges, less and less energy is contained in the outer edges—thereby making larger spot sizes less effective. Small spot sizes allow for cleaner edge definition of the weld and more intense spot energy.

The working distance can be set so that the interface for welding can lie above or below the focal point of the lens, depending on the application. Welding in the area above the intersection point may be desirable when the top material, due to its semicrystalline nature or concentration of absorbing fillers, is absorbing more light than desired and the concentration of light before the weld area will result in overheating. Overheating typically produces carbon, creating more absorption and preventing the light to pass through the top layer.

Having produced a spot of light, the next step is to move this light or move the assembly, allowing for an infinite number of weld patterns to be produced. The most common method is to incorporate an *x-y* table. Either the laser can be mounted to the *x-y* allowing for the beam to traverse across the assembly, or the clamped assembly can be mounted to the *x-y* table, allowing for the assembly to move below the laser source. The utilization of a robotic arm can allow for the welding of different planes.

In addition to optics creating a spot of light, custom optic lenses can be created that will produce rings of light. The same amount of light energy contained in the spot described above is dispersed into the ring pattern. Therefore, a standard 50-W HPDL can effectively apply a ring pattern with a maximum ring dimension of 25.0-mm O.D. × 23.0-mm I.D. The minimum ring thickness is 1.0 mm. Other custom optic lenses can allow for elliptical patterns of light to be emitted. A pattern that is approximately 1.0-mm wide and 10.0-mm long can be emitted; however, unlike the ring pattern, the light energy will not be balanced across the area. The typical weld duration for the ring or elliptical pattern is 1–2 s.

In addition to weld pattern flexibility, another major benefit of the contour spot method is the ability to incorporate a pyrometer to obtain a closed-loop temperature feedback of the weld profile. Temperature sensing can provide a complete temperature profile of the weld pattern, providing precise confirmation that the desired temperature level has been achieved throughout the weld pattern.

36.4.2 Simultaneous Welding

As discussed above, instead of optically coupling the laser light into a fiber and using optics to create an elliptical shape, the light can be immediately collimated and shaped into a fixed line. If fast cycle times are required, a series of collimated beams can be presented to the part, creating lines, squares, or rectangular shapes. A standard line dimension of 1.0-mm wide and 60 mm in length will produce a balanced deposition of laser to the work piece. Though line lengths as wide as 50 mm are achievable, it becomes difficult to balance energy across longer weld lines. The benefits of this method are consistent edge definition, a fixed pattern, and high throughput.

36.4.3 Mask Welding

Mask welding, a patented process by Leister, utilizes the same collimation technique as described in the line welding section above; however, the collimated light is presented to a mask which is placed above the work piece, allowing the laser to only pass through specific areas. The mask shape, therefore, determines the pattern of weld produced. Where the laser light is allowed to enter, welding is achieved (see Fig. 36.7).

A precise mask is produced via photolithographic removal of predetermined portions of a metallic coated glass. The pattern can be as varied as required by the application.

The main advantage of this process is that it allows for very precise and very fine weld lines. Weld lines as narrow as 100 μm have been successfully made with the mask welding process. In addition, this process allows the possibility of producing welds with elaborate structures or contours. During one weld sequence, it is plausible to weld lines with different widths and shapes, as well as whole areas of weld.

The mask process has found a strong foothold in the medical and microfluidic markets because of its ability to produce very precise and very repeatable weld shapes.

When welding microfluidic channels, alignment of the mask with the channel is critical to ensure proper location of the weld. If precise location of weld is not required, very quick cycle times are possible, as the alignment can be performed mechanically with a fixture. In situations where the position of the desired weld is required to be extremely precise, a vision system can be employed. With a vision system, fiducials of the mask can be compared with fiducials of the assembly, accurately

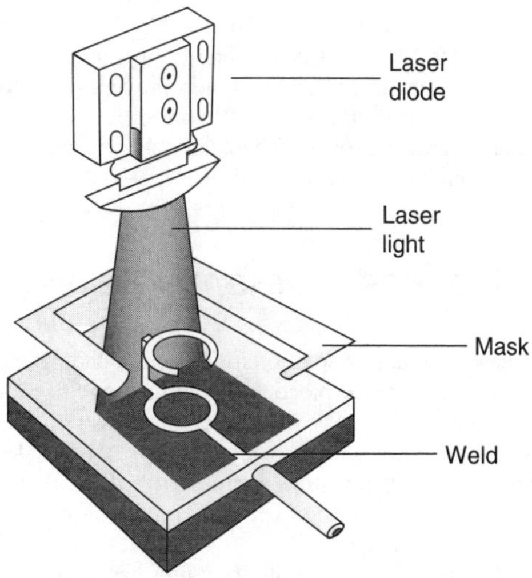

FIGURE 36.7 Mask welding technique. (*Courtesy of LEISTER Technologies.*)

positioning the mask within ±2.0 μm. The vision system for the mask alignment provides additional controlling mechanisms for the deviation of microstructured plastic parts in comparison to the microstructures on the mask. The measured tolerance during the alignment procedure can then generate error information and show influence on the accuracy of mask alignment process.[4]

After the adjustment process, scanning the laser source over the mask creates the desired bond. The concept of mask welding in micrometer range has implications relating to the rapid heating and cooling of plastic during the welding process. The key factors in establishing the desired welding accuracy are not only the quality of mask and beam shape of the laser, which mainly determine the precise heat transfer, but also the speed of heat treatment and response of material due to the melt flow.

The laser light can be applied to very specific areas and weld lines can be as narrow as 100 μm (0.004 in).

36.4.4 Quasisimultaneous Welding

Through the use of a galvo system, the laser light can be focused onto two mirrors that simultaneously move, creating an instantaneous illumination of complex shapes and patterns. Though the system utilizes one beam of light, the light can be moved in an extremely fast manner. This method can be employed in situations where the complete illumination of the pattern is required to achieve a collapse of the weld area. The system can also move more slowly, so as to emulate the function of the spot method discussed earlier.

36.5 DIODE LASER SAFETY

The United States has a *Center for Disease and Radiological Health* (CDRH) that regulates the use of lasers in the workplace. Lasers are classified as class I, class II, class III, or class IV. These are classifications relating to the power of the laser and the diffusion into the workplace. A class I laser is safe for normal operation. Laser systems are required to obtain an Accession number from the CDRH to verify its safe use in the workplace. The Accession number ensures that the specific laser system has the necessary safety features to allow for its use in a public facility. A standard high power diode laser is rated class IV. To obtain a class I rating, the diode laser system must be designed such that when the laser is turned on, no light is able to escape into the surrounding workplace. The chamber need not be air tight; it need only prevent light from escaping. If doors are lowered to close off the assembly during the process, the specific sensors that signal that the door is closed need to be documented and approved. There are also specific labeling instructions and emergency stop requirements that have to be adhered to.

36.6 ALTERNATIVE METHODS OF PLASTIC ASSEMBLY

Adhesives have long been a solution for difficult bonding applications. Hot melt and cold glue solutions can be applied to the work piece and clamped together for a seal. There are also pressure-sensitive adhesives available that can be preapplied to the work piece and clamped for an instant bond. Though adhesives can be deposited in extremely rapid cycles (less than 1 s), as the work piece becomes smaller and smaller it becomes increasingly difficult to control the deposition of the adhesive. Additionally, as polymer design becomes more complex, chemical compatibility between the plastic and the adhesive is challenged.

Since the mid-1970s, ultrasonic bonding has proven to be a very rapid and predictable method for joining plastics. The typical ultrasonic process requires one of the parts to have an energy director (typically a 90° included angle, 0.012 in tall and 0.012 in wide) on its surface. The two halves are then pressed together with a force of 1–300 lb force. With the parts clamped, the ultrasonic head, or horn, is cycled at a rate of 15–40 cycles per second. This ultrasonic vibration causes the energy director to create heat, via friction, resulting in a melt, or weld, between the two pieces. For the majority of applications this process is exceptional and cycle times can be less than 1 s. For this process to be successful, energy directors are required to be incorporated into the design of the part. Assemblies that can allow for energy directors, some degree of flash (particulate), and a slight marring of the top surface are well suited for this method, as are assemblies that do not have sensitive structures or delicate connectors within the assembly that can be fractured by the high frequency vibration. The size of the parts being assembled can be as large as 8 in × 8 in when using a standard 20-kH system.

Vibration welding is a method whereby the two pieces being assembled are vibrated or rubbed against each other at high speeds and with a given clamping force. This method of bonding is suitable for welding large parts where excess flash and a less than consistent finished stack height are not a concern.

Laser has developed as a method to bond plastics in ways that overcome the challenges confronted by other joining methods. The mask process, for instance, now allows for extremely small weld areas—down to 100 μm in width.

36.7 CONCLUSION

Diode laser welding has proved to be a viable method for joining plastics. The medical industry has exhibited great interest in this joining method due to its ability to produce precise bonds without creating particulate, contamination, or surface marring. The ability to seal films is also very desirable. As sensors and other assemblies become more compact the minimal heat-affected zone of the laser method allows for precise bonding close to sensitive structures or components. Dissimilar materials can be joined with this process, relying on overlaps of the softening characteristics of the polymers. Additionally, gas-tight welds can be achieved without the use of gaskets, adhesives, or fasteners. Applying the laser light in the myriad of shapes discussed above allow for optimization of the desired bonding area.

REFERENCES

1. J. Rotheiser, "Laser Welding of Plastics," *Plastics Decorating,* July/August 2002.
2. National Inventors Hall of Fame, http//www.invent.org, October 15, 2001.
3. Hessler, Thomas, "Promising Technology: Laser Welding of Plastics," *Euroforum,* 26, October 27, 1999.
4. J. Chen, "Laser Assembly Technology for Planar Microfluidic Devices," *60th Annual Technical Conference for the Society of Plastic Engineers,* 2002.

CHAPTER 37
ELECTRICAL DISCHARGE MACHINING

Gisbert Ledvon
Charmilles Technologies Corporation
Lincolnshire, Illinois

37.1 INTRODUCTION

The *electrical discharge machining* (EDM) process of erosion using electrical discharge was developed in 1770 by an English scientist. The Russian scientists B. and N. Lazerenco adopted the principle and developed a process to control the material removal of conductive materials. In 1952, Charmilles Technology Corporation, based in Geneva, Switzerland, began working together with Mr. Lazarenko to develop the first industrial economical die sinking EDM machine, which was introduced in 1955 at the EMO in Milan, Italy. Shortly after, the team developed a surface finish scale to communicate between designer and EDM operator—the surface finish can be achieved on an EDM machine. This surface finish scale was also adopted by the German association of engineers VDI (*Verein Deutscher Ingenuere*) and is now known as VDI 3400. This surface finish scale has numbers starting at CH0 = 0.10 μm RA = mirror finish and CH45, the roughest finish, at 18.0 μm RA.

When *numerical controls* (NC) for machine tools became more available, EDM became more popular, and an additional process, the wire EDM, was developed in early 1970. The maximum cutting speed then was about 1 in^2/h, and has increased to 40 in^2/h in 2003. EDM machines are no longer nonconventional, representing now more than 7 percent of the number of cutting tool machines sold in 2000. Job shops are discovering more and more new applications by using wire or die-sinking EDM (see Fig. 37.1).

37.2 THE PRINCIPLE OF EDM

The EDM process removes electrical conductive material using electrical discharge. The process takes place in a nonconductive dielectric fluid. The electrode (cutting tool) and the work piece are located a specified distance from each other, while a voltage and current are applied to the electrode. When a certain charge between the two surfaces is reached (like a capacitor), a discharge occurs and a spark can be seen (see Fig. 37.2).

The shape of the spark generates heat between 8000 and 20,000°C and the material on the surface melts, vaporizes, and is flushed away by the dielectric fluid.

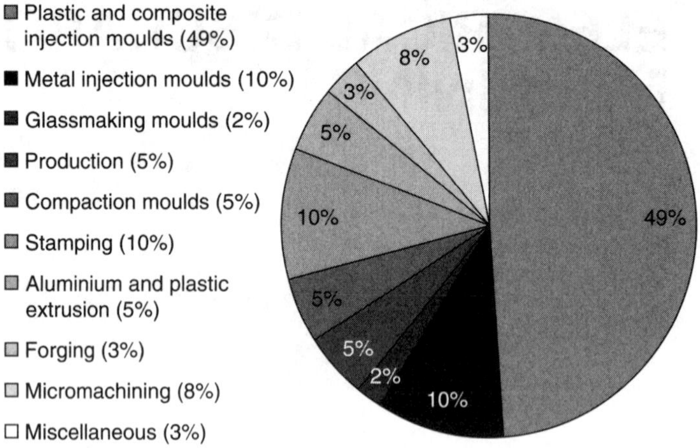

FIGURE 37.1 Industries using EDM machines.

Fundamentally, two different EDM technologies are applied. The first is the die sinking process; this process utilizes an electrode either made out of graphite copper, copper tungsten, or other conductive materials. Die sinking machines are mainly used to build molds for the plastic, cast iron parts, and various production parts as seen in Fig. 37.3.

The second, wire EDM technology, utilizes a precise wire as a cutting tool. Made originally from brass, coated wires have been developed to improve cutting speed. The most popular wire diameter is 0.25 mm (0.010 in). This diameter gives good removal properties. The wire is guided by an upper and a lower wire guide, the guides having the same diameter as the wire. In some cases, "v-guides" are used to allow for more than one wire diameter to be used with the same guide. Alternately, site v-guides can't be used to cut large taper angles.

The lower guide is mounted on the lower work head, which is mounted on the *X-Y* axes, and the upper head with wire guide on the *U-V* axes. On the same guide system, a flushing nozzle forces high-pressured (16–20 bar) dielectric fluid into the kerf to remove the eroded particles and cool the wire and the workpiece during the cutting process. All axes are controlled by a CNC system moving the axes based on a G-code generated by a CAD/CAM programming system.

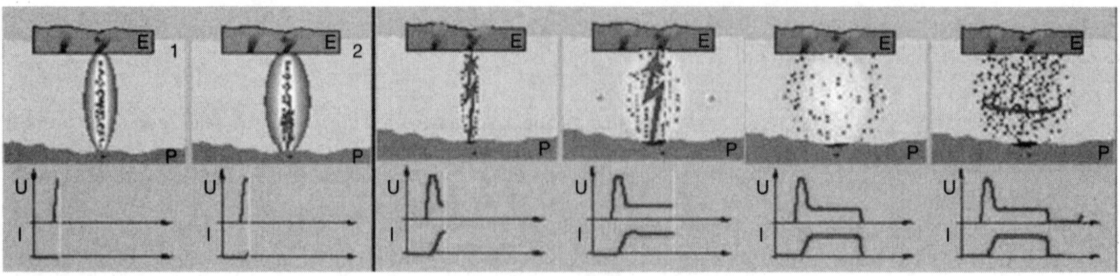

FIGURE 37.2 Tool and work reach a discharge that causes cutting operation. (E = electrode, P = workpiece, U = voltage, I = current)

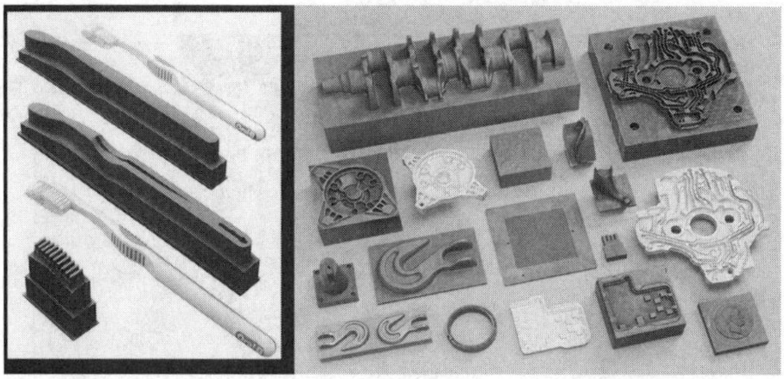

FIGURE 37.3 Die sinking machines are used to build molds for plastic and cast iron parts.

37.3 TYPES OF DIE-SINKING EDM MACHINE

Die-sinking machines have three basic designs. The first and simplest die-sinking EDM machine is known as the "small EDM hole driller." This machine is mainly used to EDM holes from 0.1 to 3 mm in diameter in any conductive material within seconds, using an aggressive spark generator and a rotating round tube electrode. Dielectric fluid, usually deionized water, is pumped through the electrode to remove the material eroded as fast as possible. While the holes are not extremely accurate (±0.1 mm), this process does provide for rapid cutting, which is the primary goal in some cases. Common applications include start holes to feed the wire for the wire EDM machine, or drill cooling holes for molds or aerospace parts, to name only some applications.

The next type is the manual or conventional type. These machines have a moving *X-Y* axis machining table, a moving *Z*-axis column, and a spark generator. The operator selects the correct generator setting based on electrode and workpiece material.

Finally, the CNC diesinker is a more sophisticated machine tool. CNC diesinkers have a fixed or moving machine table, a moving or fixed upper column housing the *X-Y-Z-C* movement, a tool changer to change electrodes, and the spark generator housing also the CNC. Programming is executed with an on-board expert system that makes the operation of the machine simple, and ensures predictable results in surface finish and accuracy on the part.

All the die sinking machines using dielectric fluid use either a mineral-based oil or synthetic oil. The dielectric system is equipped with a filter system that filters the removed material so the dielectric is clean in the work tank and in the spark gap (gap between electrode and workpiece). Surface finishes down to mirror finish can be achieved depending on electrode and workpiece material.

37.4 TYPES OF WIRE EDM MACHINE

Wire EDM machines are categorized as submerged or nonsubmerged (coaxial flushing) machines. With a submerged wire EDM, the part and the wire electrode are submerged with deionized water for the duration of the cutting process. The nonsubmerged machine operates by submerging the cutting zone around the wire (see Fig. 37.4).

Both machine types have the following components (see Fig. 37.5):

- Dielectric filter and deionization system
- Wire transportation system
- Automatic wire cutting and threading system

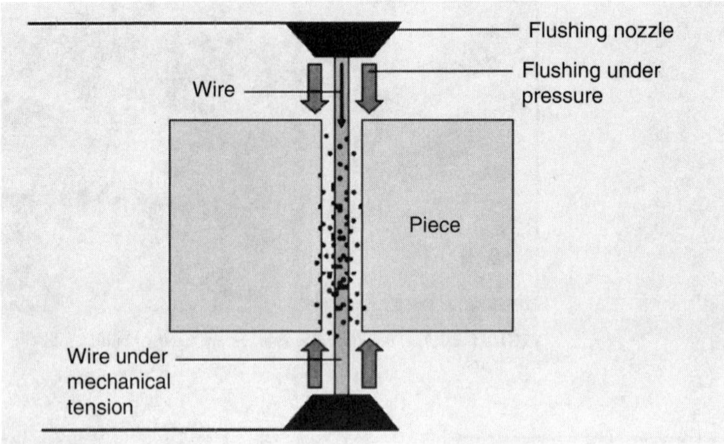

FIGURE 37.4 Nonsubmerged wire EDM.

- Slug removal system
- Fixed work piece table
- *X-Y*, *U-V*, and *Z* axes
- CNC control with spark generator

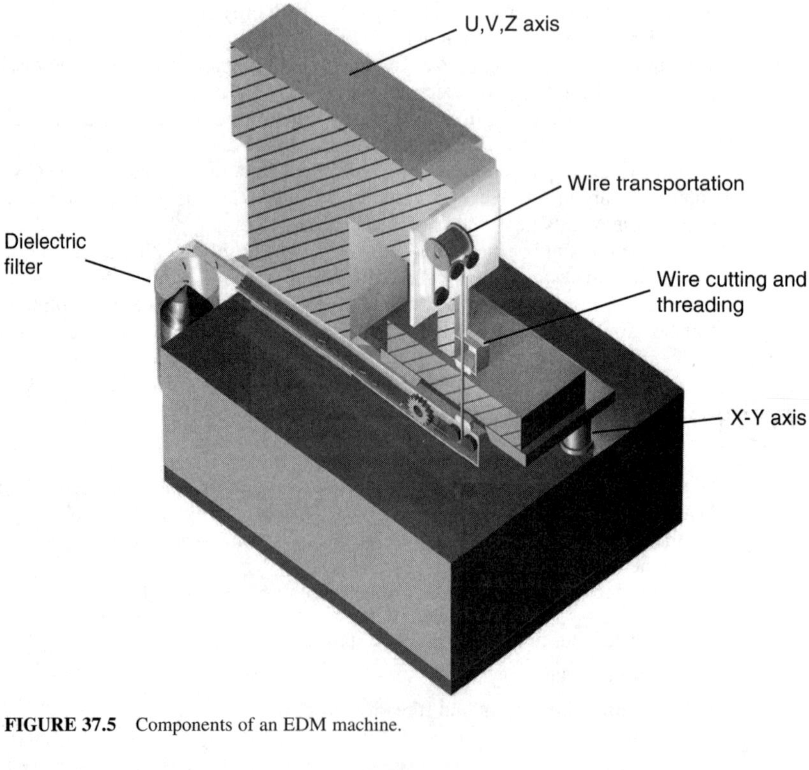

FIGURE 37.5 Components of an EDM machine.

Some wire EDM machines now have the capability to switch automatically from large (e.g., 0.3 mm) to small wire diameter (e.g., 0.1 mm) automatically to enhance cutting performance and productivity. For example, a larger wire diameter can be used to do the rough cut, and the smaller wire diameter is used to finish details, such as small inside radii.

The modern CNC-integrated expert systems allow the operator to run the machine unattended without breaking the wire. The expert systems monitor the cutting conditions and make adjustments to spark frequency and flushing pressure so the wire will not break, even in difficult flushing conditions. To ensure the desired accuracy and surface finish without reworking the part, expert systems onboard the CNC or on a PC generate a specific technology for each job. The user provides basis data including workpiece thickness, material to cut, smallest inside radii in the part, maximum taper angle, desired accuracy, and surface finish. Based on the data input the expert system generates a specific technology for the specific part. For example, assuming the part has to be cut again in a few months, the operator would just call up the same technology and would achieve the same results in terms of accuracy and surface finish. This differentiates the wire EDM from any other cutting tool technology. Since the wire EDM can cut any kind of conductive materials and achieves cutting speeds of 400 mm^2/min. (37 in^2/h) in harden tool steel, EDM has begun to replace conventional machines like mills or grinders. EDM provides greater precision and surface finishes, and can cut small slots 0.040 mm and doesn't leave burrs on the cutting edge.

37.4.1 Geometry and Surface Finish

In general, the accuracy of the part depends largely on the design of the machine and the best positioning accuracy. Part accuracy is achieved if the EDM machine is equipped with glass scale on all machine axes or at least on X and Y if no taper accuracy is required. Another important factor is the wire guide system, where closed guides are the most reliable and accurate solution. Most accurate wire EDM machines also offer corner control expert systems; these systems ensure that the wire is as straight as possible between the upper and lower wire guide before (see Figs. 37.6(*a*), 37.6(*b*)) entering or exiting a sharp corner or small geometry detail.

This is achieved by reducing the cutting speed, reducing the flushing pressure, and increasing the wire tension. Extremely accurate EDM machines offer this "corner control expert system" not only for rough cuts, but also for skim or finishing cuts.

To make wire EDM more efficient, wire-threading systems allow the machines to operate unattended or even in conjunction with an automatic loading system for parts like a robot. As the threading system cuts the wire, the machine moves to the next workpiece, stops at a hole location,

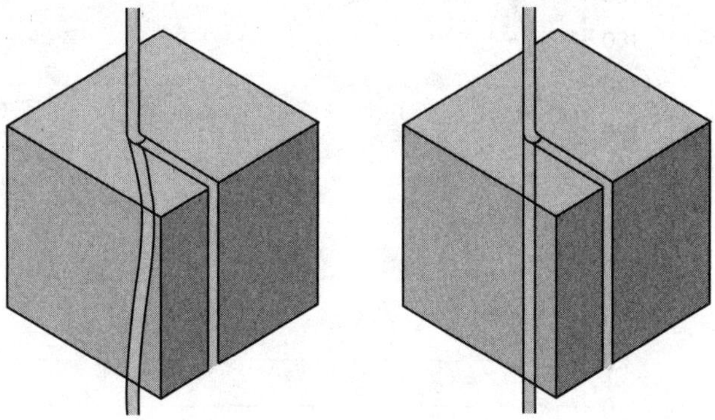

(a) (b)

FIGURE 37.6 The expert system ensures the wire is as straight as possible.

and the wire is automatically threaded. Then the machine starts cutting the next part or opening and so on. Buyers of wire EDM machines should consider the following factors when selecting a machine:

1. Submerged or not submerged?
   - If you only do punch and die work with small taper angle up to 10 degrees, take advantage of a nonsubmerged machine, but look for an enclosed machine design for maximum temperature stability in the work zone. An expert system prevents wire breakage by automatically adjusting flushing pressure and generator settings when cutting from the outside into a part, or through a part with variable height.
   - If long parts have to be handled that are larger than the worktable, consider a nonsubmerged machine for maximum flexibility.
2. For large taper applications with 15, 20, 30, or 45 degrees and tall part cutting, e.g., up to 600-mm height, choose a submerged wire EDM.
3. Job shops with a variety of applications should choose a machine with large *Z*-travel and large *U-V* axis travel. The most versatile machine would have *X-Y* = *U-V* travel.

37.4.2 Some Technological Information

Wire diameter influences cutting speed. As a rule, larger wire diameter can handle more energy, therefore more material can be removed and the cutting speed increases. The cutting or removal rate is calculated as follows: Removal rate per min = Workpiece height x linear cutting speed in min.

The most efficient cutting speed is achieved at around 50-mm workpiece height (see Fig. 37.7). In the United States, cutting speed is specified in "square inch per hour"; in the rest of the world, it is specified in mm^2/min.

The same formula applies, replacing mm with inches and minutes with hours. The second aspect of cutting speed is the material to be cut. The melting point and density of the material has a large influence on the cutting speed (e.g., aluminum can be cut faster than carbide or steel). EDM technology is very efficient in cutting alloys, titanium, or polycrystalline diamond (PCD), difficult materials to machine conventionally.

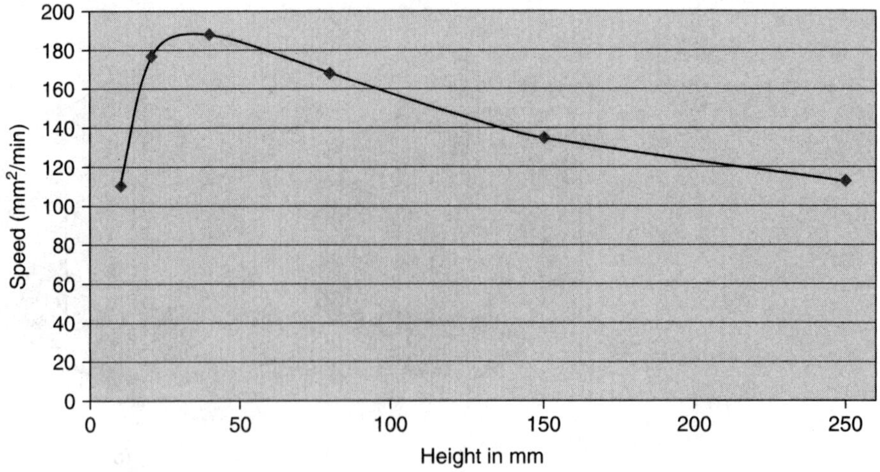

FIGURE 37.7 Optimum cutting speed is achieved at 50-mm workpiece height.

37.4.3 Workpiece Holding and Wire Alignment

The workpiece is either clamped directly on the worktable or held with a vise, preventing interference with the flushing nozzles. The flushing nozzles need to be as close as possible to the surface of the work piece to generate enough flushing pressure, which affects cutting speed.

The wire needs to be perpendicular to the workpiece table to cut a perpendicular part or exact parting line for a die or mold. Most wire EDMs have an automatic alignment function using a small device mounted on the table. The wire is fed into the loop hole of this device, and the machine then finds the center of that hole by means of electrical contact between the wire and the measuring device. When the machine finds the center, the *Z*-axis will move to maximum height position and perform the same measuring cycle function. If the first measurement differs from the second, the machine moves the *U-V* axis by the measured deviation and, at that point, the wire is perpendicular to the worktable.

37.4.4 Maintenance and Operation of a Wire EDM

A wire EDM machine needs the following components to operate. Because these parts need to be replaced as wear occurs, they are termed consumables:

- Water filters
- Power contacts
- Resin to generate deionized water
- Flushing nozzles
- EDM wire

The machine should have a chiller unit to keep the dielectric water at a constant temperature of 20°C or 68°F to ensure maximum cutting accuracy. The machine should not been placed next to a grinder, punch press, or graphite mill. Doing so will have impact on machining accuracy and the lifetime of the machine tool. The best location would be a separate room, air conditioned if possible to ensure maximum cutting accuracy. It is important to remember that some leading wire EDM machines' cutting accuracy is within microns or even a fraction of a micron. As a result, temperature has an affect on the part expansion as well as on the machine tool.

37.4.5 Programming

Some wire EDM machines offer an onboard CAM system, which makes change in the geometry or the program on the fly simple.

Programming of any CNC machine should be done on a PC and in a quiet environment like a closed office where the programmer can concentrate. If an error is made during programming, it can lead into costly mistakes. As a result, investing in a CAM system that can process DXF, IGES, etc., or other data formats from CAD systems is a wise decision. Another important factor to consider is the "postprocessor." This feature generates the G/M code for the machine. It is important that the postoutput is compatible with the machine's G/M codes.

37.5 USE OF DIE-SINKING EDM

In general, die-sinking EDM machines are used to machine a cavity or a blind hole, while wire EDMs cut through openings and any type of contour that can be drawn (see Fig. 37.8). As a result, die sinking machines are found mainly in mold shops or production applications (see Fig. 37.9).

FIGURE 37.8 Cuts done with wire EDM machines.

FIGURE 37.9 EDM machines cut molds for injection molding machines producing plastic parts.

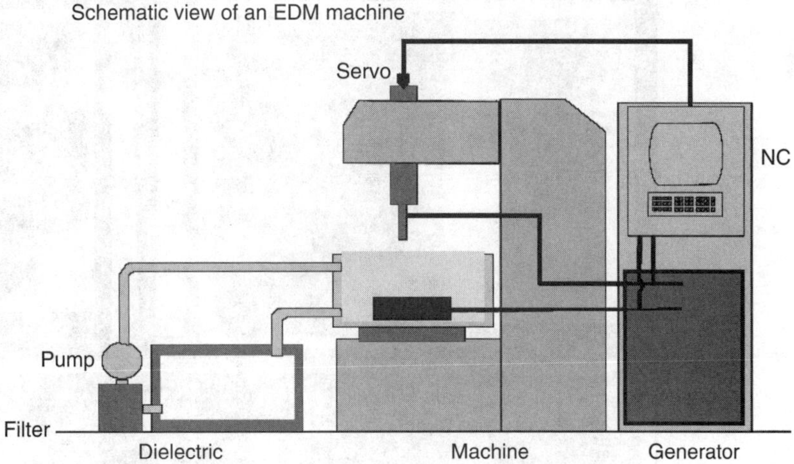

FIGURE 37.10 Construction of an EDM machine.

For the most part, the principles of wire vs. die-sinking EDM are similar, the main difference being the construction of the machine tool itself (see Fig. 37.10), which is a reproduction of a positive shaped electrode (see Fig. 37.11). In the United States, electrodes are most often made of graphite, which offers the best removal rate and is easy to manufacture, e.g., grinding or *high speed machining* (HSM).

Graphite is also an excellent material to use to cut deep ribs, which are commonly used in the mold industry (Fig. 37.12). Copper electrodes are used for finer surface finishes, and copper tungsten is

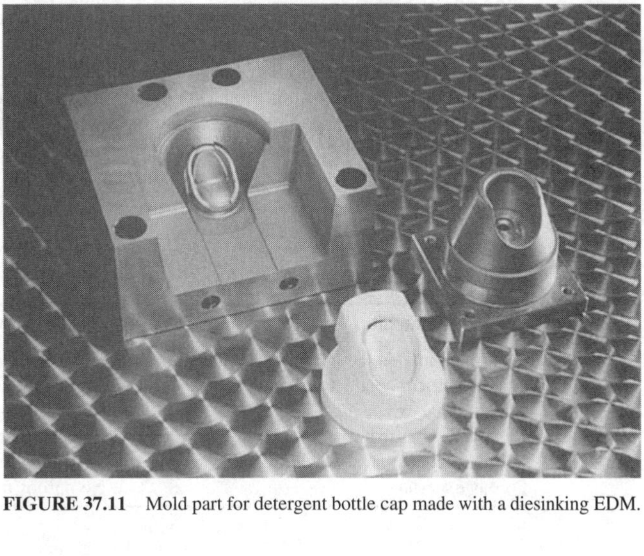

FIGURE 37.11 Mold part for detergent bottle cap made with a diesinking EDM.

FIGURE 37.12 Example of rib electrodes. (*Top*: Electrodes; *Bottom*: Finish workpiece cut open)

used to minimize electrode wear when cutting carbide. Table 37.1 below compares different characteristics of copper and graphite as electrode material (also see Table 37.2).

37.5.1 Workpiece and Electrode Holding

The workpiece is either clamped directly on the work table or mounted on a magnet. Shops who automate their EDMs utilizing pallets on the workpiece table and electrode holder systems on the ram. This allows them to mill or grind the electrodes in reference to the electrode holder, and makes it easy to transfer the electrode from the mill to the EDM machine without determining the electrode location in reference to the *x*-*y* zero position. Most CNC die-sinking EDM machines have measuring cycles to detect the exact workpiece location and electrode location. This ensures that the cavity is cut at the exact location the print or design drawing calls for in the mold.

TABLE 37.1 Different Characteristics of Copper and Graphite as Electrode Materials

Graphite	Copper
Good thermal conductivity	Excellent thermal conductivity
Good electrical conductivity	Excellent electrical conductivity
Low thermal expansion	Relatively large thermal expansion
Relatively light material	Relatively heavy material
Very easy mechanical machining	Very easy mechanical machining
Porous material	No porous material

TABLE 37.2 Recommended Electrode Materials Based on Applications

Application	Electrode material
Large electrode	Graphite because it is light
Strong roughing	Preferably graphite (higher removal rate)
Surface finish from CH 20 to CH 33	Copper or graphite
Surface finish from CH 16 to CH 22	Copper, copper-tungsten, fine grain graphite
Surface finish from CH 16 to CH 22 and complex shape	Fine grain graphite
Surface finish <CH 16	Copper, copper-tungsten
Surface finish <CH 16 and complex shape	Copper-tungsten (relatively easy to mill and to grind)
Machining of carbide	Copper-tungsten, eventually copper

37.5.2 Programming

Modern CNC die-sinking machines have onboard program expert systems in some case, also available for PC. These expert systems work interactively and, based on information the operator provides, generate a technology and a cutting program. In most cases the following data have to be provided: workpiece material, cutting depth, electrode material, and surface finish required. This software can also be used to determine the best electrode undersize. The undersize plays an important role for the cutting speed. As a rule, if the undersize is large, the eroded material can be flushed out faster. For example, a designer using this EDM expert system can determine the undersize of the electrode before it is made to insure max efficiency in the EDM process.

37.5.3 Safety for Die-Sinking EDM

The workpiece and electrode must always be fully submerged with the dielectric fluid, a minimum of 40 mm between the electrode cutting and the dielectric level; otherwise there is risk of combustion when sparks are too close to the dielectric fluid surface. If the machine is operated unattended, an automatic fire extinguisher system connected to the CNC is necessary. Do not touch an electrode and a workpiece at the same time, as this can lead to electrical injury.

37.6 CONCLUSION

In summary, EDM technology is one of the most advanced technologies to be automated for several reasons:

- Consistent results in accuracy and surface finish due to on board expert systems
- Unattended machining due to electrode changers and automatic wire-threading systems
- Available remote monitoring software so the machine can be monitored via internet from home PC or cell phone
- No cutting chips need to be removed
- Automatic optimization of cutting condition if expert systems are on board of the CNC

FURTHER READING

Carl Sommer, Steve Sommer, *Wire EDM Handbook*, 4th ed., Advance Publishing, Houston, Texas, 1999.
E. Bud Guitrau, *EDM Handbook*, Gardner Publications, Cincinatti, Ohio, 1997.

USEFUL WEBSITES

www.edmtoday.org
www.charmillesus.com

CHAPTER 38
ABRASIVE JET MACHINING

John H. Olsen
OMAX Corporation
Kent, Washington

38.1 INTRODUCTION

38.1.1 Brief Description of Process

Abrasive jet machining is a grinding process where the abrasive particles are moved through the workpiece by a stream of high speed water rather than by a solid wheel. By mass flow, the stream is approximately 90 percent water and 10 percent abrasive. This provides excellent cooling and there is no heat-affected zone at the cut surface. The jet is typically about 0.030" in diameter and virtually any two- dimensional shape that can be drawn can be cut. A typical setup is shown in Fig. 38.1.

The jet is formed by pumping clean water at 40,000–60,000 psi through a fine orifice, then entraining dry abrasive into the stream and passing the mixture through a mixing tube where the water accelerates the abrasive up to speed. In general the process is most competitive for materials between $^1/_8$" and 2" in thickness. A collection of parts made by abrasive jet machining is shown in Fig. 38.2.

38.1.2 History

High pressure water without abrasives began to be used as a cutting tool for soft materials in about 1970. The increase in cutting power by the addition of abrasives came in about 1980. The first abrasive jets were crude devices that were often sold as the tool of last resort for severing difficult material. Dimensional control was comparable with oxy-acetylene burning. Then, advances in material technology allowed the manufacture of long life nozzles that could keep their shape during a long cut. At about the same time computational power became very low cost so that advanced control systems could be built. These two advances increased the precision of the process to permit cutting within 0.001 in of the true line in some cases and within 0.005 in. in almost all cases. Moreover, the advanced controls allowed inexperienced operators to make rapid setups and produce good parts without trial and error.

38.1.3 Factors Influencing Selection of Abrasive Jet Machining

Abrasive jet machining is chosen for many reasons, but all boil down to consideration of costs. Cost savings come from many areas including

1. Low capital cost
2. High cutting speed in difficult materials

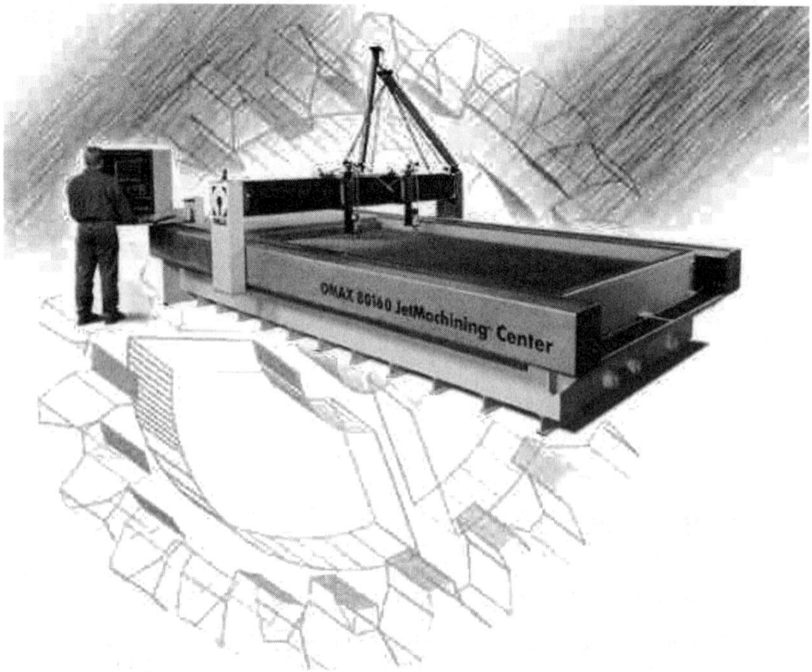

FIGURE 38.1 Precision machining for larger parts or multiple parts from stock, up to 5" × 10". (*Courtesy of OMAX corporation.*)

FIGURE 38.2 A collection of parts made by abrasive jet machining.

3. Minimal clamping and fixturing required for the workpiece
4. Rapid set up and programming
5. No heat affected zones on the cut surface
6. Waste in the form of valuable large pieces, not oily chips
7. Very thin sections can be made without deformation or melting.

Abrasive jet machining can also be compared with other processes. It is faster than wire EDM machining. It can make the same general shapes as EDM but is less accurate. The abrasive jet is not sensitive to slag and scale on or within the material that can cause problems by breaking wires in EDM machining. It can also be used on insulating materials.

The abrasive jet is also often considered as an alternate to various thermal cutting processes. Lasers are fast and accurate in thin materials where they outperform abrasive jets in terms of cutting speed. However, many materials (copper, brass, glass, ceramics, and others) can not be cut with lasers. The abrasive jet provides better precision and speed than lasers in thicker materials ($^1/_2$ in and up). Plasma cutting and oxy-acetylene burning are faster than both lasers and abrasive jets in thicker materials, but the range of materials cut by these processes is even more restricted than for lasers and these other thermal processes are, in general, less accurate than either lasers or abrasive jets. Finally, all the thermal processes leave a heat affected edge that often interferes with subsequent welding or machining while the abrasive jet does not.

Production of flat parts on machining centers requires fixturing and tooling and short runs are often impractical under CNC control. The abrasive jet can make parts from plate without tooling and fixtures directly from a CAD file. For this reason, abrasive jets are often found in prototype shops where they very effectively compete with manual machinery.

38.2 THE CUTTING PROCESS

38.2.1 Geometry of Jet and Character of the Cut Surface

In general, the jet cannot be regarded as a rigid tool. Abrasive jets bend as they pass through the cut material so that the exit point lags the entry point. At sufficiently high speeds there is a small side-to-side motion as well that produces striations on the cut surface. In Fig. 38.3 we see a part with five fingers cut at differing traverse speeds but equal times to produce each finger. In the long finger we can see the lag as recorded by the striations. The lag is usually not important for straight line cutting where maximum cutting speeds are determined by the striation that can be tolerated. In shape cutting, the lag can produce serious geometry errors on tight radii and it is the lag that sets the maximum speed for radii and corners.

FIGURE 38.3 Five fingers cut in equal times show effect of cutting speed.

The material in the long finger of Fig. 38.3 was cut at nearly the maximum possible cutting speed. Note that the top half of the cut has a much better surface than the bottom. If one were cutting a plate of half thickness at this same speed, the surface would be like the top half of the long finger in Fig. 38.3. The notion of "quality" used in jet cutting springs from this concept. A quality 1 cut will just make it through the material. For quality 2, the speed is set so that the jet could cut through a plate of double thickness, and the jet produces the better finish of the top half and so on for increasing qualities. The five fingers of Fig. 38.3 were made with qualities 1 through 5. There is no further significant surface improvement for qualities above 5. In modern controllers, the user specifies the quality and the controller chooses the speed.

Another effect of importance to the user is that the kerf left by the jet is generally tapered. With a fast moving jet the kerf is generally widest at the top and becomes more straight as the cutting speed is reduced. At very slow speeds the taper may even reverse and become widest at the bottom. In general, taper is the greatest problem in thin materials of less than 0.25in where cutting speeds are high.

38.2.2 Influence of Jet Parameters

The speed at which one can make a linear cut in a material of known machinability, M (Table 38.1), was given by Zeng[1,2] as:

$$V = \left(\frac{fa.\ M.P^{1.594}.d^{1.374}.Ma^{0.343}}{163.Q.H.Dm^{0.618}} \right)^{1.15}$$

where P = Stagnation pressure of the water jet, in thousands of psi (ksi)
d = Orifice diameter, in
Ma = Abrasive flow rate, lb/min
fa = Abrasive factor (1.0 for garnet)
Q = Quality index (1–5)
H = Material thickness, in
Dm = Mixing tube diameter, in
V = Traverse speed, in/min
M = See Table 38.1

Speeds must be slower than given above for small radii or the lag in the jet will cause serious shape errors. In contrast with oxy-acetylene burning, piercing is most effective with the jet moving. Motion during piercing allows the reflected jet to be diverted so that it does not disturb the incoming jet.

TABLE 38.1 Machinability (M) of Various Materials

Materials	M
Hardened tool steel	80
Mild steel	87
Copper	110
Titanium	115
Aluminum	213
Granite	322
Plexiglass	690
Pine wood	2637

38.3 EQUIPMENT

38.3.1 Pumping Equipment

All high pressure pumps are positive displacement pumps where a solid plunger is pushed into a closed water-filled chamber expelling the water. There are two commonly used means for moving the plunger: a crankshaft in the crank drive pump and a hydraulic cylinder in the intensifier type pump. In general, crank drive pumps have a higher operating efficiency than intensifier pumps (95 percent vs. 70 percent) and have lower energy costs. However, the crank drive pumps usually have a higher plunger speed and seal wear is greater (Fig. 38.4). The maintenance interval for a crank drive pump may be in the 200–400-h range whereas that for an intensifier may be in the 300–1000-h range. In either case, a rebuild is a 1–3-h job.

38.3.2 X-Y Cutting Equipment

The major use of abrasive jet machining is for making what may be described as flat two-dimensional parts usually from plate or sheet stock. The nozzle is carried by an X-Y mechanism under servo control (Fig. 38.5). A number of manufacturers are producing standard machines. Buyers should consider the following factors in selecting a machine:

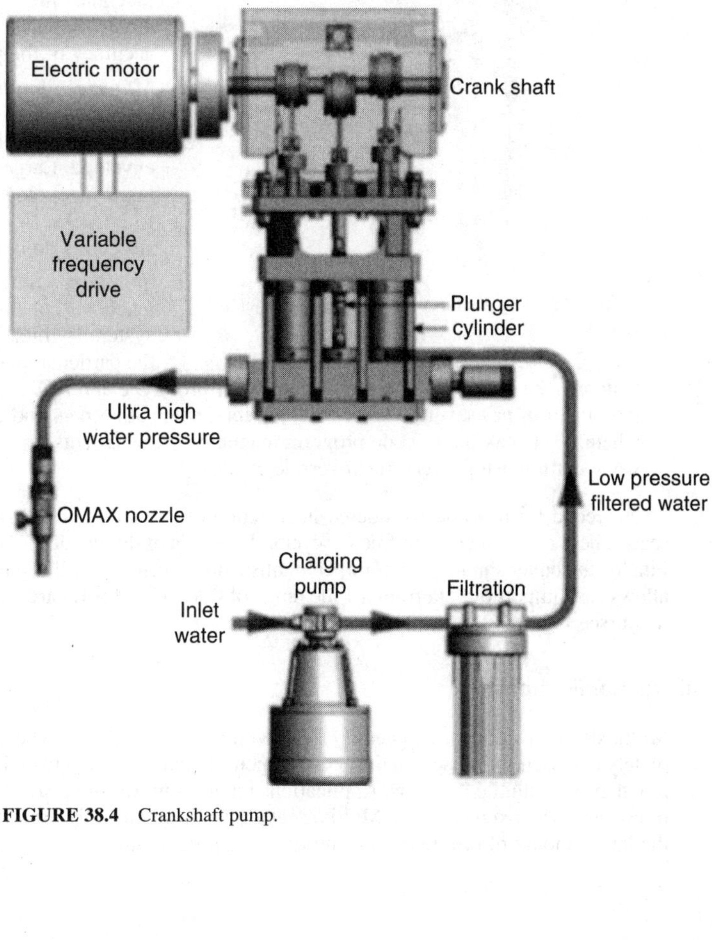

FIGURE 38.4 Crankshaft pump.

FIGURE 38.5 The architecture of an abrasive jet is fairly simple.

1. *Protection of the machine elements from water and abrasive.* Look for bellows that completely surround the machine elements. The operator should not be able to blow grit into the bearings with an air gun.
2. *Noise of the jet and pumping equipment.* Cutting should be performed under water for lowest noise. Otherwise, a separate room for the equipment should be considered.
3. *The precision of parts to be manufactured.* For loose tolerance parts to be used in fabrications savings may be obtained by buying a low precision lower cost machine. For other parts, higher precision in the abrasive machining step may allow skipping of subsequent secondary operations and resulting cost savings there. For job shop applications where the exact class of work to be done is not known, the more precise machine is recommended so as to be able to accept the widest range of work. Because the jet shape is dependent upon traverse speed, precision machines must accurately control the speed along the path as well as the positioning accuracy of the path.
4. *Material loading and removal.* Small machines are easy to load manually and can often work on plates larger than the machine envelope. Large machines are often used for large sheets loaded with overhead cranes or fork trucks. Be sure that guards and machine structures do not interfere with loading.
5. *Software.* Software for high precision machines contains a cutting model or "expert system" which sets the cutting speeds for the particular material and part geometry. This feature is essential for efficient production of prototype and low volume parts. Without this feature, the user has several choices: (1) accept defects at corners and curves, (2) cut slowly everywhere, (3) tweak the G-code program manually with trial cuts, or (4) manually adjust the speed while cutting using a feed rate override control.

A second factor to be considered in selection of software is the cost of upgrades and additional seats. These costs range from free to several thousands of dollars depending on the particular supplier. Finally, the buyer should look for open published file formats for the geometry and path storage. This allows easy import of data from a wide range of CAD-CAM software, nesting programs, and scanner input (see Fig. 38.6).

38.3.3 Multiaxis Applications

Multiaxis applications are generally done with custom machines where the cutting process is completely enclosed for noise and dust containment. Machines range from large X-Y gantries with additional axes to multiaxis robots. Applications range from trimming sheet metal and fiberglass parts to removing risers from castings. Multiaxis applications are usually production applications because of the large amount of programming and fixturing time required.

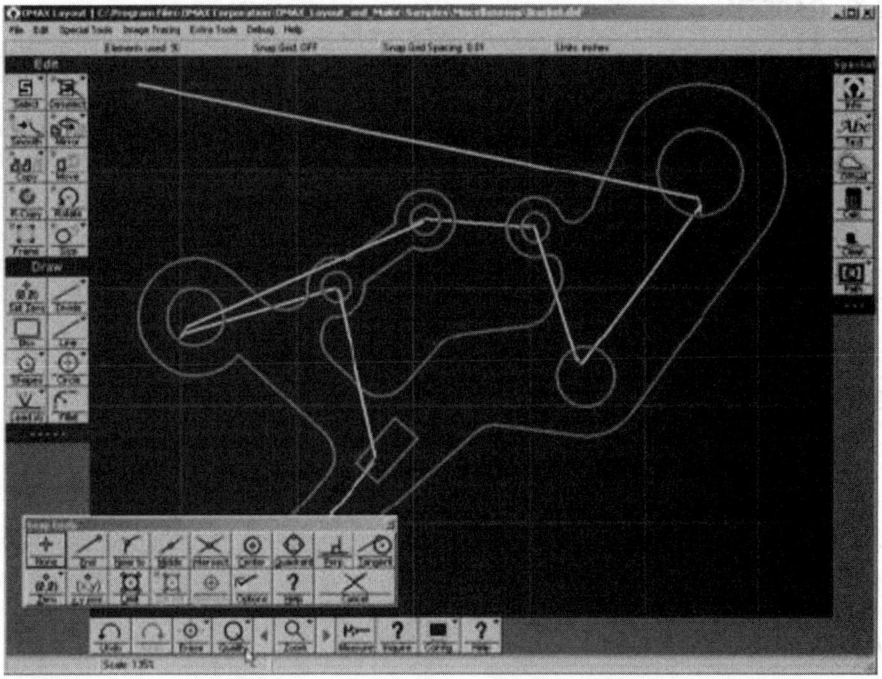

FIGURE 38.6 Layout is the OMAX program that lets you draw the part you want to make. This full-featured CAD program was created by OMAX and designed to work with the JetMachining center.

38.3.4 Utility Requirements

Clean mineral free water is required for long nozzle operation. The same elements in hard water that cause boiler scale form deposits on waterjet nozzles. Water softeners and sometimes reverse osmosis systems are often required to achieve long nozzle life. In extreme cases, hard water can also form harmful deposits within a pump. A cutting nozzle will use from 0.1 to 2 gallons per minute depending on size. In addition often 1–3 gallons per minute is required for intensifier cooling. Usual tap water pressures are sufficient for supply, but some manufacturers include a boost pump to insure that there is no loss of pumping due to unexpected supply pressure dips.

Most systems require three-phase electric power and use normal factory voltage levels. Power requirements would range from 15 kW to run a 20-hp pump up to 150 kW for a 200-hp pump. Small shops without three phase power can install a phase converter if they have sufficient single phase power. Another option, but a rarely used one for manufacturing applications, is to drive the pump with diesel power.

Often the control systems associated with jet cutting require a small supply of compressed air at normal shop pressures of 80–100 psi. This would almost certainly be a requirement if automatic on-off valves are used as they are most commonly air actuated.

The waste stream includes water, bits of the material being cut and abrasives. Usually the water is routed to a drain and the solids are removed separately. However, if the material being cut is hazardous, for example lead or beryllium, the waste water cannot be put in the drain. Closed loop water recycle systems must be used in these cases. Otherwise, the solid waste is directly put with waste routed to a land fill. The solid waste stream from abrasive jet cutting can be recycled to reclaim the spent abrasive in cases where the usage is large enough to justify the cost of the recycle system.

38.4 SAFETY

The largest danger is the jet itself. It can easily cut through human flesh and bone and an exposed jet generates noise levels that can easily harm human hearing. These dangers are most difficult to deal with in fully three-dimensional cutting in applications like removing risers from castings with abrasive jets. The solution for these applications is usually completely sealing the entire apparatus in a cutting box or room. In cutting shapes from or slitting flat materials, the problem is much less severe. The nozzle is kept very close to the material for both safety and best cutting performance and the jet is received in a catcher immediately below the material. When cutting materials where wetting is not an issue, it is a good idea to cut $^1/_2$ in or more under water. This almost totally eliminates any jet noise and makes a cleaner operation by suppressing splash back and mist that carries fine abrasive dust.

REFERENCES

1. Zeng, J., "Mechanisms of Brittle Material Erosion Associated with High Pressure Abrasive Waterjet Processing," Ph.D. diss., University of Rhode Island, Kingston, RI, 1992.
2. Zeng, J., Olsen, J., and Olsen, C., "The Abrasive Waterjet as a Precision Metal Cutting Tool," *Proceedings of the 10th American Waterjet Conference,* Houston, TX, August 14–17, 1999.

INFORMATION RESOURCE

http://www.waterjets.org.

CHAPTER 39
TOOLING MATERIALS FOR PLASTICS MOLDING APPLICATIONS

James Kaszynski
Böhler Uddeholm
Rolling Meadows, Illinois

39.1 INTRODUCTION

When it comes to the plastic injection molding process, a great deal of emphasis is placed on the mold design, tooling cost, and the amount of time required to bring the mold into a production environment. Often, the importance of selecting the optimum material from which to build the mold is not fully evaluated. There is a vast array of mold materials from which to select, and therefore it is important to fully understand what these different grades offer in terms of performance and productivity.

The selection process may also be controlled by material cost. However, the mold steel or mold material is typically estimated to be only 10–15 percent of the overall tooling cost. Most of the expense is attributed to design, labor, machining, heat treatment, and reworking. In addition, production loss because of tooling maintenance must also be considered.

To obtain optimum performance, the proper selection of the mold material is essential. This will ultimately be a multistep process. Begin by identifying the characteristics that the finished component must possess. Some of the most important things to consider are the following:

- Surface requirements of the molded part
- Complexity of the design
- Size of the mold

Next consider the type of plastic resin from which the molded part will be formed. This requires one to ask the following:

- Does the resin contain abrasive fillers or additives?
- Does it contain corrosive agents or is it likely to produce corrosive by-products that will attack the mold material?

The third set of conditions to consider is the required service life of the mold. That is, how many finished, molded components must this tool produce? The mold material selection for a short-run, prototype tool will be much different than the material chosen for a tool expected to produce a million parts. Let's consider these subjects individually and see how they are related to each other in the course of the selection process.

39.2 SURFACE FINISH OF MOLDED COMPONENT AND MOLD STEEL "POLISHABILITY"

Let's begin by examining the characteristics that the molded part must possess. It must be understood that any defect that appears on the surface of the mold will be replicated on to the surface of the molded part. If the mold surface contains scratches, pits, welds, uneven textured patterns, and the like, and they are visible on the mold surface, they most likely will appear to some degree on the molded part. In some cases the molded component may not have any cosmetic or aesthetic requirements. If so, some surface flaws on the part may be acceptable and not considered to be a cause for rejection.

In other circumstances, the finished part may need to function as a lens or may have some consumer application that requires a smooth, flawless appearance. Under these conditions a nearly perfectly polished mold surface will be necessary.

In an effort to define such a subjective parameter as mold material *polishability,* a scale defined by the Society of the Plastics Industry (SPI) was created (see Table 39.1). If the surface finish of the molded part is a high priority, one must consider the use of a mold material that will reach what is commonly referred to as a number one, or more specifically an A1-finish.

The producers of mold steels will often utilize a process known as *electroslag remelting* (ESR) to produce materials that are capable of being polished to a high surface finish. This additional step in the steel manufacturing process helps ensure that the mold steel will contain a low level of nonmetallic inclusions. Nonmetallic inclusions are brittle, constituents such as oxides and sulfides that are formed within the mold steel due to either the presence of tramp elements or the influence of the atmosphere during the mold steel melting and casting process.

An ESR furnace is basically a movable copper mold that contains a basic slag (Fig. 39.1). The heat of the slag is used to melt the as-cast ingot. The ingot melts droplet-by-droplet and the dense steel falls through the slag and resolidifies at the base of the mold. The slag acts as a filter, absorbing sulfur. In addition, the relatively fast resolidification results in a material with a relatively low level of segregation. The result is a steel with a low inclusion level and a homogeneous microstructure.

How does the presence of the nonmetallic inclusions and segregation within the mold steel limit the integrity of a mold steel? Lets refer to Fig. 39.2, which schematically represents the presence of a sulfide inclusion contained within a mold steel.

After the grinding process, any such inclusions will be nearly undetectable. However, during the stoning/polishing operation the softer steel matrix that surrounds the hard, brittle sulfide will be

TABLE 39.1 Empirical Ranking System Defining Mold Material Polishability.

Type of finish	Current SPI finish	Description	Previous finish number
	A1	Comparable to	#1
High polish	A2	Slightly finer than	#2
	A3	More imperfections than	
	B1	Finer than	#3
Paper finish	B2	Slightly finer than	
	B3	Comparable to	
	C1	Finer than	#4
Stone finish	C2	Slightly finer than	
	C3	Comparable to	
	D1	Finer than	#5
Dry blasted	D2	Slightly finer than	
	D3	Comparable to	

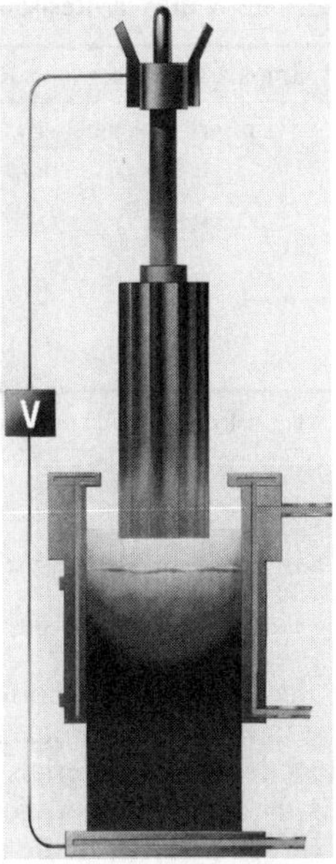

FIGURE 39.1 A schematic representation of the ESR unit (left). The slag contained with the copper mold is heated and used to remelt the as-casted ingot. As the as-casted ingot is consumed the mold moves upwards and leaving behind an ESR-ingot with superior properties. The photograph on the right is an actual ESR furnace in operation. (*Photo courtesy of Uddeholm Steel.*)

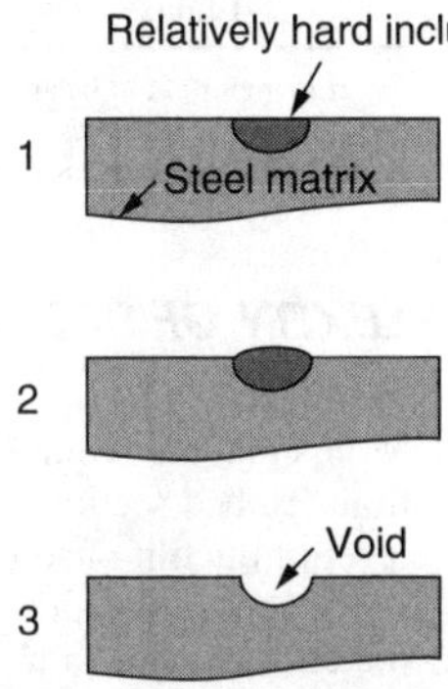

FIGURE 39.2 Schematic representation (left) of a sulfide inclusion present within the matrix of a mold steel. As the polishing process progresses there is a tendency to pull the sulfides and other hard constitutes within the mold steel (e.g., carbides) from the matrix. This defect or pit left on the core or cavity surface will be replicated on the surface of the mold part.

preferentially removed. Effectively, the inclusion is lifted from the surface. As the polishing operation continues the inclusion is pulled from the matrix of the steel, and a void or a pit is left behind on the steel surface. An actual example of a pitted surface is shown in Fig. 39.3.

Since polishing is typically the last operation performed prior to putting the mold into production, such a problem is not detected until the final stages of mold construction. Once such defects are found one must stone and repolish, in hopes of finding a plane within the steel that does not contain these inclusions. To avoid these issues it is imperative that when surface appearance is a high priority, one selects an electroslag remelted mold steel.

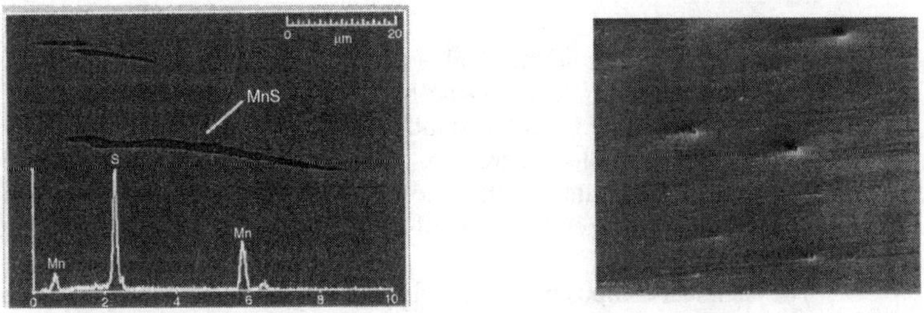

FIGURE 39.3 The photomicrograph on the left shows some actual sulfide inclusions that have formed within the mold steel matrix. Note that its orientation is running parallel to the rolling/forging direction. The photomicrograph on the right shows an actual example of a mold surface, which contains pits/voids from areas where the inclusions were torn.

TABLE 39.2 A Relative Comparison of Toughness Along With Wear Resistance.

Mold material	Hardness (HRC)	Impact toughness	Relative wear resistance
		(Empirical scale 1–10; 10 being the highest)	
P20	30	9	3
S7	56	6	6
H13	50	7	5
420 stainless	50	4	5
420F stainless	37	4	4
Elmax	58	3	9

It is important to balance toughness and wear resistance when selecting a mold material for a given application. Please note that these values are presented only as a relative comparison. Actual test values may vary based on original block size, heat treatment condition, hardness level, and sample orientation.

39.3 COMPLEXITY OF DESIGN

Molds that are required to produce plastic parts that contain sharp corners, thick-to-thin wall transitions, ribbed sections, etc. must be manufactured from mold materials that possess a relatively high level of ductility and toughness. These parameters define the materials ability to resist cracking.

A good indication of a material's ductility is the percentage of elongation that a sample of the material reaches in a standardized tensile test. However, it should be noted that the testing of tool steels hardened above a level of 55 Rockwell C (HRC) is not commonly performed. The reason is that at these hardness levels the axial alignment of the sample within the testing apparatus will strongly affect the measured values. The lack of precision prevents the measurement of reliable data and, therefore, rarely will one find this information unless the testing has been conducted at elevated temperatures when the samples become less brittle.

However, toughness is also an indication of crack resistance and it may be determined by comparing data from any number of standardized impact toughness tests (Charpy V-notch, Izod, unnotched-test samples, etc.) Table 39.2 provides a relative comparison for some commonly used mold steels based on an arbitrary scale of 1–10 (10 being the greatest).

The impact toughness test measures the amount of energy absorbed by a standard specimen as it is fractured with a pendulum. The values are reported in either foot-pounds or joules, and the higher the value the tougher the mold material. It is important to note, that the measured values will vary depending on the type of impact test, the notch geometry and the specimen size. Also, the orientation of the sample relative to the original block and the heat treated condition will all affect the measured values. Therefore, when comparing data from impact toughness testing one needs to be aware of these external conditions so that the values may be compared on a "level playing field."

39.4 WEAR RESISTANCE OF THE MOLD CAVITY/CORE

Wear resistance of a mold material is a function of its chemical composition and its hardness level. Some grades of mold steel may be heat-treated to achieve a hardness level exceeding 60 HRC. This may be desirable from the standpoint of wear resistance; however, it is likely to be too brittle for most applications. Therefore, when selecting mold materials it is necessary to balance the material's wear resistance with its level of ductility or toughness. Again, referring to Table 39.2 will provide a relative comparison of toughness and wear resistance among some commonly used mold materials.

The chemical composition of a mold steel will determine its level of wear resistance. By definition, a mold steel is comprised predominantly of iron with typically no more than a 2-percent addition of carbon. In essence, the mold steel or any alloy for that matter is a solution or more specifically a solid solution made up of iron, carbon, and any number of alloying elements.

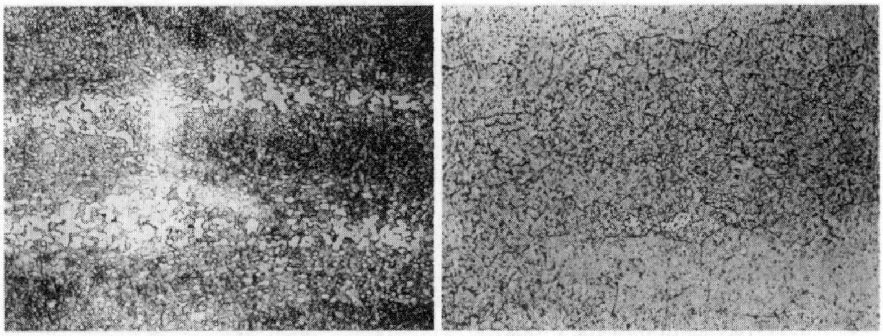

FIGURE 39.4 A comparison of two stainless mold steels with similar chemical compositions. The microstructure on the left contains large, blocky carbides (white areas) that run parallel to the rolling/forging direction of the original block. The sample on the right has undergone an additional operation known as electro-slag remelting (ESR). The carbides are still present, however, they exist as relatively fine, uniformly distributed particles.

Some common alloying elements include chromium, molybdenum, and vanadium. These elements may be defined as carbide-formers, since they react with the carbon to form a combination of chromium, molybdenum and/or vanadium carbides. The carbides that are present within the matrix provide the steel with its wear resistance. The greater the amount of carbon and alloying elements present, the greater will be the percentage of the wear resistant carbide particles that will form within the material.

To provide an understanding of this phenomenon one needs to consider the microstructure of the mold steel. The microstructure as the name implies is an examination of the material under high magnification. Basically, a sample of the material is polished to a mirror-finish, etched with an acidic solution and examined with a metallurgical microscope. An example of some mold steel microstructures is shown in Fig. 39.4.

The carbides that form within the mold steel are hard particles held within a softer, more ductile matrix. The more carbide present, the greater the wear resistance; however, their presence also reduces toughness or crack resistance. The carbides are relatively brittle and when they form as large, blocky particles their presence can lead to crack propagation and overall poor material toughness. Again, it is this balancing act that must be considered. One wants to maximize wear resistance without making the mold too susceptible to cracking or premature failure.

39.5 SIZE OF THE MOLD

The overall size of the mold will determine whether it is prudent to use a grade of mold steel that will require a heat treatment process (i.e., a quench and tempered steel or an age-hardened steel) or to select one that is delivered to the toolmaker in a pre-hardened condition (i.e., prehardened mold steel, copper, and aluminum based materials).

When using a quench and tempered mold steel, the heat treatment procedure must be incorporated into the mold building process. When the soft annealed, steel is received by the mold maker, the ideal scenario for the mold building sequence is as follows:

- Perform the bulk of the rough machining, leaving a significant amount of machine stock to allow for movement during a stress relief operation.
- Stress relieve tool at 1200°F held at temperature for 2 h; the majority of movement will occur at this stage of the mold building process.

- Continue machining until the semifinished dimensions (i.e., leave adequate machining stock to allow for growth, shrinkage, distortion, etc. that may occur during heat treating operation).
- Heat-treat steel according to manufacturers recommendation. (Many mold steel manufacturers specify procedures to ensure optimum properties for their materials.)
- Finish machine to final dimensions.

It is essential to quench the tool rapidly from the hardening temperature in order to reach a high hardness level and also to form a sound microstructure. When relatively large molds are to be constructed the heat treatment process can be a challenge. Large tooling may not be capable of being quenched quickly enough to ensure a sound microstructure or consistent hardness level throughout the cross section of the block.

When such conditions exist it becomes necessary to consider using a prehardened material. These materials are typically supplied at a hardness ranging from 28 to 40 HRC. They are soft enough to machine, yet have adequate hardness for many molding applications.

In general, these materials do not require any thermal processing. The only time when this is warranted is when a significant amount of material must be removed from a piece of mold steel. Removing excessive amounts of material will often lead to the introduction of machining stresses. These stresses may make the material dimensionally unstable and susceptible to movement during operation. Therefore, a thermal stress temper (not a stress relief) may be advisable and should be incorporated into the mold building process. This extra step will remove the stresses before the tool is put into a production environment.

A stress temper should be specified at a temperature of 50–100°F below the highest tempering temperature that was used during the hardening operation of the mold steel. This will ensure that the machining stresses are reduced without leading to softening of the material. This mold building sequence should be as follows:

- Rough machining, leaving a generous amount of machining stock to compensate for movement/distortion.
- Thermal stress temper at a temperature that is 50–100°F below the highest temperature used during the heat treat process (may have to contact steel manufacturer to determine the tempering temperature used for the particular grade in question).
- Finish machine to final dimensions.

39.6 CORROSION-RESISTANT MOLD MATERIALS

Many plastic resins contain either corrosive agents or have the potential to produce corrosive byproducts during the molding process. The corrosive agents may be additives required to impart specific properties such as UV stabilizers or flame inhibitors. However, some resins themselves may breakdown during the melting stage. For example, polyvinyl chloride (PVC) may disassociate, allowing chlorine to combine with water to form hydrochloric acid. Such conditions will lead to the attack of the mold surfaces and components.

If such conditions exist, it may be necessary to consider using a mold material that will resist this corrosion or perhaps a secondary operation may be required to provide a protective barrier. That is, some type of a surface treatment such as plating.

Lets begin by considering the use of a corrosion resistant mold material. In many circumstances this will lead to the selection of one of two types of stainless steel. The first choice is a *quench and tempered* stainless mold steel (e.g., AISI 420 stainless) and the second is an *age-hardened* stainless mold steel (e.g., 15-5 precipitation hardening).

By referring to the previous sections we know that each grade offers some unique properties that need to be considered in order to make the best selection. The 420 type materials require a heat treatment operation that brings the material to relatively high temperatures, for a designated period of time, followed by rapid cooling. This can lead to excessive amounts of distortion or possible cracking when relatively large or complex molds are involved.

However, the advantage of using the 420 type stainless is that the material is capable of reaching hardness levels in the range of 48–52 HRC. This provides a mold with the ability to withstand wear and indentation while maintaining a reasonably good resistance to corrosive attack.

The age-hardening type stainless materials use a different mechanism to reach hardness. Here the materials are provided to the mold builder in what is referred to as a solutionized condition. That is, specific alloying elements are maintained in solid solution or simply stated are dissolved within the bulk or matrix of the steel. At the appropriate time, the material is placed in a furnace at relatively low temperatures causing these alloying elements to precipitate out as a uniform, dispersion of small particles. The net result is that the steel can reach, in most cases, a hardness level of up to 42 HRC.

These age-hardening (sometimes referred to as precipitation hardening) mold steels allow for the building of large and/or complex molds with a reduced risk of cracking or excessive distortion during the heat treatment operation. The main drawback is that their hardness level does not reach that of the quench and tempered type materials. Therefore, they will not withstand the same level of wear resistance or resistance to indentation as compared to the harder mold materials. Other options for corrosion resistance include using a noncorrosion resistant material (e.g., P20, H13, S7, etc.) and coating it with a corrosion-resistant barrier. There are a multitude of choices for selecting coatings and surface treatments. This is an area that is continually being developed with the introduction of new products.

When deciding on a specific surface treatment one fundamental warning must be considered. What temperature is used to apply the surface treatment? Since this is the last operation in the mold building process, the application temperature must be below the tempering or aging temperature used during the hardening process of the mold material. This general rule must be followed. This requires knowledge of the prior heat treatment history or in the case of prehardened materials, information from the material supplier.

The reason is that if this temperature is exceeded during the surface treatment operation, the mold material will soften or distort. Each condition will lead to either a degradation of material properties or perhaps a complete loss of dimensional tolerances. Therefore, when selecting any secondary operation such as a surface treatment (plating, coating, nitriding, etc.) it is essential to adhere to this rule.

39.7 THERMALLY CONDUCTIVE MOLD MATERIALS

In plastics molding operations, the cycle time often becomes a critical factor in determining the overall costs associated with manufacturing finished components. When thousands or millions of parts must be produced, the savings of only a few seconds with each cycle can make a significant difference.

The ability to remove heat from the resin and quickly solidify it, may also affect the quality of the finished piece. Thin-walled components or parts that contain variations in wall thickness may be difficult to mold due to dimensional instability. Often times a faster cooling rate will not only improve cycle time, but will also result in finished parts that have more accurate dimensions and tighter tolerances.

When faced with such issues it becomes important to consider some alternative tooling materials from which to build the mold. Due to their outstanding thermal conductivity properties, aluminum and copper base alloys have the ability to extract heat quickly from the resin. These mold materials used independently or in conjunction with traditional mold steels may be used to build the entire core and cavity or used strategically at specific problem areas within the mold.

39.8 ALUMINUM MOLD MATERIALS

Specialized, high-strength aluminum mold materials have been developed specifically for the plastics molding industry. Supplied in the prehardened condition, there is no need for any heat treatment during the mold building process. These grades have excellent thermal conductivity properties thereby, reducing cycle times and simplifying the requirements for the mold cooling system.

The as-supplied hardness typically ranges from 160 to 170 Brinell. Although, this hardness level is significantly lower than that of a typical, prehardened mold steel (e.g., P20 type mold steel supplied at ~300 Brinell), it is still adequate for many low-wear, plastics molding applications, including

- Blow molding
- Short-to-medium run injection molds
- Prototype tooling

When producing a cavity, it is the center or core of the block that will function as the molding surface. Often times there is a significant drop in hardness from the surface-to-core, particularly in larger sized blocks of standard aluminum grades, such as Type 6061 or Type 7075. This difference in hardness becomes more pronounced for pieces that are greater than 6 in. in thickness.

A lower hardness on the mold surface will lead to more rapid wear and washout of the mold details. To help reduce the affects of wear, mold quality aluminum materials have been developed which exhibit excellent, through-hardening properties. A comparison between some commonly used grades is shown below in Fig. 39.5.

Another important factor to keep in mind, when considering an aluminum tooling material is its strength level. If the stresses applied during the molding operation exceed the material's yield strength, the mold will become permanently distorted. If the stresses exceed the mold material's tensile strength, the mold will experience a catastrophic failure.

There is a relationship between the hardness of a material and its tensile strength. These two properties are directly proportional to each other. That is, as the hardness increases, the tensile strength also increases. This property is significant from the aspect of designing a mold and affects the required wall thickness dimensions, placement of water lines, etc. An example of the tensile strength for some commonly used aluminum mold materials, as a function of plate thickness, is shown below in Fig. 39.6.

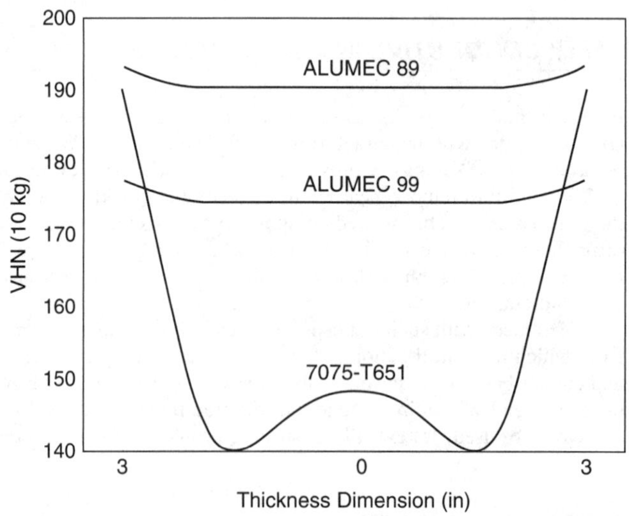

FIGURE 39.5 Example of a hardness profile for several different grades of aluminum, commonly used for plastics molding applications (6-in thick cross section). The loss in hardness near the core of some aluminum grades may result in premature wear of the mold surfaces and parting lines.

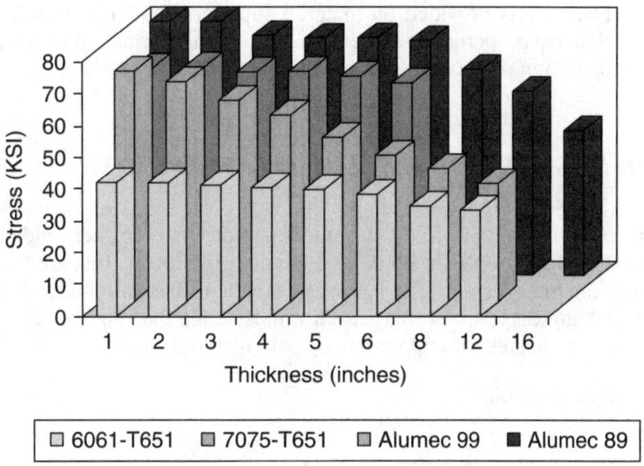

FIGURE 39.6 Average tensile strength values of several commercially available aluminum mold materials as a function of their plate thickness. The difference in tensile strength becomes more significant as the plate thickness increases.

The characteristics that make aluminum mold materials desirable for certain molding applications includes the following:

- Machinability (metal removal rates that are 8–10× faster than for a typical mold steel in conventional machining; 4–5× faster in EDM operations)
- No heat treatment required
- Thermal conductivity (3× greater than a typical mold steel)
- Low density (one third that of steel)
- Inherently corrosion resistant in many environments

These materials are readily machinable, do not require heat treatment, and are capable of being modified via welding if design changes are required. These characteristics make high-strength aluminum alloys ideal for projects that require extremely short lead times (e.g., prototype molding). In addition, the low density of aluminum makes the handling and operation of large molds considerably easier and reduces the wear on press components.

What are the limitations? Although high-strength aluminum molding materials have been developed specifically for the plastics industry, one must realize that they still do not have the strength levels of hardened mold steel such as AISI P20, H13, S7, etc. This is well understood by designers and mold builders that work with aluminum tooling materials on a regular basis and it must be taken into consideration when designing the mold.

The following points must be recognized when designing aluminum molds:

- Wall thickness of 1.5× greater than that of a mold steel is typically required.
- Very deep cavities should be interlocked at the sides to prevent flexing.
- Wall thickness should be equal to at least the cavity depth.
- Core-height should not be greater than 2× its width, unless supported.
- Avoid sharp corners; use as generous a radius as possible.
- Coolant-lines should be placed at a distance no closer than 2× the cooling-line diameter from the mold surface.

Data sheets provided by material suppliers contain a wealth of information with regard to the mechanical properties of these materials. Such information provides important data that may be used in the design phase of the mold building operation.

39.9 COPPER-BASE ALLOYS FOR MOLD APPLICATIONS

Copper-base alloys offer some unique options to toolmakers and plastic molders. These materials are often indispensable for producing quality plastic parts that are free from warpage, shrinkage marks, and internal stresses. They possess excellent thermal conductivity characteristics and a relatively high hardness level that may reach values within the range of 30–40 HRC. Applications that are often good candidates for copper-based mold materials include:

- Injection molds
- Core pins
- Blow molds (pinch-offs, neck rings)
- Injection nozzles

The ability to remove the heat from the resin will determine the cycle time and dimensional stability of a molded component. One may even consider a plastic mold to function as a heat exchanger. Due to the shrinkage of the resin during the solidification process, the majority of the heat build-up will typically occur at the core-side of the mold.

The exceptional thermal conductivity of copper-base alloys can help promote rapid and uniform cooling in these critical areas. The copper materials may be strategically inserted within a mold to provide rapid cooling at a specific location or used to build the entire core and/or cavity.

Although these materials possess thermal conductivity values that are much greater than mold steel, the heat must still be removed via a cooling system. Otherwise, the copper will function as heat sink, at first drawing heat away from the resin, but then retaining it. This heat must be transferred to a cooling system either directly via a coolant line or indirectly through the use of core pins and a chill plate.

An advantage of the copper material is that its excellent thermal conductivity properties will help simplify the cooling system. This means that perhaps fewer coolant channels will be required compared to a similar core made from a mold steel. This is an advantage to the designer, who must also plan and leave room for the ejector system on the core side of the tool.

When considering the placement of the cooling lines there are some rules of thumb that should be adhered to when using copper based materials. For example, the distance from the mold surface to any cooling line should be a minimum of two times the cooling line diameter. The distance between adjacent cooling lines should be a minimum of three diameters (see Fig. 39.7).

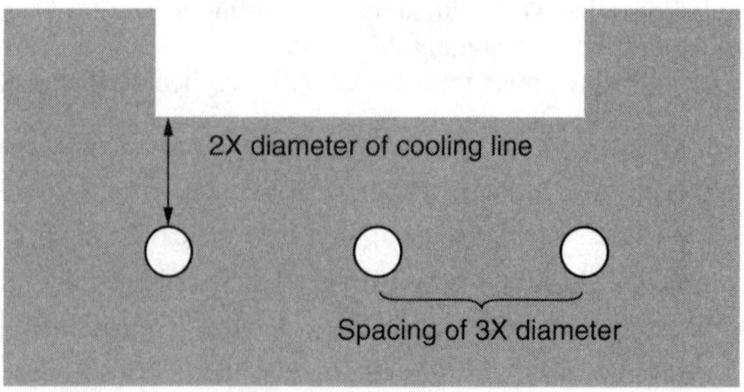

FIGURE 39.7 Schematic representation of cooling line placement.

This helps minimize stresses and still provides good proximity of the coolant to the molding surfaces. Again, the purpose of the coolant is to prevent the copper from retaining the heat that is removed from the plastic. The risk of placing cooling lines too close to the mold surfaces or to each other, adds an unnecessary risk for cracking.

The machining rates for conventional machining processes are much greater compared to a mold steel. However, the high thermal conductivity of the copper-base materials will increase the time required for *electrical discharge machining* (EDM). These affects can be minimized by performing the bulk of the machining first through conventional means before moving to the EDM process for the finished machining operations.

Copper-based mold materials can be welded without a significant loss of strength. Weld filler metals are available that match the chemical composition and hardness profile of the base metal. Many of these copper based materials lend themselves to texturing and have the ability to be polished to a mirror finish.

The wear resistance of the aluminum and copper base materials may be increased through the use of various surface treatments. Chrome plating, electroless-nickel plating, or low temperature *physical vapor deposition* (PVD) processes may be used. Before performing any type of surface treatment, it is important to make sure that the temperature used during its application does not adversely affect the hardness and strength of the mold material. The material supplier can provide the maximum temperature at which the mold material may be exposed without suffering any degradation in properties.

39.9.1 High Performance Powder Metallurgical Steels

Today, many resins contain reinforcing fibers and fillers to improve the strength of molded parts. These additives are inherently very abrasive and promote erosion and premature wear of molds and molding components. In addition to wear, many resins will become chemically aggressive at molding temperatures. These corrosive by-products are then available to attack noncorrosion resistant mold steels.

To combat such problems it becomes necessary to use highly alloyed tooling materials. If such materials are manufactured through conventional methods they are often too brittle for practical use. Segregation problems and the formation of large, blocky carbides act as stress risers that promote cracking.

However, highly alloyed mold steels can be produced via the powder metallurgical (P/M) method. These materials contain significant amounts of spherically shaped, uniformly distributed carbide particles. This allows for excellent wear properties while maintaining the mold steel's resistance to brittle failure. The composition can also be balanced to provide the mold steel with a high level of corrosion resistance. The P/M manufacturing process provides alternatives to the more traditional plastic mold steels for demanding applications.

39.10 STANDARD MOLD STEEL PRODUCTION METHODS

In order to appreciate the advantages of the P/M process, it is important to first have a fundamental understanding on how conventional mold steels are produced. The process begins by melting starting material in an *electric arc furnace* (EAF). The starting material is comprised of carefully selected, low alloy scrap steel with the lowest possible level of impurities. EAF units can melt from 10 tons, to as much as 50 tons of starting material per heat.

After the primary melting is completed, the molten steel is transferred to a ladle or refining vessel where it undergoes what is commonly referred to as secondary or ladle metallurgy. At this stage the composition of the steel is evaluated and adjusted to produce the final chemical composition. Additional steps are often involved such as slag treatments and degassing procedures used to remove undesirable elements.

Following this refining process, the molten steel is poured into large molds where it solidifies into a simple shape that is referred to as an ingot. These ingots will take several hours to solidify completely.

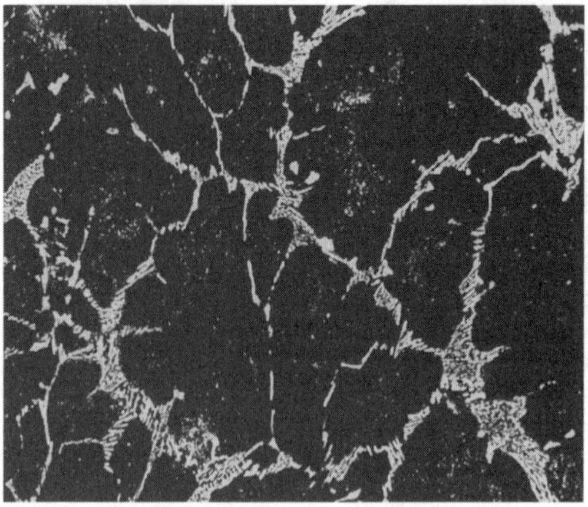

FIGURE 39.8 Example of alloy carbide networks (white, web-like areas) that form in an as-cast mold steel (500×).

This relatively long period of time will lead to significant amounts of chemical segregation. The result is a wide variation in the chemical composition and mechanical properties throughout the cross section of an ingot. These differences will be more pronounced as the alloy content of the steel increases.

It is important to note that steel is in fact a "solution" comprised predominantly of iron with additional alloying elements such as carbon, chromium, molybdenum, and vanadium. These alloying elements are more soluble when the steel is in the liquid state as opposed to when it is in the solid state. Therefore, as the molten steel begins to freeze, the alloying elements migrate to the remaining liquid, as the solidification front progresses from the surface toward the core of the ingot. Therefore, the center of the ingot will become progressively richer in alloying elements.

The alloying elements and the carbon within the steel will react with each other to form carbides. The carbides will not be evenly distributed throughout the cross section of the ingot, but instead they will also be concentrated in certain regions. These segregated regions are commonly referred to as carbide networks as shown below in Fig. 39.8.

To some extent, thermal treatments and mechanical working, such as forging and rolling, will break down the pattern of chemical and carbide segregation. However, the carbide networks will have a tendency to align themselves as stringers or banded areas that run parallel to the primary hot working direction.

Carbide particles are inherently very hard and provide the steel with its abrasive wear resistance. However, when the carbides form as long stringers they will act as a stress risers within the steel. These regions will be more susceptible to brittle failure and the overall toughness properties of the steel will decrease.

39.11 POWDER METALLURGICAL PROCESS FOR MOLD STEEL PRODUCTION

With a basic understanding of conventional mold steel manufacturing one can appreciate the benefits of producing these materials via the powder metallurgical process. The P/M process begins with the creation of very fine grains of powder that are formed through a process referred to as atomization.

The process begins by melting the starting material in an induction furnace. When the initial melting is completed the molten steel is transferred to a holding vessel known as a tundish. During this stage of production, the surface of the molten steel is covered with a slag. The molten steel is treated with the slag to remove undesirable tramp elements, thereby, providing for a very clean material.

In the P/M process the molten steel is not poured from the ladle into an ingot mold. Instead, when the ladle is tapped, the stream of liquid steel is struck by a series of nozzles that emit high-pressure nitrogen gas. As the nitrogen gas strikes the column of molten steel, the steel is dispersed into tiny particles that solidify instantaneously.

Each particle that is formed is essentially a tiny ingot. However, instead of taking several hours to solidify, as in the conventional steel making process, the powder solidifies in only a fraction of a second. This prevents the segregation problems that occur due to the long solidification rates experienced during ingot casting. Therefore, the chemical composition of the powder is consistent from particle to particle and the alloy carbides that form are small, spherical and evenly distributed.

The powder is not handled in open air, but is poured directly into steel containers under a protective atmosphere. The container is then evacuated and the top is welded shut. The sealed containers are then processed through an operation that is referred to as *hot isostatic pressing* (HIP, see Fig. 39.9). During this operation the container is subjected to an equal amount of pressure from all directions while it is held at an elevated temperature. The combination of high temperature and pressure leads to a complete bonding of the powder. The result is a fully dense piece of steel that can be forged or rolled to the required dimensions.

It is the rapid solidification which occurs in the P/M process that allows for the production of highly alloyed mold steels which are free of both chemical and carbide segregation. To understand the advantage of the P/M process it is beneficial to compare the microstructure of conventionally produced mold steel to a P/M mold steel.

As shown below in Fig. 39.10 the differences are obvious. The formation of the blocky, carbide stringers in the conventional steel are inherent stress risers that make the steel more susceptible to chipping and cracking. It is the uniform, spherical carbide particles that form in the P/M tool steels

FIGURE 39.9 The powder is placed into a capsule and welded shut (shown on the right). Under this condition the powder is only approximately 70% dense. After the hot isostatic pressing operation the entire capsule is compressed at high temperatures and extreme pressures to consolidate the powder inside into a 100% dense piece of steel (as-hipped capsule shown on the left).

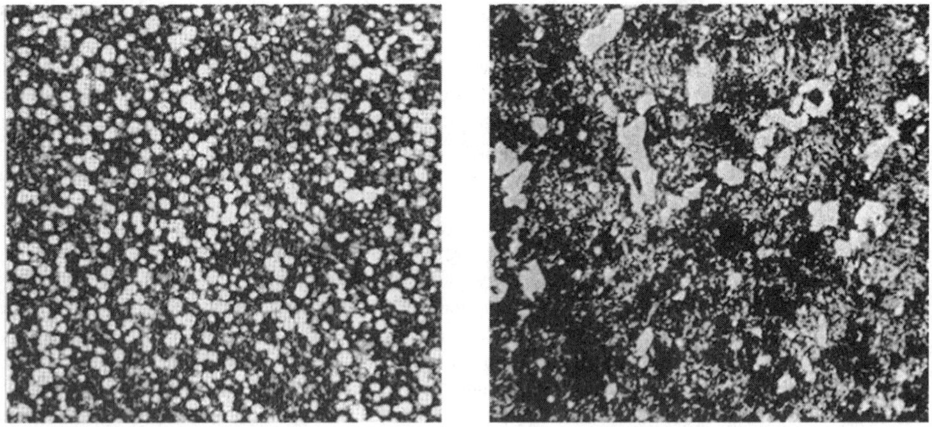

FIGURE 39.10 The microstructures shown above are for two steels of comparable chemical compositions. The difference is that the one on the left was manufactured via the P/M process. The uniformly dispersed, carbide particles (white regions) provide the mold steel with excellent wear resistance, while maintaining excellent toughness characteristics (500×).

that are responsible for providing the abrasive wear resistance without degrading the steel's toughness properties.

The carbide particles range in size from 1 to 4 μm. The additional benefit of the P/M production method is that the carbides are also spherically shaped and uniformly distributed throughout the matrix. Regardless of the size of the rolled or forged piece, the carbide distribution will remain consistent throughout the entire cross section.

39.12 SUMMARY

Selecting the best mold material for a given application requires the evaluation of many different variables. Initially, one needs to consider the properties and characteristics of the plastic resin from which the components will be molded as well as the functionality of the part. Other factors worthy of consideration are the size of the mold, required service life and the complexity of the design. Table 39.3 lists some commonly used materials for mold building; Table 39.4 compares their hardness and thermal conductivity.

Cycle time will also contribute to the overall costs per part produced. As previously discussed, the ability to remove heat quickly from the resin will shorten cycle times and often yields finished parts that have better dimensional stability. This is an additional consideration when trying to optimize mold performance.

In summary, the selection of the mold material must be a joint effort between the designer, toolmaker, and often times the heat treater and material supplier. Each group must use their expertise to determine the best balance of properties that will lead to the optimum mold material selection.

TABLE 39.3. Typical Materials for Mold Building Applications

Typical materials for mold building applications	
Quench & tempered mold steels	S7, H13, A2, D2
Pre-hardened mold steels	P20, 420F
Age-hardened mold steels	15-5 PH Stainless, Corrax, M261
Non-ferrous pre-hardened mold materials	7075 Aluminum, Alumec 89, Moldmax, Protherm

TABLE 39.4 Hardness and Thermal Conductivity Comparison of Mold Materials

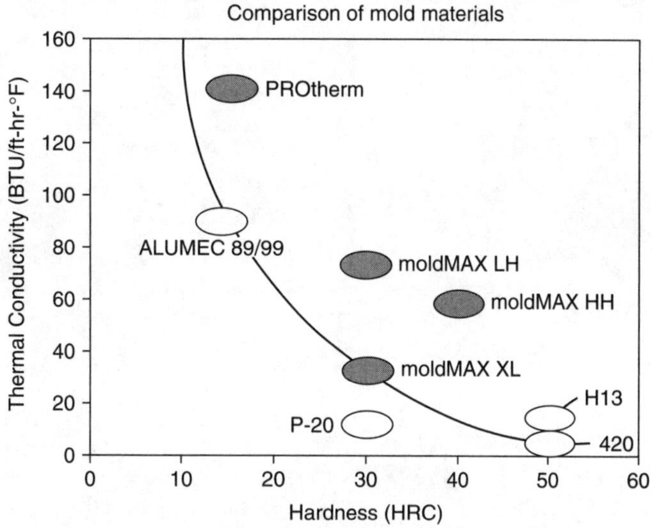

Trade name	Grade type	Typical working hardness	Thermal conductivity (BTU/ft .h .°F)
Alumec 89	High-strength aluminum	164–168 BHN	95
Alumec 99	High-strength/high corrosion resistant aluminum	158–162 BHN	95
Protherm	Beryllium copper	190 BHN	140
Moldmax—low hard	Beryllium copper	26–32 HRC	75
Moldmax—high hard	Beryllium copper	36–42 HRC	60
Moldmax XL	Large block, copper-base alloy	30 HRC	40
Typical mold steels	P20	30 HRC	17
	H13	50 HRC	15
	420 stainless	50 HRC	09

All values for thermal conductivity measured at room temperature.

CHAPTER 40
INJECTION MOLDS FOR THERMOPLASTICS

Fred G. Steil
D-M-E Company
Madison Heights, Michigan

40.1 INTRODUCTION

Injection molds for thermoplastics are typically designed and constructed of steel to fill the need for a particular part or set of plastic parts. Some short run prototype molds are made of aluminum or laser-sintered material. In order to design a particular mold, the mold designer must be given certain information. This information must contain the following: the part design, the required production rate, the amount of parts the mold may be expected to produce over its life, and any special requirements of the part. Before discussing how the part design affects the design of the mold, it will be helpful to define some terms specific to mold making (Fig. 40.1).

40.2 INJECTION MOLD COMPONENT DEFINITIONS

2^n. The numbers 1, 2, 4, 8, 16, 32, 64, etc.

Baffle. A device used to alter the flow path of coolant.

Bubbler. A tube to conduct coolant to the inside of the distal end of a core.

Cavity. The female part of the molding surface that forms the molded part. Usually found on the "A" or fixed half of the mold.

Core. The male part of the molding surface that forms the molded part. Usually found on the "B" or movable or ejection side of the mold.

Draft. The angle on the sides of cavities and cores that allow the plastic part to be readily released and removed from the mold.

Ejection system. That part of the mold that ejects the part. It may contain ejector pins, ejector sleeves, strippers, lifters, two-stage ejection, accelerated ejection mechanisms and more, including the ejector assembly and its plates.

Ejector pin. A steel pin that is attached to the ejector system of the mold and is used to eject the part from the mold.

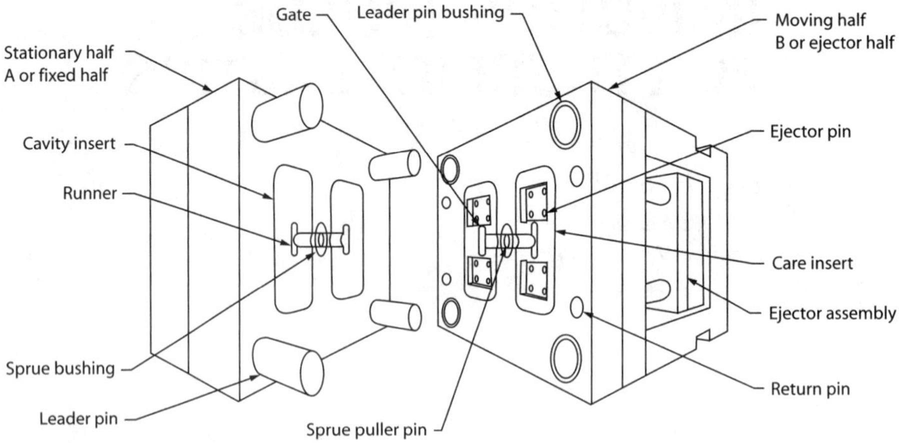

FIGURE 40.1 Mold components.

Interlock. The method of positive alignment between cavity, stripper, and core using interlocking tapers or engaging straight surfaces.

Leader pin. A steel pin generally located at the periphery of the mold, used to guide the mold halves together in the proper alignment.

Lifter. A device attached to the ejector plate and containing part of the molding surface for releasing an undercut as the ejection system moves forward.

Gate. Located at the very end of the runner, the gate is a short restriction in the flow path of the melt stream, connecting it to the cavity.

Hot runner. A method of conducting plastic from the machine nozzle to the gate where the plastic is kept in a contiguous liquid state. It is particularly suited for top gating parts.

Leader pin bushing. A steel or bronze bushing to mate with the leader pin and provide proper alignment of the mold halves. Always found on the opposite side of the mold from the leader pin.

Parting line. The place between plates, where the mold separates when it opens, often designated by (P/L).

Projected area. The size of the molding area as seen from the top (locating ring) side of the mold.

Return pin. The return pins push the ejector plates back to the molding position as the mold closes.

Runner. The flow path of the plastic from the sprue bushing to the gate(s).

Side action. A method of removing the part of the mold that forms an undercut in the molded part. Usually the side action is parallel to the parting plane of the mold.

Slide. A piece of the mold that slides a side core or undercut cavity or core component out of the way to allow ejection. The slide can be activated by angle pins in the mold or by external cylinders. A slide is held in place, during molding, by a heel block.

Splits. A pair of opposing slides, sliding parallel to the parting line.

Sprue. The portion of the runner that is formed by the sprue bushing.

Sprue bushing. Conducts the plastic from the machine nozzle to the beginning of the runner.

Stripper. A part of the ejection system where the stripper plate surrounding a core, strips the part from the core.

Sucker pin. A name given to any pin that is used to pull a plastic item across the parting line as the mold opens.

Undercut. A condition that exists in the molded part that prevents its being ejected from the mold, perpendicular to the parting line, e.g., a thread.

Unscrewing mold. Refers to a mold where the molded parts contain threads which must be unscrewed to remove them from the mold components that formed them.

40.3 PART DESIGN

The part design is perhaps the most important input to the mold designer. The shape of the part will determine the parting line of the mold. If there are undercuts, they must be considered in order to remove the part from the mold. The part must, under most conditions, allow a slight taper to its sides and its protrusions in order to remove it from the mold. This taper is known as *draft*. The amount of draft required is often determined by the plastics material selected for the part as well as any texture on the parts' surface. The design of the part may also dictate the gate location and the position in which the part will be formed in the mold. The position in which the part is formed will affect how the part can be removed or ejected from the mold.

40.4 PRODUCTION RATE

The required production rate of the mold is needed in order to choose the amount of cavities the mold will contain. The part is formed by injecting plastic into the mold at a controlled rate, time, temperature, and pressure. It is then cooled sufficiently to allow the mold to be opened and the part ejected. This cycle is repeated over and over. The time it takes to complete the cycle is called the *cycle time*. The more cavities there are in the mold, the more pieces will be produced every cycle. Multi cavity molds are generally constructed to have 2^n cavities. This allows the mold to be constructed with a geometrically balanced runner system. A geometrically *balanced runner system* is one where the distance from the end of the sprue to each and every cavity is identical.

40.4.1 Expected Volume to be Produced

The amount of parts the mold is expected to produce is an important piece of information that unfortunately is often overlooked. If the mold is needed to produce many millions of parts, it needs to be constructed much differently than if it only needs to produce several hundred parts. Molds that produce standard unscrewed closures, for instance, are generally designed to last 15 years. They undergo planned maintenance every 2 years for wear parts and a major reconditioning every 5 years. To be able to survive this amount of work, the mold plates are usually made of hardened carbon steel or hardened stainless steel. The cavities and cores are inserts also made of hardened tool steels. Moving parts run on bearing material and the mold components are usually interlocked to assure concentricity of the parts and to prevent wear from misalignment. Compare this to a mold that must only produce 300 housings. That mold may be made of aluminum. It may rely on just the leader pins to align the mold halves. The cavity and core may be cut directly into the mold plates. Of course the cost is far less, but the cost is usually amortized across the pieces produced and the machine hour rate needed to produce them.

40.4.2 Special Requirements of the Part

Many parts have special requirements. Often these requirements determine the plastics material to be molded. These requirements may include surface finish, either glossy or textured, color, strength, transparency, flexibility, electrical resistance, lubricity, or other more subtle issues. It may also be that part geometry will require the mold to be built a certain way. This might include a special ejection, unscrewing devices, top injection, side action, splits, or special cooling items.

40.5 SELECTION OF MOLDING MACHINE

In order to operate the mold, the correct size of molding machine must be determined. The molds' vertical length must fit on the machine platen. The mold width must fit between the machine's tie bars to assure the parts will drop unobstructed. The molds thickness, or *stack height,* must fit within the machine's maximum mold height dimension. The mold opening required eject the parts plus the thickness of the mold must be less than the machine's *daylight* dimension. The total volume of all of the cavities plus the volume of the cold runner must be less than 80 percent of the machine volume shot capacity. The screw recovery rate must be adequate for the estimated cycle time. The tonnage of the press must be between 2 and 3 tons/in^2 of projected molding area. The lower number is used for general purpose molding the higher number for thin wall or high pressure molding (lenses etc.).

40.6 TYPES OF MOLDS

There are many types of molds. Most molds, however, are made using standard configurations. Figure 40.2 shows various mold types and names the various plates. In order to aid the mold makers' efficiency, companies such as DME Company or National Tool supply mold bases and standard mold supplies.

40.6.1 Two-Plate Mold

In the two-plate mold (A and B series in Fig. 40.1) the cavities are mounted in the A plate and the cores in the B plate. A sprue bushing is mounted in the A or stationary side of the mold. This sprue bushing can center gate a single cavity or feed into a cold runner system of a multicavity mold. A two-plate mold can also utilize a hot runner system by adding a hot half to the top of the A plate. This hot half or package hot runner system then feeds, using nozzle drops, the individual cavities below. The B half of the mold usually contains an ejector box, which, in turn, contains the ejector and ejector retainer plates. These plates rest on *stop pins* or *buttons* in the back position. They are held against the buttons, when the mold halves are closed, by return pins that protrude through the B plate and touch the A plate. The ejector pins and other ejection devices are held in the ejector retainer plate and are moved forward when the ejector plate is moved forward by the molding machine. The return pins return the ejector assembly when the mold closes. The ejector plates can be guided by the addition of guide pins and bushings. This is done to prevent cocking of the ejector plates, particularly in those molds with asymmetric ejector patterns.

40.6.2 Stripper Plate Molds

Stripper plate molds (X and AX series in Fig. 40.1) are similar to two-plate molds with the addition of a stripper plate. The stripper plate is a floating plate that contains bushings that ride on the leader pins in the mold base. The stripper plate may contain a stripper ring made of a cavity or core steel. This ring usually forms an edge of the part and uses that edge to strip the part from the core. The stripper plate is often attached to the return pins and therefore travels with the ejector plates. Because mold design is a very creative process, many design variations are possible.

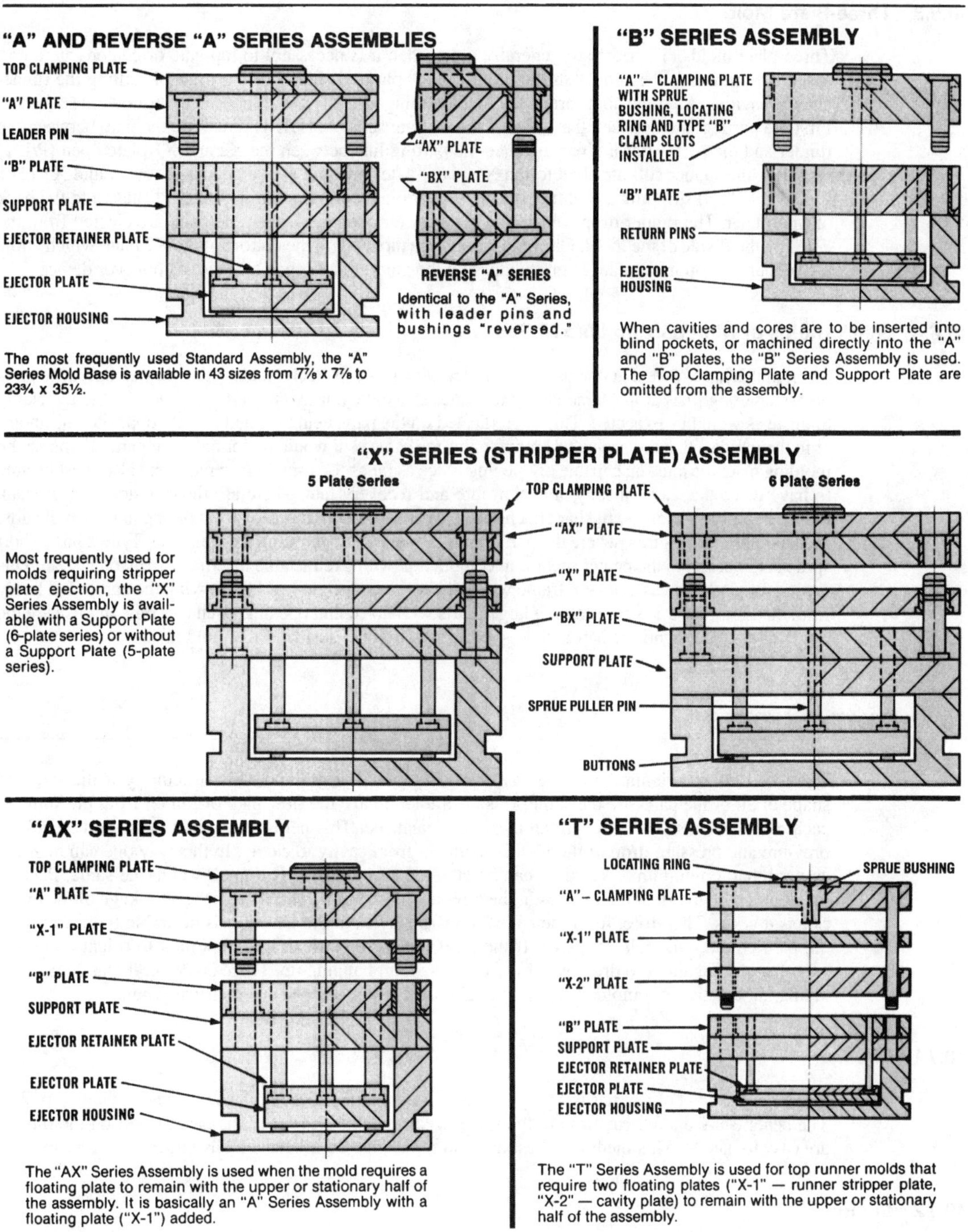

FIGURE 40.2 Standard mold base terminology.

40.6.3 Three-Plate Mold

Three-plate molds (T series) are generally used when it is necessary to top gate or center gate a part and a cold runner is desired. With the mold closed, plastic is injected into the mold filling the runner and the cavities. After the appropriate length of cooling time the mold opens in the following sequence. First the parting line between the X-2 and the B plate separate (P/L 1), breaking the gate between the runner and the part. After a given distance the parting line between the X2 and X1 plate open (P/L 2) revealing the runner still attached to the X1 plate. After opening a given distance, the A and X1 plate separate, (P/L 3) stripping the runner off the sucker pins protruding through the X1 plate into the back of the runner. The runner drops out of the mold. After opening further, the parts are ejected from the cores on the B side of the mold. Often separators are mounted to the bottom of the X2 plate which allow the runner to drop into an auger granulator and the parts to fall separately into a conveyor device.

40.6.4 Stack Molds or Multiparting Line Molds

Stack molds have more that one parting line. The advantage of stack molds is that they can produce parts on both parting lines at the same clamp tonnage. Twice the number of parts does not double the clamp tonnage as would be expected. This is predicated on the parts being located exactly opposite one another in the mold so that the projected area remains the same as it would be for only one part. Of course the molding machine must be capable of opening twice as far as for a single face mold and also must be able to have twice the shot capacity, injection rate and recovery rate. There are three basic types of stack molds. Type 1 has a long sprue bar attached to the center section. It is used for molding many small parts, such as caps and closures, where the parts can fall freely without hitting the sprue bar. Type 2 has a short sprue bar attached to the center section in conjunction with a retractable machine extension nozzle. This is used for parts such as a picture frame where the machine nozzle retracts allowing the part to fall freely from the mold. Type 3 is for molding larger parts with no central opening, such as a tray. In this case the runner is routed around the part and crosses the parting line outside of the molding area.

40.7 CAVITY LAYOUTS

The location and orientation of the cavities in a mold is very important particularly if the size and shape of all of the parts are the same. It is desirable to have the flow path or runner from the sprue to each cavity be identical in length and size, i.e. balanced. This helps assure that the fill rate, packing pressure and pressure drop of the melt is identical from cavity to cavity. In this way one can be more assured that a minimum cycle time can be attained as well as parts that all weigh the same. Plastics molecules tend to orient themselves in the direction of the flow. This affects the shrinkage of the material as it cools. Therefore, for dimensional consistency cavity to cavity, it is desirable to gate each of the parts in the same relative place. If the parts are asymmetrical, it is important to orient them in a way that permits them to drop out of the mold without hanging up. There are several classic ways of laying out cavities in a mold. They are the circular, in-line and the symmetrical groups.

40.7.1 Circular

The circular layout of cavities is a very practical system to produce in relatively small cavitation molds. The benefits are that all cavities are the same distance from the sprue. The parts, if left on the runner, are easy to handle. This method is best used on molds that do not have many slides or side actions.

40.7.2 In-Line

The in-line arrangement of cavities is very common. It is often used for molds that are produced using slides or splits. In a typical eight cavity mold, four cavities might be formed above the sprue and four below, all in one vertical line. In this way, the right and left slides can be moved apart by angle pins as the mold opens, and the parts easily ejected. The difficulty in this arrangement occurs in trying to balance the runner system. It can, however, usually be accomplished by branching the runner into two as many times as necessary to feed 2^n cavities. Often one sees molds with multiple rows of cavities. For instance consider a 48-cavity mold. It can be 4 vertical rows of 12 or 6 rows of 8 cavities. To feed these cavities is very difficult in an in-line manner. Therefore, these molds are usually fed by a three-plate or hot runner where the branching can be done on one level of the mold and then dropped to the cavities below where needed. Hence, hot runner nozzles are often called hot drops.

40.7.3 Symmetrical Groups

Symmetrical groups of cavities are the most common method of arranging large cavitation molds. These types of molds include conventional or cold runner molds where the runner branches out symmetrically to feed each part uniformly, three-plate molds and hot runner molds. In this manner the plastic can be distributed in one level and transferred to the cavity in another level in a balanced manner. The cavity pattern can be one of several. One common way is to drop into the cavity plate in several locations and then distribute in a cold runner to a circular arrays of cavities. Many hot runner manufacturers also supply small modular manifolds of 2–16 drops. These manifolds are often placed under a larger distributor manifold to feed a multicavity mold. Of course, three-plate molds and hot runners can also be arranged to feed the desired cavities directly without the modular manifolds.

40.8 GATING

40.8.1 Conventional

The easiest gate to make is a simple short land, usually ground from the runner to the cavity. Parts using this type of gate must be twisted or cut from the runner. This sometimes leaves a small vestige or blemish where the runner is cut away (see Fig. 40.3).

40.8.2 Tab Gate

This gate allows a large volume of material to flow into the part with very little shear (Fig. 40.4). It is used where a blush, splay, or other imperfection may be expected. The imperfection will often be caught in the tab, leaving clean material to flow into the cavity.

40.8.3 Subgate

A subgate (Fig. 40.5) is also called a submarine or cashew gate. It is used to minimize the gate vestige and to automatically degate the part upon ejection.

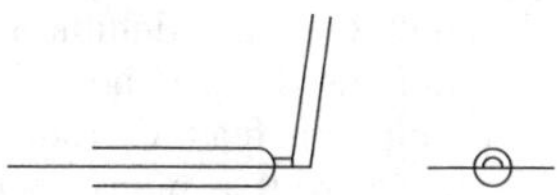

FIGURE 40.3 Conventional gating.

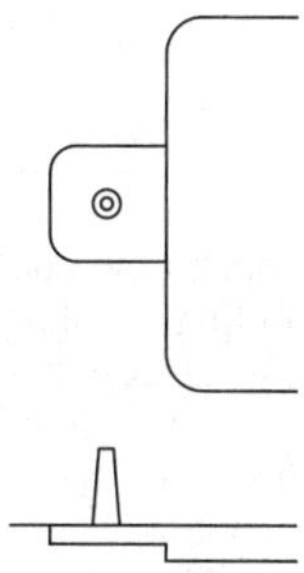

FIGURE 40.4 Tab gate.

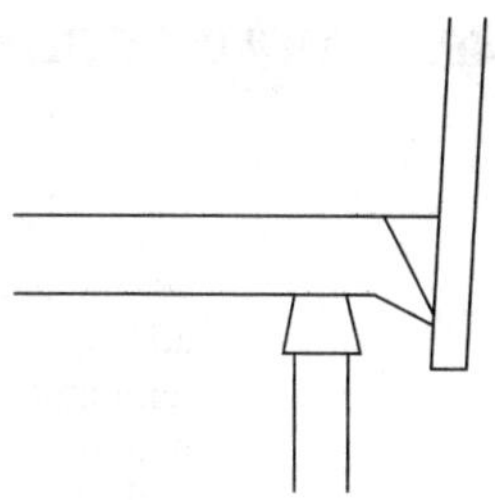

FIGURE 40.5 Subgate.

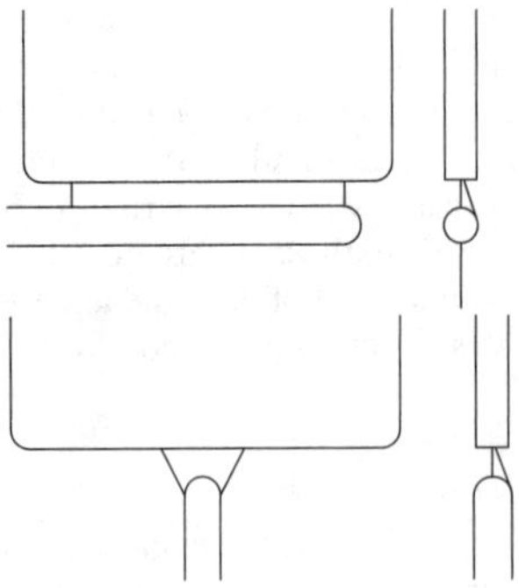

FIGURE 40.6 Flash gate.

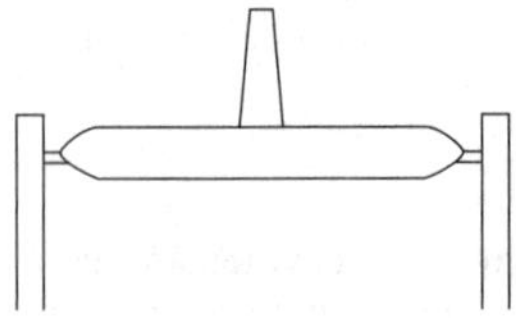

FIGURE 40.7 Disc gate.

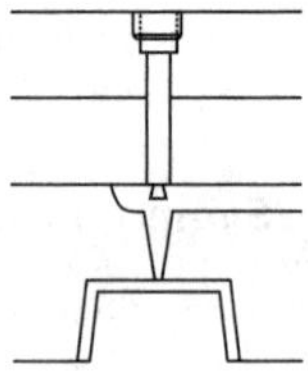

FIGURE 40.8 Point gate.

40.8.4 Flash Gate

The flash gate (Fig. 40.6) is used to fill the part evenly from one side of a flat part to the other. In this way the molecules will shrink in the same direction when the part cools, and the part will not tend to warp. A variation of the flash gate is the fan gate, where the gate is fanned out from the end of the runner to the part. The depth of the gate starts at the same depth as the runner and ends as a thin flash gate at the cavity.

40.8.5 Disc Gate

The disk gate (Fig. 40.7) is very similar to the flash gate except that it is circular and used to gate into a round hole in the part, particularly in the center hole of round parts. In this way the plastic flows radially and therefore the part will remain round after shrinkage occurs. This gate is often used in the molding of spools, hubs, or pulleys.

40.8.6 Point Gate

The point gate (Fig. 40.8) is used in conjunction with three-plate molds or hot runner molds. The runner is contained in a level of the mold behind the cavities. Therefore the parts can be gated into their center rather than peripherally as in other gating methods. This is particularly desirable in round parts, such as caps. It must be kept as small as possible, hence the name point gate or pin point gate.

40.9 MOLD COOLING

Molding thermoplastics requires the cooling of the injected plastic (Fig. 40.9). The mold must be equipped to provide this cooling. Cooling is a relative term. Depending on the polymer being molded, the mold temperature can be from below 32°F to above 300°F. The cooling medium used is usually water or water with additives as ethylene glycol and/or corrosion inhibitors. In the case of temperatures over 200°F, hot oil or thermal fluids are generally used. There are many standard components sold to facilitate cooling in many of the more common features found in molds. This includes baffles, cascades, and bubblers to reach inside cores as well as plugs to channel water and thermal pins to cool areas that are difficult to reach with water lines.

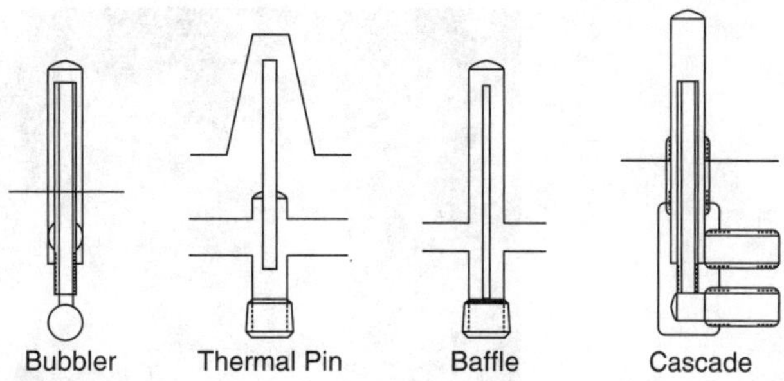

FIGURE 40.9 Mold cooling bubbler, thermal pin, baffle, and cascade.

40.10 HOT RUNNER SYSTEMS

A hot runner system (Fig. 40.10) is a means to conduct the melt in a molten state, from where it leaves the molding machine to either the cavities (direct part gating) or to a cold runner that in turn feeds the part. In order to keep the melt in the optimum condition, most hot runner systems employ a balanced, heated manifold, and subsequent hot drops to supply molten plastic to the parts.

40.11 MOLD MANUFACTURING

Mold manufacturing has changed greatly over the past decade. Most books on the subject still refer to small vertical mills, engine lathes, surface grinders, and jig bores as the staples of mold shop equipment. The rapid acceptance of two- and three-dimensional CAD, as well as CAM has changed moldmaking forever. In addition, the advent of high speed machining, hard turning, and hard milling is also having a significant influence on the speed of design to part production. The pantograph and duplicator mills have given way to *electrodischarge machines* (EDM), both sinker and wire. In order to reduce time and the time cost of money, moldmakers, rather than making them themselves, purchase custom mold bases from vendors that specialize in their manufacture, such as D-M-E Company in Madison Heights, Michigan. They are usually available within a few days. Mold bases (Fig. 40.11) can be purchased in millions of combinations of size, type, plate thicknesses, plate materials, with milled pockets, gun drilled water lines, and a large amount of installed options to aid in

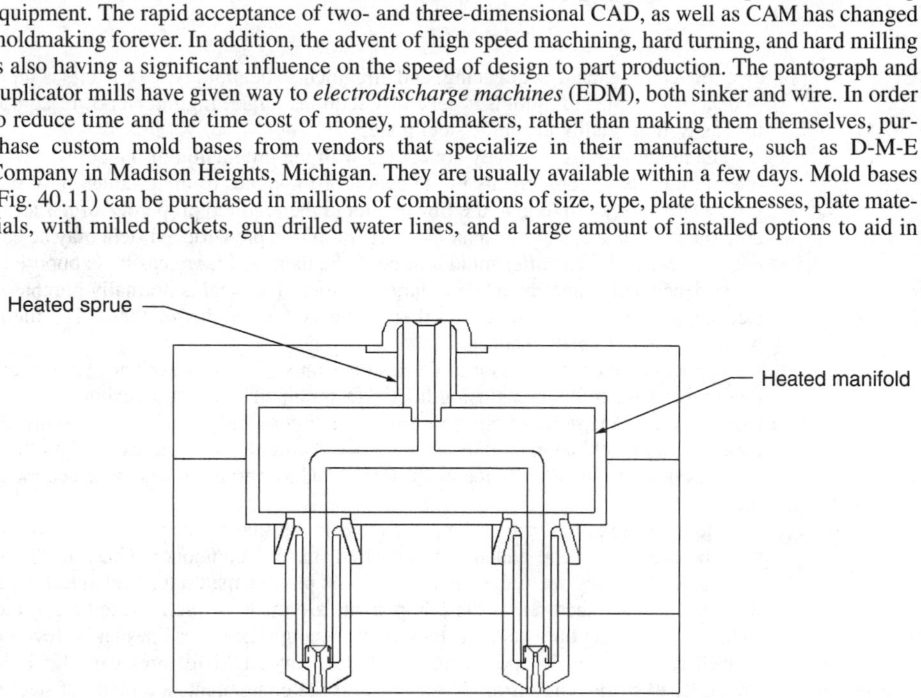

FIGURE 40.10 Hot runner system.

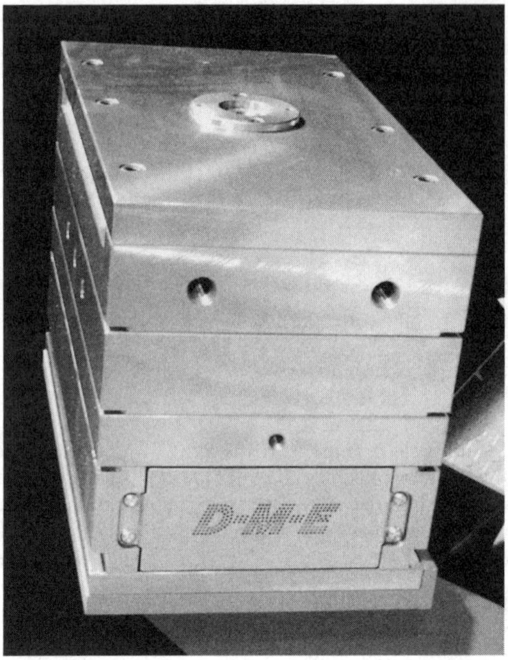

FIGURE 40.11 Standard mold base. (*Photo courtesy of D-M-E* company)

ejection, cooling, plate sequencing, and alignment. The mold base or frame is the fundamental component of the injection mold. It generally requires larger, high horsepower equipment to manufacture as well as readily available steel plate.

The moldmaker is primarily concerned with the production of the cores, cavities and the fit and operation of these components in the action of the mold. In the production of a cavity or core, the proper steel is chosen to give the finished mold the required properties of hardness, wear resistance, corrosion resistance, and life span. For larger molds a prehardened steel may be chosen knowing that the trade off will be a softer mold that could be damaged more easily as opposed to a steel that can be hardened, but cannot be made in large sections. The steel is normally purchased from a specialty steel supply house. The moldmaker then saw cuts it to rough size, then it is milled to squared blocks or rough turned to blank shapes.

The moldmaker often has a mold designer on staff. This designer has the capability to draw or model the mold components using his CAD system. These CAD designs are usually modified and translated to CAM programming for down loading to various machines. Many shops also program their various machines manually on the floor. Particularly in the case of multi cavity molds, with many cavities producing the identical item, CNC machining gives more accurate results in far less time.

The steps are to first rough the piece close to net shape. The plastics material molding shrinkage must be determined and added to the desired finished component sizes. In this way, as the plastic part cools, it will shrink to the correct size. The plastics material shrinkage should be obtained from the material manufacturer. If the components are made of prehardened steel, they can be brought right to size leaving only enough stock for polishing. The water lines and screw holes are then added as well as any core pin and ejector pin holes. Any EDM features can also be done at this time. Prehardened steels range from 34 to 44 Rc. If the components are made of steel that must be hardened, the roughed parts receive the machining for water lines, screw holes, core pins, and ejector pins. Usually 0.005–0.010-in stock is left on all areas that will be finish ground, EDM'd, lapped, etc. after hardening. The components are then sent out for heat treating (hardening). Depending on use, most mold components are hardened from 45 to 55 Rc. Once hard, the parts are brought to size by grinding or EDM. The screw hole threads must be cleaned and the pin holes lapped to fit the pins perfectly. Once the components are at this stage they are fitted into the pockets of the mold base by careful grinding, keeping in mind the position of the cavity to the core, in their respective mold halves, in order to maintain the proper wall thickness of the molded part. Slides and inserted components are also fitted at this point. Once the components are fitted to their respective pockets, screw holes and ejector pin holes are spotted and drilled into the mold base.

The components, cavities, cores, and slides are now ready for polishing. Although many aids are available for the polisher, the art of polishing is still done largely by hand. It begins by stoning out the cutter marks in the surface of the steel with ever finer grits of polishing stones. The desired finish is determined by the aesthetic requirements of the part as well as the release properties required by the plastics material that will be molded. If a very high polish is required, the final polishing is done with very fine diamond polishing compounds. If the mold finish specifies plating, the components can be sent out for plating at this stage. If a texture such as a leatherette is desired, once polished, the finish may be textured by sending it a texturing specialty company such as Mold-Tech of Youngstown Ohio.

After polishing, the mold is assembled. During the assembly process, the ejector pins are cut and ground to length, the gates are ground and finished, vents are machined and the runner path is polished and any final fits are completed. All mold actions, such as slides and ejection, are then tested. The mold is then ready for spotting. Spotting is a procedure where one side of the mold parting line is coated with a thin layer of *Prussian blue* pigment. The mold halves are pressed together and than opened. If the mold is correct, the blue will spot the uncoated half evenly. If not, the high spots must be ground or the mold adjusted until the parting line seats and seals evenly.

After all checks have been completed to assure that the mold designer's wishes have been met, the mold is ready to release for part qualification at the injection molding facility. The qualification items, such as dimension, strength, finish and fit, are based on the criteria given to the mold designer at the beginning of the process. If the part passes the qualification tests, the mold is ready for production.

FURTHER READING

DuBois, H., and Pribble, W., *Plastics Mold Engineering Handbook*, 5th ed., Chapman & Hall, New York, 1995.

Dym, Joseph, *Injection Molds and Molding*, Van Nostrand Reinhold, New York, 1987.

CHAPTER 41
MACHINE TOOL DESIGN ON FLEXIBLE MACHINING CENTERS

Mal Sudhakar
Mikron Bostomatic Corporation
Holliston, Massachussetts

41.1 INTRODUCTION

The consumption of machining centers in the United States is around 12,000 units annually at a value of over $1 billion. Of all the machine tool types, machining centers represent the single most frequently purchased and used machine tool in manufacturing today.

Machining centers are numerically controlled machines with multipurpose machining capabilities such as milling, drilling, boring, tapping, reaming, and even turning and grinding capabilities. This flexibility not only provides productivity improvements but also allows one machine to replace several single purpose machines or machining processes.

41.2 CLASSIFICATION

There is a vast array of machining centers available in the market and the selection of a machining center for an application can be a difficult and confusing process. Fundamentally, machining centers can be classified as vertical, horizontal, or universal.

Vertical Machining Center (VMC). Vertical machining centers as the name implies have a vertical spindle whereas horizontal machining centers have a horizontal spindle. Vertical machining centers are more popular than horizontals as they are lower in price for a comparable working range. With vertical machines, setup and handling are easier, but they are limited to single-sided machining unless indexers or a rotary table is used.

Horizontal Machining Centers (HMC). Horizontal machines allow four-sided machining of a workpiece, but can be more difficult to set up. As chips fall freely, chip flow is better with horizontals.

Universal Machining Centers. Universal machines generally feature a spindle that can swivel to a vertical or horizontal position, offering the best of both worlds and also the ability to machine five sides of a workpiece in a single setup.

The selection of a vertical, horizontal, or universal machine requires a careful evaluation of the application and the return on investment.

TABLE 41.1 Comparison Between Standard Machining Center vs. High-Speed Machining Center

	Standard vertical machining center	High-speed machining center
Construction	Fixed column C-frame	Bridge-type —
Guideways	Box ways/linear Roller guideways	Linear roller Guideways
Drive train	Ball screw	Ball screw/linear motor
Spindle speed	Up to 6000/8000 rpm	Up to 42,000/60,000 rpm
Spindle type	Belted or gear driven	Integral motor spindle
Spindle torque	High	Low
Feedrates	Low	High
Acceleration	0.1–0.2 g	1–2.5 g
Cutting forces	High	Low
Tool diameters	Large	Small

Multiaxis Machining Centers. A further classification of a machining center is whether it is a three-, four-, or five-axis machine. A three-axis machine features three linear axes, X, Y, and Z. A four-axis machine features a rotary table. A five-axis machine features, in addition, a tilting table or a swiveling/tilting head, i.e., it has three linear axes and two rotary axes.

Standard vs. High-Speed Machining Centers. With the advent of high-speed machining in recent years, a new classification of a machining center is whether it is a standard or high-speed machining center. Whereas a standard machining center may have a top spindle speed of 6000 or 8000 revolutions per minute (rpm), a high-speed machining center can go up to 42,000 or 60,000 rpm. Spindle speed is only one of the distinguishing characteristics.

Standard and high-speed machining centers are built differently from the ground up. Table 41.1 lists some of the major differences. If standard and high-speed machining centers represent two ends of the spectrum, there are a large number of machining centers that fall in between in their characteristics and are labeled high-performance machining centers. Next we shall examine the design and construction of the popular standard vertical machining center and the more state-of-the-art high-speed machining center.

41.3 VERTICAL MACHINING CENTERS

41.3.1 Construction

For a three-axis vertical machining center a number of different designs are available, such as the fixed column C-frame, traveling column C-frame, fixed rail bridge, gantry bridge, traveling column ram, and the like. The fixed column C-frame is the most common and popular design for a three-axis vertical machining center, which is shown in Fig. 41.1. The spindle carrier mounted on the fixed column provides the Z-axis movement and a compound table mounted on the machine bed provides the X and Y movement. The machine bed and column are of a well-ribbed cast iron construction which provides good stiffness and moderate dampening.

The fixed column C-frame is ideal for parts where the X-Y dimensions typically exceed the height of the part, such as molds and dies, flat parts, plates, etc. The stiffness of the C-frame decreases with increased Z-height so the "sweet spot" is close to the table. Stiffness is also improved by having a minimum of overhang for the compound table. The machine in Fig. 41.1 features a wide machine

FIGURE 41.1 Fixed column C-frame VMC.

bed to support the compound table and a minimum of overhang for the table in its extreme X-axis positions, which helps in providing good geometric accuracy.

Thermal stability of the C-frame, however, is average because the machine's axes and most of its structure are located within the work area, making it hard to isolate them from heat generated in the cutting process. The front of the column will be warmer than the rear, resulting in a tendency of the column to bend slightly backwards.

Manufacturing costs for the fixed column C-frame design are lower than with other designs and generally the machines provide good machining performance and accuracy for a wide range of small to medium parts while being cost-effective. However, once parts require large Y-axis travels, (above 25"), the C-frame design has weaknesses. The spindle has a large overhang from the Z-axis way system, magnifying angular errors and reducing stiffness. Also, for very tall parts, additional angular errors are created because the tool-work plane is far from the X and Y axes. A large load on the table also negatively affects the dynamic performance of the X and Y axes and generally table loads on C-frame machines do not exceed 2000 lb.

41.3.2 Guideways

Vertical machining centers were typically constructed with hardened and ground box ways which provided good stiffness and dampening under high cutting forces. However, with the requirement of high rapid traverse and feed rates, increasingly linear roller guideways are being used as depicted in Fig. 41.1. A properly selected linear guideway can provide the stiffness of a comparably sized box way. Linear roller guideways possess very low static and dynamic friction and allow very high feed rates and accelerations. These way systems are modular, facilitating machine assembly. If necessary, the guideways can be replaced in the field.

41.3.3 Drive Train

The drive train of a machining center consists of the axis motors (servomotor), coupling, ball screws, and the connection to the machine slide. The drive train plays a critical role in establishing the positioning accuracy and dynamic performance of a machining center because it links the CNC and the machine slides.

Ball screws are the principal drive mechanisms used in machining centers today. Ball screws utilize a series of steel balls and recirculators built into the nut to convert rotary motion into linear motion needed to drive machine axes. This conversion process is extremely efficient in a ball screw, requiring only about 30 percent of the torque compared to a conventional lead screw.

Feedback on the position of the slide is provided by one of two systems—a rotary encoder mounted to the rear of the motor or a direct measuring system which consists of a linear scale mounted in close proximity to the ball screw. Linear scales provide higher accuracy as they directly measure motion of the machine slide. By directly mounting the scale to the moving slide, inaccuracies in the ball screw do not affect the position data fed to the CNC as when using a rotary encoder.

41.3.4 Spindles

Spindles on machining centers are one of three types: belted, gear driven, or integral motor spindles. Typically on standard vertical machining centers belted or gear driven spindles are used while high-speed machining centers feature integral motor spindles. Gear driven spindles provide the benefit of very high torque at low speeds but are also characterized by high heat generation, transmission losses, and vibration. Top end speeds for geared spindles are 6000–8000 rpm.

FIGURE 41.2 Toolchanger.

Belted spindles fall in between the high torque/low speed characteristics of geared spindles and the low torque/high-speed characteristics of integral motor spindles. The use of a belt allows power to be smoothly and efficiently transferred from the spindle motor to the spindle. Belted spindles can achieve speeds of up to 15,000 rpm and are cost-effective compared to integral motor spindles for spindle speeds up to 15,000 rpm.

41.3.5 Toolchanger

Vertical machining centers feature an automatic toolchanger (Fig. 41.2) for changing tools automatically into the machine spindle upon command in the NC program. A side-mounted disk carries tools in multiple pockets. Upon command a new tool is presented to the intermediate exchange arm, which swaps the tool in the spindle with a new tool. A C-frame machining center requires only a Z-axis motion to change tools and can achieve fast tool change times on some machines within a few seconds.

41.4 HIGH-SPEED MACHINING CENTERS

41.4.1 Construction

With the requirements of high spindle speeds, feeds, and acceleration, the design of a high-speed machining center is quite different from that of a standard vertical machining center. It is not sufficient to equip a C-frame machine with a high-speed spindle and a fast control and expect good results. Fundamentally a high-speed machine differs from a conventional machine in that the stationary structural machine components are made as stiff and heavy as possible, whereas the moving masses (spindle carrier, machine table) are made as stiff and light as possible. With high-speed machining there are lower cutting forces but higher dynamic loads from acceleration and deceleration. The stiff and heavy stationary structure is needed to absorb these forces, and, in turn, to keep the forces as low as possible, the moving slides must have a reduced mass.

A popular method of construction for a high-speed machining center is the bridge-type construction, a modified bridge construction or an overhead gantry system. Compared to a C-frame, which has an overhanging spindle and overlapping guideways for the table, the bridge design provides an optimal distribution of the moving masses and the spindle is rigidly mounted close to the guideways.

A drawback of the bridge construction however has been accessibility to the working area both for the operator and for an automated loading device such as a palletchanger or robot. A new type of construction has solved this problem while providing many other benefits. Figure 41.3 illustrates a new bridge design in polymer concrete, which provides excellent accessibility to the working area from the front for an operator. The large opening allows good access for a palletchanger or robot. The pyramid shaped construction with three-point support provides a stiff stationary structure, which absorbs the high acceleration/deceleration forces of the moving slides.

The polymer concrete construction, which is becoming increasingly popular for high-speed machining centers, provides 6–10 times better vibration dampening than cast iron. It also offers superior thermal stability, as the thermal conductivity of polymer concrete is $^1/_{20}$th that of cast iron. Overall the results are better tool life, better accuracy, surface quality, and quieter operation.

The new polymer concrete construction also greatly reduces manufacturing costs and assembly times. Special bonding techniques enable the creation of complex structures to be created as a one-piece or monobloc structure, eliminating any mechanical joining, while providing superior stiffness. Such complex designs would be difficult to manufacture conventionally. Previously guideway surfaces required cast in steel inserts to be machined or the surfaces themselves to be ground. Newly developed casting techniques enable guideway surfaces to be prepared with a straightness and parallelism of 0.0002" in 50" for the direct mounting of linear guideways. No machining or assembly is required for any of the major structural components—truly a breakthrough in machine tool construction.

FIGURE 41.3 Bridge design high speed machining center.

The linear guideways, ball screws, slides, motors, linear scales, etc., are mounted on the monobloc structure. Other interesting features are that the linear guideways and ball screws have permanent grease lubrication. No oil lines are required, there is no mixing of the oil with coolant or no oil disposal, and the system is maintenance free. The spindle and feed motors are liquid cooled thus conducting away heat from the structure. A laser tool measurement system, Fig. 41.4, serves to measure the length/radius of tools at the rotating speed thus compensating for mechanical and thermal displacements of the spindle as well as thermal growth of the tool holder.

41.4.2 Spindles

While various types of spindles are available for high-speed machining centers, integral motor spindles with hybrid ceramic bearings remain the workhorse for high-speed machining (Fig. 41.5). Hybrid bearings feature ceramic balls in a steel race. Ceramic balls provide lower centrifugal forces, higher stiffness, less uncontrolled thermal effects, lower friction, and require less lubrication. The benefits are that hybrid bearings stand up to higher speed than all-steel bearings, they are more rigid, they run quieter due to less vibration, and they last longer. The spindle is cooled by a recirculating water jacket.

FIGURE 41.4 Laser tool measurement system.

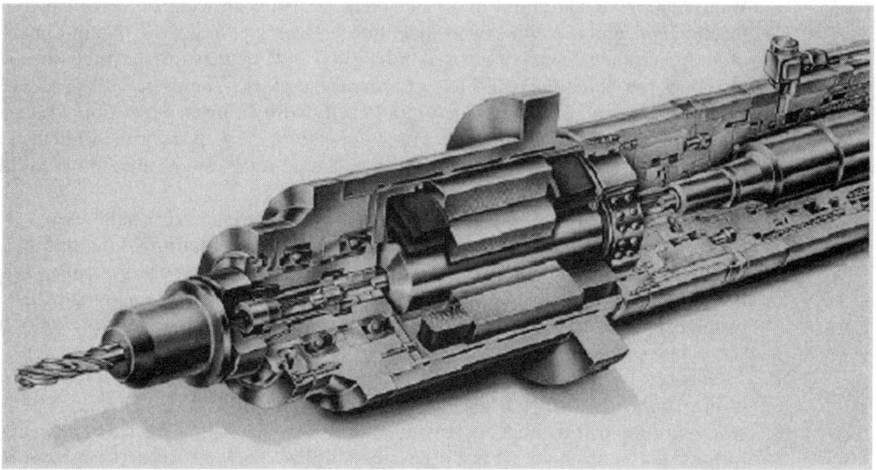

FIGURE 41.5 Workhorse of high speed machining.

Vector control is the underlying technology of the new high-performance motor spindles. It provides several advantages including load-independent speed, a wide range of speeds, high torque at low speeds for larger diameter tools, extremely fast run-up, and braking times and rigid tapping.

For instance, the spindle shown in Fig. 41.5 has a speed range of 100–42,000 rpm and 19 HP, can handle tools up to 5/8" in diameter, and can accelerate from 0 to 42,000 rpm within 3 s. The spindle has excellent characteristics for roughing and finishing hardened steel and for machining a variety of other materials such as aluminum, copper, and graphite.

For high-speed machining with high-rotational speeds, the interface between the spindle and the toolholder and in turn between the toolholder and the tool takes on a special significance. When running a spindle beyond 13,000 rpm, balancing of the toolholder/tool becomes indispensable and the use of the HSK interface is strongly recommended.

41.4.3 Controls

High-speed machining places high demands on CNC controls. Cutting of complex three-dimensional surfaces at very high feed rates and contouring accuracies poses special requirements on the CNC control. Some of these include:

- Large memory capacity and the ability to edit large programs at the control. Controls today have a hard disk capacity of several gigabytes.
- High-speed communication capabilities—both external and internal. Large programs have to be downloaded quickly and today with the Ethernet network connection transmission rates can be up to 100 Mbits/s.
- Short block processing times to allow machining at high feed rates. The required block processing time is related to the acceleration characteristics of the machine, i.e., the higher the acceleration, the shorter the block processing time required.
- Sufficient look ahead. The control must anticipate changes in the contour ahead and must correspondingly accelerate and decelerate the axes to machine a contour accurately. High-speed controls today have a look-ahead capability of several hundred blocks.

41.4.4 Automation

With 8760 available hours in a year, there is also a growing requirement to automate high-speed machining centers, as with other machines in the shop, to get the maximum capacity utilization out of the equipment. With improved reliability and security of the high-speed machining process, machines are being designed to run around the clock. There are still processes that are better run attended, for example, roughing a mold cavity with an insert-type tool, while others such as finishing operation can run unattended. Finishing with a fine stepover can eliminate or reduce the requirement for polishing operations. There is thus a requirement to be able to run multiple jobs through the same machine.

The first step toward this is a quick-change setup system or a palletizing system on the machine. The pallets can be changed quickly in and out of the machine enabling multiple jobs to be processed through the machine. The pallets also have a common reference system as they can be transferred from machine to machine or process to process. The next step is to automate the loading and unloading of the pallets with a mechanical arm, either a palletchanger or a robot, when the requirement is to serve multiple machines.

Figure 41.6 illustrates a high-speed machining center that offers a table chuck and pallet system in lieu of the conventional machine table as a standard feature. The machine is also offered with a seven-position palletchanger as a standard feature. No longer is the palletchanger or a robot is a feature for production machines. It is a necessity to utilize machines around the clock and get the maximum return on investment.

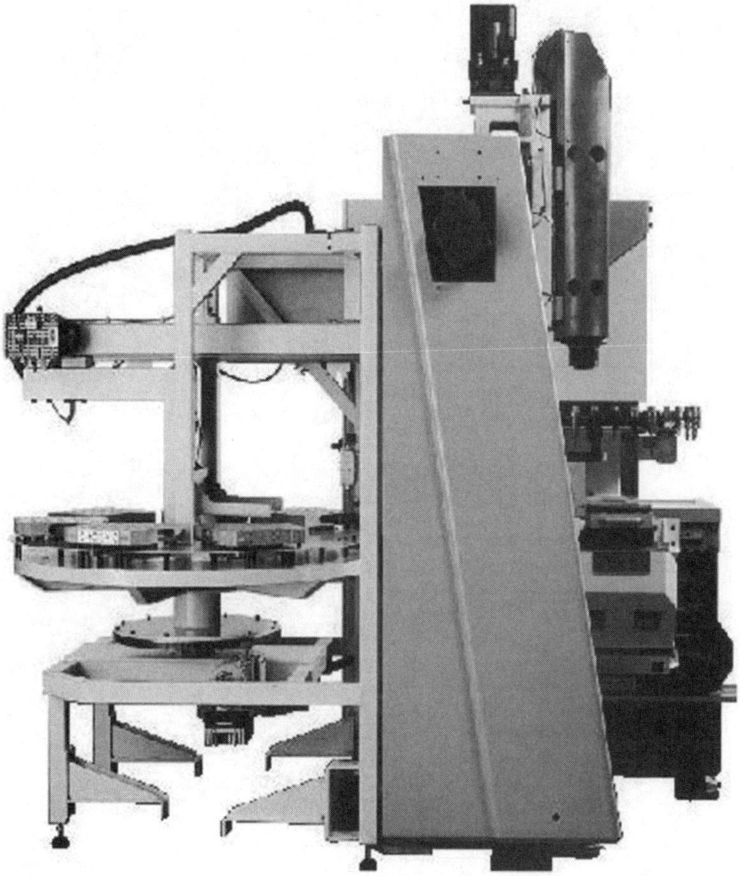

FIGURE 41.6 Table chuck and pallet system.

41.5 FUTURE TRENDS

Increasingly, machining centers with larger axis travels or requiring high-speed positioning capabilities are being equipped with linear motor drive trains. Linear motors are essentially rotary motors unrolled into a flat form. A linear motor can eliminate a wide range of components typically associated with a ball screw drive—the ball screw itself, support bearings, bearing blocks, the motor, and coupling.

Linear motors have the benefits of having no wear parts and can provide very high feed rates up to 5000 or 6000 in/min and accelerations of 3–5 g. There is also no backlash and consequently, high motion accuracy. However, the drawbacks are high costs, approximately 30 percent more than ball screw drive trains, high heat development which requires elaborate cooling systems, and high energy consumption. The linear motors, which have strong magnetic forces, have to be well-protected against chips.

Linear motors are therefore used in large travel machines or in horizontal machining centers used in automotive production applications where the benefits of higher productivity outweigh the drawbacks. In time, however, as the drawbacks are overcome and costs come down, linear motors will be more prevalent on machining centers.

CHAPTER 42
LUBRICATION DEVICES AND SYSTEMS

Peter M. Sweeney
Bijur Lubricating Corporation
Morrisville, North Carolina

42.1 INTRODUCTION

This chapter highlights the various automatic lubrication devices incorporated into industrial machinery to maintain production efficiency. Simple and sophisticated devices are detailed as well as practices available to plant maintenance personnel.

42.1.1 Lubricating Devices and Systems

The need to deliver correct amounts of lubricant to friction points of production machinery is critical in meeting industrial output. Automatic lubrication devices eliminate the inherent human error of manual lubrication and deliver controlled accurate lubricant amounts to machinery friction points in a timely manner.

Oil and grease lubrication devices and centralized distribution systems play a critical role in maintaining the performance of industrial production machines. Automatic lubricators reduce friction between moving mechanical friction points, conserve energy, and prevent or minimize bearing wear, as well as maintain peak operating machine efficiency. In addition machine downtime and operating expenses are reduced.

Lubrication devices and systems can be adapted to various types of friction points encountered throughout industry; typical applications include chain-drives, slides, plain bearings, and antifriction-type bearings, including roller and needle bearings. Specialty lubrication devices are available for many unique applications including high-speed spindle applications and special purpose conveyors, and the like.

Various types of lubricating systems offer individual characteristics and advantages. Factors affecting selection of a lubricating system can vary widely: environment, number of application points and placement, type and quantity of lubricant delivered, as well as adaptability to machinery.

Basic lubricating devices and systems can be categorized under the following headings:

42.1.2 Single Point Lubrication Grease Feeders and Oilers

A broad selection of easy-to-install single point lubricators and accessories is available. Units are offered in various reservoir capacities with accessories to satisfy a broad spectrum of bearing point requirements for machine tools, grease- or oil-lubricated bearings, gearboxes, motors drives, and chains.

Gravity Feed Oilers. Act as a constant filled reservoir in oil-lubricated bearings. Units can also be used as an oil level indicator, reserve reservoir, or to relieve backpressure on machine start-up.

Drip-Feed Oilers. A popular single point oiling device, the operator sets the drip rate needle valve by viewing the oil drop delivery rate to the application point. Units are supplied with a built-in manual shut-off valve.

Chain Oilers. Similar to drip feed oilers, units are supplied with various oil distribution brush assemblies to feed lubricant to critical friction areas on roller chain link pins, and the like. The operator sets the lubricant feed rate to satisfy individual applications. Models are available with various brush configurations to disperse oil to chain pins and bushings. Proper lubrication also prevents chain failure by protecting against rust and corrosion. The brush bristles help remove dirt and contamination.

Wick Feed Oilers. Wick feed oilers are similar to drip devices. They are ideally suited for applications where precise oiling is required. Units are easily adapted to all classes of machinery. The characteristics of the wick material control the oil feed rate—the wick operates through capillary action providing a predetermined flow based on oil grade; a selection of wicks are offered to achieve desired oil feed rates.

Grease Feeders

On-Demand Feeders. These devices offer simple economical methods of dispensing industrial greases to a wide variety of applications. Units are normally refillable and have the ability to operate in various industrial environments, including dynamic applications where shock and vibration are encountered. Individual grease feeders usually are screwed into grease fitting connections on bearing housings.

On-demand spring feed devices typically incorporate a spring-loaded piston, which forces lubricant at low pressure through a graduated grease-metering rod to provide a balanced grease delivery during operation. A variety of spring rates are offered to handle various types of grease (Fig. 42.1).

Time-Release Gas Feeders. These are disposable (nonrefillable) grease filled reservoir devices with built-in gas generators that operate over a broad temperature range. An inert gas is created in an expandable sealed chamber from a chemical reaction between an activator and an electrolyte. Pressure build-up in the housing forces a piston forward to discharge lubricant from the reservoir over a fixed time period.

Models are offered in variety of preset timed discharge periods. Battery-operated programmable electrochemical models are also offered.

Positive-Displacement Feeders. This compact self-contained assembly utilizes a small electric motor to actuate piston-to-discharge volumetric lubricant outputs over set time periods. Controlled positive displacement piston movement provides operators with stable outputs under all operating conditions—this feature provides positive control of lubricant deliveries.

A built-in electronic programmable timer controls operating cycles. This unit has a broad range of adjustable time settings to satisfy most industrial applications.

Individual application selection guides for all grease feeder models are available. A broad selection of reservoir capacities is also offered.

42.1.3 Centralized Lubrication Systems

A centralized lubrication system configuration utilizes a lubricant distribution network through which lubricant is fed from a central point or a common lubricant source. Typically, lubricant meters are used throughout the system to feed various deliveries to application points connected to the network.

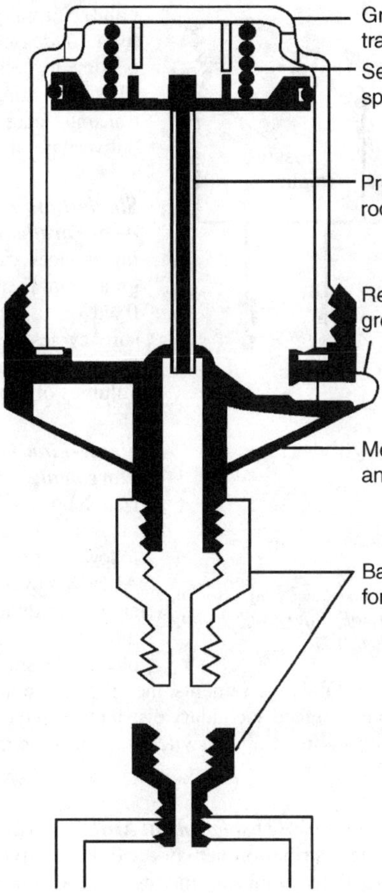

FIGURE 42.1 Spring-fed single-point grease feeder. (*Reprinted with permission of LubeSite Systems, Inc.*)

A centralized system may be installed on an individual machine or on a group of machines. Operation can be manual or automatic.

Zerk Fed (Grouped Nipple) Systems. A simple distribution network comprises a manually operated system in which individual points are fed from a central battery of lubricant nipples. This simple layout permits lubrication of a limited number of machinery friction points.

Single-Line Feed Systems—Resistance Type for Oil. A lubricator forces a predetermined amount of oil into the restrictor-type distribution system. During the discharge operation, pressure is built up in the system and oil is apportioned to each lube point. Typical industrial applications call for cyclic operated lubricators operating at preset intervals. A fixed orifice resistance fitting (flow-unit) meters oil to each individual point. A broad selection of flow-unit rates offer a wide range of delivery to satisfy diverse applications (see Fig. 42.2).

Single-Line Feed Systems—Oil Mist Resistance Systems. Distribution principle is similar to restrictor-type metering systems; oil is atomized at the pump and mixed with the operating air supply in the distribution lines. The oil mist is carried along the distribution lines in mist form to the lube

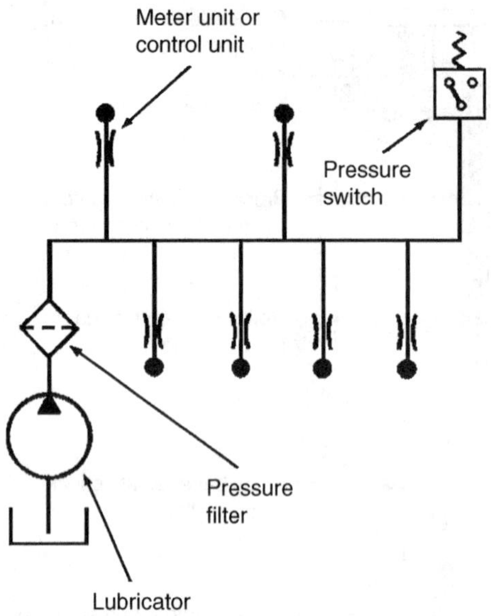

FIGURE 42.2 Single-line resistance oil system. *(Reprinted with permission of Bijur Lubricating Corporation.)*

point. Normally the delivered mist is reclassified by a condenser-type apportioning device at the bearing to deliver mist, oil drops, or oil spray. Mist lubricating systems offer users the ability to transmit accurate quantities of lubricant to the delivery points on a continuous basis.

Single-Line Feed Systems—Positive Displacement Parallel Distributors. Volumetric metering devices (injectors) are offered for oil and grease lubricants. Models are available with fixed or adjustable outputs. Programmed pressure cycles in the distribution network create forward piston movements to discharge fixed volumes of lubricant to each outlet.

Single-Line Feed Systems—Progressive Series Movement. Pressure build-up from a centralized lubricator feeds lubricant into a distribution block; in operation, a prearranged number of pistons follow a series of movements through the divider. As each movement takes place, lubricant is delivered. The divider must complete a prearranged movement to enable each point connected to the block to be successfully lubricated (see Fig. 42.3).

To meet a broad variety of lube requirements, the progressive movement system typically utilizes a primary or master distributor to feed secondary distributors in the system network. The cyclic operated system is monitored for malfunction by switches monitoring movement as well as over-pressure indicators.

Single-Line Feed Systems—Positive Displacement Air/Oil Mixing Distributors. Lubricant is distributed through a single-line distribution network and parallel type or series movement dispensing devices are incorporated into a special lube air mixing valve. In a typical installation, a separate mixing valve serves each lubricant point to deliver spray droplets to the application point.

Air/oil mix distribution systems offer great flexibility and control over a broad range of viscous fluids and temperature ranges. The system is suited to lubricate both slow and high velocity moving surfaces (see Fig. 42.4).

Dual-Line Systems—Positive Displacement Metering Blocks. These heavy-duty systems are typically used to dispense grease lubricants to large machine elements with an extensive number of lube points. The system consists of twin delivery feed lines from the centralized lubricator connected to all metering blocks comprising the system network as well as two outlets from each block section (see Fig. 42.5).

Each dispensing block utilizes two pistons, a control pilot piston and a metering discharge piston. In operation, pressure build-up in one feed line from the pump causes the pilot piston to act as a "slide valve" to direct the pressurized lubricant to actuate the discharge piston to discharge through one outlet. Alternating the pressure from one feed line to the other, the slide valve changes direction and metered lubricant is obtained through the second block outlet. A pressure-reversing valve is typically mounted on the centralized pump.

Multiline Systems. The system comprises multiple feed lines from a centralized pump to handle the number of points connected to the system. Each line has a dedicated pumping unit or metered outlet for each application point. The discharged lubricant from each pumping unit is delivered to

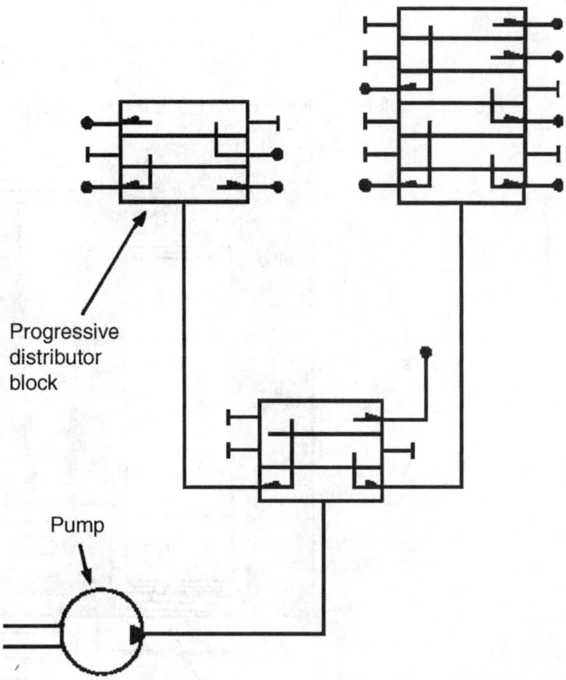

FIGURE 42.3 Single-line progressive series movement system, oil and grease. *(Reprinted with permission of Farval Lubrication Systems, Inc.)*

the lube point by a direct feed line. The system is ideally suited to machinery having a limited numbers of application points. Oil and grease systems are available.

Recirculating Oil Systems. Incorporated into a broad range of machinery, these systems supply lubricant continuously in small and large amounts. System output deliveries to bearings can comprise a few drops/min to gallons/min. Adjustable individual flow meters can be located at or close to bearing entry points.

Continuous Feed Oil Systems. In these systems, the operating principle is similar to recirculating systems; typical layouts incorporate single line feed systems using restrictors or piston distributors and continuously supply oil in closely controlled small quantities. These systems can operate in a total-loss situation without delivering excess quantities of oil at bearing points.

System Control and Monitoring. During an operational cycle, an automatic lubricator delivers outputs at preset internals. An electric timer normally controls this function. A system controller offers adjustable programming as well as monitoring the system network for "Fail-safe" operation, including pump function and low system operating pressure. Programmable Logic Controllers (PLCs) offer additional system control by monitoring pump failure, low oil level in reservoir, clogged filter, and low operating pressures as well as detecting piston movement in series progressive systems through monitoring switches. PLC outputs can be incorporated into the controls of the machine being lubricated to provide automatic "read-outs" of the lubrication system's operating functions.

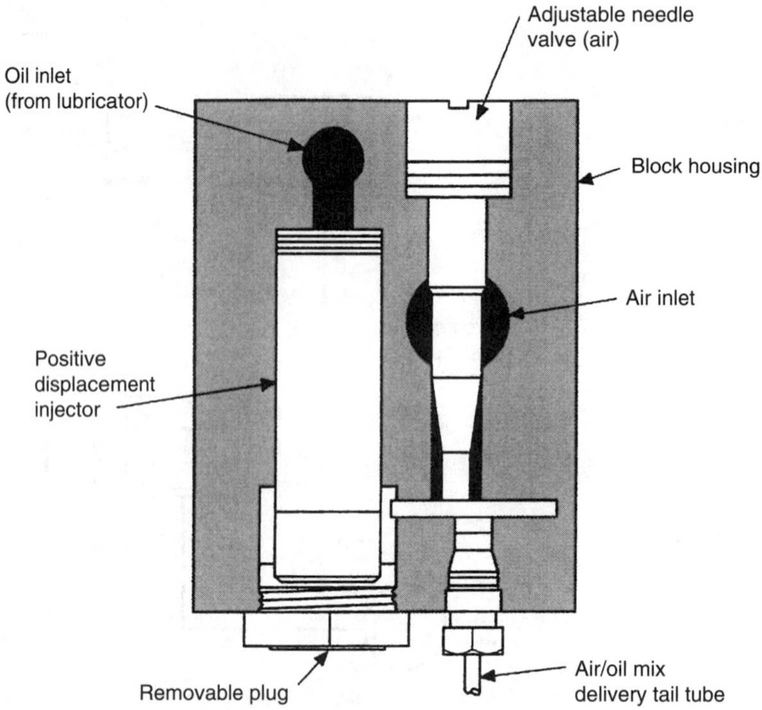

FIGURE 42.4 Positive displacement air/oil mixing distributor. (*Reprinted with permission of Bijur Lubricating Corporation.*)

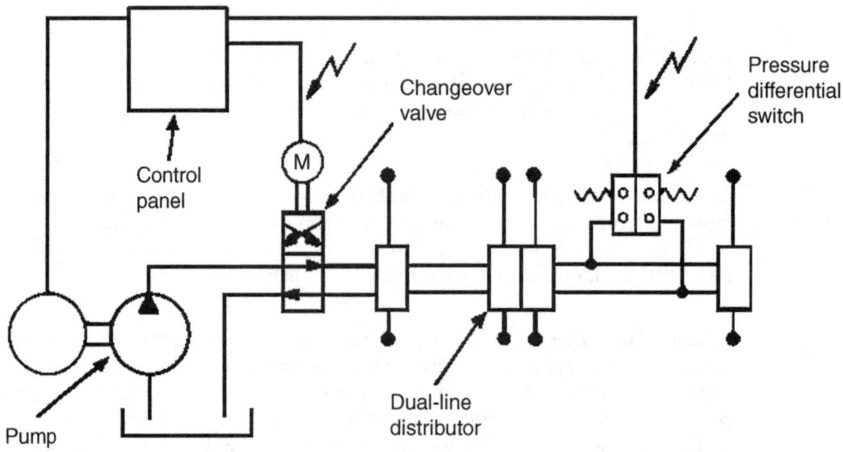

FIGURE 42.5 Dual-line system, oil and grease. (*Reprinted with permission of Delimon GmbH.*)

Preventative Maintenance. Engineers incorporate procedures to follow a simple routine of service and inspection of the machinery's lubrication system to ensure optimum performance from the system and the machinery. System maintenance practices must ensure controlled clean lubricant is always used in systems. A regular service routine of filter inspection/replacement and proper lubricant connections to application points eliminates problems before they gain a foothold and grow into larger problems. Lubrication equipment and lubricant manufacturers' offer user guidelines to obtain maximum performance from effective lubrication programs.

Lubrication Management Programs. Computerized lubrication programs are ideal for large and medium-large manufacturing facilities with production machinery incorporating hundreds of lubrication points with a broad range delivery requirements and use of various lubricants. Custom-designed plant lubrication management software programs are the nucleus of plant-wide preventative maintenance practices.

These programs lay the groundwork to check and maintain the effectiveness of lubrication systems and proper lubricant usage plant-wide. A database of lubrication needs by individual application points establishes the necessary procedures and practices based on an individual machine's operation. Adherence to the program's guidelines and procedures by plant personnel assure maximum results are obtained from a planned lubrication program that reduces plant downtime expense as well as conserving use of lubricant.

42.2 CONCLUDING COMMENTS

Lubricating devices play an increasingly important role in maintaining a successful plant operation. Careful selection of lubricating equipment and lubricating programs result in wide-ranging benefits in industrial outputs and machine operating efficiency.

INFORMATION RESOURCES

Bijur Lubricating Corporation, www.bijur.com.

Delimon GmbH, www.delimon.com.

Farval Lubricating Systems, Inc., www.farval.com.

LubeSite Systems, Inc., www.lubesite.com.

CHAPTER 43
CHIP PROCESSING AND FILTRATION

Kenneth F. Smith
Chief Engineer
Mayfran International
Cleveland, Ohio

43.1 INTRODUCTION

Just a few decades ago, machine tools were all manual, with an operator for every machine, who calculated tool movements and positions, turned cranks, turned on power feeds, and swept chips off machine ways, fixtures, and tools. Speeds and feeds were slow, coolant and cutting fluids were flooded onto parts and tools under low pressure, and material removal rates were anemic. There was plenty of time and labor available to keep machines and work areas clean. Environmental issues were virtually nonexistent.

Then, with the advent of the first NC machines, gradual changes began. Operators, though still required, were no longer making key decisions and typically could tend to more than one machine. Speeds and feeds began to increase slightly. And the labor force, both skilled and nonskilled, started to shrink. More and more, people became concerned with pollution and the environment. Chip and coolant handling requirements became more important as a production, quality, maintenance, and cost element.

43.2 CHALLENGES OF CHIP AND COOLANT HANDLING

Today, machine tool advancements—higher spindle speeds and feed rates, conversational programming, fast tool changes, automated load/unload—along with new tooling systems and inserts, have lead to increased spindle uptime, reduced direct labor involvement, and tighter tolerance and repeatability performance. With these advancements, a machine may possibly be running cast iron parts in the morning, aluminum through the afternoon, and steel, untended, over the night shift. While producing more parts faster, these new machines are also producing more chips faster. For example, consider a 30-HP machining center that can produce up to 25 ft^3 of chips and 150 in^3 of sludge in a 24-h time period machining cast iron parts. As machine performance is improved, the work asked of the center increases. Such is the case of a five-axis machining center producing aerospace fan blades. The fan blades start out as nearly 600 lb of solid forged titanium. They finish as 40 lb of components—and almost 560 lb of chips.

In many of today's applications, the new machines run into trouble too frequently because processing and handling of chips and coolant is insufficient. The consequences can be devastating and in direct contrast to those anticipated: increased downtime cleaning out sumps, increased downtime for changing tools as they dull and wear more quickly, increased direct labor for maintenance, a degrading of part tolerances and surface finishes, and higher coolant costs and subsequent disposal costs.

As older machines were much more tolerant to chip migration (higher particle content in coolant), simple separating conveyors were usually adequate for removing most of the larger particles. The smaller fines had little realized effect. Now, however, the very sophistication of today's machines, that makes them more productive, also makes them more susceptible to the problems of chip migration and coolant contamination.

In the future, these trends will continue, requiring even greater chip handling and coolant cleaning capabilities. Growing environmental concerns will drive up disposal costs, while expenses for clean up of spills and overflows of fluids will skyrocket. This in turn will make coolant cleaning and reclamation critical, and put a large emphasis on high-pressure, targeted coolant use. Health and safety considerations will require cleaner working areas and a contaminate-free environment. A decreasing dependence upon human intervention will require additional optics and sensors to monitor machine, tool, and part conditions. Therefore cleaner machine conditions will be necessary. As manufacturers become leaner and more agile, levels of flexibility rise, creating a more diverse array of chip styles and range of materials to be processed. Labor costs and availability will drive machine systems to greater degrees of automation with less need for human intervention. The expanding role of lightweight materials such as aluminum, titanium, and magnesium creating large volume, low-density chips can quickly overwhelm conventional chip handling mechanisms as well as increase handling and transportation costs. And, though current efforts at "dry" machining may seem like tomorrow's panacea, coolant is still used, typically turning to an airborne mist that may raise health issues, while the dry chips, no longer flushed by fluids, usually require vacuum systems to remove them from machines, parts, and workholding.

Current machine tool technologies, and those predicted for the near future, are challenging the chip handling and coolant processing systems. Here are practical solutions, systems, and strategies to keep chip and coolant processing on pace with machine tool productivity.

43.3 CENTRAL AND INDIVIDUAL SEPARATION SYSTEMS

There are currently two philosophies that govern how most machining operations are set up for effective chip, swarf, and coolant handling within a manufacturing environment. One method is the use of central systems, where both chips and coolant from many machines (can be multiple machine cells, transfer lines, or entire plants) are automatically transported to and filtered through a centralized unit, with the clean coolant returned to the machines. The second strategy involves stand alone, individual separation/filtration stations located at each machine. Both have their advantages, disadvantages, and within each group are technologies that can categorize their effectiveness and appropriateness for applications.

43.4 CENTRAL SYSTEM AND TRANSPORT METHODS

A centralized system consists of two major components: the conveyance system carrying chips and coolants from the machines to the filter, and the filtering device itself. The conveyance system is generally divided into two subclassifications: in-the-floor systems and overhead systems. The older of these two and, therefore, the most recognized is the in-the-floor system.

The concept is simple—trenches are constructed within the floor, following plant layout guidelines for machine types and locations. Coolant and chips from each machine then flow into the trench and are transported to the central filter unit by using one of five usual methods: hinged belt conveyor, drag conveyor, push-bar system, simple sloped velocity flume type trench, or on overhead style conveyance system.

The selection of which equipment to use will depend upon the application. Hinged-belt conveyors provide the most versatility in carrying different types of chips, turnings, and tramp metal. Drag conveyors are limited to broken chips that flow with limited tramp metal.

Reciprocating push-bar troughs are capable of conveying different types of chips and turnings with limited tramp metal. Sloped flumes are limited to broken chips with limited amounts of tramp metal and need high coolant flow rates to keep material moving.

43.4.1 Advantages and Disadvantages of a Central System

The advantages of in-the-floor systems are obvious, because they are beneath the floor, they are out of the way and do not take up valuable manufacturing floor space. Also, in-floor systems consist of known technology, i.e., time and field proven equipment with established performance parameters, specifications, and well-recognized installation protocols. Typically, in-floor systems can be designed and sized to handle a wide range of load capacities, with changes in trench depths and widths.

The in-the-floor style of chip and coolant handling, in addition, provides an emergency reservoir in the event of a machine malfunction or blockage, where excess coolant from spills or overflows will have a place to safely runoff. This can be a significant factor, helping to minimize clean up time and exposure to the problem.

What are the downsides to in-floor conveyance systems? First, for installation in new facilities or new machining lines, trench systems will add time and costs to construction independent of the cost of equipment. To retrofit existing plants with an in-floor system will usually require disconnecting and moving machines, causing production downtime. Second, in-floor units become virtual nonflexible monuments once completed, so thorough advanced planning is critical. They can be costly and time consuming to modify in cases of adding machines or altering plant layouts. Third, since the system is out of sight, environmental problems may be difficult to detect, resulting in greater damage and requiring more extensive corrective measures. Even small leaks, if left unchecked for too long, can seep into soils and ground waters, requiring expensive clean ups and potential fines.

A fourth drawback to the in-floor type transport system is its undesirable and oftentimes unpleasant environment—its "grunge" factor—when maintenance is required. Being below floor level, the trenches and flumes will not only be a collection area for chips and coolant to traverse, but also for all types of dirt, debris and refuse from machines, employees, and material handling equipment. Surfaces will typically be dirty, oily/greasy, and present a health and safety risk to maintenance personnel. This hostile environment may also increase the frequency of needed maintenance procedures, diminish optimum performance of equipment components sooner than their expected service life, and depending upon access, repairs may take longer than if the units were installed above floor and easily accessible.

Now, attention is given to overhead style conveyance systems. These types of chip and coolant systems fall into three basic categories or transfer methods. These are pressurized-pump type operation, a sloped flume style, and, third, the horizontal pipe with vertical, gravity-induced suction-like operation. All three methods require pump stations at each machine tool to provide the necessary pressure to push chips and coolant up through a vertical riser section of piping where it joins with horizontal runs suspended in ceiling trusses. The horizontal run(s) of pipe, with an additional vertical riser lifts from other pump stations added along the way, travel to the central filter unit where the chips are separated from the coolant and discharged into hoppers. The coolant is filtered, and then returned to the machines.

The pressurized pump-style equipment requires lower coolant volume for transportation purposes, but does offset these accomplishments with high-energy consumption and, therefore, higher operational cost since the pump is sized larger. Also, installation cost can be higher because of the need for individual return pipes per pump station.

The sloping pipe flume technique, like the subfloor flumes, has the appearance of the traditional older technology that requires a volume of coolant to flush the sloped pipe to keep the material flowing smoothly and to keep the pipes clear of blockage and backups. This also requires increased headroom due to the sloped pipe.

The horizontal pipe/gravity induced flushing system is one of the newest technologies available and requires some explanation as to its operating principles. This system, instead of flushing with additional coolant volume, creates a turbulent flow to dislodge chips and particles and keep pipes clear of build up (Fig. 43.1). The flow is only as great as the sum of coolant used by all of the machine tools, so pipe diameters can typically be smaller than sloped flush pipe systems.

As an example, but again dependent on the individual application, the vertical rise discharge pipe may be of 2-in diameter and flows into a horizontal pipe run of initially 5-in diameter. As the horizontal run continues and other vertical discharge connections from various machines are added, the horizontal pipe diameter may increase several times before its descent to the central chip separation unit and filtering device. Less obtrusive than other overhead systems, operating expenses are minimal since the turbulent flow is created during a programmed (only as-needed) purge cycle, the frequency of which is determined by the amount and type of chips produced by each application. The simplicity of this type chip/coolant transport system allows for a brief description of its operation. All overhead type systems start at the machine tools with a pump station. The pump station and all connections to the machine are typically at floor level, and all chips, strings, fines and the coolant are removed from the machine(s) bed to these stations. Pump stations connected to machining operations that provide long, stringy chips may require a mechanical shredder device to reduce the stringy chips to smaller, more uniform chips that can pass through the pump station and the piping without clogging. From the pumpback station's outlet, the vertical rise discharge pipe joins the horizontal run that is routed to the central filter.

The unique feature of this type of system is its flush or purge method. At predetermined time intervals, the system's PLC type controller signals a discharge valve located just prior to the separation/filter unit to close, in effect allowing the system's pipes to fill with chips and coolant. Sensors monitor the level of coolant, and once the correct height is reached, a second signal is sent to open the discharge valve (foot valve). At that time basic laws of physics create a siphon effect that flushes coolant, chips, and sediment through the pipes and to the filter unit.

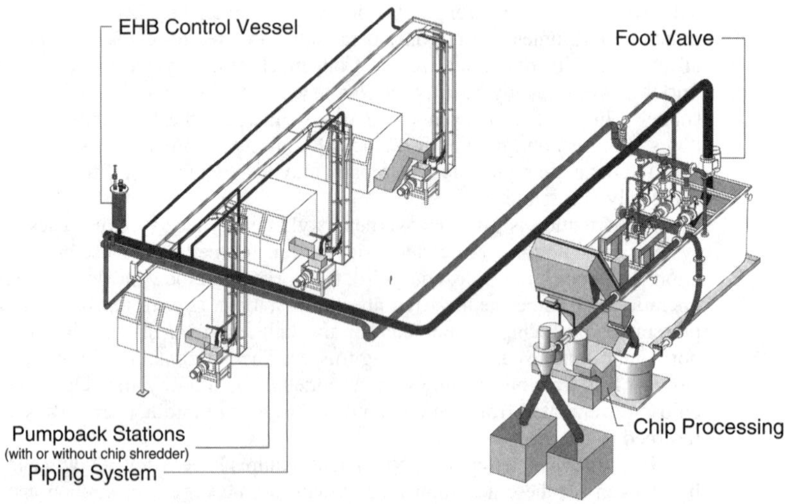

FIGURE 43.1 Mayfran International manufactures an alternative to the subfloor trenches or flumes with its Viavent system that puts chips and coolant transport above the shop floor. Using a network of pipes, it allows more flexibility for reconfiguration at lower costs when changes are needed. (*Illustration courtesy of Mayfran International.*)

Key to the system is the controlled discharge valve or foot valve. It allows the horizontal pipe to accumulate used machining fluids and creates the turbulent flow when the valve opens. Therefore, it does not require additional high pressure, energy-consuming pumps to create that flow, nor is the introduction of large amounts of additional coolant needed to clean out the lines. This turbulent flow purge cycle also eliminates the potential in the overhead pipe for fungus build-up.

The benefits inherent with the overhead type central chip and coolant transportation systems, as compared to subfloor style conveyor styles, include their relative ease of installation and minimal interruption to ongoing building construction or to production.

Time for project completion, whether in a new facility or an existing plant, is greatly reduced. The overhead system enhances flexibility of the centralized configuration, allowing for modifications of runs, additions, or deletions of machines accomplished with basic pipe fitting skills. There is no demolition of floors, large equipment to move around, or disruption to normal plant operations. The overhead pipes provide a personnel-friendly, clean environment for any maintenance that may be required, while the open routing of the pipes provides visual confirmation of tight seal throughout the coolant path. Any leaks can be quickly identified and corrected before major contamination occurs. What's more, the closed-loop construction of the overhead systems can help minimize coolant evaporation and mist in the air, improving health conditions and reducing coolant loss and consumption.

Overhead systems do have shortcomings. Among them is the fact that the technology, and especially the vertical induced suction, turbulent flow systems, are relatively new and are slow to be embraced as reliable, practical alternatives. Used extensively in Europe, these types of systems are just now making inroads into North American operations.

In addition, a machine that puts out a nonanticipated chip size or style, say a long string, could be a problem with in-floor systems, and may require that some type of chip processing device, such as a shredder or crusher, be installed before the pumpback station to avoid lines becoming clogged. Also, there is the additional investment, maintenance and operation costs of the pump stations. The fact that the interior of the pipes is visually inaccessible (except with fiber optic cameras) for inspection can make troubleshooting and finding blockages and other causes of slow flow rates more difficult.

43.5 COOLANT FILTRATION FOR A CENTRAL SYSTEM

Once chips and coolant have been transported, either by in-the-floor or overhead system, from the machines, they reach the central filter unit. In almost every application, the filter device has a vacuum type operation, with a pump creating a low-pressure vacuum on one side, the "clean" side, of the filter media. Many of these filters have a conveyor that removes larger chips. The filter media may be a disposable material, or permanent with an integral backwash cycle (a periodic pressure wash, using clean coolant, to remove filtered particles from the media surface), or a combination of the two different types. Disposable filter materials may include nonwoven fiber and paper, which may be surface treated with materials that further aid in filtration. Permanent media is usually woven synthetic fibers, wedge-wire or micro screen. Typically, central filters can provide coolant cleanliness from 50 μm to 5 μm, average.

Regardless of the style of system chosen—in the floor or overhead—the centralized chip and coolant handling option has advantages and disadvantages that must be weighed before any decision is made. On the positive side, a central system eliminates equipment duplication that will usually save floor space and be less intrusive to the manufacturing environment in and around machine tools. With all chips transported to a single area, wringing and removal/disposal operations are convenient, efficient, and economical, and usually the immediate area at the machines will stay cleaner and safer due to the fact that chip discharge and handling is now performed at a distant location. Because fewer pieces of equipment are needed, overall maintenance and repair expenses for central systems may be at a lower level. Central filters also allow for simpler management of the various types of soluble cutting lubricants (coolants). Uniform concentrations and pH levels can be maintained easier in central filter units opposed to multiple, stand-alone filter units.

The drawbacks of using central systems include the initial investment. Also consider that operational flexibility is thwarted since all of the machines connected to the system will have to use the same coolant, so materials and processes will have to be similar. If scrap value of the reclaimed chips is important, all of the machines will have to run closely related materials, such as cast irons, steels, alloys, etc., since chips cannot be kept segregated and will be mixed together within the central transport and single filtration system. And, even though the overhead-type systems provide an enhanced agility, there are still substantial costs, time, and planning required in making a good system. Then, there is the unthinkable—if the central system malfunctions and goes down, it can bring production to a halt at all of its online machine tools.

As a reference, today's typical machine tool using high pressure circulation will require particles no larger than 10–15 μm nominal and lower if possible. Machine tools not equipped with high pressure systems may operate satisfactorily with particle sizes of 50 μm. However, the cleaner the coolant the better since dirty coolant can dull cutting tools faster, create premature and excessive wear on machine components such as bearings, and develop quality problems with part dimensions and surface finishes. High concentrations of particles can also adversely affect the stability of the various soluble cutting lubricants used in today's metal cutting processes. Unstable coolant can become difficult to filter, become contaminated with bacteria, thus resulting in total replenishment of the coolant.

The centralized filtering technique offers flow rates that can exceed 15,000 gallons of coolant per minute, so even large, multiple machine stations and plant-wide operations can utilize this cleaning method. Generally, central units are of two design principles—one being a conveyor-style bed type where the liquids flow down through permanent media. The second design concept is that of a rotary drum filter that draws the fluids through the center of the drum.

Both designs provide similar filtration capacity and performance. One of the only differences is that, in most cases, the rotary drum cannot operate with disposable media except as an add-on filter (cartridge or bag) down the line. A major consideration when selecting a system, however, is the choice between permanent and disposable filtration media. The permanent system with the additional backwash mechanisms will generally be a more expensive initial investment, while the disposable style will require continual media purchase and disposal costs over its lifetime.

43.6 STAND-ALONE CHIP COOLANT SYSTEM

The stand-alone (at-the-machine) strategy for chip handling and coolant processing also has its pluses and minuses. There is a wide range of equipment and performance levels offered, depending on application needs, to meet most manufacturing and machining operations. Because these systems of chip and coolant control process the coolant at the machine, there is an additional concern, one not usually associated with centralized systems that must be addressed when deciding upon stand-alone units.

That problem is when coolant, acting as a heat sink, begins to lose its ability to cool. This happens when chips, fines, and super fines migrate to and accumulate in the "clean side" reservoir, to the point where their volume displaces too much of the fluid. And, since travel to a central point is not necessary, there is little time for the fluid to cool down, and the temperature of the coolant begins to rise. Also, due to the decreases in coolant volume, a constant pressure may be difficult to maintain. Once this happens, temperature fluctuations can occur at the point of machining, resulting in inconsistent part dimensions and varying finish characteristics. This situation is exacerbated when cast iron is being machined, as the dust-like super fines, billions of particles, create coolant slurry and then sludge. As sludge builds-up in the reservoir, it has to be cleaned sooner, but temperatures remain high as even more particles remain in suspension (instead of settling to the bottom with time). This abrasive mixture will quickly coat parts, tools, workholding, and machine components. Surface finishes of machined parts will deteriorate while maintenance requirements skyrocket. In addition, environmental concerns of coolant overflowing, spilling onto floors and runoff, have to be considered as the reservoir fills with chips and swarf.

43.7 STAND-ALONE TRANSPORT AND FILTRATION SYSTEM

With the above points to consider, a review of available stand-alone systems reveals three levels of performance offered. First there are the basic and simple conveyors: hinged-belt, drag and/or scraper types, and some magnetic units—which all offer very basic, large particle separation, and no actual filtration. The larger contaminates are carried or pushed to hoppers while gravity and openings in the machinery allow coolant to drain. Mesh screen or wedge-wire in the coolant reservoir may provide additional separation, but particles of up to 1/4 in. in size may also pass through the openings and remain in the coolant. This level of chip separation can only be used when high-pressure circulation and clean coolant are not required in the machining process. Of course, magnetic conveyors will not be effective for separating chips from coolant when the application is machining nonferrous materials.

The second level of at-the-machine chip and coolant handling is a chip conveyor (as above) used in combination with a filter. In the past, this usually meant two separate units joined together to act as one unit. Now, a new technology that integrates the conveyor and filter into a single, compact machine, saving floor space and simplifying overall operation, is being employed. These combination units will typically offer mechanical separation through the conveyor system, plus a secondary, permanent media filter that can remove particles of 50 μm (nominal) or larger. These machines can handle virtually all types of chips and are ideal for applications, such as the machining of cast iron and die cast aluminum. The single-unit systems are designed to replace existing systems, or as specified equipment for new machine tools, and with modifications for custom machine tool arrangements.

A typical unit will consist of a single scraper conveyor with or without a hinged steel belt conveyor over top for segregation of larger chips, then an indexing, permanent media filter drum for fine particulate filtration. A pressurized wash system is included that means virtually no downtime for manual cleaning. The process can be viewed through a clear access panel for visual inspection and verification of the cleaning action. The filtered coolant is held in a clean coolant holding tank for return to the machine tool (Fig. 43.2).

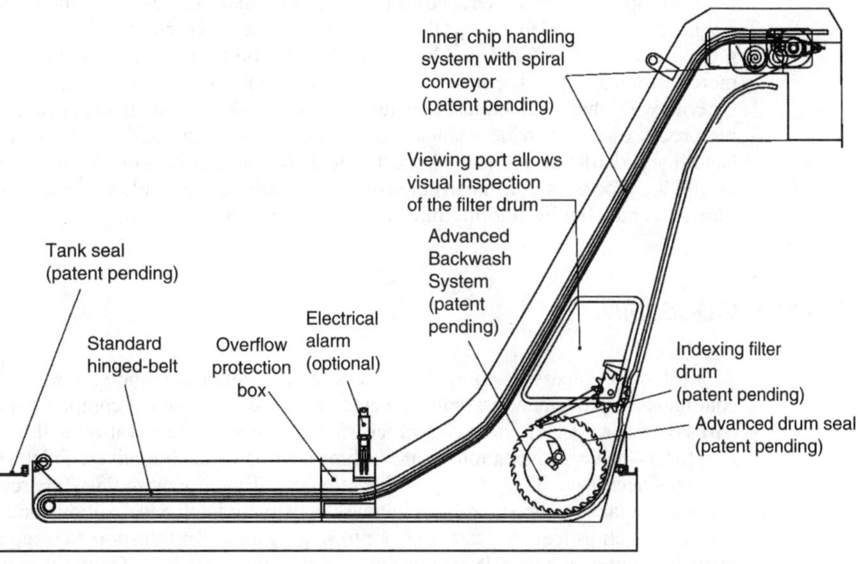

FIGURE 43.2 The ConSep 2000 II is the latest development of Mayfran International for at-the-machine chip handling and coolant cleaning all in one unit. (*Illustration courtesy of Mayfran International.*)

The third and highest level of stand-alone coolant cleaning employs the chip conveyor and filtration drum system as described immediately above, combined with an additional (nondisposable media) filter/cleaner that can clean coolant to a 10–15 nominal level. The first stage of this system's conveying and cleaning process is identical to the conveyor and filter drum unit, and then takes the coolant through additional cleaning process step. Typically a cyclonic media-free system is used at this stage to provide excellent coolant cleaning while preserving a low need for maintenance and human intervention in the cleaning process. With the addition of auxiliary bag style filters or other types of disposable media, coolant may be cleaned to 5-μm levels or less.

This level of coolant filtration is ideal for machine tools where high coolant circulation pressures of up to 1000 psi may be necessary, when very tight tolerances on part features and finishes are demanded, and when exacting, vulnerable machine components could be compromised by contaminates.

43.7.1 Advantages and Disadvantages of a Stand-Alone System

The list of advantages afforded by the stand-alone units, as compared to centralized systems, includes the high level of flexibility that allows machines to be moved freely, machines to be added and deleted from the plant without any expensive alterations, and long-term construction and installation considerations. The agility of the at-the-machine chip handling also means that each cutting machine can work with different coolants chosen to provide optimum results for its application, and each machine can work with different materials without worry of cross-contamination of chips. Likewise, separate machine systems means that any malfunction or repair service keeps downtime confined to just one machine tool, and, similarly, contaminated coolant in one machine will not spread throughout all machines, requiring a complete and costly coolant replacement procedure. This strategy of chip handling at individual machines also means that, since the machine tools with their chip handling equipment may be acquired in stages, there is no single, large investment to justify and fund.

Among the disadvantages of going this route in chip handling is the need for an increased amount of time and labor to tend the numerous and separated chip hoppers and coolant sumps at each machine. This argument also applies to the fact that each machine has a sump pump and tank that will require attention from time to time for cleaning. Then too, the greater number of mechanical units will likely increase demands on maintenance and increase the odds of breakdowns and repair frequency. There is also the fact that, as each machine tool is equipped with the necessary chip equipment, floor space may become limited and accessibility in, around, and between machines may be more restricted than if the machines were linked to a centralized system.

Following these guidelines, plus factoring in application elements such as material and machining processes will provide a sound, initial starting point in the decision of which chip and coolant handling and filtration system is best suited for an organization. Elements such as material and machining processes, estimated chip configuration style, size, volume, and weight, as well as coolant formulas and labor availability must also be considered.

43.8 CHIP PROCESSING

To most, chip-processing equipment is a necessary option; to others, they are a luxury. However, as continuous improvement programs go enterprise-wide, and global competition makes even fractional cost savings critical, efficient chip processing and coolant conservation will become more important. Add to this the implementation of far-reaching environmental initiatives, and the processing equipment becomes necessary. That is why devices that can dry chips with 98–99 percent (by weight) of the residual coolant removed, and can increase density and decrease volume, and automatically handle and remove chips from the factory floor to staging areas for transport to scrap dealers, smelters, and recycling centers are becoming more critical to plant operations. Configurations include continuous feed units for use after discharge from central filters or batch loaded processing machines for applications with at-the-machine coolant and chip separation.

Various processing equipment are available and include gravity draining of fluids, drying by spinning or wringing, and making chip volumes denser through compression and compacting. With these auxiliary machines, it may sometimes be difficult to justify the financial investment in such equipment because the goal may be long term. Organizations should, however, expect more efficient chip handling and transport, reduced storage and labor needs, and also safer operations and lower environmental concerns when chip processing equipment is employed.

43.8.1 Chip Processing Equipment

The first step in most chip processing, coolant recovery requirements, and material recycling efforts could be some type of chip shredding or crushing. As was mentioned earlier, shredding may be necessary in overhead, centralized chip transport systems where long strings become tangled nest that could cause blockage of pipe runs. But a more common reward offered by chip shredding is the high density, volume reduction benefits while allowing the material to be processed further (Fig. 43.3).

43.8.2 Crusher and Shredder

A typical chip crusher usually consists of a set of hardened cutters rotating in opposite directions. They can reduce volumes of chips by as much as 50 percent, converting turnings, balls of turnings, and strings, into smaller granular chips.

What does this mean in terms of density and why is it important? For example, assume that a container will hold up to 75 ft^3 of chips and only one container can be taken to a metal scrap dealer per trip. If that container is loaded with loose, light turnings from an aluminum machining application, the nominal weight of which is approximately 10 lb/ft^3, the scrap load will be about 750 lb. If the same container is then filled with loose turnings processed through a chip crusher and converted to

FIGURE 43.3 Heavy-duty wringers that have continuous feed systems are recommended for large volumes of chip processing. (*Photo courtesy of Mayfran International.*)

fine chips, the per cubic foot weight increases to a range of 20–30 lb, thus the load weight would, on the conservative end, approach 1500 pounds. If a scrap dealer pays $0.25 per pound of aluminum, one trip would pay $187.50 while the denser load would result in a $365.00 payback for the same transport costs.

Crushers and shredders are available in sizes and capacities suitable for integration at the individual machine and pumpback stations application, for batch loading in a centralized chip staging area, or for plant-wide automatic infeed and discharge applications.

Typically, the metal chips and swarf will be more valuable, the drier the material is, when received. The less coolant remaining on the chips reduces the risks of contamination of the environment, and reduces the wet chip penalty cost. Plus, the more coolant that can be reclaimed at the machining operation, the more can be reused, reducing acquisition costs. Probably the simplest drying method is gravity and the easiest piece of equipment to use for this is a conveyor or hopper with a grate in the bottom for draining. On the conveyor that chips and swarf travel some coolant will drain off. As it inclines, additional amounts of fluid may drain. But, depending on the time factor, coolant may still remain.

43.8.3 Chip Wringing System

The next method is chip wringing using centrifugal force where chips are either batch loaded or continuous fed into a spinning drum, like the spin cycle of a clothes washing machine. The g-forces throw the coolant off the chips and through the wedge wire screen or perforation for reclamation. As more chips are introduced into the drum, chips will be forced up the sides of the drum, passing the wedge wire screen, over the rim into waiting hoppers, or slung through pipes to empty bins or onto conveyors for transfer. Although these types of wringers can be used with most materials the higher payback usually comes from aluminum, titanium, and exotic type metals. A third style of wringer has what is called a live bottom where a plate mechanism at the bottom of the drum is programmed to periodically rise and force chips over the wedge wire screen and out of the spinning drum. This style is best for high volumes of heavier chips and fine, sludge-type swarf when the weight of the solids would make lifting and ejecting the material by spinning action alone difficult or impossible.

43.8.4 Compactors

A final technology for drying chips does double duty and can be included with the devices for reducing volume. These machines are known as compactors or briquetting machines that compress loose chip materials into briquettes, pucks, or bricks (Figs. 43.4 and 43.5). Capacities often range from 50 kg/h up to 3000 kg (110–6,614 lbs.) for processing from a single machine, a cell, system, or an entire machining facility. Importantly, during the compression operation, these machines can squeeze dry the chips, removing up to 98–99 percent of the residual coolant and cutting oils for reclamation.

With the chips dried and, through compression or shredding, made denser, the material can be sent, via heavy-duty conveyors from the final processing step directly to disposal staging points. Incline conveyors can lift the chips or briquettes to heights for deposit into dumpsters, truck trailers or rail cars.

43.8.5 Advantages of Chip Processing

The advantages of using chip processing equipment, such as wringer and briquetting systems, will usually include higher value for scrap, more reusable coolant recovered therefore lowering purchases, labor savings, lower removal and transportation costs, and reduce risks of environmental accidents and costly cleanups. Accidents from unruly chips or slips on coolant-wet floors, along with health concerns may be alleviated with proper and thorough chip processing and load out methods in place.

FIGURE 43.4 Aluminum compressed into small, dense cylinders. (*Photo courtesy of Mayfran International.*)

The major problem with chip processing is the difficulty manufacturing and maintenance personnel may have in convincing financial departments the cost is justified. Return on investment figures can extend over long periods and many costs savings are hidden or based on "what ifs"—if there is no coolant spill, how much were the cleanup savings and what was the increase in uptime and what is the coolant savings? Only estimates can be used.

FIGURE 43.5 Large briquetting equipment can compress a variety of ferrous and nonferrous metals. (*Photo courtesy of Mayfran International.*)

A second, less prevalent but possible problem is that scrap dealers, smelters, and recyclers may not automatically increase scrap values, especially with briquette-type compacting. Negotiations may be required to obtain their acceptance and buy-in of the improved scrap condition.

43.9 THE FUTURE

As the machine tools of today evolve to meet the next generation of challenges, the need for cleaner coolants to improve part quality, protect tools, parts, and machines will continue to increase. Flexibility will claim a greater foothold in fast-response manufacturing. Demands for safe, clean work and living environments will expand. Can new cutting tool technologies make coolant-free dry machining more practical on the factory floor? Will a decreasing labor force necessitate even more automation and artificial intelligence? What is the next lightweight material to find a place in machining?

Present day filtration systems, chip drying or briquetting equipment, combined with chip and coolant overhead transport systems are keeping pace now. And they will continue to evolve, stride for stride, side-by-side with the machine tools, to afford maximum production, savings and safety as these questions and others are answered.

CHAPTER 44
DIRECT NUMERICAL CONTROL

Keith Frantz
VP of Development, Cimnet, Inc.
Robesonia, Pennsylvania

44.1 INTRODUCTION

The most obvious failure of numerical control (NC) file management is the crash of a machine tool into its fixturing, its workpiece, or worse—its operator—because the wrong part program was run. This is an attention-getting event that brings a shop to a halt. Fortunately, though it's one of the most costly single events, it's also one of the rarer failures of NC file management. The more subtle shortcomings that show up all the time are the ones that really combine to steal productivity and profits from a facility.

This article introduces the problems that challenge today's metal fabrication manufacturer and how *direct numerical control* (DNC) systems will help to solve these problems. The benefits of DNC systems are widely understood and acknowledged to improve factory productivity and product quality.

44.2 WHAT IS DNC?

Direct numerical control is a data communications system connecting a group of NC machines to a common computer or network of computers. A DNC system manages the storage of numerous NC part programs and has provision for on-demand distribution (downloading) of NC programs to the machines. A DNC system usually has additional functionality for the management, display, and editing of NC part programs.

44.3 INVESTING IN DNC

Calculating the cost of managing your DNC system is not easy. There are many activities involved with an NC machining operation, some of which are easily overlooked. Some occur in the ordinary course of events; others happen infrequently and on an irregular basis.

The cost of the recurring activities can at least be estimated by taking a daily average of the number of programs worked on, and adding up the times spent on each activity involved for a single program, multiplying these two numbers together, and multiplying the result by an average hourly cost. Creation of new part programs will usually have to be costed separately from the running of existing part programs because most shops reuse files as parts that were made previously are rerun for new jobs.

Costing nonroutine activities is much harder, because both their frequency and the cost of each occurrence are usually guesses, and thus can vary widely. One approach would be to make a list of each event involving NC part programs that has been known to occur and get a number of knowledgeable individuals to make independent estimates of how often they happen and what they cost when they do. Don't forget to include direct costs as well as labor and overhead when costing sporadic occurrences—wasted stock, replacement parts, extra supplies, etc.

If your facility is like most, you will be astounded at the costs involved if a thorough analysis is made. *At almost every point, however, there are actions you may be able to take that will make an incremental improvement,* depending on how things are currently working. These fall into the same general categories as most manufacturing improvements—saving time, reducing costs, and reducing waste and scrap.

44.4 IMPROVING YOUR DNC SYSTEM

In order to see where improvements can be made in the processes relating to NC files, let's first consider the routine activities from beginning to end, and then look at things that don't happen all the time.

44.4.1 Programming

The first thing that happens to an NC file is, of course, that it is created. The actual time required to produce the post-processed part program is largely dependent on the CAM system and the programmer's skill in using it. However, his job is not finished when the posted file is written. It needs to be identified with the correct part, operation, and machine, and associated with the right tooling, fixturing, and any special instructions. If the file is to be stored on physical media for transfer to the machine, the media must be punched or written, and then stored in the right location with appropriate identification. Even if there is a DNC system so that the transfer will be done electronically, the file must be put in the appropriate place and identified.

Where Programming Improvements can be Made

Use Electronic Storage. If the file does not need to reside on physical media, the time required transferring it to that media and the cost of the media itself can be eliminated. This is especially true with paper or Mylar tape media, and, to a lesser extent, with floppy disks or magnetic tape. The time required to identify and store the file will be much less if it is kept in electronic form rather than being transferred to a physical media.

Efficient File Management. Even if the file is stored electronically, the job of properly identifying it, associating it with the other information about tooling, fixturing, etc., and putting them all in the right place can be simplified if there is computerized support for NC (and other) file management. The NC file may internally denote what it is for, what tooling it uses, etc., but these are not apparent when looking at a directory of file names. Descriptive directory and file names help, and this is a lot easier if names are not limited to the DOS 8.3 format. Having a regular file (and directory) naming convention can extend the usability of a flat file-based system by insuring that a new file is not inadvertently given the name of an existing file. In practice, the applicability of any naming scheme depends on the regularity of the part names, operation numbers, etc., that go into it. As a simple example, creating a part program for an operation with a three-digit operation number where only two were provided for in the file naming scheme may cause an ambiguity—is program 300120 for part 3001, operation 20, or part 300, operation 120? This is actually a variation on the intelligent vs. nonintelligent part numbering debate. Just as any intelligent part numbering scheme will eventually break and need to be reworked, so also any file naming scheme will eventually break down. A system that provides for storage of information other than in the file's path will eventually prove more flexible. The most straightforward place to store this information is in a database, which also makes the information available to other systems and makes it easy to maintain. By storing the program file's "metadata"—data about the file's data, in this case, to what it pertains—in the database separately

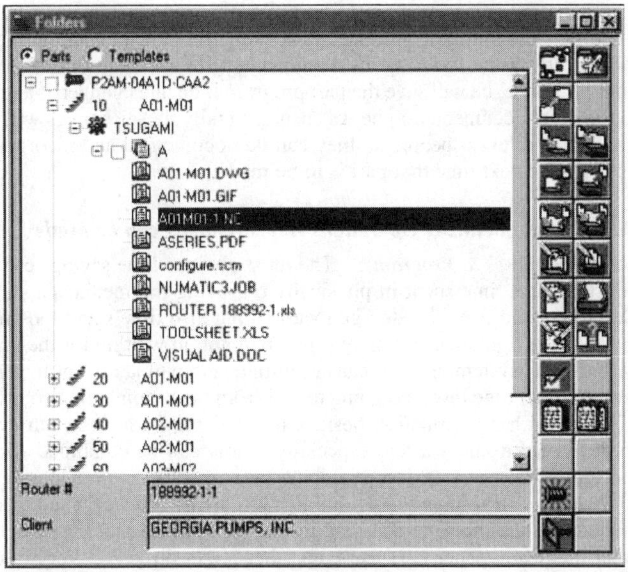

FIGURE 44.1 Folders of NC programs and drawings.

from the file itself, it becomes possible to identify multiple contexts for the same file, e.g., if the same part program is used for a particular operation on several parts, the file only needs to be stored once, yet can be identified as being used for each part (Fig. 44.1).

44.4.2 Shop Supervision

The next thing that happens to an NC program is that it is identified as one of the inputs needed to run a particular operation on the factory floor. This is a specific case of the more general need for supervisors to instruct operators what to do. Since the order entry and job scheduling systems are not necessarily tied to programming, the need for a particular NC file might arise before the file is ready for manufacturing.

Where Shop Supervision Improvements can be Made

Automate Transfer of Work Instructions to Operators. How shop supervisors know which jobs to tell operators to do is beyond the scope of this document, but since their instructions to the operator may involve specifying the NC file(s) to use, automating the process of getting work instructions to the operator is going to save time and eliminate one possible source for confusion. Automated instructions are almost always written and are less prone to miscommunication than verbal instructions. Unlike pieces of paper, they also cannot be misplaced or destroyed.

Have a System to List All Information Required. If there is a system in place for supervisors to know whether the information the operator will require for an operation (such as the part program) is available, the supervisor will not waste the operator's time on a wild goose chase if it isn't.

44.4.3 Manufacturing Operations

When an operator knows he has a specific job to do, the steps he must take with respect to the NC part program begin with retrieving the part program from wherever it is stored, and loading it into the control. He will almost certainly check it out somehow—inspecting the code, dry running the machine, etc. He may need to edit it, due to a change in tooling or fixturing not anticipated by programming, a last-minute revision of the part, or just because he has a better way to do the job. In doing these things he will need to consult other documentation for the current task—a blueprint, a tool sheet, special fixturing instructions, etc. He will then run the parts and if all goes well, usually after he's done, he will save the part program if he had changed it at all, along with any changes made to the other documentation he was using. Finally these changes will need to be brought to the attention of the relevant people so they can be documented and incorporated into the information to be used for the next time that part is to be made.

Where Manufacturing Operations Improvements can be Made

Local Access to Programs. The most obvious time savings can be had by eliminating sneakernet time—the time spent in physically retrieving the media. This involves not only the time spent walking to and from the storage location, but also time spent looking up the right one, of verifying that it is the right one, and more often than not, in waiting for the media to be made available. Even with a DNC system, if the means to initiate a download is not locally available at the control, there is sneakernet time involved going to and from the terminal that controls communications. Also, there is usually other information besides the part program to be retrieved. If these documents can be retrieved electronically, and especially if this can be combined with the fetching of the program, more time can be saved (see Fig. 44.2).

Download Program Electronically. If using physical media for part transfer, it needs to be loaded into the machine and read in. Physical media and the devices used to read it are subject to wear and tear—paper tape tears, read heads get dirty, etc. Very often the reading process must be

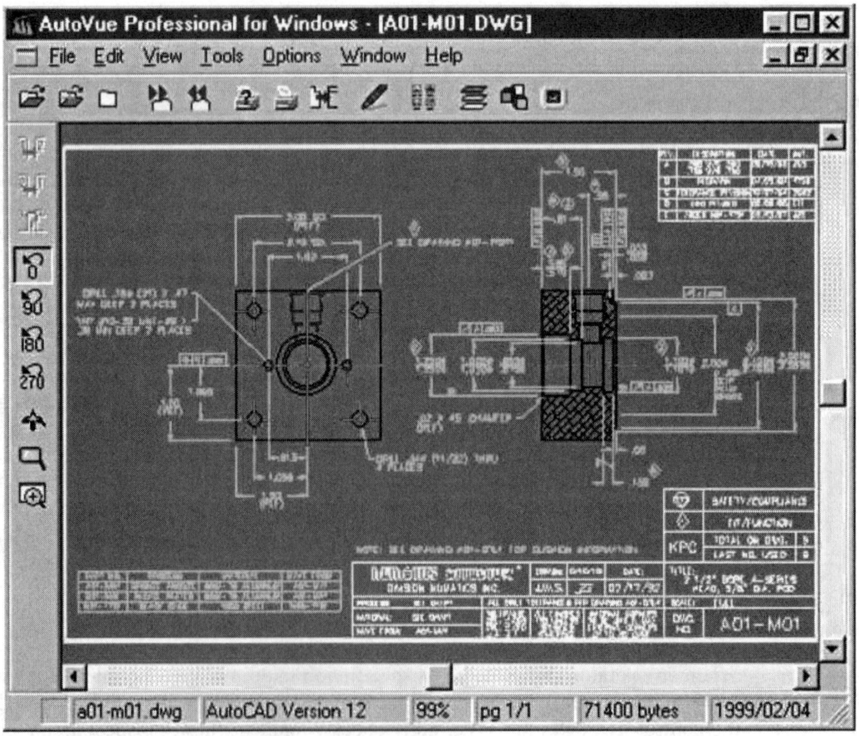

FIGURE 44.2 View drawings online.

repeated, or some maintenance must occur before the reading can be successfully completed. Cutting down on the use of mechanical devices in this process not only saves a little time every time a new program is read into the control, but occasionally saves quite a bit of time when the process doesn't work the first time.

Verification Program With Solid Modeling Software. If the program can be verified electronically, this will be faster than dry-running (at a reduced feed rate) or visually inspecting the program. To be most effective, more than the tool center line path must be shown—the actual tool and stock geometry must be included in the model (see Fig. 44.3).

Modify Program Locally. If editing is needed, and it cannot effectively be done on the control, the operator must get programming to make the change or find a terminal to make the change himself. Having a complete part program editor available locally could end up saving a great deal of time (see Fig. 44.4).

Manage Toolbreak Restarts. An unfortunate fact of life in metal cutting is that tools sometimes break. This will mean backing up to some point in the NC program, and restarting again from that point. While straightforward if the entire part program resides in the control's memory, recovering from this can be a time consuming activity if "drip-feeding" from an external source. This usually

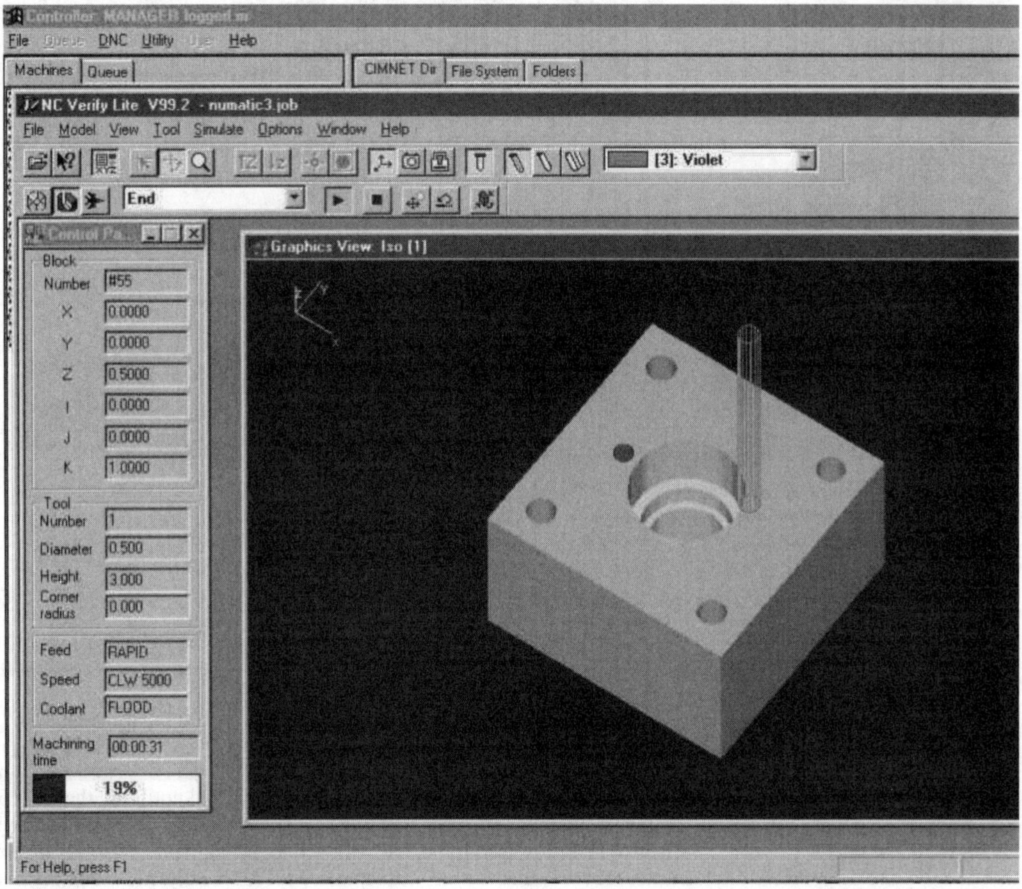

FIGURE 44.3 Viewing virtual tool path.

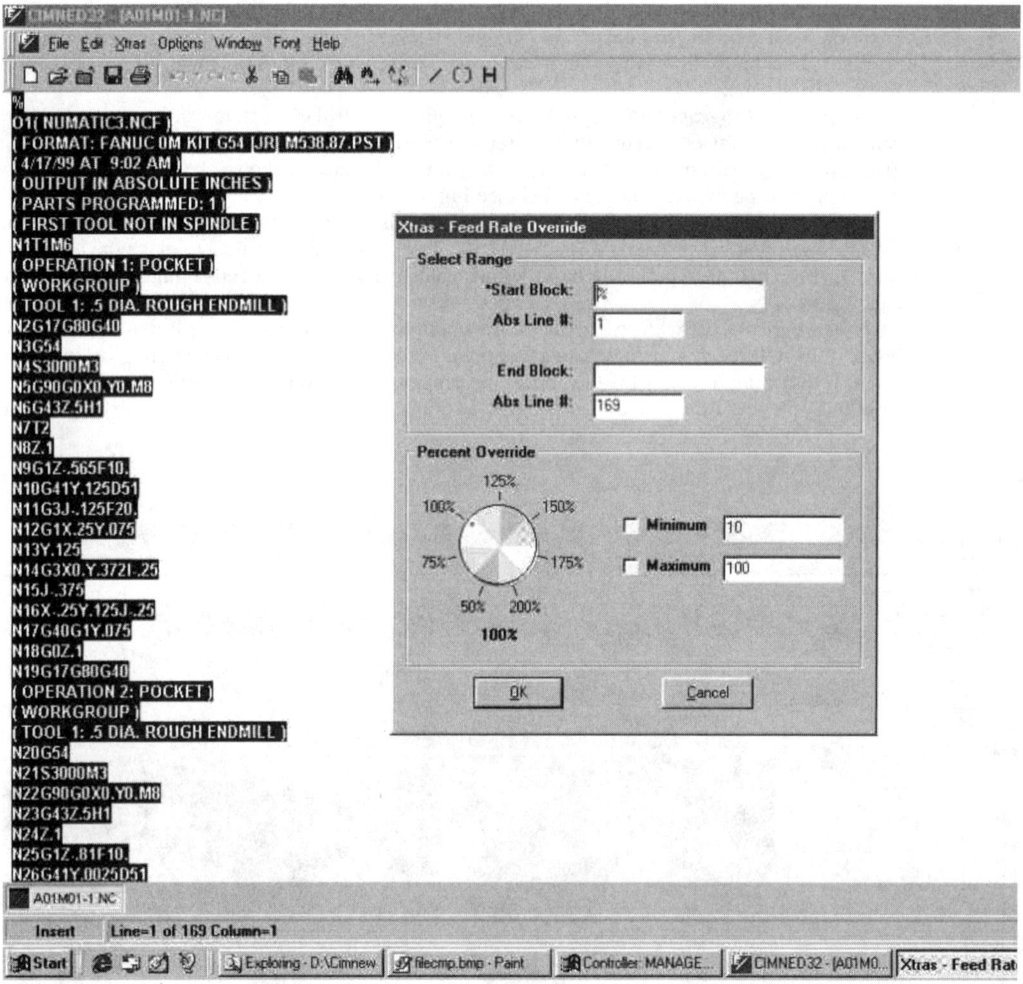

FIGURE 44.4 Editing NC program.

also means manually inputting MDI setup information. A system that can electronically provide just the portion of the program yet to be run, with the requisite header, tool change, etc. can save a lot of work.

Save Changes to Programs Electronically. If the program is changed, it needs to be saved for future use, or at least reference. If it needs to be stored on physical media (by punching a tape or writing a disk) there is time that could be saved both in the actual transfer itself (especially punching a long tape), and in storing, labeling, and logging the revised version, and notifying the right people about it. The same is true of drawings or other paper documents if they are marked up—while the operator is not a draftsman, markups need to be disseminated to the right people so the originals can be updated. Any of these activities that can be accomplished electronically, including the notification of people, will save time, and potentially reduce errors.

44.4.4 Nonrecurring Activities

Other activities which touch on the handling of NC files happen less frequently—perhaps only a few times—yet if not handled well they can take a long time.

Where Improvements can be Made

Log the Program Used Electronically. Finding out what program was used to run a particular lot or batch may not even be possible unless a log of files is maintained. If a paper log is used, it may not only be time-consuming to locate a particular entry, but also a little time will be required to record each program run.

Use Generic Hardware. Repairing or replacing the equipment used to transfer programs will be more difficult, time-consuming, and expensive, the older and more specialized the equipment is. Paper tape readers and punchers, custom disk or tape drives, and even proprietary DNC hardware are always relatively expensive pieces of equipment because they were never produced in large volumes, and in the future this will only get worse as devices are discontinued. While generic hardware cannot be used for everything, it should be used wherever possible because it is low cost and comparatively easy to repair or replace. Even software can become obsolete—not that the programs themselves degrade, but they may not any longer be compatible with currently-available replacement hardware. For example, many DOS-based communications programs cannot be run on Windows NT or Windows 2000 because these operating systems do not allow application programs to directly access the computer's hardware, and this was the norm for most DOS programs. They require the use of a software driver, and few DOS programs are written to use such a driver.

Expand With Generic Hardware. Expanding to new machines or facilities will involve some of the same considerations as equipment replacement. In addition, there are often different ways of accomplishing the same goal. For example, if a machine needs to be upgraded to run longer programs, this can be accomplished by either adding memory to it (an expensive proposition, not to mention that there is a limit as to how much memory can be added) or by allowing the machine to drip feed from an external source such as a DNC system. For a single machine, with a finite memory requirement, the former may be cheaper, but the latter can be applied to other machines for little or no incremental cost—and there is no practical upper limit to program size. Similarly, expanding a system based on proprietary hardware may be less costly in the short term than replacing it with one using ordinary PCs, but the incremental cost of further expansions will always be greater in the long run.

Reduce Training With Familiar Interfaces. Training of new operators can take a lot of time. Using systems having interfaces with which the operators are already familiar can minimize this.

Recover From Disasters. Catastrophic failures can occur due to either human error or natural causes—floods, storms, etc. Backup and recovery planning must be part of any enterprise that intends to survive such an occurrence. The impact of such a failure can range from an inconvenience—a single machine goes down for a while, requiring production to be rerouted—to a disaster. The extent to which the NC file management system allows for quickly adapting to whatever happens can be a major cost savings.

44.5 DNC COMMUNICATIONS

Factories using NC machines vary from small job shops up to large multinational manufacturers. In all cases, most DNC systems have the communications capabilities to expand from small installations with a single computer up to site-wide installations with servers and networked computers. The method of communicating with the NC machines is similar in all cases.

44.5.1 Serial or Ethernet

There are two mainstream methods of communicating to your NC machines from DNC systems. One is via the RS232 serial port on your computer and the other is via Ethernet. The method depends on the communications capabilities of the controller on the machine itself. The more modern machines support Ethernet.

44.5.2 Hubs and BTR

In the case of serial communications, many DNC systems have multiport serial devices to communicate with many NC machines as seen in drawing below. On older machines, it was common to use paper tape to load a program into a machine. This too can be incorporated into a DNC system using a *behind the reader device* or BTR. A BTR solution involves putting a computer serial port in series with the old paper system to simulate the loading of the program into the machine tool (see Fig. 44.5).

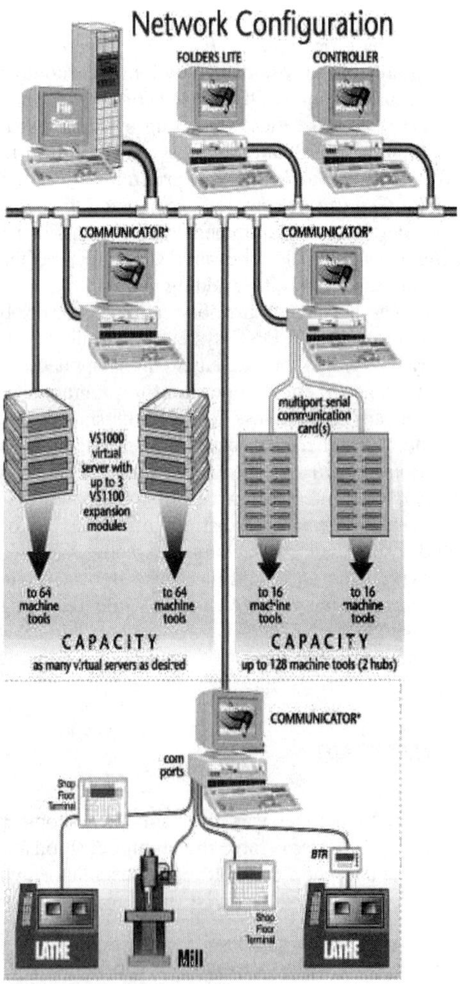

FIGURE 44.5 DNC control network.

44.6 CONCLUSION

By this point it will be apparent that the elimination of physical media using software running on a common PC will in time save many times the cost of whatever equipment is required to accomplish it. DNC systems are the rule rather than the exception, and for good reason—they're cost-effective. But what characteristics should the DNC system possess to optimize the flow of part programs and their related information? To summarize the points made in the preceding sections:

1. Support for long file and directory names
2. Grouping of related information with the part program
3. A system for identifying the files according to their part, operation, machine, and other relevant information, preferably in a database
4. A way for supervisors to organize the operators' work and let them know what they should work on
5. A way for the operator to transfer part programs right from the machine without having to go somewhere else
6. Electronic verification of the part program
7. A good G-code editor if the controls themselves are not capable of editing
8. A way to view and edit or at least mark up other documentation
9. A way to download a properly-formatted partial program to pick up from a tool break
10. A way to segregate uploaded files from others and properly identify them
11. A way for operators to notify others of changes or problems
12. A log of part program transfers—uploads and downloads
13. Uses standard, up-to-date PC hardware and system software
14. Support for multiple simultaneous drip-feed downloads
15. A conventional user interface
16. Allowance for flexibility in routings
17. Easy backup and recovery

In addition there are a few other points to consider in deciding whether an upgrade to a modern DNC system is worthwhile:

1. Is the system upgradeable to allow it to eventually be tied in to the rest of the company's information systems? Can it be integrated with data collection, monitoring, document management, and scheduling?
2. Does the DNC system support special communications protocols that were not present in the existing system? It may be possible to bring machines into the system that previously operated as independent islands.
3. To what extent is security implemented in the system? Even if the facility's work does not deal with sensitive information, a good security system can prevent mistakes, sabotage, and just plain old "messing around." It should be possible to limit who can access which files, for example to differentiate proven from unproven programs.

In summary, a modern, flexible, expandable DNC system can go a long way toward saving time and reducing errors in your facility. Ordinary day-to-day activities will be simplified, and the impact of extraordinary events will be minimized.

INFORMATION RESOURCES

www.cimnet.com
www.mmsonline.com
www.ndx.com
www.e4production.net
www.modernapplicationsnews.com
www.metalworkingdigest.com

P · A · R · T · 5

ROBOTICS, MACHINE VISION, AND SURFACE PREPARATION

CHAPTER 45
FUNDAMENTALS AND TRENDS IN ROBOTIC AUTOMATION

Charles E. Boyer
ABB Inc.
Fort Collins, Colorado

45.1 INTRODUCTION

Automation has come a long way since Henry Ford's mass production assembly lines. Ford helped to cut the cost of factory produced goods, while improving the working conditions for the factory workers. By simplifying each worker's task and decreasing the demand for skilled workers, the assembly line contributed to a reduction in the price of a T-Ford automobile. The work, however, was boring and led to increased workforce turnover. Still it would be several years before machines would begin to replace workers who no longer wanted to do the repetitive, heavy, and occasionally dangerous operations where robots are now utilized.

Automated welding with robots is a mature technology that is still growing. Once an amazing capability that increased the speed of automobile assembly lines, welding robots are now feasible for many other uses. Two trends brought us to this point—the development of flexible, portable robot cells; and computers that have become steadily faster, smaller, and cheaper. Because of these trends, a business that buys a robot can expect an acceptable return on investment in less than a year.

45.2 DESIGNS: CARTESIAN, SCARA, CYLINDRICAL, POLAR, REVOLUTE, ARTICULATED

There are several types of robotic arms on the market, and selecting the correct configuration will depend on the operation to be performed. Among the numerous styles available the most popular include the Cartesian, SCARA, and Articulated. There are several advantages and disadvantages to each of these.

Cartesian robots move in three coordinate directions X, Y, and Z at right angles to each other (Fig. 45.1). The primary advantage of Cartesians is that they are capable of moving in multiple linear directions and have the most rigid robotic structure for a given length, since the axes are supported at both ends. Due to their rigid structure these robots can manipulate relatively high loads, so they are often used for pick-and-place applications, machine tool loading, and stacking parts in bins. Cartesians can also be used for assembly and even genetic research or any task that is high volume and repetitive. The main disadvantage of Cartesians is that they require a large amount of space in which to operate and are sometimes referred to as a gantry robot.

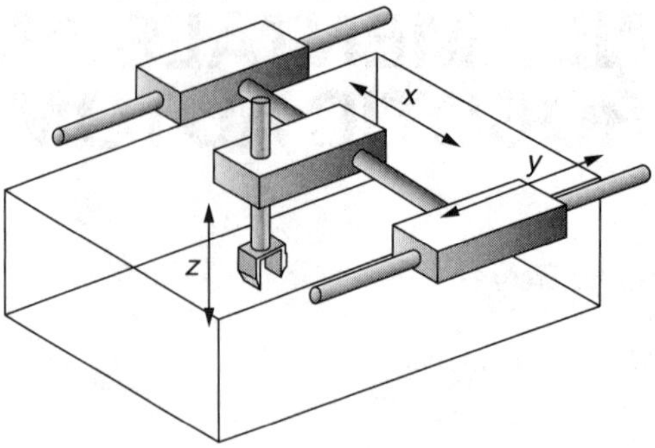

FIGURE 45.1 Cartesian style (gantry) robot.

SCARA robots are used primarily for pick and place, and general-purpose type applications such as sealing, assembly operations, and handling machine tools. SCARA robots are repeatable and ideal for palletizing and simple welding applications. SCARAS have two parallel rotary joints to maintain compliance in a plane (Fig. 45.2).

Articulated robots have motions similar to those of a human arm. These machines comprise both a "shoulder" and "elbow" joint along with a "wrist" axis and vertical motion (Fig. 45.3). Articulated robots are ideal for general-purpose applications requiring fast, repeatable, and articulate point-to-point movements such as palletizing, machine loading/unloading, and assembly. Other tasks that are ideal include spray painting, welding, and sealing applications. Articulateds are rigid, but less so than Cartesians because all joints are at the end of the arm, resulting in unsupported mass and in more deflection.

The advantages of Articulate/Jointed Arm/Revolute robots include flexibility to reach over obstacles, compatibility with other equipment, and good reach. However, large and variable torques on joints present counter-balance problems. Poor resolution and position errors are other disadvantages.

Two linear axes make the mechanical design of cylindrical robots less complex than Cartesians. The cylindrical robots also provide moderate collision-free movements and good accuracy, although they are less accurate and achieve less resolution than Cartesians. Cylindrical robots are useful in radar systems for tracking flying objects and in the design and analysis of revolute robots (see Fig. 45.4).

Spherical, or polar, robots have minimal structural complexities (Fig. 45.5). They are compatible with other robots and machines in common work spaces, and achieve good resolution because errors in position are perpendicular. They also have minimal structural complexities. Counter-balance problems and limited collision avoidance are disadvantages of the polar robots.

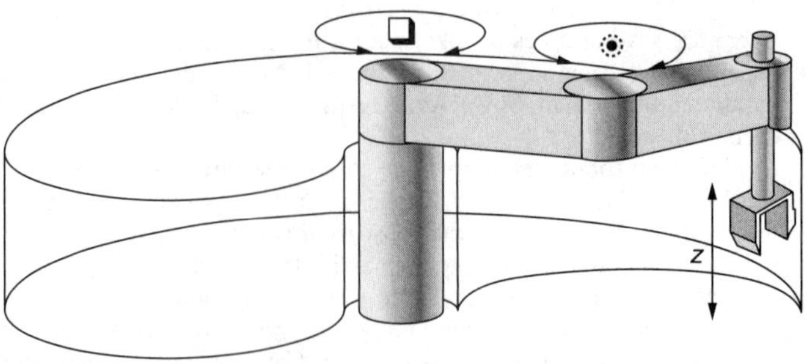

FIGURE 45.2 SCARA style robot.

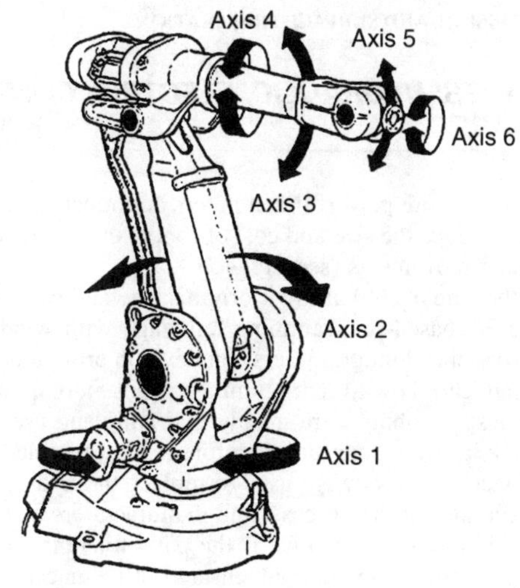

FIGURE 45.3 Articulated style robot.

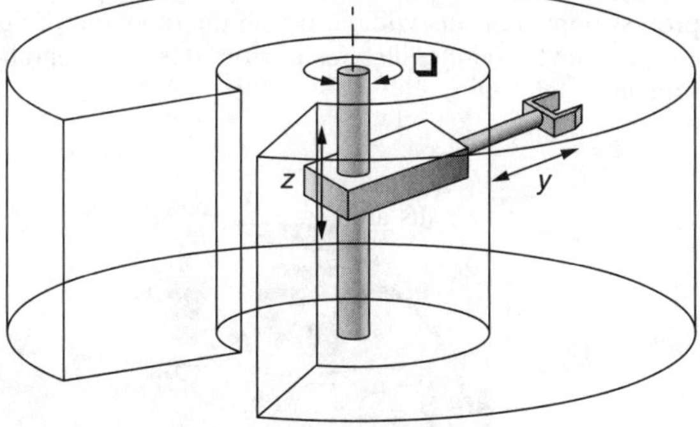

FIGURE 45.4 Cylindrical style robot.

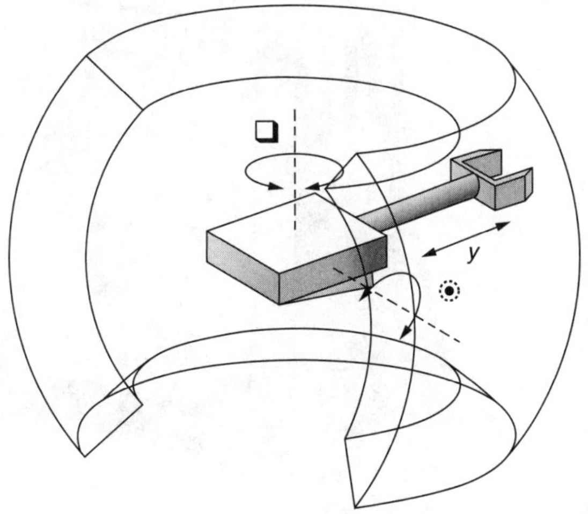

FIGURE 45.5 Polar style (spherical) robot.

45.3 EQUIPMENT TYPES: HYDRAULIC, ELECTRIC, CONTROLLER EVOLUTION, SOFTWARE

Faster, cheaper, more powerful computers and much smaller electronic devices allowed robot manufacturers to reduce the size and cost of robot controllers. Controllers are computers that direct robot and positioner movements (see Fig. 45.6).

Today, they are about half their original size and can do much more. Programming, too, is easier. Most use a PC-based programming language with windows, drop-down menus, and cutting and pasting. Experienced programmers can develop programs faster and more easily, and new robot cell operators can learn how to reprogram the robot more quickly with the new controllers.

The earliest of robots were mostly hydraulic and used to lift heavy objects. However, accuracy and repeatability were poor and therefore not acceptable for precision operations.

When electric motors were used to replace the hydraulics a new problem arose. Now the robots were too accurate for the process, and manufacturers had to improve the quality of the parts before they arrived at the robot stations. If the parts tolerance was within a specified amount, sensor systems were sometimes used to compensate for the inaccuracies. This, however, worked only within a small number of operations and presented new problems to solve as well.

Offline programming software is the wave of the future. This will allow the robot to continually work while engineers develop new designs, improve current components, do time studies, and process improvements without taking the robot out of operation. The new software will also make it possible to switch much faster from one job to another, keeping production downtime to a minimum.

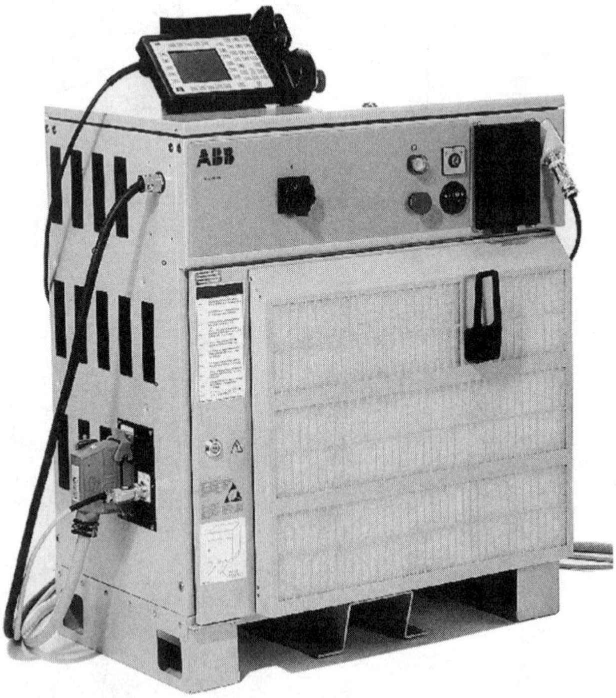

FIGURE 45.6 Robot controller.

45.4 APPLICATIONS

Applications that utilize robotics include: welding, foundry, painting, machine tending, material handling, packaging, palletizing, pick and place, press brake operations, and cutting processes, such as plasma, laser, and water jet (see Fig. 45.7). All of these operations involve undesirable or tedious activities that have become unpopular with the skilled worker, who is therefore increasingly harder to replace. From the undesirable environments of welding, foundries, painting, and cutting with plasma or lasers to the repetitive operations of machine tending, pick and place, and press brake operations to the back-breaking, heavy lifting required in palletizing and material handling, robots have filled in when humans would no longer accept working under these types of conditions.

45.4.1 Welding

Robots work essentially like humans. In welding, a robot arm strikes an arc, holds a torch at the right distance from a joint, and moves the torch along the joint. The robot then pulls the torch back, moves to the start of the next weld, and repeats the operation. However, the robot does not have to raise a hood to see the next joint, and the robot never gets tired or distracted. The robot can move quickly between welds and produce consistently high quality welds 24 h a day.

These advantages make robot welders ideal for assembly lines. The robot can be mounted beside the line and make the same welds repeatedly as long as the line is running. Assembly line automation made consistently good welds possible at an acceptable cost. Since assembly line robot welders worked so well, the next logical step was to use robots to weld parts, as well as the assemblies they

FIGURE 45.7 Typical packaging robot application.

go on. Robot manufacturers designed robot stations, called *cells*, that contained a welding robot and a platform to hold the part. These robot cells could sit on a factory floor and weld parts brought to them. The robot cell consisted of the robot, a fixture to hold the part, a positioner to turn it, and perimeter walls to create a safety barrier. Once the part was fastened in the fixture, the robot could automatically make all the welds the part needed. If the part required welds that the robot could not reach, the positioner could rotate or turn the part to bring those weld locations within reach.

This worked well for companies that needed to weld large numbers of one part. Manufacturers designed a cell specifically for that part. They placed a standard robot and a standard positioner in a cell and designed a fixture for the part. Over time, the company could recover the cost of the robot cell. If the part design changed, however, the company might have to scrap that cell and order another one. Only the biggest shops could afford to do that.

To increase the useful life of cells, robot manufacturers began designing flexible, portable cells that companies could change in the field to accommodate part design changes. Much of the flexibility was in the positioner. More flexible positioners could adapt easily to part design changes and different but similar parts. Flexibility increased until modularized systems could accommodate 70 or 80 percent of the typical welding applications. A company did not need to have a large order for one part to justify a robot welding cell (Fig. 45.8). The robot cell market expanded to include smaller businesses that might only weld several hundred parts for one order, and then switch to a different part.

Modular Robot Cells. A portable welding cell does not have to occupy valuable floor space when it is not in use. A forklift can move cells from one part of the shop to another. If a company has a small parts run, rather than move all the parts to the robot cell, they can move the cell to the parts. The weld station does not tie up some part of the shop for long periods, and shop traffic is reduced (see Fig. 45.9).

For example, several robot manufacturers offer modular welding packages. Their customers like the fact that the cells provide a complete, portable turnkey system. All of the components, including the standard six-axis robot, sophisticated programmable controller, positioner, welding power supply,

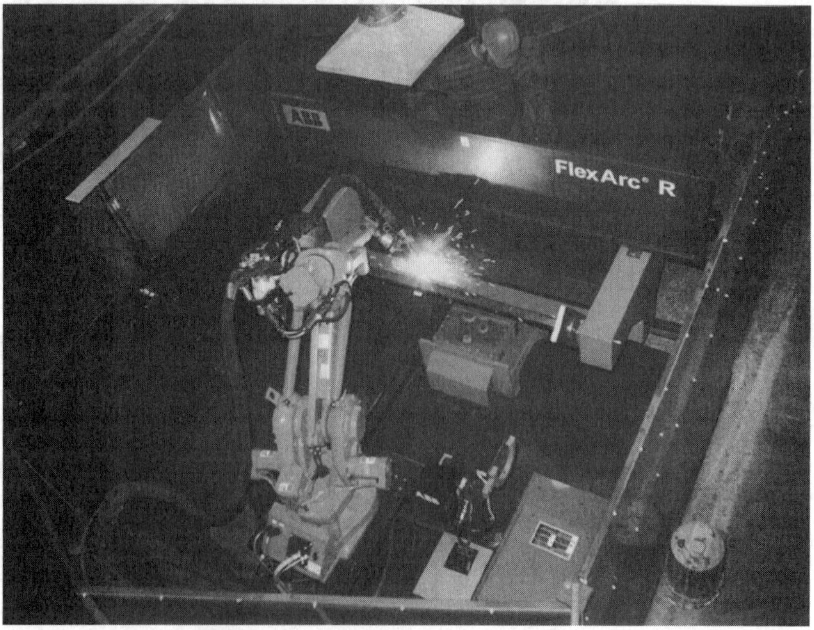

FIGURE 45.8 Standard self-contained welding cell.

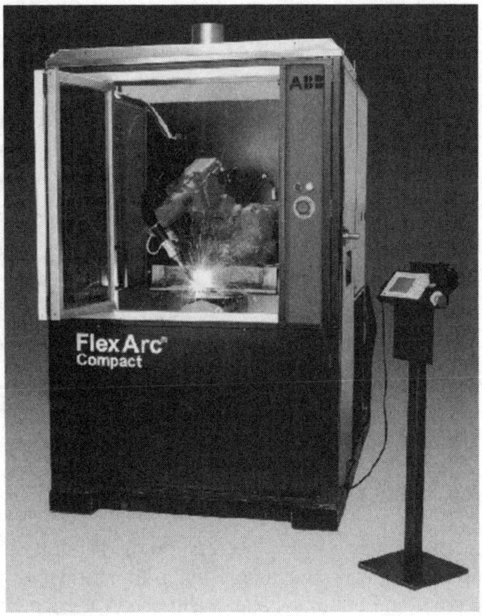

FIGURE 45.9 Modular welding cell.

welding torch, and perimeter fencing is mounted on a common base for easy shipment and quick setup on the factory floor.

Years ago a typical robot customer would have had its components put together by a third party engineering company. Currently, cell standardization makes it possible to build cells ahead of time and keep them in stock, bringing robot manufacturing costs down and enabling faster delivery. These factors are important to smaller businesses that need the flexibility to accept work on short notice.

Changing the cell to weld a new part used to require changing the fixture and reprogramming the robot. The operator had to stop the robot to do this, and a stopped robot was not making money for the company. Manufacturers began looking for faster ways to change the fixture and reprogram the robot. One way to make the fixture changeover faster was to include locating pins in the positioner. Locating pins ensured that the fixture could only attach to the positioner the correct way. Manufacturers built modular fixtures so that although the fundamental mount was the same, the components could be swapped with less setup time.

45.4.2 Painting

The environment in a painting operation can be a health hazard to people. Additionally, it's typically noisy, uses chemicals that are carcinogenic, requires handling of bulky paint guns and hoses, and requires performing repetitive tasks. In order to protect the human painter from this environment, additional safety equipment such as masks and suits must be worn, further adding to the difficulty of repeatedly producing a quality finish.

Manual painting requires additional capital and operating costs for larger air handling equipment to meet safety regulations, and for training and maintaining skilled painters, as well as all the administrative costs associated with any human worker.

Robots can also handle more weight than dedicated equipment or humans, thus allowing them to carry paint delivery equipment, such as color changing valves on the robot arm, which is close to the applicator (Fig. 45.10). The shorter the distance between the color change valves and the spray gun,

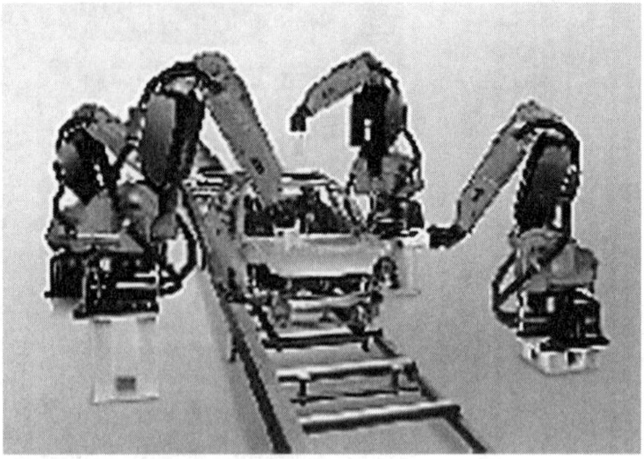

FIGURE 45.10 Painting robot application.

the quicker the system can replace one color of paint at the gun or applicator with another—thus minimizing cycle times and increasing production rate capability. This also helps minimize the amount of paint and purge solvent waste used for each color change. The latest paint robots on the market today have payload capacities that can accommodate as many as 32 different color valves, gear meters for two component materials, and associated paint lines.

45.4.3 Material Handling

Repetitive, heavy, monotonous, and boring operations are usually involved in material handling operations. Short cycle times are also usually critical in the process. Robots used in these processes are large enough to lift substantial weights including the grippers and end effectors required to manipulate the products, i.e., lifting heavy fixtures and engines, turning car bodies, grinding and polishing parts, and the like.

Today there is a growing trend to extend automation beyond the cell—using material-handling robots to deliver parts to the cell and take them away. A material-handling robot (Fig. 45.11) can get a part from a rack or conveyor and place it in the cell fixture. After processing, the material-handling robot can remove the finished part and put it on a conveyor belt. The material-handling robot and process robot work in tandem, their controllers coordinated by a master controller through a local area network (LAN). In addition to saving labor costs, the material-handling robot works with precision that a human could not match. It can quickly pick up a part and place it exactly in the fixture every time. It also avoids the safety concerns that always come with a human working near machinery.

A transfer line extends the advantage of material-handling robots. A transfer line is a series of cells. Each cell has a fixture to handle different processes on the same part. Instead of repositioning the part in one cell, the material-handling robot moves it from cell to cell and puts it in each fixture in the correct position. The robot can quickly and precisely remove the part from one cell and place it in the next, and can bring new parts to the first cell and take them away from the last. For example, if you are welding the four corners of a box-shaped object, you can do it with four welding cells. The welding robots in each cell weld a different corner.

A material-handling robot can also act as a fixture and hold the part in position, or after placing the part in a fixture it can hold a bracket in place while the robot welds it to the part.

Material-handling robots can go beyond the welding operation. They make a "lights out" fabricating shop possible, one where no human is involved on the shop floor. Robots might pick up one of several sizes of sheet metal from a rack, put it on a machine that washes the sheet, then another

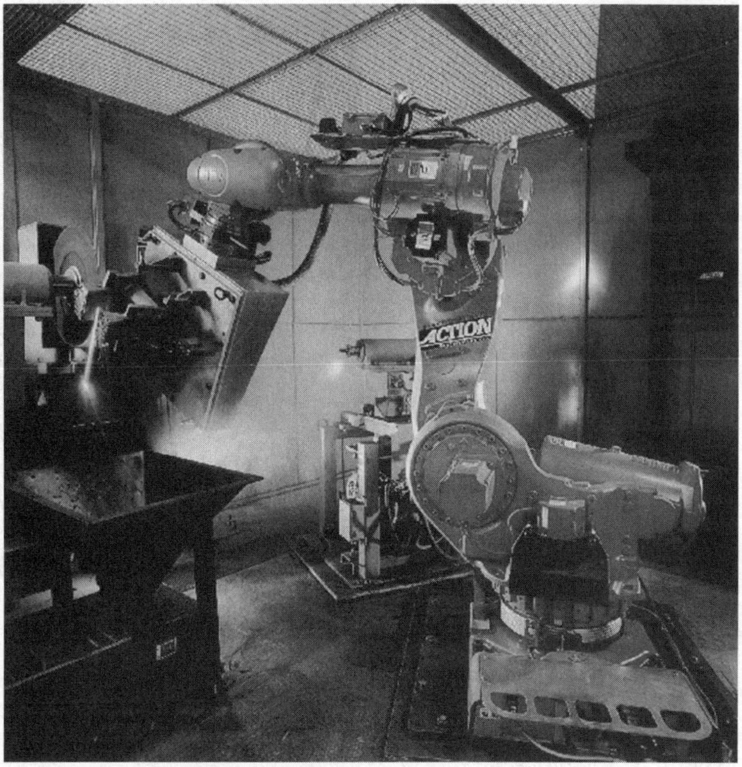

FIGURE 45.11 Material handling robot.

machine that cuts the blank from the sheet, followed by a forming brake that forms the blank to a certain shape, then another machine that cuts holes in it, then a series of welding stations, and finally place it on a conveyor to be delivered for testing.

45.4.4 Machine and Press Tending

Reduced floor space and labor costs along with the critical timing of parts production are just some of the reasons for using robots to tend machines in the manufacturing process. Robots can keep manufacturing cost down, allowing companies to compete with the cheaper overseas labor rates. Robots' ability to perform tasks in undesirable and dangerous environments is another reason for automating the process with robotics, i.e., foundry or forges, deburring castings, feeding press brakes, and the like.

Picking, Packing, and Palletizing. These operations require robots with speed and accuracy, usually tied to a vision system for picking items from a conveyor. This can include anything from food items (i.e., candy, cheese, frozen food, and the like) to larger boxes of staples (i.e., diapers, bottled products, etc.). The items are then placed in an open slot on another conveyor or stacked on pallets in preparation for shipment. Robots for these operations vary in design from high speed, light weight versions to larger robots required to pick up heavier weights for palletizing (Fig. 45.12). Robots that work in the food industry have special requirements for corrosion prevention and are sealed to prevent contamination.

FIGURE 45.12 Palletizing robot application.

45.5 OPERATION CONCERNS

45.5.1 Quality/ISO Requirements

There has been a move in recent years to an improved sense of quality awareness. Manufacturers are being pressured more and more to produce a consistent quality product at a reduced price, in lower quantities on a just-in-time schedule. When coupled with the labor shortage and improved working conditions demanded by today's workers, automation has become the major result.

To meet ISO 9001 standards, welds must consistently achieve the necessary bond strength. Weld penetration is a major factor in bond strength. A robot has a steady hand, and it can be controlled to consistently achieve the desired weld penetration. If something happens when a robot is welding—an arc isn't quite right or something has failed—the robot stops and warns the operator that there is a problem.

A robot performing electrical assembly or painting also controls the process to a point where it can sense if an error has occurred or the process is out of tolerance and will signal the operator that steps need to be taken to bring it back into compliance.

45.5.2 Positioning and Accuracy

The early hydraulic robots were capable of lifting large amounts of weight but their movement and positioning were only approximate at best. In order for robots to be productive, their motions had to be more controlled and consistent. Enter the electric motorized robot. Even though their weight capacity was less, they were more accurate and positioning improved tenfold. Now the robot could move between points continually within a prescribed tolerance that was more acceptable.

This also caused some problems. Now the parts being handled, welded, and the like had to be more consistent in their manufacture so they would work in the robot cell. Quality became a necessity for improving the process. As positioning capabilities improve, robot speeds are increasing.

When moving with electric motors at a high rate, the motors will actually allow the robot to overshoot its path. If there is an obstacle in its path it could hit and damage the robot. Today's robots take that into consideration and offer dynamic braking to help control the actual path traveled.

Safety. While one of the main selling points for robotics was that it got the operator out of dangerous environments and away from repetitive activities, other concerns emerged that had to be addressed. Chief among these is keeping the worker from accidentally walking into the area where a robot is operating and getting injured. Unfortunately, with the cost of robotics some manufacturers are willing to cut corners in this area at the expense of the workers' safety.

Robot manufacturers are trying to make this as simple as possible by building safety into the cell where possible (Fig. 45.13). This can be as simple as fencing in the area, to installing sensing mats or sophisticated light beams or curtains that identify worker activity and shut down the robot immediately. Vacuum systems can be installed to remove the welding smoke and keep the shop clean.

Offline Part Programming. Robot manufacturers are always looking for ways to program the robot controller faster. The standard way to program the controller for a part has been with a *teach pendant*. This is a hand-held box connected by a cable to the robot controller that lets an operator manually control the robot's movements. The operator uses buttons or a joy stick on the teach pendant to move the robot arm in different directions. Programmers use the teach pendant to steer the robot to each point where a process begins and ends and enter that point in the robot's program (Fig. 45.14).

A robot manufacturer just recently introduced a software product that provides a 3D simulation of a real production cell. When a customer orders the system, the company delivers a virtual copy of the cell that contains the exact system in 3D simulation. The customer runs the software on their own computer. The company provides a model of the cell configuration for the customer to take back and put on its PC. They can then start programming parts, designing fixtures, and conducting time studies to get ready for the real cell.

Customers can use the software model to create a program for the robot. Customers can also create 3D solid models of the part, the torch, and fixture tooling, and import these models into the software in the IGES, STET, or ACIS format.

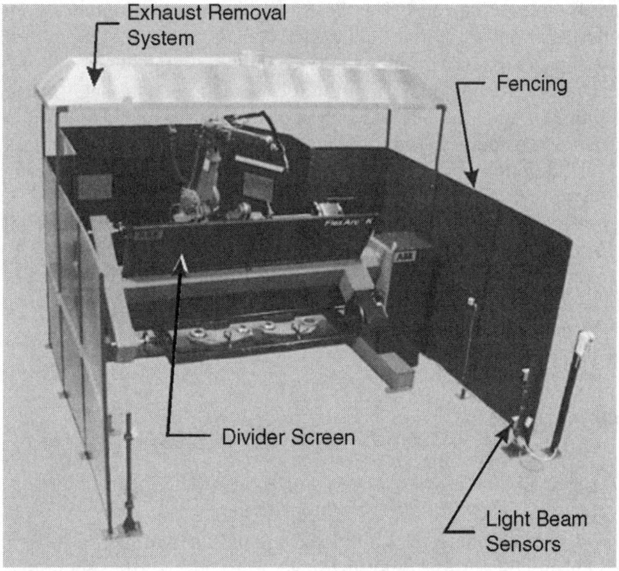

FIGURE 45.13 Typical welding cell safety components.

FIGURE 45.14 Offline part programming software.

The robot 3D simulation lets customers fully evaluate the new system's performance before delivery. The software displays a full spatial environment where customers can rotate the part, zoom in, and pan around the cell. They can evaluate different solutions for a part and check the robot's reach in different positions. They can create motion paths and lay down a virtual weld to determine the torch angle and other parameters. Once they have created a motion path, they can make welding cycle time estimates.

One manufacturer offers a programming language which runs on any computer platform. When copied to the robot controller, exactly the same code that is running in the computer will be running on the robot controller. The *virtual controller* even includes a *virtual teach pendant* that lets users interface with the virtual controller the same way they would with a robot controller. The virtual teach pendant has a graphic interface with the same push buttons or joystick, the same windows, and the same error messages that the robot teach pendant has. New operators can train on the software simulation with the virtual teach pendant. Virtual teach pendant instructions become part of the program downloaded to the robot controller. No further program translation is needed.

Traditionally, fixture tooling designers designed tooling around a part before the robot program was written. The part had to be in place before the programmer could create the program. Then the programmer might discover that a fixture clamp prevents the robot from reaching a specific location. With today's software programs, the programmer can create the basic robot program without fixture tooling. He or she can put the part in position and determine part positioning requirements in the simulation. Then, the tooling manufacturer can use the simulation to build tooling that will not interfere with the robot.

45.6 JUSTIFICATIONS

45.6.1 Qualitative

From the beginning, flexibility and minimal downtime have been the key goals in automation. Two recent trends outside robot manufacturing have further increased the usefulness of robots: just-in-time inventory control; and faster, cheaper, more powerful computers.

All these trends have made automated welding more attractive to small shops. If a small company wins a large contract and needs to increase production in its welding shop, it has two choices. It can try to find more welders, or automate with a robot. The welding industry is experiencing a labor shortage. A small, flexible, portable robot cell might be all the company needs to handle the increased production. If business keeps growing, it can automate in stages.

Robot manufacturers now offer numerous configurations to meet and fulfill the majority of customer requirements. Utilizing standard components, such as the robot, each cell features differently configured process stations, from a simple no frills offering to multiaxis sophisticated designs, in order to meet the variety of application needs. For welding applications a company can buy a two-dimensional robot cell configuration with only a worktable to weld a simple part that the robot can access without repositioning. If the company gets a new part that they need to rotate, they can add a positioner that is just the size they need.

For example, a bicycle frame is large, but needs few welds. However, it must be turned in order to reach all the welds, requiring a large but lightweight positioner. A caterpillar heavy equipment part will be heavy but not necessarily large. Because of the weight, it needs a large fixture and a heavy-duty positioner.

The welding industry is moving toward consistently higher manufacturing quality to meet ISO 9001 standards. Automation has made consistently good welds possible at an acceptable cost. Whether a weld is done at 8 a.m. or 5 p.m. or on a Monday or Friday, a robot does it with the same quality. A manual welder might get tired at the end of the day and not be moving as quickly, or may not be feeling well, and the consistency can vary. At companies where welders rotate to different jobs, quality can even vary from one welder to another. In fact, years ago the attitude was that it was best not to buy a car built on Monday or Friday because those were days workers were likely to be most distracted or absent and quality would suffer. One manufacturer has reduced product rejects from 20 to 25 percent, using manual welding, to less than one percent using a robotic welding cell (see Fig. 45.15).

45.6.2 Quantitative

Adopting the just-in-time inventory philosophy, companies no longer want to produce large quantities of many parts to keep in stock. They want to produce only the parts they need when they need them. With just-in-time inventory control, a company only produces the quantity of a part it needs to fill an order. It doesn't need to produce extra parts to be sure there's enough on hand for future orders. With flexible, portable robot cells, a company can quickly retool to produce a part to fill an order.

FIGURE 45.15 Manual welder.

45.7 CONCLUSIONS AND THE FUTURE

Accessing real-time production data from off-site locations is a developing trend in automation. When problems occur, key pieces of data can speed the diagnosis, but monitoring production and retrieving reliable diagnostic information from on-site operators is time-consuming. New comprehensive software platforms have been developed to monitor plant-wide production data, statistics, and diagnostic information remotely using a standard Internet browser. Authorized users can monitor production data from anywhere, at any time—even from the golf course! The software programs continuously track the status of connected robot controllers and transmit live updates. Users can collect data for a single work cell, for a production line, or for an entire plant. They can analyze recorded data on-line with ad hoc queries, embed custom reports linked to process data, or download data for further analysis off-line.

What will the future bring? We will see continuing improvement in robot cells, but at a slower rate. Existing systems can handle a large majority of parts. Robots in the cells are not going to change much. Manufacturers started with five-axis robots and now are producing six-axis robots, which can do any job that can be automated.

The biggest changes in the future will be in controller electronics. Controllers will become even smaller and smarter, and they will be able to do more; cabling will become nonexistent. Some six-axis robot controllers can already control 12 or more axes, including axes from the positioners and track systems. We are starting to put multiple robots into a cell using one controller. As electronic miniaturization continues, this trend will grow.

Robots will be able to see and feel through sensory feedback, allowing them to address applications where the process or parts are not predictable or where they vary in dimension, location, or application.

Robot manufacturers are also looking for new and exciting areas of use outside the industry. Service robots for the home consumer are starting to get attention. Robots are now being asked to do vacuum cleaning or lawn mowing, retrieve the mail, do the dishes, or patrol a yard or structure to watch for open windows, smoke in the air, and even intruders. You'll find them delivering the mail or even assisting nurses with medical supplies.

The future of service robots is still unknown but the industrial robot has become an important part of the manufacturing industry. Robots can enhance the quality of work life but the next generation of workers will require advanced cognitive skills for the jobs available after robots take over the more low-level activities.

FURTHER READING

ABB Robotics press archive.

Charles E. Boyer, "Trends in Welding Automation," *Welding Design & Fabrication,* June 2001.

Lars Westerlund, *The Extended Arm of Man,* Stockholm, 2000.

CHAPTER 46
MACHINE VISION

Nello Zuech
President, Vision Systems International, Inc.
Yardley, Pennsylvania

46.1 INTRODUCTION

Machine vision is a technology used in manufacturing industries as a substitute for the human vision function. Machine vision systems embody techniques leading to decisions based on the equivalent functionality of vision, but without operator intervention.

Characteristics of these techniques include noncontact sensing of electromagnetic radiation, direct or indirect operation on an image, the use of a computer to process the sensor data and analyze that data to reach a decision with regard to the scene being examined, and an understanding that the ultimate function of the system's performance is control (process control, quality control, machine control, or robot control).

Machine vision technology is not a single technical approach. Rather, there are many techniques and implementations. Machine vision involves three distinct activities and each has multiple approaches to accomplish the result desired for an application: image formation/acquisition, image processing, and image analysis/decision-making action (Fig. 46.1).

Image formation/acquisition (Fig. 46.2) involves the stage of the process that couples object space to image space and results in the quantized and sampled data formatted in a way that makes it possible for a computer or dedicated computerlike circuitry to operate on the data. This stage itself typically includes lighting, optics, sensor, analog-to-digital converter, and frame buffer memory to store the image data, as well as application-specific material handling and complementary sensors such as presence triggers.

Each of these can be executed in any number of ways, sometimes dictated by the application specifics. Lighting can be active or passive. Passive lighting refers to those lighting techniques designed to provide illumination (ideally as uniform as possible) of the object so that the sensor receives a sufficient amount of light to yield a reliable electron image by the photon to electron imaging transducer. Again, there are specific passive lighting techniques—fluorescent, incandescent, fiber optics, and projected.

Active lighting involves those arrangements that operate on the image of the light itself. "Structured" light is one such popular arrangement. Typically, a line of light is projected across the object and the image of the line of light, as deformed by the geometry of the object, is acquired and analyzed. There are many forms of active lighting arrangements.

Similarly, there are different sensors that can be used to acquire an image. In a flying spot scanning arrangement, the image is acquired from a single element point detector as a time-dependent signal representative of the image data. Linear arrays are also used and are especially beneficial if an object is consistently moving past the scanner station. The image data are acquired one line at a time.

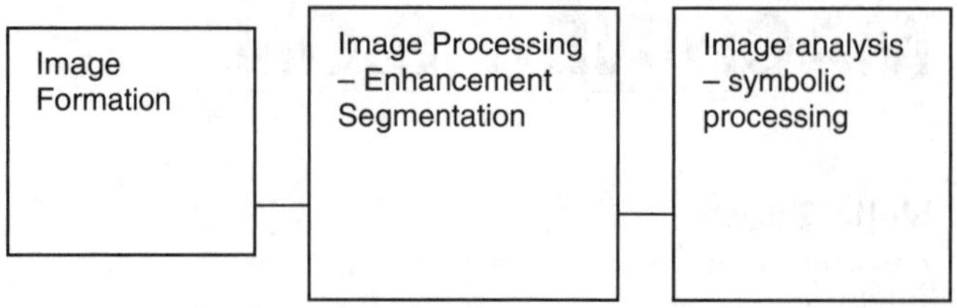

FIGURE 46.1 Machine vision block diagram.

The sensor most frequently identified with machine vision is the area array; however, it is important to understand that machine vision is not restricted to only those systems that acquire images using area arrays.

Beyond differences in sensing or converting the optical image into an electronic image, there are differences in how the analog signal is handled, converted to a digital signal, and stored. Some executions do some signal enhancement processing in the analog domain. Some operate on the digitized image in real time without storing a frame. Cameras may themselves be digital. A number of standards have emerged (IEEE-1394 and CameraLink) designed to input digital data into a personal computer making a *vision engine* based on host-based processing possible.

Processing on the image also varies with each execution. Some vision platforms and system implementations have extensive computing capacity to enhance and segment images, some less so. Different executions base analysis and decisions on different routines.

This discussion on differences is meant solely to emphasize that they exist. True general-purpose machine vision does not exist. Rather, there are many embodiments of the complement of techniques together representing machine vision.

In addition to recognizing that there are variations in machine vision technology, it is important to understand that machine vision companies have made certain business decisions that also dictate the segment of the market in which they compete. At least four business strategies have emerged, influencing product design implementations.

Some companies are offering vision engine components or vision toolkits. These still have a long way to go before they truly imitate the human eye–brain function, but they generally have some robustness, making them sufficiently flexible to handle a variety of applications under certain constraints. There are several versions of these.

46.1.1 Frame Grabber

A board that includes A/D, LUTs, memory to store one or more frames, D/As, but does not include any on-board processing capability. In some cases, they merely condition the image data out of a

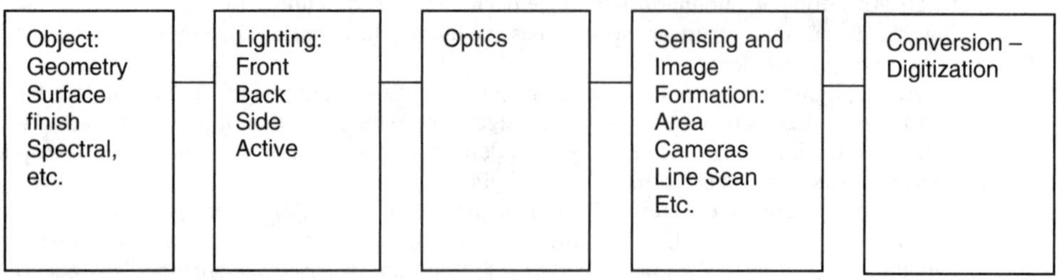

FIGURE 46.2 Image formation.

camera making it compatible with processing by a personal computer. These boards can operate with either digital or analog cameras.

46.1.2 Image Processing Board

A board product, such as a frame grabber with firmware that performs certain image processing algorithmic primitives at real time rates, and off-loading the computer requirements to the firmware from the computer itself. The interface supplied generally requires a familiarity with image processing and analysis, since one will generally start at the algorithm level to develop an application. These boards can operate with either digital or analog cameras.

46.1.3 Vision Processor

A vision processor is a board or box that in addition to including frame grabber functionality also includes on-board computer processing capability. This could be in the form of an embedded microprocessor, FPGA, DSP, array processor, ASIC, proprietary hardware, and the like. The graphic user interface is such that little or no reference is made to image processing and analysis. Rather, the interface refers to generic machine vision applications (flaw inspection, gauging, assembly verification, find/locate, *optical character recognition* (OCR), *optical character verification* (OCV), and the like) and walks the user through an application setup via menus or icons.

These systems may or may not have the ability to get into refining specific algorithms for the more sophisticated user. Vision processor systems are sold to application-specific machine vision system builders, merchant system integrators, OEMs, or end users.

46.1.4 Smart Camera Vendor

A smart camera vendor is a company that offers a camera with embedded intelligence, such as a microprocessor or DSP, which can be programmed to make the camera behave like a vision processor-based system. The graphic user interface is such that little or no reference is made to image processing and analysis. Rather, the interface refers to generic machine vision applications (flaw inspection, gauging, assembly verification, find/locate, OCR, OCV, and the like), and walks the user through an application setup via menus or icons. Cameras may or may not have the ability to get into refining specific algorithms for the more sophisticated user.

The second segment of the market includes those companies that have adapted the vision engines for use in a specific application niche. Generally, complementary material handling is what adapts these systems to the dedicated application. Should the application disappear, the company that bought the product could unbundle the vision platform and, with appropriate application engineering, configure the system for another application. Examples of these systems include systems specifically for ball grid array inspection, compact disk/optical media inspection, and LED inspection. Typically these really represent canned applications of the multipurpose platform, not a turnkey system.

The third segment includes those companies that offer application-specific machine vision systems or turnkey machine vision for dedicated tasks. These are typically for industry-specific applications although there are some more generic systems that address applications that cut across manufacturing industries. An example of the latter are the machine vision-based gauging systems used to make measurements on formed parts regardless of the materials or industry involved.

Interface refers specifically to the application itself, not generic machine vision applications or imaging functions. In other words, machine vision technology is virtually transparent to the user. The vision hardware can be either based on commercially available vision processors (image board level products, general-purpose machine vision systems), or proprietary vision computer products. ASMV systems are typically sold directly to end users.

The fourth segment includes those companies that offer customized machine vision systems generally built around a common technology base. An example of this would be systems found in the automotive industry for gap and flushness measurements in sheet metal assemblies based on a family of lighting/sensor probes or in the wood products industry where they collect 3D volume data to optimize the yield from a log.

On top of this, one can also find machine vision software suppliers targeting those who want to take complete ownership of the vision technology. These companies supply generic software libraries that can be adapted to many different applications and/or a supplier of software tools for specific applications, for example: OCR, OCV, alignment, robot guidance, BGA inspection, LCD inspection, IC package inspection, component alignment, and the like.

An infrastructure has also developed to support the installation of machine vision technology. There are numerous merchant system integrators. These are companies that provide a machine vision system with integration services, adapting the vision system to a specific customer's requirements. A system integrator is project oriented. They provide a turnkey system based on commercially available vision engine products plus value added—application engineering, GUI, material handling, and the like.

A major challenge, however, is that the machine vision industry is forever changing. The neat pigeonholes that have been historically used to classify certain companies no longer apply. Some companies that had traditionally been board level producers, today offer products that look more like general-purpose machine vision systems or vision processors. In some cases they even offer application-specific software and in some other cases they deliver turnkey application-specific machine vision systems.

Companies that had traditionally offered general-purpose machine vision products are also migrating into application-specific markets. Mergers and acquisitions have resulted in companies that now offer a range of products that cut across classes.

Several companies offer vision processors/general-purpose vision platforms that have sufficient functionality permitting them to be configured for a variety of applications. Some of these same companies are suppliers of products that address a specific set of applications such as OCR and OCV. Some companies are suppliers of vision processors/image processing board sets that also offer the functionality of a vision platform and can be utilized to address many applications like the general-purpose vision platforms.

The vast majority of the suppliers that make up the machine vision industry are suppliers of industry-specific niche application products. There is often as much value associated with peripheral equipment necessary to provide a turnkey solution, as there is value of the machine vision content in the system.

46.2 MACHINE VISION TECHNOLOGY

Machine vision is conceptually a relatively simple technology. There are many different executions, but for the most part machine vision involves combining television and computers. In its most straightforward form, it is the analysis of television pictures by computer. This concept transcends what is commonly referred to as machine vision and is generally associated with the broader field of electronic imaging.

In other words, machine vision is a subset of the field of electronic imaging, and specifically the subset associated with the application of electronic imaging technology or techniques in industrial manufacturing settings. In manufacturing, machine vision is employed for the purpose of control: quality control, process control, machine tool control, or robot control. As much as anything, the application of the technology of electronic imaging is what distinguishes machine vision from other subsets of the fields, such as applications in medicine, offices, document scanning, and the like.

Machine vision begins with an image—a picture. In many ways the issues associated with a quality image in machine vision are similar to the issues associated with obtaining a quality image in a photograph. In the first place, quality lighting is required in order to obtain a bright enough reflected image of the object. Lighting should be uniformly distributed over the object. Nonuniform lighting will affect the distribution of brightness values that will be picked up by the television camera.

As is the case in photography, lighting tricks can be used in order to exaggerate certain conditions in the scene being viewed. For example, it is possible that shadows can in effect include high contrast information that can be used to make a decision about the scene being viewed.

The types of lamps that are used to provide illumination may also influence the quality of the image. For example, fluorescent lamps have a higher blue spectral output than that of incandescent lamps. While the blue spectral output is more consistent with the spectral sensitivity of the eye, higher infrared output is typically more compatible with the spectral sensitivity of solid-state sensors that are used in machine vision.

It has been found that the sensitivity of human inspectors can be enhanced by using softer lighting or fluorescent lamps with gases that provide more red spectral output. It may also be the case with machine vision. That is, the lamp's spectral output may influence the contrast associated with the specific feature one is attempting to analyze.

As in photography, machine vision uses a lens to capture a picture of the object and focus it onto a sensor plane. The quality of the lens will influence the quality of the image. Distortions and aberrations could affect the size of features in image space. Vignetting in a lens can affect the distribution of light across the image plane. Magnification of the lens has to be appropriate for the application. As much as possible the image of the object should fill the image plane of the sensor. Allowances have to be made for any registration errors associated with the position of the object and the repeatability of that positioning. The focal length and aperture have to be optimized in order to handle the depth of field associated with the object.

The imaging sensor that is used in the machine vision system will basically dictate the limit of discrimination of detail that will be experienced with the system. Imaging sensors have a finite number of discrete detectors and this number limits the number of spatial data elements that can be processed or into which the image will be dissected. In a typical television-based machine vision system today the number of spatial data points is on the order of 500 to 1000 horizontal × 500 to 1000 vertical.

What this means basically is that the smallest piece of information that can be discriminated is going to be a function of the field of view. Just like in photography one can use panoramic optics to take a view of a mountain range, although a family might be in the picture in the foothills of the mountains—it is unlikely that you would be able to discriminate the family in the picture. On the other hand, using a different lens and moving closer to the family, one would be able to capture the facial expressions of each member, but the resulting picture would not include the peaks of the mountains.

So, for example, given that an application requires a one-in field of view and a sensor with the equivalent of 500 spatial data points is used, one would have a spatial data point that would be approximately 0.002 in on the side. Significantly, the ability of machine vision today to discriminate details in a scene is generally better than the size of a spatial data point.

In a manner basically analogous to how an eye can see stars in a night sky because of the contrast associated with the star light, so too in machine vision techniques exist which allow systems to be able to discriminate details smaller than a spatial data element. Again, contrast is critical. The claims for subpixel sensitivity vary from vendor to vendor and depend very much on their execution and the application.

In all machine vision systems up until this point in our discussion, the information or the image has been in an analog format. For a computer to operate on the picture the analog image must be digitized. This operation basically consists of sampling at discrete locations along the analog signal that corresponds to a plot of time vs. brightness, and quantizing the brightness at that sample point.

The actual brightness value is dependent on the lighting, the reflective property of the object, conditions in the atmosphere between the lighting and the object and between the object and the camera, and the specific detector sensitivity in the imaging sensor. Most vision systems today characterize the brightness into a value of between 0 and 255. The brightness so characterized is generally referred to as a shade of gray.

For the most part today machine vision systems are monochromatic. Consequently, the color may also be a factor in the brightness value. That is, it is possible to have a shade of red and a shade of green (and so on) all of which would have the same brightness value. In many cases where color issues are a concern, filters are used in order to eliminate all colors that are not of interest to the particular application. In this way the gray shades are an indicator of the saturation level associated with a specific color.

At last we have a picture that has been prepared for a computer. In most machine vision systems today, the digitized image is stored in memory that is separated from the computer memory. This dedicated memory is referred to as a frame store—where frame is synonymous with the term used in television to describe a single picture. In some cases the dedicated hardware that includes the frame store also includes the analog-to-digital converter as well as other electronics to permit one to view images after processing steps have been conducted on the image to view the effects of these processing procedures.

Now the computer can operate on the image. The operation of the computer on the image is generally referred to as image processing. In addition to operating on the image, the computer is also used to analyze the image and make a decision on the basis of the analyzed image and perform an operation accordingly. What is typically referred to as the machine vision system is the combination of image processing, analysis, and decision-making techniques that are embodied in the computer.

A good analogy can be made to a toolbox. Virtually all machine vision systems today include certain fundamental tools, much like a hammer, screwdriver or pliers. Beyond these, different suppliers have developed additional tools, more often than not driven by a specific class of applications. Consequently the description frequently given for machine vision as being an "idiot savant" is quite apropos. That is, most of the platforms are brilliant on one set of applications, but "idiots" or truly not the optimal for other applications.

It is important, therefore, to select the vision platform or toolbox with the most appropriate tools for an application. Significantly, no machine vision systems exist today that come anywhere near simulating the comprehensive image understanding capabilities that people have. It is noted that for many applications many different tools will actually do the job and in many cases without sacrificing performance. On the other hand, in some cases while the tools appear to do the job, performance might be marginal, in a manner analogous to when we attempt to use a flat head screwdriver in order to turn a screw with a Phillips head.

Image processing is generally performed on most images for basically two reasons: to improve or enhance the image and, therefore, make the decision associated with the image more reliable; and to segment the image or to separate the features of importance from those that are unimportant. Enhancement might be performed, for example, to correct for the nonuniformity in sensitivity from photo site to photo site in the imaging sensor, correct for distortion, correct for nonuniformity of illumination, to enhance the contrast in the scene, correct for perspective, and the like.

These enhancement steps could be as simple as adding or subtracting a specific value to each shade of gray or can involve a variety of logical operations on the picture. There are many such routines. One routine that is commonly found as a tool for image processing in most vision platforms today is a histogram routine. This involves developing a frequency distribution associated with the number of times a given gray shade is determined.

One use of histograms is to improve contrast. This involves mathematically redistributing the histogram so that pixels are assigned to gray shades covering 0 to 255, for example. In an image with this type of contrast enhancement it could be easier to establish boundaries or easier to establish a specific gray shade level or threshold to use to binarize the image. Binarizing an image, or segmenting an image based on a threshold, above which all pixels are turned on and below which all pixels are turned off, is a conventional segmentation tool included in most vision platforms and can be effective where high contrast exists.

Where contrast in a scene is not substantial, segmentation based on edges may be more appropriate. Edges can be characterized as locations where gradients or gray shade changes take place. Both the gradient as well as the direction of change can be used as properties to characterize an edge. Significantly edges can be caused by shadows as well as reflectance changes on the surface in addition to the boundaries of the object itself. Artifacts in the image may also contribute to edges. For example, unwanted porosity may also be characterized by increased edges.

There are many different ways edges are characterized. One of the simplest is just using the fact that there are sharp gray scale changes at an edge. Significantly, however, edges in fact appear across several neighboring pixels and what one has is in fact a profile of an edge across the pixels. Because of this there are ways to mathematically discriminate the physical position of an edge to a value less than the size of the pixel. Again, there are many ways that these subpixel calculations have been made and the results are very application dependent. Consequently although claims are made of one part in ten or better subpixelling capability, it is important to understand that the properties of a given application can reduce the effectiveness of subpixelling techniques.

Having performed image-processing routines to enhance and segment an image, the computer is now used to analyze the image. The specific analysis conducted is again going to be very application dependent. In the case of a robot guidance application, for example, a geometric analysis would typically be conducted on the segmented image. Looking at the thresholded-segmented image or edge segmented image one would be able to calculate the centroid property and furnish this as a coordinate in space for the robot to pick up an object, for example.

In the case of using vision systems to perform inspections of one type or another, there are literally hundreds of different types of analysis techniques that have emerged. The number of pixels associated with the binarized or thresholded picture, for example, could be counted. This could be a relatively simple measure of the completeness of an object. The number of transitions or times that one goes from black to white can be counted. The distance between transitions can be counted and can serve as a measurement between boundaries of an object. The number of pixels that are associated with an edge can be counted. Vectors associated with the direction of the gradient at an edge can be used as the analysis features. A model based on the edges can be derived where the edges can be characterized as vectors of a certain length and angle. Geometric features can be extracted from the enhanced image and used as the basis of decisions.

These same techniques can be used in conjunction with pattern recognition applications. In each case one can define a pattern or more of the above mentioned features extracted from the image. For example, maybe a combination of the transition counts and edge pixels would be sufficient to make a judgment about patterns, where that combination is sufficient to distinguish between the patterns. Another approach might be to use geometric properties to distinguish patterns. These might include length and width ratios, perimeter, and the like.

The computer having reduced the image to a set of features used as the basis of analysis would then typically use a deterministic or probabilistic approach to analyze the features. A probabilistic approach is one that basically suggests that given a certain property associated with a feature, there is a high probability that the object is in fact good. So, for example, using the total number of pixels as an indication of the completeness of an object one would be able to suggest that if the total number of pixels exceeded say 10,000 there is a high probability that the object is complete. If less than 10,000 the object should be rejected because it would be characterized as incomplete. Some refer to this as goodness-of-fit criteria. It is also possible to set a boundary around this criteria. That is, it should fall between 10,000 and 10,500. An indication of a pixel count greater than 10,500 could be an indication, for example, of excess flashing.

A deterministic approach is one that will use physical feature properties as the criteria. For example, the distance between two boundaries has to be 1 ± 0.005 in. The perimeter of the object must fall within 12 ± 0.020 in. The pattern must match the following criteria in order to be considered a match: length/width ratio of a certain value, perimeter of a certain value, centroid of a given calculated value, and the like.

In a deterministic mode each of the features can be associated with a vector in decision space. In a pattern recognition application, the combined feature vector or the shortest distance to the known feature set for each of the patterns is the one that would be selected. This type of evaluation is referred to as decision theoretic. Another type of analysis is one based on syntactic techniques. In these cases, primitives associated with pieces of the image are extracted and the relationship between them is compared to a known database associated with the image.

In other words, the primitives and their relationship to each other have to abide by a set of rules. Using syntactic techniques one may be able to infer certain primitives and their position knowing something about other primitives in the image and their position with respect to each other.

This could be a technique to handle parts that might be overlapping and still be able to make certain decisions associated with those parts, even though one cannot see them entirely.

As you can see there are many vision tools that are available and the specific tools that one requires are application dependent. Today one can find machine vision type technology in virtually every manufacturing industry. The largest adopter by far is the electronics industry. In microelectronics, machine vision techniques are used to automatically perform inspections throughout the integrated circuit manufacturing process: photomask fabrication, post die slicing inspection, precap inspection and final package inspection for mark integrity.

46.3 RULES OF THUMB FOR EVALUATING MACHINE VISION APPLICATIONS

So you think you have a machine vision application. Do you want to somehow determine if the project is at least remotely feasible and you don't want to use a company's salesman to do the evaluation? How does one go about doing that? Well there are several rules of thumb that can be used to get at least some measure of feasibility. These all start with having at least a fundamental understanding of how a computer operates on a television image to sample and quantize the data. Understanding what happens is relatively straightforward if one understands that the TV image is very analogous to a photograph.

The computer operating on the television image in effect samples the data in object space into a finite number of spatial (2D) data points which are called pixels. Each pixel is assigned an address in the computer and a quantized value, which can vary from 0 to 63 in some machine vision systems or from 0 to 255 in others. The actual number of sampled data points is going to be dictated by the camera properties, the analog to digital converter sampling rate, and the memory format of the picture buffer or frame buffer as it is called.

Today more often than not the limiting factor is the television camera that is being used. Since most machine vision vendors today are using cameras that have solid-state photo sensor arrays on the order of 500 or so, by 500 or so one can make certain judgments about an application just knowing this figure and assuming each pixel is approximately square. For example, given that the object you are viewing is going to take up a 1-in field of view, the size of the smallest piece of spatial data in object space will be on the order of 2 mils, or 1 in divided by 500. In other words, the data associated with a pixel in the computer will reflect a geographic region on the object on the order of 2 mils by 2 mils.

One can so establish what the smallest spatial data point in object space will be very quickly for any application: X (mils) = largest dimension/500. Significantly this may not be the size of the smallest detail a machine vision system can observe in conjunction with the application. The nature of the application, contrast associated with the detail that you want to detect, and positional repeatability are the principal factors that will also contribute to the size of the smallest detail that can be seen by the machine vision system.

The nature of the application refers to exactly what you want to do with the vision system: verify that an assembly is correct, make a dimensional measurement on the object, locate the object in space, detect flaws or cosmetic defects on the object, read characters, or recognize the object. Contrast has to do with the difference in shade of gray between what you want to discriminate and the background, for example.

The organizational repeatability is in effect just that—how repeatable will the object be positioned in front of the camera. If it can not be positioned precisely the same way each time then it means that the field of view will have to be opened up to include the entire area in which one can expect to find the object. This will in turn mean that there will be fewer pixels covering the object itself. Vibration is another issue which can impact the size of a pixel as does typically motion in the direction of the camera itself since then optical magnification may become a factor—increasing or decreasing the size of the spatial data point in object space.

Let's take the generic applications one at a time. You want to verify that all of the features on an assembly are in place. If you can perceive that there is a high contrast between each of the features and the background or when a feature is in place or not in place, then the smallest feature one can expect to be able to detect would have to cover a 2 pixel by 2 pixel or so area. If, on the other hand, the contrast is relatively low then a good rule of thumb is that the feature should cover at least 1 percent of the field of view, or in the case of 500 by 500 pixels, a total of some 2500 pixels. So knowing the size of a pixel in object space one can multiply that value 2 or 2500, depending on contrast, to determine the area of the smallest detectable feature.

In the case of making dimensional measurements with a machine vision system one can consider the 500 pixels in each direction as if they were 500 marks as on a ruler. Significantly, just as in making measurements with a ruler, a person can interpolate where the edge of a feature falls within lines on a ruler, so, too, can a machine vision system. This ability to interpolate, however, is very application dependent. Today the claims of vision companies vary all the way from one-third of a pixel to one-tenth or one-fifteenth of a pixel. For purposes of a rule of thumb, you can use one-tenth of a pixel.

What will this mean in conjunction with a dimensional measuring application? Metrologists have used a number of rules of thumb themselves in conjunction with measuring instruments. For example, the accuracy and repeatability of the measurement instrument itself should be 10 times better than the tolerance associated with the dimension being checked. Today this figure is frequently modified to one-fourth of the tolerance. The other rule of thumb that is often used by metrologists is that the sum of repeatability and accuracy should be a factor of three or one-third the tolerance.

So how does one establish what the repeatability of a vision system should be? Given the subpixel capability of one-tenth of a pixel mentioned above and as in the example, an object that is 1 in on a side, the discrimination (the smallest change in dimension detectable with the measuring instrument) associated with the machine vision system as a measuring tool would be one-tenth of the smallest spatial data point or two mils or 0.0002 in. Repeatability will be typically ± the discrimination value or 0.0002 in.

Accuracy, which is determined by calibration against a standard, can be expected to run about the same. Hence, the sum of accuracy and repeatability in this example would be 0.0004 in. Using the three-to-one rule, the part tolerance should be no tighter than 0.0012 in for machine vision to be a reliable metrology tool. In other words, if your part tolerance for this size part is on the order of ±0.001 in or greater, the vision system would be suitable for making the dimensional check.

As you can see, as the parts become larger and with the same type tolerances, machine vision might not be an appropriate means for making the dimensional check that is, based on the use of area cameras that only have 500 × 500 discrete photosites. Conversely, if the tolerances were tighter the same would be true.

Using machine vision to perform a part location function one can expect to achieve basically the same results as making dimensional checks. That is, most vendors whose systems are suitable for performing part location claim an ability to perform that function to a repeatability and accuracy of ± one-tenth of a pixel. Using our example again, namely a 1-in part, one would be able to use a vision system to find the position of that part to within ±0.0002 in.

For applications involving flaw detection, contrast is especially critical in determining what can be detected. Where contrast is extremely high, virtually white on black, it is possible to detect flaws that are on the order of one-third of a pixel. Significantly, one can detect these flaws but not actually measure them or classify them. When detecting flaws that are characterized as geometric in nature, for example, scratches or porosity, it is noted that creative lighting and staging techniques can frequently exaggerate the presence of such flaws. So if those were the only flaws one wanted to detect and detection was all that was necessary, a rule of thumb would be that the flaw has to be greater than one-third of a pixel in size.

Where contrast is moderate, the rule of thumb associated with assembly verification, namely that the flaw cover an area of 2 by 2 pixels would be appropriate. Classifying a flaw with moderate contrast would require that it cover a larger area, on the order of 25 pixels or so. Again, where contrast associated with a flaw is relatively low as is the case with many stains, the 1 percent of the field of view rule would hold or it should cover 2500 or so pixels. Significantly, if it is a question that one is trying to detect flaws in a background that is itself a varying pattern (stains on a printed fabric, for example), the chances are that one would only be able to detect very high contrast flaws.

For applications involving OCR or OCV, the rule of thumb is that the stroke width of the smallest character should be at least 3-pixels wide. A typical character should cover an area on the order of 20–25 pixels by 20–25 pixels. The critical issue here then is the length of the string of characters that one wants to read. At 20 pixels across a character and two pixels spacing between characters, the maximum length of the character string would be on the order of 22 characters in order to fit into a camera with a 500-photosite arrangement. In optical character recognition/verification applications, a bold font style is desirable. In general it is also true that only one font style can be handled at a given time.

Another rule of thumb is that the best OCR systems have a correct read rate on the order of 99.9 percent. In other words, one out of every thousand characters will be either misread or a "no-read." The impact of this should be evaluated. For example, if 300 objects per minute are to be read, and 0.1 percent are sorted as "no-reads," in 1 h you would have approximately 20 products to be read manually. Is this acceptable? This is the best-case scenario. The worst case would be if they were misread.

When it comes to pattern recognition applications, a reasonable rule of thumb is that the differences between the patterns should be characterized by something on the object that is greater than 1 percent of the field of view or again on the order of 2500 pixels. Significantly, the gray shade pattern can be a major factor in making it possible to see pattern differences or to recognize patterns that have differences of far less than 2500 pixels. This would be the case, for example, where both geometry and color are factors.

Significantly, where more than one generic application is involved in the actual application, the worst-case scenario should be determined and used as the criteria to establish feasibility. Throughout this rule of thumb analysis, the dictating factor has been the number of photosites in the camera. Significantly, today solid-state cameras do exist that have up to 1000 × 1000 photo sites. These cameras, however, are not cheap. It is even possible that they would be more expensive than the vision system itself. Furthermore, few commercialized machine vision systems have the capacity to process so many pixels and make vision/decisions at anywhere near real time rates.

An alternative, however, to capturing images of an object with an area camera would be to use a linear array camera. There are several vision companies who offer linear array based vision systems where the linear arrays have up to 2000 photosites. Using a linear array one would have to move an object under the camera or move the camera over the object in order to capture a two dimensional image. Significantly, if the object is going to be moved under the camera, the speed with which it passes must be well regulated and the operating speed of the camera in combination with the speed of the object as it passes underneath the camera will in effect dictate the size of the pixel in the direction of travel.

Typically vision systems that use these principles will operate at up to 2-MHz rates. For a 2000-element array that means that you will be scanning 1000 lines per second (2,000,000/2000) in the direction of travel. For example, given an object speed of 10 in/s (10,000 mils/s), at a sample rate of 1000 lines/s, the effective pixel size in the direction of travel will be 10 mils (10,000/1000). So when evaluating machine vision applications, you may want to consider the possibility that the application can be addressed with a linear array based technique. In these instances all of the size details one can discriminate in object space would be proportionally better. For example, with a 2000-element linear array, everything would be four times better than using an area camera with 500 × 500 photosites.

Significantly, these are meant to be rules of thumb and should be only used as such in the evaluation of an application. Having performed this type of evaluation, however, it would be more reasonable for you to decide whether or not to pursue an application. It will avoid your wasting time with sales people trying to convince you that your application is "a piece of cake."

46.4 APPLICATIONS

Throughout the manufacturing process, machine vision is also used to provide feedback for position correction in conjunction with a variety of manufacturing processes such as die slicing and bonding and wire bonding. In the macro electronic industry machine vision is being used to inspect printed circuit boards for conductor width spacing, populated printed circuit boards for completeness, and post solder inspection for solder integrity.

As in microelectronics it is also being used to perform positional feedback in conjunction with component placement. It has become an integral part of the manufacturing process associated with the placement of chip carriers with relatively high-density pin counts.

In industries that produce products on a continuous web, such as the paper, plastic, and textile industries, machine vision techniques are being used to perform an inspection of the integrity of the product being produced. Where coatings are applied to such products, machine vision is also being used to guaranty the coverage and quality of coverage. In the printing industry one finds machine vision being used in conjunction with registration.

The food industry finds machine vision being used in the process end to inspect products for sorting purposes, that is, sorting out defective conditions or misshapen product or undersize/oversize product and the like. At the packaging end it is being used to verify the size and shape of contents, such as candy bars and cookies to make sure they will fit in their respective packages.

Throughout the consumer manufacturing industries one will find machine vision in various applications. These include label verification, that is, verifying the position, quality and correctness of the label. In the pharmaceutical industry one finds it being used to perform character verification, that is, verifying the correctness as well as the integrity of the character sets corresponding to date and lot code.

The automotive industry finds itself using machine vision for many applications. These include looking at the flushness and fit of sheet metal assemblies, including the final car assembly; looking at paint qualities, such as gloss; inspecting for flaws on sheet metal stampings; verifying the completeness of a variety of assemblies from ball bearings to transmissions; etc., used in conjunction with robots to provide visual feedback for: sealant applications, windshield insertion applications, robotic hydro piercing operations, robotic seam tracking operations, and the like.

Virtually every industry has seen the adoption of machine vision in some way or another. The toothbrush industry, for example, has vision systems that are used to verify the integrity of the toothbrush. The plastics industry looks at empty mold cavities to make sure that they are empty before filling them again. The container industry is using machine vision techniques widely. In metal cans they look at the quality of the can ends for cosmetic flaws, presence of compound, score depth on converted ends, and the like. The can itself is examined to inspect it for defective conditions internally.

The glass container industry uses machine vision widely to inspect for sidewall defects, mouth defects, and empty bottle states as well as dimensions and shapes. In these cases vision techniques have proven to be able to handle 1800 to 2000 objects/min.

46.5 DEVELOPING A MACHINE VISION PROJECT

How do I know what machine vision techniques are most suitable for my application? A studied approach is usually required unless the application is one that has a system that has been widely deployed throughout an industry. In that case the pioneering work has already been done. Adaption to one's own situation while not trivial may have little risk. To find out if your application has been solved, ask around. Today most machine vision companies when contacted and when asked, if they do not offer the specific solution, if they know of any other company that does, will generally respond with candor and advise accordingly. Consultants may also be able to identify sources of specific solutions.

Having identified those sources they should be contacted to identify their referenced accounts and these in turn should be contacted to determine—why they were selected, what has been the experience, service, and the like, and would they purchase the same product? This should help to narrow down the number of companies to be solicited for the project. In this case the ultimate selection will no doubt be largely based on price, though policies such as training, warranty, service, spare parts, and the like should also be considered as they will impact the life cycle cost.

What do you do if you find your application is not a proliferation of someone else's success? In this case a detailed application description and functional specification should be prepared. This means really getting to know the application—what are all the exceptions and variables? The most critical ones are position and appearance. These must be understood and described comprehensively.

What are the specific requirements of the application? Will the system first have to find the object—even minor translation due to vibration can be a problem for some machine vision executions. In addition to translation, will the part be presented in different rotations? Are different colors, shades, specular properties, finishes, and the like anticipated? Does the application require recognition? Is it gauging? What are the part tolerances? What percent of the tolerance band would it be acceptable to discriminate? If flaw detection, what size flaw is a flaw? Is the flaw characterized by reflectance change, by geometric change, and the like.

Having prepared the spec, at least a preliminary acceptance test for system buy-off should be prepared and solicitations should be forwarded to potential suppliers. How do you identify those suppliers? A telephone survey of the 150 or so companies is one approach. Again, use of a consultant can greatly accelerate the search. In any event, the leading question should be whether or not they have successfully delivered systems that address similar requirements.

Since we have already established that the application does not represent the proliferation of an existing system solution, the best one can expect is to find a number of companies that have been successful in delivering systems that address needs similar to yours and seemed to have been able to handle similar complexities. So, for example, if the application is flaw detection—are the type and size flaws similar? Are the part, size and geometric complexity, and material similar? Is part positioning similar, and so forth?

This survey should narrow the number of companies to be solicited to four to six. The solicitation package should demand a certain proposal response. It is important to get a response that reflects that the application has truly been thought about. It is not sufficient to get a quotation and cover letter that basically says "trust me" and "when I get the order I will think about how I'm going to handle it." The proposal should give system details. What lighting will be used and why was that arrangement selected? How about the camera properties, have they been thought through? How about timing, resolution, sampling considerations, and the like?

Most importantly, does the proposal reflect an understanding of how the properties of the vision platform will be applied and can it defend that those properties are appropriate for the application? How will location analysis be handled? What image processing routines will be enabled specifically to address the application? A litany of the image processing routines inherent in the platform is not the issue. Rather what preprocessing is being recommended, if any? What analysis routines, and the like? Along with this an estimate should be prepared of the timing associated with the execution from snapping a picture to signal availability reflecting the results of a decision. This should be consistent with your throughput requirements.

When a vendor has thought through the application this way and conducted a rather comprehensive analysis, he is in a good position to provide both a schedule of project development tasks and a good estimate of the project cost. By insisting on this type analysis in the proposal, both vendor and buyer should avoid surprises. Among other things it will give the buyer a sense that the application is understood. Those proposals responsive in this manner should be further evaluated using a systematic procedure such as Kepnor-Tregoe decision-making techniques. These involve establishing criteria to use as the basis of the evaluation, applying a weighting factor to the criterion and then evaluating each of the responses against each weighted criterion to come up with a value. This value represents a measure of how a company satisfies the criterion along with the relative importance of that criterion to the project. In some cases, the score given should be 0 if the approach fails to satisfy one of the absolute requirements of the application.

Having made a decision on a vendor, justifying the project may be the next issue. Significantly, justification based solely on labor displacement is unlikely to satisfy the ROI requirements. Quantifying additional savings is more difficult but in reality may yield an even greater impact than labor savings. Product returns and warranty cost should be evaluated to assess how much they will be reduced by the machine vision system. The cost of rework should be a matter of record. In addition, however, a value can be calculated for the space associated with rework inventory. The rework inventory itself should be included. The cost of rejects and related material costs, the cost of waste disposal associated with rejects, and the cost of freight costs on returns are all very tangible quantifiable costs.

There are other savings, through less tangible, which should be estimated and quantified. These include items such as:

1. The cost of overruns to compensate for yield
2. The avoidance of inspection bottlenecks and impact on inventory income and inventory turnover
3. The elimination of adding value to scrap conditions
4. The potential for increased machine uptime and productivity accordingly
5. The elimination of schedule upsets due to the production of items that require rework

Another observation is that when considering the savings due to labor displacement, it is important to include all the savings. These include the following:

1. Recruiting
2. Training
3. Scrap rework created while learning a new job
4. Average workers compensation paid for injuries
5. Average educational grant per employee
6. Personnel/payroll department costs per employee

Overall the deployment of machine vision will result in improved and predictable quality. This in turn will yield improved customer satisfaction and an opportunity to increase market share—the biggest payback of all.

FURTHER READING

Automated Imaging Association (the North American trade association of the machine vision industry): www.machinevisiononline.org.

Machine Vision Association, Society of Manufacturing Engineers: www.sme.org.

Nello Zuech, *Understanding and Applying Machine Vision*, Marcel Dekker, New York, 1999.

Vision System Design magazine: www.vision-systems.com.

CHAPTER 47
AUTOMATED ASSEMBLY SYSTEMS

Steve Benedict
Com Tal Machine & Engineering, Inc.
St. Paul, Minnesota

47.1 INTRODUCTION

Automated assembly is a natural outgrowth of the Industrial Revolution and represents one of the most efficient and cost-effective methods of manufacturing modern consumer and industrial products.

Early manufacturing is often referred to as the "craftsman" era in which an individual would design and build an entire article, usually one at a time, for his or her own use. Eventually some folks began specializing in building specific products for sale to the community, but the manufacturing methods were still relatively crude by modern standards.

Eli Whitney is recognized as a pioneer in the use of interchangeable parts in manufacturing. This concept was largely driven by the Colonial firearms industry and gave birth to the American concept of mass production.

The Industrial Revolution ushered in the "assembly line" era which greatly increased production and efficiency, but was still relatively labor-intensive. This system, in which a worker might fabricate only a small piece of the whole product and never see the completed product, often resulted in a lower worker satisfaction and a disconnection from the craftsman concept of pride in workmanship.

Although Ransome Eli Olds invented the basic concept of the assembly line in 1901 with the first mass-produced automobile (the "curved-dash" Oldsmobile), it was Henry Ford who invented an improved assembly line and installed the first conveyor belt–based assembly line in his factory around 1913–14. The assembly line reduced production costs for cars by reducing assembly time, allowing Ford's famous Model T to be assembled in ninety-three minutes.

The increasing sophistication of machine tools, controls, and processes in the mid to late 1900s, as well as ever-increasing labor costs, fueled the interest in automated assembly. The increasing availability, sophistication, and processing power of modern programmable logic controllers, solid-state machine control systems, robotics, and vision systems provide manufacturers with cost-effective manufacturing methods unheard of only a few decades ago.

Some argue that modern automation eliminates assembly jobs and has a negative effect on job creation, but one must keep three important factors in mind:

- Automation can enhance worker and plant safety by performing many hazardous operations to which workers should not be exposed.
- Many of the jobs performed by automation are tedious, repetitive operations that offer workers little in terms of job satisfaction, interest, or variety.

- As technological advancements continue to create the ability to engineer and manufacture more compact medical and electrical devices, automation can create a precise and statistically controlled repetitive environment based on quality.

Many manufacturers retrain workers who formerly performed these tedious tasks to operate and maintain the automation equipment. This results in a higher accomplishment level job and increased job satisfaction for the worker, and decreased costs and higher productivity for the manufacturer.

With greater pressure in today's society to bring products to market faster and more economically, it is important for manufacturers to investigate and evaluate the need to automate their production. Some products justify automation much easier than others. For example, a precision medical device that needs to be produced by the millions, and that requires statistical tracking of product codes, inspection results, and product quality issues would probably be a good candidate for automation. In contrast, a low-quantity product that requires manual assembly and/or inspections might not justify the expense of automation. One must keep in mind that automated assembly does not need to be an "all or nothing" approach and it is quite common to have a semi-automated production line.

The next time you walk though your favorite store, look at the vast variety of products ranging from toothbrushes to children's toys. Someone or something has manufactured the item, formed the box or container, placed the item in the container, applied a label or printed a code onto the container, loaded the container into a shipping carton, then taped or shrink-wrapped the carton and placed it on a pallet or skid. Most of these steps were performed by automated equipment.

What makes automation projects interesting for engineers and designers is the staggering variety to which ever-changing technologies can be applied to the manufacture and packaging of the products we all use everyday.

47.2 ELEMENTS OF MODERN AUTOMATION SYSTEMS

47.2.1 Intermittent Motion Versus Continuous Motion

The question of whether to use intermittent or continuous motion is usually determined by the required production rate and the complexity of the assembly process.

In an intermittent-motion system, the components are transferred to an assembly station; it remains stationary during the assembly process. Intermittent-motion part transfer systems have an indexing (moving) motion and a dwell (stationary) period. Intermittent-motion systems are generally less expensive than continuous-motion systems, but have a practical production rate limit of about 40 parts per minute.

In a continuous-motion system, the part transfer systems and components remain in continuous motion and the parts are assembled on the fly. Continuous-motion systems are usually more expensive owing to the faster part handling requirements and the greater quantity of parts that need to be provided to the machine, but can achieve maximum production rates of 500 parts or higher. Continuous-motion systems are particularly well-suited to multicomponent products that are consumed in high volume such as data cartridges, diskettes, tape dispensers, trinkets, pens, pencils, disposable razors and razor blades, and other low-cost disposable/consumable products.

47.2.2 Main Part Transfer Systems

One of the first factors to consider when designing an automation system is how the product will be transferred through the machine.

Common methods of intermittent part transfer include indexing dial plates, linear translators, walking beams, indexing conveyors and floating-puck conveyors. Component parts are most often assembled in fixtures (sometimes referred to as *nests*) on the main part transfer. In many circumstances machines can be equipped with interchangeable nests and suitable changeover tooling to allow the machine to run several different part sizes or configurations (see Fig. 47.1).

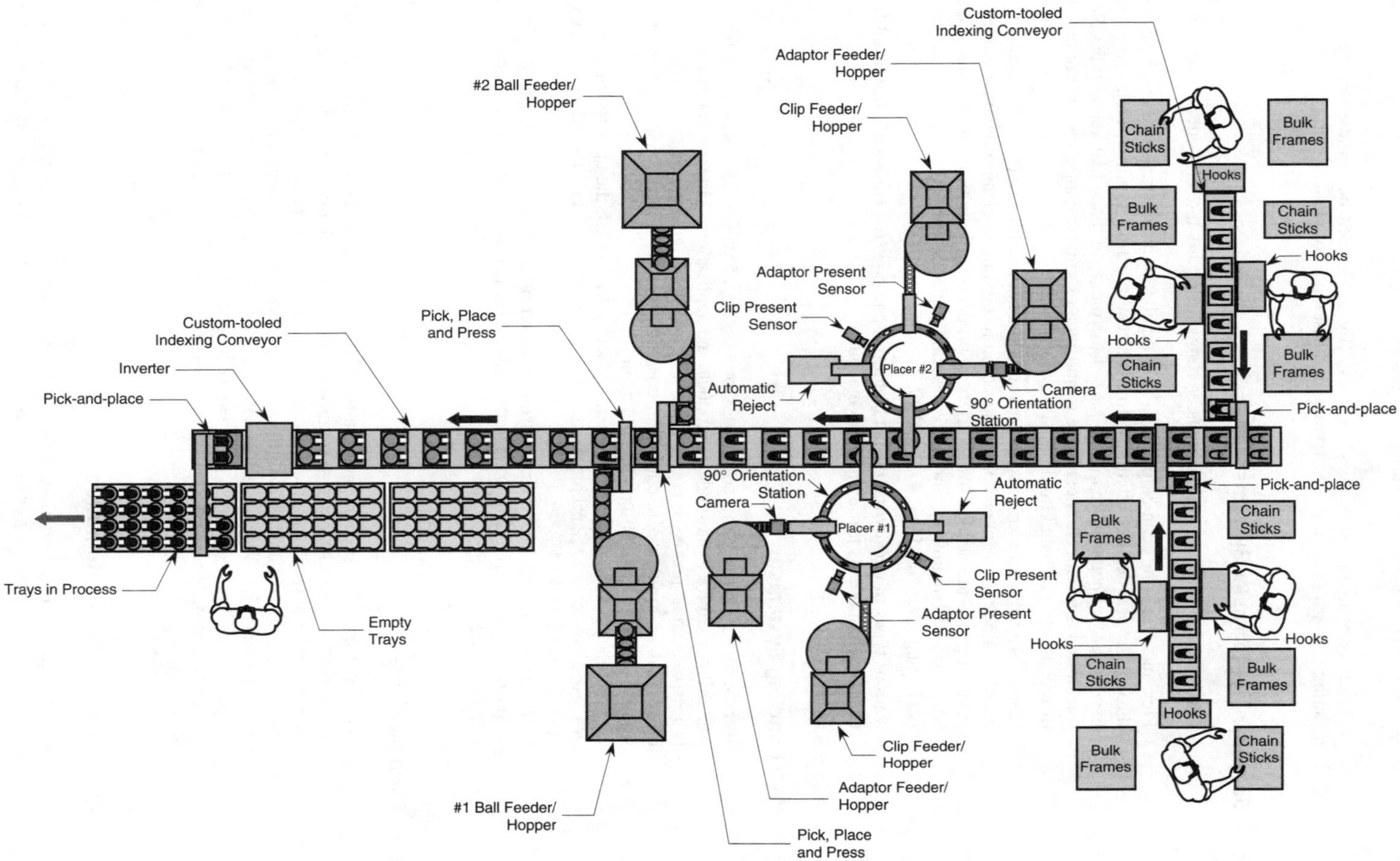

FIGURE 47.1 Assembly system concept drawing. (*Courtesy of ComTal Machine & Engineering*)

Continuous-motion machines often use flighted/lugged belt conveyors, lead screws and intermeshing *star wheels* to transfer, fixture, and assemble component parts.

47.2.3 Secondary Part Transfer Systems

Various mechanisms are used to transfer component parts into and out of assembly stations.

- Pick-and-place units can be pneumatic, hydraulic, or mechanical, and can be designed to function through several degrees of motion to match the function required. Pick-and-place units typically have an end-effector, which is the part of the pick-and-place that actually handles the parts to be transferred. End-effectors can have a multitude of gripper configurations, or can use vacuum cups or magnets to grasp and hold parts.
- Robots are similar in function to pick-and-place units, but are typically used when more sophisticated assembly movements are required, in hazardous or clean room environments, or when multiple programs are needed for a variety of product configurations. Robots are used extensively for welding and painting in automotive assembly plants, and for palletizing cartons at the ends of packaging lines (see Figs. 47.2 and 47.3).
- Various types of conveyors, pushers, shuttles and powered rollers are also used to manipulate and transfer components throughout the machine.

47.2.4 Web Handling Systems

Web handling involves the continuous processing of woven and nonwoven fibers, plastic and/or paper films used for a multitude of products and applications. Modern web handling involves multiple processes such as precision registration, alignment and tension control, laminating, coating, curing, rotary and intermittent die-cutting, ultrasonic welding, slitting, winding and rewinding, on-line inspection, sorting, collating, and packaging.

Web lines are most often controlled by servo drives with the appropriate master/slave hierarchy to ensure precision registration from layer-to-layer and/or end-to-end (i.e., flexographic printing presses) (see Fig. 47.4).

Modern web lines are typically used for paper products such as napkins, medical diagnostic test strips, transdermal drug delivery patches, labels, bandages, gauzes and wound dressings of all types, and surgical drapes.

47.2.5 Machine Vision

The past decade has seen an increase in the use of vision systems for product registration, location, and inspection. Modern vision systems are capable of resolution in the single-digit micron range, but a large part of a system's effectiveness and applicability depends on correct camera mounting and placement (angles/distances), lighting, and developing the appropriate software algorithms and machine integration. While machine vision can provide statistical accountability for quality assurance, the parameters that determine a good part or a bad part must be clearly defined. With the accuracy and resolution of contemporary vision systems, these systems can essentially reject every part going through the machine.

47.2.6 Controls and System Integration

One big change in automation in the last 15 years or so has been the proliferation of *programmable logic controllers* (PLCs), *personal computers* (PCs) and touch-screen operator interfaces. Hard-wired

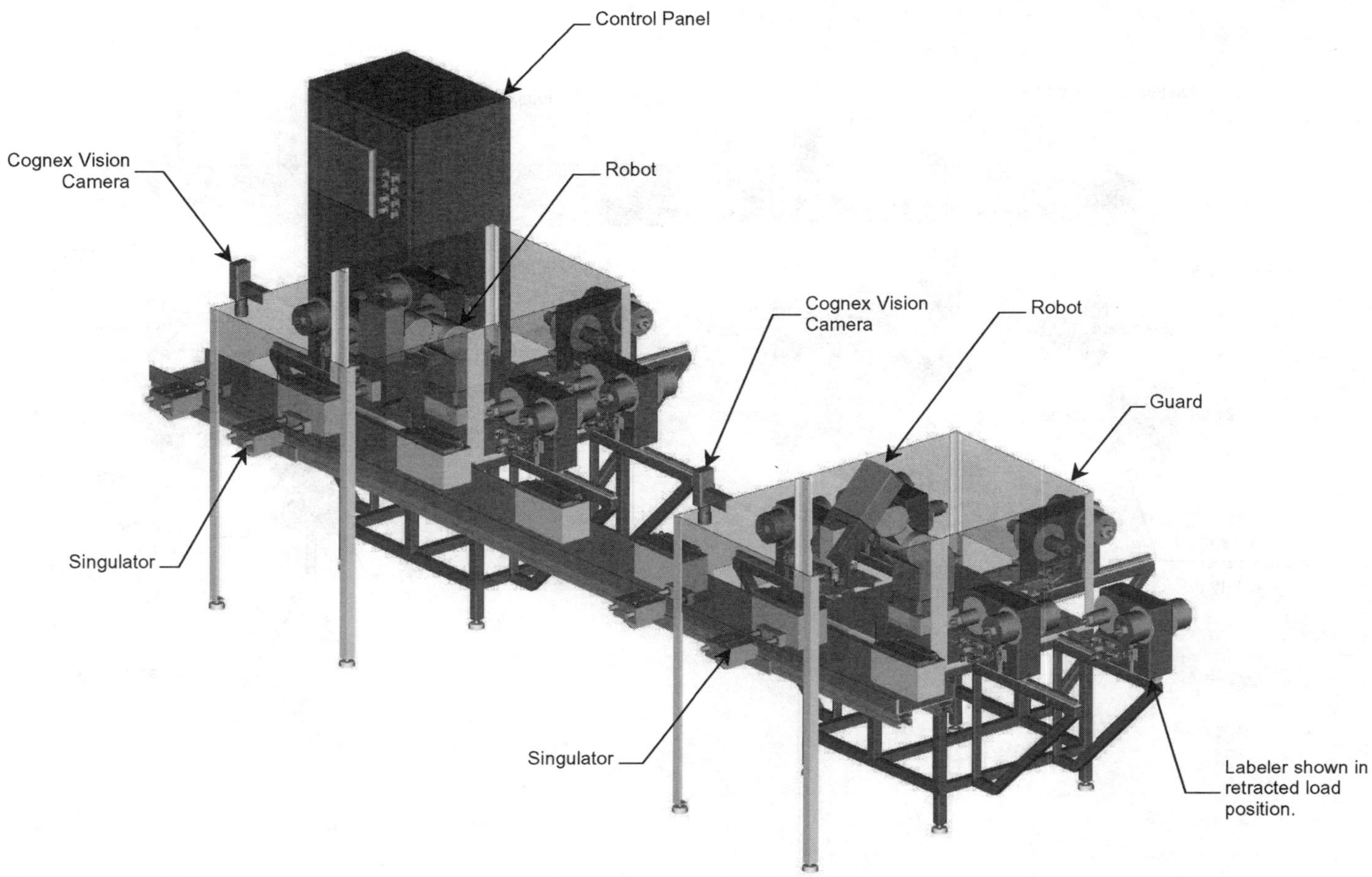

FIGURE 47.2 Battery labeling system with robotic assembly. (*Courtesy of ComTal Machine & Engineering*)

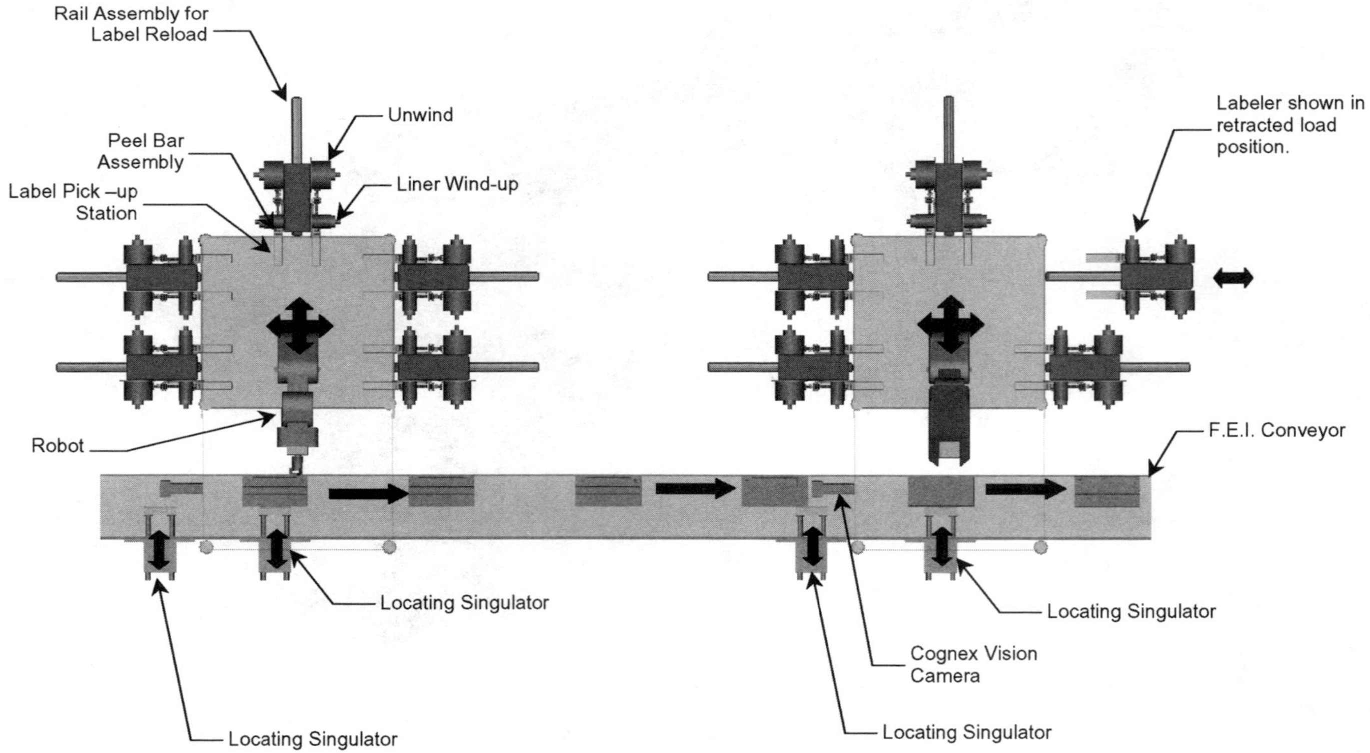

FIGURE 47.3 Battery labeling system in 2-D. (*Courtesy of ComTal Machine & Engineering*)

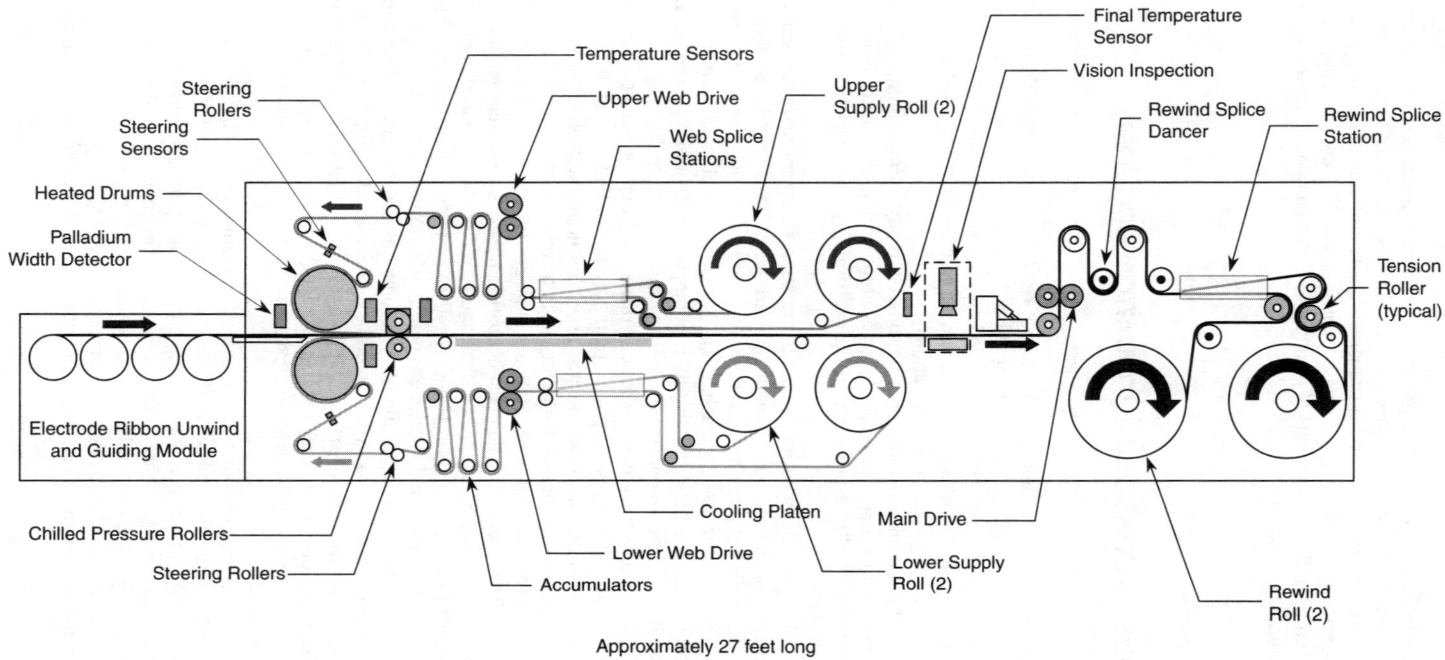

FIGURE 47.4 Sample web line concept drawing. (*Courtesy of ComTal Machine & Engineering*)

controls have pretty much given way to more sophisticated input/output devices that are much easier to integrate into complex manufacturing systems. These modern control systems make it much easier and less expensive to implement more complex machine logic. PLCs have proven to be better for binary/digital data and single inputs/outputs. PCs are more appropriate for machine control when dealing with large amounts of analog data, large amounts of operator input, complex data analysis and complex algorithms.

47.3 REASONS TO CONSIDER AUTOMATION: ECONOMY AND BENEFITS

Many countries realized that they not only must compete with one another, but with the enormous pools of inexpensive labor in the third world. Mechanized assembly offers not only competition to inexpensive labor, but allows a product with verified quality to be manufactured in close proximity to major marketplaces. It often means that manufacturing can be done very profitably in areas with a greater degree of political stability.

The rate of increase in productivity in many industrial countries is declining. The largest percentage of remaining direct labor content in manufacturing is in assembly. It exceeds all direct labor costs for fabrication, forming, molding, plating, finishing, and metallurgical treatment combined.

It is important to involve an automation company or expert early in the evaluation process to determine the risks and benefits that are related to a specific product. Minor product modifications can be the determining factor whether automation is feasible. For example, certain markings on a product can be used by machine vision to orient or align placement, as well as chamfers or tolerance build-ups. It is important that all the components that make up the final assembly or product be held within specific tolerances to avoid any interference problems.

Management and/or ownership understands that certain processes require equipment, like injection molders to produce plastic parts, presses to bend material or form parts and other specialized fabrication machines. Automated assembly is an alternative that has to weigh against the costs of capital investment, the availability of relatively inexperienced and unskilled labor.

Engineers, plant personnel, and managers must carefully analyze their manufacturing facility to see where automation can best be justified. As discussed earlier, automation presents a viable solution for parts assembled in a chemical or clean room environment, or for components that are too small to hand-assemble. However, when considering easily assembled, high volume components such as consumer disposables, potential automation purchasers must evaluate their production needs over the expected life of the product and of the automation equipment, and the cost savings realized from increased production and decreased labor. The time it takes for the cost savings to equal the cost of the automation equipment is often referred to as the "break-even" point or payback period. Assuming a stable product demand, the decreased cost of manufacture after the "break-even" point can produce significant profit as the automation equipment has essentially paid for itself.

Consider a hypothetical example of a manufacturer that assembles a consumer-disposable product by hand. Their customers' sales have skyrocketed and production requirements have gone from one million units per year to five million units per year. Although the evaluation process has many variables (market share, technology, etc.), the underlying objective is to make profit. If automated assembly cannot result in a reasonable payback period, it simply does not make sense. While company owners, shareholders and managers are obviously concerned about safety and product quality, in today's competitive environment the question to engineering or manufacturing to justify factory automation is simply, will it make money? To accurately determine if automation is viable, the current assembly process must be evaluated in terms of:

- How long does it take to assemble each unit?
- What is our current reject rate?
- Can we fulfill our orders at this production rate?
- What is the cost to assemble each unit manually?
- How long would it reasonably take to assemble each unit using automation?
- What would the expected reject rate be using automation?
- What would the cost be to assemble each unit using automation?

The increased productivity per hour, in conjunction with the increased profit per part, will determine the payback period. While this may seem elementary, the key is to evaluate the expected machine rate over the available run time. The decision must be made whether to purchase a $1 million continuous-motion machine that assembles 200 parts per minute in an eight-hour shift, or a $500,000 machine that assembles 100 parts per minute requiring two shifts. The cheaper machine may be easier to justify, and also presents the option of purchasing a second machine (less the non-recurring engineering costs) as production needs increase. Establishing a realistic internal budget will help establish the path for the best automated or semi-automated approach.

47.4 WHAT TO EXPECT FROM A REPUTABLE AUTOMATION COMPANY

When evaluating companies to fulfill your automation needs, it is important to look for companies with a broad base of expertise in many different industrial fields, and companies capable of providing turnkey system integration services for complete manufacturing lines.

An able automation partner will work with you to develop and document specifications for all required equipment, as well as suggest potential sources of that equipment whether available as standard commercial equipment or completely custom. The specifications should detail all operations that each machine will be required to perform.

Automation suppliers should assist customers in the development and selection of preferred brands of components, and should help determine control system and documentation requirements, materials, inspection procedures, acceptance and performance criteria, project reporting needs, installation assistance, maintenance, start-up, training, warranty, and spare parts requirements.

Automation companies can provide CAD-generated concept layout drawings to illustrate proposed product flow and full material handling requirements, locations of all equipment, footprint size requirements for the complete line, and the locations of all utilities and operator workstations.

By providing estimates for each phase of a project, projected costs for each machine, and detailed milestones for the entire project or for each individual machine, a reputable automation supplier can help ensure the success of the customer's project.

47.4.1 Request for Quote Information Form

Company name: ________________
Project name: ________________

- Price: Firm ______ Budgetary ______
- Project type: Standalone ______ Tabletop ______ Turnkey______ Modify existing______
- Customer's preferred concept: Continuous motion rotary______ Indexing dial plate______ Inline walking beam Conveyor______ Palletized Conveyor______ Other______
- Concept drawing required: Yes______ No______
- Operator load: Operator unload______ Auto load______ Auto unload______
- Current production rate: Parts/minute ______ parts/hour ______ shifts/week ______
- Requested rate: parts/minute______ parts/hour ___ annual output______
- How many operators are they trying to eliminate per shift? ______ How many operators do they want to have assigned to this equipment?______ Per shift ______ Shifts/week____
- The customer's average annual cost per operator: ______
- Direction of product flow, from operator side of equipment: ______
- Foot print size space available for this equipment: ______
- Include in pricing: Mechanical design______ Electrical design______Controls installation and startup______ Manuals______
- CAD preference: Solid works______ AutoCAD______ Unigraphics______ Pro/E______
- Is the Acceptance Test Criteria included in the specification: Yes______ No______

  If not, describe here: ________________
- Compliance/validation requirements: CE______ UL______ FDA/CDRH______ CGMP______ Other______
- Clean room: Yes______ No______ Class ______
- Inspection requirements: Operator______ Vision system______ Other______
- Definition of rejects: ________________
- Particular materials that will withstand wear from the product:
- Minimum door/path size from dock to installation point:
- Does customer have Preferred Brands List for commercial items? Yes (see attached) No
- Preferred brands: PLC: ______ Servos ______Solenoid valves ______
- Human machine interface preferences: Pushbuttons______ Touch-screen______Panelmate______ Panel view______ Color______ Monochrome______ Display size:
- Guarding preference: Perimeter______ Interlocks______Light curtain______Touch-switches______
- Power: ______VAC
- Plant air supply: ______ psi ______ scfm cap. ______ In-plant vacuum? Yes______ No______

47.4.2 Product Design for Automated Assembly

Assemblies should be designed to facilitate the monitoring of the assembly process as it occurs in each incremental step of the machine sequence.

Access should be provided in the design for sensors to determine presence and correct relative position of each component immediately after each insertion or joining operation.

Reference locations or surfaces should be included in the assembly design whenever inserted components have less thickness or size than possible height stack-up of dimensional tolerances in parts previously assembled.

When designing products, one must always be mindful of the fact that components may need to be fed and oriented by automation whether through vibratory or centrifugal feeders, tape and reel, web registration, vision guided, etc.

47.4.3 Operation and Service Manual

An *operation and service manual* (OSM) is an essential part of machine. A typical OSM should cover the following sections:

1. *Introduction*

- *Machine description.* Gives a brief description of the machine listing the major components and station functions.
- *Sequence of operation.* Gives a step-by-step description of how the machine runs. Note that we can often get most of this information from the quote.
- *About this manual.* Standard boilerplate; references technical manuals for purchased/commercial components.
- *Safety.* Includes the compressed air and electrical disconnects, and any additional safety hazards or concerns (fumes, UV light, heat sources, and the like). We also include photos and explanations of all safety warning stickers on the machine.

2. *Installation* included only if the customer is going to install the machine; usually includes adjusting leveling pads, connecting main power and air, and attaching and connecting any peripheral equipment such as conveyors, feeder bowls, and hoppers.

3. *Machine controls* include descriptions and graphics of all control panels and operator interface screens.

4. *Set up* includes items like machine threading, loading product into feeder bowls or hoppers, minor adjustments to stations, and the like. If the machine requires extensive set up to run a different product configuration we usually add a Machine Changeover section.

5. *Operation before you operate the machine* includes items like how to load/stage product, power-up the machine, switch on the compressed air, power-up any peripheral equipment, and the like.

- *To operate the machine in the auto mode.* Instructs the operator how to run the machine in the normal/automatic mode. If the machine is equipped with a touch-screen, we show the displays and screens at each step of operation. If the operators are required to perform manual assembly tasks during semiautomatic operation, we will describe and show an illustration of each task.
- *To jog the machine in the manual mode.* Instructs the operator and/or Maintenance Technician how to jog the machine components; includes any relevant screens and instructions on how to reset the machine for automatic operation after jogging.
- *Shutdown procedures.* Instruct the operator how to cycle any remaining product through the machine and shut the machine down.
- *To clear errors/jams.* Instructs the operator how to clear jams, clear error messages and restart the machine.
- *To restart the machine after an e-stop.* Instructs the operator how to clear and restart the machine after an e-stop/alarm.

6. *Maintenance* lists the preventive maintenance procedures by order of frequency (every shift, daily, weekly, and the like); also includes any common maintenance items that may need to be addressed as required (adjusting belt tensions, replacing wear items such as knife blades).

7. *Adjustments and repair* lists any items that could need replacement or adjustment that don't qualify as normal maintenance items.

8. *Troubleshooting* lists likely problems, their cause, and correction.

9. *Recommended spare parts* recommended by project manager/engineers.

47.5 FUTURE TRENDS AND CONCLUSION

The area of assembly and related testing is the only remaining area of manufacturing that is capable of broad-scale significant reductions in direct labor cost. The product consumer is demanding more quality while pride in work seems to be on the decline.

With greater pressure on manufacturers to deliver cheaper products more quickly while still controlling inventory, there is a greater demand for automation concepts that emphasize modularity and flexibility. With shorter product life cycles comes a narrower window for payback on thc dollars invested in capital equipment. The automation that manufacturers purchase needs to be more adaptable for the next new and improved model or next generation of products.

As the suppliers of automation respond to these needs, modular bases or chassis for automated cells continue to evolve. There are many gimmicks on the market today. They simply get more attention when the terms "modular" and "flexible" are used. The current state of custom automation has precluded the development of a single concept that can be applied across the board for all potential types of custom automation. The race for the ultimate solution will continue at a feverish pitch.

While there will always be a need for certain products to be assembled on dedicated lines, flexible automation will continue to evolve into the future. When the total commercial cost of an automated assembly machine can exceed 40 percent of the overall quoted price, it is important to try to get the most out of the purchased equipment. Utilizing flexible components such as servo-controlled indexers, interchangeable nests on dial tables, programmable pick-and-place units, robots, visions systems, industrial personal computers and PLCs provides versatility, and allows equipment to be modified and adapted to new products while retaining the value of the initial commercial component cost.

Ingenuity in design, keeping it simple, and providing flexibility and value for the long term will be the keys to successful automation solutions in the future.

INFORMATION RESOURCES

Organizations:

Society of Manufacturing Engineers: www.sme.org.
Society of Plastic Engineers: www.4spe.org

Periodicals:

Assembly Magazine: www. assemblymag.com
MachineDesign: www.machinedesign.com
Design News: designnews.com
Mediacal Device and Diagnostic Industry: www.device.link
Control Engineering: www.controleng.com

CHAPTER 48
FINISHING METAL SURFACES

Leslie W. Flott
Summit Process Consultant, Inc.
Wabash, Indiana

48.1 INTRODUCTION

The term finish, or finish process, refers to any process or course of action by which the surface of a material is altered. The surface may be altered by mechanical, chemical, electrochemical, or photoelectric means to produce a surface that has a more desired texture, overall appearance, corrosion resistance, or solderability. The finish that is applied to a basis material is most often the last surface modification performed before the product is expected to carry out its end use. The following is intended to introduce the reader to the design for metal finishing, cleaning, and electrochemical processes.

48.2 DESIGNING FOR FINISHING

The metal finisher is limited as to what one can do by certain basic principles of mechanical finishing and of electroplating. The design engineer should understand the limitations imposed by type, shape, and size of components to facilitate quality finishing at an acceptable cost. The designer can exert as much, or more, influence on the quality attainable in finishing a part as can the metal finisher.

Designers of goods to be finished need a broader understanding of the processes used. Finishers must know what kind of metal they are finishing. Every metal requires a somewhat different pretreatment. Even variation between alloys may require modification of the pretreatment process. An incorrect identification can result in damage or the destruction of the parts. Metal preparation affects adhesion and appearance, as well as composition and properties of the final deposit. Providing particulars about preexisting soils helps the customer to receive the metal finish desired. Environmental regulations have severely restricted cleaning options, but most soils can be safely removed if the finishers know what they're dealing with.

Before parts can be processed, the finishers must know what kind of metal they are dealing with. Many metals require a special pretreatment. Aluminum, for example, is dissolved in solutions formulated for cleaning steel. Even different alloys may require different pretreatment. A wrought aluminum alloy, such as 1100, must be cleaned differently than 380 diecast alloy. Equally important, especially for anodizing, is not to mix alloys in the same lot or within a single weldment. Mixing alloys will always result in a less uniform finish within a lot, or a single weldment, and may result in some parts being destroyed.

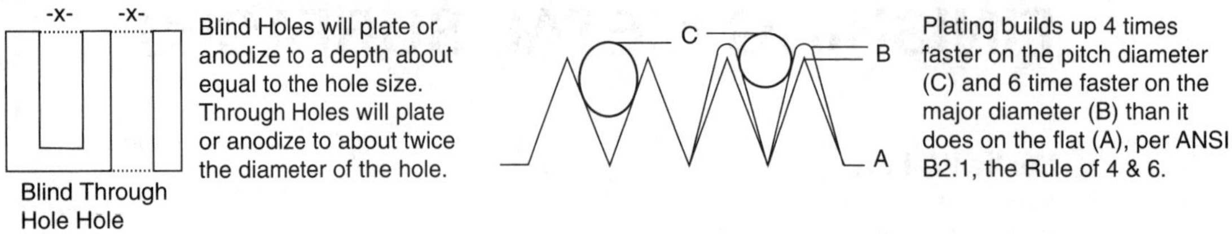

FIGURE 48.1 Plating and anodizing on threads and holes.

The following items must be considered when requesting any metal finishing services:

Base metal. Is the metal to be plated stainless steel, high nickel content, cold roll, hot roll, high carbon, aluminum, copper, or die cast? Which alloy? What heat treat? What gage metal, and the like?

Specification. What is the minimum thickness required? Is there a maximum? What type of finish? What specification is called for? Is the part automotive, military, aerospace, and the like?

Print requirements. Do the parts have a critical micro finish? Is masking required? Is etch cleaning allowed? Are the areas where rack marks cannot be allowed? Does the 3/4-in ball rule apply?

Testing and certification. What *acceptable quality level* is acceptable? What tests are requested? Are special tests or unusual documentation required, and the like?

Dimensions and tolerances. Does the design allow for build-up on threads or outside diameters? What is the weight per part, weight per 1000, and the like? (see Fig. 48.1)

Heat treat prior to plating. Are the parts hardened? Is baking for hydrogen embrittlement relief before or after plating required? If hydrogen embrittlement is a problem, will the standard 375°F bake for 3 h be sufficient? Are there any special baking instruction, etc.?

Cosmetic concerns. What affect will minor blemishes have on acceptance? Do cosmetic or functional concerns override, etc.?

An important consideration in specifying metal finishes is significant surfaces. In most cases the same quality or thickness is not essential to the appearance and serviceability when the parts are assembled in the normal position. Certain surfaces, those that cannot be touched by a 3/4-in ball for example, are excluded in many specifications.

48.3 DESIGN FOR PLATING

Coating distribution can be estimated from models and computer simulations, but practical experience provides clues for designers and finishers. Listed below are some simple empirical rules for improving coating uniformity:

1. All significant surfaces should be clearly marked on the drawings.
2. Round all edges to a radius of at least 0.01625 to 0.030125, preferably the latter.
3. All angles and corners, inside and out, should be provided with a generous radius.
4. Avoid both concave and very flat surfaces. A convex or crowned area, with a minimum 0.015-in per inch crown, will receive more uniform coatings.

5. The width of blind holes, recesses, grooves, and slots should not be less than half the depth.
6. Minimize the number of blind holes, exempt them from minimum thickness requirements, and their diameters must never be less than 0.200 in.
7. All threaded holes should be countersunk to prevent build up on their edges.
8. Use broad low ribs when needed using a minimum height and radius all sides and corners by 0.0625 in.
9. Use recessed letters rather than raised ones and round off all edges.
10. Studs for fasteners should be as short as possible with rounded inside angles and tips.
11. All bosses and studs should face the same way and should have hollow centers angled 90° to the plane of the part.
12. Avoid using more than one basis metal in any one part or order as dissimilar metals can cause interference with coverage and adhesion.
13. Small parts, to be barrel processed, must be sturdy enough to withstand the multiple impacts of barrel rotation and not entangle so as to cause damage or incomplete finishing.
14. Small flat parts, which tend to nest together, should be provided with ridges or dimples to prevent such nesting.

48.3.1 Significant Surfaces

Significant surfaces are often the subject of agreement between customer and manufacturer and shall be indicated on the drawings of the parts, or by the provision of suitably marked samples. The designers must understand that one of the most important factors determining the cost of a coating is the thickness required on significant surfaces. Fundamental laws of electrochemistry (current distribution) operate to prevent perfectly uniform deposition of an electrodeposited coating on a cathode of any practical shape and size. Portions of the work that are nearer the anodes tend to receive a heavier deposit. Sharp edges or protrusions at all current densities tend to steal a disproportionate share of the current. The goal of the designer and the plater is to make thickness variations as small as possible. At the same time, uneconomical wastage of metal by excessive build-up on both nonsignificant and significant areas must be avoided. The same difference in plated thickness found within a plated article also exists from piece to piece on the rack of plated work and within the barrel load. Metal distribution is of particular importance in precious metal plating owing to both cost sensitivity from over-plating and the lack of adequate corrosion protection in underplated areas.

General principles and sketches from many sources help to teach from practical experience. ASTM Standard B-5O7 also provides the designer with helpful information.

48.3.2 Throwing and the 3/4-in Rule

Plating into corners and other tight places is not just difficult; it is often not possible. Even where plating is deposited into recesses it is generally much thinner in the recess or corner than it is on a flat surface. The plating typically thins in the final 3/4-in. A thickness of 0.0002 in at a distance of 1 in from the corner may well be 0.0000050 in or less in the corner. Additionally, not all plating baths deposit into corners as well as others, i.e., with the same efficiency. This phenomenon of depositing or plating metals into recesses is called *throw*. In fact, many plating specifications include what is called the 3/4-in rule (Fig. 48.2). Simply stated the rule says that the thickness requirements do not normally apply to any area of the part being plated that cannot be touched with a 3/4-in diameter ball.

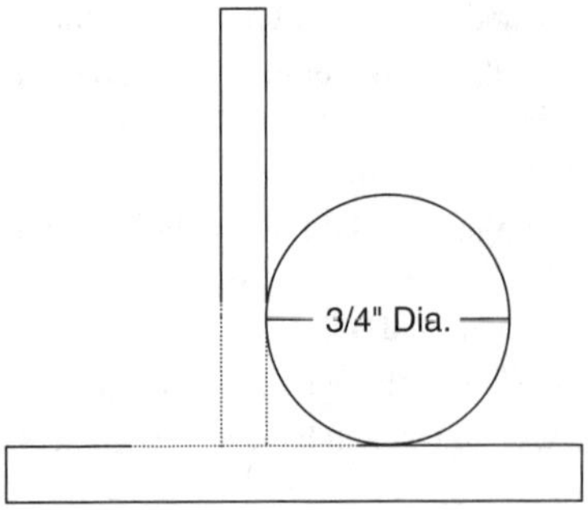

FIGURE 48.2 Illustration of the $^3/_4$-in rule.

Throw is defined as "The improvement of the coating (usually metal) distribution over the current distribution on an electrode (usually the cathode) in a given solution, under specific conditions. The term is also used for anodic processes (anodizing) for which the definition is analogous."

In Fig. 48.3, the areas shown in gray will not plate sufficiently. The $^3/_4$-in rule applies. The walls around the recess are analogous to a building protected by a lightning rod. The rod, or the sharp edged and more accessible surfaces near the recess more readily attracts the charges metal ions, which in turn means less plating in the recess. The phenomenon is often referred to as a Faraday Cage. In essence, the projections act very much like a lightning rod, drawing the current away from the tight inside corners.

The same problem exists for holes, both those going completely through the part and blind holes. Typically plating will throw approximately $1^1/_2$ times the diameter of the hole. Again chrome is especially hard to throw into holes. Chrome may only throw half the diameter into the hole, while tin and zinc will throw slightly better that $1^1/_2$ diameters.

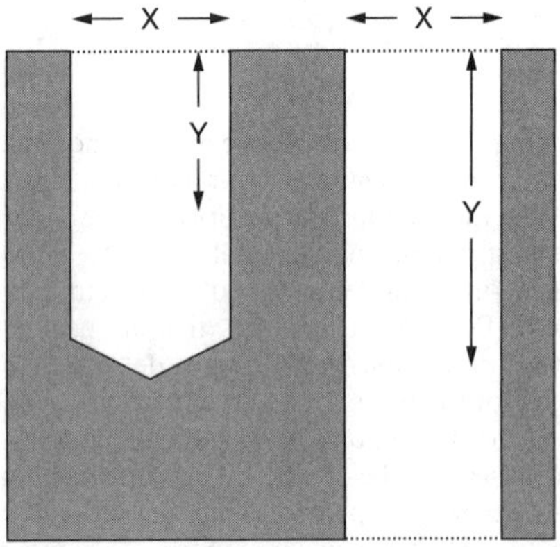

FIGURE 48.3 Illustrating throw into blind and through holes. "Y" indicates the plating depth.

However, the problem is not without possible cure. The simple fact is, if the customer insists and is willing to pay the difference the parts can be plated using conforming anodes. This will commonly increase the plating cost by as much as 10 to 100 times the original cost, depending on the part topology and is therefore seldom requested.

Chrome over nickel is almost a special case. Nickel has a yellow cast and chrome is more blue white, therefore the contrast in corners is especially obvious. In addition, nickel plating is porous and the protection it offers is mechanical not sacrificial as is the case with metals like zinc. Nickel over steel offers only 24 h salt spray resistance compared to 98 h for yellow chromate. Alternate choices of plating, where the special wear characteristics of chrome are not an issue may be substituted. Tin over nickel offers exceptionally good corrosion resistance (as much as 1000 h or more), if the customer allows the substitution.

48.3.3 Design for Barrel Plating

Not all parts to be plated or anodized need to be polished before further finishing. If they need to be smoothed or cleaned they are usually done in a barrel. They may also be tumbled or vibrated with or without special media. When designing to improve appearance, or quality in general, certain rules apply. The following applies typically to parts intended to be barrel finished:

1. Blind holes, recesses, and other crevices should be avoided whenever possible since they may trap tumbling compounds, liquids, or metal residue.
2. Barrel plating rubs parts against, thus intricate patterns, which could be blurred or otherwise marred, are to be avoided if the parts are to be barrel finished.
3. Fragile, or delicate, will almost assuredly be damaged for the reasons above. Barrel finishing demands sturdy parts.

4. Small flat parts tend to stick, or nest, together. Incorporating dimples or ridges into the parts avoids the problem.
5. Design the parts so that liquids can easily enter and drain from them. Threaded blind holes will often draw liquids into them by capillary action. The liquids tend to leak out later marring the finish.
6. Significant surfaces must be external for barrel plated both for proper cleaning and for applying the finish itself. Wherever possible, making the inside angle concave rather than sharply angled will greatly aid in obtaining a more uniformed finish.

Barrel may be as small as 3 to 5 in. in diameter and 10 in long to a couple of feet or more in diameter and 5- or 6-ft long. Barrels are generally made of perforated polyethylene, polypropylene, or Lucite. The smaller barrels will have a single bar running through its axis often with thin chains attached that make contact with the parts. There are also two chains on the outside end of the barrel, which serve to conduct current into the interior of the barrel. The barrel usually has a series of baffles or plates on the interior, which cause the parts to tumble so that all parts ultimately have an even chance of being plated. Larger barrel have danglers that connect from a contact bar into the interior of the barrel. These danglers are various sized as needed and look like battery cables. Usually a tin, or solder weight is attached to the inside end of the dangler so that the danglers from the two end cross. With many types of plating, for example, the plating builds up during plating and is stripped when the parts go through an acid bath during the cleaning cycle. In those baths where the acid is insufficient to strip them, the danglers will gain considerable weight and may actually damage the parts as the danglers grow beyond a certain size. In these cases the danglers must periodically have to be cut off. The excess weight may also cause the danglers to twist so that they remove themselves.

Some other special purpose barrels are available with button contacts or other special features depending on the parts being plated. Some barrels are designed not to rotate at all. These are generally intended for more delicate parts. These barrels that oscillate rather than rotate are referred to as *horizontal oscillating barrels* (HOB). Barrels are typically loaded by weight. The plater calculates or estimates the surface area of one part in grams and calculates how many of a specific part may be loaded at one time. Depending on packing fraction and other factors, it may take several attempts to arrive at the ideal load size.

48.3.4 Designing for Racking, Draining, and Air Entrapment

Design always has a considerable effect on the efficacy and permanence of the finish to be applied. Designers and finishers alike are encouraged to research the effects that design may have on the finished parts.

Most metal parts weighing more than a few ounces or that require a high microfinish, jewelry, for example, cannot be barrel plated, but are rather rack plated. Most difficult racking problems can be solved by design modification. Rack plating is, however, normally more expensive than barrel plating. The cost depends on the number of parts that can be run per hour, and labor costs. Designer note: Specially designed conforming anodes can reduce the natural variance within a part, but this means that the parts and the anodes must be wired individually. This process, while possible, is usually extremely expensive. Parts that might be plated for fractions of a cent in a barrel, or for a penny or two on a rack, might well cost 10 cents or more a piece because this is a very labor-intense process. The designer is often better off redesigning the part.

Of primary importance is electrical contact; without it the parts either will not plate at all or they may double plate (especially nickel), which causes the parts to peel and/or blister. The caution applies to both plated parts and anodized parts, but for different reasons. The anodized film is a nonconductor; therefore when a part being anodized breaks contact the finishing process is done. This results in a thin anodize and may cause other parts to burn.

Another important consideration is drainage. Trapped solutions will carry over into subsequent tanks causing contamination, possible corrosion, and wasted chemicals. Designers should avoid rolled edges, blind holes, and welded joints wherever possible. Drain holes should be provided, especially for irregular or tubular products. Even parts that appear to have been designed to drain well may have air pockets on the inside corners that allow the air to corrode the basis metal. Good drainage also means avoiding shapes that can trap air. Inside corners can trap air blocking access of solution to critical areas. Trapped hydrogen and oxygen from cleaning and plating process can also build up inside parts.

48.3.5 Basic Metals

A wide variety of metals are used as the substrates for mechanical, chemical, and electrochemical finishes and painting. The existing metal surfaces of these substrates are in many, if not most, instances unsuitable for the applications for which they have been chosen. Oxides, scales, nonmetal inclusions, and organic compounds must in most cases be removed. The basis metals are chosen for specific properties that they have, but each metal also has limitations that the designer wishes to overcome, or of which he or she may be unaware; the finisher attempts to compensate. Nearly any metal may be plated. The following are among the most popular.

48.3.6 Aluminum and Its Alloys

Aluminum is a silvery-white metal with many valuable properties. It is light (density 2.70 g/cm^3), nontoxic, and can be easily machined or cast. Aluminum has an electrical conductivity 60 percent that of copper and a much lower density; it is used extensively for electrical transmission lines. Pure aluminum is soft, but can be strengthened by alloying with small amounts of copper, magnesium, and silicon.

Based on its chemical reactivity, aluminum should not be very useful at all. Its standard reduction potential is −1.66 V, indicating that it is a very good reducing agent. It is far more active than iron, which has a reduction potential of –0.44 V. Aluminum weathers far better than iron, however, because the product of its corrosion, Al_2O_3, adheres strongly to the metal's surface, protecting it from further reaction. This is quite different from the behavior of iron's corrosion product, rust. Rust flakes off the surface of iron, exposing the surface to further corrosion.

The process of anodizing frequently enhances the protective oxide coating on aluminum. Aluminum metal is made the anode in an electrolytic cell, where aluminum metal is oxidized and its protective oxide coating is thickened. The oxide coating on aluminum renders it impervious to most acids. However, any chemicals that form a strong complex with aluminum ions, e.g., OH^- and Cl^- ions, will react with the oxide coating. Therefore, aluminum will dissolve in hydrochloric acid and sodium hydroxide solutions.

48.3.7 Polishing

It is not unusual for the grinding of metals to leave a somewhat rough surface. Further mechanical finishing may be required depending on the needs of the finished product—for example the high gloss common to chrome-plated bumpers. Thus the grinding operation is most often followed by a polishing operation, which in turn is followed by a buffing step.

The purpose of polishing is to remove a significant amount of metal resulting in a smoother surface preparatory to more refined finishing procedures. Polishing wheels are typically made of canvas, muslin, felt, or leather. The physical construction of the wheels varies depending on the shape of the parts being polished. The most common wheels in use are made of woven cotton fabric. The most durable wheels in use are made of individual disks of canvas glued together. The softest wheels are made of muslin sewn together.

No hard and fast rules exist for selecting polishing wheels. In general the more rigid wheels are used for coarse polishing and softer wheels for irregular surfaces and fine polishing. Very small parts are sometimes tumbled to both remove burrs and produce a more uniform surface finish.

48.4 CHEMICAL FINISHES

The finishes applied to metal surfaces may be cosmetic or serve more practical purposes, such as corrosion protection. These finishes are achievable through purely chemical means and include pickling, bright dipping, and conversion coatings.

48.4.1 Passivation of Stainless Steel

Passivation of stainless steel is a nonelectrical process whereby the free iron is chemically removed from the surface of stainless steel. This prevents the formation of possible corrosion sites and the development of tightly adhering oxides. The 300 series alloys are generally preferred for passivation, as the passivation process will actually discolor some of the 400 series alloys.

Passivation imparts a limited *neutral salt spray* (NSS) corrosion protection to the stainless steel, lasting usually not much over 2 h. It is critical when making assemblies to be passivated that all component parts be made of the same alloy; different alloys may be indistinguishable before passivation but may have a different appearance afterward. Since different solutions are used to passivate different alloys, they must be properly identified. Mixing alloys may not only result in differences in appearance but may also result in some parts being destroyed. Either QQ-P-35 or ASTM A-380 are generally referenced for the passivation of stainless steel.

48.4.2 Chromate Conversion Coatings

The chromate conversion coatings and process are known by several names: chromating, chromate dipping, bright dipping, passivating, and dichromating. These terms are all interchangeable terms and refer to the process of dipping, or spraying, a solution, which leaves a very thin film of hevavalent or trivalent, and an assortment of related compounds on an active metal surface.

Chromate conversion coatings are complex chemical coatings designed to retard the formation of white corrosion (zinc oxide) products and thus provide additional corrosion protection. The substrate metal participates in the coating reaction and becomes a component of the coating. The collaboration has a profound effect on the properties of the coating. Among the metals commonly chromated are zinc, zinc die castings, hot-dipped galvanized steel, cadmium, aluminum and sometimes copper and silver. Chromate films are typically very thin, on the order of 0.0000001 in and contribute no measurable thickness to the overall coating.

The chemistry involves a reaction between the metal surface and an aqueous solution containing chromates (chromium salts, either hexavalent or trivalent) and certain activators or film formers. Activators may include sulfates, fluorides, phosphates, and sometimes complex cyanides. Normally a given chromate is designed to work on a particular metal; but in a few cases will work on two or more. The chromate solution is normally acidic, pH typically about 1.5 to 2.0. About 0.00005 in of plating thickness is dissolved during the chromating process.

> **Note 1: Precautions**
>
> Neither plating nor chromate conversion coatings will correct problems that are inherent in the base metal. Plating may in fact make such surface defects as discoloration, roughness or pits due to welding, brazing or rust more prominent than they would appear on the original surface. Plating will not fill in or cover over existing surface flaws.

Chromates come in a wide variety of colors. The color is largely owing to the relative thickness of the chromate film. A difference of even a few angstroms can cause color variation. The color may also vary with the total immersion time in the bath and other factors. Therefore, variation within a single order and even on a single part is not remarkable, nor is this necessarily a sign of poor workmanship. The colors generally available are: clear (or blue), yellow, black, and olive drab. Most chromates require a time to set up. The fresh chromate will easily scratch and if it becomes damp may rub off. The typical time it takes for a chromate to set up is 24 h. After that the film is generally hard to remove. Corrosion resistance for chromates and most other metal finishes is measured using ASTM B117, The Neutral Salt Spray Test (NSS). Those requesting an NSS are cautioned, within the text of the test itself that the test does not correlate with corrosion resistance in the world in general. A difference of 100 percent in salt spray results does not guarantee a 100 percent improvement is the atmosphere where the part is used. This author has always used a rule of thumb as far as salt spray resistance, based on observation not scientific tests. Twenty-four hours of NSS is roughly equivalent to one year in a noncorrosive environment.

Clear chromate offers from 8 to 12 h to white corrosion (zinc oxide or *white rust*), and has a clear to slightly iridescent blue appearance. This is one of the earliest types of chromating solutions and is still used today. Other terms used to describe *clear chromate* include: blue, blue-bright, colorless, and clear silver-white, all of which terms are used interchangeably for the same color chromate. Older style clear chromates often deposited as a golden-yellow coating and were bleached by immersion in a dilute alkali solution to obtain the clear appearance. More up-to-date blue bright chromates are single dip solutions; often using only trivalent chrome salts and are therefore more environmentally friendly. Some varieties are capable of achieving NSS results of 96 h. Some varieties of this type of bath may be dyed an assortment of colors. These colors are typically used for identification purposes.

Gold, or yellow, chromate coatings are applied from baths containing chromates with nitrate, sulfate, or chloride activators and produce a distinct iridescent golden-yellow color. Typically *yellow chromates* are not pure yellow, but usually contain patches of red and green iridescence, owing to minute differences in the thickness of the chromate film and how it refracts visible light. Prolonged immersion in a yellow chromate bath will produce a brownish film that is often powdery. If the color is a smutty brown there is a chance the chromate will rub off. Yellow chromate solutions contain more hexavalent chromium than clear films, which accounts for their color. Yellow chromate offers in excess of 96 h NSS corrosion protection and is an excellent paint base.

GM 4252-M is typical of specifications calling for yellow chromate. In paragraph 4.1 (Appearance) is the following note: "In the case of zinc plus chromate and cadmium plus chromate finishes listed herein, a clean, commercial finish is required; *range of color, iridescence and sheen are not critical.*" (emphasis added)

Olive drab (OD), or forest green, chromate is the ultimate in commonly available conversion coatings, with NSS resistance in excess of 150 h. OD chromate is most commonly specified for military applications. It is generally not feasible to apply OD chromate to barrel plate work. The color and the corrosion resistance are owing to the inclusion of an organic acid modifier to the chromate formula. Many customers find the color is not especially pleasing but specify OD (as forest green) for nonmilitary applications for its functional rather than its decorative value.

Black chromate is usually achieved by incorporating a soluble silver salt in a yellow chromate formula, which produces a deposit of black silver chromate. This coating offers excellent corrosion resistance and a jet black semimatte to matte appearance. Black chromate has even found some applications in the space program where it is used on solar collectors. The use of silver salts makes black chromate coatings rather more expensive and sensitive to chloride contamination. This is especially a problem when black chromate is to be applied over parts plated in a chloride bath.

48.4.3 Pickling

Pickling is the process by which the surface of iron or steel (or other metals) is treated with acid to remove scale, rust, and dirt, preparatory to further processing such as cold rolling, tinning, galvanizing, and polishing. The most common impurities are iron oxides and scale. Scale is the oxides of iron that form on the surface of steel after heating.

The oxide or scale removal processes are major users of certain chemicals. In a typical year 25 percent of all hydrochloric acid and 5 percent of all sulfuric acid produced are consumed by the metal finishing industry to remove scale and oxides from steel parts. Hydrochloric acid is an excellent choice for removing surface oxidation, rust. While it is often more costly than sulfuric acid, many smaller shops prefer to use hydrochloric acid because it may be used at room temperature. Hydrochloric acid tends to emit chlorine fumes at temperatures above 130°F and is therefore seldom heated without considerable ventilation. Hydrochloric acid emits fumes at lower temperatures and higher concentrations, so pickling solutions are usually made up at 10 to 40 percent—rarely higher. Even shops that do very little pickling can usually keep a crock around for occasional use.

Sulfuric acid, while it makes up into an excellent pickling, usually has to be heated to be effective. Sulfuric acid is used mostly for economic reasons. The procedure is rather simple. The steel is immersed in a tank large enough so that the acid can touch all surfaces. The acid attacks the oxide and/or scale first and exposes an oxide free surface.

An inhibitor is commonly added to sulfuric acid pickling baths to reduce the rate of attack on the basis-metal. The inhibitor does not affect the scale removal rate. There are a wide variety of commercial inhibitors on the market, and finishers are advised to check out newly submitted parts against more that one inhibitor. Some inhibitors work very well on low carbon steel and poorly on high carbon steel, and some may actually attack hot-rolled steel and not affect cold-rolled steel at all. It is incumbent on the designer to tell the finisher what steel is being used.

The amount of time in the pickle depends on the uniformity and thickness of the scale or oxide coat. If the rust is light or superficial, it may take only a few seconds to expose a clean surface. When the surface rust is deeper, the removal often leaves a rough or pitted surface. Pickling will not in general improve the surface uniformity of the parts. If the parts have been allowed to sit for an extended period before they are scheduled for finishing so that thick or nonuniform rust forms, pickling may render them unusable for some purposes. This is because the acid attacks the rust preferentially to the steel.

Some parts are very difficult to plate and some will not plate at all. The old time platers would allow these parts to sit for a short time until they developed a thin film of rust. The acid pickle removes the rust film exposing a clean surface on the parts, which usually plates quite well. Great caution is required when using this trick, since allowing the parts to sit too long may allow them to start pitting; pickling cannot correct this condition.

48.4.4 Bright Dips

Some mild steels are commonly bright dipped to brighten the surface. The degree of surface shine achievable depends on such variables as the carbon content and the alloying ingredients. One solution useful for producing a luster on common steels is made up of 2.5 percent oxalic acid, 1.3 percent hydrogen peroxide, and a trace of sulfuric acid (0.007 to 0.010 percent), at a nominal 65°F. Compared to some other bright dips this is a slow action formula, but generally a safe one. Metal is removed at a rate of approximately 0.0001 in every 15 min. This rate is increased a bit by agitating the bath or raising the temperature to 105°F.

48.4.5 Dipping Copper Alloys

Copper alloys offer a different problem for metal finishers. Copper rich materials, like brass and bronze, are resistant to sulfuric acid while copper oxide is readily attacked. A popular bright dip for copper and its alloys uses 5 to 10 percent sulfuric acid solution at room temperature to remove scale. If the scale is especially heavy, heating the bath to about 170–180°F may be necessary. Pickling copper alloys can usually be done in 1 to 15 min. In cases where organic contaminates are involved the addition of a small amount of sodium dichromate will increase the speed and effectiveness of the pickling solution.

Most pickling baths do not produce a smooth or bright finish. Bright dipping of copper and its alloys requires the addition of an oxidizing agent to the sulfuric acid solution; many of these are proprietary and there will be some attack on the basis-metal as well.

Common bright-dip additives include chromic acid, dichromates, ferric sulfate, and nitric acid. The rate of attack depends on the relative amounts of the sulfuric acid and the oxidizing agent being used. The rate is generally fast. The rate can be further increased by raising the temperature of the bath and may be slowed down by the addition of inhibitors, i.e., 10 gm/l of sodium nitrate. The addition of up to 50 percent phosphoric acid tends to smooth better and to prevent the tendency to tarnish. These bright dip baths are usually limited to racked parts. For dipping part in bulk it is standard practice to dilute the bath with water. Some finishers prefer to bright dip bulk parts in a tank equipped with cooling coils to prevent the temperature from rising.

48.4.6 Dipping Aluminum Alloys

Aluminum is different from other common metals in that it is inert to oxidizing solutions. The more popular pickling solutions listed above meet these criteria. Mixtures of phosphoric acid, nitric acid, and glacial acetic acid work very well to remove oxide from aluminum; this includes anodize coatings. The thickness removed varies from 0.00004 to 0.0002 in/min depending on the alloy, the solution chosen, and the temperature.

48.4.7 Cleaning Metal Surfaces

In the preparation of metal parts for plating or anodizing, by far the most critical step is how to clean the parts. Cleaning affects the adhesion, appearance, composition, and corrosion resistance of the final deposit. It is therefore vital that the plater be given as much information about possible preexisting contaminations, such as inclusions in the base metal, the types and nature of the lubricants and cutting oils used in metalworking, the use or presence of synthetic organics or other coating such as cured paints, and the likely existence of inorganic films such as heat treat scale. Environmental regulations have severely restricted the cleaning options available to metal finishers, and procedures that were common a few years ago may now be illegal or very severely restricted. While the choices for cleaning materials are more limited, most soils and contaminates can be safely removed if the platers know what they are dealing with. Information supplied by the customer is the key to most successful metal finishing. Common contaminates include oxides and scale, buffing compounds, cutting oils, greases, machine oils, and the like. The finishes to be applied are not likely to adhere if the parts are not surgically clean.

In a majority of cases ferrous metals can be adequately cleaned with uninhibited alkaline compounds at fairly high concentrations. These cleaners are commonly used at a 6 to 8 oz/gal concentration, and at temperatures of 140 to 180°F. Copper alloys may be cleaned in the same solution but generally at a lower concentration and often with an inhibitor. It is critical for the metal finisher to be informed as to the nature of the soils expected to be found. Different classes of contaminants commonly require different cleaning solutions. Contaminants may be organic (man-made or natural), or inorganic.

Natural organic soils are regularly found on metal parts. Organic soils are mostly processing oils and greases, buffing compounds, cutting oils, protective oils, and mold lubricants or releases; some man-made and some not. In the pure, or nearly pure state most of these are easily removed. Most animal and vegetable oils are saponifiable, although the reaction time tends to be slow. These oils are, however, usually insoluble in water, but are soluble in polar solvents. Past practice was to remove organic greases and oils with solvents, commonly trichloroethylene—now largely illegal because of environmental concerns. Many cleaners available today are based on citric acid that seem to do a good job on this type of soil.

Inorganic soils may result from pickling, or other pretreatment and can be very difficult to remove. Inorganic soils also include metal chips and other debris left after machining, polishing, or buffing operations.

Compounded oils and greases, such as sulfurized mineral oil, sulfurized lard oil, synthetic moor oils, and extreme pressure lubricants are initially selected because they have the ability to adhere to the metal surface during extreme conditions. They are therefore hard to remove. The finisher may need to experiment to find the ideal cleaner for these materials, which will vary depending on the nature of the soils.

There is a simple test for cleanliness that anyone can perform called the *water break test*. This test is based on the ability of a clean metal surface to form and sustain an unbroken film of water. The clean part is immersed in clear, cold water and removed; if the surface shows and retains for 30 s a uniform film of water the part is deemed to be clean.

A modification of this test the *spray test* using pure deionized water in a light spray. If the surface forms a film instantly, flashes over, it suggests that some detergent or cleaner still remains on the surface. If a uniform film forms after 5 s it is assumed that the part is clean. If the water beads up on the surface the parts are still dirty.

48.5 ELECTROCHEMICAL PROCESSES

Other than mechanical finishes, electrochemical processes are probably the most common ways of finishing metal parts. The most common electrochemical processes are anodizing, electroplating, and autocatalytic (also called electroless) plating. Anodizing involves nothing more elaborate than inserting oxygen into the interstices of the basis metal. The latter two processes involve adding a generally thin coating of one metal on top of a basis metal surface. These include zinc on steel, copper-nickel-chrome on steel or a copper alloy, and the like. The section above may help the reader to better understand what is happening the plating process.

48.6 ANODIZING

The chemistry of aluminum (and its alloys) is fascinating. Since this metal is more reactive than iron one would expect it to corrode quickly. Surprisingly it doesn't, instead it forms an initial oxide layer that protects the metal underneath. Aluminum oxide is a tough material second only to diamond for hardness. The oxide layer formed in air is amorphous with only token strength. The white powder can easily be scraped off leading to further oxidation. Anodizing on the other hand is a process that forms a particularly structured, and dense oxide layer, which resists abrasion and thus protects the underlying metal. See Fig. 48.4 below. This layer is colorless, but it's possible to introduce a dye at one stage in the process to permanently color the surface. Not all aluminum alloys can be easily anodized; cast aluminum (usually with a high silica content) is particularly difficult.

It is important for the people doing the anodizing to know what alloy was used to make the part to be anodized. The nature of the aluminum alloy is important because of such factors as alloying elements, and their concentration or content. The more iron, magnesium, or silicon, the less likely the process is to work. Also, the product should be newly machined (for a shinier, more professional finish) or sandblasted (for a flatter, though less even, finish). All oil, dirt, glass beads, and the like, must be removed prior to the anodizing process, which can be done by thoroughly cleaning with special inhibited cleaners. After this bath, a smutty brown to black finish is obtained. To remove, rinse the part, an effective method is to pickle in 12 percent HNO_3 or 1 percent HF at ambient temperature for 5 min followed by a water rinse. Specimens known to have had embedded iron particles were found

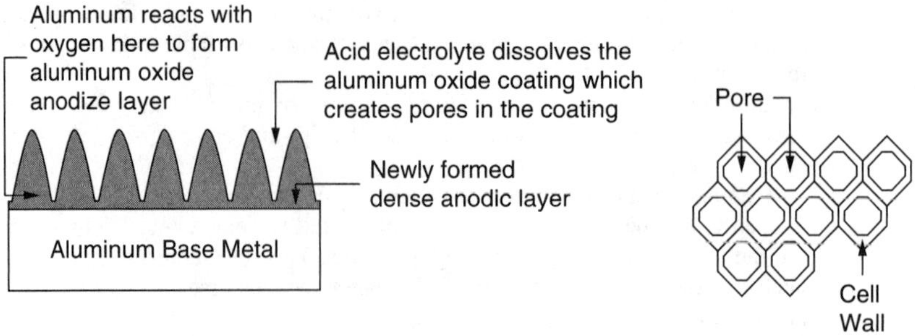

FIGURE 48.4 How anodize film is formed.

to be completely free of any surface iron contamination by the scanning electron microscope following this procedure.

Anodizing is an environmentally safe process producing a tightly adhering layer of oxide of aluminum 0.0005 to 0.003 in thick, resistant to corrosion and abrasion. In anodizing, two competing processes occur at once. At the positive anode, a tough oxide coating is formed which also dissolves in the electrolyte making it porous. The pores in anodize coating absorb dyes producing an integral color unlikely to chip or peel. With care, anodized articles can last a lifetime. Anodizing is corrosion resistance, (336 h per ASTM B117) and with a near diamond hard surface. Anodize hardness abrasion resistance is tested by the *Tabor Abrasion Test.*

48.6.1 Anodize Film Formation

Anodizing changes the dimensions of a part, so when designing parts to be anodized, allow for this change and specify coating thickness on blueprints and purchase orders (Fig. 48.5). Anodizing grows outward and also penetrates the aluminum, each by roughly one-half of the overall film thickness. The part grows by an amount equal to the total anodize film thickness. Anodizing is also nonconductive.

The nature of the anodize film, i.e., penetration, and build up, make stripping the film a very serious matter. A film thickness of 0.002 in on a cubic part exactly 1.000 in will leave a block of approximately 0.998 in after stripping. The coating can be stripped quickly in a solution of caustic soda or other alkalis, but the surface may be left pitted and etched. A safer, if slower, strip process is to use a solution of 40 g/L of phosphoric acid (75 percent) and 20–25 g/L of chromic oxide (CrO_3) at 190°F. This strip formula also works when it is necessary to determine the exact weigh of the coating. A density of 3 g/mL can then be used to calculate the film thickness.

48.6.2 Affect of Alloy Ingredients

Since the anodize film is the oxide of the basis metal, it follows that alloying ingredients should have an affect on the properties of these ingredients. The current efficiency for pure or nearly pure

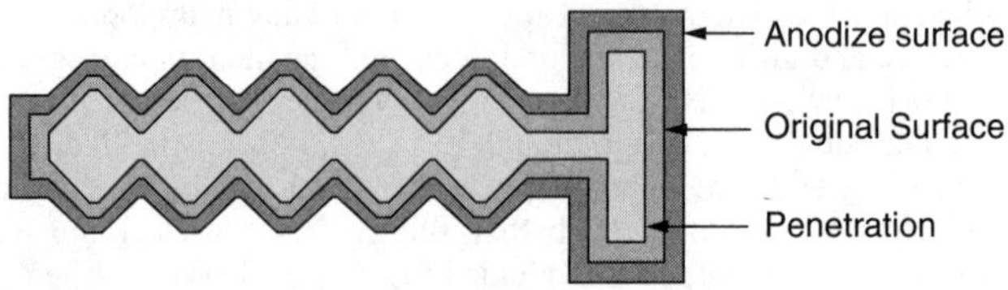

FIGURE 48.5 Anodize film build-up.

aluminum, 1100 for example, is nearly 100 percent. This current efficiency drops when alloying ingredients are added to create special need alloys.

Zinc, magnesium, copper, and manganese are common alloying elements. Alloy content of these metals at or about 1 1/2 percent have no great effect on the anodize process. Coppers at approximately 4 percent (2024) causes a drop in anodize efficiency by as much as 20 percent. In addition, when 2024 is used in a weldment with other alloys, it seems to draw current disproportionately. The 2024 will often burn up completely destroying the part. Silicon can be a serious problem if the parts are to be dyed. Since the dye forms *lakes* in the anodize cells and the silicon doesn't anodize the part will show white blotches wherever free silicon is on the surface. Dipping the parts in a bath of concentrated nitric acid with approximately 3 to 4 percent sodium fluoride added can solve this latter problem. The sodium fluoride removes only the free silicon from the surface.

48.6.3 Clear Anodizing of Aluminum

Aluminum oxidizes very quickly, forming an aluminum oxide coating that inhibits further oxidization. This coating is useless as is, in terms of protecting the metal, because it is so thin. Immersing the part in an electrolytic solution (165 to 180 g/L of sulfuric acid at 60 to 80°F) and passing an electrical current (at 10 to 15 A/ft^2) through it can produce a thicker coating. The resulting film is nearly colorless on some alloys and can be easily dyed because it is very porous at the molecular level. Then, by placing the part in boiling water, the film's pores can be sealed; the oxide changes to a hydrated structure as a result.

Sulfuric, or clear anodizing, has a silver, bronze, tan, or gray color, depending on the alloy. Sulfuric anodizing coating thickness runs from 0.0005 to 0.0015 in for wrought alloys and a bit less for cast alloys.

48.6.4 Hard Anodizing, "Hardcoat"

Sulfuric acid at 13 percent by volume, at a temperature of approximately 3°C, (50°F) applies a thicker, denser, and much harder anodize film. Note that while the temperature is not critical, it is important; it should be closely monitored. The current density should be about 15 A/ft^2. The aluminum piece should stay in a bath without additives for 2 to 4 h. This time can be drastically reduced by the addition of commercial hardcoat additives available from a number of suppliers.

Hard anodizing leaves the material more absorptive. For hardcoat anodizing the cathode, or negative lead, should be lead, with the anode-to-cathode proportion being about approximately 1:1. Hard anodize coatings should not ordinarily be sealed because sealing has a deleterious effect on the abrasion resistance of the anodize film. Hardcoat anodizing is more abrasion resistant than tool steel with an Rc in the 65–70 range, developing a thickness from 0.0020 to 0.0030 in on wrought materials and less on cast alloys. For maximum wear resistance hardcoat is not sealed. Teflon may be applied to hardcoat for a low friction surface.

48.6.5 Coloring Anodize Coatings

Dyeing is the next step, if desired. Since the pores are extremely small, many common dyes will not work. Some wool dyes are known to work, or commercial anodizing dyes are available from industrial suppliers. Typically, the dyeing process involves immersion in the dye solution, which is heated to improve their effectiveness. The commercial dyes are specifically designed for coloring anodize films and have much superior resistance to ultraviolet light.

As with many other features of plating and anodizing, some alloys (6062) dye better because the anodize film itself is colorless. In addition, essentially all chromic anodize film have a yellowish cast that affects the final color. Likewise, hardcoat tends to darken the anodize film on most alloys, and is often a poor candidate for dyeing.

Color matching is not work for a beginner. The exact shade produced is influenced by more than just the time, temperature and concentration of the dye. The specific alloy chosen and the condition of the anodize process as well. When all pieces of a set cannot be anodized at one time, or if some parts of the final assembly are die cast and others are wrought, or when different alloys are used, the possible problems multiply.

Iron in the alloy causes the colors to be duller; magnesium in the alloy can cause the parts to be streaky. Copper can make the colors more intense but can also dull some colors. Alloys with 5 percent or more silicon can only be dyed successfully with darker colors. Alloys, like 328, with 12 percent or more silicon are dark gray after anodizing and suitable only for being dyed black, and as mentioned above all silicon alloys should be dipped in a nitricbifluoride solution to avoid light spots.

48.6.6 Black MLW Dye

Black MLW dye is perhaps the most commonly used dye for anodized parts. This along with other dyes intended for dyeing anodize parts may be purchased from several commercial sources. Price varies with supplier and quantity purchased. Small samples or test quantities are usually available. Black MLW is favorable because it is both light fast in the visible spectrum and a rather flat black. Its drawbacks are that it is fairly expensive and a minimum order is typically 5 lbs. This dye requires special working procedures (the dust is hazardous to the lungs). Black MLW is used at a temperature between 51 and 60°C, for 5–10 min depending on quality of blackness desired. The concentration is 10 g/L.

48.6.7 Other Anodize Colors

The process of coloring anodized film is fairly simple. There are three fundamental methods for using organic dyes: first, the parts may be dyed cold, followed by sealing in hot (nearly boiling deionized water) water or steam. Second, the parts may be dyed at moderate temperatures (140 to 180°F) and then sealed in approximately 5 percent nickel acetate or nickel chloride solutions. The third method has fallen out of favor in the past few years. This method dyes in a boiling solution which ideally dyes and seals in one operation. The only problem with organic dyes is that the color fades when exposed to ultraviolet light, including both fluorescent and natural sunlight.

Some anodizers prefer to use pigments to color the anodized film. This generally involves using ferrous ammonium oxalate, which imparts a yellowing color to the parts. Another method is called the two-step process. The two-step process is an electrolytic process. A variety of color can be achieved ranging from a light champagne to a deep black. Most two-step color processes are proprietary. Other colors are achievable from specialty suppliers. The two-step color when done properly can achieve ultraviolet light stability as much as an order of magnitude greater than what is generally achievable using organic dyes. For outdoor applications this can be a critical advantage.

48.6.8 Sealing

Sealing is then done by putting the part in boiling water (preferable deionized water), which changes the film from gamma aluminum oxide into a hydrated form called boehmite. After anodizing and dyeing, if necessary, boil the parts for about 20 min. This seal can be improved through that addition of a trace (0.5 percent or less) of sodium dichromate. Other seals include nickel chloride, nickel acetate, and other commercial seals.

The sealing process hydrates the porous anodic layer, closing it, protecting it, and sealing in the dye. For most applications, sealing is desirable. The sealing process can be done with boiling water. Two to five percent nickel acetate work better than a hot water seal. The temperature of the seal bath should be between 70 and 87°C, with duration of 5–20 min, although for thicker coatings longer may help depending on how porous the coating is. Note that after each step, the object should be rinsed with deionized, or distilled water, both of which will work fine.

48.7 ELECTROPLATING PROCESS

The following is intended as an introduction to electroplating. It attempts to explain some of the basic science of the electroplating process for design engineers and others who may be less than totally familiar with the principles behind electroplating. It is not intended to be a manual for doing practical industrial electroplating, nor does it attempt to discuss all of the intricacies of real-world operations.

Electroplating is the deposition of a metallic coating onto an object by putting a negative charge onto the object and immersing it into a solution, which contains a salt of the metal to be deposited. The metallic ions of the salt carry a positive charge and are attracted to the part. When they reach it, the negatively charged part provides the electrons to reduce the positively charged ions to metallic form. Most plating problems that occur aren't plating problems at all; they are people problems. Plating bath chemistry is carefully controlled, as are load size, current, and consistency. Those who design parts to be plated (or anodized) must seek to broaden their own knowledge of the plating process so as to make informed decisions about finishing specifications.

48.7.1 Typical Electroplating Process

Platers immerse objects into a series of chemical baths in order to change their surfaces conditions. The number of tanks and their chemical make-up differs based on the desired result and every plating plant is different.

Electroplating protects, beautifies, insulates, or increases the corrosion resistance, conductivity, or solderability of metal objects, generally of iron or copper alloys, but sometimes other metals as well. The process involves dissolving metal at the anode and depositing it at the cathode. Direct current applied between the electrodes is conducted through the solution containing metallic salts. Because the ions depleted by the plating process at the cathode are replenished at the anode, the process is essentially self maintaining.

48.7.2 How Does It Work?

Designers must remember that things like polishing, pretreatment, and posttreatments may often be more critical than the electroplating step itself. In the real world, it is often not possible to get bright and acceptable plated coatings without the use of patented organic additives from commercial sources. Safety issues must be given due consideration, along with proper waste treatment and disposal, and a host of other issues (Fig. 48.6).

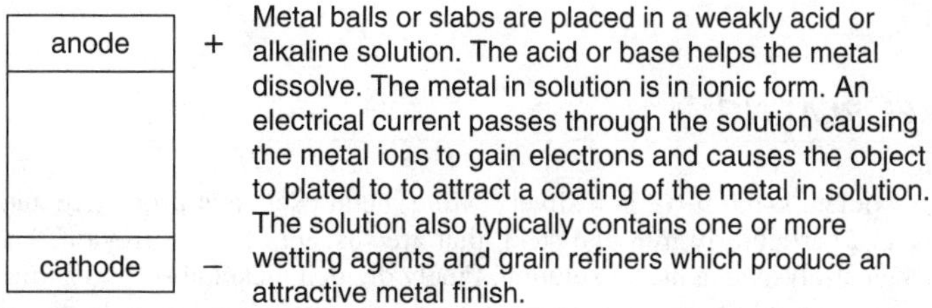

FIGURE 48.6 How electroplating works.

TABLE 48.1 Electrical and Magnetic Fields of Selected Elements Important to Electroplating

Ion	Half-cell potential
Mg^{++}	2.40
Al^{+++}	1.70
Zn^{++}	0.827
Cr^{++}	0.762
Fe^{++}	0.441
Ni^{++}	0.231
Sn^{++}	0.136
Fe^{+++}	0.045
H^2	0.000
Cu^{++}	–0.344
Ag^{++}	–0.798
Au^{++}	–1.360

Any element will replace any metal below in the series. Elements top to bottom are progressively more noble.

Imagine that an object made of steel has been appropriately cleaned, and that it is to be plated with copper. A wire is attached to the object, and the other end of the wire is attached to the negative pole of a battery. To the positive pole of the battery we connect another wire; the other end of this wire is connected to a rod made of copper. Now the cell is filled with a solution of a salt of the metal to be plated. It is theoretically possible to use a molten salt, and in rare cases that is done, but most of the time the salt is simply dissolved in water. The $CuCl_2$ ionizes in water to Cu^{++} and two parts of Cl^-.

The object to be plated is negatively charged, it attracts the positively charged Cu^{++}. The Cu^{++} reaches the object, and electrons flow from the object to the Cu^{++}. For each atom of Cu^{++}, two electrons are required to neutralize it or reduce it to metallic form. Meanwhile, the negatively charged Cl^- ions are attracted to the positively charged anode. At the anode, electrons are removed from the copper metal, oxidizing it to the Cu^{++} state. Thus the copper metal dissolves as Cu^{++} into the solution, supplying replacement copper for that which has been plated out, and we retain a solution of copper chloride in the cell.

Copper chloride is used in this example for reasons of simplicity. It is not recommended that copper be used for experimentation or demonstration because some people are quite allergic to it; nor is it recommended that chloride salts be used, because it is possible to release chlorine gas from them. See the material that follows. Many of the items that are precious metal plated first had their surfaces mechanical polishing or mass finished. The first step after polishing is to clean and activate the basis metal so proper adhesion and defect-free appearance may be realized. Cleaning, degreasing, and activation are carried out in aqueous solutions, utilizing a variety of alkalis, surfactants, detergents, and other, usually commercial, cleaning agents. The surfaces are normally activated with dilute mineral acids (see Table 48.1).

After cleaning, most of the items that are to be precious metal plated are first coated with a strike of copper or nickel.

48.8 NICKEL PLATING

Nickel, symbol Ni, is a silvery white metal used as a protective and ornamental coating for metals, particularly iron and steel, that are susceptible to corrosion. The nickel plate is deposited by electrolysis in a nickel solution. Finely divided nickel absorbs 17 times its own volume of hydrogen and is used as a catalyst in many processes, including the hydrogenation of oils.

Nickel plating is one of the most versatile of all finishes. Its applications encompass decorative, engineering, and electroforming. Nickel plating is a yellowish-white, hard, reflective finish—used for wear resistance, solderability or dimensional restoration. Nickel plate is often applied over copper and under chrome for a decorative finish. Nickel is a very hard metal with relatively poor ductility, consequently, parts to be nickel plated should be bent into final shape before plating whenever possible. Nickel may be specified by QQ-N-290-Nickel Plate, class 1–"Decorative", types I to III are for steel parts and Types V to VII for copper base parts and B 456 , Copper/Nickel/Chrome for a variety of *service classes* (SCs) which vary with alloy and base metal.

Bright nickel plating is a highly reflective finish, which often eliminates the need for subsequent polishing. The brightest finishes result, however, from first polishing the base metal substrate. The brighter the ensuing nickel plating the greater the internal stresses and less ductility. It is best to avoid specifying bright nickel if the parts are to be bent or crimped after plating.

Semibright nickel has a more satiny finish than bright nickel and may be marginally more ductile. If heat shock or minor bending of the parts is anticipated, it would be better to specify semibright nickel so as to reduce the risk of the plating flaking off.

Nickel plating is characteristically brittle. When brighteners are added to make the finish brighter, the finish also becomes more brittle and flakes off when bent, heated, or otherwise stressed. The conventional bath with very low brighteners loses somewhat in brightness but is less brittle. The watts

	Nickel plating baths		
	Bright nickel	Sulfamate nickel	Watts nickel
Nickel sulfate			225–300 g/L
Nickel ammonium sulfate	315–450 g/L	500–600 g/L	
Nickel chloride	5–15 g/L	0–22 g/L	37–45 g/L
Boric acid	30–45 g/L	30–45 g/L	30–45 g/L
Temperature	32–60°C	60–70°C	44–66°C
Agitation	Air or Mechanical	Air or Mechanical	Air or Mechanical
Current density	up to 90 A/dm^2	0.5–32 A/dm^2	3–11 A/dm^2
Anodes	Nickel	Nickel	Nickel
pH	3.5–4.5	3.5–4.5	3.0–4.2

nickel tends to be less bright and also somewhat more flexible. The sulfamate nickel bath is very low stress and therefore is used on parts that may need to be bent after they are plated. Extremes of heat or mechanical stress may still result in flaking and peeling.

Nickel plating is, strictly speaking, mechanical in its protection. If the coating is scratched or punctured the basis metal is immediately susceptible to oxidation, or corrosion. While the nickel provides good corrosion resistance, once the coating is breached, the basis metal will begin to oxidize. In addition, nickel plate tends to contain many microcracks, which also tend to lower its corrosion resistance. Applying a good strike or multiple layers of nickel tends to seal many of these microcracks, extending the life of the coating.

48.8.1 Decorative Nickel Plate

Decorative bright nickel is frequently used as an undercoat for precious metal coatings. The nickel undercoat provides brightness, enhanced reflectivity, leveling, better corrosion resistance, reduced porosity, and acts as a diffusion barrier to prevent the base metal from migrating into the precious metal topcoat.

48.8.2 Plating on Stainless Steel and High Nickel Alloys

Plating on stainless steel may be done, after suitable pretreatment, by any method listed here. Stainless steels and high nickel alloys form a tightly adhering passive oxide film within minutes of being plated.

Stainless parts can be plated with other metals if a fresh flash of wood's nickel is applied, providing an active surface for subsequent plating. ASTM B-254 addresses plating on stainless steel.

A *wood's nickel strike*, the exact solution used may vary considerably, what is important is that the ration of nickel chloride to hydrochloric acid remains at essentially a 1:1 ratio.

Wood's Nickel Strike. (Nickel chloride16–64 oz/gal, Hydrochloric acid 8–32 fl. oz/gal). This strike is best applied at room temperature using nickel anodes and a current density of 20–200 A/ft^2. Other methods of activation will work, but the wood's nickel strike is generally considered to be foolproof.

48.8.3 Nickel Contact Dermatitis

Many people are allergic to nickel. They develop to varying degrees a form of contact dermatitis. As a result attempts have been made to limit the use of nickel in products that will come in contact with the skin. The European Community (EC) has led the way in this effort.

The use of nickel in components that come in contact with the skin, or are inserted into the human body have been reviewed and subjected to legislation in the EC. Early in 1994, the EC Council of Research Ministers adopted Directive 94/27/EC, limiting the use of nickel. The Directive was implemented at the end of 1997. Every member of the EC has adopted the Directive. The Directive forbids the use of nickel in the following applications:

1. Post assemblies that are inserted into pierced ears or other human body parts, unless the nickel content is less than 0.05 percent.
2. In products intended to come into direct contact with the skin, e.g., earrings, necklaces, bracelets, anklets, snaps, fasteners, zippers, if the rate of release is more than 0.5 μg/cm^2 per week.
3. In products such as those that have a nonnickel coating (a lacquer or heavy gold electroplate for example) whose rate of nickel release will not exceed 0.5 μg/cm^2 per week for a period of at least 2 years of normal wear.

48.9 ZINC PLATING

Zinc Plating is a soft ductile, decorative, marginally solderable, corrosion resistant finish. Even before zinc was plated onto other metals it was widely used as a coating on wire, roofing, and other materials for corrosion protection. The metal parts were, and still are, dipped into a bath of molten zinc. The process is called galvanizing. Some people refer to zinc plating as electrogalvanizing.

Zinc, because it is the most reactive of all metals commonly plated, is not in the strictest terms a protective coating. It is rather a sacrificial coating. When exposed to a corrosive atmosphere, the zinc coating corrodes preferentially to the ferrous basis metal on which it has been applied. Ideally as long as there is some zinc left the coating corrodes rather that the basis metal. Unlike most other commonly plated metals, zinc protects the substrate by sacrificing itself and thus corrodes before the base metal. This means that zinc will protect even if the zinc coating sustains minor damage, such as scratches of small punctures, which is often an advantage that other types of plating do not offer. Zinc is the most reactive of all common metals; however, any metal may be attacked or dissolved by ordinary liquids such as soft drinks and vinegar. The ultimate corrosion resistance of zinc is a function of the plating thickness. To increase the corrosion resistance of zinc a conversion coating is usually added.

Electroplaters have historically been able to produce bright zinc finisher with baths that were in chemical balance. There have been chiefly three main types of zinc plating baths. The oldest of these has been the *zinc cyanide bath*. Historically, most plating solutions, including zinc, contained cyanide. In 1990 environmental regulations made it advisable, if not mandatory, to find alternative plating baths without cyanide.

Chloride zinc has been in use since about 1980, when environmental pressures began to strongly suggest or even demanded the replacement of cyanide baths with noncyanide baths. *Chloride baths* have high cathode efficiency and excellent throwing power, because of the high cathode efficiency over the entire current density range.

The three most common types of zinc plating baths are listed below.

	Bath composition in oz/gal						
	Zinc metal	Sodium hydroxide	Sodium cyanide	Ammonium chloride	Potassium chloride	Boric acid	pH
Cyanide zinc	1–1.5	10–12	1.5–2.5				
Ammonium chloride	2–4.0			16–20			5–6.0
Alkaline noncyanide	.08–3.0	10–20					
Nonammonium all potassium chloride	3.0–5.0				25–30	3–5	5.0–5.5

These baths are typical but not necessarily ideal. The various suppliers of plating materials offer more specific recommendations. Wetting agents, brighteners, grain refiners, and other additives can modify the baths and improve the overall quality of the finish. It is recommended that the user at the least explore what the suppliers have to offer. Each of the classes of baths has some common advantages and some limitations. The plater should be willing to examine carefully what is available.

These baths, with or without the proprietary additives, differ in certain special ways. Cyanide zinc baths tend, in general, to be more forgiving in terms of cleanliness of the parts themselves. The cyanide dissolves most common metals and even gold, but it is limited somewhat be carbonitrite heat treat scale. It also does a poor job of plating over a steel part that has been decarburized. Decarburization occurs when the heat treat furnace is opened before the steel has cooled down sufficiently. The inert atmosphere escapes and oxygen enters the furnace and reacts with the carbon in the steel.

Acid chloride zinc throws somewhat better than cyanide zinc and therefore covers better, especially in crevices and recesses. This makes for slightly higher cathode efficiency and works rather well in barrel plating. The solution seems to be wetter than water and therefore there is a bit more drag-out with a chloride than with cyanide. Plater can use a drip tank and/or an evaporator to minimize the problem of excess solution build-up. The higher metal content of the bath means that there may be more metal in the waste treatment system.

Alkaline noncyanide zinc baths throw very well and have a high cathode compared to the anode efficiency. This means that in many installations the zinc tends to deplete. In some shops this means making more frequent additions, or having an offline zinc generator. Once the plater learns to balance the system and maintains it properly, the system works very well.

Ammonium chloride zinc plating baths, generally, work well and have excellent throw which would seem to make them the zinc bath of choice. The only problem is that ammonium forms complexes with many different metals and so its presence complicates waste-treatment.

Almost all zinc plating is treated with a posttreatment of one sort or another. When the part is to be painted, it is common to add a posttreatment of iron phosphate which not only helps to protect the base metal and the zinc coating, but also makes an excellent base for the paint. If the parts are not to be painted, a coating of chromate is usually added. The specific chromate used is a function of the end use, the environment in which the parts are expected to function, and to a degree cosmetic.

Zinc without a chromate is typically good for a couple of hours before white corrosion appears (zinc oxide) and as much as 8 h before *red rust* appears (iron oxide). A clear or blue chromate is good for about 12 h to white and 24 h salt spray resistance to red rust compared to 98 h for yellow chromate and 150 h for black or olive drab chromates. Yellow is the most common chromate applied, but most yellow chromates contain hexavalent chrome which is a known carcinogen.

Many new developments have come along in recent years, at least one commercial blue chromate has been marketed recently that does not contain hexavalent chrome and show promise of better corrosion resistance that the yellow, black, or olive drab chromates. There are also several post chromate dips available which offer even better corrosion resistance. Some of these products are solvent based, which may be unacceptable for their release of volatile organic compounds (VOCs). Other that are water based add some protection but generally not as much as the solvent based coating.

48.9.1 Plating on Aluminum

Since aluminum surfaces oxidize almost immediately, the first step in plating aluminum involves removing that oxide layer. This entails etching the surface and providing a fine etch finish. Because of the resistance of aluminum to acids, the etchant is alkaline. Immerse the aluminum part for 1–3 min in a hot (160–180°F) solution of 3 oz/gal of sodium carbonate and 3 oz/gal of trisodium phosphate, or a 30-to 60-s dip in a 5 percent sodium hydroxide solution at 120–140°F.

The caustic dip leaves a smut on most alloys. Parts made of 6061 alloy produce practically no visible smut, while the smut on 7075 is deep black. The finisher should note that in weldments this color discrepancy is a sure fired sign poof mixed alloys, which could cause burning during the anodize process.

To remove this smut and any other metallic impurities on the surface left by the caustic, an acid dip follows. The acid dip removes both the smut and any metallic surface contaminants. The make up of the dip depending on the alloy being used, as follows:

	Acid dips for aluminum (Desmuts)				
	Sulfuric acid (%)	Nitric acid (%)	Hydrofluoric acid (%)	Time in seconds	Temp.
Pure aluminum		50		30–60	Ambient
Series 3000 & 5000		50		30–60	Ambient
Series 2000 & 6000	25			60–120	180–190°F
Rinse & follow with		50		30–60	Ambient
Series 7000 & high					
Silicon alloys		75	25	30–60	Ambient

After the acid dip the surface must be protected from oxidation until the plating can be accomplished. The most popular way of doing this is to use a zincate, which applies an immersion coating of zinc on the clean surface.

	Zincate solutions (oz/gal)					
	Sodium hydroxide	Zinc oxide	Ferric chloride hexahydrate	Sodium potassium tartrate	Sodium nitrate	Immersion time
Type 1	16	2.7		6.7	0.13	30 s
Type 2	67	13	0.3	1.3		30–60 s

There are also several good zincates on the market for those who prefer, commercial bronze solutions and even a dip into a noncyanide bath are reportedly successful.

There are two approaches to using a zincate immersion coating. The first is to employ a single dip. The parts are dipped into the zincate, briefly dipped in clean, clear running water and then dipped into a copper strike. It is important that the copper strike have a pH of less than 10.5. Above this pH the copper bath dissolves the zinc faster that it can plate by the strike, and so it replaces the zincate with an immersion copper which itself has poor adhesion. The second method uses a double dip in the zincate. The first dip is applied as usual, but is then dipped into a 50 percent nitric acid bath, followed by a second zincate dip. When nothing else works to get a good surface for the copper strike, the two-step method works.

After the copper strike any other type of plating can be applied. Another caution, when the overplate is tin it is important to know if the part will subsequently be soldered or spot-welded. In the case of welding a thickness of tin should be between 0.0001 and 0.00025 in for good solderability, for spot welding, however, it is better for the tin to be on the thin side. Because of the high heat conductivity of aluminum a heavy tin can cause the welding tip to float this causing a poor weld.

BIBLIOGRAPHY

Dini, J. W., *Electrodeposition: The Materials Science of Coatings and Substrates*, Noyes, Park Ridge, NJ, 1993.

Durney, L. J., *Electroplating Engineering Handbook*, 4th ed., Van Nostrand Reinhold, New York, 1996.

Metal Finishing Magazine, *Metal Finishing Handbook and Directory*, published annualy.

Schlesinger, M., and Paunovic, M. (eds.), *Modern Electroplating*, 4th ed., Wiley, New York, 2000.

Wernick, S., Sheasby, P. G., and Pinner, R., *The Surface Treatment and Finishing of Aluminum and Its Alloys*, 2 vols., 5th ed., ASM International, Materials Park, OH, 1987.

CHAPTER 49
COATING PROCESSES

Rodger Talbert
R. Talbert Consulting, Inc.
Grand Rapids, Michigan

49.1 INTRODUCTION

Virtually every durable good product manufactured has a coating on its surface: automobiles, computer cabinets, televisions, bicycles, furniture, toys, and thousands of other products. A coating provides a decorative or protective finish on a metal, plastic, or wood product. Coatings are also applied for insulation, to reflect light, or conduct electricity.

An average person can see the decorative value of the finish and many times a product is purchased for its color. Resistance to the elements is a less obvious property of the coating but consumers do appreciate that a product stays cleaner, lasts longer, and does not rust because it is protected by an industrial coating. There are many types of coatings manufactured to provide unique properties to a specific product. Some coatings are pigmented to provide special colors and some are clear, applied to provide high gloss or impact resistance. Proper selection, handling, and application of a coating require understanding of the different materials available and their unique qualities.

49.2 COATING CLASSIFICATION

Coatings may be classified by their function, curing or application characteristics, the resin system used to formulate them, or their composition.
Coatings classified by function:

1. *Undercoat.* A coating film beneath the topcoat.
2. *Primer.* An undercoat applied directly to the surface being prior to application of a topcoat. Primers are usually used to increase the adhesion of the next coat or to increase corrosion resistance, or both.
3. *Sealer.* Usually an undercoat that enhances the adhesion between any two coatings. Sometimes a sealer is used to prevent movement of pigment or oils into an upper coat.
4. *Surfacer.* An undercoat used to fill surface imperfections. Surfacers are usually designed for easy sanding.
5. *Topcoat.* The final and actual coat that meets the eye.

Coatings classified by curing or application characteristics:

1. *Air-drying.* Liquid coatings that do not need heat to cure. House paints and some automotive repair products are examples.
2. *Thermal cure coatings.* Coatings that must be heated to cure.
3. *Electrodeposition.* A liquid water-based coating applied by dipping the part into a tank and applying voltage to the solution.
4. *Autodeposition.* A liquid water-based coating applied by dipping the part into a tank where a chemical reaction takes place and deposits the coating.

Coatings classified by resin system:

1. *Two component liquid paints.* Paints with two reactive components that are mixed just prior to use.
2. *Lacquers.* Coatings that dry by evaporative loss of solvent. No reaction takes place.
3. *Enamels.* These are coatings that involve a chemical reaction between binder materials resulting in the formation of a solid film. Most high gloss coatings are called enamels.
4. *Latex.* A liquid paint that is applied as a waterborne emulsion. Drying involves evaporative loss of water and the merging, or coalescence, of the paint particles to form a film.

Coatings classified by their composition:

1. *Conventional solvent based liquid paint.* Paints with less than 50 percent solids by weight, more than 50 percent solvents. All liquid industrial coatings were made this way in the past.
2. *High solids liquid paint.* Paints that contain more than 60 percent solids by weight. The lower solvent content helps the coater meet environmental regulations.
3. *Waterborne paint.* Paints that are dissolved or suspended in water. Like high solids, these coatings are designed to reduce solvent content.
4. *Plural component paints.* Paints that have two or more ingredients blended with a catalyst just before application.
5. *Powder coatings.* Dry-blended coatings that do not use solvent or water in the formula.

There are a variety of different finishing processes used by industry to provide protection for a product, a decorative appearance, or both. An understanding of the coatings, the application process, the basic equipment, the characteristics of the raw material, the finished paint film, and the advantages and limitations of each coating is important to manufacturers.

In addition to understanding the coating process, it is necessary to know the environmental issues that govern the finishing industry. Most manufacturers have always been concerned with the performance and cost of the coatings they applied. In today's environmentally sensitive society, they must be equally aware of the compliance issue. Environmental regulation has a major impact on the development and application of coatings for industry.

No two finishing systems are exactly alike. As a result, each time a system is built, the people involved must make decisions about the type of coating to be used, the particular equipment to be used, and the design features. To make sure that the system works as expected, the factors that affect the operation must be carefully defined and evaluated.

49.3 FINISHING SYSTEM PROCESSES AND DEFINITIONS

The fundamental components for finishing include some method of part preparation, an application system, and a method of drying or curing the coating.

49.3.1 Part Preparation

Preparation of the part surface is critical to coating adhesion and performance. Surface preparation involves filling joints or holes, smoothing the surface, removing dirt and grease, and sometimes doing surface conversions such as phosphating. The surface preparation process prepares the surface for good coating adhesion and enhances the appearance and performance of the finish. Mechanical methods include sanding, grinding, blasting, and tumbling. Chemical methods include solvent wiping, vapor degreasing, dip washing, aqueous cleaning, and spray washing.

49.3.2 Application Systems

Coatings can be applied by dipping the part into a tank or by spraying the coating onto the part. Some operations dip the part into a tank of paint, remove the part and allow it to air-dry. These are parts that do not require particularly high quality coatings. Dip systems for higher quality primer or finish coat include electrocoat and autodeposition. Spray application is the most common because it provides a high quality finish with flexibility for multiple colors.

49.3.3 Drying and Curing

Some solvent-based and waterborne liquid coating materials, referred to as air-dry coatings, will dry without being exposed to heat or light energy. Some liquid coatings require exposure to heat or light energy to chemically react and cure. Systems designed for coatings that require thermal or light energy to cure will include a cure oven. Air-dry coatings may also use an oven for accelerated drying or more consistent process times.

The coatings and process methods used are determined by the type and volume of product, the cost per applied square foot, and the quality requirements for the product.

49.3.4 Product Definition

The size, shape, and volume of parts to be coated will influence the type of coating to be used, the size of the equipment, and degree of automation used. Different part preparation methods, coating technologies and types of systems can be configured for a specific situation.

The product quality is usually related to the end use. The expected quality of the part in the field will influence the type of pretreatment used, filtration, and many other factors.

The product size and volume usually defines the scope of the finish system. Once the clearance requirements, production rates, and quality standards are defined, the size and nature of the system can be determined.

49.3.5 Process Definition

A finishing system is made up of several discreet subsystems. The steps that make up a typical process are the following:

- Load and unload space
- Cleaning system
- Dry-off oven (if the part is cleaned by an aqueous process)
- Cool down area (if the part is subjected to a dry-off oven)
- Application area
- Cure oven

49.3.6 Load and Unload Space

The area allotted for loading and unloading the parts must have adequate time and space. A comfortable amount of time must be allowed for load and unload and the staging area must be large enough for parts, racks, loaders, and free movement in and out of the area. It is also critical to consider ergonomics and worker safety. Lifting heavy loads onto a moving conveyor is challenging and manufacturers need to provide appropriate devices and enough people to avoid damage to products personal injury.

49.3.7 Cleaning

The part surface has to be clean and dry for good coating adhesion and performance. Mechanical cleaning methods remove dirt and scale by using abrasion. For example, a batch pressure blast system is commonly used for removal of old paint, grease, and mill scale for steel parts prior to finishing. The blast media removes organic and inorganic soils and the roughened surface provides a good anchor pattern for paint adhesion.

Chemical cleaning uses an alkaline or acidic product in a dilute aqueous solution to loosen and remove organic soils. Chemical cleaning is often done in multiple stage spray washers that clean and rinse the part. A conversion coating is often used to alter the normally reactive surface of metal parts, applying a thin, nonmetallic layer on the surface that enhances adhesion and performance. Conversion coatings are usually iron or zinc phosphate materials.

49.3.8 Dry-Off Oven

Drying a freshly cleaned part must be done quickly and efficiently. Drying is accomplished by the use of air and heat for a certain time in a dry-off oven. Air can be directed to the part with a fan in the form of a blow-off or as a part of the dry-off oven. The air will break up pockets of water and improve the drying process.

49.3.9 Cool Down Area

Between the dry-off oven and the application area the part must be cooled to near ambient temperature. This can be accomplished by adequate conveyor travel or forced cool down. Determining how to cool the part depends on the part mass, temperature of the part exiting the dry-off oven, and space available for cooling.

49.3.10 Application Area

Design of the application area begins with the coating supplier. They supply the necessary information about the number of coats required, flash times, and characteristics that determine how the application area will be laid out. The equipment suppliers can determine how to accomplish the process.

49.3.11 Cure Oven

The cure oven design is related to the required time at temperature as supplied by the coating supplier. Once the length of time in the oven has been established, the configuration will be determined by the part size and available space.

49.4 FINISHING SYSTEM DESIGN CONSIDERATIONS

The design considerations for a finishing system include environmental considerations, racking, controls, supply of coating materials, and ventilation.

49.4.1 Environmental Considerations

In today's environmentally sensitive atmosphere, air quality, waste water, and paint overspray are important considerations in the design and layout of a finishing system. Waste treatment and abatement equipment are a normal part of today's finishing system. The customer will have to determine what types of wastes will be generated and in what volume. Next, they must consult with the Department of Natural Resources (DNR) to determine what abatement requirements they must meet. With this information, the equipment supplier can determine the pieces of equipment required and their arrangement.

49.4.2 Racking

How the parts will be racked is important to the designer. The system designer must be sure that the rack will fit through the system and that parts will not be blown off in the washer or dry-off. If the part is double hung, the designer must make sure that it will make it through the conveyor turns.

49.4.3 Controls

Together the system designer and the customer will decide what type of controls is required. Controls should be simple, effective, and safe. Many times it is useful to have output from the controls such as temperature readouts, pressures, cycle times, and cycle recorders. In all cases, controls must be reliable and safe.

Programmable Logic Controller (PLC) controls can be useful for tight control, documentation, and to integrate operation features of the system. Changes to operating parameters and system monitoring can be done from one central location. PLCs can provide flexible control and they may not cost much more than relay control. The decision to use a PLC should be on real need for the benefits. On smaller, more basic systems, relay control is usually less expensive, simpler to operate, and perfectly reliable.

49.4.4 Supply of Coating Material

The paint storage location and method of delivery to the application area is part of the design process. Later in the text, we will deal in depth with paint storage and circulation. The method of delivery and its impact on production should be considered very carefully.

49.4.5 Ventilation

Washers and ovens will produce heat gain in the building. Part of the system plan should be how to evacuate the heat and make up the air that is exhausted from the system equipment. If the system is for powder coating, dip, or similar nonexhausting application area, some ventilation fans should be installed and a properly sized air make up heater installed to create enough air turnovers to relieve heat gain. If the system is a liquid spray, the exhaust volume and air make up will provide adequate ventilation.

Keep the system as simple as possible. Make sure it will handle requirements; fit into the budget; fit into the building; and be accessible, maintainable, and a good value. Do not add controls that complicate the operation without adding obvious benefit. Make sure the system is flexible but keep it sized to the primary product.

49.5 COATING METHODS

There are many different types of coating and many different ways to apply them. Industrial coating use is heavily influenced by quality, cost, the end use of the product, and environmental issues.

The most common coating types and processes include conventional solvent-based coatings, high solids, water-based, plural-component coatings, powder coating, electrodeposition, and autodeposition.

49.5.1 Conventional Solvent-Based Paint

Historically, liquid paint contained over 50 percent solvent. The solvent is used to blend the ingredients and keep them wetted during application and flow-out. Using a variety of different solvents the paint formulator can create different appearance and performance characteristics to meet a specific application.

These coatings are usually applied by spray equipment and curing may be air-dry or thermal. Pretreatment standards are not as high as they are with some coatings because the solvent can dissolve some soils and still provide good adhesion. Solvent-based, air-dry coatings are often used for assembled parts that are heat sensitive or field painting where an oven is impractical.

Although solvents are useful for the paint manufacturer and applicator, they have been linked to pollution issues. The pollution of air from combustion and industrial activities has created a need to control emissions of solvents. Certain solvents, because of their molecular structure are highly reactive and contribute to smog and related problems. These problems include eye irritation, irritation of the respiratory tract, vegetation damage, and reduced visibility. Legislation has been created to reduce the impact of solvents on the atmosphere. The West Coast, with Los Angeles (Rule 66k) being the forerunner, has started a trend that has spread across the country and exists in every major metropolitan area through state and local regulations. The original air quality classification for organic solvents is photochemically reactive or nonphotochemically reactive.

The current Environmental Protection Agency (EPA) regulations address air quality by defining allowable emission levels for variouFs industries and finishing operations. *Volatile organic compounds* (VOC) measured in pounds per gallon of coating and tons of emissions in a unit of time, is used in the current EPA regulations. In addition, *hazardous air pollutants* (HAPS) are listed solvents facing additional restrictions.

The Federal Environmental Protection Agency has established national emission standards for HAPs as part of the Clean Air Act.

Conventional solvent-based paints are still used in limited volume situations or field painting in North America and elsewhere. In some areas of the world where raw materials are harder to come by or pollution pressure is modest they may still be the coatings of choice.

49.5.2 High Solids

High-solids materials are solvent-based paints that contain a higher volume of solids, and therefore less solvent, than traditional solvent based formulas. High-solids coatings are typically applied by spray guns in a finishing system that features a spray-washer, dry-off oven, application area, cure oven, conveyor, and a load/unload area to stage the parts in and out of the system.

The spray washer is usually at least five stages of process. The standard five-stage washer includes a cleaner, rinse, phosphate solution, and rinse/deal. Additional stages are often required to meet the finished product performance standards and to assure removal of all surface contaminants.

These coatings are applied at relatively thin film thickness (0.75 to 1.5 mils) and they are intolerant of surface contamination. Deionized water is often used for the final rinse to produce a super clean part. After the parts are washed they are dried in an oven.

Application of high solids varies widely. The application area may be two booths, one for the "A" side and one for the "B" side of the part, with one manual sprayer in each booth, or it may involve as many as 10 or 11 booths with a combination of manual spray, automatic spray, and robotic spray. The variables considered include:

- Does the process require one, two, three coats, or more?
- How much flash time is required between coats?
- Is it necessary to cure between coats?

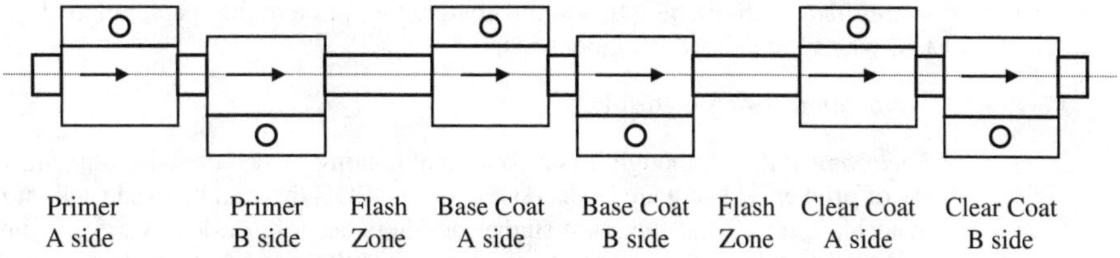

FIGURE 49.1 Base-coat/clear-coat application area.

The behavior of the coating and the requirements of the customer will impact the design of the application area. The coatings supplier should be consulted when designing the application area. A typical base-coat/clear-coat application area is shown in Fig. 49.1 below.

The cure oven for a high-solids line is usually a convection type. Cure cycles vary widely dependent on the material being applied and the substrate. For a steel substrate, a typical cure cycle would probably fall between 15 and 30 min, and 250 and 350°F. A typical time/temperature curve for high-solids is shown in graph below (Fig. 49.2).

The VOC content of high-solids paints is defined in pounds per gallon. For example, 3.0 VOC paint contains 3 lb of solvent in one gallon of paint. Coatings manufacturers are continually developing materials with higher-solids content, working to keep ahead of the regulatory requirements. Materials with 75 to 80 percent solids are common and testing is being conducted on 100 percent solid materials.

For the company that is currently applying liquid coatings, high-solids provides an excellent short-term compliance alternative. Some changes must be made to deal with the lower solvent content, but the basic technology is still the same.

Spray equipment has been developed, new guns, gun tips, delivery systems, and paint heaters to help the applicator deal with the viscosity of the high-solids materials.

Advantages using high solids:

- *Color.* Color variety is virtually unlimited and color match is relatively simple and quick. Color changes can be made in seconds simply by flushing the fluid path with solvent and compressed air.
- *Appearance.* Solvent based coatings provide a very smooth, glossy, and attractive surface at thin film thickness.
- *Function.* Masking of surfaces for decorative coatings is relatively simple; touch-up and repair of defects is fairly simple.

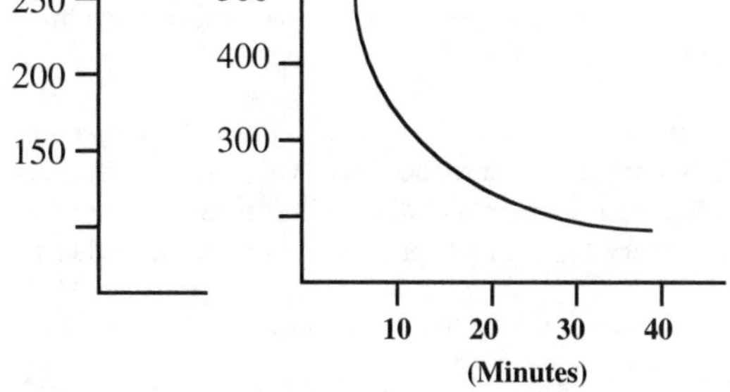

FIGURE 49.2 Time/temperature curve for cure of high solids.

- *Compliance.* Switching to a liquid coating that has a higher percentage of solids can be an excellent way of meeting EPA guidelines.

Limitations using high solids:

- *Environmental.* Although it is a compliant coating, it still contains solvent, which creates a variety of problems. There are limits on the tons of VOC that can be exhausted into the atmosphere, the material is toxic, and the paint sludge or filters are a hazardous waste. A limit on emissions can eventually lead to a limit on production or the purchase of abatement equipment to stay within emission standards. Toxic materials and hazardous wastes are becoming increasingly more expensive to deal with. The "cradle to grave" concept makes the disposal of hazardous wastes very undesirable.
- *Air make-up.* The air that is exhausted to atmosphere from the spray booths must be replaced in the plant with tempered air to provide proper ventilation. Also, the oven exhaust volume is greater than required for lower VOC coatings such as waterborne or powder coating. Air make-up is a significant cost.
- *Handling.* In general, the higher solids content makes these materials more difficult to handle, harder to pump, harder to atomize, and harder to clean up after than paints with a lower solvent content. To overcome the poor flow characteristics, some changes are required in the types of guns and other application equipment; and heaters are often required to achieve proper spray viscosity. Since the material does not air dry, the guns, booth interior, and filters are very messy and difficult to clean up.

49.5.3 Waterborne Coatings

Waterborne coatings are similar to solvent-based coatings in some ways but they substitute water for a large percentage of the organic solvent. They are usually spray applied, although some are applied by dip. They were originally developed to reduce the solvent content of liquid coatings. Waterborne coatings contain less solvent and produce a lower volume of VOC in the exhaust stream than High Solids. In many ways, they provide the same advantages of other liquid spray applications. Many coaters have successfully converted to water-based materials, especially in automotive base coats. They still use some solvent in the formulation and they require some special features in the application area and the fluid handling equipment. Flash times are longer due to the slower evaporation rate of water as opposed to solvent. The fluid delivery system must be all stainless steel to avoid corrosion. Film build characteristics are unique and may require different booth and spray equipment arrangements than solvent materials. The same basic system equipment components are required as used for solvent-based materials.

Waterborne materials are offered to assist liquid coaters with compliance issues and satisfy regulations governing the application of coatings. With regulations evolving to new standards that are making it tougher to comply with High Solids materials, and advances in performance and appearance of water-based paints, more and more operations are using them. Waterborne coatings do have some organic solvents in the paint formula. They are not 100 percent VOC free, so some manufacturers have chosen other technologies that will provide more long term compliance.

Advantages using waterborne coatings:

- *Environmental.* Water-based liquid formulas are more environmentally friendly than solvent based materials. They produce less VOC, are less toxic, and less of a fire hazard than solvent materials. Sludge treatment and disposal costs are relatively low. The use of DI water for purging the fluid delivery system and spray guns provides a further reduction in solvent use.

Limitations using waterborne coatings:

- *Fluid delivery.* Circulation systems and fittings must be stainless steel to protect from rust. Because the waterborne paint is a very good conductor, the fluid supply source must be isolated from ground to avoid the loss of electrostatic charge from the gun tip back through the fluid

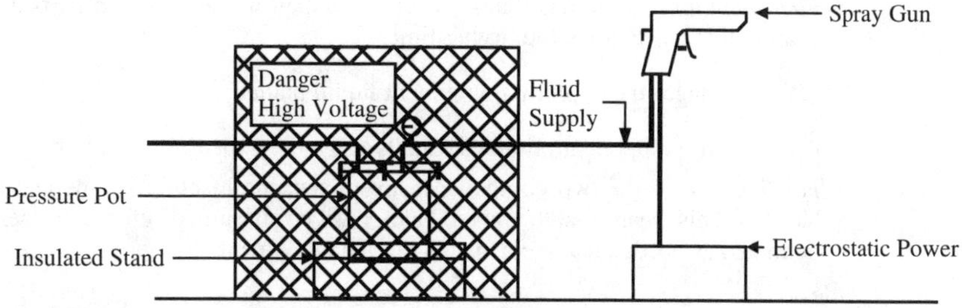

FIGURE 49.3 Electrostatic spray of waterborne coating.

supply hose. When the fluid supply source is isolated from ground, it becomes charged during spraying, and it must be enclosed to protect personnel from dangerously high voltage levels (Fig. 49.3). Since electrostatic spray is commonly required as the *best available control technology* (BACT), this can be a cost factor and an operational headache. Application equipment companies are now offering equipment that provides a voltage block in the pumping system. Some companies have had success with these systems, especially for manual spray.

- *Controls.* The temperature and humidity must be carefully controlled because the material is sensitive to any climatic change. The film is more sensitive to dirt and oil than solvent materials because the solvent has a tendency to dissolve and neutralize minor soils. Longer flash time (possibly heated) is required than solvent-based materials, and it is more sensitive to improper warm up or cure cycles.

49.5.4 Plural Component Liquid Paints

A two-component product uses catalyst and resin that react when mixed to form a paint film. Multicomponent urethanes and epoxies are used for a variety of industrial and automotive coatings. A urethane is formed when an isocyanate chemically reacts with an alcohol. They form excellent high gloss films that are flexible and resistant to outdoor weathering. They are very popular for plastic parts that cannot take the higher cure temperature required for typical thermal-set products.

Two-part epoxies provide the durable finish and excellent chemical resistance of one-part epoxies with the added advantage that they do not need a high temperature cure. This makes them very popular for large parts with heat sensitive components. The combination of high performance and ambient drying makes these coatings very useful for the marine industry. Of course epoxies do not resist sunlight very well so outdoor applications that require good appearance over time are not good candidates for epoxy systems.

Application equipment is similar to other liquid technologies. Mixing is required and there are a variety of methods that can be used. Batch mixing is common for lower volume operations. Once catalyzed, the material has a limited pot life so automated mixing is more popular in larger production situations. Automatic mixing systems may be as simple as a piston pump or more sophisticated metering systems.

Automated finishing systems are built for two-component urethanes on plastic parts for automotive and other industries. Washers are sometimes used but many times plastic parts do not go through a power washer. The cure oven for these systems operates at around 1800°F and is used primarily to accelerate the cross-linking of the coating and provide a consistent drying cycle.

Advantages using plural component liquid paints:

- *Quality.* These products typically provide exceptional quality.
- *Ambient cure.* On of the chief advantages of catalyzed materials is good performance without thermal curing. Products that are heat sensitive or cannot be put through an oven can still have a very high quality finish.

- *Field painting.* Products that need to be painted in the field can be coated with a high quality coating that has predictable drying time.

Disadvantages using plural component liquid paints:

- *Cost.* Two-component products usually are more expensive then baking enamels.
- *Pot life.* Once the two components are mixed, the reaction begins and the product eventually hardens. This means that the mixed batch must be consumed before it hardens, or the material must be mixed as needed using expensive mixing equipment.

49.5.5 Powder Coatings

Thermoset powder coatings are typically applied by electrostatic spray as a dry film. The charged material clings to the part surface due to the electrostatic charge. When the dry powder coat is introduced into an oven, it will melt and cross-link into a continuous film. The system components are very similar to a liquid system. Cure cycles for powder tend to be a little shorter and hotter than their liquid counterparts. One element of planning a powder system that is different is that there is no need to consider abatement due to the lack of hydrocarbon emissions.

The chief advantage of powder is that it is nearly 100 percent solids so it is one of the best compliance options. Material utilization can be very high because overspray can be captured and reused. There is no mixing and there are no hazardous waste products.

Other than the environmental issues, the primary difference between powder and liquid paints is the spray booth. They both have an enclosure for the spray process but the powder booth (Fig. 49.4) uses a recovery system to capture the overspray for reuse. The air used to contain the powder in the booth is filtered and returned to the plant, so no air make-up is required for the booth.

Powder coating was developed to offer an environmentally safe alternative to solvent materials. The raw material is literally a powder, mixed dry, extruded, and ground into a final product that contains virtually no VOC. Like liquid coatings, it can be formulated to provide different properties and in almost any color. Color matching is more time consuming and expensive than with liquid coatings. If a batch of powder is not correct, it is discarded, unlike liquid paints, which can be modified to make the color correct.

Thermoset powders are widely accepted as the best compliance alternative. They are tough, chemically resistant, and they have excellent appearance characteristics. The fact that the oversprayed powder is collected for reuse is a major environmental plus and the high material utilization percentage makes powder cost-effective.

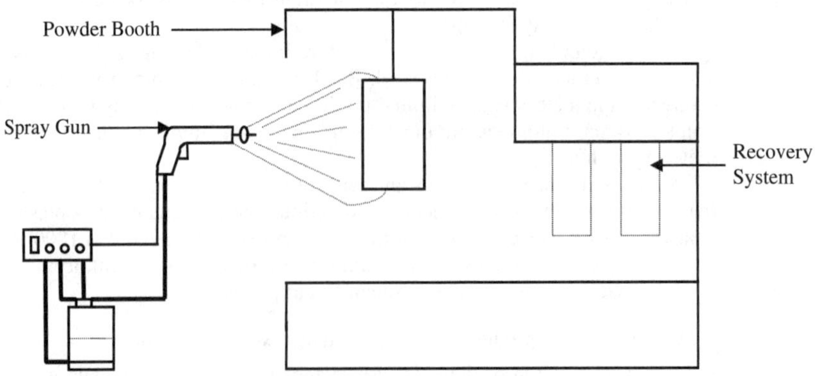

FIGURE 49.4 Powder coating booth.

Advantages using powder coating booth:

- *Material utilization.* Because the overspray is collected for reuse, material utilization of 95 to 98 percent is possible. Powder that is cleaned up from the floor, obsolete material, and recovered powder that is unfit due to repeated cycling through the recovery system, must be disposed of in a normal landfill. Recovery does require the correct equipment and it takes more time than spray-to-waste.
- *Environmental.* The major advantage of powder is the fact that it does not contribute to pollution in any significant way. There is no VOC, it is not toxic, low risk of fire hazard and no hazardous waste.
- *Air make-up.* Air make-up costs are lower because the air that contains the powder in the spray booth can be filtered and returned to the plant and a powder cure oven requires less exhaust than a solvent or a waterborne coating.
- *Function.* There is no solvent to buy, mix, or store. Expensive liquid circulation systems are not required. Fewer surface defects are produced because the coating is less sensitive to shrinkage stresses, less likely to trap airborne dirt, does not solvent pop and is more forgiving of higher film build. Because it is a one-coat process and does not require a flash zone, powder systems may require less space. Manual spray technique is easier to teach and in some ways not as sensitive as liquid coating application.
- *Film characteristics.* Although liquid coatings can be formulated to provide desired abrasion resistance, chemical resistance, and impact resistance, on average, powder provides a tougher and higher quality surface.
- *Specialty finishes.* Powders can be formulated to provide a variety of unique appearance characteristics (stone look, wrinkles, textures, vanes, etc.) in a one-coat process.

Limitations using powder coating booth:

- *Film thickness.* Because it is applied dry, powder cannot be controlled at mil thickness comparable to liquid coatings. Thicker films mean slightly more orange peel, difficulty with masking and repair, and difficulty holding tight tolerances.
- *Color.* The major problem with powder is color. Any color or gloss can be formulated, but the process is more expensive and time consuming than liquid paint. A typical liquid paint can be color matched in 4 hours. A typical powder coat color match takes 72 hours or more. An incorrect batch of powder must be scrapped. Metallic coatings are difficult to apply with current technology.
- When a color change takes place, the entire system, including the booth walls and recovery system, must be purged of the previous color before the new color can be introduced. This can be time consuming and result in loss of production time. Many times, duplicate equipment is required to reduce color change time.
- *Substrate.* The cure temperature of powder coatings is typically between 325 and 400°F. This makes it very difficult to coat plastics or wood. Some nonmetal substrates are powder coated and cured by Infrared energy. Another complication for nonmetal substrates is that powder is applied electrostatically and these surfaces are nonconductive. When powder is applied to nonconductive substrates, this is overcome by first applying a conductive prep-coat or by preheating the part and letting the powder gel on contact.
- Also, very large products can present a problem because powder overspray must be contained and it does not air dry. This makes it almost impossible to use on airplanes, tank cars, and the like, because of impractical size requirements of the booth and oven.

49.5.6 Electrodeposition

Electrodeposition, or E-coat, is applied by submerging the parts into a water-based material in a dip tank (Fig. 49.5). The parts are grounded and the paint film is deposited through an electrification

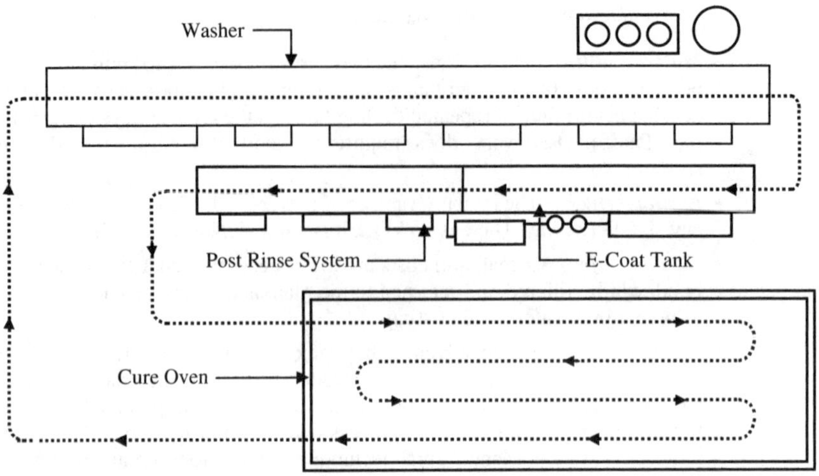

FIGURE 49.5 Electrodeposition system, plan view.

process using a rectifier. Following the dip tank, the parts are conveyed through a series of postrinses. The postrinse process will remove excess paint solids that cling to the part as drag-out. The postrinse is usually a closed loop system; the DI water rinses the parts removing paint solids, the rinse solution is circulated through an ultra-filter that separates the recovered paint solids and the DI water, the paint solids are returned to the dip tank and the recovered DI water, called permeate, is used for the postrinse. After the postrinse the parts are allowed 5 to 10 minutes of flash and then transported into an oven to be cured.

The washer will usually be a seven-stage zinc phosphate system with a DI final rinse. The parts can enter the dip tank wet, so there is usually no dry-off oven. The cure oven is 30 minutes or longer, operating at 350 to 400°F. A low-temperature dehydration zone of 1 to 3 min is sometimes used at the oven entrance to evaporate excess moisture. In some cases the dehydration zone may be as long as 10 to 15 min.

Overhead monorail is the most commonly used conveyor, but some e-coat systems use a programmed hoist or other means of transportation.

Since the solution is water-based and there is no exhaust from the application area, e-coat is a very good compliance alternative. Sometimes, the oven exhaust is smoky and has a strong odor, so even though it has very little VOC, it may require an afterburner.

E-coat was created as a primer and is still used primarily for parts that require corrosion resistance but do not need a decorative finish in a variety of colors. It does provide very good corrosion resistance and is used for many automobile parts as a primer for body components, or as a topcoat for parts that are hidden in the final assembly, such as oil pans, under the seat or under the dashboard.

Advantages using electrodeposition:

- *Environmental.* In terms of air quality, e-coat produces very little VOC. Some systems discharge a percentage of the overflowed permeate from the postrinses but a properly designed and installed system recovers the overflow. The e-coat bath is typically not a fire hazard and is low in toxicity.
- *Finish quality.* Because e-coat is a dip process, it penetrates holes and weld seams that are not accessible to spray coatings. This provides excellent corrosion resistance in harsh environments. The surface is a very uniform thin film with no runs or sags.
- *Production.* The application of the coating is done in the dip tank so skilled sprayers are not required and parts can be racked very densely, allowing the system to produce large volumes with great consistency and low rejects. It does not require compressed air and it is not sensitive to temperature or humidity. Also, it provides very high material utilization.

Limitations using electrodeposition:

- *Color.* With the large dip tank and postrinse arrangement, changing colors is not very practical. To have more than one color requires a separate e-coat tank and postrinse. For color change, a link in the conveyor chain is detached and hooked to a loop that passes through another tank. At best, e-coat is a two-color system. Any more than two colors would require too much space and capital expense. Most e-coat systems are one color. This means that it is not a viable alternative for multicolor products. The available colors and glosses are limited.
- *Capital equipment expense.* When compared to other coating equipments, an e-coat system is very expensive to install and it requires a lot of power to operate the rectifier that supports the electrodeposition process.

49.5.7 Autodeposition

Autodeposition is an alternative to e-coat that contains low or no VOC. It is a waterborne material that is deposited by a chemical process. The bath is composed of a mildly acidic latex emulsion polymer, DI water, and some proprietary ingredients. When the parts are dipped into the bath a chemical reaction takes place that deposits a film that penetrates all surfaces and provides excellent corrosion resistance.

An immersion of 60 to 120 s will produce a film of 0.5 to 1.0 mil. The film will continue to build if the immersion time is increased but it is eventually self-limiting as the available sites for chemical activity are covered. Cure cycles vary from 20 to 30 min at 212 to 350°F.

Since it does not depend on an electrical process for deposition, autophoretic coating will provide very good penetration into cracks and seams.

A typical autophoretic system would include an overhead conveyor or programmed hoist and the process sequence the following process cycle:

Stage	Process
1	Alkaline spray clean
–2	Alkaline immersion clean
–3	Tap water immersion rinse
–4	DI water immersion rinse
–4a	Virgin DI water spray rinse
–5	Autodeposition coating immersion
–5a	DI water mist
–6	Tap water immersion rinse
–6a	DI spray mist
–7	Reaction immersion rinse
8	DI mist tunnel
9	Cure oven

Advantages using autodeposition:

- *Environment.* The process is very friendly to the environment, producing little if any VOC.
- *Cost.* Because the deposition process is chemical rather than electrical, operating cost is modest. The applied cost is competitive with other compliance coatings and a conversion coating is not required. Capital equipment can be lower than some of the alternatives.

Limitations using autodeposition:

- *Color.* Like e-coat, autodeposition is strictly for operations where multiple colors are not required. It does not have the appearance characteristics required for a topcoat.
- *Product mass.* Very heavy parts may require unacceptably long cure cycles. Testing should be done for heavy parts.

49.5.8 Compliance Coatings in North America

Through the Clean Air Act Amendments (CAAA), the EPA has established air quality standards and required improvements in the emission levels of VOC. The term VOC includes almost all common solvents used in industry and contributes to smog at lower levels of the atmosphere, or the depletion of the earth's protective ozone layer in the upper atmosphere. States are required to develop a plan for achieving the specified reductions of VOC.

The acceptable solvent content of compliant coatings and the levels of hydrocarbons that can be exhausted varies for different geographic areas (states, counties, cites) and industries. The coatings industry has developed a variety of products that reduce VOC and meet federal guidelines for compliance. The most common of these compliant coatings are high solids, waterborne coatings, powder coatings, electrodeposition, and autodeposition. Each of these materials has operational as well as compliance strengths and limitations as described. A full comparison needs to be done to balance the needs of quality, color, and cost.

Summary. The primary issues in the comparison of coatings are performance, appearance, cost, and compliance. To determine the correct process, the needs of the product and the company must be defined and prioritized and then used to judge the alternatives.

All coatings have strengths and weaknesses. What is best for one application may not work at all for another. To determine the most suitable coating requires an analysis of the materials and processes available, compared to the requirements of the end product. In many situations this may mean a combination of two different technologies. For example, powder for standard, large volume colors, and high solids for the short run, odd colors, or an e-coat primer and powder top coat.

The first step in the selection process is an evaluation of the coatings that meet the required performance and appearance characteristics for the product.

The compliance issue should be a major part of the coatings system decision. One caution that must be considered is that environmental regulation is an evolving situation. This means that what is acceptable today, may not be tomorrow. For this reason, large capital expenditures should be considered carefully. For example, a $500,000 investment to convert from high solids to waterborne may only buy a facility temporary compliance. It may be better to focus on improving transfer efficiency and using higher solids materials and consider other alternatives for the future, such as powder.

When the options have been narrowed down by consideration of the coating performance, appearance, and compliance, the cost factor will lead to the ultimate choice. When cost is considered, it is very important to consider the operating cost as well as the original equipment cost.

Another consideration is abatement. If a solvent based paint is chosen, it may be necessary to incinerate the emissions from the spray booths and/or ovens and flash tunnels. Although the equipment is expensive, certain types of oxidizers offer a long-term answer to the compliance issue without changes to the application process. Also, it may be possible to use a secondary heat exchanger and recover the heat from the incineration process for use as plant or process heat.

49.6 PAINT APPLICATION

Liquid paint can be applied in many different ways. Each method has some advantages and some limitations. The needs of the product and the manufacturing process will help determine the correct application method. An understanding of the different application processes is necessary in order to evaluate them for a specific task.

49.6.1 Dip Coating

Dipping simply involves immersing the part in a tank of paint. Excess paint drains back into the tank once the part is removed. Parts may be handled in batches or with a conveyor.

Dip tanks range in size from a few to several thousand gallons. Provisions must be made for circulation, filtering, and control of temperatures and viscosity. Because of the fire hazard involved, dip operations using flammable paints must be equipped with CO_2 fire extinguishing systems and dump tanks.

Advantages of dip coating:

1. Simple equipment
2. Low skilled labor requirement
3. Easily automated
4. Close racking of parts
5. Various part shapes can be mixed
6. Little paint is wasted if the part is allowed to drip over the tank

Limitations of dip coating:

1. Viscosity control is critical. If the paint is too viscous, too much paint is used (wasteful) and film thickness is too high. Low viscosity results in paint films which are too thin. Consequently, corrosion resistance may be too low for primers and hiding power may be inadequate for topcoats.
2. Possible settling of the paint in the circulation system.
3. Gradual tank contamination is possible. Parts entering the tank must be clean and dry to avoid carrying in contaminants.
4. Some parts are hard to immerse because trapped air keeps them afloat.
5. Tanks must be large enough to immerse the entire part.
6. There is a fire and toxicity hazard associated with large volumes of paint solvents.
7. Film thickness control is a problem. Paint may become trapped in recesses and drain slowly or not at all. On vertical surfaces the paint tends to be thicker at the bottom of a panel than at the top. This is caused by an increase in viscosity as the paint loses solvent while draining. It often results in fatty edges that can later break off under thermal or mechanical shock.
8. The drain-off area must be cleaned regularly. Since this is usually done manually, some dried paint may fall into the circulation system and clog the filters, pumps, and nozzles.

49.6.2 Flow Coating

With flow coating the part is held over a tank and a stream of paint is pumped over the part. It is much like hosing off a car. The excess paint drains into the tank and is recirculated over the parts. The variables that are important for flow coating include pumping, filtering, viscosity, temperature, ventilation and fire control (see Fig. 49.6).

Advantages of flow coating:

1. Simple equipment
2. Little labor required
3. Easily automated
4. Dead air spaces can be coated if a stream of paint can be directed upward into the recess.
5. Smaller volumes of paint are required because the parts need not be immersed.
6. Little paint is wasted if the parts are allowed to drain over the tank.

Limitations of flow coating:

1. High solvent loss by evaporation. Solvent must be replenished on a regular basis.
2. Possible settling of the paint in the circulation systems
3. Fire and toxicity hazards associated with the high solvent loss
4. Inability to coat certain recesses and inside surfaces

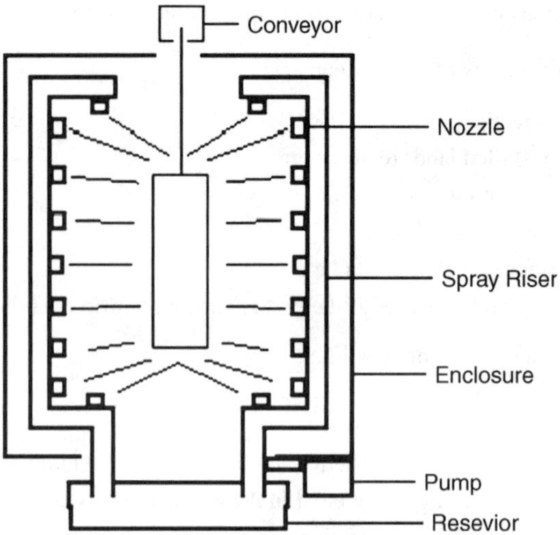

FIGURE 49.6 Flow coating.

5. Gradual paint contamination if the paint rinses dirt, salts, oils, etc. from the parts.
6. Film thickness control is a problem. Paint may become trapped in recesses and drain slowly or not at all. On vertical surfaces wedging may occur resulting in fatty edges.
7. The drain-off area must be cleaned regularly. Since this is usually done manually, some dried paint may fall into the circulation system and clog the filters, pumps, and nozzles.

Dipping and flow coating methods should be considered when high production painting of relatively simple shapes is required and quality appearance and corrosion resistance is not important. Viscosity and film thickness control are the major problems to be expected. Efficient use of paint and low labor rates are the primary advantages.

These processes are limited to waterborne coatings because of environmental considerations. High solids coatings do not lend themselves to these types of application because of the higher viscosities that cause poorer flow characteristics and high film builds.

49.6.3 Spray Application

Paint spray guns are available in many different types that have their own unique features. Commonly used spray guns include air spray, airless spray, and electrostatic spray.

Conventional Air Spray Application. A conventional air spray gun is a tool using compressed air to atomize a fluid paint or lacquer and applying it to a surface. The compressed air and paint enter the gun through separate passages and are mixed and ejected in a controlled pattern.

Air spray guns can be fed from an attached cup or a pressure feed system that uses a pressure pot and compressed air to supply paint through a fluid hose.

A cup gun is referred to as a suction or siphon feed system. A stream of compressed air creates a vacuum allowing atmospheric pressure to force the fluid material from an attached container to the spray head. The advantage of this system is the ability for quick color change where small amounts of material are used. It is limited to low viscosity materials, such as furniture lacquers, and used with quart size containers.

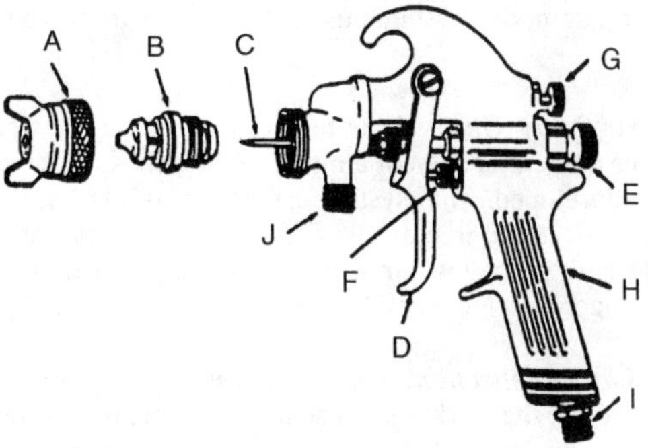

FIGURE 49.7 Compressed air spray gun.

A pressure feed gun applies paint that has been forced to the gun by pressure, either from compressed air or a pump (Fig. 49.7). This type of gun is used when the paint viscosity is too high.

A. *Air cap.* The part that directs compressed air into the fluid stream for atomizing and the formation of a spray pattern.

B. *Fluid tip.* The fluid nozzle located behind the air cap that regulates and directs fluid into the air streams created by the air cap.

C. *Fluid needle.* The part that seats on the fluid tip opening and opens and closes with action of the trigger. The fluid tip and needle are matched and should be replaced as a unit for best results.

D. *Trigger.* The part that is pulled to draw the fluid needle back, allowing fluid to pass through the nozzle. When released, the passage of fluid is stopped. For complete fluid flow, the trigger must be pulled all the way back.

E. *Fluid adjustment.* The fluid knob controls the travel of the fluid needle. When the travel of the needle is restricted the amount of fluid going through the fluid nozzle is limited. For pressure systems the normal adjustment is to screw out until a thread or two shows on the screw to assure full fluid flow. Fluid volume can be regulated with this knob for siphon systems.

F. *Air valve.* The air valve is activated when the trigger is pulled to release air through the gun body to the air cap.

G. *Spreader.* The spreader regulates air to the holes in the horn of the air cap to regulate the size of the spray pattern. Turning the adjustment clockwise will reduce the fan pattern, counter-clockwise will increase the fan pattern.

H. *Gun body.* The handle and main body of the gun.

I. *Air inlet.* The main airline from paint pot or container is attached to this fitting. This fitting is generally smaller in diameter than the fluid inlet.

J. *Fluid inlet.* The fluid line from a pressure pot, paint tank, or from a paint pump is attached to this fitting which is normally larger in diameter than the air inlet.

Automatic Air Spray. Automatic spray guns are used for higher volume operations. They are similar to hand spray guns but they are usually triggered on and off automatically as parts are carried past on a moving conveyor. Automatic gun triggering improves material efficiency.

Automatic guns can be mounted in a stationary position or on a gun mover. Horizontal or vertical reciprocators are often used to move the gun across the surface being painted. The movement of the reciprocator is related to the part surface and position. Flat panels or similar parts are usually painted with a machine that traverses over them in a horizontal position. Parts that are hung vertically from an overhead conveyor use a vertical machine. Some systems use rotary mover systems and most automotive systems use robotic arms to move the spray gun into the contours of the car body.

HVLP Air-Spray Gun. In order to improve the transfer efficiency of conventional coatings and reduce the amount of solvent emitted to the atmosphere the *high volume low pressure* (HVLP) spray gun was developed. This system essentially mixes large volumes of air at low pressures with the paint to break it into fine droplets. The fluid delivery pressure is lower than conventional spray systems. Because of the low air pressures (6–10 psi) there is very little bounce back of paint as it hits a part and much less tendency for the paint to blow by the part.

Airless Spray Application. An Airless spraying system utilizes hydraulic pressure to atomize paint and deliver it to the workpiece. The hydraulic pressure is released through an orifice in the spray nozzle. At high pressures, the fluid is separated into small droplets resulting in a very fine atomized spray. Because compressed air is not used to form the spray pattern, the term *airless* is used.

In conventional spray application the paint is atomized by compressed air at the nozzle. The more viscous the material, the greater the air pressure required for atomization. Fluid particles are blown at high speed toward the work and *overspray* and *rebound* carry paint away from the target. This combination can result in a paint loss as high as 50 percent.

Like other spray technologies, airless requires more pressure to move higher viscosity materials. Unlike air atomization, increased hydraulic pressure does not create excessive velocity of the atomized particles. Because the atomized fluid moves forward by its own momentum and meets high resistance in the air, it is slowed down and lands softly on the surface without appreciable rebound. The almost complete absence of rebound combined with very little overspray can result in very high transfer efficiency, more paint deposited on the surface at a higher production speed.

Less solvent is required because higher viscosity materials can be used and less overspray results in lower booth maintenance costs.

One coat finishing is common with the airless spray method because higher viscosity and solids materials may be applied, resulting in thicker dry films. Airless spraying penetrates porous surfaces more readily than conventional spray, resulting in faster application. Because less solvent content is required in airless spraying, there is a lower concentration of flammable vapors.

There is improved adhesion through better wetting of the surface and smoother paint films are attained due to the wetness of the coating that permits better flow-out. The absence of air impingement results in smoother coatings when applying heavier films.

The primary disadvantages of airless are the lack of control over the shape of the spray and film build characteristics. Without compressed air, fine atomization is not possible, resulting in heavier films and more risk of runs or sags. Airless systems are best suited to applications that require heavy coats without intense quality standards, such as the application of air-dry materials on large pieces of equipment.

Electrostatic Spraying. In electrostatic spray, a corona field of free ions is generated at the gun tip from an electrical source, through a cable, to a discharge electrode. Paint particles of neutral polarity are passed through this field of free ions where they pick up extra electrons. The paint atoms become negatively charged due to the excess electrons and they are attracted to the grounded substrate, moving along lines of force created by the discharge of voltage at the gun tip. The resulting electrostatic charge increases paint transfer efficiency.

Electrostatic charge can be used with any of the various technologies described for hand application. It is often used with automatic spray guns or rotary atomizers. A rotary atomizer is an entirely different way of atomizing paint. It uses a rotating electrically charged disk, bell, or cone to break up the paint. Atomization is achieved by a combination of centrifugal and electrostatic force. Rotary atomizers can atomize high viscosity paints with minimum turbulence that reduces overspray and increases paint efficiency. Rotational speeds greater than 35,000 RPM are common. Charging voltages are in the 50–60 kV range.

Electrostatic spray is not a cure-all for every production line. In some cases it can lead to substantial savings and in some cases no savings at all. Usually more parts can be coated per gallon with electrostatic equipment than with conventional spray because paint is attracted to the part and overspray is reduced.

The amount of paint actually applied to the parts may increase from about 40 percent to as high as 70 percent under favorable conditions when conventional systems are converted to electrostatic. Some high-speed rotary systems may approach 90 percent efficiency. The shape of the parts has a lot to do with the efficiency. Large flat parts are more efficiently coated than bedsprings or bicycle frames, for example.

Spray Booths. Spray booths are enclosures around the painting area. Their job is to keep paint overspray out of the plant area and sometimes dirt out of the wet paint. Booth airflow may be downdraft or sidedraft, and overspray collection may be water wash or dry filter.

Downdraft booths are usually used for large objects so that they can be painted from two sides without turning the part. Side-draft booths are generally used for smaller objects sprayed from one side only. Both downdraft and side-draft booths have air ducts for removing the solvent-laden air so that it does not enter the plant. The exhaust air is routed to atmosphere or some type of abatement arrangement.

49.6.4 Other Paint Application Methods

Liquid paint is sometimes applied with roller coating operations or in curtain coating operations. These systems are often high volume production lines that run on flat stock or coil stock.

Roller Coating. A roller coating operation is capable of coating large areas in the form of sheet or strip at high speed and with a continuous and even thickness of paint or varnish. The rollers can also be used to laminate plastic film to metal substrates.

In a three-roller operation (Fig. 49.8), the coating is fed between the feed and applicator rollers rotating in opposite directions, which in turn is fed to the sheet. The third roller acts as a pressure roller to ensure that the sheet comes into correct and even contact with the wet applicator rolls.

1. Feed
2. Application
3. Pressure

Curtain Coating. Curtain coating is a sophisticated method of pouring a continuous film of lacquer or paint over flat sheets. The sheets are conveyed under a trough with an adjustable slot. The trough acts as a pouring head and control of the gap regulates the volume of material, controlling the film thickness deposited on the sheets. Curtain coating works well for coating flat sheets of metal or wood (Fig. 49.9).

A second conveyor moves the coated panel away from the gap. Very little waste is incurred and excess falls down into a reservoir below the conveyor line and is re-circulated back to the trough by means of a pump.

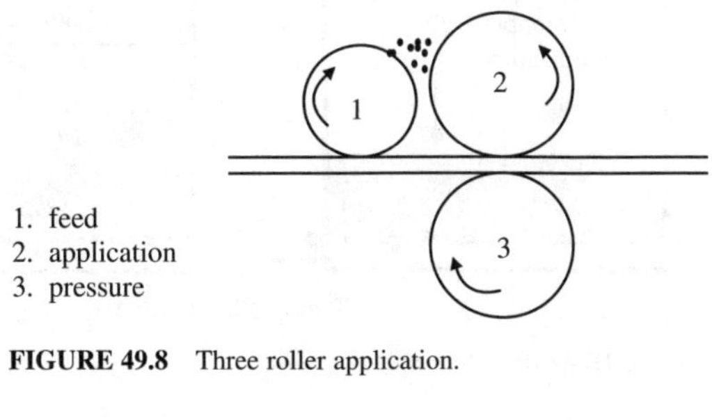

FIGURE 49.8 Three roller application.

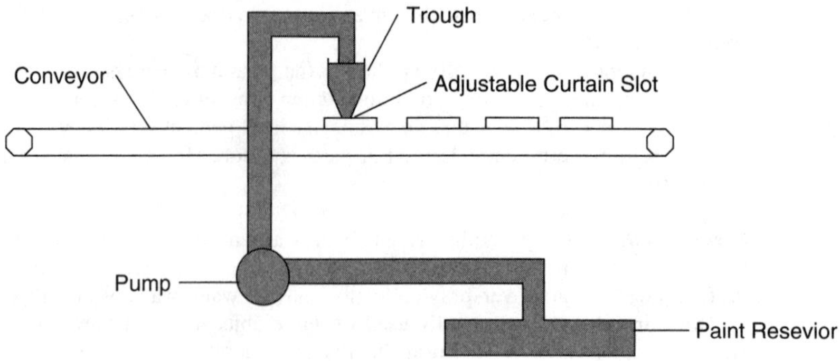

FIGURE 49.9 Curtain coating.

The most important part of curtain coater is the adjustment slot (or gap) in the trough. The gap can be accurately controlled to develop predetermined film weights. The trough and slot must be kept scrupulously clean because any dirt or solid material will break the curtain and cause a surface defect.

Summary. There are many devices available for application of liquid coatings. The correct equipment for a particular application is usually based on the size, volume, and specific characteristics of the parts to be painted and the paint material. Testing is useful to determine the correct system for a given application.

49.7 POWDER COATING APPLICATION

49.7.1 Fluidized Bed Application

Early in the history of powder coatings, thermoplastic materials were applied by the fluidized bed dip process (Fig. 49.10). The fluidized bed is a container that holds the powder material with an air chamber at the bottom referred to as an inlet plenum. The container and the plenum are separated by a membrane that is porous enough for air to pass through but not porous enough for solids to pass through. Compressed air is introduced into the plenum and up through the fluidizing membrane. As the compressed air passes up through the container, the powder particles are suspended

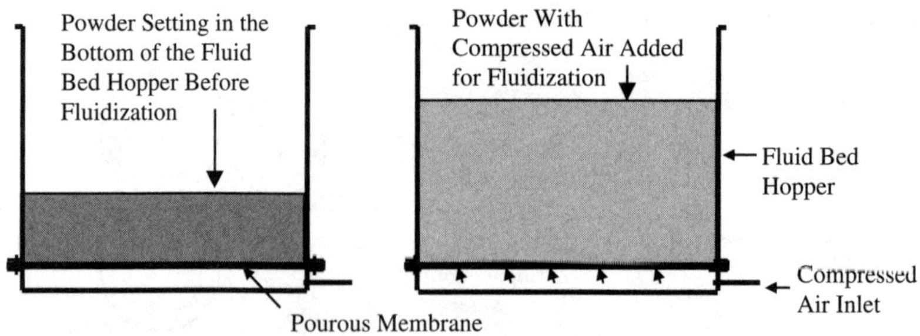

FIGURE 49.10 Fluidization of powder.

in the air-stream. In this suspended state, referred to as fluidization, the powder/air mixture behaves somewhat like a liquid.

For fluidized bed application the part is preheated and dipped into the fluidized bed of powder. The powder material will fuse upon contact with the hot part, creating a thick continuous film (10–20 mils) on the metal surface. In cases where the part does not have sufficient mass to completely fuse the powder, the part will be put through a short post-cure cycle, typically 3–5 min at 400 to 500°F (204 to 260°C).

49.7.2 Electrostatic Fluidized Bed Coating

Electrostatic fluidized bed application uses the same fluidizing technique and the conventional fluidized bed dip process but with much less powder depth in the bed. An electrostatic charging media is placed inside the bed so that the powder material becomes charged as the fluidizing air lifts it up. Charged particles of powder move upward and form a cloud of charged powder above the fluid bed. When a grounded part is passed through the charged cloud the particles will be attracted to its surface. The parts are not preheated as they are for the conventional fluidized bed dip process.

49.7.3 Electrostatic Spray Application

Electrostatic spray application uses a fluidized bed as a feed hopper to hold the powder and fluidize it so that it can be pumped to the tip of a spray gun using compressed air for transport from the feed hopper to the gun tip. The spray gun is designed to impart an electrostatic charge to the powder material and direct it toward a grounded workpiece.

This process makes it possible to apply much thinner coatings with a wide variety of decorative and protective features.

The electrostatic charge can be imparted with voltage, called corona charging, or by frictional contact with the inside of the gun barrel, called tribo charging.

In a corona charging system, a voltage source generates current through a voltage cable that carries it to the powder gun tip. Powder is pumped through the gun and out of the gun tip using compressed air. As the powder passes through the electrostatic field at the gun tip, it picks up a charge and is attracted to a grounded substrate. The part is then conveyed into an oven for curing of the powder. In the cure oven, the powder melts and cross-links to a hard film to complete the process.

Electrostatic spray application of powder is the most common application method. The parts to be coated are cleaned, dried, and cooled; the coating is applied and cured at the required temperature for the required time; and then cooled for removal from the line.

An electrostatic spray application system includes a delivery system and a charging system (Fig. 49.11). The basic pieces of equipment that make up an application system include the following:

- The feed hopper
- The powder spray gun
- The electrostatic power source
- The controls

49.7.4 Corona Charging

Corona charging systems are the most widely used type of electrostatic spray equipment. Corona equipment uses voltage to supply a charge to the powder particles.

Powder is pumped from the feed hopper through a hose to the tip of the spray gun by the delivery system. A charging electrode at the gun tip is connected to a high voltage generator. The generator can be located inside the control panel and connected to the electrode by a high voltage cable or it can be located inside the gun barrel. High voltage multipliers that are inside the gun barrel are connected to the control panel by a low voltage cable.

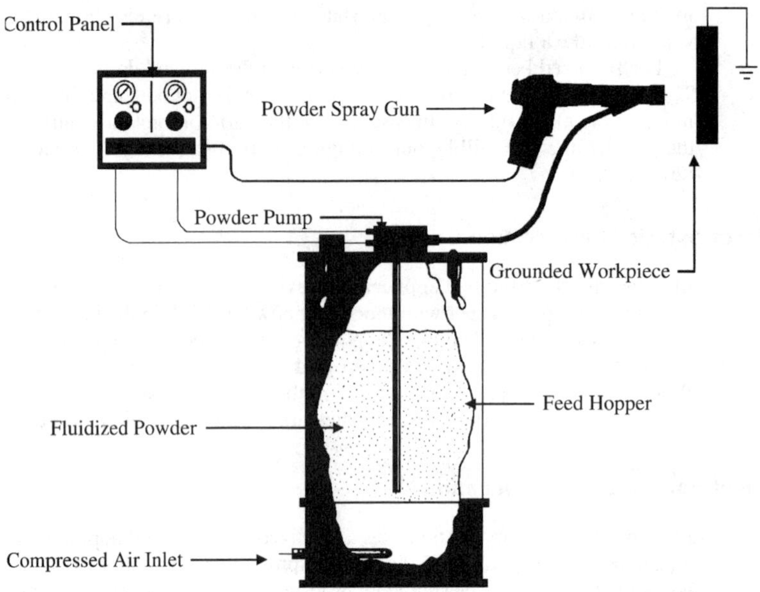

FIGURE 49.11 Powder coating spray application unit.

When a corona-charging gun is triggered it will create a sharply nonuniform electric field between the gun tip and a nearby, grounded object. Field lines will develop between the sharp tip of the gun electrode and the surface of a grounded part.

Free electrons or ions are always present in the air. If these free ions pass through the strong electric field at the gun tip, they will begin to move along the field lines and get accelerated by the field force. As electrons accelerate along the field lines they will collide with air molecules.

When the combination of applied voltage and electrode geometry creates an electric field in excess of the breakdown strength of air, a corona discharge is generated. The impact of the free ions on the air molecule will be strong enough to split it, forming two secondary negative ions and one positive ion. The secondary electrons will get accelerated in the electric field and repeat the process of splitting air molecules to form more free ions. The resulting negative ions will accelerate along the field lines to the part while the left over positive ions will accelerate along the field lines in the opposite direction, toward the gun electrode. The space between the spray gun and the grounded part becomes filled with millions of free electrons and positive ions.

The grounded, conductive parts provide a receiving electrode to form an electrostatic field between the gun and the part. As powder particles pass through this corona, they become charged, usually with negative ions, and follow the electric field and the air currents to the earth grounded workpiece. As air brings the particles within a few centimeters of the part, Coulombic attraction pulls the particles toward the part surface and causes them to adhere there until some other force interferes or they are cured by the introduction of heat or radiant energy.

49.7.5 Tribo Charging

Another device that can help to deal with Faraday cage effect is the tribo-charging gun. In a tribo-charging gun, the powder particles are charged by frictional contact with the material inside the gun body. A powder and air mixture enters the gun and passes through a tubular section that is made of a material that is known to be a good acceptor of electrons, such as Teflon. As the powder particles collide with the walls of the tube, they pick up a positive charge by giving up electrons to the tube, causing the tube to become negatively charged. The negative charge is then passed from the gun barrel to ground through a cable.

The powder material must be a formula that is a good electron donor, such as nylon or epoxy. Not all powder materials can be used effectively in a tribo-charging gun. In addition to the powder chemistry, tribo guns are more sensitive to particle size than corona guns. Generally, larger particles will be more effective in a tribo gun. Powder material suppliers can work with the applicator to make sure that the material is suitable for tribo application.

49.7.6 Powder Bells and Discs

The powder bell is fashioned with some of the same principles as rotary atomizers made for the application of liquid coatings.

The powder bell is mounted in a horizontal position on a stationary or reciprocating gun mount. A turbine rotates a bell shaped disc at the gun tip. Powder is delivered to the bell by compressed air. The centrifugal force of the rotating bell ejects the powder in an even pattern from small holes or slots around the edge of the disc. The charging electrode is disk or needle shaped and located at the gun tip. The corona field and charging method is the same as a standard air spray gun. The pattern and velocity can be controlled by the rotational speed of the bell and a shaping air supply.

The bell can deliver a large volume of powder over a large surface area with a high level of charging efficiency. Large surfaces without significant Faraday areas can be coated with excellent film thickness control and high *first pass transfer efficiency* (FPTE).

The pattern of powder delivered from the bell is broader than a typical spray gun and more concentrated around the edges than in the center. In systems where gun triggering is used the bell may not be as efficient as a spray gun because of this large spray pattern. The bell needs to be triggered before the part arrives and remains triggered until the part passes. Since the bell pattern may be as much as 18-in wide a lot of overspray is generated during the lead into the spray zone and the lag out of the spray zone.

Summary. Application of powder is predominantly accomplished with corona spray systems. Any number of spray guns can be used for different applications and volumes. Many systems will use a combination of automatic and manual application in one booth. Gun movers can also be used to provide a single- or multiple-axis motion to the guns for coverage on complex shapes.

49.8 FUTURE TRENDS IN COATINGS

The same issues that affect the selection of a particular coating option also affect the trends in development and use: quality, cost, and environmental compliance. Manufacturers are looking for a coating material that offers them a broad variety of colors and glosses, is reasonable to apply, matches their particular performance requirements, helps them meet regulatory guidelines and comes at a reasonable cost.

With the compliance issue in mind, much of the development of coatings technology is in products that do not contain solvents. These products include ultra-high solids liquid coatings, UV curable coatings, and powder coating. In addition, newer electrocoat materials are very low VOC and no-lead formulas. A brief look at these technologies will provide some insight into the current situation and outlook for the future.

49.8.1 Ultra-High Solids

Certain resin systems have been used to formulate paints with very low solvent content, and, in some cases, 100 percent solids content. In most cases, color flexibility and performance has not been comparable to more traditional liquid coating formulas but the research continues. Some paint manufacturers are marketing zero VOC coatings now that have successful applications.

Zero VOC polyurethanes are being used in several applications, such as aerospace. They are formulated for areas where flammability, solvent exposure, and emissions of volatile organic compounds are a concern. Water is used as a carrier to substitute organic solvents. These products have superior gloss and gloss retention when compared to traditional water borne polyurethanes, and are comparable to conventional solvent based polyurethanes. They also have excellent resistance to chemicals, good dry film appearance, and excellent dry rates at ambient room temperatures.

They are designed for use over properly prepared and primed surfaces in aerospace and general industrial applications. They are available in gloss, semigloss, and many colors.

49.8.2 Radiation Curable Coatings

Laboratory development of UV cured coatings began in the '40s and the first commercial applications for the technology occurred in the mid '60s for coating wood products in Western Europe. In 1971 the first commercial application for UV cured particleboard filler was reported.

The energy crisis and increasing concerns over environmental issues accelerated development of UV coatings during the '70s. Radiation cured liquid coatings have several advantages that have stimulated interest over the years.

1. The film properties are very durable due to high crosslink density.
2. Very fast cure cycles
3. Very low temperature for curing
4. Zero or low solvent levels that reduce volatile emissions

Like any coating technology, liquid UV coatings also have disadvantages:

1. Uniform dispersion of pigment can be difficult and curing highly pigmented coatings may be a challenge due to the dependence on light energy dispersion.
2. The volatility of low-molecular weight radiation cure materials requires caution in handling to avoid skin irritation and sensitization.
3. Hundred percent solids liquid materials can be often difficult to atomize and spray uniformly. They may also be sensitive to differences in temperature and humidity in the application process.
4. High shrinkage of coatings when cured may cause challenges in edge coverage.
5. Some materials produce objectionable odors because of uncured monomers, and other raw materials such as photoinitiators.

Products that require a tough surface of a clear, scratch resistant coating are possible candidates for UV coatings. The substrates are often heat sensitive materials such as wood, paper, or plastic run at high production speeds. Because of the limited line-of-sight curing of UV materials the products are often flat such as boards, sheets, and rolls of materials. There has been limited use of UV coatings to date in pigmented systems, metal applications, or applications for three-dimensional construction. However, the compliance issue may lead to further development of UV applications.

49.8.3 UV Powder Development

For the past several years there has been a continuing effort to develop a market for application of powder coatings using UV cure technology. The use of UV lamps and the presence of a reactive photoiniator in the coating make it possible to cure powder on heat sensitive substrates.

In 1998 the first commercial UV powder was installed at Baldor Electric Motors in Westville, Oklahoma. The line was used to powder coat fully assembled electric motors. UV powder made it possible to coat these motors that already contained heat sensitive wiring, lubricants, and lacquers. In 1999 the first commercial UV powder line for wood was installed in Wales at Stilexo Products, a manufacturer of ready to assemble furniture. In 1999 the first U.S. installation for UV powder coating wood was installed at Decorative Veneer.

Powders are formulated in a dry mix process that makes it easier to develop different colors and looks than liquid UV coatings. And powder overspray can be recovered for reuse, offering some economic advantages.

UV powder materials are available and functional but they do cost more than their thermosetting counterparts. At the present time they are not widely used but the future growth is likely.

49.8.4 Powder Coating

There is still a lot of potential growth in thermoset powder coatings. Traditional barriers to growth included slower color change time (compared to liquid coatings) and high cure temperatures. Manufacturers of equipment and materials have work diligently to shorten color change time and provide materials with lower cure temperatures.

Powder offers very good film performance quality and a variety of colors and specialty looks in different chemistries. With exceptional environmental characteristics and the ability to reuse overspray, the future of powder coatings looks good.

49.8.5 Electrocoat

With very low or no VOC and HAPs, electrocoat offers an excellent compliance option. Suppliers continue to work on advances in corrosion protection, lower cure capabilities, and enhanced physical properties. Many ecoat systems offer tailored formulas for specific situations where low cure is needed or exceptional corrosion protection.

The future of electrocoat technology will continue to focus on enhanced performance with minimal environmental emissions and practical applied cost. Eventually, all electrocoat systems will be HAPs- and VOC-free without sacrificing quality performance.

The one limiting factor with ecoat is the inability to run a variety of colors. Still, new applications will be developed and the increases demand for better performance will extend ecoat into other markets for undercoat corrosion resistance.

Summary. Industrial coating development and selection is driven by ease of application, performance and appearance, capital equipment cost, applied cost, and environmental compliance. Growth in the future will depend on how well a coating meets a manufacturer's needs and concerns.

Traditional coatings depend on solvent to provide the fluidity needed for blending and flow during application. Future coatings will continue to move away from organic solvents and find new ways to reduce emissions.

CHAPTER 50
ADHESIVE BONDING AND SEALING

David J. Dunn
F.L.D. Enterprises
Aurora, Ohio

50.1 INTRODUCTION

Adhesives are materials designed to hold materials together by surface attraction, often as alternatives to mechanical fastening systems. Sealing has been defined as "the art and science of preventing leaks."[1] A sealant is normally used either to prevent a fluid or gas from escaping from an assembly or to prevent the ingress of components of the atmosphere into the assembly. The distinction between adhesives and sealants is not always very clear. Many sealants are adhesives but their primary function is to seal a joint with adhesion merely being one important property. Some systems are true adhesive-sealants and fulfill the dual role of bonding and sealing a joint.

50.2 ADHESIVES

Adhesives come in several forms—thin liquids, thick pastes, films, powders, preapplied on tapes, or solids that must be melted. Adhesives can be designed with a wide range of strengths, all the way from weak temporary adhesives for holding papers in place to high strength structural systems that bond cars and airplanes. In many industries, adhesives compete with mechanical fastening systems such as nuts, bolts, and rivets, or welding and soldering.[2]

Adhesives offer several benefits for the joining of materials:

- They distribute loads across the entire joint area.
- They show excellent fatigue properties.
- They attenuate mechanical vibrations and sound.
- Adhesives often fulfill a dual role of acting as both adhesive and sealing a joint against ingress of water or other fluids.
- Because of their low electrical conductivity, adhesives significantly reduce galvanic corrosion between dissimilar metals.
- Adhesive assembly is often much faster and cost-effective than mechanical methods.

Important properties that must be considered when choosing an adhesive include the following:

- Rate of cure
- Gap filling capability
- Tensile shear strength
- Peel strength
- Impact resistance
- Load bearing capability
- Heat and cold resistance
- Fluid resistance
- Health and safety issues

50.3 TYPES OF ADHESIVES

The number of adhesive technologies for load bearing applications is somewhat limited. Most thermoplastic systems, for example thermoplastic hot melts, are totally excluded because they will creep under load, particularly at elevated temperatures. Highly cross linked structures are necessary to support loads.

The most important adhesives for engineering applications are as follows:

- Anaerobics
- Epoxies
- Reactive acrylics
- Polyurethanes
- Special formulations of cyanoacrylates
- Reactive hot melt polyurethanes

Anaerobics are one-component liquids that cure rapidly when confined between two surfaces. Their most successful application is to replace lock washers in threaded fastener assemblies (e.g. nuts and bolts), where they fulfill the dual function of locking and sealing the assembly. Frequently referred to as liquid lock washers, anaerobics adhere strongly to the metal surfaces and prevent loosening from vibration. They also completely fill the void space in a threaded assembly and protect it against corrosion. This is still the most distinctive application for anaerobics because they do not merely act as a liquid alternative to lock washers but truly augment the assembly, particularly in situations under vibration, such as in vehicle applications. To remain effective, assembled fasteners must maintain under tension and the adhesion provided by an anaerobic increase the forces required to loosen the assembly. Furthermore, even if an assembly should loosen, the fastener will not disassemble completely (such as a screw falling out or a nut coming off) because the cured anaerobic completely fills the void spaces between threads—thus preventing catastrophic failures.

Threadlockers can be formulated to have very low strengths, such as for small adjustment screws, medium strengths for assemblies requiring disassembly, or high strength for permanent assemblies.

In addition to liquid anaerobics, threadlockers are supplied in a solid form that can be preapplied to a fastener. In these systems, the anaerobic adhesives or a catalyst are microencapsulated in a thin shell. This preapplied system remains stable until a nut is applied to the fastener when the shearing action of engagement breaks the microcapsules and the adhesive cures, locking the fastener.

Other important applications for anaerobics include the sealing of threaded pipe joints, gasketing, structural adhesives, and the bonding of slip fitted cylindrical parts (often called "retaining" compounds).

Epoxy adhesives have been the major adhesive family used for structural bonding of metals and composites in the aerospace, industrial, and automotive industries. They are characterized by curing to hard infusible resins that bond to a wide range of metals and have excellent heat and environmental resistance. Epoxies have been used since the 1940s and have an excellent track record of successful structural bonding. Both two component and heat cured versions are available and literally thousands of formulations have been developed over the years for specific applications. One of the traditional weaknesses of epoxies, namely, their brittleness, has been the focus of research attention for the last 20 years or so and significant breakthroughs have been made to produce tougher epoxies by the incorporation of rubbers such as *carboxy terminated butadiene-acrylonitrile* (CTBN), polysulfides, and urethanes. The necessity for long cycle times for heat curing has been mitigated in many automotive and industrial applications—first, by using current paint bake ovens in the automotive industry to cure the adhesives; and second, by using technologies like induction heating where parts can be fixtured in seconds.

Reactive acrylic adhesives are two-component 100 percent reactive systems, based on methacrylate and acrylate monomers that are toughened with dissolved rubbers and are cured either by using a surface primer, or by mixing two components together. These adhesives have proved to be extremely useful in demanding plastic and metal bonding applications where fixture times of a few minutes and strong, impact resistant bonds are required. They often compete very effectively with epoxy adhesives. An attractive feature of acrylic adhesives is that with very few exceptions, they are formulated from a wide range of low cost monomers. Acrylic adhesives do have the inconvenience of being two-part systems and some have unpleasant odors and are flammable. However the vast range of available raw materials makes them very attractive to the adhesives formulator.

Polyurethane adhesives are one or two component systems that cure to very tough flexible adhesives. One-component systems use atmospheric moisture (or can be misted with sprayed water), whereas two component systems must be mixed thoroughly before use. The largest single application for polyurethanes is the structural bonding and sealing of automotive windshields and they have also proven to be very effective in large area panel bonding of metals and composites.

Cyanoacrylates, the so-called one-component instant adhesives, were traditionally not considered as structural adhesives because of their thermoplastic nature, brittleness and limited thermal resistance. However, new formulations toughened with elastomers have true structural properties but their use is limited to relatively small parts because of their high cost.

Reactive hot melt polyurethanes (RHMUs) are a relatively new technology, and are single component adhesives that are applied molten like conventional hot melt adhesives, with virtually instant fixturing, and then cure by reaction with atmospheric moisture to give a cross-linked adhesive. They are very versatile adhesives whose major limitation is the expensive equipment required for storage and application. Typical applications are large area panel bonding, high performance bookbinding adhesives, and bonding automotive interior trim. Table 50.1 compares the properties of these different types of adhesives.

TABLE 50.1 Comparison of Adhesives

	Anaerobics	Epoxies	Reactive acrylics	Polyurethanes	Cyanoacrylates	RHMU
Relative rate of cure	Medium	Slow	Fast	Slow	Very fast	Fast fixture slow cure
Gap filling	0.5 mm	No limit	1 mm	No limit	0.5 mm	3–5 mm
Tensile shear strength	3000–4000 psi (21–28 Mpa)	3000–5000 psi (21–35 Mpa)	3000–4000 psi (21–28 Mpa)	2000 psi (14 Mpa)	2000–3000 psi (14–21 Mpa)	1000–2000 psi (7–14 Mpa)
Peel strength	Very low	Low	High	Very high	Low	High
Impact resistance	Very low	Low	High	High	Low	High
High temp. resistance	200°C	200°C	150°C	180°C	100°C	150°C
Fluid resistance	Excellent	Excellent	Good	Fair	Fair	Fair
Cost	High	Low	Medium	Medium	Very high	High

50.4 TYPICAL APPLICATIONS FOR ADHESIVES

Adhesives are used in a wide range of industries, most frequently to bond metals or plastics.

50.4.1 Bonding of Metals

Although new materials like plastics and composites have replaced metals in many industrial and household products, metals will continue to be used where their unique combination of properties such as high strength and high temperature resistance are coupled with low cost.

Epoxy adhesives have traditionally been the most commonly used adhesives for metals in industrial and aerospace applications. Reactive acrylics have been shown to be very effective in bonding unprepared surfaces such as those contaminated with drawing or cutting oils. Researchers at Henkel-Loctite have recently shown that by adding activators to common cutting oils, reactive acrylics can be cured by metal surfaces fabricated with these lubricants.[3] Cyanoacrylate adhesives are very effective in the rapid assembly of small metal components.

In addition to the joining of iron and steel, technologies for the joining of other metals are being researched in order to produce assemblies of exceptional strength and durability under high stress situations and in hostile environments. One driving force for using alternative metals is the tremendous weight savings that can be achieved through the use of so-called light metals such as aluminum and magnesium. This is of particular interest to the automotive industry where weight savings can be translated into considerable gains in both fuel economy and performance.

Aluminum is 60 percent lighter than steel and has a very good strength to weight ratio. The metal has excellent corrosion resistance and is used widely in industrial, construction and consumer goods.

Magnesium is the lightest metal that is stable under normal ambient conditions and produced in quantity. It is 33 percent lighter than aluminum and 75 percent lighter than steel and iron.

Nonferrous metals like aluminum, magnesium and titanium have been bonded using high performance epoxies and nitrile-phenolics in the construction of aircraft, often in combination with other fastening methods such as rivets. Lengthy and expensive pretreatment and curing processes have been designed to ensure absolute reliability in bonded joints, both in terms of structural strength and resistance to corrosion. Although this experience gives confidence to the nonaerospace design engineer, most of the assembly processes used in aerospace are totally incompatible with the rapid production line assembly operations practiced by industrial and automotive manufacturers.

The use of aluminum sheet offers considerable potential to reduce the weight of an automobile body. The mass production of cars with aluminum and plastics replacing steel in body structures is still some years away for both cost and performance reasons but many components such as doors, fenders, hoods, and trunk lids are now manufactured out of materials like thermoset polyurethane, unsaturated polyester *sheet molding compounds* (SMC) and aluminum. Mechanical fastening and heat curing cannot be used with these systems and adhesive bonding with two-component epoxies and polyurethanes is employed. H. Flegel of Daimler-Chrysler has described the use of structural adhesives in autobody construction and concluded that modern lightweight design, safety, and modular concepts can no longer do without adhesively bonded joints and the strength they provide in a crash scenario.[4] With reference to the Mercedes S-Class coupe, he describes that it is a multimaterial design using steel in areas such as roof pillars, longitudinal members, and cross members that are subjected to high loads during a crash; aluminum for the manufacture of large-area parts such as the hood, roof, tailboard and rear fender; magnesium for the interiors of the doors; and plastic for attachments such as the trunk lid, bumper, and front fender. The body-in-white is adhesively bonded and reinforced using self-piercing rivets. The combined joining method considerably increases the energy absorption capacity and hence the crash resistance, in contrast to spot-welded joints. In addition the structural rigidity of the body can be increased, depending on the vehicle design, by about 15 to 30 percent compared with welded components. Some 90 m of adhesively bonded joints are used in this vehicle.

Adhesive bonding is usually not the only possible joining method available to the design engineer. Table 50.2 shows the strengths and weaknesses of adhesive bonding of metals compared to other methods.

TABLE 50.2 Comparison of Assembly Methods for Metals

	Adhesive bonding	Mechanical fastening	Soldering/ brazing	Arc welding	Spot welding	Riveting
Joint stresses	Uniform distribution	Point of high stress at fasteners	Fairly uniform distribution	Fairly uniform distribution structure	Local stress points in structure	Local stress points in structure
Fatigue resistance	Excellent	Need to prevent vibration loosening	Good	Special methods often necessary to enhance resistance	Often poor	Good
Use for dissimilar metals?	Excellent for most combinations	Usually good-some problems with soft metals	Some capability	Limited capability	Limited capability	Excellent
Joint preparation	Cleaning and pre-treatment often necessary	Hole drilling or tapping required	Pre-fluxing necessary	Little or none required	Little or none required	Holes when non-piercing rivets used
Temperature resistance	Limited	Very high	Limited by softening point of metals used	Very high	Very high	Very high
Joint appearance	Invisible bondlines	Surface discontinuities, often show	Usually acceptable	Usually acceptable	Surface discontinuities, often show	Surface discontinuities, often show
Production speeds	Slow to fast depending on adhesive type	Slow to medium	Can be very fast	Can be very fast	Can be very fast	Can be very fast
Material costs	Low	High	Medium	Low	Low	High
Operating costs	Low	Medium	Medium	High	Medium	Medium
Capital costs	Low	Low	Low	High	High	Low

Other technologies and combinations of technologies which can be used are

Mechanical Clinching. This is a form of mechanical fastening which involves the joining of two sheets of metal by deep drawing the sheets into a cup shape using punch and die tooling.

Clinch Bonding. This technique combines mechanical clinching and adhesive bonding.

Weld Bonding. This is a process that combines adhesive bonding with resistance spot welding and is carried out by either spot welding through an uncured adhesive joint or by flowing adhesive into a spot welded joint. The spot-welding is often used to fixture the assembly while the adhesive cures.

Rivet Bonding. The combination of a small number of rivets and a structural adhesive can usually give a much stronger assembly than a large number of rivets used alone. This process has been widely used in the assembly of school buses for many years.

50.4.2 Bonding of Plastics and Composites

There has been a tremendous increase in recent years in the use of plastics to replace metals in component design. Since plastics do not always lend themselves to traditional methods of fastening, there have been increasing demands for adhesive bonding of these materials.

Three types of plastics-base materials can be categorized for bonding:

Thermoplastics, e.g., polystyrene or polyethylene

Thermosets, e.g., phenolics or epoxies

Composites: Plastics that are reinforced with mineral or metallic filler

From the adhesives technologist's point of view, plastics are somewhat of an anomaly: on the one hand bonded joints can be made where the plastic falls before the adhesive; on the other hand certain plastics are very difficult to bond. Although plastics often present difficulties in adhesive bonding (Table 50.3), all can be joined successfully with the correct choice of adhesive, joint design, and surface preparation.

Often compromises in the choice of an adhesive have to be made, for example one may use a very expensive fast-curing adhesive but make tremendous gains in productivity. In other cases a less expensive adhesive may be acceptable but necessitate long cure times in ovens. Modern adhesives for plastics typically are epoxies, cyanoacrylates, polyurethanes, or reactive acrylics, although hot melts can often give good bonds. Solvent cements are still used but mainly on household plumbing fixtures.

The major factor that has limited the penetration of epoxies into plastics-bonding applications has been brittleness. Rubber toughened systems have improved this problem to a certain extent. Two-component polyurethanes are extremely versatile in plastics bonding and are used widely for polyester *sheet-molding compound* (SMC) bonding where their good adhesion and high flexibility are key properties. Limitations include the moisture-sensitivity of the isocyanates leading to foam

TABLE 50.3 Problems in Bonding Plastics

Inherent flexibility of plastics—induces peel and cleavage stresses in bonds
High thermal expansion coefficients compared to metals
Stress cracking from liquids
Presence of mold-release agents
Low polarity surfaces

formation, the need for primers for some surfaces, and some health and safety concerns concerning the toxicity and irritancy of isocyanates. Cyanoacrylates are probably the most versatile adhesives for plastics, with excellent adhesion to most plastics. Limitations include gap filling only to 0.5 mm, temperature resistance to about 100°C, and poor impact resistance. Reactive acrylics have excellent plastics-bonding abilities due in part to their solvency for many plastics. However this can cause stress-cracking in some cases. The major limitation of this class of adhesive is limited gap-filling to about 1 mm. Two-part mixable versions of these adhesives have greater gap filling ability, leading to even greater capabilities for structural bonding and repair of plastics and composites. UV acrylics are also available and can be cured in 1–30 s using UV or visible light. These are used successfully on transparent substrates like PVC and polycarbonate.

When attempts are made to bond nonpolar plastics such as polyethylene, polypropylene, and *thermoplastic polyolefins* (TPOs), very weak assemblies are obtained with tensile shear strengths of less than 0.7 Mpa (100 psi) and which readily fail under slight impact or exposure to moisture.

A significant breakthrough was the discovery by Loctite Corporation that certain surface primers can give considerable enhancements to adhesive strength with cyanoacrylate adhesives.[5] These primers give bonds so strong that failure often occurs in the polyolefin substrates (shear strengths of over 7 MPa [1000 psi] are consistently obtained). 3M has recently introduced two-part reactive acrylic adhesives that can bond many low surface energy plastics, including many grades of polypropylene, polyethylene, and TPOs without special surface preparation. This is a very versatile, tough adhesive system with good tensile shear strengths on a wide range of substrates plus peel strengths of 28–32 N/cm (16–18 pli). The major limitation of these adhesives appears to be their cure time, with time to reach handling strength being 2 to 3 h at room temperature and full cure in 8 to 24 h—although heat can accelerate the reactions.

Plasma treatment has always been an option for treating polyolefin surfaces to improve bondability but has traditionally been a difficult and expensive process. However, new equipment such as Openair plasma technology[6] is allowing cost effective treatments.

50.5 SEALANTS

Although most sealants are highly viscous liquids, some, such as porosity sealants, can be very low viscosity "water-thin" liquids. Liquid sealants compete with preformed seals and gaskets based on elastomers, plastics, or composite materials.

There are usually several types of sealant available for particular application. The product designer has to be able to choose the most cost-effective sealant that will perform in the joint to be sealed and have the requisite durability for the expected lifetime of the product.[7]

All sealants must fulfill three basic functions:

1. Fill the space to create a seal.
2. Form an impervious barrier to fluid flow.
3. Maintain the seal in the operating environment.

First, the sealant must flow and totally fill the space between the surfaces, and it is very important that it be able to conform to surface irregularities. Strong adhesion to the joint surfaces is also a necessary requirement for most sealant applications.

Second, the sealant must be totally impervious to fluid flow. Some older traditional sealing materials such as fiber sheet packing and cork are impermeable only when sufficiently compressed to close their inherent porosity. Most elastomers used in sealants are inherently impermeable to liquids, but care must be taken when gases or vapors are involved. For example, silicones are excellent sealants for liquid water but have high moisture vapor transmission rates.

Third, the sealant must maintain the seal throughout the life expectancy of the joint, often under severe operating and environmental conditions. Aircraft, automotive, and industrial sealed assemblies are often exposed to severe movements caused by vibration, mechanical strain, and changes in temperature, pressure, and velocity. The sealants in these systems must also be inert to the chemical effects of sealed liquids, which can encompass a wide range of materials, including aqueous solutions, hydraulic fluids, gasoline, organic chemicals, motor oil, jet fuel, liquid oxygen, strong acids, and alkalis. Aggressive liquids can cause the swelling, chemical degradation, and debonding of a sealant. Although a small amount of swelling can help to seal a joint, severe swelling is often the first symptom of the onset of chemical degradation.

Important properties that must be considered when choosing a sealant include the following:

- Rate of cure
- Depth of cure
- Shrinkage on curing
- Adhesion
- Hardness and flexibility
- Tensile strength
- Compressive strength and compression set
- Stress relaxation
- Creep
- Heat and cold resistance
- Fluid resistance
- Moisture and gas permeability

50.6 TYPES OF SEALANTS

Sealants can be liquids that cure in place, or preformed seals and gaskets that are usually made from high performance elastomers. Table 50.4 shows the main types of sealing materials.

Figure 50.1 shows selected high performance elastomers in the classic ASTM format, representing them in terms of their temperature and oil swell resistance performance.

TABLE 50.4 Common Sealing Materials

Cure in-place sealants
Anaerobics
Silicones
Polysulfides
Pre-formed sealants
Acrylic (ACM)
Ethylene Acrylic (AEM)
Epichlorohydrin (ECO)
Ethylene/Vinyl Acetate (EVM)
Chlorinated polyethylene (CPE)
Chlorosulphonated ethylene (CSM)
Fluoroelastomers (FKM)
Perfluoroelastomers (FFKM)
Hydrogenated Nitrile (HNBR)
Silicone (VMQ)
Fluorosilicone (FVMQ)

( ) indicates standard ASTM abbreviations for elastomers.

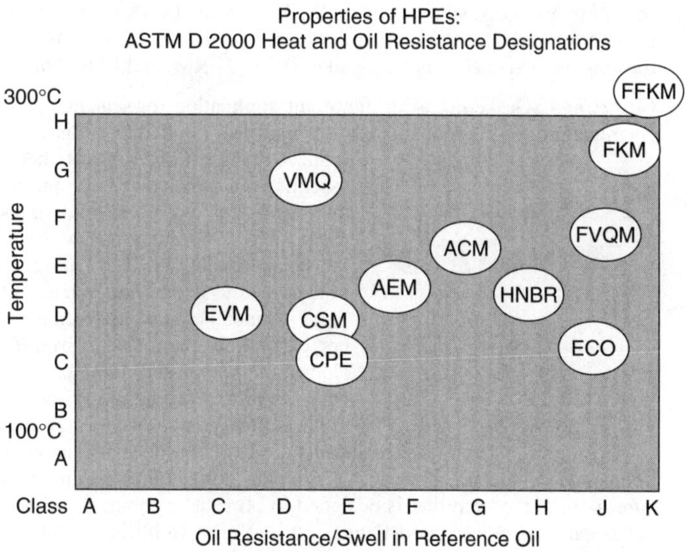

FIGURE 50.1 Temperature and oil swell resistance of high performance elastomers.

50.7 TYPICAL APPLICATIONS FOR SEALANTS

The major applications for sealants are:

- Threaded pipe joints
- Dynamic and static seals
- Gasketing
- Potting and encapsulation
- Porosity sealing
- Aircraft joint sealing

Threaded Pipe Joints. These are used in automotive, industrial, and chemical processing. Anaerobic sealants have gained wide acceptance for sealing metal threaded joints because of their ease of application and high performance. They are much more versatile than polytetrafluoroethylene (PTFE) tape or pipe dopes and can be formulated to give varying levels of adhesion, allowing for permanent assemblies or those that might need disassembly for maintenance operations. The only application where anaerobics are not recommended is for sealing liquid oxygen systems, because anaerobics are organic in nature and could act as fuel for a fire. High quality PTFE tape or paste or gel sealants based on mixtures of polychlorotrifluoroethylene and PTFE are the preferred sealants (see Fig. 50.2).

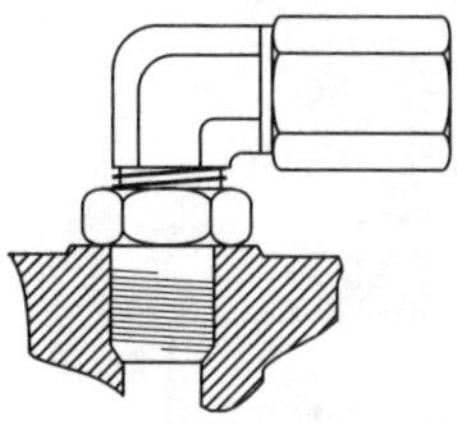

FIGURE 50.2 Threaded pipe joint.

Dynamic and Static Seals. With the exception of gasketing where liquid cure in place sealants are sometimes used, most dynamic and static seals are based on preformed O-rings and other seal designs made out of high performance elastomers. Automotive applications in engine, transmission, and axle sealing frequently use large volumes of ACM, AEM, and HNBR rubbers with smaller amounts of FKM and FVMQ for specialized applications such as resistance to fuel. In aerospace joints, the highest performing and expensive materials like FVMQ, FKM, and FFKM are commonplace.

Gasketing. Gasketing is an important application for sealants, particularly in the automotive and chemical processing industries. A conventional molded or cut gasket creates a seal by compression, induced by the tightening of the fasteners (see Fig. 50.3). It is essentially a compressed spring, comprising flanges, gasket, and fasteners. Any loss of tension in the system caused by stretching of the fasteners, vibration loosening, or creep or relaxation of the gasket will ultimately lead to failure. In contrast, a liquid formed-in-place-gasket does not cure until after assembly. It has metal-to-metal contact between the flanges, which, with the fasteners, carry all of the tension. Even if some tension is lost, the adhesion of the sealant will often prevent the failure of the system. It is very important that flange systems be specifically designed for or modified for formed-in-place-gaskets.

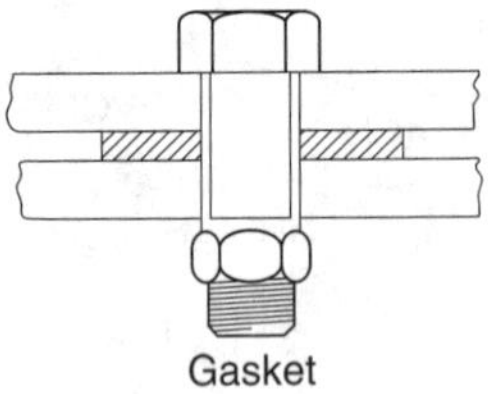

FIGURE 50.3 Gasket.

Two types of liquid gasket sealants are commonly used: anaerobics and medium-modulus Room Temperature Vulcanization (RTV) silicones. Anaerobics require rigid flanges and small gaps and cure to a rigid system that is designed to prevent any joint movement. Medium-modulus RTV silicones can be used on rigid flanges but are also useful in flexible flange situations in which joint movement is expected. In this case, it is very important that the flange design allows a sufficient gap, primarily to permit atmospheric moisture to cure the silicone and secondarily to maximize the joint movement capability of the sealant. RTV silicones have been used in the United States since 1971 for formed-in-place automotive gasketing in engine, transmission, and rear axle sealing.

RTV silicones face strong competition from molded silicone gaskets used by U.S. automotive manufacturers, and European manufacturers have always preferred to use molded rubber gaskets in most applications. Furthermore, many applications are developing for liquid injection molded gaskets that can be injected and cured into preformed grooves in an assembled component. A new development is the formation of gaskets on flanges by applying a liquid silicone or acrylic and curing it by UV radiation.

Potting and Encapsulation. These are terms used in the protective sealing of components on printed circuit boards for the electronics industry. Potting (Fig. 50.4) involves enclosing a component in a reservoir and filling with a so-called potting compound. Development of adhesives with controlled flow characteristics can often replace potting with encapsulation, thus removing the necessity for the reservoir. This type of encapsulant is frequently called a "glob top" adhesive. Epoxies, acrylics, and silicones are the most common materials used.

Porosity Sealing. Most castings and powdered metal parts have porosity, inherent in the powder metal compacting and sintering process and caused by volatilization of organic impurities in castings. With castings such as iron or aluminum, this porosity allows fluid leakage and with powdered metal parts, can cause severe plating problems. Pores can be quite large, designated as macroporosity,

FIGURE 50.4 Potting and encapsulation.

or very small, called microporosity. Sealing of the pores is a routine necessity and is carried out by impregnation with a liquid sealant followed by a curing step. The modern way of sealing this porosity is to use anaerobic impregnants. They are low viscosity impregnants and have revolutionized porosity sealing because of their ability to penetrate microporosity and cure very quickly at room temperature.

Aircraft Joint Sealing. Sealants are used widely in aircraft to maintain pressurization in cabin areas, to retain fuel in storage areas, to achieve exterior surface aerodynamic smoothness, and to weatherproof the airframe. They need to have special properties for aerospace applications, including high and low temperature resistance (–60 to 120°C), resistance to jet fuel, adhesion to aluminum, stainless steel, titanium and composites, and have the ability to maintain electrical continuity. Two-part polysulphides are the most widely used products because of their outstanding fuel resistance and flexibility. Epoxy cured polythiothers are used for faster curing. Noncuring polysulphide pastes are often used as channel sealants round fuel tanks. Fluorosilicones and cyanosilicones have also been used for this application.

50.8 APPLYING AND CURING OF ADHESIVES AND SEALANTS

Adhesives can be applied as shots and beads, roll coated, or sprayed. Application equipment ranges from simple bottles and tubes to sophisticated robotic automatic systems. The simplest application techniques are airless spraying or roll coating—these are used frequently in the bonding of large panels, e.g., structural insulated panels or insulated garage doors. However, most adhesives and sealants need to be applied in small amounts and with some precision. This can be done with manual applicators such as syringes or cartridges or can be automated.

Important factors to consider when choosing dispensing equipment include the following:

- Nature of the product—single or dual component
- Viscosity and rheology
- Precision required during application
- Possible interaction of the product with materials in the dispenser
- Pot life issues in the dispenser due to heat or atmospheric contamination

In general, single component materials are the easiest to handle but there are several potential problems that may arise:

- Anaerobic adhesives should be kept in contact with atmospheric oxygen to maximize pot-life and care should be taken to exclude transition metals in the dispensing lines and valves, which can act as accelerators for the adhesives. Stainless steel and polyethylene are usually the best materials for dispensing lines and pots.
- Hot melt polyurethanes dispensers must be purged with dry air or nitrogen to prevent premature curing and since their viscosities tend to increase rapidly on prolonged storage at high temperatures, this must be checked frequently and appropriate timers or temperature controllers should be installed to prevent overheating during line shutdowns.
- Cyanoacrylates are quite sensitive to acidic or basic impurities and their stability and reactivity should be checked after storage in dispensers. Clean up can often be a problem with cyanoacrylates where the adhesive often cures in, for example, drip trays or on the outside of joints. Precise application is the best solution—otherwise disposable drip trays or trays coated with PTFE should be used.
- Seals in equipment can be swollen, destroyed, or bonded. Usually, constructing seals from high performance materials like PTFE or fluoroelastomers will mitigate these problems.

Two-component adhesives and sealants have some similar potential handling problems:

- Epoxy adhesives are sometimes highly filled with abrasive particles, which must be taken into account in the design of dispensers and seals.
- Two-component urethanes are very sensitive to contamination by moisture, which can cause premature curing or gas bubbles in the adhesive.

Two-component materials can be premixed and applied within the specific pot-life of the mixed material or can be stored separately and then be metered and mixed. In recent years many types of two-component syringes and cartridges have become available with varying mix ratios and fitted with static mixing nozzles.

Fully automated systems are available for both single and two component adhesives and sealants, and there are a number of different options, degrees of sophistication and price levels. An automated system comprises the following:

Delivery System. Materials can be supplied in bulk, in pails, drums, large totes or tanks, or in small containers that include cartridges, syringes, or cans. Depending mainly on the viscosity of the material, a delivery system can use gravity feed or be pressurized. Peristaltic pumps or gear pumps can also be used for delivery.

Metering. The materials are applied by measured volume shots. The most common metering systems are positive displacement piston fillers and timed shot dispensers.

Mixing. Mixing of two component materials can be done using a static (sometimes called motionless) mixer where the materials are forced through a series of right- and left-handed helical elements at 90° with no moving parts. Dynamic mixers are also available that use high shear to mix the two components. Pin/blade and helical designs are common and plastic mixers with rotating mix elements become common in recent years.

Dispensing. Valves range from simple ball valves to pinch valves and needle valves. The "suck-back" or snuffer valve is invaluable for dripless dispensing. The liquid flow commences on the forward stroke of the valve spool. When the spool retracts, a vacuum is created and an adjustable, dripless snuff-back occurs at the dispense-nozzle outlet.

Automation. Automation offers many benefits from increased production rates, accuracy and consistency to decreased labor costs. Most automated systems still use indexing rotary tables or conveyors and apply adhesives and sealants using a "Z" lift to move the dispense valve vertically to a preset height. Once the valve is elevated, a new part is indexed under the valve nozzle, the valve is lowered and dispensing starts. Programmable XYZ tables can handle applications requiring several different dispense patterns. The most sophisticated motion device is a multiaxis robot and these are now being used quite widely in a wide range of industries.

50.9 HEALTH AND SAFETY ISSUES

Many health and safety issues with adhesives and sealants have been mitigated with the movement away from solvent-based systems. Some activators and primers are still solvent based but have been changed from chlorinated solvents to safer ones such as alcohols, acetone, and hydrocarbons. Some reactive acrylic activators are used undiluted or dissolved in monomers.

However, there are still some concerns, even with 100 percent solids systems. Examples are

Epoxy hardeners. Some amines are quite toxic.

Cyanoacrylates. Highly irritant vapors.

Polyurethanes. Some isocyanates in these systems are irritating and toxic.

Anaerobics. A small percentage of people can suffer skin sensitization.

Reactive acrylics. Systems based on methyl methacrylate have noxious irritating odors and are highly flammable.

Reactive hot melt polyurethanes. Danger of burns from molten adhesive.

50.10 FUTURE TRENDS

The increasing use of plastics and composites will continue to fuel demand for high performance adhesives and they will continue to replace other methods of fastening. Sealants will need improvements in performance and will find new applications.

In particular:

- The increasing use of polyester SMC, other composites, and light metals in automotive panels will increase demand for polyurethanes and epoxies.
- Better and faster curing adhesives for plastics like polyethylene and TPOs will lead to increased usage.
- The auto industry is increasingly demanding better products to meet emissions legislation, increase fuel economy, and lengthen warranties. They will use lower viscosity lubricants and aggressive oil additives, and will challenge the sealant industry to develop higher performing and cost-effective products.
- Liquid cure-in place gaskets will decline in the auto industry and be replaced by UV cured form-in-place systems and liquid injection molded systems. Currently limited exclusively to silicones, these technologies will be expanded to use alternative high performance elastomers.
- The increase in the use of electronics in automotive will enable many new applications for RTV silicones to seal and weatherproof components.
- High performance liquid and reactive hot melt urethanes will continue to expand their market share in automotive and in the bonding of large area panels.

REFERENCES

1. G.S. Haviland, *Machinery Adhesives for Locking Retaining and Sealing,* Marcel Dekker, New York, 1986.
2. D.J. Dunn, *Adhesives and Sealants,* Rapra Technology, UK, 2003.
3. S. Morrill, J. Woods, and T. LaBelle, *26th Annual Meeting of the Adhesion Society,* Myrtle Beach, SC, February 2003.
4. H.A. Flegel, *Auto Technology,* Vol. 5, No. 64, 2002.
5. Y. Okamoto and P.T. Klemarczyk, *J. Adhesion,* Vol. 40, No. 81, 1993.
6. www.plasmatreat.com
7. D.J. Dunn, *Engineered Material Handbook,* Vol. 3, ASM International, 1990, p. 48.

P • A • R • T • 6

MANUFACTURING PROCESSES DESIGN

CHAPTER 51
LEAN MANUFACTURING

Takashi Asano
Japan Management Association Consultants, Inc.
Cincinnati, Ohio

51.1 INTRODUCTION

Lean manufacturing is a system whose techniques aim to significantly eliminate waste in the manufacturing process. Japan had learned management and improvement techniques/methods such as *industrial engineering* (IE) and *quality control* (QC) from Europe and America. Those methods were further developed in Japan and resulted in the technique known throughout the world as KAIZEN. During this movement, Toyota developed their own unique methods in manufacturing. The concept was completely different from the mass production method (represented by Ford's T-type). In Ford, they produced one type of automobile as a batch. That's why it is referred to as "mass production." However, Toyota started to produce automobiles one by one when it was actually needed. Moreover, suppliers started to produce parts only when they were needed (producing to actual demand). In contrast to the *push production* used by almost all other manufacturers, Toyota's concept was referred to as *pull production*. Moreover, they tended to expose problems by reducing inventory for materials, work-in-process, and then solved those remaining problems thoroughly. Lean manufacturing is a name applied to these ideas and methods. Toyota used its own term, *Toyota Production System* (TPS), instead of using lean manufacturing.

Producing products without waste and providing them to customers is one of the essential functions in manufacturing companies. In the current economic situation, many manufacturers in North America have moved/have been moving to Mexico and to China. Therefore, manufacturers especially in North America must maintain/achieve cost competitiveness by completely eliminating waste from their manufacturing processes and gaining the competence to achieve these processes. In this chapter, I will explain the concepts, the tools, and procedures of lean manufacturing from a high-level view. Moreover, I will explain the KANBAN system in detail, as it is one of the core parts of lean manufacturing. Finally, I will explain the analysis and improvement tools for stagnation of materials. As the procedure for lean manufacturing, I will introduce the technique of *value stream mapping*.

51.2 CONCEPT OF LEAN MANUFACTURING

Although there are several ways to categorize wastes, I simply categorized them into the following three:

- Material waste
- Equipment waste
- Operation waste

Finally, these are all related to the waste of money. In short, you need to remove waste from these main production factors.

51.2.1 About Material Waste

Toyota paid special attention to the waste of overproduction. Products or parts produced above the necessarily required are mainly categorized into *material waste*. However, costs of equipments and operations are added to material costs corresponding to the progress of production processes. When the product or the part is not used, all costs incurred by the product become wasteful. Following these often-held beliefs causes the waste of over production.

- Production with a large amount of the same item reduces cost. (mass production)
- Continual production despite many work-in-processes at the down stream process is beneficial. (partial optimization)
- There is a limit in production capacity. Moreover, the production method does not change at once. (abandonment)
- There is a change in demand, or, because the procurement goods are not always available when required it is necessary to stock the items just in case. (safety stock)

Toyota completely changed the current ideas of the time, and achieved the fastest materials flow. *Material* mentioned here is a generic name which includes raw materials, material, parts, work-in-process, components, subassembly parts, and final products. The first fundamental principle behind the transformation from conventional material flow to the ideal material flow is *pull production*. The second fundamental principle is the leveling of production (HEIJUNKA). Pull production is a method which makes only the amount required by the following process. Leveling of production aims to constantly make different kinds of parts in a short cycle and not use a large batch for production. Both methods are combined to give you the fastest material flow and minimum materials waste.

In addition to the above, it is important to look into the lead-time from receiving materials to shipping parts postproduction. This is an important element as inventory costs increase and cash flow suffers if this period is too long. This production period can be broken down into the following four elements.

- Processing period
- Movement period
- Inspection period
- Stagnation period

An important period is the processing of materials (physical transformation, changing in quality, assembly, and decomposition) and it is regarded as *value added time*. The movement period and the inspection period are necessary for the production of the materials in some degree, and finally the stagnation period is waste of time. The ratio of the value added time to the production period can be measured by picking a specific part. Generally speaking, the production period is hundreds to several thousand times greater than the value added time, although this differs by the product type.

When you think of a stagnation place for two or more materials it is possible to divide the character of the stagnation place into three types. This is based on the different status of materials when they come to a place and when they go out of a place. These statuses are the differences in time (timing), amount (quantity), and order (sequence). This will be described in detail later.

51.2.2 About Equipment Waste

It is necessary to consider two points when discussing the investment required, and the operations necessary for eliminating equipment waste. The first point is to only introduce/use equipment, which has the sole function of making the products that are selling. (Do not invest in excessive plant and equipment although it is a difficult management decision.) The production facility engineer should always consider what occurs on the site and how the machine is actually handled from a viewpoint of material flow. The second point is to use current equipment efficiently.

To use it efficiently does not mean to use the maximum capacity of the equipment to make the products. Rather, efficiently means to produce products based on actual demand without wasting equipment time. In this case the index of *overall equipment efficiency* (OEE) in *total productive maintenance* (TPM) is used as one of the ratios of value added time to available time of equipment.

51.2.3 About Operation Waste

To reduce/remove waste from operations, the first thing to do is to breakdown operation time into three elements. There is the time used for basic functions, the time for assisting basic functions, and then the remaining time. The next step is to increase the ratio of time for basic functions. Basic function means value added time. When discussing the removal of operation waste, there are many people who think that it simply means cutting head count. However, this is far from the truth. The words of Mr. Eiji Toyoda (former president of Toyota) support this point: "A person's life is an accumulation of time, just one hour is equivalent to a person's life. Employees provide their precious hours of life to the company, so we have to use it effectively, otherwise, we are wasting their life."

51.3 LEAN PRODUCTION AS A CORPORATE CULTURE

When introducing lean manufacturing in an organization, it is important to also introduce the concept of lean manufacturing as a corporate culture. The philosophy of making a product without waste is the fundamental belief behind lean manufacturing and should be held strongly by the shop floor worker all the way up to the president of the company. GEMBAism, practicality and trust are also important elements of the culture.

GEMBA means the actual shop floor in Japanese. It is the factory site where the raw materials are actually turned into products. For instance, controlling the quality of product is to control the processes at the factory site (GEMBA). It does not mean just checking the quality documentation at the office. If there is something wrong with the quality of the product, one must go to GEMBA to see how to fix the problem. Mr. Kiichiro Toyoda (former president of Toyota) established this culture in Toyota. Moreover, to a certain degree this example is indicative of the manufacturing culture in Japan. When you introduce lean manufacturing, you must go to GEMBA rather than just sit at the office. This is a principle that must not be forgotten.

The principle of practicality is as follows. Although strategy and high-level plans are important, the most important aspect is implementation. The most difficult phase is not in planning but in implementation. The results depend on how the implementation occurs.

As stated by Mr. Eiji Toyoda, trust/respect of all employees is one of the fundamental ideas behind the philosophy of lean production. If there is no such philosophy, no employee would be willing to improve the company.

This philosophy will never change even if the techniques within lean manufacturing change in the future. In other words, it can be said that lean manufacturing is a culture where the company culture is more important than the techniques which comprise it.

51.4 METHODOLOGY AND TOOLS

There are many tools that can be used when introducing lean manufacturing. These are tools introduced from the fields of industrial engineering, QC, KAIZEN, TPM and other tools/methodologies that Toyota originally developed.

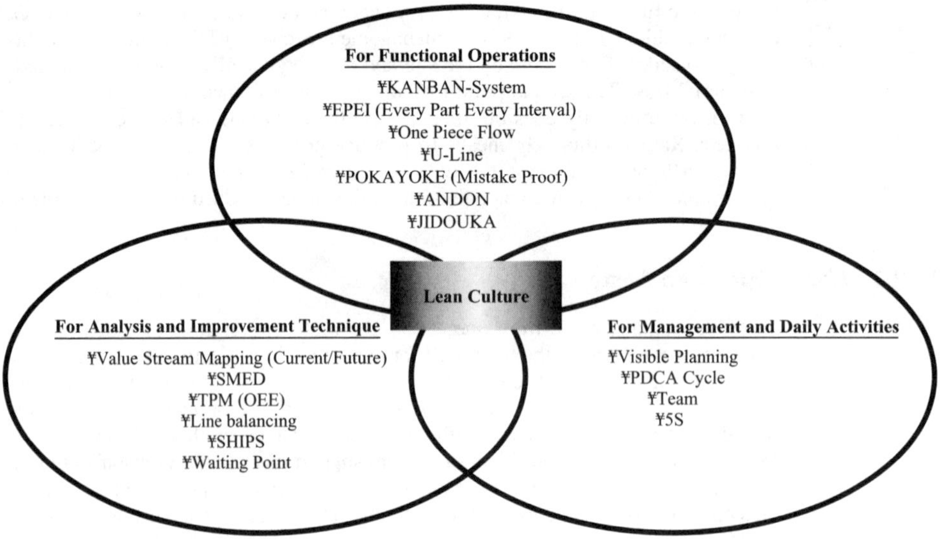

FIGURE 51.1 Tools for lean manufacturing.

These tools can be roughly divided into three areas (Fig. 51.1). First of all, there is a group of tools that are used as solutions such as the KANBAN method or one-piece flow (IKKO-NAGASHI), etc. The second group of tools is used for analysis and improvement and include methodologies/techniques such as value stream mapping, single minutes exchange of die (SMED), TPM, etc. The third group of tools is used for management purposes. Examples of these include PDCA-cycle, visible management, team activity, etc. I will now explain these tools one by one.

51.4.1 KANBAN System

What is KANBAN System? The KANBAN system originated from the production system developed by Toyota Motor. Another name of the system is the *just-in-time* (JIT) system. Waste of all kinds is generated in the production of any product. Toyota classifies these wastes into seven categories: overproduction, waiting time, conveyance, processes, inventory, motion, and defects. Placing a special emphasis on waste arising from overproduction, Toyota developed the KANBAN system as a measure aimed at obtaining long-term improvements in its production system.

Just-in-time began to be noticed from around the latter half of the 1960s—Toyota's significant growth and the worldwide attention that came with it fueled this new interest.

Despite the fame of the KANBAN system, there are a considerable number of people who do not have an accurate understanding of the system. Let's take a look at a few common misconceptions.

1. KANBAN system does not require the use of a computer to calculate the required materials.
2. Because of the JIT nature of the system, there is no inventory.
3. KANBAN system is aimed exclusively at just reducing inventories, etc. Unfortunately, these conceptions are all off the mark. In the KANBAN system, the amount of materials, parts and components needed for a set time cycle—a month or ten days, for example—is calculated using a computer. Inventory, of course, is kept, and is linked into the system. Also, the KANBAN system is not used merely to reduce inventories kept on hand; it has a much broader aim of improving the level of the overall production system. A more in-depth knowledge of the system, therefore, is necessary to avoid and to correct such misconceptions.

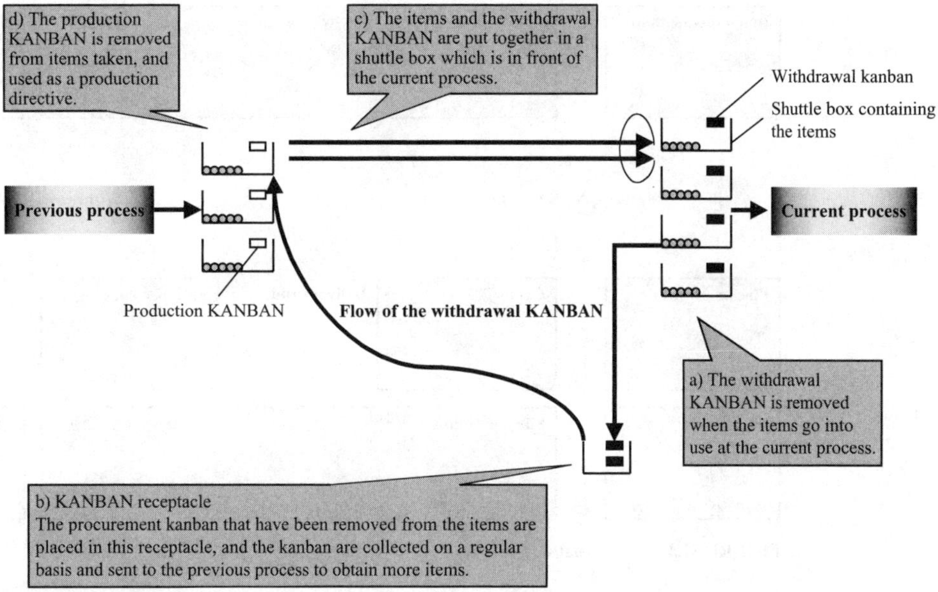

FIGURE 51.2 Procedural flow of the withdrawal KANBAN.

The Format of KANBAN System. The KANBAN system makes use of cards, called KANBAN, on which there is data used as instructions for the ordering, movement, storage and production of materials, parts or components, and semifinished goods. These KANBAN can be divided into two general categories: withdrawal KANBAN and production KANBAN.

Withdrawal KANBAN. The withdrawal KANBAN is used to have parts, components, etc., procured and delivered from a previous process, another factory, or from an outside supplier. Its usage is illustrated in Fig. 51.2. Before the process begins, the necessary parts/components are acquired on the basis of this KANBAN and placed in shuttle boxes. One withdrawal KANBAN is placed into the shuttle box along with the articles. When the parts go into use at the main process, the withdrawal KANBAN is removed and placed into the proper KANBAN receptacle ((a) in Fig. 51.2). The KANBAN that is gathered in this receptacle is collected on a regular basis, and parts corresponding in volume to the number of KANBAN collected are procured form the previous process (b). The withdrawal KANBAN is attached to the relevant parts stocked at the previous process and are sent together to the main process and stocked there (c). The shuttle boxes stacked behind the previous process contain a production KANBAN, which is removed upon receipt of a withdrawal KANBAN and used as a production directive at that process. The repetition of this withdrawal KANBAN cycle maintains the inventories between two processes.

For the sake of clarity, the explanation shows shuttle boxes only at the process before the current one; but it should be pointed out that in reality shuttle boxes containing withdrawal KANBAN are also present midstream. In this case, the withdrawal KANBAN maintains the connections between the processes. Should the previous process be an outside supplier, the KANBAN may be called procurement KANBAN instead of a withdrawal KANBAN. A sample of an actual withdrawal KANBAN is shown in Fig. 51.3.

Production KANBAN. Let's move on now to an explanation of the production KANBAN. This is a work commencement order, and serves as a directive for production to begin. The procedural flow is shown in Fig. 51.4. Here, items containing a production KANBAN are placed behind the relevant process. When these items are taken away as a result of the withdrawal KANBAN discussed in the previous section, their production KANBAN is removed and placed in a receptacle ((a) in Fig. 51.4).

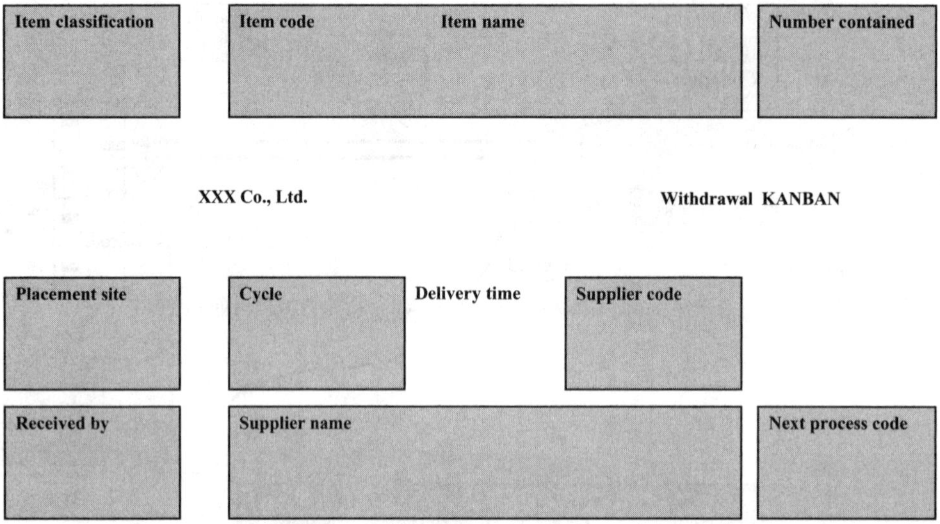

FIGURE 51.3 Withdrawal KANBAN.

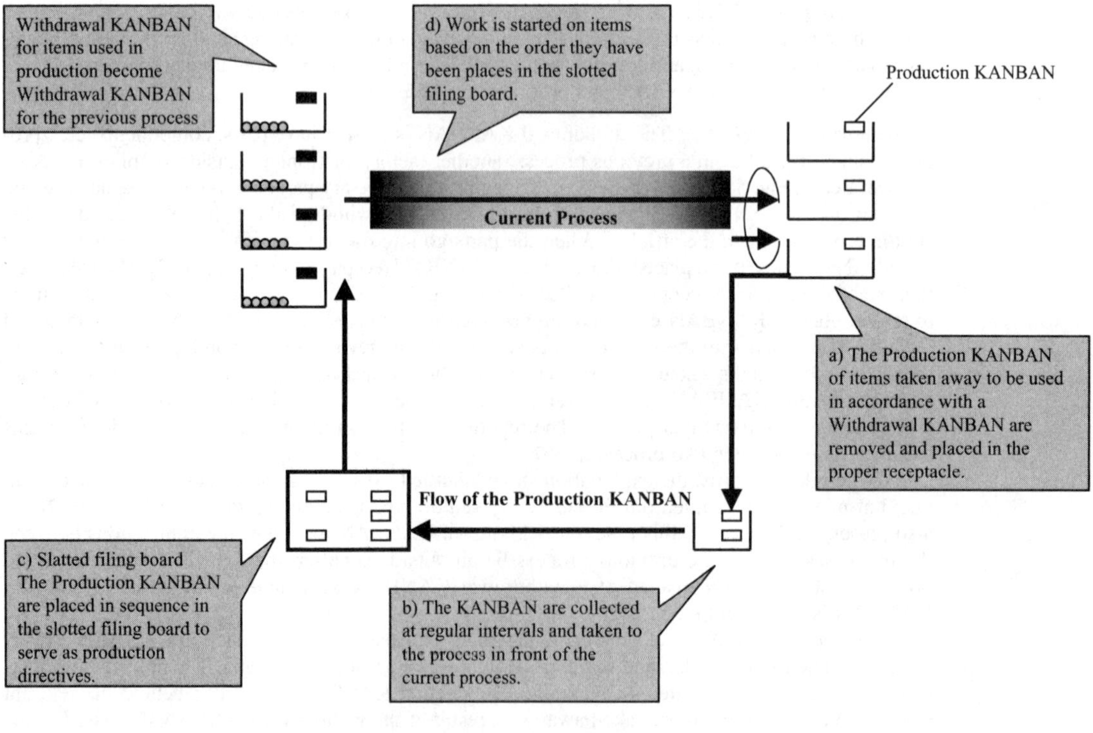

FIGURE 51.4 Procedural flow of the production KANBAN.

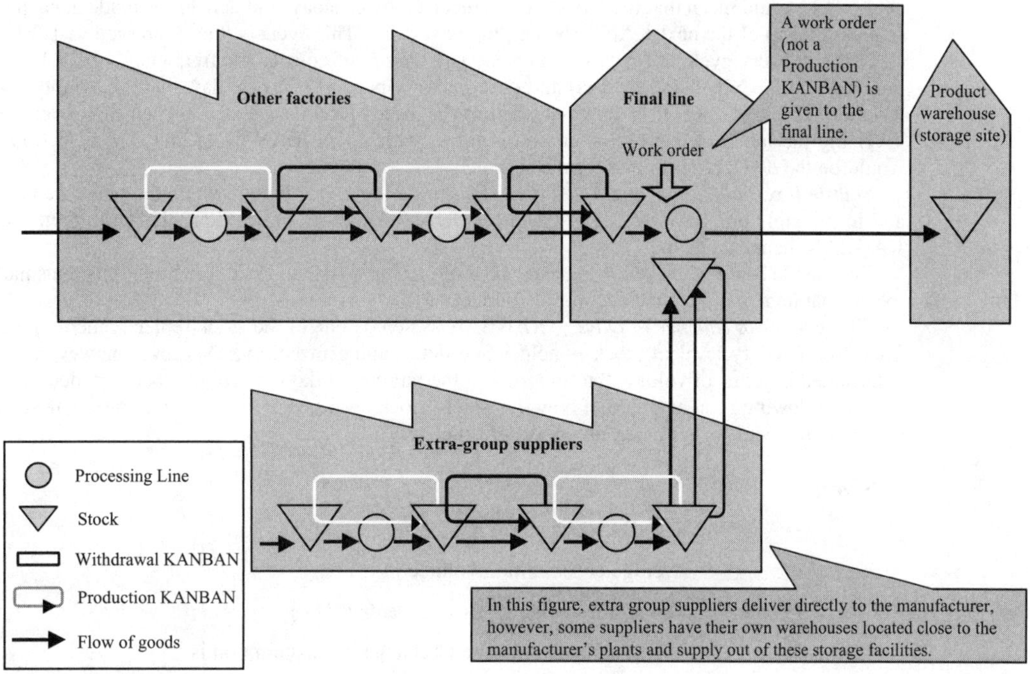

FIGURE 51.5 Overall image of KANBAN system.

These KANBAN are collected at regular intervals, taken to the previous process (b), and placed in sequence in a slot of a filing board (c). The person doing the work proceeds by producing the items in accordance with the production KANBAN in the order they are filed (d). The finished items are placed into a shuttle box together with a production KANBAN, and placed behind the relevant current process. In this manner, the production KANBAN functions as a production directive aimed at having the same amount of an item made as is taken away to be used.

Overall Image of the KANBAN System. The KANBAN system uses the two main types of KANBAN described previously to keep production moving while maintaining a close linkage and connection between outside suppliers, and intracompany factories and processes. The overall image of this system is shown in Fig. 51.5.

To explain briefly, the manufacturer's final line is given a work order—not a production KANBAN—based on the company's production schedule and plan. All processes and outside suppliers in front of this line are linked by the procurement and production KANBAN. This enables the movement of goods and production to proceed in a chain reaction effect with just this one work order.

Terms Used in the KANBAN System. Now that we have understood the basic framework of the KANBAN system, let's move on to a discussion of the various terms used with it in order to gain an even greater insight into the system and its workings. The actual terms used may differ from company to company; however, the basic meanings and their implications should be fairly consistent.

KANBAN Cycle. This refers to the cycle that KANBAN completes in making its rounds, and is usually expressed using three numbers.

Let's use an example here of a procurement KANBAN given to an outside supplier for our explanation, and say that the KANBAN is presented to the supplier at a cycle of twice a day (at 10:00 and 15:00, for example) and that delivery for each KANBAN's worth of goods is to be made at the next cycle. This would mean that the KANBAN is received twice a day, and delivery is made at each subsequent pickup of the next KANBAN, i.e., the next cycle. This cycle is thus expressed as 1:2:1.

If the delivery cycle is four times per day (matching, of course, the frequency with which the procurement KANBAN is received) and each delivery period is 1/4 of a day, then we would have a KANBAN cycle of 1:4:1. In order words, delivery is made four times per day, at each subsequent time.

Using the mode of expression common today, with a KANBAN cycle of $l:m:n$, then $(l/m)*n$ would be the delivery lead-time for the supplier.

Shuttle Box and Shuttle Box Amount. The shuttle box is the box in which the parts are placed, and in principle one KANBAN is used for one box. It shuttles around along the same route as the KANBAN, hence its name.

The shuttle box amount refers to the number of parts in the shuttle box, and represents the smallest lot for that item and the smallest unit of production.

Safety Stock in Number of Days. KANBAN system is one of the stock replenishment systems. Therefore a safety level of stock is held when determining inventory. This level, however, is not determined in terms of volume, but in terms of the number of days' worth of stocks needed.

The following is an example of how this safety stock level is determined for an item whose consumption rate varies over a certain period of time.

Given

KANBAN cycle: $l:m:n$

Average consumption volume/day: D

Maximum consumption volume/day: D_{max}

Then the maximum rate of change in consumption is

$v = (D_{max} f D)/D$

To find the safe level in number of days to ensure that shortages do not occur even if maximum consumption occurs on consecutive days, we need to perform the following calculation:

$$\text{Safe level in days } a = v \times (l/m) \times (n + 1)$$
$$= \{vl(n + 1)\}/m$$

Per-Day Volume. This is the number of parts required for a certain period of time—normally, the amount required per day over a month's time. This is derived by dividing the total volume required over the period by the number of operating days. This per-day volume is obtained for each inventory point by computing the net volume requirements using the figures contained in the final product production plan projections.

Current Numbers of Kanban Cards. The number of KANBAN cards currently required is termed the current numbers of KANBAN card. The calculation is based on the per-day volume figures for one month, or for one ten-day period (in Japan), and the number of actual KANBAN cards is increased or reduced in accordance with the calculation.

Computing Current Numbers of Kanban Cards. Let's take a closer look at how the current numbers of KANBAN cards is computed.

Example

Given

KANBAN cycle, $l:m:n = l$ days, m times, delivered after the nth time

Per-day volume, $D = D$ items/day

Safe stock in days, $a = a$ days' worth

Number contained, $L = L$ units

(1) Current numbers of KANBAN cards is the total of the following:

Delivery time period consumption

$$= \text{delivery time period (days)} \times \text{consumption volume/(days)}$$

$$= (l/m) \times n \times D$$

(2) KANBAN holding period consumption

$$= \text{KANBAN holding period (days)} \times \text{consumption (days)}$$

$$= (l/m) \times D$$

(3) Safety stock days' volume

$$= \text{safe stock days} \times \text{consumption volume (days)}$$

$$= a \times D$$

Current card number = {(1) + (2) + (3)}/number contained

$$= \frac{\{(l/m) \times n \times D\} + \{(l/m) \times D\} + (a \times D)}{L}$$

$$= \frac{\text{D} \times \{(l/m) \times (n+1) + a\}}{L}$$

(round off to the next highest number)

Position in Terms of Production Control

Prototype of the KANBAN System. The history of production control techniques is not complete without examples of systems which resemble the KANBAN system.

The double bin system for ordering parts is just one such technique, and in fact, it can be viewed as a prototype of the KANBAN system. Here, two bins (containers corresponding to shuttle boxes in KANBAN system) are used. When one bin becomes empty, an order is placed for that amount to replenish it, with delivery set for before the other box will be emptied. The container KANBAN is quite similar to those used in the double bin system. The container KANBAN is for the most part a variation of the withdrawal KANBAN. Here, item name, number of items contained, and other information is attached directly to the container. Thus, the container itself is used as a KANBAN. In contrast with the double bin system, we could call this the *n* bin (multiple bin) system.

Pull Production System. As a production control technique, the KANBAN system can be classified as a pull production system, as opposed to the conventional push systems. In push systems, production directives based on scheduling plans are given to each control point, and goods are pushed through from the upstream processes to downstream processes. In the KANBAN system, on the other hand, a production directive is given to the final process along, with only current card number change directives being given to the various control points. The goods used by an upstream process are replenished during the course of daily productions, making this a pull system.

With these features in mind, let us look at aims of the KANBAN system and then move on to explore some future trends.

Definition of KANBAN System. A simple definition and explanation of the KANBAN system is as follows.

Stocks are held between factories (intracompany and outside) and between processes, and are used in accordance with production directives for the final product. As this takes place, production and delivery directives are transmitted in the opposite direction to the flow of goods through the use of cards, known as KANBAN, causing goods to be made or to be directed in a specific flow. The only control functions are the production directive given to the final process in short cycles (up to one-day units), and stock level directives (current card number increase or decrease) given to each

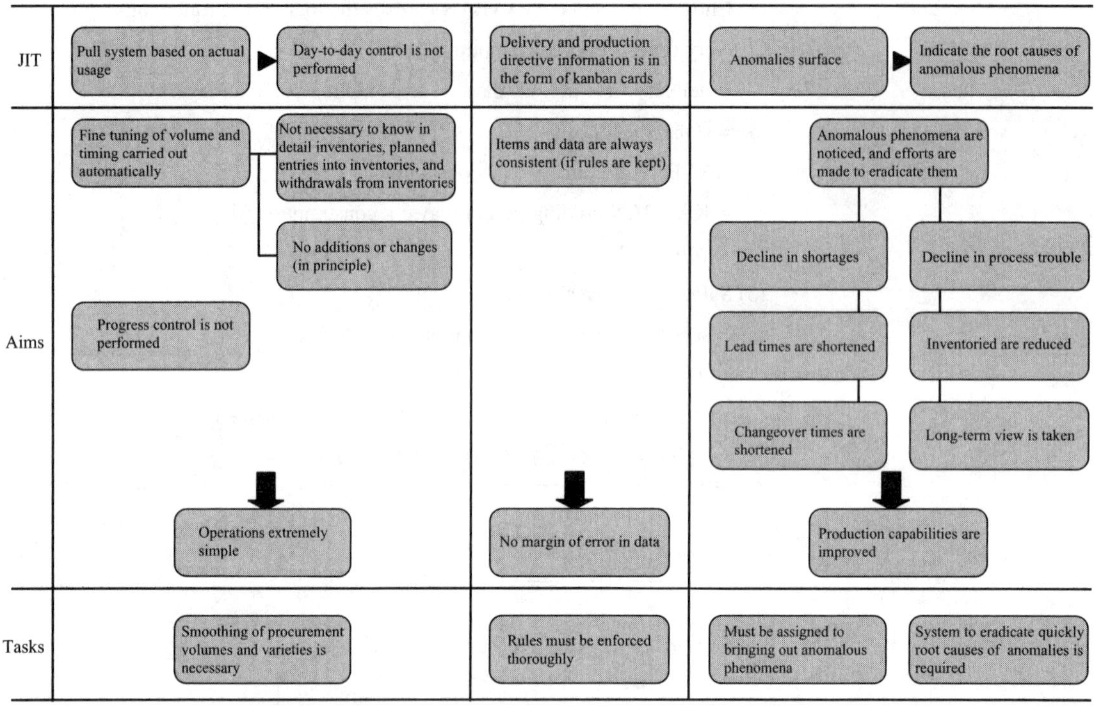

FIGURE 51.6 Goals and tasks of JIT.

control point in long cycles (one month, or one ten-day period). Anomalies such as troubles in the process or delays in supply are discovered immediately in the form of line stoppages. The discovery of such anomalous phenomena is used to identify root causes and to promote modifications, with the ultimate goal of long-term improvement in production capabilities.

The goals and tasks of the *just in time* system based on this definition are listed in (Fig. 51.5).

Aims of the KANBAN System. The aims of the KANBAN system are to simplify daily control for production and to improve production capabilities from a long-range point of view. Toyota engaged in an extreme pursuit of the latter, using the somewhat contrary slogan of "stop the lines." When a line no longer was brought to a halt by a current card number, a further shortening of lead-time and reduction of inventory were sought, gradually lessening the current card number. There is a tendency on the shop floor for the foremen and workers to make an excess volume in order to prevent anomalies and problems from surfacing. This produces the waste of over production. KANBAN system seeks to eliminate this waste by having items made only in accordance with KANBAN, and by baring all problems on the shop floor by controlling the current numbers of KANBAN card. The other aims of a KANBAN system can be seen in Fig. 51.6.

Next, let us touch upon the tasks associated with the KANBAN system. I would like to discuss two of the most significant of these tasks, drawing on my experience with actually working to introduce the system.

Tasks Associated with Introducing the KANBAN System EPEI: Every Part Every Interval (HEIJUNKA). Because stocks are held, procurement of all items and the issuance of production directives are based on actual use, shortages quickly result when the dispersion of actual use is large. The only way to prevent this is to increase the size of the inventories held. If one seeks to avoid both shortages and overstocking, then it becomes necessary to apply EPEI for the production directive

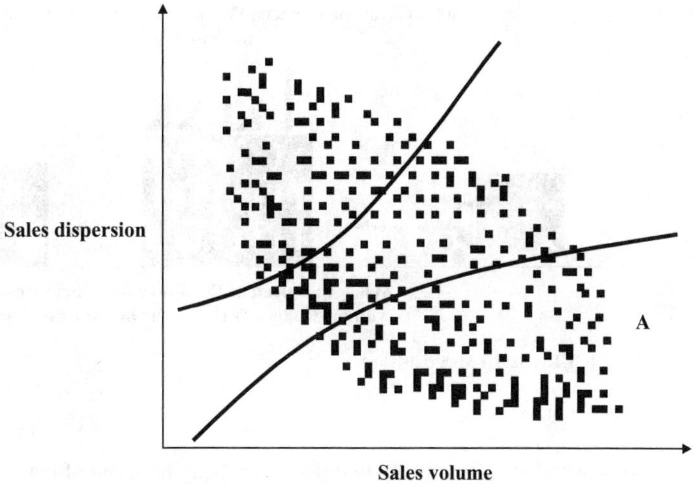

FIGURE 51.7 Sales dispersion vs. sales volume.

given to the final process in terms of its varieties and volumes. EPEI will be explained in detail, later. Achieving EPEI is no simple task. For example, if there were a wide variance in product sales, or sales were exceptionally high during the bonus season, due to production capacity limitations—there is a need to produce the volumes required in advance in anticipation of the inordinate sales peaks. This would in turn mean greater product inventories.

This would seem to indicate that JIT is most suited for introduction into areas in which demand swings are relatively small and for which demand volume is large. The area indicated by A in Fig. 51.7 corresponds to these conditions.

Surfacing of Anomalous Phenomena. The other critical task is accepting the fact that anomalous phenomena will surface. This is an extremely difficult point when introducing JIT, as we naturally have the tendency to work to prevent anomalies from occurring.

We may, for example, remove the KANBAN from a still-unused shuttle box to send a production directive for an item that appears in danger of running short (because stocks are extremely low, etc.). Or we may tend to overuse reserve KANBAN. Should this occur, we would become unable to grasp the current numbers of KANBAN cards and to control the flow of goods. It would also greatly increase our work. And of course, the aims of the JIT systems would not have been met.

In JIT, we must resign ourselves to spilling a little blood, and to become determined to play strictly by the rules. To achieve the kind of production improvement attained by Toyota, it is essential for top management and those in charge at the working level to accept this. A system for making improvements must also be instituted to ensure that an anomalous phenomenon that has surfaced never does again. It is only when JIT is teamed with improvement activities that its full benefits can be obtained. Failure to recognize this can only lead to failure of the system.

I think that you have understood the KANBAN system which is the one of core parts of lean manufacturing.

51.4.2 Analysis for Stagnation of Materials

I have explained how to make materials flow. The next step is to explain techniques used for the analysis of the stagnation of materials.

In manufacturing, the material, parts, half-finished goods, and the final product flow form the *value stream* of a material factory, a parts plant, the final assembly plant, the wholesaler, and retailer

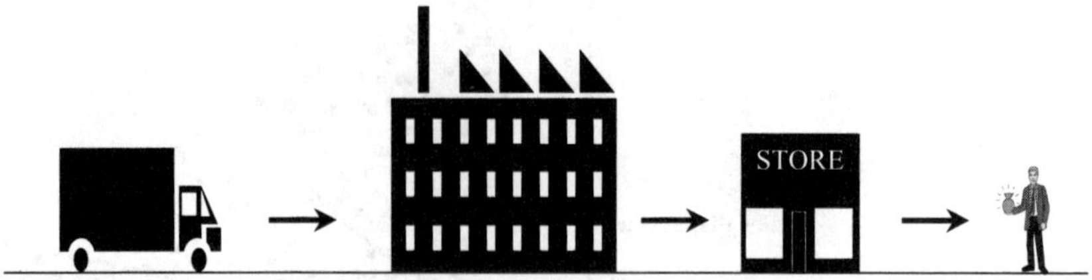

FIGURE 51.8 Materials flow.

ending at end user (Fig. 51.8). In order to heighten the speed of material flow, it is effective to look into the stagnation of material flow.

The conventional method for analyzing processes is basically comprised of just four elements. As previously mentioned, these are processing, movement, inspection, and stagnation. Regarding stagnation, there is just one status to be analyzed. The reason for this is that it is an analysis technique focusing on just one kind of material/part, not many materials or parts at the same time. Generally speaking, when you place a material in a location, you put many kinds of materials/parts not just one kind of material/part. When you analyze the meaning of stagnation you need consider this viewpoint. You can understand there is a difference in the types of stagnation. For instance, there is a case of that 10 pieces of parts A, 30 pieces of parts B, and 5 pieces of parts C are put at the place in the order (In), and 3 days later 3 pieces of parts B, 2 pieces of parts C and 10 pieces of parts A are picked up and shipped (Out). The other case might be that 3 pieces of parts A, 2 pieces of parts B, and 1 piece of parts C are put (In) and taken (Out) immediately in the same sequence and in the same numbers of parts as it were put.

The elements that form differences between (In) and (Out) of stagnation are the timing, the amount of parts and the sequence. Using these three elements, the pattern of stagnation is divided into three areas (Fig. 51.9). They are the *stock point,* the *distribution point,* and the *waiting point.*

At the stock point, when materials are received—the timing, the amount of parts, and the sequence of demand are not known. For instance, imagine the parts warehouse of an automotive dealership. Stock parts are stocked without knowing when they are used. Products produced based on demand forecasting are put into this type of stagnation place.

To understand the distribution point, imagine a distribution trader's transshipment base. When the trader places an item at the distribution center, the timing, the amount of parts, and the shipping sequence has already been decided. They are taken and shipped with different amounts (kind) and sequence (combination). Waiting point is the place where materials are taken in the same amount and in the same order as they were put in a short time. For instance, the place is the stagnation of the seats at the final assembly line of the car.

A theoretical method of making the material flow fast is to change the stock point to the distribution point, distribution point to waiting point, waiting point to transferring and finally removing the transferring area. The ideal stagnation shown in Fig. 51.9 is the perfect achievement of JIT.

There are three stock points in the supplier and the assembly line where JIT has not been introduced (Fig. 51.10). Let's analyze the stagnation of the materials in the automobile company where JIT system is introduced. Generally, there are three kinds of methods of supplying parts to the automobile assembly plant. They are the stock replenishment supply methods, the sequential supply methods, and the synchronized supply methods.

Figure 51.11 shows you stock replenishment supply methods. When parts are used at the final assembly line, they are replenished using the KANBAN system. Compared to the conventional

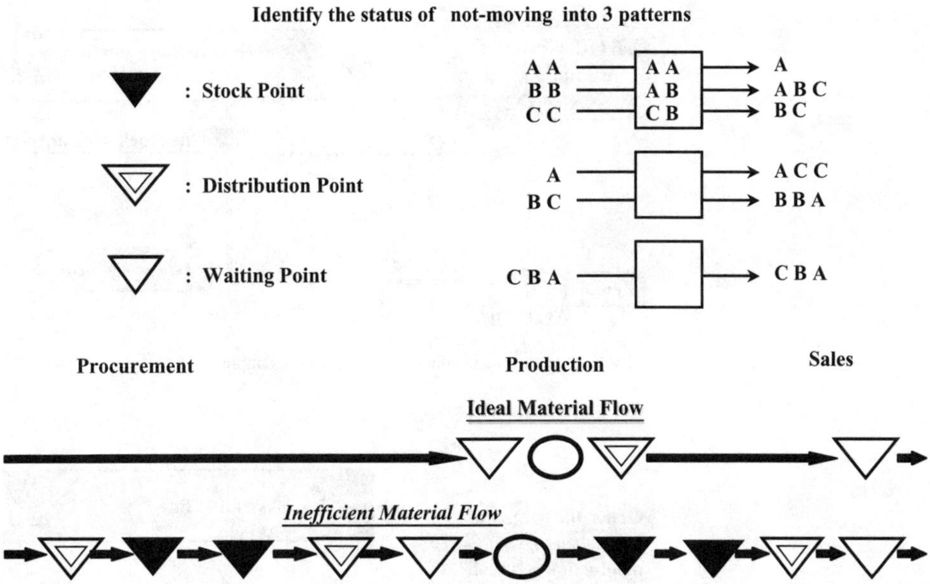

FIGURE 51.9 Three patterns and ideal material flow.

method in Fig. 51.10, one stock point has decreased. Using sequential supply method, parts are delivered at the same sequence as it used at the assembly line (Fig. 51.12).

With this method, the waiting point is placed next to the assembly line and the distribution point is placed on the supplier side while the stock point is reduced by one. With the synchronized supply method, shown in Fig. 51.13, that has been introduced when considering big parts such as doors and seats at car assembly line there is only one small stock point and one distribution point in the supplier and one waiting point at the assembly line. I think you understand that stock decreases substantially when compared with conventional method shown in first example, and the speed of material-flow accelerates.

It is powerful to use these three symbols for mapping patterns of stagnation to tie to the improvement activities.

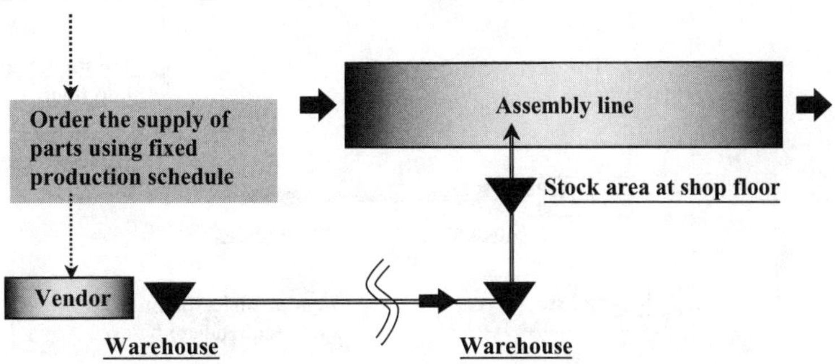

FIGURE 51.10 Conventional parts (lot) supply—example.

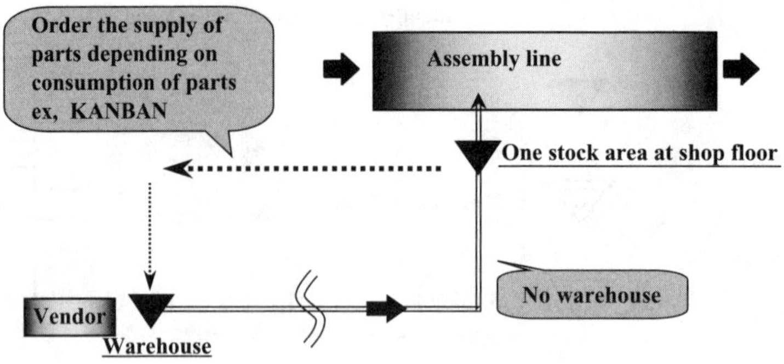

FIGURE 51.11 Replenishment parts supply—example.

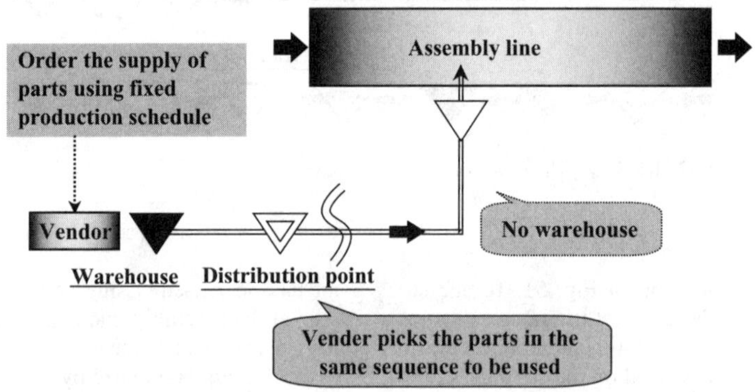

FIGURE 51.12 Sequential parts supply—example.

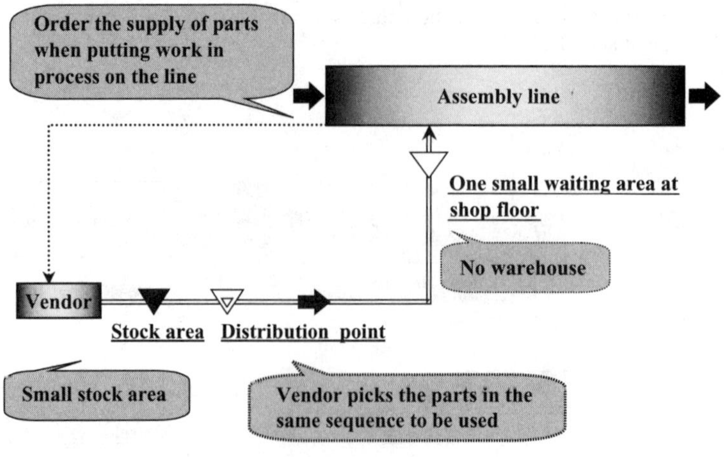

FIGURE 51.13 Synchronized parts supply—example.

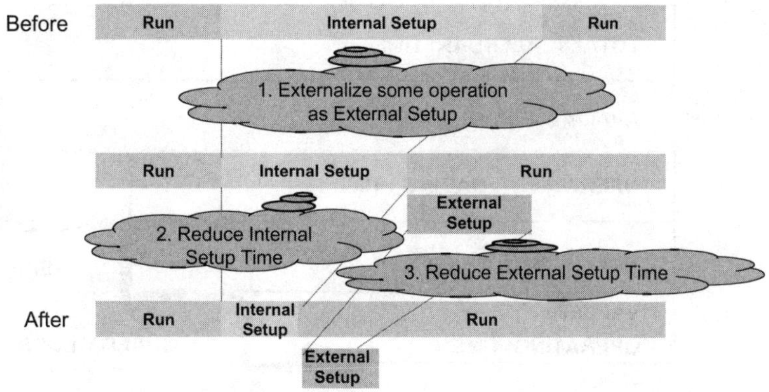

FIGURE 51.14 Single minute exchange of die.

51.4.3 Single Minute Exchange of Die

The word "Single" (Single Digit) is symbolically used as the name for the method of decreasing the setup time of the machine/equipment to less than 10 min. The method allows for having more setups at a process. This means that one can produce smaller numbers of a certain product per one setup/lot resulting in low level of stocks.

There is a famous method named *economical order quantity* (EOQ) which determines the purchase/production batch economically. This method works even today with the hidden conditions that all products produced are sold with the targeted price at once and the setup cost of competitors is the same or higher than your company. Actually, products produced based on forecasting cannot exactly be sold. And if competitors drastically shorten the time of setup, they can use this advantage for lead-time competitiveness. Activities in JIT system aimed at how EOQ can be reduced by shortening the time of setup while conventional methods just calculated EOQ using constant conditions of set up times and other variables. The concrete method is *single minute exchange of die* (SMED). It is a method developed with Toyota by Mr. Shigeo Shingo, a former senior consultant in JMAC.

SMED classifies the time necessary for the setup process into two elements. One is the internal setup time and the other is the external setup time. Internal setup is the setup operation while the machine is stopping and the external setup is the setup operation while the machine is working.

The procedure of SMED is as shown in (Fig. 51.14). You need to map the current operation of the setup process. Next, change some operations into external setup. Then, reduce the internal setup time. Finally, reduce the external setup time. A small-batch production or one-piece flow is enforceable by this shortened setup time and this results in a faster material flow.

51.4.4 TPM: Reduction of Waste of Equipment

The method and idea to operate the equipment efficiently is as follows.

First, you need to recognize that 100% utilization of equipments/machines is almost impossible and not good for your business when you plan/operate machines. Because the demand in the future when the mechanical equipment is introduced cannot be accurately estimated, it is common that the capacity of equipment is a little more than actual demand. And to bridge the difference between machine capability and actual demand, the master operation plan (operating shift, days, and time for each machine) is made. Then, daily efforts will be made to increase productivity (remove waste) in this available time.

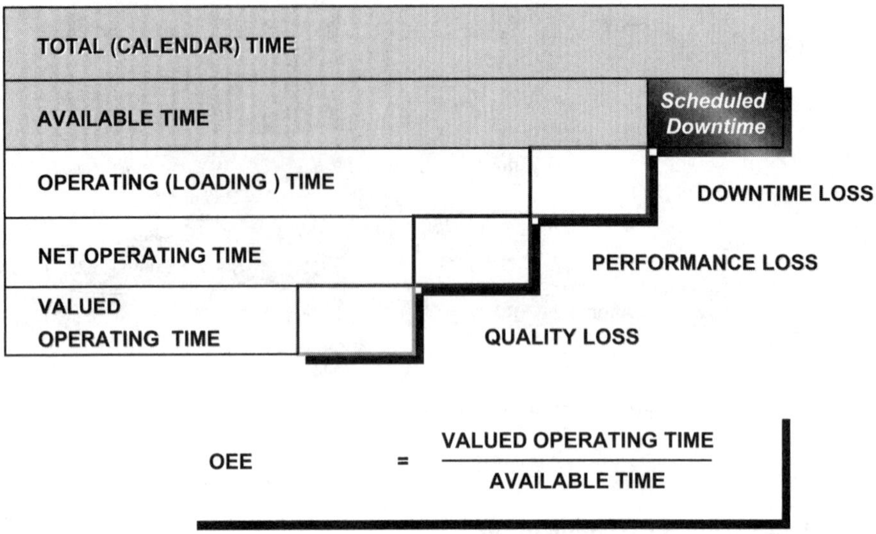

FIGURE 51.15 Overall equipment efficiency.

Figure 51.15 shows you a definition of time. There is *calendar time*. When you subtract scheduled *down time* from calendar time you can get *available time*. *Operating time* is the difference between available time and nonscheduled down time. When you subtract performance loss from *operating time* you can get *net operating time*. Finally, when you subtract quality loss from net operating time, you can get *valued operating time*. The ratio of valued operating time over available time is defined as *overall equipment efficiency* (OEE) in *total productive maintenance* (TPM). One of the core activities in TPM is to remove those losses/wastes that are between available time and valued operating time. See the losses which TPM defines.

- Down time loss
  1. Failure loss
  2. Set up/adjustment loss
  3. Cutting blade change loss
  4. Start up loss
- Performance loss
  1. Minor stoppage loss
  2. Speed loss
- Quality loss
  1. Defect/rework loss

In TPM there are eight pillars those are core activities including the above. Japan Institute of Plant Maintenance (JIPM), which is one of JMA group, has developed TPM with Denso.

51.4.5 5S

5S is a program or a series of activities that aim to change the workplace into a well-organized environment, which fosters good work. This allows employees to achieve higher morale and job satisfaction.

This results in better quality, productivity, and safety. The origin of the name of 5S is from five Japanese words starting at S. They are the following:

Seiri: Sort
Seiton: Store (Organize)
Seiso: Shine (Clean)
Seiketsu: Standardize
Shitsuke: Sustain

51.4.6 Value Stream Mapping

Value stream mapping (VSM) aims to map the relations among entity, material, and information flow. There is the current VSM (Fig. 51.16) and the future VSM (Fig. 51.17). In the future VSM you need to identify the result target. In this case, the future does not mean long-term, rather the future state should be what you would like to achieve in three months to a year. According to the product and the lean manufacturing enterprise, sometimes it is easy to develop the future VSM, but when difficulties arise, it is advised to retain help from outside your firm. As I described in KANBAN system, it is relatively easy to develop a future VSM for automobile or electric suppliers that are producing

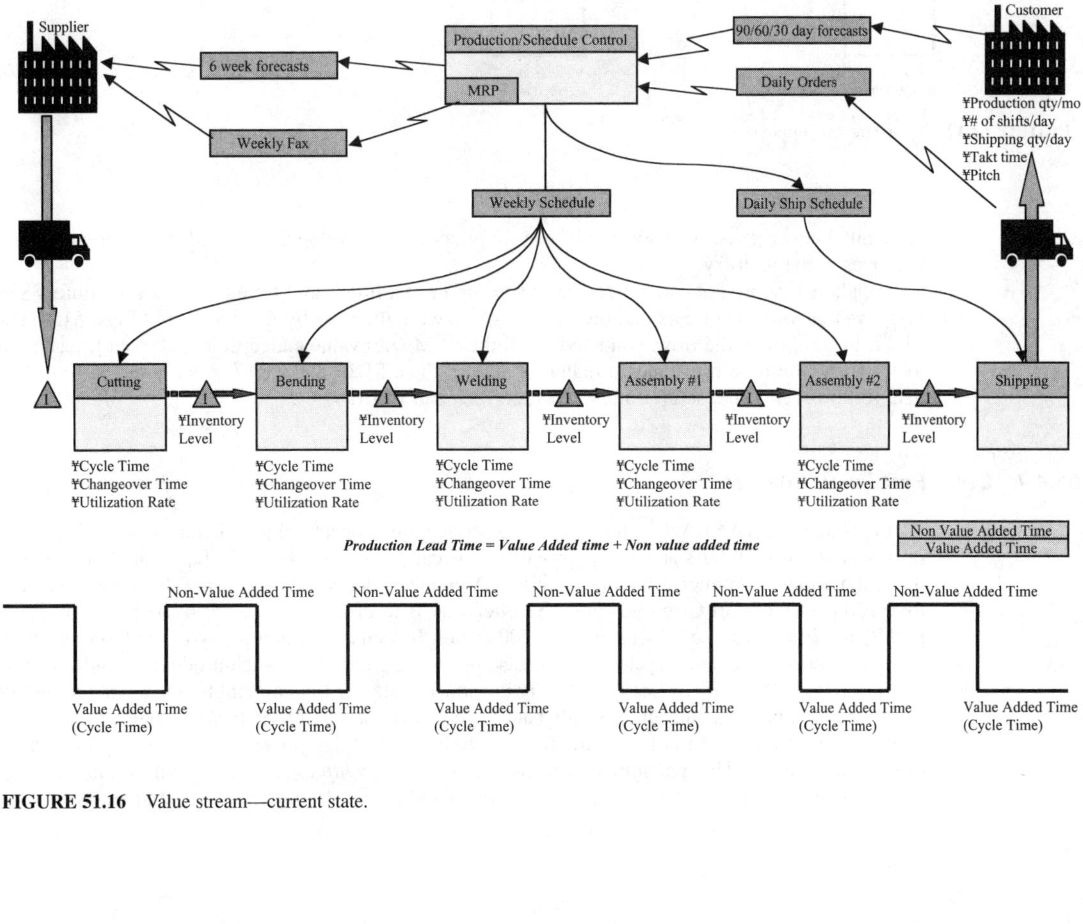

FIGURE 51.16 Value stream—current state.

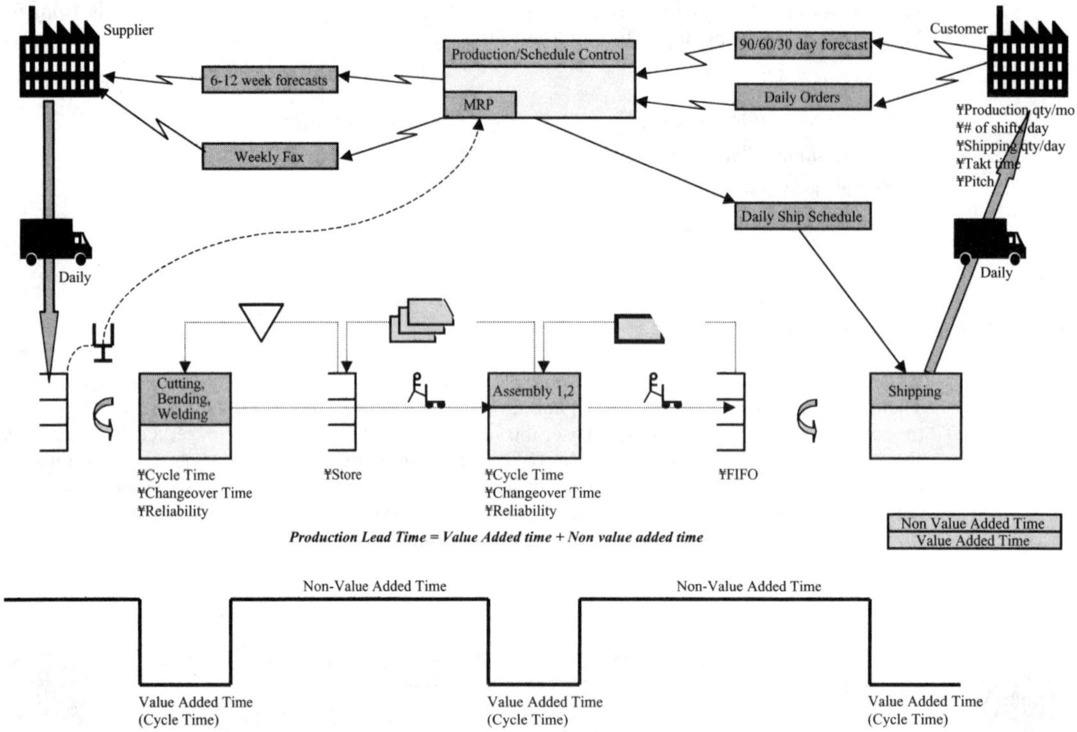

FIGURE 51.17 Value stream—future state.

large numbers of parts. A majority of these suppliers could develop these VSMs by referencing best practices in the industry.

Suppliers that are not categorized in this group need a little more time to develop a future VSM. I believe that you can understand the method of drawing the flow by just looking at Figs. 51.16 and 51.17. In addition to the flow, you need to fill the VSM with value added time and other times in the total production time for objected materials like in Figs. 51.16 and 51.17. Also, you need to fill in data for the cycle time, setup time, utilization rate, and EPEI.

51.4.7 Every Part Every Interval

As I explained in KANBAN system EPEI is a production concept, which intends to make one cycle of production for all parts at short as possible. It is called "HEIJUNKA" in Japanese. For example, it is not optimal to produce 1000 pieces of part A in one cycle, 500 pieces of part B in one cycle, and then 200 pieces of part C in one cycle. A better way is to produce 10 pieces of part A, 5 pieces of part B, and 2 pieces of part C—each using 100 cycles. In lean manufacturing, you need to keep shortening the period of this one cycle. This means approaching the small batch production and finally to one piece flow. When EPEI has been firmly introduced into the final assembly line at an automobile plant, the supplier can introduce small batch production or one-piece flow in their organization. When the degree of EPEI is low at the final assembly line, the negative effects are amplified at the upstream processes. This phenomenon is known as the *bull whip effect*. As a result of this, suppliers will stock many parts as safety stocks to be supplied. Otherwise, frequent shortage of parts will occur.

Let me explain a simple case to calculate the period of EPEI. Assumption: there are three types of parts—1, 2, and 3. Each part has the same processing time and the same setup time. The pattern of the production cycle is part 1, 2, 3, 1, 2, 3, 1, 2, 3, . . .

Production amount	Aa, Ab, Ac
Processing (cycle) time	P
Changeover time	C
Available time	A
Utilization rate	u%

With the above assumptions, the shortest EPEI becomes the following:

Operating time	O = A*u%
Total production time	TP = P*(Aa + Ab + Ac)
Total changeover time	TC = O*f*TP
Number of total changeover	NC = TC/C
Number of cycle	NCL = NC/3
Time of one cycle of EPEI	EPEI = O/NCL

A summary of the objectives of EPEI is as follows:

- Use smaller batch sizes for production. Reduce over production caused by large batch production
- Avoid high number of defects caused by large batch production
- Reduce fluctuation of parts demand to upstream processes
- Absorb fluctuation of the operating time caused by each model at the final assembly line
- Create flexible production system to fit the diversity of the product line

51.4.8 Other Tools

In addition to the above, there are various tools shown in Fig. 51.1. There is the U-line technique, which intends to heighten the speed of material flow and productivity by using U shape lines and not straight lines. ANDON (Lamp) uses a warning light to inform of an abnormality or a line stoppage etc. JIDOUKA means that machines are used effectively when people work efficiently with them. POKAYOKE (Mistake Proof) aims to prevent human error by using physical mechanism such as an automatic kicker for the stamping machine to detect defect parts. PDCA cycle is used for management purposes and team activities are more effective than working alone. *Strategic high productivity system* (SHIPS) is a powerful program which aims to increase productivity for operations.

51.5 PROCEDURE FOR IMPLEMENTATION OF LEAN PRODUCTION

I will explain one of the basic procedures of lean production implementation, but of course, the implementation would vary depending on the characteristics of the processes and the products. There are six steps as shown in (Fig. 51.18).

1. Selection of pilot site
2. Value stream mapping: current
3. Value stream mapping: future (target setting)

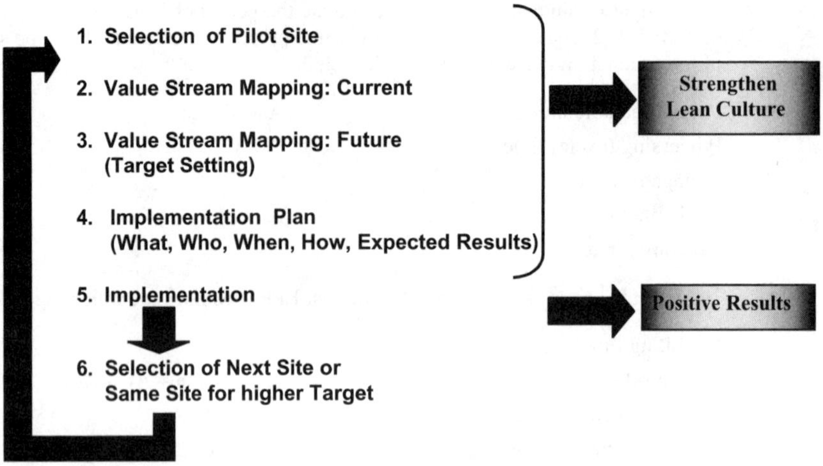

FIGURE 51.18 Lean manufacturing step.

4. Implementation plan (what, who, when, how, expected results)
5. Implementation
6. Selection of next site or same site for higher target

Each Step

1. Selection of Pilot Site. This is the step to decide your target processes and part groups. It is a good idea to select easy processes and part groups to improve in order to encourage people by achieving success. As an extreme example, it is OK to select just one stamping processes as the first pilot. When you use an outside expert, a good idea is to select easy pilot site to show positive results. This will help future improvement activities to function more easily.

2. Value Stream Mapping: Current. Create a *value stream mapping* (VSM) from suppliers to customers and include processes which you select. There is no need to spend too much time in order to draw the map.

3. Value Stream Mapping: Future. Develop VSM for the near future with the assumption that you can realize these targets within 3 months to 1 year. You need to develop targeted cycle times, setup times, utilization rates, and EPEI for the future flow of materials and information. When you draw a future material flow, you can apply three patterns of stagnation which I have explained before. If there is a stock point, try to change it to the distribution point. If there is a distribution point, try to change it to the waiting point (first-in first-out). Try to connect processes if it is already a waiting point. If the processes are linked with a push concept, change it to pull. If there is a KANBAN link, try to reduce the number of KANBANs. You can use the tools as possible solutions as shown in Fig. 51.1.

4. Implementation Plan (What, Who, When, How, Expected Results). Develop an implementation plan, which includes the problems to solve, responsibilities, the time frame, methodologies, and the expected results. The tools to be used are shown in Fig. 51.1 as analysis and improvement tools. Do not spend much time in steps 1–4. While the planning stage is important, there are many cases that spend all their time for planning but not for implementation. In lean manufacturing activities, there are many trial and error attempts at implementation and it works well.

5. Implementation. This is the most important stage in lean manufacturing. The methods for job instruction will change, the way of connecting the upstream process with the downstream process will change, processes will change, and machines will change. Moreover, the management side will change using visible management, PDCA cycle, team activities, and so on. The lean manufacturing concept and tools, which I have described, will all be used in this phase. Then the performance targets must be achieved.

When things are going wrong you need check the implementation by asking the following questions:

1. Is there sufficient motivation?

2. Is the concept of lean manufacturing understood? Is each technique understood?

3. Is it really being used?

If the answers to A, B, and C are yes, but there are still no positive results, then you need to find the bottleneck to the improvement activities. Sometimes it will be a long setup time or unstable machine times. In this scenario, you need to place an expert in the problem area. When the problem is not identified, an expert opinion should be brought in for assistance. One cycle of the first pilot site ends by 1–5.

6. Selection of Next Site or Same Site for Higher Target. Lean manufacturing is a continuous activity. When one cycle of 1–5 ends, a new site is chosen and the activities repeated. By following this cycle, the level of lean implementation will continuously improve. Moreover, when the level of lean manufacturing is raised, the skills and competence of the leaders and the members of lean activities will be strengthened. Their competence becomes wider and deeper in such areas as distribution, mechanical engineering, product development, and sales.

51.6 FUTURE

Lean manufacturing was one of the unique methods in the automotive industry. However, there are many companies those have implemented it or have started to implement. Firms that have firmly understood lean manufacturing and truly believe the methods and culture have been achieving success. For all manufacturers, it is essential to produce products without waste, and this is a core concept of lean manufacturing. Lean manufacturing is now at the stage of becoming a basic concept, not just for automotive industry but also for all manufacturers. Even if the tools and methods are changed, the concept of continuous activities to remove waste at the GEMBA (actual site, down floor) will remain. Ultimately, the lean culture will become the culture for all manufacturers.

CHAPTER 52
WORK CELL DESIGN

H. Lee Hales
Bruce J. Andersen
William E. Fillmore
Richard Muther and Associates
Marietta, Georgia

52.1 OVERVIEW

This chapter explains step by step how to plan a manufacturing cell. It discusses the information and analyses required at each step and the outputs achieved. Several types of cells are discussed. Special issues related to automation are also discussed. A comprehensive checklist is provided covering all major aspects of cell planning and operation, including physical arrangement, operating procedures, organization, and training.

52.2 BACKGROUND

52.2.1 Definition of a Manufacturing Cell

A manufacturing cell consists of two or more operations, work stations, or machines dedicated to processing one or a limited number of parts or products. A cell has a defined working area and is scheduled, managed, and measured as a single unit of production facilities. Typically, a cell is relatively small, and may be virtually self-managed. Usually, the outputs of a cell are more-or-less complete parts or assemblies, ready for use by a downstream operation or cell, or for shipment to a customer.

Three aspects—physical, procedural, and personal—must be addressed when planning a manufacturing cell. Cells consist of *physical facilities* such as layout, material handling, machinery, and utilities. Cells also require *operating procedures* for quality, engineering, materials management, maintenance, and accounting. And because cells employ *personnel* in various jobs and capacities, they also require policies, organizational structure, leadership, and training.

A cell is essentially a production line (or layout by product) for a group or family of similar items. It is an alternative to layout and organization by process, in which materials typically move through successive departments of similar processes or operations. This layout by process generally leads to higher inventories as parts wait between departmental operations, especially if larger batches or lots are produced. There is more material handling required to move between departments, and overall processing time is longer. Exposure to quality problems is greater, since more time may pass and more nonconforming parts may be produced before the downstream department notices a problem.

52.2.2 Benefits of Cells

The principal physical change made with a manufacturing cell is to reduce the distance between operations. In turn, this reduces material handling, cycle times, inventory, quality problems, and space requirements. Plants installing cells consistently report the following benefits when compared to process-oriented layouts and organizations:

- *Reduced materials handling.* Reductions of 67 to 90 percent in distance traveled are not uncommon, since operations are adjacent within a dedicated area.
- *Reduced inventory in process.* Reductions of 50 to 90 percent are common, since material is not waiting ahead of distant processing operations. Also, within the cell, smaller lots or single-piece flow is used, further reducing the amount of material in process.
- *Shorter time in production.* From days to hours or minutes, since parts and products can flow quickly between adjacent operations.

In addition to these primary, quantifiable benefits, companies using cells also report:

- Easier production control
- Greater operator productivity
- Quicker action on quality problems
- More effective training
- Better utilization of personnel
- Better handling of engineering changes

These secondary benefits result from the smaller, more-focused, and simplified nature of cellular operations.

52.2.3 Difficulties in Planning and Managing Cells

To obtain the benefits of cells, planners and managers must overcome the following difficulties:

- *Worker rejection or lack of acceptance.* Often due to lack of operator involvement in planning the cell, or to insufficient motivation and explanation by management, especially if the outcome is perceived to be a work force reduction.
- *Lack of support or opposition by support staff in production planning, inventory control and/or cost accounting.* Usually when creation of the cell causes changes in procedures and practices, or reduces the amount of detail reported from the plant floor.
- *Reduced machine utilization.* Due to dedication of equipment to cells and to families of parts. In some cases, additional, duplicated machinery may be required. In other cases, large and high-speed equipment that may be appropriate in a process-oriented department or job shop must be replaced by slower and usually smaller, lower-capacity machines that are more appropriate to the volumes of a cell.
- *Need to train or retrain operators.* Often for a wider range of duties and responsibilities.
- *Wage and performance measurement problems.* Especially when individual and piece-rate incentives are in use. The team-oriented nature of the typical cell and the goals of inventory reduction may work against traditional incentives and measures.

52.3 TYPES OF MANUFACTURING CELLS

Cells take different forms based upon the characteristics of the *parts* (P) and *quantities* (Q) produced, and the nature of the process sequence or *routing* (R) employed. The relationship of these characteristics—P, Q, and R—and their influence on manufacturing cells can be seen in Fig. 52.1.

52.3.1 Production Line, Group of Parts, and Functional Cells

Cells are typically used to serve the broad middle range of a *product–quantity* (P–Q) *distribution*. Very high quantities of a part or product—typically above 1 million units per year—lend themselves to dedicated mass-production techniques such as high-speed automation, progressive assembly lines, or transfer machines. At the other extreme, very low quantities and intermittent production are insufficient to justify the dedicated resources of a cell. Items at this end of the P–Q curve are best produced in a general-purpose job shop. In between these quantity extremes are the many items, parts or products that may be grouped or combined in some way to justify the formation of one or more manufacturing cells.

Within the middle range, a *production line cell* may be dedicated to one or few high-volume items. This type of cell will have many of the attributes of a traditional progressive line, but is usually less mechanized or automated.

Medium and lower production quantities are typically manufactured in *group technology* or *group-of-parts cells*. These are the most common types of cells. They exhibit progressive flow, but the variety of parts and routings works against a production line.

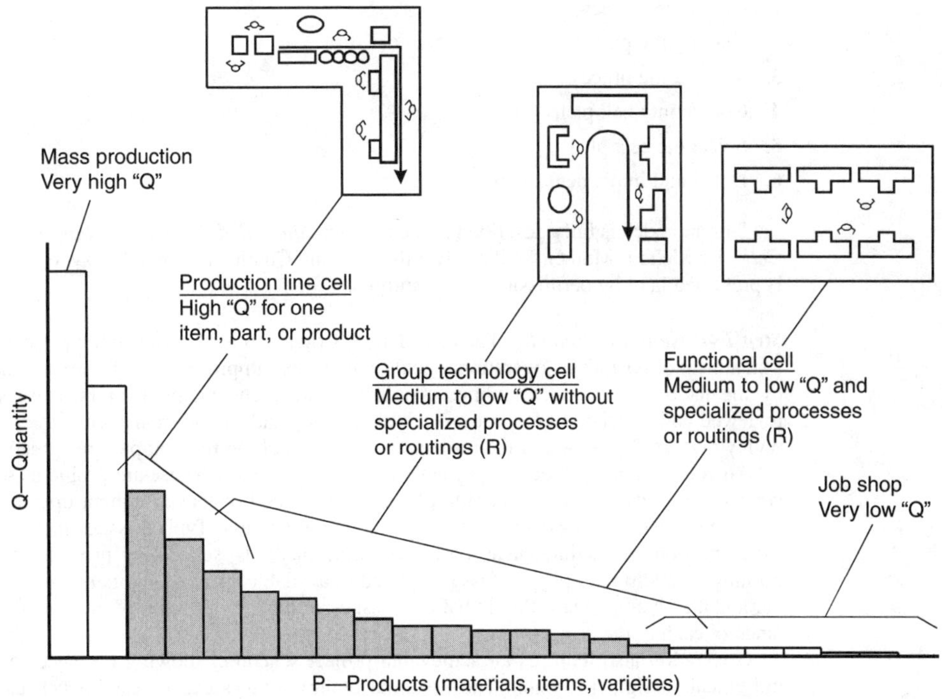

FIGURE 52.1 Key considerations and types of manufacturing cells. (*From Planning Manufacturing Cells, © Copyright 2002 Richard Muther & Associates.*)

If the processing steps are specialized in some way, requiring special machinery and utilities, or special enclosures of some kind, then a *functional cell* may be appropriate. Functional cells are often used for painting, plating, heat treating, specialized cleaning, and similar batch or environmentally sensitive operations. If the functional cell processes parts for other group-of-parts or production line cells it will introduce extra handling, cycle time, and inventory, since parts must be transported and held ahead of and behind the functional cell. For this reason, planners should first examine the practicality of decentralizing or duplicating the specialized process(es) into group-of-parts or production line cells.

The steps required to plan a manufacturing cell are the same for all three types of cells—production line, group technology, and functional. However, the emphasis and specific techniques used will vary somewhat based upon the physical nature of the manufacturing processes involved. For example, when planning for machining and fabrication cells the capacity of key machines is critical and may be relatively fixed. The time required to change from one part or item to another is also critical. Allowances for set-up and capacity losses to changeovers are very important. Manpower planning may be of secondary importance, after the number of machines has been determined. In contrast, when planning for progressive assembly, the variability of operation times must be understood, and the work must be balanced among the operators to assure good utilization of labor. In such assembly cells, utilization of equipment may be a secondary issue.

52.4 HOW TO PLAN A MANUFACTURING CELL

Most cells can be planned using a simple six-step approach:

1. Orient the project
2. Classify the parts
3. Analyze the process
4. Couple into cell plans
5. Select the best plan
6. Detail and implement the plan

This approach is fully described in the booklet, *Simplified Systematic Planning of Manufacturing Cells,* by Richard Muther, William E. Fillmore, and Charles P. Rome.[1] A synopsis of this approach is presented here by permission of the authors.

Step 1—Orient the Project. The cell planner's first step is to organize the project, beginning with a statement of objectives, operational goals, and desired improvements. External conditions imposed by the facility or the surroundings should be noted. The planning or business situation is also reviewed and understood for issues such as urgency and timing, management constraints, or other policy matters. The scope of the project and the form of the final output are agreed upon.

All cell-planning projects begin with a set of open issues. These are problems, opportunities, or simply questions that will affect the planning of the cell or its subsequent operation. These issues must be resolved and answered during the planning process. Typical issues include responsibilities for inspection and maintenance, cost accounting methods, scheduling procedures, job design and training, in addition to physical issues related to available space, equipment, and utilities. The planner and the planning team should list their issues at the first opportunity, and rate the relative importance of each to the project.

Orientation also requires an achievable project schedule, showing the necessary tasks and the individuals assigned to each. The essential planning tasks can be established using the six-step procedure outlined above, adapted to the specifics of the project at hand. The final output of step 1, Orientation, can be summarized on a simple worksheet or form like that shown in Fig. 52.2.

ORIENTATION & ISSUES WORKSHEET

Project Name *Oven Assembly Cell* Project No. *99509*
By *Team* With *DM*
Date *10/8* Sheet *1* of *1*

1 Objectives *Reduce material handling; accommodate plant ordering procedures; minimize throughput time; attain desired output rate, no more, no less.*

2 External Conditions *Locate in old Receiving area. Moves of baskets and totes-on-pallets by fork truck.*

3 Situation *Quick start up required to meet customer demand. Use available equipment.*

4 Scope and Form of Output *Cell must start deliveries by 11/15.*

	PLANNING ISSUES			Action to Resolve	Resp	Proposed Resolution	X
1	*Will team be cross trained?*	*I*					
2	*What will project life be?*	*E*					
3	*Can takt time goal be met?*	*A*					
4	*Must use avail. Equipment*	*A*					
5	*Must start delivery by 11/15*	*A*					
6							
7							
8							
9							
10							

Dominance/Importance Rating ↑ ↑ Mark "X" if beyond control of company/plant/project

Notes:

Distribution *Team* By *Team* With *DM* Status as of

PROJECT SCHEDULE

No.	Action Required	Who	10/8	10/15	10/22	10/29	11/5	Notes
1	*Define & schedule projetct*	*Team*	■					
2	*Classify parts*	*BG*	■					
3	*Analyze & document process*	*PH*	■					
4	*Balance operations*	*PH*	■					
5	*Equipment & Flow Diagram*	*PH*	■					
6	*Develop cell plans*	*Team*	■	■				
7	*Evaluation meeting*	*All*			■			
8	*Make implementation plan*	*DM*			■			
9	*Install available equipment*	*DM*				■		
10	*Complete implementation*	*All*	→	→	→	→	→	*Target 11/15*

Notes:

FIGURE 52.2 Cell planning orientation and issues worksheet for a small cell assembling sheet metal ovens. (*© Copyright 1995 Richard Muther.*)

Step 2—Classify the Parts. Most projects have a candidate list of potential parts that could be made in the cell. These parts typically have the same or similar routings. The planner must still clarify and confirm that these candidate parts do belong in the cell, and identify those that do not. Classifying the parts also simplifies the analysis and design of the cell. The first cut at classification usually involves the physical characteristics of the candidate parts. These include the following:

- Basic material type
- Quality level, tolerance, or finish
- Size
- Weight or density
- Shape
- Risk of damage

Additional common considerations for classification include the following:

- Quantity or volume of demand
- Routing or process sequence (and any special or dominant considerations)
- Service or utility requirements (related to the process equipment required)
- Timing (may be demand-related, e.g., seasonality; schedule-related peaks or valleys; shift-related; or possibly related to processing time if some parts have very long or very short processing times)

Less common but occasionally significant classification factors include building features, safety issues, regulatory considerations, marketing-related considerations, and even organizational factors that may be reflected in the way that a specific part is scheduled and produced.

All of these factors can be tied together into a worksheet like the one shown in Fig. 52.3.

The planner identifies and records the physical characteristics and other considerations for each part or item. If it seems awkward to record or rate each amount or specific dimension, one can rate

PRODUCT/PART CLASSIFICATION WORKSHEET

Project *Oven Assembly Cell* Project No. *99509*
By *BG* With *Team* Date *10/9* Sheet *1* of *1*

Item No.	Name	Basic Material	Physical Characteristics: Quality Level	Size	Weight-Density	Shape	Risk of Damage	Value			Other Considerations*: Q	R	S	T	B	F	L	K	O		Further Explanation	Class. Identif.
A	*Top Half*	*Steel*	*O*	*E*	*8*	*U*	*O*				*1*	*Assembly*	*O*	*To be calculated*	*U*	*O*	*U*	*U*	*U*			*b*
B	*Lower Half*	*Steel*	*O*	*E*	*8*	*U*	*O*				*1*		*O*		*U*	*O*	*U*	*U*	*U*			*b*
C	*Coil Holders*	*Steel*	*O*	*O*	*4*	*I*	*O*				*2*		*O*		*U*	*O*	*U*	*U*	*U*			*d*
D	*Thermo Bracket*	*Steel*	*O*	*O*	*4*	*I*	*O*				*1*		*O*		*U*	*O*	*U*	*U*	*U*			*d*
E	*Nut*	*Steel*	*O*	*O*	*2*	*H*	*O*				*1*		*O*		*U*	*O*	*U*	*U*	*U*			*d*
F	*Back*	*Steel*	*O*	*I*	*8*	*F*	*O*				*1*		*O*		*U*	*O*	*U*	*U*	*U*			*c*
G	*Reinforcements*	*Steel*	*O*	*O*	*3*	*FL*	*O*				*2*		*O*		*U*	*O*	*U*	*U*	*U*			*d*
-	*Oven Assembly*	*Steel*	*O*	*A*	*44*	*R*	*I*				*1*		*O*		*U*	*O*	*U*	*U*	*U*			*a*

*** Other Considerations**
Q Quantity/Volume/Demand
R Routing/Process domination
S Service/Utility requirements
T Timing/Complementary seasons
B Building features
F Safety problems
L Legal/Regulatory problems
K Market togetherness
O Others/Operators/Organization

Reference Notations
a. *Weight shown in ounces*
b. *Finished assembly: 24" x 19" x 16"*
c. *Number of pieces per unit*
d.
e.

FIGURE 52.3 Product/part classification worksheet for the parts sketched in Fig. 52.5. (*© Copyright 1995 Richard Muther.*)

the importance or significance of each characteristic as to its contrast or dissimilarity with the other parts. Use the vowel-letter, order-of-magnitude rating code illustrated in Fig. 52.3 and defined below.

A—Abnormally great

E—Especially significant

I—Important

O—Ordinary

U—Unimportant

After recording or rating the physical characteristics and other considerations for each part or item, note those parts that have similar characteristics—that is, classify the parts according to the most important characteristics and considerations. Assign a class code letter to each class, group, or combination of meaningful similarities. Enter the appropriate class letter code for each part or item in the class-identification column.

When a large number of different parts will be produced, planners should place special emphasis on sorting the parts into groups or subgroups with similar operational sequences or routings. Those assigned to a class will all go through the same operations. Generally, for cells producing many parts different parts, this is the most useful type of classification for the subsequent steps in cell planning. The final output of step 2, Classify the Parts, is a clear listing of the classes or groups of parts to be produced in the cell.

Step 3—Analyze the Process. In step 3, the planner uses charts and diagrams to visualize the routings for each class or subgroup of parts, and then calculates the numbers of machines and/or operators and workplaces that will be required to satisfy the target production rates and quantities.

If the plan is for an assembly cell, the preferred way to visualize the process is with an operation process chart like that shown in Fig. 52.4. This example visualizes the assembly of the sheet metal oven pictured in Fig. 52.5. The oven consists of a top half, a bottom half, and a back, to which smaller parts are added.

In addition to showing the progressive assembly of the finished item, the process chart also shows the labor time at each step. Given a target production rate and the number of working hours available, the planner calculates the work content of the process and breaks it into meaningful work assignments. In this way, the required number of operators and workplaces is determined, along with the flow of materials between them. Assumptions or calculations must be made to establish the time that will be lost to breaks and to nonvalue-adding tasks such as material handling, housekeeping, and the like. The formal name for this process is *line balancing*. A good line balance achieves the desired production rate with the minimum number of operators and minimal idle time.

Once balanced to the planning team's satisfaction, workplaces and equipment are defined and represented in an equipment-and-flow diagram. This is the final output of step 3 (see Fig. 52.6). In this example, scaled templates represent the equipment. Such graphic detail is useful but not mandatory. A simple square symbol can be used to represent each operator workstation, or each machine. Numbers of lines and lower-case letters designate the flow of parts and materials.

When planning a machining or fabrication cell, the Group-of-Parts process chart is used to illustrate the sequence of operations for each class of parts (see Fig. 52.7).

The group-of-parts process chart must be accompanies by a capacity analysis showing the types and quantities of machines required by the cell. A simple form of this capacity analysis is shown in Fig. 52.8.

When calculating the number of machines, planners must be sure to add allowances to downtime, schedule interference, and changeovers between individual parts and groups of parts.

A good cell capacity plan meets the desired production rate with an appropriate number of machines and level of utilization. Usually, the analysis will reveal over and under-utilization of some equipment planned for the cell. If the analysis reveals overutilization, the planner may choose to:

- Remove parts from the cell to reduce utilization of the equipment.
- Purchase more equipment.
- Reduce process, changeover, or maintenance times.

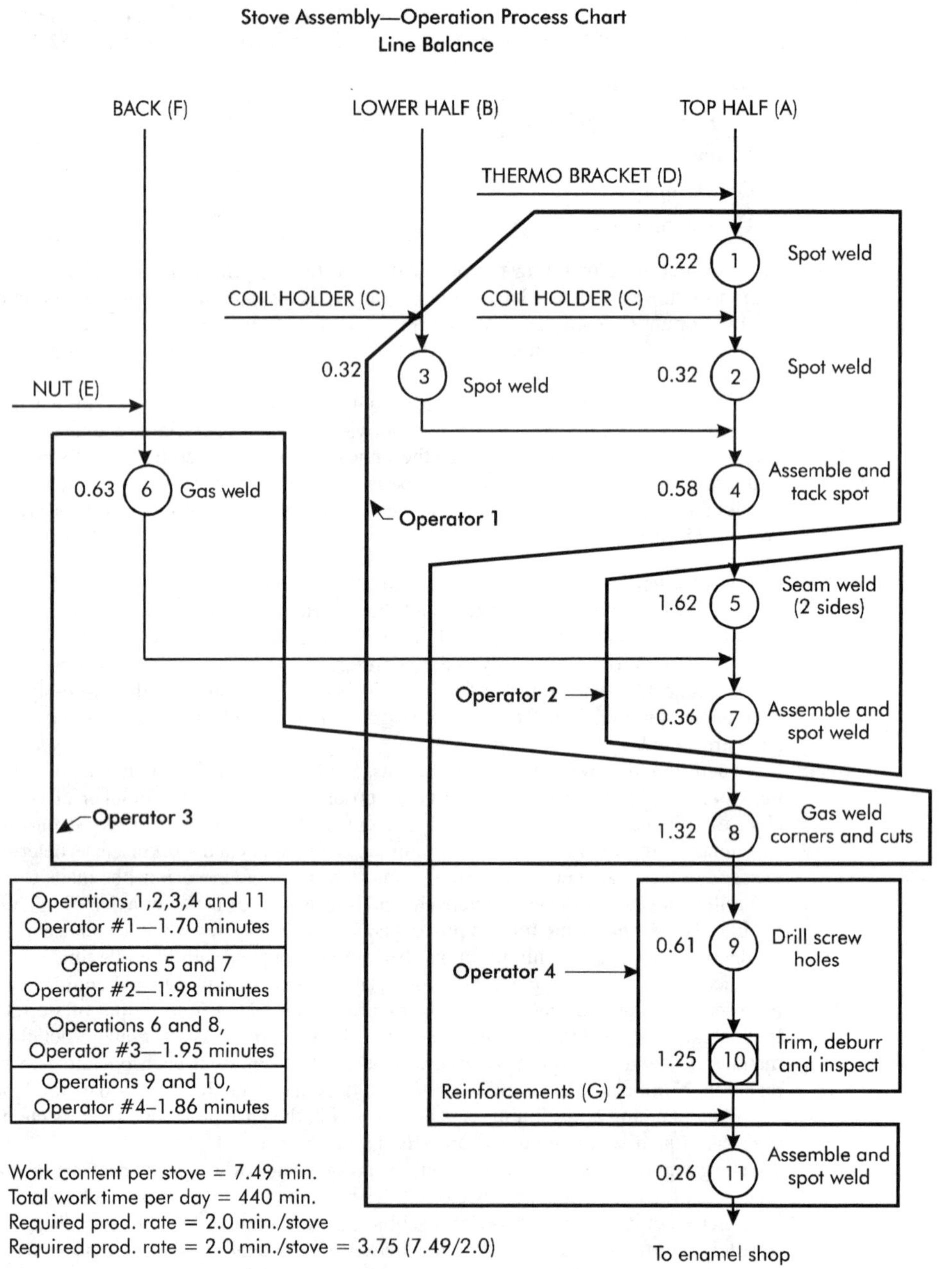

FIGURE 52.4 Operation process chart. *(From Planning Manufacturing Cells, © Copyright 2002 Richard Muther & Associates.)*

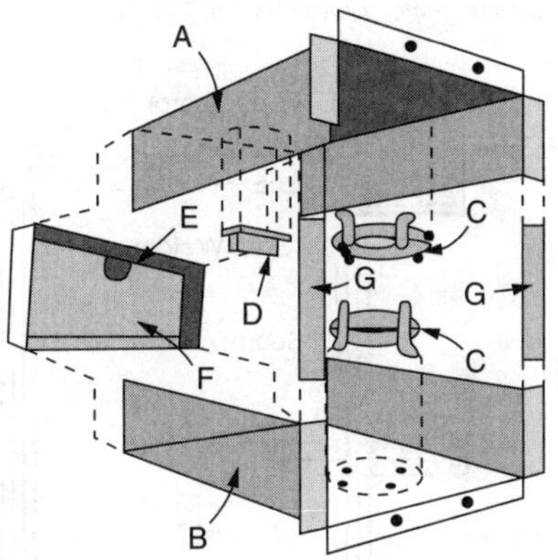

FIGURE 52.5 Sketch of the sheet metal oven charted in Fig. 52.4.

If the analysis reveals underutilization, the planner may choose to:

- Add parts to the cell to increase the utilization of equipment.
- Remove parts from the cell to eliminate the need for the equipment.
- Change the manufacturing process to eliminate the need for the equipment.
- Leave the equipment external to the cell and route parts to it.

In machining or fabrication cells where the throughput is paced more by the operators than the machines, the cell planner may need to conduct a line balancing exercise, in addition to rough capacity and utilization analyses. In some cases, computer simulation may be useful to examine the implications of changes to product mix and peaks in production volume.

Once the number of machines has been determined, an equipment-and-flow diagram is prepared, similar to that shown earlier in Fig. 52.5. The standardized work form shown in Fig. 52.9 is a similar type of conceptual visualization that also illustrates operator times and cycles.

Step 4—Couple into Cell Plans. A cell plan is a coupling of parts and process into an effective arrangement and operating plan. It should include:

- The layout of operating equipment (physical)
- The method(s) of moving or handling parts and materials (physical)
- The procedures or methods of scheduling, operating, and supporting the cell (procedural)
- The policies, organizational structure, and training required to make the cell work (personal)

The best way to begin this step is by sketching a layout from the equipment-and-flow diagram developed in step 3. Once the machinery and workplace layout is visualized, the material handling and any storage methods are determined. Material handling equipment, containers and storage, or parts-feeding equipment are added to the layout. The planner also adds any support equipment not already visualized in the workplaces, such as tool and die storage, fixture storage, gage tables and tool set-up, inspection areas, supply storage, trash bins, desks, computer terminals and printers, display boards, meeting areas, etc.

Once the layout and handling methods—the physical aspects of cell planning—have been determined, the planning team turns its attention to the procedural and personal aspects. In our experience, the procedural and personal aspects are often more important than the layout and handling in assuring a successful manufacturing cell. These aspects include the procedures and policies for staffing, scheduling, maintenance, quality, training, production reporting, performance measurement, and compensation. In practice, some of these will have already been determined during the layout and handling discussion; the remainder should be clearly defined by the team and approved by management. The documentation of a viable cell plan will also require the resolution of any remaining planning issues listed earlier in step 1. The final output of step 4 is one or more documented cell plans. These will take the form of a layout with associated policies and operating procedures like that shown in Fig. 52.10.

Step 5—Select the Best Plan. In step 5, the planning team and other decision-makers will evaluate the alternatives prepared in step 4 and select the best plan. Typically, this selection will be based upon comparisons of costs and intangible factors. Typical considerations include the following:

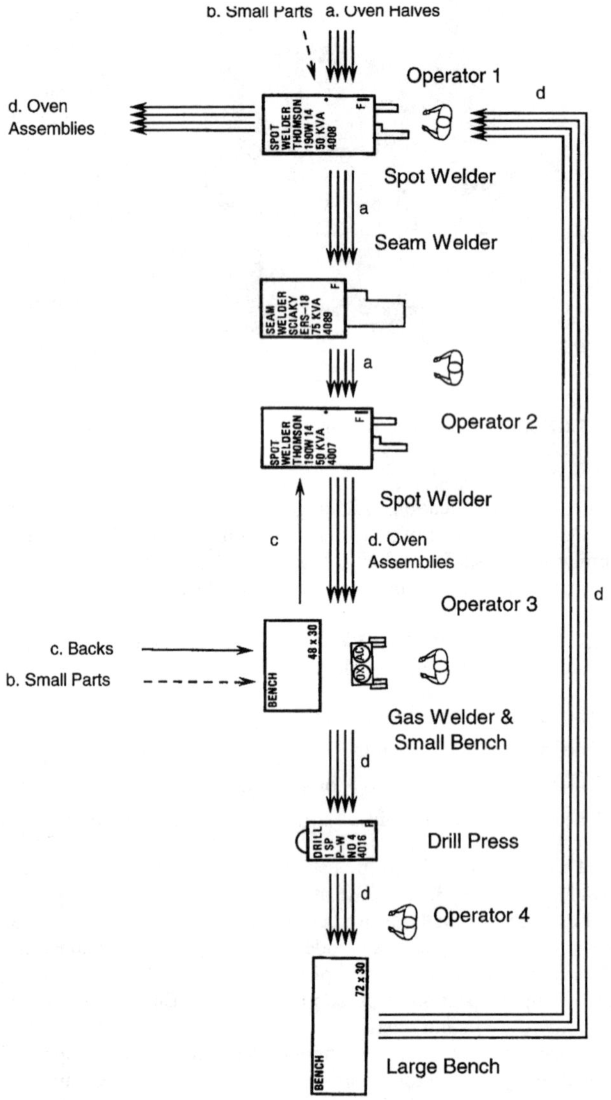

FIGURE 52.6 Equipment-and-flow diagram for the sheet metal oven charted in Fig. 52.4.

Investment Costs (and Savings or Avoidance)

- New production machinery
- Material handling equipment
- Pallets, containers, and storage equipment
- Auxiliary or support equipment
- Building or area preparation
- One-time move costs, including overtime

GROUP-OF-PARTS PROCESS CHART

Plant *Beeville* Project *Misc. Shafts* Date *9/2*
By *L.P.* With *B.H.* Sheet *1* of *1*

Parts and Quantity → / Process ↓ / Sequence ↘	a	b	c	d	e	f	Operating Times Hours per Year	Number of Machines Required
P Q	1 2750 6 2750 3 5500	2 300 2 650 3 650	6 3800 2 3150	5 2750 3 3000 3 2500	8 3000			
Op. Centered, Mch. 1	1	1	1	1	1		1,100	1
Op. Turn Contour, Mch. 20	2	2	2	2	2		3,972	2
Op. Finish Turn, Mch. 30	3	3	3	3	3		3,967	2
Op. Mill Key, Mch. 50		4 @					132	1
Op. Mill Shoulder, Mch. 12	4	4 @					366	1
Op. Mill Spline, Mch. 40			4	4	4		3,780	2
Op. Mill Thread, Mch. 60	5	5	5	5	5		1,599	1
Op. Deburr, Mch. 70	6		6	6	6		2,112	2
Op. Cyl. Grind, Mch. 80		6		7	7		2,263	2
Op. Mill Gear, Mch. 90		7		8			1,926	1
Op. Cut Gear, Mch. 10					8		2,812	1
Op. Inspect, Mch. 14	7	8	7	9	9		As needed	1

Group Description:
- a. Threaded Shafts
- b. Gear and Thread
- c. Spline and Thread
- d. Gear and Spline – 1
- e. Gear and Spline – 2
- f.
- g.

Referenced Notations:
- a. @ Alternate routing, either – or
- b.
- c.
- d.
- e.
- f.
- g.

FIGURE 52.7 Group-of-parts process chart for a family of steel shafts. (*From Planning Manufacturing Cells © Copyright 2002 Richard Muther & Associates.*)

- Training and run-in
- Engineering services
- Permits, taxes, freight, or other miscellaneous costs
- Inventory increases or reductions (one-time basis)

PRODUCT CLASS/DESCRIPTION	Work Center Number>>			MACHINE TYPE										
				1	20	30	50	12	40	60	70	80	90	10
	Part No.	Model	Pieces per Year	Center'g Lathe	Contour Lathe	Engine Lathe	Key Mill	Univ. Mill	Spline Mill	Thread Mill	Drill Press	Cylinder Grinder	Gear Mill	Gear Cutter
Carried forward from Page 2 (hours/year)				750	2192	2775	0	0	3521	718	838	1585	1563	1750
(b) Shaft with gears and threads														
Rear axle gear shaft	47345 53049	LS 20 B 30	300	8	55	28	---	15	---	15	---	123	15	---
Power take-off shaft	36456 70459	L 10 L 20	650	13	108	32	---	18	---	43	---	132	39	---
Auxiliary gear shaft	56097 78905 76890	LS 20 B 20 B 30	650	13	135	74	120	---	---	67	---	149	32	---
(a) Shaft with threads														
Injection pump drive shaft	46785	M 85	2750	42	335	255	---	22	---	205	275	---	---	---
Steering knuckle arm	46056 45159 45907 45650 45432 45329	L 10 BH 10 L 20 BH 20 BV 20 B 30	2750	46	160	225	---	168	---	182	463	---	---	---
King pin	46554 56354 10101	L-B 10 L-B 20 B 30	5500	92	350	275	---	92	---	228	245	---	---	---
Machine Run Time (hours/year, incl. load & unload)				964	3335	3664	120	315	3521	1458	1821	1989	1649	1750
Setups (estimated hours per year)				100	600	260	3	30	210	111	21	200	200	522
Scheduled Maintenance (est. hrs per year)				24	25	28	6	14	33	20	180	50	52	360
Allowance for Unplanned Downtime (est. hrs per year)				12	12	15	3	7	16	10	90	24	25	180
Total Machine Hours Per Year				**1100**	**3972**	**3967**	**132**	**366**	**3780**	**1599**	**2112**	**2263**	**1926**	**2812**
No. of Machines Required				0.6	2.0	2.0	0.1	0.2	1.9	0.8	1.1	1.1	1.0	1.4
No. of Machines Available				1	2	2	1	1	2	1	2	2	1	1
Capacity Utilization				**55%**	**99%**	**99%**	**7%**	**18%**	**95%**	**80%**	**53%**	**57%**	**96%**	**141%**

FIGURE 52.8 Capacity utilization worksheet for the family of steel shafts charted in Fig. 52.7. *(From Planning Manufacturing Cells © Copyright 2002 Richard Muther & Associates.)*

Operating Costs

- Direct labor
- Fringe benefits and other personnel-related costs
- Indirect labor
- Maintenance
- Rental of equipment or space
- Utilities
- Inventory increases or decreases (annual carrying cost)
- Scrap and rework

Intangible Factors

- Flexibility
- Response time to changing production demand
- Ease of supervision
- Ease of material handling
- Utilization of floor space

STANDARDIZED WORK

Manual Work ——— / Machine Time ········ / Walking ∧∧∧∧∧∧	Company/Plant *Standard Automotive*	Part/Product *Model J Gear - #366675AH*
	Operation/Process *Gear Machining Cell*	By *WEF* With *JJ* Issue Date *19-May*
	Production Rate *1100/day* Takt Time *55 seconds*	Sheet *1* of *2* Rev. Date *6-Jun*

Ref. Symbol	Cyclical Work Element	Manual	Machine	Walking
1	*Pick blank from basket in*	*1*	*0*	*2*
2	*Load L-1 & start*	*1.5*	*45*	*0*
3	*Pick part from L-1 eject*	*1.5*	*0*	*3*
4	*Load T-1 & start*	*2*	*21*	*1.5*
5	*Pick from T-1 eject*	*2*	*0*	*3*
6	*Load RG-1 & start*	*3*	*42*	*0*
7	*Pick from RG-1 eject*	*1.5*	*0*	*3*
8	*Load Rolmac & start*	*2*	*14*	*0*
9	*Pick from Rolmac eject*	*1.5*	*0*	*2*
10	*Load washer basket*	*2*	*12*	*0*
11	*Pick blank from basket in*	*1*	*0*	*3*
12	*Load L-2 & start*	*1.5*	*40*	*0*
13	*Pick from L-2 eject*	*1.5*	*0*	*2.5*
	Continued on Sheet 2			
	Sub-Total	22	174	20
	Total Operator Cycle Time			

(Element Time (Sec.): Manual, Machine, Walking)

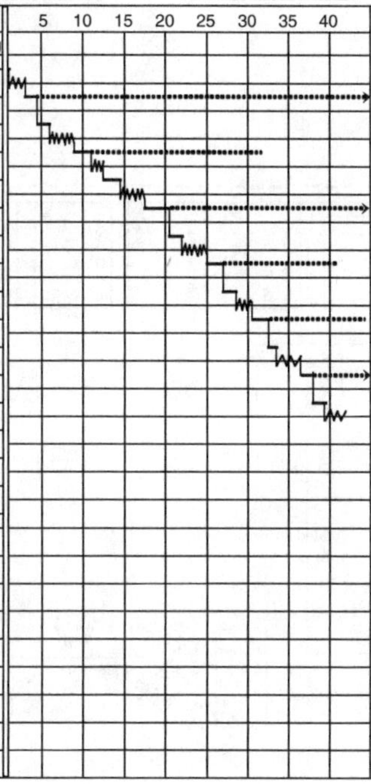

No.	Non-Cyclical Work Element	Time	Freq.	Total
a	*Team meeting*		*Daily*	
b	*Houskeeping*		*Daily*	
c	*Machine maintenance*		*Daily*	
	Total Operator Time			

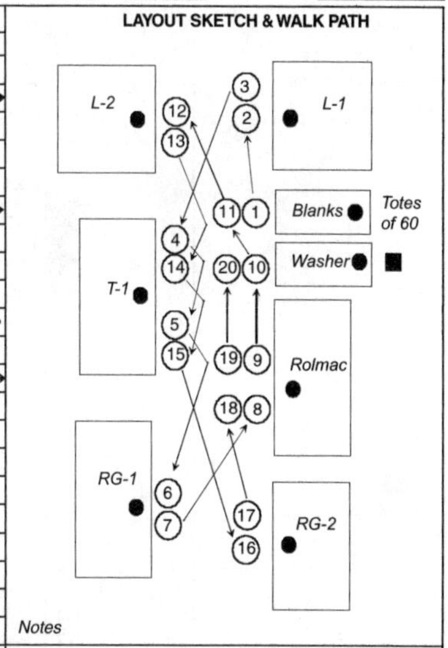

Notes

● *Work-in-Process* ■ *Inspection*

One or two operators in alternating walk path between L-1 & L-2; R-1 & R-2. Gage & inspection table after washer not shown.

Notes

FIGURE 52.9 Standardized work for a gear machining cell. (*From Planning Manufacturing Cells © Copyright 2002 Richard Muther & Associates.*)

- Ease of installation (avoidance of disruption)
- Acceptance by key employees
- Effect on quality

Costs are rarely sufficient for selecting the best cell plan. There are typically too many intangible considerations involved. And in many cases, the costs of alternative plans fall within a relatively narrow range. In practice, the final selection often rests on intangibles.

The weighted-factor method is the most effective way to make selections based upon intangible factors. After making a list of the relevant factors, weights should be assigned to indicate their relative importance. An effective scale is 1 to 10—with 10 being most important. Next, the cell operating team should rate the performance or effectiveness of each alternative on each weighted factor. It is important that ratings be made by cell operators and the appropriate plant support personnel—those closest to the action and responsible for making the selected plan work.

Since ratings are subjective they are best made with a simple vowel-code scale, and converted later into numerical values and scores. The following scale and values are effective:

A—Almost perfect results (excellent) 4 points

E—Especially good results (very good) 3 points

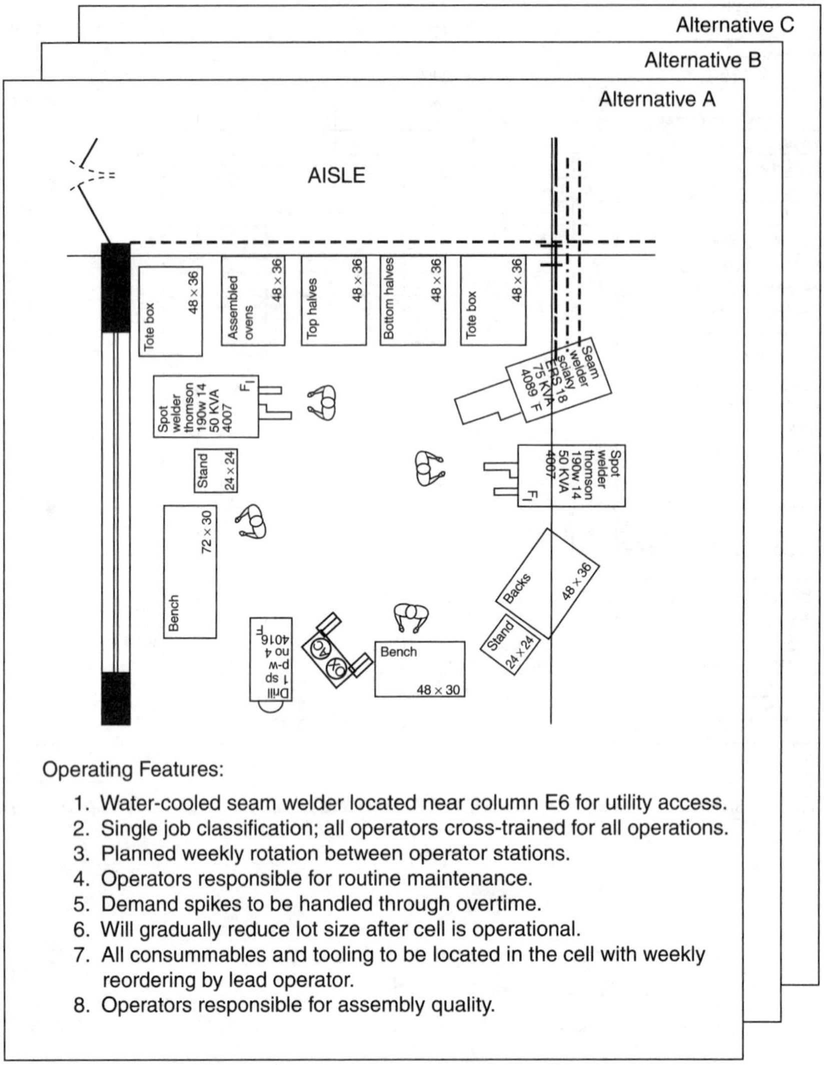

FIGURE 52.10 Alternative cell plans for the oven assembly cell diagrammed in Fig. 52.6.

I—Important results (good) 2 points

O—Ordinary results (fair) 1 point

U—Unimportant Results (poor) 0 points

X—Not Acceptable Results fix or remove from consideration

Rating values are multiplied by factor weights and down totaled to arrive at a score for each alternative plan. If one plan scores 15 to 20 percent higher than the rest, it is probably the better plan. If the costs are acceptable, this plan should be selected. If no plan scores significantly better than any other, then pick the least expensive, or consider additional factors. The final output of Step 5 is a selected cell plan.

Step 6—Detail and Implement the Plan. Once selected, details must still be worked out and preparations made to implement the cell plan. Detailing should begin with an updated, dimensioned drawing of the selected cell layout—typically at a fairly large scale, say 1:50. The detailing step should produce a scaled plan view of each work place showing the following:

- Normal operator work position
- Location of tooling, gauges, and controls
- Parts containers, fixtures, and workplace-handling devices
- Utility drops and connection points
- Door swings and access points on control panels and machinery
- Position of overhead lighting

In some cases, an elevation sketch may be useful, showing vertical placement of work surfaces, fixtures, containers, etc. Where highly fixed machinery is tied together with fixed conveyors, or where robots are used, it may also be necessary to develop a three-dimensional computer model of the cell in order to simulate or test for interference and proper placement.

In light manufacturing cells—machining or assembly—where equipment is easily adjusted during installation, this sophistication is typically unnecessary. Conventional plan views are usually sufficient.

If space is available and time permits, great insight can often be gained by creating a life-size mock-up of the cell, using cardboard, wood, light metal, or plastic tubing (see Fig. 52.11). By involving the cell operators in this mock-up, a great deal of useful detailing can be accomplished in a very short time. In our experience, mock-ups provide two significant benefits: (1) They uncover overlooked details that may be expensive to change later in implementation, and (2) they obtain a much greater level of operator involvement and interaction than is possible with an on-screen computer model, or with a two-dimensional plan view of the layout.

FIGURE 52.11 Construction of a physical mock-up for the gear machining cell charted in Fig. 52.9. *(From Planning Manufacturing Cells © Copyright 2002 Richard Muther & Associates.)*

Implementing a cell is an opportunity on a relatively small scale to make progress on plant-wide improvement initiatives. The cell implementation plan should include tasks, time, and money for the following common improvements:

- *Housekeeping and Safety.* Disposition of unnecessary items, fixing of leaks, cleaning and painting of machines, floors, ceilings, machine guards, aisleway guard rails, posts, etc.
- *Visual Control.* Marking and striping, signs for machines and workplaces, labeling for tool and fixture storage and containers, signal lights, performance displays.
- *Quality Management.* Certification of machine and process capabilities, tool and gauge lists and calibration plans, mistake-proofing and failure analyses, control plans, training, etc.
- *Maintenance.* Repair and rebuilding of machines, replacement of worn-out equipment, preventive maintenance schedules, operator maintenance procedures, etc.
- *Set-Up Reduction.* Video-taping, time- study, and methods analysis; redesign of fixtures, tools, and machines; duplication of key equipment, gauges, and fixtures, redefinition of responsibilities; training, etc.

Once the necessary tasks have been defined, they should be assigned to the appropriate individuals, estimated in terms of time and resources, and placed into a schedule, recognizing any dependencies between the tasks. The final output of step 6 is the selected cell plan detailed and ready for implementation.

52.5 MORE COMPLEX CELLS

Several considerations can complicate cell-planning projects. Chief among these is the question of how many cells are needed. Given a set of candidate parts and their desired production rates or quantities, the planner must occasionally decide whether a single cell is appropriate, or whether the work should be spread across multiple cells.

When one or more cells feed others, the project becomes one like planning a "mini factory." (See Fig. 52.12). Additional analysis is required to agree upon the material-handling methods and scheduling procedures for moving parts *between* the cells. It may also be necessary to share personnel or equipment capacity *across* the cells being planned. And, the project may also have to decide on common policies (across all of the cells) for organization, supervision, and performance measurement.

Even when planning a single cell, complications can be introduced if there is a wide range of possible locations for the cell, the appropriate level of automation is unclear, or there is the potential for radical organizational change, such as a move from traditional supervision to self-directed teams.

52.5.1 Four Phased Approach

Complex cell planning projects are best planned in four overlapping phases.

1. Orientation
2. Overall cell plan
3. Detailed cell plans
4. Implementation

The scope of these phases is illustrated in Fig. 52.13. Specific steps within each phase are fully described in Hales and Andersen, *Planning Manufacturing Cells.*[2] A synopsis of each phase is presented here.

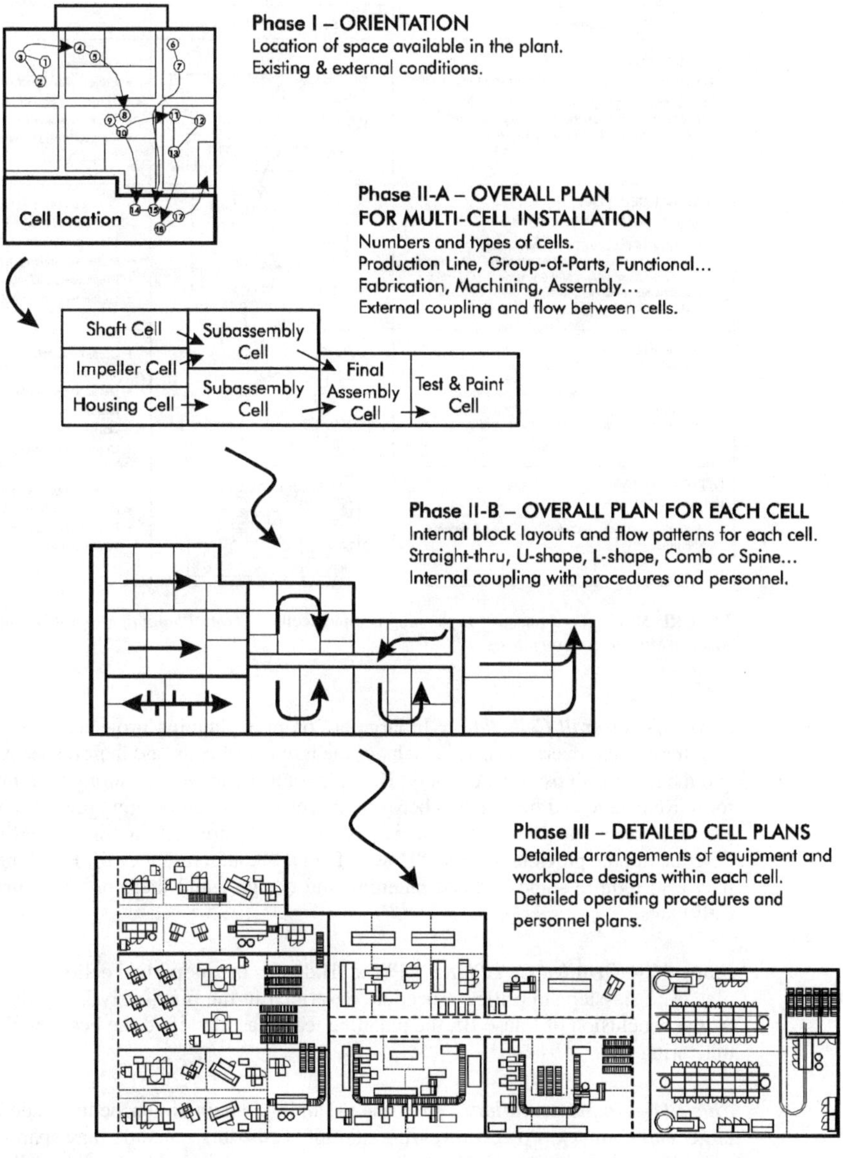

FIGURE 52.12 Complex, multicell planning. *(From Planning Manufacturing Cells © Copyright 2002 Richard Muther & Associates.)*

Phase I—Orientation. Complex projects or those with very large scope may need an entire phase to determine the best location(s) for the prospective cell(s), the handling to and from, the issues involved, and the plan for planning the cell. Reaching sound decisions may require creating or updating a master plan for the total facility. It may also require some conceptual planning of the prospective cells themselves—hence the overlap with phase II, Overall Cell Plan.

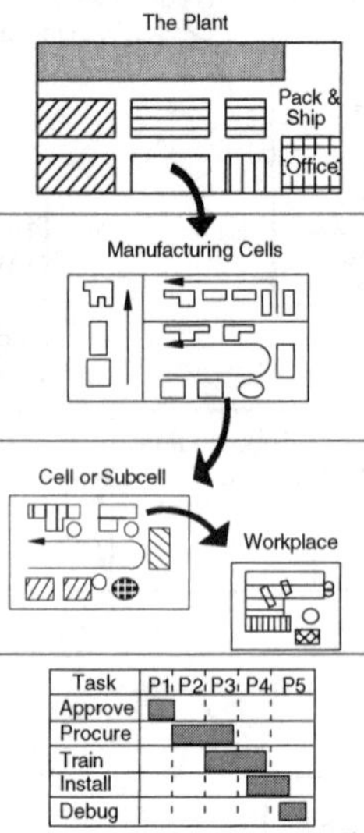

Planning Phases	Scope & Visualization	Outputs
I. ORIENTATION Location, external conditions and planning the planning. The surrounding physical and non-physical conditions/considerations; plus the objective, situation and plan for planning the cell.	The Plant	• Objectives • Project plan & schedule • Targeted parts or operational area • Planned location • External conditions • Scope; "givens", or limits of team's authority
II. OVERALL CELL PLAN The general plan for cellular operations. Definition of the cell or cells: how many; what parts, what processes and equipment. Alternative block layouts of space and equipment. Operating relationships and integration between cells and the rest of the plant.	Manufacturing Cells	• Definition of cells: what parts, what operations, what equipment • Block layout of cells and supporting areas • Material handling and coupling between cells • Preliminary cell layouts • General operating policies and practices
III. DETAILED CELL PLANS Detail plans and designs for each cell. The detailed arrangements of equipment and workplace design. Specification of detailed operating procedures and methods within each cell.	Cell or Subcell Workplace	• Confirmed definition of cell: parts, operations, equipment • Detailed layout for each cell • Material handling within the cells (between work stations) • Detailed operating policies and procedures
IV. IMPLEMENTATION Do; take action on the plans. Schedule, provide, procure, train, install, cleanup, debug...	Task: Approve, Procure, Train, Install, Debug (P1 P2 P3 P4 P5)	• Project approval and funding • Detailed implementation plan • Procured equipment • Working cell

FIGURE 52.13 Four phases for planning complex cells. *(From Planning Manufacturing Cells © Copyright 2002 Richard Muther & Associates.)*

Phase II—Overall Cell Plan. In large or complex planning projects, phase II defines the general plan for cellular operations. This includes the number of cells and their respective parts and processes, and the relationships between them. Block layouts are developed along with material handling plans for movement to and from and in between the cells. General operating practices or policies are decided. These planning activities and decisions are not addressed in the simplified six-step approach described in the previous section, "How to Plan a Manufacturing Cell." Reaching decisions in Phase II may also require some detailed planning and design—and therefore overlaps Phase III—Detailed Cell Plans.

Phase III—Detailed Cell Plans. Phase III details the individual cell(s) within the selected overall plan. The six-step simplified procedure described in the previously is highly effective for purpose. At the conclusion of Phase III, the planning team has identified the best detailed plan for each manufacturing cell.

Phase IV—Implementation. In Phase IV, an implementation schedule is defined for each cell. On larger and complex projects involving multiple cells, this schedule may span several months. It will typically include many interdependencies between the individual cell installations, and changes to the surrounding facilities, organization, and management systems. The team then obtains approval and funding, procures necessary equipment and services, directs the physical and procedural implementation, and debugs and releases the cell for production.

52.5.2 Impact of Automation and Technology

Most manufacturing cells consist of conventional, operator-controlled machinery and equipment. However, in some industries and processing situations, highly automated machinery may be used. For example, cells for high volume, repetitive welding may use robots instead of human operators.

The same is true with other hazardous operations such as forging. In high volume assembly of many small or precision components, cells may consist of automated assembly machines, and pick-and-place robots, often connected by conveyors. The entire cell may be computer controlled, with operators providing service and material handling to and from the cell. In some cases, the material handling may be automated using automated guided vehicles.

In between the extremes of all manual or fully automatic operations, cells may include some limited automation for parts feeding and loading, or material handling between operations. A common example is the use of automatic ejection or unloading devices on machine tools. These are typically used to present a completed part to the operator during the same cycle used to manually load the machine. Other examples of selective automation include conveyorized parts washers, curing tunnels, or similar process equipment. Typically the conveyor automates movement between sequential operations without operator intervention or effort, and may drop finished parts into a container.

If an automated cell with computerized controls is planned, extra attention should be given to estimating the costs of equipment, software, systems integration, and to the ongoing maintenance of the system. Adherence to sound technical standards and thorough documentation of all computerized systems will help to keep these costs down. As noted earlier, advanced visualization with 3-dimensional computer models is often valuable. Computer simulation may also be used.

Use of automation and advanced technology in a manufacturing cell is most appropriate when the following conditions apply:

- Production volumes are very high, typically above 500,000 units per year, and predictable or steady.
- Product lives are relatively long (before extensive changes or reconfigurations are required).
- Product designs are relative stable.
- Labor is expensive.
- The company or plant has prior successful experience with automated systems.
- The processes are hazardous or unsafe for human operators.
- Very high repeatability and precision are required.
- The processing technology is stable.

When several of these conditions are met, then at least one alternative cell plan should make use of automation, to be sure that good opportunities are not overlooked.

52.6 CHECKLIST FOR CELL PLANNING AND DESIGN

The paragraphs below contain a checklist and brief discussion of the most common choices or decisions to be made when planning or designing a cell. The topics are organized around the three aspects of cell planning discussed earlier: physical, procedural, and personal. The order of presentation follows roughly the order in which the choices and decisions should be made during a planning project—starting with the physical, followed by the procedural, and finally the personal or personnel-related.

52.6.1 Physical Questions

Layout and Flow Patterns

1. Which material flow pattern should be used within the cell?

- Straight-through
- U-shaped
- L-shaped
- Comb or spine

Basic Cell Flow Patterns and Their Benefits

Pattern	Benefits
Straight Thru 	1. Easy to understand, follow, schedule, and control. 2. Allows straight, inexpensive handling methods. 3. Easy access on two sides. 4. Avoids congestion at point of delivery and take away.
U-Shape or Circular 	1. Automatically returns product, holding fixtures, and mobile handling equipment to cell entrance. 2. Delivery and take away point are the same; allows convenient handling to and from the cell. 3. Workers in center can assist one another more readily. 4. Easier to assign multiple operations to an operator. Allows easier line balancing.
L-Shape 	1. Allows fitting lengthy series of operations into limited space. 2. Lets feeding cells start on an aisle and end at point of use. 3. May allow isolation of dangerous or costly-to-move equipment in the elbow, with savings in implementation cost and/or two directions for expansion. 4. Easy to segregate in-flow and out-flow of physically different materials, products, supplies, and special services.
Comb or Spine 	1. Lends itself to two-way flow. 2. Well suited to cells with highly variable routings. 3. Allows "fingers/teeth" to be segregated for special requirements. 4. Well-suited to functional cells.

FIGURE 52.14 Basic cell flow patterns. *(From Planning Manufacturing Cells © Copyright 2002 Richard Muther & Associates.)*

2. Cells may be physically arranged into one of four basic flow patterns. (See Fig. 52.14.) While the U-shape is frequently advocated and very common, the other patterns do have occasional advantages and appropriate uses. Your authors believe that the best cell layouts are achieved when the planning team forces itself to consider at least two alternative flow patterns, if only to stimulate discussion.
3. How does the choice of flow pattern fit with the overall plant layout and material flow? When deciding on the internal flow pattern for each cell, do not overlook its relationship to and impact upon the overall plant layout. The layout of aisles and general flow pattern in the factory may favor or even force a particular flow pattern within the cell.

Handling and Storage

1. What are the groups or classes of material to be moved? General categories to be examined include:
   - Incoming parts and materials to the cell
   - Work in process between workstations within the cell
   - Outgoing parts and materials leaving the cell

2. Classes should be defined with an eye toward common handling and storage methods. The classes for work-in-process should have already been defined through parts classification and analysis of the process. But a review of incoming and outgoing parts and materials may introduce additional classes, not yet identified or considered.
3. What handling equipment should be used for each class of material? Typical choices include: forklifts, tugs and carts, push-carts, walkie pallet jacks, conveyors, slides, chutes, overhead handling devices, or simply the operators themselves, hand carrying parts or materials.
4. What containers or transport units will be used? Typical choices include pallets, skids, bulk or large containers, small containers and totes, cartons, or in some cases, the items themselves.
5. Where and how will materials be stored or staged? What equipment will be used? Typical choices include: the floor, flow racks, shelves, pallet racks, cabinets, cartridges or magazines integrated directly into machines, or directly on work benches themselves.
6. How much material will be staged or stored? Typically expressed in minutes, hours, or days of coverage at an expected production rate.
7. How much staging or storage space will be required? And where should it be placed in the layout?

Supporting Services and Utilities

1. What process-related supporting services are required? Space and equipment are often required for tool & die storage, fixture storage, gage tables and benches, tool set-up, inspection areas, supplies, trash, and empty containers.
2. What personnel-related supporting services are required? Typicals include: shop desks and work areas; team meeting area; computer terminals and printers, telephones and public address speakers, document storage, bulletin boards, etc.
3. What special utilities are required? Water and drains? Special electrification? Special ventilation or exhausts? Lighting?

52.6.2 Procedural Questions

Quality Assurance/Control

1. Who will be responsible for quality? Will operators inspect their own work? Each other's work? Or, will dedicated inspectors be used? From within the cell or from outside?
2. What techniques will be used? e.g., visual, statistical process control, mistake proofing, etc.
3. Will special equipment be required?
4. What specifications or procedures are relevant or should be incorporated into the plan?

Engineering

1. Who is responsible for engineering the parts and processes involved?
   - Product engineering
   - Manufacturing/process/industrial engineering
2. How will tooling be managed? Externally by a central organization, or internally within the cell? Will tools be shared or dedicated to each machine?
3. Where will tools be stored? External to the cell? Internal? Centrally or at each work place or machine?
4. Who will be responsible for set-up? External or internal specialists? Operators themselves? Teams?

Materials Management

1. How will production be reported? Aggregate or total units only? Each unit as completed? First and/or last operation as performed? At the completion of each operation?
2. How will reporting be accomplished? Using paper forms? Key entry? Bar code scanning or other electronic method?
3. How will the cell be scheduled and by whom?
4. How will specific parts and jobs be sequenced? Who will be responsible?
5. Is line balancing needed?
6. What is the strategy for workload and capacity management? How will the cell respond to changes in product mix, bottlenecks and peaks?
   - With extra, idle machine capacity?
   - With extra labor or floating personnel?
   - With overtime?
   - With help from adjacent cells?
   - By off-loading work?
   - By building ahead to inventory?
   - By rebalancing or reassigning operators?

Maintenance

1. Who will be responsible for maintaining machinery and equipment? External by a central organization, or internally by cell operators?
2. Have specific maintenance duties and frequencies been defined?
3. Who will be responsible for housekeeping? Cell operators or external personnel?
4. Are preventive maintenance procedures required?
5. Are statistical or predictive maintenance procedures appropriate or necessary?
6. Will the cell require special equipment or services to hold or recycle waste, chips, oils, coolant, scrap, etc.?

Accounting

1. Will new accounting or reporting procedures be required? For costs? For labor? For material usage?
2. Will the cell be treated as a single work or cost center for reporting purposes? Or, will costs be charged to specific operations within the cell?
3. Will labor reporting distinguish between direct and indirect activities?
4. Will labor be reported against specific jobs or orders?
5. Will inventory be charged or assigned to the cell? Will the cell be a controlled inventory location?
6. How will scrap and rework be tracked and reported?

52.6.3 Personnel-Related Questions

Supervision and Performance Measurement

1. Will the cell have a designated supervisor or team leader? Or will the cell operate as a self-directed team?
2. How will cell performance be measured and reported, and to whom?

Job Definitions and Assignments

1. Will new positions be required? Have they been defined?
2. Will cell operators be specialists or cross-trained to work anywhere in the cell?
3. Will operators rotate assignments on a regular basis?
4. How will the initial cell operators be recruited or assigned? Will the opportunity be posted plant-wide?
5. How will future operators be assigned?

Compensation and Incentives

1. Will operators be compensated on a different basis from other parts of the plant?
2. Will operators be paid for skills or cross training?
3. Will a group incentive be used? How will it be calculated?

52.6.4 Systematic Planning and Involvement

The cell planner can generally achieve a good result by planning each individual cell with the six-step procedure outlined above. If the project is large or complex, and involves multiple cells, the additional structure of four overlapping phases will be helpful. As each cell is planned, look first at the physical, then the procedural, and finally at the personal aspects of the project. At every step, the planner should involve prospective operating personnel and others from the relevant supporting groups. Working as a group to answer the questions above will assure that the final, selected plan will have a smooth implementation and deliver the benefits desired.

52.6.5 Role of the Manufacturing Engineer

In many cases, the manufacturing engineer may serve as the primary cell planner. But for best results, production personnel should play leading roles in planning their own cells. When operators and first-line supervisors lead the project, the manufacturing engineer plays an important supporting role, typically focused on the analytical steps and physical aspects of the cell plan. The engineer will often lead or perform much of the work on:

- Classification of parts
- Definition of processes and routings
- Capacity analysis
- Layout planning

The manufacturing engineer may also assist in developing operating procedures, and will often provide cost estimates and comparisons of costs and savings among alternative plans.

52.7 CONCLUSIONS AND FUTURE TRENDS

By moving operations closer together, manufacturing cells reduce material handling, cycle times, inventory, quality problems, and space requirements. In addition to these primarily quantifiable benefits, the focused nature, and typically small size of cells also leads to:

- Easier production control
- Greater operator productivity

- Quicker action on quality problems
- More effective training
- Better utilization of personnel
- Better handling of engineering changes

Because of these benefits, cells provide a focused and practical way to implement the principles of The Toyota Production System, Lean Manufacturing, World Class Manufacturing, Just-In-Time, and other forms of plant-wide productivity improvement. Visual management and control, elimination of waste, set-up reductions, pull signals and continuous flow are all easier to achieve when implemented through individual manufacturing cells. The popularity of plant- and company-wide improvement programs will continue to expand the use of manufacturing cells.

Because they are relatively quick and easy to reconfigure, cells have become the preferred manufacturing model for high-variety, medium-to-low volume, short-life-cycle products. Increasing marketing emphasis on highly tailored, relatively low-volume products will also expand the use of manufacturing cells.

REFERENCES

1. Muther, Richard, William E. Fillmore, and Charles P. Rome, *Simplified Systematic Planning of Manufacturing Cells,* Management and Industrial Research Publications, Kansas City, Missouri, 1996. (booklet)
2. Hales, H. Lee, and Bruce J. Andersen, *Planning Manufacturing Cells,* Society of Manufacturing Engineers, Dearborn, Michigan, 2002. (textbook and videotapes)

CHAPTER 53
WORK MEASUREMENT

Lawrence S. Aft
Aft Systems, Inc.
Roswell, Georgia

53.1 INTRODUCTION

Work measurement is performed to develop the standard time to perform operations. Time standards have traditionally been defined as the time required by a typical operator, working at a normal pace, to perform a specified task using a prescribed method, with time for personal matters, fatigue, and delay allowed. Time standards, work standards, or just standards, are critical pieces of management information that apply to all kinds of work, including manufacturing, assembly, and clerical.

Standards provide information essential for the successful operation of an organization. This information includes:

- *Data for scheduling.* Production schedules cannot be set unless times for all operations are known. Delivery dates cannot be promised unless times for all operations are known.
- *Data for staffing.* The number of workers required cannot accurately be determined unless the time required to process the existing work is known. Continuing management of the work force requires the use of labor variance reports. Labor variance reports are also most useful for determining changes in work methods, especially the subtle or incremental changes.
- *Data for line balancing.* The correct number of workstations for optimum workflow depends on the processing time, or standard, at each workstation. Operation times and set-up times are key pieces of this information.
- *Data for materials requirement planning.* MRP systems cannot operate properly without accurate work standards.
- *Data for system simulation.* Simulation models cannot accurately simulate operation unless times for all operations are known.
- *Data for wage payment.* To be equitable, wages generally must be related to productivity. Expected productivity when compared with actual productivity can only be determined through the use of work standards.
- *Data for costing.* Ultimately the profitability of an organization lies in its ability to sell products for more than it costs to produce them. Work standards are necessary for determining not only the labor component of costs, but also the correct allocation of production costs to specific products.
- *Data for employee evaluation.* In order to assess whether or not individual employees are performing as well as they should, a performance standard is necessary for measurement. Without a standard it is impossible to determine the level of performance.

53.2 TIME STANDARDS

As defined above, the standard time is the time required by a typical operator, working at a normal pace, to perform a specified task using a prescribed method, with time for personal, fatigue, and delay allowed. Some of the key factors of this definition are the understanding of a typical operator, the concept of normal pace, the reliance on prescribed method, and the designation of the allowance.

The definition of standard time specifies a typical operator. The typical operator is an operator who is representative of the people performing the task. The typical operator is not the best nor the worst. The typical operator is skilled in the job and can perform it consistently throughout the entire workday.

The normal pace is a rate of work that can be maintained for an entire workday. It is neither too fast nor too slow. It is the pace of a typical worker. Rarely will any worker perform at the normal pace for an entire workday. The normal pace represents an ideal that the industrial engineer judges the average worker should be able to maintain day in and day out.

A key part of the definition is the phrase relating to prescribed method. Work standards measure the time required to correctly perform defined tasks. Part of the definition must include a statement regarding the quality of the work performed.

All workers have personal needs that must be attended to. Workers also tend to become tired as the workday progresses. When developing a time standard these factors must be included and allowed for. Additionally, there will be occasional unexpected and often uncontrollable delays, such as material shortages or equipment breakdowns that also must be allowed for. The *personal, fatigue, and delay* (PFD) factors, depending on the nature of the work being performed, can be a significant part of the work standard.

53.2.1 Measuring Work

Standards have traditionally been developed in one of three major ways.

1. The first of these is *estimation.* There are two ways that estimation is done. Sometimes the time required is provided via a SWAG.* An individual who is believed to be knowledgeable about the task examines the work to be completed and then states, "It ought to take about that many hours to get all the pieces run." Sometimes it does. Sometimes it does not. Sometimes work gets completed early, other times bottlenecks develop and schedules are missed. The other commonly used method of estimation involves the use of historical data. Prior runs are examined and actual times and production quantities are used to develop a historical standard. The danger with historical standards lies in Parkinson's Rule† as applied to industrial engineering.
2. Standards are also set using direct observation and measurement. There are three common methods used for setting standards using direct observation—time study, work sampling, and physiological work measurement. Time study is defined as follows:
   - Time study is the analysis of a given operation to determine the elements of work required to perform it, the order in which these elements occur, and the times which are required to perform them effectively.†† A work sampling study consists of a large number of observations taken at random intervals; in taking the observations, the state or condition of the object of study is

* SWAG is an acronym for Scientific Wild Ass Guess.

† Parkinson's Rule as applied to Industrial Engineering states that the amount of time required to complete a task is directly proportional to the time available. The more time available, the longer each individual item will require. In other words, the standard expands or contracts based upon the availability of time to complete the work.

†† Maynard, 1956.

noted, and this state is classified into predefined categories of activity pertinent to the particular work situation. From the proportions of observations in each category, inferences are drawn concerning the total work activity under study.*

- Another way to directly measure work performed is by physiological means. This is based on the fact that work is equal to force times distance. Energy is required to perform work.
- Physical work results in changes in oxygen consumption, heart rate, pulmonary ventilation, body temperature, and lactic acid concentration in the blood. Although some of these factors are only slightly affected by muscular activity, there is a linear correlation between heart rate, oxygen consumption, and total ventilation, and the physical work performed by an individual. Of these three, the first two—heart rate and oxygen consumption—are most widely used for measuring the physiological cost of human work.†
- Many studies have shown that the difference between well trained workers and beginners on a job is significant. The physiological cost to the beginner would be greater when the beginner attempts to produce at the normal pace. Physiological measurements are used to compare the cost to the worker for performing varying tasks.††

3. The third general way of setting work standards is through the use of standard data systems. Mil-Std (1567) defined standard data as "a compilation of all the elements that are used for performing a given class of work with normal elemental time values for each element. The data are used as a basis for determining time standards on work similar to that from which the data were determined without making actual time studies."

Standard data is the term used to describe time data for groups of motions rather than single motions. Such data are used to set standard times for new work without having to take complete and detailed studies of the work. They are compiled from existing detailed studies of manual work and are arranged in the form of tables, graphs, and formulas for ready use. Knowledge of how the new job must be done makes it possible to select the appropriate time data from these records to obtain the proper standard time for the job.§

- There are two types of standard data. One is what is often referred to as macroscopic standard data.

  Many operations in a given plant have several common elements. The element "walking," for example, is a component of many different jobs. Diverse activities such as painting, handling, or working on a site invariably involve an element of "walking." When these activities are times, the same common element is in fact timed again and again. The job of the work study analyst would therefore be made much easier if he had at his disposal a set of data from which he could readily derive standard times for these common work elements with necessarily going into the process of timing each one.¶

  Macroscopic standard data takes advantages of similarities of activities within like families of operations and uses those similarities to develop standards for related activities. Standard data can reduce the time and labor required to set standards.‖
- The other type of standard data is what might be called microscopic standard data. This type of standard data is also often referred to as predetermined time systems. It is a motion based method of work measurement.

  By carefully describing all of the motions required to perform a particular job, the analyst will have to carefully study the method being used to perform the job. When the motions required to complete the work have been identified, the standard can be set. In predetermined time systems, each motion that is described and coded has a specific time allowed for its completion. By completely identifying all of the motions required the entire time for a sequence of motions or for an entire operation can be synthesized. Once the allowance is applied, an accurate time standard can be prepared. This procedure, of course,

* Heiland and Richardson, 1957. † Barnes, 1980. †† Brouha, 1960. § Bailey and Presgrave, 1958.
¶ International Labour Office, 1979. ‖ Aft and Merritt, 1984

is based on the assumption that the correct motions have been identified before the times are assigned.*

A wide variety of predetermined time systems exist. Some will be described in more detail in subsequent sections. Regardless of the specific system selected, they all are used in a similar fashion. Initially the task being studied has to be precisely defined in terms of the motions involved. This requires a complete understanding of the operation. Once the motions are defined then times for individual motions are retrieved from the system's database. The individual motion times are combined and an appropriate allowance is incorporated resulting in a time standard for the task.

53.3 TIME STUDY

Time study involves the use of a timing device, study of the existing work method, recording observed times, rating the subject's performance compared with normal pace, and adding the PFD allowance. Time study is most effective for developing standards for highly repetitive tasks which have relatively short cycle times. When work is nonrepetitive and has relatively long cycle times, such as some clerical and maintenance tasks, then work sampling is an appropriate method for setting standards.

The first requirement for making a time study is having a timing device. Traditionally this tool has been a stopwatch, however, electronic devices such as the PDA are now being used to collect this data. Section 53.8 illustrates the use of such devices. Second, the analyst must describe the method used. Typically this involves a detailed description of key subdivisions of the work being performed called elements. These are created after a "best" method is established. The best method is a productive and ergonomically sound approach to the tasks required in the job being performed. Elements are components of a job or task that is logical divisions of the job, have easily identifiable starting and stopping points, and repeat on a regular basis throughout the work day. Each element must have an easily identifiable starting and stopping point. When completed there can be no discontinuities in the work performed.†

Once elements are identified then the elapsed times for each element are measured. Sufficient repetitions of the job are measured until the desired accuracy, e.g., ±5 percent and the desired statistical confidence,†† e.g., 95 percent are achieved. While the data is being collected, the subject of the study is rated. This process will be discussed in the next section.

Once the data has been collected the analyst will average the observed times for each element, prorate the averages for elements that don't occur every cycle based on their relative frequency of occurrence, apply the rating factor, and sum these values creating the *normal time*. After the personal, fatigue, and delay allowance is applied the result is the standard time. The process for determining the allowance will also be covered in another section of this chapter.

53.3.1 Performance Rating

The normal pace, regardless of the standard used, is an idealized pace. No individual is capable of consistently working at the normal pace for the entire work day. Some people work faster than normal and some work slower than normal. Sometimes the process of being time studied causes the subject to change the work pace to try to conform with what is believed to be normal or to conform to a peer group concept of normal. The industrial engineer must be aware of this tactic, whether or not it is done consciously, and compensate for it by rating the performance of the job being studied.

---

* Aft, 2000. † Aft, *Work Measurement and Methods Improvement,* Chapter 5.

†† We are after all measuring a sample from all possible repetitions of this job.

Rating is the process of comparing the actual work being performed with the analyst's concept of normal pace and evaluating the observed performance quantitatively. Rating often uses the normal pace as a base of 100 percent. A worker performing at a pace 20 percent faster than normal would be rated at 120 percent.

There is no magic to rating. To rate effectively, the analyst must be able to compare the observed activity to the predetermined concept of standard or normal pace. It is most important that every analyst in the organization have the same basic standard firmly entrenched in his or her mind. Consistency is of utmost importance. Analysts must work to calibrate themselves and the organization's view of normal performance. Companies use a variety of techniques to help achieve this consistency. Some purchase commercially available rating films, videotapes, or CD's, while others produce their own.

Generally, experienced anallysts will be able to rate every job within ±5 percent of every other analyst within the organization. This skill is developed only through practice and experience. A new analyst often will work in conjunction with an experienced one to achieve the consistency that is necessary to set fair and effective time standards.

53.3.2 Personal, Fatigue, and Delay Allowance

In addition to acknowledging that workers can work at speeds other than the normal pace, it must be acknowledged that workers cannot work for an entire work day without some rest. Sometimes, rest is the worker's own doing, such as a visit to the water fountain or restroom. Sometimes it is beyond the workers' control, such as when a machine breaks down or when the line runs out of inventory. As the day progresses, workers become tired and are less able to perform as they did early in the shift.* The standard must be adjusted to reflect the personal, fatigue, and delays that are part of every job.

Some typical allowance factors that many jobs experience include the following:

- Tool changes and adjustments
- Stock handling
- Servicing equipment
- Job preparation and cleanup
- Inspection
- Interruption by supervisor
- Personal time

The personal, fatigue, and delay (PFD) allowance is usually expressed as a percentage of the standard time and added to the time allowed to complete the particular task being studied. The amount of time provided for the allowance should be based on actual observation of the work performed as measured during an all-day production study. Such a study identifies and measures actual delays and personal time. By comparing performance from hour to hour during the work day the impact of fatigue can be determined.

53.3.3 Sample Size Estimation

The number of cycles to time in a time study depends. It depends on two things. First of all it depends on how much confidence we want in our result. Remembering that the time standard represents an average time, we first introduce the concept of statistical confidence. It is defined as "the probability

* Aft, *Work Measurement and Methods Improvement,* Chapter 5.

TABLE 53.1 Z Value at Various Confidence Levels

Confidence level (percent)	Z value (approximate)
50	0.67
60	0.84
70	1.04
75	1.15
80	1.28
85	1.44
90	1.645
95	1.96
99	2.575

that an interval about a sample statistic actually includes the population parameter."* Confidence is usually expressed as a percentage, e.g., 95 percent confidence. The second "it depends" is the concept of statistical accuracy. Accuracy is "the closeness of agreement between an observed value and an accepted reference value."† Accuracy is often stated in terms of a percentage, as in ±5 percent.

Thus the number of cycles to time depends on the confidence and accuracy that is desired. There are two ways to approach this. We can either specify the confidence and accuracy we desire ahead of time or we can collect our data and then specify either the confidence or the accuracy and calculate the other.

Specifying the number of cycles to time for a specified confidence and accuracy level will be addressed first. A key to the number required in this calculation is the variability of the elemental times. In order to determine the number of cycles to time a small sample is actually timed. In addition to the average the sample standard deviation is calculated. These values are then substituted into an equation that will calculate the required number of samples. This equation also requires the specification of the confidence level that translates into a value from the standard normal probability distribution. The equation to determine the sample size is:

$$n = \frac{s^2 z^2}{k^2 \bar{x}^2}$$

where z comes from the standard normal table. Some typical values of z are shown in Table 53.1. s is the sample standard deviation k is the accuracy $\bar{x}$ is the average.

The standard deviation is calculated from the observed elemental times. The formula for the sample standard deviation is

$$\sqrt{\frac{\sum (x - \bar{x})^2}{n - 1}}$$

where x = each observed elemental time
$\bar{x}$ = average of the observed elemental times
n = the number of cycles initially timed

* ASQ's *Glossary and Tables for Statistical Quality Control,* American Society for Quality, Milwaukee, 1983, p. 2.
† ASQ's *Glossary and Tables for Statistical Quality Control,* American Society for Quality, Milwaukee, 1983, p. 1.

The calculation for number of cycles is performed for each element. The number of cycles to time is the largest of these numbers.

Example

An analyst studies a job. This particular job has but one element. A preliminary study of 10 cycles had the following results:

$$\bar{x} = 2.8 \text{ minutes}$$

$$s = 0.6 \text{ minutes}$$

How many cycles should be studied for 95 percent confidence and 5 percent* accuracy? To answer this question the following formula is used:

$$n = \frac{s^2 z^2}{k^2 \bar{x}^2}$$

The z value corresponding to 95 percent confidence is read from Table 53.1 as 1.96. The standard deviation, s, is 0.6 min.

The sample size is then calculated: $n = 70.56$ or 71 total cycles should be timed. Based on the initial 10 this leaves 61 cycles remaining to be timed in order for the analyst to state that he or she is 95 percent confident that the measured time is within 5 percent of the true time required for the task to be performed.

53.4 PREDETERMINED TIME SYSTEMS

There are times when it is neither possible nor practical to set time standards using stopwatch time study. There are also times when it might not be possible to use a company-developed data base to synthesize time standards. When this situation arises and when the organization desires to have a measure of time required, the use of predetermined time standards might be appropriated.

These are standard data systems that are designed to be used in a wide variety of product and process applications. They are sometimes viewed as general purpose systems, designed to work anywhere, or special purpose systems, designed for a specific process or product, such as sewing.

The use of predetermined time systems requires special training and skill. It is not the intent of this chapter to create expertise in the systems described below, but rather to describe their features.

53.4.1 Advantages and Limitations

Predetermined time systems have four major advantages as well as some limitations. The benefits include the following:

1. All predetermined time systems require a complete methods analysis prior to the setting of the standard. Each motion must be identified. Obvious methods problems and other inefficiencies are readily identified when this much attention is paid to the work method being used. The resulting analysis also yields a well-documented procedure for performing the task. For new jobs they force the establishment of a sound, well thought out method.
2. Predetermined time systems do not require the analyst to perform performance rating. This eliminates some subjectivity from the resulting standard and provides a more consistent standard.

---

* 5 percent accuracy means that the true value is within 5 percent of the average. This is considered to be excellent accuracy.

3. In order to develop work standards using a direct observation method, the work must be performed and measured while it is being performed. Predetermined time systems allow the analyst to visualize the work and synthesize the standard even if the task is still in the planning phase.
4. Predetermined time systems provide information about learning time. The development of learning curves and their subsequent application is an essential part of determining the cost of a new product or service.

Although there are significant benefits associated with predetermined time standards, there are also some limitations. A major disadvantage is the difficulty encountered with machine-paced operations. Most of the predetermined systems were designed for human motion times, not machine times. Some of the systems have been designed for specific types of work, such as clerical operations or sewing operations and the motions defined within the systems do not transfer well to other types of work. Predetermined time systems have many definitions and rules associated with the proper application of times. Whether or not it is a disadvantage is debatable, but a significant amount of training is required prior to individuals being competent to apply most of the systems.

53.4.2 Example Systems

Predetermined time systems provide information about manual work cycles in terms of basic human motions. There are differences between the criteria adopted for the classification of these motions. Broadly speaking, there are two main sets:

1. Object related classification
2. Behavior related classification

In an object related system reference may be made to characteristics of parts or to the nature of the surround conditions. Behavior related systems classify motions according to what they look like to an observer.*

Another way to look at the classification of predetermined time systems is as motion based, action based, and activity based. *Motion-based* encompasses all those systems that are made up of "basic motions"—time elements that cannot be broken down into smaller elements. *Action-based* are such systems that consist of elements combined of basic motions into "actions." *Activity-based* are systems also consisting of elements that are combinations of basic motions or (in most cases) action elements. Activity-based elements are then put together in a sequence representing a complete activity such as "move object from A to B" or "fasten screw with screwdriver."

Some examples of motion based predetermined time systems are presented first.

Methods Time Measurement (MTM-1).† The most widely publicized system of performance rating ever developed was presented in Lowry, Maynard, and Stegemerten (1940). The basis of the rating was the use of four factors: skill, effort, consistency, and performance. Maynard and Steemerten teamed with John Schwab to expand this idea into Methods Time Measurement or MTM.†† (This is now known as MTM-1.) According to Robert Rice this method is the most widely used system of predetermined times.§ Maynard and associates performed many micromotion studies to come up with their standard elements and times. Because MTM was and is readily available, it is not surprising that it is the most frequently used of all the systems and it is also the most frequently imitated. Standard MTM-1 data is shown in Fig. 53.1. MTM-1 is defined as a procedure that analyzes any manual operation or method into the basic operations required to perform it and assigns to each motion a predetermined time standard which is determined by the nature and the condition under

* International Labour Office, 1979.

† MTM-1, MTM-2, MTM-3, and MTM-MEK are copyrighted and are the property of the MTM Association for Standards and Research and as such cannot be reproduced without the written authorization of the MTM Association for Standards and Research.

†† Maynard, Schwab, and Stegemerten, 1948. § Rice, 1977.

TABLE I – REACH – R

Distance Moved Inches	Time TMU A	Time TMU B	Time TMU C or D	Time TMU E	Hand In Motion A	Hand In Motion B	CASE AND DESCRIPTION
3/4 or less	2.0	2.0	2.0	2.0	1.6	1.6	A Reach to object in fixed location, or to object in other hand or on which other hand rests.
1	2.5	2.5	3.6	2.4	2.3	2.3	
2	4.0	4.0	5.9	3.8	3.5	2.7	
3	5.3	5.3	7.3	5.3	4.5	3.6	B Reach to single object in location which may vary slightly from cycle to cycle.
4	6.1	6.4	8.4	6.8	4.9	4.3	
5	6.5	7.8	9.4	7.4	5.3	5.0	
6	7.0	8.6	10.1	8.0	5.7	5.7	
7	7.4	9.3	10.8	8.7	6.1	6.5	C Reach to object jumbled with other objects in a group so that search and select occur.
8	7.9	10.1	11.5	9.3	6.5	7.2	
9	8.3	10.8	12.2	9.9	6.9	7.9	
10	8.7	11.5	12.9	10.5	7.3	8.6	
12	9.6	12.9	14.2	11.8	8.1	10.1	
14	10.5	14.4	15.6	13.0	8.9	11.5	D Reach to a very small object or where accurate grasp is required.
16	11.4	15.8	17.0	14.2	9.7	12.9	
18	12.3	17.2	18.4	15.5	10.5	14.4	
20	13.1	18.6	19.8	16.7	11.3	15.8	
22	14.0	20.1	21.2	18.0	12.1	17.3	E Reach to indefinite location to get hand in position for body balance or next motion or out of way.
24	14.9	21.5	22.5	19.2	12.9	18.8	
26	15.8	22.9	23.9	20.4	13.7	20.2	
28	16.7	24.4	25.3	21.7	14.5	21.7	
30	17.5	25.8	26.7	22.9	15.3	23.2	
Additional	0.4	0.7	0.7	0.6			TMU per inch over 30 inches

TABLE II – MOVE – M

Distance Moved Inches	Time TMU A	Time TMU B	Time TMU C	Hand In Motion B	Wt. (lb.) Up to	Dynamic Factor	Static Constant TMU	CASE AND DESCRIPTION
3/4 or less	2.0	2.0	2.0	1.7				
1	2.5	2.9	3.4	2.3	2.5	1.00	0	A Move object to other hand or against stop.
2	3.6	4.6	5.2	2.9				
3	4.9	5.7	6.7	3.6	7.5	1.06	2.2	
4	6.1	6.9	8.0	4.3				
5	7.3	8.0	9.2	5.0	12.5	1.11	3.9	
6	8.1	8.9	10.3	5.7				
7	8.9	9.7	11.1	6.5	17.5	1.17	5.6	B Move object to approximate or indefinite location.
8	9.7	10.6	11.8	7.2				
9	10.5	11.5	12.7	7.9	22.5	1.22	7.4	
10	11.3	12.2	13.5	8.6				
12	12.9	13.4	15.2	10.0	27.5	1.28	9.1	
14	14.4	14.6	16.9	11.4				
16	16.0	15.8	18.7	12.8	32.5	1.33	10.8	
18	17.6	17.0	20.4	14.2				
20	19.2	18.2	22.1	15.6	37.5	1.39	12.5	C Move object to exact location.
22	20.8	19.4	23.8	17.0				
24	22.4	20.6	25.5	18.4	42.5	1.44	14.3	
26	24.0	21.8	27.3	19.8				
28	25.5	23.1	29.0	21.2	47.5	1.50	16.0	
30	27.1	24.3	30.7	22.7				
Additional	0.8	0.6	0.85		TMU per inch over 30 inches			

TABLE III A – TURN – T

Weight	30°	45°	60°	75°	90°	105°	120°	135°	150°	165°	180°
	Time TMU for Degrees Turned										
Small – 0 to 2 Pounds	2.8	3.5	4.1	4.8	5.4	6.1	6.8	7.4	8.1	8.7	9.4
Medium – 2.1 to 10 Pounds	4.4	5.5	6.5	7.5	8.5	9.6	10.6	11.6	12.7	13.7	14.8
Large – 10.1 to 35 Pounds	8.4	10.5	12.3	14.4	16.2	18.3	20.4	22.2	24.3	26.1	28.2

TABLE III B – APPLY PRESSURE – AP

FULL CYCLE			COMPONENTS		
SYMBOL	TMU	DESCRIPTION	SYMBOL	TMU	DESCRIPTION
APA	10.6	AF + DM + RLF	AF	3.4	Apply Force
			DM	4.2	Dwell, Minimum
APB	16.2	APA + G2	RLF	3.0	Release Force

TABLE IV – GRASP – G

TYPE OF GRASP	Case	Time TMU	DESCRIPTION	
PICK-UP	1A	2.0	Any size object by itself, easily grasped	
	1B	3.5	Object very small or lying close against a flat surface	
	1C1	7.3	Diameter larger than 1/2"	Interference with Grasp on bottom and one side of nearly cylindrical object.
	1C2	8.7	Diameter 1/4" to 1/2"	
	1C3	10.8	Diameter less than 1/4"	
REGRASP	2	5.6	Change grasp without relinquishing control	
TRANSFER	3	5.6	Control transferred from one hand to the other.	
SELECT	4A	7.3	Larger than 1" x 1" x 1"	Object jumbled with other objects so that search and select occur.
	4B	9.1	1/4" x 1/4" x 1/8" to 1" x 1" x 1"	
	4C	12.9	Smaller than 1/4" x 1/4" x 1/8"	
CONTACT	5	0	Contact, Sliding, or Hook Grasp.	

EFFECTIVE NET WEIGHT			
Effective Net Weight (ENW)	No. of Hands	Spatial	Sliding
	1	W	$W \times F_c$
	2	W/2	$W/2 \times F_c$

W = Weight in pounds
F_c = Coefficient of Friction

TABLE V – POSITION* – P

CLASS OF FIT		Symmetry	Easy To Handle	Difficult To Handle
1—Loose	No pressure required	S	5.6	11.2
		SS	9.1	14.7
		NS	10.4	16.0
2—Close	Light pressure required	S	16.2	21.8
		SS	19.7	25.3
		NS	21.0	26.6
3—Exact	Heavy pressure required.	S	43.0	48.6
		SS	46.5	52.1
		NS	47.8	53.4
SUPPLEMENTARY RULE FOR SURFACE ALIGNMENT				
P1SE per alignment: >1/16 ≤1/4"		P2SE per alignment: ≤1/16"		

*Distance moved to engage—1" or less.

TABLE VI – RELEASE – RL

Case	Time TMU	DESCRIPTION
1	2.0	Normal release performed by opening fingers as independent motion.
2	0	Contact Release

TABLE VII – DISENGAGE – D

CLASS OF FIT	HEIGHT OF RECOIL	EASY TO HANDLE	DIFFICULT TO HANDLE
1—LOOSE—Very slight effort, blends with subsequent move.	Up to 1"	4.0	5.7
2—CLOSE—Normal effort, slight recoil.	Over 1" to 5"	7.5	11.8
3—TIGHT—Considerable effort, hand recoils markedly.	Over 5" to 12"	22.9	34.7

SUPPLEMENTARY

CLASS OF FIT	CARE IN HANDLING	BINDING
1—LOOSE	Allow Class 2	—
2—CLOSE	Allow Class 3	One G2 per Bind
3—TIGHT	Change Method	One APB per Bind

TABLE VIII – EYE TRAVEL AND EYE FOCUS – ET AND EF

Eye Travel Time = $15.2 \times \frac{T}{D}$ TMU, with a maximum value of 20 TMU.

where T = the distance between points from and to which the eye travels.
D = the perpendicular distance from the eye to the line of travel T.

Eye Focus Time = 7.3 TMU.

SUPPLEMENTARY INFORMATION

– Area of Normal Vision = Circle 4" in Diameter 16" from Eyes

– Reading Formula = 5.05 N Where N = The Number of Words.

MTM ASSOCIATION FOR STANDARDS AND RESEARCH

MTM association

1111 E. Touhy Avenue
Des Plaines, IL 60018
Phone: 847/299-1111
Fax: 847/299-3509

FIGURE 53.1 MTM-1. (Reproduced by permission of MTM Association.)

which it is made.* Reach is the most common or basic MTM-1 motion. The other motions involved include the following:

Move—In which the predominant purpose is to transport an object to a destination.

Turn—In which the hand is turned or rotated about the long axis of the forearm.

Position—The motion employed to align, orient, and engage one object with another.

Grasp—Where the main purpose is to secure sufficient control of one or more objects with the fingers or the hand.

Release—Motions that identify the operator relinquishing control of an object.

Disengage—The motion is used to identify when contact between two objects is broken.

Eye Times—Used when the eyes direct hand or body motions.

Body Motions—Motions are made by the entire body, not just the hands, fingers, or arms.

MODAPTS. MODAPTS is a relatively easy-to-use predetermined time system. MODAPTS stands for *modular arrangement of predetermined time standards.*

MODAPTS is an Australian–developed time system that is based on the premise that larger body sections take longer to move than that of smaller sections. For example in this system it takes twice as long to move a hand as it does to move a finger. It takes three times as long to move the forearm as it does a finger, and it takes four times as long to move the whole arm outward. From this simple framework, MODAPTS has built an entire system of predetermined macro time standards.†

Because it describes work in human rather than mechanical terms, it has many more potential applications than earlier work analysis systems. The application is integrated with desktop computer processing capabilities, which simplifies its use.

MODAPTS is a recognized industrial engineering technique, meeting all criteria of the U.S. Defense Department and Department of Labor to develop industrial standards. Performance times are based on the premise that motions will be carried out at the most energy-efficient speed.

MODAPTS is used to analyze all types of industrial, office, and materials handling tasks. Data from MODAPTS studies are used for planning and scheduling, cost estimating and analysis, ergonomic evaluation of manual tasks, and the development of labor standards.††

BasicMOST. MOST§ concentrates on the movement of objects. Efficient, smooth, productive work is performed when the basic motion patterns are tactically arranged and smoothly choreographed. This provides the basis for the MOST sequence models (see Table 53.2). The primary work units are no longer basic motions, but fundamental activities (collections of basic motions) dealing with moving objects. These activities are described in terms of subactivities fixed in sequence. In other words, to move an object a standard sequence of events occurs.

Objects can be moved in only one of two ways: either they are picked up and moved freely through space, or they are moved and maintain contact with another surface. The use of tools is analyzed through a separate activity sequence model which allows the analyst the opportunity to follow the movement of a hand tool through a standard sequence of events, which, in fact, is a combination of the two basic sequence models.

Consequently, only three activity sequences are needed for describing manual work.¶ The MOST technique there is comprised of the following basic sequence models:

- The general move sequence—for the spatial movement of an object freely through the air.
- The controlled move sequence—for the movement of an object when it remains in contact with a surface or is attached to another object during the movement.
- The tool use sequence—for the use of common hand tools.

---

* Karger and Bayh, 1987. † Masud and Malzahn, 1985.

†† Additional information about MODAPTS is available from the International MODAPTS Association, Inc. The material presented here is reproduced with permission.

§ Zandin, 1980.

¶ Information about MOST reproduced with permission of Marcel Dekker.

TABLE 53.2 Basic Most Sequence Models

Manual handling			Equipment handling		
Activity	Sequence model	Parameters	Activity	Sequence model	Parameters
General move	ABGABPA	A - Action distance B - Body motion G - Gain control P - Place	Move with manual crane-jib type	ATKFVLVPTA	A - Action distance T - Transport empty K - Hook up and unhook F - Free object V - Vertical move L - Loaded move P - Place
Controlled move	ABGMXIA	M - Move controlled X - Process time I - Align			
Tool use	ABGABP ABPA	F - Fasten L - Loosen C - Cut S - Surface treat M - Measure R - Record T - Think	Move with powered crane-bridge type	ATKTPTA	A - Action distance T - Transport K - Hook up and unhook P - Place
			Move with truck	ASTLTLTA	A - Action distance S - Start and park T - Transport L - Load or unload

General move is defined as moving objects manually from one location to another freely through the air. To account for the various ways in which a general move can occur, the activity sequence is made up of four subactivities:

A—Action distance (mainly horizontal)

B—Body motion (mainly vertical)

G—Gain control

P—Place

Controlled move sequence is used to cover such activities as operating a lever or crank, activating a button or switch, or simply sliding an object over a surface. In addition to the A, B, and G parameters from the general move sequence, the sequence model for controlled move contains the following subactivities:

M—Move controlled

X—Process time

I—Align

Tool use (equipment use) sequence covers the use of hand tools for such activities as fastening or loosening, cutting, cleaning, gauging, and writing. Also, certain activities requiring the use of the brain for mental processes can be classified as *tool use*. The tool use sequence model is a combination of general move and controlled move activities.

Table 53.2 shows the sequence models comprising the basic MOST techniques. While the three manual sequences comprise the basic MOST technique, three other sequence models were designed to simplify the work measurement procedure for dealing with heavy objects.

Manual Crane Sequence covers the use of a manually traversed jib, crane, monorail crane, or bridge crane for moving heavier objects.

Powered Crane Sequence covers the use powered cranes for moving heavier objects.

Truck Sequence covers the transportation of objects using riding or walking equipment such as a forklift, stacker, pallet lift, or hand truck.

MOST is appropriate for any manual work that contains variation from one cycle to another. MOST should not be used in situations in which a cycle is repeated identically over a long period of time. In these situations, which, by the way, do not occur very often, a more detailed system should be chosen as the analytical tool.

53.5 WORK SAMPLING

For many jobs it is impractical to set performance standards or productivity measures by the methods described thus far in this chapter. There are many jobs where time study, standard data, and predetermined time standards just are not useful, productive ways to set standards. When jobs do not have short cycle times or high repetition rates, such as those in maintenance, material handling, clerical and the like, these methods are not effective. Jobs with long cycle times and low frequency of repetition require the use of work sampling.

Work sampling is the process of making sufficient random observations of an operator's activities to determine the relative amount of time the operator spends on the various activities associated with the job. Although it is not the express purpose of work sampling to determine how long specific tasks should take, work sampling data, when coupled with historical production data for key volume indicators and performance levelling, can provide information that can be used to establish standards. The major goal of work sampling, however, is to determine the proportion of the workday spent on certain types of work.

53.5.1 Methodology

The method to conduct a work sampling study is as follows:

- Establish the purpose of the study. Is the intent to set a standard to drive staffing levels or is it to identify the proportion of time spent on each major category of work?
- Identify the subjects. Who is doing the work within the organization?
- Identify the measures of output. This is essential if the objective of the study is to develop standards. For example, in an insurance office, the number of claims processed might be such a key volume indicator.
- Establish a time period for the study. The period of the study should be long enough to be representative of normal operations.
- Define the activities. This includes a definition of the major tasks performed. An example of some work definitions is included in a case study included at the end of this section.
- Determine the number of observations needed. After the work elements are defined, the number of observations for the desired accuracy and confidence level must be determined.
- Schedule the observations. If an analyst will record the data, use a random number table or random number generator. Devices such as the PDA and Divilbiss' JD7 Random Reminder are most useful for developing the random observation schedule.
- Inform the personnel involved. As in any productivity measurement study this part of the procedure is important. Workers and their supervisors might think that they personally are being measured rather than the work they are doing.
- Record the raw data.
- Summarize the data. Determine the proportion of time spent on each activity.

53.5.2 Application—Case Study

Background. The CFO of a large collection agency contacted the author (and his colleagues) to study some of the back office operations at the agency. The CFO was very concerned that labor costs and head count were considerably larger than required to operate the business profitably. Two departments were the focus of his immediate attention. First was the new business department. Second was payment processing.

New business was responsible for entering all new accounts into the agency's computer database. The new accounts were generated by the sales staff and typically consisted of one of the following types of accounts:

1. Major credit card such as MasterCard, Visa, or American Express
2. Major retail account such as a department store
3. Utility
4. Small retail account
5. Health care related account
6. Student loan account

Each of these six accounts received different information from the original credit grantor, which meant that each type of account required a different set of information in the agency's database.

When the CFO called the headcount in the new business operation had just reached 16 direct employees. When benefits and overtime were considered, the average annual cost of one of these employees was $30,000. With an annual departmental cost of close to half a million dollars, the CFO was justified in his concern that the staffing level was appropriate.

Investigation. In order to determine the staffing levels required it was necessary to first of all determine the work currently being performed; second, ascertain whether or not the work being performed was appropriate and necessary; third, verify that an appropriate or "best" method was being used; and fourth, determine the time required to perform the work. Discussions with the department manager indicated that a number of different activities were required to enter any new business into the system. Furthermore, these activities were somewhat specialized. For example, an employee entering major retail accounts could not enter health care accounts. Any questions regarding the appropriateness of the work being performed or the work method used were met with significant resistance. Because of this and because the collection agency was very wary of industrial engineering work, the CFO decided that the traditional improvement aspects of the study would be best delayed and that the primary issue of managing the staffing levels be addressed with the initial study. Future improvements could be compared with the existing base line data that would be developed during this study.

Due to the fact that six different items were "being produced" in the department, it was decided that work sampling would provide the best estimate of the time required to complete the activities. This time, when coupled with measures of volume processed by the department, would result in a staffing chart for each type of new business entered into the data base.

Discussions with the manager and employees processing the accounts led to the following set of common work definitions being established for any type of new business (Table 53.3):

These definitions were agreed upon by all parties involved. The work sampling study was then conducted. Initial studies by the industrial engineers indicated that the shortest of the work elements defined required 7 percent of the workday. The sample size calculation for 95 percent confidence and 5 percent accuracy indicated a total of 21,300 observations* were required. In order to economically collect this amount of data random time generators known as random reminders were purchased from Divilbiss Electronics and given to each of the personnel in the new business department. Using the definitions above, each person was asked to complete the form shown in Table 53.4 each time the random time generator signaled. Forms were submitted each day and compiled by the industrial engineers.

In addition to the tally marks indicating the work performed, additional information was collected each day. This included information on the computer system the data was entered into. Due to rapid

* Barnes, *Motion and Time Study,* page 414.

TABLE 53.3 Activity Definitions for New Business

Batch and prepare includes the following activities: prepare check client, separate inpatient from outpatient, separate agency number from client number, run tape or batch media, complete batch sheet, and prepare logs.

Data entry involves any direct entering of information into the system.

Acknowledgments includes verification, error correction on the system and acknowledgment, and mailing file to client and/or branch.

Returns and changes refers to processing mail returns and address changes in the system.

Sample books includes any activities performed while updating sample (training) books.

Running reports includes pulling electronic and manual numbers.

Miscellaneous refers to any activity performed that is not listed above. Please list this with a description someplace on the form.

Breaks/personal/lunch refers to any time spent on those activities.

growth there were several different databases being used for some of the specific types of new businesses entered. The branch information indicated which office would ultimately be responsible for working the accounts, e.g., whether Dallas or Chicago would be contacting the debtors. The client was the specific name of the original credit grantor. The hours worked were clock in time until clock out time. The volume was the operator's count (provided by her terminal) of the number of accounts added to the database on the given day. This information provided useful information for analysis during the standards setting process.

Analysis. The information gathered during the work sampling study was compiled for each of the six types of business.

1. Major credit card such as MasterCard, Visa, or American Express
2. Major retail account such as a department store

TABLE 53.4 Work Sampling Data Collection Form

Activity	Tally	Total
Data entry		
Returns and changes		
Acknowledgements		
Sample books		
Running reports		
Breaks/personal/lunch		
Batch and prepare		
Miscellaneous		
Comments	System	
	Branch	
	Client	
	Hours worked	
	Volume	
	Date	
	Name	

TABLE 53.5 Sample Staffing Table

Volume	Hours	Volume	Hours	Volume	Hours
100	0.83	525	4.36	950	7.89
125	1.04	550	4.57	975	8.09
150	1.25	575	4.77	1000	8.30
175	1.45	600	4.98	1025	8.51
200	1.66	625	5.19	1050	8.72
225	1.87	650	5.40	1075	8.92
250	2.08	675	5.60	1100	9.13
275	2.28	700	5.81	1125	9.34
300	2.49	725	6.02	1150	9.55
325	2.70	750	6.23	1175	9.75
350	2.91	775	6.43	1200	9.96
375	3.11	800	6.64	1225	10.17
400	3.32	825	6.85	1250	10.38
425	3.53	850	7.06	1275	10.58
450	3.74	875	7.26	1300	10.79
475	3.94	900	7.47	1325	11.00
500	4.15	925	7.68	1350	11.21

3. Utility
4. Small retail account
5. Health care related account
6. Student loan account

The productive work elements were identified and the proportion of time spent on each of these activities was identified. The performance of the operators was evaluated and a leveling factor was applied. This information was then combined with the hours worked information supplied and the volume of new business posted yielding a standard hours per item entered. As might be expected the standards for each of the different types of business was significantly different. The standards were then used to develop a staffing table for each of the different types of business. Table 53.5 shows a portion of a staffing table for one of the new business categories.

Tables like the one shown in Table 53.5 were developed for each of the six types of new business. The tables, when combined with available information on volume, permitted the department manager to schedule the operators in order to complete the required work.

53.6 Learning Curve

Before a production or manufacturing plan can be translated into a production schedule, the times required for the various operations must be determined. Normally this is a matter of applying the time standards developed by industrial engineering to the various operations. However, when the product to be produced is new, employees must learn their jobs. The learning curve attempts to compensate for this effect. This is especially noticeable for large scale products with low production quantities.

Learning curves are used primarily to estimate costs. Based on the concept that less labor will be required for each succeeding unit, certain key costs can be determined or estimated.

Mathematical relationships have been developed that will determine the following:

- Average hours/unit
- Expected hours for a given unit based on the fact that each successive unit requires less time, but not at a uniformly decreasing rate

TABLE 53.6 Sample Values for Learning Curve Constants

Learning percentage	Constant (n)
95	0.074
90	0.152
85	0.235
80	0.322
75	0.415
70	0.515

Based on these estimates a number of other useful pieces of information can be determined. These include, for each potential production quantity, the following:

- Average hours per unit based on production to date
- Cumulative hours based on production to date
- Labor cost per unit as a function of the time required

This information can be used to calculate a breakeven production quantity, since most products sell for the same uniform price. Generally an organization will lose money on the first unit produced but make a significant profit on the last one delivered.

If we let y be the average hours per unit, x the number of units produced, c the time required for the first unit, and n a constant, depending on the slope of the learning curve (an 80 percent learning curve implies that when production is doubled only 80 percent of labor time required for the prior production will be required to produce the doubled production), the value for y is calculated with the equation:

$$y = cx^{-n}$$

Sample values for learning curve constants appear in Table 53.6.

The learning curve equation can be used to calculate time and cost information. The total time (TT) required is the product of the average hours per unit and the number of units to be produced.

$$TT = xy$$

The time for the last unit, or the mth unit, is calculated using the following relationship.

$$T_m = (1 - n)y$$

53.6.1 Sample Calculations

An organization plans to produce 50 units to fulfill a contract. Based on the best engineering estimates they anticipate 350,000 h will be required to produce unit 1. They also believe that an 80-percent learning curve will be in effect.

The average time per unit is calculated

$$y = cx^{-n} = (350{,}000)(50)^{-.322} = 99{,}312 \text{ h}$$

The total time required to produce all fifty units

$$TT = xy = (50)(99{,}312) = 4{,}965{,}600 \text{ h}$$

The time required to produce the last or 50th unit

$$T_m = (1 - n)y = (1 - 0.322)(99{,}312) = 67{,}344 \text{ h}$$

53.7 PERFORMING STUDIES

Regardless of the methodology used, all work measurement studies should include the following key events. First of all the objective for the study should be stated. Then specific jobs to be studied must be identified. Next, the current method should be documented. Documentation includes tools, equipment, layout, equipment, inspection status, and environmental information that might impact on the performance of the task. If it is a brand new, never-been-performed job, this information should still be available.

After the initial documentation is complete a methods analysis should be performed to identify obvious and necessary improvements. Once these are completed and documented, the actual study begins. Permission from the supervisor, operator, and if applicable, the union should be sought. While not necessary, it is a common courtesy and will normally make the data collection process easier.

During the study itself the data is collected and the method used verified for conformance to the published procedure. The data is analyzed using the appropriate methodology. After the analysis is completed, the results should be verified to make sure they are reasonable and make sense for the operation.

When that is complete the standard should be published so that this useful management information can be used. All documentation should be saved in case any questions arise. The documentation will also prove beneficial whenever the operation is audited, to assure that the standard time remains accurate.

53.8 CURRENT COMPUTER APPLICATIONS

Computers have entered the work measurement arena. Their primary use, as of this writing, is in the data collection and analysis process. Organizations such as Quetech, Ltd. have adapted the personal digitital assistant for data collection and developed excel applications to work in concert with these.

WorkStudy+, developed by Quetech Ltd., is a portable software solution for performing time studies and work sampling. It runs on Palm OS compatible handheld devices, and replaces the traditional stopwatch, study board, and pen method.*

WorkStudy+ closely mimics the traditional approach of conducting time studies, but allows the time study practitioner to operate more discreetly and efficiently by eliminating all manual data entry tasks typically associated with time studies.

WorkStudy+ offers several advantages over the traditional method of conducting time studies.

1. *Elimination of manual data entry tasks*. With WorkStudy+, time study and sampling data is captured electronically into a handheld computer, and can easily be transferred to a PC for further analysis. This eliminates all manual data entry, and provides a quick, accurate, and consistent account of all studies data.
2. *Hierarchical element list*. Another important advantage is the extension of the traditional element list into a hierarchically organized element tree. This allows a more detailed elemental breakdown of the job being studied, and makes studies with even hundreds of elements manageable.

---

* Palm OS is a registered trademark of Palm, Inc. Excel is a trademark of Microsoft Corporation.

3. *Immediate access to statistical information.* WorkStudy+ contains functionality that allows the observer to view detailed statistical information such as elapsed time, avoidable and unavoidable delay time, standard deviation, and sample count needed to reach confidence interval for each element. Since this information is available to the observer immediately upon completion of a study, decisions about the validity of data or the need to capture additional observations can be made before even leaving the study area.
4. *Ease, flexibility, and accuracy of data capture.* Many aspects of the data capture process can be automated in WorkStudy+. Of particular importance is the ability to preconfigure an element sequence for each study. With the sequence specified ahead of time, the observer only needs to press one button for each observation (equivalent to pressing the stopwatch timer button). The software automatically selects the proper element and records the observation. This allows accurate recording of observations even if they are as short as 1–2 s.
5. *Generation of statistical reports.* Completed time and sampling studies can be imported into Excel, where they can be analyzed with WorkStudy+'s report generating functionality. The captured study data can be turned into finished reports in a matter of seconds.

As previously mentioned, collecting time study data with WorkStudy+ closely resembles the traditional stopwatch method.

The first step consists of creating a new study file and populating it with the elements that will be observed. Working on the PC or directly on the handheld, these elements can be descriptively named (such as "Get part" or "Process order") and organized hierarchically into element groups. Although it is advisable to have the elements well defined before initiating any study, WorkStudy+ allows for the addition or editing of elements at any time. This could prove handy should unforeseen elements be encountered when collecting data.

During the collection phase, WorkStudy+ displays information about elemental observations in real time while allowing the observer to performance rate, track cycles, take notes and more. Recording an observation with WorkStudy+ is very similar to the stopwatch method; the user simply presses a "start" button, at which point the timer turns on. The user then chooses the observed element from the element list and is free to edit the performance rating of that element or add any comments to it. WorkStudy+ allows for up to five studies to be performed simultaneously. (This is often referred to as multisubject studies, where each subject is timed by a separate clock.)

Once a study is completed, all the data can be transferred directly into Microsoft Excel for easy summarization. WorkStudy+ provides sophisticated Excel-based statistical generating tools with which users can custom-create chart templates with standard headings, logos, or aesthetics to meet their exact needs. With the click of a button studies can be directly imported into the custom made spreadsheet.

FURTHER READING

Aft, Lawrence, and Thomas Merritt, "Meeting the Requirements of MIL-STD 1567 and Lockheed-Georgia Computerized Standard Data Development System," *Annual International Industrial Engineering Conference Proceedings,* 1984.

Aft, Lawrence, *Productivity, Measurement and Improvement,* Prentice Hall, Englewood Cliffs, New Jersey, 1992.

Aft, Lawrence, *Work Measurement and Methods Improvement,* Wiley, New York, 2000.

Aft, Lawrence, *Production and Inventory Control,* Technology Publications, 1987.

Barnes, Ralph, *Motion and Time Study,* Wiley, New York, 1980.

Brouha, Lucien, *Physiology in Industry,* Pergamon, 1960.

Heiland and Richardson, *Work Sampling,* McGraw-Hill, New York, 1957.

International Labour Office, *Introduction to Work Study,* ILO, Geneva, Switzerland, 1997.

Karger and Bayh, *Engineered Work Measurement,* Industrial Press, New York, 1987.

Masud, Abu, Don Malzahn, and Scott Singleton, "A High Level Predetermined Time Standard System and Short Cycle Tasks," *Annual International Industrial Engineering Conference Proceedings*, 1985.

Rice, R.S., "Survey of Work Measurement and Wage Incentives," *Industrial Engineering,* July, 1977.

Zandin, Kjell, *MOST Work Measurement Systems,* Marcel Dekker, New York, 1980.

INFORMATION RESOURCES

Books

Maynard, H.B. (ed.), *Industrial Engineering Handbook,* McGraw-Hill, New York, 2001.

Maynard, H.B., Stegemerten, G.J., and Schwab, J.L., *Methods Time Measurement,* McGraw-Hill, New York, 1948.

Training Manuals

MTM Association for Standards and Research, *Various Training Manuals.*

Methods Workshop Limited, *General Sewing Data Users Manual.*

Articles

Brisley, Chester L., "Comparison of Predetermined Time Systems," *Fall Industrial Engineering Conference Proceedings,* 1978.

CHAPTER 54
ENGINEERING ECONOMICS

Gerald A. Fleischer
Industrial and Systems Engineering
University of Southern California

As is the case with other types of capital allocation decisions, *engineering* economy rests on the proposition that refusal to expend scarce resources is rarely, if ever, the most prudent course of action. Rather, the problem is one of choosing from among a variety of investment alternatives in order to satisfy the decision makers' intermediate and longer term objectives. The operative word is *economy,* and the essential ingredient in economy is consideration of the economic consequences of alternatives over a measured period of time, the planning horizon. This chapter is dedicated to the principles and procedures for evaluating the economic consequences of engineering plans, programs, designs, policies, and the like. The effects of income taxes and relative price changes ("inflation") are also considered.

54.1 FUNDAMENTAL PRINCIPLES

Before developing the mathematical models appropriate to evaluating capital proposals, it will be useful to identify the fundamental principles that give rise to the rationale of capital allocation. Moreover, some of these principles lead directly to the quantitative techniques developed subsequently.

1. *Only feasible alternatives should be considered.* The capital budgeting analysis begins with determination of all feasible alternatives, since courses of action that are not feasible, because of certain contractual or technological considerations, are properly excluded.
2. *Using a common unit of measurement (a common denominator) makes consequences commensurable.* All decisions are made in a single dimension, and money units—dollars, francs, pesos, yen, and so forth—seem to be most generally suitable. Of course, not all consequences may be evaluated in money terms. (See principle 9 below.)
3. *Only differences are relevant.* The prospective consequences that are common to all contending alternatives need not be considered in an analysis, because including them affects all alternatives equally.
4. *All sunk costs are irrelevant to an economic choice.* A *sunk cost* is an expense or a revenue that has occurred before the decision. All events that take place before a decision are common to all the alternatives, so sunk costs are not differences among alternatives.
5. *All alternatives must be examined over a common planning horizon.* The *planning horizon* is the period of time over which the prospective consequences of various alternatives are assessed. (The planning horizon is often referred to as the *study period* or *period of analysis.*)
6. *Criteria for investment decisions should include the time value of money and related problems of capital rationing.*

7. *Separable decisions should be made separately.* This principle requires the careful evaluation of all capital-allocation problems to determine the number and type of decisions to be made.
8. *The relative degrees of uncertainty associated with various forecasts should be considered.* Because estimates are only predictions of future events, it is probable that the actual outcomes will differ to a greater or lesser degree from the original estimates. Formal consideration of the type and degree of uncertainty ensures that the quality of the solution is evident to those responsible for capital-allocation decisions.
9. *Decisions should give weight to consequences that are not reducible to monetary units.* The irreducible as well as monetary consequences of proposed alternatives should be clearly specified in order to give managers of capital all reasonable data on which to base their decisions.

54.2 EQUIVALENCE AND THE MATHEMATICS OF COMPOUND INTERESTS

A central notion in engineering economy is that cash flows (that is, the receipt or payment of an amount of money) that differ in magnitude but that occur at different points in time may be *equivalent*. This equivalence is a function of the appropriate interest rate per unit time and the relevant time interval. Mathematical relationships describing the equivalence property under a variety of conditions are described in the remainder of this section.

Useful Conventions. The following conventions will be used in this chapter.

Cash Flow Diagrams. In the literature of engineering economy, cash flow diagrams are frequently used to illustrate the amount and timing of cash flows. Generally, a horizontal bar or line is used to represent time, and vertical vectors (arrows) are used to represent positive or negative cash flows at the appropriate points in time. These cash flow diagrams are illustrated later in Fig. 54.1. The "shaded" arrows in the right-hand portion of the figure represent cash flowing continuously and uniformly throughout the indicated period(s).

Functional Notation. As the algebraic form of the various equivalence factors can be complex, it is useful to adopt a standardized format which is easily learned and has a mnemonic connotation. The format which is in general use* is of the form

$$(X|Y, i, N)$$

which is read as "to find the equivalent amount X given amount Y, the interest rate i and the number of compounding or discounting periods N."

Discrete Cash Flows—End of-Period Compounding. Assume that a cash flow A_j occurs at end of period j. Interest is compounded or discounted at the end of each period at rate i period. The interest rate i is constant over $j = 1, 2, \ldots, N$. The periods are of equal duration.

Single Cash Flows. Consider a single cash flow P to be invested at the beginning of a time series of exactly N periods. Let F represent the *equivalent future value* of P as measured at the end of N periods hence, assuming that interest is compounded at the end of each and every period at interest rate i. Then,

$$F = P(1 + i)^N = P(F/P, i, N) \tag{54.1}$$

It follows immediately that, given a future amount F flowing at the end of N periods hence, the *equivalent present value* P is given by

$$P = F(1 + i)^{-N} = F(P/F, i, N) \tag{54.2}$$

---

* This is the functional notation recommended in *Industrial Engineering Terminology,* revised edition, Industrial Engineering and Management Press, Industrial Engineering Institute, Norcross, GA, 1991.

The growth multiplier as shown in Eq. (54.1), $(1 + i)^N$, is known in the literature of engineering economy as the (*single payment*) *compound amount factor.* The discounting multiplier shown in Eq. (54.2) is known as the (*single payment*) *present worth factor*.

The cash flow diagrams, algebraic forms, and functional forms for these two factors are shown later in Fig. 54.1. Tabulated values of $i = 10$ percent and for various values of N are given in Table 54.6.

Examples. A sum of \$1000 is invested in a fund which earns interest at the rate of 1 percent per month, compounded monthly. To determine the value of the fund after 24 months, using Eq. (54.1):

$$F = \$1000(1.01)^{24} = \$1269.73$$

A certain investment is expected to yield a return of \$100,000 exactly 8 years in the future. Assuming a discount rate of 10 percent per year, what is the equivalent present value? Using Eq. (54.2):

$$P = \$100{,}000(1.10)^{-8} = \$46{,}651$$

How long will it take a sum of money to double if interest is earned at the rate of 8 percent per period? From Eq. (54.1):

$$\$2 = \$1(0.8)^{-N}$$

$$N = \ln 2/\ln 1.08 \simeq 9 \text{ periods}$$

An investment of \$10,000 yields a return of \$20,000 five years later. What (annual) rate of return was earned? From Eq. (54.1):

$$\$20{,}000 = \$10{,}000(1 + i)^5$$

$$i = (\$20{,}000/\$10{,}000)^{1/5} - 1 = 14.87\%$$

Uniform Series (Annuity). Consider a uniform series of cash flows A occurring at the end of each of N consecutive periods. That is, $A_j = A$ for $j = 1, 2, \ldots, N$. The equivalent future value F at the end of N periods is given by

$$F = A\left[\frac{(1+i)^N - 1}{i}\right] = A(F/A, i, N) \qquad (54.3)$$

The factor in brackets is known as the (*uniform series*) *compound amount factor*. To find A given F:

$$A = F\left[\frac{i}{(1+i)^N - 1}\right] = F(A/F, i, N) \qquad (54.4)$$

The factor in brackets is known as the *sinking fund factor*.

The equivalent present value of this uniform series is given by

$$P = A\left[\frac{(1+i)^N - 1}{i(1+i)^N}\right] = A(P/A, i, N) \qquad (54.5)$$

The factor in brackets is known as the (*uniform series*) *present worth factor*. To find A given P:

$$A = P\left[\frac{i(1+i)^N}{(1+i)^N - 1}\right] = P(A/P, i, N) \qquad (54.6)$$

The factor in brackets is known as the *capital recovery factor*.

As before, the appropriate cash flow diagrams, algebraic forms, and functional forms are shown in Fig. 54.1. Tabulated values for $i = 10$ percent are given in Table 54.6.

Examples. (A 10 percent interest rate is assumed for all the following examples.) A sum of \$1000 is invested at the end of each period for 15 periods. What is the amount in the fund after the 15th payment has been made? From Eq. (54.3):

$$F = \$10{,}000(F/A, 10\%, 15)$$
$$= \$10{,}000(31.772) = \$317{,}720$$

(Note that the value for the compound amount factor has been taken from Table 54.6.) How much must be invested at the end of each year for 15 years in order to have \$20,000 in the fund after the 15th payment? From Eq. (54.4):

$$A = \$20{,}000(A/F, 10\%, 15)$$
$$= \$20{,}000(0.0315) = \$630$$

How much must be invested today in order to yield returns of \$2500 at the end of each and every year for 8 years? From Eq. (54.5):

$$P = \$2500(P/A, 10\%, 8)$$
$$= \$2500(5.335) = \$13{,}337$$

Certain equipment costs, \$50,000, will be used for 5 years, and will have no value at the end of 5 years. What is the equivalent annual (end of year) cost? From Eq. (54.6):

$$A = \$50{,}000(A/P, 10\%, 5)$$
$$= \$50{,}000(0.2638) = \$13{,}190$$

Arithmetic Gradient Series. Let $A_j = (j - 1)G$ for $j = 1, 2, \ldots, N$, where G represents the amount of increase or decrease in cash flow from one period to the next. This results in an arithmetic series of the cash flows of the form $0, G, 2G, \ldots, (N - 1)G$ for periods $1, 2, \ldots, N$, respectively. Given the gradient G, the equivalent present value is given by

$$P = G\left[\frac{(1+i)^N - iN - 1}{i^2(1+i)^N}\right] = G(P/G, i, N) \tag{54.7}$$

and the equivalent uniform series is given by

$$A = G\left[\frac{(1+i)^N - iN - 1}{i(1+i)^N - i}\right] = G(A/G, i, N) \tag{54.8}$$

Again, the appropriate cash flow diagrams, algebraic forms, and functional forms are shown later in Fig. 54.1. Representative tabulated values are given in Table 54.6 for $(P/G, 10\%, N)$ and $(A/G, 10\%, N)$.

Example. Costs of manufacturing are assumed to be \$100,000 the first year and to increase by \$10,000 in each of the years 2 through 7. If interest is at 10 percent per year, determine the equivalent present value of these costs. Using Eqs. (54.5) and (54.7) and taking the appropriate factor values from Table 54.6:

$$P = \$100{,}000(P/A, 10\%, 7) + \$10{,}000(P/G, 10\%, 7)$$
$$= \$100{,}000(4.868) + \$10{,}000(12.763) = \$614{,}430$$

Note: This analysis assumes that all cash flows occur at end of year.

END-OF-PERIOD COMPOUNDING EFFECTIVE INTEREST RATE (i)		CONTINUOUS COMPOUNDING NOMINAL INTEREST RATE (r)
1* $P = P\,(F/P, i, N) = P\,(1 + i)^N$ 3* $P = F\,(P/F, i, N) = F\,(1 + i)^{-N}$	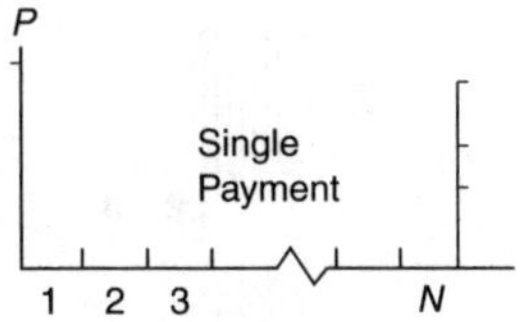	4* $P = F\,(P/F, r, N) = Fe^{-rN}$
5* $F = A\,(F/A, i, N) = A\left[\frac{(1 + i)^N - 1}{i}\right]$ 7* $A = F\,(A/F, i, N) = F\left[\frac{i}{(1 + i)^N - 1}\right]$	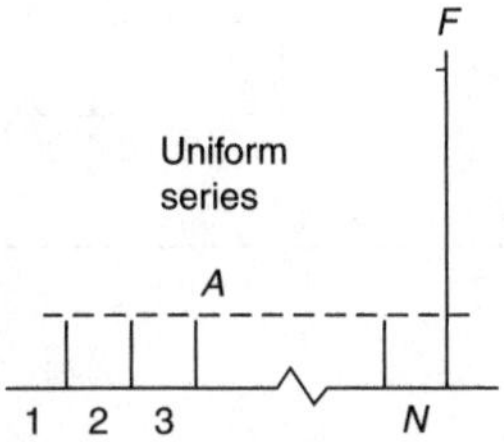	6 $P = A\,(F/A, r, N) = A\left[\frac{e^{rN} - 1}{e^r - 1}\right]$ 8 $A = F\,(A/F, r, N) = F\left[\frac{e^r - 1}{e^{rN} - 1}\right]$
9* $P = A\,(P/A, i, N) = A\left[\frac{(1 + i)^N - 1}{i(1 + i)^N}\right]$ 11* $A = P\,(A/P, i, N) = P\left[\frac{i\,(1 + i)^N}{(1 + i)^N - 1}\right]$	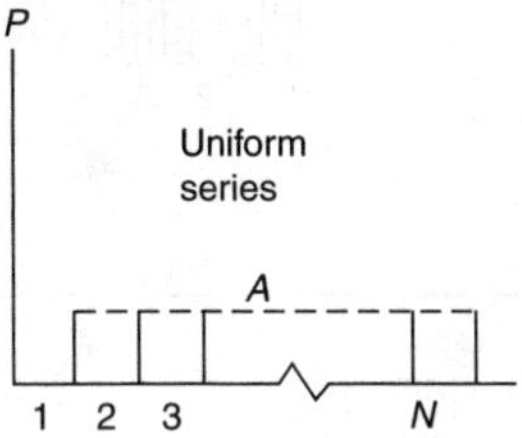	10 $P = A\,(P/A, r, N) = A\left[\frac{e^{rN} - 1}{e^{rN}(e^r - 1)}\right]$ 12 $A = P\,(A/P, r, N) = P\left[\frac{e^{rN}(e^r - 1)}{e^{rN} - 1}\right]$
13* $P = G\,(P/G, i, N) = G\left[\frac{(1 + i)^N - iN - 1}{i^2(1 + i)^N}\right]$ 15* $A = G\,(A/G, i, N) = G\left[\frac{(1 + i)^N - iN - 1}{i\,(1 + i)^N - 1}\right]$	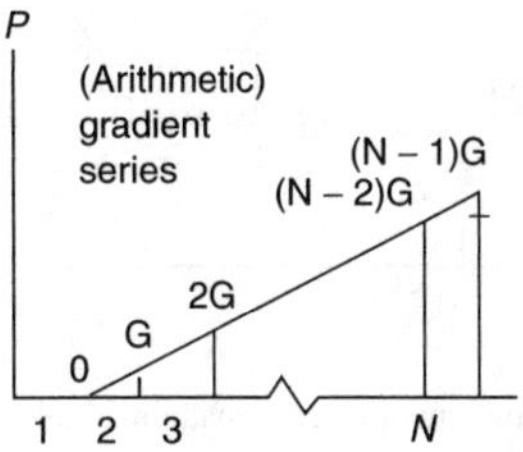	14* $P = G\,(P/G, r, N) = G\left[\frac{e^r - Ne^r + N - 1}{e^{rN}(e^r - 1)^2}\right]$ 16* $A = G\,(A/G, i, N) = G\left[\frac{e^r - Ne^r + N - 1}{(e^r - 1)(e^N - 1)}\right]$

(a)

FIGURE 54.1 Cash flow models and mathematical models for selected compound interest factors: (*a*) Discrete cash flows.

CONTINUOUS COMPOUNDING EFFECTIVE INTEREST RATE (i)		CONTINUOUS COMPOUNDING NOMINAL INTEREST RATE (r)
17 $F = \bar{P}(F/\bar{P}, i, N)$ $= \bar{P}\left[\frac{i(1+i)^{N-1}}{\ln(1+i)}\right]$	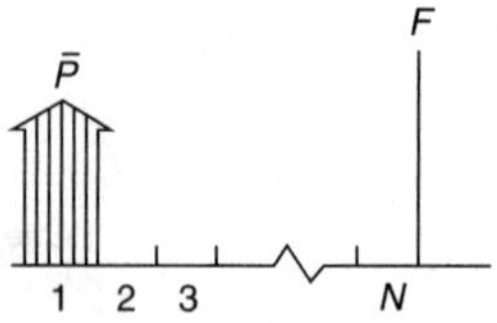	
19* $P = \bar{F}(P/\bar{F}, i, N)$ $= \bar{F}\left[\frac{i(1+i)^{-N}}{\ln(1+i)}\right]$	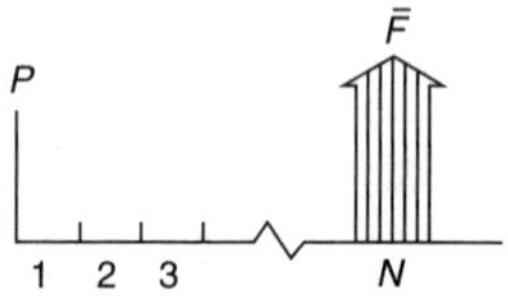	20 $P = \bar{F}(P/\bar{F}, r, N)$ $= \bar{F}\left[\frac{e^r - 1}{re^N}\right]$
21* $F = \bar{A}(F/\bar{A}, i, N)$ $= \bar{A}\left[\frac{(1+i)^N - 1}{\ln(1+i)}\right]$ 23 $\bar{A} = F(\bar{A}/F, i, N)$ $= F\left[\frac{\ln(1+i)}{(1+i)^N - 1}\right]$	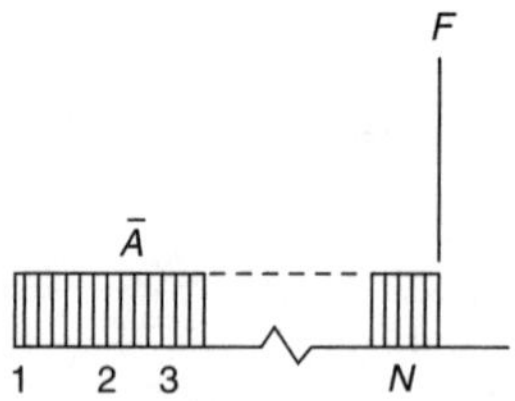	22 $F = \bar{A}(F/\bar{A}, r, N)$ $= \bar{A}\left[\frac{e^{rN} - 1}{r}\right]$ 24 $\bar{A} = F(\bar{A}/F, r, N)$ $= F\left[\frac{r}{e^{rN} - 1}\right]$
25* $P = \bar{A}(P/\bar{A}, i, N)$ $= \bar{A}\left[\frac{(1+i)^N - 1}{(1+i)^N \ln(1+i)}\right]$ 27 $\bar{A} = P(\bar{A}/P, i, N)$ $= P\left[\frac{(1+i)^N \ln(1+i)}{(1+i)^N - 1}\right]$	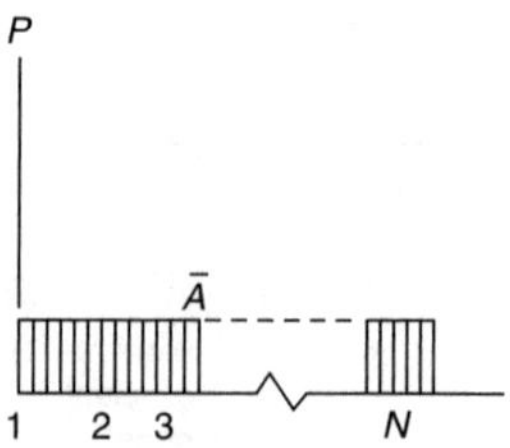	26 $P = \bar{A}(P/\bar{A}, r, N)$ $= \bar{A}\left[\frac{e^{rN} - 1}{re^{rN}}\right]$ 28 $\bar{A} = P(\bar{A}/P, r, N)$ $= P\left[\frac{re^{rN}}{e^{rN} - 1}\right]$

(b)

FIGURE 54.1 Cash flow models and mathematical models for selected compound interest factors. (*b*) Continuous cash flows.

Geometric Gradient Series. Consider a series of cash flows $A_1, A_2, \ldots, A_N$ where the A_j's are related as follows:

$$A_j = A_{j-1}(1 + g) = A_1(1 + g)^{j-1} \tag{54.9}$$

where g represents the rate of increase or decrease in cash flows from one period to the next. With cash flows discounted at rate i per period, the equivalent present value of the geometric series is given by

$$P = A_1\left[\frac{1-(1+g)^N(1+i)^{-N}}{i-g}\right] \tag{54.10}$$

As N approaches infinity, this series is convergent if $g < i$. Otherwise ($g \geq i$) the series is divergent.

Example. Manufacturing costs are expected to be \$100,000 in the first year, increasing by 5 percent each year over a 7-year period. Find the equivalent present value of these cash flows assuming a 10 percent discount rate (per year) and end-of-year cash flows. Using Eq. (54.10):

$$P = \$100{,}000(P/A_1, 10\%, 5\%, 7)$$

$$= \$100{,}000(5.5587) = \$555{,}870$$

Effective and Nominal Interest Rates. An interest rate is meaningful only if it is related to a particular period of time. Nevertheless, the "time tag" is frequently omitted in speech because it is usually understood in context. If someone reports earnings of 6 percent on investments, for example, it is implied that the rate of return is *per year*. However, in many cases the interest-rate period is a week, a month, or some other interval of time, rather than the more usual year (per annum). At this point it would be useful to examine the process whereby interest rates and their respective time tags are made commensurate.

As before, let i represents the *effective* interest rate per period. Let the period be divided into M subperiods of equal length. If interest is compounded at the end of each subperiod at rate i_s per subperiod, then the relationship between the effective interest rates per period and per subperiod is given by

$$i = (1 + i_s)^M - 1 \tag{54.11}$$

The *nominal* interest rate per period, r, is simply the effective interest rate per subperiod times the number of subperiods, or

$$r = Mi_s \tag{54.12}$$

Periods and Subperiods. An Example. It is often necessary to compare interest rates over a common time interval. Consider, for example, the case of consumer credit, say, a major oil company or bank "charge card" for which interest is compounded monthly at rate 1.5 percent of the unpaid balance. Here $i_s = 0.015$ and $M = 12$. The *nominal* rate per annum, by Eq. (54.12) is $12 \times 0.015 = 0.18$. The *effective* rate per annum, by Eq. (54.11) is

$$i = (1.015)^{12} - 1 = 0.1956$$

Periods and Superperiods. Consider a uniform series of cash flows A occurring at regular intervals. Specifically, the cash flows occur every M periods, with the first cash flow occurring at the end of period m and the last cash flow occurring at the end of period n, where $1 \leq m \leq n \leq N$. There are exactly $[(n - m)/M] + 1$ cash flows in M periods, where M is integer-valued, with the start of the first superperiod at the end of period $m - M$. The equivalent present value of this uniform series of cash flows is given by

$$P = A\left\{\frac{(1+i)^{n-m+M} - 1}{(1+i)^n[(1+i)^M - 1]}\right\} \tag{54.13}$$

For example, consider major overhaul expenses of \$20,000 each occurring at the end of year 5 and continuing, every 2 years, up to and including year 13. (A_j = –\$20,000 for j = 5, 7, 9, 11, 13.) Assuming a 10 percent discount rate:

$$P = \$20{,}000\left\{\frac{(1.10)^{13-5+2}-1}{(1+1)^{13}[(1.10)^2-1]}\right\}$$

$$= \$20{,}000(1.19834) = \$43{,}967$$

Continuous Cash Flows—Continuous Compounding. Assuming that the number of subperiods at the end of which interest is compounded or discounted becomes infinitely large, the effective interest rate per period is

$$i = \lim_{M\to\infty}\left\{\left[\left(1+\frac{1}{M/r}\right)^{M/r}\right]^r - 1\right\} \tag{54.14}$$

where e is the base of the natural (napierian) logarithm system and is approximately equal to 2.71828.

Assume that a total of $\bar{A}$ dollars flows over one interest period, with $\bar{A}/M$ flowing at the end of each and every one of the M subperiods within the period. As before, the *effective* interest rate is i per period and the nominal rate is r per period. Interest is compounded at effective rate $i_s = r/M$ per subperiod. Let A represent the equivalent value at the end of the period:

$$A = (\bar{A}/M)(F/A, i_s, M)$$

As the number of subperiods (M) becomes infinitely large, it may be shown that

$$A = \bar{A}\left[\frac{e^r-1}{r}\right] = \bar{A}\left[\frac{i}{\ln(1+i)}\right] \tag{54.15}$$

The value in brackets is known as the *funds flow conversion factor* because it has the effect of converting a continuous cash flow (during the period) to a discrete cash flow (at the end of the period). The funds flow conversion factor is useful in modifying the end-of-period factors, previously discussed, to accommodate the continuous assumptions. To illustrate, consider the factor for determining the equivalent present value of a cash flow $(\bar{F})$ flowing continuously and uniformly during the Nth period. Hence, combining Eqs. (54.2) and (54.15), we get

$$P = \bar{F}\left[\frac{i}{\ln(1+i)}\right](1+i)^{-N}$$

$$= \bar{F}(P/\bar{F}, i, N) \tag{54.16}$$

Similarly,

$$P = \bar{A}\left[\frac{i}{\ln(1+i)}\right]\left[\frac{(1+i)^N-1}{i(1+i)^N}\right]$$

$$= \bar{A}(P/\bar{A}, i, N) \tag{54.17}$$

For ease of reference, all of the equivalence models described above are summarized in Fig. 54.1. Models for *discrete cash flows* are shown in Fig. 54.1(*a*) under the two compounding conventions: (1) *end-of-period* compounding at effective interest rate i and (2) *continuous* compounding at nominal interest rate r. Models for *continuous cash flows* are shown in Fig. 54.1(*b*) under the assumption of continuous compounding at (1) effective interest rate i and (2) nominal interest rate r.

54.3 METHODS FOR SELECTING AMONG ALTERNATIVES

A variety of methods are used to evaluate alternative investments. Associated with each is a *statistic,* that is, a "figure of merit," and a *decision rule* which is used to select from among alternatives on the basis of the statistics. These are presented briefly here for a set of evaluation methods which are most commonly used in engineering economy.

Present Worth (Net Present Value). *Present worth* (PW) and *net present value* (NPV) are equivalent terms. The former is widely used in the literature of engineering economy; the latter is common to the literature of finance and accounting.

Present Worth of a Proposed Investment. The *present worth* is the equivalent present value of the cash flows generated by the proposed investment over a specified time interval (planning horizon N) with discounting at a specified interest rate i. One of several algebraic expressions for PW, assuming end-of-period cash flows A_j and end-of-period discounting at rate i, is

$$PW = \sum_{j=0}^{N} A_j (1+i)^{-j} \tag{54.18}$$

The *planning horizon* represents that period of time over which the proposed project is to be evaluated. It should be of sufficient duration to reflect all significant differences between the project and alternative investments. The discount rate i is the *minimum attractive rate of return*, that is, the rate of return which could be expected if the funds to be invested in the proposed project were to be invested elsewhere.

Present Worth of the "Do Nothing" Alternative. Let P represent the initial investment in a proposed project. If P were to be invested elsewhere, rather than in the proposed project, then this do nothing alternative would yield $P(1 + i)^N$, assuming compounding at rate i for N periods. The present worth of this course of action is zero, as can be seen by

$$\begin{aligned} \text{PW} &= A_0 + A_N (1 + i)^{-N} \\ &= -P + P(1 + i)^N (1 + i)^{-N} = 0 \end{aligned}$$

Comparing the proposed investment with the do nothing alternative, it follows that the investment is economically attractive (preferred to doing nothing) if its PW > 0.

The do nothing alternative is sometimes known as *alternative zero*. Here, PW(∅) = 0.

Multiple (More Than Two) Alternatives. We have seen that given two alternatives, (1) the proposed project and (2) do nothing, the "invest" decision is indicated if PW > 0. But suppose that there are more than two alternatives under consideration. In this case, the PWs of each of the alternatives are rank-ordered, and *that alternative yielding the maximum PW is to be preferred* (in an economic sense only, of course). To illustrate, consider the four mutually exclusive alternatives summarized in Table 54.1. Present worths have been determined using Eq. (54.18) and assuming $i = 20$ percent. As noted in the table, the correct rank ordering of the set of alternatives is IV > II > III > ∅ > I.

It is not necessary to adjust the PW statistic for differences in initial cost. This is so because any funds invested elsewhere yield a PW of zero. In our example, consider alternatives II and III. Initial costs are \$1000 and \$1100, respectively. Alternative II may be viewed as requiring \$1000 in

TABLE 54.1 Cash Flows for Four Mutually Exclusive Alternatives

	Assume $i = 20\%$			
End of period	Alternative I	Alternative II	Alternative III	Alternative IV
0	–$1000	–$1000	–$1100	–$2000
1–10	0	300	320	520
10	4000	0	0	0
Net cash flow	$3000	$2000	$2100	$3500
PW	–$354	$258	$242	$306
AW	–$85	$62	$58	$73
FW	–$2192	$1596	$1496	$1894

the project (yielding PW of $258) and $100 elsewhere (yielding PW of $0). The total PW(II) = $258. This may now be compared directly with alternative III: PW(III) = $242. Each alternative accounts for a total investment of $1100.

Annual Worth (Equivalent Uniform Annual Cost). The *annual worth* (AW) is the uniform series over N periods equivalent to the present worth at interest rate i. It is a weighted average periodic worth, weighted by the interest rate. Mathematically,

$$AW = (PW)(A/P, i, N) \tag{54.19}$$

If $i = 0$ percent, then AW is simply the average cash flow per period, that is,

$$AW = (1/N)\sum_{j=0}^{N} A_j$$

By convention, this is known as the *annual* method, although the period may be week, month, or the like. This method is most often used with respect to *costs,* and in such cases it is known as the *equivalent uniform annual cost* (EUAC) method.

The decision rule applicable to PW is also applicable for AW (and EUAC). That is, a proposal is preferred to the do nothing alternative if AW > 0, and multiple alternatives may be rank-ordered on the basis of declining AW (or increasing EUAC). Given any pair of alternatives, say, X and Y, if PW(X) > PW(Y), then AW(X) > AW(Y). This is so because (A/P, i, N) is a constant for all alternatives as long as i and N remain constant.

The annual worth method is illustrated in Table 54.1. Note that the ranking of alternatives is consistent with that of the PW method: IV > II > III > ∅ > I.

Future Worth. In the *future worth* (FW) method, all cash flows are converted to a single equivalent value at the end of the planning horizon, period N. Mathematically:

$$FW = (PW)(F/P, i, N)$$

The decision rule applicable to PW is also applicable to FW. A set of mutually exclusive investment opportunities may be rank-ordered by using either PW, AW, or FW. The results will be consistent. The future worth method is illustrated in Table 54.1.

Rate of Return

Internal Rate of Return. The internal rate of return (IRR), often known simply as the *rate of return* (RoR), is that interest rate i^* for which the net present value of all project cash flows is zero. When all cash flows are discounted at rate i^*, the equivalent present value of all project benefits

TABLE 54.2 (Internal) Rate of Return Analysis of Alternatives from Table 54.1

Step	Comparison of alternatives	Cash flows (A_j) A_0	A_1–A_{10}	A_{10}	Incremental Rate of return (%)	Conclusion (MARR = 20%)
1	∅ → I	–\$1000	0	4000	14.9	I < ∅
2	∅ → II	–1000	300	0	27.3	II > ∅
3	∅ → III	–1100	320	0	26.3	III > ∅
4	∅ → IV	–2000	550	0	24.4	IV > ∅
5	II → III	–100	20	0	15.1	III < ∅
6	II → IV	–1000	250	0	21.4	IV > II

exactly equals the equivalent present value of all project costs. One mathematical definition of the IRR is that rate i^* that satisfies the equation

$$\sum_{j=0}^{N} A_j (1+i^*)^{-j} \equiv 0 \tag{54.20}$$

This formula assumes discrete cash flows A_j and end-of-period discounting in periods $j = 1, 2, \ldots, N$.

The discount rate used in present worth calculations is the opportunity cost—a measure of the return that could be earned on capital if it were invested elsewhere. Thus a given proposed project should be economically attractive if and only if its IRR exceeds the cost of opportunities forgone as measured by the firm's *minimum attractive rate of return* (MARR). That is, an increment of investment is justified if, for that proposal, IRR > MARR.

Multiple Alternatives. Unlike the PW/AW/FW methods, mutually exclusive projects may *not* be rank-ordered on the basis of their respective IRRs. Rather, an incremental procedure must be implemented. Alternatives must be considered pairwise, with decisions made about the attractiveness of each increment of investment. As shown in Table 54.2, we conclude that IV > II > ∅ > I. These results are consistent with those found by the PW/AW/FW methods.

Multiple Solutions. Consider the end-of-period model described by Eq. (54.20):

$$\sum_{j=0}^{N} A_j (1+i^*)^{-j} \equiv 0$$

This expression may also be written as

$$A_0 + A_1 x + A_2 x^2 + \cdots + A_N x^N = 0 \tag{54.21}$$

where $x = (1 + i^*)^{-1}$. Solving for x leads to i^*, so we want to find the roots x of this Nth-order polynomial expression. Only the real, positive roots are of interest, of course, because any meaningful values of i^* must be real and positive. There are many possible solutions for x, however, depending upon the signs and magnitudes of the cash flows A_j. Multiple solutions for x—and, by extension, i^*—are possible. In those instances where multiple IRRs are obtained, it is recommended that the PW method, rather than the rate of return method, be used.

Benefit–Cost Ratio. The benefit–cost ratio method is widely used in the public sector.

Benefit–Cost Ratio and Acceptance Criterion. The essential element of the benefit–cost ratio method is almost trivial, but it can be misleading in its simplicity. An investment is justified only if the incremental benefits B resulting from it exceed the resulting incremental costs C. Of course, all benefits and costs must be stated in equivalent terms, that is, with measurement at the same point(s) in time. Normally, both benefits and costs are stated as "present value" or are "annualized" by using compound interest factors as appropriate. Thus,

$$B{:}C = \frac{\text{PW (or AW) of all “benefits”}}{\text{PW (or AW) of all “cost”}} \tag{54.22}$$

Clearly, if benefits must exceed costs, the ratio of benefits to costs must exceed unity. That is, if $B > C$, then $B{:}C > 1.0$. This statement of the *acceptance criterion* is true only if the incremental costs C are positive. It is possible, when evaluating certain alternatives, for the incremental costs to be negative, that is, for the proposed project to result in a reduction of costs. Negative benefits arise when the incremental effect is a reduction in benefits. In summary,

For $C > 0$, if $B{:}C > 1.0$, accept; otherwise reject.

For $C < 0{:}C > 1.0$, reject; otherwise accept.

Multiple Alternatives. Like the rate of return method, the proper use of the benefit–cost ratio method requires incremental analysis. Mutually exclusive alternatives *should not be rank-ordered on the basis of benefit–cost ratios*. Pairwise comparisons are necessary to test whether increments of costs are justified by increments of benefits.

To illustrate, consider two alternative projects U and T.

Present Worths

Comparison	Benefits, \$	Costs, \$	$B{:}C$	Conclusion
$\varnothing \rightarrow T$	700,000	200,000	3.50	$T > \varnothing$
$\varnothing \rightarrow U$	1,200,000	600,000	2.00	$U > \varnothing$
$\varnothing \rightarrow U$	500,000	400,000	1.25	$U > T$

On the basis of benefit–cost ratios, it is clear that both T and U are preferred to the do nothing alternative (∅). Moreover, the *incremental* analysis indicates that U is preferred to T since the incremental $B{:}C$ (=1.25) exceeds unity.

It will be noted here that PW analysis would yield the same result: PW(T) = \$500,000 and PW(U) = \$600,000. It may be shown that this result obtains in general. That is, for any number of mutually exclusive alternatives, ranking based on *proper* use of the benefit–cost ratio method using *incremental* analysis will always yield the same rank order resulting from proper use of the present worth method.

Payback. The payback method is widely used in industry to determine the relative attractiveness of investment proposals. The essence of this technique is the determination of the *number of periods required to recover an initial investment.* Once this has been done for all alternatives under consideration, a comparison is made on the basis of respective payback periods.

Payback, or *payout*, as it is sometimes known, is the number of periods required for cumulative benefits to equal cumulative costs. Costs and benefits are usually expressed as cash flows, although discounted present values of cash flows may be used. In either case, the payback method is based on the assumption that the relative merit of a proposed investment is measured by this statistic. The smaller the payback (period), the better the proposal.

(*Undiscounted*) *payback* is that value of N^* such that

$$P = \sum_{j=1}^{N^*} A_j \tag{54.23}$$

where P is the initial investment and A_j is the cash flow in period j. *Discounted payback,* used much less frequently, is that value of N^* such that

$$P = \sum_{j=1}^{N^*} A_j(1+i)^{-j} \tag{54.24}$$

The principal objection to the use of payback as a primary figure of merit is that all consequences beyond the end of the payback periods are ignored. This may be illustrated by a simple example. Consider two alternatives *V* and *W*. The discount rate is 10 percent and the planning horizon is 5 years. Cash flows and the relevant results are as follows:

End of year	Alternative *V*	Alternative *W*
0 (initial cost)	–$8000	–$9000
1–5 (net revenues)	4000	3000
5 (salvage value)	0	8000
Undiscounted payback	2 years	3 years
PW at 10%	$7163	$7339

Alt. V has the shorter payback period, but Alt. W has the larger PW.

Payback is a useful measure to the extent that it provides some indication of how long it might take before the initial investment is recovered. It is a helpful *supplementary* measure of the attractiveness of an investment, but it should never be used as the sole measure of quality.

Return on Investment. There are a number of approaches, widely used in industry, that use accounting data (income and expenses) rather than cash flows to determine "rate of return," where income and expense are reflected in the firm's accounting statements. Although there is no universally accepted terminology, this accounting-based approach is generally known as *return on investment* (ROI), whereas the cash flow approach results in IRR or RoR.

One formulation of the RoI is the ratio of the average annual accounting profit to the *original book value* of the asset. Another variation is the ratio of the average annual accounting profit to the *average book value* of the asset over its service life. In any event, such computations are based on depreciation expense, an accounting item which is not a cash flow and which is affected by relevant tax regulations. (See "Depreciation" below.) Therefore, the use of RoI is not recommended as an appropriate figure of merit.

Unequal Service Lives. One of the fundamental principles, noted earlier, is that alternative investment proposals must be evaluated over a common planning horizon. Unequal service lives among competing feasible alternatives complicate this analysis. For example, consider two alternatives: one has life of N_1, the other has life of N_2, and $N_1 < N_2$.

Repeatability (Identical Replication) Assumption. One approach, widely used in engineering economy textbooks, is to assume that (1) each alternative will be replaced at the end of its service life by an identical replacement, that is, the amounts and timing of all cash flows in the first and all succeeding replacements will be identical to the initial alternative; and (2) the planning horizon is at least as long as the common multiple of the lives of the alternatives.

Under these assumptions, the planning horizon is the least common multiple of N_1 and N_2. The *annual worth* method may be used directly since the AW for alternative 1 over N_1 periods is the same as the AW for alternative 1 over the planning horizon.

Specified Planning Horizon. Although commonly used in the literature of engineering economy, the repeatability assumption is rarely appropriate in real-world applications. In such cases, it is generally more reasonable to define the planning horizon *N* on some basis other than the service lives of the competing alternatives. Equipment under consideration may be related to a certain product, for example, which will be manufactured over a specified time period.

If the planning horizon is *longer* than the service life of one or more of the alternatives, it will be necessary to estimate the cash flow consequences, if any, during the interval(s) between the service life (or lives) and the end of the planning horizon. If the planning horizon is *shorter* than the service lives of one or more of the alternatives, all cash flows beyond the end of the planning horizon are irrelevant. In the latter case it will be necessary to estimate the salvage value of the "truncated" proposal at the end of the planning horizon.

54.4 AFTER-TAX ECONOMY STUDIES

Most individuals and business firms are directly influenced by taxation. Cash flows resulting from taxes paid (or avoided) must be included in evaluation models, along with cash flows from investment, maintenance, operations, and so on. Thus decision makers have a clear interest in cash flows for taxes and related topics.

Depreciation. There is a good deal of misunderstanding about the precise meaning of *depreciation*. In economic analysis, depreciation is not a measure of the loss in market value or equipment, land, buildings, and the like. It is *not* a measure of reduced serviceability. Depreciation is strictly an *accounting* concept. Perhaps the best definitions provided by the Committee on Terminology of the American Institute of Certified Public Accountants:

Depreciation accounting is a system of accounting which aims to distribute the cost or other basic value of tangible capital assets, less salvage (if any), over the estimated life of the unit (which may be a group of assets) in a systematic and rational manner. It is a process of allocation, not of valuation. Depreciation for the year is the portion of the total charge under such a system that is allocated to the year.*

Depreciable property may be tangible or intangible. *Tangible* property is any property that can be seen or touched. *Intangible* property is any other property, for example, a copyright or franchise.

Depreciable property may be real or personal. *Real* property is land and generally anything erected on, growing on, or attached to the land. *Personal* property is any other property, for example, machinery or equipment. (*Note:* Land is never depreciable as it has no determinable life.)

To be depreciable, property must meet three requirements: (1) it must be used in business or held for the production of income; (2) it must have a determinable life longer than 1 year; and (3) it must be something that wears out, decays, gets used up, becomes obsolete, or loses value from natural causes.

Depreciation begins when the property is placed in service; it ends when the property is removed from service.

For the purpose of computing taxable income on income tax returns, the rules for computing allowable depreciation are governed by the relevant taxing authority. An excellent reference for federal income taxes is *How to Depreciate Property, Publication 946*, published by the Internal Revenue Service (IRS), U.S. Department of the Treasury. Publication 946 is updated annually.†

A variety of depreciation methods have been and are currently permitted by taxing authorities in the United States and other countries. The discussion which follows is limited to the three methods which are of most interest at the present time. The straight line and declining balance methods are used mainly outside the United States. The Modified Accelerated Cost Recovery System (MACRS) is used currently by the federal government as well as by most states in the United States. Moreover, as will be shown, the straight line and declining balance methods are imbedded within the MACRS method, and it is for this reason that the straight line and declining balance methods are included here.

---

* American Institute of Certified Public Accountants, *Accounting Research Bulletin No. 22* (American Institute of Certified Public Accountants, New York, 1944) and American Institute of Certified Public Accountants, *Accounting Terminology Bulletin No. 1* (American Institute of Certified Public Accountants, New York, 1953).

† The discussion of depreciation accounting is necessarily abbreviated in this handbook. The reader is encouraged to consult competent tax professionals and/or relevant publications of the Internal Revenue Service for more thorough treatment of this complex topic.

1. *Straight-line method.* In general, the allowable depreciation in tax year j, D_j, is given by

$$D_j = \frac{B-S}{N} \quad \text{for } j=1,\ldots,N \tag{54.25}$$

where B is the adjusted cost basis, S is the estimated salvage value, and N is the depreciable life.

Allowable depreciation must be prorated on the basis of the period of service for the tax year in which the property is placed in service and the year in which it is removed from service. For example, suppose that $B = \$90{,}000$, $N = 6$ years, $S = \$18{,}000$ after 6 years, and the property is to be placed in service at midyear. In this case,

$$D_j = \frac{\$90{,}000 - \$18{,}000}{6} = \$12{,}000 \qquad \text{for } j = 2,\ldots,6$$

$$D_1 = D_7 = (6/12)(\$12{,}000) = \$6000$$

The *book value* of the property at any point in the time is the initial cost less the accumulated depreciation. In the numerical example above, the book value at the start of the third tax year would be \$90,000 – \$6000 – \$12,000 = \$72,000.

2. *Declining balance method.* The amount of depreciation taken each year is subtracted from the book value before the following year's depreciation is computed. A constant depreciation rate a applies to a smaller, or declining, balance each year. In general,

$$D_j = \begin{cases} \pi_1 aB & \text{for } j = 1 \\ aB_j & \text{for } j = 2,3,\ldots,N+1 \end{cases} \tag{54.26}$$

where π_1 = portion of the first tax year in which the property is placed in service

$$(0 < \pi_1 \leq 1)$$

B_j = book value in year j prior to determining the allowable depreciation

Assuming that the property is placed in service at the start of the tax year (π_1 = 1.00), it may be shown that

$$D_j = Ba(1 - a)^{j-1} \tag{54.27}$$

When $a = 2/N$, the depreciation scheme is known as the *double declining balance method,* or simply DDB.

To illustrate using the previous example, suppose that we have DDB with $a = 2/6 = 0.333$. Since $\pi_1 = 6/12 = 0.5$,

$$D_1 = \pi_1 aB = 0.5(0.333)(\$90{,}000) = \$15{,}000$$

$$D_2 = a(B - D_1) = 0.333(\$90{,}000 - \$15{,}000) = \$25{,}000$$

Salvage value is not deducted from the cost or other basis in determining the annual depreciation allowance, but *the asset cannot be depreciated below the expected salvage value.* In other words, once book value equals salvage value, no further depreciation may be claimed.

3. *MACRS (GDS and ADS).* Under the 1986 Tax Reform Act, the *modified accelerated cost recovery system* (MACRS, pronounced "makers") is permitted for the purpose of determining taxable income on federal income tax returns. MACRS consists of two systems that determine how qualified property is depreciated. The main system is called the *general depreciation system* (GDS) and the other is called the *alternative depreciation system* (ADS). MACRS applies to most depreciable property placed in service after December 31, 1986.

- *Class Lives and Property Classes.* Both GDS and ADS have preestablished class lives for most property. These are summarized in a Table of Class Lives and Recovery Periods at the back of IRS *Publication 946*. There are eight recovery periods based on these class lives: 3-, 5-, 7-, 10-, 15-, and 20-year properties, as well two additional real property classes, *nonresidential real* property and *residential rental* property.
- *Depreciation Methods.* There are a number of ways to depreciate property under MACRS, depending upon the property class, the way the property is used, and the taxpayer's election to use either GDS or ADS. These are summarized below:

Property class	Primary GDS method	Optional method
3-, 5-, 7-, 10-year (nonfarm)	200% DB over GDS recovery period	Straight line over GDS recovery period or 150% DB over ADS recovery period
15-, 20-year (nonfarm) *or* property used in farming, except real property	150% DB over GDS recovery period	Straight line over GDS recovery period or Straight line over ADS recovery period
Nonresidential real and residential rental property	Straight line over GDS recovery period	Straight line over fixed ADS recovery period

Where the declining balance method is used, switch to the straight line method occurs in the first tax year for which the SL method, when applied to the adjusted basis at the beginning of the year, will yield a larger deduction than had the DB method been continued. Zero salvage value is assumed for the purpose of computing allowable depreciation expense.

The Placed-in-Service Convention. With certain exceptions, MACRS assumes that all property placed in service (or disposed of) during a tax year is placed in service (or disposed of) at the midpoint of that year. This is the *half-year* convention.

Depreciation Percentages. The annual depreciation percentages under GDS, assuming the half-year convention, are summarized in Table 54.3.

For 3-, 5-, 7-, 10-, 15-, and 20-year properties, the depreciation percentage in year j for property class k under ADS is given by

$$p_j = \begin{cases} 0.5/k, & j=1 \\ 1.0/k, & j=2,3,\ldots,k \\ 0.5/k, & j=k+1 \end{cases} \tag{54.28}$$

Other Deductions from Taxable Income. In addition to depreciation, there are several other ways in which the cost of certain assets may be recovered over time.

Amortization. Amortization permits the taxpayer to recover certain capital expenditures in a way that is like straight-line depreciation. Qualifying expenditures include certain costs incurred in setting up a business (for example, survey of potential markets, analysis of available facilities), the cost of a certified pollution control facility, bond premiums, and the costs of trademarks and trade names. Expenditures are amortized on a straight-line basis over a 60-month period or more.

Depletion. Depletion is similar to depreciation and amortization. It is a deduction from taxable income applicable to a mineral property, an oil, gas, or geothermal well, or standing timber. There are two ways to figure depletion: cost depletion and percentage depletion. With certain restrictions, the taxpayer may choose either method.

Section 179 Expense. The taxpayer may elect to treat the cost of certain qualifying property as an expense rather than as a capital expenditure in the year the property is placed in service.

TABLE 54.3 Annual Depreciation Percentages Under MACRS (Half-Year Convention)

Recovery year	Recovery period (k)					
	3-year	5-year	7-year	10-year	15-year	20-year
1	33.33	20.00	14.29	10.00	5.00	3.750
2	44.45	32.00	24.49	18.00	9.50	7.219
3	14.81	19.20	17.49	14.40	8.55	6.677
4	7.41	11.52	12.49	11.52	7.70	6.177
5		11.52	8.93	9.22	6.93	5.713
6		5.76	8.92	7.37	6.23	5.285
7			8.93	6.55	5.90	4.888
8			4.46	6.55	5.90	4.522
9				6.56	5.91	4.462
10				6.55	5.90	4.461
11				3.28	5.91	4.462
12					5.90	4.461
13					5.91	4.462
14					5.90	4.461
15					5.91	4.462
16					2.95	4.461
17						4.462
18						4.461
19						4.462
20						4.461
21						2.231

Qualifying property is "Section 38 property"—generally, property used in the trade or business with a useful life of 3 years or more for which depreciation or amortization is allowable, with certain limitations—and that is purchased for use in the active conduct of the trade or business.

The total cost that may be deducted for a tax year may not exceed some maximum amount M^*. The expense deduction is further limited by the taxpayer's total investment during the year in Section 179 property: the maximum M is reduced by \$1 for each dollar of cost in excess of \$200,000. That is, no Section 179 expense deduction may be used if total investment in Section 179 property during the tax year exceeds \$200,000 + M. Moreover, the total cost that may be deducted is also limited to the taxable income which is from the active conduct of any trade or business of the taxpayer during the tax year. See IRS Publication 946 for more information.

The cost basis of the property must be reduced by the amount of the Section 179 expense deduction, if any, before the allowable depreciation expense is determined.

Gains and Losses on Disposal of Depreciable Assets. The value of an asset on disposal is rarely equal to its book value at the time of sale or other disposition. When this inequality occurs, a gain or loss on disposal is established.

In general, the *gain* on disposition of depreciable property is the net salvage value minus the adjusted basis of the property (its book value) at the time of disposal. The *adjusted basis* is the original cost basis less any accumulated depreciation, amortization, Section 179 expense deduction, and where appropriate, any basis adjustments due to the investment credit claimed on the property. A negative gain is considered a *loss* on disposal.

---

*M = \$18,500 in 1998, \$19,000 in 1999, \$20,000 in 2000, \$24,000 in 2001 and 2002, and \$25,000 after 2002.

All gains and losses on disposal are treated as *ordinary* gains or losses, *capital* gains or losses, or some combination of the two. The rules for determining these amounts are too complex to be discussed adequately here. Interested readers should therefore consult a competent expert and/or read the appropriate sections in *Tax Guide for Small Business* (IRS Publication 334) or a similar reference.

Federal Income Tax Rates for Corporations. Income tax rates for corporations are adjusted from time to time, largely in order to affect the level of economic activity. Currently the *marginal* federal income tax rate for corporations is given below.

If the taxable income is:

At least	But not more than	Marginal tax rate is
\$0	\$50,000	0.15
\$50,000	\$75,000	0.25
\$75,000	\$100,000	0.34
\$100,000	\$335,000	0.39
\$335,000	\$10 million	0.34
\$10 million	\$15 million	0.35
\$15 million	\$18 1/3 million	0.38
\$18 1/3 million	and over	0.35

It may be shown that the *average* tax rate is 35 percent if the total taxable income is at least \$18 1/3 million.

When income is taxed by more than one jurisdiction, the appropriate tax rate for economy studies is a combination of the rates imposed by the jurisdictions. If these rates are independent, they may simply be added. But the combinatorial rule is not quite so simple when there is interdependence. Income taxes paid to local and state governments, for example, are deductible from taxable income on federal income tax returns, but the reverse is not true; federal income taxes are not deductible from local returns. Thus, considering only state (t_s) and federal (t_f) income tax rates, the *combined incremental tax rate* (t) for economy studies is given by

$$t = t_s + t_f(1 - t_s) \tag{54.29}$$

Timing of Cash Flows for Income Taxes. The equivalent present value of tax consequences requires estimates of the *timing* of cash flows for taxes. A variety of operating conditions affect the timing of income tax payments. It is neither feasible nor desirable to catalog all such conditions here. In most cases, however, the following assumptions will serve as a reasonable approximation.

1. Income taxes are paid quarterly at the end of each quarter of the tax year.
2. Ninety percent of the firm's income tax liability is paid in the tax year in which the expense occurs; the remaining 10 percent is paid in the first quarter of the following tax year.
3. The four quarterly tax payments during the tax year are uniform.

The timing of these cash flows can be approximated by a weighted average of quarter-ending dates.

$$0.225(1/4 + 2/4 + 3/4 + 4/4) + 0.1(5/4) = 0.6875$$

That is, the cash flow for income taxes in a given tax year can be assumed to be concentrated at a point 0.6875 into the tax year. (An alternative approach is to assume that cash flows for income taxes occur at the end of the tax year.)

After-Tax Analysis. The following procedures are followed to prepare an after tax analysis.

1. Specify the assumptions and principal parameter values, including the following:
   - Tax rates (federal and other taxing jurisdictions, as appropriate)

- Relevant methods related to depreciation, amortization, depletion, investment tax credit, and Section 179 expense deduction
- Length of planning horizon
- Minimum attractive rate return—the interest rate to be used for discounting cash flows
  Caution: This rate should represent the after-tax opportunity cost to the taxpayer. It will almost always be lower than the pretax MARR. The same discounting rate should not be used for both before-tax and after-tax analyses.

2. Estimate the amounts and timing of cash flows other than income taxes. It will be useful to separate these cash flows into three categories:
- Cash flows that have a *direct* effect on taxable income, as either income or expense. Examples: sales receipts, direct labor savings, material costs, property taxes, interest payments, state and local income taxes (on federal returns)
- Cash flows that have an *indirect* effect on taxable income through depreciation, amortization, depletion, Section 179 expense deduction, and gain or loss on disposal. Examples: initial cost of depreciable property, salvage value
- Cash flows that do not affect taxable income. Examples: working capital and that portion of loan repayments that represents payment of principal

3. Determine the amounts and timing of cash flows for income taxes.

4. Find the equivalent present value of cash flows for income taxes at the *beginning of the first tax year.* To that end, let P_j denote the equivalent value of the cash flow for taxes in year *j*, as measured at the start of tax year *j*.

$$P_j \simeq T_j(1+i)^{-0.6875} \qquad j = 1, 2, \ldots, N+1 \tag{54.30}$$

where *i* is the effective annual discount rate and *N* is the number of years in the planning horizon.

The equivalent present value of all the cash flows for taxes, as measured at the start of the first tax year, is given by

$$P(T) = \sum_{j=1}^{N+1} P_j(1+i)^{-j+1} = \sum_{j=1}^{N+1} T_j(1+i)^{0.3125-j} \tag{54.31}$$

5. Find the equivalent present value of the cash flows for taxes, where "present" is defined as the start of the planning horizon. For example, if the property is placed in service at the end of the third month of the tax year, the present value adjustment is $P(T) \times (1 + \mathrm{i})^{3/12}$.

6. Find the equivalent present value of all other cash flows estimated in step 2 above. Use the after-tax MARR. Here the "present" is defined as the start of the planning horizon.

7. Combine (54.5) and (54.6) to yield the total *net present value* (NPV), or present worth (PW).

Note: If it is desired to determine the *after-tax rate of return* rather than PW (or FW, EUAC, and so on), steps 4 to 7 above must be modified. With the appropriate present worth equation for all cash flows, set the equation equal to zero and find the value of the interest rate i^* such that PW = 0. This is the after-tax IRR for the proposed investment.

Example. Consider the possible acquisition of certain manufacturing equipment with *initial* cost of $400,000. The equipment is expected to be kept in service for 6 years and then sold for an estimated $40,000 *salvage value. Working capital* of $50,000 will be required at the start of the 6-year period; the working capital will be recovered intact at the end of 6 years. If acquired, this equipment is expected to result in *savings* of $100,000 each year. The timing of these savings is such that the continuous cash flow assumption will be adopted throughout each year. The firm's after-tax MARR is 10 percent per year. The present worth of these cash flows, other than

TABLE 54.4 Cash Flows for Income Taxes—Numerical Example

					Tax rate = 0.35			
Tax year j	Depreciation rate p_j (5)	Depreciation D_j	Gain G_N	Other revenue R_j	Taxable income $R_j - D_j + G_N$	Income taxes T_j	PW factor $(1.10)^{0.3125-j}$	PW @10% p_j
1	0.200	$80,000		$40,000	$(40,000)	$(14,000)	0.93657	$(13,112)
2	0.3200	128,000		80,000	(48,000)	(16,800)	0.85143	(14,304)
3	0.1920	76,800		80,000	3,200	1,120	0.77403	867
4	0.1152	46,080		80,000	33,920	11,872	0.70366	8,354
5	0.1152	46,080		80,000	33,920	11,872	0.63969	7.594
6	0.0576	23,040		80,000	56,960	19,936	0.58154	11,594
7	0.0000	—	$40,000	40,000	80,000	28,000	0.52867	14,803
	Cost basis = $400,000			PW measured at start of 1st tax year				$15,796
				Adjustment factor ($^1/_2$ year)				$\times (1.10)^{0.5}$
				PW measured at start of planning horizon				$16,566

income taxes, is

$$\begin{aligned} \text{PW} &= -\$400{,}000 + \$40{,}000(P/F, 10\%, 6) \\ &\quad - \$50{,}000 + \$50{,}000(P/F, 10\%, 6) \\ &\quad + \$1000{,}000(P/\bar{A}, 10\%, 6) \\ &= \$57{,}800 \end{aligned}$$

Assume that there is no Section 179 expense deduction. The equipment will be placed in service at the middle of the tax year and depreciated under MACRS as a 5-year recovery property using the half-year convention. The incremental federal income tax rate is 0.35; there are no other relevant income tax affected by this proposed investment. The PW of the effects of cash flows due to income taxes is summarized in Table 54.4. The total PW for this proposed project is as follows:

Cash flows other than income taxes	$57,759
Effect on cash flows due to income taxes	−16,566
Net present worth	$41,193

Spreadsheet Analyses. A wide variety of computer programs are available for before-tax and/or after-tax analyses of investment programs. (Relevant computer software is discussed from time to time in the journal, *The Engineering Economist.*) In addition, any of several spreadsheet programs currently available may be readily adapted to economic analyses, usually with very little additional programming. For example, Lotus and Excel include financial functions to find the present and future values of a single payment and a uniform series (annuity), as well as to find the IRR of a series of cash flows.

Tables 54.4 and 54.5 are illustrations of computer-generated spreadsheets.

54.5 INCORPORATING PRICE LEVEL CHANGES INTO THE ANALYSIS

The effects of price level changes can be significant to the analysis. Cash flows, proxy measures of goods and services received and expended, are affected by both the *quantities* of goods and services as well as their *prices*. Thus, to the extent that changes in price levels affect cash flows, these changes must be incorporated into the analysis.

TABLE 54.5 Spreadsheet Analysis—Numerical Example

				MARR = 10%		
Project year *j*	Investment and salvage value	Working capital	Savings during year *j*	PW of discrete cash flows	PW of continuous cash flows	Total present value
0	($400,000)	($50,000)		($450,000)		($450,000)
1			$100,000		$95,382	$95,382
2			$100,000		$86,711	$86,711
3			$100,000		$78,828	$78,828
4			$100,000		$71,662	$71,662
5			$100,000		$65,147	$65,147
6	$40,000	$50,000	$100,000	$50,803	$59,225	$110,028
Total	$360,000)	$0	$600,000	($399,197)	$456,957	$57,759
			Present Worth (NPV) of Cash Flows for Taxes			($16,566)
			Net Present Worth			$41,193

The *Consumer Price Index* (CPI) is but one of a large number of indexes that are regularly used to monitor and report for specific economic analyses. Analysts should be interested in relative price changes of goods and service that are germane to the particular investment alternatives under consideration. The appropriate indexes are those that are related, say, to construction materials, costs of certain labor skills, energy, and other cost and revenue factors.

General Concepts and Notation. Let p_1 and p_2 represent the prices of a certain good of service at two points in time t_1 *and* t_2, and let $n = t_2 - t_1$. The relative rate of price changes between t_1 and t_2, the average per period, is given by

$$g = \sqrt[n]{p_2 / p_1} - 1 \tag{54.32}$$

We have *inflation* when $g > 0$ and *disinflation* when $g < 0$.

Let A_j = cash flow resulting from the exchange of certain goods or services, at end of period *j*, stated in terms of *constant* dollars. (Analogous terms are *now* or *real* dollars.) Let A_j^* = *cash flows for those same goods or services in actual* dollars. (Analogous terms are *then* or *current* dollars.) Then

$$A_j^* = A_j (1 + g)^j \tag{54.33}$$

where *g* is the periodic rate of increase or decrease in relative price (the *inflation rate*).

As before, let i = the MARR in the absence of inflation, that is, the real MARR. Let i^* = the MARR required taking into consideration inflation, that is, the *nominal* MARR. The periodic rate of increase or decrease in the MARR due to inflation *f* is given by

$$f = \left(\frac{1 + i^*}{1 + i} \right) - 1 = \frac{i^* - 1}{1 + i} \tag{54.34}$$

Other relationships of interest are

$$i^* = (1 + i)(1 + f) - 1 = i + f + if \tag{54.35}$$

and

$$i = \left(\frac{1 + i^*}{1 + f} \right) - 1 = \frac{i^* - f}{1 + f} \tag{54.36}$$

Models for Analysis. It may be shown that the *future worth* of a series of cash flows A_j^* ($j = 1, 2, \ldots, N$) is given by

$$FW = (1 + i^*)^N \sum_{j=0}^{N} A_j (1 + d) \qquad (54.37)$$

where

$$d = \frac{(1 + i)(1 + f)}{1 + g} - 1 \qquad (54.38)$$

and i, f, and g are as defined previously. From Eq. (54.37) it follows that the *present worth* is given by

$$PW = \sum_{j=0}^{N} A_j (1 + d)^{-j} \qquad (54.39)$$

Note: In these models it is assumed that both the cash flows and the MARR are affected by inflation, the former by g and the latter by f, and $f \neq g$. If it is assumed that both i and A_j are affected by the same rate, that is, $f = g$, then Eq. (54.39) reduces to

$$PW = \sum_{j=0}^{N} A_j (1 + i)^{-j} \qquad (54.40)$$

which is the same as the PW model ignoring inflation.

To illustrate, consider cash flows in constant dollars (A_j) of \$80,000 at the end of each year for 8 years. The inflation rate for the cash flows (g) is 6 percent per year, the nominal MARR (i^*) is 9 percent per year, and the inflationary effect on the MARR (f) is 4.6 percent per year. Then

$$d = \frac{1 + i^*}{1 + g} - 1 = \frac{1.09}{1.09} - 1 = 0.0283$$

and

$$PW = \sum_{j=1}^{8} A_j (1 + d)^{-j} = \$80{,}000(P/A, 2.83\%, 8) = \$565{,}000 \qquad (54.41)$$

Multiple Factors Affected Differently by Inflation. In the preceding section it is assumed that the project consists of a single price component affected by rate g per period. But most investments consist of a variety of components, among which rates of price changes may be expected to differ significantly. For example, the price of the labor component may be expected to increase at the rate of 7 percent per year, and the price of the materials component is expected to decrease at the rate of 5 percent per year. The appropriate analysis in such cases is an extension of Eqs. (54.37) through (54.39).

Consider a project consisting of two factors, and let A_{j1} and a_{j2} represent the cash flows associated with each of these factors. Let g_1 and g_2 represent the relevant inflation rates, so that

$$A_j^* = A_{j1}(1 + g_j)^j + A_{j2}(1 + g_2)^j$$

It follows that

$$FW = (1 + i^*)^N \left\{ \left[\sum_{j=1}^{N} A_{j1} (1 + d_1)^{-j} \right] + \left[\sum_{j=1}^{N} A_{j2} (1 + d_2)^{-j} \right] \right. \qquad (54.42)$$

and

$$PW = \left\{ \left[\sum_{j=1}^{N} A_{j1} (1 + d_1)^{-j} \right] + \left[\sum_{j=1}^{N} A_{j2} (1 + d_2)^{-j} \right] \right\} \qquad (54.43)$$

where

$$d_1 = (1 + i^*)/(1 + g_1) \quad d_2 = (1 + i^*)/(1 + g_2) \tag{54.44}$$

Interpretation of IRR Under Inflation. If *constant* dollars (A_j) are used to determine the internal rate of return, then the *inflation-free* IRR is that value of ρ such that

$$\sum_{j=0}^{N} A_j (1 + \rho)^{-j} = 0 \tag{54.45}$$

The project is acceptable if $\rho > i$, where i is the inflation-free MARR as in the preceding section.

If *actual* dollars (A_{j^*}) are used to determine the internal rate of return, then the *inflation-adjusted IRR* is that value of ρ^* such that

$$\sum_{j=0}^{N} A_j^* (1 + \rho^*) = 0 \tag{54.46}$$

To illustrate, consider a project which requires an initial investment of \$100,000 and for which a salvage value of \$20,000 is expected after 5 years. If accepted, this project will result in annual savings of \$30,000 at the end of each year over the 5-year period. All cash flow estimates are based on constant dollars. It may be shown that, based on these assumptions, $\rho \simeq 19$ percent.

It is assumed that the cash flows for this proposal will be affected by an inflation rate (g) of 10 percent per year. Thus $A_j^* = A_j (1.10)_j$, and from Eq. (54.45), $\rho^* \simeq 31$ percent.

The investor's inflation-free MARR (i) is assumed to be 25 percent. If it is assumed that the MARR is affected by an inflation rate (g) of 10 percent per year, then $i^* = 1.10(1.25) - 1 = 0.375$.

Each of the two comparisons indicates that the proposed project is not acceptable: $\rho(19\%) < i\ (25\%)$ and $\rho^*\ (31\%) < i^*\ (37.5\%)$.

54.6 TREATING RISK AND UNCERTAINTY IN THE ANALYSIS

It is imperative that the analyst recognize the uncertainty inherent in all economy studies. The past is irrelevant, except when it helps predict the future. Only the future is relevant, and the future is inherently uncertain.

At this point it will be useful to distinguish between risk and uncertainty, two terms widely used when dealing with the noncertain future. *Risk* refers to situations in which a probability distribution underlies future events and the characteristics of this distribution are known or can be estimated. Decisions involving *uncertainty* occur when nothing is known or can be assumed about the relative likelihood, or probability, of future events. Uncertainty situations may arise when the relative attractiveness of various alternatives is a function of the outcome of pending labor negotiations or local elections, or when permit applications are being considered by a government planning commission.

A wide spectrum of analytical procedures is available for the formal consideration of risk and uncertainty in analyses. Space does not permit a comprehensive review of all these procedures. The reader is referred to any of the general references included in suggestions for further reading for discussion of one or more of the following:

- Sensitivity analysis
- Risk analysis
- Decision theory applications
- Digital computer (Monte Carlo) simulation
- Decision trees

TABLE 54.6 Compound Interest Tables for i = 10%

(10 Percent)												
	Single payment			Uniform series				Uniform series		Gradient series		
	Compound amount	Present worth		Compound amount		Present worth		Sinking fund	Capital recovery	Uniform series	Present worth	
N	F/P	P/F	$P/\bar{F}$	F/A	$F/\bar{A}$	P/A	$P/\bar{A}$	A/F	A/P	A/G	P/G	N
1	1.100	0.9091	0.9538	1.000	1.049	0.909	0.954	1.0000	1.1000	0.000	0.000	1
2	1.210	0.8264	0.8671	2.100	2.203	1.736	1.821	0.4762	0.5762	0.476	0.826	2
3	1.331	0.7513	0.7883	3.310	3.473	2.487	2.609	0.3021	0.4021	0.937	2.329	3
4	1.464	0.6830	0.7166	4.641	4.869	3.170	3.326	0.2155	0.3155	1.381	4.378	4
5	1.611	0.6209	0.6515	6.105	6.406	3.791	3.977	0.1638	0.2638	1.810	6.862	5
6	1.772	0.5645	0.5922	7.716	8.095	4.355	4.570	0.1296	0.2296	2.224	9.684	6
7	1.949	0.5132	0.5384	9.487	9.954	4.868	5.108	0.1054	0.2054	2.622	12.763	7
8	2.144	0.4665	0.4895	11.436	11.999	5.335	5.597	0.0874	0.1874	3.004	16.029	8
9	2.358	0.4241	0.4450	13.579	14.248	5.759	6.042	0.0736	0.1736	3.372	19.421	9
10	2.594	0.3855	0.4045	15.937	16.722	6.145	6.447	0.0627	0.1627	3.725	22.891	10
11	2.853	0.3505	0.3677	18.531	19.443	6.495	6.815	0.0540	0.1540	4.064	26.396	11
12	3.138	0.3186	0.3343	21.384	22.437	6.814	7.149	0.0468	0.1468	4.388	29.901	12
13	3.452	0.2897	0.3039	24.523	25.729	7.103	7.453	0.0408	0.1408	4.699	33.377	13
14	3.797	0.2633	0.2763	27.975	29.352	7.367	7.729	0.0357	0.1357	4.996	36.801	14
15	4.177	0.2394	0.2512	31.772	33.336	7.606	7.980	0.0315	0.1315	5.279	40.152	15
16	4.595	0.2176	0.2283	35.950	37.719	7.824	8.209	0.0278	0.1278	5.549	43.416	16
17	5.054	0.1978	0.2076	40.545	42.540	8.022	8.416	0.0247	0.1247	5.807	46.582	17
18	5.560	0.1799	0.1887	45.599	47.843	8.201	8.605	0.0219	0.1219	6.053	49.640	18
19	6.116	0.1635	0.1716	51.159	53.676	8.365	8.777	0.0195	0.1195	6.286	52.583	19
20	6.728	0.1486	0.1560	57.275	60.093	8.514	8.932	0.0175	0.1175	6.508	55.407	20
21	7.400	0.1351	0.1418	64,003	67,152	8.649	9.074	0.0156	0.1156	6.719	58.110	21
22	8.140	0.1228	0.1289	71,403	74,916	8.772	9.203	0.0140	0.1140	6.919	60.689	22
23	8.954	0.1117	0.1172	79,543	83,457	8.883	9.320	0.0126	0.1126	7.108	63.146	23
24	9.850	0.1015	0.1065	88,497	92,852	8.985	9.427	0.0113	0.1113	7.288	65.481	24
25	10.835	0.0923	0.0968	98,347	103,186	9.077	9.524	0.0102	0.1102	7.458	67.696	25
26	11.918	0.0839	0.0880	109,182	114,554	9.161	9.612	0.0092	0.1092	7.619	69.794	26
27	13.110	0.0763	0.0800	121,100	127,059	9.237	9.692	0.0083	0.1083	7.770	71.777	27
28	14.421	0.0693	0.0728	134,210	140,814	9.307	9.765	0.0075	0.1075	7.914	73.650	28
29	15.863	0.0630	0.0661	148,631	155,945	9.370	9.831	0.0067	0.1067	8.049	75.415	29
30	17.449	0.0573	0.0601	164,494	172,588	9.427	9.891	0.0061	0.1061	8.176	77.077	30
31	19.194	0.0521	0.0547	181,944	190,896	9.479	9.945	0.0055	0.1055	8.296	78.640	31
32	21.114	0.0474	0.0497	201,138	211,035	9.526	9.995	0.0050	0.1050	8.409	80.108	32
33	23.225	0.0431	0.0452	222,252	233,188	9.569	10.040	0.0045	0.1045	8.515	81.486	33
34	25.548	0.0391	0.0411	245,477	257,556	9.609	10.081	0.0041	0.1041	8.615	82.777	34
35	28.102	0.0356	0.0373	271,025	284,361	9.644	10.119	0.0037	0.1037	8.709	83.987	35
40	45.259	0.0221	0.0232	442,593	464,371	9.779	10.260	0.0023	0.1023	9.096	88.953	40
45	72.891	0.0137	0.0144	718,906	754,280	9.863	10.348	0.0014	0.1014	9.374	92.454	45
50	117.391	0.0085	0.0089	11,163.91	1,221.181	9.915	10.403	0.0009	0.1009	9.570	94.889	50
55	189.059	0.0053	0.0055	1,880.594	11,973.13	9.947	10.437	0.0005	0.1005	9.708	96.562	55
60	304.482	0.0033	0.0034	3,034.821	3,184.151	9.967	10.458	0.0003	0.1003	9.802	97.701	60
65	490.372	0.0020	0.0021	4,893.72	5,134.51	9.980	10.471	0.0002	0.1002	9.867	98.471	65
70	789.748	0.0013	0.0013	7,887.48	8,275.592	9.987	10.479	0.0001	0.1001	9.911	98.987	70
80	2,048.41	0.0005	0.0005	20,474.05	21,481,484	9.995	10.487	0.0000	0.1000	9.961	99.561	80
90	5,313.04	0.0002	0.0002	53,120,348	55,734.17	9.998	10.490	0.0000	0.1000	9.983	99.812	90

Some of these procedures can be found elsewhere in this handbook. Other procedures widely used in industry include the following:

- *Increasing the minimum attractive rate of return.* Some analysts advocate adjusting the minimum attractive rate of return to compensate for risky investments, suggesting that, since some investments will not turn out as well as expected, they will be compensated for by an incremental *safety margin*, Δi. This approach, however, fails to come to grips with the risk or uncertainty associated with estimates for specific alternatives, and thus an element Δi in the minimum attractive rate of return penalizes all alternatives equally.
- *Differentiating rates of return by risk class.* Rather than building a safety margin into a single minimum attractive rate of return, some firms establish several risk classes with separate standards for each class. For example, a firm may require low-risk investments to yield at least 15 percent and medium-risk investments to yield at least 20 percent, and it may define a minimum attractive rate of return of 25 percent for high-risk proposals. The analyst then judges to which class a specific proposal belongs, and the relevant minimum attractive rate of return is used in the analysis. Although this approach is a step away from treating all alternatives equally, it is less than satisfactory in that it fails to focus attention on the uncertainty associated with the individual proposals. No two proposals have precisely the same degree of risk, and grouping alternatives by class obscures this point. Moreover, the attention of the decision maker should be directed to the causes of uncertainty, that is, to the individual estimates.
- *Decreasing the expected project life.* Still another measure frequently employed to compensate for uncertainty is to decrease the expected project life. It is argued that estimates become less and less reliable as they occur further and further into the future; thus shortening project life is equivalent to ignoring those distant, unreliable estimates. Furthermore, distant consequences are more likely to be favorable than unfavorable. That is, distant estimated cash flows are generally positive (resulting from net revenues) and estimated cash flows near date zero are more likely to be negative (resulting from startup costs). Reducing expected project life, however, has the effect of penalizing the proposal by precluding possible future benefits, thereby allowing for risk in much the same way that increasing the minimum attractive rate of return penalizes marginally attractive proposals. Again, this procedure is to be criticized on the basis that it obscures uncertain estimates.

54.7 COMPOUND INTEREST TABLES (10 PERCENT)

Table 54.6 presents compound interest tables for the single payment, the uniform series, and the gradient series.

FURTHER READING

Books

Blank, Leland T., and Anthony J. Tarquin, *Engineering Economy,* 4th ed., McGraw-Hill, New York, 1997.

Fleischer, Gerald A., *Introduction to Engineering Economy,* PWS Pub. Co., 1994.

Newnan, Donald G., *Engineering Economic Analysis,* 6th ed., Engineering Press, San Jose, CA, 1996.

Park, Chan S., and G. P. Sharp-Bette, *Advanced Engineering Economics,* Wiley, New York, 1990.

Park, Chan S., *Contemporary Engineering Economics,* 2nd ed., Addison-Wesley, Reading, MA, 1996.

Sullivan, W. G., G. J. Bontadell, and E. Wicks, *Engineering Economy*, 11th ed, Prentice-Hall, Upper Saddle River, NJ, 2000.

Thuesen, H. G., and W. J. Fabrycky, *Engineering Economy,* 8th ed., Prentice-Hall, Englewood Cliffs, NJ, 1993.

Wellington, Arthur M., *The Economic Theory of Railway Location,* 2d ed., Wiley, New York, 1887. (This book is of historical importance; it was the first to address the issue of economic evaluation of capital investments due to engineering design decisions. Wellington is widely considered to be the "father of engineering economy.")

Journals

Decision Science
The Engineering Economist
Financial Management
Harvard Business Review
IIE Transactions
Industrial Engineering
Journal of Business
Journal of Finance
Journal of Finance & Quantitative Analysis
Management Science

CHAPTER 55
MRP AND ERP

F. Robert Jacobs
Indiana University
Bloomington, Indiana

Kevin J. Gaudette
Indiana University
Bloomington, Indiana

55.1 MATERIAL REQUIREMENTS PLANNING

Material requirements planning (MRP) and its descendents have had a pervasive effect on production and inventory management over the past three decades. In fact, the majority of manufacturing companies in the United States use (or have used) an MRP-based system for planning production and ordering parts. This section discusses the theory and evolution of MRP, its inputs and system logic, and a variety of relevant topics of concern to users. It also describes other related systems in the MRP family of systems.

55.1.1 BACKGROUND

Independent vs. Dependent Demand. The key to understanding the theoretical basis of MRP is to first understand the two types of item demand: independent and dependent. Independent demand refers to the case where demand for each item is assumed to be unrelated to demand for other items. For example, if a firm manufactures bicycles, independent demand techniques for managing inventory ignore the fact that there are exactly two tires, one seat, one handlebar assembly, and the like on each bicycle. Instead, historical or forecasted demands are used for each part independently to determine the timing and quantity of orders. Although there are admittedly examples in which the independence assumption holds true, in most manufacturing environments demand for parts and assemblies are directly related to demands for finished goods, as illustrated by the simple bicycle example.

In contrast, dependent demand techniques recognize the direct link (dependence) between the production schedule for an end item and the need for parts and assemblies to support this schedule. Again using the simple example of a bicycle, if 10 bicycles are scheduled to be produced next Monday we know that we will need 20 tires, 10 seats, and so on. Although conceptually simple, data requirements and calculations can be substantial for complex products. Computations that took hours or even days 30 years ago therefore limited MRP's use until computing power began to catch up.

Historical Development. Prior to the late 1960s, most manufacturing firms used independent demand techniques applied to each part (in fact, many still do!). The most common of these techniques is known as the *reorder point* (ROP), which is a stock level that when reached triggers the placement of an order. Dependent demand systems such as MRP had difficulty in the early days of computing due to their large data requirements, and so these independent systems were for many years the most viable technique for managing inventory. As computing power became cheaper and more accessible, MRP became more and more prevalent. Since the early 1970s, MRP and its descendents have grown to dominate production and inventory management in manufacturing settings.

Lumpy Demand. Aside from the intuitive appeal of MRP stemming from its dependent demand logic, there are several reasons why ROP techniques have gradually fallen out of favor in the manufacturing industry. One of the most important is the issue of demand variability. A fundamental assumption of ROP techniques is that demands are consistent through time. If we project based on demand that we will need 200 widgets over the next 200 production days, the ROP assumption is that we will need one per day. In practice, however, we may use 40 today and then go several weeks without using a single widget. This condition of highly variable demand is known as demand "lumpiness," and is not handled well with an ROP system. It results in excess inventory being carried at times, and at other times results in shortages that stop production.

For complex products with multiple assembly levels, the problem of lumpy demand gets worse. The demands for components that are lower in the assembly structure are subject not only to their own demand "lumps" but also to those of the higher assemblies in which they are used. This amplifies the amount of lumpiness down through the levels of assembly.

Why MRP? MRP is designed to work well in manufacturing environments where production schedules are developed to meet customer demand. MRP calculates the exact requirements over time for the subassemblies, parts, and raw materials needed to support this schedule. The system uses a bill of materials that corresponds to how the product is put together. The current inventory status of items and lead times are considered in the calculations made by the system.

In the following sections, we begin to look at the major inputs to an MRP system, followed by a discussion of the system logic.

55.1.2 System Inputs

There are just three primary inputs to any MRP system: the *master production schedule* (MPS), the *bill of material* (BOM), and inventory status data. The MPS drives the requirements, the BOM defines the product assembly structure, and inventory status data tells the system what will be available to meet requirements in each period. Each plays a critical role in the logic of the system, and data accuracy in each is a prerequisite for effective MRP use. All three are discussed in more detail in the paragraphs that follow.

Master Production Schedule. Since the fundamental assumption of dependent demand systems is that demand for all components is dependent on demand for the end item, it should come as no surprise that the end-item production schedule drives the system. This production schedule for end items is called the master production schedule, or MPS, and is the first of three inputs to the MRP system. The MPS drives the calculations that determine the materials required for production, so it can be considered the heart of the MRP system.

The MPS is a simple concept, albeit one that is critical to the function of MRP. Production requirements for each end item are grouped into time buckets, or time periods. Time buckets can be anywhere from a day to a quarter or longer, but the most widely used time interval is one week. At the time of an MRP run, the bucket that is about to begin is defined as period 1 by convention, followed by period 2, period 3, and so on. Figure 55.1 shows an example MPS for three products (A, B, and C).

The number of periods included in the plan defines the planning horizon (16 weeks in the Fig. 55.1 example). The appropriate length of the planning horizon varies by company, and is driven in part

	PLANNING HORIZON															
	FIRM								TENTATIVE							
Period / Product	1	2	3	4	5	6	7	8	9	10	11	12	13	14	15	16
A	25			40	30		30		50	10	30	50	25	60	10	
B		40	40		30	20	30	60	25	90	80	10	10		30	30
C	90	75	80	80	55	95	60	20	40			50	80	60	75	85

FIGURE 55.1 Master production schedule example.

by external vendor lead times, internal production lead times, and the uncertainty of demand. In cases where vendor lead times are long, the planning horizon must obviously be extended so that orders can be placed well in advance of production. Likewise in the case where production lead times are long. Conversely, for companies operating in a just-in-time environment lead times are typically shorter, so the planning horizon may be kept relatively short as well.

High levels of demand uncertainty make long-range planning ineffective, so again short planning horizons may be desirable in this case. A common alternative is to divide the planning horizon into two parts: firm and tentative. The firm portion covers the cumulative lead time of the entire process, and therefore ends with finished goods that must be acted upon starting in period 1. The tentative portion extends beyond the firm portion, and is used for general planning purposes only.

The MPS, then, is simply comprised of scheduled production requirements for each product, by period, for the entire planning horizon. Two major considerations go into the generation of the MPS that are critical to the proper functioning of the MRP system in which it is used. First, the requirements must be accurate. Although seemingly obvious, inaccurate requirements flow down to the component levels in the MRP system and therefore affect each and every component in the product. Inaccuracies therefore can (and do) cause both inventory shortages and excess on a broad scale, along with their associated costs and production delays. Second, the requirements must be feasible. Even accurate demand forecasts are of limited value if the plant capacity is inadequate to meet their associated levels and timing of production. For this reason, most modern MRP systems are capacity constrained, meaning they contain a *capacity requirements planning* (CRP) module that adjusts the MPS to a feasible level before the final material plan is executed. Capacity planning is discussed in more detail later in the chapter.

Bill of Material. If the MPS is the heart of the MRP system, the BOM is the skeleton that holds it all together. At its most basic level, the BOM is a database containing all parts. More importantly, it defines the sequential material and assembly relationships required to produce an end item. It therefore links each part to its next-higher assembly, which is then linked to its next-higher assembly, and so on all the way up to the end item. This system of linking is called "pegging," and is critical to the basic logic of MRP. Using the simple bicycle example, the spokes, rim, hub, tire tube, and tire are all pegged to the wheel assembly, which is then pegged to the bicycle assembly.

When we talk about the hierarchy of parts and assemblies that make up the finished product, it is common to describe it in terms of parent-child relationships. At the highest level, the bicycle is the parent and the wheel assembly, frame, seat assembly, and the like are its "children." Likewise, at the next level the wheel assembly becomes the parent of the hub assembly, spoke, rim, and tire children. This relationship between each constituent component (child) and its next higher assembly (parent) forms the basic structure of the BOM database.

Consider a realistic case with multiple products, each containing hundreds of parts. Many of the parts may be used on several different products and are therefore referred to as common parts. Further complicating the situation is the case where a particular part is used in different subassemblies of the same product (i.e., it has multiple parents). Clearly, a comprehensive database structure is required to maintain the bookkeeping on all of the parts in the BOM and their relationships. To accomplish this task, several techniques are employed. One such technique, parent-child relationships, has already been discussed. Another is a system of levels to identify where in the production

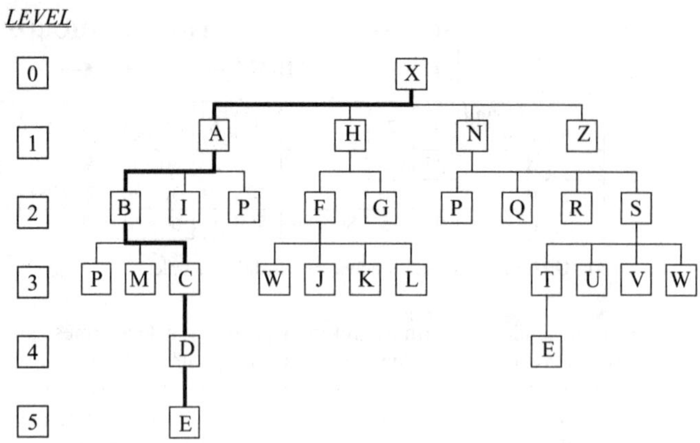

FIGURE 55.2 Example multilevel product structure.

process each part falls. In the United States, the accepted convention is to assign a level code of 0 to the end item, 1 to its direct children, and so on, as illustrated in Fig. 55.2.

Finally, low-level coding is used to ensure accurate and efficient calculations for parts or components that are used in more than one assembly level throughout the product structure. A low-level code identifies the lowest level in the BOM in which a component appears. Part E in Fig. 55.2, for example, is a level-5 child of part D and a level-4 child of part T. Its low-level code is therefore 5, the lowest level in which it is used. Low-level codes are used by the MRP system to postpone processing of component requirements until the lowest level for a part is reached. By doing this, the system avoids having to recalculate requirements several times for the same item.

Inventory Status Data. The third major input to MRP is inventory status data. These data ensure the currency of the requirements calculations, since they tell the system the quantities that are on-hand or expected to arrive for each item in each time bucket. The inventory status data is critical to MRP logic. Net requirements are calculated by subtracting the current inventory available from the total or gross requirements. These calculations are performed over time, thus giving an accurate picture of future requirements.

In addition to the on-hand and expected quantities, inventory status data generally include some indicative data elements for each item. These elements include general descriptive fields like part number, name, and description. They also contain stocking data needed in the calculation of material requirements, such as lead time, safety stock level, and lot size. Some of these data elements will be discussed in more detail later.

55.1.3 System Logic

Now that we have talked about why MRP is used in lieu of ROP techniques and have identified its three major inputs (the MPS, BOM, and inventory status data), we can begin to describe the system logic inside the "black box." As noted earlier, the logic is not conceptually or mathematically difficult. In very basic terms, MRP schedules the required quantities of parts to arrive at the correct times. We therefore begin our discussion of system logic with two concepts that are a fundamental part of the MRP calculation: requirements explosion and timing.

Requirements Explosion. Recall that MRP is a dependent demand system, meaning that requirements for all parts and components are dependent on those of the end item. To calculate these requirements, MRP uses a relatively straightforward technique known as requirements explosion. Again we refer to our simple bicycle example to illustrate the concept (Fig. 55.3). Assume we have

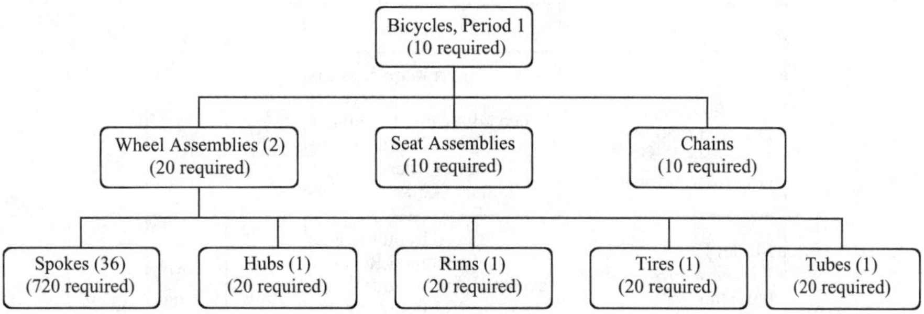

FIGURE 55.3 Graphical representation of requirements explosion.

a requirement to build 10 bicycles in the current week (period 1). By using the quantity per next higher assembly, shown beside each component in parentheses, we can easily calculate the requirements of the next level of parts and components. In this case, we need 20 wheel assemblies (2 per assembly), 10 seat assemblies, 10 chains, and so on. Focusing now on the wheel assemblies, we can likewise calculate the requirements for its children parts. For our 20 wheel assemblies, we need 20 rims, 20 tires and tubes, 20 hub assemblies, and 720 spokes (36 per wheel × 20 wheels). If we are assembling our own hubs, we would also need to calculate the requirements for hub shells, bearings, axles, and caps as well.

Even for this simple (and partial) example, it is obvious that the complexity of the requirements explosion can grow very quickly. For plants that produce multiple complex products, the data requirements can be significant. This is especially so when configuration and technical data change over time, underscoring the need for a current and accurate bill of materials. We discuss this challenge more in the section on the BOM. Although the calculations are theoretically straightforward, in practice the requirements explosion can be quite complex.

Timing. The second concept that is integral to MRP logic is that of timing. Just as the required quantities must be calculated, the timing of those requirements must be determined so that parts and assemblies arrive when needed for the production process. If parts arrive too early, excess inventory results. This excess costs money to store, ties up cash that could be invested elsewhere, and adversely affects production times and quality. Conversely, if parts do not arrive on time, production delays result. So timing is a critical component of an effective MRP system.

The correct timing of production and material orders is based on lead times. An item's lead time is simply the amount of time it typically takes either to receive an order from a vendor (purchasing lead time) or to produce an item in-house (production lead time). Material and production orders must be offset by the lead time so that they arrive when needed. This concept is explained in more detail in subsequent paragraphs, as are the problems that arise when lead times vary.

The Time-Phased Record. Before moving on to the individual elements of the MRP calculation, it is helpful to visualize them as a whole in a structured way that shows their relationships. This structure is the basis of the time-phased record. The MRP time-phased record is a collection of relevant data for each raw material, part, assembly, and finished good. In its most basic form, the record shows the gross requirements, scheduled receipts, projected balance, net requirements, and planned order releases for a single item for each period in a planning horizon. It also contains item-specific data like the lead time, quantity on hand, safety stock level, and order quantity. To facilitate the discussions that follow, a more appropriate example is now introduced to replace our simple bicycle. The example will be used for the remainder of the MRP discussion to illustrate specific topics.

Figure 55.4 illustrates the product structures and time-phased records for electric meters used to measure power consumption in residential buildings. Two types of meters are used, model A and model B, depending on the ranges of voltage and amperage used in the building. Model A consists

Item	Data Fields	Week 4	5	6	7	8	9
Model A	Gross Requirements						1250
LT = 2 weeks	Scheduled Receipts						
On hand = 50	Projected Available Balance	50	50	50	50	50	50
Safety stock = 0	Net Requirements						1200
Order qty. = lot-for-lot	Planned Order Receipts						1200
	Planned Order Releases				1200		
Model B	Gross Requirements						470
LT = 2 weeks	Scheduled Receipts		10				
On hand = 60	Projected Available Balance	60	60	70	70	70	70
Safety stock = 0	Net Requirements						400
Order qty. = lot-for-lot	Planned Order Receipts						400
	Planned Order Releases				400		
Subassembly C	Gross Requirements				400 +1200		
LT = 1 week	Scheduled Receipts						
On hand = 40	Projected Available Balance	35	35	35	35	435	435
Safety stock = 5	Net Requirements				1565		
Order qty. = 2000	Planned Order Receipts				2000		
	Planned Order Releases			2000			
Subassembly D	Gross Requirements			4000	1200		270
LT = 1 week	Scheduled Receipts	100					
On hand = 200	Projected Available Balance	180	280	280	1280	80	80
Safety stock = 20	Net Requirements			3720			190
Order qty. = 5000	Planned Order Receipts			5000			5000
	Planned Order Releases		5000			5000	

FIGURE 55.4 (A) Time-phased record. (B) Example product structure.

of two subassemblies, C and D, while its subassembly C also contains two additional units of subassembly D (a transformer). Model B uses only subassembly C (and by extension, two "D" transformers). In addition to complete meters, the subassemblies are sold separately for repairs or changeovers to a different voltage or power load.

The time-phased record is the vehicle through which the quantity and timing calculations discussed in the previous two sections are made. In fact, it may be helpful to think of it as a group of cells in a spreadsheet. Each field in the record plays a specific role in the calculation, as will be discussed in the following sections.

Gross Requirements. The first data field in the time-phased record (i.e., the top row of the table in Fig. 55.4) is the gross requirements field. The gross requirements are simply the total scheduled requirements for each period, and emanate directly from the MPS. It is here that the exploded, time-phased requirements first manifest themselves. For end items, the gross requirements are simply the scheduled production quantities from the MPS (i.e., no explosion or time-phasing needed). For lower level parts they are the exploded, time-phased requirements resulting from the MRP calculations. Because the timing of the gross requirements for component parts depends on when their parents actually start assembly, the entire time-phased record must be calculated sequentially level by level, beginning with end items and working down through levels 1, 2, and so on.

On-Hand, Scheduled Receipts, and Net Requirements. Once we've determined the gross requirements and their associated times, we must account for inventory status to determine our net requirements. At the start of the planning horizon, there may be some inventory in stock which can be applied to meet the gross requirements. This inventory is known as the on-hand quantity in the MRP time-phased record. There may also be an existing order or orders that have not yet been received. These scheduled receipts must also be taken into consideration when determining the net requirements for each period. The net requirements, then, are the gross requirements in each period, less any on-hand inventory and scheduled receipts for that period.

Lead Time Offset and Planned Order Releases. The net requirements are determined for each part and for each period in the planning horizon. The MRP system must account for the *lead time* to determine when the order or production should be initiated, if necessary, so that each item will be available at the appropriate time. This is known as a lead time offset, and is a fundamental part of the MRP calculation. For example, there is a net requirement for 1200 units of model A in week 9 (Fig. 55.4), and its lead time is 2 weeks. The lead time offset therefore plans an order release to begin assembly in week 7. This also means that 1200 units each of assemblies C and D must be available at that time, so there is a gross requirement for period 7 for these components. The order releases for C and D are then offset by their lead times, one week to period 6. By backing up the net requirements for each part by their associated lead times, we can determine when to release the order so that it is available when needed. This is known as a planned order release.

55.1.4 Advanced Concepts

Until now, the assumption is that demand forecasts are always correct. We can order exactly what we need to arrive exactly when we need it; lead times are precisely the same each time we place an order. Unfortunately this is not always the case in practice, so many techniques have been developed over the years to deal with demand and lead time uncertainty and production economics. The discussion that follows describes the two most important: lot sizes and buffering.

Lot Sizing. So far our discussion has assumed that exact requirements are calculated (by explosion) and time-phased, so that the exact quantity arrives in exactly the right period. In an MRP setting, we use lot sizing to determine the actual size of the order. There are several reasons why it is sometimes wise to batch manufacturing and purchase orders. For example, ordering or setup costs may be high enough that ordering precisely what is needed on a continual basis becomes unnecessarily expensive. Also, it is common to receive quantity discounts, particularly for smaller, cheaper items, so it may make sense to batch orders or manufacturing runs to save money on the unit cost. A great deal of research has focused on developing lot sizing techniques; we present four of the most common here and offer some guidelines for choosing between them.

Economic Order Quantity. One of the simplest lot sizing techniques (aside from lot-for-lot, which involves ordering the exact quantity needed for each period) is known as the *economic order quantity*, or EOQ. It is also the most experienced, having been originally developed in 1915 by F. W. Harris. In 1934, a consultant by the name of Wilson coupled the Harris technique with an ROP methodology, publicized it, and the EOQ became known as "Wilson's EOQ Model." The example in Fig. 55.5 illustrates one of EOQ's major limitations, which is discussed below.

For our sample problem, the *average weekly demand* (D) is 92.1 units, the *ordering cost* (C_O) is \$300 per order, and the *holding cost* (C_H) is \$2 per unit per week. The order quantity is calculated using Wilson's famous formula given below.

$$Q^* = \sqrt{\frac{2C_O\overline{D}}{C_H}} = \sqrt{\frac{2(300)(1105)}{2}} = 166$$

Wilson's formula assumes that demand is constant over time. In our case, however, demand varies from 0 (week 11) all the way up to 270 (week 8). Our first order of 166 units in period 1 therefore lasts until period 6, so we carry excess inventory for 5 weeks. Going into periods 7, 8, and 9, our order quantity will not cover the demand, so we have to increase the order quantities to match demand as shown in Fig. 55.5a. The actual holding costs and ordering costs over the 12-week period total \$4865. A better alternative is to order upcoming requirements on a regular, periodic basis, which is the topic of the following section.

(a)

Week	*1*	*2*	*3*	*4*	*5*	*6*	*7*	*8*	*9*	*10*	*11*	*12*
Requirements	10	10	15	20	70	180	250	270	230	40	0	10
EOQ Order	166					166	223	270	230	166		
Beg. Inventory	166	156	146	131	111	207	250	270	230	166	126	126
End Inventory	156	146	131	111	41	27	0	0	0	126	126	116

Ordering Cost	$1,800
Inventory Carrying Cost	3,065
Total	$4,865

(b)

Week	*1*	*2*	*3*	*4*	*5*	*6*	*7*	*8*	*9*	*10*	*11*	*12*
Requirements	10	10	15	20	70	180	250	270	230	40	0	10
POQ Order	20		35		250		520		270			10
Beg. Inventory	20	10	35	20	250	180	520	270	270	40	0	10
End Inventory	10	0	20	0	180	0	270	0	40	0	0	0

Ordering Cost	$1,800
Inventory Carrying Cost	2,145
Total	$3,945

(c)

Week	*1*	*2*	*3*	*4*	*5*	*6*	*7*	*8*	*9*	*10*	*11*	*12*
Requirements	10	10	15	20	70	180	250	270	230	40	0	10
PPB Order	55				70	180	250	270	270			10
Beg. Inventory	55	45	35	20	70	180	250	270	270	40	0	10
End Inventory	45	35	20	0	0	0	0	0	40	0	0	0

Ordering Cost	$2,100
Inventory Carrying Cost	1,385
Total	$3,485

(d)

Week	*1*	*2*	*3*	*4*	*5*	*6*	*7*	*8*	*9*	*10*	*11*	*12*
Requirements	10	10	15	20	70	180	250	270	230	40	0	10
WW Order	55				70	180	250	270	280			
Beg. Inventory	55	45	35	20	70	180	250	270	280	50	10	10
End Inventory	45	35	20	0	0	0	0	0	50	10	10	0

Ordering Cost	$1,800
Inventory Carrying Cost	1,445
Total	$3,245

FIGURE 55.5 Lot sizing techniques.

Periodic Order Quantity. As shown in the previous section, the EOQ model performs poorly when requirements vary from period to period. An alternative to EOQ's fixed order quantities is to calculate the best *time between orders* (TBO). To do this, we simply calculate the EOQ as before and divide it by the average demand rate. In our example, the time interval is about 2 weeks (166/92.1 = 1.8). Every two weeks, then, we would order the exact number of units needed until the next order. The total number of orders remains at 6 as in the EOQ case, but since we are now ordering our exact requirements, the inventory carrying cost is reduced by about 30 percent. The resulting total cost is now only $3945, a 19 percent improvement over the EOQ solution. Still, we can see that the *periodic order quantity* (POQ) solution can easily be improved. For example, pooling the orders for the first 4 weeks into a single order of 55 reduces the cost by an additional $160. The following section presents a technique that considers more of the information in the schedule to arrive at a better solution.

Part Period Balancing. *Part period balancing* (PPB) equates the total cost of placing orders with the cost of carrying inventory, and in doing so arrives at a low-cost (although not always the *lowest* cost) solution. Starting with week 1, we have many alternatives. We can order for week 1's requirements only, for week 1 and week 2, for weeks 1 through 3, and so on. In any of the alternatives, we start with a single order costing \$300 (our ordering cost per order). PPB requires the calculation of carrying costs for all of the alternatives, and selection of the one that is closest to the order cost, therefore "balancing" the order and carrying costs. The calculations are as follows:

1. $(\$2) \times [(1/2) \times 10] = \$10$
2. $(\$2) \times [(1/2) \times 10] + [(3/2) \times 10] = \$40$
3. $(\$2) \times [(1/2) \times 10] + [(3/2) \times 10] + [(5/2) \times 15] = \$115$
4. $(\$2) \times [(1/2) \times 10] + [(3/2) \times 10] + [(5/2) \times 15] + [(7/2) \times 20] = \$255$
5. $(\$2) \times [(1/2) \times 10] + [(3/2) \times 10] + [(5/2) \times 15] + [(7/2) \times 20] + [(9/2) \times 70] = \$885$

In this case, alternative 4 (ordering 55 units to cover the first 4 weeks) is closest to our ordering cost of \$300, so that is the one we choose. We then move on to week 5, our next decision point. Again, we can order for week 5 only, weeks 5 and 6, and so on. The calculations this time are as follows:

1. $(\$2) \times [(1/2) \times 70] = \$70$
2. $(\$2) \times [(1/2) \times 70] + [(3/2) \times 180] = \$610$

Alternative 1 is closest to \$300 without exceeding it, so we order 70 units to cover week 5 on this round. Carrying through with this procedure, we get the final plan in Fig. 55.5c. The total inventory carrying cost is reduced by \$760. Even though a seventh order is required, the total cost drops by \$460, a reduction of about 13 percent from the POQ solution. Clearly, the PPB procedure is an improvement over the EOQ and POQ approaches, but it still does not guarantee the lowest cost, since it does not explore every potential ordering scheme. To do that, we turn to the optimizing Wagner-Whitin (WW) algorithm.

Wagner-Whitin Algorithm. Whereas the EOQ, POQ, and PPB procedures are approximating heuristics that are simple to implement but do not guaranty an optimal solution, the WW algorithm yields the optimal (lowest cost) solution. The WW algorithm is complex, however, and so its details are omitted here for brevity. In fact its complexity has limited its use mainly to academic research, where it is used as an optimal baseline for comparing other techniques. We focus instead on the results when WW is applied to our example. Fig. 55.5d shows the WW-generated ordering plan, which reduces the total cost by \$240 over the PPB solution. Notice that the only difference between the two is that the WW algorithm orders the 10 units for week 12 in week 9. This results in an additional carrying cost of \$60 (10 units × 3 weeks/unit × \$2/week), but avoids a \$300 ordering cost by eliminating the seventh order. The net reduction is therefore \$240. By exploring all possible solutions, WW found a better solution.

Buffering Against Uncertainty. While lot sizing techniques help to deal with varying demands, they unfortunately do very little to help with uncertainty. In fact, research has repeatedly confirmed one thing with regard to uncertainty and lot sizing: as uncertainty rises, the differences in performance of different lot sizing techniques decrease. In other words, in a highly uncertain environment it makes little difference which technique is used. As a result, reducing uncertainty should be the primary focus of any manufacturer before lot sizing techniques are tested. Uncertainty is a fact of life, however, and it is always present to some degree. Buffers are therefore used to mitigate its negative effects.

Sources and Types of Uncertainty. Before discussing the two types of buffers used to counteract uncertainty, it is important to understand its sources and types. There are two types of uncertainty affecting both supply and demand: quantity uncertainty and timing uncertainty. On the demand end, quantity uncertainty simply describes the fact that actual requirements are higher or lower than forecasted, while timing uncertainty refers to the case where requirements shift back and forth between periods. Likewise on the supply side, orders can arrive early or late (timing uncertainty) or can contain more or less than we planned for (quantity uncertainty).

Order Quantity = 50 Lead Time = 2 weeks		Period				
No Buffering		*1*	*2*	*3*	*4*	*5*
Gross requirements		20	40	20	0	30
Scheduled receipts			50			50
Projected available balance	40	20	30	10	10	30
Planned order release				50		

Safety stock = 20 units						
Gross requirements		20	40	20	0	30
Scheduled receipts			50	50		
Projected available balance	40	20	30	60	60	30
Planned order release		50				

Safety lead time = 1 week						
Gross requirements		20	40	20	0	30
Scheduled receipts			50		50	
Projected available balance	40	20	30	10	60	30
Planned order release			50			

FIGURE 55.6 Safety stock and safety lead time.

Safety Stock and Safety Lead Time. As mentioned in the previous section, there are two ways to buffer against uncertainty. The first is to carry additional stock in case requirements are higher or earlier than expected. This additional stock is known as buffer or safety stock. The level of safety stock is calculated using the standard deviation of demand. To illustrate, suppose we have a part for which weekly demand averages 50 units. The demand varies according to a normal distribution with a standard deviation of five units. Holding two standard deviations of safety stock will give us 95 percent confidence that we will have enough extra inventory to meet demand. This is called a 95 percent service level. So in this example we order what we need each week, as described earlier in the paper, and in addition carry 10 units of safety stock to buffer against higher or earlier demands than anticipated.

The second way to buffer against uncertainty is to release orders earlier than planned. This method is called safety lead time. Although it also artificially raises the amount of inventory held, safety lead time does so on a set schedule with varying quantities equal to requirements. In contrast, safety stock raises the inventory by a set quantity independent of the time-phased requirements. In practice, safety stock and safety lead time can affect ordering quite differently, as shown in Fig. 55.6.

In the first case, with no buffering at all, an order is released for the order quantity of 50 units in week 3 to arrive in week 5. This leaves no room for error, because if requirements in week 4 rise above 10 units the inventory will be short. The second case illustrates the same problem with a safety stock of 20 units. Here, an order is released in week 1 when the projected balance would have dropped below the safety stock level, to arrive in week 3. Finally, the last case shows a safety lead time of 1 week. This case is identical to the first, except that the order release in week 3 is now released in week 2, to arrive in week 4. In both cases, sufficient stock is on hand to cover a change in requirements in weeks 3 or 4.

Performance Characteristics. Given the preceding discussion, the question concerning which is better—safety stock or safety lead time—can be addressed. This question has been addressed by researchers over the past couple of decades. The following list provides general guidelines that have been tested and validated under a variety of conditions, both analytically and through the use of simulation experiments.

- For timing uncertainty (demand or supply), use safety lead time.
- For quantity uncertainty (demand or supply), use safety stock.
- As either type of uncertainty increases, the choice of the correct buffering mechanism becomes more critical.

Although we have discussed two alternative ways of mitigating uncertainty, the primary focus of any manufacturer should be to reduce uncertainty at both sources and of both types. It should be clear that both of the buffering mechanisms increase inventory levels and, therefore, inventory cost. High levels of inventory can also mask production and quality problems that would otherwise be obvious to managers, and can create additional problems that result from an overdependence on inventory. By addressing uncertainty at its sources, the dependence on buffering mechanisms is reduced and many problems are avoided.

55.1.5 Other MRP Concepts and Terms

To this point we have given a very general overview of MRP, its components, its logic, and a few select advanced topics. For completeness, we now add some concepts and terms that are commonly used in MRP environments, but that did not fit into the context of our earlier discussions. The list is by no means all-inclusive, but provides a quick reference.

Rough-Cut Capacity Planning. In order to ensure that the final material plan is feasible, capacity must be considered. This is typically done at two points in the process. The first look at capacity follows the generation of an initial MPS, and is called *rough-cut capacity planning* (RCCP). Rough-cut capacity planning is simply a validation of the MPS with respect to available resources such as labor and machines. If the MPS is determined to be feasible, it is validated and sent to the MRP system for requirements explosion and time-phasing. If, however, the MPS is infeasible, either an adjustment must be made to the MPS itself or the capacity must be adjusted, e.g., by scheduling overtime.

RCCP draws on bills of resources to estimate feasibility. These bills are similar to the BOM conceptually, but instead of simply listing required materials they list all required resources for each process. To complete a valid RCCP, time buckets should be the same duration as those used in the MPS. In this way, the detailed resource requirements for each bucket can be directly matched and compared.

The second capacity check is performed by CRP following the generation of a material plan by the MRP system. CRP is discussed in more detail later in the chapter.

Replanning. An important topic not yet discussed is that of replanning. In simple terms, replanning refers to an MRP run in which the material plan is generated. Two issues must be addressed when deciding on a replanning strategy: type and frequency. Two basic types of replanning are used in MRP systems. The first is called regenerative, and uses a complete recalculation (explosion) of all requirements. The existing plan is erased from the system and replaced by a completely new one, which includes all current information. Since the entire plan is regenerated, the processing time can be significant. As a result, many companies complete the regenerated plan off-line before overwriting the existing plan. Others process the plan on-line, but during off-peak hours or weekends. Both methods help to avoid degradation of system performance.

The second approach to replanning, called net-change, recalculates only those items that are affected by changes that have occurred since the last plan was generated. Net-change replanning requires considerably less computer time, making it possible to replan more frequently without disrupting the system. But data records must be meticulously maintained through strict review and transaction processing procedures, since only a fraction of records are reviewed in each cycle. Without such oversight, data errors can linger undetected in the system for weeks or even months.

In addition to determining the type of replanning that is most appropriate, the replanning frequency must be considered. The appropriate frequency varies by firm, and depends in part upon the stability of its product data, requirements, and supply chain. In general, the more things change, the more frequently planning should be done. Replanning too frequently can have serious side effects, however, as discussed in the following section.

Nervousness. Nervousness is a term that describes the degree of instability of orders planned in the system. More specifically, it describes how changes in inputs (i.e., the MPS, inventory data, or BOM) manifest themselves in the outputs (i.e., the material plan). There are a number of sources of system nervousness, such as changes to the quantities or timing of orders and adjustments to safety stock levels. In fact, virtually any changes in the data or MPS can impact the existing plan, sometimes significantly. This nervousness can also be magnified by such things as the choice of lot sizes, frequent replanning, and a multilevel product structure.

Several techniques are generally used to reduce system nervousness in practice. Here we describe two of the most effective. The first, and most obvious, is to address the sources of change. As mentioned in the previous section, there are many small system and data changes that can lead to significant changes in the material plan. To the maximum extent possible, quantities and lead times should be stabilized. Quantity changes can be reduced, for example, by including spare parts requirements in the MRP gross requirements. If not included in the plan, these requirements will show up as unplanned demands and require changes to the planned orders. Another method used to stabilize demand quantities is to freeze the MPS for the current period(s). On the lead time side, close cooperation with suppliers can help to reduce variability, thereby reducing the need to replan. Finally, changes in parameters such as safety stock should be minimized. By addressing these common sources of change, the overall system nervousness can be dampened significantly.

A second technique used to reduce nervousness is the selection of appropriate lot-sizing methods. Typically, it makes sense to use a fixed order quantity approach at the end item level, a fixed order quantity or lot-for-lot approach at the intermediate levels, and period order quantities at the lower levels. This approach tends to limit end-item quantity changes and, by extension, the "ripples" they cause down through the BOM.

Bucketless MRP System. As computing power has increased over the past few decades, the necessity to mass requirements into long time buckets has decreased. In place of weekly or monthly requirements, we gradually gained the capability to plan in terms of daily or hourly requirements. Taken to the extreme, buckets have been eliminated in bucketless MRP systems. These systems have been extended further into real-time systems, which are run daily and updated as needed in real time.

Phantom Assemblies. From a planning perspective, it is important to reduce the number of assemblies to the minimum possible to simplify the BOM. Elimination of some assemblies may be technically feasible, but it may still be necessary to retain assembly stock numbers to allow for occasional stocking. In this case, phantom assemblies are used in the BOM. These assemblies are items that are physically built, but rarely stocked. To illustrate, consider a gearbox assembly in a transmission. The gearbox is assembled from its parts on a feeder line that moves directly into the final assembly line. The gearbox does not physically move into stock and back to the main assembly line, but it does exist as an entity. In fact, occasional overproduction or a market for service parts may require that we stock the gearbox as a unique component. Still, for the MRP run we do not need to complicate the explosion by adding an additional layer. The phantom assembly allows the MRP system to ignore the gearbox in terms of planning, while still retaining it in the BOM. Phantom assemblies are also known by the terms phantom bill of material, transient bill of material, and blowthrough (since the MRP system blows through the phantom assembly when planning).

Yield and Scrap Factors. In many production environments, scrap is a fact of life. MRP systems therefore use yield and scrap factors to account for scrap loss in the planning process. For example, if 100 computer chips are required to meet customer demand in a given week, and the typical yield is 70 percent, then the MRP system adjusts the gross requirement to 143 ($^{100}/_{0.70}$). In doing so, the final output after scrap is enough to meet requirements.

Replacement Factors. Similar to the yield factor, the replacement factor is used to plan component requirements in repair or remanufacturing operations. The replacement factor is calculated by dividing the total use of a component by the total number of end items repaired or remanufactured. It is then applied in the MRP system to plan component requirements. For example, if 100 transmissions are to be rebuilt in a given week and the replacement factor for torque converters is 0.8, then a requirement for 80 torque converters is used in the MRP explosion.

55.1.6 Related Systems

As computing power has continued to grow at an exponential rate over the years, MRP systems have capitalized by evolving more and more capability. This final section chronicles that evolution, ultimately leading into the next major topic: enterprise resource planning. Also discussed are three other systems that were developed to meet specific needs: CRP, *distribution requirements planning* (DRP), and *distribution resource planning* (DRP II). These systems actually became common modules in many advanced MRP systems, further expanding their capabilities and moving them closer to total enterprise planning. Figure. 55.7 illustrates the functionality of MRP, closed-loop MRP, MRP II, DRP, and DRP II.

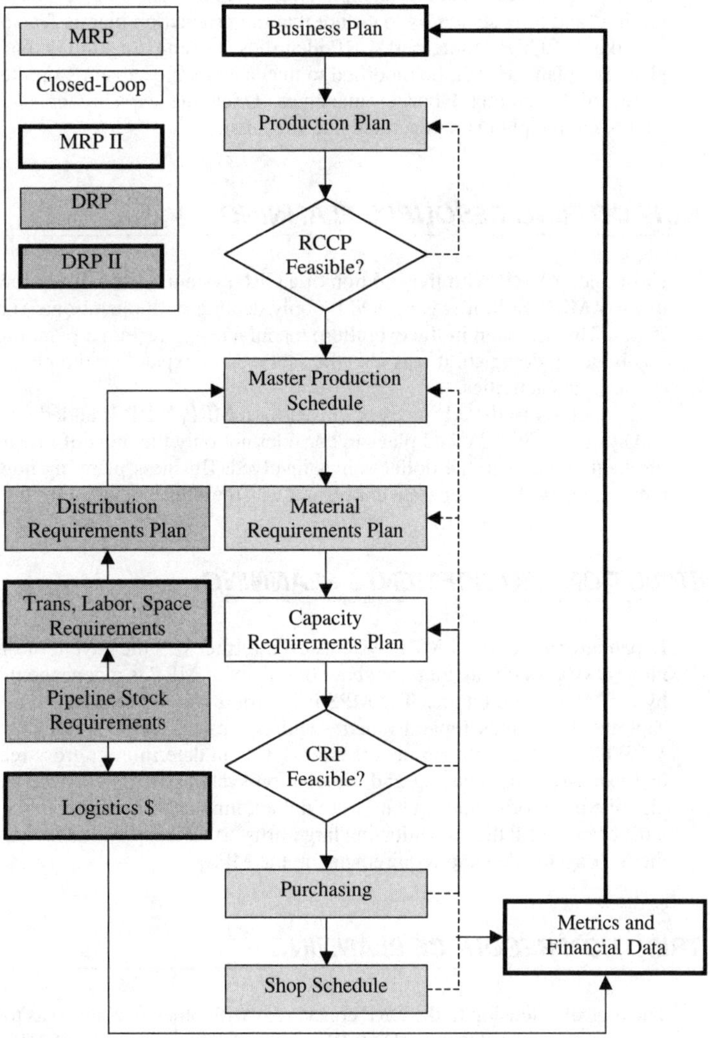

FIGURE 55.7 MRP and related systems.

Closed-Loop MRP. In the early days of MRP, the material plan was generated without regard to resource feasibility. The plan often created as many problems as it solved, leading to the establishment of dedicated expeditors to solve problems. Subsequently, MRP systems gradually evolved to include a capacity check feedback loop. It was at this time that CRP first came into use.

55.2 CAPACITY REQUIREMENTS PLANNING

While RCCP provides a capacity check of the MPS, it is, as its name implies, very rough. Often an MPS that is deemed feasible by RCCP becomes infeasible at the more detailed shop level. MRP's answer to this problem is capacity requirements planning (CRP). CRP is a sub-system that provides a final check of the material plan prior to its execution. This check includes the detailed labor and material resources needed to complete production, right down to the work center level. It uses part routings and time standards to ensure that the production plan is feasible in every time period.

Like its RCCP counterpart, CRP identifies capacity constraints that will prevent completion of a plan. The plan can then be modified to match available capacity, or alternatively the capacity can be increased by adding labor or machines. Once the capacity constraints identified by CRP are addressed, the plan is regenerated and executed.

55.3 MANUFACTURING RESOURCE PLANNING

Closed-loop MRP, with the addition of a CRP system and feedback loop, made great improvements to early MRP. Still, it was limited by only dealing with operational (i.e., production and inventory) issues. The next step in the evolution, manufacturing resource planning (MRP II), began to address this issue. In doing so, it was also the first step in expanding the planning function to include more of the firm's activities.

In addition to the advances of closed-loop MRP, MRP II added connectivity to the firm's financial systems. This allowed plans to consider not only the units of raw materials and parts needed for production, but also the dollar values involved. Business planning now became internal to the system, as opposed to a separate entity, and was therefore included in the feedback loop.

55.4 DISTRIBUTION REQUIREMENTS PLANNING

Expanding the scope of MRP systems even further was the advent of distribution requirements planning (DRP). As discussed throughout this chapter, MRP is a dependent demand system that is driven by the MPS for end items. The MPS is in turn driven by forecasted requirements. DRP, by contrast, captures those stock replenishment requirements that were previously external to the MPS. While MRP had always included on-site stock levels in determining gross requirements, many large firms had extensive warehousing and distribution systems that went well beyond the main plant. These pipeline inventory requirements were not automatically included in the MPS. As a result, DRP systems began to fill the gap, allowing large firms to collect pipeline inventory requirements and include them alongside demand requirements in the MPS.

55.5 DISTRIBUTION RESOURCE PLANNING

The logical extension to the DRP concept, as with other systems, was to broaden its scope. Thus, distribution resource planning (DRP II) was born. The enhancements offered by DRP II parallel those of MRP II. In addition to the consideration of pipeline stock requirements, DRP II also plans such logistics requirements as warehouse space, labor, transportation, and cash.

55.6 ENTERPRISE RESOURCE PLANNING

From the preceding discussion, it may appear that the evolution of MRP-based systems has been a continual patchwork project. To a large extent it has, paced by the evolution of computing power. Despite the hurdles, the ongoing efforts to improve and expand MRP over the course of over three decades yielded amazing results in productivity. Similar evolutionary efforts were happening during the same period in accounting, human resources, and sales and marketing systems. Ultimately, the result in most firms was a set of very capable systems that did not communicate. The promise of an integrated approach gave rise to a new concept: *enterprise resource planning*, or ERP. This section looks at ERP in detail.

55.6.1 Overview

The term "enterprise resource planning" has been used in many ways over the past decade, so we begin with a definition to lay the groundwork. APICS, The Educational Society for Resource Management, defines enterprise resource planning as follows:

> A method for the effective planning and control of all resources needed to take, make, ship, and account for customer orders in a manufacturing, distribution, or service company.

From the definition above, ERP may appear to be very similar to the MRP II systems discussed previously. There are a few major philosophical differences, however, that are identified in the definition. First, ERP encompasses *all* resources, while MRP II is limited to those used for manufacturing activities. Second, ERP includes the processes of taking, making, shipping, and accounting for customer orders. MRP II, by contrast, focuses on making. Finally, ERP is used by manufacturing, distribution, and service sector firms, while MRP II is a manufacturing system. Clearly, the scope of ERP extends well beyond that of its MRP-based ancestors. There are also important system differences that are illustrated in the APICS definition of ERP systems:

> An accounting-oriented information system for identifying and planning the enterprise-wide resources needed to take, make, ship, and account for customer orders. An ERP system differs from the typical MRP II system in technical requirements such as graphical user interface, relational database, use of fourth-generation language, and computer-assisted software engineering tools in development, client/server architecture, and open-system portability.

Certainly, many MRP II systems had *graphical user interfaces* (GUIs) and used relational databases. Still, the definition implicitly addresses the two major system differences between ERP and its predecessors: a *common database* and *integration*. Figure. 55.8 illustrates the integrated nature of

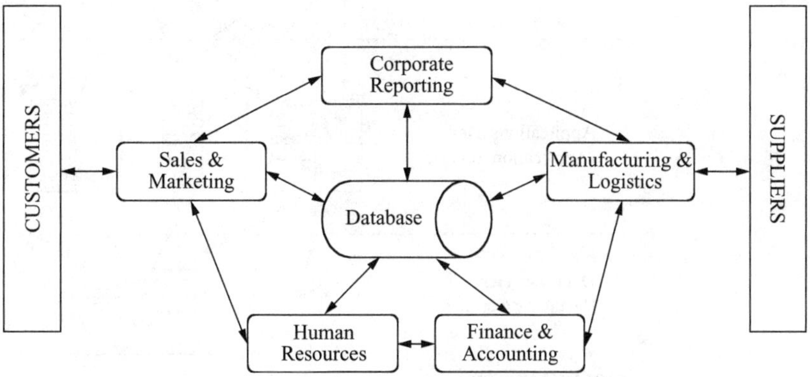

FIGURE 55.8 ERP integration with a common database.

the ERP functional application modules with a common database, as well as the interfaces with customers and suppliers.

55.6.2 Common Database

Although it is a seemingly straightforward concept, using a common relational database for an entire enterprise was nevertheless very rare prior to ERP. Typical legacy systems evolved within the confines of individual functions like production, inventory management, accounting, finance, human resources, and so on. Until the early 1990s, these systems usually used dedicated flat files to store data, a relic of punch cards that made storage and retrieval difficult at best. It was typical to have tens or even hundreds of these systems using common data fields, but since the systems were independent the data rarely matched. For example, an order-entry system used by customer service employees would contain a set of forecasted demands and firm orders. The system used to generate the MPS would have its own numbers, which may or may not have agreed with those in the customer service system. The accounting system would likewise contain its own set of numbers.

Compounding the problem of multiple databases was the differences in their configuration. An exception code, for example, may have been in column 68 of the manufacturing database and column 25 of the accounting database. Reconciling data therefore became very labor and system intensive, and required complicated schemas to keep the bookkeeping straight. Most firms developed and/or purchased more advanced systems over time, only to find that they needed complex programs to transfer data between them—systems that were difficult (and expensive) to maintain. The result was a complicated web of systems. ERP solves this data problem by using a common, relational database that is used by all applications in the enterprise. The database is updated real-time, so that it is always current and is accessible to all functional components that have a need for it.

To operationalize the concept of a shared database, ERP systems use a three-tiered architecture (Fig. 55.9). At the core of the enterprise is the database layer. The database server (or server*s*, for large firms) houses all of the enterprise's data, which is shared by all of the ERP system's programs in the application layer. The application servers run the actual programs, called modules, which support the work of the firm. Modules will be discussed in more detail later. These application modules are accessible by users in the presentation layer. The presentation layer is comprised of client workstations throughout the organization. At this layer, each user sees the same graphic user interface, although access to some data may be restricted to certain users.

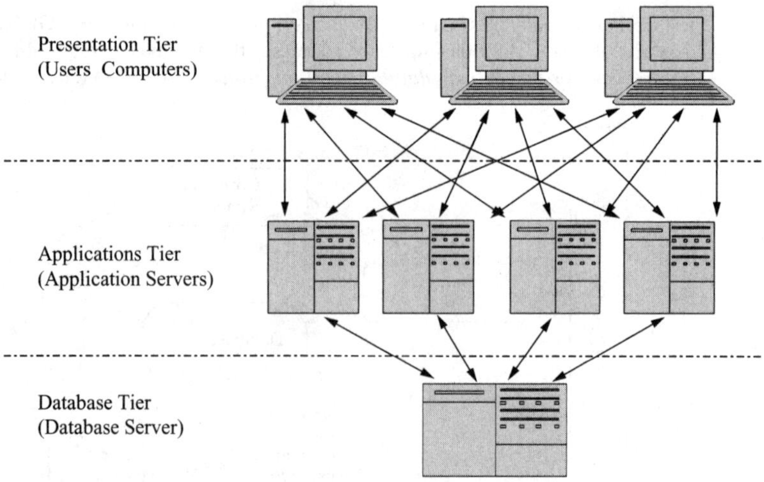

FIGURE 55.9 ERP architecture.

The concept of a common database is critical to the function of any ERP system. It means that every employee of the firm, regardless of their job or location, has access to exactly the same data. The forecasted and firm requirements seen by the sales force are identical to those seen by the shop floor. The accounting department likewise sees the same numbers. Instead of needing a large number of custom programs to retrieve and transfer data from and to multiple databases, all programs draw from the same source. In fact, without a common database the ERP concept would not be possible. Common data only solves part of the problem, however, since a system must be in place to ensure that the common data are used by all relevant functions for all relevant tasks. In short, the various functions must be integrated to capitalize on the potential benefit of a common database.

55.6.3 Integration

Just as the old legacy systems relied on specialized programs to transfer data between systems, they also relied on such programs to integrate all of the different functions of the firm. In most cases, these programs were augmented by a great deal of manual data entry. For example, customer service takes a phone order and enters the data into its system. An order form is printed and copies are sent via office mail to production scheduling, materials management, and accounting. The production is scheduled via the MRP system (more input), which generates orders for parts. Upon completion of production, more paperwork travels to the accounting department where billing invoices can be generated and mailed to the customer (again, after some data entry).

ERP systems address these inefficiencies by controlling all of the individual tasks that are required for a given process, instead of focusing on the functions of the firms. This is critical, since most processes cut across multiple functional areas (Fig. 55.10). Using our example from the previous paragraph, a simple customer order requires processes to be completed by nearly every functional area of the firm. When the order is entered via the customer service screen by an employee, the information is immediately transferred to the database. More importantly, it is automatically transmitted to the production, accounting, distribution, and inventory application systems in the second layer. Once there, the production schedule is updated, material orders are generated, accounting systems are alerted, and transportation is scheduled for delivery. In other words, ERP looks at the entire process from end to end instead of dealing with all of its individual tasks separately.

Through the use of a three-tiered system architecture with a common database, and by integrating all of the functional areas of the firm around that database, ERP has solved most of the system problems that previously plagued business. So how does ERP serve the functional areas of a business? In the next section we answer this question by taking a detailed look at the functional components, or modules, that comprise most modern ERP systems.

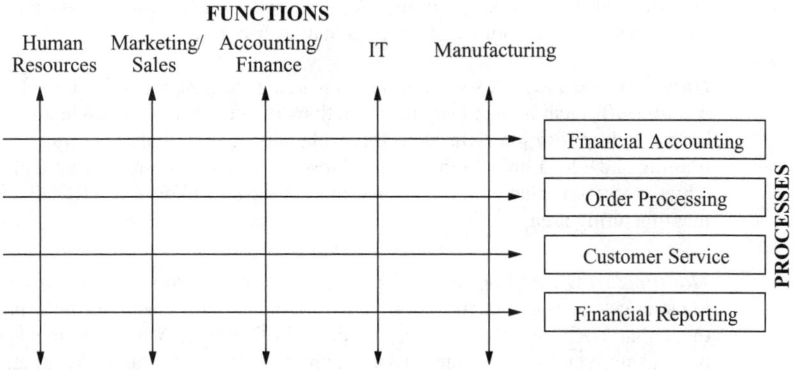

FIGURE 55.10 Functional integration.

55.6.4 Functional Components

Generally speaking, an ERP system can be any collection of programs that are designed to work together (i.e., they are integrated) and share a common database. More commonly, they are off-the-shelf software packages that have been developed by a single vendor. In either case, four requirements must be met in order for the system to be considered a true ERP system. First, the system must be multifunctional. This means that it performs most or all of the functions of the firm under one umbrella system, as opposed to limiting itself to a single functional area. Second, it must be integrated, as discussed in the previous paragraph. Third, an ERP system must be modular. Modularity means that the system is programmed in blocks that can be used as a whole or as a subset of individual modules. This building block approach allows ERP systems to be tailored to the needs of individual firms. It also allows firms to implement ERP on an incremental basis, gradually adding more modules as needed until the full suite is implemented. Finally, an ERP system must facilitate planning and control, to include such activities as forecasting, planning, and production and inventory management.

ERP systems perform two distinct types of roles. The first, transaction processing, involves posting and managing all of the routine activities that make the business run. Order processing, receipt of material, invoicing, and payment are typical transactions that are processed in the manner defined by the firm. The second, decision support, relates to the management decisions that make the firm operate more efficiently. Since the decision support role is less reliant on up-to-the-minute data, many large firms maintain a duplicate repository of historical data called a data warehouse on a separate server. This allows managers and analysts within the company to access data and make decisions without tying up the system. Some companies, like Wal-Mart for example, even allow its suppliers access to some of the data so that they can manage their own inventory levels in accordance with actual usage.

Moving now from the conceptual to the practical, ERP systems are composed of a number of application programs called modules. Although the actual modules vary by vendor, they typically include at least the following areas: financial accounting, human resources, manufacturing and logistics, and sales and distribution.

Financial Accounting. At the heart of almost any task in a firm, there is an accounting transaction. An order for parts or raw material generates accounts payable transactions. A completed customer order generates an accounts receivable transaction. For inventory pulled from stock, there is an inventory adjustment on the balance sheet, and the list goes on. It should come as no surprise, then, that financial accounting systems plan an important role in modern ERP. Since financial accounting is almost completely reliant on current and accurate data, it is also one area with a large potential benefit.

The benefits are even greater for large firms with multiple holdings. ERP systems span not only functional areas, but also company and geographic lines. This makes them ideally suited for capturing financial data and accounting transactions in real time from all business units in an enterprise. Decisions at the firm level can therefore be made much more quickly and with a higher degree of confidence in the reliability of the data than ever before.

Human Resources. A second functional area benefiting from the advent of ERP systems is human resources. Typical human resources functions include payroll and benefits administration, employee hiring and application data, workforce planning and scheduling, travel expense accounting, and training and education administration. Most of these functions overlap with the accounting systems, while workforce planning and scheduling comes into play in the RCCP and CRP processes in the manufacturing area.

Manufacturing and Logistics. The largest number of modules falls under the manufacturing and logistics category. In fact, most of the MRP-based systems discussed in the previous section, as well as those that MRP interacts with, are modules in this area. A material management module manages purchasing, vendor selection, ongoing contracts, invoicing, asset tracking, receipt, storage assignment, and inventory control—among other tasks. The production planning and control module performs master production scheduling, RCCP, MRP (requirements explosion, e.g.), CRP, and product costing functions. It may also design machine and work-station routings for job shop operations. The plant maintenance module handles the planning, scheduling, and performance of preventative maintenance and unscheduled repairs. This information then folds into the capacity calculations used by RCCP and CRP modules, and alerts the sales force to potential delays in filling orders. The quality management module plans and implements procedures for inspection and quality assurance. For example, it may indicate to production workers which units are to be pulled for inspection and log the results when the inspection is complete. This information is tracked over time to alert management to quality problems. A project management module allows users to plan large, complex jobs using industry-standard techniques like *critical path method* (CPM), *program evaluation and review technique* (PERT), and Gantt charts.

Sales and Distribution. The sales and distribution area is the main interface with the customer. Typically its modules draw upon a wide range of data from all of the other functional areas, and allow the sales force to provide real-time information to customers. The customer management module manages the customer database and aids the marketing department in designing offerings, setting prices, and targeting advertising efforts. Sales order modules are the main interface between the sales force and the rest of the firm, passing order information to all activities instantly. The distribution module handles shipping and transportation, and for international firms may also include import and export functions as well. The sales and distribution modules may also include billing, invoicing, and rebate functions that interact closely with financial accounting modules.

Bolt-On Modules. In addition to its standard packages, many ERP vendors also offer customized modules that either cater to a specialized industry or perform specialized functions. These modules, like those in the standard suite, are designed to interact seamlessly with its counterparts in the ERP system. Third-party vendors have also developed an extensive market for these bolt-on modules, offering a wide range of planning and decision support functions to the user that were previously unavailable. Many firms also develop, or contract the development of, tailored applications. Needless to say, the compatibility of bolt-on modules is always a concern, since the ERP vendor does not typically support products other than its own. Still, these external products can provide specific capabilities to the firm that may otherwise be unavailable in the core ERP package.

55.7 ENTERPRISE PERFORMANCE MEASURES

To this point, our discussion has focused on the operational advantages of ERP. Another benefit of common, integrated data is that it allows the firm to measure performance in new ways that reflect the performance of the entire enterprise. This is particularly so for large firms with multiple business units, which can use the database to perform strategic analysis and make long-range decisions. This capability avoids the pitfalls of traditional lagging measures like income statements and cash flows, which look at past performance rather than toward the future.

The ERP framework also allows a more holistic approach to measuring performance than did the nonintegrated systems of the past. The latter approach typically led to local optimization, rather than every functional unit acting together in the best interest of the firm. For example, consider a typical flow of material and product through a manufacturing firm. At the front end, sales and distribution is concerned with minimizing the cost of distributing product. It is encouraged by way of the firm's metrics to use full truck loads to minimize transportation costs, and will usually avoid higher-cost options like air and *less-than-truckload* (LTL) shipments. Manufacturing is usually measured by, among other things, utilization rates. It is given incentive to keep all machines and labor busy for the maximum amount of time possible. At the back end of the flow, the purchasing department is likewise given incentive to minimize its costs. This leads to large lot sizes when ordering, and important concerns such as quality, delivery, and payment terms take a back seat in the decision process.

The general example from the previous paragraph illustrates the ills of using a disjointed approach to measuring performance. The purchasing group has ordered in large lots to reduce unit costs, building up excess raw materials and parts in the warehouse; the manufacturing group has set production levels to maximize utilization and has likewise built up excess *work-in-process* (WIP)

Measure	Description	Best-in-Class	Average or Median
Delivery Performance	Percentage of orders shipped according to schedule	93%	69%
Fill rate by line item	Percentage of actual line items filled (considering orders with multiple line items)	97%	88%
Perfect Order Fulfillment	Percentage of complete orders filled and shipped on time	92.4%	65.7%
Order Fulfillment Lead Time	Time between initial order entry and customer receipt	135 days	225 days
Warranty Cost as a Percent of Revenue	Actual warranty cost divided by revenue	1.2%	2.4%
Inventory Days of Supply	Number of days worth of inventory carried in stock	55 days	84 days
Cash-to-Cash Cycle Time	Time it takes to turn cash used to purchase materials into cash from a customer	35.6 days	99.4 days
Asset Turns	Number of times the same assets can be used to generate revenue and profit	4.7 turns	1.7 turns

FIGURE 55.11 Integrated ERP metrics.

and finished goods inventories; and the sales and distribution group has built up excess inventory in its distribution warehouses, and has probably been late on some deliveries as well. With ERP, a system-wide set of performance metrics is possible that transcends the individual groups and optimizes the enterprise's performance.

The Supply Chain Council (www.supply-chain.org) has proposed a number of supply chain metrics that encompass the actions of all parts of the enterprise. These sets of metrics are focused on the different categories of companies to which they best apply. Figure 55.11 shows an example set of metrics for large industrial product manufacturers, and includes industry average and "best-in-class" benchmarks.

Cash-to-Cash Cycle Time. Cash-to-cash cycle time provides an excellent illustration of a holistic performance measure that spans the enterprise, so we look at it in more detail to illustrate the power of integrated metrics. At a top level, cash-to-cash cycle time measures the amount of time it takes to turn expenditures for raw materials and parts into cash from the customer. More specifically, it is calculated as follows:

$$\begin{aligned}\text{Cash-to-cash cycle time} &= (\text{average days of inventory})\\ &\quad + (\text{average days of accounts receivable})\\ &\quad - (\text{days of accounts payable})\end{aligned}$$

Average Days of Inventory. We now need to calculate the three elements of the cash-to-cash cycle time. We begin by calculating the average daily cost of sales C_d:

$$C_d = S_d CS$$

where, C_d = Average daily cost of sales
CS = Cost of sales (percent)

Now the average days of inventory I_d is calculated by dividing the current value of inventory (I) by the average daily cost of sales (C_d) from above:

$$I_d = \frac{I}{C_d}$$

where, I_d = Average days of inventory
I = Current value of inventory

Average Days of Accounts Receivable. The average days of accounts receivable measures the days of sales outstanding. It shows the average time it takes for customers to pay once an order is shipped. First we determine average daily sales S_d as follows:

$$S_d = \frac{S}{d}$$

where, S_d = Average daily sales
S = Sales over d days

The average days of accounts receivable AR_d is then calculated by dividing the current accounts receivable AR by the average daily sales S_d:

$$AR_d = \frac{AR}{S_d}$$

where, AR_d = Average days of accounts receivable
AR = Accounts receivable

Days of Accounts Payable. The final component of cash-to-cash cycle time is the average days of accounts payable. It measures the level of accounts payable relative to the cost of sales:

$$AP_d = \frac{AP}{C_d}$$

where, AP_d = Average days of accounts payable
AP = Accounts payable

Combining the components above, we can now calculate the cash-to-cash cycle time:

$$\text{Cash-to-cash cycle time} = AR_d + I_d - AP_d$$

The example below demonstrates the calculations.

Data:

Sales over last 30 days = \$1,020,000

Accounts receivable at end of month = \$200,000

Inventory value at end of month = \$400,000

Cost of sales = 60 percent of total sales

Accounts payable at end of month = \$160,000

$$S_d = \frac{S}{d} = \frac{1{,}020{,}000}{30} = 34{,}000$$

$$AR_d = \frac{AR}{S_d} = \frac{200,000}{34,000} = 5.88 \text{ days}$$

$$C_d = S_d CS = 34,000(0.6) = 20,400$$

$$I_d = \frac{I}{C_d} = \frac{400,000}{20,400} = 19.6 \text{ days}$$

$$AP_d = \frac{AP}{C_d} = \frac{160,000}{20,400} = 7.84 \text{ days}$$

$$\text{Cash-to-cash cycle time} = AR_d + I_d - AP_d = 5.88 + 19.6 - 7.84 = 17.64 \text{ days}$$

By using system-wide metrics like the cash-to-cash cycle time, all functions are given incentive to improve the same performance measures. Looking at individual elements (the average days inventory or days of accounts payable, e.g.) can then pinpoint the specific problems that need to be addressed.

55.7.1 Evaluating, Selecting, and Implementing an ERP Package

From the preceding discussion, it may seem as though ERP is the answer to manufacturing's dreams. Although it does represent a major improvement over the ways of the past, it is certainly not a panacea. In fact, the manufacturing industry is littered with examples of ERP implementation efforts that have not lived up to their billings. The reasons are varied, but in most cases they stem from one important issue: information systems do not run businesses, people do. This may seem like an obvious statement, but in most failed ERP efforts, the company fails to recognize this important point.

Failure to understand the strengths, weaknesses, and processes of the company manifests itself in several ways. Most ERP systems are structured, to varying degrees depending on the vendor, around industry "best practices." This means that firms must change their practices to some extent to match those of the system. In some cases, this degrades the competitive edge of the firm. For example, if a firm's market niche involves a highly-customized and flexible product line, its BOM will likely be very dynamic. There may even be the need to alter the BOM during actual production and assembly to match the changing desires of the customer. Many ERP systems do not accommodate this capability, so embracing the embedded "best practices" in the system may actually hinder the ability of the firm to meet its customers' needs. Clearly this is not the desired outcome.

Another potential barrier to a successful implementation is the cost. For smaller companies, the benefits simply may not justify the hefty price tag of most ERP systems. Implementation can take months or even years and can cost millions of dollars. In addition to the actual hardware and software, consultants are usually needed to aid in the implementation efforts, raising the price tag even further. As with any capital investment, the benefits must outweigh the costs. With these potential pitfalls in mind, the remainder of this section offers suggestions for the evaluation and selection of an ERP system.

Evaluating Alternatives. There are many issues to consider when evaluating ERP alternatives, which can be broadly categorized into four areas: business environment, internal processes, cost, and information technology. We discuss each category separately below.

Business Environment. From the perspective of the business environment, three factors dominate the ERP decision: firm size, complexity, and industry. To date, ERP has been widely implemented by larger firms where the payoff will justify the high cost. Increasingly, small- to mid-size firms are following suit, but their needs are clearly different from those of large firms. Smaller firms typically require lower-cost ERP solutions that are much less restrictive, since their resources are limited and they rely on agile operations to stay competitive. Large firms, by contrast, typically have a more traditional structure that lends itself well to a more expensive, established ERP system.

The complexity of the business is also an important consideration. Firms with a complex group of business units spanning the globe are more likely to benefit from the integration of an ERP system, while less complex operations may not need such additional help. Finally, the specific industry plays a role in the decision. In many industries, ERP systems have become standard practice and therefore have become almost a prerequisite to stay competitive. Specific ERP systems have also come to dominate certain industries, which adds further pressure to adopt the "industry standard" in order to seamlessly interact with suppliers and customers.

Internal Processes. The internal processes of a firm also impact the ERP decision. Companies must first carefully consider the scope of functionality needed from the ERP system before committing to a solution. If only a few modules are needed for a small operation, custom systems may be the best choice. Manufacturing firms are generally organized as either a job shop (discrete manufacturing) or as a flow shop, although some do a hybrid of both. The two processes have very different requirements, particularly with respect to the manufacturing modules, so the type of process becomes a major consideration in the ERP decision. Beyond the designation as a job shop or process shop, some firms have unique or sophisticated processes that require specialized systems. Again, this becomes a major factor to analyze. Finally, the methods used for inventory management, forecasting, material management, and the like must be compatible with those offered by the ERP system. If not, the firm will be forced to change its processes and may lose competitive advantage as a result.

Cost. There are many financial considerations involved in the evaluation of ERP systems, but the most important is the cost-benefit ratio. Before committing to a particular application, the benefits must be listed and quantified wherever possible. These benefits must then be weighed against a realistic estimate of the cost of implementation. Typical costs include the software license, hardware upgrades, telecommunications systems like EDI, training, and consulting fees. Others may include such intangible costs as lost sales during implementation. For large companies, the cost of the software license alone can run into the tens of millions, with consulting fees about equal to the software expense. As with any capital investment, ERP implementation must justify the cost and provide a reasonable return on investment.

Cost-benefit analysis techniques vary by firm, but research indicates that nearly 90 percent use either *return on investment* (ROI) or payback period. *Earned value analysis* (EVA) is also used, albeit to a lesser extent. Returns also vary, but the highest percentage fall into the 15 to 25 percent range. That said, it is not uncommon for firms to show a negative return on an ERP investment. Smaller firms have a particularly daunting task to show a positive return, since a substantially larger portion of their revenues are invested in the ERP implementation. In fact, firms with $50 million or less in sales spend over 13 percent of annual revenues on the process, while large firms spend less than 5 percent of revenues on average. Firms with revenues exceeding $5 billion spend less than 1 percent. It should be clear that careful selection and implementation becomes more critical the smaller the firm.

Information Technology. The final area to consider in evaluating ERP systems is information technology. Implementing and maintaining an ERP system requires an enormous amount of IT expertise. A firm that does not have this expertise in-house may be forced to either develop the capability or contract it out to a third party. Either way, the costs can be substantial. In addition, the system requirements may require major upgrades to the hardware and telecommunications infrastructure of the firm, especially if the existing systems are outdated to begin with.

Selecting an ERP Solution. The preceding section offered many points to consider when evaluating alternatives. This section briefly describes some of those alternatives. To begin with, we discuss the five largest commercial packages available today: SAP, Oracle, PeopleSoft, Baan, and J.D. Edwards.

SAP R/3. SAP AG is a German company founded in 1972 that has been a pioneer in the development and evolution of ERP. Although the company name is often used synonymously with its flagship ERP system, the latter is actually called R/3. SAP R/3 has been so successful that by 1999 SAP AG became the third largest software vendor in the world, trailing only Microsoft and Oracle. It also garners more than 35 percent of the ERP market. In terms of modules, R/3 has one of the broadest sets of product offerings available. It has also been at the forefront of developing internet-based ERP systems with its recent launch of mySAP.com.

Oracle Applications. Although newer to the ERP market than SAP, database software giant Oracle has quickly become second only to SAP in ERP market share. With its dominance in database applications as a foundation, it parlayed its core expertise into the ERP realm in 1987 with Oracle *Applications*. *Applications* offers over 50 modules to its customers, a broad product set that rivals that of SAP. In addition to its ERP offerings, Oracle also provides the core database application behind many of the industry leaders, making it a partner to some of its major competitors. Like SAP, Oracle has recently developed web-based versions of its ERP software.

PeopleSoft8. PeopleSoft is a relative newcomer to the ERP market. From its inception in 1987, it has specialized in human resource management and financial services modules. It has since expanded its offerings to include manufacturing, materials management, distribution, and supply chain planning to its *PeopleSoft8* ERP system. Although it has captured only 10 percent of the market share, making it third behind SAP AG and Oracle, PeopleSoft has developed a loyal customer base due to its flexibility. In fact, it is widely regarded by its customers as a collaborative company that caters to unique individual requirements. Like SAP and Oracle, it offers an e-business application.

BaanERP. The fourth largest ERP vendor in the world, with a market share of about 5 percent, is Baan. Baan began in the manufacturing software market before expanding its scope to ERP with its *Triton* application. This was later renamed *Baan IV*, which later became *BaanERP*, its current designation. *BaanERP* is used throughout the aerospace, automotive, defense, and electronics industries.

J.D. Edwards OneWorld. Rounding out the top five ERP vendors is J.D. Edwards & Company with its *OneWorld* application. *OneWorld* is regarded as a more flexible system than its competitors, making it a viable option for smaller firms with unique requirements. It also offers an internet version called *OneWorld XE*.

Other Options. Beyond the major vendors discussed above, there are dozens of smaller companies that have established themselves in the ERP market. Most of these companies focus on narrower market segments and specialize in certain areas. In addition, there are many third-party vendors that offer specialized modules and bolt-on applications that are compatible with most existing database applications. As a result, the competition has increased as small- to mid-size firms have begun to implement ERP systems. Many such companies have limited resources, and as a result are forced to implement ERP incrementally. They also often have unique requirements that may not be met by larger packages that are structured around industry best practices.

There are six general options when considering an ERP solution. The first is to select a single ERP package in entirety, SAP R/3 for example. Second, a single package can be used alongside other internal, legacy, and bolt-on systems. The third option is to select the best modules from several different packages and use them together atop the common database. Fourth, several different ERP packages can be used with additional internal, legacy, and bolt-on systems. Fifth, a firm can opt to design its own ERP system in-house. And finally, a firm can design its own system but use some external packages for specialized functions. Surveys have shown that about 40 percent of all firms choose to implement a single package, with another 50 percent implementing a single package with other systems added on. Most of the remainder use modules from different vendors, with only about 1.5 percent developing their own systems internally.

Implementation. We close the topic of ERP with a discussion of implementation, since it can be the most problematic phase of the process. This is the time when the reality of the "people side" of information systems comes through. Typical hurdles to implementation include organizational resistance to process changes, training the workforce on a completely new and foreign system, and consulting costs that often grow over time as unanticipated problems arise. To overcome these hurdles, it is imperative that the entire organization be involved from the beginning, including the evaluation and selection of alternatives. Only with the buy-in from those that will ultimately use the system can an ERP implementation have a chance of success.

Implementation can take many months to several years. Over a third of firms responding to one survey completed the process in less than a year, and about 80 percent in 2 years or less. Only a handful (2.1 percent) took more than 4 years (see Mabert, Soni, and Venkataramanan, 2000). The duration depends on several factors, including the implementation strategy, the aggressiveness of the schedule, the scope of the effort, and the size and resources of the firm. Implementation strategy is perhaps the most significant driver that is within the company's span of control. Five strategies are commonly used in the process: big bang, mini big bang, module phasing, site phasing, and module and site phasing.

Big Bang. Big Bang, as its name implies, is the most aggressive approach to implementation. It involves a complete cutover to the new system at a predetermined (and well-publicized) point in time. This is not to say that it is at all sudden, since a great deal of planning, training, and testing typically precedes the flipping of the switch. Still, it is by far the most extreme strategy. It is also the most popular with regard to ERP implementation, with over 40 percent of firms choosing this option.

There are several advantages to this strategy. First, the implementation duration is typically much shorter, about 15 months on average. Second, largely because of the shorter duration the implementation cost is reduced. Finally, the firm is able to reap the benefits of the ERP system much more quickly. The advantages come with risks, however. Big Bang, unless it is done very carefully, can seriously disrupt operations for a period of time while users adjust to the new system. Planning and testing are therefore absolutely critical when using this approach.

Mini Big Bang. The second strategy, Mini Big Bang, offers a somewhat more conservative alternative to the first. Like Big Bang, there is a cutover on a given date. The difference is that the switch is limited to a subset of modules at a time, typically a functional group. For example, the financial modules may be switched on a certain date, followed by the manufacturing modules 2 weeks later, followed by other groups of modules at later dates. Although this approach increases the duration of the effort over the Big Bang method by 2 months on average, the risks of disruption are greatly reduced. Still, only about 17 percent of firms choose this strategy.

Phased Implementation. The remaining 43 percent of firms choose to implement in phases, a strategy which has several advantages. First, the changes are gradually introduced over time, so the risk of disruption associated with Big Bang is greatly reduced. Second, the effort consumes fewer resources in the planning and training phases since these phases become a bit less critical to overall success. Third, the organization can learn from mistakes along the way, making each successive step smoother. And finally, the organizational momentum can build on early successes, moving the implementation team forward in a gradual procession to the common goal.

The decrease in risk comes with a price, however. A phased approach can take more than twice as long to complete as the Big Bang alternative, which often takes the edge off of the excitement to implement. Support from the users can easily wane during this extended period. Phasing also requires temporary patches to allow newly-implemented modules to interact with legacy systems.

Three types of phased implementation are commonly used for ERP projects. The most popular is site phasing, in which the entire system is implemented at one location at a time. A second phasing option is module phasing, where modules are implemented one at a time. This is similar to the Mini Big Bang approach, yet less aggressive. The difference is that module phasing implements a single module at a time, while Mini Big Bang involves functionally related groups of modules. The third, and slowest to implement, is phasing by module and site. This approach involves implementing ERP site by site, with modules implemented one at a time at each site.

WEBSITES

The following list of websites, although far from comprehensive, provides a good start for additional on-line information on MRP, MRP II, ERP, and related topics. A brief description of the site contents is included with each listing.

http://members.apics.org/publications APICS, The Educational Society for Resource Management, offers an excellent selection of on-line articles and research on MRP, MRP II, ERP, and other materials management topics. Membership is required to access most publications, and is available for a modest fee.

http://www.erpforum.be/ ERP Forum is an online resource offering news, market information, and literature on a variety of ERP and related topics, such as database management, business-to-business (B2B), supply chain management (SCM), and data warehousing.

http://www.softselect.com Soft Select offers an online database of ERP systems, listing over 130 vendors. It has partnered with APICS to provide research on system vendors, and its scorecards are available at the APICS website (see above).

http://www.erpfans.com This independent user group site acts as a clearing house for ERP information, and includes news, resource listings, and a searchable chat room forum.

http://www.erpassist.com This site is similar to erpfans.com, and offers on-line newsletters, news, documents, searchable Q&A, discussion groups, and links related to ERP.

http://www.brint.com Brint.com is an on-line information portal covering a variety of business topics, including MRP, MRP II, and ERP.

http://www.purchasingresearchservice.com This site offers various news, articles, publications, an events listing, groups, and product reviews.

http://www.tangram.co.uk/GI.html Tangram is a United Kingdom-based website with general information pages on MRP, MRP II, and ERP.

http://www.engineering-group.com/ENGINEERSMALL/mrp.htm This site offers general information on MRP, MRP II, and ERP, as well as a list of links to on-line articles.

REFERENCE

Supply Chain Council, Inc. www.supply-chain.org

FURTHER READING

Chase, Richard B., Nicholas J. Aquilano, and F. Robert Jacobs, *Production and Operations Management: Manufacturing and Services*. McGraw-Hill, New York, 2002.

Cox, James F., and John H. Blackstone, eds., *APICS Dictionary,* 10th ed., APICS—The Educational Society for Resource Management, Alexandria, VA, 2002.

MRP & MRP II

Orlicky, Joseph, *Material Requirements Planning*. McGraw-Hill, New York, 1975.

Toomey, John W., *MRP II: Planning for Manufacturing Excellence*. Chapman & Hall, New York, 1996.

ERP

Hossain, Liaquat, Jon David Patrick, and M. A. Rashid, *Enterprise Resource Planning: Global Opportunities and Challenges,* Idea Group Publishing, Hershey, PA, 2002.

Jacobs, F. Robert, and D. Clay Whybark, *Why ERP? A Primer on SAP Implementation,* McGraw-Hill, New York, 2000.

Mabert, Vincent A., Ashok Soni, and M. A. Venkataramanan, "Enterprise Resource Planning Survey of U.S. Manufacturing Firms," *Production and Inventory Management Journal,* Vol. 41, pp. 52–58, 2000.

Mabert, Vincent A., Ashok Soni, and M. A. Venkataramanan. "Enterprise Resource Planning: Measuring Value," *Production and Inventory Management Journal,* Vol. 42, pp. 46–51, 2001.

Nah, Fiona Fui-Hoon. *Enterprise Resource Planning Solutions and Management*. IRM Press, Hershey, PA, 2002.

CHAPTER 56
SIX SIGMA AND LEAN MANUFACTURING

Sophronia Ward
Pinnacle Partners, Inc.
Oak Ridge, Tennessee

Sheila R. Poling
Pinnacle Partners, Inc.
Oak Ridge, Tennessee

56.1 OVERVIEW

Six Sigma is a business process that allows companies to drastically improve their bottom line by designing and monitoring everyday business activities in ways that minimize waste and resources while increasing customer satisfaction.[1] It is a strategy driven, process focused, and project enabled improvement discipline with the goal of defect-free output.

Six Sigma evolved from a quality initiative at Motorola, Inc. in the mid-1980s. Observations showed that early field failures of products could be traced to those which had been reworked while those that were produced defect free did not exhibit early failures. Based on these findings Motorola launched an effort to reduce defects by preventing them from happening in the first place. This required that they focus on the processes that produced the products as well as the design of the product itself. There were many ways for products to fail. In order to assure that products could be made defect free, the chance of each possible type of defect had to be reduced almost to zero. Defects were associated with customer *critical-to-quality* (CTQ) characteristics or CTQs. Motorola realized that each CTQ would need to have a frequency of defects in the range of 3.4 *defects per million opportunities* (DPMO) to assure that a product with multiple CTQs would work correctly and not fail in the field. Characteristics that achieved a DPMO of 3.4 were considered to be at a quality level of Six Sigma (see Fig. 56.1).

But the emphasis was not on quality alone. Early on, the Six Sigma focus deliberately connected improving quality and reducing costs. Higher quality at lower costs was the goal. This new linkage became the foundation of the Six Sigma initiative at Motorola. Since the late 1980s, numerous companies such as Motorola, Allied Signal, and General Electric have made quantum leaps in quality as well as major impacts to the bottom-line profits of their organizations using the Six Sigma approach. Six Sigma is a rigorous discipline that ties improved quality, reduced costs, and greater customer satisfaction together to accomplish the strategic goals of an organization.

56.2 CONCEPT AND PHILOSOPHY OF SIX SIGMA

Six Sigma incorporates many of the elements from early quality efforts, integrating them into a comprehensive initiative which focuses on defect prevention and bottom-line results to achieve the strategic

Sigma Threshold	% Conformance	Defects per Million Opportunities (DPMO)
3	93.3193	66,807
4	99.379	6,210
5	99.9767	233
6	99.99966	3.4

FIGURE 56.1 Six Sigma for a single CTQ.

goals of an organization. Projects are chartered with both quality and financial goals. Each project has a *champion* who is usually the owner of the process involved and a *Black Belt* who is trained in the skills and tools necessary to manage the project to a successful completion. As the process owner, the Champion is accountable for the project's success.

Project work focuses on improving a process or processes that are central to the project. Processes that are directly involved in producing a product or service to a customer as well as those that are integral to running the business can all be improved using the Six Sigma methodology. Each process is studied thoroughly to understand why defects or defective results happen. Once the causes of the defects are isolated, prevention is possible. Preventing causes of defects is the core of the Six Sigma work.

The philosophy of preventing defects and defective results has two benefits: a higher level of quality and reduced costs. Higher quality comes from designing both the products and processes so that defects can't happen or the incidence is greatly reduced. The reduced costs are a direct result of the savings achieved because there is essentially no rework involved and products and services perform as intended. The saying, "There is never enough time to do it right, but all the time in the world to do it over," no longer applies. The goal is for every CTQ in every process to operate at the Six Sigma level.

Each CTQ is evaluated at the beginning of a project using current data to determine the sigma level. As the process is improved, the sigma level of each CTQ is reassessed to show progress toward the Six Sigma goal.

But the Six Sigma philosophy goes beyond improving the current processes in an organization. Six Sigma is the framework within which an organization looks to the future. An organization's strategic objectives can be achieved if the Six Sigma philosophy is directed to designing and producing products and services that meet the needs of an evolving customer. All processes in marketing and sales, customer services, and research and development are included in the Six Sigma initiative.

56.3 THE HISTORY OF SIX SIGMA

56.3.1 Motorola Story

The origin of Six Sigma is found at Motorola. In a study of the field life of a product, a Motorola engineer observed that products that were defect free at manufacturing rarely failed in early use by a customer. However, products which had failed during production and undergone rework were more susceptible to early failures in the hands of the customer. It was as if those products that were originally made well worked well, while those that originally had a defect might have other defects as well. Catching and fixing one defect did not assure a good product. It was not enough to find and fix defects, they had to be prevented.

Preventing defects is possible only when the processes that produce the product are set up so that defects can't or don't happen. Motorola focused on "how" the work was done in each and every process. If defects were prevented, then there would be no need for rework and quality would be higher. Since there were lots of processes, Motorola developed a measure for the Six Sigma work that could be applied to every process.

The measure Motorola developed was called the *sigma level* or *sigma value* for each CTQ associated with a process. The sigma level is based on a statistical analysis linking the DPMO and the capability of the CTQ with respect to customer requirements. A CTQ with 3.4 DPMO is associated with capability values of $C_p = 2.0$ and $C_{pk} = 1.5$ and is considered to have achieved the Six Sigma level.

Not all processes are alike, but the output of every process can be inspected for defects. Each opportunity for a defect can be associated with a CTQ characteristic for the output of that process. Using data on defects for CTQ characteristics collected from the outputs of a process, a sigma value can be determined for the average opportunity of a defect. The sigma level of the CTQs for each process provides a method of comparing the performance of a wide variety of processes.

Motorola successfully applied the concepts, philosophy, and techniques of Six Sigma to the design, development, and production of the Bandit pager. The quality level of this pager was unsurpassed. Motorola showed that the traditional approach of detecting and fixing defects resulted in CTQs at the four sigma level of quality. Four sigma quality is associated with 6210 DPMO. Six Sigma quality, or 3.4 DPMO, led to elimination of costly inspection and rework, which in turn led to decreases in manufacturing time and increases in customer satisfaction. Customers were happy and Motorola reaped staggering financial savings.

56.3.2 General Electric Story

In 1995, Larry Bossidy of AlliedSignal introduced General Electric (GE) top management to Six Sigma. Bossidy's account of the benefits realized at AlliedSignal led Jack Welch, then CEO of GE, to launch a Six Sigma initiative which would bring all of GE's products to a Six Sigma level by 2000. At the time, GE's processes were operating between three and four sigma, which would mean an average of 35,000 DPMO. To go from three or four sigma to Six Sigma would require a monumental training and education effort.

Fortunately GE could build on its previous initiative called *work-out*. The work-out program had prepared GE employees to speak out without fear of reprisal, take on more responsibility for their jobs, eliminate waste, and work toward common goals. GE employees were primed for Six Sigma. Nevertheless they needed rigorous training in the Six Sigma methodology with a heavy emphasis on statistical methods. The rollout was expensive, but the rewards were overwhelming. From 1996 through 1998, GE's training investment approached $1 billion and the returns on that investment were already keeping pace. By 1999 they expected $1.5 billion in savings.

GE's businesses range from a variety of products to various services including GE Capital. In all areas, Six Sigma projects were adding to the bottom line while improving the delivery of services. Streamlining processes, preventing failures, and eliminating rework made way for a faster response time and increased productivity. These all combined to pave the way for a greater market share.

Throughout GE, the Six Sigma initiative returned benefits to the organization's bottom line and to its ability to serve its customers.

56.4 THE STRATEGIC CONCEPT FOR SUCCESSFUL SIX SIGMA

Six Sigma began as an effort to eliminate early failures associated with products once they reached the customer's hands. Many organizations see this benefit of preventing defects during manufacturing and launch numerous Six Sigma projects under the guidance of trained experts known as Black Belts. Even when such projects are successful, an organization can fail to reap the maximum benefits of Six Sigma if the project work is not tied to the organization's strategy. This was the Achilles heel of previous quality efforts and will be so for Six Sigma unless an organization embraces Six Sigma from a strategic context.

56.4.1 Foundation for Success

The foundation for success as a Six Sigma organization is to link projects to the overall strategic goals of an organization. This is critical to realize the full potential of a Six Sigma initiative. Every organization will launch numerous projects, but not all projects will meet the criteria of a Six Sigma project. Only those projects supported by top management to further the strategic objectives of the organization and have the potential to return substantial savings to the bottom line will have the designation of a Six Sigma project. Such projects will be high profile, will require dedicated efforts, and will go beyond the business as usual syndrome. Six Sigma projects should advance an organization toward its strategic goals and return a streamlined process that prevents defects.

In order to manage Six Sigma projects successfully, an organization needs to create a strategic council to oversee the selecting and monitoring of projects. This Strategic Council will consist of upper level managers who are knowledgeable about the strategic goals of the organization and can evaluate the potential of a proposed project. The strategic council will also make recommendations as to the number of Six Sigma projects that are feasible to pursue at any one time.

56.4.2 Metrics for Management

Every organization is run by the numbers. This means that decision makers at all levels use numbers, or data, as the basis for decision. Financial values, productivity and efficiency values, and customer data are all used to make daily, weekly, monthly, quarterly, and annual decisions. The familiar data values currently in use reflect a particular philosophy of running a business.

Since a Six Sigma initiative puts the emphasis on prevention of defects rather than the detection of them after the fact, some of the current measures that managers use will need to be changed. There will still be a focus on the bottom-line profitability of the organization and financial measures will be important. Some of the efficiency or productivity measures will need review to assure that they reflect the focus on process improvement. In addition, there will need to be measures of the Six Sigma initiative that assure the new sigma level and defects per million opportunities of the CTQ characteristics is being maintained.

The new set of metrics, commonly called the scoreboard, must be linked at every level of the organization and also to the strategic goals. Six Sigma projects will be selected based on the strategic goals and the values of the metrics on the scoreboard. Improvement of the metrics on the scoreboard must tie directly to achieving the strategic goals.

56.4.3 Leadership for the Future

A Six Sigma company focuses on improving processes throughout the organization. Top management provides the leadership for the entire organization to achieve the breakthroughs possible under a Six Sigma initiative. As Larry Bossidy of AlliedSignal and Jack Welch of GE so ably demonstrated, the leadership at the top is critical. Six Sigma requires a discipline throughout an organization that can only be successfully deployed as a result of top management's leadership.

Six Sigma requires leadership first for the strategic direction of the organization and then for the initiative that will accomplish the strategic goals. An organization that does not intend to improve its financial well-being as well as its ability to serve customers will not benefit from Six Sigma. A Six Sigma effort without a strategic vision will fall short of what is possible. Thus, the benefits of a Six Sigma implementation are focused on the future. Leadership is a critical element to envision the future and set the activities in motion to accomplish that future. Six Sigma is the methodology to achieve the vision set by the leadership.

56.4.4 Culture and Mindset of the Six Sigma Organization

In a Six Sigma organization, everyone focuses on improving a process. The emphasis is on improving "how" things are done. This is a different mindset than the one currently in practice. There are two different versions of current practice. One is widespread in the manufacturing arena. It consists of inspecting everything after it is done to see what works and what doesn't. Rework is done where possible and the parts that can't be reworked are discarded.

The second version of current practice is common in service processes. The focus is on measuring the performance of a service to a goal. All of the results that do not meet the goal are subjected to investigation. The results of the investigations often lead to contingencies that are employed. One such action is that of adding people to eliminate a backlog. Once the people go back to their regular jobs, the backlog grows again.

Current reality in most organizations is "do it" and then "fix it." All such efforts are doomed to continue on the same path. The only way out, and the Six Sigma way, is to concentrate on the process rather than the result. Only when the process is able to deliver what is required according to the CTQ characteristics will an organization be able to reap the benefits of high quality and low cost. The perception that such processes are forbiddingly expensive is not borne out in reality. Typically the best processes are less expensive to run.

A culture that focuses on the process is different from one that focuses on results. The two cannot exist simultaneously. And the key is that the way to achieve superior results most economically comes from focusing on the process. An old adage says, "You can't inspect quality into a product." Regardless of the truth, people continue to do what they have done for years—burn the toast and scrape it.

56.4.5 Choosing the Six Sigma Projects

Projects are the backbone of the tactical side of Six Sigma. They are specifically selected and chartered to forward the strategic goals of an organization. The chartering process for each project outlines the scope of the project, how it supports the strategic goals of the organization, the process to be improved, and the scorecard for the project. Improvement goals for the process and financial goals are included.

The Six Sigma strategic council maintains oversight on all Six Sigma projects chartered. Reviews on the progress of current projects are conducted at least quarterly. Projects should have a scope that is aggressive, but can be accomplished in a 4- to 6-month time frame. The strategic council continually updates the need for additional projects as resources become available.

It is easy to fall into a trap of thinking that all projects should be Six Sigma projects. This is not the case. There are numerous projects in every organization that will be set up and completed without requiring the Six Sigma framework. Six Sigma projects are time consuming, of strategic importance, and require substantial resources. They are expected to accomplish significant results, including bottom-line savings and increased customer satisfaction.

56.5 ROLES AND ACCOUNTABILITIES IN A SIX SIGMA ORGANIZATION

The roles and accountabilities in a Six Sigma organization are divided into the strategic and tactical. The strategic council is formed at the upper management level and manages the selection of projects. Members of the Strategic Council must be knowledgeable about the vision and strategic goals of the organization for the future so appropriate projects can be chosen.

Some members of the strategic council may also serve as champions for the Six Sigma effort. Champions promote the Six Sigma initiative at several levels. Organizational Champions are actively involved in promoting the Six Sigma initiative as well as assessing the potential of projects. Project champions own the process that is integral to the project. They support the work of the projects by removing roadblocks that can undermine the success of the project. Such roadblocks may include a lack of availability of needed resources or current practices that conflict with the Six Sigma focus on improving a process. Every organization has systems in place to assure that the organization functions. Many of these systems, such as data systems or accounting systems, may actually prevent the progress possible from a Six Sigma initiative because they were set up to support a different way to manage the company. Champions may be involved in reviewing these systems and recommending revisions as necessary to support the Six Sigma initiative.

There are several roles at the tactical level for working on the projects. The Six Sigma Black Belt is the person assigned to run the project. This person is specially trained and educated to manage the project, lead a project team through the Six Sigma methodology, and bring it to a successful conclusion. A Black Belt will work closely with a Champion on the project. The Champion will provide a connection to the organization and the strategic goals while the Black Belt runs the day-to-day work.

In addition to Black Belts, other specially trained individuals work on project teams. Green Belts are trained in teamwork and some of the Six Sigma methodology. They are able to assist the Black Belt on the project. Other individuals may be trained as Yellow or Brown Belts to work on special Six Sigma projects. The amount of expertise required for a successful Six Sigma project can appear staggering. This is the nature of a Six Sigma project. They are about business as *unusual*. Fundamental change is going on.

56.6 THE TACTICAL APPROACH FOR SIX SIGMA

The tactical approach for Six Sigma begins with the careful selection of projects and the assignment of a project champion and Black Belt. This is the *recognition phase* of a Six Sigma project. The project champion has ownership of the project while the Black Belt leads and runs the day-to-day activities of the project team. Working together these two people assure that a Six Sigma project is focused, makes progress, and has the necessary resources to achieve success.

Project activities are guided by the DMAIC methodology. The five phases of DMAIC are *define, measure, analyze, improve,* and *control*. Activities in each phase are set out specifically and are designed for success of the project. Following the DMAIC model in a disciplined fashion is a key to successful projects. At the end of the project, there are two additional phases—standardize and institutionalize—that focus on turning the newly improved process back to the process owners who will take responsibility of maintaining the improvements.

56.6.1 Chartering Six Sigma Projects

In a Six Sigma organization, the strategic council will review, evaluate, and prioritize proposed projects for consideration as Six Sigma projects. As resources become available, in particular a Champion and Black Belt, an approved project can be chartered. The chartering process includes summarizing the reasons for selecting the project, identifying the process involved to be improved, final selection of the Black Belt and Champion for the project, identifying team members, setting the goal for the project, identifying resources needed, and outlining a time frame for the project. If sufficient resources are available, Six Sigma projects will typically take approximately 4 to 6 months from start to finish.

The charter for a Six Sigma project is critical. It is one of the mechanisms that sets Six Sigma projects apart from the rest. All pieces of the charter need to be in place so that the Black Belt, Champion, and team members will be clear on the project, the process, and the goal for improvement as well as the expected timing.

56.6.2 Project Teams

Most organizations have a wealth of people experienced in working on project teams. Six Sigma project teams consist of specially trained people and those who work in the process associated with the project. The core team should consist of 6 to 10 members with extra resources available as needed. Some resources may be needed in the measure and analyze phases of the project work and not required in the other phases. Having these resources available will make a big difference in the success of the project.

Six Sigma Projects follow a systematic discipline called the DMAIC Model. This model consists of 5 major components:

Within each step there are various activities to be completed by the Project Team members and/or their support resources.

FIGURE 56.2 The DMAIC model.

Each project will have many stakeholders, those who have a stake in the improvement of the process. All stakeholders will need to be informed regularly as to the progress of the project.

Project teams meet at least once a week with the Black Belt in charge of setting the meeting and the agenda. Outside the actual meeting, every team member will have assignments for work to complete before the next meeting. All team members must be aware and plan for meeting time as well as work between meetings

56.6.3 DMAIC Model

The five phases of the DMAIC model are *define, measure, analyze, improve,* and *control* (Fig. 56.2). While this model suggests a linear progression with each phase leading to the next, there will always be some iterative work between the phases. Each phase involves a certain amount of work to accomplish which is integral to the successful completion of a Six Sigma project.

The Phases of DMAIC. The *define phase* is for focusing the Six Sigma project and creating a roadmap with time line to guide all project activities toward a successful conclusion. Under the leadership of the Black Belt, the project team works on mapping the process associated with the project and setting up communication with all stakeholders of the process. Logistical decisions about team members and meeting times are decided. The charter is finalized and signed off by everyone. If questions arise about availability of resources such as time or people, the Champion helps resolve these issues.

The *measure phase* concentrates on collecting data to assess current process performance. It involves deciding all of the measures that are pertinent to the project. These include the scorecard metrics that tie to the scoreboard of the organization as well as all CTQ characteristics for the process. All measures must be clearly identified and data collection plans for each one set up. Then data can be collected regularly for the analyze phase.

In the *analyze phase,* the various techniques are used to determine how the current process performance compares to the performance goals in the project charter. Data are analyzed with the express purpose of understanding why the current process works the way it does. Specifically, it is important to determine if the process has a predictable or unpredictable behavior. Predictable processes have routine sources of variation that are always present and impact the process day in and day out. Unpredictable processes have both routine sources of variation plus special, unpredictable sources of variation that can knock a process off track.

Activities in the *improve phase* are directed toward finding out how to improve the process. This phase involves investigations to see what changes to the process will make the process better to meet the project goals. The project team may conduct experiments to see what can be achieved and the associated costs. Once several solutions are identified, the project team can evaluate and pilot them for their potential benefits. Before a final improvement solution is selected and implemented, the project team needs to assess any potential problems with all solutions. Resolution of all potential problems will assure that the solution will not be undermined in the future.

Finally, the *control phase* in the DMAIC model must focus on making the changes permanent for sustained process improvement. This means that the project team will work on those activities that are critical for the project team to turn over an improved process to the process owners. To reap the benefits of any Six Sigma project, the improvements must be sustainable by the organization, not by the project team. Elements included in the control phase are training of employees, control plans, and process measures that will keep attention on the process. The transition from the control phase of the DMAIC model used by the project team back to the organization comes with the standardize and institutionalize phases.

Tools and Techniques. In each phase of the DMAIC model there are goals, with activities and deliverables. A variety of tools and techniques are available to accomplish the goals. Also, there are tools and techniques for effective meetings and project management to keep the team members on track and working together productively.

The deliverables for the define phase are the charter with the scope and goals of the project, a process map, a timeline for the project, and a communication plan for project progress. Tools that are used in this phase include process mapping techniques, customer surveys to establish CTQs, worksheets to complete the elements of the project charter, Pert chart, Gantt chart, a stakeholder analysis to assess the level of support for the project, and a scoreboard with the measures that tie the project to the strategic goals of the organization. Graphical tools, such as Pareto charts, scatter diagrams, and run charts may be used in the define phase to support the need for the project. Finally, relevant financial data and reports can be useful to support the case for the project.

Data collection and graphical summaries are critical deliverables of the measure phase. Some of the tools from which to choose are Pareto charts, cause and effect diagrams, brainstorming, histograms, multivari charts, check sheets, bar charts, run charts, scatter diagrams, quality function deployment (QFD), data collection plans, and failure mode and effects analysis (FMEA). All of the data collected and summarized will reveal current process behavior. It is critical to have data and summaries of all CTQs and the process settings or variables that drive or control the CTQs. One helpful way to organize the information is to add these summaries to the process map. The information will be readily available for use by the project team and the process map will become a living document for the current process. Finally, the sigma of all CTQs needs to be evaluated.

In the analyze phase, numerous graphical and statistical analysis techniques are utilized. The analysis of current process behavior is conducted with control chart or process behavior chart techniques. Average and range charts, individual and moving range charts, three-way charts, difference charts, Z-charts, and cusum charts are some of the ones that are often used. To find specific causes of variation, summaries such as Pareto chart and investigative techniques such as root cause analysis are very useful. Additional analysis techniques are histograms, capability analysis, confidence intervals, regression analysis, and measurement process analysis. Finally the potential of the current process can be determined using Taguchi's loss function.

Many of the activities in the Improve phase involve some experimenting and analysis of the experimental data. Statistical techniques of regression and design of experiments are essential in this phase. Some specific techniques include hypothesis testing, confidence intervals, regression analysis and correlation analysis, analysis of means, analysis of mean ranges, analysis of variance, factorial designs, fractional factorial designs, screening designs, Plackett-Burman designs, response surface analysis, and reliability testing. Techniques of generating ideas for possible solutions to improve the process are valuable. Nominal group technique along with criteria ranking can yield solution ideas. These must be evaluated and prioritized. Piloting a proposed improvement solution is critical and risk assessments are needed for each proposed improvement.

In the control phase the activities are directed to assuring that process improvement will be sustained. Control plans which include standard operating procedures, training plans, data collection and monitoring plans, reporting plans, Poka-Yoke, and stakeholder analysis are all needed in this phase. These documents are included as part of the project completion package to give to the process owners for use when the project team is no longer involved. This will assure that the process continues to operate in its improved state.

Standardize and Institutionalize Process Improvements. At the end of the control phase, the Six Sigma project team must turn the process back over to the process owners. The process owners are ultimately responsible for maintaining process improvements. The new process must be institutionalized in such a way that it cannot revert to its former state.

56.7 SIX SIGMA AND LEAN MANUFACTURING

There are many parallels between the Six Sigma and lean philosophies. Both are focused on the process and process improvements. The techniques of each one compliment the other and provide enhanced results.

56.7.1 Comparing the Two Methodologies

Six Sigma focuses on improving existing processes to give results that are essentially defect free. A CTQ characteristic or CTQ that reaches the Six Sigma level will have only 3.4 DPMO. There are two ideas at work here: first, preventing defects is more cost effective and reduces the need for inspection and rework; second, all products and services have multiple CTQ characteristics. If each CTQ reaches the six sigma level, then the product or service will work as the customer expects.

The reality is that Six Sigma can lead to an improved process with mechanisms to maintain the improvements but there are still inefficiencies in the process. All of the benefits of eliminating non-value-added steps may not yet be evaluated or addressed. Also, the flow throughout an organization that is integral to the pull systems of lean may not have been considered.

What we know today as lean manufacturing began with the concepts that Ford Motor Company pioneered at the Rouge facilities in Michigan. Based on the idea of raw steel coming in and finished cars going out, the Rouge facility had taken production to new heights. Then Toyota took this idea a step further. The brilliance of lean was to move beyond craft and mass production techniques to short runs with essentially no set up time required. With short runs and small batch sizes, quality problems would be noticed immediately and there would be no backlog of inferior parts or materials. Small batches made quality immediate visible and led to quality improvements (see Fig. 56.3).

Waste of time, waste of materials, and waste of money—all drive the lean concepts. A number of offshoots of the original lean ideas have emerged. Eliminating wasted efforts by removing non-value-added steps in a process has led to *value stream mapping*. Eliminating excess inventory and small batches has led to a "pull" system of production. All of the lean efforts are directed toward

	Objectives	Focus	What's missing
Six sigma	• Improved customer satisfaction • Improved quality • Increased profitability	• Reducing variation in products and services • Eliminating defects to achieve a six sigma level of quality	• Speed and efficiency • Streamlined process
Lean manufacturing	• Reduced waste • Decreased inventory • Reduced costs • Increased speed	• Streamlining processes • Improving efficiency	• Reduction in variation • Prevention of defects

FIGURE 56.3 Comparing the two methodologies.

removing wastes of various types in the production systems. A lean approach opens up the opportunity to achieve reductions in time to market and in costs while improving product quality.

56.7.2 Benefits of a Synchronized Approach

Both concepts focus on the process. Lean techniques may well be the ones to use to achieve the improvements in quality and cost reduction that are the objectives of the Six Sigma initiative. Six Sigma may well provide the statistical techniques that will take the lean initiatives to the next level. They both are critical and if orchestrated in concert can be much more effective than either one individually.

56.8 OBSTACLES IN SIX SIGMA IMPLEMENTATION

Many companies are already reaping the benefits of a Six Sigma initiative. The two most celebrated ones are Motorola and GE. Six Sigma was developed at Motorola and GE has taken it to astonishing levels of success, particularly outside the traditional realm of manufacturing. It is important to realize what made these successes happen so that organizations can avoid obstacles that can impede the Six Sigma initiative.

The biggest obstacle for Six Sigma success is "business as usual." Many of the quality initiatives of the 1980s failed to live up to expectation because they were voluntary, haphazard, and lacked committed leadership. Six Sigma can fail for these same reasons. One look at the successful Six Sigma companies reveals that the senior executives in the organization are heavily involved and lead the initiative. In order for Six Sigma to succeed, many of the organization's systems and structures, such as information systems, reporting systems, hiring and reward practices, and financial and accounting systems, will need to be reviewed, and possibly modified or completely changed. These systems maintain the organization in its current state. A Six Sigma initiative must ultimately become the foundation of the way to do business for the future.

A second obstacle is recognizing the magnitude of the potential of Six Sigma. An organization that tries to implement Six Sigma without sufficient resources will almost certainly fall short of its potential. At GE, Jack Welch committed millions of dollars to training and additional resources to support Six Sigma projects. The results were staggering. You don't get staggering results without a monumental commitment.

Finally, guidance from experts in Six Sigma initiatives is essential. The investment in training and coaching throughout the first several years of a Six Sigma initiative will pay back immeasurably in early successes. The best way to go down a new road is to look ahead, out of the windshield, and not continually stare in the rear view mirror. In an organization with a rich and successful history, taking a new road will not be easy no matter what the rewards. Just sticking to the path can be difficult. Guidance from experts will help make the journey less difficult and eliminate most of the detours and short cuts that people are tempted to make.

56.9 OPPORTUNITIES WITH SUCCESSFUL SIX SIGMA

The opportunities that a success Six Sigma initiative will bring appear unbounded. Improved processes run smoother, at lower cost, with higher quality, and better service to the customer. Plus, everyone in the organization can put their various experiences and expertise to better use when they don't have to inspect and rework or scrap what has been done. Time is available to move to the next level of achievement. Fire fighting is no longer required, but the expertise that was used in fire fighting can be focused on much tougher issues, such as making the processes even better. Forward momentum fosters forward momentum. The potential is unlimited, the benefits are astounding. The most difficult challenge for those seeking to go down the Six Sigma path is letting go of the past. Knowledge from the past was sufficient to bring you to the current state of success that you enjoy. New knowledge is required to take you to the future successes you envision. Six Sigma provides a method of systematically building new knowledge and using it to improve processes to benefit all.

REFERENCES

Harry, Mikel and Schroeder, Richard, *Six Sigma, The Breakthrough Management Strategy.* Doubleday, New York, 2000.

Snee, Ronald and Hoerl, Roger. *Leading Six Sigma: A Step-by-Step Guide based on Experience with GE and Other Six sigma Companies.* Financial Times Prentice Hall, Upper Saddle River, New Jersey, 2003.

Womack, James P. and Jones, Daniel T. *Lean Thinking: Banish Waste and Create Wealth in Your Corporation.* Simon and Schuster, New York, 1996.

Womack, James P., Jones, Daniel T., and Roos, Daniel, *The Machine that Changed the World: The Story of Lean Production.* HarperCollins, New York, 1990.

FURTHER READING

Collins, Jim. *Good to Great: Why Some Companies Make the Leap… and Others Don't.* HarperCollins Publishers, New York, 2001.

De Feo, Joseph, Barnard, William, *Juran Institute's Six Sigma Breakthrough and Beyond,* McGraw-Hill, New York, 2004.

Pyzdek, Thomas, *The Six Sigma Handbook,* McGraw-Hill, New York, 2003.

Ward, Dr. Sophronia. (2000–2004) *Brain Teasers.* A series of manufacturing case studies published by Quality Magazine. Bensenville, IL. WWW.qualitymag.com

Yang, Kai, Ei-Haik, Basem, *Design for Six Sigma,* McGraw-Hill, New York, 2003.

CHAPTER 57
STATISTICAL PROCESS CONTROL

Roderick A. Munro
RAM Q Universe, Inc.
Reno, Nevada

57.1 INTRODUCTION

SPC can be defined as the use of statistical techniques to depict, in real time, how a process is performing through the use of graphical tools. Many engineers have heard about the Shewhart control charts, now being called process behavior charts, which were first developed in the late 1920s. However, there are many other tools available that we will briefly review in this chapter with references to where more information can be found if needed. It is recommended that the manufacturing engineer use this as an overview of the topic of *statistical process control* (SPC), which should not be confused with *statistical process display* (SPD). Many organizations get into the habit of posting graphs around the organization without the benefits of using what the graphs are intended to tell the operator and supervisors. Thus SPC is a real time graphical tool process that gives insight into the process behavior that is being studied.

The tools are listed alphabetically to allow for ease of finding the reference quickly. This should in no way be taken to indicate a sequence or importance of use. All of these tools are useful for their intended usage and the manufacturing engineer is encouraged to become aware of as many of these as possible. This will prevent the old adage of everything starting to look like a nail if you only have a hammer in your toolbox.

Tip. In this section, you will note a number of "tips" listed after the discussion of the various tools. These are listed to give additional insight into the use of the SPC tools. To ensure that SPC works within your organization, you *must* ensure that the gages are performing properly. Use of *measurement system analysis* (MSA) or *gage repeatability and reproducibility* (GR&R) studies are strongly recommended.

57.2 SPC PRINCIPLE AND TECHNOLOGIES

Variation is the basic law of nature that no two things are exactly alike. There are usually many reasons for things not being constant. We have procedures for how to do the work, but small things can and will change causing the output of the process to be different. A common way to describe this today is with the formula: $Y = f(x)$ and called out as Y equals the function of the xs. Graphically this is most easily seen when using the cause-and-effect diagram. The effect is the Y of the formula and the causes are the xs (see Fig. 57.1).

The traditional view of quality (sometimes called the goal post mentality) depicts that some parts are clearly made within specifications while others are outside of specifications. There is no

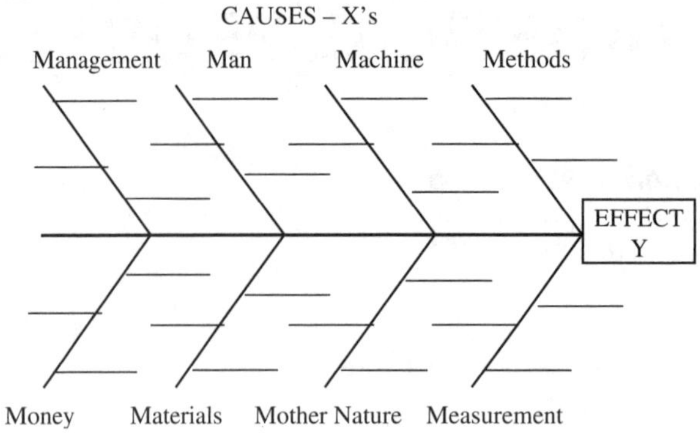

FIGURE 57.1 Cause and effect diagram.

relationship called out for $Y = f(x)$, but the question that should be asked here is what is the real difference between parts if one is just inside the spec and another is just outside the spec? Put together, these two parts are very close and will probably function equally well, or poorly, when used by the customer. That is one of the reasons that people who use this traditional model will tend to ship the part at the spec limit (even if just outside the spec limits) because they think they can get a few more sales to the customer that will not be noticed. This usually happened at the end of the shipping periods, e.g., the end of each month.

The change in view that has occurred (called the Taguchi Loss Function) states that all characteristics, which are measured by the "*xs*," should be aimed for a target value that is aimed at the middle of the specification limits. In this case, parts that are just in or out of specification have nearly the same loss to the customer and will not be accepted very well. As parts move away from the target value, the cost to the customer, and thus society, increases as issues or problems with the use of those parts increase. The goal today is that we need to reduce variation (both common cause and special cause) so that the customer sees more parts that are closer to the target value of what they want—not what we can produce.

57.3 APPLICATIONS

The eight most commonly used tools in SPC include cause-and-effect diagrams, checksheets, flow charts, histograms, Pareto diagrams, process behavior charts, run charts and scatter diagrams. Each of these tools is designed to show a different view of the process variation and to assist the manufacturing engineering in identifying the common and special causes of the variation. Generally (Deming 1992), you will find that to reduce common cause variation, capital investment and other factor requirement management support will be needed; while operators and supervisors can usually handle the special cause variation on the shop floor.

57.4 PLANNING AND IMPLEMENTATION

The remainder of this section will discuss the use of some of the common SPC tools found in manufacturing operations today.

57.4.1 Cause-and-Effects Diagram (CE Diagram)

Also called the *Ishikawa diagram* or the *fishbone diagram,* this tool was first developed in Japan to help improve quality by studying the causes (factors) and effects of a process in greater detail to illustrate their relationships on a diagram and make them more useful to production. "CE Diagrams are drawn to clearly illustrate the various causes affecting product quality by sorting out and relating the causes. Therefore a good CE Diagram is one that fits the purpose, and there is no one definite form" (Ishikawa 1971).

Tips. Remember to ask the five Ws and H (what, why, when, where, who, and how) when identifying the causes to be used in the diagram. The effect can be any outcome of the process, sometimes a positive situation (think prevention vs. detection). The people doing the work should be involved with creating the diagram. The causes should include topics in the basic five Ms (man, machine, methods, materials, mother nature, measurement, money, or management). Note that the five Ms really have eight items and that they are generic and no offense is meant toward women. Use appropriate language for your organization.

Sample. The author once used a CE diagram to help a group of academics visualize a problem statement that they had struggle with for many hours. Using the basic frame of the diagram, he was able to focus the group's attention on each of the stems to develop a single picture, in less than thirty minutes, of the factors in play that would eventually lead to their solution.

57.4.2 Check Sheets

A check sheet can be any set of words, tally lists, or graphics designed to assist in conducting a planned review or observation of a process on the shop floor. They are commonly used in many areas of our society to help ensure that something is done in a certain sequence (airplane pilots use checksheets to take off and land aircraft) or to tally information in a sequence that becomes useful in real time information (incoming inspection).

Tips. Pretest a check sheets before full use to ensure that it collects the desired information and that users understand how the information is to be displayed. Using pictures of the product and allowing operations to make a mark on the picture whenever something is not according to specifications makes for a very simple collection technique.

Sample. In one plant the author worked with, we created a profile of the product and tapped a copy of the paper to each unit going down the line. After working with the operators, each person marked on each unit anything they noted that was not exactly the way it should have been. At the end of the line, inspectors collected the paper and keep a single page showing all the issues for that day's production. The inspectors actually created a pictorial Pareto diagram from the individual pictorial check sheets.

57.4.3 Flow Charts

Flow charts (aka process maps, flow maps, and process flow diagrams) are a pictorial representation of the process flow or sequence of events on the shop floor. You are creating a representation of the steps in a process or system as it actually operates or is supposed to operate. Many software programs are available to assist the manufacturing engineer in creating the flow charts.

Tips. These are very common in many organizations and the primary challenge is to use similar figures and symbols to represent the same items throughout the organization. The two common ways to produce these is either by working with the various people in the system to identify the actual process or to create a "should be" flow map of what the process should do. The common challenge with asking people in the process is that the odds are that you will get very different views of what is happening by the different functions within the organization. Thus time is needed to work out the differences.

Sample. The author has seen many "as is," "should be," "could be," and "what if" process flow maps used in any number of situations from the boardroom to the shop floor. This tool is very useful in getting common agreement with a group of people of what is or is supposed to be happening in a process.

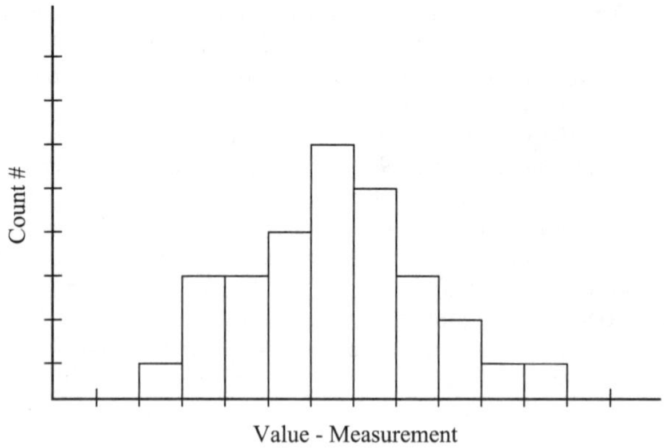

FIGURE 57.2 Histogram.

57.4.4 Histogram

A histogram is a frequency distribution (usually shown horizontally) that graphically displays the measurements taken from a process and shows how those data points are distributed and centered over a measurement scale (see Fig. 57.2). Histograms give a picture of what the process is producing over a specified period of time, although not sequentially (see run charts or process behavior charts). A clear picture of the process variation for the specified time frame become evident and comparisons can be made against the expected process output vs. the actual production.

Tip. Whenever measurements are being made, ensure that you can trust those measurements through GR&R studies. Sometimes drawing a bell shaped curve (many software programs do this automatically) can help show how normal the process is behaving. Watch for bimodal and multimodal distributions, which could be a sign that the various machines, shifts, or people are operating slightly differently.

Sample. A bimodal distribution indicates that something in the process is not quite the same. If two machines and two operators are involved, have the operators switch machines and compare the before and after results. Is it the machines or is it the people causing the bimodal distribution? Many times you find that it is one of the machines (maintenance hates this as many times the machine that is not the same is the one that they just rebuilt or refurbished). This test is called the "old switcheroo"!

57.4.5 Pareto Diagram

The Pareto principle basically states that 80 percent of the effect is caused by 20 percent of the causes (commonly called the $^{80}/_{20}$ rule). The Pareto chart organizes data to show which items or issues have the biggest impact on the process or system (see Fig. 57.3). Then, on the chart, we stratify the data to show the groups, starting with the largest and working down to the lowest number of items in each group. The idea is that by organizing the data in this format, we can develop a plan to work on problems that will give us the biggest return for our process improvement efforts.

Tip. Pick a specific time frame to collect the attribute data for the chart. Ensure that the operators and/or inspectors are viewing the process in a similar manner to allow for consistency in data collection. A Pareto chart is easy to develop by hand; however, use of computers makes for easier manipulation of the data as things change.

Sample. A group of executives once scoffed at this basic concept of $^{80}/_{20}$. They challenged the author to prove that this concept worked and how it might relate to them. Having received information about the company ahead of the engagement, the author was able to point out to them that 80 percent of their total sales volume was directly related to 20 percent of their customer base!

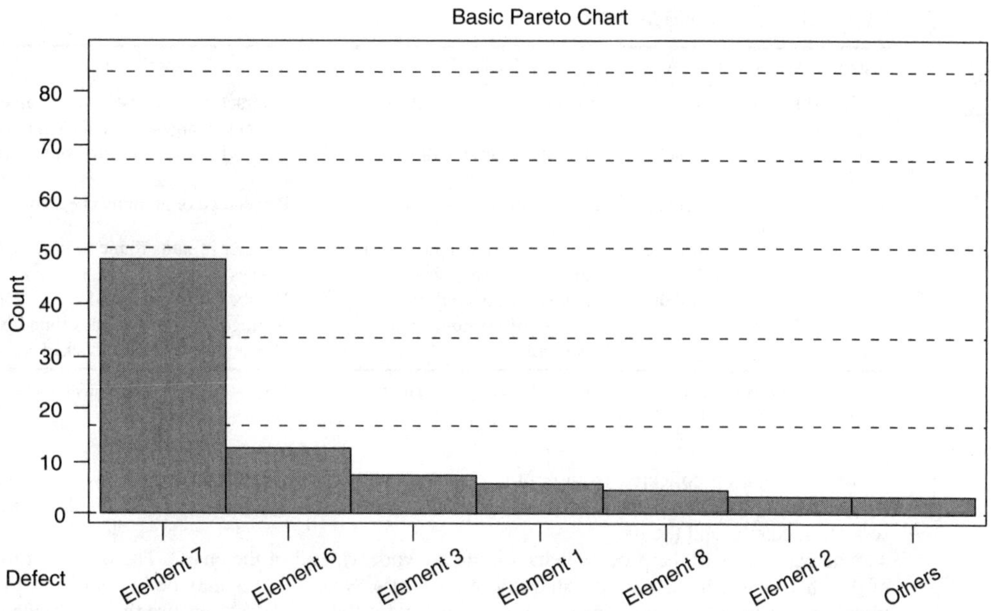

FIGURE 57.3 Pareto chart.

57.4.6 Paynter Charts

A suppler quality assurance engineer working with electrical suppliers developed the Paynter chart at Ford Motor Company. It in essence combines the ideas of the run chart and the Pareto chart into a single numerical table of information. This can then show how the distribution of issues (starting from the $^{80}/_{20}$ rule) is changing over the time of the study. Usually only shown in numerical values (vs. graphically), this time sequence is very good at viewing overall detail of improvement in the process which is usually hidden in attribute process behavior charts.

Tip. When dealing with the need to show quick process changes or the improvement that is being made on a daily basis, this tool works very well in gathering all the data in one location for engineering management review. Also useful when you decide on an engineering change in a process and want to view the output on a daily basis to see if the production is producing what was expected and how the process is maturing.

Sample. The original use of this chart was to view the number of issues found on wiring harnesses, and how the supplier's effort to improve the process was actually playing out at the assembly plants that used the product. Over the period of time from problem identification to resolution, management was able to track the supplier progress on a daily basis until the problem solving process was complete.

57.4.7 Process Behavior Charts (Control Charts)

The *process behavior charts* developed by Walter Shewhart were used primarily for long production runs of similar parts (Shewhart 1932). The 30 plus different charts that are available today were originally called *control charts* (many books still use this term) or *statistical process control* (limiting SPC to only the basic *process behavior charts*). In this section we will focus on only the six most commonly used charts, they are—Xbar and R; individual and moving range; p; np; c; and u (see Table 57.1). The primary distinguisher is the type of data measurements that are collected. Variable

TABLE 57.1 Process Behavior Charts

Chart name	Data type	Measure*	Description
X-bar and R	Variable	Averages of variable data	Subtract the smallest sample value from the largest to identify the range.
Individual and moving range	Variable	Individual variable data	Used when averages are not available
p	Attribute	Fraction of nonconforming units	Percentage of all units checked
np	Attribute	Number of total nonconforming units	Number of units found to have issues
c	Attribute	Number of nonconforming	Number of issues found
u	Attribute	Number of nonconforming per unit	Average number of issues found per the number of units checked

*Some references refer to nonconforming as "defect." Your industry may have a product liability issue with the term "defect" so "nonconforming" is used in this section.

data is information collected from continuous measurement devices, e.g., length, weight, volume, roughness, and the like. Attribute data is ordinal information, e.g., go/no go, good/bad, blemishes, scratches, counts, and the like.

The basic rules of the process behavior charts work with all of the charts. The primary function of the chart is to demonstrate the stability of a process. Note that this may be in conflict with continual improvement, but you must have a starting point (benchmark) to ensure that you have made improvements. Without a stable process behavior chart, you are unable to calculate capability of the process and you will be forever guessing on what factors are causing variation within your system. The charts will distinguish the differences between special (assignable) and common (random) cause variation and give the manufacturing engineer the evidence needed to make process improvements.

57.4.8 X-Bar and R

The X-bar and R (sometime sigma is used instead of the range, thus transforming the chart into the X-bar and S chart) was the first chart developed (see Fig. 57.4). It was used extensively through WWII because it is easy for operators to use without the need of a calculator or computer. If sample size five is chosen, simply add up the five numbers, double the value, and move the theoretical decimal point one place to the left! You now have the mathematical average of the five numbers. This only works with sample size of five and this is why many textbooks suggest five for the sample size.

57.4.9 Individual and Moving Range

When destructive testing or high cost measurement is involved, it is usually impractical to test more than one part or process parameter (see Fig. 57.5). Thus an *individuals* chart can be used to monitor the process behavior for patterns, trends, or runs. As in all variables charts, start by observing the range chart for stability and then study the actual measurements.

57.4.10 Attribute Charts (p, np, c, and u)

The attribute charts are not usually considered as robust as the variables charts but are still highly prized for their ability to monitor a process and show stability when variable data is not available (see Figs. 57.6, 57.7, 57.8, and 57.9). One note here for the manufacturing engineer is that as the process improvements are made in the process, larger and larger sample sizes will be needed to detect nonconforming rates and patterns in the process. The need for very large sample sizes is one

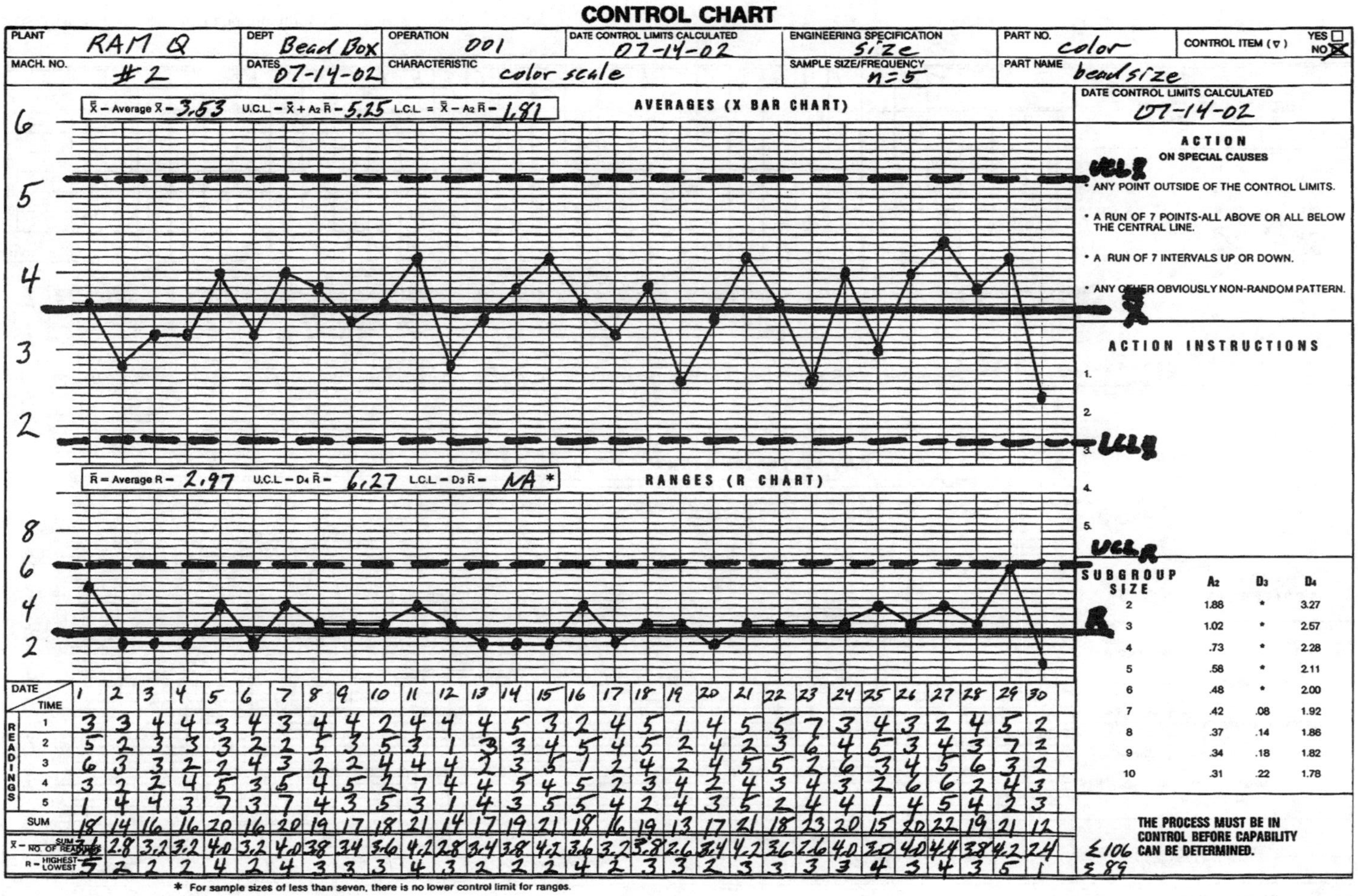

FIGURE 57.4 X-bar and R chart.

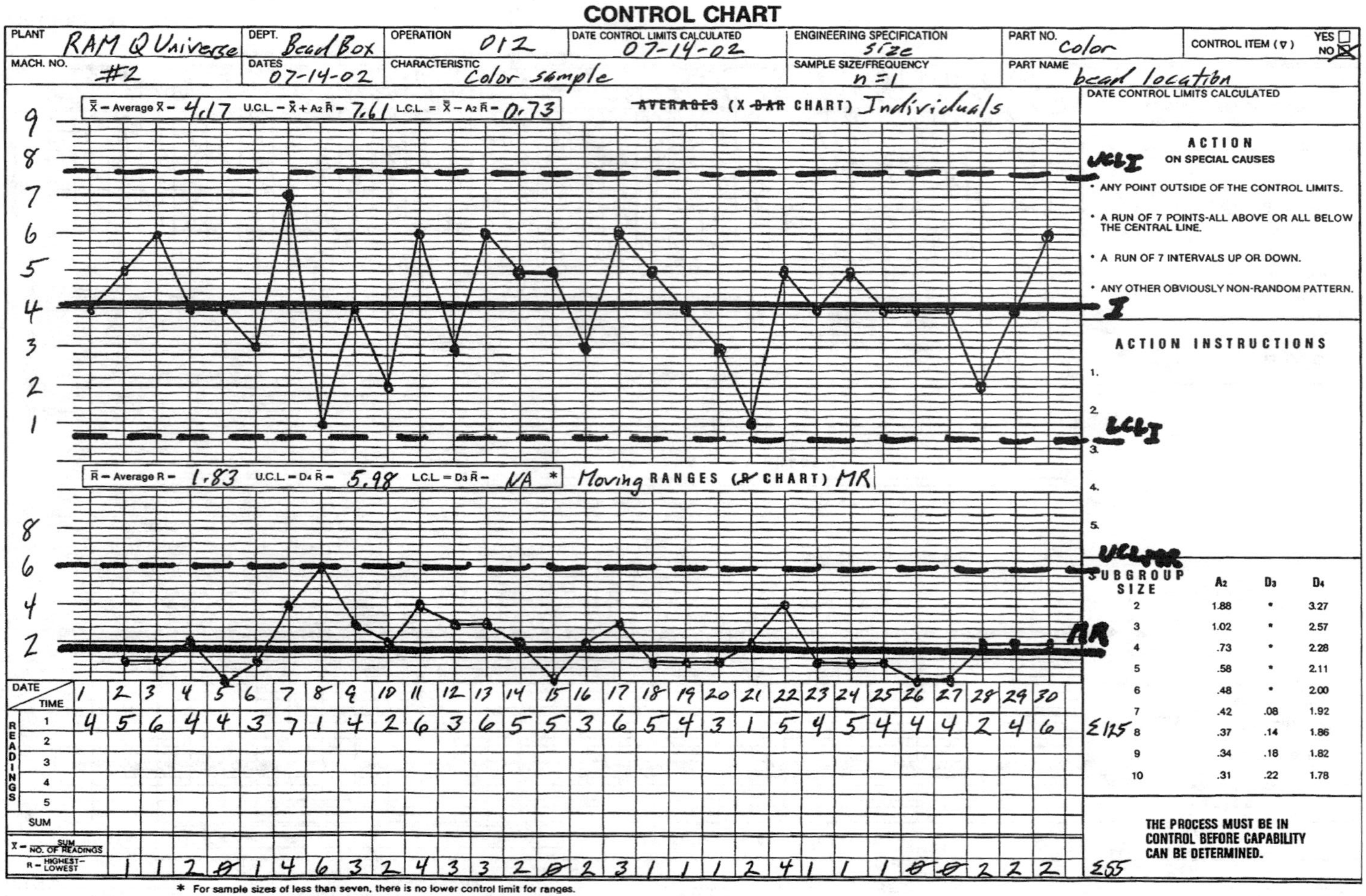

FIGURE 57.5 Individual and moving range chart.

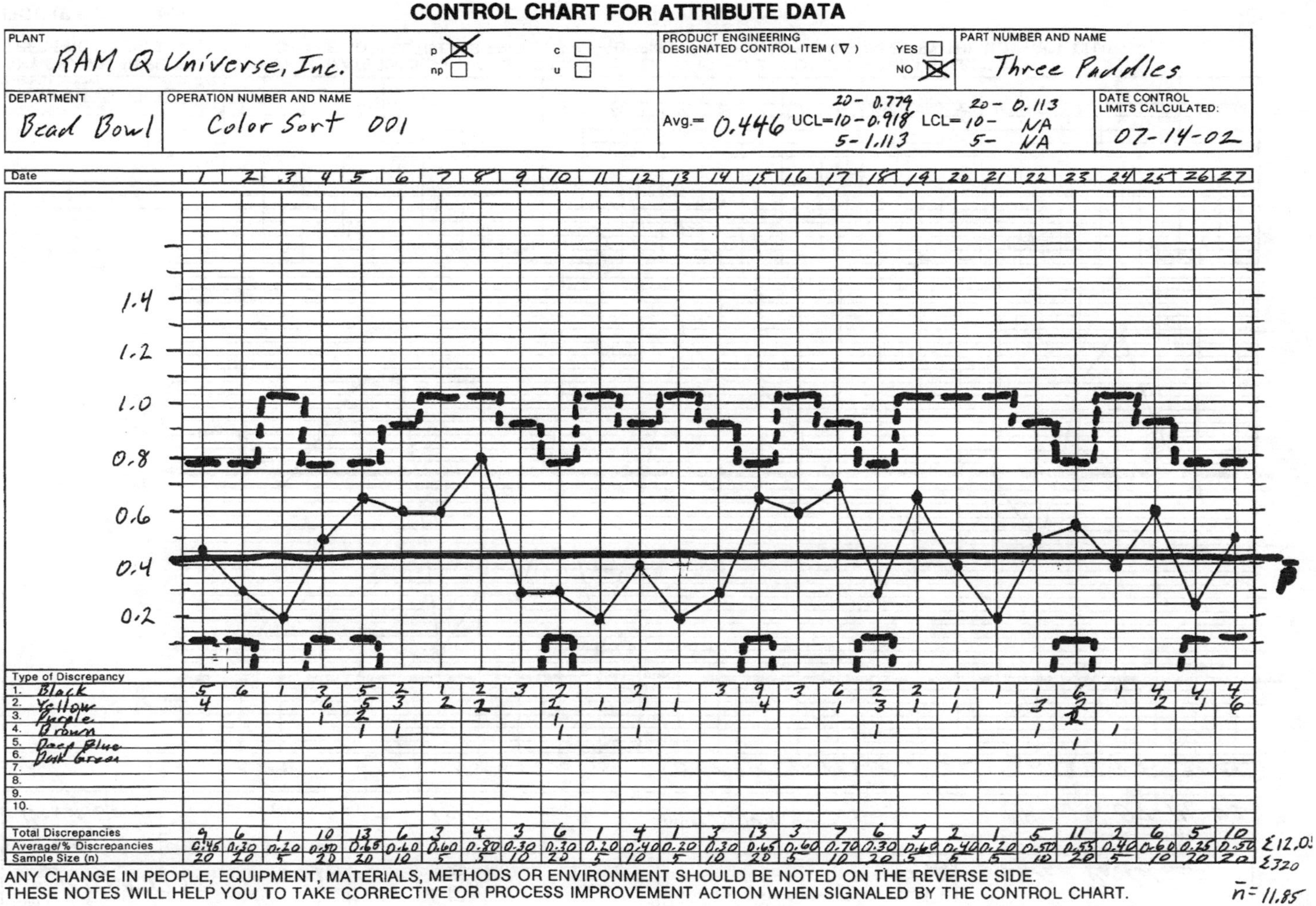

FIGURE 57.6 p chart.

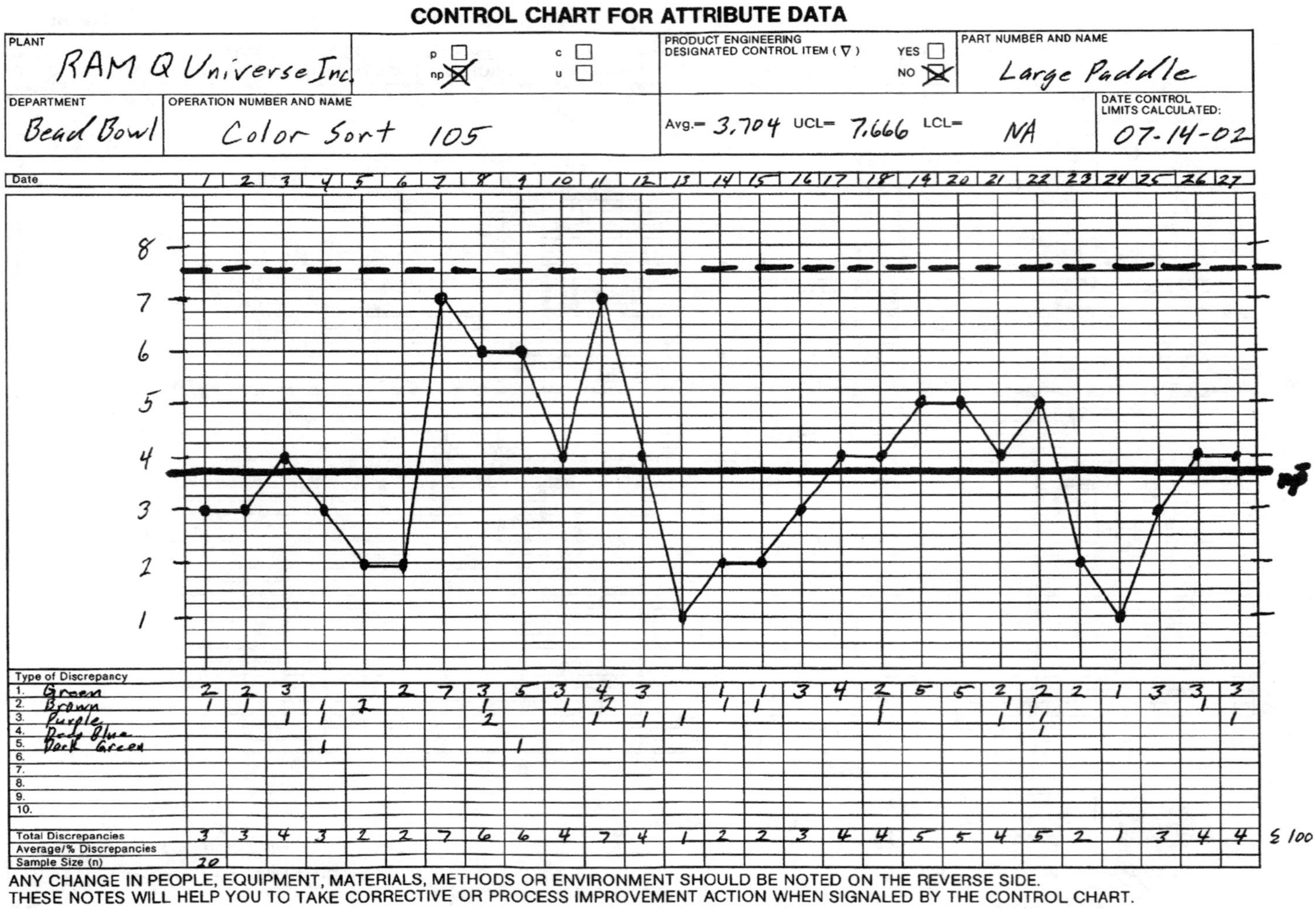

FIGURE 57.7 np chart.

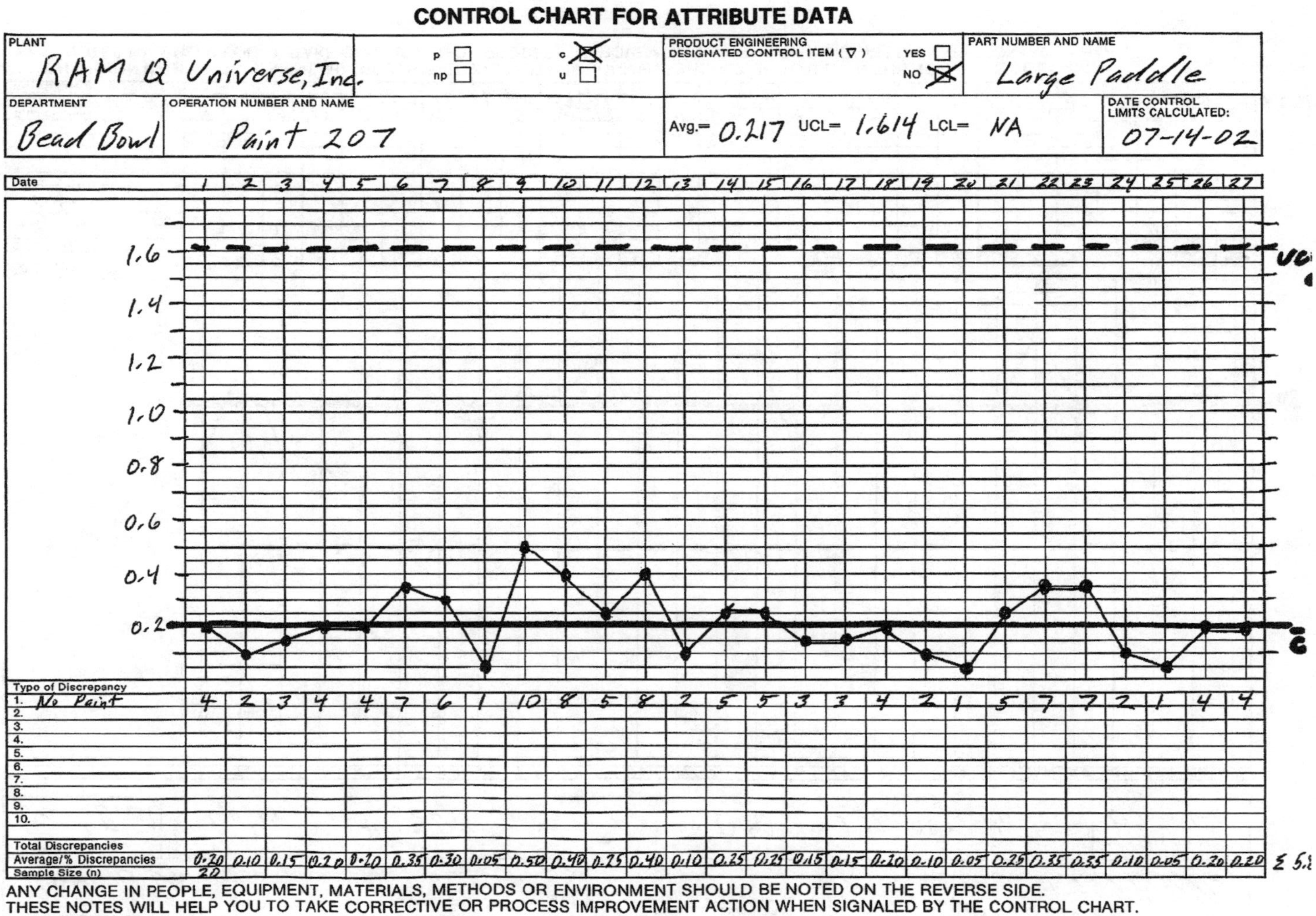

FIGURE 57.8 c chart.

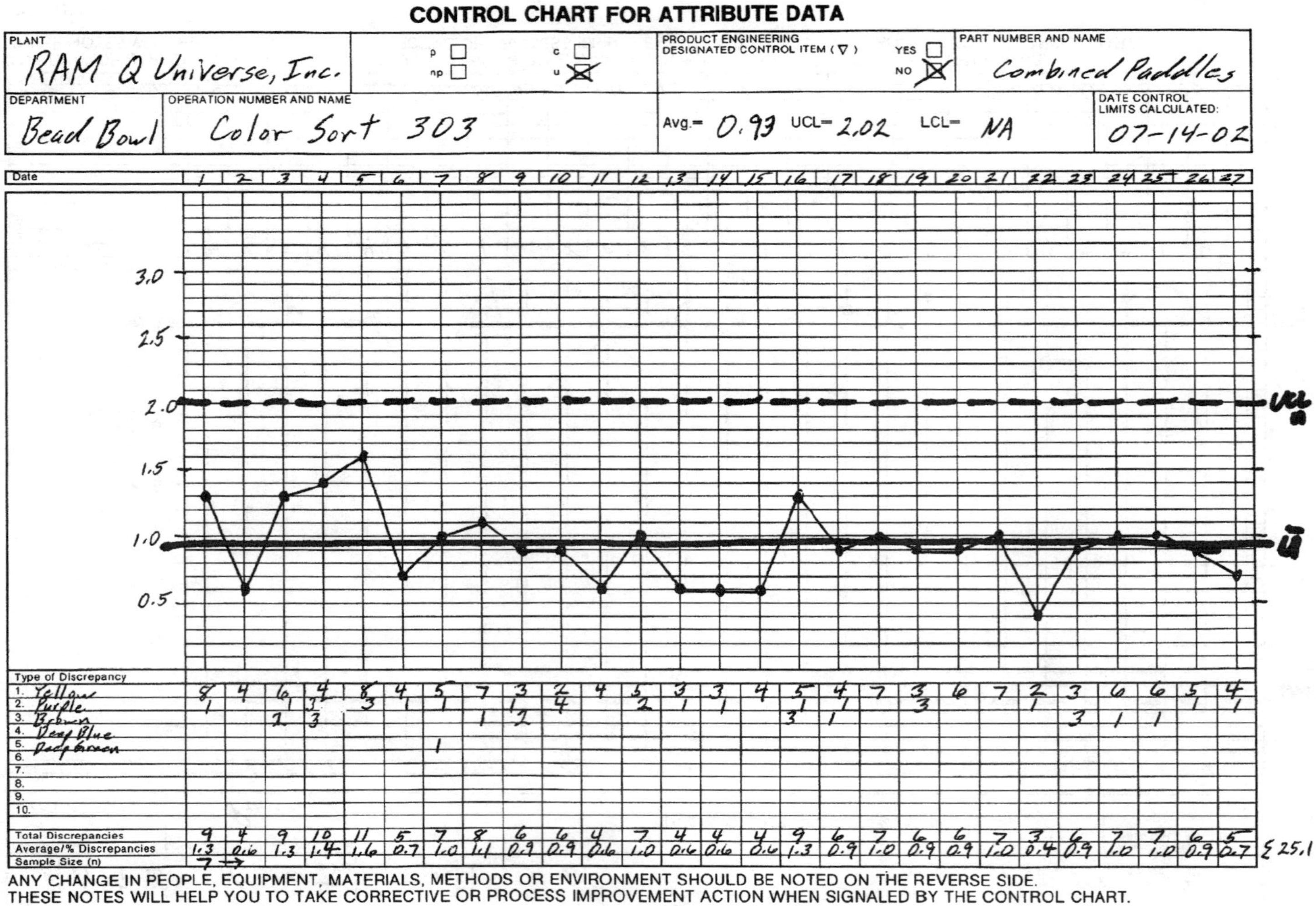

FIGURE 57.9 u chart.

of the primary reasons that many textbooks strongly suggest finding a variable measure in the process to monitor.

Tip. There is far too much material here to cover in a couple of pages; thus the list in the reference section. These books (Ishikawa 1971, Juran 1999, Munro 2002, Stamatis 1997, Wheeler 2001, and AT&T 1956) have a wealth of information on the application and use of these and other charts. The manufacturing engineer may also want to discuss the use of these charts with the quality office in your organization as they may have other applications in the company that you will be able to get ideas from.

Sample. The author's first use of one of these charts was the *individual and moving range* used on a late 1970s model vehicle to monitor gas mileage (this was before computer controls). By using the chart as a prevention tool, the author saved over a thousand dollars over a 3-year period on maintenance and other costs related to the use of the car.

57.4.11 Run Charts

A run chart is a line graph that shows measurements from a process, system, or machine in relationship to time (see Fig. 57.10). Virtually no calculations are required for this chart and it is very useful in monitoring process variation for patterns, trends, and shifts. The run chart (aka trend chart or line graph) can be used almost anywhere there is attribute or variable data.

Tip. Very simple chart to construct by hand, however, when comparing charts, ensure that the scales are the same! Many times computers will change the scale to make the chart fit the available space without notifying the user. Many false readings or interpretations have resulted from not watching the scale shift.

Sample. As with many of these tools, the run chart can be used at home as well as in the production process. The author has monitored home utility usage of water, gas and electric to look for ways of energy conservation and to monitor the processes.

57.4.12 Scatter Diagram

Scatter diagrams (aka correlation charts, regression plots) are pictorial ways of showing the relationships between two factors (see Figs. 57.11, 57.12, and 57.13). The base diagram lists each factor on one of the axes of the graph and plots the paired measured information. Patters in the data plots can show how much, if any, relationship there is and the strength of the relationship.

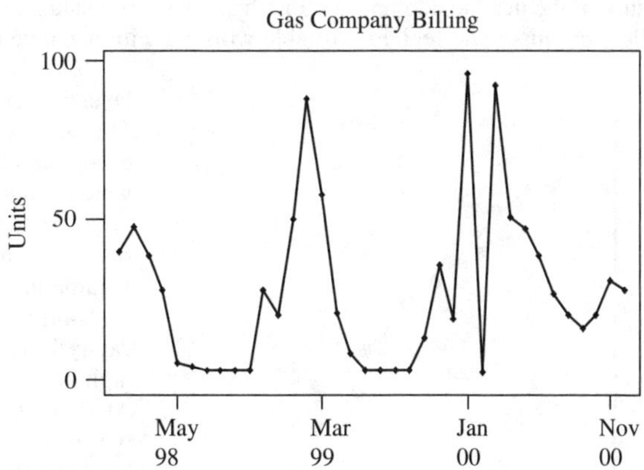

FIGURE 57.10 Run chart.

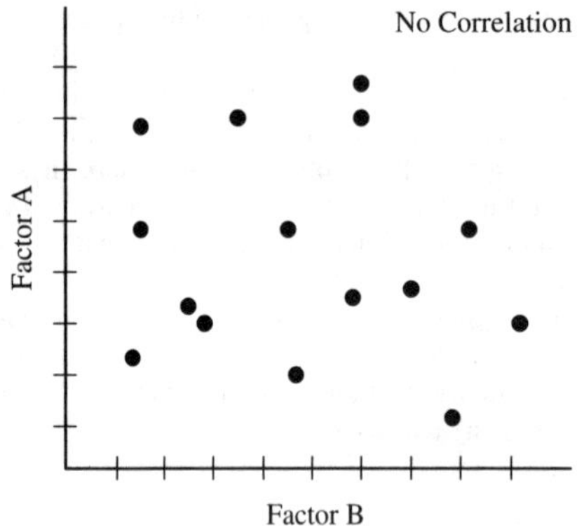

FIGURE 57.11 Scatter diagram—no correlation.

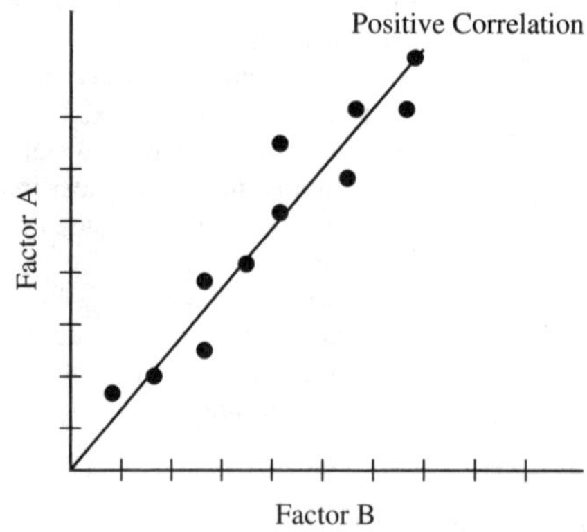

FIGURE 57.12 Scatter diagram—positive correlation.

Tip. Sometimes things may seem to be related while a third factor is actually the controlling element. (For example, you can prove with a scatter diagram that ice cream causes drowning! As ice cream sales go up, so do swimming accidents. The hidden factor is that it is summer.) Look for the causality of the factors that you are planning to study.

Sample. The author used this tool on one study to look at relationships between air temperature and liquid volume in a large chemical storage take. Management had supposed that the operators were overusing the chemical when in reality the temperature of the outside air variation caused the variation of usage. No one had taken this into consideration when the chemical mixing process was developed.

57.4.13 Short Run SPC

The short run SPC technique has been developed to use the same process behavior charts when frequent changeovers occur or short production runs are the norm. All of the same rules and charts apply with the one exception of how the data is plotted. Instead of plotting the actual measured data, a conversion of the data is made from the target value or nominal value for that specific process. Because of the need to add and subtract from the target value, operators will have to be able to handle a little more math and feel comfortable working with negative numbers.

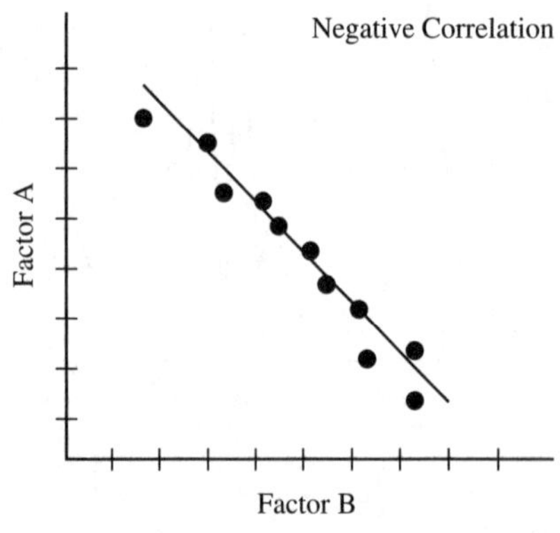

FIGURE 57.13 Scatter diagram—negative correlation.

Tip. Note that we are plotting the process behavior and not specific part measurements. This allows the short run SPC technique to work exceptionally well in a number of applications were changeover occur frequently and/or normal production consists of relatively short production runs, e.g., a machine shop, mold building, low volume industry as aerospace, and the like.

Sample. An injection machine with a large cavity low cycle time mold is able to produce a high number of parts with a short period of time. After studying the mold to ensure that each cavity is statistically capable, the engineer identified a cavity that is nearest the nominal value for each mold that is typically used in this machine. As each mold is set up for that days run, that one cavity is plotted on an Xbar and R chart using four consecutive shots of the same cavity once every 45 min of production time. This frequency and sample size were determined by the manufacturing engineer given the past history of the speed of the system and how often the process can change.

Process Capability. Process capability is a mathematical calculation to determine how the manufacturing process (*voice of the process*, VOP) compares with the engineering specifications. The intention is that the engineering specification will match the wants and needs of the customers who use the products and services we produce (*voice of the customer,* VOC). There are several different calculations that have been developed over the years with the most popular ones being the Cp and Cpk. This section will only deal with these two ratios.

Capability Ratio $\text{Cp} = \dfrac{\text{USL} - \text{LSL}}{6\text{s}}$

FIGURE 57.14 Cp calculation.

Cp is the process potential calculation. Cp looks at the engineering specification limits as the numerator and the manufacturing process spread (descriptive statistics six standard deviation value) as the denominator (see Fig. 57.14). If variable process behavior charts are being used in the manufacturing process, then the range value can be used to estimate the six standard deviations value and the engineering prints will contain the specifications that manufacturing is suppose to work with.

The numerical value can never be negative. A value of Cp = 1.0 indicates that the tails of the six standard deviation calculation meet the engineering specification width. (Note that this calculation does not take into consideration the location of the process to the engineering specifications.)

Process Capability $\text{Cpk} = \dfrac{\text{USL} - \overline{\overline{\text{X}}}}{3\text{s}}$

lesser value of: $\text{Cpk} = \dfrac{\overline{\overline{\text{X}}} - \text{LSL}}{3\text{s}}$

FIGURE 57.15 Cpk calculation.

Cpk give the process location in relationship to the engineering specifications. In this case, there are actually two calculations that need to be done with the resulting value being the lesser of—*upper specification limit* (USL) minus the process average (*X*) as the numerator divided by three standard deviations as the denominator, or the process average (*X*) minus the *lower specification limit* (LSL) as the numerator divided by three standard deviations as the denominator (see Fig. 57.15).

For Cpk, if the value is negative, that means that the manufacturing process average is outside one of the engineering specification limits (obviously not a good situation). The Cpk value can only be equal to or lower than the Cp value and gives a centering value of the manufacturing process to the engineering specifications.

Tip. We are not talking about the buzz around Six Sigma in this section. However, conversion charts are available (see Fig. 57.16) to look at how what is being called Six Sigma in industry today compares with process capability. Please note that many of the Six Sigma practitioners use a 1.5-sigma shift factor in calculating their values.

Cpk	Six Sigma	DPMO	Yield
2.00	6.0	3.4	99.99966
1.67	5.0	230	99.977
1.33	4.0	6210	99.379
1.00	3.0	66800	93.329
0.67	2.0	308000	69.2
0.33	1.0	690000	31

FIGURE 57.16 Six Sigma comparison chart.

Sample. The author was once called in to arbitrate a situation between a large company Automotive Original Equipment Operator (OEM) and one of their suppliers around the capability of a part being supplied. The customer wanted a Cp and Cpk of 1.33 minimum (±4 standard deviations). This was during the late 1980s when initial production Cp and Cpk were to be at 1.67 and ongoing production was to be 1.33. The part had been designed and tools cut in the very early 1960s and the part was designed to be at the low end of the specification (note that today we want engineering to design things to nominal—the middle of the specification). The tools and resulting parts had been produced for nearly 20 years with no issues at the assembly plants and never any warranty issues. The Cp was 20.0 but the Cpk was 0.5! The OEM wanted the supplier to fix the problem!

The solution given by the author was for the OEM engineering to change the specifications, but they did not want to do that because of the cost involved with hanging prints. The answer to this was to tell management that the engineering specification was going to be cut in half (resulting in a Cp of 10—still very good) and at the same time to center the specifications on the process average (resulting in a Cpk of 10). This would give the supplier relief of the new customer mandates that were not in place when the part was designed.

57.4.14 Other Tools

Other tools (Munro 2002) that could be used on the production floor include—*advance quality planning* (AQP), benchmarking, brainstorming, control plan, *cost of quality* (cost of poor quality), employee involvement, *failure mode and effects analysis* (FMEA), five Ss, lean manufacturing, *measurement system analysis* (MSA), process capability, *plan-do-study-act* (PDSA), sampling plan, *and standardize-do-check-act* (SDCA).

57.5 CONCLUSION

As we have seen, SPC is far more than just the traditional process behavior charts that are referenced in some books and articles. There are a number of statistical tools that can be grouped under the umbrella of SPC. Many of these tools have been used very successfully for many decades in production operations to help monitor and improve the process and parts in the plant. These are the same tools that are used in the Six Sigma methodology and many other quality programs that have been touted over the years. These tools work very well and are only limited by your imagination.

Cp and Cpk should always be used together to get an idea of the process potential and the process capability when you have a stable, in control process. The same application can be made to the measurement system that checks the process. Knowing how much measurement error might be in the process capability is especially important as manufacturing looks for ways to demonstrate the continual improvement requirements of many customers today. There are other ratios used for non-stable or not in control applications and the reader is directed to look at the Further Reading list for more detail.

REFERENCES

Deming, W. E. (1992). *Quality, Productivity, and Competitive Position.* MIT, Boston, MA.

Ishikawa, K. (1971). *Guide to Quality Control.* Asian Productively Organization: Kraus International Publications, White Plains, NY.

Shewhart, W. A. (1932). *Economic Control of Quality of Manufactured Product.* Van Nostrand Company, New York, NY.

FURTHER READING

AT&T (1956). *Statistical Quality Control Handbook.* AT&T Technologies, Indianapolis, IN.

Juran, J. M. (1999). *Juran's Quality Control Handbook,* 5th ed. McGraw-Hill, New York, NY.

Measurement Systems Analysis Work Group (2002). *Measurement System Analysis (MSA),* 3rd ed. Automotive Industry Action Group (AIAG), Southfield, MI.

Munro, R. A. *Quality Digest: Using Capability Indexes in Your Shop*. May 1992, Vol. 05, No. 12.

Munro, R. A. (2002). *Six Sigma for the Shop Floor: A Pocket Guide.* American Society for Quality, Milwaukee, WI.

Smith, R. D., Munro, R. A., and Bowen, R. (2004). *The ISO/TS 16949:2002.* Paton Press.

Stamatis, D. H. (2003). *Failure Mode and Effect Analysis: FMEA from Theory to Execution,* 2d ed., Revised and Expanded. American Society for Quality, Milwaukee, WI.

Stamatis, D. H. (1997). *TQM Engineering Handbook.* Marcel Dekker, New York, NY.

Stamatis, D. H. (2002). *Six Sigma and Beyond: Statistics and Probability.* St. Lucie Press, Boca Raton, FL.

Wheeler, D. J. and Poling, S. R. (2001). *Building Continual Improvement: SPC for the Service Sector,* 2d ed. SPC Press: Knoxville, TN.

CHAPTER 58
ERGONOMICS

David Curry
Packer Engineering
Naperville, Illinois

Albert Karvelis
Packer Engineering
Naperville, Illinois

58.1 INTRODUCTION

The word *ergonomics* (sometimes known in the United States as *human factors*) stems from the Greek words *ergon* (work) and *nomos* (laws) and is basically the science of analyzing work and then designing jobs, equipment, tools, and methods to most appropriately fit the capabilities of the worker. Its primary focus within the workplace lies on the prevention of injuries and the improvement of worker efficiency. Leaving aside the ethical issue of the employers responsibility to minimize potential injuries to their workforce, ergonomics can be a positive force in the workplace from an economic standpoint in at least two different ways: reduction in costs associated with work-related injuries (e.g., lost workdays, workman's comp costs, and associated medical costs) and increased profits through improvements in overall worker productivity. Excuses for ignoring ergonomics range from such old standbys as "this is the way we've always done it" to blaming problems on workers or unions to assertions that ergonomics is not an "exact science." Most (if not all) such platitudes are wishful thinking at best, and willful ignorance at worst. The fact of the matter is that ergonomic intervention in the workplace does not have to be complicated, and can potentially pay considerable dividends when skillfully employed. In one survey by Joyce Institute/Arthur D. Little, 92 percent of the respondents reported decreases in worker compensation costs of greater than 20 percent, 72 percent reported productivity gains in excess of 20 percent, and half reported quality increases exceeding 20 percent (Joyce, 2001). To put it briefly, the potential payoff in focusing on ergonomic considerations within the workplace is high, often paying back the effort and expenditure involved in the short-range time frame, while benefits continue to accrue over the long-term.

Examples of increased productivity are not difficult to find. One logging firm performed a simple ergonomic improvement to the seating and visibility in 23 tractor-trailer units at a cost of $300 per unit. As a result, downtime owing to accidental damage dropped by over $2000 per year per unit, and productivity increased by one extra load per day. This resulted in a cost savings of $65,000 per year for an total investment of $6900, an almost 10:1 payoff in a single year. In another case in Sweden, a steel mill ergonomically redesigned a semiautomated materials handling system. Overall noise level was reduced from 96 to 78 dB, production increased by 10 percent, and rejection rates were reduced by 60 percent. The system costs, including design and development, were paid back within the first 15 months.

Savings through reductions in worker injuries and absenteeism are also easy to document. In 1979, Deere and Company (the agricultural equipment manufacturer) implemented a company-wide ergonomics program involving extensive employee participation. By 1984, the firm had reduced workers compensation costs by a total of 32 percent; by 1996, they had recorded an 83 percent reduction in back-related injuries. In the early 1980s, the Union Pacific Railroad's Palestine Car Shop had the worst safety statistics among all of the firm's shop operations, with 579 lost and 194 restricted/limited workdays and 4 percent absenteeism among workers. After ergonomically redesigning the tasks and tools involved, the figures for 1988 showed a reduction in total injuries of almost two-thirds, in lost days from 579 to 0, in restricted days of almost 80 percent, and in absenteeism of 75 percent. Actual work performed in the same shop during the same period almost doubled, going from 1564 to 2900 cars per year (Warkotsch 1994; Hendrick 1996).

This chapter presents guidance on ergonomics in the working environment, the tasks being performed, and work methodologies.

58.2 THE WORKING ENVIRONMENT

58.2.1 Illumination

One of the most critical components of workplace design in terms of both productivity and worker comfort is that of adequate lighting. While the human visual system is functional across a range of 10^{16} levels of illumination, this does not imply that all result in equal performance!

Illuminance is the amount of light falling on a surface, while *luminance* is the amount of light reflected from a surface. To some degree, the suggested level of illumination within the working environment varies with such variables as the age of the worker, the nature of the task being performed, and the reflectance of the background, but *general* guidance is available (see Table 58.1).

Too high a level of illumination will result in unacceptable levels of *glare* (excessive brightness that exceeds the adaptation level of the eyes) and unacceptable levels of shadow, often obscuring critical detail. There are two types of glare which must be taken into consideration within the manufacturing environment. The first is *direct glare,* caused by having the source of illumination within the visual field of the employee. A number of methods can be used to control this problem (Rea 2000), such as

- Decreasing the luminance of the light sources
- Reducing the area of high luminance causing the glare
- Increasing the angle between the glare source and the line of vision

TABLE 58.1 Recommended Level of Illumination for Various Types of Tasks

Activity type or area	Range of illumination (lux)	Range of illumination (fcd)
Public areas with dark surroundings	20–50	2–5
Simple orientation for short visits	50–100	5–10
Working spaces where visual tasks are only occasionally performed	100–200	10–20
Performing visual tasks of high contrast or large size	200–500	20–50
Performing visual tasks of medium contrast or small size	500–1000	50–100
Performing visual tasks of low contrast or very small size	1000–2000	100–200
Performance of visual tasks of low contrast and very small size for prolonged periods of time	2000–5000	200–500
Performance of very prolonged and exacting visual tasks	5000–10,000	500–1000
Performance of very special visual tasks of extremely low contrast and small size	10,000–20,000	1000–2000

Source: Sanders and McCormick (1993).

- Increasing the level of luminance around the glare source
- Placing something between the glare source and the line of sight

In practice, lighting sources within the normal field of view should be shielded to at least 25° from the horizontal, with 45° being preferable to minimize direct glare.

Reflected glare is caused by the reflection of sources of illumination from shiny surfaces and can most easily be minimized by either utilizing less powerful sources of illumination or by reorienting work so that the light is not reflected into the worker's normal line of vision. *Discomfort glare* is a sensation of annoyance or pain caused by differences in brightness within the visual field, while *disability glare* is glare which interferes with visual performance. Disability glare (though not discomfort glare) appears to be strongly related to age, with older workers suffering more than younger under the same condition (Sanders and McCormick 1993).

There are a number of other important issues that must be considered with respect to lighting in addition to simple illumination and glare. Color perception, for example, is directly affected by the level of illumination present—below a certain level of illumination, the color receptors of the eye are nonfunctional. Further, the human eye sees largely based on differences in contrast between an object and its background (both color and brightness contrasts). Up to a certain point (about a 10:1 ratio), the higher the relative brightness contrast between the two (the *luminance ratio*), the greater the level of detail that can be perceived by the individual. It is also true, however, that the eye functions best where luminance levels are more or less constant throughout the rest of the workplace. There must be a trade-off with respect to this issue in the manufacturing environment, since it is not usually possible to maintain the same level of illumination throughout the entire area that might be possible within a smaller or more controlled environment. Recommendations for maximum luminance ratios for normal working environments (Rea 2000) are as follows:

- Between tasks and adjacent darker surroundings: 3 to 1
- Between tasks and adjacent lighter surroundings: 1 to 3
- Between tasks and remote darker areas: 10 to 1
- Between tasks and remote lighter areas: 1 to 10

Another issue which may be of concern is that of flicker. Most industrial lighting is provided by fluorescent fixtures which use magnetic ballasts and are connected to 60 Hz ac power systems. This results in the lights flickering at 120 times/s. While this is normally above the level of perception for most people and most tasks, it can present problems in some situations (notably visual inspection). Problems may include distraction, eyestrain, nausea, and increased visual fatigue. To alleviate this, the use of high-frequency electronic ballasts (10 to 50 kHz) or three phase lighting should be considered. (Rea 2000)

Finally, the type of illumination provided (direct or indirect) must be considered. For most tasks, indirect illumination is preferred in order to prevent objectionable shadowed areas; however, some tasks, such as fine visual inspection, may benefit from use of more direct lighting techniques to highlight imperfections.

58.2.2 Temperature

Another important environmental factor within the workplace is temperature. The body strives to maintain a consistent core temperature of approximately 98.6°F (37°C), and the less effort that is required to do this, the more comfortable the working environment is perceived to be. Optimal comfort conditions are those acceptable to 95 percent of the population, leaving 2.5 percent at either extreme. Research has shown that for 8-h exposures, temperatures between 66 and 79°F (19 to 26°C) are considered comfortable, provided that the humidity in the upper part of the range and the air velocity at the lower are not extreme. Temperatures within the range of 68 to 78°F (20 to 25.5°C) are generally considered more acceptable by most workers (Eastman Kodak 1983). When the environment

TABLE 58.2 Maximum Recommended Work Loads, Heat Discomfort Zone

		Maximum recommended work load*			
Temperature		Relative humidity			
°C	°F	20%	40%	60%	80%
27	80	Very heavy	Very heavy	Very heavy	Heavy
32	90	Very heavy	Heavy	Moderate	Light
38	100	Heavy	Moderate	Light	Not recommended
43	110	Moderate	Light	Not recommended	Not recommended
49	120	Light	Not recommended	Not recommended	Not recommended

Source: Eastman Kodak Company (1983).

is such that either heat is carried away from the body too rapidly or if excess heat cannot be removed fast enough, the result is (at least) discomfort for the worker. The body does have the ability to regulate its internal thermal environment within a limited range; however, temperature extremes can lead to a number of potential problems. High heat and humidity conditions result in increased worker fatigue and can lead to potential health hazards, while low temperatures may lead to decreased productivity owing to loss of hand and finger flexibility. Both conditions may result in increased worker distraction.

Heat transfer to or from the body comes about through two principal mechanisms: a change in the level of energy expenditure or a change in the flow of blood to the body's surface areas. *Vasodilation* is a process through which blood flow to the skin area is increased, leading to more rapid heat loss to the environment through both radiation and convection. If the core body temperature is still too high, sweating occurs in an attempt to maintain heat balance. *Vasoconstriction* involves a lessening of the blood flow to the skin area, decreasing its temperature and increasing shell insulation and thus reducing heat loss. In more extreme conditions, shivering (rapid muscle contractions) occurs to increase heat production.

Heat loss through the evaporation of sweat is limited by the level of moisture already existing in the air, and thus humidity may have a large affect on the subjective perception of discomfort at higher temperature level. Research has shown that an increase in humidity from 50 to 90 percent at a temperature of 79°F has been linked to up to a fourfold increase in discomfort level (Fanger 1977). Humidity levels of less than 70 percent are preferable during warmer seasons, and those above 20 percent are recommended for all exposures exceeding 2 h (ASHRAE 1974). Table 58.2 below presents recommended maximum workloads for 2-hour exposures at various heat and humidity levels.

Air velocity greater than 1.65 ft/s or durations less than 2 h will allow for heavier work within each condition. Examples for each of the work categories mentioned in the table are as follows:

- *Light.* Small parts assembly, milling machine or drill press operation, small parts finishing
- *Medium.* Bench work, lathe or medium-sized press operation, machining, bricklaying
- *Heavy.* Cement making, industrial cleaning, large-sized packing, moving light cases to and from a pallet
- *Very heavy.* Shoveling or ditch digging, handling moderately heavy cases (>15 lb) to and from a pallet, lifting 45 lb cases 10 times/min.

Several potential disorders may stem from severe or prolonged heat stress. Among them in order of severity are: *heat rash,* a rash on the skin resulting from blocked sweat glands, sweat retention, and inflammation; *heat cramps*, muscles spasms commonly in the arms, legs, and abdomen resulting from salt deprivation caused by excessive sweating; *heat exhaustion,* weakness, nausea, vomiting, dizziness, and possibly fainting owing to dehydration; and *heat stroke,* as the result of an excessive rise in body temperature and characterized by nausea, headaches, cerebral dysfunction, and possibly unconsciousness or death.

Several personal factors may play a large role in determining an individual's reaction to heat stress. Physically fit individuals perform work tasks with lower increases in heart rate and heat build up. Increasing age normally results in more sluggish activity by the sweat glands and less total body water, leading to lower thermoregulatory efficiency. Some studies have indicated that men are less susceptible to heat-related maladies than are women, but this may primarily be owing to a generally higher level of fitness.

Several performance-related problems may be associated with work performed under less than ideal temperature conditions. Reductions in body core temperatures to less than 96.8°F (36°C) normally result in decreased vigilance, while reductions below 95°F (35°C) are associated with central nervous system coordination reductions. For tasks involving manual manipulation, joint temperatures of less than 75°F (24°C) and nerve temperatures of less than 68°F (20°C) result in severe reductions in the ability to perform fine motor tasks. Finger skin temperatures of below 59°F (15°C) result in a loss in manual dexterity. As a rule, mental performance begins to deteriorate slightly with room temperatures above 77°C (25°C) for the unacclimated worker, though workers who have had to acclimatize themselves to higher temperatures may show no degradation in performance until temperatures above 86 to 95°F (30 to 35°C) are reached. Short-term, maximum strength tasks are typically not affected by high heat levels, but extended high-intensity work suffers greatly until workers become acclimatized to higher heat levels (up to 2 weeks) (Kroemer et al. 1997).

58.2.3 Vibration

Whole-Body Vibration. Evidence suggests that short-term exposure to whole-body vibration has limited physiological effects of negligible significance. For longer-term exposure, whole-body vibration effects tend to be more pronounced, both in the operator performance and the physiological areas—particularly to the lumbar spine. It should be noted, however, that the physiological affects of vibration are, in most cases, extremely difficult to isolate from those associated with awkward postures and extended sitting. Areas of particular concern to industry include such activities as vehicle, production, and power tool operations. Reported physiological effects for vibration in the 2 to 20 Hz range include abdominal pain, loss of equilibrium, nausea, muscle contractions, chest pain and shortness of breath (Eastman Kodak 1983). Loss of visual acuity owing to blurring occurs primarily in the 10 to 30 Hz range depending on the amplitude; some degradation in manual precision has been shown to occur with vibration in the 5 to 25 Hz range (Grandjean 1988). Tasks involving primarily mental activity (e.g., reaction time, pattern recognition, and monitoring) appear to be affected very little by whole-body vibration (Sanders and McCormick 1993).

Heavy vehicles, such as buses, commercial trucks, or construction equipment produce vibration in the 0.1 to 20 Hz frequency range with accelerations up to 0.4 g (about 13 ft/s^2 or 3.9 m/s^2), but predominantly less than 0.2 g (6.4 ft/s^2 or 1.9 m/s^2). Many sources (e.g., *The Occupational Ergonomics Handbook*) caution against prolonged exposure to vibration in vehicular environments, but the evidence supporting injury from prolonged vibration exposure in this environment is mixed and is confounded with such factors as awkward sitting postures, and lack of movement, and prolonged isometric and isotonic contraction of the back muscles. A recent study (Battie et al. 2002) employing monozygotic (identical) twins with large differences in lifetime driving exposure revealed no difference in disc degeneration levels in the lumbar spine as a function of occupational driving.

Most such vibration is oriented in the vertical direction (Wasserman and Badger 1974). The current U.S. standard with respect to whole body vibration is ANSI S3.18-2002 (which is identical to ISO 2631-1:1997). The standard provides a methodology for measuring vibration based on root-mean-squared averaging within defined frequency bands; measured vibration levels are then modified by a weighting function that is a function of the vibration frequency and the orientation of the vibration (e.g., x, y, or z dimensions). Figure 58.1 shows a diagram relating the weighted acceleration values to exposure times. The shaded area corresponds to a caution zone in which potential health risks exist; the area above the zone corresponds to exposure levels where health risks are likely to occur. According to the standard, health risks below the cautionary zone have either not been clearly documented or not been objectively observed.

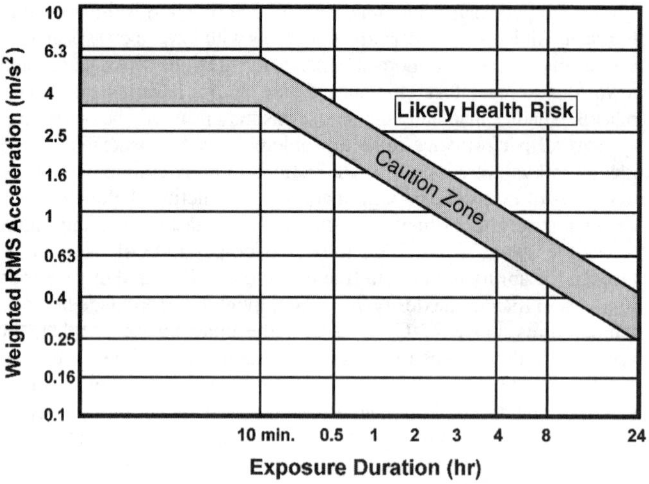

FIGURE 58.1 Health guidance zones. (*Source: International Standard Organization 2631-1:1997, Annex B*).

The standard also supplies general guidance with regard to operator comfort for use in public transportation. These values are presented in Table 58.3.

TABLE 58.3 Comfort Assessments of Vibration Environments

Vibration level (m/s^2)	Rider perception
Less than 0.315	Not uncomfortable
0.315 to 0.63	A little uncomfortable
0.5 to 1	Fairly uncomfortable
0.8 to 1.6	Uncomfortable
1.25 to 2.5	Very uncomfortable
Greater than 2	Extremely uncomfortable

Source: ISO 2631-1:1997, Annex C.

Segmental Vibration. Most structures within the human body have resonant bands in the 4 to 8 Hz range and long-term exposure to vibration in this band has been shown to produce negative effects in some cases. In the head and spinal regions, vibration between 2.5 and 5 Hz have an effect on the vertebrae of the neck and lumbar regions, while those between 4 and 6 Hz set up resonances between 4 and 6 Hz (Grandjean 1988). For the hands, vibration in the range from 8 to 500 Hz and from 1.5 to 80 g are of particular concern. Segmental vibration varies with the particular type of tool in question, as well as characteristics such as its weight, size, and design. Prolonged use of tools such as jack-hammers, drills, and riveters has been associated with the disease *Reynaud's Syndrome* which will be discussed later in this chapter. This disease is characterized by numbness in the hands or feet, finger cramps, loss of tactile sensitivity, and increased sensitivity to cold (Hutchingson 1981).

58.2.4 Noise

Sound can be defined as vibration in a medium that stimulate the auditory nerves, while *noise* is sound that is in some sense objectionable. Sound intensity is defined in terms of power per unit area, and is commonly expressed using the decibel scale, a logarithmic scale for comparing multiple sound pressure levels. The baseline for this scale, 0 dB, is defined as being 20 microbars of pressure per square meter, which represents the lowest level at which a 1000 Hz pure tone can be heard under ideal conditions by the average adult. In general, the ear is less sensitive to frequencies below 500 Hz and to those above 5000 Hz, and thus a sound of equal intensity in the 500 to 5000 Hz range is perceived by the listener as louder than one falling outside this range. There are a number of different sound pressure weighting scales that adjust the straight decibel scale to make bring it more in line with the human auditory system, the most common of which is A-weighting (expressed in dBA) which adds or subtracts up to 39 dB from the unweighted sound pressure level.

The decibel scale is a logarithmic one, thus an increase of 10 dB represents a tenfold increase in sound power; a 3 dB change represents a doubling of sound power, and a 6 dB change represents a doubling in the sound pressure level. The ratio between competing sound sources is obtained by subtracting the value for the quieter source from the louder. Since this scale is not linear in nature, simple addition or subtraction of noise contribution values from different sources does *not* represent of the final sound level (e.g., taking one machine producing 95 dB of noise from a pair of identical machines does not reduce the overall sound pressure level to 0 dB).

Figure 58.2 and the mechanics of computing sound pressure levels arising from multiple sources is borrowed from Peterson and Gross, Jr. (1972). Use of the chart below is relatively straightforward. To add two sound pressure levels, one first determines the difference in decibels between two noise sources, then examines the curved side of the graph to find the appropriate value, and finally reads the appropriate value off the left hand side of the chart to determine the value that should be added to the larger of the two components to obtain the total. Example: Find the total sound pressure level from combining a 90 and a 98 dB source. The difference between the two is 8 dB. Reading off the left side of the chart, one obtains a value of 0.6. The total value then is 98 + 0.6 or 98.6 dB.

The process is slightly different for subtraction. In this case, one finds the difference between the total sound pressure level and the contribution from the source to be subtracted. If this difference is less than three, one finds the appropriate value off the left-hand side of the graph and follows the appropriate line rightward across to the curved section. The value from the curved section is then

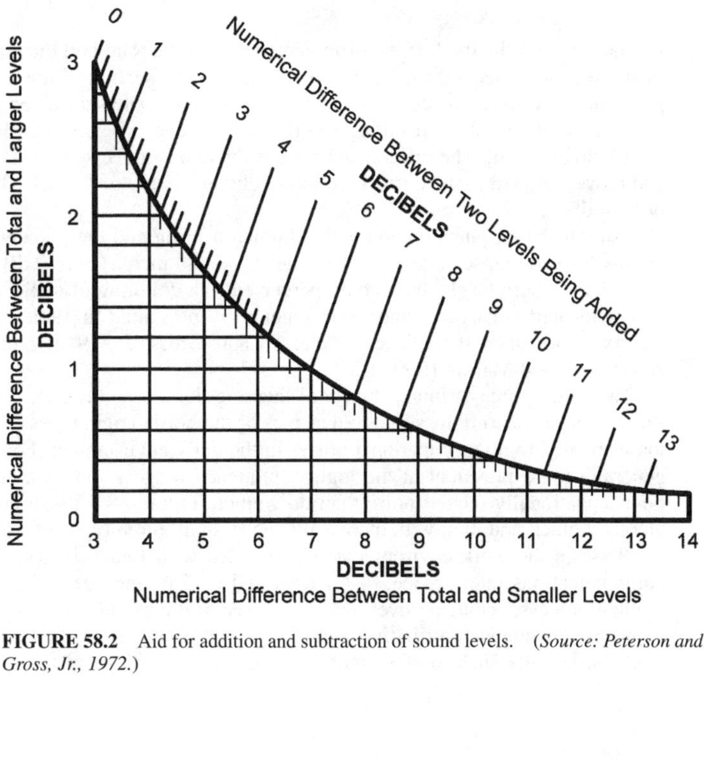

FIGURE 58.2 Aid for addition and subtraction of sound levels. (*Source: Peterson and Gross, Jr., 1972.*)

TABLE 58.4 Representative Sound Levels and Associated Hearing Risks

Environment/task	Normal sound level (dB)	Hearing risk
Wristwatch at arms length	10	—
Quiet whisper at ear	20	—
Quiet bedroom	40	—
Average conversation at 3 ft	60	—
Average automobile, Freight train @100 ft, Vacuum cleaner	70	—
Airline cabin, Pneumatic drill @ 50 ft	85	Damage risk limit
Textile weaving plant, Boiler room, Print press plant, Power mower with muffler	90	—
Electric furnace area	100	—
Casting shakeout area, Riveting machine, Cutoff saw, Chainsaw	110	—
Propeller aircraft @ 100 ft, Rock concert 6 ft from speakers, Jackhammer	120 140	Threshold of discomfort
Jet takeoff @ 100 ft	160	Threshold of pain Eardrum ruptures

subtracted from the total. If the difference between the total and the source to be subtracted is more than three, one selects the difference from the bottom section of the graphic and follows the appropriate line upward to the curve to determine the value to be subtracted from the total sound pressure level. Example: Find the resulting sound pressure level when an 85 dB machine is removed from a 90 dB environment. The difference between the two is 5 dB, so one starts at the bottom of the graph and moves upward to the curved section to obtain a value of 1.6 dB. The total value is then 90 – 1.6 or 88.4 dB.

An interesting phenomenon is that human hearing and the perception of loudness operate on a largely logarithmic scale as well. This means that an increase of 10 dB equates to a tenfold increase in sound pressure levels, but is only perceived as a doubling of subjective loudness. Representative values for particular tasks and environments are presented in Table 58.4 along with physiological effects of sound at these levels (Peterson and Gross, Jr. 1972; Sanders and McCormick 1993; Karwowski and Marras 1999).

There are three primary types of hearing loss: *presbycusis*, *sociocusis,* and occupational. Presbycusis is hearing loss incurred as part of the normal aging process, while sociocusis is hearing losses related to nonoccupational noises in the daily environment. For the most part, presbycusis is generally more prevalent at the higher frequency ranges, and men suffer from more exposure to nonoccupationally related noise than do women. Figure 58.3 illustrates the average threshold shift among women and men with increasing age from all three types of exposure combined.

Noise in the work environment is associated with four primary negative effects: hearing loss, communications interference, distraction, and performance degradation. Hearing loss is usually a gradual process, occurring over a period of years of exposure. Intensity, frequency, and duration of exposure are major contributing factors to such losses, as well as individual differences between workers. Usually such loss occurs first in the upper portions of the frequency range, resulting in a

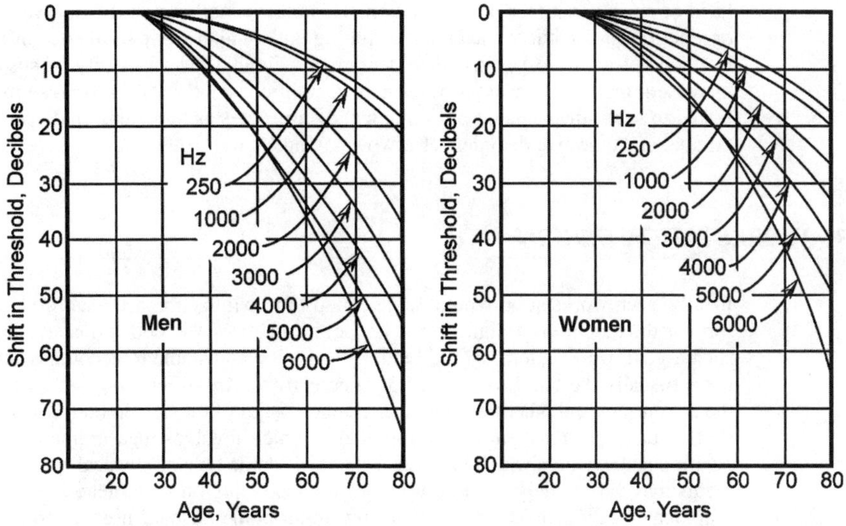

FIGURE 58.3 Average shift in hearing threshold with age. (*Source: Peterson and Gross, Jr., 1972.*)

loss of in the clarity and fidelity of sounds. A 1990 estimate by the National Institute of Health was that over 10 million people in the United States alone have significant noise-related hearing loss.

Research has shown that noise levels below 85 dB are usually not associated with ear damage, though they may contribute substantially to distraction and loss of productivity (particularly those in excess of 95 dB). Current OSHA regulations mandate that employers institute a hearing conservation program when employee noise exposures equal or exceed an 8-h time weighted average of 85 dBA. If daily exposure exceeds 90 dBA, steps must be taken to reduce exposure either through work scheduling or the institution of engineering controls. Hearing protection is mandated when scheduling or engineering controls fail to reduce exposure below the permissible level. Table 58.5 details the maximum OSHA-permissible exposure to continuous noise at various intensity levels.

Performance wise, information transfer tasks involving high detail and complexity are the first to show degradation in response to noise—noise which is variable or intermittent in nature or at frequencies above 2000 Hz generally most likely to interfere with work (Eastman Kodak 1983). To obtain reliable effects on performance, it is usually necessary for noise levels to exceed 95 dB in intensity. Broadbent (1976) identified three psychological effects of noise in the workplace. First, confidence in decisions is increased, though such confidence may or may not be justifiable. Second, attention is more focused almost exclusively on the most critical aspects of the task at

TABLE 58.5 Maximum Permissible Noise Exposure

Duration per day (h)	Sound level (dBA)
8	90
6	92
4	95
3	97
2	100
1–1/2	102
1	105
1/2	110
1/4 or less	115

Source: 29 CFR, 1910.95.

hand or primary sources of information. This may actually serve to improve performance on simple or repetitive tasks, but can lead to the "tuning out" of alternative sources of information or other critical tasks, leading to performance decrement. Finally, the variability of sustained performance is increased, though the average level may remain constant. Prolonged noise exposure has also been linked to such stress-related physiological disturbances as hypertension, heart irregularities, extreme fatigue, and digestive disorders (Karwowski and Marras 1999).

58.3 WORKSTATION DESIGN

Ideally, each workstation would be designed to maximize the accessability, productivity, and comfort for the individual assigned to it. Since, in practice, it is both expensive and difficult to achieve such a goal, workstations should be designed to be adjustable to accommodate a range of potential users (usually the 5th through the 95th percentiles). *Anthropometry* deals with the measurement of the dimensions and other physiological characteristics of the human body. Such measurement focuses on two primary areas: *static* anthropometry which involves measurements of the body in a fixed position and *dynamic* which involves the body is some type of physical activity. Since such measurements may vary widely from group to group depending on such factors age, gender, race, and geographical location, it is critical that any anthropometric data used in the design of products or workplaces be appropriate to the population that will be employing them.

Table 58.6 provides some important workstation design–related dimensions for 5th, 50th, and 95th percentile humans (Kroemer 1983; Panero and Zelnick 1979; Das 1998). In practice, designing for the range from a 5th percentile female to a 95th percentile male will accommodate 95 percent of the total population, owing to the overlap between the two groups. Standing dimensions assume a relaxed (slumped) posture. It is important to note that since there is normal variation across body dimensions across individuals, that the simple addition of dimensional values within a given percentile range for all the body elements will not produce a composite at the same percentile value (i.e., addition of 5th percentile legs, trunk, head, and the like will not produce a representative composite 5th percentile body). One study has shown that taking a composite of 5th percentile values for all body components, for example, produces a composite approximately 6 in shorter in height than an actual 5th percentile human (Robinette and McConville 1981).

58.3.1 Types of Workstations

Generally, workers either sit or stand while performing work and the choice between the two types of workstation should be based on the nature of the job being performed. Standing workstations are appropriate in cases where a large work area is required, heavy or frequent lifting or the moving of large or heavy objects is required, or large forces must be exerted by the hands and arms. Seated workstations are appropriate for jobs requiring extended periods of time, since they involve less fatigue and generally less stress on the body. Sit/stand workstations in which employees are provided with a work surface high enough to do standing tasks as well as an appropriately designed elevated seat may be a good compromise for tasks involving both sitting and standing operations (Wickens at al. 1998).

Each of these types of workstation involves some negative consequences. Prolonged standing work may lead to both excessive fatigue and fluids pooling in the legs, so the use of mats and frequent rest breaks is recommended for worker comfort. Studies have indicated that, since the human spine and pelvis rotate when sitting, loads on the lumbar spine are up to twice that for a standing posture (Grandjean 1988). This increases the likelihood of backaches or other spinal problems with prolonged sitting. Research also indicates that the load on the lumbar vertebrae is increased even farther if the feet are not planted firmly, allowing the legs to take some of the strain off of the spine itself. This means that the use of a proper seat, with adequate foot support, is critical for seated use of sit/stand workstations.

TABLE 58.6 Selected Anthropometric Dimensions

Dimension	Gender	Percentiles 5	50	95
Standing			**in (cm)**	
Height (standing)	Male	65.4 (166.2)	69.3 (176.1)	73.3 (186.3)
	Female	60.6 (153.8)	64.3 (163.2)	68.4 (173.8)
Eye height (standing)	Male	63.0 (160.0)	64.9 (164.9)	68.8 (174.8)
	Female	56.5 (143.6)	60.4 (153.5)	64.3 (163.4)
Shoulder height (standing)	Male	54.3 (138.0)	58.1 (147.7)	61.7 (156.8)
	Female	50.4 (127.9)	54.1 (137.5)	57.7 (146.6)
Elbow height	Male	41.6 (105.6)	44.5 (113.0)	47.4 (120.4)
	Female	39.0 (99.0)	41.4 (105.1)	43.8 (111.2)
Elbow–elbow breadth	Male	15.7 (40.0)	17.8 (45.1)	20.4 (51.7)
	Female	14.1 (35.7)	15.0 (38.2)	17.2 (43.8)
Body depth	Male	10.5 (26.7)	11.9 (30.2)	13.4 (34.0)
	Female	8.6 (21.8)	9.7 (24.6)	10.9 (27.6)
Arm length	Male	26.9 (68.3)	29.6 (75.2)	32.3 (82.0)
	Female	23.7 (60.2)	26.0 (66.0)	28.5 (72.4)
Forearm length	Male	14.6 (37.1)	15.9 (40.4)	17.2 (43.7)
	Female	12.8 (32.5)	14.4 (36.6)	16.0 (40.7)
Seated				
Height	Male	32.3 (82.1)	34.5 (87.6)	36.5 (92.7)
	Female	30.9 (78.5)	32.6 (82.8)	34.3 (87.1)
Eye height	Male	27.9 (70.9)	30.0 (76.2)	32.0 (81.3)
	Female	27.0 (68.6)	29.6 (75.2)	32.3 (82.0)
Elbow rest height (seat to elbow)	Male	7.4 (18.8)	9.1 (23.1)	10.8 (27.4)
	Female	7.4 (18.8)	9.1 (23.1)	10.8 (27.4)
Thigh clearance height (seat to top of thigh)	Male	5.6 (14.2).	6.4 (16.2)	7.3 (18.5)
	Female	4.9 (12.4)	5.7 (14.4)	6.5 (16.5)
Knee height	Male	19.3 (49.0)	21.4 (54.4)	23.4 (59.4)
	Female	17.9 (45.5)	19.6 (49.8)	21.5 (54.6)
Buttock to knee distance	Male	21.3 (54.1)	23.3 (59.2)	25.2 (64.0)
	Female	20.4 (51.8)	22.4 (56.9)	24.6 (62.5)
Popliteal height (floor to bottom of thigh)	Male	16.7 (42.4)	18.0 (45.7)	19.2 (48.7)
	Female	14.7 (37.3)	16.0 (40.6)	17.2 (43.6)
Hip breadth	Male	12.2 (31.0)	14.0 (35.6)	15.9 (40.4)
	Female	12.3 (31.2)	14.3 (36.3)	17.1 (43.4)

Source: Kroemer (1989), Panero and Zelnick (1979), Das (1998).

Standing Workstations. Two factors are critical for determining work surface height for any type of work being performed: elbow height and the type of work to be performed. For normal handwork, the optimal height is normally between 2 and 4 in (5 to 10 cm) below standing elbow height, with the arms bent at right angles to the floor. For more precise handwork, it may be desirable to support the elbow itself and to raise working materials closer to the eyes; in such cases, surface heights will need to be slightly higher. Table 58.7 provides an illustration of appropriate heights and ranges of adjustments for standing workstations. Values for fixed heights are appropriate to accommodate taller workers and assume that platforms will be available to lift smaller workers to the appropriate heights.

TABLE 58.7 Recommended Standing Work-Surface Heights for Three Types of Tasks

		Fixed height		Adjustable height	
Type of task	Gender	in	cm	in	cm
Precision work (elbows supported)	Male	49.5	126	42.0 to 49.5	107 to 126
	Female	45.5	116	37.0 to 45.5	94 to 116
Light assembly	Male	42.0	107	34.5 to 42.0	88 to 107
	Female	38.0	96	32.0 to 38.0	81 to 96
Heavy work	Male	39.0	99	31.5 to 39.0	80 to 99
	Female	35.0	89	29.0 to 35.0	74 to 89

Source: Sanders and McCormick (1993).

Horizontal work surface size is based on the concepts of normal and maximal work areas for standing or sit-stand workers, originally proposed by Barnes (1963) and Farley (1955). The normal work area is defined as that which can be conveniently reached by the worker with a sweep of the forearm while the upper arm remains stationary in a relaxed downward position. The maximal work area is that which can be reached by extending the arm from the shoulder without flexing the torso. The values for these areas were further modified by Squires (1956) to account for the dynamic interaction of the forearm and moving elbow. All of these areas are illustrated in Fig. 58.4. Items which are frequently used in normal work (particularly for repetitive tasks) should be located within the normal working area, while those that are used only infrequently, may be placed farther out within the maximum working area. For lateral size, one formula for obtaining minimum clearance at the

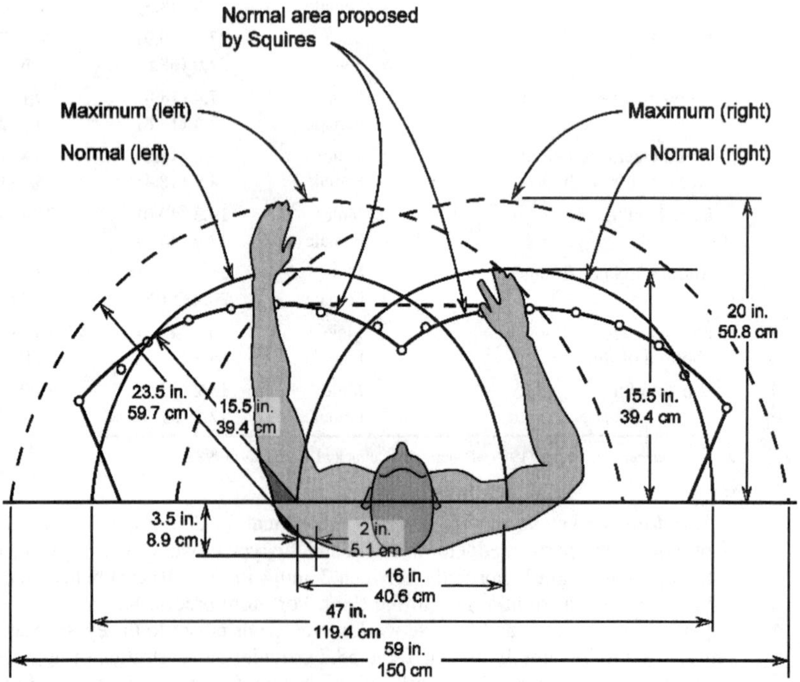

FIGURE 58.4 Normal and maximum working areas (in inches and centimeters) proposed by Barnes and normal work area proposed by Squires. (*Sanders and McCormick 1993.*)

waist level is to add 4 in (10 cm) to the waist dimensions of the worker. At the elbow level, an equivalent formula would be to add the same value to the sum of the elbow-to-elbow value plus the body depth.

To minimize fatigue and discomfort, workers should try to adopt as near to a neutral standing position as possible, by standing with ears, shoulders, and hips in the same plane and the spine erect. Arms should remain close to the body, with the elbows at the sides. When working in a standing position for extended periods, either one foot or the other should be supported on a short footstool, antifatigue mats should be employed, and the weight should be shifted often to reduce prolonged standing in a static position.

Seated Workstations. Two of the most critical issues regarding seated workstations are those of work-surface height and sitting position. To minimize stress on the body, it is preferable to place the body in as close as possible to a neutral posture in order to minimize the amount of stress on the skeletal and muscle system (think of an astronaut floating vertically in space). In order to accomplish this, seating should be adjustable so that the angles assumed by the knees and hips are approximately 90°, with the feet planted flat on the floor. Elbows should be at approximately 90°or inclined slightly downward for keyboard work, with the arms fully supported by armrests, wrists flat, and the shoulders falling freely (not pushed upwards by armrests or muscular effort). The spine should be supported in its natural curvature through lumbar support on the chair. Given that most work surfaces have only limited vertical adjustment yet still must accommodate taller workers, this sitting position may require the employment of some type of floor-mounted footrest for shorter workers.

With regard to the work surface itself, dimensions should correspond to those of the standing workstation described earlier, but in no case less than 20 in. in width. The area beneath this surface should be large enough to provide adequate thigh, leg, and foot clearance for even the tallest workers. The Human Factors Society recommends a minimum depth of 15 in (38 cm) at knee level and 23.5 in (59.0 cm) at toe level, a minimum width of 20 in (50.8 cm), and a minimum height of 26.2 in for nonadjustable surfaces and from 20.2 to 26.2 in for adjustable surfaces (51.3 to 66.5 cm) (ANSI/HFS, 1988). The latter measurement is from the floor to the bottom of the working surface.

Visual Displays. For visual displays, the American National Standards Institute (ANSI) specifies that text size shall be a minimum of 16 min of visual angle for reading tasks in which legibility is important and recommends a character height of 20 to 22 min for such tasks (HFES 1988). For the normal user, the typical eye to screen distance for seated work is approximately 20 in, making the minimum text height 0.09 in (2.3 mm) and the preferred height between 0.116 and 0.128 in (2.9 and 3.3 mm). Maximum character height should be no more than 24 min of visual angle (0.14 in or 3.6 mm) to minimize reading time. For longer reading distances, these values should be increased accordingly.

The effects of display polarity (dark text on light background or light text on dark background) on readability are mixed, though some do show performance superiority for the use of a light background. Almost all current generation monitors allow for the adjustment of display refresh speed. This speed should be set at no less than 72 Hz, and preferably as high as possible to avoid perception of flicker, particularly when using a very large monitor or a light background to display text. The monitor should be placed directly in front of the user, requiring no rotation of the torso.

For seated workstations where computer operations are conducted, the monitor should be positioned at a height where the topmost line of text is at or slightly below the worker's eye level. For bifocal wearers, such a placement may require an uncomfortable positioning of the head and neck to place the text within the close-focus (bottom region) of the eyeglasses; in such cases, the use of single-vision reading glasses rather than bifocal lenses is recommended. Keyboards should be placed directly in front of the body, and their vertical position should be such that the forearms form an angle from 90°to 110° with the upper arms while typing. Traditional keyboard designs normally require that the arms be angled inward while the wrists are angled outward for proper keying positions. This position can cause a variety of musculoskeletal disorders over periods of extended use. When possible, the use of split keyboards should be considered in order to allow the arms to conform to a more natural position and reduce the impact of extended use on the hands, wrists, and arms. When typing for more than 1 or 2 h at a time, periodic short rests and stretching exercises are strongly encouraged.

58.4 WORK DESIGN

58.4.1 Work/Rest Cycles

Heavy work has been defined as "any activity that calls for great physical exertion, and is characterized by high energy consumption and severe stresses on the heart and lungs" (Grandjean 1988). Modern tools and equipment have reduced the energy expenditure involved in a great number of tasks, but workers, like any other mechanism still expend energy while involved in work no matter what the nature of the task might be. In practice, at a maximum, the large muscles of the human body convert a maximum of about 25 percent of the energy that enters the body in terms of fuel into useful work (Helander 1995), making them about equivalent to an internal combustion engine (Brown and Brengelmann 1965).

Work can be broken down into a number of categories based on the energy expenditure involved in its accomplishment. Energy expenditure in humans is normally measured in kilocalories, and there have been a number of recommendations regarding the appropriate upper limit for work-related energy expenditures. Lehmann (1958) estimated that maximum sustained level of energy expenditure for healthy adult male was approximately 4800 kcal/day. After subtracting energy required for basic body functions and leisure activities, this left a maximum of 2500 kcal available for work, or 5.2 kcal/min for an 8-h day. A maximum of 5.0 kcal/min for men and 3.35 kcal/min over an 8-h workday was recommended by Ayoub and Mital (1989), with higher values for 4 h of work (6.25 and 4.20, respectively) (5 kcal/min is approximately the energy expenditure of the average man walking at 3.5 mph). The maximum recommended by the National Institute for Occupational Safety and Health was 5.2 kcal/min for a healthy 35-year-old male worker, though in practice they advocate a value of 3.5 kcal/min so as to accommodate more female or older workers. Table 58.8 below shows a categorization for work levels and some sample energy expenditures for selected types of tasks.

Since these recommended maximum expenditure rates above represent averages over extended periods of time, for work which exceed these levels—periodic breaks for rest are necessary during

TABLE 58.8 Energy Expenditure Rates for Typical Tasks

	Energy expenditure per minute (kcal/min)		
Grade of work	Whole body	Upper body only	Examples
Light work	1.0 to 2.5	Less than 1.8	Lecturing, nursing, light assembly, printing, bench soldering
Moderate work	2.5 to 3.8	1.8 to 2.6	Brick laying, press operation, machine sawing
Heavy work	3.8 to 6.0	2.6 to 4.2	Carpentry, digging, pushing wheelbarrows, lathe operation, most agricultural work
Very heavy work	6.0 to 10.0	4.2 to 7.0	Jackhammer operation, chopping wood, shoveling (7 kg weight), medium-sized press operation, sawing wood by hand
Extremely heavy work	Over 10.0	Over 7.0	Firefighting, repeated heavy lifting (30 kg boxes, 10/minute, 0 to150 cm)

Source: Eastman Kodak Company (1986).

the course of the day. Length and scheduling of such rest periods is often of major concern. One simple method for calculating required rest is to use the equation below (Murrell 1965).

$$R = \frac{T(W - S)}{W - BM}$$

Where R = rest required in minutes
T = total work time in minutes
W = average energy consumption of work in kcal/min
S = recommended average energy expenditure in kcal/min
BM = the basal metabolic rate in kcal/min (Men = 1.7, women = 1.4, general population = 1.5)

Using the equation above, if a worker were engaged in work at a level of 6.0 kcal/min for a period of 15 min and assuming a maximum recommended level of 5.0, then the appropriate rest break would be about 3.5 min. In practice, the general values for basal metabolic rate (the energy expenditure rate of the body at rest) above can be used, or an individual estimate can be made using the equation $70W^{0.75}$, where W is the body weight in kilograms.

Another method for calculating the appropriate working period uses the equation

$$TW = \frac{25}{x - 5}$$

Where TW is length of working time and x represents the level of energy expenditure in kilocalories. For the example above, the optimum work period would be 25 min. Length of the recovery period is then calculated by the equation

$$TR = \frac{25}{5 - a}$$

Where a is the basal metabolic rate (i.e., 1.5 kcal/min). The appropriate rest period in this instance would be approximately 7 min. Length of rest periods and their frequency should be adjusted upward for temperatures above normal.

It is important to recognize that both of these methods deal with average expenditures over extended working periods, followed by periods of rest involving little or no activity. In practice, it is often possible to rest workers by interspersing work tasks with lower energy expenditure levels with those with greater levels. Work involving levels of 5 kcal/min or less can be performed for extended periods with no rest without risk workers experiencing undue fatigue.

58.4.2 Manual Materials Handling

Manual materials handling involving handling, lifting, or dragging loads often involve sufficient effort as to be classified as heavy work. The primary concern surrounding such activities, however, is not that of energy expenditure, but that of the load they impose on the intervertebral discs of the spine and their potential for causing back problems (Grandjean 1988). According to the National Institute for Occupational Safety and Health (NIOSH 2001), in 1996 there were a total over 525,000 overexertion injuries in the private sector in the United States, with almost 60 percent of these related to lifting activities (2001). The median number of days away from work was 5, though 23 percent were related to over 20 days work loss. The Department of Labor reports that nearly 20 percent of all occupational injuries are to the back and that these account for roughly 25 percent of all workers' compensation costs.

The lower back, particularly the spinal disc between the fifth lumbar and first sacral vertebrae (colloquially known as L5/S1), is one of the most vulnerable points in the entire human musculoskeletal system (the L4/L5 disc is also a problem area). When bending and lifting, the horizontal distance between this disc and the point where the weight is applied (the shoulders) acts as a lever arm, multiplying the forces that are applied at this point in the lower back. The forward movement imposed by the weight of the torso and load have to be counterbalanced by muscle force exerted in the opposite direction by the lower spinal muscles. Since their attachment point is much closer to the point of rotation than that of the load being lifted (about 2 in vs. the distance from the lower back to the shoulders), they have far less leverage and commensurately more force must be exerted. Even worse, this load is not equally distributed over the entire surface of the disc, being concentrated primarily on the front side. This is why individuals are often exhorted to keep their back as vertical as possible when lifting by keeping the load close to the body trunk as possible—the horizontal distance between the lower spine and the shoulders is minimized in such a posture.

NIOSH has developed an equation for evaluating two-handed lifting tasks for healthy adult workers. The *recommended weight limit* (RWL) was developed for a specific set of task conditions and represents the weight load that nearly all healthy workers could lift over a substantial work period (e.g., up to 8 h) without increased risk of lifting-related low back pain. The RWL is used along with the weight actually lifted (*L*) to calculate the hazard level or *lifting index* (*LI*) using the equation

$$LI = \frac{L}{RWL}$$

In instances where the *LI* is greater than 1.0, the task *may* pose a risk for some workers and a redesign of the lifting task is recommended. In cases where $LI > 3$, however, many or most workers exposed to the task *are* at high risk of developing low-back pain or injury (Wickens et al. 1998). The RWL is designed to protect about 95 percent of male workers and 85 percent of female workers.

The equation for calculating the *RWL* (Waters et al. 1994) is expressed as

$$RWL = LC \times HM \times VM \times DM \times AM \times FM \times CM$$

where

LC is the load constant. This value is 51 lb in U.S. units or 23 kg in metric units and represents the maximum weight recommended for lifting.

HM is a horizontal multiplier. *HM* is 10/*H* in inches or 25/*H* in centimeters, where *H* is the horizontal distance of the hands when measured from a midpoint between both ankles. For lifts in which precision is required for final placement, *H* should be measured at both the origin and endpoint of the lift. *H* assumes a minimum value of 10 in (25 cm) and a maximum value of 25 in (63 cm). For values higher than the maximum, this value should be set to 0.

VM is a vertical multiplier. *V* represents the vertical height of the hands from the floor and should be measured from the middle knuckle at both the origin and the destination of the lift. *VM* is based on the deviation of *V* from the optimal height of 30 in (75 cm), and is calculated using the formula

$$1 - (.0075(|V - 30|))$$

when *V* is measured in inches, and

$$1 - (.003(|V - 75|))$$

when *V* is measured in centimeters. While there is no minimum value, a maximum value is set based on reach; *VM* is 0 for values in excess of 70 in or 175 cm.

DM is a distance multiplier and is calculated using the equation

$$.82 + \frac{1.8}{D}$$

when measured in inches and

$$.82 + \frac{45}{D}$$

when measured in centimeters. D in these equations is the vertical travel distance between the origin and destination of the lift. It assumes a minimum value of 10 in (25 cm) and a maximum of 70 in (175 cm). For distances greater than 70 in, DM should be set to 0.

AM is an asymmetric multiplier. Ideally, all lifts should begin and end without requiring the body to be rotated from its normal position, but asymmetric lifts occur when the origin and destination are oriented at an angle from one another, lifting across the body occurs, or lifting is done to maintain body balance in obstructed areas. The asymmetry angle (A) is defined as the angle between the asymmetry line and the mid-sagittal line. The asymmetry line is a horizontal line midway between the ankle bones and a point on the load directly between the midpoint of the gripping points of the load. The sagittal line is defined as being the line passing through the midpoint between the ankle bones and the mid-sagittal plane (one that splits the body into two equal right and left halves) of the body while in a neutral posture (i.e., no twisting of the torso, legs, or shoulders, and the hands directly in front of the body). This angle is assumed to be between 0° and 135° and is set to 0 for values in excess of 135° (i.e., no load should be lifted). The AM is calculated from the equation

$$1 - .0032A$$

CM is a coupling multiplier. This value represents an index which evaluates the effectiveness of the contact between the hand and the object being lifted (i.e., how good of a grip the lifter can obtain on the object). A good coupling will decrease the maximum grasp forces which must be exerted and increase the maximum weight which can be lifted, while a poor coupling increases the grasp force required, and lowers the maximum weight which can be lifted. Coupling quality is assessed based on the criteria presented in Table 58.9. Values for the CM itself can be obtained from Table 58.10.

TABLE 58.9 Coupling Assessment for the Revised NIOSH Lifting Equation

Good	Fair	Poor
1. For containers of optimal design, such as some boxes, crates, and the like a "good" hand-to-object coupling would be defined as handles or handhold cutouts of optimal design.[a,b,c]	**1.** For containers of optimal design, such as some boxes, crates, and the like a "fair" hand-to-object coupling would be defined as handles or handhold cutouts of less than optimal design.[a,b,c,d]	**1.** Containers of less than optimal design or loose parts or irregular objects that are bulky, hard to handle, or have sharp edges.[e]
2. For loose parts or objects that are not usually containerized, such as casting, stocks, and supply materials a "good" hand-to-object coupling would be defined as a comfortable grip in which the hand can be easily wrapped around the object.[f]	**2.** For containers of optimal design with no handles or handhold cutouts or for loose parts or irregular objects, a "fair" hand-to-object coupling is defined as a grip in which the hand can be flexed about 90°.[d]	**2.** Lifting nonrigid bags (i.e., bags that sag in the middle)

[a]An optimal handle design has 0.75–1.5 in (1.9 to 3.8 cm) diameter, ≥4.5 in (11.5 cm) length, 2 in (5cm) clearance, cylindrical shape, and a smooth, nonslip surface.

[b]An optimal handhold cutout has the following approximate characteristics: ≥1.5 in (3.8 cm) height, 4.5 in (11.5 cm) length, semioval shape, ≥2 in (5 cm) clearance, smooth nonslip surface, and ≥0.25 in (0.60 cm) container thickness (e.g., double thickness cardboard).

[c]An optimal container design has ≤16 in (40 cm) frontal length, ≤12 in (30 cm) height, and a smooth nonslip surface.

[d]A worker should be capable of clamping the fingers at nearly 90° under the container, such as required when lifting a cardboard box from the floor.

[e]A container is considered less than optimal if it has a frontal length >16 in (40 cm), height >12 in (30 cm), rough or slippery surfaces, sharp edges, asymmetric center of mass, unstable contents, or requires the use of gloves. A loose object is considered bulky if the load cannot easily be balanced between the hand grasps.

[f]A worker should be able to comfortably wrap the hand around the object without causing excessive wrist deviations or awkward postures, and the grip should not require excessive force.

Source: Waters et al. (1994).

TABLE 58.10 Coupling Multipliers

Coupling Type	Coupling multiplier V < 30 in (75 cm)	Coupling multiplier V ≥ 30 in (75 cm)
Good	1.0	1.0
Fair	0.95	1.0
Poor	0.90	0.90

Source: Waters et al. (1994).

FM is a frequency multiplier, which is a function of the average number of lifts per minute, the vertical position of the hands at the origin, and the duration of lifting. Lifts per minute is usually based on an average taken over a 15 min lifting period. For values under 0.2 lifts/min, a value of 0.2 lifts/min should be used. Duration of lifting is broken into three categories based on the patterns of continuous work time and recovery time. Continuous work time is defined as a period of uninterrupted work, while recovery time means time spent involved in light work (e.g., light assembly, deskwork) following a period of continuous work. Short work durations are those lasting for 1 h or less, followed by a recovery period of at least 1.2 times the length of the continuous work. Moderate durations are those lasting between 1 and 2 h, followed by a recovery period of at least 0.3 times the working period. Long durations are those between 2 and 8 h with standard industrial rest allowances (lunch, morning, and afternoon breaks). For infrequent lifting (values under 0.1 lifts/min), the rest intervals between lifts are usually long enough to consider the task "short" duration, no matter how long the activity actually goes on. Appropriate values for *FM* can be found in Table 58.11.

Since this explanation is necessarily rather complex, the following practical example will be used to demonstrate the employment of the equation.

TABLE 58.11 Frequency Multiplier Values

Frequency	Work duration ≤1 h		>1 but ≤2 h		>2 but ≤8 h	
lifts\min	V < 30 in	V ≥ 30 in	V < 30 in	V ≥ 30 in	V < 30 in	V ≥ 30 in
≤0.2	1.00	1.00	0.95	0.95	0.85	0.85
0.5	0.97	0.97	0.92	0.92	0.81	0.81
1	0.94	0.94	0.88	0.88	0.75	0.75
2	0.91	0.91	0.84	0.84	0.65	0.65
3	0.88	0.88	0.79	0.79	0.55	0.55
4	0.84	0.84	0.72	0.72	0.45	0.45
5	0.80	0.80	0.60	0.60	0.35	0.35
6	0.75	0.75	0.50	0.50	0.27	0.27
7	0.70	0.70	0.42	0.42	0.22	0.22
8	0.60	0.60	0.35	0.35	0.18	0.18
9	0.52	0.52	0.30	0.30	0.00	0.15
10	0.45	0.45	0.26	0.26	0.00	0.13
11	0.41	0.41	0.00	0.23	0.00	0.00
12	0.37	0.37	0.00	0.21	0.00	0.00
13	0.00	0.34	0.00	0.00	0.00	0.00
14	0.00	0.31	0.00	0.00	0.00	0.00
15	0.00	0.28	0.00	0.00	0.00	0.00
>15	0.00	0.00	0.00	0.00	0.00	0.00

Source: Waters et al. (1994).

Example: A worker is lifting 18-lb cartons from the floor, rotating his torso, and then stacking the cartons on 24 in high carts. Each carton measures 18 in. in width, 8 in. in height, and is equipped with well-designed handles in the middle of each side. There are 200 cartons to task and he completes the task in 40 min by working continuously. Evaluate the task.

Since it is necessary to pause during the course of the lift to stack the cartons, the lifting task will be evaluated at both the origin and destination.

Hand Location: Origin ($H = 18$ in, $V = 44$ in), Destination ($H = 18$ in, $V = 7$ in)

$HM = 10/H = 10/18 = 0.556$ both origin and destination

$VM = 1 - (0.0075| V - 30|$; $VM = 0.805$ for origin and 0.97 for destination

Vertical Distance $= D = 24$ in; $DM = 0.82 + 1.8/D$; $DM = 0.895$

Asymmetric Angle: $A = 30°$; $AM = 1 - (0.0032A)$; $AM = 0.904$

Frequency: 200/40; $F = 5$ lifts/min

Duration: 1 h ($FM = 0.8$ from Table 58.11

Coupling: Well designed handle, so "Good" from Table 58.10; $CM = 1.00$

$RWL_{Origin} = (51)(0.556)(0.805)(0.895)(1.00)(0.8)(1.00) = 16.34$ lbs; $LI = 18/16.34 = 1.1$

$RWL_{Destination} = (51)(0.556)(0.97)(0.895)(.904)(0.8)(1.00) = 17.80$ lbs; $LI = 18/17.80 = 1.0$

Since *LI* for the task at the origin slightly exceeds 1.0, task redesign should be considered. In this case, something as simple as placing a pallet under the cartons on the floor might be appropriate. Studies have indicated that *LI*s in excess of 2.0 have been associated with higher incidence of back injuries.

There are a number of caveats to using the NIOSH Lifting Equation. The equation assumes that tasks other than lifting involve minimal expenditure of energy on the part of the worker and do not include such activities as pushing, pulling, walking, carrying, or climbing. If such activities make up more than 10 percent of the employee's activities, then other job evaluation methods may be necessary. Further, the equation assumes that all tasks are carried out under reasonable working conditions with regard to both temperature and humidity (19 to 26°C, or 66 to 79°F, and 35 to 50 percent). Also, the equation is unsuitable for examining one-handed lifting, or those performed in unusual or awkward postures (seated, kneeling, or in confined workspaces). Finally, the equation is also not designed for use in high speed lifting tasks (over 30 in/s), with low friction working surfaces (under 0.40 coefficient of friction between shoes and flooring), or for evaluating work with wheelbarrows or shovels. If any of these conditions apply, other methods of evaluating the work task should be employed.

Tables 58.12a through 58.12c list maximum acceptable loads based on psychophysical considerations (acceptability) and may provide some guidance for lifting tasks not covered by the NIOSH equation. It should be noted that these tables represent subsets of more extensive data sets contained in the original source.

Pulling and pushing activities can also lead to potential risk of back injuries as they also impose compressive loads on the spine. Table 58.13 presents data on recommended upper limits for horizontal pushing and pulling tasks. Limits are set to allow the majority of the workforce to perform the tasks and are for tasks involving forces applied with the arms between the waist and shoulders. Tasks performed in higher or lower positions prevent the arms from being properly positioned to exert maximal force and limits must be reduced accordingly. For maximum acceptable pushing and pulling loads based on psychophysical studies, please see Table 58.14a and 58.14b. Again, it should be noted that these tables are subsets of larger data sets contained in the original publication (Snook and Ciriello 1991).

For vertical pulling and pushing tasks, values are somewhat higher since the weight of the body can be used in the former and the muscles of the torso and legs in the latter. Table 58.15 presents recommended maximum limits for vertical operations. Again, these limits are set to allow the majority of the working population to perform these tasks.

TABLE 58.12*a* Maximum Acceptable Weights for Lifting Task for Males (kg)

			Floor level to knuckle height one lift every								Knuckle height to shoulder height one lift every								Shoulder height to arm reach one lift every							
			5	9	14	1	2	5	30	8	5	9	14	1	2	5	30	8	5	9	14	1	2	5	30	8
Width*	Distance†	%‡		(s)			(min)			(h)		(s)			(min)			(h)		(s)			(min)			(h)
34	25	90	10	12	14	18	20	20	23	27	11	14	16	20	20	21	23	26	10	13	15	19	19	19	22	24
		75	15	18	21	26	30	28	33	38	14	18	21	26	27	28	31	34	13	17	20	24	25	26	29	31
		50	20	24	28	35	40	38	44	52	18	23	27	33	34	35	39	43	16	22	25	31	31	33	36	40
		25	26	30	35	44	50	48	55	65	21	28	32	40	41	42	47	52	20	26	30	37	38	39	44	46
		10	29	35	41	52	59	57	66	76	25	33	37	47	47	49	55	60	23	30	35	43	44	45	51	55
	51	90	9	10	12	16	18	20	23	24	9	12	14	17	17	18	20	22	8	11	13	16	16	17	18	20
		75	12	15	18	23	26	29	33	34	12	16	18	22	23	23	26	29	11	14	17	21	21	22	24	26
		50	17	20	24	31	35	39	44	46	15	20	23	28	29	30	33	36	14	18	32	36	27	28	31	34
		25	21	25	30	39	44	49	55	57	18	24	27	34	35	36	40	44	17	22	25	32	32	33	37	41
		10	25	30	35	46	52	58	66	68	21	28	32	40	40	42	46	51	19	26	29	37	37	39	43	47
49	25	90	8	10	12	16	18	19	20	23	10	13	15	18	18	19	21	23	9	11	12	16	16	17	19	21
		75	12	15	17	23	26	28	29	33	13	17	20	23	24	25	27	30	11	14	16	21	21	22	25	27
		50	16	20	23	30	34	37	38	45	17	22	25	30	30	31	35	38	14	18	21	27	27	28	32	35
		25	21	25	29	38	43	47	48	56	20	27	30	36	36	38	42	46	16	22	25	33	33	34	38	42
		10	24	29	34	45	51	56	57	67	23	31	35	42	42	44	49	53	19	25	29	38	38	40	44	48
	51	90	7	9	10	14	16	17	18	20	8	11	13	15	15	16	18	19	7	9	11	14	14	14	16	18
		75	10	13	15	20	23	25	25	30	11	15	17	20	20	21	23	25	9	12	14	18	18	19	21	23
		50	14	17	20	27	30	33	34	40	14	19	21	25	25	26	29	32	12	15	18	23	23	24	27	29
		25	18	21	25	34	38	42	43	50	17	23	26	30	31	32	36	39	14	19	21	28	28	29	32	35
		10	21	25	29	40	45	49	50	59	20	26	30	35	36	37	41	45	16	22	25	32	32	34	37	41

*Box width (the dimension away from the body). This value is based on the position of the hands in front of the body while lifting and not the true width of the box. To obtain "box width," a value of 1/2 the actual dimension of the box is used under the assumption that the hands are located around the centerpoint of the box.

†Vertical distance of lift (cm)

‡Percentage of industrial population

Source: Snook and Ciriello (1991).

TABLE 58.12*b* Maximum Acceptable Weights for Lifting Task for Females (kg)

			Floor level to knuckle height one lift every								Knuckle height to shoulder height one lift every								Shoulder height to arm reach one lift every							
			5	9	14	1	2	5	30	8	5	9	14	1	2	5	30	8	5	9	14	1	2	5	30	8
Width*	Distance†	%‡		(s)			(min)			(h)		(s)			(min)			(h)		(s)			(min)			(h)
34	25	90	8	10	11	11	12	12	14	19	8	8	9	12	12	12	14	16	7	7	8	10	11	11	12	14
		75	10	12	13	14	15	15	17	23	9	10	11	13	14	14	16	18	8	8	9	12	12	12	14	16
		50	12	15	16	17	18	19	21	28	10	11	13	16	17	17	18	21	9	10	11	13	14	14	16	18
		25	14	17	19	20	22	22	24	33	12	13	14	18	19	19	21	24	10	11	12	15	16	16	18	21
		10	16	20	21	23	25	25	28	38	13	14	16	19	21	21	23	27	11	12	14	17	18	18	20	23
	51	90	7	9	9	11	12	12	14	19	8	8	9	12	12	12	14	16	7	7	8	9	10	10	11	12
		75	9	11	12	14	15	15	16	22	9	10	11	12	13	13	14	17	8	8	9	11	11	11	12	14
		50	11	13	14	16	18	18	20	27	10	11	13	14	15	15	17	19	9	10	11	12	13	13	14	17
		25	13	15	17	19	21	21	24	32	12	13	14	16	17	17	19	22	10	11	12	14	15	15	16	19
		10	14	18	19	22	24	24	27	36	13	14	16	18	19	19	21	24	11	12	14	15	16	16	18	21
49	25	90	6	8	8	9	10	10	11	15	6	7	8	10	11	11	12	14	5	6	7	8	9	9	10	11
		75	8	10	11	12	12	13	14	19	7	8	9	12	13	13	14	17	6	7	8	9	10	10	11	13
		50	10	12	13	14	15	15	17	23	9	10	11	14	15	15	16	19	7	8	9	11	12	12	13	15
		25	11	14	15	16	18	18	20	27	10	11	12	16	17	17	19	22	8	9	10	12	13	13	15	17
		10	13	16	17	19	20	21	23	31	11	12	14	18	19	19	21	24	9	10	11	14	15	15	16	19
	51	90	6	7	8	9	10	10	11	15	6	7	8	9	10	10	11	13	5	6	7	7	8	8	9	10
		75	7	9	9	11	12	12	14	18	7	8	9	11	12	12	13	15	6	7	8	9	9	9	10	12
		50	9	10	11	13	15	15	16	22	9	9	11	13	14	14	15	17	7	8	9	10	11	11	12	14
		25	10	12	13	16	17	17	19	26	10	11	12	14	16	16	17	20	8	9	10	11	12	12	13	15
		10	11	14	15	18	19	20	22	30	11	12	14	16	17	17	19	22	9	10	11	13	14	14	15	17

*Box width (the dimension away from the body). This value is based on the position of the hands in front of the body while lifting and not the true width of the box. To obtain "box width," a value of 1/2 the actual dimension of the box is used under the assumption that the hands are located around the centerpoint of the box.

†Vertical distance of lift (cm)

‡Percentage of industrial population

Source: Snook and Ciriello (1991)

TABLE 58.12c Maximum Acceptable Weight to Carry (kg) for Males and Females

			2.1 m Carry once carry every							4.3 m Carry once carry every							8.5 m Carry once carry every						
	Height (cm)	%	6 (s)	12 (s)	1 (min)	2 (min)	5 (min)	30 (min)	8 (h)	10 (s)	16 (s)	1 (min)	2 (min)	5 (min)	30 (min)	8 (h)	18 (s)	24 (s)	1 (min)	2 (min)	5 (min)	30 (min)	8 (h)
Males	79	90	13	17	21	21	23	26	31	11	14	18	19	21	23	27	13	15	17	18	20	22	26
		75	18	23	28	29	32	36	42	16	19	25	25	28	32	37	17	20	24	24	27	30	35
		50	23	30	37	37	41	46	54	20	25	32	33	36	41	48	22	26	31	31	35	39	46
		25	28	37	45	46	51	57	67	25	30	40	40	45	50	59	27	32	38	38	42	48	56
		10	33	43	53	53	59	66	78	29	35	47	47	52	59	69	32	38	44	45	50	56	65
	111	90	10	14	17	17	19	21	25	9	11	15	15	17	19	22	10	11	13	13	15	17	20
		75	14	19	23	23	26	29	34	13	16	21	21	23	26	30	13	15	18	18	20	23	27
		50	19	25	30	30	33	38	44	17	20	27	27	30	34	39	17	19	23	24	26	29	35
		25	23	30	37	37	41	46	54	20	25	33	33	37	41	48	21	24	29	29	32	36	43
		10	27	35	43	43	48	54	63	24	29	38	39	43	48	57	24	28	34	34	38	42	50
Females	72	90	13	14	16	16	16	16	22	10	11	14	14	14	14	20	12	12	14	14	14	14	19
		75	15	17	18	18	19	19	25	11	13	16	16	17	17	23	14	15	16	16	17	17	23
		50	17	19	21	21	22	22	29	13	15	19	19	20	20	26	16	17	19	19	20	20	26
		25	20	22	24	24	25	25	33	15	17	22	22	22	22	30	18	19	21	22	22	22	30
		10	22	24	27	27	28	28	37	17	19	24	24	25	25	33	20	21	24	24	25	25	33
	105	90	11	12	13	13	13	13	18	9	10	13	13	13	13	18	10	11	12	12	12	12	16
		75	13	14	15	15	16	16	21	11	12	15	15	16	16	21	12	13	14	14	14	14	19
		50	15	16	18	18	18	18	25	12	13	18	18	18	18	24	14	15	16	16	16	16	22
		25	17	18	20	20	21	21	28	14	15	20	20	21	21	28	15	17	18	18	19	18	25
		10	19	20	22	22	23	23	31	16	17	22	22	23	23	31	17	19	20	20	21	21	28

Source: Snook and Ciriello (1991).

TABLE 58.13 Recommended Upper Force Limits for Horizontal Pushing and Pulling Tasks

Condition	Forces that should not be exceeded, in newtons (lbf)	Examples of activities
Standing		
Whole body involved	225 (50)	Truck/cart handling; moving wheeled or castered equipment; sliding rolls on shafts
Primarily arm and shoulder muscles, arms fully extended	110 (24)	Leaning over obstacle to move objects; pushing at or above shoulder height
Kneeling	188 (42)	Removing or replacing components from equipment; handling in confined work environments
Seated	130 (29)	Operating a vertical lever such as a floor shift on heavy equipment; moving trays or products on and off conveyors

Source: Eastman Kodak Company (1986).

TABLE 58.14a Maximum Acceptable Sustained Push Forces for Males and Females (kg)

	Height (cm)	%	2.1 m push one push every							7.6 m push one push every							15.2 m push one push every							30.5 m push one push every					45.7 m push one push every					61.0 m push one push every			
			6	12	1	2	5	30	8	15	22	1	2	5	30	8	25	35	1	2	5	30	8	1	2	5	30	8	1	2	5	30	8	2	5	30	8
			(s)		(min)				(h)	(s)		(min)				(h)	(s)		(min)				(h)	(min)				(h)	(min)				(h)	(min)			(h)
Males	95	90	10	13	16	17	19	19	23	8	10	13	13	15	15	18	8	10	11	12	13	13	16	8	10	12	13	16	7	8	9	11	13	7	8	9	11
		75	14	18	22	22	25	26	31	11	13	17	18	20	21	25	11	13	15	16	18	18	21	11	13	16	18	21	9	11	13	15	18	9	11	12	15
		50	18	23	28	29	33	34	40	14	17	22	23	26	27	32	14	17	19	20	23	23	28	15	17	20	23	27	12	14	17	19	23	12	14	16	19
		25	22	28	34	35	40	41	49	17	21	27	29	32	33	39	18	21	24	25	28	29	34	18	21	25	28	33	15	18	21	24	28	15	17	20	23
		10	26	33	40	41	46	48	57	20	24	32	33	37	38	45	20	25	28	29	32	33	40	21	25	29	33	39	17	20	24	27	32	17	20	23	27
	144	90	10	13	15	16	18	18	22	8	9	13	13	15	16	18	8	9	11	12	13	14	16	8	10	12	13	16	7	8	10	11	13	7	8	9	11
		75	13	17	21	22	24	25	30	10	13	17	18	20	21	25	11	13	15	16	18	18	22	11	13	16	18	21	10	11	13	15	18	9	11	13	15
		50	17	22	27	28	31	32	38	13	16	22	23	26	27	32	14	17	20	20	23	24	28	15	17	20	23	28	12	14	17	19	23	12	14	16	19
		25	21	27	33	34	38	40	47	16	20	28	29	32	33	39	17	20	24	25	28	29	34	18	21	25	29	34	15	18	21	24	28	15	17	20	24
		10	25	31	38	40	45	46	54	19	23	32	33	38	39	46	20	24	28	29	33	34	40	21	25	29	33	39	18	21	24	28	33	17	20	23	28
Females	89	90	6	7	9	9	10	11	13	6	7	8	8	9	9	11	5	6	6	7	7	8	10	5	6	6	7	9	5	6	6	6	8	4	4	5	6
		75	8	11	13	13	15	16	19	9	10	11	11	13	13	17	7	8	9	10	11	11	14	8	9	9	10	13	7	8	8	9	12	6	6	7	9
		50	11	15	18	18	20	21	26	12	13	15	15	17	18	22	9	11	13	13	14	15	19	10	12	12	13	17	10	11	11	12	16	8	9	9	12
		25	14	18	22	23	25	27	33	15	17	19	19	21	23	28	12	14	16	16	18	19	24	13	15	15	16	22	12	14	14	15	20	11	11	12	15
		10	17	22	26	27	30	32	39	17	20	22	23	25	27	33	14	17	19	19	21	23	28	16	18	18	19	26	14	16	17	18	24	13	13	14	18
	135	90	6	8	10	10	11	12	14	6	7	7	7	8	9	11	5	6	6	6	7	7	9	5	6	6	6	8	5	5	5	6	8	4	4	4	6
		75	9	12	14	14	16	17	21	9	10	11	11	12	13	16	7	8	9	9	10	11	13	7	8	9	9	12	7	8	8	8	11	6	6	6	9
		50	12	16	19	20	21	23	28	12	14	14	15	16	17	21	10	11	12	12	14	14	18	10	11	12	12	16	9	10	11	11	15	8	8	9	12
		25	16	20	24	25	27	29	36	15	17	18	18	20	22	27	12	14	15	16	17	18	22	13	14	15	15	21	11	13	13	14	19	10	10	11	15
		10	18	23	28	29	32	34	42	18	20	21	22	24	26	32	14	17	18	18	20	22	27	15	17	17	18	25	14	15	16	17	22	12	12	13	17

Source: Snook and Ciriello (1991).

TABLE 58.14*b* Maximum Acceptable Sustained Pull Forces for Males and Females (kg)

	Height (cm)	%	2.1 m pull one pull every							7.6 m pull one pull every							15.2 m pull one pull every						
			6 (s)	12 (s)	1 (min)	2 (min)	5 (min)	30 (min)	8 (h)	15 (s)	22 (s)	1 (min)	2 (min)	5 (min)	30 (min)	8 (h)	25 (s)	35 (s)	1 (min)	2 (min)	5 (min)	30 (min)	8 (h)
Males	64	90	11	14	17	18	20	21	25	9	11	14	15	17	17	20	9	11	12	13	15	15	18
		75	14	19	23	23	26	27	32	11	14	19	19	22	22	26	12	14	16	17	19	19	23
		50	17	23	28	29	32	34	40	14	18	23	24	27	28	33	15	18	20	21	23	24	28
		25	20	27	33	35	39	40	48	17	21	27	28	32	33	39	18	21	24	25	28	29	34
		10	23	31	38	40	45	46	54	19	24	31	32	37	38	45	20	24	27	28	32	33	39
	95	90	10	13	16	17	19	20	24	8	10	13	14	16	16	19	9	10	12	12	14	14	17
		75	13	17	21	22	25	26	30	11	13	17	18	20	21	25	11	14	15	15	18	18	22
		50	16	21	26	27	31	32	37	13	17	21	22	25	26	31	14	17	19	19	22	23	27
		25	19	26	31	33	37	38	45	16	20	26	27	30	31	37	17	20	22	23	26	27	32
		10	22	29	36	37	42	43	51	18	23	29	31	34	36	42	19	23	26	27	30	31	37
Females	57	90	5	8	9	9	10	11	13	6	7	8	8	9	10	12	5	6	7	7	7	8	10
		75	7	11	12	12	13	14	18	8	9	11	11	12	13	16	7	8	9	9	10	11	13
		50	9	14	15	16	17	18	23	10	12	13	14	15	16	20	8	10	11	12	13	14	17
		25	11	17	18	19	21	22	27	13	15	16	17	19	20	24	10	12	14	14	16	17	21
		10	13	20	21	22	24	26	32	15	17	19	20	22	23	28	12	14	16	16	18	19	24
	89	90	6	9	10	10	11	12	14	7	8	9	9	10	10	13	5	6	7	7	8	9	11
		75	8	12	13	13	15	16	19	9	10	11	12	13	14	17	7	8	10	10	11	12	14
		50	10	15	16	17	19	20	25	11	13	15	15	16	18	22	9	11	12	13	14	15	18
		25	12	18	20	21	23	24	30	14	16	18	18	20	22	27	11	13	15	15	17	18	22
		10	14	21	23	24	26	28	35	16	18	21	21	23	25	31	13	15	17	18	20	21	26

	Height (cm)	%	30.5 m pull one pull every					45.7 m pull one pull every					61.0 m pull one pull every			
			1 (min)	2 (min)	5 (min)	30 (min)	8 (h)	1 (min)	2 (min)	5 (min)	30 (min)	8 (h)	2 (min)	5 (min)	30 (min)	8 (h)
Males	64	90	9	11	13	15	18	8	9	11	12	15	8	9	10	12
		75	12	14	17	19	23	10	12	14	16	19	10	12	13	16
		50	15	18	21	24	27	13	15	17	20	23	12	14	16	20
		25	18	21	25	28	33	15	18	21	24	28	15	17	20	23
		10	21	24	28	32	38	17	20	24	27	32	17	20	23	27
	95	90	9	10	12	14	17	7	9	10	12	14	7	9	10	12
		75	12	13	16	18	21	10	11	13	15	18	9	11	13	15
		50	14	17	19	22	26	12	14	16	19	22	12	14	16	18
		25	17	20	23	27	32	14	17	19	22	26	14	16	19	22
		10	19	23	27	31	36	16	19	22	25	30	16	19	21	25
Females	57	90	6	6	6	7	9	5	6	6	6	8	4	5	5	6
		75	7	8	9	9	12	7	8	8	8	11	6	6	6	9
		50	9	11	11	12	16	9	10	10	11	14	8	8	8	11
		25	11	13	13	14	19	11	12	12	13	17	9	10	10	13
		10	13	15	16	16	22	12	14	14	15	20	11	11	12	16
	89	90	6	7	7	7	10	5	6	6	7	9	5	5	5	7
		75	8	9	9	10	13	7	8	9	9	12	6	7	7	9
		50	10	12	12	13	17	9	11	11	12	15	8	8	9	12
		25	12	14	15	15	21	11	13	13	14	19	10	10	11	15
		10	15	16	17	18	24	13	15	16	16	22	12	12	13	17

Source: Snook and Ciriello (1991).

TABLE 58.15 Recommended Upper Limits for Vertical Pushing and Pulling Forces in Standing Tasks

Conditions	Upper limit of force for design in newtons (lbf)	Examples of activities
Pull down, above head height	540 (120)	Activating a control, hook grip; safety shower handle or manual control
	200 (45)	Activating a control, power grip; <5 cm (2 in) diameter grip surface
Pull down, shoulder level	315 (70)	Activating a control, hook grip. Threading up operations as in paper manufacturing or cable stringing
Pull up, 25 cm (10 in) above the floor	315 (70)	Lifting an object with one hand
Elbow height	148 (33)	Raising a lid or access port cover, palm up
Shoulder height	75 (17)	
Push down, elbow height	287 (64)	Wrapping, packing, sealing cases.
Push up, shoulder height (boosting)	202 (45)	Raising a corner or end of an object, like a pipe or beam. Lifting an object onto a high shelf.

Source: Eastman Kodak Company (1986).

The best solution for reducing the risk of lifting-related manual material handling injuries is, of course, to eliminate such tasks. Since this is often impractical, the following job redesign techniques may be employed (Ayoub and Mital 1989):

1. Decrease the weight of objects being handled.
2. Use two or more people to move heavy or large objects.
3. Change the activity: pushing or pulling is better than carrying.
4. Reduce the horizontal distance from the beginning and endpoint of lifts.
5. Do not stack materials at more than shoulder height.
6. Keep heavy objects at knuckle height.
7. Reduce the required frequency of lifting.
8. Incorporate rest periods.
9. Incorporate job rotation to less strenuous jobs.
10. Design containers with handles that can be held close to the body.

Many employers attempt to provide employees with backbelts for use by their employees rather than redesign work tasks. NIOSH recently completed a study evaluating the effectiveness of using such devices for employees involved in materials handling tasks in the retail environment (Wassell et al. 2000). Their results showed that use of such devices had no effect on the incidence of back injury claims or instances of low back pain.

58.4.3 Design and Selection of Hand-tools

Hand-tools have been with us since the dawn of time, but during the last century the introduction of powered tools has added new areas of concern owing to the increased forces (and increased possibility for injury) involved. It is important that tool design take this issue as well as proper tool usage into consideration during the design phase. Proper selection of both powered and unpowered hand-tools can be a strong factor in both the reduction of injuries and the reduction in worker fatigue.

There are two primary grips used in tool operation: the *power grip* and the *precision grip*. The power grip involves making a fist with the hand by reaching around the tools with the forefingers and overlapping them with the thumb, with the handle of the tool positioned perpendicularly to the axis of the hand. The power grip can be divided into three primary subdivisions: (1) where force is exerted parallel to the forearm, e.g., sawing; (2) where force is exerted at an angle to the forearm, e.g., hammering; and (3) where torque is exerted around the forearm, e.g., using a screwdriver (Konz 1990). The precision grip employs either an internal grip where the tool is held within the hand (e.g., a knife) or an external grip where the tool is pinched between the thumb and index (and possibly middle) finger (e.g., a pen). It has been estimated that the precision grip allows for only about 20 percent of the strength found in a power grip, and is used primarily in tasks involving higher levels of control than can be exerted with the power grip.

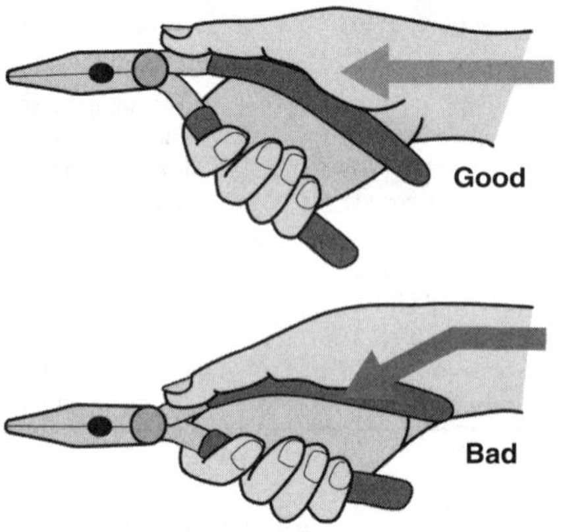

FIGURE 58.5 Proper and improper wrist positions in tool use.

One of the main occupational hazards stemming from the design of hand tools is that of cumulative trauma to muscles and tissues. Two of the simplest guidelines for tool design and selection are that they should minimize required movements of the hand and that the hand and wrist should remain as close as possible to their neutral positions (i.e., keeping the wrist straight—see Fig. 58.5). Whenever the wrist is moved from its neutral position, there is a loss of grip strength of up to 25 percent (Terrell and Purswell 1976). Some of the occupational hazards associated with prolonged maintenance of the hand in awkward postures include pain, reduced grip strength, and possible cumulative trauma disorders. Further, extended gripping of hand tools may cause obstruction of blood flow to areas of the hands and fingers, leading to tingling, numbness, and eventual tissue damage. Ideally, tool handles should have as large a gripping surface as practical to spread the compressive force over a greater area. The determination of the appropriate size of such a surface is, however, an issue of considerable complexity (Freivalds 1999).

Anthropometrically speaking, there are few areas of the body with as great a difference in size and strength between the genders as in the human hand. Female maximum grip strength is approximately 60 percent of male strength, while the average female hand length is approximately the same size as that of a 5th percentile male (50th percentile male, 7.6 in; 50th percentile female, 6.9 in) (Kamon and Goldfuss 1978; Woodson et al. 1992). Proper tool design can minimize the force required to properly operate tools as well as the detrimental effect on the body structure from their use.

Handle Diameter. Handle diameter is a major determinant of maximum grip strength, the stresses placed on the hand during tool use, and fatigue, thus considerable research has been focused on determination of optimal handle dimensions across a variety of applications. Ideally, in order to minimize muscle fatigue and stress-related disorders, a tool should be designed to maximize the amount of force that *can* be applied, while simultaneously minimizing the amount of force that *must* be applied. In terms of size, the proper power grip for a cylindrical handle should be such that the fingers and thumb surround the handle and barely touch. This is difficult to accomplish in practice, since preferred handle size is thus largely a function of operator hand size. A grip diameter of approximately 1.6 in (40 mm) appears to be most appropriate for a power grip. Studies have shown that grip forces with handles of approximately 2.0 in (50 cm) are 5% lower and those at approximately 2.5 in (65 mm) are 30% lower than those at the 1.6 in diameter (Fraser 1989).

For precision grips, recommended sizes are much smaller. Experiments involving screwdriver handles have recommended diameters of 8 mm based on speed of operation (Hunt 1934). Recommendations for pen sizes have ranged from 13 to 30 mm, with the latter a function of reduced fatigue and strain and greater maximal force capability (Kao 1974; Sperling 1986). Diameters of less than 6 mm should be avoided, as sizes in this range tend to produce hand pain if much force is required (Konz 1990).

Handle Shape. Optimal handle shape has been shown to be largely a function of the nature of the task being performed. A comprehensive study performed by Cochrane and Riley (1986), involved testing 36 handles on six common types of tasks: thrusting and pulling in the plane of the handle (e.g., stabbing and withdrawing a knife), pulling and pushing perpendicular to the plane of the handle (e.g., sawing), and twisting around the long axis of the handle (e.g., using a screwdriver) to determine the configuration commensurate with maximum force (power grip). Results indicated that for thrusting and pulling types of tasks, a triangular handle with a circumference of 11 cm was best for mixed-gender populations. For orthogonal push/pull activities, rectangular handles with width-to-height ratios of 1:1.25 and circumferences of 9 cm were superior to other configurations. For tasks involving wrist extensions/flexions, rectangular handles with width-to-height ratios of up to 1:2.0 were best, with larger handles (up to 13 cm) better than smaller ones. For a composite tool used in all tasks, a rectangular handle with a height-to-width ratio between 1:1.25 and 1:1.5 was recommended as the best compromise shape.

Handle Length. Optimal handle length has received less experimental study, but a reasonable goal in design is to at least allow for enough space to accommodate all four fingers for a power grip. Since hand breadth at the palm ranges from 71 mm (5th percentile female) to 97 mm (95th percentile male) (Garrett 1971), 100 mm is a practical minimum dimension, though 120 mm is more practical (Eastman Kodak 1983). Even larger values should be considered if gloves are commonly used. For external precision grips, the shaft of the tool must be long enough to be supported by the area of the hand between the thumb and index finger (at least 100 mm). For external precision grips, the tool must be long enough to extend past the palm area, while not being so long as to extend past the wrist in order to minimize potential tissue compression (Konz 1990).

An important factor in the design of powered hand tools is the ability of the operator to optimally support and control the tool, as well as being able to react to the forces exerted by the tool in order to prevent loss of control. This issue is largely a function of the weight and mass distribution of the tool itself (this consideration will also serve to minimize the potential for the operator to lose control of the tool, potentially leading to injury). Ideally, all powered tools should be neutrally balanced with all necessary attachments having their center of mass aligned with that of the operator's hand (Radwin 1999). Subjective studies have indicated an operator preference for power tools with masses between 0.9 and 1.75 kg, while others have indicated that tools weighing less than 1 kg were perceived as requiring less effort to use than those weighing between 2 and 3 kg (Ulin et al., 1993).

Handle material selection should largely be based on the nature of the task being performed. In general, preference should be given to grip surfaces that are slightly compressible, nonconductive of heat and electricity, and smooth. Compressability reduces both tool vibration and hand slippage, while smooth surfaces reduce the potential for hand pain owing to localized high pressure areas (Konz 1990).

58.5 CUMULATIVE TRAUMA DISORDERS

Cumulative trauma disorders (CTDs), also known as *musculoskeletal disorders* (MSDs), or *repetitive stress injuries* (RSI), are on the rise primarily owing to the pace of modern work. Assembly line production techniques emphasize increased worker speed, often on simple repetitive tasks, with relatively little time available for rest or recovery. In some cases, the exact same movement may be repeated as many as 25,000 times per day (Luopajarvi et al. 1979). According to the Bureau of Labor Statistics (2001), in 1996 there were over 2.8 million reported workplace injuries associated with repeated trauma in private industry in the United States, making up almost 65 percent of all reported illnesses. In terms of total number of cases, the automotive industry reported almost 44,000 injuries alone, while meat packing and poultry processing added an additional 27,400 cases. In terms of incident rates, meat packing plants and knit underwear mills lead other industries with over 900 incidents per 10,000 workers annually. Almost 74,000 injuries owing to repetitive motion were reported, with almost half of these occurring within the manufacturing sector.

These numbers, as large as they are, may represent only a fraction of the actual number of cases. It is often difficult to determine the cause of CTD, as they normally only develop over periods of extended exposure (usually months or years). Causal linkages are also hard to determine, since the onset of such injuries is gradual in nature, and the injured individual's current job may only have contributed to or exacerbated an existing condition rather than having been the sole cause of it. Table 58.16 shows a variety of manufacturing operations, details the occupational factors that may lead to CTDs, and lists the particular CTDs which could result from their performance.

TABLE 58.16 Job, Identified Disorder, and Occupational Risk Factors

Job	Occupational factors	Possible CTDs
Buffing/grinding	Repetitive motions, prolonged flexed shoulders, vibration, forceful ulnar deviation, repetitive forearm pronation	Tenosynovitis, thoracic outlet, carpal tunnel, De Quervain's disease, pronator teres
Punch press operation	Repetitive forceful wrist extension/ flexion, repetitive shoulder abduction/ flexion, forearm supination. Repetitive ulnar deviation in pushing controls.	Tendinitis of wrists and shoulders
Overhead assembly	Sustained hyperextension of arms. Hands above shoulders.	Thoracic outlet syndrome, shoulder tendinitis
Belt conveyor assembly	Arms extended, abducted, or flexed more than 60°, repetitive, forceful wrist motions	Tendinitis of shoulder and wrist, carpal tunnel, thoracic outlet syndrome
Typing/keypunching	Static, restricted posture, arms abducted/ flexed high speed finger movement, palmar base pressure, ulnar deviation	Tension neck, thoracic outlet syndrome, carpal tunnel syndrome
Sewing and cutting	Repetitive shoulder flexion, repetitive ulnar deviation; Repetitive wrist flexion/ extension, palmar base pressure	Thoracic outlet syndrome, De Quervain's disease, carpal tunnel
Small parts assembly	Prolonged restricted posture, forceful ulnar deviation and thumb pressure, repetitive wrist motion, forceful wrist extension and pronation	Tension neck, thoracic outlet syndrome, wrist tendinitis, epicondylitis
Bench work	Sustained elbow flexion with pressure on the ulnar groove	Ulnar nerve entrapment
Packing	Prolonged load on shoulders, repetitive wrist motions, overexertion, forceful ulnar deviations	Tendinitis of shoulder and wrist, tension neck, carpal tunnel, De Quervain's disease
Truck driving	Prolonged shoulder abduction and flexion	Thoracic outlet syndrome
Core making	Repetitive wrist motions	Tendinitis of the wrist
Carpentry/bricklaying	Hammering, pressure on the palmar base	Carpal tunnel, Guyon tunnel
Stockroom/shipping	Prolonged load on shoulder in unnatural position	Thoracic outlet syndrome, shoulder tendinitis
Materials handling	Carrying heavy loads on shoulders	Thoracic outlet syndrome, shoulder tendinitis
Lumber/construction	Repetitive throwing of heavy loads	Shoulder tendinitis, epicondylitis
Butcher/meat packing	Ulnar deviation, flexed wrist with exertion	De Quervain's disease, carpal tunnel

Source: Putz-Anderson (1988).

The particular CTDs mentioned above will be discussed in more detail later in this section, but some basic vocabulary is necessary to understand the nature of the risks detailed above.

Abduction. Movement of a joint sideways away from the body (e.g., extending the arm outward to the side)

Adduction. Movement of a joint sideways toward the body (e.g., lowering an arm extended sideways from the body)

Flexion. Movement involving two ventral surfaces of the body moving closer together (e.g., when flexing your arm at the elbow joint, your forearm approaches your upper arm)

Extension. This movement is the opposite of flexion. When two dorsal surfaces are moved away from each other (e.g., straightening the fingers or straightening the arm at the elbow joint)

Pronation/supination. The forearm is able to rotate about its longitudinal axis, so that with the elbow flexed to a right angle it is possible for the hand to be placed palm up or palm down. With the palm up the forearm is supinated and with the palm down the forearm is pronated.

Ulnar/Planar deviation. Rotation of the hand at the wrist toward the thumb and smallest finger respectively

Conditions which lead to MSDs include—decreased blood flow to muscles, joints, and nerves; nerve compression; tendon and/or joint damage; and muscle strain. There are several work factors which contribute to the development of these types of maladies:

1. *Awkward or deviated postures.* All joints within the body have an optimal position and range of motion. Awkward postures lead to reduced blood flow, while placing undue stress on the tissues. Some of the most common causes of inappropriate work postures include improper workspace design, poor seating, inadequate tool design, and bad lifting technique.
2. *Excessive force.* Requirements for high levels of force often place undue stresses on the muscles, tendons, and joints, a problem that is often compounded by awkward postures. Even relatively low force levels may contribute to CTDs when they are applied by small muscle groups (e.g., tightly gripping pens, pencils, or precision hand tools). Gloves further exacerbate this by requiring the application of even more force to overcome their attendant decrease in sensitivity or lack of good contact with the tool or work piece.
3. *Repetitive work.* Repetitive tasks when performed at high rates of speed often prevent muscles from recuperating during work operations.
4. *Extended work durations.* The constant expansion and contraction of the muscles results in a pumping action which facilitates blood flow through the muscle tissues. This pumping action is required to carry oxygen and nutrients to the muscles and waste products (primarily lactic acid and carbon dioxide) away from them. Some work tasks may require constant exertion of the muscle in a more or less static contraction, which restricts blood flow and allows for a build up of waste products. The former reduces the capacity of the muscle to do useful work, while the latter can actively deteriorate the structure of the muscle itself.
5. *Mechanical trauma.* The design of many tools brings sharp edges or other limited contact surfaces into contact with the hands. This results in all of the force exerted by the hand being applied over a relatively small area, potentially causing high levels of compression and damaging the body tissues involved.
6. *Vibration.* Segmental vibration occurs when a part of the body is in contact with a vibrating surface, but this vibration is not transferred to the body as a whole. High frequency segmental vibration has been shown to interfere with blood flow to the extremities and even lower frequencies have been linked to increased grip forces (required to control the tool), leading to restriction of blood flow to the fingers or nerve compression.
7. *Temperature extremes.* Cold temperatures result in a shifting of blood flow from the extremities and away from the surface of the skin. This in turn may compound problems caused by the other factors listed above.

CTD injuries have a number of common characteristics (Armstrong 2002):

1. Mechanical and physiological processes
2. Relationship with work intensity and duration
3. Development normally requires weeks, months, or years
4. Recovery time requires weeks, months, or years
5. Symptoms are often poorly localized, nonspecific, and episodic
6. Often unreported
7. Stem from more than one causal factor

58.5.1 Types of MSDs

In general, there are three types of CTDs: those involving either the tendons and their sheaths, the nerves or the neurovascular system. Tendons are connective tissues that attach the muscles to the bones with which they are associated (think of them as the cables in a winch), while ligaments connect bones to one another. Tendon sheaths surround tendons much in the manner that rubber covers surround bicycle cables. Nerves are pathways that carry commands from the brain to the effectors of the body and sensation back to the brain (control and feedback circuits).

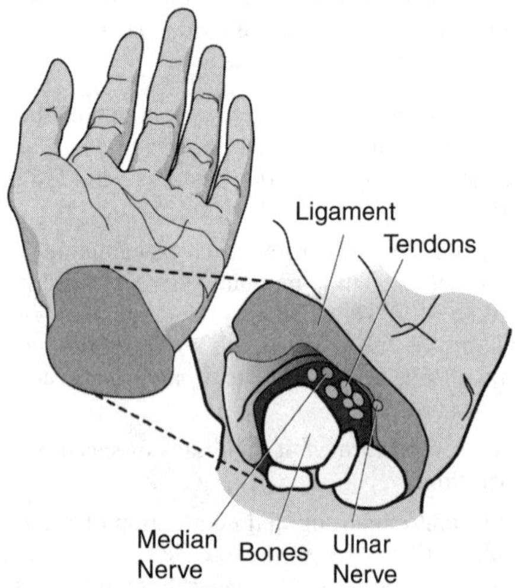

FIGURE 58.6 Human wrist.

Possibly the most well known single CTD is *carpal tunnel syndrome*. The carpal tunnel is an opening in the wrist bounded on one side by the carpal bones on the bottom and by carpal ligament on the top. Through this opening pass a number of tendons and blood vessels, as well as the median nerve which enervates the palm, thumb, and first two fingers of the hand as well as the radial side of the ring finger (see Fig. 58.6). Even when the wrist is in a neutral posture, the carpal tunnel is extremely crowded, and becomes even more reduced in size when the hand or fingers are flexed/extended or the wrist undergoes either radial or ulnar deviation. Problems also develop when this area is compressed from outside by continued pressure on the tunnel ligament itself (such as when the wrist is rested on the edge of a desk for long periods while typing). All of these elements irritate the tendons and cause them to swell, further reducing the area of the carpal tunnel. This compression can lead to impaired blood flow to the median nerve and symptoms such as numbness, lack of strength in a power grip, burning, prickling, tingling, and diminished sweat function in the affected area. In time, the pain may also radiate up the arm to the shoulder and sometimes the neck, and if the condition is allowed to progress, weakness of the thenar muscles may result. This hinders one from being able to bring the thumb into opposition with the other fingers, and thus hinders one's ability to grasp objects. Over time, the nerve can become permanently damaged.

Guyon canal syndrome is similar to carpal tunnel syndrome, except that it affects the little finger and the outside of the ring finger, and is caused by the compression of the ulnar nerve as it passes through Guyon's canal in the wrist (same cause, different nerve). Contributory conditions involve overuse of the wrist, particularly tasks which involve bending the wrist down and out, or which put constant pressure on the palm. Another potential, non-work-related contributor is arthritis involving the wrist bones and joints.

It is important to note that these disabilities are not restricted only to work-related causes; they can result from other medical conditions, such as pregnancy. Occupational risk factors that have been associated with the development of carpal tunnel include: highly repetitive work, rapid and repetitive finger movements, bent wrist postures, excessive force exerted in repetitive work or awkward wrist postures, cold temperatures, mechanical trauma, and high vibration levels. Personal risk factors include pregnancy or the use of oral contraceptives, menopause, hobbies that place excessive strain on the wrist (e.g., racquetball, computer use, hand sewing), a previous wrist fracture, rheumatoid arthritis, diabetes, smoking, obesity, diabetes, and alcoholism. Carpal tunnel syndrome commonly affects both men and women (predominantly women), usually in the 40- to 60-year-old age range (Moore 1998).

The best method of prevention of this particular CTD is through the elimination or reduction of work involving awkward wrist postures, repetitive motion, or excessive vibration. One common approach is the redesign of handtools to allow for a more neutral wrist posture during their employment. Another helpful technique is to redesign tasks to allow for frequent rest breaks where the hands and wrists are allowed to return to neutral postures or to rotate workers through a number of different job tasks throughout the work cycle to reduce the amount of time spent working in awkward postures or performing repetitive tasks.

Tendinitis is traditionally defined as an inflammation of the tendons caused by repeated tensing of the muscle/tendon unit. If such exertions are further continued, the fibers that make up the tendon then begin to break apart. *Lateral epicondylitis* is such an inflammation at the point where the tendons of the lower arm attach to the small bony point on the outside of the elbow (the lateral epicondyle). This condition is directly related to the motions that tense the extensor and supinator muscles of the wrist. These are the same muscles which are most frequently used when a tennis player hits a backhand (hence the nickname "tennis elbow"), though less than 5 percent of those who suffer from this condition actually play tennis. Men and women are equally affected, with most cases occurring between the ages of 35 and 55 (Moore 1998). Up to one-third of such cases may be the result of blunt trauma to the affected area, but repeated pronation and supination of the forearm against resistance or when the hand is maintaining a grip, overexertion, and forced extension of the wrist appear to be major contributing factors (Moore 1998). A related injury is *medial epicondylitis* (also known as "golfer's elbow"), which affects the tendon attaching the wrist flexor muscles to the inside of the elbow.

A condition occasionally confused with lateral epicondylitis is *radial tunnel syndrome*. Though the pain is experienced in the elbow area, unlike epicondylitis, this condition is more similar to that of carpal tunnel in that it is caused by compression of the radial nerve. This nerve passes from the neck, through the arm, through a tunnel across the outside of the elbow (where it may become constricted), and down to the hand. This condition is associated with overuse of the arm for twisting activities, repetitive forceful pushing and pulling, bending of the wrist, and gripping and pinching movements of the arm—any of which may cause irritation of the radial nerve.

Tenosynovitis is a repetitive tendon injury involving the synovial sheath surrounding the tendons. These sheaths are filled with a lubricant called synovial fluid, which eases the movement of the tendon. When overused, the sheath secretes excess fluid and begins to swell, thus restricting movement of the tendon and becoming painful (Putz-Anderson 1988; Helander 1995). Tasks which exceed 1500 to 2000 repetitions per hour have been shown to produce this condition in the hands (Curwin and Stanish 1984).

De Quervain's disease (stenosing tenosynovitis) is a special case of tenosynovitis which affects the tendons at the base of the thumb at the side of the wrist. Excessive friction between the two tendons which pull the thumb out and back from the hand causes a thickening of their sheaths and consequent constriction of the tendons inside (Helander 1995; Putz-Anderson 1988). In some cases, this condition may be the result of acute trauma; however, a more often cited cause is overuse of the thumb. Tasks related to the condition commonly involve the rotation of the wrist while exerting a strong grip, such as when wringing clothing or operating a manual screwdriver. This condition affects primarily women (10 times as frequent as men) in the 35 to 55 age group (Moore 1998).

Trigger finger or *trigger thumb* (stenosing tenosynovitis of the digits) is a condition in which the sheath around the tendons of the fingers are so swollen that the tendon becomes locked in place, resulting in snapping or locking movements and pain when attempts are made to extend the affected digit (since the flexor muscles are stronger than the flexors, flexion is usually still possible) (Moore 1998). Though usually this condition develops gradually, it can also result from acute trauma. In the workplace, this condition is commonly associated with overuse of tool handles with hard or sharp edges, extended use of pistol-gripped power tools, or long periods gripping a steering wheel. These conditions primarily affect the thumb, middle and ring fingers of the dominant hand; the index and little finger are usually unaffected (Putz-Anderson 1988; Karwowski and Marras 1999; Moore 1998).

Thoracic Outlet Syndrome involves the compression of the three nerves of the arm and blood vessels between the neck and shoulder. Symptoms are similar to those of carpal tunnel, and involve numbness in the hands and fingers. It is commonly caused by the adoption of postures which restrict blood flow through the arms, depriving the associated muscles, tendons, and ligaments of both oxygen and nutrients. Examples of such postures include those that require the shoulders to be rotated back and down or that of working overhead.

Neuritis is the inflammation of the nerves caused by stretching or rubbing of them against bone, leading to nerve damage. This is commonly caused by repeated use of the upper extremities in awkward postures and is characterized by tingling and numbness in affected areas (Wickens et al. 1998). Ischemia is the sensation of tingling or numbness which occurs as the result of restricted localized blood flow, such as that which occurs in the fingers with excessive palmar pressures while using a handtool.

Reynaud's Disease or "vibration induced white finger syndrome"also involves a thickening of the tendon sheaths. It is commonly associated with handtool vibration in the 50 to 100 Hz range, which leads to constriction of the blood vessels in the hand and reduced blood flow to the fingers. It is characterized by tingling or numbness, pale/cold/ashen skin, and eventual loss of sensation and control in both the fingers and hand, and may be compounded by other factors causing vasoconstriction such as smoking or low temperatures (Helander 1995; Putz-Anderson 1998).

58.6 WORKPLACE SAFETY

Workplace safety should be of paramount importance at all stages of the life cycle in the manufacturing process—it is required by law (OSHA), it is required by society and professional ethics, and it is also good business practice. Workplace safety is often an industrial hygiene issue and as such an extremely brcad field. The following sections will only highlight key regulations and concepts which are related to safe workplace practices, particularly with respect to stationary equipment. The focus on this section is the *system model* composed of three interrelated elements: (1) the worker, (2) the machine, and (3) the environment. These components provide the structure for performing a job hazard analysis of the production worker.

58.6.1 Federal Legislation

The Williams-Steiger Occupational Safety and Health Act of 1970 requires that every employer covered under the Act must provide "a place of employment which is free from recognized hazards that are causing or likely to cause death or physical harm to their employees." Workplaces covered include most private manufacturing concerns, as well as the recently added U.S. postal service. Other similar federal regulations cover workplaces in mining-related industries. It further requires that "employees comply with standards, rules, regulations, and orders issued under the Act which are applicable to their own actions and conduct."(29 CFR §1903.1)

The Act encourages and gives individual states the opportunity to develop their own safety and health programs; however, such programs must be approved by the federal OSHA and must be at least as effective when compared to the federal standards.

A listing of the detailed requirements of all of the OSHA standards embodied in 29 CFR 1910 is obviously outside the scope of this chapter, but some selected issues will be addressed. It is important, however, for the manufacturing engineer to become at least familiar with the requirements contained in all sections of this document, since even a single piece of production machinery may be affected by requirements listed in such diverse segments as subpart D (Walking-Working Surfaces), subpart O (Machinery and Machine Guarding), subpart S (Electrical), or subpart M (Compressed Gas). Compliance with the Act is mandatory.

Of particular significance to workplace safety in the manufacturing environment is the section dealing with "Lockout/Tagout"(29 CFR §1910.47). This portion of the regulations identifies the practices and procedures necessary to shut down and lock out or tag out machines and equipment, and mandates training to employees in their role in lockout and further requires employers to monitor their employees to insure that lockout/tagout is being used. OSHA has estimated that failure to control energy when performing maintenance, servicing or cleaning by itself accounts for nearly 10 percent of the serious accidents in many industries (*Federal Register,* 1989).

Service and maintenance activities are necessary components of the industrial process, as are lubricating, cleaning, unjamming, and making minor adjustments and simple tool changes to equipment. When a machine is being used in normal production, Section 1910.212 dictates what safeguards must be in place. According to OSHA, if a jam were to occur during operation that would require removal of guarding, the machinery must be shut down and locked out before such "servicing" may take place. The fact that such actions are not part of normal production is quite explicitly stated in the *Federal Register* (Vol. 54, No. 169), ". . . Although this action takes place during normal production operations, it is not actually production, but is servicing of the equipment to perform its production function..." At this point, the provisions of Section 1910.147 (The control of hazardous energy—lockout/tagout) come into play.

If a servicing activity (such as lubricating, cleaning, or unjamming) takes place during production, the employee performing this activity may become exposed to hazards that are not normally encountered as part of the production operation itself. Workers must then deenergize the equipment and lockout/tagout the machine before performing such servicing operations to prevent potential exposure to moving parts. There are very few exceptions to this requirement. The printing industry, for example, attempted to obtain relief from these requirements in order to allow for the use of inching/jogging methods while cleaning print rollers. The final ruling by OSHA was relaxed only minimally; cleaning could only take place after jogging was complete and the machinery was stopped. This "service only when it is safe" philosophy maximizes the safety to the employee, usually at only a minimal cost in terms of time and convenience.

58.6.2 National Consensus Standards

ANSI, founded in 1918, has served as the administrator and coordinator of the United States private sector voluntary standardization system for more than 80 years. ANSI does not itself develop standards; rather it facilitates their development by promoting and publishing standards, such as those involving safety issues.

ANSI safety standards for machinery and equipment are voluntary standards, which generally reflect national consensus by a wide and balanced group of interested parties. Such standards are typically drafted by professional engineering society members, and (at least for review purposes) government representatives, manufacturers of the type equipment under consideration, independent safety consultants, users of the equipment, and union personnel. In some cases, certain of these standards are directly incorporated by reference into OSHA regulations. Such ANSI standards are a valuable resource for providing safety information and requirements that represent the consensus of good practice.

ANSI safety standards often provide a breakdown of responsibilities between: (1) the manufacturer of equipment, (2) the installer, and (3) the end user and/or operator. Such a breakdown of responsibilities helps in structuring specification documents for equipment. Owing to the large number of ANSI safety related standards, when requiring standards compliance in either purchase order or request for quote documents, it is important to cite by name which particular ANSI standards are expected to be complied with by the vendor. Nevertheless, the ultimate responsibility for ensuring workplace safety rests with the end user/operator.

Examples of ANSI standards which may be relevant to maintaining a safe manufacturing environment include, but are not limited to the following:

ANSI A14.3-1992 "Requirements for Fixed Ladders"

ANSI/ASME B15.1 1994 "Safety Standard for Power Transmission . . . "

ANSI A10.5 1992 "Requirements for Material Hoists"

ANSI/ASME B20.1 "Safety Standards for Conveyors and Related Equipment"

Other safety standards also exist which are not published by ANSI but nevertheless represent a national consensus regarding practices which promote workplace safety. The American Petroleum Institute (API), for example, produces RP 2003.82 "protection against ignitions arising out of static lightning and stray currents" and the National Fire Protection Association (NFPA) puts out NFPA 77 "static electricity."

In an effort to provide for the safe workplace, it is important for the user to became aware of and apply all the appropriate consensus standards to the manufacturing equipment and environment.

58.6.3 Machinery Hazard Control

Machinery accident prevention must be part of an ongoing program. The process cannot develop on its own—it must be planned, monitored, and controlled. Safety professionals have long recognized that the manufacturing process involves a system. This system is predicated on the concept that the performance of any manufacturing task requires an interaction between the worker, the equipment and materials (machine), and the environment. This interrelationship is diagrammatically shown in Fig. 58.7.

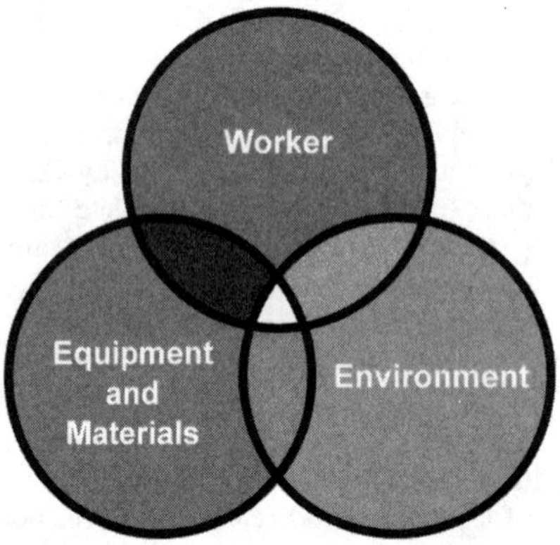

FIGURE 58.7 Task performance interaction. (*Source: National Safety Council.*)

According to the National Safety Council, accidents occurring in the workplace can be attributed directly or indirectly to an oversight, omission, or malfunction of the management system regarding one or more of the following items:

1. *Human factors.* Either the worker or another person
2. *Situational factors.* Such as tools, facilities, equipment, or materials
3. *Environmental factors.* Such as noise, vibration, temperature extremes, and visibility

If an adequate management hazard control system is properly designed, monitored, and controlled for the specific worker system (worker, machine, and environment) the likelihood of accidents occurring in the workplace is greatly reduced.

An effective hazard control program has six steps or processes (Firenze 1978).

1. Hazard identification
2. Risk assessment
3. Management decision-making
4. Establishing preventative and corrective measures
5. Monitoring
6. Evaluating program effectiveness

Several resources are readily available to assist in the establishment of the six-step hazard control program listed above. The National Safety Council's *Accident Prevention Manual: Administration & Programs* provides a very comprehensive listing, addresses, and telephone numbers of resources to assist in the establishment of hazard control programs, such as the American Society of Safety Engineers, the American Society of Mechanical Engineers, and others.

58.6.4 Hazard Identification and Risk Assessment

Every job should be examined for its potential to result in an injury and the potential severity of the injury in light of the entire system (worker–machine–environment). There is no uniformly accepted procedure, "recipe," or checklist that is appropriate to all situations. Firenze (1978) provides three different hazard assessment schemes noting that all are essentially qualitative in nature. One commonly employed approach is to use a *task analysis* such as the one detailed by the National Safety Council (1992). The first stage of such an exercise is to break the operation of a piece of machinery or a production task into its component subelements. Examples of a set of task categories for production machinery might include (but would not be limited to) the following:

1. Set up and try out
2. Operation in a production mode (reaching in, feeding material, removing material, and the like)
3. Tool change
4. Planned maintenance
5. Unplanned maintenance (unjamming, cleaning, and the like)
6. Major/minor repair
7. Recovery from a "crash"
8. Trouble shooting
9. Housekeeping
10. Emergency shutdown

It is important to assess the potential worker involvement (if any) in these tasks and to identify potential hazards owing to such causes as pinching, shearing, crushing, impact, thermal/chemical exposure, or electrocution in the interaction. Each step of an operation must be assessed individually while keeping in mind the relationship between the task and the workers and equipment, materials, environment, and other workers that may not be directly involved.

The next step is to independently identify both the consequences of potential injuries and their probabilities of occurrence. Consequences fall into one of four possible categories (National Safety Council, 1992):

- *Category 1: Catastrophic.* May cause death or loss of a facility
- *Category 2: Critical.* May cause severe injury, severe occupational illness, or major property damage
- *Category 3: Marginal.* May cause minor injury or minor occupational illness resulting in lost workdays or minor property damage
- *Category 4: Negligible.* Probably would not affect personnel safety or health and thus, less than a lost workday, but nevertheless is in violation of specific criteria

The probability of occurrence is estimated by taking into account frequency, duration and extent of exposure, training, and awareness of the hazard. Some examples of factors which may be considered in estimating the probability of harm are as follows:

Machine/task history (reliability, history of accidents/near misses, and the like)

Workplace environment

Human factors/ergonomics

Reliability of safety equipment in place

Protective measures and their maintenance

Training

Probabilities are assigned based on four categories (National Safety Council, 1992):

- *Category 1.* Likely to occur immediately or within a short period of time when exposed to the hazard
- *Category 2.* Probably will occur in time
- *Category 3.* Possible to occur in time
- *Category 4.* Unlikely to occur

For example, if material is hand-fed into a press or into rollers as part of task no. 2 above (Operation in a Production Mode), a potential hazard might be a crushing injury involving getting a hand caught in the machine, depending on the configuration of the machine, the characteristics of the material being fed, and the location of the worker. The consequences in this case might be *critical* (severe injury) and the probability such an occurrence might be "C." An alternative possible injury might involve the dropping of feedstock, which could result in only a *marginal* crushing injury, but have a higher probability of occurrence.

An exposure assessment is typically also conducted after the severity and probability analysis. A typical scheme is to assign a value ranging from 1 to 4, as presented below (National Safety Council, 1992):

- *Category 1.* More than 50 different people are regularly exposed to the hazard.
- *Category 2.* Between 10 and 49 different people are regularly exposed to the hazard.
- *Category 3.* Between five and nine different people are regularly exposed to the hazard.
- *Category 4.* Less than five different people are regularly exposed to the hazard.

By examining the combination of severity, probability, and exposure to injury, judgment can then made regarding the overall risk level of the task hazard. Risk levels are typically rated on a 5-point scale, with a high of 1 (critical) to a low of 5 (negligible).

58.6.5 Management Decision Making and Risk Reduction

Once a risk level assessment is complete, a decision by management must be made to provide for risk reduction where the original risk level is determined to be unacceptable. As noted by Firenze (1978), risk assessment is qualitative in nature, hence an unacceptable level of risk is qualitatively and situationally dependent. The selection and implementation of risk reduction controls involve risk-utility considerations:

Technological feasibility

Economic impact

Ergonomic impact

Productivity

Potential for the creation of new hazards

Corrective measures can be of three types: (1) administrative (training, education, housekeeping, and the like), (2) engineering (design, lockout procedures, process changes, warning equipment, and the like), or (3) protective equipment and tools.

It has long been recognized that the preferred control alternative is typically to attack the hazard at the source through design or substitution of a less harmful agent than the one posing the hazard (subject to the risk-utility considerations above). The second alternative is to control the hazardous task by providing a physical barrier between the worker and the hazard. Examples of such controls might be machine guards, protective curtains for shielding eyes from welding arcs, chemical hoods, and the like. The third alternative is to direct the worker activity through the use of personal protective tools, devices, and training (National Safety Council, 1992).

The final selection of the most appropriate control means or combination of control means must also be considered in light of the worker–machine–environment system in order to be effective. Additional consideration should also be given to any task which may require the removal, modification, or disablement of one or more of the selected safeguards or controls—in such cases, OHSA-mandated lockout/tagout procedures must be employed. It is important to note that the manuals provided with purchased machinery often contain valuable instruction on safe operating procedures and practices. These materials should be considered by management in the implementation of the optimal control measures in order to reduce unacceptable risks.

Selecting a particular safeguarding means or relying on a machine supplier's original safeguarding is often insufficient. According to the occupational safety and health administration: " . . . Even the most elaborate safeguarding system cannot offer effective protection unless the worker knows how to use it and why. Specific and detailed training is therefore a crucial part of any effort to provide safeguarding against machine-related hazards . . . " (U.S. Department of Labor, 1992).

Thorough operator training should involve instruction and/or hands on training in the following:

Familiarization with the hazards associated with the machinery in the workers job space

Explanation of what the selected safeguarding means are, how and when to use them, and how the safeguarding reduces the risks

Instruction on what action to take when the safeguarding means is found altered or inoperative or in some way compromised

Instruction in lockout/tagout

Procedures for cleaning, adjusting, maintenance, repair, or unjamming of machinery

58.6.6 Monitoring and Evaluating Effectiveness

A key aspect of monitoring is to make certain that the chosen safety strategy is in place and is working properly over time. Safeguarding controls may be rendered ineffective by the workers or their supervisors, therefore periodic verification of their continued presence and effectiveness must take place. Another aspect of monitoring is the need to reperform the hazard analysis as the production process is modified, changed, or ages, not only to maintain the desired level of safety, but also to provide the opportunity to uncover hazards that were previously overlooked.

Traditional means of evaluating effectiveness are monitoring and studying the accident statistics within a production facility or a similar facility. A program that actively solicits data from "near misses" as well as actual accidents/injuries and analyzes this information is a superior method of monitoring and evaluating a production process.

REFERENCES

ANSI/HFS 100-1988, American National Standard for Human Factors Engineering of Visual Display Terminal Workstations. The Human Factors Society, Santa Monica, CA, 1988.

Armstrong, T. J., *Work related upper limb disorders,* http://www-personal.engin.umich.edu/~tja/CTD1.html, 2002.

ASHRAE, *Handbook of Fundamentals,* American Society of Heating, Refrigeration, and Air-Conditioning Engineers, New York, 1974.

Ayoub, M., and A. Mital, *Manual Materials Handling,* Taylor & Francis, London, 1989.

Barnes, R. M., *Motion and Time Study*, 5th ed., Wiley, New York, 1963.

Battie, M. C., T. Videman, L. E. Gibbons, Hannu Manninen, Kevin Gill, Malcolm Oppe, and Jaakko Kaprio, "Occupational Driving and Lumbar Disc Degeneration: A Case-Control Study," *The Lancet,* October 15, 2002. Available at: http://image.thelancet.com/extras/01art9329web.pdf

Broadbent, D., "Noise and the Details of Experiments: A Reply to Poulton," *Applied Ergonomics,* 7, 231–235, 1976.

Brown, A., and G. Brengelmann, "Energy Metabolism," in *Physiology and Biophysics,* T. Ruch and H. Patton, (eds.), Saunders, Philadelphia, 1965.

Cochrane, D., and M. Riley, "The Effects of Handle Shape and Size on Exerted Forces," *Human Factors,* 28(3), 253–265, 1986.

Curwin, S., and W. Stanish, *Tendinitis: Its Etiology and Treatment,* Collamore Press, Lexington, KY, 1984.

Das, B., "Manufacturing Workstation Design," in W. Karwowski and G. Salvendy (eds.), *Ergonomics in Manufacturing: Raising Productivity through Workplace Improvement,* Society of Manufacturing Engineers, Dearborn, MI, 1998.

Eastman Kodak, *Ergonomic Design for People at Work,* Vol. 1, Van Nostrand Reinhold, New York, 1983.

Eastman Kodak, *Ergonomic Design for People at Work,* Vol. 2. Van Nostrand Reinhold, New York, 1986.

Fanger, P., *Thermal Comfort, Analyses and Applications in Environmental Engineering,* Danish Technical Press, Copenhagen, 1977.

Farley, R. R. "Some Principles of Methods and Motion Study as Used in Development Work," *General Motors Engineering Journal,* 2(6): 20–25, 1955.

Control of Hazardous Energy Sources, *Federal Register*, 54(169) 36644–36696, 1989.

Fraser, T. M., *The Worker at Work,* Taylor & Francis, Bristol, PA, 1989.

Firenze, R. J., *The Process of Hazard Control,* Kendall/Hunt Publishing Company, Dubuque, IA, 1978.

Freivalds, A., "Ergonomics of Hand Tools," in W. Karwowski and W. Marras (eds.), *The Occupational Ergonomics Handbook,* CRC Press, Boca Raton, FL, 1999.

Garrett, J., "The adult human hand: some anthropometric and biomechanical considerations," *Human Factors,* 13:117–131, 1971.

Grandjean, E., *Fitting the Task to the Man: A Textbook of Occupational Ergonomics,* 4th ed., Taylor & Francis, London, 1988.

Helander, M., *A Guide to the Ergonomics of Manufacturing,* Taylor & Francis, Bristol, PA, 1995.

Hendrick, H., "Good Ergonomics is Good Economics," Presidential Address at the 1996 Human Factors and Ergonomics Society Annual Meeting; http://hfes.org/publications/goodergo.pdf

Hunt, L., "A Study of Screwdrivers for Small Assembly Work," *The Human Factor,* 9(2), 70–73, 1934.

Hutchingson, R. D., *New Horizons for Human Factors in Design*. McGraw-Hill, New York, 1981.

ANSI/HFS 100-1988, American National Standard for Human Factors Engineering of Visual Display Terminal Workstations, Human Factors Society, Santa Monica, CA, 1988.

ISO 2631-1, Mechanical Vibration and Shock—Evaluation of Human Exposure to Whole-Body Vibration, *General Requirements,* Volume 1, International Standards Organization, Geneva, 1997.

Joyce, M., "The Business Case for Ergonomics/Human Factors—Bottom Line: Incorporating Ergonomics into Your Business Plan Adds Value and Stimulates Profitability," 2001, www.meadhatcher.com/human.html

Kamon, E. and A. Goldfuss, "In-Plant Evaluation of the Muscle Strength of Workers," *American Industrial Hygiene Association Journal,* 43: 853–857, 1978.

Kao, H., "Human Factors Design of Writing Instruments for Children: The Effect of Pen Size Variations," *Proceedings of the 18th Annual Meeting of the Human Factors Society,* Santa Monika, CA, 1974.

Karwowski, W., and W. Marras, *The Occupational Ergonomics Handbook,* CRC Press, Boca Raton, FL, 1999.

Konz, S., *Work Design: Industrial Ergonomics,* 3rd ed., Horizon Publishing, Scottsdale, AZ, 1990.

Kroemer, K., "Engineering Anthropometry," *Ergonomics,* 32(7), 767–784, 1983.

Kroemer, K. H., H. J. Kroemer, and K. E. Kroemer-Elbert, K. E. *Engineering Physiology: Bases of Human Factors/Ergonomics,* 3rd ed, Van Nostrand Reinhold, New York, 1997.

Lehmann, G., "Physiological Measurements as a Basis of Work Organization in Industry. *Ergonomics,* 1, 328–344, 1958.

Luopajarvi, T., I. Kuorinka, M. Virolainen, and M. Holmberg, "Prevalence of Tenosynovitis and Other Injuries of the Upper Extremities in Repetitive Work," *Scan. J. Work Environ. and Health,* 5(3), 48–55, 1979.

Moore, S. J., "An Overview of Upper Extremity Disorders," in W. Karwowski and G. Salvendy (eds.), *Ergonomics in Manufacturing*: *Raising Productivity through Workplace Improvement,* Society of Manufacturing Engineers, Dearborn, MI, 1998.

Murrell, K., *Human Performance in Industry,* Reinhold, New York, 1965.

National Safety Council (1992). Accident Prevention Manual for Business and Industry: Administration & Programs, Chicago, IL, 10th Edition.

NIOSH, *A NIOSH Look at Data From the Bureau of Labor Statistics—Worker Health by Industry and Occupation: Musculoskeletal Disorders, Anxiety Disorders, Dermatitis, Hernia,* U.S. Department of Health and Human Services, Public Health Service, Centers for Disease Control and Prevention, National Institute for Occupational Safety and Health, 2001.

Oh, S., and R. G. Radwin, "Pistol Grip Power Tool Handle and Trigger Size Effects on Grip Exertions and Operator Preference," *Human Factors,* 35(3), 551–569, 1993.

Panero, J. and M. Zelnik, *Human Dimension and Interior Space*: *A Source Book of Design Reference Standards,* Watson-Guptill Publications, New York, 1979.

Peterson, A., and E. Gross, Jr., *Handbook of Noise Measurement*, 7th ed., General Radio Co., New Concord, MA, 1972.

Putz-Anderson, Vern, *Cumulative Trauma Disorders*: *A Manual for Musculoskeletal Diseases of the Upper Limbs,* Taylor & Francis, Bristol, PA, 1988.

Radwin, R., "Hand Tools: Design and Evaluation," in W. Karwowski and W. Marras (eds.), *The Occupational Ergonomics Handbook,* CRC Press, Boca Raton, FL, 1999.

Rea, Mark S., *The IESNA Lighting Handbook*: *Reference and Application,* 9th ed., IESNA, New York, 2000.

Robinette, K. and J. McConville, "An alternative to percentile models," SAE Technical Paper Series no. 810217, Society of Automotive Engineers, Warrendale, PA, 1981.

Sanders, M., and E. McCormick, *Human Factors in Engineering and Design,* 7th ed., McGraw-Hill, New York, 1993.

Snook, S. H., and V. M. Ciriello, "The design of manual handling tasks: revised tables of maximum acceptable weights and forces," *Ergonomics,* 34(9), 1197–1213, 1991.

Sperling, L., *Work with Hand Tools: Development of Force, Exertion, and Discomfort in Work with Different Grips and Grip Dimensions,* Report in Swedish with English summary, Arbetarskyddsstyrelsen, Undersokninsrapport, 25, 1986.

Squires, P. C., *The Shape of the Normal Work Area,* Report No. 275, Navy Department, Bureau of Medicine and Surgery, Medical Research Laboratory, New London, CT, 1956.

Terrel, R., and J. Purswell, "The influence of forearm and wrist orientation on static grip strength as a design criterion for hand tools," *Proceedings of the Human Factors Society,* 20: 28–32, Santa Monica, CA, 1976.

U.S. Department of Labor, Occupational Safety and Health Administration, "OSHA Publication 3067: Concepts and Techniques of Machine Safeguarding", 1992. Available at: http://www.osha.gov/Publications/Mach_SafeGuard/chapt1.html

Ulin, S., T. Armstrong, S. Snook, and W. Keyserling. "Examination of the Effect of Tool Mass and Work Posture on Ratings of Perceived Exertion for a Screw Driving Task", Int. J. of Indust. Ergonomics, 12:105–115, 1993.

Warkotsch, W., "Ergonomic Research in South African Forestry," *Suid-Afrikaanse Bosboutydskrif,* 171, 53–62, 1994.

Wassell, J., L. Gardner, D. Landsittel, and J. Johnston, "A prospective study of back belts for prevention of back pain and injury," *Journal of the American Medical Association*, 284(21), 2000.

Wasserman, D., D. Badger, T. Doyle and L. Margolies, "Industrial Vibration—An Overview," *Journal of the American Society of Safety Engineers,* 10, 38–40.

Waters, T. R., V. Putz-Anderson, and A. Garg, *Applications Manual for the Revised NIOSH Lifting Equation,* DHHS (NIOSH) Publication No. 94-110, U.S. Department of Health and Human Services, Public Health Service, Centers for Disease Control and Prevention, National Institute for Occupational Safety and Health, 1994.

Wickens, C. D., S. E. Gordon, and Y. Liu, *An Introduction to Human Factors Engineering,* Longman, New York, 1998.

Woodson, W., B. Tillman, and P. Tillman, *Human Factors Design Handbook: Information and Guidelines for the Design of Systems Facilities, Equipment and Products for Human Use,* 2nd ed. McGraw-Hill, New York, 1992.

FURTHER READING

Flynn, J. E., "The IES approach to recommendations regarding levels of illumination," *Lighting Design and Application,* 9 (9):74–77, 1979.

Village, J., and M. Lott, "Work-relatedness and threshold limit values for back disorders owing to exposure to whole body vibration," *Proceedings of the IEA 2000/HFES2000 Conference,* Human Factors and Ergonomics Society, Santa Monica, CA, 2000.

CHAPTER 59
TOTAL PRODUCTIVE MAINTENANCE

Atsushi Terada
JMA Consultants America, Inc.
Arlington Heights, Illinois

59.1 INTRODUCTION

It is necessary to link cost reduction with the elimination of various kinds of waste in order for a firm to survive the current severe economic environment. Therefore, the need to use equipment to its utmost capacity has become paramount. As quality demands have become more and more stringent, it has become necessary to establish and maintain conditions that prevent equipment from creating defects and thus to avoid shipping defects. Requirements for production of small lots with a wide range of product variety, *just-in-time* (JIT) production, and cost reductions have imposed severe conditions on the manufacturing industry.

Faced with a situation where equipment (even if new and purchased at great expense) breaks down frequently, has many short stoppages and produces numerous defects, it is impossible to implement a flexible production system and/or JIT. Therefore, the elimination of equipment losses has become an urgent issue.

As the operation and functional level of equipment advances, so does the necessary skill level of operators and maintenance personnel on the production floor. In other words, today's operators must not think of themselves as just using or driving a machine from day to day, but must be able to retain a sense of ownership, protecting and caring for their machines. As for maintenance personnel, the prevention of equipment malfunctions needs to take precedence over "fire-fighting."

In addition, a worker-friendly environment in the workplace has become more and more important. This means that the prevention of oil leaks, machining debris (shavings and the like) and other dirt coming from equipment has also become very important. Although responsibility for the elimination of equipment losses has previously been solely the domain of maintenance and production people, it has gradually become recognized that *all* employees need to be involved in this process.

As a program for promoting better equipment usage, *total productive maintenance* (TPM) has been advocated in Japan from the 1970s. TPM is a form of *productive maintenance* (PM) involving all employees.

59.2 TRANSITION OF EQUIPMENT MANAGEMENT TECHNOLOGY

59.2.1 Breakdown Maintenance

The first stage involved the notion of *breakdown maintenance* (BM) which had been done since before the 1940s. This is the idea of fixing something when it is broken. It is not easy to schedule BM before the actual occurrence and this results in lower efficiency in regards to workers, materials, and machine parts arrangement. This method is still adopted today if the stop loss of the production facility can be disregarded. In a word, when mean time between failures is not constant, mean time to repair is short and the part cost to exchange regularly is high, this method is adopted.

59.2.2 Preventive Maintenance

Preventive maintenance, introduced in the United States in the 1940s, emerged as the second stage in equipment management. The fundamental idea behind preventive maintenance is to maintain equipment before it breaks down. Preventive maintenance intends to exchange parts and units at economical intervals of time to prevent sudden breakdowns and equipment stoppages. The interval for preventive maintenance is based on the scale and expected lifetime of the equipment. Typically such intervals are yearly, weekly, or daily periods; however, it should be noted that preventive maintenance is not economical if it is done excessively. The plan of preventive maintenance is made after consideration of the cost of breakdown maintenance, cost of preventive maintenance, and cost of stop losses owing to unplanned breakdowns.

59.2.3 Productive Maintenance

The third stage in equipment management is PM, which was introduced in the 1960s. This is an excellent maintenance method in terms of economy. The purpose of this maintenance method is to improve the productivity of the firm by reducing total cost such as equipment cost, operation cost, and maintenance cost. Reducing losses associated with the deterioration of the equipment is another important objective.

PM consists of two ideas. The first idea is corrective maintenance. This involves improving equipment by making it easy to maintain and repair, and ultimately getting to the point where maintenance is no longer required. In addition, corrective maintenance aims to improve equipment to increase its productivity. The second idea involves maintenance prevention. To reduce the maintenance cost of equipment fundamentally, devise the maintenance method ahead of time and purchase equipment which does not require maintenance. Procuring and maintaining equipment which is consistent with this principle fulfills the objective of maintenance prevention.

59.2.4 Total Productive Maintenance

The fourth stage is TPM *total productive maintenance*, which includes autonomous small group activities on the part of operators. It began in the 1970s. When Nippon Denso captured the PM prize awarded by JIPE (currently JIPM, a nonprofit organization, Japan Institute of Plant Maintenance), the examiners visiting the factory floor were impressed with the "productivity maintenance with the involvement of all employees" which Denso had developed. This later came to be known as TPM and, in the history of Japanese progress in the field of PM, Denso's receipt of this prize is considered to be the birth of TPM. Conventional versions of PM did not cover the entire workplace and were focused on factory staff and maintenance people, without the involvement of top management or all operators. The T in TPM refers to "total" and TPM is a revolutionary step forward in solving this problem. With TPM top management takes responsibility for the project and for promotion of total productive maintenance, and all operators are involved in project activities with a high level of enthusiasm and desire. Even though this was an early and not fully developed version of TPM, it marked the birth of a revolutionary new approach.

59.2.5 Predictive Maintenance

Finally, the fifth stage of equipment management is the concept of *predictive maintenance* and this spread throughout the industry in the 1980s.

The concept of predictive maintenance is to diagnose the deterioration situation and the performance situation of the equipment, and to develop maintenance activities based on the results of the diagnosis. To do this, one must understand the deterioration level of the equipment accurately and precisely. This is the concept of condition-based maintenance which maintains equipment when deterioration is observed and maintenance is really necessary.

Development of equipment diagnosis technology allowed equipment to be quantitatively evaluated and its condition assessed. As a result, the check-inspect-replace process for equipment moved from time-based (known as time-based maintenance) to condition-based.

59.3 OUTLINE OF TPM

59.3.1 Definition of TPM

The definition of TPM is as follows according to JIPM.

1. Aims at building up a corporate culture that thoroughly pursues production system efficiency improvement or *overall equipment efficiency* (OEE)
2. Creates a system to prevent every kind of loss, for example, to achieve zero accidents, zero defects, and zero failures, based on "actual site" and "actual object" over the entire life cycle of a production system
3. Covers all departments including production, product and process development, and marketing and administration
4. Requires full involvement from top management across the board to frontline employees
5. Achieves zero losses by overlapping small-group activities

59.3.2 TPM Award

When firms first introduce TPM into their organization, many of them have the ultimate goal of capturing the TPM award. JIPM introduced the TPM award in 1971.

In Japan, the Deming Prize for quality was established in 1950s, and the fact that many firms aimed to improve their quality in order to win this prize is well known.

The TPM award was established from this Deming Prize about 20 years later. Each enterprise in Japan soon began to aim at winning this TPM award, recognizing that it came next to the Deming Prize.

At first, only those firms in Japan were considered for the TPM prize; however, from the 1990s firms outside Japan have vied for the TPM award as well. In fact, these days one-third of all plants to capture the prize are outside of Japan. The TPM prize is based on six criteria. Award winners are selected through the process of preliminary assessment, document assessment, and final (on-site) assessment.

For further details, please refer to the JIPM homepage.

59.4 EIGHT PILLARS OF TPM

In TPM, to achieve targets effectively and efficiently, eight pillars have been adopted. These pillars are activities that are cross-functional.

1. Focused improvement
- ***a.*** The aim of this activity
  - **(1)** Acquire the ability to discover and solve problems of equipment losses.
- ***b.*** Important subjects
  - **(1)** Elimination of the six major chronic losses
  - **(2)** Substantial cost reduction
  - **(3)** The improvement of management techniques and engineering techniques

2. Autonomous maintenance
- ***a.*** The aim of this activity
  - **(1)** Restore deteriorated equipment and establish a system whereby the operators think and act in terms of "We take the responsibility to protect our own equipment."
- ***b.*** Important subjects
  - **(1)** Establishment of 2S
    - (*a*) *Sort.* Eliminate the unnecessary
    - (*b*) *Stabilize.* Establish permanent locations for the essential
  - **(2)** Steady development of the 7 *steps* of autonomous maintenance
  - **(3)** Improvement of the maintenance skill of operators

3. Planned maintenance
- ***a.*** The aim of this activity
  - **(1)** Improvement of conventional maintenance capabilities, preventing breakdowns, and reducing maintenance costs
- ***b.*** Important subjects
  - **(1)** Support for autonomous maintenance and focused improvement activities
  - **(2)** Realization of "no sudden malfunctions" by promoting corrective maintenance/periodic maintenance
  - **(3)** Enforcement of predictive maintenance through development of diagnostic technology

4. Education and training
- ***a.*** The aim of this activity
  - **(1)** Use skilled people to handle equipment and operations
- ***b.*** Important subjects
  - **(1)** Promotion of maintenance skill training and office automation skill education
  - **(2)** Establishment and promotion of a skill development program
  - **(3)** Upgrading and improvement of the education and training system

5. Initial management
- ***a.*** The aim of this activity
  - **(1)** Based on *maintenance prevention* (MP) design, establish a highly reliable, maintainable, and economical system. MP is defined as "activities to implement design that features high degrees of reliability, maintainability, safety, and flexibility, in consideration of maintenance information and new technologies, when planning or constructing new equipment, and to thereby reduce maintenance cost or deterioration losses" by TPM Encyclopedia published by JIPM.
- ***b.*** Important subjects
  - **(1)** Collection and usage of MP information
  - **(2)** Development of a "concurrent operation" system
  - **(3)** Promotion of equipment development which fully considers *life cycle cost*

6. Quality maintenance
- ***a.*** The aim of this activity
  - **(1)** Acquire the control capability of a "cause system" which is superior to that of a "results system," and thus achieve "zero defects"

b. Important subjects
(1) Elimination of customers' claims
(2) Realization of no defects originating from equipment or people
(3) Establishment and maintenance of optimal 4M conditions (man, machine, materials, and methods), including selection of settings

7. TPM for indirect office work
a. The aim of activity
(1) Reduction of indirect work and its cost
(2) Improvement of quality of service to customers and/or the production division
b. Important subjects
(1) *Clarify purpose.* Elimination of method losses and human losses
(2) Simplification of operations, and sharing of knowledge
(3) Improvement of service to customers and/or the production division

8. Management of safety, health, and environment
(1) Establishment of a safe and sanitary environment
a. The aim of this activity
(1) Achievement of a workplace environment which is safe, bright, and easy-to-work-in
b. Important subjects
(1) Realization of no accidents and no pollution
(2) Reduction of noise, humidity, and the like
(3) Achieve a complete "healthy workplace" environment

59.5 O.E.E. AND LOSSES

59.5.1 The Structure of Equipment Usage Loss

Using equipment efficiently means to use equipment which has the best possible performance, function, and specifications; however, in reality, obstacles will hinder equipment usage from attaining its optimal performance.

We need to eliminate these obstacles to high usage in order to achieve the optimal condition for equipment.

The loss structure is shown in Fig. 59.1.

1. *Breakdown loss.* These losses contribute to the deterioration of the original equipment capability and equipment productivity.
These losses can be classified into two types.

- *Overall functional stoppage.* Even though the initial problem is often only partial functional stoppage, such partial functional stoppage results in overall functional stoppage. This type of stoppage happens suddenly.
- *Function reduction.* This partial function reduction does not inevitably lead to total functional stoppage. It does, however, result in equipment speed loss, defects, and/or lower yield.

2. *Setup/adjustment loss (Change-over loss).* Equipment stoppage owing to setup/adjustment falls under this type.

Setup time is defined as the time between the stoppage of the previous product/model production and the next good product/model production start time. In other words, setup time is the time from when previous production is finished to the time when the production preparation is completed and the next perfect product is produced.

Some people do not realize that setup/adjustment loss is lost time. Furthermore, some firms regard only the physical die change as the setup. All elements of setup/adjustment need to be considered as equipment loss. These losses should be minimized as much as possible. As a first step, the target time for setup/adjustment should be in the single digits (less than 10 min).

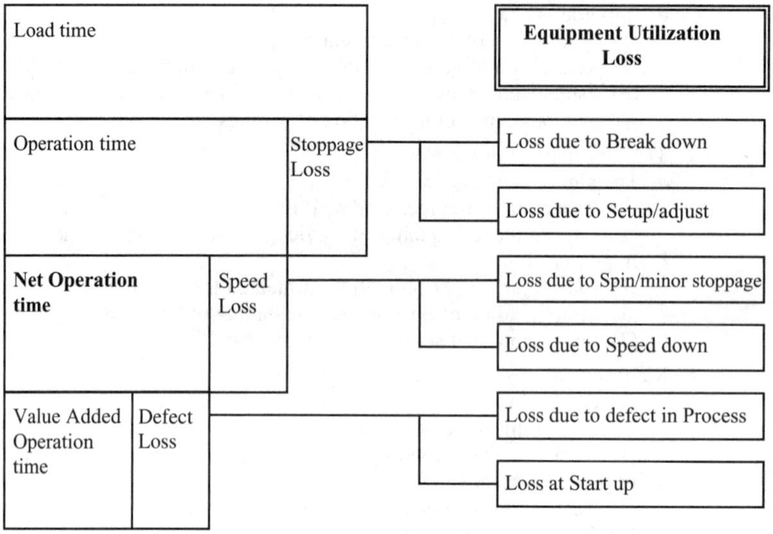

FIGURE 59.1 Equipment usage loss.

3. *Minor stoppage (short stoppage) loss.* Minor stoppage (short stoppage) is different from normal machine breakdown. This stoppage loss occurs when a machine is stopped owing to a temporary problem or a part is just spinning (equipment is running, but not passing on parts). For example, a machine may be spinning because a chute is clogged with parts. Alternatively, spinning may occur when a sensor/switch finds or picks out defect parts, temporarily stopping equipment. In the above case, the equipment normally resumes after the parts that caused the clogged chute are removed, or the sensor/switch is reset.

   Therefore, this short stoppage is different from normal machine breakdown (functional stoppage).

4. *Speed loss.* This is the difference between design/original speed and actual speed. For example, when companies try to run equipment at the designed speed, the defect ratio (and/or mechanical problems) often increases. To offset this, the speed is slowed down in order to decrease the defect ratio. This slow down is called speed loss.

5. *Defect/rework loss.* This is the loss owing to defects or rework. In general, defects are counted only as scrapped products; however, the parts that need to be reworked also need to be considered because additional wasted operation time is required to rework these subpar products.

6. *Start up/yield loss.* This loss is the time from production start time to the time when good products can be produced consistently. Few people realize that this loss is often very large. If any conditions such as unstable processing standards, lack of maintenance of tools/jigs or die, cutting blade ineffectiveness, or poor skill level of operators exist, the result is a loss. This loss sometimes is very difficult to find.

The six-equipment losses have been explained above. These losses are factors that prevent equipment from achieving its intended productivity/efficiency. Therefore, reducing and even eliminating these losses is essential to increase equipment productivity/efficiency.

Control Indicator for Equipment Loss: Overall Equipment Efficiency. The contents of these losses are explained on previous pages. These losses need first to be quantified in order to attack and

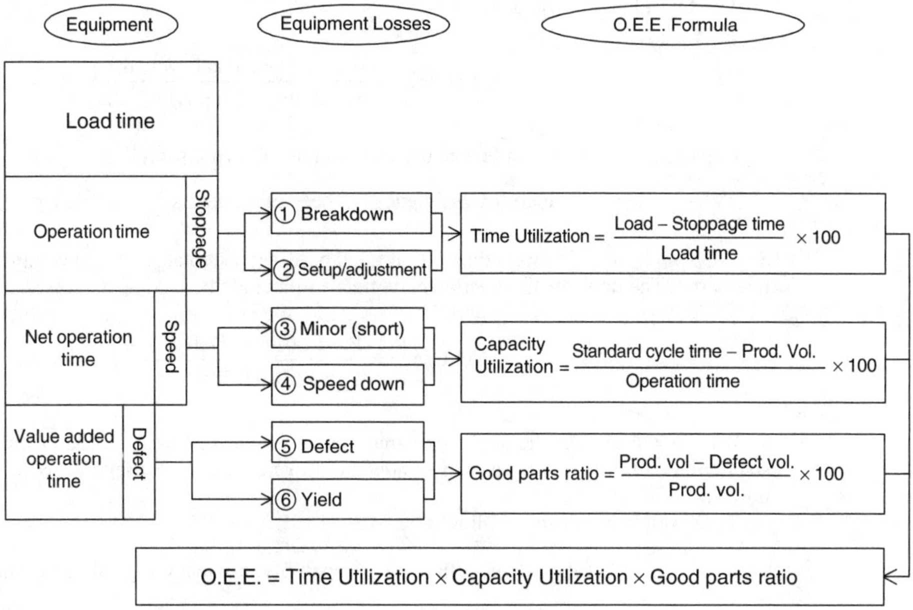

FIGURE 59.2 Overall equipment efficiency.

reduce the losses. The quantified loss structure is called *overall equipment efficiency* (O.E.E.). The formula to quantify these losses is explained below (Fig. 59.2).

1. *Time usage.* Time usage is the ratio of actual operation time against load time (equipment running time to produce the required production volume.) The formula is shown as follows:

$$\text{Time Usage} = \frac{\text{Load time} - \text{Stoppage time}}{\text{Load time}}$$

- *Load time.* Total time minus planned nonproduction time (planned maintenance time or meetings, and the like) during a day's operation time
- *Stoppage time.* Machine stoppage time owing to machine breakdowns, setups, adjustments, changing blades, and the like

2. *Capability usage.* Capability usage consists of speed usage and net usage.

Speed usage means the ratio of actual speed to original/designed speed. In short, it is an indicator of whether a machine is operating at the decided/set speed (standard cycle time). If the actual speed is lower than the original/design speed, then this is known as speed loss.

The formula is shown as follows.

$$\text{Speed Usage} = \frac{\text{Standard cycle time}}{\text{Actual cycle time}}$$

Net usage shows the degree to which a machine is properly operating over a specific period of time. This is not an indicator of machine speed but is an indicator which shows whether a machine is consistently operated or not. Based on this calculation, short stoppage and trouble loss can be calculated.

The formula is shown as follows:

$$\text{Net Usage} = \frac{\text{Prod. Vol.} \times \text{Actual cycle time}}{\text{Load time} - \text{Stoppage time}}$$

Capability efficiency can be calculated with the following formula:

$$\text{Capability Efficiency} = \text{Speed efficiency} \times \text{Net efficiency}$$

3. *Good parts ratio.* This is the ratio of good parts against total produced volume (including defective parts). The number of defective parts needs to include both scrap and repaired parts.

$$\text{Net Usage} = \frac{\text{Prod. Vol.} - \text{Defect Vol.}}{\text{Prod. Vol.}}$$

4. *What is OEE?* As the above explanations have shown, losses can be calculated from several standpoints. O.E.E. is a measure to understand overall machine usage based on all of the above points.

The formula is shown as follows:

$$\text{O.E.E.} = \text{Time efficiency} \times \text{capability efficiency} \times \text{good parts ratio}$$

The purpose of classifying the losses into these categories is to define the losses and to more clearly understand and discover the weak points of the equipment

59.6 ACTIVITY OF EACH PILLAR

59.6.1 Focused Improvement

The theme of focused improvement activities is divided roughly into three subthemes: The theme of improvement of the model line through the combined efforts of managers and staff, the *project theme* which involves a cross-functional staff organization, and the *circle theme* led by managers.

Also, problems which occur chronically and could not be solved in the past must be thoroughly examined and improved. For example, reducing standard tooling time and eliminating idle time are often chronic problems.

Up to now the reasons why these chronic losses have remained are as follows:

- Weak phenomenon analysis
- Inappropriate methods for listing up possible causal factors
- Insufficient check-ups on defects
- Inability to see losses

There are focused improvement activities to eliminate these problems.

With individual TPM improvements, the objective is always "zero"—total elimination of defects and waste. Observing the three key factors of actual work site, actual items, and actual phenomena (in Japanese, the philosophy of the "3 Gens"), PM analysis is done following basic principles and rules and steps are taken to find root causes. If, using principles of physics and chemistry, true causes can be identified, it is truly possible to achieve "zeros" (zero problems, zero waste, and the like).

Another aspect of TPM is to develop a horizontal rollout manual based on the results achieved on the model line and thoroughly roll out improvements to other lines. To eliminate the recurrence of problems, it is necessary to pursue root causes through thorough analysis, divide countermeasures into the following three categories and make maintenance management easy:

1. *Fixed factor.* Factors which, once fixed, do not revert back
2. *Semifixed factor.* Factors which degrade over a certain period, finally moving beyond the reference value
3. *Variable factors.* Factors which are always varying, so that reference values may be exceeded

Fixed factors can be removed from the list of maintenance management items, semifixed and variable factors, however, must be addressed in written PM standards, autonomous PM standards, and standard operating procedures and must be the subject of strict maintenance management. In focused improvement, selecting the model line(s), model process(s), and/or model equipment which contributes directly to the plant results is very important. One may think about the selection and the allotment of the improvement themes in the model process (equipment) as follows.

First of all, the six major losses must be identified. For these losses, it is necessary to decompose these structures and quantify the losses. In respect to loss structures which can be quantified, the rule is start improvement projects (themes) from those losses which create the true bottlenecks. Moreover, in each theme, a good way is to start from the loss with a high contribution rate to the O.E.E.

In addition, the cost structure of the model line (manufacturing cost structure) is clarified, and the relationship between the six major losses and the cost structure is understood. It is important to select as improvement themes those losses which, based on cost structure, are having the most impact.

Once an improvement theme is selected, classify it and then decide on the subthemes. At this time, pursuing the subthemes with a zero-defect oriented approach becomes the key to achieving zero problems; however, when the theme is too large, the challenge to zero might become difficult because it takes too much time to find the solution, and the analysis is too wide-ranging. For instance, assume that "reduce minor stoppages to zero" was selected as a focused improvement theme.

In general, there is not only one, but actually a lot of minor stoppage phenomena. In such a case, it is advisable to focus on one phenomenon initially and reduce it to zero. This also helps build confidence among participants as to the effectiveness of the approach.

59.6.2 Autonomous Maintenance

Equipment maintenance programs and their operation gave birth to the need to differentiate functions (division of labor) and to establish a special post for equipment maintenance. In this scenario, a dedicated maintenance man did the maintenance job and the operator was only responsible for the operation of his/her equipment. As a result, operator's concern for equipment was lost, even in the case of a small breakdown, and the maintenance man was called; however, the one who understands the condition of the equipment the best is the one who actually operates it.

If the worker acquires a basic knowledge of their equipment and the methods for proper operation and daily maintenance, such as cleaning, lubrication, and tightening, significant results in improving breakdown detection and prevention can be expected at an early stage.

TPM emphasizes this principle. Thus, this kind of equipment maintenance activity is known as autonomous maintenance. When introducing autonomous maintenance, equipment should be cleaned thoroughly, and the dust and dirt that has piled up for many years needs to be removed. Moreover, the objective of this cleaning is not simply to clean the exterior of the equipment. Rather, the true objective is to use the cleaning activity as a way to discover and check potential process problems with the equipment (parts wearing out, clogging, looseness, cracks, shortage of oil, and the like). In other words, the concept is "cleaning as inspection." When this is practiced, a tag is attached to the equipment, indicating the detected trouble, and once the problem is corrected, the tag is removed. Through these activities, the employees' level of concern for the equipment will be improved.

Once a habit of maintaining equipment cleanness is solidly in place, those places which soon again become oily or where shavings accumulate, even though routine cleaning is done, will become apparent. Then, because the source of contamination is now understood, measures can be taken to make sure it does not occur again. For instance, a partial cover which confines the generation of chips and splashing of oil can be conceived of as a corrective measure. In this way, difficult parts of daily cleaning and lubrication can be improved, and inspection time and maintenance time will be shortened.

For example, it becomes considerably easier to inspect the equipment by adding a transparent cover and peephole. It is also advantageous to raise equipment for easy access for cleaning, and to centralize the locations of check points. When this stage is reached, reliability and maintainability of equipment increase, and the beneficial effects of autonomous maintenance can be actually felt. The associate is made to understand the structure and the function of equipment deeply, and he or she gains the skill to inspect the equipment effectively.

If autonomous maintenance reaches this level, the cleaning and inspection standards can be built from previous experience and training, and it becomes possible for operators to maintain those standards surely.

When operators keep standards which they made by themselves and revise the standards by themselves, then autonomous management has been achieved.

59.6.3 Planned Maintenance

All maintenance is not done based on plan even if it is known as planned maintenance. First of all, maintenance is classified into preventive maintenance, breakdown maintenance, and corrective maintenance. So even if using the term "planned maintenance," not all maintenance is done based on a plan.

In the evolution of maintenance methods, there have been a variety of names for maintenance methods. Current classifications of maintenance are as follows.

Maintenance is classified into preventive maintenance, BM, and corrective maintenance. Usually when we chose a maintenance method, either preventive maintenance or BM will be selected.

We should base our decision to select preventive maintenance or BM on factors such as cost and/or technical feasibility. According to the type of the breakdown, it may be impossible to take action beforehand. Such breakdowns are classified assuming that they will be dealt with through breakdown maintenance—in other words, maintenance will be performed after the fact.

In addition, preventive maintenance can be classified into time-based maintenance and condition-based maintenance. This condition-based maintenance is called predictive maintenance. To apply predictive maintenance, it is necessary to predict breakdowns appropriately in view of cost, time, and technology. The information necessary to predict breakdowns can be obtained by daily checkups, regular service, and monitoring the current condition with diagnostic equipment. Corrective maintenance means the activity which improves equipment (parts, components, and the like) to enable preventive maintenance to be implemented surely. Moreover, it also consists of activities which reduce the maintenance time when breakdowns do occur.

To implement this planned maintenance, autonomous maintenance activities conducted by the operating department and conventional maintenance activities conducted by the maintenance department should work together, maintenance activities being performed according to plan. Furthermore, continuous improvement is essential.

To advance planned maintenance activities, the maintenance department has to develop activities which serve the dual purposes of improving the equipment itself and improving the maintenance techniques and skills. Therefore, it is necessary to introduce the appropriate training and new diagnostic technology and the like. Moreover, it is necessary to have an appropriate information management system to manage the spare parts and to establish appropriate maintenance plans. The areas which this information management system should cover include the tracking of breakdown and repair times, calculation of required spare parts, and inventory control for spare parts.

59.6.4 Education and Training

A firm can achieve its full potential by fostering its human resources and receiving the best effort from its employees. With TPM activities the objectives are changing the economic environment, advancing technology, upgrading equipment, and modernization of management. To achieve these objectives, an important requirement is to apply human resources (associates) who are knowledgeable and conscientious in regard to equipment and their work.

Employees who work with the production equipment on a daily basis need to adopt the mentality that we need to look after our equipment ourselves. Employees who work in the maintenance department should acquire the skills and techniques necessary to work in a professional manner, and employees who work in the equipment design and production engineering department should acquire professional knowledge in technology, managerial techniques, and management acumen; however, the actual situation on the site is typically as follows.

The production department makes their direct employees concentrate only on manufacturing operations. The maintenance department is overwhelmed with repair requests so they neglect to train their employees on maintenance techniques and to improve their employees' skill level. These factors contribute to lower efficiency, lower productivity, and reduced safety and morale in the plant.

Furthermore, the production engineering department struggles with equipment that is not suitable for the realities of the shop floor and not easy to maintain.

For TPM activities to eliminate this vicious circle from its root, all members participating in and promoting the PM activities need to improve their individual skill level in regard to management, technology, and know-how; however, it is up to company management to see that this improvement in individual skills is achieved.

Significant results cannot be obtained if the firm does not raise the level of individual management acumen, technology, and employee skills. Therefore, any firm which implements TPM must make a significant effort to provide skill training and management skill training. This can be done by developing an internal training system which focuses on individual capability development and emphasizes training on maintenance skills and management techniques.

59.6.5 Initial Management

With the growth of product diversification and shortened product life cycles, it is important to be more efficient in product and equipment development. The objective should be to shorten the product development period and the mass-production start-up activities (such as equipment investment and the like). To achieve these objectives, initial management is included as part of the TPM activities.

The following activities are included in initial management:

- Product initial management
- Equipment initial management
- Commissioning control

To support these activities, product design technology information management and the MP design technology information management are necessary.

Product Initial Management. As consumer needs diversify, the competitiveness of the market regarding quality, price, and delivery increases significantly. Therefore, it is very important to plan, develop, design, and make products that satisfy customers' quality requirements cheaply and quickly.

It is said that 80 percent of the product cost is established during the product design stage, and 80 percent of the costs associated with problems is decided at the product and equipment design stage.

Therefore, during the planning and design stage, it is necessary to consider quality, ease of manufacturing, and ease of performing quality assurance. The activities to achieve this objective are called product initial management.

Equipment Initial Management. When equipment is remodeled or newly installed, problems arising at the commissioning stage often delay equipment startup and reduce overall effectiveness. This occurs when project teams fail to consider reliability, maintainability, operability, economy, safety, and ease of performing quality assurance during the early management phase. To shorten test operation and commissioning and to achieve immediate, problem-free start-up, all available technical capabilities must be used to weed out potential problems during the planning and design stage.

Commissioning Control. Commissioning control is an activity which aims to detect problems and debug processes during actual production after the equipment is installed and the trial run is completed. By doing this, stable operations are established at an early stage. Furthermore, commissioning control is also relevant when existing equipment is used for new products. As for the objects of debugging at the commissioning control, priority will be given to problems with process capability and/or quality, to problems with material handling, and to minor stoppages. Moreover, it is important to establish standards for operations, changeovers, lubrication, and inspection. Finally, it is important to train the operators and maintenance employees.

Management of Product Design Technology Information and MP Information. Appropriate information management is indispensable to execute the three activities described previously. The necessary information can be divided into product design technology information and MP information (equipment design technology information). To ensure equipment installation and planning for maintenance are done efficiently and efforts are made to improve systems, it is important to implement an information management system which can classify, accumulate, and reuse the information coming from the problems which occur during the product development, equipment development, commissioning control, and mass-production stages.

59.6.6 Quality Maintenance

As products become more sophisticated and precise, customers expect a high level of quality assurance. For this reason, consistency of product quality and continuous quality improvement become important issues for the manufacturer.

To improve the level of quality assurance through each stage of product planning, designing, production preparation, and production, conditions for zero defects need to be set. By making sure these conditions are maintained throughout the production process, a manufacturer can be assured that customer quality requirements will be met.

In the past, the general method for quality assurance was to check the quality at the end of each process.

However, it is difficult to achieve a high level of quality assurance from a system which monitors only results. In other words, it is important to control the conditions in the upstream processes. In addition, it is necessary to clarify the causal relationship among quality, equipment accuracy, die accuracy, tool accuracy, manufacturing conditions, operation methods, and the like. Furthermore, the implementation systems managing these conditions should also be clarified.

In the current manufacturing environment, the core of production has shifted from labor to equipment. Automation and unmanned operations are now more common.

Therefore, the idea of "building quality into the equipment" has become more important and in many ways has replaced the previous notion that "quality is determined in each process."

In other words, assuring a high level of quality is increasingly determined by the ability to control the condition of the equipment. Moreover, when quality is controlled by the state of equipment in this way, it is necessary to implement preventive measures to find equipment and process conditions that produce no defects, monitor variations in those conditions, and eliminate ahead of time any possibility of producing defects.

Equipment maintenance which implements such preventive measures is called quality maintenance, and its concept is the following: Quality maintenance consists of activities which establish the proper equipment conditions. The proper conditions do not allow quality defects to be produced and have the goal of maintaining the equipment in perfect condition to produce defect-free products. Quality defects are prevented by checking and measuring equipment conditions periodically and verifying that the measured values lie within the specified range. Potential quality defects are predicted by examining trends in the measured values, and are prevented by taking measures in advance.

59.6.7 TPM for Indirect Office Work

The TPM activity started originally as an activity in the manufacturing department; however, TPM is becoming the activity of the reformation of corporate culture now, it is necessary and indispensable to involve the indirect department. The role of the indirect department is in the support of the direct department. Therefore, when the theme of the improvement is selected, the following two views are necessary.

One is "How do we support the TPM activities of the direct department?" and another one is "What issues must we address to maximize our own efficiency?"

Autonomous maintenance, one of TPM's unique characters, is the key to successful TPM in indirect departments. Developing a program of administrative autonomous maintenance is essential for efficient, trouble-free work execution.

In indirect department, 5S is essential for workplace management as well as direct department. Through autonomous maintenance, it is expected apply step-by-step methodology and advance the improvement steadily.

Sort. Eliminate the unnecessary

Stabilize. Establish permanent locations for the essential

Shine. Find ways to keep things clean and inspect through cleaning

Standardize. Make adherence easy

Sustain. Self-discipline

59.6.8 Management of Safety, Health and Environment

It is assumed that zero accidents and zero pollution are the essential requirements to win the TPM award. In fact, by introducing the TPM activities, safety in the workplace and the work environment have improved considerably. This is owing to the following factors.

By thoroughly implementing 5S, visibility and cleanliness improves in the shop floor. An unsafe place decreases by implementing self-imposed improvement and focused improvement. The process for detecting equipment abnormalities is improved. Therefore, abnormalities can be detected early, and appropriate measures can be taken. It comes to be able to follow the rule that decided by them. Equipment trouble often signifies an unsafe environment. Therefore, achieving zero breakdowns and zero defects improves the safety level in the shop floor.

TPM activities are activities which aim to change the culture in the whole company. So an action plan focusing on safety and environmental issues needs to be created.

59.7 RESULT OF TPM ACTIVITY

Results at firms which have won TPM awards are significant. Some examples of results are as follows: 1.5–2 times more productivity of added-value per employee, process defective rate 1/10, a 30 percent reduction in manufacturing costs, zero accidents, zero pollution, and 5–10 times more suggestions.

Two results of TPM:

59.7.1. TPM Quantitative Results

Productivity

Equipment productivity	1.5~2 or more times
Labor productivity	1.5~2 or more times

Equipment overall efficiency	85 percent or more (an assembly system), 90 percent or more (a process system)
Number of sudden failures	1/5~1/100 or less
In-process defect rate	1/5~1/10 or less
Cost reduction	20~30 percent or more
Service Levels	
Delivery claim	1/3~1/or less 5
Delivery achievement rate	99 percent or more
Inventory	
Finished good/ WIP	1/2 or less
Safety	
Accident	zero
Moral	
Number of suggestions	5 per person per month
Return on Investment	
	5 or more times

59.7.2. TPM Qualitative Results

With improvement of safety consciousness based on correct work owing to activities conducted by all employees together, a feeling of solidarity exists and teamwork improves. A sense of trying to discover abnormalities is supported and an awareness of problems improves. By using various techniques, employees throughout the firm are able to improve their problem solving abilities. By significantly changing the activities conducted and the work environment, the latent talent and ability of the employees can be brought out. This results in an exciting atmosphere.

The barriers within an organization (the sectionalism) disappear and communication within the organization is enhanced. An organization's weak points become clear, and the organization's ability to attack these weak points is improved.

The company environment is better and impressions of guests visiting the firm improve. Moreover, the confidence of the firm rises. The awareness that you should "protect your equipment by yourself" grows. Goals such as "zero malfunctions" and "zero defects" are achieved, and everybody's self confidence increases.

REFERENCES

TPM Encyclopedia, Japan Institute of Plant Maintenance (1996)
TPM in Process Industries, Productivity Press (1994)

INFORMATION RESOURCES

JMA Consultants, Inc., www.jmac.co.jp
TPM in Process Industries, Productivity Press www.jipm.or.jp

CHAPTER 60
PROJECT MANAGEMENT IN MANUFACTURING

Kevin D. Creehan
Center for High Performance, Virginia Polytechnic Institute and State University
Blacksburg, Virginia

60.1 INTRODUCTION

As companies have moved into the twenty-first century, the philosophy of project management has played an increasing role in obtaining their global objectives. Global competition, along with the influx of computer technology, has created an environment of change in the modern workplace that is occurring faster than ever before, and at an increasing level of complexity. The discipline of project management has served, and will continue to serve as the fundamental philosophy to deal with product and manufacturing process changes as projects become the foremost mechanism for creating new products and processes. This change is occurring, however, because organizations have had the willingness and strength to adopt project management as a body of knowledge rather than as a set of tools or a special case of traditional management. As product life cycles grow shorter and shorter, the use of projects as a fundamental means of instituting change in the enterprise becomes more necessary. And as the constant need for project implementation grows, the need to establish project management as a strategy within the enterprise becomes obvious. This manner of thinking will serve as the enterprise's primary means to cope with ongoing change existing in today's markets.

Many manufacturing industries have long ignored the project management philosophy despite the innumerable successes witnessed through its employment. These companies will inevitably be left behind, if they have not already, as this manner of thinking grows in popularity and is generally accepted as the best way to deal with changes in manufacturing industries. Several examples can be cited to demonstrate this approach to be true and will be discussed in detail as this chapter progresses.

In order to understand the importance of project management, it is essential to first fully understand what a project is. The Project Management Institute (PMI) defines a project as "a temporary endeavor undertaken to create a unique product or service."[1] It can be noted that this definition implies that every project must have a clear beginning and end, and its end result must create something that did not previously exist. In the manufacturing environment, a project could be the construction of a new facility, new product or process development, transferring technology to the manufacturing operation, commercializing a new manufacturing process, reorganizing the shop floor, or otherwise adding to the competence of the business. At any rate, a project is always a complex undertaking that pulls together multiple organizational resources to accomplish unique objectives and, thus, is the primary instrument of change in the organization.

Consequently, project management is the primary strategy used to cope with change in an organization. It allows members of the organization to apply "knowledge, skills, tools, and techniques to project activities in order to meet stakeholder needs and expectations from a project."[2] Most succinctly, "project management is a series of activities embodied in a process of getting things done on a project by working with members of the project team and with other people in order to reach the project schedule, cost, and technical performance objectives."[3] The origins of project management, elements of which can be traced back to ancient civilizations, have been studied and disseminated for only a few decades. It is therefore widely considered a modern discipline because it has led to modern innovations in management structure and management techniques, the purposes of which have been to better use existing resources; however, its impact in this relatively small amount of time has been immeasurable. It has spanned across all industries and contributed to revolutionary results, such as the atomic bomb, space exploration, and the explosion of computer technology in today's world. Project management has truly developed a new vision in the management of today's organizations.

An indication of project management in the modern manufacturing world is the *product realization process*. Quite simply, this interdepartmental process seeks to use a team-based, interactive environment to develop new products, keeping the team intact from product concept, to market and customer need determination, to product design, to manufacturing, to product introduction. The success of this modern organizational arrangement contradicts the established belief that revolutionary ideas are only developed through entrepreneurship. Rather, by utilizing the skills and talents of several members of a diverse group, a product realization team has a broader scope of expertise and a wider range of personal preferences, experiences, and skills. This methodology has led to noteworthy results in several areas of the manufacturing world, including the automotive, aerospace, computer, chemical, electronics, communications, and consumer products industries.

Further, the modern concept of *continuous improvement* in the workplace as a manufacturing philosophy is greatly facilitated by the project management philosophy and the tools that it provides. The need for continuous improvement has generated multiple contemporary techniques, such as *just-in-time* (JIT) inventory management, *materials requirements planning* (MRP), *total quality management* (TQM), *computer integrated manufacturing* (CIM), and *flexible manufacturing systems* (FMS). As Cleland outlines, the two strategies that companies can use to advance their technologies are the continuous improvement of the existing products and processes, and a push for a breakthrough technology.[4] Both of these strategies are aided by the use of the project management, the philosophy of which instills several common concepts into the daily operations of the organization.

60.1.1 Example of Project Management in Manufacturing

Leading automobile manufacturers, such as Ford Motor Company, Daimler Chrysler, General Motors, and Toyota, have successfully implemented a multi-project management philosophy into their product and manufacturing process development. By allowing their engineers the opportunity to contribute to multiple projects at any given time, while creating products that share components, these corporations have witnessed tremendous savings in their product and process development costs, and therefore in their overall production costs as well. The goal of their philosophy is to develop new products and processes that share components but maintain unique development teams, which ensures that each product will differ sufficiently to attract different customers.[5] By adopting cross-functional teams as well as other philosophical changes, the organizations have been able to shorten lead times and improve product concepts. They introduced a matrix business structure, which combines functional departments and cross-functional teams in the form of projects, rather than eliminating functional departments completely. Thus, the projects integrate across functions but maintain functional expertise within the organization. From this philosophy, these matrix organizations that are based on project work have discovered the following advantages:[6]

- Barriers between functional departments have virtually disappeared as different departments work together toward a common product concept and common project goals.
- Project engineers have the opportunity to influence the whole product concept or project effort, rather than only their individual components.

- Engineers from different specialties combine to make product or process development more innovative.
- Because engineers remain part of a functional department, they retain the ability to accumulate special technical expertise, which aids in the production of radical innovations in particular technologies.

60.2 PROJECT MANAGEMENT INSTITUTE

Project management continues to grow in popularity worldwide. The PMI was founded in 1969 and has grown to almost 90,000 members throughout the world. As the leading nonprofit professional association in the area of project management, PMI has established standards, provided seminars, developed educational programs, designed a certification program, and conducted research into current project management methodology and techniques. The PMI vision statement reads: "To be respected throughout the world for advancing project management excellence through professionalism."

PMI has further expanded the dissemination of project management practices through the publication of the PMI Standard, *A Guide to the Project Management Body of Knowledge* (PMBOK Guide)—2000 edition. This document, which identifies the elements of the *project management body of knowledge* (PMBOK) that are generally accepted, describes nine knowledge areas:

- Project management integration
- Project scope management
- Project time management
- Project cost management
- Project quality management
- Project human resource management
- Project communications management
- Project risk management
- Project procurement management

60.3 FUNDAMENTALS OF PROJECT MANAGEMENT

60.3.1 Overview of Projects

A project is "any undertaking that has definite, final objectives representing specified values to be used in the satisfaction of some need or desire."[7] As the fundamental means of change in the organization, projects have become building blocks in the execution of organizational strategies. In the manufacturing environment, they are used in the development of new products or processes, and lead to superior market position, growth opportunity, and ultimately survival. Every project has several common, essential characteristics that distinguish them from other activities performed within an organization:

- Projects are temporary, planned activities with defined beginning and end points.
- Projects have objectives that are designed to meet time, cost, and technical performance specifications.
- Projects are complex, and the scope requires the crossing of functional and often organizational boundaries.
- Projects play a vital role in preparing the organization for its future, and contribute directly to its future success or failure.
- Projects contribute to the organization something that has never previously been accomplished.
- Projects consume resources, such as people, equipment, and funds.

Although different people may conclude differently whether a project has been a success or a failure, generally its success is defined by, at least, the achievement of the project objectives on time and within budget, but more importantly, by its positive impact on the sponsoring organization; however, if the user deems the results acceptable or outstanding, a project may also be considered successful despite late delivery or overspent budget. In this respect, project success and failure is often qualified differently for each project stakeholder, individuals with a vested interest in the project. That being said, the ability to identify in advance that a project is headed toward failure is an essential attribute to any project team and any project manager. By grasping project management as a philosophy, the project team will develop the ability to recognize the matters that determine project success or failure, and thus will improve their own probability of success by possessing the knowledge to positively resolve those issues.

Projects vary greatly depending on the type of organization conducting the project. In manufacturing, a project can be as small as reorganizing the shop floor to as large as the construction of a new manufacturing facility. In either case, the projects deliver something to the organization that did not previously exist. Moylan wrote that there are two broad classifications of manufacturing projects: the *engineer-to-order* (ETO) project, in which each new product requires a major project to introduce it, and the *capital project*, which does not relate to new product development, but rather involves an improvement in manufacturing capability and capacity via a facility infrastructure implementation.[8] In any case, the project is the operator of change within the organization. This change, regardless of the form in which it presents itself, needs to be managed.

60.3.2 When to Use Project Management

Project management, then, is "the art of directing and coordinating human and material resources throughout the life cycle of a project by using modern management techniques to achieve predetermined objectives of scope, cost, time, quality, and participant satisfaction."[9] Its need arises because the traditional methods of business organization do not provide the dynamics required to successfully manage project work, which by definition are complex, out of the ordinary, unfamiliar endeavors. The need to share resources between functional units of the organization to give single-minded attention to a specific customer or specific problem requires the tearing down of traditional functional barriers within the organization, and therefore of traditional management practices as well. Cleland outlined the primary motives for employing a project management framework, and these include, but are not limited to the following:[10]

- To share resources across organizational units
- To focus attention on specific customers in specific market segments
- To integrate systems and subsystems simultaneously or in parallel within independent organizations
- To manage focused interorganizational efforts from a profit-center perspective
- To deal with specific planned or unplanned problems and opportunities
- To manage an organization's crisis control plan
- To expedite responses to new events in the organization or its environment
- To accommodate the inherent interdependency within an organizational system
- To combine several proven methods of organizational design, such as product, functional, and geographic
- To bring a wide range of expertise and viewpoints into focus on tasks, opportunities, and problems
- To formalize an informal management process, such as project engineering
- To deal with the magnitude of an undertaking requiring massive input of capital, technology, skills, and resources

- To manage unique or rare activity
- To facilitate the participation of organizational members in the management process of the organization
- To deal with a new technology that requires pooling existing resources and capacities
- To deal with a task that is bigger than anything the organization is accustomed to handling
- To meet competition

Example of When to Use Project Management. Genentech, Inc. implemented a project management philosophy to develop a world-class flexible manufacturing system in its Vacaville, California plant for pharmaceutical manufacturing. The facility has multiproduct, multistream process capability that allows the plant to perform complex pharmaceutical batching operations. The project was 6 years in duration, and cost over $250 m.

Genentech began planning the project for additional production capacity in 1994 to accommodate several new products; however, the ability to produce such a variety of complex biotechnology products in the same facility is challenging. Moreover, pharmaceutical producers are regulated by strict Food and Drug Administration (FDA) regulations intended to make sure that all processes and systems will consistently produce products that meet predetermined specifications and quality attributes.

The objectives of the project were to build a flexible, multiproduct batch manufacturing plant with the automation ability to easily introduce new products and facilitate throughput, verify that the automation systems provide repeatable operational results, and integrate the systems to collect the necessary process data to minimize production cycle times. With hundreds of product recipes involved in producing a single final product, efficient project management and project control is essential. The project team developed a customized information and control system that receives immediate product life cycle reports and process data.[11]

In the 2000, Genentech's state-of-the-art Vacaville manufacturing facility received FDA licensure as a multiproduct facility.[12] The success of this massive project has been directly attributed to superior teamwork functioning toward a clear vision.

60.3.3 The Project Life Cycle

Managing a project is a complex undertaking. So much so that understanding or even predicting all of the necessary activities required throughout the project is exceedingly complicated. Consequently, the need for a model of the project's life cycle arises to facilitate the understanding of the project's complexity, duration, and its associated management functions.

A project's life cycle can range in duration from a matter of days or weeks to a matter of years. But every project, regardless of its duration, goes through a series of phases that separate the project tasks and the managerial approach required. Between each of these phases are decision points, which are used to determine when or if the next phase of the project should be carried out. These end-of-phase decision points are often referred to as stage gates, kill points, phase exits, go/no-go points, and the like. Several diverse models have been put forward to represent the project life cycle, with varying degrees of detail. Some models demonstrate significant detail, and depict the project proceeding through several major phases, while others remain very broad. The wide diversity among these models is expected and necessary owing to the complexity and diversity of modern-day projects. Nevertheless, despite the varying numbers of stages put forth by each of these models, all of them at some level depict projects going sequentially through four fundamental, generic phases:

1. Concept development and project definition
   - This phase begins as a vision in someone's mind and is put in motion via a project definition. In this phase, the need for the project and the work required to accomplish the objectives of the project are analyzed, evaluated, and promoted to the organization. Once a project proposal is

developed and accepted, the primary stakeholders agree on technical performance specifications, and broad estimates are developed for the schedule and budget.

2. Project planning

- In this phase, a detailed project plan is laid out—describing how and by whom, specific project tasks will be performed. Included in this plan are the project team members, preliminary studies and analyses, project targets, and a detailed schedule for the execution phase.

3. Project execution or operation

- This phase consists of the daily management and operations of the project, monitoring the performance, updating the project plan, and examining potential areas of improvement.

4. Project delivery and termination

- In this phase, the project deliverables are transferred as specified in the project plan, the project is terminated, the project team is disbanded, rewarded, and reassigned, and any postimplementation review processes, including the recording of lessons-learned, are performed.

The project life cycle mirrors that of many other items in the world in that they are created, prepared for full existence, exist, and then die. Generally the project life cycle will define, or help to define, the work that is required in each phase and the participants who should be involved.

One of the first tasks performed in a project is to develop a rough estimate of the major tasks to be performed in each of these phases. As more is learned about the project, these estimates, or work packages, must be reconsidered and updated. Figure 60.1 illustrates the level of effort/man-hours

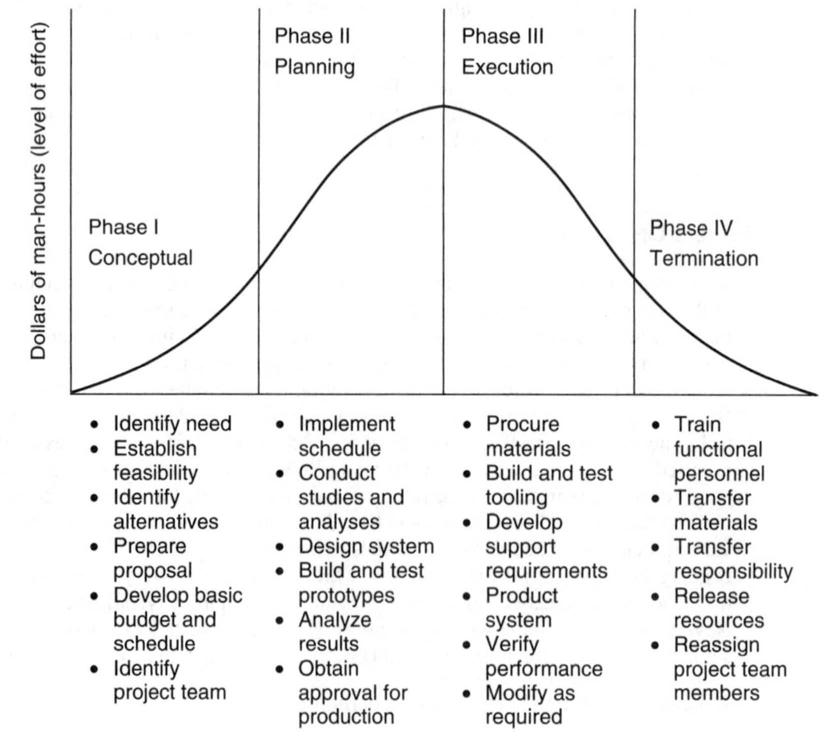

FIGURE 60.1 Tasks accomplished by project phase.[13]

required within each of the four fundamental phases of the life cycle, along with a list of typical tasks associated with the phase.

Further, as the life cycle progresses, the cost, time, and performance parameters will fluctuate, and as such, the resources allotted to each must be constantly managed.

The project life cycle, and the typical tasks associated with each phase, provides a starting point template for the management of projects. The life cycle can be used as a general focal point to which a project manager can constantly refer, so to ensure that the project achieve its time, cost, and technical performance objectives; however, project management cannot be approached as the management of a series of individual tasks. For this reason, a systems approach to project management is required.

60.3.4 The Project Management System

In management literature, a management system is characterized as a group of elements, either human or nonhuman, that is organized and arranged in such a way that the elements can act as a whole toward achieving some common goal.[14] The complex and continuous nature of projects requires that a system be in place that addresses the manner in which project management will be applied. By establishing a project management system, the overwhelming task of managing a project can be separated into more definite subsystems that address all of the essential variables in the project management process. As a fundamental philosophical model of project management, the systems approach facilitates the management and execution of the major activities throughout the life cycle of the project.

A project management system supplies the team with a theoretical model to use as a guide on how to manage the project from strategic planning through implementation. The principle subsystems of a project management system include the following:[15]

- *Facilitative organizational subsystem.* This subsystem is needed to allocate the resources to the project. It is in this subsystem that project authority, responsibility, and accountability are designated, and individual and collective roles are established.
- *Project control subsystem.* This subsystem is needed to establish the performance standards for the project. Further, in this subsystem the actual progress is compared to the planned progress, so that corrective action may be taken. To accomplish this, regular review of project progress is discussed among project stakeholders.
- *Project management information subsystem.* This subsystem is needed to provide the necessary formal (information retrieval system) and informal data (conversation among project team members) to allow the team to plan, oversee, and make decisions in the management of the project.
- *Techniques and methodologies.* This subsystem contains the necessary techniques and methodologies to facilitate the planning, oversight, and decision-making in the management of the project. Such systems include the *program evaluation and review technique* (PERT), the *critical path method* (CPM), project cost estimators, and technical performance assessments.
- *Cultural ambience subsystem.* This subsystem consists of the cultural environment in which the project takes place, including the attitudes, perceptions, prejudices, assumptions, and values of the people associated with the project. The cultural ambience affects how people act, react, think, and feel, as well as their actions and words.
- *Planning subsystem.* This subsystem provides the means to determine the resources required in support of the project, and how they will be used. An integrated project plan will provide the project a performance criterion to monitor, evaluate, and control the activities of the project. In this plan, a work breakdown structure is developed to demonstrate how the project will be broken down into its component parts.

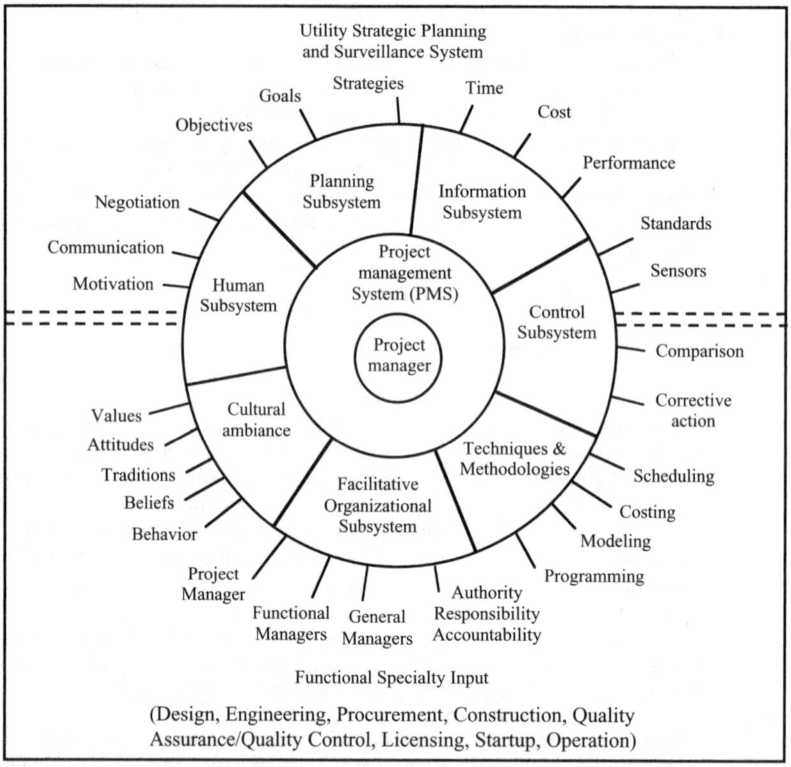

FIGURE 60.2 Project management system.[16]

- *Human subsystem.* This subsystem considers the sociology, psychology, anthropology, communications, semantics, leadership, and motivation of the project participants. It requires that the management and leadership style lead to trust, loyalty, and commitment among the project team.

Figure 60.2 illustrates a project management system for a public utility organization. It demonstrates how the use of a project management system transcends all functional departments within the organization while identifying the primary areas of focus within each of the subsystems.

The systems approach is a logical, disciplined, and ongoing process of problem solving, and is vital to project success. A project management system brings with it an ability to analyze the total project in the following ways:[17]

- It forces review of the interrelationship of the various subsystems.
- It is a dynamic process that integrates all activities into a meaningful total system.
- It systematically assembles and matches the parts of the system into a unified whole.
- It seeks an optimal solution or strategy in solving a problem.

60.3.5 Project Management Functions

Traditional management is broadly considered to contain five core functions: planning, organizing, motivating, directing, and controlling. The project management process consists of the same core functions of management, but they are executed through altered approaches. The knowledge, skills, tools, and techniques of project management are most easily described in terms of these major activities, which are outlined below, as defined by Cleland and Ireland:[18]

- *Planning—What are we aiming for and why?* Develop the objectives, goals, and strategies to provide for the commitment of resources to support the project.
- *Organizing—What is involved and why?* Identify the human and nonhuman resources required, provide a suitable layout for these resources, and establish individual and collective roles of the members of the project team, who serve as a focal point for the use of resources to support the project.
- *Motivation—What motivates people to do their best work?* Establish a cultural system that brings out the best of people in their project work.
- *Directing—Who decides what and when?* Provide for the leadership competency necessary to ensure the making and execution of decisions involving the project.
- *Control—Who judges the results and by what standards?* Monitor, evaluate, and control the use of resources on the project consistent with project and organizational plans.

The continuous process, that is project management, requires numerous activities within each of these functions, as is discussed as the chapter progresses. Although they are identified as independent functions, throughout the duration of a project these functions remain interrelated. Clearly the effective execution of any significant project activity, say the development of a project information system, requires a *motivated* workforce that has been unambiguously *directed* in its task, and steered in the right direction when swayed off course (*controlled*). As always, meticulous *planning* within a committed, well-*organized* team in advance of the execution is crucial.

The PMI defines the five group processes of project management as initiating, planning, executing, controlling, and closing.[19] They characterize these groups as linked, or interdependent, because the results produced by one process often becomes the input of another.

The continuous challenges presented by projects create interdependence between the core functions. Effective management of each function relies on the effective management of the others. The interdependence between the core functions re-emphasizes the benefits of developing a system of project management, which institutes a logical process of problem solving. Project management systems are most effective when weaved into the strategic management of the organization.

60.3.6 The Strategic Context of Project Management

The importance of projects to the survival and growth of manufacturing organizations mandates that senior management support and keep close watch over them. As the fundamental means of change within a manufacturing organization, future strategies depend on a selection of projects that, when successful, lead to new products or manufacturing processes, and inevitable growth of the organization. As a result, organizations depend on the success of projects, via successful project management, for their survival. And as such, project management must fit into the strategic management of the manufacturing organization. So the strategic planning process now extends to participants at all levels of the organization, rather than only senior management.

Consequently, project management has become more and more involved in the strategic planning and project selection processes of the manufacturing organization. As such, project management is being identified as a core competency of organizations and provides competitive advantage. This involvement provides several advantages to all levels of the organization:

- Improved communication within the organization allows all members to become intimately familiar with, and initiate meaningful dialog about the organizational mission, objectives, goals, and strategies that shape the future direction of the organization.
- All members of the organization may develop an overall perspective of the organizational focus, and are involved in shaping it.

- The organization and senior management gain insight from project management regarding the organization's resource capabilities and constraints.
- All members of the enterprise understand from the outset the scope and the strategic context of their projects in relation to the other projects within the organization.
- Project managers become aware of project selection criteria and the selection process that facilitates the reassignment of resources and priorities among projects.

60.4 ORGANIZATIONAL DESIGN

60.4.1 Project Organization

The traditional concept of organizational design distributes authority, responsibility, and accountability by dividing the organization into units—usually according to functionality. When projects arise in an organization of this type, they are coordinated among the high levels of management and their tasks are divided and assigned to the corresponding functional units. This approach tends to greatly slow the completion of projects because the assigned responsibility for project activities is spread across several departments, limiting the communication channels and eliminating any momentum that may be created through the completion of work packages.

The desire for hybrid structures arises from the common barriers that exist horizontally within a traditional organizational design, such as departmentalization across functional, product, customer, territorial, and process units. Moreover, barriers exist vertically as well, dividing each of the horizontal units by position or rank. The hierarchies that exist within these traditional organizations tend to induce long lead times, and fail to develop an organization-wide focus on projects.

On the other hand, purely project team organizations (sometimes called a projectized organization) bring together a group of personnel from many functional areas on a full-time basis. In such an organization, the project manager maintains complete authority over the entire project. Consequently, effective channels of communication develop that results in quick response time, and effectively eliminates long lead times. A wide body of research has demonstrated that this approach delivers project results in a much more efficient manner than do functional organizations, and, as a result, companies are continuing to grow in this direction. The disadvantages of these organizational designs, which distribute the authority, responsibility, and accountability into strictly project units, are that sometimes costs increase to the organization because the walls that exist between project teams inhibits the sharing of personnel resources between projects.[20] Further, owing to the lack of opportunities for dialog between the members of the same functional units, specific technical knowledge within each unit of the organization begins to suffer as time goes on and technology changes.[20]

For these reasons, the project environment and the functional environment often coexist. Ordinarily, today's organizations that make use of a project approach organize the project teams according to some form of a matrix organizational philosophy. This modification of the traditional and projectized organizational designs temporarily brings together personnel from several of the organization's functional units, while the project team and functional units share the authority, responsibility, and accountability of each task. In this way, a "horizontal" project organizational structure is superimposed onto the traditional "vertical" structure of a functional organization. As such, within the matrix organization there are typically two chains of command, one along functional lines and one along project lines. Typically, the project managers have responsibility and accountability for project success while the functional managers have the responsibility of sustaining technical excellence throughout each activity. Figure 60.3 demonstrates a sample matrix organizational structure. This illustration clearly demonstrates how personnel are drawn from the organization's functional units to perform specific tasks within a temporary project organization.

Companies apply the matrix concept in various ways. Different models of the matrix organization assign the authority, responsibility, and accountability to the project manager and functional managers in varying degree. The *functional matrix* assigns the project manager with the authority to

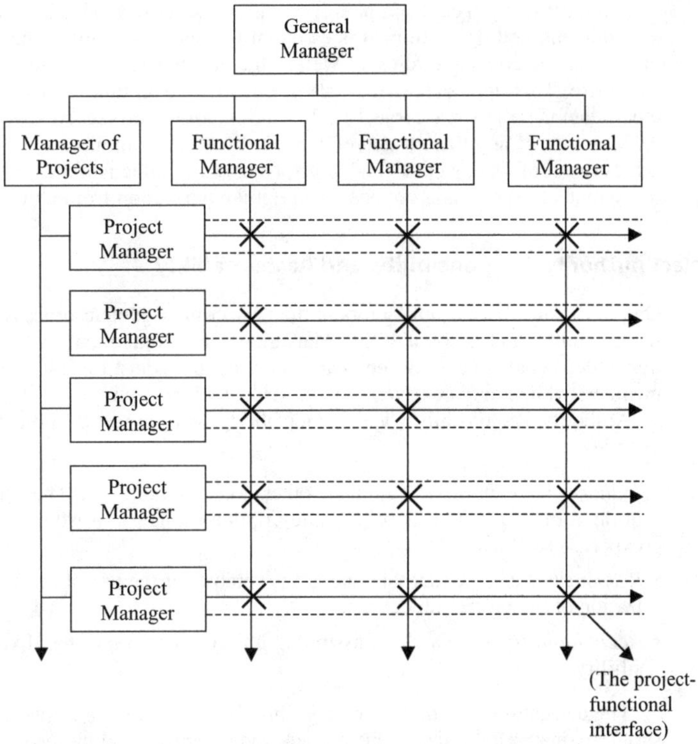

FIGURE 60.3 A basic project management matrix.[21]

manage a project across the functional units of the organization, but the balance of authority gives preferentiality to the functional managers. Conversely, the *project matrix* tips the balance of authority to the project manger by assigning the project manager total responsibility for the completion of the project. It follows then that the *balanced matrix* assigns the project and functional managers equal authority, responsibility, and accountability. The *project matrix* has become the design of choice among many organizations owing to its superior performance with respect to achieving time, cost, and technical performance objectives. It has been shown that projects relying on a functional organization or a functional matrix were substantially less successful than those in a projectized structure. Moreover, the project matrix has been shown to outperform the balanced matrix in terms of schedule objectives, while outperforming projectized organizations in terms of cost objectives.[22]

60.4.2 Project Teams

There are increasing numbers of companies that use project teams effectively to improve their performance. A matrix organizational design, which facilitates the use of teams, defines how authority, responsibility, and accountability are to be assigned to the individual members, including clarification of how the team members are to collaborate with other team stakeholders.

Studies have repeatedly demonstrated that people are more motivated to perform well in the work environment if a culture exists in which they feel they belong, make a difference in work that has meaning, and have opportunities for growth. Active membership on a project team can fulfill this need and result in a more energized work force, so long as the managers are willing to share the authority, responsibility, and accountability for the project, while maintaining their leadership role. This culture must to be actively developed, as it does not appear automatically. An organization-wide commitment to a project team philosophy is necessary to establish a culture of trust, respect, loyalty, commitment, and dedication. It is essential that everyone within the organization understand their individual and collective roles, as well as the objectives, goals, and strategies of the project and the enterprise. This approach will result in effective communication between, and a commitment to, other team members, the project leader, and important project stakeholders.

Members of an efficient and effective team feel a commitment to the team, the project, and the achievement of its objectives and goals. Constant, productive interaction produces a synergy within the team that expects and obtains meaningful contribution from all of its members.

60.4.3 Project Authority, Responsibility, and Accountability

One of the most important factors in the project management process is the allocation of authority, responsibility, and accountability. Managing across functional and organizational lines presents a tremendous challenge to project teams, and the team must be able to weigh the input of other team members while being decisive as an individual when required.

Authority, responsibility, and accountability with respect to project management can be defined as follows:[23]

- *Authority* is the power to command others to act or not to act. This authority can be derived legally, in the form of a policy or position description (de jure), or from a person's knowledge, skill, and expertise (de facto).
- *Responsibility* is the fact that one is answerable for the use of resources on a project and the realization of the project objectives.
- *Accountability* is the state of assuming liability for something of value because of one's responsibility.

The delegation of authority, responsibility, and accountability varies based on the organizational structure (matrix), the size of the project, and organizational philosophy; however, before instituting a project management system into an organization, it is imperative that the roles of authority be clearly defined in writing, and understood by the general, project, and functional managers.

60.4.4 Project Leadership

Project success cannot be obtained simply through increased knowledge, implementation of a high-quality process, or the use of specialized techniques and tools. Rather, because it depends on a cohesive unit of human resources, motivation and leadership from the senior level of the organization play a vital role in project success and failure. Effective leadership is displayed in many ways, but there are several characteristics that remain constant among all leaders. All successful leaders have the following attributes in common:

- Have confidence in their own abilities as well as the team's collective abilities
- Attack each problem with an infectious, positive energy that spreads throughout the team
- Are the hardest working members of their team
- Do not need to be motivated by an outside force; they are self-starters and highly motivated
- Are honest, truthful, and deal in a straightforward manner with every person, under all circumstances
- Have an undying commitment to get the job done
- Exhibit fairness and patience with other members of the team, which adds to their existing credibility
- Have the uncanny ability to surround themselves with great people, and have the common sense to stay out of the way and allow them to do their jobs

- Precisely communicate their vision and inspire those around them toward the achievement of that vision
- Deal with the world as it is, not as they wish it would be
- Seek to be educated in unfamiliar matters that are essential to reaching the objective
- Are strong and resilient in the face of adversity
- Are committed to continuous improvement of themselves, their team, and their organization
- Are team players

Although all great leaders have these, and many more, attributes in common, this is not to suggest that they lead in similar manners. In fact, the truth is quite the opposite. Wideman and Shenhar identified four distinct types of project leaders, and identified their similarities and differences:[24]

- *The explorer.* Bold, courageous, imaginative entrepreneur-type project leaders have a vision of the future, and projects are the stepping-stones. They are salesmen, and are confident and charismatic. They delegate day-to-day problems to others.
- *The coordinator.* Humble, sensitive, and responsive to the views of team members who then take responsibility for their decisions. They ensure that team issues surface, which are then discussed and resolved to the team's satisfaction. They have the ability to facilitate compromise within the team.
- *The driver.* Action-oriented, realistic, and resourceful, they are always well prepared and maintain focus on the project mission. People with similar traits find them easy to work with, but those who do not have similar traits do not.
- *The administrator.* Highly organized, they insist that the work be carefully scheduled and procedural. They demonstrate intellectual logic while recognizing the need for stability.

While it is rare to find a leader that exhibits all four of these traits, effective leadership can be attained through any of these avenues, as long as the project manager has been appropriately selected, matching the personality traits to the situation.

60.4.5 Linear Responsibility Chart

The linear responsibility chart is an easy-to-read innovation that details the work package/organizational position pairings in the organization. It goes beyond the traditional charting techniques that outline the formal hierarchy of the organization to include the degree of participation of all members of the organization for all activities performed. It lays out a detailed description of the authority and responsibility relationships for organizational activities.

The organizational activities included in the linear responsibility chart are called work packages. A work package is a discrete unit of work, clearly distinguishable from all other work packages, at the appropriate level of the organization where work is assigned. A work package can be integrated with other work packages to support higher-level work packages.

The linear responsibility chart is developed by identifying the responsible members for each work package. These responsibilities are defined at the work package/organizational position interface on the chart, by using symbols or letters to represent the appropriate responsibility. As a result, the authority/responsibility patterns and the organizational interdependencies can be read directly from the chart. The following symbols demonstrate sample responsibilities:

- *A—Approval.* This person approves the work package.
- *P—Primary responsibility.* This person has primary authority and responsibility for accomplishing the work package.
- *R—Review.* This person reviews the output of the work package.
- *N—Notification.* This person is notified of the output of the work package.

- *O—Output.* This person receives the output of the work package.
- *I—Input.* This person contributes input to the work package.
- *W—Work is done.* This person performs the work for the work package.
- *I*—Initiation.* This person initiates the work package.

Linear responsibility charts may be used within the context of a project as well as within the context of the organization as a whole. Project managers can use them to establish and understand the authority relationships among their project team members and to clarify the roles of the project team. Furthermore, the development of an LRC allows each individual the opportunity to discuss their role as well as the roles of their colleagues.[25]

60.5 STAKEHOLDER MANAGEMENT

60.5.1 Identifying Project Stakeholders

Stakeholders are individuals with a vested interest in the completion of the project, but not necessarily in support of its completion. Project managers must be able to recognize, network, and cooperate with all of the institutions and individuals in the project's environment. These stakeholders extend far beyond the project team, senior management, and the customer. In fact, project stakeholders can include organizational shareholders, the local community, government agencies, suppliers, customers, employees, labor unions, creditors, competitors, contractors, and the general public, to name a few.

These stakeholders have the ability to affect the outcome of a project, sometimes to its detriment. For this reason, all potential stakeholders in the project must be identified initially, and their interest in the project must be recognized so that the project team may predict their behavior when strategies are implemented. This formidable, ongoing task must be constantly managed. And a task of such magnitude requires a system to facilitate its accomplishment.

60.5.2 Project Stakeholder Management Process

The need for such a system arises from the realization that key external stakeholders maintain an immense influence on the attainment of project objectives. Cleland identified the following basic premises, which serve as guides for the development of a project stakeholder management process:[26]

- Stakeholder management is essential for ensuring success in managing projects.
- A formal approach is required for performing a stakeholder management process. Multiyear projects are subject to so much change that informal means of stakeholder management are inadequate.
- Project stakeholder management should provide the project team with adequate intelligence for the selection of realistic options in the management of project stakeholders.
- Information on project stakeholders can be gained from a variety of sources, some of which might superficially seem to be unprofitable.

The project stakeholder management process consists of executing the management functions of planning, organizing, motivating, directing, and controlling the resources used to cope with external stakeholders' strategies. The continuous stakeholder management process consists of the following phases, which are executed in relation to the aforementioned management functions:

- Identify project stakeholders
- Gather information on the stakeholders
- Identify the stakeholders' missions
- Determine stakeholder strengths and weaknesses

- Identify stakeholder strategy
- Predict stakeholder behavior
- Implement stakeholder management strategy

The project stakeholder management process aids in the attainment of reliable information about the strengths, weaknesses, capabilities, and the options open to each project stakeholder. Once these options and capabilities have been identified, the project team may better predict the stakeholder's behavior in the face of project activities, and how this behavior will influence the project's end result. At this point, the project team will have a better idea of how to develop its own strategies to best manage the stakeholder.

60.6 PROJECT OPERATIONS

60.6.1 Project Planning

Without question, one of the most important, yet most overlooked aspect of the project team's job is project planning. Project planning is the logical determination of how to begin, execute, and end a project in a forecasted environment. The most challenging activity for the project leader, project planning begins as a vision, and involves determining the likely events of the future, and then developing a strategy for how the organizational resources will be allocated to take advantage of these future conditions. But when most effective, project planning is performed through an iterative process throughout the entire life cycle of the project. To that end, insufficient planning will inevitably lead to decreased execution of project objectives.

Project planning begins within the context of the strategic plan of the organization. Projects are often years in duration. As such, the project plan must contain both short-term operational and long-term strategic thinking, and must be contributed to by many members of the organization. The strategic roles of key individuals involved in project planning are as follows:[27]

- *Board of directors* review and approve key project plans and maintain surveillance over the implementation of the plans.
- *Senior management* direct the design, development, and implementation of a strategic planning system and a project planning philosophy and process.
- *Functional managers* integrate modern functional technology into the project plans.
- *Project managers* integrate and coordinate the project planning activity.
- *Work package managers* provide input to the project plans.

Project planning includes a wide scope of activity. Of primary concern to the project leader are long-range considerations such as objective and goal setting, budgeting and cost estimation, resource usage estimation, and deliverable requirements. The leader must deal with determining those activities and those resources that are needed to make certain that the project will be satisfactorily completed.

The project team must be able to balance the long-term strategic thinking with a short-term implementation plan that will direct them through the execution of the project. To do so, a few detailed tasks may be performed up-front that simplify the operational requirements of the project.

60.6.2 Work Breakdown Structure

One of the most fundamental and vital objectives of project planning is the detailed definition of the work required. A *work breakdown structure* (WBS) divides the overall project into work packages that represent singular units of work, at the level where the work is performed. Each project is subdivided into tasks that are assigned and accomplished by one individual or organizational unit. As a result of the creation of a WBS, the fundamental units of work can be assigned to members of the team who then become accountable for the completion of the work.

A work breakdown structure has several characteristics that provide a common framework from which to move forward:

- Provides work packages that are clearly distinguishable from one another
- Helps determine project schedule
- Aids in determination of budget and resource requirements
- Limits work-in-process by reducing work to short packages
- Defines scope and duration of work packages
- Summarizes all products and services comprising the project
- Displays the interrelationships of the work packages to each other, to the total project, and other activities in the organization
- Helps establish authority-–responsibility relationships
- Provides a basis for controlling the allocation of resources within the project

The work breakdown structure is constructed by decomposing the project into major initiatives, then dividing the major initiatives into tasks, dividing the tasks into subtasks, and so forth into small, manageable components. The WBS may be depicted using a hierarchical pictorial representation or by using an ordered numbering system. By using an ordered numbering system, following a general outline scheme, the action items show themselves. The outline format allows for improved communication and understanding of the total project and its integral subsystems, for example:

1. Project
  a. Major initiative 1
  b. Major initiative 2
    (**1**) Task 1
      (***a***) Subtask 1
      (***b***) Subtask 2
    (**2**) Task 2
  c. Major initiative 3

Projects are planned, organized, and controlled around the lowest level of the work breakdown structure. Each descending level represents an increased level of detail within the project. It is only through the completion of the lower level work packages that the upper levels, major initiatives, and eventually the project can be completed.

The work breakdown structure provides a tool with which to manage the innumerable components of a project. Each work package is assigned to a work package manager, who is then held responsible for achieving a specific, measurable objective within a designated schedule and budget, by both the project and functional managers.

60.6.3 Scheduling

Another important output of project planning is the project master schedule, which visually represents the timeframe for all of the necessary project activities. The project schedule establishes realistic beginning and ending points for each project activity. By establishing the time parameters, project teams can coordinate their efforts during the entire life cycle of the project, as well as monitor and control the progress of the project.

Useful project schedules allow the project team to identify critical work packages and tasks, and are based on probable time estimates. Owing to this uncertainty, the schedule development process is iterative and flexible in nature. Cleland identified the following sequence of steps that are required to develop the project schedule:[28]

1. Define the project objectives, goals, and overall strategies.
2. Develop the project work breakdown structure with associated work packages.
3. Sequence the project work packages and tasks.
4. Estimate the time and cost elements.
5. Review the master schedule with time constraints.
6. Reconcile the schedule with organizational resource constraints.
7. Review the schedule for its consistency with project costs and with technical performance objectives.
8. Senior managers approve the schedule.

Several tools exist that facilitate the development of the project schedule, the most popular of which are the *critical path method* (CPM), *graphical evaluation and review technique* (GERT), *program evaluation and review technique* (PERT), and project management software.

60.6.4 Project Management Information System

The *project management information system* (PMIS), a subsystem of the project management system described above, keeps track of the essential information for planning, organizing, directing, and controlling of the use of resources on the project. Accurate information is required in a timely manner so that the project team can continuously monitor, evaluate, and control the resources used on the project. Further, the information needs to be user-friendly, so that it can be passed on to key stakeholders and senior management for review.

An effective PMIS provides the project team with an understanding of the current status of the project with respect to the time, cost, and technical performance objectives. It should make the team aware of variances to the project plan, and provide insight into the specific corrective actions that can be implemented as a result. To do so, it is essential that the PMIS focus on critical areas needing attention rather than reporting on all areas of the project.

The PMIS is not limited to the formal information that displays project performance information. Rather, because the PMIS is essentially a communication device, it also includes the informal information that arises from discussion among the team, with project stakeholders, and through many other sources. This information gives the project team an indication of how those outside the project team feel about the project. The data received through these channels are often every bit as important as those received through formal information retrieval.

60.6.5 A Project Management Information System Application

Nabisco[29], one of America's largest snack food companies, has implemented sound project management philosophy throughout the enterprise to improve communication and, through its project management information system implementation, is saving millions of dollars each year by eliminating unsuccessful product development projects and expensive product-specific equipment.

In the snack food industry, the need to constantly develop and bring to market new products is accelerated by the fact that many snack foods purchases are impulse purchases. Moreover, the fine line between those new product concepts to pursue and those to discard is narrowed even further by the loss of millions of dollars in research and development, manufacturing, and marketing that is incurred as a result of a single poor decision.

To consistently create successful new products and eliminate unsuccessful products, Nabisco built a unique project management information system to assist their new product development process by coordinating all communication within the project and establishing go/no-go decision points for every project. The process provides a means for cross-functional project teams to maintain constant communication and enhances communication among team members by providing a framework for all project milestones.

After only seven months, Nabisco was able to reduce its new product trials by a third, allowing them to concentrate fully on those products that have a higher probability of success. In addition, they save millions of dollars each year by avoiding those projects that require large capital investments in specialized equipment that cannot be reused with other products. In total, Nabisco is better meeting its customers' requests, and thus, increasing sales.

60.6.6 Project Monitoring, Evaluation, and Control

During the execution phase of a project, the most important management function is controlling, or ensuring that the project proceeds as planned (Fig. 60.4). More specifically, the project manager monitors and evaluates the actual progress of the project to the planned progress, with respect to the time, cost, and technical performance objectives, so that corrective action (control) may be taken if the actual progress and planned progress differ.

The control of a project is an ongoing process throughout the life cycle of the project, but has four distinct steps: developing performance standards (planning), observing performance (monitoring), comparing actual and planned performance (evaluation), and taking corrective action (control). These steps demonstrate that the planning and controlling functions of project management are linked. The monitoring and evaluating of a project is the link between the planning and controlling functions.

The process of establishing standards begins in the planning phase of the project. During this period, among other goals, the scope of the project is outlined, the work packages are identified, the budget and schedule are estimated, and stakeholder requirements are established. From these items, coupled with the organizational plan, performance standards are derived.

Once the execution phase of the project begins, the performance must be observed. To observe performance is to receive sufficient information to compare the actual and planned progress in the next step of the control process. The amount of information that constitutes "sufficient information" varies, and is subject to personal opinion. Nevertheless, it is at this point that an effective project management information system is necessary. Both formal and informal information need to be collected through formal progress reports, conversations, briefings, internal audits, and stakeholder interaction. These data will provide the basis for evaluation of the project's progress.

By continually comparing actual progress and planned progress, the project team is able to keep a scorecard regarding the evolution of the project. They can determine if there are deviations from the plan, the degree of any deviations, and what could be done to rectify the problems. Furthermore,

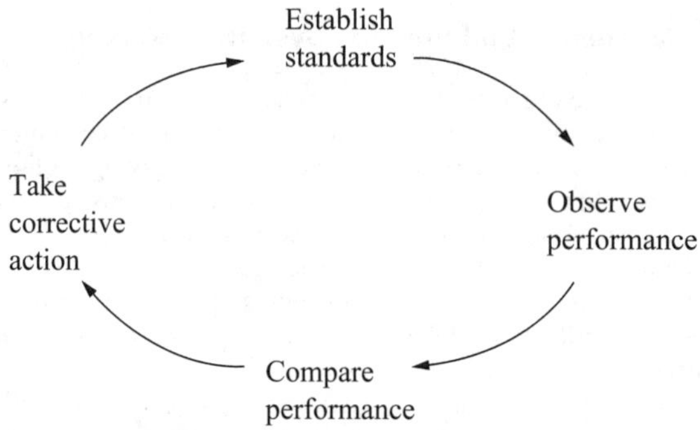

FIGURE 60.4 The control system.[30]

they can foresee potential future problems and determine if anything is performing better than expected, and in that case, if and how that overperformance can be better exploited.

The monitoring and evaluation of project activities provide the team the intelligence needed to make informed decisions regarding any corrective action that may be necessary. This corrective action could be in the form of replanning, reallocating resources, or changing the way the project is managed. In any case, the changes made must be in direct result of timely, accurate information gathered, and their effect on the time, cost, and technical performance objectives.

60.6.7 Project Termination

Projects must end at some point, by definition. But the termination of a project is not an instantaneous phenomenon at the completion of the execution phase. Several important tasks must be performed to close out the project life cycle. In the case of a successful project, before it can be terminated, the technical capabilities produced by the project need to be turned over to those who will execute the day-to-day operations of the deliverables. In other words, operational personnel who will be working with the deliverable regularly, whether it is a product, service, or process, must be trained and the responsibility for the deliverable must be transferred to them. Furthermore, learning from the project must be documented, the resources must be released, and the project team members must be reassigned.

There are many reasons to terminate a project. Most often these reasons are simple; the project was successful or the project failed. A successful project is characterized by delivering the agreed-on technical performance capability to the customer, hopefully on time and within budget. A project failure is characterized by overrun cost and schedule objectives, and/or a failure to meet technical performance specifications. Other reasons to terminate a project include: a change in strategy so that the project is no longer a strategic fit, an external environmental change renders the project useless, or the project's importance has diminished within the organization.

As the Nabisco example indicated, often the termination of a project, which usually implies failure, actually implies success. If substantial amounts of money or man-hours can be saved as a result of terminating a project that is destined to fail, one could look at the project as a success. Hence, a manager's ability to recognize that a project is failing is a valuable attribute.

60.7 PRODUCT DEVELOPMENT PROJECT MANAGEMENT

The continuous reduction of the life cycles of modern-day products is a trend that has inspired and continues to stimulate the need for project management today; however, applying the project management philosophy to new product development projects is difficult owing to the unpredictability of the creative process. Nevertheless, more manufacturing and product development companies are implementing project management to condense development cycles, reduce costs, and make business processes more efficient as they move to provide customer-driven results and systems solutions.

Time in modern product development is critical. A six-month delay in getting a product to market can cost 33 percent of its life cycle profits, and as much as 80 percent of a product's costs are incurred during its development.[31] These enormous up-front costs, and the speed at which they must be implemented, dictate that project management be used to manage the product development process.

Project management helps direct the strategic planning of the organization and aids in the management of organizational resources, but many companies shy from project management because it requires a cultural change by the employees and organization as a whole; however, it can be shown that the high rate of project failures is often a direct result of the lack of a structured approach to managing organizational resources. Furthermore, structured lines of communication, and sharing of information that is widely considered to be essential to successful product development are facilitated through the use of project management.

Without question, project management supports the move toward more integrated, rapid part and product development approaches. By bringing together all of the information about a product cycle's resources, time, and costs, project management facilitates the attainment of profit-bearing new products in a minimal time period.

60.7.1 A Product Development Project Management Example

Boeing, Inc. has incorporated a project management philosophy into its new product design process. Beginning with the innovative breakthrough model 777 jet, both the company culture and design technologies experienced a massive overhaul. Seeking integration, design teams replaced the long-established procedure of dividing the work among different engineering departments.

The goal of the overhaul was to be able to prebuild the entire airplane in a sophisticated computer-aided design program. If successful, all design conflicts could be resolved before the physical assembly took place.

The radical change in philosophy led to increased communication and collaboration throughout the organization. The less threatening, more exploratory conversations spurred discussion on future design issues among various engineering design groups within the company. The resulting conversations resolved anticipatory design conflicts before more serious conflicts could emerge.

The philosophical change at Boeing led to new interaction between previously isolated disciplines, tore down the walls of the functional organization, and inevitably created new value to the organization.[32]

Summary. Global competition, along with the influx of computer technology, has created an environment of change in the modern workplace that is occurring faster than ever before. The philosophy of project management has served, and will continue to serve as the fundamental way to deal with product and manufacturing process changes, as projects become the primary mechanism for creating new products and processes. Its impact on the modern world is clear and widespread, and the discipline continues to grow. The key to success in the twenty-first century will be the organization's ability to adapt to new opportunities in a constantly changing, increasingly accessible global environment. Project management, the most efficient methodology to deal with change, appears to be the pathway to this impending success.

REFERENCES

1. Project Management Institute, *A Guide to the Project Management Body of Knowledge,* 2000 ed., Project Management Institute, Newtown Square, PA, 2000, p. 4.
2. Project Management Institute, *A Guide to the Project Management Body of Knowledge,* 2000 ed., Project Management Institute, Newtown Square, PA, 2000, p. 6.
3. David I. Cleland, *Project Management: Strategic Design and Implementation,* 2d ed., McGraw-Hill, New York,1994, p. 39.
4. David I. Cleland, *Project Management: Strategic Design and Implementation,* 2d ed., McGraw-Hill, New York, 1994, p. 441.
5. Michael Cusumano and Kentaro Nobeoka, *Thinking Beyond Lean: How Multi-Project Management Is Transforming Product Development at Toyota and Other Companies,* The Free Press, New York, 1998, p. 1.
6. Michael Cusumano and Kentaro Nobeoka, *Thinking Beyond Lean: How Multi-Project Management Is Transforming Product Development at Toyota and Other Companies,* The Free Press, New York, 1998, pp. 159–160.
7. Ralph Currier Davis, *The Fundamentals of Top Management,* Harper and Brothers, New York, 1951, p. 256.
8. William A. Moylan, "Managing Manufacturing by Project," *Proceedings of the Project Management Institute 2001 Seminars and Symposium,* Nashville, TN, November 5, 2001.

9. R. Max Wideman, "Tl e PMBOK Report—PMI Body of Knowledge Standard," *Project Management Journal*, Vol. 17, No. 3, August 1986, pp. 15–24.
10. David I. Cleland, *Project Management: Strategic Design and Implementation,* 2d ed., McGraw-Hill, New York, 1994, pp. 57–74.
11. Dave Harrold, "Project Success Requires Vision and Attention to Detail," *Control Engineering,* May 2000.
12. Genentech, Inc., "About Genentech—Our History," Genentech, Inc.; http://www.gene.com/gene/about_genentech/history/#2000, 2000.
13. John R. Adams and Stephen E. Barndt, "Behavioral Implications of the Project Life Cycle," in David I. Cleland and William R. King (eds.), *Project Management Handbook,* Van Nostrand Reinhold, New York, 1983.
14. Harold Kerzner, *Project Management*, 3rd ed., Van Nostrand Reinhold, New York, 1989, p. 72.
15. David I. Cleland and Lewis R. Ireland, *Project Manager's Portable Handbook,* McGraw-Hill, New York, 2000, pp. 7.66–7.68.
16. David I. Cleland, "Defining a Project Management System," *Project Management Quarterly,* Vol. 10, No. 4, p. 39.
17. Harold Kerzner, *Project Management*, 3rd ed., Van Nostrand Reinhold, New York, 1989, p. 86.
18. David I. Cleland and Lewis R. Ireland, *Project Manager's Portable Handbook,* McGraw-Hill, New York, 2000, pp. 1.16–1.17.
19. Project Management Institute, *A Guide to the Project Management Body of Knowledge,* 2000 ed., Project Management Institute, Newtown Square, PA, 2000, p. 30.
20. Harold Kerzner, *Project Management,* 3rd ed., Van Nostrand Reinhold. New York, 1989, p. 114.
21 David I. Cleland, *Project Management: Strategic Design and Implementation,* 2d ed., McGraw-Hill, New York, 1994, p. 187.
22. David I. Cleland, *Project Management: Strategic Design and Implementation,* 2d ed., McGraw-Hill, New York, 1994, p. 188.
23. David I. Cleland and Lewis R. Ireland, *Project Manager's Portable Handbook,* McGraw-Hill, New York, 2000, pp. 2.12--2.15.
24. R. Max Wideman and Aaron J. Shenhar, "Professional and Personal Development Management: A Practical Approach to Education and Training," Chap. 20 in Joan Knutson (ed), *Project Management for Business Professionals,* John Wiley & Sons, New York, 2001, pp. 365–366.
25. David I. Cleland, *Project Management: Strategic Design and Implementation,* 2d ed., McGraw-Hill, New York, 1994, pp. 207–216.
26. Ibid, pp. 144–157.
27. Ibid, p. 247.
28. Ibid, p. 253.
29. Microsoft Business, "Nabisco's Journey Halts Half-Baked Ideas, Saves Dough," http://www.microsoft.com/business/casestudies/productivity/microsoft_nabisco.asp, Sept. 19, 2000.
30. David I. Cleland, *Project Management: Strategic Design and Implementations,* 2d ed., McGraw-Hill, New York, 1994, p. 285.
31. Lisa Kempfer, "First to market: The power of managing products," *Computer-Aided Engineering*, Cleveland, OH, February 1999.
32. Michael Schrage, The Proto Project, *Fast Company—The Magazine,* http://www.fastcompany.com/online/24/schrage.html, May 1999.

FURTHER READING

Cleland, David I., and William R. King, *Systems Analysis and Project Management,* 3rd ed., McGraw-Hill, New York, 1983.

Cleland, David I., *Strategic Management of Teams,* John Wiley & Sons, New York, 1996.

Davis, Ralph Currier, *The Fundamentals of Top Management,* Harper and Brothers, New York, 1951.

Gray, Clifford F., and Erik W. Larson, *Project Management: The Managerial Process,* McGraw-Hill, New York, 2000.

Kliem, Ralph L., and Irwin S. Ludin, *Project Management Practitioner's Handbook,* AMACOM, New York, 1998.

Lewis, James P., *Project Planning, Scheduling, & Control,* rev. ed., Irwin Professional Publishing, Chicago,1995.

Wheelwright, Steven C., and Kim B. Clark, *Revolutionizing Product Development: Quantum Leaps in Speed, Efficiency, and Quality*, The Free Press, New York, 1992.

INFORMATION RESOURCE

For more information on the PMI, visit www.pmi.org, email pmihq@pmi.org, or telephone +610-356-4600.

CHAPTER 61
POLLUTION PREVENTION AND THE ENVIRONMENTAL PROTECTION SYSTEM

Nicholas P. Cheremisinoff
Clean Technologies and Pollution Prevention Programs
Princeton Energy Resources International, LLC
Rockville, Maryland

61.1 INTRODUCTION

Prevention refers to taking actions that do not allow pollution or waste to occur. For example, if we are using a hazardous solvent like trichloroethylene (a human carcinogen) in a vapor degreasing operation for metal cleaning, we need be concerned with several forms of pollution and waste, as well as possible environmental damages that create long-term liabilities. To name a few: workers may be exposed to hazardous air emissions, soil and groundwater could become contaminated from an inadvertent spill from the mishandling of this material, a regulated hazardous sludge will have to be disposed of, and we will have to replace the solvent in the degreaser over time (thereby facing the problem of having to dispose of a hazardous liquid waste, as well as deal with the ongoing costs of supplying fresh chemical feedstock). To minimize some of these wastes and pollution, we could substitute another chemical that is more environmentally friendly. We may also prevent these wastes and pollution by changing the technology altogether, for example, by switching from solvent degreasing to a water soluble or emulsion-based metal cleaning operation. In addition, there may be a variety of process operating conditions and good housekeeping practices to choose from, which achieve varying degrees of pollution reduction and waste minimization.

From a business standpoint we should ask ourselves whether alternative approaches to managing environmental problems and achieving compliance are more cost-effective than the current methods used in the operation. In many situations, alternatives that result in equivalent or lower levels of pollution and waste are seemingly more costly than less environmentally friendly operations that meet the legal emission standards; however, when we consider the liabilities associated with less environmentally friendly operations, a more costly initial investment into a P2 technology can be justified. In many situations it really reduces to how much risk a company is willing to accept in managing certain types of wastes and pollution.

61.2 HIERARCHY OF POLLUTION MANAGEMENT APPROACHES

Pollution management approaches to a company's environmental issues can be described as strategies. These strategies traditionally have relied on so-called *end-of-pipe treatment* technologies, which have two disadvantages when compared to P2:

1. They require on-going costs that are associated with operations and maintenance, use of energy, and they carry many hidden and indirect costs.
2. There is liability that continues forever, because when we rely only on controls, pollution and waste ultimately must be disposed of. These wastes are not destroyed and simply persist as an ongoing threat and potential liability.

There is a hierarchy to pollution management strategies. Strategies that reduce or eliminate wastes before they are even created are preferable to those that incur ongoing costs for treating and disposing of wastes that are generated continuously. These strategies reduce the costs associated with the first disadvantage of end-of-pipe technologies, and they minimize or can even eliminate the second disadvantage. Strategies ranked in order of preference (from highest to lowest) are illustrated in Fig. 61.1 and are summarized as follows:

- *Prevention.* This strategy prevents wastes or pollution from ever being formed in the first place.
- *Recycling.* In the majority of manufacturing processes waste generation is unavoidable. Therefore, operations should pursue those strategies that minimize these inefficiencies to the greatest extent possible. Recycling and reuse of spec materials and recovery of certain pollutants for reuse (known as *resource recovery*) are common strategies to achieving this goal.
- *Treatment.* When wastes cannot be prevented or minimized through reuse or recycling, then we need to pursue strategies aimed at reducing volumes and/or toxicity. Engineers often confuse

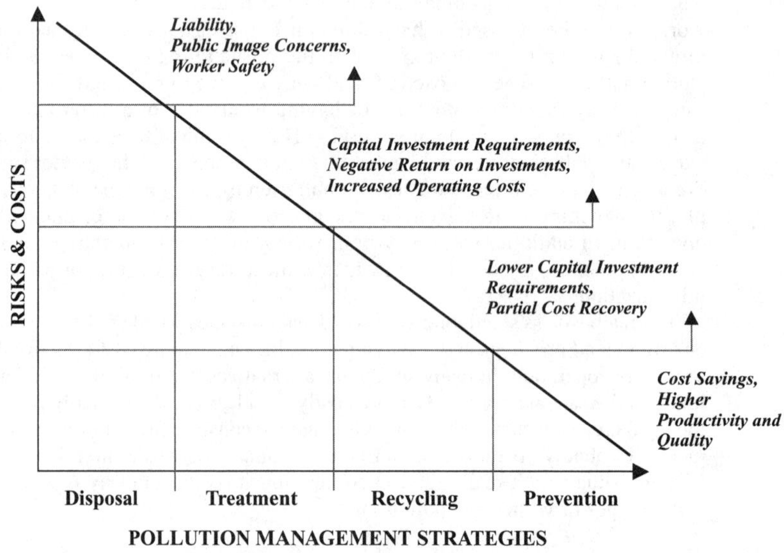

FIGURE 61.1 Hierarchy of pollution management strategies.

TABLE 61.1 Examples of Pollution Management Strategies

Priority	Strategy	Example	Specific applications
1	Prevention (source reduction)	Process changes Designing products that minimize environmental impacts Source elimination	Modifying processes to avoid or reduce use of solvents Modify products to extend coating life
2	Recycling	Reuse Reclamation	Solvent recycling Metal recovery from spent surface treatment baths Recovery of VOCs
3	Treatment	Stabilization Neutralization Precipitation Evaporation Incineration Gas scrubbing	Thermal destruction of organic solvents Precipitation of heavy metals from spent plating baths
4	Disposal	Off-site disposal	Landfill

end-of-pipe controls with treatment technologies. In fact, for most situations they are not treatment technologies. Treatment technologies often are add-on processes that focus on stabilization of wastes, reducing toxicity, reducing volume before ultimate disposal, or in some cases creating limited-use by-products (an example would be the vitrification of a toxic solid waste in a cement kiln and then using the ash as an extender in an aggregate like cement for a construction industry application).

- *Disposal.* The only other strategy available to us is disposal. Waste disposal practices are integrated into the environmental management strategies of most manufacturing operations, and quite often are among the highest direct cost components. From a business standpoint, it is the least desirable strategy and one that can be directly addressed by P2 and waste minimization practices.

Some examples of each of the strategies are given in Table 61.1.

When considering strategies, we should look closely at the complexity, the effectiveness, and costs of the technologies being considered. Table 61.2 provides a relative comparison between different waste and pollution reduction technologies.

61.3 FOUR TIERS OF POLLUTION COSTS

Decisions on investing in pollution management strategies that rely heavily on end-of-pipe treatment technologies are often the result of ignoring or not recognizing the true or total costs associated with pollution control and post waste handling. To account for all of the costs we need to recognize that they fall into four distinct categories or tiers. McHugh (1990) originally introduced this concept:

- *Tier 1 costs.* Defined as usual or normal costs
- *Tier 2 costs.* Hidden or direct costs
- *Tier 3 costs.* Future liability costs
- *Tier 4 costs.* Less tangible costs

Examples of each category follow.

TABLE 61.2 Comparison of Different P2 and Waste Minimization Technologies

Preventive technique or technology	Relative cost	Complexity	Effectiveness
Class 1 options			
Pollution detection, monitoring and taking corrective actions	Low to high	Low to high	High
P2 and waste reduction audit	Low	Low	High
Class 2 options			
Improved management techniques	Low	Low	High
Substitution of raw materials	Low to high	Low	High
Product redesign and reformulation	High	High	Medium
Equipment technology alterations	High	High	Medium
Procedure modifications	Low to high	Low to high	Medium
Class 3 options			
Materials reclamation	Payback	Medium	High
Off-site waste recycling	Payback	Medium	High

61.3.1 Tier 1: Usual or Normal Costs

These include such costs to the operation as:

- Pollution fee, i.e., fee paid for discharging pollutants within legally allowable emissions standards
- Direct labor costs
- Raw materials, e.g., feedstock chemicals and water
- Energy
- Capital equipment items
- Site preparation for pollution control equipment
- Equipment tie-ins and process modifications to accommodate pollution controls
- Employee training
- Permits to construct

61.3.2 Tier 2: Hidden or Indirect Costs

These include such costs to the operation as:

- Expenses associated with monitoring
- Permitting fee, e.g., permits to operate, permit renewals, and the like
- Environmental transformation costs, e.g., those costs associated with transforming a water pollution problem into a solid waste disposal issue—an example is a dewatering operation such as a filter

press to treat suspended solids in a wastewater stream. This strategy generates a sludge that must be disposed of. Another example is a wet scrubber used to capture particulates in order to address an air pollution problem. In this example, although we eliminate or reduce an air emission problem, we generate a water pollution problem with the use of this technology. The water pollution problem may also have to be addressed by separate treatment technologies, resulting in a sludge disposal problem.

- Environmental impact statements—depending on the nature of the project, EISs can be many thousands of dollars and require long lead times.
- Health and safety assessments
- Service agreements
- Legal costs
- Control instrumentation
- Maintenance and replacement costs
- Reporting and recordkeeping

61.3.3 Tier 3: Future Liabilities

These include such costs to the operation as the following:

- Remedial action costs—site cleanup plus cleanup costs associated with third-party damages
- Personal injury to workers
- Health risks and injury to the public
- More stringent compliance—environmental compliance is often a moving target, and hence a control technology used today may become obsolete a few years down the road
- Inflation, e.g., this could impact on higher feedstock costs and energy, as well as waste disposal services

61.3.4 Tier 4: Less Tangible Costs

These include such costs to the operation as the following:

- *Consumer response and loss of confidence owing to perceived poor environmental management.* It is well recognized in technologically advanced societies that consumers and investors favor those companies that are environmentally conscious of their actions and have proactive environmental management programs.
- *Employee relations.* Poor management of environmental issues places workers at health risks.
- *Establishing and extending lines of financial credit.* Lending institutions will not extend loans to, or provide favorable terms for credit lines if a company runs a high risk operation from an environmental management standpoint.
- *Adverse impacts on sales of property, mergers, and acquisitions.* Companies that are looking to partner or merge with others are not willing to take on the other firm's environmental liabilities. Poor environmental management can bring enormous liabilities to investors and partner organizations.
- *Higher insurance premiums.* Those companies that have a poor environmental track record represent a risk, and therefore the potential loss of assets from fire or explosion, or health risks to workers and the public are high. Insurers will not take these risl:s, or they will impose high premiums with very limited coverage.

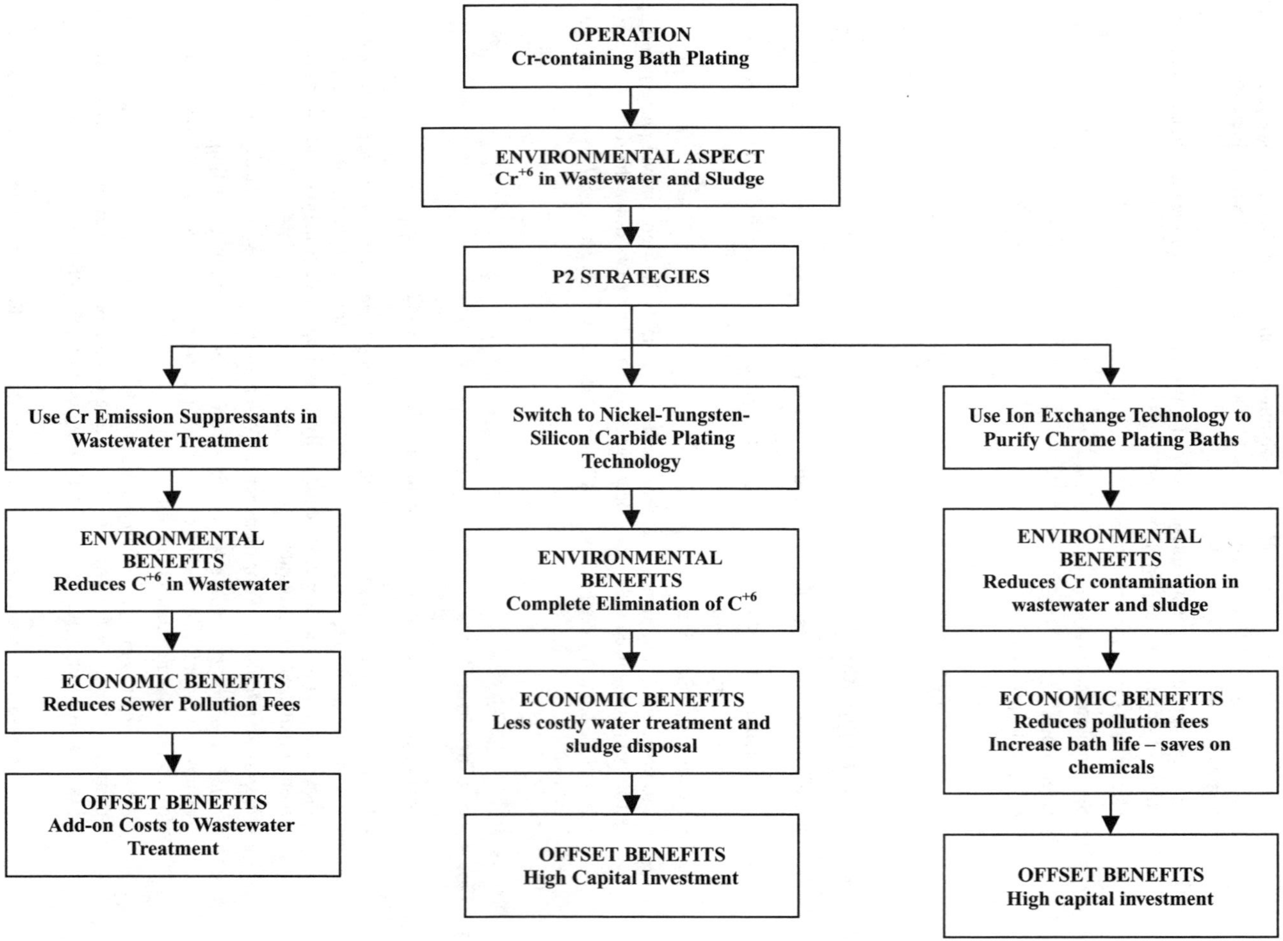

FIGURE 61.2 Three alternative P2 approaches.

- *Becoming the focus of frequent environmental inspections and corrective actions.* If environmental regulators recognize a recurring theme of noncompliance or accidents, then the facility will be singled out for numerous violations, fines, inspections, and corrective actions. All of these can result in significant costs to an operation, including interruptions in production schedules and legal costs.
- *Less leverage in negotiating environmental fines and penalties.* The environmental laws in technologically advanced nations are so complex, and operations are often fragmented, leading to innocent violations of environmental laws. If a company has had a history of poor environmental management, then it has little basis to negotiate for leniency on innocent violations. The problem is perceived as systematic.

Tier 1 and 2 cost components are usually identified in selecting pollution management strategies. If we look at these tier cost components alone, then it is easy to justify investments into control-based technologies. In contrast, Tier 3 and 4 cost tiers are often not given a sufficient weighting, i.e., they are considered less important than more direct and tangible costs, usually because components are more difficult to assign values to; or they are simply not recognized. This is a mistake, because by attaching the proper importance and most probable costs to these components, other strategies become apparent and more attractive.

Proper evaluation of Tier 3 and 4 cost components in assessing total costs and then applying the results as a basis to selecting proper pollution management strategies requires the application of risk assessment principles. Since the ultimate goal is to arrive at cost effective strategies that meet a compliance target, or to go beyond that target, risk assessment should be considered as a part of *life cycle costing analysis* (LCCA).

Understanding and dissecting the cost components within each tier are critical to making decisions on investing into the most economical pollution management strategies. When companies only focus their attention on the obvious costs (Tiers 1 and 2), critical factors that could alter the decision on investments into pollution prevention may be missed. As an example, consider an electroplating operation that relies on the use of chromium solutions. This type of operation produces highly toxic sludge and wastewaters that contain hexavalent chromium, a confirmed human carcinogen. Stringent environmental regulations demand the applications of *best available technologies* (BATs) to control wastewater emissions and to dispose of the toxic sludge in a safe manner. But BATs generally tend to be expensive, especially for regulated hazardous wastes. Figure 61.2 summarizes three alternative P2 approaches to the conventional control technologies. By only focusing on Tier 1 and 2 cost components, it's quite possible, depending on the size of the operation, that the least cost strategy is to continue with BATs (i.e., end-of-pipe). When the analysis focuses on such issues as worker health risks, cumulative sewer discharge fees, on-site wastewater treatment costs, and the cumulative costs associated with sludge stabilization, on-site staging of wastes, recordkeeping and waste reporting requirements, and transport and disposal, then other alternatives that displace or eliminate the generation of hexavalent chromium wastes as a part of the process become much more attractive. When we consider also the potential future liabilities when the facility is sold, such as the costs that may be faced in remediating the property, then higher investments into strategies that eliminate chrome as a part of the operation are justifiable. This is the basis under which life cycle costing techniques are applied to making logical investment decisions.

61.4 IMPORTANCE OF P2 TO YOUR BUSINESS

The first overall objective of any pollution management strategy, and indeed an *environmental management system* (EMS), is to bring a facility into compliance with environmental regulations. Simply stated, there are only two choices any company faces:

- Comply with environmental regulations and remain in business.
- Don't comply, and face heavy fines, penalties, costly interruptions, or in the extreme, go out of business.

When a company approaches this objective with strategies based largely on end-of-pipe treatment technologies, it is addressing only the legal requirements for staying in business, and not growth. By repeatedly trying to meet more stringent effluent emission standards through the use of more advanced controls, a company is always on the defensive, i.e., playing catch-up with compliance requirements.

P2 strategies achieve the same end results as the application of control technologies (i.e., they help meet a compliance requirement), but P2 is more cost effective. The reason for this is that P2 strategies reduce Tier 1 and 2 costs through savings not achievable through control technologies, which are simply add-on process components that impose additional demands on energy, materials, and labor. The savings can be applied toward reinvesting into modernization and improvements to operations, or in the financing of the implementation of an EMS. This enables a company to grow its operations.

P2 strategies also reduce or eliminate costs associated with future liabilities (Tier 3) and less tangible costs (Tier 4). In the ideal case, prevention eliminates waste and pollution altogether. Hence, issues like third-party damages, or joint and several liabilities for off-site damages are no longer a concern.

When P2 strategies are employed, the owners/operators of facilities can exceed compliance requirements, because in the ideal case zero waste and pollution is the ultimate goal. Another way of looking at this is that when an organization focuses on continual reductions of waste and pollution, operating costs are steadily reduced over time, and profitability is maximized.

61.5 P2 IN THE CONTEXT OF AN EMS

P2 programs work best within the framework of an EMS. As noted by Cheremisinoff and Bendavid-Val (2001) :

> An environmental management system is an approach … a tool … a set of procedures … a planned and organized way of doing things …

In other words, an EMS is a systematic approach for a company to manage its environmental concerns. Most people tend to separate the functions of an EMS and pollution prevention in their minds, or think of these as related only in the sense that P2 is the preferred way of addressing environmental aspects of a company's operations that have been identified through an EMS. But in fact, a pollution prevention program must be fully integrated with an EMS in order to maximize the benefits a company can achieve through its EMS.

A P2 Program consists of a systematic approach to identifying more cost effective strategies and technologies for waste handling, then implementing those strategies and tracking their environmental and economic performances for the purposes of establishing new targets and goals, starting with the highest environmental priorities first, and then iteratively repeating the process for, and applying lessons learned to, successive pollution problems through the entire operation. Through the application of this iterative process, incremental savings are achieved with each new P2 activity. The savings are cumulative and hence, over time, represent significant capital returns that can be reinvested into a company's operations. Successive rounds of continual P2 improvements are what constitute the basis for an EMS.

A policy commitment to pollution prevention (P2) is one of the very first things the ISO 14001 EMS standard requires. We all know why: end-of-pipe approaches to dealing with pollution and waste tend only to alter the medium, method, and location of disposal; generally consume even more energy and other resources in the cleanup process; and because of that, add to the costs of production. Moreover, unless a regulated hazardous waste is completely immobilized or destroyed, it remains a liability forever. By contrast P2, as ISO 14001 notes in its definitions section, tends to reduce adverse environmental impacts while improving production efficiency and reducing costs; and because pollution is reduced or eliminated at the source, liabilities are substantially reduced.

Indeed, arguably most of the environmental and economic performance benefits, reduction in risk of environmental liabilities, and hence the sustainability benefits of an EMS derive from the P2 activities and investments that it generates.

There's no such thing as a meaningful EMS without P2 at its core. P2 is not just a concept, an idea, an approach, or a mindset; it requires detailed analysis and planning that incorporate a rigorous methodology and a set of engineering, management, and life-cycle costing tools. True, you can walk around the shop floor and identify many P2 housekeeping and relatively low-cost investment opportunities—leaking valves, lights left on unnecessarily, heat losses from poor insulation, materials that could be recycled internally, waste that could be reduced with automated controls, and so on. But to identify the bigger-ticket P2 opportunities that yield major production process efficiencies and cost savings, properly evaluate those opportunities from a bottom-line point of view and then comparatively evaluate them and devise a P2 investment strategy, and to do this on a continual improvement basis, requires a well thought-out P2 program with a structure like the one illustrated in Fig. 61.3, similar to an EMS.

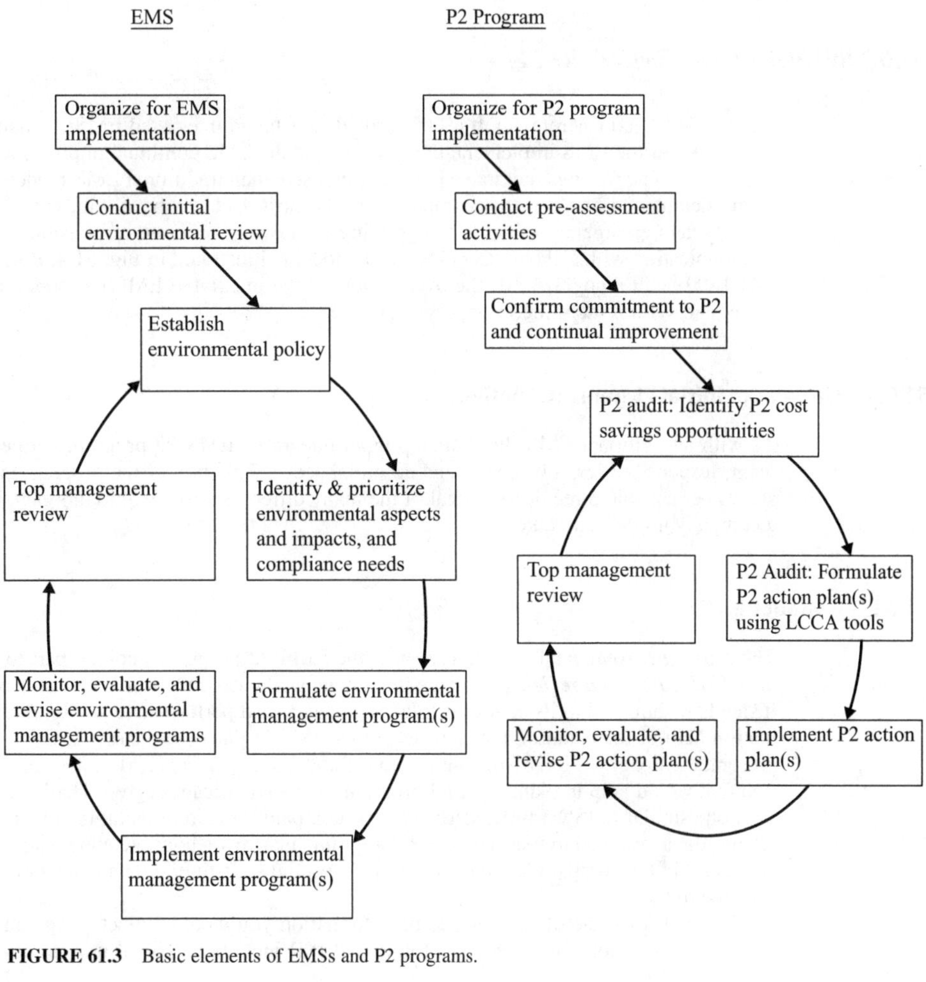

FIGURE 61.3 Basic elements of EMSs and P2 programs.

In Chapter 7 of *Green Profits* (2001), a 21-step model for a P2 audit is provided. Figure 61.3 has adapted the outlines of that model to a continual improvement program framework. The P2 program begins with three steps that need to be taken as part of setting it up the first time: organizing to put the program in place, gathering and interpreting basic production process information, and confirming top management's commitment to a program of continual improvement in preventing pollution. Next there are five steps that constitute the ongoing P2 continual improvement program—they amount to an adaptation of the famous plan-do-check-act management system model.

The necessarily similar structures of EMSs and P2 programs make it possible to integrate the elements of a P2 program smoothly into an EMS. When this is done, P2 is raised from being a mere commitment within the EMS to being a major operational component of the EMS. Instead of looking just at the regulatory compliance status and environmental aspects of your company's operations, and then at ways to improve on these, you would also be specifically and systematically looking for high-return P2 opportunities from the outset, and doing so continuously. An integrated EMS/P2 program reduces operating costs and the risks of environmental liability more, better, and faster than a traditional EMS. The earlier major cost savings can be used to help finance investments, further accelerating improvements in business and environmental performance for the company. How an integrated EMS/P2 program would work is worth examining.

61.6 INTEGRATING EMS AND P2

Bendavid-Val and Cheremisinoff (2003) provide a detailed integrated EMS/P2 model that moves P2 to the front of the EMS implementation process and the EMS continual improvement cycle. In addition, it has some features that make it more responsive than traditional EMS models, like ISO 14001, to the heightened concern that companies in the post-9/11 era have with environmental liability, security, and emergencies, and to the growing interest in corporate responsibility and constructive communication with stakeholders. The basic model is illustrated in Fig. 61.4. Readers familiar with ISO 14001 will recognize that the overall flow of the integrated EMS/P2 model in Fig. 61.4 is the same; and yet not the same, as in a typical EMS.

61.6.1 Undertake Initial Organizing Activities

As with a traditional EMS the first step in an integrated EMS/P2 program is to establish commitment, responsibilities, awareness, and an initial scope. This box of the program model in Fig. 61.4 shows perhaps the three most critical of the many different possible initial organizing activities that a company might undertake.

61.6.2 Conduct IE/P2R

The initial environmental and P2 review is the EMS/P2 program's counterpart to the EMS-related *initial environmental review* (IER). A company may conduct an IER to establish a baseline on which it stands with regard to its environmental compliance and performance and to develop related information helpful for planning the steps to put its EMS in place. Typically an IER, including the one presented on pages 179–185 of Cheremisinoff and Bendavid-Val (2001), does not develop information that would help in planning a P2 program to any significant degree. That's because an EMS, at least one similar to ISO 14001, starts with the company's environmental interactions, "aspects," and so the IER is oriented to that; while a P2 program, apart from housekeeping, starts with inefficiency and waste in the company's production processes, and so its initial data collection would need to be oriented to that.

Examples of some of the P2-specific information you would collect during an IE/P2R that you would normally not collect during a traditional IER include details of unit processes, an initial list

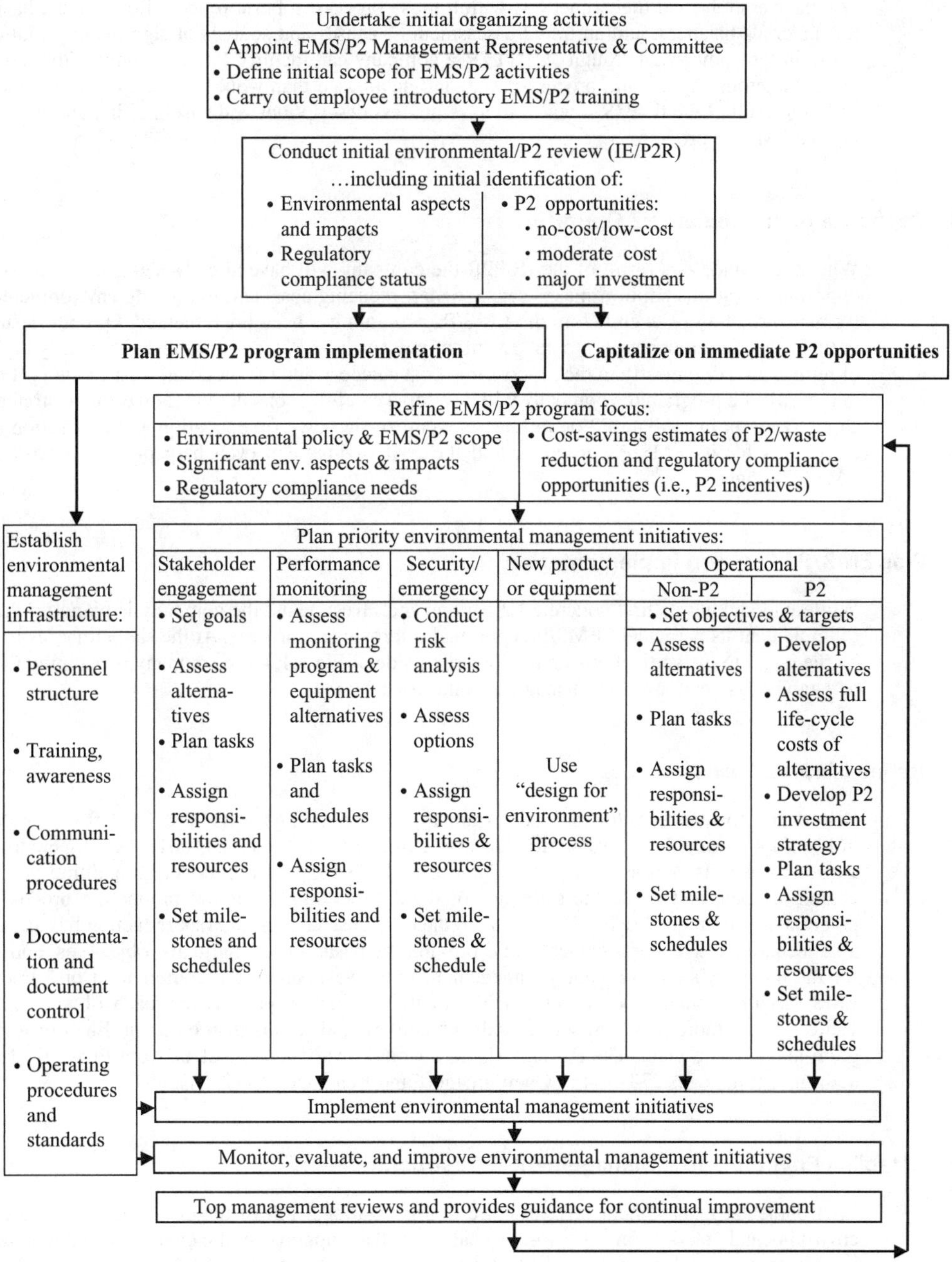

FIGURE 61.4 An integrated EMS/P2 program.

of unit operations and their functions within those processes, basic process flow sheets, basic data for materials balances, and an initial assessment of causes and sources of significant pollution and waste in the company. Through an IE/P2R a company can identify P2 opportunities that are much more numerous, significant, process-based, and data-based than would be the case through an IER; and as a result of the IE/P2R it will also have process descriptions and baselines in hand for use during the EMS/P2 program.

61.6.3 Capitalize on Immediate P2 Opportunities

Why wait? Since as a result of the IE/P2R the company will have already identified some solid P2 opportunities at this point, it makes sense to start reducing costs and improving environmental performance right away, even before the EMS/P2 program has been implemented. Of course, jumping in this way your company has to be careful to invest only in P2 opportunities with clearly high rates of return, probably mostly in the no-cost/low-cost category and the moderate cost category. Later on in the EMS/P2 program the company will develop a carefully considered P2 investment strategy. But capitalizing on immediate P2 opportunities can provide powerful education and motivation among employees for its EMS/P2 program, and the cost savings can even help offset the costs of the EMS/P2 program.

61.6.4 Plan EMS/P2 Program Implementation

While undertaking its first concrete P2 actions and investments, the company is planning the steps to implement its permanent EMS/P2 continual improvement process. At the same time, as is shown by the arrow to the vertical box on the lower left side of Fig. 61.4, the company is planning the steps to establish its environmental management infrastructure.

61.6.5 Refine EMS/P2 Focus

Here the company refines the initial determinations made during the IE/P2R. The left side of this box in Fig. 61.4 is roughly equivalent to clauses 4.2 through 4.3.2 of ISO 14001, covering environmental policy, aspects, and legal requirements. In the EMS/P2 program model, in addition you would extend, deepen, and refine the estimates of regulatory compliance and production process costs, potential environmental liabilities, and the related P2 cost savings and risk reduction based on available technologies. These together would provide the basis for setting goals, objectives, and targets for the company's environmental management initiatives. Again, the combination of a traditional EMS "environmental aspects" approach with the production process approach of a P2 program yields a much more powerful set of goals, objectives, and targets than either an EMS or a P2 program alone would yield. The P2 analysis here would involve certain steps from Phase II of the P2 audit model in pages 273–304 of Cheremisinoff and Bendavid-Val (2001).

61.6.6 Establish Environmental Management Infrastructure

Environmental management infrastructure is organizational infrastructure meant to support sound environmental practices on a day-to-day basis as well as support specific environmental management initiatives generated through the EMS/P2 program. This box in Fig. 61.4 is roughly equivalent to clause 4.4 (Implementation and Operation) of the ISO 14001 EMS standard.

However, the model in Fig. 61.4 reflects belief that environmental security and emergency preparedness should be a type of environmental management initiative rather than a part of the supporting infrastructure. This would mean that under the EMS/P2 program the company would always be looking for initiatives that would help it improve in this area, and therefore environmental security and emergency concerns would receive more deliberate, continuous, and systematic scrutiny.

61.6.5 Plan Priority Environmental Management Initiatives

This box in Fig. 61.4 is the counterpart of ISO 14001's clauses 4.3.3 (Objectives and Targets) and 4.3.4 (Environmental Management Programs). By "environmental management initiatives" we mean carefully planned activities designed to address the company's needs for improvement that have been identified, refined, and prioritized in earlier phases of EMS/P2 program work. These initiatives are basically the same as what ISO 14001 calls environmental management programs, and what the Cheremisinoff and Bendavid-Val (2001) P2 audit model calls action plans. The prominence of this box in Fig. 61.4 is meant to emphasize that environmental management initiatives are the centerpiece of an EMS/P2 program.

Figure 61.4 shows six possible types of environmental management initiatives. Each type of initiative requires its own planning process, as suggested in short-hand in the column for each type. The types, or categories, of environmental management initiatives reflect the areas of environmental management concern that the company's management wants to be sure are the primary focus of its continual improvement efforts. We do not have space in this chapter to dwell on what is encompassed by each type of environmental management initiative shown in Fig. 61.4 and the unique planning requirements of each, much as we would like to. But we do need to look for a moment at the types of initiatives shown in the two columns at the right of this box. These are environmental management initiatives aimed at achieving specific performance targets, incidentally the only kind of environmental management program explicitly required in ISO 14001 and most other EMS models. As the illustration suggests, you can employ a fairly routine planning process when for whatever reason a nonP2 approach must be used to achieve a particular performance target. For the collection of performance targets that will be achieved through P2 techniques and technologies (the rightmost column) you will need to perform extensive economic analysis, including life-cycle costing, and develop a full P2 investment strategy. The investment strategy may even be structured so that the economic benefits of earlier P2 investments help finance subsequent P2 investments. In any case, the P2 planning work at this point would involve many of the steps in Phases II and III of the P2 audit model presented by Cheremisinoff and Bendavid-Val (2001) (pages 273–316).

Planning for EMS/P2 environmental management initiatives is followed, as shown in Fig. 61.4, by the familiar steps of implementation; monitoring, evaluation, and revision; top management review; and the next cycle of continual improvement. As with ISO 14001, once the pieces are in place the integrated EMS/P2 program does not really operate as a cycle with discrete sequential elements: rather, all its elements should be operating more-or-less all the time.

61.7 CLOSING REMARKS

The integrated EMS/P2 program described here provides a framework that accommodates all the requirements of ISO 14001 and therefore can be used to pursue ISO 14001 certification, if that's what your company wishes to do. But an integrated EMS/P2 program can yield environmental and economic performance benefits faster than typical ISO 14001-type EMSs, and the model recommended offers other advantages as well. In Bendavid-Val and Cheremisinoff (2003), the reader will find further elaboration, including the following:

- Methodological detail specifically for the P2 elements of the integrated EMS/P2 program
- The content and planning process for each type of environmental management initiative
- An approach to establishing environmental management infrastructure in a company

- Converting a traditional EMS into an integrated EMS/P2 program without disruption
- Building an integrated EMS/P2 program on the base of a quality management program
- Case illustrations

REFERENCES

Bendavid-Val, A., and N. P. Cheremisinoff, *Achieving Environmental Excellence: Integrating P2 and EMS to Increase Profits,* Government Institutes, Rockville, MD, 2003.

Cheremisinoff, N. P., and A. Bendavid-Val, *Green Profits: The Manager's Handbook for ISO 14001 and Pollution Prevention,* Butterworth-Heinemann Publishers, UK, 2001.

Cheremisinoff, N. P., and A. Bendavid-Val, *Green Profits,* Chap. 7, http://www.bh.com/engineering/default.asp?isbn=0750674016

McHugh, R. T., *The Economics of Waste Minimization,* McGraw-Hill, New York, 1990.

FURTHER READING

American Conference of Governmental Industrial Hygienists, *TLVs and BEIs: Threshold Limit Values for Chemical Substances and Physical Agents,* Biological Exposure Indices, ACGIH, Cincinnati, OH, 1996.

Berner, Elizabeth K., and R. A. Berner, *The Global Water Cycle: Geochemistry and Environment,* Prentice-Hall, Englewood Cliffs, NJ, 1987.

Cheremisinoff, Nicholas P., *Calculating and Reporting Toxic Chemical Releases for Pollution Control,* SciTech Publishers, Matewan, NJ, 1990.

Cheremisinoff, Nicholas P., *Handbook of Industrial Toxicology and Hazardous Materials,* Marcel Dekker, New York, 1999.

Cheremisinoff, Nicholas P., *Handbook of Pollution Prevention Practices,* Marcel Dekker, New York, 2001.

EPA-625-7-91-017, *Guides to Pollution Prevention: The Pharmaceutical Industry,* Risk Reduction Engineering Laboratory and Center for Environmental Research and Development, USEPA, Cincinnati, OH, Oct. 1991.

Nelson, Michael J. K., S.O.Montgomery, E.J. O'Neill, and P.H. Pritchard, "Aerobic metabolism of trichloroethylene by a bacterial isolate," *Applied and Environmental Microbiology,* 52:383–384, 1986.

Parsons, Frances, P. R. Wood, and J. DeMarco, "Transformations of tetrachloroethene and trichloroethene in microcosms and groundwater," *Journal of the American Water Works Association,* 56–59, February 1984.

Schwille, Friedrich, *Dense Chlorinated Solvents in Porous and Fractured Media,* Lewis, New York, 1985. (James F. Pankow, trans.)

Whitten, Kenneth W., K. D. Gailey, and R. E. Davis, *General Chemistry With Quantitative Analysis,* Saunders College Publishing, New York, 1992.

Wilson, John T., and B. H. Wilson, "Biotransformation of trichloroethylene in soil," *Applied and Environmental Microbiology,* 49:242–243,1985.

INFORMATION RESOURCE

U.S. Environmental Protection Agency, 63 FR 34338, June 24, 1998.

INDEX

Z

ABOUT THE EDITOR

Hwaiyu Geng, CMfgE, PE, is a Project Manager in Corporate Project Planning and Management at Hewlett-Packard Company, Palo Alto, California. While in charge of planning and implementing Hewlett-Packard's $51-million inkjet plant in Shanghai, China, he charted and directed the design, construction, and manufacturing plan, and completed the installation and startup of the manufacturing plant. Also, as Hewlett-Packard's Seismic Program Manager, Mr. Geng established safety and quality control procedures for the corporation's seismic assurance and mitigation projects in Shanghai and elsewhere.